Tunnel Engineering Handbook

Tunnel Engineering Handbook

Second Edition

Edited by

JOHN O. BICKEL

THOMAS R. KUESEL

ELWYN H. KING

KLUWER ACADEMIC PUBLISHERS
BOSTON/DORDRECHT/LONDON

Distributors for North, Central and South America:
Kluwer Academic Publishers
101 Philip Drive
Assinippi Park
Norwell, Massachusetts 02061 USA
Telephone (781) 871-6600
Fax (781) 871-6528
E-Mail <kluwer@wkap.com>

Distributors for all other countries:
Kluwer Academic Publishers Group
Distribution Centre
Post Office Box 322
3300 AH Dordrecht, THE NETHERLANDS
Telephone 31 78 6576 000
Fax 31 78 6576 474
E-Mail services@wkap.nl>

Electronic Services <http://www.wkap.nl>

Library of Congress Cataloging-in-Publication

Tunnel engineering handbook / edited by John O. Bickel, Thomas R.
Kuesel and Elwyn H. King. -- 2nd ed.
p. cm.

ISBN-13: 978-1-4613-8053-5 e-ISBN-13: 978-1-4613-0449-4
DOI: 10.1007/978-1-4613-0449-4

1. Tunneling. I. Kuesel, T.R. II. King, Elwyn H.
TA805.T82 1996 96-17571
624.1'93--dc20 CIP

Cover photo of Glenwood Canyon Tunnels, Colorado:
© David Sailors, courtesy of Parsons Brinckerhoff
Cover design: Curtis Tow Graphics

Printed on acid-free paper.

This printing is a digital duplication of the original edition.

Dedicated in memory of

John O. Bickel
1896–1991

Parsons Brinckerhoff Quade & Douglas, Inc.

Partner 1954–1968
Associated Consultant 1968–1991
Principal Tunnel Engineer 1932–1991

Contents

Preface

The first edition of the *Tunnel Engineering Handbook* was conceived and personally guided by John O. Bickel. It was a labor of love and perseverance, distilling the experience of a 50-year career in tunnel engineering. When it was published in 1984, John was keenly aware of its deficiencies and imperfections. Nonetheless, he was impelled to release it for publication by the knowledge that it filled a vacuum in engineering literature. At that time, no text covered planning, design, construction, and operation of all types of tunnels—soft ground, hard rock, cut-and-cover, and immersed tubes (or "sunken tubes" in the convention of the time).

Almost from the time of original publication, John set out to organize a second edition, which could correct the deficiencies of the first. In the 10 years that this project has gestated, there have been many advances in tunneling. It is the intent of this edition to reflect these advances, as well as to amplify the coverage of areas that were omitted or slighted in the first edition and to update the previous material that remains pertinent.

Accordingly, the second edition includes eight completely new chapters—Tunnel Stabilization and Lining, Tunneling in Difficult Ground, Deep Shafts, Water Conveyance Tunnels, Small-Diameter Tunnels, Fire Life Safety, Tunnel Rehabilitation, and Tunnel Construction Contracting. The original two chapters on soft ground tunneling and shield tunnels have been merged into one, as have the two on cut-and-cover and subway construction. All the remaining chapters have been updated, and most have been extensively rewritten.

The title remains *Tunnel Engineering Handbook,* but John always recognized that you could not "engineer" a tunnel properly without considering how it might be constructed and for what purpose it was intended. So the first five chapters cover matters that are primarily the concern of the tunnel designer. The next twelve treat the wide spectrum of tunnel construction methods, but all with relevance to the matters a tunnel engineer needs to understand and consider in the layout and design of a tunnel project. The next seven chapters deal with the operating systems for transportation tunnels—all the things needed to transform a hole in the ground into a useful, convenient, and safe public facility.

Tunnels age, even as do tunnel engineers. But the life of a tunnel frequently extends beyond a human life span, and so a chapter has been added on tunnel rehabilitation, to discuss how old tunnels may be rejuvenated, and how their useful lives may be extended.

Finally, although this book is not about tunnel construction contracting, a short chapter has been added to explain the evolution and philosophic basis for some unique provisions of modern tunnel construction contracts.

This book was written for a broad spectrum of readers, ranging from engineers seeking technical guidance to owners and other decision makers hoping to glean a better understanding of alternatives. To serve this audience better, we have opted not to include an index in the second edition. Instead, we have prepared an annotated table of contents, which provides a detailed guide to the many subjects contained herein. In the text itself, the authors have included cross-references to others chapters as appropriate. It is out hope that this combination will help readers locate information more quickly than a traditional, keyword-based index would.

The preparation of this book has involved the dedication and perseverance of many individuals. It could not have been completed without the unflagging support of Parsons Brinckerhoff, and especially the encouragement and patience of its president, James L. Lammie, who stuck with the enterprise when it seemed becalmed or lost. We appreciate his generating the wind that filled our sails and finally brought us into port. But we would never have made it without the tireless and skillful production editing of Nellie Negrin Finnegan and Karen Tongish, who corralled our distracted and procrastinating chapter authors, coaxed and badgered their manuscripts, and converted a huge pile of raw drafts into a coherent, readable text.

John Bickel did not survive to complete the work on the second edition, but he left a strong beacon that has lighted the way for his successors. All of us who have labored on John's legacy have striven to uphold the high standards to which he held us. This book is a memorial to his inspiration, and a tribute to his vision.

Tunneling brings man into confrontation with the infinite variety and complexity of nature. A professional career in tunneling leads to appreciation of several aphorisms:

- Nature is always smarter than some of us, and sometimes smarter than all of us.

- A little learning is a dangerous thing.

and, from a Chinese fortune cookie:

- Listen to advice, but make your own decisions.

This handbook endeavors to collect the best advice currently available from the most experienced professionals in the field of tunneling. It is in no sense to be treated as a cookbook or to replace the judgment of knowledgeable engineers with regard to specific applications to specific projects. Neither the editors, the chapter authors, nor the publishers assume any liability for application or misappl cation of any material in this book to any public or privat undertaking, nor do they warrant or guarantee the "correc ness" of any statements or opinions expressed herein, in an specific situation.

Caveat emptor.

Thomas R. Kuese
Charlottesville, Virgini

Elwyn H. Kin
San Francisco, Californi

1

An Introduction to Tunnel Engineering

Elwyn H. King
Principal Professional Associate (retired),
Parsons Brinckerhoff Quade & Douglas, Inc.
Thomas R. Kuesel
Consulting Engineer, Chairman Emeritus,
Parsons Brinckerhoff Quade & Douglas, Inc.

A tunnel is much more than just a tunnel. It serves any of myriad functions—highway, railroad, or rapid transit artery; pedestrian passageway; fresh water conveyance, cooling water supply, wastewater collector or transport; hydropower generator; or utility corridor. Tunnels are constructed by cut-and-cover methods; in long, prefabricated sections sunk in place as in immersed tubes; in short prefabricated sections pushed into place from jacking pits; by drilling and blasting; by mechanized means such as tunnel boring machines or continuous miners (roadheaders), with the aid of a protective shield in free or compressed air; and they will eventually be constructed in ways now existing only in our imaginations. In cross section it takes one of several shapes—circular, multicurve, horseshoe, cathedral arch, arched, or flat-roofed, and with clear spans of from a few feet to more than 50 ft and, in cavern form, much wider. Its length can vary from less than 100 ft to more than 30 miles. A tunnel can be located in any of a variety of places—under mountains, cities, rivers, lakes, sea estuaries, straits, or bays. Finally, a tunnel is constructed in one of innumerable media—soft ground, mixed face, rock, uniform, jumbled, layered, dry, wet, stable, flowing, squeezing.

Most of all, a tunnel exists because there is demonstrated need—to move people or material where no other means is practical or adequate, or to accomplish the required movement more directly, more quickly, or less obtrusively. The need may be for storage, either short term as for storage of stormwater flows to reduce the otherwise high peak capacities required of wastewater treatment plants, or longer term as for storage of vital raw materials or products.

TUNNEL ELEMENTS

A tunnel also is more than just a hole in the ground to provide for a desired movement of people or material. To accomplish the movement satisfactorily, one or more of a variety of facilities in simple or complex form must be provided in addition to the continuous space. The most obvious need for highway tunnels is ventilation (Chapter 20). Fresh air must be supplied in proportion to tunnel usage. The air also must be of reasonable purity; this requires constant monitoring for pollutants and consequent adjustment of the air supply and exhaust.

Another essential for highway tunnels is some form and degree of lighting (Chapter 21). In low-use rural tunnels, the vehicles' headlights may suffice. In high-use (generally urban) tunnels, a sophisticated system of high-level, adjustable lighting is necessary both for safety and for ensuring maximum appropriate speeds.

While ventilation and lighting are the most obvious of day-to-day tunnel operational needs, there is a greater need. That this need may arise only once every 10, 25, or 50 years is unimportant because of the possible consequences when it does. All tunnels used for transport of people must have adequate fire life safety provisions (Chapter 19). The need has always been obvious in rapid transit tunnels, but it has only recently been recognized for more spacious highway tunnels.

Fire life safety, ventilation, and lighting are also important to railroad and transit tunnels, although in quite different degree. Lighting requirements are much reduced, needed more for orientation and maintenance than for passage. However, fire life safety requirements are more rigorous due to the greater number of people transported. BART (the San Francisco Bay Area Rapid Transit System) gained recognition for the design of close, parallel twin tunnels with appropriately spaced and equipped connecting passageways, now accepted as the best possible layout for providing refuge from a raging fire or deadly smoke. Ventilation takes a somewhat different form because a train's piston effect can handle a large part of the normal ventilation requirement, especially for transit tunnels and short rail tunnels. For transit tunnels, then, emergency ventilation becomes more important. On long railroad tunnels with diesel locomotives and a high volume of freight traffic, the tunnel ventilation capacity may constrain the tunnel's hourly traffic capacity. The ventilation air flow rate governs the speed the locomotives can attain without overheating and the time required to purge diesel exhaust fumes before the next train can enter the tunnel. In some long tunnels, in order to increase the air flow past the train, unorthodox construction may be required, such as the trackway doors installed at portals at the Cascade, Flathead, and Moffet Tunnels, and at portals and at an intermediate shaft at the Mt. Macdonald Tunnel (Chapter 20).

Modern highway tunnels include elaborate traffic surveillance and control systems (Chapter 24), coordinated with emergency ventilation and lighting systems, and provisions for protected egress of motorists in the event of a fire and access for fire-fighting personnel. The types of surveillance and control for a highway tunnel depend to a considerable extent on the approach roads. Because a tunnel is confined space and presents psychological hazards, there is a need for more detailed knowledge of traffic conditions, better communication with drivers, and faster and more stylized response to emergencies than on the open road or on bridges.

DETAILS

Having considered the principal elements involved in a tunnel project, it is now appropriate to consider how a tunnel is actually designed and constructed. "Design" includes layout, location, ground conditions, and the theoretical and practical consideration of how the tunnel hole is kept open, temporarily during construction and permanently during operation and use.

The need for a well-defined and inclusive tunnel cross section with adequate spatial allowance for all necessary functions is obvious (Chapter 2). While most tunnels are constructed by one primary method, in some situations the best layout is a combination of different construction methods, and for long water crossings sometimes a combination of tunnels and bridges. The controls exerted on transportation tunnel layouts by operating considerations are discussed in Chapter 2, which also gives examples of hybrid tunnel and bridge–tunnel layouts. Determining the precise location of where a tunnel should be sited beneath the earth's surface and where it will actually be constructed (ideally, these are identical) is also an obvious need (Chapter 3). While most elements of tunnel design and construction can be the subject of lively arguments, there can be no argument about the need to determine the tunneling medium as accurately as practical (Chapter 4). The only limitation, given the tunnel engineers' insatiable appetite for more and better information, is the availability of time and money.

While the basic principles of tunnel engineering have remained unchanged, improvements in equipment and field techniques have taken place. Tunnel design procedures have changed dramatically in the last 30 years. Current or desirable philosophies and approaches are set forth in Chapter 5, which illustrates the great variety of tunnel structure types in common use and how their selection depends on ground conditions, construction methods, and end use. Chapter 5 also discusses how the design of tunnel structures differs fundamentally from that of aboveground structures, and it examines the pitfalls of mathematical analysis of tunnels.

Chapters 6 and 7 cover tunnel construction in soft ground and hard rock; the extremes and mixtures of these two basic types are frequently labeled "difficult ground," and they are covered in Chapter 8.

Chapters 9 and 10 cover access shafts for relatively shallow tunnels, and deep shafts for mining, water power, or storage projects.

The most revolutionary development in modern tunnel construction has been the rise of the tunnel boring machine (TBM). Chapter 11 gives a brief history of this development, and discusses how TBMs work and the special considerations required if a tunnel is to be constructed with one. The other major development of modern tunnel construction is the widespread use of shotcrete, which is covered in Chapter 12.

Most tunnel designers are quite innocent of any awareness of how tunnels are actually built. Chapter 13 seeks to illuminate this dark hole with a discussion of materials handling and the construction plant. Mastery of this field is acquired only by long practical experience, but all tunnel engineers should have some familiarity with the practical problems confronting a tunnel construction contractor.

The next three chapters cover three special classes of tunnels—immersed tubes (Chapter 14), water conveyance (Chapter 15), and small diameter (Chapter 16)—and the special considerations involved in their design and construction.

A wholly different approach to tunneling, to dig the tunnel out from the surface (cut-and-cover construction), is discussed in Chapter 17. Although much recent emphasis has been on trying to construct tunnels without disturbing the ground surface and the existing structures thereon, there are still many situations in which cut-and-cover is the appropriate technology, and sometimes the only feasible method. Many advances in this ancient technology have been made in recent years.

Tunneling makes use of high-technology equipment and advanced theory, but it is ultimately performed by human beings. No treatment of tunneling would be complete without special emphasis on safety, both during construction and in operation and public use. These subjects are covered in Chapter 18 (Safety Provisions) and in Chapter 19 (Fire Life Safety).

The next seven chapters are devoted to the operating systems that are needed to convert a hole in the ground into a transportation tunnel—ventilation, lighting, electric power supply and distribution, water supply and drainage, surveillance and control, finishes, and service buildings and ancillary spaces.

The first 26 chapters deal with new tunnels. But tunnels characteristically have very long service lives, and they are rarely abandoned because they give out. (A rare exception is the Baytown Tunnel under the Houston Ship Channel, which was replaced by a bridge after 40 years. Area subsidence resulting from the pumping of groundwater and hydrocarbons caused a relative rise in sea level, which threatened to flood the portals.) Chapter 27 treats tunnel rehabilitation, a developing art that bears witness to the enduring nature of the work of tunnel engineers.

It is appropriate to end this introduction by stating that one book cannot cover all aspects of tunneling in depth. Readers are unlikely to find new material in their own areas of expertise, but may well gain in understanding of a fellow engineer's field. While the heavy construction items are generally confined to basics, the subjects relating to operations are frequently treated in their most sophisticated aspect so that the reader will know what is possible. It is hoped that potential tunnel owners will gain an appreciation of the diversity of problems that tunnel engineers must address, and that potential tunnel engineers will obtain an overall view of the field.

2 Tunnel Layout

Elwyn H. King,
Principal Professional Associate (retired),
Parsons Brinckerhoff Quade & Douglas, Inc.
Thomas R. Kuesel
Consulting Engineer, Chairman Emeritus,
Parsons Brinckerhoff Quade & Douglas, Inc.

This chapter covers considerations of internal clearances and overall alignment that are common to all transportation tunnels, including highway, railway, and rapid transit. Water tunnels are governed by different considerations, discussed in Chapter 15. Also covered here are limitations on the tunnel layout imposed by operating requirements and by factors inherent in certain construction methods. Finally, alternative layout concepts for underwater tunnels, which generally admit a greater variety of approaches than land tunnels, are discussed.

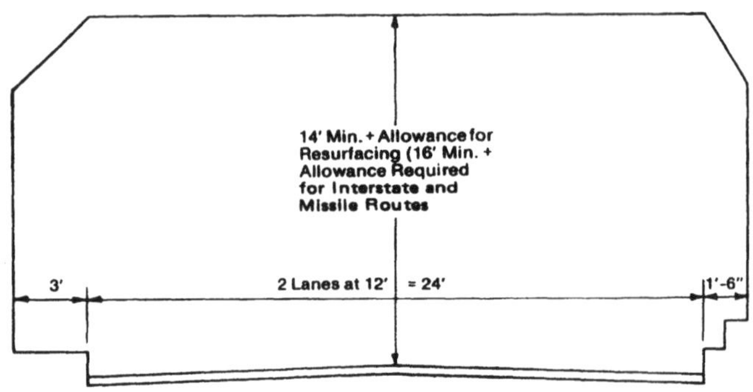

Fig. 2-1. ASHTO clearances for a two-lane primary highway.

CLEARANCES FOR HIGHWAY TUNNELS

Standards for lane and shoulder width and vertical clearance for highways have been established by the Federal Highway Administration (FHWA) and by the American Association of State Highway and Transportation Officials (AASHTO) according to classification. Figure 2-1 shows AASHTO tunnel clearances for a two-lane primary highway. For an additional lane, the width should be increased by at least 10 ft, preferably by 12 ft. For state trunk highways and interstate systems, the vertical clearance should be at least 16 ft over the entire width of roadway, plus an allowance for resurfacing. For curved alignments with superelevations, clearances may be increased to provide for overhang.

In addition to the width of traveled lanes, it is customary to provide left and right shoulders, flush with the pavement surface. The left shoulder primarily provides a "shy distance" to prevent motorists from moving away from the tunnel side wall. The right shoulder may also provide a refuge area for disabled vehicles. Shoulder widths in tunnels are generally narrower than on open highways, both to conserve costs and in recognition that emergency towing service is generally available on heavily used tunnels. Horizontal clearances on curved tunnels may need to be increased to provide sight distance past the tunnel wall. In many older tunnels, particularly two-lane sections with two-way traffic, raised maintenance walks were provided for tunnel operation personnel. As closed-circuit television (CCTV) camera surveillance has replaced foot patrols, maintenance walks have generally been eliminated, and lanes are closed when maintenance access is required.

In addition to the prescribed vertical clearance for vehicle operation, vertical space must frequently be provided for tunnel lighting, traffic lights, signs, and sometimes ventilation fans.

In horizontally curved tunnels, provisions must be made to accommodate superelevation of the roadway, and superelevation transitions at the ends of alignment curves.

ALIGNMENT AND GRADES FOR HIGHWAY TUNNELS

Alignment

Alignment should be straight, if possible. If curves are required, the minimum radius is determined by stopping sight distances and acceptable superelevation in relation to design speed (Table 2-1). Where shoulders are narrow, horizontal sight distance may be restricted by the proximity of the tunnel sidewall.

Table 2-1. Sight Distances for Stopping

Design speed (mph)	Sight distance (ft)
30	200
40	275
50	400
60	525
70	625

Passing distances do not apply, since passing in tunnels is not permitted.

Tunnels on Interstate Highways

Design speed cannot be less than 60 mph unless otherwise restricted in urban areas; the minimum radius of curvature preferably should not be less than 1,500 ft. Other tunnels should be designed for the speeds governing the approach highways according to state or local regulations.

Curvature and superelevation should be correlated in accordance with AASHTO "Policy on Geometric Design of Rural Highways" (AASHTO, 1989). Superelevations should not exceed 10 to 12% on sharply curved tunnels, but a maximum of 6% is preferred for 60-mph operation. For wide roadways, high superelevation rates lead to long superelevation transition runouts, which are undesirable.

Grades

Upgrades in tunnels carrying heavy traffic are preferably limited to 3.5% to reduce ventilation requirements. For long two-lane tunnels with two-way traffic, a maximum grade of 3% is desirable to maintain reasonable truck speed. For downgrade traffic, 4% or more is acceptable. For lighter traffic volumes, grades up to 5% or even 6% have been used for economy's sake. In immersed tube tunnels, grades between channel lines controlling navigation width and depth are at a minimum adequate for drainage to a low point, preferably not less than 0.5%. Lengths of vertical curves at grade changes are governed by design speed and stopping sight distance. Sight distance on sag vertical curves may be controlled by the tunnel ceiling.

CLEARANCES FOR RAILROAD TUNNELS

Although clearances may vary for individual railroads to suit the dimensions of their equipment, the minimum clearances for tunnels are identical to those recommended by the American Railway Engineering Association (AREA) for bridges on tangent tracks (Figure 2-2).

On curved track, the lateral clearances should be increased for the midordinate and overhang of a car 88 ft long and 62 ft between the centers of trucks, equivalent to 1 in. per degree of curvature. Clearances for superelevation are governed by AREA recommendations.

Additional clearance requirements for third rail, catenary, or pantographs should conform to the diagrams issued by the Electrical Section, Engineering Division, of the Association of American Railroads. For 25 kV catenaries, the control height is increased to 25 ft; for 50 kV lines, use 26 ft.

In all cases, the latest standards of AREA should be used for new construction, and special legal requirements and standards of using railroads should govern if exceeding these clearances.

Circular tunnels are fitted to the clearance diagrams with such modifications as may be acceptable.

ALIGNMENT AND GRADES FOR RAILROAD TUNNELS

Alignment

Where possible, from the standpoint of general alignment and cost, tunnel alignment should be straight to facilitate construction. Curved tunnels have been used on many railroads, particularly in mountainous areas, and railroad curves frequently have spiral transitions at their ends. (Both portals of the Mt. Macdonald Tunnel in British Columbia are located in the middle of spiral curves.)

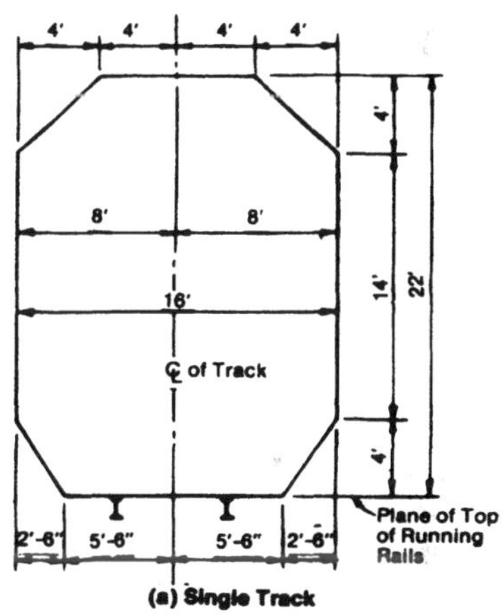

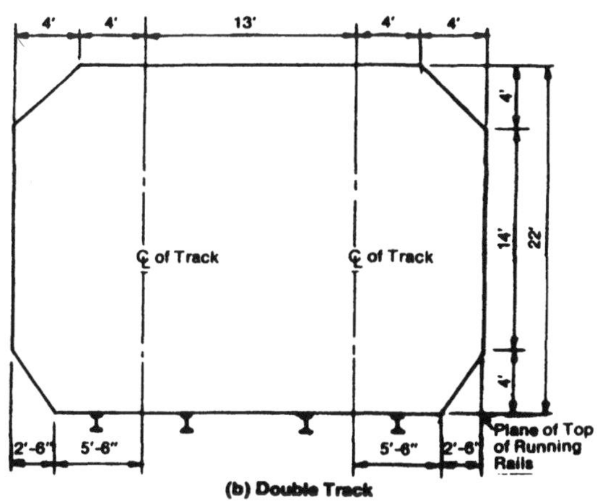

Fig. 2-2. Minimum clearances recommended for railroad tunnels (American Railway Engineering Association).

Radii of curvature and superelevation of track are governed by maximum train speeds of the particular railroad.

Grades

If possible, maximum grades in straight tunnels should not exceed 75% of the ruling grade of the line. This grade should extend about 3,000 ft before and 1,000 ft beyond the tunnel. Grades in curved tunnels should be compensated for curvature in the same manner as for open lines.

CLEARANCES FOR RAPID TRANSIT TUNNELS

Each rapid transit system establishes its own rolling stock, power supply system, and signal space.

Figure 2-3 shows the normal clearance diagram of the New York City Independent Subway System.

Figure 2-4 shows the clearances for the San Francisco Bay Area Rapid Transit System in the shield-driven circular tunnels for tangent line, and with superelevation on curves. The cars are 10 ft wide and 70 ft long on a 5-ft, 6-in. gauge track. Clearances allow for overhang of cars, for tilting due to superelevation and sway, and for a broken spring or defective suspension.

ALIGNMENT AND GRADES FOR RAPID TRANSIT TUNNELS

Operating requirements of the particular system govern the curvature and limiting grades of its lines. The New York City Independent Subway has a minimum radius of 350 ft with spiral transition curves for radii below 2,300 ft. The maximum grades for this system are 3% between stations and 1.5% for crossovers and turnouts.

The San Francisco Bay Area Rapid Transit System, generally designed for train speeds of 80 mph on its 5-ft, 6-in. gauge, but constructed with several curves with a restricted radius of 500 ft, determines the relationship among speed, radius, and superelevation of horizontal curves by:

$$E = \frac{4.65V^2}{R} - U \tag{2-1}$$

where

E = superelevation (in.)
R = radius (ft)
V = train speed (mph)
U = unbalanced superelevation, which should not exceed 2-3/4 in. optimum, or 4 in. as an absolute maximum

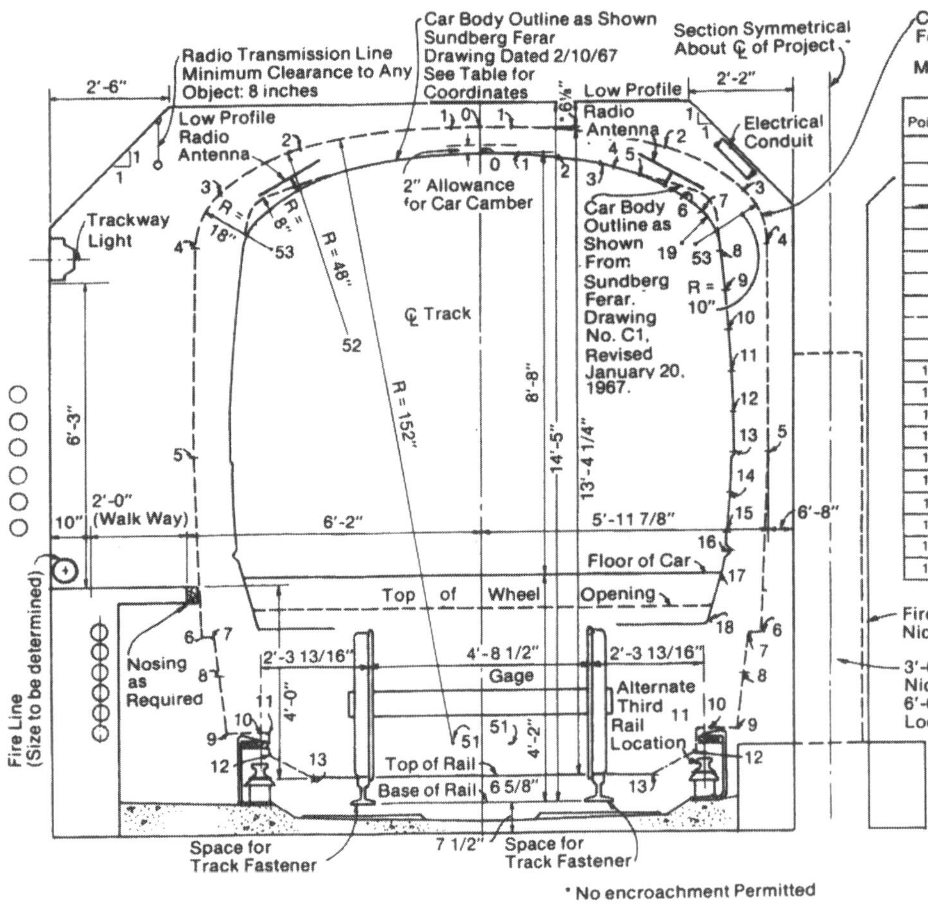

M.C.T.A. Car Outline Coordinates

Point	"X" From ℄ of Car	"Y" From Top of Rail
0	0.	154.00
1	10.00	153.56*
2	20.00	152.63
3	30.00	150.94
4	33.70	150.00
5	40.00	148.06
6	50.00	143.31
7	54.55	140.00
8	59.75	130.00
9	60.56	120.00
10	61.19	110.00
11	61.84	100.00
12	62.47	90.00
13	63.00	80.00
14	62.38	70.00
15	61.56	60.00
16	61.80	56.00
17	60.00	50.00
18	56.00	38.50
19	49.35	130.75

M.C.T.A. Car Clearance Line Coordinates

Point	"X" From ℄ of Car	"Y" From Top of Rail
0	0.	160.23
1	7.12	160.23
2	45.03	155.42
3	65.73	144.11
4	71.48	130.92
5	71.86	79.83
6	69.97	35.93
7	66.97	35.93
8	66.00	25.37
9	64.00	11.00
10	56.25	11.00
11	53.25	11.00
12	53.25	5.48
13	42.00	0.
[1] 51	7.12	8.23
[2] 52	33.06	108.94
[3] 53	53.48	130.92

1 Center of Arc From Point 1 to Point 2, Radius = 152"

2 Center of Arc From Point 2 to Point 3, Radius = 48"

3 Center of Arc From Point 3 to Point 4, Radius = 18"

All Dimensions Are in Inches

Fig. 2-3. Normal clearance diagram of the New York City Independent Subway System.

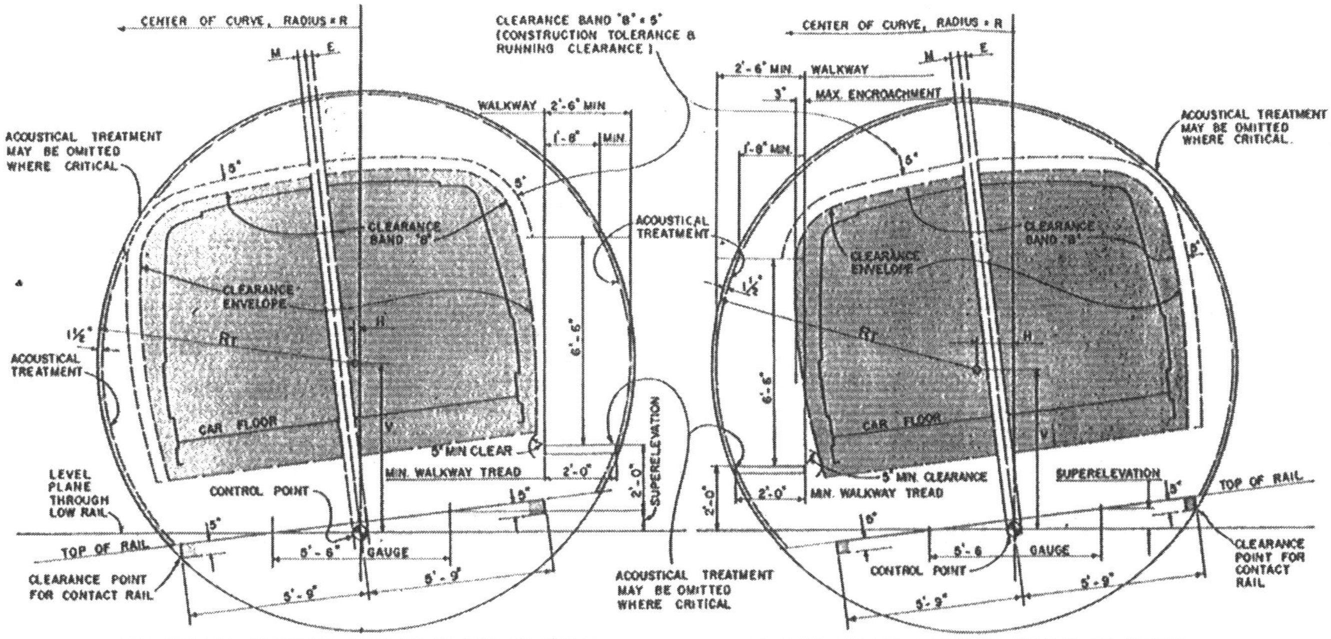

WALKWAY ON OPPOSITE SIDE TO CENTER OF CURVE

R TRACK RADIUS OF CURVATURE ⇨		TANGENT TO 1500'		1499' TO 500'	
RT RADIUS TO STRUCTURAL SURFACE ⇨		8'-3"		8'-6"	
Es SUPERELEVATION ⇨		0	6¾"	0	6¾"
TUNNEL CENTER LOCATION WITH RESPECT TO CONTROL POINT	H	→6"	←2"	→8"	0
	V	4'-11"	5'-3"	5'-1"	5'-6"

⟶ ARROW INDICATES DIRECTION THAT TUNNEL CENTER IS OFFSET FROM TRACK CENTERLINE.

WALKWAY ON SAME SIDE AS CENTER OF CURVE

R TRACK RADIUS OF CURVATURE ⇨		TANGENT TO 1500'		1499' TO 500'	
RT RADIUS TO STRUCTURAL SURFACE ⇨		8'-3"		8'-6"	
Es SUPERELEVATION ⇨		0	6¾"	0	6¾"
TUNNEL CENTER LOCATION WITH RESPECT TO CONTROL POINT	H	←6"	←1'-2"	←8"	←1'-4"
	V	4'-11"	5'-2"	5'-1"	5'-6"

CLEARANCE DIAGRAMS FOR TUNNEL SECTIONS

(a)

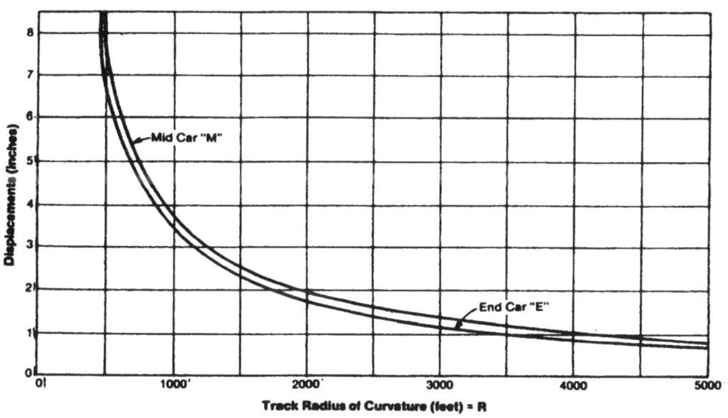

(b)

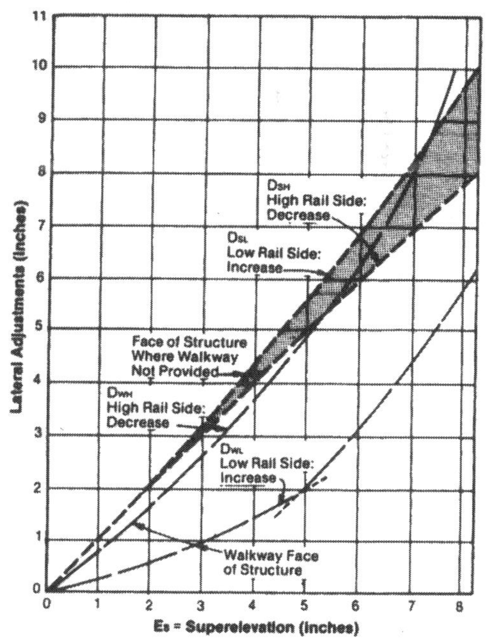

Note
Curves for D_{WH} & D_{WL} are for walkway tread 2'-0" above adjacent rail. Where walkway tread is at the level of adjacent rail or below, the curves for D_{SH} & D_{SL} shall apply.

(c)

Fig. 2-4. Clearances for the San Francisco Bay Area Rapid Transit System in the shield-driven circular tunnels.

At 80 mph design speed, the radius with an optimum superelevation would be 5,000 ft. For a maximum permissible superelevation of 8-1/4 in., the minimum radius required would be 3,600 ft.

For standard 4-ft, 8-1/2 in. gauge track, such as the Southern California Rapid Transit District's Metro Rail Project, the superelevation formula coefficient becomes 4.011, the absolute maximum unbalance for that system is 4.5 in., the desired maximum actual superelevation is 4 in., and the absolute maximum 6 in. Other Metro criteria require that for mainline tracks, whenever possible, the geometry is to accommodate the maximum design speed of 75 mph, the minimum to be 45 mph; desired minimum curve radius is 1,000 ft, absolute minimum 750 ft; desired minimum tangent constant profile grade, or circular curve length (ft) of 3 times the minimum design speed (mph) with absolute minimum of 75 ft; and 450-ft station platforms centered on 600-ft tangents. The maximum sustained grade is 3.0%; for short lengths, this can be increased but should not exceed 6.0%.

CONTROLS ON LAYOUT OF UNDERWATER TRANSPORTATION TUNNELS

The material in the following section has been adapted from a paper presented at a 1986 conference on undersea tunnels in Tokyo (Kuesel, 1986). The objectives of this section are (1) to consider alternative configurations and construction methods suitable for underwater transportation tunnels; (2) to examine the limitations on these configurations imposed by design and operating considerations, and by present construction capabilities; and (3) to illustrate the variety of conditions encountered through examples of completed and proposed undersea tunnel crossings.

The first underwater tunnel on record was built in Babylon, by simply diverting the river and building the tunnel on top of the dry river bed. If this is considered overly ambitious for most underwater crossings, three practical construction methods remain:

- To use mining methods (generally in rock), taking measures to control water inflow so that the work can proceed under normal atmospheric pressure
- To use shield methods (generally in soft ground), stabilizing the working face by various measures
- To use immersed tube methods, with prefabricated tube sections lowered onto a prepared foundation and joined underwater

The relative suitability of each of these methods will depend primarily on the hydrographic and geotechnical conditions of the project site.

Before considering construction limitations, it is appropriate to examine the limitations imposed by operating and other design considerations. These may be differentiated between rail and highway considerations.

Design Limitations

Rail Tunnels. For a rail facility, the primary limit is the allowable gradient of the profile. For surface mainline freight operations, the limit is frequently set at 1%. Maximum gradients for major rail tunnel projects are given in Table 2-2.

The limiting gradient depends on the length of the grade, the volume and schedule of freight and passenger traffic, and the economics of railroad operations. The Dutch have found it feasible to construct and operate rail tunnels under shallow canals at gradients up to 2.5%. Rapid transit tunnels can generally accommodate gradients of 4%.

For electric power operation, ventilation of railway tunnels is not generally a design control. For diesel operations, it is necessary to purge the diesel exhaust fumes after passage of each train, and also to ensure adequate air flow past the locomotives to prevent overheating. This can be accomplished with a system of fans and tunnel doors that permits control of the volume and direction of air flow when the tunnel is occupied by a train. Diesel ventilation considerations impose a somewhat flexible limit on the spacing of shafts. For the Mt. Macdonald Tunnel, a maximum spacing of 5.5 mi (8.9 km) is provided.

Highway Tunnels. For highway tunnels, gradients up to 6% have been used successfully. For heavy traffic volumes, it is desirable to adopt a limit of 4% upgrade because the ventilation requirements rise rapidly for higher gradients.

Ventilation is a prime consideration in the layout of highway tunnels. The longest existing lengths between ventilation points pertain to mountain tunnels with generally free-flowing and light to moderate traffic volumes (see Table 2-3).

Proposals for mined highway tunnels for the English Channel Crossing included the Channel Expressway project, with a length of 11.6 mi (18.75 km) between ventilation shafts, and the Euroroute project, with a length of 6.4 mi (10.3 km) between shafts.

The Kan-Etsu and Enasan Tunnels in Japan, which have relatively heavy traffic, both have intermediate recirculation stations between ventilation shafts, where particulates are removed from the air by electrostatic precipitators. In general, tunnels with lengths between ventilation points exceeding about 2 mi (3 km) require special ventilation measures if they carry substantial traffic volumes.

Table 2-2. Maximum gradients for major rail tunnel projects

Tunnel	Gradient (%)
Seikan Tunnel, Japan	1.2
Kanmon Tunnel, Japan	2.2
Shin-Kanmon Tunnel, Japan	1.8
English Channel Tunnel	1.1
Mersey Tunnel, England	3.7
Severn Tunnel, England	1.1
Mt. Macdonald Tunnel, Canada	0.7
Bosphorus Tunnel, Turkey (proposed)	1.8

Table 2-3. Maximum distance between ventilation points in mountain highway tunnels

Tunnel	Location	Maximum length between vent points (ft)	Total length portal to portal (ft)
Mont Blanc	France-Italy	38,050	38,050
St. Gotthard	Switzerland-Italy	18,560	53,540
Arlberg	Austria	14,760	45,850
Frejus	France-Italy	14,210	42,020
Kan-Etsu	Japan	12,250	35,840
Seelisberg	Switzerland	12,140	30,180
Enasen	Japan	10,500	28,010
Eisenhower	United States	8,950	8,950

For heavily traveled urban highway tunnels, ventilation requirements become more onerous. Examples of long lengths between ventilation points on existing underwater urban highway tunnels are shown in Table 2-4.

The Tokyo Bay Tunnel is to have one intermediate ventilation point in a length of 5.9 mi (9.5 km).

The electric power required to ventilate a tunnel of given total length varies in proportion to the sums of the fourth power of the distances between vent points. Therefore, the maximum distance between vent points can be an important factor in the evaluation of operating costs and economic feasibility. The power requirement also varies with the cube of the traffic volume, which is why ventilation lengths can be longer for rural tunnels than for urban tunnels.

The governing consideration in the design of tunnel ventilation systems (particularly for highway tunnels) is likely to be provisions for control of tunnel fires. This matter also involves provisions for emergency access and egress, and it will weigh heavily in the layout and design of any undersea tunnel. Historically, greater concern has been given to tunnel fire provisions in the United States than in other countries. However, in recent years the matter has received increased attention in Japan and Europe.

Highway tunnel ventilation is discussed in depth in Chapter 20.

Construction Limitations

Mined Tunnels. Turning to construction limitations, mining methods generally require that the work be executed in free air. The most important limitation on mining methods is, therefore, the capability to control the inflow of groundwater into the tunnel heading. This is accomplished first by seeking the most favorable geology for the tunnel alignment and profile. Because it is exceptionally difficult to determine from the surface the location and extent of local geological defects buried beneath the sea, mined tunnels frequently include pilot headings and extensive probing ahead of the working face to locate leaking joints and pervious zones. Groundwater flowing into undersea tunnels is generally under high pressure and has an inexhaustible reservoir. For these reasons, control by drainage and pumping or by raising the internal air pressure is not effective; therefore, the primary method of groundwater control is grouting. To be effective, grouting should be performed ahead of the pilot tunnel face, but this interferes with the progress of tunnel excavation and greatly increases its duration and cost. The Seikan Tunnel, which connects the Japanese islands of Honshu and Hokkaido, illustrates both the great range of geologic defects that may be dealt with by these methods and the great cost of doing so.

Where geological conditions are relatively well known and the rock is sufficiently impermeable that a major grouting program is not required, mining by full-face rotary tunnel boring machines may be considered. The advantages are more rapid progress and reduced cost if favorable ground conditions persist; the major disadvantage is vulnerability to major delays if serious water-bearing zones are encountered.

To minimize the risks of unexpected encounters with major water flows, exploratory or pilot tunnels may be driven in advance of the main headings. These may also serve as grouting galleries during construction, and as service, ventilation, and emergency access/egress galleries in the completed tunnel. The Mersey Road Tunnel in England used a pilot tunnel, which was enlarged to full section with a boring machine.

Where short lengths of difficult ground are encountered, they may be dealt with by freezing the ground. The time and cost required to implement a freezing program usually make the method uneconomical as a general construction system.

Other than economic considerations, there are no inherent limits on the depth of mined rock tunnels. At great depth and in some geologies, ground heat may become a problem, but this can generally be controlled by increased ventilation, at an appropriate cost. Although sedimentary rocks bearing small amounts of gas may be dealt with similarly, major gas-bearing formations are likely to preclude the economical use of mining methods.

Table 2-4. Maximum distance between ventilation points in underwater urban highway tunnels

Tunnel	Location	Maximum length between vent points (ft)	Total length portal to portal (ft)
Hampton Roads	Norfolk	7,150	7,500
Baltimore Harbor	Baltimore	6,300	7,640
Fort McHenry	Baltimore	5,710	7,200
Lincoln	New York	5,410	8,220
Mersey Kingsway	Liverpool	4,300	7,380
Brooklyn-Battery	New York	3,940	9,120
Holland	New York	3,380	8,300

Shield Tunnels. Shield methods for subaqueous tunnel construction are generally limited to shallow depths under estuaries and coastal sites. They are most suited to sites with deep deposits of moderately soft, impervious clay. Under such conditions the excavation may be accomplished in free air, although it may be necessary to pressurize the face of the tunnel heading if the clay is very soft and highly stressed and tends to squeeze into the tunnel.

Shield tunnels through pervious ground, or through generally impervious ground in which pervious lenses or zones may be encountered, are generally limited to the depth at which the groundwater pressure may be balanced by internal air pressure. Medical limits for worker safety give an absolute limit of about 3 atm, corresponding to 90 ft (30 m) below the water surface. However, if these conditions exist over a substantial length, the economic limit is likely to be nearer 60 ft (20 m).

Development of pressurized-face tunnel boring machines has made use of general compressed-air tunneling largely obsolete. However, the present limits on the capacity of shield tail seals, and the need to provide access to the face to deal with obstructions or to effect repairs, indicate that the above limits on shield tunnels in pervious ground cannot be greatly extended. Nonetheless, it should be noted that the Tokyo Bay tunnel construction is using pressurized face shield tunnels at a depth of 160 ft (50 m), and Denmark's Great Belt Railway Tunnel boring machines reached a depth of 240 ft (75 m) below the sea surface. Both of these tunnels are situated in relatively impervious strata at their deepest points.

Immersed Tubes. Immersed tubes are generally suited to crossings of soft-bottomed estuaries in which trenches may be excavated by floating equipment. The deepest trench for an existing tunnel extends 135 ft (41 m) below sea level, for the San Francisco Trans-Bay Tube (Figure 2-5). A hard rock bottom, particularly if it has an irregular or steep profile, will increase the cost of trenching substantially, but short sections of rock trench may readily be accommodated.

Immersed tubes are usually buried beneath the level of the existing sea bed, although short projections above the bottom across deep natural trenches may be mounded over for protection.

In a departure from precedent, the proposed Euroroute bridge–tunnel crossing of the English Channel would have only partially embedded the tubes in a shallow trench dredged in soft chalk rock. The maximum trench depth proposed was 200 ft (60 m) below sea level.

Studies for an immersed tube crossing of Denmark's Great Belt proposed a trench extending to 160 ft (50 m) below sea level.

Other limitations on immersed tube construction include the following:

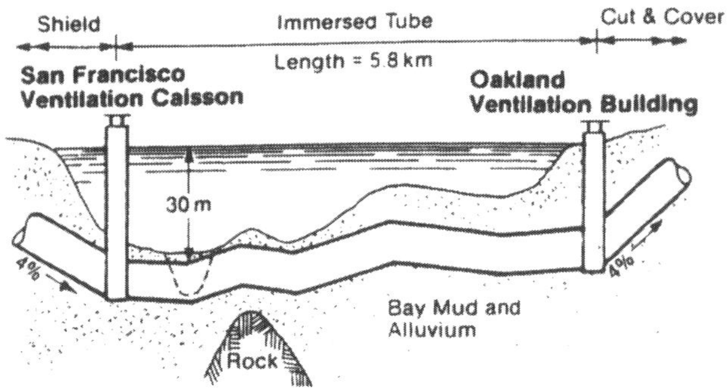

Fig. 2-5. San Francisco Trans-Bay Tube.

- There must be sufficient duration of slack tidal current to permit lowering the tube—preferably less than 3 ft/sec (1 m/sec) over a duration of 2 hours.
- The bottom must not be so soft and unstable that the trench cannot be kept open.
- The site must be reasonably free from rapid deposition of fluid silts, which can alter the density of water in the trench and affect the balance of buoyancy at tube placement.

Floating Tunnels. Although floating tunnels have been proposed for several deep water crossings, none have been constructed. The concept is similar to that of the immersed tube, but in this case the tube sections have positive buoyancy and are tethered at an appropriate depth below the sea surface by cables anchored into the sea bed.

The first serious proposal for a floating tunnel was made in 1940 by Charles Andrew, who made a preliminary design for a crossing of the 800-ft-deep (250 m) Puget Sound at Seattle, Washington (see Figure 2-6). A similar proposal was studied for the proposed Messina Straits crossing in 1993, and a somewhat different layout was seriously considered for a crossing of the 1,600-m-deep Eidfjord in Norway in 1979.

Tubes Supported Above the Sea Bed. Designs of conventional negatively buoyant immersed tubes on pile or pier foundations projecting above the sea bed have also been proposed. All designs of tubes above the sea bed are exposed to the hazards of currents and storm waves and the risk of sinking vessels and dragging anchors or tackle.

Floating tunnels risk the additional hazards of loss of buoyancy through marine growth accretions, and of anchor cable corrosion and anchorage block scour.

Representative Tunnels—Existing and Proposed

The choice between alternative methods is greatly dependent on hydrographic and geological site conditions. At most sites, the conditions vary considerably along the length of the crossing, and a layout involving combinations of different methods may be advantageous. These may include different tunneling methods or bridge–tunnel combinations. The possibilities are best illustrated by examples of existing works.

Tunnels wholly in rock are exemplified by the Seikan Tunnel (Figure 2-7). Other notable examples are the Kanmon Tunnel in Japan, the English Channel Tunnel, and the

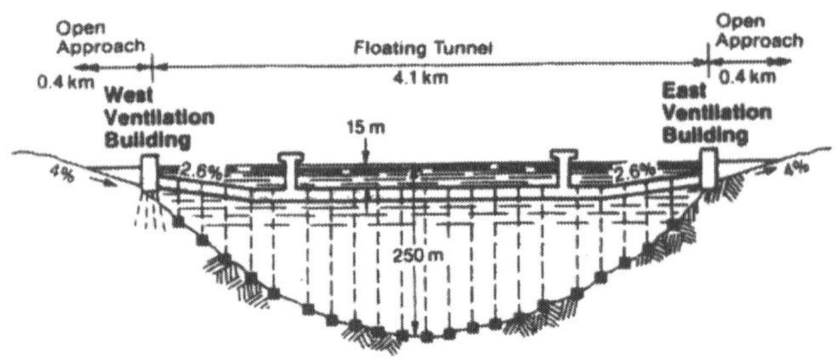

Fig. 2-6. Puget Sound floating tunnel (proposed).

Mersey and Severn Tunnels in England. There are many shield tunnels under shallow estuaries and harbors. Perhaps the greatest collection exists in New York City, where there are 39 individual shield-driven bores under the Hudson, East, and Harlem Rivers. Figure 2-8 shows the locations of the transportation tunnels connecting to Manhattan Island in New York City. Many of the shield tunnels connect to rock tunnel approaches and involved difficult mixed-face work, generally accomplished under compressed air.

The use of immersed tubes is exemplified by the San Francisco Trans-Bay Tube (Figure 2-7). There are many other examples, including several in Tokyo, four in Hong Kong, and one in Sydney, Australia, as well as numerous projects in northern Europe.

As an example of floating tunnel proposals, Figure 2-8 shows Andrew's 1940 layout for a crossing of Puget Sound. As mentioned, similar proposals have been made for the Messina Straits and Norwegian fjords.

Figure 2-9 illustrates a hybrid tunnel, the 63rd Street Tunnel in New York City, in which two deeply eroded valleys in a rock region were accommodated by two immersed tube sections, joined to rock tunnels at both ends and through an intervening rock ridge island. A similar layout was developed for the proposed Bosphorus Rail Tunnel in Istanbul, Turkey.

A combination of immersed tubes and shield tunnels is illustrated by the Detroit–Windsor Tunnel between the United States and Canada (Figure 2-10). This combination was facilitated by the deep deposits of impermeable clay that characterize the site.

A combination of bored tunnels and bridges, connected by a portal island, is provided by the Trans-Tokyo Bay Tunnel (Figure 2-11). The first bridge–tunnel crossing was the Hampton Roads Crossing in Virginia (Figure 2-12), completed in 1957. By building two portal islands at the edges of the deep central channel, and connecting them to shore with fixed bridges, the tunnel length was reduced from 3.5 mi (6 km) to a little more than 1.4 mi (2 km), with consequent great reductions in ventilation problems and project costs. A similar layout was subsequently used for the 17-mi (28-km) Outer Chesapeake Bay Crossing.

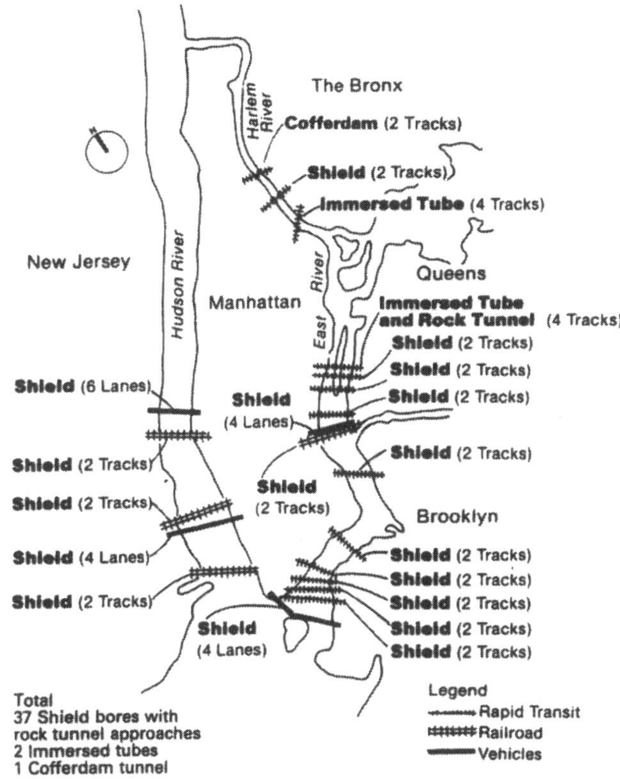

Fig. 2-8. New York City subaqueous tunnels.

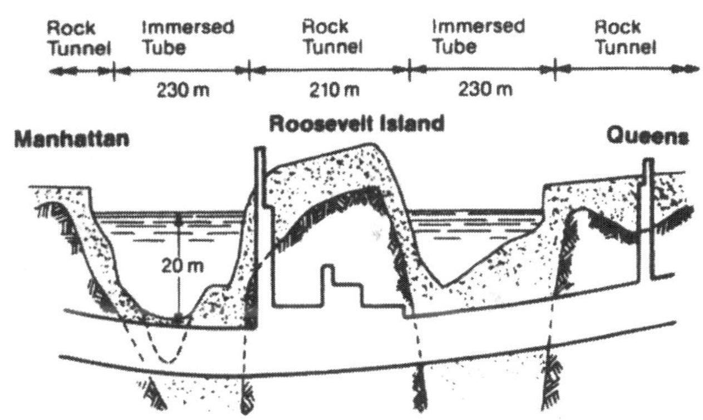

Fig. 2-9. New York City 63rd Street Tunnel.

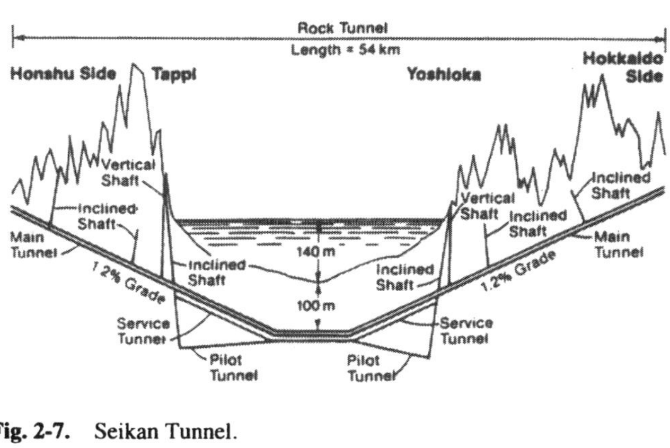

Fig. 2-7. Seikan Tunnel.

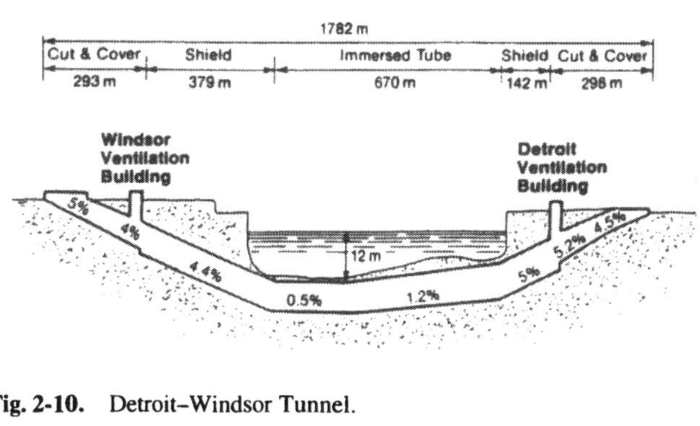

Fig. 2-10. Detroit–Windsor Tunnel.

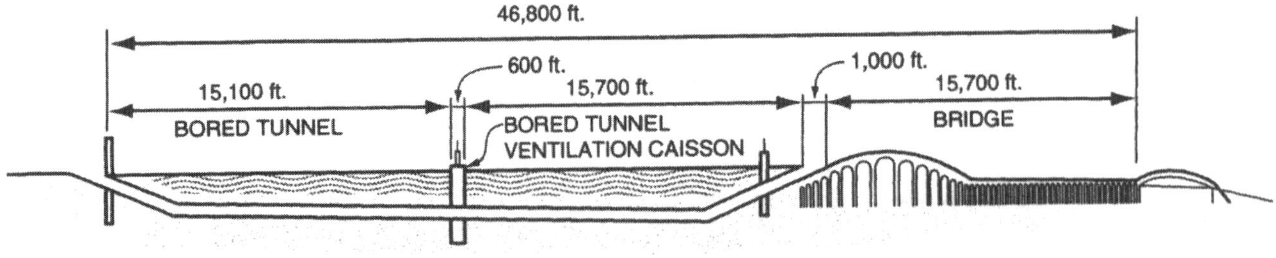

Fig. 2-11. Trans-Tokyo Bay Tunnel.

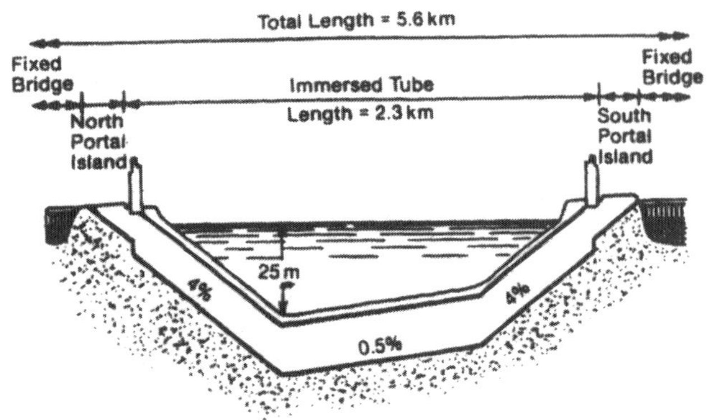

Fig. 2-12. Hampton Roads Bridge–Tunnel.

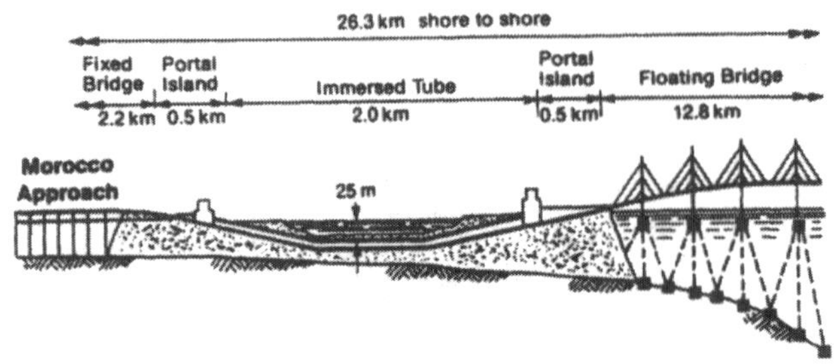

Fig. 2-13. Gibraltar Bridge–Tunnel (proposed). The Moroccan side is shown; the Spanish side is similar.

A different concept for a bridge–tunnel crossing was presented to the Madrid Colloquium on the proposed Gibraltar crossing in 1982 (Kuesel, 1982). Adapted to the particular site conditions of this project, the concept is shown in Figure 2-13. Two immersed tube tunnels, catering to separate European and African shipping channels, together with four portal islands, are provided in moderately deep water. The shallow-depth sections of the crossing adjacent to both shores are covered by fixed bridges, and the central deep gorge by a floating bridge structure.

As can be seen from this brief catalog of existing and proposed projects, the variety of solutions to undersea tunnel projects is matched only by the variety of site conditions to which they must be adapted. No one concept is superior for all conditions. For each new proposed project, the full range of possible alternatives should be considered in order to develop the best solution.

REFERENCES

AASHTO (1989) *Policy on Geometric Design of Rural Highways*, American Association of State Highway Transportation Officials, Baltimore, Md.

KUESEL, T.R. (1982) "A Bridge–Tunnel Crossing for the Strait of Gibraltar," *Proceedings of the Coloquio Internaçional sobre le factilibad de una comunicaçión fija a través del Estrecho de Gibraltar, Madrid 9 al 13 Noviembre, 1982* (ISBN 84-500-8985-7, p. 505).

KUESEL, T.R. (1986) "Alternative Concepts for Undersea Tunnels," *Tunneling and Underground Space Technology*, Vol. 1, No. 2/4.

3 | Tunnel Surveys and Alignment Control

Peter K. Frobenius,
Project Manager, Bechtel Corp.
William S. Robinson,
President (Retired), Towill, Inc.

Sputnik 1, launched October 4, 1957, heralded a revolution in technology that has continued to expand at an ever-increasing rate. The development of electronic computation and miniaturization of computer components, together with advances in communication and satellite technology, have significantly affected the surveyor, providing instruments and techniques that greatly enhance the accuracy and production rate of surveys for all aspects of planning, design, and construction. In the recent past, requirements for field and office work often had to be tempered by the cost and time needed to accomplish the survey and prepare computations and maps, because of the amount of labor needed to accomplish the work. Today, the most sophisticated technologies of the "Space Age" have been directed toward the surveying industry, resulting in the ability to precisely determine survey positions by satellite and computer technology; acquire and record survey data with minimal chance of human error; conduct sophisticated calculations to determine the most probable answer to complex data sets; and to plot, store, retrieve, transmit, and interpret survey information to a unprecedented level of efficiency and accuracy. The ability of the engineer, surveyor, and constructor to utilize this technology has resulted in dramatic advances in the surveying technology needed for all phases of tunnel design and construction.

Surveying is an integral part of a tunnel project from early in the conceptual stage to the completion of the as-built drawings. A preliminary survey, expanded from the existing data but including limited field work, is necessary and must be provided for planning and concept development. As the planned project develops, photogrammetric mapping, recording of seismic activity (if any), and geophysical profiling become necessary, as does accurate location of bore holes for the geotechnical investigations.

CURRENT STATE OF SURVEYING TECHNOLOGY

Significant advances have been made in surveying technology since the first edition of this book was published.

Theodolites

The "Total Station" concept is prevalent in present theodolite design. This concept embodies the following features in one compact instrument with superior inherent accuracy and greatly improved operating characteristics:

- Infrared laser ranging system that operates parallel to or through the axis of the theodolite telescope. Distance accuracy up to 1 mm plus 1 ppm is attainable, and distance can be determined within 2 to 4 after acquiring the target. Most instruments can operate in a tracking mode for staking or topographic surveys, requiring about 1 to determine distance.
- Horizontal and vertical angular measurements are read by electronic micrometer capable of reading to 0.1 of arc in the most accurate models.
- Electronic leveling system that corrects horizontal and vertical readings for instrument "out-of-level" attitude.
- Electronic display system and storage to display and record keyboard entry data, angles, distances, point identity, etc.
- Internal computer providing the ability to convert slope distance to level distance and height difference, curvature and refraction corrections, traverse computations, stakeout computations, and cut sheets.

- Ability to store all survey data and download data for final computation and plan drafting via "field-to-finish" software.
- Vertical angular accuracy sufficient to attain third-order level accuracy by trigonometric leveling (Figure 3-1).

Global Positioning System (GPS)

GPS surveying has preempted triangulation and traversing as a means for conducting long-range surveys, and it is rapidly encroaching on shorter-range surveys for cadastral, topographic, and staking surveys. As it requires at least four satellites in its visible horizontal, it is not suitable for tunnel interiors, downtown locations, forested areas, or other locations where the GPS antenna cannot be exposed to the satellites via direct line of sight. GPS features include

- Automatic acquisition of satellites as they come into view.
- Static accuracy of 2–4 mm ± 1 ppm (average).
- Automatic timing and recording of satellite data over extended time periods.
- Rapid acquisition of static coordinates.
- Ability to acquire third-order position data while traveling along an unobstructed survey route using rapid static techniques and software.
- Ability to provide real-time positioning of moving survey vessels, vehicles, etc.
- Ability to determine elevation relative to the spheroid.
- Software for baseline reduction and network adjustment is commercially available (Figure 3-2).

Leveling Instruments

Both the precision and utility of level instruments have been improved significantly in the last 20 years. Features of modern level instruments include:

- Automatic reading and recording of the level rod via a "bar code" type rod pattern and optical scanning level. These systems are available for both first- and second-order leveling (Figure 3-3).
- Reversible pendulum, nonmagnetic levels for first-order leveling.
- Accessory data collectors and software that record automatic or manual level readings, check each reading for wire spread and balanced sight lines, and provide go/no-go instructions to reread or proceed to the next setup.
- Laptop, handheld, and desktop computers with specialized survey software for adjustment and plotting of survey data.

Fig. 3-1. Leica/Wild T2002 first-order theodolite with DI2000 distance meter (Total Station configuration).

Fig. 3-2. Global Positioning System (GPS) receiver (Trimble Model 4000 ST).

Fig. 3-3. Leica/Wild NA3000 first-order self-reading level with bar code rod.

Computer-Assisted Drafting (CAD)

CAD is in common use throughout the engineering and construction industries. Its capabilities include

- Fast, accurate drafting
- Control of line weights, symbols, and so on
- Uniformity of drafted sheets
- Ability to transfer data by wire to remote plotting station
- Ability to change scale and orientation
- Ability to draft in color

Geographic Information System (GIS)

GIS is designed for managing data in a complex environment, and it is a capable tool for project management. It will accept all types of data, such as digital, text, graphic, tabular, imagery, etc., and organize this data in a series of interrelated layers for ready recovery. Information stored in the system can be selectively retrieved, compared, overlain on other data, composited with several other data layers, updated, removed, revised, plotted, transmitted, etc.

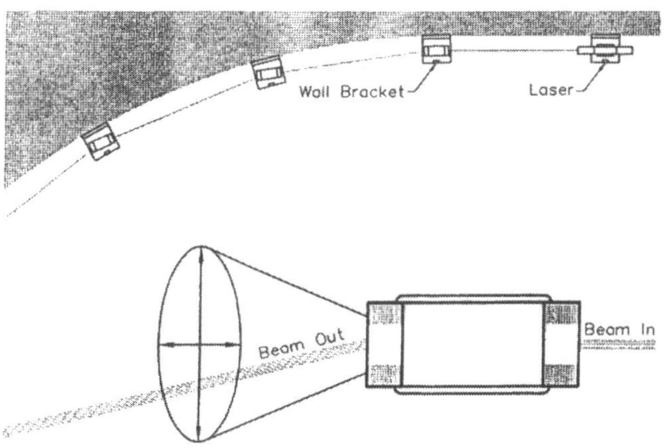

Fig. 3-4. Line and grade diverter.

Tunnel As-Built Surveys

In railway tunnels, a specially constructed measuring carriage on which flexible touch sensors are mounted can be pulled through the tunnel. The sensors indicate and record areas of insufficient clearance. Cross-section measurements can also be carried out by rotating the beam of a high-energy electronic distance measuring instrument in a plane normal to the tunnel axis, and recording the distance to the tunnel wall and the station at each measuring step (*World Tunneling*, 1980).

Line and Grade Diverters

Line and grade diverters with adjustable horizontal and vertical beam deflection can deflect a tunnel control laser beam in a series of tangents around horizontal and vertical curves. Several diverters can be used in series to provide tunneling machine control around curves without moving or repointing the laser (Figure 3-4).

Gyro Azimuth Theodolites

Modern Gyro Azimuth theodolites are capable of defining azimuth in a tunnel environment to less than 3 sec of arc (Figure 3-5).

Taylor Hobson Spheres, etc.

Taylor Hobson spheres and precise optical plummet instruments facilitate transfer of surface horizontal control lines to tunnel control traverse with superior accuracy (Figure 3-6).

Robotic Survey Instruments

Robotic survey instruments such as the Wild TM series and Geodimeter Slope Monitoring System are motorized electronic distance meter (EDM) instruments that can search out and measure horizontal and vertical angles and distance to reflectorized targets. These instruments can be programmed for continuous or intermittent operation, or they can be activated

Fig. 3-5. DMT Gryomat 200. Gyrotheodolite mounted on primary horizontal control monument, Superconducting Super Collider project. (Photo courtesy Measurement Science, Inc.)

Fig. 3-6. The Taylor Hobson sphere.

remotely upon command. Survey data is transmitted to a central computer for recording and processing. Robotic survey instruments are useful for remote monitoring of mining and construction sites, slope stability, glacier and ice field movement, and structure deformation and movement.

GENERAL SURVEYING REQUIREMENTS AND PROCEDURES

Accuracy

Requirements for survey accuracy are dictated by the type of project, length of the tunnel, and other factors. Conventional tunnels less than 10,000 ft long can be reliably controlled and monitored using horizontal and vertical surveying techniques that were in use 50–100 years ago. In contrast, specifications for the Department of Energy (DOE) Superconducting Super Collider in Waxahachie, Texas, required such precise tunneling and construction that the highest order of geodetic control and construction surveys was essential. Standards for all orders of horizontal and vertical surveys, together with field and office procedures, network design, instrument calibration, monumentation, etc., are described in the following publications:

- "Standards and Specifications for Geodetic Control Networks," Federal Geodetic Control Committee, Rockville, Maryland, September 1984 (reprinted August 1993)
- "Geometric Geodetic Accuracy Standards and Specifications for Using GPS Relative Positioning Techniques," Federal Geodetic Control Committee, Version 5.0, May 1988

Standards for distance and elevation accuracy are given in Tables 3-1 and 3-2.

The following survey specifications are recommended for major projects:

- Primary Horizontal Control: Second Order Class 1 plus closing error 1:70,000

Table 3-1. Distance Accuracy Standards

Classification	Minimum Distance Accuracy
First Order	1:100,1000
Second Order, Class I	1:50,000
Second Order, Class II	1:20,000
Third Order, Class I	1:10,000
Third Order, Class II	1:5,000

- Primary Vertical Control: Second Order Class 1
- Tunnel Control Traverse: Second Order Class 1
- Tunnel Control Benchmarks: Second Order Class 1

High-speed rail and specialized projects may require first-order or higher horizontal and vertical accuracy.

Survey instruments and methods used to transfer working line and elevation underground and to set monuments for construction line and grade should provide for the following:

1. Angular measurements to the nearest 1 sec of arc
2. Stationing to the nearest 1/1,000 ft
3. Benchmark elevations to the nearest 1/1000 ft
4. Adjusted coordinates to the nearest 1/1,000 ft

The precision of the target readings of the tunneling machine control system and tunnel ring measurements as performed after each shove should be in the range of one to two hundredths of a foot. The short time available for the performance of these measurements explains the lesser precision required.

Instrument Adjustments

Theodolites, Total Stations, EDMs, and GPS units cannot be adjusted or calibrated in the field. This work must be done in a competent service facility. Level instruments, however, require regular testing to assure that the horizontal crosshair defines a true level plane, and field adjustments are required if the "peg test" or other testing techniques indicate that the horizontal crosshair does not define a level plane.

On continuing projects where leveling is a daily task, time can be saved by replacing routine "peg testing" with the following procedure:

- Consign one automatic level for office calibration only. This instrument should be in perfect adjustment (preferably by in-

Table 3-2. Elevation Accuracy Standards

Classification	Maximum Elevation Difference Accuracy
First Order	0.5mm/km
Second Order, Class I	0.7mm/km
Second Order, Class II	1.0mm/km
Third Order, Class I	1.3mm/km
Third Order, Class II	2.0mm/km

strument shop) and should not be used for any other purpose but adjusting field levels.

- The level should be tripod-mounted on a stable foundation, or mounted on a steel bracket fixed to a concrete wall or column. It should be protected from abuse, dust, rain, and the like.
- Focus the eyepiece to infinity, and point telescope toward the level to be adjusted.
- Set up flashlight, mirror, or other light source behind eyepiece, with white paper between the light and the eyepiece to diffuse the light.
- Set up level to be tested on line with calibration level and at approximately the same level (within ±1/2 in.).
- Focus level to be tested at infinity, and point in direction of calibration level.
- Observe through eyepiece of level being tested. Adjust direction of both levels until crosshairs of calibration level are seen superimposed on the crosshairs of the level being tested.
- Refer to Figure 3-7 and, if needed, adjust horizontal crosshair of field level until horizontal crosshairs of both instruments coincide.

Preliminary Surveys

A preliminary horizontal and vertical control survey is required to obtain general site data for route selection and for design. This survey should be expanded from existing records and monuments that are based on the same horizontal and vertical datum that will be used for final design of the structures. Additional temporary monuments and benchmarks are placed as needed to support field investigations, mapping, environmental studies, and route selection.

U.S. Geological Survey topographic maps (1:24,000 series with 10-ft or 20-ft contours) may be used for preliminary route selection, but when the project corridor has been defined, new aerial photography should be obtained and photogrammetric maps should be prepared to facilitate portal design, access, right of way, drainage, depth of cover, geologic, seismic, and other studies. The scale and contour interval selected for these maps may be influenced by siting (urban or rural), drainage, and critical depth of cover. Scale and contour selection would include consideration of

1 in. = 50 ft or 100 ft scale with 1-ft, 2-ft or 5-ft contour interval
1 in. = 100 ft or 200 ft scale with 5-ft or 10-ft contour interval
1 in. = 400 ft or 500 ft scale with 10-ft or 20-ft contour interval

The cost of preparing photogrammetric maps is influenced predominantly by the contour interval, with the smallest interval being most costly. This should not, however, result in a decision to use a larger interval map solely to reduce cost when a more accurate map is needed.

Equipment and Techniques. Modern mapping equipment and techniques provide a wide range of products and services to support planning and design, and ongoing construction management, including

- Measurement of earthwork pay quantities.
- Digital ortho mapping, wherein the aerial photographic image is digitized in true plan position and scale, and can be inserted into the project Geographic Information System (GIS) or database.
- Digital topographic mapping, wherein contours and planimetric features are directly digitized during the map compilation process and can be CAD-plotted and/or inserted into the project GIS.
- Software enabling manipulation of digital map and survey data to extract profiles, cross sections, spot elevations, etc., and to superimpose this data selectively with design, right of way, geologic, and other data sets that have been digitized into the GIS/database.
- CAD plotting of selected layers, or combination of layers of map data, such as contours, building outlines, road outlines, vegetation, utilities, drainage, fences, walls, and so on. This data can also be selectively plotted with property and boundary information, design information, land ownership data, utility size, type, age, etc., and demographic and other information that has been entered into the GIS/database.

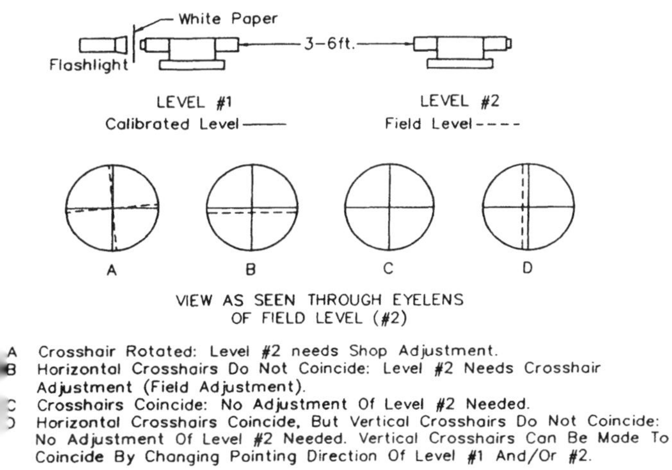

Fig. 3-7. Adjustment of level instruments.

Hydrographic Surveys

Where construction of a submersed tunnel or tube is planned, hydrographic surveys to determine bottom topography will be needed, together with current direction and velocity studies if reliable data is not available from NOAA, USGS, USCE, State Department of Water Resources, or some other source. In planning the hydrographic survey, an investigation should be made to determine the existence and location of submarine pipelines, cables, cathodic devices, etc., that may affect design or construction of the submersed tunnel or tube. Additional surveys such as magnetometer, seismic sub-bottom, electromagnetic toning, side scan sonar, and the like, may be required to detect and locate these features. These additional surveys may be done simultaneously or sequentially with the basic hydrographic survey (Figures 3-8–3-10). Data generated in the hydrographic survey should be based on the same horizontal datum as project control

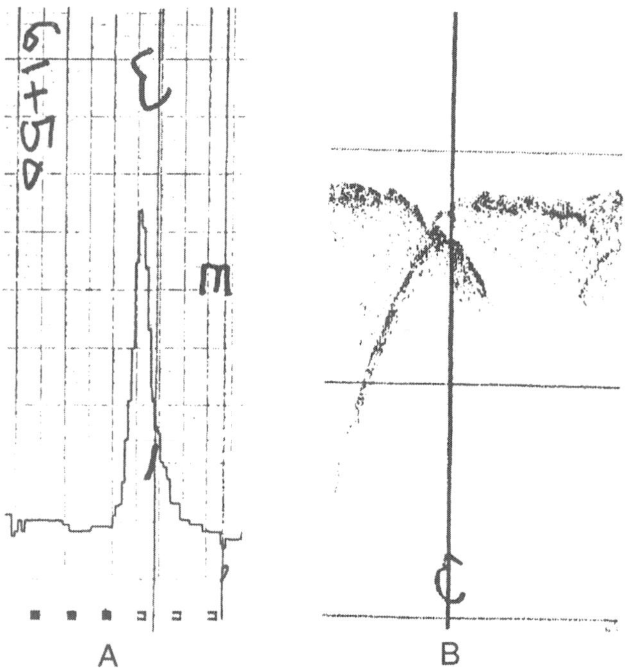

Fig. 3-8. (a) Marine magnetometer chart showing magnetic anomaly indicating location of 12-in. submersed fuel line in approximate water depth of 25 ft (Geometrics Model 826 marine magnetometer with side-mounted sensor). (b) Seismic subbottom chart showing typical signature of submersed pipeline; depth of cover approximately 3 ft (ORE Pipeliner 3.5–7.0 KHz).

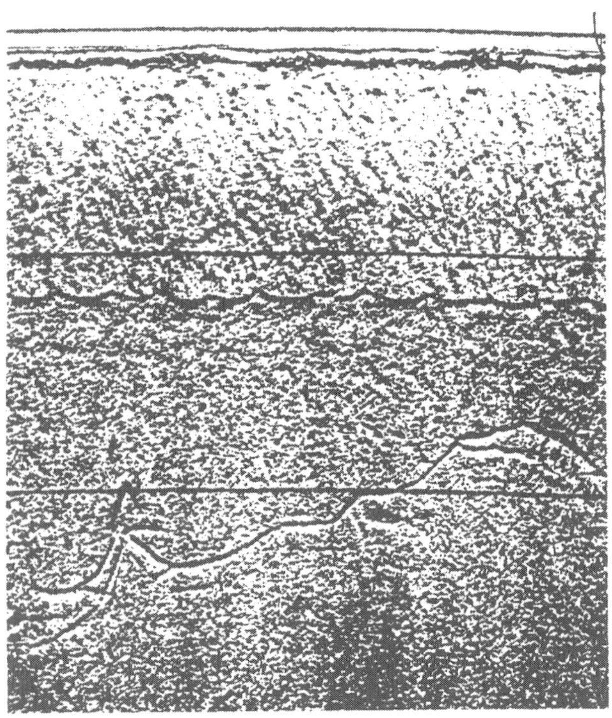

Fig. 3-9. Sonar side scan chart showing lost anchor chair and submerged buoy on bay bottom; note double chain and harness (Klein Model 260; 500 KHz).

surveys, and they should be compatible with the project GIS/database. Vertical datum selected for the hydrographic survey should be based on the primary monument elevations, but it may be expressed in terms of National Geodetic Vertical Datum of 1929 (NGVD), Mean Lower Low Water Datum, or other datum selected by the designer.

Utility Surveys

Utility information is required for preliminary and final route selection and to determine the type and extent of utility protection, relocation, or reconstruction needed. This information is obtained from surveys commissioned for the project, and from existing utility maps and records.

Utility surveys are needed to collect new data, corroborate existing data, and composite all data in maps and reports that will enable the designer to prepare tunnel plans knowing that utilities have been accurately located, typed, and identified. Utility surveys, like all other surveys on the project, must be based on the primary horizontal and vertical control network, and they must be sufficiently accurate to ensure that all utility features are located within design tolerances. In most cases, accuracy of 0.10 ft is acceptable, although specialized situations may require other standards of accuracy.

Utility surveys must be based on a written scope of work that defines the level of detail and survey accuracy needed by the designer. Derivation of this scope is the responsibility of the designer. Obtaining and organizing the survey data and preparing maps and plans illustrating such data is the responsibility of the utility surveyor. Responsibility for reconciling differences that inevitably occur between existing utility maps and commissioned survey data should be decided and included as a statement in the scope of work.

Maps showing the location and type of both aboveground and buried utilities are normally available from the owners

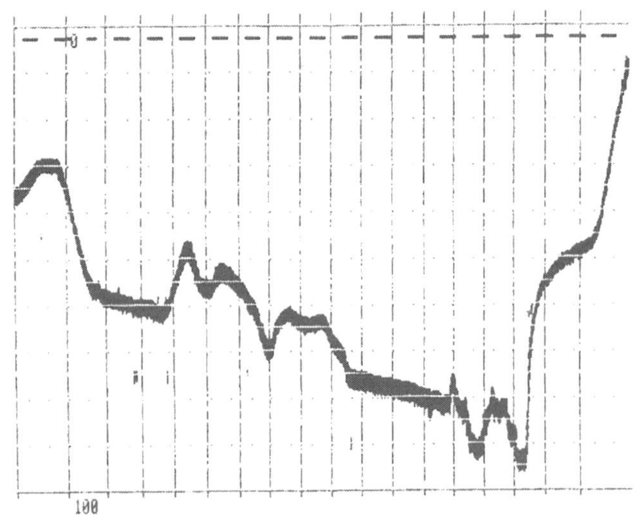

Fig. 3-10. Depth recorder chart showing bay bottom in water depth ranging from 28 to 92 ft; vertical lines indicate 100-ft increments traveled by the survey vessel (Odom Echotrac 200 KHz).

of the utilities (utility companies, cities, utility districts, etc.). These maps are for informational purposes and generally do not warranty that the utility features shown actually exist, that they are in the exact location shown on the map, or that np additional features exist that are not shown. In general, surface features tend to be reasonably well positioned on utility maps, but underground connections (pipes, conduits, cables, etc.) are usually shown as straight lines connecting the surface features. During original construction of such utilities, trenching may have been designed as a series of straight lines, but in actuality, buried obstructions such as boulders, unstable soil, or unmapped existing utilities necessitated deviation from the designed trench alignment. In many instances, as-built surveys were never done after construction, and the design map, without any notation of as-constructed alignment changes, became the only map recording the location of the constructed utilities.

Although generally less troublesome than buried utilities, surface features, such as poles, valves, hydrants, drop inlets, manholes, are often incorrectly placed or omitted from utility maps. Manholes and valves are often paved over during street improvements, and features may be relocated (without remapping) to accommodate improvements and expansion. In areas of intense development, it is not unusual to find 6 to 10 major surface utility features serving underground utilities at a street intersection, and ground inspection of each feature may be needed to ascertain which features on the ground relate to particular symbols on the map. Prior to starting surveys, local utility companies and agencies should be contacted to locate and mark underground utilities on the site, and to pothole to determine exact location and depth in critical areas.

The perceived reliability of utility data is often reflected in construction bid costs. If the reliability of utility information is questionable, bids will be inflated. If the owner or designer accepts responsibility for the accuracy and completeness of utility information, the bids may be lower, but the owner or designer assumes responsibility for additional costs or damages resulting from faulty utility data. For this reason, it is not advisable for the surveyor to undertake a survey based on a scope of work that contains generalized phrases such as "all utilities shall be located" or "surveyor shall warrant accuracy of all utility features." It is impossible to know beforehand the status and recoverability of underground utilities, or that all underground utilities can be located, or that underground pipes, cables, conduits, and the like are in the locations shown by existing utility plans. It may be necessary to undertake extensive subsurface excavation and surveying that are beyond the time and cost limitations of the designer's budget. On some sites, it is particularly difficult to affirm that all utilities have been located and identified without a prohibitive amount of investigation. Excavations in urban sites routinely uncover active or inactive pipes, cables, etc., that are not shown on utility plans, and which cannot be detected or located with any degree of certainty by indirect surveying techniques.

The requirement for utility information varies with tunneling methods and the site. Urban sites house many more utility features, and property values are higher than at rural sites. Urban sites are generally more restrictive, and routing changes may require utility rerouting. Deep tunnel construction may not pass through any utility systems, but vibration, blasting shock, and settlement may affect surface and underground utilities in the project corridor. Cut-and-cover construction, particularly in urban areas, extensively affects overlying and adjacent utilities. Gas, water, sewerage, stormwater, electrical, telephone, and other utility mains and distribution systems may require excavation, rerouting, strengthening, or reconstruction, and they may also require positional monitoring during construction.

Equipment and Techniques. Instruments and systems available for locating utilities include

- Photogrammetric mapping: routinely used to document the location of prepainted surface features such as manholes, valves, inlets, hydrants, etc. This is normally done during the photogrammetric mapping phase of the survey work. Horizontal accuracy of well-defined utility features is usually within 0.6% of original map scale, e.g., 0.6 ft on 1 in. = 100 ft scale map.
- Magnetic surveys: ferrous bodies such as iron and steel pipes, barrels, piles, etc., induce anomalies in the earth's magnetic field. Land and marine magnetometers detect the anomalies, whose amplitude is a function of the ferrous mass and the distance from the surface.
- Electromagnetic toning: a low-frequency AC current is conducted into linear metal features such as pipelines, cables, cable jackets, etc., by connecting an AC tone generator to an exposed section of the feature. A handheld receiver detects the feature by electromagnetic signals whose magnitudes are a function of the strength of induced AC current, distance between tone generator and mobile receiver, depth of cover over the feature, electrical conductivity of the feature, and electrical insulation between the feature and its burial medium (earth, water). Operating AC electrical cables may also be detected by electromagnetic toning. In situations where the toning signal is clear and concise, depth of the feature may be determined within ±2 ft by vertical triangulation using two receivers (Figure 3-11) or one receiver at two triangulation positions.
- Ground-penetrating radar: a portable instrument that emits radar frequency signals vertically downward and plots energy pulses reflected by buried objects. The system performs best in low conductive soils, and it becomes inoperative in highly conductive soils. Terracotta, transite, and wood pipes, as well as metal pipes and objects, have been detected to depths of 20 ft or more under ideal conditions.
- Air/vacuum excavating systems: equipment includes pump, hose, and dirt storage tank. A small (±10-in. diameter) hole is excavated down to the utility by vacuum excavating without the potential for damage associated with traditional excavation methods.
- Geographic information systems provide a means to enter and retrieve a wide range of utility information, including

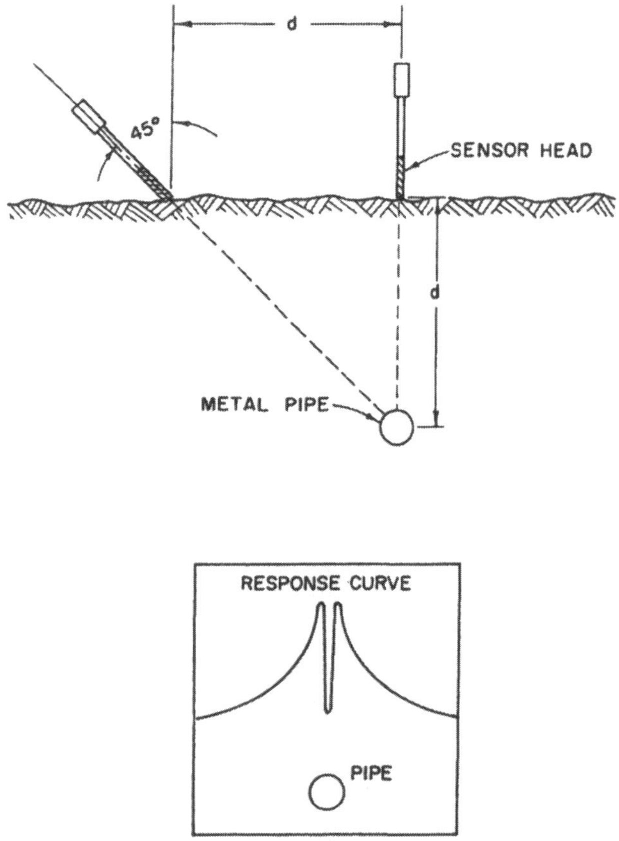

Fig. 3-11. Locating buried metal pipe by vertical triangulation using low-frequency electromagnetic toning. Pipe is first located using sensor in vertical mode. Depth of burial is then determined by locating pipe with sensor 45° to vertical. Burial depth is equal to distance betweeen sensor positions. Typical response curve indicates null when sensor is pointing at center of pipe.

location, type, date of construction and repair, ownership, right of way, etc. This information is stored in dedicated data layers, and it can be readily accessed to display or plot both technical and demographic information.

- "Smart pigs" for oil and gas lines are presently in prototype stage. These are instrument packages that, when forced by pressure or traction to traverse through the pipe, transmit coordinate information defining their depth and location.

Primary Survey Network

Primary surveys are the basic positional reference for the project. These surveys must be founded on stable and accessible monuments, and they must be conducted to a high degree of accuracy to meet project needs. The survey work, computations, adjustment, and data recording must be accurate and reliable so that design and construction can proceed with absolute confidence in the credibility of the survey data.

Survey Control. Primary horizontal surveys are conducted using triangulation, electronic distance meter (EDM) traverse, trilateration, Global Positioning System (GPS) surveys, or a combination of these methods. Horizontal surveys must contain sufficient redundancy for quality assurance, and they should be adjusted by least square methods.

Primary vertical surveys are closed level circuits run from existing first-order NGS (or other acceptable agency) benchmarks through benchmarks set for the project. Elevations are usually adjusted to the National Geodetic Vertical Datum of 1929 (NGVD29) and are based on the most recently published elevations of the original benchmarks. Survey instruments must be maintained in proper adjustment, and rods should be calibrated prior to starting the work. Instruments and rods must meet specifications for the work to be done, and procedures should comply with NGS standards for geodetic surveys.

After completion of route selection, a horizontal and vertical survey of high accuracy is conducted, with permanent monuments installed near portals, adits, and other selected locations in the project corridor. Design and execution of the survey must be done with the objective of establishing a singular and authoritative survey system that is based on securely founded monuments and meets the accuracy standards required for the project. All subsequent surveys and construction work must be based solely on the control survey network, and the project plans and specifications should contain specific statements affirming this.

Although most major projects in the United States are tied by survey to the North American Datum (NAD) horizontally, and vertically to the NGVD29, the primary control survey network and the coordinates and elevations derived from it may be based on any selected datum. In selecting a coordinate system on which to base the primary control horizontal survey system, persuasive arguments are often advanced by local government, utility companies, and the like in favor of using existing monuments in the job site and adjusting project surveys to coordinates already established for these monuments. It is imprudent to agree to this, because older surveys, even first-order traverse or triangulation, do not usually meet the attainable accuracy of present surveying instruments and techniques, and because existing coordinate sets can only be held by distorting the primary survey adjustment at the expense of primary survey accuracy.

A commonly used computational technique, particularly for Global Positioning System surveys, is to hold the existing published coordinates of one existing NGS monument as the basis for project coordinates, and refer to the published coordinates of one or more additional NGS monuments to determine azimuth for the project. This removes artificial constraints from the adjustment process, and it ensures that coordinates will reflect only the undistorted survey results. Existing monuments in the project site may be used if they are physically stable, suitably located, and accessible; however, to avoid confusion, existing coordinates and elevations of these monuments should not be recorded nor used in project records.

Electronic Distance Measuring. Modern EDM instruments (Total Stations) combine accurate measurement of angles and distances, internal computer logic, computer processing of data, and storage of observed angle and distance data. Range of distance measurement depends upon type of EDM used, number of reflective prisms, and clarity of the air. Typical range is 2,000–3,000 m, with some specialized instruments ranging in excess of 7,000 m. Standard deviation of angle and distance measurements vary with the various models and makes of EDMs available. Angular standard deviation ranges from 0.5 to 6.0 secs of arc. Distance standard deviation ranges from 1 mm plus 1 ppm to 5 mm plus 3 ppm (Figure 3-1). (See *P.O.B. Magazine*, April–May, 1994.)

EDMs with data collectors can download survey data to be processed and plotted using specialized "field-to-finish" software.

Global Positioning System (GPS). Coordinate positioning of widely spaced control monuments is usually accomplished by GPS surveys, which utilize the signal transit time from ground station to satellites to determine the relative position of monuments in a control network. The accuracy of measurement is dependent upon the number of satellites observed, configuration of the satellite group observed, elapsed time of observation, quality of transmission, type of GPS receiver, and other factors including network design and techniques used to process data. Under normal conditions, GPS surveying can attain position accuracy of 2–4 mm (0.006–0.013 ft), ±1 ppm of the distance to the base station, using dual-frequency receivers (Figure 3-2).

GPS surveying requires the simultaneous operation of several receiving instruments located at different stations throughout the survey network, and the success of an observing session depends upon each instrument being in place and operating at a predetermined time. This requires detailed advance planning, which includes the need to schedule access, travel routes, right of entry, traffic conditions, etc. Night operations substantially increase the planning effort needed to ensure successful GPS operations; however, with a full galaxy of satellites now available, night operations are usually not necessary.

Although intervisibility between GPS monuments is not required to complete a GPS survey, it is desirable so that subsequent surveys will have reference backsights for azimuth control. Intervisible monuments must be sufficiently distant from each other so that combined GPS position error does not result in excessive azimuth error between monuments. GPS data is usually postprocessed using baseline reduction and least square adjustment algorithms for determination of coordinates, and elevations referred to the spheroid.

Although GPS surveying is now considered "conventional," high-order GPS surveys entail extremely sophisticated procedures for both field and office work. Accordingly, the work should be planned and executed under the direction of a qualified GPS specialist with strong credentials in the application of advanced geodesy to design and construction.

Monument Construction and Records. Permanent monuments and benchmarks are constructed of brass or aluminum disks secured in cast-in-place concrete posts or grouted into concrete structures or exposed bedrock. Concrete posts bearing benchmarks or control monuments should not be bedded in loose shale, clay, peat, or other expansive or compressive soils; they should be anchored in stable underlying soil or rock. Monuments should be constructed to withstand effects of freezing and thawing. During installation of monuments, a survey monument record should be prepared describing type and location of monument, reference ties to nearby features, and a description of the location and route to be followed to find the monument. Upon completion and adjustment of the primary survey, coordinates, elevations, azimuth, and distance to intervisible monuments, date of survey, type of survey, and identification of survey agency should be added to complete the monument record form. This information should be embodied in final tunnel alignment plan and profile drawings (Figures 3-12 and 3-13).

On projects requiring extreme precision, stability of survey monuments is of prime importance, and requires specialized design and construction to detect and resist earth forces (Figure 3-5).

Published Geodetic Data. Geodetic control data and cartographic information that pertain to the National Networks

Fig. 3-12. Bennchmark monument with 1/2-in. driven steel rod, PVC casing, and cast aluminum collar and cover (NSG 3D Monument).

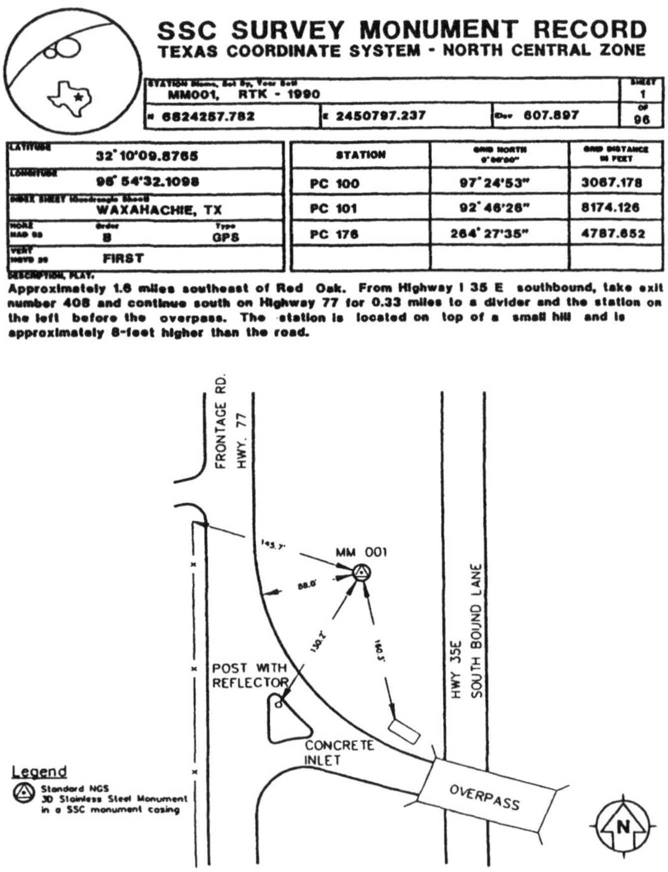

Fig. 3-13. Monument record form (Superconducting Super Collider project).

of Geodetic Control are widely distributed by the NGS National Geodetic Information Center (NGIC) to scientific and surveying–mapping communities.

Geodetic control data for the national networks are primarily published as standard quadrangles of 30 ft in latitude by 30 ft in longitude. However, in congested areas, the standard quadrangles are 15 ft in latitude by 15 ft in longitude. In most areas of Alaska, because of the sparseness of control, quadrangle units are 1 ft in latitude by 1 ft in longitude.

NOAA provides other related geodetic data, e.g., calibration baseline data, gravity values, astronomic positions, preliminary adjusted horizontal positions, horizontal and vertical data for crustal movement studies, Universal Transverse Mercator, (UTM) coordinate data, and other information.

The NGIC receives data from all NOAA field operations and mark recovery programs. In addition, other federal, state, and local governments, and private organizations contribute survey data from their field operations. These are incorporated into the NGIC control files. NOAA has entered into formal agreements with several government agencies whereby NGIC publishes, maintains, and distributes geodetic data received from these organizations.

New micropublishing techniques have been introduced in the form of computer-generated microforms. Some geodetic data are available on magnetic tape, microfilm, and microfiche, and horizontal and vertical geodetic control data for the entire United States are available for purchase on a set of 5 CD-ROMs. Charges for automated data are determined on the basis of the individual requests, and reflect processing time, materials, and postage. For additional information, contact NOAA, National Geodetic Survey N/CG174, 1314 E. West Highway, Station 9202, Silver Springs, MD 20910; Telephone: (301) 713-3242, FAX: (301) 713-4172.

Grid Factor and Sea Level Reduction Factor. When construction plans are being prepared, each plan (or set of plans) should include a statement defining the relationship between distances derived from grid coordinates and corresponding distances on the ground. This relationship is a function of grid scale factor and the elevation above sea level of the project at the center of each sheet. Grid scale factor varies in a north–south direction on Lambert Conformal Projection grids, and in an east–west direction on Transverse Mercator Projection grids. The National Geodetic Survey has prepared and published tables for each projection, covering all State Plane Coordinate zones in each state plus Puerto Rico, Virgin Islands, St. Croix, and American Samoa.

An elevation scale factor, or sea level reduction factor, is used to convert ground measured distances at project elevation to the corresponding distances at sea level elevation. Assuming the earth's average radius to be 20,906,000 feet, the sea level reduction factor is determined as follows:

$$\text{Sea level reduction factor} = 1 - \frac{\text{project elevation above sea level}}{20{,}906{,}000}$$

Example:

Length of line measured on the ground = 10,000,000 ft
Elevation of measured line = 2,000 ft above sea level
Grid scale factor at project = 0.999994 (from NGS tables)

1. Sea level factor $= \left(1 - \dfrac{2{,}000}{20{,}906{,}000}\right)$
2. Length of line at sea level = 10,000,000 ft × 0.9999044 = 9999.044 ft
3. Grid length = 9999.044 ft × 0.999994 = 9998.984 ft

From this it is apparent that grid lengths calculated from coordinates will differ from lengths measured on the ground, and a statement should be included in each plan defining the relationship between grid and ground lengths. In the foregoing example, the following statement would be appropriate: "To convert grid lengths to ground lengths, multiply grid lengths by 1.0001016," or conversely, "To convert ground lengths to grid lengths, multiply ground lengths by 0.9998984."

On major projects, it may be advisable to design a special coordinate grid for horizontal control to minimize the effects of sea level reduction and grid scale factor, thus eliminating the need to convert from ground length to grid length, and vice versa.

TRACK CENTERLINE

Division of Responsibility between Resident Engineer and Contractor

Survey costs are small in comparison with the expenditures involved in tunnel driving. Nevertheless, if tunneling is held up because of faulty survey work or because of interference by the engineer's survey crew with driving operations, the resulting losses can be considerable. For this reason, specifications relating to tunnel driving accuracy should be written as a performance specification, and the contractor should have full responsibility for transferring line and grade from the primary surface control into the tunnel and for development of tunnel construction control procedures. If delays or rework are caused by errors in the basic survey data furnished by the engineer, or by unwarranted interference during the engineer's check survey operations, contractor claims for additional compensation are inevitable.

Before the start of construction, the engineer's surveyor performs all survey work such as preliminary surveys and primary control surveys on the surface. During construction, the engineer's survey responsibility should be limited to maintaining the basic survey network, monitoring existing structures, and checking results of the contractor's work. This includes making sure that underground survey control of adjoining contractors agrees at the contract interface. Check work should be done by the engineer's quality control surveyor on a defined schedule, with both field and office work completed and reviewed by the engineer as soon as practicable to detect any errors in the contractor's survey work, and to limit the impact on construction that such errors may cause. Any out-of-tolerance differences with the contractor's surveys or deviation from construction plans should be verified and brought to the contractor's attention without delay.

Both space and time for surveys are usually limited in tunneling projects. Although it is essential that the engineer's and contractor's surveyors maintain independence in their field and office operations, it may be feasible to combine forces when time and/or space are critically limited. This can be achieved by assembling a composite survey team and equally sharing the task of making survey measurements and observations, with each party independently recording all measurements and completing computations. If data collectors are used, the data log can be copied, or the original log can be used by both parties for their computations and preparation of plans.

Information concerning groundwater level as obtained from observation well readings is of vital importance for the contractor's tunneling operation. It is, therefore, reasonable to include installation of observation wells, maintenance of the wells, and periodic reading of water levels in the contractor's contractual obligation. Water level records should be made available to the engineer at the time of recording.

Level readings of surface settlement points, which serve as indication of construction problems at the tunnel heading, are not in the immediate practical interest of the contractor. As a matter of fact, the chance of inaccurate level readings during times when heading problems are encountered is greater than during normal operation. The contractor is preoccupied with the construction problems at the heading during times of trouble and, therefore, spends minimal time on required surface level readings. For this reason, surface levels over the tunnels should be run and evaluated by the quality control surveyor, and the results should be made available to the contractor. Installation and monitoring of special recording devices, such as subsurface settlement points, inclinometers, and strain gauges, should also be the quality control surveyor's responsibility.

TUNNEL GEOMETRY

Relationship of Centerline Track to Centerline Tunnel

On a rapid transit system, centerline of track and centerline of tunnel are normally not identical because of clearance requirements. Centerline of track is the basic control during layout of the system. During construction of the tunnel, ~~however, it is desirable~~ from a practical standpoint that the contractor's and the engineer's field personnel use centerline of tunnel rather than centerline of track as the basis of tunnel control.

The vertical and horizontal offset from centerline of track to centerline of tunnel varies with the superelevation of track. The resulting tunnel centerline is a curve of complex mathematical definition (Figure 3-14). Therefore, a tunnel centerline should be developed that is composed of tangent, circular, and transition spiral sections and approximates the complex theoretical tunnel centerline within a specified tolerance (0.25 in.). This centerline should be incorporated into the contract drawings of the tunnel contract, and all tunnel control should be based on this curve.

A computer printout listing coordinates of points, tangent bearing, and elevation of points and slope at 5-ft intervals on the tunnel centerline should also be incorporated into the contract documents. Since stationing of centerline tunnel and centerline track will not agree because of different curve radii, station equations between centerline tunnel and centerline track should be incorporated at the beginning and the end of each construction contract, at TS (tangent spiral) points, at SC (spiral curve) points, and at such points as vent shafts or cross passages. Stationing for tunnel centerline should start at station 0 + 00 for each tunnel contract. Stationing of track proceeds through the entire system, which, generally, is made up of several tunnel contracts.

If the rapid transit system has a natural center point from which several lines branch out in different directions, the station 0 + 00 should be assigned to this point. Stationing then proceeds to the outlying areas, and future extensions of the system can be added without upsetting the stationing sequence.

Working Line. The working line is the survey line used by the contractor's field personnel to establish shield or

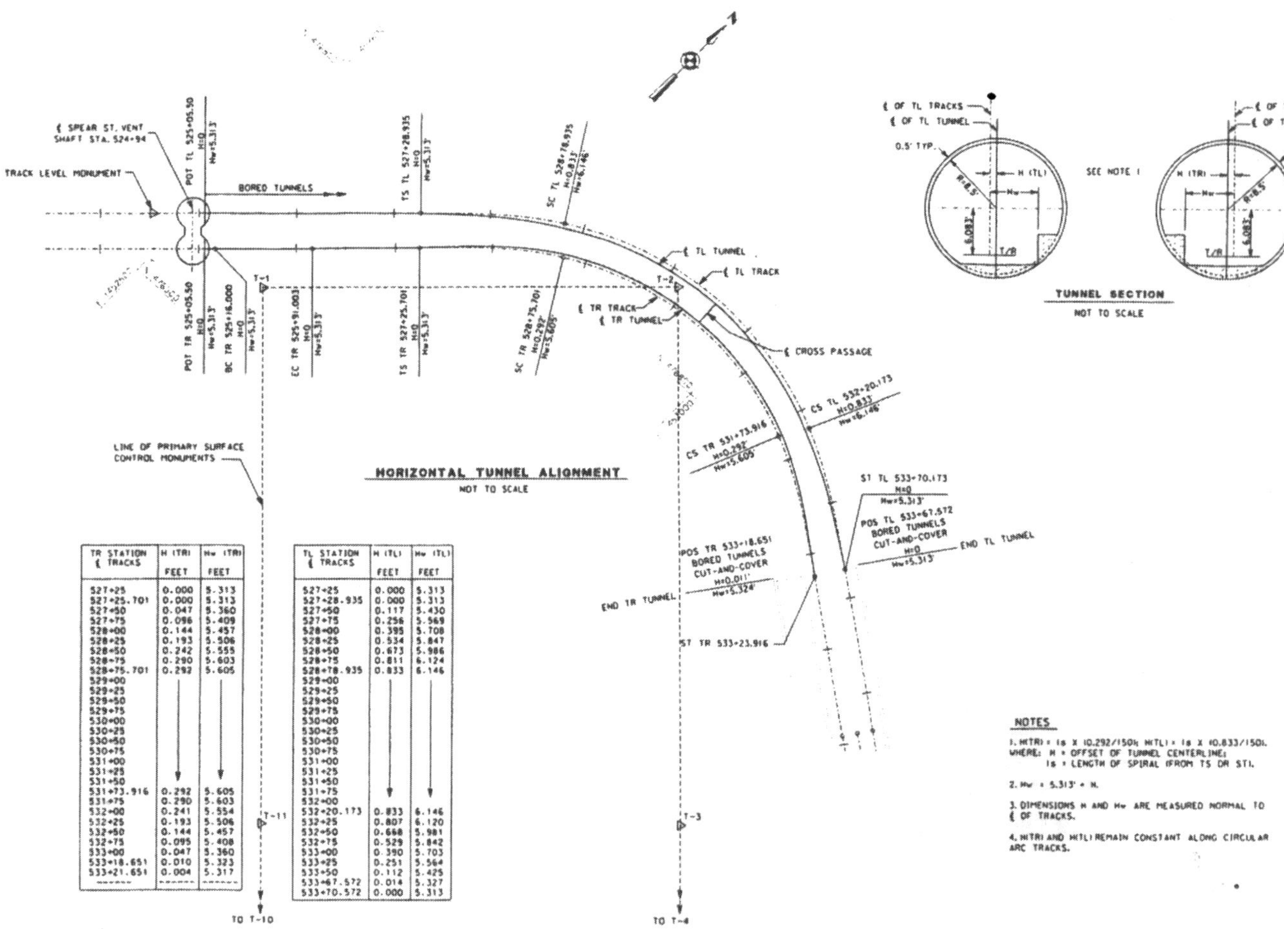

TR STATION ℄ TRACKS	H (TR) FEET	Hw (TR) FEET
527+25	0.000	5.313
527+25.701	0.000	5.313
527+50	0.047	5.360
527+75	0.096	5.409
528+00	0.144	5.457
528+25	0.193	5.506
528+50	0.242	5.555
528+75	0.290	5.603
528+75.701	0.292	5.605
529+00		
529+25		
529+50		
529+75		
530+00		
530+25		
530+50		
530+75		
531+00		
531+25		
531+50		
531+73.916	0.292	5.605
531+75	0.290	5.603
532+00	0.241	5.554
532+25	0.193	5.506
532+50	0.144	5.457
532+75	0.095	5.408
533+00	0.047	5.360
533+18.651	0.010	5.323
533+21.651	0.004	5.317
------	------	------

TL STATION ℄ TRACKS	H (TL) FEET	Hw (TL) FEET
527+25	0.000	5.313
527+28.935	0.000	5.313
527+50	0.117	5.430
527+75	0.256	5.569
528+00	0.395	5.708
528+25	0.534	5.847
528+50	0.673	5.986
528+75	0.811	6.124
528+78.935	0.833	6.146
529+00		
529+25		
529+50		
529+75		
530+00		
530+25		
530+50		
530+75		
531+00		
531+25		
531+50		
531+75		
532+00		
532+20.173	0.833	6.146
532+25	0.807	6.120
532+50	0.668	5.981
532+75	0.529	5.842
533+00	0.390	5.703
533+25	0.251	5.564
533+50	0.112	5.425
533+67.572	0.014	5.327
533+70.572	0.000	5.313

Fig. 3-14. Relationship between centerline of track, centerline of tunnels, and primary surface control monuments.

tunneling machine guidance in the tunnel. The working line may coincide with the tunnel centerline or may run through the laser position points for the laser setup. The selection of the working line must be left to the contractor to suit his tunnel equipment and methods. The working line is usually established by traverse survey methods.

Tunnel Lining Geometry and Taper Rings. Steel, cast iron, or precast lining rings are installed in tunnels to support the ground. The lining used for the construction of the soft ground tunnels of the San Francisco BART system is typical. The lining rings were composed of six welded steel segments and a small key segment (see Figure 3-15). The segments were bolted together and to the flange of the last-erected ring in the tail of the shield to form a standard lining ring of 2.5-ft width and 17.5-ft outer diameter. The standard rings were intended for use on tunnel sections with curve radii over 1,500 ft. Special rings of 18.0-ft outer diameter were installed in tunnel sections where curve radii are shorter than 1,500 ft. On these curves, the overhang of subway cars in curves required additional clearance.

If straight lining rings are erected in sequence, they will form a straight tunnel section, except for deflections of rings to one side or the other caused by ring distortion. Therefore, where the tunnel was constructed on a vertical or horizontal curve, taper rings were required to change the horizontal or vertical direction of the tunnel. The width of these rings tapered from the standard 30-in. width to reduced widths of 29.25, 28.75, and 28 in., as shown in Figure 3-15 and Table 3-3. The radius of the tunnel curve determines the type and number of taper rings required. Table 3-3 shows that taper rings of 17.5-ft outer diameter that taper 0.75 in. form a tunnel curve of 691.25-ft radius. For a longer curve radius, alternating straight and taper rings are installed.

Usually the contractor has to make an estimate of the number of straight and taper rings required before the tunnel is started. Based on the radius and length of curves involved and the geometry of the taper ring, he makes an estimate of the theoretical number of taper and straight rings. The actual number of taper rings required will always be greater than the theoretical number because some of the taper rings are used up to correct tunnel driving deviations. To develop a feeling for the actual number of taper rings required, the number of installed rings is compared with the theoretical number of taper rings for tangent and curved tunnel sections in Table 3-4. Extra taper rings can always be installed back

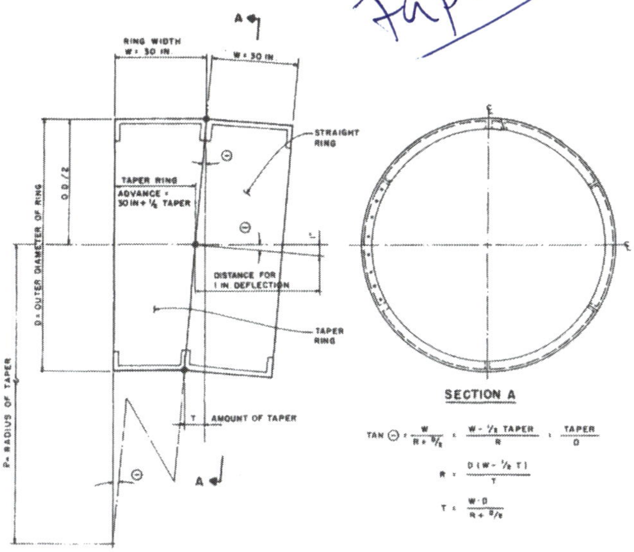

Fig. 3-15. Taper ring geometry.

Table 3-4. Straight and Taper Rings Required for Varying Curve Radii

			Line	Tunnel Lined with 2.5-ft. Wide Steel Lining Rings: 17.5-ft Diameter, Straight Tunnel	17.5-ft Diameter, 1,785-ft Radius Tunnel	18.0-ft Diameter, 655-ft Radius Tunnel
Theoretical number of rings required	Straight Rings		1	660	234	111
	Taper Rings	0.75-inch taper	2	—	148	—
		1.25-inch taper	3	—	—	119
		2.0-inch taper	4	—	—	37
Theoretical footage			5	1650.00	995.0	667.50
Actual Footage*			6	1649.10	955.0	656.85
Actual number of Rings Installed	Straight Rings		7	621	204	85
	Taper Rings	0.75-inch taper	8	39	178	—
		1.25-inch taper	9	—	—	138
		2.0-inch taper	10	—	—	44
		0.75-inch taper	11	6%	8%	—
		1.25-inch taper	12	—	—	7%
		2.0-inch taper	13	—	—	3%

* The difference between theoretical and actual footage is caused by the fact that each taper ring is about one-half a taper shorter at the centerline tunnel than the straight ring.

to back to form a double straight ring. A comparison of lines 3 and 4 of Table 3-4 shows that the theoretical tunnel footage, as computed by multiplying the number of rings installed by the width of tunnel rings, has to be reduced by a fraction of the taper for each taper ring installed.

When a taper ring is erected so that its widest point lies exactly on a vertical or horizontal diameter, the change of direction is entirely vertical or horizontal. Quite often, however, the widest point of the taper ring must be erected at an intermediate point because of the need to break ring joints. When this is done, changes of direction take place in both vertical and horizontal planes. Taper clocks, as shown in Figure 3-16, are convenient for visualizing the relationship between vertical and horizontal change of direction for various positions of the taper ring. In order to use the taper clock, a reproduction of this clock is mounted on cardboard, and the inner ring is separated from the outside ring. The outside ring is rotated in position to resemble the ring installed before the taper ring under consideration. The taper ring in the position shown in Figure 3-15 will produce an overhang of 0.15 ft and a lead of 0.057 ft on the left, which will throw the tunnel to the right. If elevating the tunnel is not desired, the taper ring could be rotated two bolt holes clockwise, resulting in zero overhang and a lead of 0.06 ft on the left. This would, however, cause the ring joints to line up with the joints of the previously erected ring, which adversely affects lining strength.

SURVEY WORK DURING CONSTRUCTION

Tunnels Driven from Portals

Where tunnels are driven from portals, one work point at the portal and a backsight, both on the working line, are sufficient to extend the working line into the tunnels. The theodolite is set up over the work point at the portal. A backsight is taken to the previously established backsight on the working line. Then the telescope is plunged, and work points are set on the working line in the tunnel.

Tunnels Driven from Work Shafts

In many cases, tunnel work is conducted from work shafts. Two methods of transferring line and levels from the surface to the bottom of the shafts are described. The first

Table 3-3. Taper Ring Data

Item	Ring O.D. 17 ft., 6 in.	Ring O.D. 18 ft., 0 in.	Ring O.D. 18 ft., 0 in.
Ring O.D. (inches)	210.0	216.0	216.0
Amount of Taper (inches)	0.75	1.25	2.00
Angle θ (degrees)	0°, 12', 16.65"	0°, 19', 53.65"	0°, 31', 49.81"
Deflection in 30 inches (inches)	0.107	0.174	0.278
Distance to Deflect 1 inch (feet)	23.33	14.44	9.10
Radius of Continuous taper rings (feet)	691.25	423.0	261.0

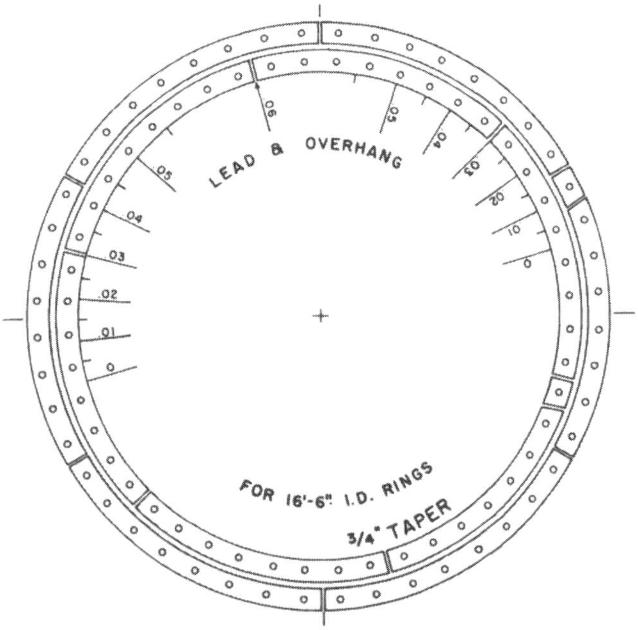

Fig. 3-16. Taper clock.

method is by transit sights. Two work points located on the working line are set at opposite edges of the shaft (see Figure 3-17). The theodolite is set up over one of the points (WP1), and a backsight is taken to a target on the extension of the working line (WP2). Then the scope is plunged, and the point on the opposite edge of the shaft (WP3) is sighted to ensure the tangent line. After the theodolite is thus aligned, the line is extended down and across to the bottom of the shaft, where a work point is established (WP4). Then the theodolite is set up over the work point at the opposite edge of the shaft (WP3), and the same procedure is repeated, resulting in two work points on the working line at the bottom of the shaft.

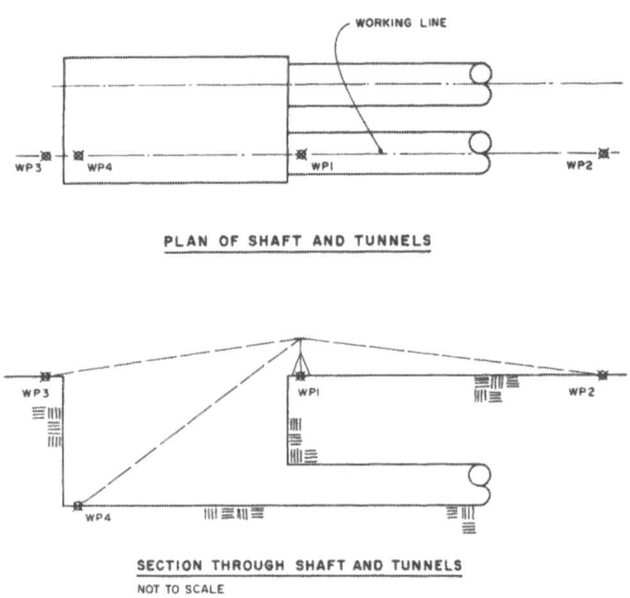

Fig. 3-17. Transfer of working lines in shaft.

Another method of transferring the line into a shaft is by means of steel wires supporting heavy weights hung in pails filled with oil. The working line on the surface is marked by two work points at opposite edges of the shaft. A theodolite is set over one work point and sighted on the work point at the opposite edge of the shaft. Two steel wires, each supporting heavy weights hanging down the shaft, are brought in line with the theodolite sight. An instrument set up in the shaft can be moved (wiggled in) to be in line with both wires. This instrument is then on working line and can be used to establish work points at the bottom of the shaft. In lieu of the wire weights suspended from the surface, an optical plummet (optical plumbing device) may be set up in the shaft under a surface target on the working line to establish a work point at the bottom of the shaft. Azimuth of the working line in the tunnel can be validated by using a gyrotheodolite. Stationing is transferred into the shaft by the methods described earlier.

Levels are transferred down by suspending a standard steel tape into the shaft, supporting a weight producing standard tension. On the surface, a tape reading is made from a known height of the instrument. Then the level is set up on the bottom of the shaft, and a reading is made on the suspended tape. The height of the instrument at the bottom of the shaft is obtained by subtracting the difference in tape readings from the height of the instrument at the upper level and correcting for temperature. Levels may also be transferred using an electronic distance measuring (EDM) instrument in vertical mode (see Figure 3-18).

Carrying Working Lines through a Compressed Air Lock

When the working line is extended through a pressurized air lock, care must be taken to position the instrument and work points at locations in the lock where the distortion that

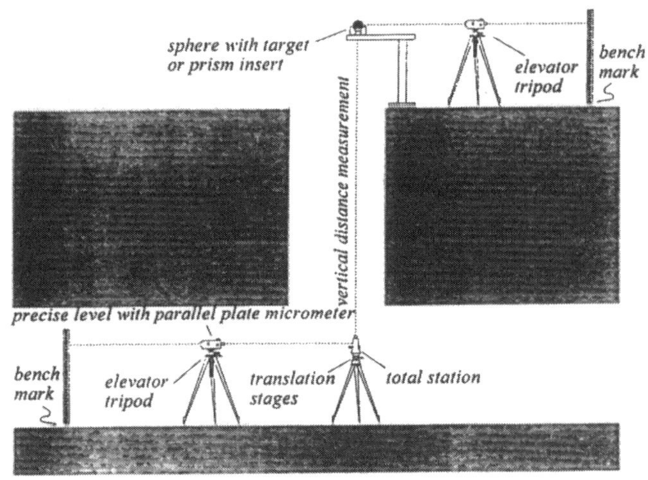

Fig. 3-18. Method used to transfer surface elevation from surface benchmarks down-shaft to tunnel control benchmarks using Taylor Hobson sphere and total station, Super Collider Project. (Courtesy, Measurement Science, Inc.)

takes place during pressure changes inside the lock is at a minimum. The instrument or work points should be located as near to the bulkhead as possible. The lock that has the best foundation should be used for locking through. A muck lock supported on concrete foundation is better than a man lock supported by steel framing.

Two methods of transferring line through the air lock are described. In the first method, the theodolite is set over the work point (WP5) in the shaft adjacent to the air lock (see Figures 3-19 and 3-20). A backsight is taken to the work point at the far end of the shaft (WP4). Then the scope of the theodolite is plunged, and three points are established in the bottom of the air lock in line with the working line in the shaft. During this operation, the free air door of the air lock is open and the air lock is in free air. Then the air lock is pressurized, and the compressed air side door of the air lock is opened. The theodolite is set up in the tunnel and is moved (wiggled in) until it is in line with the three points in the air lock. Then the scope of the theodolite is plunged, and working points are established in the tunnel.

Another method of transferring line through the air lock involves setting up the theodolite in the air lock over a working point (WP8), which was established in the air lock from the free air side (see Figure 3-21). After the theodolite is set up over the point in the air lock, a backsight is taken to the work points previously installed in the shaft on the working line (WP4 and WP5). Then the telescope of the theodolite is plunged, and the lock is pressurized. The door on the compressed air side is opened, a foresight is established in the tunnel, and work points are set on the working line. Level and stationing are carried through the air locks by similar procedures.

In pressurized tunnels, survey distances obtained by EDM instruments must be adjusted to compensate for the effects of temperature and pressure. This is done by calculating the parts per million (PPM) correction appropriate for observed temperature and pressure, and dialing this into the EDM during the measurement process. This PPM value, together with observed temperature and pressure, should be recorded in field book or data log. If the measurements are made with EDMs not equipped to dial in PPM values, temperature and pressure are recorded, and appropriate PPM factors are applied during data reduction and computations (Tables 3-5 and 3-6).

Tables 3-5 and 3-6 are derived from the following:

$$\text{PPM} = 281.8 - \frac{\text{Total pressure in inches of Hg} \times 25.4 \times 105.85}{\left[(\text{Temp } ^\circ\text{F} - 32) \times \frac{5}{9}\right] + 273.2}$$

where:

PPM = parts per million correction to measured length
Air pressure at sea level = 14.70 psi
Temperature = °F in tunnel
1 psi pressure = 2.0354 inches of mercury (Hg)

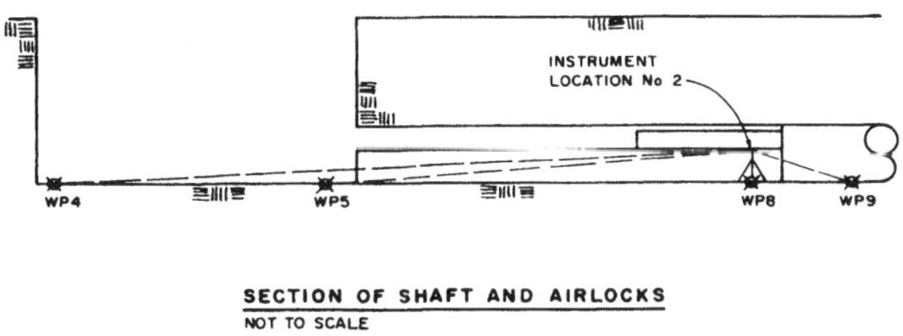

Fig. 3-19. Extending horizontal control through air locks—method 1.

Fig. 3-20. Extending line through the muck lock (San Francisco Muni Metro Turnaround Project).

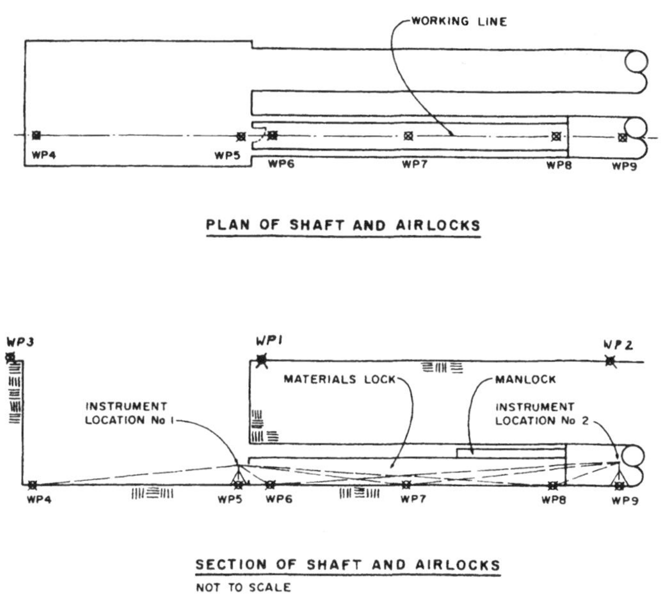

Fig. 3-21. Extending horizontal control through air locks—method 2.

Table 3-5. Parts per Million Correction (PPM) for EDM Measurements in Pressurized Tunnels

Gage	Air		PPM			
psi	psi	in./Hg	at 48°F	at 58°F	at 68°F	at 78°F
0.00	14.70	29.92	-3.4	+2.1	+7.4	+12.5
2.00	16.70	33.99	-42.2	-35.9	-29.9	-24.1
4.00	18.70	38.06	-81.0	-74.0	-67.2	-60.7
6.00	20.70	42.13	-119.8	-112.0	-104.6	-97.4
8.00	22.70	46.20	-158.6	-150.1	-141.9	-134.0
10.00	24.70	50.28	-197.4	-188.1	-179.2	-170.6
12.00	26.70	54.35	-236.2	-226.2	-216.5	-207.3
14.00	28.70	54.82	-275.0	-264.2	-253.9	-243.9

Note that positive PPM will increase measured length and negative PPM will decrease measured length. For example:

EDM measured length = 1,000.000 ft

Gage pressure = 3.5 psi (total pressure = 14.70 + .5 = 18.20 lb/in.)

Temperature = 48°F

Total pressure = 18.20 psi × 2.0359 = 37.05 inches of Hg

$$\text{PPM} = 281.8 - \frac{37.04 \times 25.4 \times 105.85}{\left[(48 - 32) \times \frac{5}{9}\right] + 273.2}$$

$$\text{PPM Factor} = \frac{-71.23}{1{,}000{,}000} = -0.0000712$$

Length correction = 1,000.000 × 0.0000712 = −0.071

Corrected length = 1,000.000 − 0.071 = 999.929 ft

Table 3-6. PPM Corrections for EDM Measurements

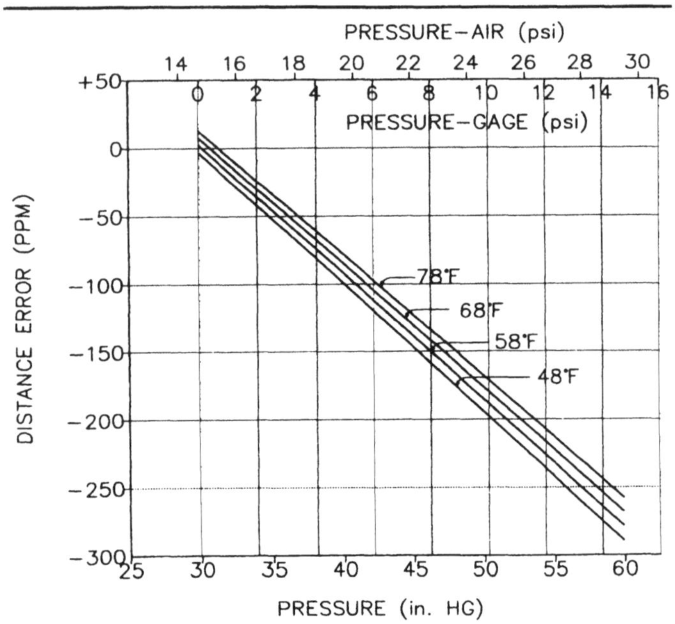

Survey Pipes from the Surface. In some compressed air tunnel installations, compressors are located close to the access shafts, and compressor vibrations, combined with the short backsight available in the access shaft and obstruction by construction equipment, make the repeated transfer of working lines through the shaft into the tunnel extremely difficult. In this case, it is advisable to bring control into the tunnel through survey holes sunk from the surface (see Figure 3-22). Two holes at a distance of about 200 ft on the working line are sufficient to transfer two work points from the surface into the tunnel and thereby establish working line and stationing in the tunnel.

A target on the working line is set in the upper part of the survey pipe. Then the point is transferred into the tunnel either by wire weights or by the use of an optical plummet.

Survey holes can also be sunk before the tunnel reaches the receiving chamber to allow a final check in time to make necessary corrections in the direction of driving before holing through. Survey pipes shown in Figure 3-22 are for compressed air tunnels. Free air tunnels simply require a drill hole from the surface of adequate diameter to drop a line from the surface.

Maintenance of Line and Grade in the Tunnel. From the work shafts or tunnel portals, the working line is carried ahead through work points; stationing is established by EDM measurement or tension chaining. If the tunnel is in stable ground, pins are driven into the crown of the tunnel to serve as work points. The theodolite is centered under a plumb bob suspended from the pin. To establish a work point in soft ground where tunnel movement is anticipated, a plumbline and target are attached to movable slides bolted to ring flanges or tunnel ribs in the crown of the tunnel (Figures 3-23 and 3-24). A survey platform is mounted below the slide, and plumb bob positions for each survey run are recorded on waterproof graph paper attached to the face of the platform. The transit is set up on the survey platform under the plumb bob suspended from the slide marking the work point. With this arrangement, surveyors can work in the crown of the tunnel without interfering with the passage of muck trains and other equipment (see Figure 3-23). Although normal sight conditions usually permit sights around

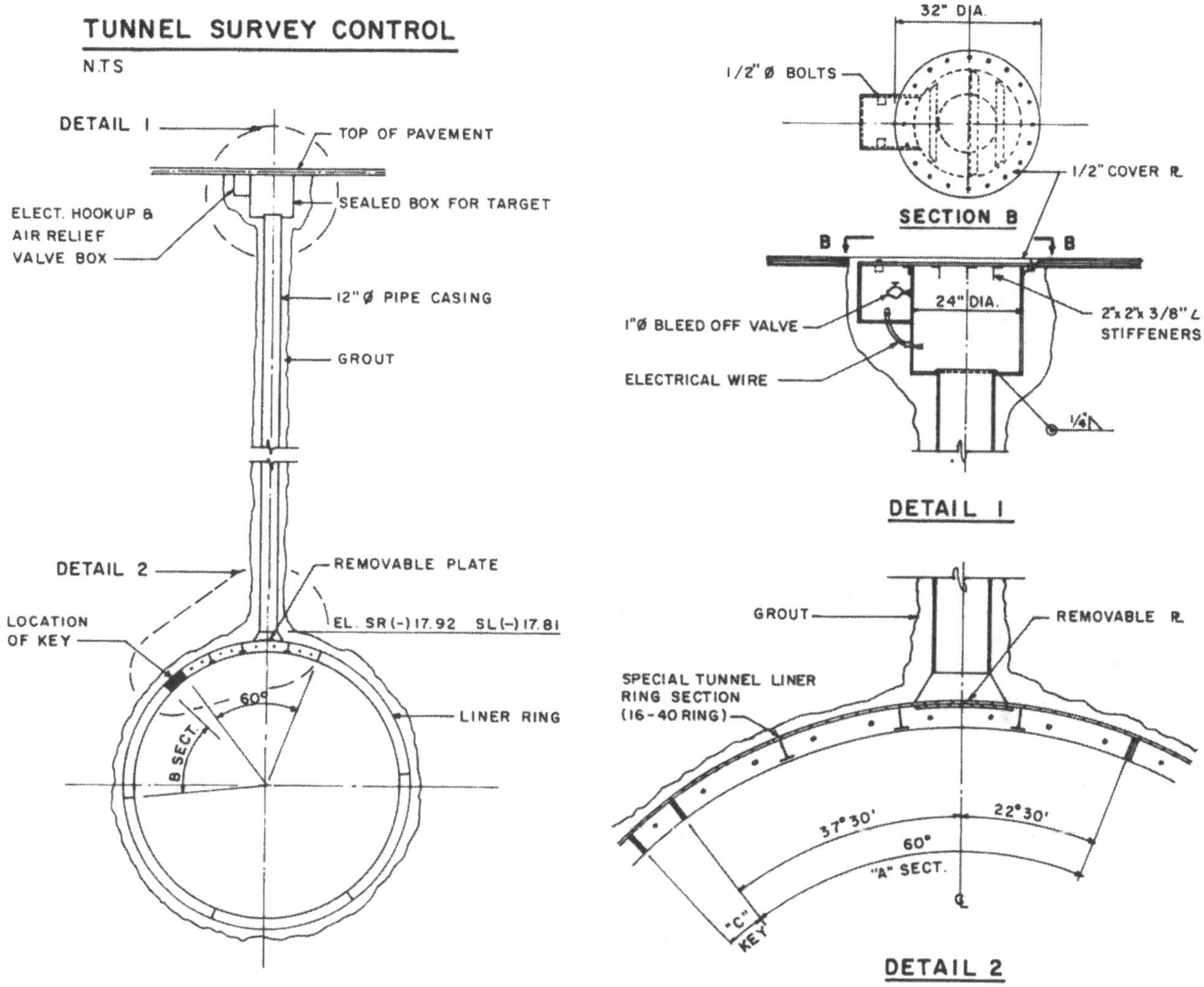

Fig. 3-22. Survey hole from surface.

600 ft, it is advisable to set work points at a distance of not more than 300 ft to allow for bad sight conditions due to smoke or fog. Levels are carried ahead by similar methods, with benchmarks set at approximately 300-ft intervals.

It is important to keep clear, concise records of survey work. Every time the line is rerun, a record of scale settings should be entered on a summary sheet on which the scale settings of previous survey runs are recorded. Chaining differences for distances between scales are shown on the same summary sheet. This survey record makes it possible to analyze the results of survey runs and to differentiate between survey errors and a pattern of tunnel movement. In soft ground, the lines are rerun at least once every week until good agreement of scale readings indicates that the location of a given control point or a section of the line is stable.

Construction Control for Drill-and-Blast Methods. Where the tunnel is excavated by drill-and-blast methods, the centerline is extended to the face before drilling for the next round starts. The centerline location is marked on the face, and the drill pattern is centered on that mark. Surveyors also give centerline location for the setting of steel sets.

Construction Control for Shields or TBMs. A shield or tunneling machine progresses in a sequence of shoves. After each shove, the shield or machine is stopped and its location and attitude are determined.

If the shield or machine is found to be off-line, adjustments of the steering mechanism are made to guide it back to its desired location.

Where tunnel lining is erected in the tail of the shield, its location and attitude are determined and recorded. The decision of whether to install standard or tapered lining sections after the next shove is based on this record.

The most practical method of shield or machine control is by laser beam and double target (for a description of the laser principle, see the *Encyclopedia of Science Technology*). A laser (see Figure 3-25) is set up at a distance behind the shield or tunneling machine to emit a laser beam from a predetermined point of origin along a predetermined line to

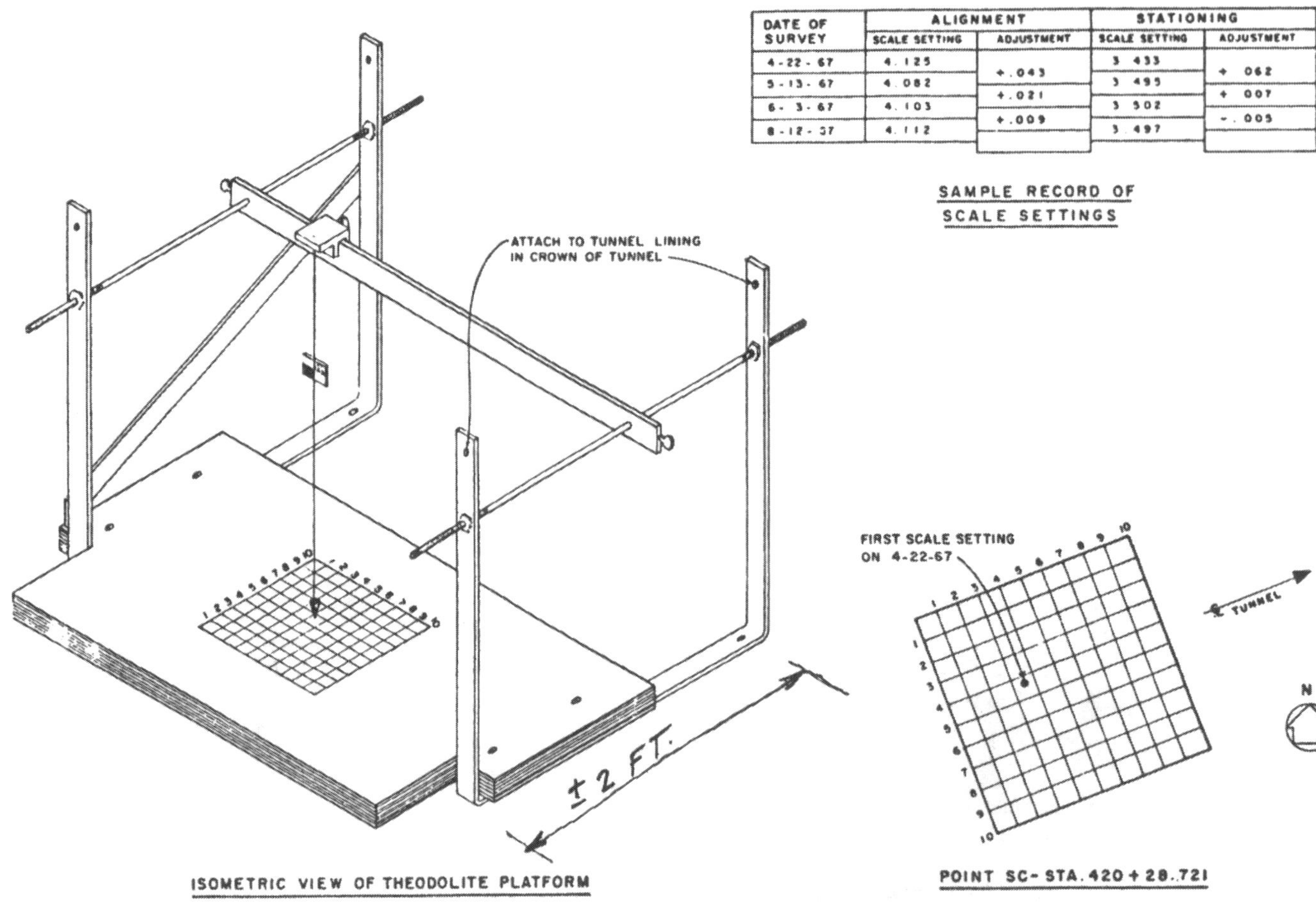

DATE OF SURVEY	ALIGNMENT		STATIONING	
	SCALE SETTING	ADJUSTMENT	SCALE SETTING	ADJUSTMENT
4-22-67	4.125	+.043	3.433	+.062
5-13-67	4.082	+.021	3.495	+.007
6-3-67	4.103	+.009	3.502	-.005
8-12-37	4.112		3.497	

Fig. 3-23. Survey platform.

Fig. 3-24. Instrumentation perched on hanging platform, making theodolite observations to centerline and springline; note instrument stand temporarily bolted to tunnel crown (Washington, D.C., Metropolitan Area Transit Authority).

the targets mounted on the shield or tunneling machine (see Figures 3-26 and 27). In the horizontal plane, the laser line is a chord line or a tangent to the tunnel centerline (see Figure 3-28). In the vertical plane, the laser line approximates the slope of the tunnel centerline (see Figure 3-29).

After the tunnel is driven to the end of one laser beam line, the laser is moved to the next laser position point, and the laser tube is set to emit the laser beam along the next pre-

Fig. 3-25. Laser setup in tunnel.

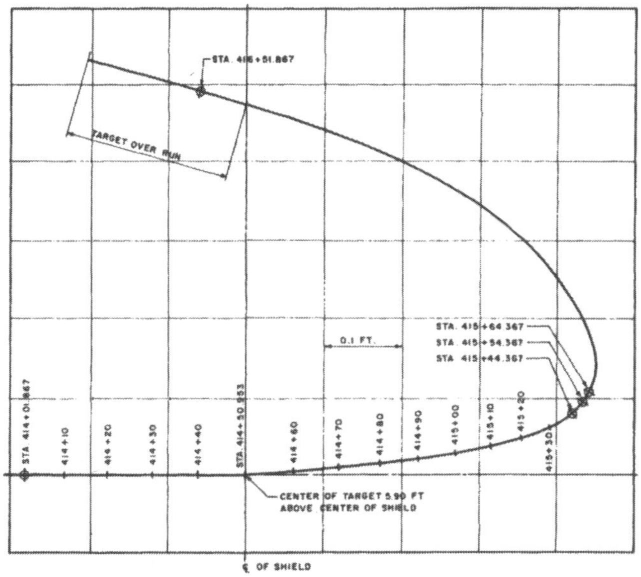

Fig. 3-26. Laser target.

determined control line. In lieu of moving and repointing the laser, beam diverter prisms can be used to define the next control line (see Figure 3-4).

Two targets, called the front target and the rear target, are mounted on the shield or tunneling machine, centered on a line parallel to the longitudinal axis and from 4 to 10 ft apart. The rear target is transparent and the leading target opaque. The targets are intersected by the laser beam, which produces a bright red spot on the target. Theoretical points of intersection between laser beam line and targets are calculated in advance for each shield location. The theoretical points of intersection are plotted on the targets and connected by a curved line (see Figure 3-26).

Fig. 3-27. Laser target mounted on shield.

The shield or tunneling machine is guided by attempting to maintain coincidence of the actual laser line intersection points with the predetermined intersection points on the target.

Calculation of Offsets to Laser Line from Tunnel Centerline. The laser line and centerline of the tunnel are plotted in plan and elevation (see Figures 3-28 and 3-29). Several trials may be necessary to find the best location of the laser line, and the following guidelines are observed in locating the laser line:

1. Find the longest unobstructed line of sight to reduce the required number of changes of laser positions.

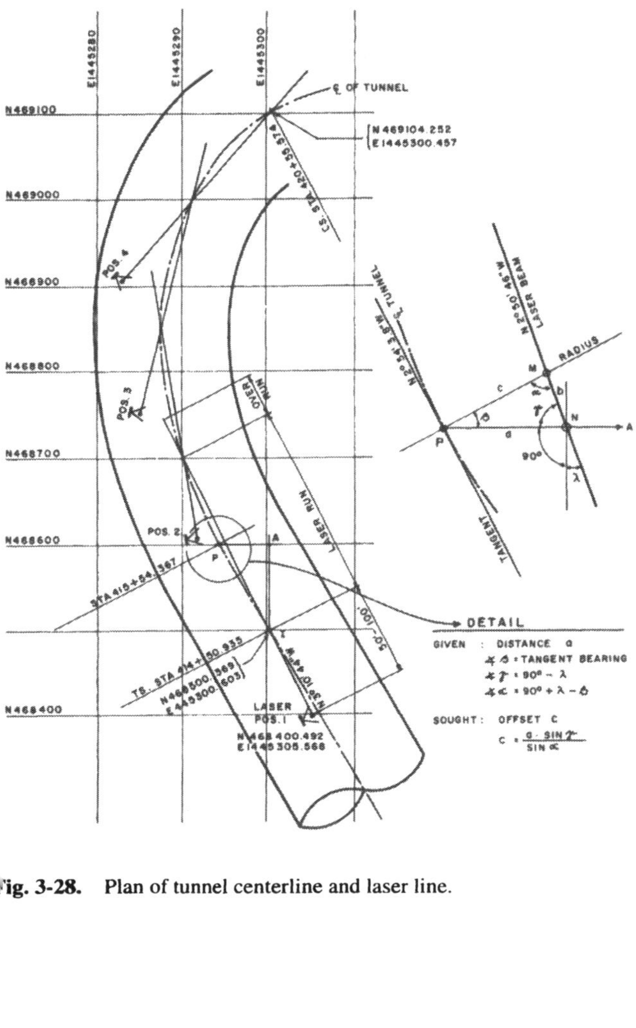

Fig. 3-28. Plan of tunnel centerline and laser line.

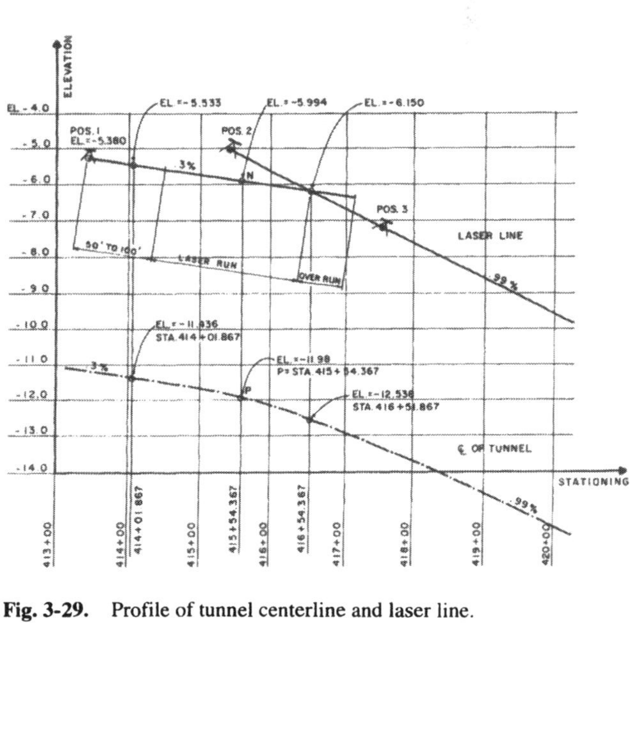

Fig. 3-29. Profile of tunnel centerline and laser line.

2. Select a laser position that is out of the way of passing tunneling equipment.
3. On tangent and flat curves, the length of laser line is limited by diffusion of the laser beam (practical limit about 1,000 ft). The beam may be focused at any point from 50 ft to infinity. At 1,000 ft, the spot can be concentrated to 1-in. diameter.
4. The target offsets for the projected laser line cannot exceed the size of the laser targets, which is often restricted due to space requirements for other equipment.
5. At the end of each laser line, an overrun is provided. This gives the heading engineer the opportunity to make the change of laser positions and targets at any convenient time while the shield or tunneling machine operates in the tunnel section covered by the overrun.

Once the coordinates and the elevation of the laser position, as well as the lateral bearing and slope of the laser line, have been determined, horizontal and vertical target offsets are calculated as shown on Table 3-7 and Figures 3-28 and 3-29. Then the target offsets are plotted on the target, as shown on Figure 3-27. The calculations are performed under the assumptions that coordinates and elevation of points on the tunnel centerline are given at 5-ft intervals. Also given are the tangent bearing and the slope of the tunnel centerline at each point. If they are not given, they must be determined by standard survey methods.

Positioning the Laser. The laser has to be positioned in its *X*, *Y*, and *Z* coordinate positions at the laser position point, and its laser beam has to be set to beam along the predetermined laser line. Several ways of mounting the laser tube are in use. Surveying lasers are equipped with a standard hub for tripod or bracket mounting (see Figure 3-26). The base is leveled by leveling screws and circular levels. In addition, coarse and fine adjustment of azimuth and elevation of the laser beam are provided.

Another way of mounting the laser tube is to secure it inside a pipe with adjustable wingbolts (see Figure 3-19). The pipe, in turn, is mounted by adjustable brackets to the tunnel lining.

Regardless of the method of mounting the laser, safety checks have to be installed to alert the surveyor if the laser drifts off alignment or is hit by construction equipment. A good method of checking utilizes a control target made of a piece of metal with a hole just large enough for the laser beam to pass through (see Figure 3-19). The control target is set on the laser beam line between the laser and the shield and the laser beam passes through the hole of the control target. Should the laser move, the disturbance is quickly noticed.

Automatic Tunnel Boring Machine (TBM) Guidance

Since the late 1970s, tunnel guidance systems have been developed that make use of computers and light-sensitive target screens to automatically indicate all tunnel boring machine (TBM) coordinates and attitude angles. This information can be displayed digitally. For example, TBMs used on the Channel Tunnel project were capable of continually measuring TBM position and predicting misalignment at the mining face (ZED 260 system).

Table 3-7. Horizontal and Vertical Offsets from Centerline Tunnel to Laser Line

		Column 1	Column 2	Column 3	Column 4	Column 5
Horizontal Offset	1. Point P at station	414 + 01.867	415 + 44.367	415 + 54.367	415 + 64.367	416 + 51.867
	2. Tangent bearing	N3 10' 44"W	N2 57' 7.9"W	N2 54' 3.8"W	N2 50' 41"W	N2 7' 48.7"W
	3. Coordinates of point P	N 468 451.359 E1 445 303.325	N 468 593.646 E1 445 295.546	N 468 603.633 E1 445 295.035	N 468 613.620 E1 445 294.534	N 468 701.035 E1 445 290.686
	4. Coordinates of intersection laser line and curve	N 468 500.369 E1 445 300.603	N 468 500.369 E1 445 300.603	N 468 500.369 E1 445 300.603	N 468 500.369 E1 445 300.603	N 468 500.369 E1 445 300.603
	5. Distance P – A = Line 4 – Line 3	2.722	5.057	5.568	6.069	9.917
	6. Distance I – A = Line 3 – Line 4	49.010	93.277	103.264	113.251	200.666
	7. Distance A – N = I A X tan λ	2.437	4.637	5.134	5.630	9.976
	8. Distance P - N = Line 5 – 7	0.285	0.420	0.434	0.439	0.059
	9. Horizontal laser line offset Distance P – M = $\frac{P - \sin \gamma}{\sin \alpha}$	0.285	0.420	0.434	0.439	0.059
Vertical Offset						
	10. Elevation point P	-11.436		-11.987		-12.538
	11. Slope of centerline tunnel at point P	0.30%		0.48%		0.64%
	12. Elevation of laser line at point P	-5.533		-5.994		-6.150
	13. Slope of laser line	0.30%		0.30%		0.30%
	14. Distance P - N = Line 10 – Line 12	5.903		5.993		6.388
	15. Vertical laser line offset: Distance P – M = $\frac{P - \sin \gamma}{\sin \sigma}$	5.900		5.991		6.388

Note: All dimensions are in feet.

The positional information is described in X and Y coordinates of the survey coordinate system (left/right and up/down in the tunnel) and the distance (Z coordinate) in the direction of tunnel centerline. The angle measurements required may be interpreted as roll (rotation about the tunnel axis), look-up or overhang (deviation from horizontal), and lead (rotation about a vertical axis). Of the angle measurements, roll and lock-up are obtained by the use of inclinometers, measuring an angle relative to the plumbline.

The X and Y measurements at the target plane and the horizontal angle between the target and the laser beam are measured by a specially fabricated target. The target intercepts the laser beam and measures the position of the laser relative to the center of the target. The beam strikes a screen at the front of the target, and the spot so formed is imaged onto two linear photodiode arrays. This provides the basis for the position measurement on the X and Y coordinates. The screen allows some of the laser beam to pass through unhindered to a third photodiode array, to measure the angle between the target and the laser beam.

The tunneling environment with its atmospheric turbulence and dust requires the target to cope with varying laser beam intensities and varying spot sizes. The target unit has its own dedicated signal processing computer, allowing the incorporation of such features as autoranging to compensate for varying signal gain and averaging techniques to enhance laser spot resolution.

In curved tunnels, a laser theodolite equipped with servo motors driving horizontal and vertical motions, encoders to read vertical and horizontal angles, and distance-measuring capability is used. Using a data transmission module on the laser theodolite, laser direction can be adjusted remotely by an operator in the on-site office, and the new distance and direction is then automatically transmitted to the site-office computer. Measurements from the inclinometers and target unit determine the position and attitude of the TBM. The calculation that is required has been outlined earlier under "Calculation of Offsets to Laser Line from Tunnel Centerline." To perform offset calculations, a site-office computer, complete with display unit and printer, is usually connected to the underground system by cable. The coordinates of the theoretical tunnel axis are compared with the TBM position and attitude data to determine shield deviation from theoretical values.

A control unit in the machine operator's cab displays all deviations of the TBM from the theoretical tunnel axis; these deviations include horizontal and vertical displacement X and Y in the target location, the predicted X and Y position at an optional drive distance, the shield lead, and the shield pitch. The machine operator can steer the TBM to conform with the theoretical tunnel centerline using the information displayed in the control unit. The machine operator can change the direction of the TBM by actuating the appropriate shove jacks and increasing or decreasing jacking pressure of the shove jacks. These are arranged around the perimeter of the shield tail to push the shield forward by extending the jack shoes against the tunnel lining erected in the tail of the shield.

Robotic Self-Tracking Total Station System for Tunnel Guidance. Instead of using a fixed laser beam for shield position measurement, a robotic Total Station instrument can be mounted on the tunnel wall with a prism reflector mounted on the TBM behind the cutting head. The total station is servo-driven to focus on the prism reflector. At each shove interval, it measures vertical and horizontal angles, and the distance to the prism. Two inclinometers measure pitch and roll of the TBM. The data are transmitted to a computer in the TBM operator's office, where X and Y deviations of the TBM and attitude angles can be determined and displayed for the TBM steering.

Articulated Shields. On some articulated shields, it is impractical to position the target unit on the front shield. This has necessitated the development of an articulation interface unit that can monitor the extensions of the articulation rams and use these values in determining the position and attitude of the front shield in relation to the tail shield where the laser target is mounted.

Tunneling Accuracy and Correction of Deviation. Typical tunnel driving specifications require that the as-built centerline of the tunnel should be located within the limits of a 3-in. diameter bull's-eye centered on the theoretical tunnel centerline. Thus, for a tunnel heading of about 5,000-ft length, the ratio of precision is

$$\frac{0.125 \text{ ft}}{5{,}000 \text{ ft}} = 1:40{,}000$$

The following alignment errors must be absorbed by the permissible tolerance:

1. Survey errors of the primary surface survey
2. Errors encountered during transfer of line and grade from the surface to the heading
3. Inability of the construction forces to keep the tunneling equipment on the indicated alignment

Experience shows that a deviation of the tunnel alignment (items 1 and 2 above) of magnitude of 0.1 ft has to be expected for a heading of 5,000-ft length. Proportionally larger or smaller deviations have to be anticipated for longer or shorter headings.

The deviation caused by the inability of the construction forces to keep the tunneling equipment on line and grade (item 3 above) depends on the equipment operation and ground conditions. The greatest deviations seem to occur in areas where ground conditions vary, e.g., where soil lenses occur. Under normal ground conditions, deviation is kept to less than 0.1 ft. In soft soil, however, the deviation may reach 0.5 ft. Therefore, it is apparent that the required accuracy may not be obtained in areas where difficult tunneling conditions exist.

In rock tunnels, it is possible to reset steel sets in order to meet the specified tolerance requirements. Since it is virtually impossible to reset lining in soft ground tunnels after the tunnel is driven, it is recommended that some additional vehicle clearance beyond the specified tolerance be incorporated into the tunnel design to absorb deviations of the magnitude outlined above. If clearance is not provided, realignment of the track or roadway centerline may be necessary to fit the as-built conditions.

SURVEY FOR CONSTRUCTION OF IMMERSED TUBES

Horizontal and Vertical Control

All tubes and bridges have different conditions for survey and construction control due to terrain, climatic conditions, reach of water crossing, vessel anchorage, vessel traffic, density of shoreline development, restrictions imposed by military reservations, parks, penal institutions, etc. Short tubes, less than a mile long, usually need three intervisible primary monuments at each end, with the tube centerline being defined, if possible, by the line between one centerline monument on each shore. Tubes longer than a mile may need additional monuments on islands, piers, bridge footings, or other sites near the tube centerline to serve as additional control during construction, when construction equipment may impede line of sight along centerline.

These monuments should be tied to National Geodetic Survey control stations or other primary monuments selected for the project, using dual-frequency GPS to attain 1:70,000 horizontal accuracy. Elevation of all monuments should be determined to Second Order Class 1 accuracy, based on NGS survey monuments whose historical record confirms little or no settlement. To ensure that the elevation of monuments at each end of the centerline is nominally correct, a level survey should be conducted between the two monuments by "Valley Crossing" methods, if site conditions permit. (This assumes that it is not feasible to conduct a level survey directly between end monuments.) The Valley Crossing method entails two calibrated level instruments or first-order theodolites sighting simultaneously in each direction to determine elevation difference between instruments.

If site conditions render Valley Crossing levels infeasible, closed level circuits should be run from each end monument to a temporary benchmark established on the shoreline. Then, during a windless period of slack tide, the elevation difference between the temporary benchmark and water surface should be measured simultaneously at both ends. This observation series should start one hour before predicted slack tide, and continue with measurements repeated at 15-min intervals until one hour after apparent slack tide. Unless the tube alignment crosses an area of excessive currents, elevation difference derived from the water transfer measurements should agree with direct level elevation of the TBMs within 0.2 ft. A third option to confirm the agreement in elevation between controlling end monuments is to determine the elevation (referred to the spheroid) of each end monument and each NGS benchmark using GPS survey and computational procedures. A disagreement larger than 0.3 ft in any of these procedures may indicate possible error in elevation of the primary benchmarks, or errors in leveling from the primary benchmarks to the controlling end monuments.

Mapping. 1 in. = 40 ft or 50 ft photogrammetric maps with 2-ft contour intervals are prepared over the terminal sites, and hydrographic surveys of a 2,000-ft wide corridor centered on the tube alignment are conducted along cross sections at nominal 200-ft intervals.

Subbottom, electromagnetic toning, magnetometer, and sonar side scan surveys are conducted at this time if needed to locate pipelines, cables, and the like. Surveys may also be needed to position rigs for geotechnical surveys and borings.

The foregoing photogrammetric mapping and hydrographic survey data is composited into mylar map sheets covering the project corridor at 1 in. = 100 ft or other suitable scale with coordinate grid, monument locations, hydrographic spot elevations and/or contours, boring locations, notations indicating horizontal and vertical datum, monument coordinates and elevations, scale bar, date of survey, north arrow, etc.

Shipyard Survey of Tube Sections. As the final horizontal alignment of a tube being laid is solely governed by the tube geometry and the relationship of its inboard end with the outboard end of the adjoining tube in place, a mathematical model of key points on each constructed tube section is needed to determine fit and angular relationship between the ends of adjoining tube sections. This model is constructed by first establishing a precise reference baseline in the concrete ways of the shipyard, affixing temporary survey targets to key points on the tube (before launching), and conducting a survey to determine local *XYZ* coordinates of each key point relative to the shipyard baseline. This survey can be done either by triangulation or by Total Station survey using reflective targets at the tube key points. In both cases, the reference baseline monuments serve as origin for coordinates and elevations. Each key point is observed from at least two baseline monuments, reading three sets of horizontal and vertical angles using a 1-sec theodolite or Total Station. Height of instrument should be measured to 0.001 ft at each instrument setup (Figure 3-30).

The *XYZ* coordinates derived from this survey describe the relationship between key points on an inclined model, because the tube is in an inclined attitude on the ways. These coordinates are then rotated to describe the model as it would be when the tube is laid on its design slope. The key point coordinates of the inboard end of the tube can then be compared with the outboard end coordinates of the adjoining tube in place. With this data (especially coordinates of the two key points on the lateral axis of the tubes), the final location of the outboard end of the tube to be laid can be projected, and shims for the joint can be designed if needed.

During steel construction of the tube in the shipyard ways, the contractor's surveyors will conduct surveys and layout out-tube centerline and other points controlling construction.

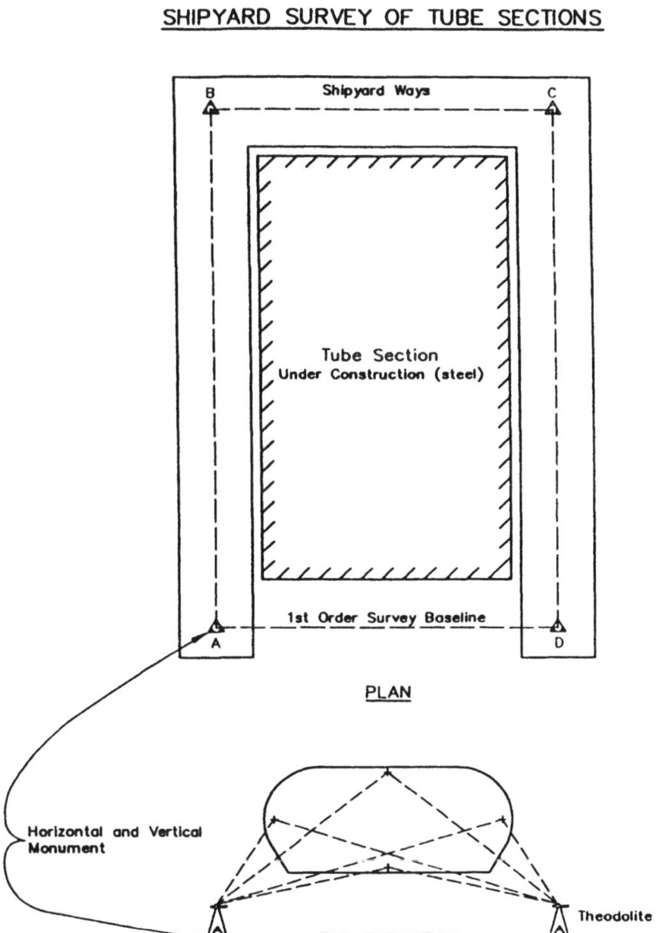

Fig. 3-30. Survey baseline and procedure used to validate tube sections before launching (San Francisco BART, Trans-Bay Tube).

These surveys may be by conventional angle-and-distance methods, with theodolite plumbed so that all surveys are referred to a horizontal plane and basic survey positions within the structure. Or shipyard methods may be used where a plane is established, generally parallel with the slope of the ways without regard to the plumbline. This plane serves as the basic reference for all surveys and construction done in the tube. In either case, there must be coordination between the contractor and engineer's survey so that the location of key points on the structure, specifically points defining the tube centerline (or control line) and points defining the attitude of the lateral and vertical axis of each end, can be measured and compared prior to launching.

In conducting surveys during tube construction, unequal heating of the sides of the steel tube structure due to sunlight can create differences in the attitude of tube elements, particularly the angle the ends make with tube centerline. A 15° F difference in temperature along one side of the tube can cause linear expansion of about 0.03 ft on the length of one side of a 300-ft- long tube. Assuming a 30-ft-wide tube, this could cause a deviation from alignment design at the outboard end of the tube (when laid) of about 0.3 ft. Surveys made to determine critical dimensions prior to launching should be scheduled to minimize the effects of differential expansion.

Use of Laser Beams to Align Dredges and Screeding Barges. A trench of 60-ft width was dredged in the bay, and a 2-ft gravel foundation course was placed in the bottom of the trench to serve as the foundation of the tube sections (see Figure 3-31). Both the dredge excavating the trench and the screeding barge that placed and screeded the foundation course were guided by laser beams. Dredge heading accuracy of 3 in. was accomplished, and the foundation course was screeded to an accuracy of 1.8 in. The principle of dredge control is illustrated in Figure 3-32. The circular beam spot of the alignment laser was converted into a line beam shape by a fan beam accessory lens. Dimensions of the beam were

4 to 6 in./mi	narrow dimension
850 ft/mi	wide dimension

The operator sights back on the narrow laser line to align the dredge or barge. He assures that the laser line has not shifted from its predetermined bearing by turning around to see the beam reflected by a retroreflector mounted on the other shore.

Tube Sections. Between ventilation buildings, the 19,113-ft length of tube was composed of 57 individually constructed reinforced concrete sections, each with a continuous exterior steel shell. The steel shells were fabricated in a shipyard and launched prior to installation of the reinforced concrete. The sections ranged in length from 273 to 336 ft and were 24 ft high and 47 ft wide. The tube contained both horizontal and vertical curves, requiring design of a number of curved tube sections. Of the total of 57 sections, 15 were curved horizontally, 4 were curved vertically, and 2 sections had vertical and horizontal curves built in (see Figure 3-33). The remaining 36 sections were straight.

Surveys During Placement. Tubes were fitted with "gland" or "snorkel" type pipes and survey towers extending from the tube at the bottom of the bay to the water surface (see Figure 3-34). Control of the tube position during placement was accomplished by using primary control monuments as reference and triangulation to the survey towers. On some tubes, the survey pipes penetrated into the tube, so that the survey tower location mark in the gallery floor inside the tube could be observed directly from the top of the survey tower above the water surface. Other tubes were fitted with survey marks in the top skin of the tube, and this also could be sighted from the top of the survey tower. Prior to placement, the survey tower and sight pipe were installed on the floating tube with sufficient inclination of the sight pipe so that, when the tube was placed on its prepared foundation, the 18-in. sight pipe would be approximately vertical. A vertical collimator (optical plummet) was mounted on a stand on top of the sight pipe, with lateral motion provided

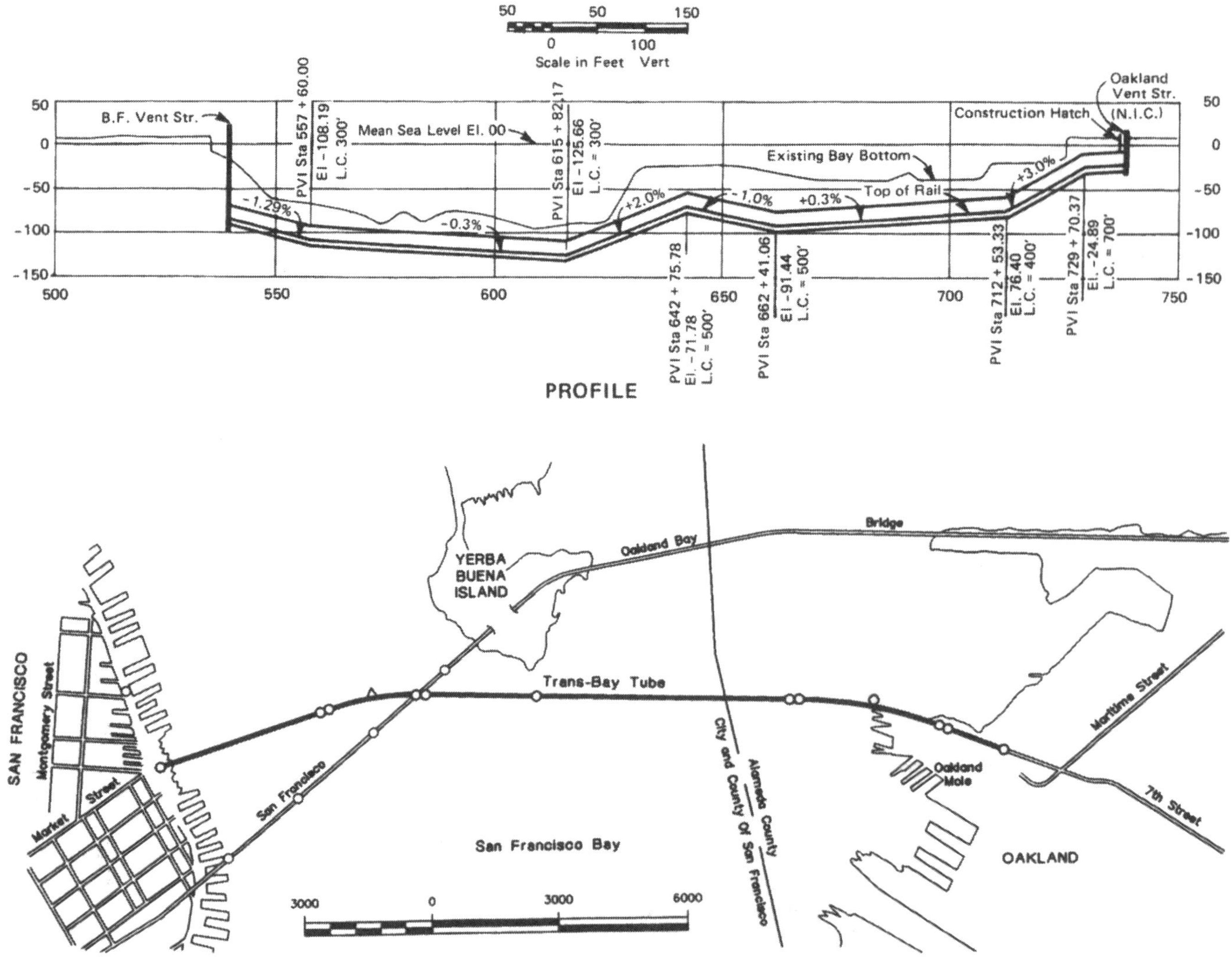

Fig. 3-31. Plan and profile of Trans-Bay Tube.

to ensure vertical centering over the survey mark at the bottom of the pipe, or over the gallery floor mark in the case of "gland" type pipes. A light was lowered into the pipe for illumination. During placement, it was necessary to relevel and recenter the collimator frequently so that the alignment surveyors could accurately observe the placement line. After the joint was made and dewatered, the collimator was again leveled and centered, and alignment surveyors made the final observation on as-placed line.

As-placed grade of the tube was observed by taking rod readings to the top of the sight pipe from level instruments set up on shore or bridge footing control monuments, or from survey towers mounted on in-place tube sections.

TUNNEL MONITORING SURVEYS

Repetitive surveys to monitor the stability of existing tunnel or tube sections are needed where a new structure is being built in the vicinity of the existing tunnel. These surveys are conducted at frequent intervals, usually at off-use hours, and the survey data is immediately processed to determine if any attitude or dimension changes are occurring in the existing structures as a result of the nearby tunnel operations. Survey requirements include establishment and/or verification of primary horizontal and vertical reference monuments; horizontal and vertical surveys to transfer coordinates and elevations from primary control into the tunnel; surveys to monitor changes in alignment, elevation, and shape in the tunnel; and photography of tunnel walls to document preconstruction condition.

Level Surveys

To monitor absolute elevation changes in an existing tunnel, precise leveling is required to periodically transfer elevation from stable benchmarks situated outside of the work area down to the monuments or markers set to monitor vertical movement in the tunnel. At least two outside primary bench-

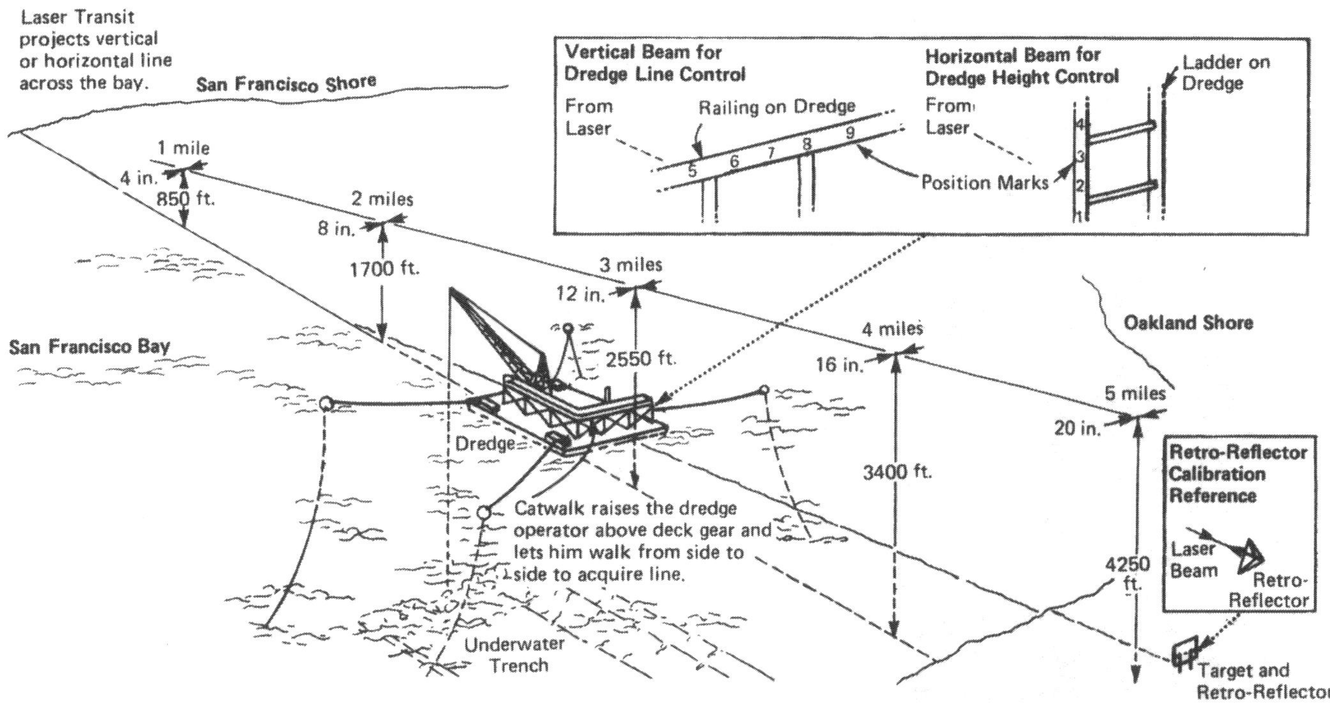

Fig. 3-32. Principle of dredge and screeding barge control.

marks should be used to serve as basic elevation reference, and the record elevation difference between these two monuments should be confirmed by direct leveling prior to leveling down into the tunnel. Disagreement between the elevation of reference benchmarks indicates instability in one or both benchmarks, and this requires selection of new reference benchmarks to ensure accurate level surveys in the tunnel.

Surveys to monitor relative elevation changes between adjacent tunnel segments require benchmarks or markers set in the crown or invert at frequent intervals (10, 20, 50 ft, etc., as determined by the engineer) and precise levels conducted at frequent intervals through these markers, recording elevations to the nearest 0.001 ft (Figure 3-35). Level instruments must be accurately adjusted, and all markers should be measured at least twice to confirm the survey. Relative elevation changes along the tunnel may also be monitored by installing, for the duration of the project, a stepped series of water level gauges using an open system of PVC pipes and upright sight tubes. This system enables accurate and rapid reading of the water level at each sight tube to determine relative changes in the elevation of the structure.

Electrolevel beams installed along the wall or crown of a tunnel electronically detect changes in slope in a 5-ft segment

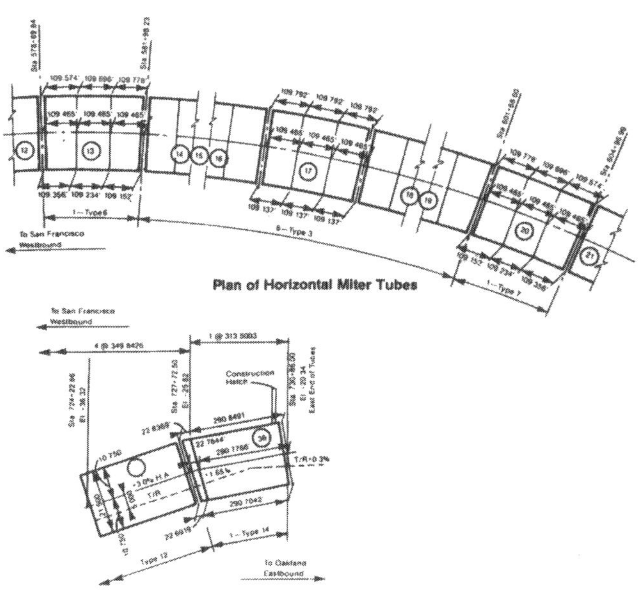

Fig. 3-33. Plan and elevation of tube sections with built-in horizontal and vertical curves.

Fig. 3-34. Tube placement; survey tower and sight pipe.

Fig. 3-35. Determining elevation of crown during deformation survey; note "bar-code" rod for automatic reading using Leica/Wild NA3000 level.

of the tunnel and transmit this information by wire, together with time of measurement, to a central data processor. A string of beams can be installed to monitor slope data over any length of tube segment, with data acquisition and processing timed to meet inspection requirements. Processing of the electrolevel data provides continual surveillance of slope data converted to relative elevations, with recording and alarm capabilities (Figure 3-36).

To measure rotation of the tunnel, difference in relative elevation is measured by direct leveling between semipermanent markers epoxied or bolted to tunnel walls at instrument height. The markers have an identity number and a clearly scribed horizontal line or vertical graduated scale to serve as a reading point. Difference in elevation is determined initially by setting up a second-order level at the approximate tunnel centerline, midway between the two markers, and observing the level readings at each wall. Changes in the relative elevation of the markers, as determined by subsequent measurements, indicate rotation. The most accurate results are obtained by using wall markers with a scribed horizontal line, and level instruments with plane parallel plate micrometers.

Alignment Surveys

Relative alignment of the tunnel is surveyed by installing alignment markers in a line in the invert or crown of the tunnel, and periodically resurveying this line by offset procedures to determine and measure lateral deformation. Distances between markers are initially measured and tied to stationing to facilitate identification and reference to plans. Distances are not normally remeasured during resurvey.

Deformation

Deformation of the tunnel cross section is directly measured by repetitive extensometer readings between markers affixed to the tunnel wall or lining. The extensometer measures changes in tunnel diameter on selected axes of the tunnel cross section (Figures 3-37 and 3-38).

Changes in cross section can also be monitored using repetitive theodolite survey of targets installed in the tunnel wall. A survey baseline consisting of markers set at approximately 100 ft intervals is established in the invert or crown and tied by precise horizontal and vertical survey to the tunnel baseline and vertical datum. Double-faced, reflectorized survey markers are established in the tunnel wall to define the horizontal axis, and in the crown to define the vertical axis of each selected cross section. The initial survey consists of a precise EDM traverse through the baseline markers, using Wild T2002 or an equivalent instrument with data collector, precision surveying tribrachs, and targets. During this survey, horizontal and vertical angles are observed, and distances are measured to all baseline stations and to all reflectorized targets and monuments in the walls, crown, and invert. These measurements are made in both plus station and minus station direction so that each wall marker is ob-

Fig. 3-36. Horizontal electrolevel beams (El Beam Sensors) installed in BART subway tunnel to monitor vertical movement during tunneling operations for San Francisco Muni Metro Turnback Project.

Fig. 3-37. Using tape extensometer to measure springline during deformation survey for San Francisco Muni Metro Turnaround Project.

served by the baseline station before it and the one after it, providing redundancy to minimize undetected errors. Distances are measured only during the initial survey and are not needed during subsequent surveys (Figure 3-39).

The survey data is processed by computer program designed to accept survey data and compute the following:

- Coordinates of baseline monuments, adjusted by least squares
- Elevation of baseline monuments
- Alignment marker offsets
- Mean value and spread of elevations and coordinates of reflectorized wall targets using redundant data
- Slope and length of the horizontal line joining opposite wall targets
- Length of the vertical line joining crown target to invert monument

The computer program must be capable of efficiently directing survey data into storage, retrieving data from storage, and organizing repetitive survey data to enable plotting, screen viewing, and printout of coordinates, elevations, lengths, and slopes, together with numerical differences so that various sets of survey data can be readily compared.

Photography. Photography of tunnel walls is needed to document preconstruction conditions. This task is facilitated by constructing a movable dolly that can be transported easily down the tunnel alignment and is fitted with a camera mount that positions the camera at approximate tunnel centerline, and enables it to rotate segmentally. This allows it to take a full round of pictures, making a panorama of the tun-

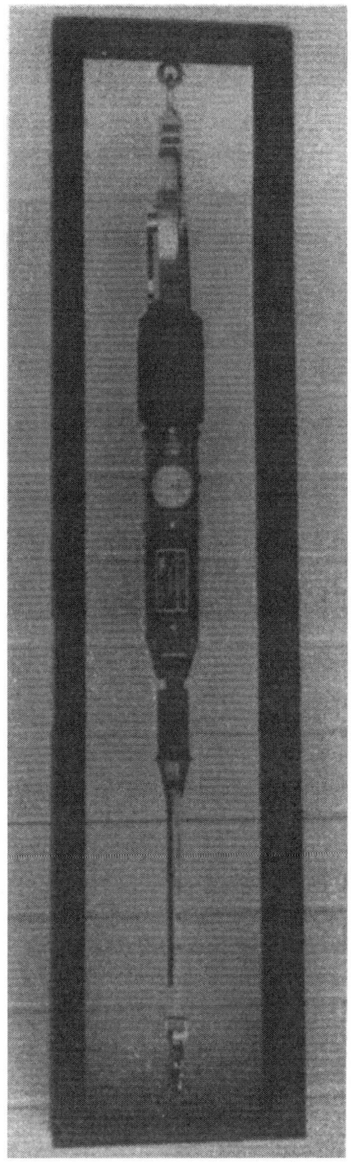

Fig. 3-38. Tape extensometer in calibration frame (Slope Indicator Model No. 51811500).

nel wall. Station difference (SD) between camera setups, in feet, is determined by the following:

$$\mathrm{SD} = \frac{0.8 \times \text{camera frame size (mm)} \times \text{radius of tunnel (ft)}}{\text{focal length of camera (mm)}}$$

Example:

35-mm camera
15-ft radius tunnel
41-mm focal length

$$\text{Distance between camera stations} = \frac{0.8 \times 35 \times 15}{41} = 10.2 \text{ ft}$$

At each exposure the camera is rotated 60°, requiring six exposures per station. Each frame should be adequately lighted. The starting frame of each round should include a

Fig. 3-39. Reflectorized survey targets epoxied to tunnel liner ring.

sign indicating station number, scale bar, and arrow showing direction of increased stationing. After completing the photography, the film is developed and check prints are made to ensure that image quality, scale, station identification, etc., are satisfactory. A full set of prints may be made later if needed. Black and white photography is generally suitable, but in older tunnels that display rust, water stains, corrosion, etc., color photographs may provide better documentation (Figure 3-40).

REPRESENTATIVE PROJECTS

WMATA

The Washington (D.C.) Area Metropolitan Transit Authority project is a 70-mile plus system with an additional 19 miles of track and 14 stations under construction. Two-thirds of the system is either tunnel or cut-and-cover construction.

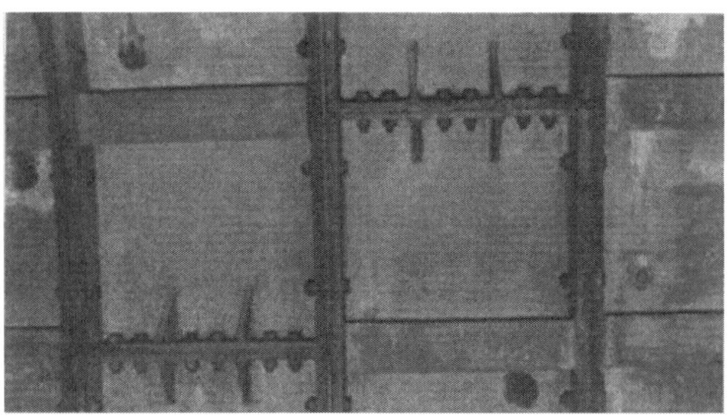

Fig. 3-40. Photograph documenting visible condition of tunnel lining prior to start of adjacent construction.

Survey services during construction are obtained by contract, with selection of the contractor based on qualifications. The contract surveyors, under the supervision of the WMATA Survey Coordinator, provide support as needed to the WMATA Resident Engineers on each construction segment. Specific survey support includes but is not limited to the following tasks: (1) Primary and secondary control surveys; (2) Verification of the structural contractors' surveys; (3) As-built quality control surveys to document construction compliance to design; (4) Earthwork and miscellaneous quantity surveys; (5) As-built trackway surveys; (6) Tunnel cross sections for determining best-fit track alignments; (7) Final track monumentation surveys; (8) Settlement and deformation surveys, etc.

Field and office techniques were developed to accelerate and simplify the task of surveying tunnel cross sections and appurtenances. High precision Total Station instruments linked to data collectors and mounted on brackets installed in the tunnel walls were used to measure angles and distances to reflective targets set at critical points defining the as-built tunnel sections. Specialized data collection, operating procedures, and software were developed to facilitate the processing of survey data into three-dimensional coordinates for rapid comparison with plan values and prior survey data. The software interfaced directly with data storage, and CAD plotting enabling rapid production of summaries, plans, and charts and selective comparison of recorded data surveyed over several time frames.

In summary, the application of specialized surveying techniques and development of unique software made it possible to automate data collection and postprocessing. This resulted in substantial increases in survey production and significant decreases in the time required for processing tunnel as-builts and quality control surveys while maintaining established survey accuracy standards. Due to the digital format of the as-built data, best-fit realignment calculations were performed more efficiently, reducing delays in the construction schedule without compromising the quality of the finished product.

Superconducting Super Collider Project (SSC)

The SSC entailed 54 miles of main tunnel and 62 shafts to house this planned Department of Energy project in Waxahachie, Texas. Because of the extreme precision required in collider physics, tunneling and construction tolerances were unusually demanding and surveying accuracy requirements challenged the outer limits of surveying skill and technology. Surveys for the primary surface networks, and for all tunneling control including quality control surveys, were the responsibility of the Architect-Engineer/Construction Management Consultant. Surveys to direct tunneling machines were the responsibility of the contractor.

In 1990, prior to final siting of the collider, a high order survey was conducted to serve as a basis for locating the final footprint of the collider, and for use in design of the primary control survey network. This preliminary survey included establishment of 180 monuments, 184 miles of double-run, first-order levels, and observation and adjustment of 12 master GPS positions to Order B procedures and accuracy.

After the survey control network was designed, specialized monuments were constructed with their foundation penetrating bedrock. Instruments designed to detect and measure monument movement due to earth pressures were imbedded in key horizontal monuments, whose locations were determined by geological studies and the geometry of the GPS survey planned for the project (Figure 3-41). The horizontal and vertical control networks would serve as a basis for all surveying and construction on the project, including tunneling, and assembly and alignment of accelerator components. GPS surveys were conducted between the horizontal control monuments using dual frequency full wavelength receivers, following procedures slightly more accurate than those specified for NGS Order A surveys. A comprehensive level network was conducted through the vertical control monuments, using instruments and procedures designed to attain better than first order class I elevations for project control (Figure 3-42).

Transfer of horizontal and vertical control from the surface primary monuments into the shafts and tunnels required innovative procedures and specialized instruments to achieve required accuracy, and to complete this work within a compressed time frame (Figures 3-18 and 3-43).

A primary underground horizontal control network was designed to minimize adverse effects of atmospheric refraction in the tunnel. Horizontal control was extended into the tunnel by forced-center Total Station traverses on relocatable survey brackets. Azimuth was periodically reinforced by reciprocal gyrotheodolite observations on selected legs of the tunnel traverse.

Surveys for control of boring operations were based on the primary tunnel traverse, and they were carried forward to the boring machines by Total Station traverse along the tunnel centerline. First Order Class I levels were extended into the tunnel from the shaft transfer points, with benchmarks installed at approximate 50-meter intervals. Accuracy of the underground surveys was ensured by the network designs, which included many redundant measurements, and by verification surveys, which were carried out at regular intervals and confirmed by closure of horizontal and vertical circuits in completed tunnel segments.

In October 1993, funding for the SSC was discontinued by vote of the U.S. Senate and House Conference Committee, thus ending one of the most challenging surveying projects in history. Although the accuracy requirements for the project were substantially more rigorous than requirements for tunnels for transportation, water, etc., the experience gained, the lessons learned, and the attainment of unheard-of construction accuracies may translate into cost and time economies in the future. Also, it may provoke a realization by designers, owners, and constructors that investment in quality surveying can yield rewards in increased production, and reduced construction claims and rework.

San Franscisco Muni Metro Turnaround (MMT)

The MMT project is a subway extension of the Muni Metro Subway under Market Street. The MMT facility includes an 840-ft twin tunnel section that connects to the

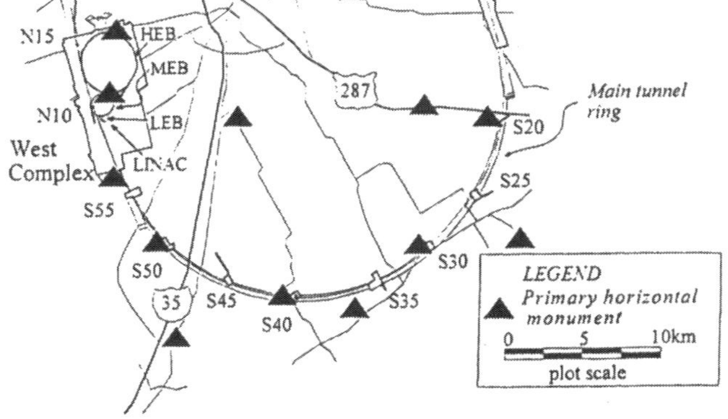

Fig. 3-41. Primary horizontal monuments, Superconducting Super Collider Project, Waxahachie, Texas. (Courtesy, Measurement Science, Inc.)

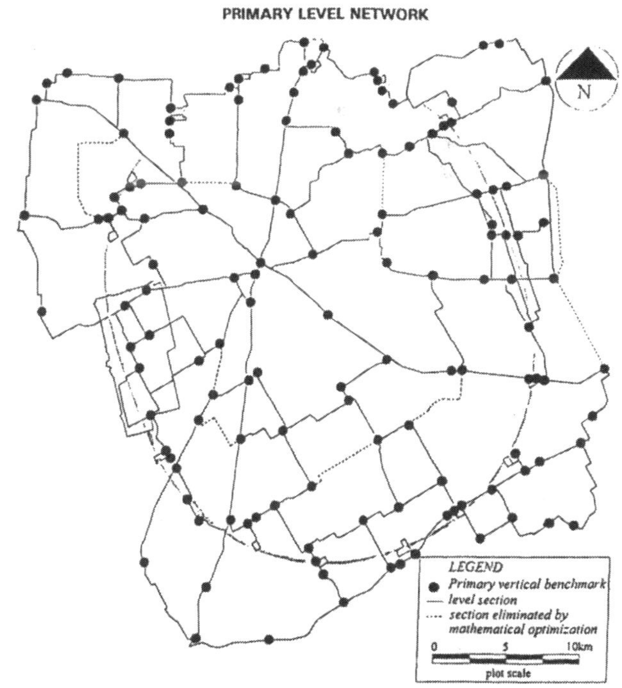

Fig. 3-42. Primary level network, Superconducting Super Collider Project, Waxahachie, Texas. (Courtesy, Measurement Science, Inc.)

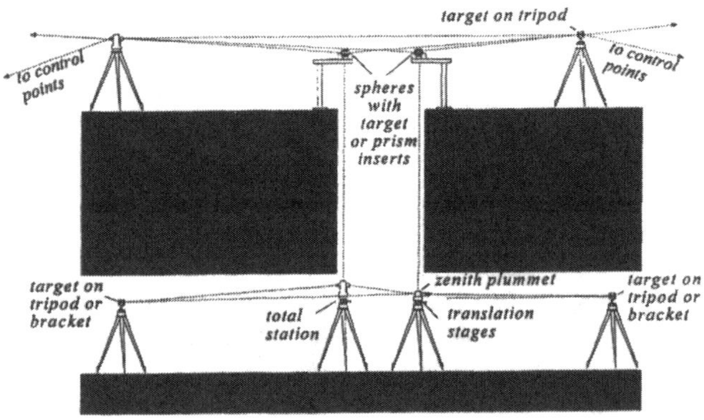

Fig. 3-43. Method used to transfer surface horizontal control down-shaft to tunnel control line, using Taylor Hobson spheres and zenith plummet, Super Collider Project. (Courtesy of Measurement Science, Inc.)

Embarcadero subway station, runs under Market Street, and turns south under the Justin Herman Plaza to the Embarcadero area (Figure 3-44). The section of the tunnel under Market Street overlays the San Francisco Bay Area Rapid Transit twin tunnels (Figure 3-45). The Muni Metro subway tracks are routed to a turnback facility in a 1,120-ft cut-and-cover section and then to a 370-ft U-wall structure, and they gradually ascend to the surface. The purpose of the project is to provide a more efficient track configuration to handle the expected growth in patronage over the next 20 years, and bring the main tracks to the surface in the median of the new Embarcadero Parkway for a future connection to a Muni Metro extension to Mission Bay.

All surveys for design and construction management were done by subcontract under the direction of the design/construction management consultant. These surveys include first-order primary horizontal and vertical control surveys, surveys to monitor stability of roads, walks, and buildings, deformation surveys in the lower level BART subway tunnels, photo documentation of buildings and BART tunnels, and as-built surveys during construction. As the entire site is in an area of imported fill, primary survey monuments were epoxied into existing sidewalks because it was felt that sidewalk monuments would be less susceptible to horizontal movement than concrete posts or rods driven into the fill material. Benchmarks were located on concrete building foundations and on the foundation piers of the nearby Embarcadero Freeway. A first order EDM traverse connected all horizontal monuments tying into existing city monuments and NGS horizontal control monuments in the area. First order levels were run through all monuments originating from existing BART benchmarks that have demonstrated vertical stability over many years.

After completion of the primary surveys, the October 1989 Loma Prieta earthquake affected the project site sufficiently to require a complete resurvey of the primary horizontal and vertical networks, and later demolition of the Embarcadero Freeway. The resurvey of the primary monuments indicated movement of up to 2 in. both horizontally and vertically on the sidewalk monuments, but benchmarks on building foundations and freeway piers showed no significant elevation changes.

Prior to starting construction, primary surveys were reconfirmed by a second resurvey of the horizontal and vertical networks. Line and grade were transferred into the BART subway tunnels; monuments, benchmarks, tiltmeters, extensometer mounts, and reflective targets were established in the tunnels; and initial surveys were made to serve as a baseline for deformation surveys during construction. Settlement points were located on the surface in roadways, sidewalks, and buildings to monitor stability during tunneling operations. BART subway walls and adjacent buildings were photographed to document preconstruction conditions.

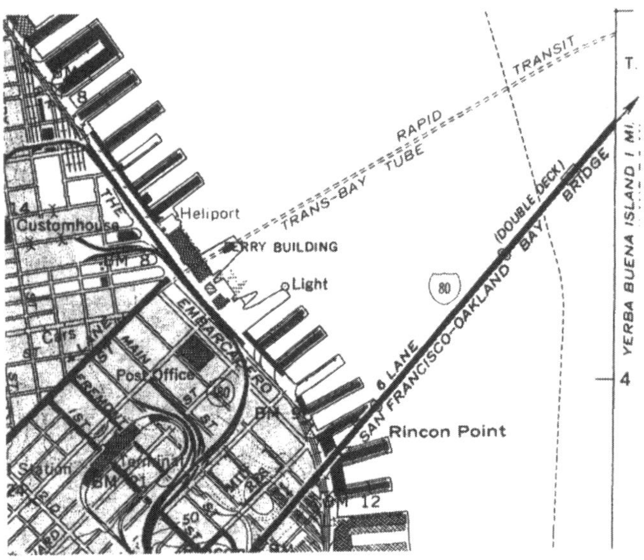

Fig. 3-44. San Francisco Muni Metro turnaround facility.

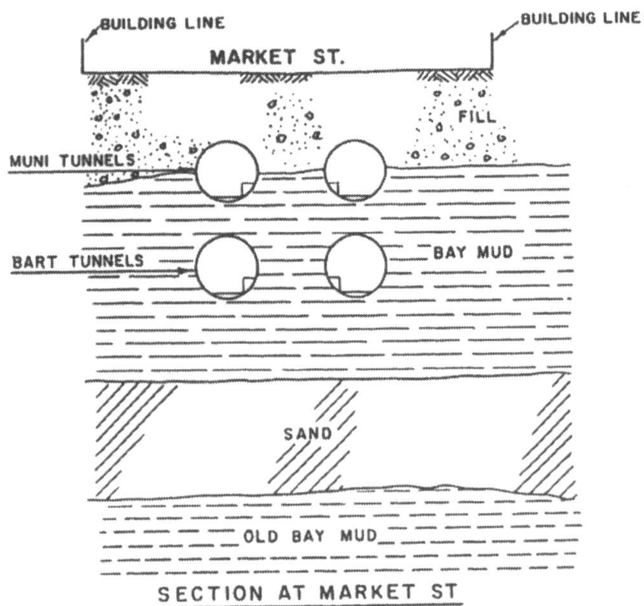

Fig. 3-45. San Francisco Muni Metro turnaround project.

During construction, the contractor's surveys extended line and grade from the primary monuments into the tunneling site, using EDM traverse to transfer control through the pressure lock into the tunnel. From this survey, the contractor established line and grade control for ring placement and for laser control of the tunnel boring machine. Deformation surveys in the BART subway tunnels and surveys to monitor roads and buildings were conducted daily by surveyors working under the direction of the design/construction management consultant. Quality control surveys were conducted weekly to verify line and grade and placement of structural elements.

Surveys for design and construction of the English Channel tunnel between England and France took place over many years, mainly from 1960 onward. Preliminary engineering work commenced after the project was authorized by a 1971 agreement between the French and British governments, and proceeded through 1974 until the project was abandoned when the final agreement between the two governments was not ratified. In 1981, Britain and France jointly issued a statement announcing their intention to proceed with the project, and in 1986 the Eurotunnel plan for a twin bore rail tunnel was selected by French and British governments. Tunneling began in December 1987 and was completed in 1991. "Fitting out," including rail tracks and walkways, power, ventilation, drainage, cooling, and servicing systems, continued to 1994.

The project consists of two railway tunnels and a service tunnel 31 mi long with terminal facilities and yards. Rail tunnels are 7.6-m (24.9-ft) diameter, and the service tunnel is 4.8-m (15.8-ft) diameter. Preliminary survey work for design and construction included connection of the main triangulation network between Britain and France, topographic and detailed surveys of land terminal sites, geological and geophysical surveys on land and subsea, offshore and coastal hydrographic surveys, and study of cross-channel level datum relationships. Global Positioning System (GPS) surveys were conducted in 1987 to verify and improve the control network (see Figure 3-46).

As a basis for computations, neither the French nor British National Grid nor UTM projections were suitable for

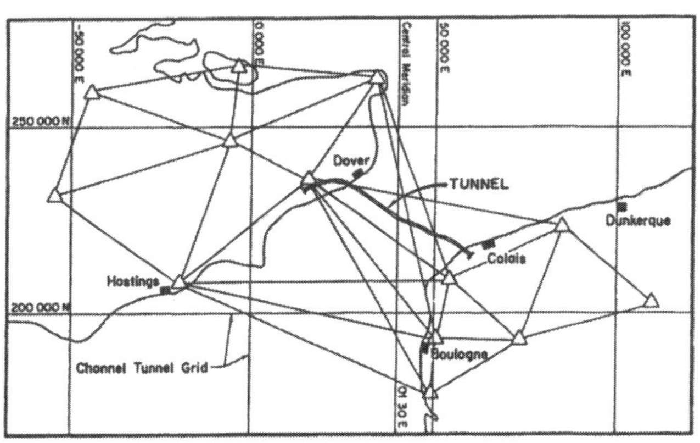

Fig. 3-46. Main control network and grid: Channel Tunnel survey.

the project because of inherent scale differences between ground and grid distances. A special grid (RTM87) was created using a Transverse Mercator projection with the Central Meridian passing near the center of the project corridor. The selection of this grid resulted in a grid scale factor of 1.000000 along the central meridian, thus eliminating the need to apply scale factor to convert from ground to grid distances and vice versa. A vertical datum (NTM88) was adopted 200 m below UK National Vertical Datum to eliminate the inconvenience of using negative elevations.

Detailed quality control procedures were developed specifying exact routines for setting up and operating instruments, and recording and processing survey data. Survey and office work was independently verified at each stage. Primary control and main control traverses were conducted to first order standards, and surveys for TBM guidance were to second order. Surveys in the tunnel were initially affected by lateral curvature of sight lines close to the tunnel walls, due to refraction caused by temperature gradients across the lines of sight. These effects were minimized in the tunnel control traverse by alternating each Total Station position to the opposite tunnel wall from its neighbor, and by frequent gyrotheodolite observations to confirm azimuth. Brackets bolted to the tunnel lining provided forced-centering capability for targets, theodolites, and gyrotheodolites to ensure exact repositioning during the tunnel control traverse.

Horizontal control was extended from the surface control network into the tunnel by EDM traverse via inclined adits, using forced-center pillars to ensure exact positioning of targets and surveying instruments. This traverse tied into the tunnel baseline and was confirmed by circuit closure through a companion adit. The underground baseline established by survey between the adits then served as control for both landward and seaward tunneling of the service and railway bores.

Elevations were transferred from surface benchmarks to tunnel monuments by precise leveling through both adits. Because of the steep gradient of the adits, precise leveling procedures could not be strictly adhered to without excessive instrument setups to maintain equality of foresights and backsights. Elevations were later verified by vertical EDM measurements from the surface upon completion of an access and ventilation shaft.

Eleven tunnel boring machines (TBMs) were used to excavate 3 landward and 3 seaward tunnel sections on each side of the channel, making a total of 12 tunnel sections. Surveys for guiding the TBMs were based on the main control survey monuments in the service tunnel. Each railway tunnel drive was controlled by its own independent traverse, and connected to the Service Tunnel control via the cross-passages at occasional intervals and adjusted where necessary.

Steering of the TBMs was controlled by a computerized guidance system based on a laser beam projected to a screen mounted on the TBM cutter head. Coordinates defining the position and alignment of the laser beam at each reset were established from the main control and loaded into an onboard computer. This established, by reference to the beam, the real-time position and attitude of the TBM, which was compared with the previously stored Design Tunnel Geometry. Variances between actual cutterhead position and design position were displayed digitally to the TBM operator, who made corrections to maintain zero variance.

Metropolitan Water Reclamation District for Greater Chicago TARP Project

The Tunnel and Reservoir Plan (TARP) was conceived by the Metropolitan Water Reclamation District of Greater Chicago (MWRDGC) as the solution to the problems of pollution and flooding caused by combined sewer overflows. The plan was the result of a study conducted by a committee comprising representatives from the State of Illinois, Cook County, the city of Chicago, and the MWRDGC.

TARP is divided into Phase I and Phase II. Phase I focuses on pollution control and consists of tunnels and pumping stations, while Phase II concentrates on flood control and consists of tunnels and reservoirs with a capacity of 41 billion gallons. Construction of Phase I began in 1975 and continues to this date. Phase I totals 109.2 mi of tunnels with diameters ranging from 8 to 33 ft. Currently, 63.9 mi of Phase I tunnels are complete and operational, 20.8 mi are under construction, and 24.5 mi remain to be constructed. To date, a total of 29 projects have been completed for Phase I of the TARP System, at a cost of more than $1.8 billion.

In the early stages of design, the Engineering Department of the MWRDGC awarded contracts to map the tunnel corridor in both topographic and planimetric formats. In addition, second-order monuments based on Illinois State Plane Coordinates and Chicago City Datum were set by surveyors contracted by the mapping consultants. Each series of monuments originated at NGS monuments and was situated near the centerline of the proposed tunnel. While some of the monuments were destroyed over time, a sufficient number remained to furnish adequate control for tunnel construction. The final design and contract drawing preparation were done by various consultants selected by the MWRDGC for their expertise. Property line information and Illinois State Plane coordinates were added to the planimetric mapping by the surveyors for use in right-of-way acquisition.

The acquisition of all right-of-way and permit applications were the responsibility of the right-of-way section of the Engineering Department, with assistance from the Law Department and special attorneys as needed. The preparation of legal descriptions and exhibits was done in house or by the design consultant, but always processed through the MWRDGC. Easements were acquired for temporary or permanent construction access as needed, and for subterranean passage.

On-site construction management was either by MWRDGC staff or consultants hired by the MWRDGC to act as field inspection teams. All layout and construction surveys were the responsibility of the contractors who were to mine the tunnel. The starting points for their surveys were the monuments set during the mapping portion of the project. These monuments were used by all of the contractors to establish their own control nets. In order to bring control into the tunnel, the contractors used similar, but different, methods to keep the mining of the tunnel within the tolerances set by the contract specifications.

One method for bringing horizontal and vertical control to the base of a tunnel currently under construction was for the contractor to plumb the construction shaft as it was being dug. Once the shaft was completed and the inspectors confirmed the plumb of the shaft, a starter tunnel was dug to allow for a longer baseline at the bottom of the shaft. When the starter tunnel was completed, horizontal control was established by the use of a theodolite fitted with a Wild GAK1 Gyro Attachment, from which true north could be determined. With horizontal control set, vertical control was brought into the tunnel by extending a tape down the shaft, a distance of some 300 ft below ground, from a benchmark on the surface. A Total Station Instrument was used to extend the control throughout the tunnel. As tunneling proceeded, the alignment was transferred to a Laserline T-2000 mounted to the tunnel wall. The laser supplied both horizontal and vertical direction to the tunnel boring machine, by being directed at a target mounted on the machine. Curves were negotiated by the use of diverters for the laser beam, which were mounted on the tunnel wall (Figures 3-4 and 3-47). MWRDGC inspectors monitor the alignment by checking into each dropshaft as it is encountered during the tunneling process. Since the coordinates of each dropshaft are known, they can be compared with the coordinates computed from the control system established in the tunnel, and adjustments can then be made to the path of the tunnel boring machine.

A different method was used by a contractor mining a tunnel under another contract. After establishing their control at the shaft sites, using a Lietz 4-B Total Station with a data collector and a Lietz Auto Leveling Engineer's Level, the construction shaft was dug. From their monuments, the

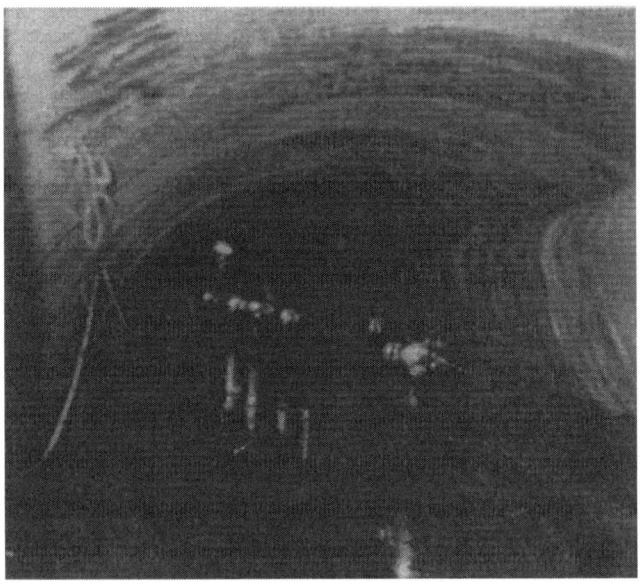

Fig. 3-47. Laser and beam diverter mounted on tunnel wall for tunnel boring machine gudiance, MWRDGC TARP project.

contractor made four notches on the top shaft ring, two each on the tunnel centerline and two on its perpendicular. Wires with flagging were then stretched between the notches, and an optical plumb bob set near the ends of each wire to transfer the centerline and its perpendicular to the tunnel floor. The benchmark was transferred down using a 300-ft steel tape. The starting station was established as the perpendicular, and the centerline was transferred to the shaft wall above the tunnel crown. The tunnel grade was transferred to the tunnel wall, 5 ft above the invert, to three pairs of tunnel spads. This enabled tunnel workers to do line and grade control with strings and plumb-bobs. The transfer of control from the surface into the tunnel was accomplished using a theodolite and a Leica engineer's level adapted for vertical sighting (auto plumbing).

The tunnel boring machine was guided by a laser beam mounted on the tunnel wall. Transparent targets were mounted about 45 ft apart on the front and rear of the machine. The use of double targets allowed the operator to check the position and attitude of the machine as it bored though the rock.

A procedure was developed to negotiate curves in the mining process using prisms and chord offsets. The curve was divided into 100-ft segments. This produced a small series of small curve portions, except for the beginning and end of the curve. Prisms were set for the 100-ft segments, and the chord offsets computed. Tapes were placed on each of the targets, and the operator kept the laser on the offset mark while mining the curve. Distortions caused by radial offset in the laser are compensated for in the laser adjustment, as it was moved and adjusted every 500 ft.

A Total Station and engineer's level were used to extend the tunnel traverse monuments through the tunnel. A north-seeking gyro was used at 1,250-ft intervals to check the line of the tunnel. As the mining proceeded past dropshaft sites, checks of shaft locations developed from the tunnel traverse were compared with the coordinates of the dropshaft sites developed from monuments at the surface. Adjustments to the tunnel monuments were made as needed using the compass rule.

The contract specifications stipulated the tolerances allowed in the tunneling process. One contract stated that an error of only 24 in. would be allowed in the lateral alignment over a tunnel length of 9.3 miles. The contractor was able to maintain tunneling accuracies well within the specifications. On a recently completed project, the largest error was 2.5 in. over a distance of 4,000 ft of tunnel. Since the tunnel accuracy was checked at dropshaft sites, the alignment was corrected before proceeding. Most tunneling deviations were attributed to the guiding of the boring machine or variations in the rock rather than the accuracy of the surveys.

REFERENCES

American Society of Civil Engineers (1982) *Engineering Surveying Manual* (Revised). Manuals and Reports on Engineering Practice No. 64, New York.

Anspach, J.H. (1994) "Integrating Subsurface Utility Engineering into Damage Prevention Programs," *Proceedings of the 1994 Excavation Damage Prevention Workshop*. National Transportation Safety Board.

Chrzanowski, A. (1993) *Modern Surveying Techniques for Mining and Civil Engineering*. Pergamon Press New York.

Chrzanowski, A., Greening, T., and Robinson, G.L. (1994) *Geodetic Control Survey for the Superconducting Super Collider Project*. Survey World (UK).

Falk, M.O., and Robinson, G.L. (1994) "Survey Engineering on the Super Collider," *P.O.B. Magazine*, August–September.

Falken, D.R., Jr. (1993) "Taking Survey Automation Underground," *P.O.B. Magazine*, April–May.

Federal Geodetic Control Committee (1974) Classification, Standards of Accuracy, and General Specification of Geodetic Control Surveys, NOAA, February (reprinted, January 1976).

Federal Geodetic Control Committee (1984) Standards and Specifications for Geodetic Control Networks, NOAA, September, (reprinted, August 1993).

Federal Geodetic Control Committee (1988) Geometric Geodetic Accuracy Standards and Specifications for Using GPS Positioning Techniques, NOAA. Version 5.0, May.

Kor, J.S., and Scapuzzi, D. (1991) "Performing the Master Surveys for the Superconducting Super Collider," *P.O.B. Magazine*, August–September.

Leick, A. (1994) *GPS Satellite Surveying*. Wiley-Interscience, New York.

Manual of Photogrammetry, 4th ed. (1980) American Society for Photogrammetry and Remote Sensing.

Mayo, R.S., and Richardson, H.W. (1980) *Practical Tunnel Driving*. McGraw-Hill, New York.

Moffitt, F.H., and Bouchard, H. (1987) *Surveying* 8th ed. Harper and Row, New York.

Price, A. (1994) "The Anglo-French Tunnel—Main Control," *Surveying World (SW) Journal for Land Survey, Hydrographic Survey and Land Information Management*, Vol. 3, No. 1.

Radcliffe, E.F. (1991) "Beneath the Sea: Channel Tunnel," *P.O.B. Magazine*, June–July.

The Institution of Civil Engineers (London) Proceedings (1992). *The Channel Tunnel. Part 1: Tunnels*.

Wells, D., et al. (1987) *Guide to GPS Positioning*. Canadian GPS Associates, University of New Brunswick Graphic Services.

World Tunneling (1980) Vol. 3, No. 2., Automatic Electronic Profile Measuring, London.

1994 Data Collector Survey, *P.O.B. Magazine*, August–September.

1994 GPS Equipment Survey, *P.O.B. Magazine* June –July.

1994 Level Instrument Survey, *P.O.B. Magazine*, December–January.

1994 Software Survey, *P.O.B. Magazine*, October–November.

1994 Total Station Survey, *P.O.B. Magazine*, April–May.

4 | Geotechnical Investigations

Harvey W. Parker
Senior Vice President, Shannon and Wilson, Inc.

GEOTECHNICAL APPROACH TO TUNNEL DESIGN

Geology plays a dominant role in many major decisions made in designing and constructing a tunnel, from determining its feasibility and cost to assessing its performance. In tunnels, unlike other structures, the ground acts not only as the loading mechanism, but as the primary supporting medium as well. When the excavation is made, the strength of the ground keeps the hole open until supports are installed. Even after supports are in place, the ground provides a substantial percentage of the load-carrying capacity. Thus, for the tunnel designer and builder, the rock or soil surrounding a tunnel is a construction material. Its engineering characteristics are as important as those of the concrete or steel used in other aspects of the work.

Because every project is unique, this chapter sets forth broad guidelines and an overall perspective or approach rather than cookbook solutions. This chapter concentrates primarily on the aspects of geotechnical issues and investigative methods that are important to tunneling. A definitive treatise on the fundamentals of geotechnical exploration can be found in many geotechnical textbooks, reports, and journals. Two comprehensive references on the state of the practice of geotechnical investigative techniques include AASHTO (1988) and Hunt (1984). Some that address tunnel issues in the United States include ASCE (1974), Ash et al. (1974), Schmidt (1974), Schmidt et al. (1976), Thompson (1980), and West et al. (1980). The philosophy of tunnel design is set forth in Chapter 5 and is augmented in Chapters 6 and 7.

It is the unanticipated problems that can create costly delays and disputes during tunnel construction. It has been shown that the more thoroughly investigated tunnels have fewer cost overruns and fewer disputes. Explorations help evaluate the feasibility, safety, design, and economics of a tunnel project by

- Developing sufficient understanding of regional geology and hydrogeology for project design and construction
- Defining the physical characteristics of the materials that will govern the behavior of the tunnel
- Helping define the feasibility of the project and alerting the engineer and contractor to conditions that may arise during construction for the preparation of contingency plans
- Providing data for selecting alternative excavation and support methods and, where project status permits, determining the most economical alignment and depth
- Providing specific rock, soil, and hydrogeologic design parameters
- Minimizing uncertainties of physical conditions for the bidder
- Predicting how the ground and groundwater will behave when excavated and supported by various methods
- Establishing a definitive design condition (geotechnical basis for the bid) so a "changed condition" can be fairly determined and administered during construction
- Improving the safety of the work
- When project funds permit, providing experience working with the specific ground at the project site through large-scale tests or test explorations, which in turn will improve the quality of design and field decisions made during construction
- Providing specific data needed to support the preparation of cost, productivity, and schedule estimates for design decisions, and for cost estimates by the owner and bidders

There is a world of difference between the geotechnical requirements for underground projects and those for surface projects. Not only are underground projects geotechnically intensive, but the nature of the ground changes both horizontally and vertically. The type of excavation method and support that is used also affect ground behavior; even the quality and timing of the construction affects how the soil or rock behaves. Every geotechnical program must predict ground behavior so that it can be translated into the selection of construction methods and support systems and the estimation of costs. The determination of stratigraphy and the elevation of the groundwater table alone are no longer sufficient for an underground project. Rather, these descriptions are just the beginning of what is needed from investigation programs.

The earlier definitive exploration is made, the more flexible the owner and designer can be in their selection of alignment and construction methods, and thus the greater the potential for savings. This is a fundamental aspect of tunnel exploration. It is also why exploration is conducted in phases. Emphasis should be placed first on defining the regional geology, and then on increasingly greater quantification of the detailed characterization of the subsurface conditions and predicted behavior. Experience has established that few projects are constructed precisely on or along the alignment established at the time the initial boring program is laid out. This should be taken into account when developing and budgeting for geotechnical investigations.

In complex geology, initial explorations frequently disclose unexpected conditions and bring up questions that must be pursued during later stages of exploration and testing. Failure to resolve these issues early on will merely defer them to the construction disputes stage.

The degree of exploration required to establish the most economical and expeditious design and construction program is in no way related to the funds available for the work. While funding may control the work accomplished, it does not determine the work actually required. It is dangerous to go below certain levels of exploration. In fact, in some cases, it may be better to have no data and acknowledge the lack of information than to have inconclusive or misleading data. In one case, a major hydroelectric cavern was relocated on the basis of one isolated boring that did not disclose adverse geologic fractures encountered during construction. Several companies went bankrupt as a result.

While general guidelines are provided here and in many other handbooks, textbooks, and journal articles, every project demands and deserves individualized, project-specific attention. This presents an enormous challenge to the geotechnical/design team, which must constantly evaluate proposed changes and suggest compromises in the scope and cost of geotechnical investigation and design. There is a lot of room for creativity and innovation, and much remains to be done in the area of underground exploration methods, interpretation, and presentation of data and alternatives to owners and contractors.

GEOTECHNICAL CHALLENGES OF THE UNDERGROUND

The underground poses some formidable, but not impossible, challenges to the geotechnical and tunnel design teams. Some of these challenges are

- There is vast uncertainty in all underground projects.
- The cost and feasibility of the project is dominated by geology.
- Every feature of geologic investigation is more demanding than traditional foundation engineering projects.
- The regional geology must be known.
- Engineering properties change with a wide range of conditions, such as time, season, rate and direction of loading, etc.—sometimes drastically.
- Groundwater is the most difficult condition/parameter to predict and the most troublesome during construction.
- Even comprehensive exploration programs recover a relatively minuscule drill core volume, less than 0.0005% of the excavated volume of the tunnel.
- It is guaranteed that the actual stratigraphy, groundwater flow, and behavior encountered during construction will be compared with the geotechnical team's predictions.

The civil engineer works with manmade materials that are subjected to frequent inspections and tests. In a concrete structure such as a tunnel lining, one cylinder is frequently taken and tested on the order of every 20 yd^3 (a volumetric sample to production ratio of 1:2,700), and the slump and consistency are at least visually observed continuously. The geology to be tunneled is, however, never seen at full scale until the ground is exposed; until then it is evaluated by observation of extremely infrequent samples and even fewer tests. For instance, if borings are drilled on 300-ft centers and continuously sampled through the tunnel horizon, the amount of material recovered would only be on the order of 0.005% of the tunnel volume. Interpretation is complicated because the characteristics of the soil or rock frequently change both horizontally and vertically every few feet.

In spite of these challenges, geotechnical explorations are largely successful, but the owner and designer must appreciate this imprecise nature of geotechnical predictions. At the same time, the geotechnical engineer must appreciate that such imprecision is contrary to the customary data precision designers, particularly structural engineers, deal with unless they are experienced in tunneling. This is a challenge to communication and mutual understanding.

A major challenge of the underground is that the predicted geology will be seen and the predicted behavior will be compared with the actual behavior during construction. It is guaranteed that both the contractor and the owner will compare the stratigraphy predicted from the boring logs with that disclosed by the geology exposed during construction. This alone is an enormous challenge, especially since most of the predicted stratigraphy will be extrapolated from borings nearby but not directly through the tunneled section. However, the challenge of the underground is even more significant since the stratigraphy alone is not the only prediction that will be checked. The predicted behavior will also be compared with actual behavior. This comparison will likely be complicated by the actual construction conditions, which may not even be similar to those assumed by the geotechnical team when the behavior was predicted.

Engineering properties are affected by many factors that change with time, with the seasons, and by the rate, degree and, direction of loading, the degree of disturbance, and other external conditions. These factors, listed in Table 4-1, are crucial to the prediction of field behavior from laboratory

Table 4-1. Variability of Geotechnical Data

Seasonal and Tidal Fluctuations in Groundwater Levels and Flows	• Seasonal fluctuations of groundwater levels and flows can be very significant
	• Groundwater levels and flows should be determined during each season
	  • Preferably over several years
	• Tidal fluctuations in soils
	  • May be smaller than tide tables predict but still significant
	  • Tidal effects are delayed in time
	• Temperature Effects
	  • Can be significant to groundwater inflow and pressure estimates
	  • Temperature sometimes strongly affects interpretation of field data

Significant Difference in Rate of Loading Between Field and Lab	• After initial excavation, rate of loading in lab far exceeds field rate
Effects of Strain Rate	• Generally, lower strain rates result in lower laboratory strength and moduli
Effects of State of Stress	• Higher levels of state of stress result in higher strength and moduli • Both in lab and field

Effects of Anisotropy	• Horizontal permeability in soil is often 10+ times higher than vertical
	• Rock and stiff soil are strongly affected by anisotropy
	  • Both strength and modulus affected
	  • Especially sedimentary and stiff fissured or jointed soil
	• Generally higher strength and lower modulus perpendicular to beds/joints

Effects of Disturbance on Soil	• Drastic effect on soils, especially soft soils • Disturbance generally results in lower strength, modulus, and compressibility • Applies to disturbance to soil samples or in field due to tunneling • Disturbance generally results in lower permeability
Significant Differences in Scale Between Lab and Field	• Generally results in lower field strength and mass modulus than lab values • Particularly significant in fissured/jointed stiff soils and jointed rock • Expect significant scale differences between lab tests, large scale tests, and full-scale behavior

Deterioration with Time	• Some soil and soft rock deteriorate rapidly when exposed
	  • Some deteriorate in presence of water
	  • Some deteriorate just in air (must log core immediately)
	• May need to protect core samples and exposed ground in tunnel

or field tests and demonstrate the complexity and uncertainty of geological data.

IMPORTANCE OF GEOLOGY

Geology affects every major decision that must be made in designing and constructing a tunnel, determining its cost, and, to a degree, even the performance of the final product. The one-to-one relationship between geology and cost is so dominant that all parties involved in the planning and design of tunnels must consider the geology and hydrogeology of the site. Decisions such as the general alignment and depth affect a host of decisions and issues because they may place the tunnel in or out of an adverse geological feature. This not only determines construction cost, it can affect long-term maintenance problems such as groundwater leakage. For this reason, the earlier exploration is made, the greater the potential for savings.

The regional geology must be understood as fully as the specific geology within the tunnel corridor. In some cases, it may even be desirable to conduct explorations, including borings, well off the site to gain an understanding of the overall geologic and hydrogeologic picture. This might aid in determining the regional strike and dip of a particular geologic feature or explain a complex groundwater regime, which in turn could be extrapolated into the project area. Such exploration can also help determine whether a certain adverse feature is local or pervasive.

Geology provides crucial insight into the third dimension, which is frequently overlooked, especially on route-type projects. Geotechnical conditions are usually presented as cross sections, or profiles, which are usually a two-dimensional representation of some idealized profile obtained by projecting off-alignment boring data and/or outcrop information to the plane of the alignment. Naturally, geology varies in all three dimensions, and a knowledge of the regional and local geology can allow one to estimate how the geology varies, not only along the alignment but perpendicular to it as well. This knowledge gives valuable insight to the relative location of adverse geological features, the extent to which the tunnel might be affected by them, and the likelihood that a different alignment may be better.

Geology also plays an important role in evaluating geotechnical conditions between borings. Significant changes can occur between adjacent borings, which, initially, are typically spaced at 500 to 1,000 ft, and it is important to use geology to interpolate between widely spaced borings to predict how long the tunnel will be in any one given ground condition.

In addition to the groundwater levels and existence of perched or artesian water, a study of the hydrogeology of the area can provide information about the flow patterns of groundwater. This is of particular interest when hazardous waste is encountered or when trying to determine groundwater conditions and behavior.

Geology also provides clues to problems that may be encountered when the tunnel is constructed. For example, deposits of glacial till frequently have boulders associated with them. Accordingly, boulders should be expected even if the actual borings did not appear to encounter boulders. Geology can also provide some insight as to how certain materials will behave during construction. For instance, on the Mt. Baker Ridge project in Seattle, it was known from previous construction in the Lawton Clay that it could be expected to be fissured and slickensided pervasively, thus changing the soil mass properties from that of an otherwise very hard clay.

Thus, geology provides a rational means of correlating particular tunneling conditions, types of ground, and case histories. It can provide a menu of potential problems as well as their solutions and a means for predicting tunnel behavior in similar geologic materials. This can be very beneficial to the owner if it is done during the initial segment of a major undertaking, such as a subway system that will be developed in stages.

PHASING AND TIMING

Need for Geotechnical Services Throughout Project Life

Geotechnical services, data collection, and evaluation should begin very early in the conceptual planning of any project and should continue through construction and even after construction to document the as-built behavior of the tunnel, as shown in Table 4-2. Naturally, the level of geotechnical services provided will change throughout the life of the project. For instance, geotechnical services might be limited at first to the collection of available geological data, then increase in level to the collection and evaluation of existing geotechnical engineering data. The level of effort may then cease temporarily while planning and financial decisions are made.

Subsequently, significant geotechnical work will be necessary during exploration, analysis, and geotechnical design during the early portions of preliminary and of final design, interspersed with relatively low levels of effort. Finally, during the latter stages of final design, when contract documents are finalized, there will be a relatively significant geotechnical effort to support the preparation of the Geotechnical Design Summary Report (GDSR) and the rest of the contract documents. A genuine need for geotechnical input extends into the bidding, construction, and postconstruction phases.

Many projects are doomed at the outset because owners (and some engineers) who are inexperienced in underground projects do not understand the vital importance of geotechnical services to underground projects. Ideally, the exploration program should be developed with input from the tunnel designer. It is essential to budget each phase properly and to budget for and make provisions for the entire geotechnical budget as a line item and not something lumped into "design." This actually allows greater control on the cost and cost-effectiveness of the geotechnical program. Guidelines for estimating geotechnical budgets are given later in this chapter.

General Geotechnical Phasing

Almost always, the sooner geotechnical information is obtained and evaluated, the greater the potential for optimization of the alignment and profile and greater cost savings. The abundant geotechnical uncertainty requires tunnel exploration and design to be iterative. The planning of each exploration phase should be based on the results of the previous phase. Most importantly, the geotechnical exploration, including evaluation and report, must be available to the decision makers on the design team in a timely manner.

Ideally, the phases should follow in immediate succession or overlap just slightly so as to maintain uniform geotechnical staffing levels. Unfortunately, this is not always possible, especially when public work is done by a multidisciplinary team. On fast-track projects, the phases may overlap or require concurrent work because of accelerated schedules. On some projects, there are large gaps between phases—the geotechnical team must be disbanded and reassembled, or a new team must be acquainted with the details of the previous phases.

Both scenarios introduce drastic inefficiencies and the potential for at least temporarily misinterpreting data. Whenever large gaps between phases are anticipated, the geotechnical team should budget for and then write a detailed report that summarizes the status and the significance of the data at the time of preparation, in case different personnel are assigned to the geotechnical team, the rest of the multidisciplinary team, or the owner during the next phase.

Table 4-2. Phases Benefiting from Geotechnical Input

Proposal and Scoping Process
Existing Data Collection/Evaluation
Geologic Reconnaissance
Feasibility Investigation and Design
Preliminary Investigation and Design
Final Investigation and Design Exploration Evaluation of Results and Geotechnical Design Identification of Cost Issues Requiring Additional Exploration Additional Exploration to Support Cost Estimates Final Design Interaction with Designers and Cost Estimators
Bidding Assistance Prebid Meetings Evaluation of Bids
Construction Services Pre-construction briefings and periodic meetings Review of submittals Construction inspection Confirmation of actual geotechnical conditions and behavior Compare actual to predicted conditions and behavior Ground Groundwater Exploration during construction to solve specific geologic needs Additional borings to more accurately predict extent of adverse geology Mapping of tunnel walls and face Probes drilled ahead of face Instrumentation to confirm design and for early warning of problems Problem solving Comprehensive as-built report
Postconstruction Services Evaluation of in-service behavior Instrumentation for in-service behavior Evaluation of in-service behavior Comparison to predicted behavior Problem solving (especially leakage) Annual or other regularly scheduled inspections Repair and Rehabilitation

Collect and Organize Phase

Spending the proper amount of time on this phase is the first step to a cost-effective investigation. Before any time is spent in the field, the published information about the geology, soils, groundwater, seismic history, and performance of structures in the project area should be collected, organized, and evaluated. In urban areas, the history of the site is also important, because it can identify old landfills or alterations to drainage patterns that may affect the project. Geologic and soils maps are especially useful, as are aerial photographs and satellite imagery. A list of various sources of geotechnical data that might be available is given in Table 4-3.

Regional and Site Reconnaissance

Reconnaissance should cover both the immediate site as well as a much larger area so that regional geologic and hydrogeologic trends can be determined or verified. Geomorphology or geologic interpretation of the landforms in an area often reveal data that points to a particular geology or adverse geological features that will influence the type and amount of exploration that should be conducted.

It is seldom possible to thoroughly explore a tunnel alignment directly during the reconnaissance phase. The geologic conditions predicted along the alignment are based on detailed knowledge of selected local areas, connected by the interpretive skills of the engineering geologist. These essential interpretations are based on a knowledge of the soil and rock types, their macro- and microscopic, mineralogic, and structural features, and the geomorphic situation and agents at work.

Table 4-3. Collection and Organization of Existing Data

Collect existing data	Current maps and charts and existing geotechnical information Historical maps and charts Geotechnical databases or files	Local agencies	Dept. of transportation (large cities) City water/sewer dept. City engineering dept.
		State and federal agencies	U.S. Bureau of Mines State & U.S. Dept. of Transportation U.S. Dept. of Energy State Dept. of Geology State Dept. of Mines & Geology U.S. Geological Survey U.S. Dept. of Agriculture Soil Conservation Service U.S. Forest Service (aerial photos)
Assemble data	Verify original source Evaluate and verify quality of data Evaluate suitability for tunnel design	• Separate fact from opinion • Coverage of area • Regional • Detailed geology • Depth • Applicability to tunnel projects • Coverage for selection of vertical and horizontal alignment • Coverage for selection of construction method • Coverage for design • Coverage for prediction of behavior • Coverage for estimating costs	
Organize and Evaluate	Plot, tabulate, and evaluate data	• Geotechnical issues • Design issues • Cost issues • Constructibility issues	

Geologic reconnaissance includes a search of available literature, the study of aerial photographs and satellite imagery, and surface geologic mapping. In developing an investigation program, it should be remembered that the answers to some geologic questions are not always found at the work site; relationships between rock units or structural features that are obscured at the work site may be obvious from some distance away. Aerial photographs are particularly helpful in geomorphic analysis and in evaluating the best alignment and in choosing or evaluating portal locations. Considerable insight into the engineering properties of rock is available from the skilled evaluation of the soil or rock's response to the natural testing laboratory of its environment.

Feasibility and Corridor Site Investigations

The feasibility study must develop enough reliable information to determine technical feasibility and provide reasonable cost estimates to determine financial feasibility and, frequently, to compare alternative route alignments. Actual borings and test pits are desirable and are frequently required to provide a sound basis for cost estimates.

Nothing will be more valuable for a tunnel corridor study than relevant, reliable, and up-to-date geology and good-quality, up-to-date aerial photo coverage. For corridor studies, general classical geological exploration with special emphasis on engineering-geology features is invaluable. Features such as general stratigraphy; fault zones; landslide hazard areas for portal evaluation; and groundwater information, such as aquifer and acquiclude identification, locations of springs, and seasonal variation of springs are geological items of great significance to tunnels.

For urban projects, identification of significant manmade features is also essential. Particular attention must be given to existing sewer lines or other buried obstacles that any future tunnel must pass over or beneath, piles for pier structures, and any potential for hazardous waste. Geologic maps should be augmented by actual geologic inspection by a qualified engineering-geologist who is well versed in the engineering and geological needs and sensitivities of tunnels to adverse geological features. This should be done personally by an experienced engineering-geologist or geological engineer who walks the corridor or spends a considerable amount of time on the ground of the corridor at the appropriate locations. Long corridors may be more amenable to geologic helicopter flights, which can land wherever desired for mapping of specific areas of interest. However, it is essential that considerable verification of the geology on the ground be conducted for any corridor reconnaissance.

Preliminary Site Investigation

A substantial portion of the geotechnical effort that goes into a project is expended during preliminary design to gather the information and to conduct the analysis needed to support preliminary engineering. Alignments may change as a result of adverse geology discovered during preliminary investigation of the site, possibly leading to additional exploration needs late in the phase.

Final Design Investigation

This is the comprehensive exploration that is planned to fill in the gaps between preliminary design borings and to confirm earlier geotechnical assumptions. The exploration program may include pump tests, in situ tests, exploratory shafts, adits or pilot tunnels, and other large-scale tests. This phase also includes confirmation of geologic profiles and engineering design parameters, analyses, detailed geotechnical evaluation of geotechnical conditions and anticipated behavior, and preparation of reports.

Final Cost Data Investigation

There is often a need for additional explorations after the major final design decisions have been made. This is intended to fill in any gaps from the final design explorations with special emphasis on obtaining the data estimators and contractors need to make a reliable cost estimate. This step has not normally been formally identified in the past, but should be in the future.

Construction and Postconstruction Investigations

Investigation continues as the ground is exposed and its behavior observed and measured.

TEAMWORK, COMMUNICATIONS, AND TRAINING

Underground projects involve a number of technical specialties. The common thread is the geology and its geotechnical aspects as ascertained by the geotechnical investigation. Geotechnical results and their significance must be understood and used properly by the entire team, not just the geotechnical engineer. Each specialist must understand the concerns of the other team members, their special needs, and how his work impacts the others' (and vice versa). More information, rather than less, should be transmitted, and a lack of need-to-know should never be presumed.

Geotechnical and Geological Specialists

Because tunnel projects are geotechnically intensive, a well-trained geotechnical engineer, geological engineer, or engineering geologist must conduct the exploration and evaluation. This geospecialist should be well versed in tunneling and should have spent significant time actually underground during construction.

Drilling, Test Pit, and Other Exploration Subcontractors

It is essential that the project manager of the drilling company as well as the driller and the driller's helper understand the special needs of your project. This will require face-to-face briefings every time a new project or phase starts, supplemented by frequent reminders in the field. In particular, precise knowledge of the depth at which geologic units are encountered is essential in tunnel projects because the actual strata changes will be exposed in the tunnel excavation. When drilling a 100-ft-deep boring for a building foundation, it may not matter whether a given stratum changes at, say, 79 or 81 ft. However, if the crown of the tunnel is planned to be at 80 ft, it is important to know whether a change occurs at 79, 80, or 81 ft.

It is also important to tell the driller whether you wish to excavate as fast as you can or if you are more interested in making sure you are not missing vital information. This requires special briefings to the driller as well as care in preparing the drilling contract. It is possible to make hybrid contracts whereby the driller is on a footage basis with special provisions for hourly payment by force account when the geotechnical team needs to go slower to improve quality of data.

The driller should also be briefed on the purpose of the boring/field test and on the type of information being sought. Groundwater information (e.g., first encounter of groundwater, any caving of holes, changes in drilling behavior or cuttings, etc.) is particularly important and the driller should know of the extreme importance of such observations. This briefing should be repeated frequently in the field.

The Designer

It is important for the designer and the geotechnical staff to be well versed in each other's concerns and issues. The designer must understand the risk associated with the limitations of the geological interpretations as well as the technical issues. Most geotechnical data is extrapolated rather than interpolated, and the interpretation of the geology and geotechnical analyses are vastly different from calculations of stresses in a structure.

The Owner

The owner and its legal staff must also be aware of the uncertainties and imprecision of geotechnical predictions. These uncertainties not only result in larger geotechnical budgets than most non–tunnel owners might expect, they also create the need for special types of specifications and risk-sharing contract provisions. Legal staff must be aware of and trained in all the new changes in contracting practices, such as Differing Site Conditions, Full Disclosure, Disputes Review Board, Geotechnical Design Summary Report (GDSR) in contract documents, etc. The role of geotechnical uncertainties in the development of these new legal trends should be well understood by the owner and its legal staff.

Construction Parties

The tunneling contractor will already be painfully aware of the hazards associated with geotechnical predictions. Nevertheless, it is important to be realistic about what can be predicted reliably and to be constantly on the lookout for unexpected conditions that need changes in construction methods or may merit filing of a changed-condition claim.

The construction manager gets essential data from the contract documents, especially the GDSR and final geotechnical report. Again, the construction manager should be highly experienced in underground construction and should be thoroughly briefed on the important geotechnical requirements and uncertainties anticipated in each particular project.

Governmental agencies and their legal staffs must be aware of the uncertainties and imprecision of geotechnical predictions, especially if the design must be reviewed and approved by a third party giving an easement to the project.

Training

Training for everyone involved in a tunnel project is absolutely essential. With proper training, team members can perform a wide variety of tasks if given the opportunity. At the minimum, some clients/owners may be oriented to the uncertainties. In particular, it is essential that the important geotechnical aspects of the project be clear to all staff as a result of training—which should be broad and conducted frequently.

SOIL CLASSIFICATION FOR TUNNELS

Soil and/or rock classifications can either be descriptive of the materials themselves or be based on how they behave during tunnel construction. The classification systems take on different characteristics depending whether they are describing soil or rock. In the following paragraphs, emphasis will be given to systems that result in a better understanding of how the material behaves in tunnels.

Unified Soil Classification System

Soils engineers in the United States have adopted the Unified Soil Classification (UC) System, which is based on the description of the soil particles themselves (see Table 4-4). In the UC system, soils are classified primarily according to the grain size that comprises more than 50 percent of the particles in any given sample, as illustrated in Table 4-3. The size gradations start with gravels (G), sands (S), silts (M), and finally, clays (C), which are the finest grain size. This broad range is divided into two major categories, coarse-grained soils (gravels and sands), which include all those with grains visible to the naked eye, and fine-grained soils (silts and clays), which are finer than the No. 200 sieve. To put this in perspective, a silt particle is the finest particle that can be seen by the naked eye; ordinary kitchen flour or Portland cement will pass the No. 200 sieve.

Coarse-grained soils are further divided into groups that describe the distribution of soil grain sizes in the sample and the nature of the smaller particles in the sample. For instance, the group symbol GW is used for well-graded gravels containing little or no fines, and the symbol SM is used for silty sands since M is the symbol for silts. The remaining group symbols are essentially self-explanatory.

In tunneling, two other categories should be given extra attention: cobbles (3 to 12 in.) and boulders (greater than 12 in.). Although these two categories are relegated to a secondary role by the UC system, they are extremely important to any tunnel project.

Fine-grained soils are classified according to their plasticity. Silts are generally nonplastic, and clays are generally plastic to very plastic. For instance, ML is a silt of low plasticity, while MH is a high-plasticity silt. Another category describing the amount of organic content is used in the fine-grained soils; i.e., OL (organic soils of low plasticity) or OH (highly plastic organic soils) and PT (peat).

Engineering Soil Properties for Tunneling

For tunnel work, the UC system must be supplemented by other descriptions such as geologic stratigraphy and with the engineering properties of the soil, strength, modulus, and permeability. Many tunneling problems are governed by the geology such that one might begin to predict the potential for some types of ground behavior just on the basis of the geologic formation.

It is important to distinguish between the engineering parameters of an intact undisturbed soil sample from the engineering properties of the mass itself, which can be characterized by the following soil mass physical properties:

- Soil mass strength
- Soil mass modulus
- Soil mass permeability

Table 4-4. Unified Soil Classification System

Criteria for Assigning Group Symbols and Group Names Using Laboratory Tests			Soil Classification	
			Group Symbol	Group Name
Coarse-Grained Soils (More than 50% retained on No. 200 sieve)	Gravels (More than 50% of coarse fraction retained on No. 4 sieve)	Clean gravels (Less than 5% fines)	GW	Well-Graded Gravel
			GP	Poorly-Graded Gravel
		Gravels with fines (More than 12% fines)	GM	Silty Gravel
			GC	Clayey Gravel
	Sands (50% or more of coarse fraction passes No. 4 sieve)	Clean sands (Less than 5% fines)	SW	Well-Graded Sand
			SP	Poorly-Graded Sand
		Sands with fines (More than 12% fines)	SM	Silty Sand
			SC	Clayey Sand
Fine-Grained Soils (50% or more passes the No. 200 sieve)	Silts and Clays (Liquid limit less than 50)	Inorganic	CL	Lean Clay
			ML	Silt
		Organic	OL	Organic Clay Organic Silt
	Silts and Clays (Liquid limit 50 or more)	Inorganic	CH	Fat Clay
			MH	Elastic Silt
		Organic	OH	Organic Clay Organic Silt
Highly Organic Soils	Primarily Organic Matter; dark in color; organic odor		PI	Peat

Note: Cobbles are defined as 3 to 12 inches. Boulders are defined as greater than 12 inches.

In coarse-grained (i.e., sandy) soils, soil mass strength is frequently estimated by the Standard Penetration Test (SPT), a standardized method to estimate the relative density by driving the SPT sampler into the soil deposit with a 140-lb hammer. The relative density of the soil is roughly correlated to the "*N*" value or blows per foot of the Standard Penetration Test as shown in Table 4-5.

For fine-grained soils, soil mass strength is described in terms of consistency or undrained shear strength as determined by estimates or actual strength tests on the soil. The generally accepted classification system for fine-grained soils is also given in Table 4-5.

Generally, lab tests give results that more closely relate to intact samples. Both soil strength and soil modulus are strongly influenced by the in situ conditions and by conditions affecting the mass properties. Soil modulus of coarse-grained soils is strongly influenced by the effective stress acting on the soil mass at and around the tunnel. Also, stiff to very stiff clays are frequently fissured and slickensided, drastically reducing the soil mass strength and modulus parameters below that of an intact sample.

Soil modulus (*E*) is a measure of the stiffness of the soil mass or its resistance to undrained deformation by external forces. It is used in simple calculations of the predicted deformations of underground openings as well as the more comprehensive analyses of the behavior of soils during tunnel construction, such as by finite element methods. The determination and use of modulus for these analyses is complex and very sensitive to many factors to a degree that cannot be given in the limited space in this chapter. However, it is important to realize that soil modulus is extremely sensitive to sampling and other forms of soil disturbance, and quality control of sampling, transportation, and testing is vital if moduli must be used in analyses.

Table 4-5. Terminology Used to Describe Soils Relative Density for Granular Soils

Relative Density	Standard Penetration Resistance (*N*-values) blows/ft	
very loose	0 to 4	Easily penetrated with 1/2-in. rebar pushed by hand
loose	4 to 10	Easily penetrated several inches with 1/2-in. rebar pushed by hand
medium dense	10 to 30	Easily to moderately penetrated with 1/2-in. rebar driven by 5-lb hammer
dense	30 to 50	Penetrated 1 ft. with 1/2-in. rebar driven by 5-lb. hammer
very dense	over 50	Penetrated only a few inches with 1/2-in. rebar driven by 5-lb hammer

Consistency For Fine-Grained (Cohesive) Soils

Consistency	Standard Penetration Resistance (N-values) blows/ft.	Approximate Unconfined Compressive Strength, tons/sq. ft.	Field Identification
very soft	less than 2	less than 0.25	Squeezes between fingers when fist is closed; easily penetrated several inches by fist
soft	2 to 4	0.25 to 0.5	Easily molded by fingers; easily penetrated several inches by thumb
medium stiff	4 to 8	0.5 to 1.0	Molded by strong pressure of fingers; can be penetrated several inches by thumb with moderate effort
stiff	8 to 15	1.0 to 2.0	Dented by strong pressures of fingers; readily indented by thumb, can be penetrated only with great effort
very stiff	15to 30	2.0 to 4.0	Readily indented by thumb nail
hard	over 30	over 4.0	Indented with difficulty by thumb nail

andy silt to noncohesive silt soils which exhibit properties of granular soils are given relative density escriptions. (After Peck et al., 1974, and NAVFAC, 1986)

Permeability is a difficult parameter to measure reliably since the results are strongly dependent on sample disturbance. However, since groundwater conditions are extremely important to tunnel behavior, permeability must sometimes be estimated on the basis of grain size or other means, or measured in situ by simple water tests in boreholes or by special full-scale pump tests. An approximate correlation between permeability and soil type and grain size is given in Figure 4-1.

Behavioral Classification of Soils for Tunneling

Terzaghi, in 1950, published the Tunnelman's Ground Classification System, a classification system of the reaction of soil to tunneling operations. Terzaghi described the representative soil types and the predicted behavior of these ground types to the tunneling construction methods in use in the 1950s. They are still useful today to describe soil behavior provided allowances are made for new technology. The reader is referred to Terzaghi (1950) and Proctor and White (1977) for details of the development and use of this system by Terzaghi. Heuer (1974) modified the Tunnelman's Ground Classification System, as shown in Table 4-6, to present the classification system in engineering terms that reflect current terminology and usage.

Sands and silty sands that have little to no cohesive strength cannot be classified according to a measurable "strength." Their behavior is a function of their relative density or compactness, grain size and amount of fines, groundwater conditions, and a whole host of other factors including angularity of grains. An approximate correlation of tunnel behavior developed by Heuer and Virgins (1987) is given in Figure 4-2, which is for dense soils (SPT $N > 30$) above the water table. It should be noted that these are very approximate correlations to illustrate broad trends. Heuer and Virgins report that very loose soils ($N < 10$) may behave one or two classes poorer, while very angular particles or some cementation would make the soils behave one or two classes better.

Silts and fine sands behave much differently from clayey soils and provide potentially difficult ground conditions, as described by Heuer and Virgins (1987). Some trends regard-

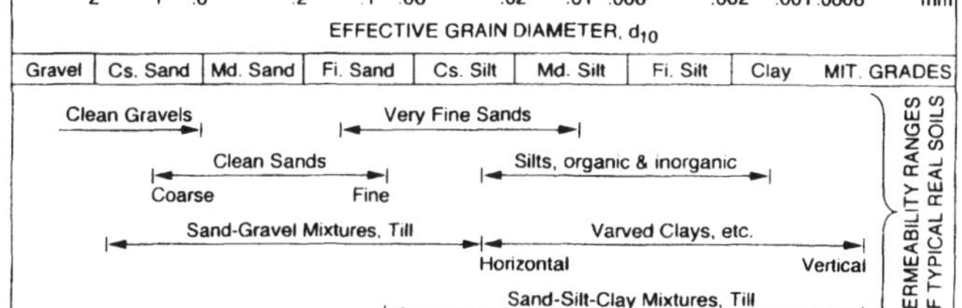

Fig. 4-1. Approximate correlations of soil type and permeability (Schmidt, 1974).

Table 4-6. Tunnelman's Ground Classification for Soils

Classification		Behavior	Typical Soil Types
Firm		Heading can be advanced without initial support, and final lining can be constructed before ground starts to move.	Loess above water table; hard clay, marl, cement sand and gravel when not highly overstressed.
Raveling	Slow raveling Fast raveling	Chunks or flakes of material begin to drop out of the arch or walls sometime after the ground has been exposed, due to loosening or to overstress and "brittle" fracture (ground separates or breaks along distinct surfaces, opposed to squeezing ground). In fast raveling ground, the process starts within a few minutes, otherwise the ground is slow raveling.	Residual soils or sand with small amounts of binder may be fast raveling below the water tale, slow raveling above. Stiff fissured clays may be slow or fast raveling depending upon degree of overstress.
Squeezing		Ground squeezes or extrudes plastically into tunnel, without visible fracturing or loss of continuity, and without perceptible increase in water content. Ductile, plastic yield and flow due to overstress.	Ground with low frictional strength. Rate of squeeze depends on degree of overstress. Occurs at shallow to medium depth in clay of very soft to medium consistency. Stiff to hard clay under high cover may move in combination of raveling at execution surface and squeezing at depth behind surface.
Running	Cohesive, running Running	Granular materials without cohesion are unstable at a slope greater than their angle of repose (±30–35). When exposed at steeper slopes they run like granulated sugar or dune sand until the slope flattens to the angle of repose.	Clean, dry granular materials. Apparent cohesion in moist sand, or weak cementation in any granular soil, may allow the material to stand for a brief period of raveling before it breaks down and runs. Such behavior is cohesive-running.
Flowing		A mixture of soil and water flows into the tunnel like a viscous fluid. The material can enter the tunnel from the invert as well as from the face, crown, and walls, and can flow for great distances, completely filling the tunnel in some cases.	Below the water table in silt, sand, or gravel without enough clay content to give significant cohesion and plasticity. May also occur in highly sensitive clay when such material is disturbed.
Swelling		Ground absorbs water, increases in volume, and expands slowly into the tunnel.	Highly preconsolidated clay with plasticity index in excess of about 30, generally containing significant percentages of montmorillonite.

ing the behavior of silty sands during tunneling as observed by Heuer and Virgins are

- Larger fines content (smaller D_{10} size)
- Wider range of grain sizes (higher uniformity coefficient C_u)
- More angular grains with interlocking structure
- Higher relative density (SPT blow count, N)

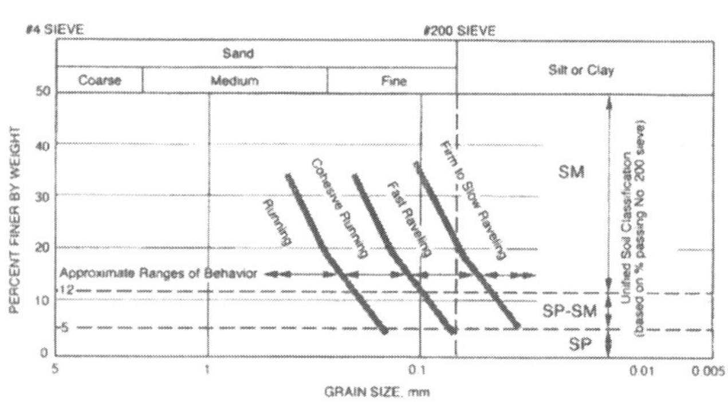

NOTES

1. Based on D_{10} size shown for dense soil, N>30, above water table developed from Terzaghi (Proctor and White, 1977).
2. Very loose soils (N<10) or rounded particles may behave 1 or 2 classes poorer.
3. Very angular sands, bonds, or cementation may behave 1 or 2 classes better.
4. Behavior below water table may be flowing and is a function of water head and permeability and other factors.

Fig. 4-2. Approximate ground behavior trends of dense silty sands above water table (Heuer and Virgens, 1987).

- More plastic fines
- More bonding due to chemical cementation or relict bonds
- Greater previous overburden loading
- Least amount of water transport during formation of deposit

Deere et al. (1969) developed a behavioristic diagram (Figure 4-3), which roughly correlates soil conditions as described by the Unified Soil Classification System with Terzaghi's Tunnelman's Ground Classification System.

There is, of course, a general relationship between behavior of soil in a tunnel and strength of fine-grained soils. As discussed in Chapter 7, shield tunneling in clay soil can be correlated with the simple Overload Ratio (OFS), which is the net overburden pressure at springline (after accounting for any air pressure) divided by undrained shear strength of the clay. This is a valuable relationship, as ground conditions are favorable for tunneling so long as OFS < 3 and the shield becomes unmanageable when OFS $= 6$. This relationship is shown graphically in Figure 4-4 (Schmidt, 1974b). These and other general tunneling relationships in soil were developed by Peck (1969), and are discussed in Chapter 6.

ROCK CLASSIFICATION

Physical Rock Descriptions

Rock types are classified according to their origin and physical characteristics as sedimentary, igneous, or metamorphic as given in Tables 4-7, 4-8, and 4-9, respectively. Terminology used to describe intact rock for strength and weathering are given in Tables 4-10 and 4-11, respectively.

The importance of geology gains even more prominence in rock classification. Since many rock formations seem to pose similar tunneling problems, it is important to describe and evaluate the stratigraphy of the rock. Since many tun-

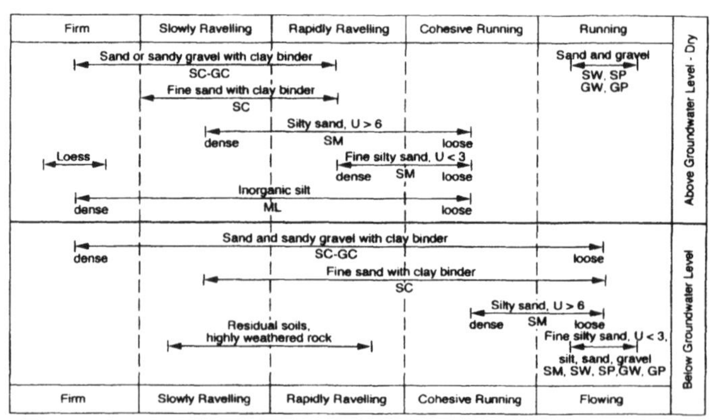

NOTES

1. Air loss (in tunnelling under compressed air) and water inflow is governed by the permeabililty, largely a function of D_{10}.
2. Behavior below groundwater table under suitable air pressure is approximately the same as above groundwater level.
3. Loose is here defined by N < 10 (standard penetration test), dense by N > 30.
4. Descriptive terms of materials according to the Unified Soil Classifiction.
5. Behavior may be somewhat better than shown above groundwater level, if material is moist and fine or silty.

Fig. 4-3. Approximate ground behavior trends of various soils (Deere et al., 1969).

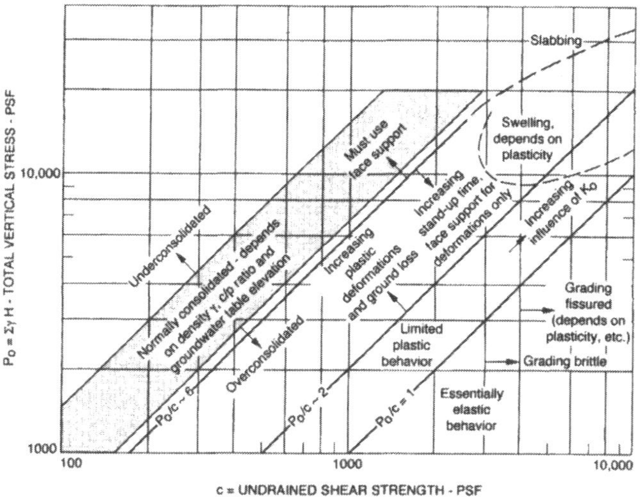

Fig. 4-4. Approximate relationship between undrained shear strength of fine-grained soils and tunnel behavior (Schmidt, 1974).

Table 4-7. Sedimentary Rock Classification

GroupGrain Size		Composition		Name
Clastic	Mostly Coarse Grains	Rounded pebbles in medium-grained matrix		Conglomerate
		Angular coarse fragments, often quite variable		Breccia
	More than 50 percent of medium grains	Medium quartz grains	Less than 10 percent of other minerals	Siliceous sandstone
			Appreciable quantity of clay minerals	Argillaceous sandstone
			Appreciable quantity of calcite	Calcareous sandstone
			Over 25 percent feldspar	Arkose
			25 to 50 percent feldspar and darker minerals	Graywacke
	More than 50 percent fine grain size	Fine to very fine quartz grains with clay minerals		Siltstone (if laminated, shale)
		Microscopic clay minerals	<10 other minerals	Shale
			Appreciable calcite	Calcareous shale
			Appreciable carbonaceous material	Carbonaceous shale
			Appreciable iron oxide cement	Ferruginous shale
Organic	Variable	Calcite and fossils		Fossiliferous limestone
	Medium to microscopic	Calcite and appreciable dolomite		Dolomite limestone or dolomite
	Variable	Carbonaceous material		Bituminous coal
Chemical	Microscopic	Calcite		Limestone
		Dolomite		Dolomite
		Quartz		Chert, Flint, etc.
		Iron compounds with quartz		Iron formation
		Halite		Rock salt
		Gypsum		Rock gypsum

After NAVFAC (1986).

neling problems are governed by the geology, one might begin to predict the potential for some types of ground behavior just on the basis of the geologic formation. For instance, limestone formations frequently have solution joints or even karst features. Also, basalt formations have distinctive cooling joints and are susceptible to lava tubes. Whenever the indicated geology is present, such features may be expected, and a potential effect on tunneling may be expected; in the above cases, this might be a potential for large water inflows.

Table 4-8. Igneous Rock Classification

Color	Light	Intermediate		Dark	
Principal Mineral	Quartz & Feldspar Other Minerals Minor	Feldspar	Feldspar & Hornblende	Augite and Feldspar	Augite Hornblende, Olivine
Texture					
Coarse, Irregular, Crystalline	Pegmatite	Syenite pegmatite	Diorite pegmatite	Gabbro pegmatite	
Coarse and Medium Crystalline	Granite	Syenite	Diorite	Gabbro	Peridotite
			Dolerite		
Fine Crystalline	Aplite			Diabase	
Aphanitic	Felsite			Basalt	
Glassy	Volcanic glass			Obsidian	
Porous (Gas Openings)	Pumice		Scoria or vesicular basalt		
Fragmental	Tuff (fine), breccia (coarse), cinders (variable)				

After NAVFAC (1986).

Table 4-9. Metamorphic Rock Classification

Texture	Structure	
Coarse Cyrstalline	Foliated	Massive
	Gneiss	Metaquartzite
Medium Crystalline	(Sericite) (Mica) Schist (Talc) (Chlorite) (etc.)	Marble Quartzite Serpentine Soapstone
Fine to Microscopic	Phyllite Slate	Hornfels Anthracite coal

After NAVFAC (1986).

Table 4-10. Scale of Intact Rock Strength

Descriptive Terminology	Strength Designation	Approximate Range of Unconfined Compressive Strength, psi	Field Identification
Extremely Soft	R0	28–100	Indented by thumbnail.
Very low strength	R1	100–1,000	Crumbles under firm blows with point of geology pick, can be peeled by a pocket knife, breakable by finger pressure with difficulty.
Low strength	R2	1,000–4,000	Can be peeled by a pocket knife with difficulty, shallow indentation made by firm blows of geology pick, craters under point impact.
Moderate strength	R3	4,000–8,000	Cannot be scraped or peeled with a pocket knife, specimen can be fractured with a single firm blow of geology hammer, and will develop a smooth pit or dent under point impact.
Medium high strength	R4	8,000–16,000	Specimen requires more than one blow with a geology hammer to fracture it; explosively dents or pits under geology hammer point impact.
High strength	R5	16,000–32,000	Specimen requires many blows of geology hammer to fracture it, does not indent.
Very high strength	R6	32,000	Specimen can only be chipped with geology pick.

After Deere and Miller (1966) and ISRM (1981).

Table 4-11. Scale of Rock Weathering

Descriptive Terminology	Symbol	Description
Fresh	W1	No visible sign of rock material weathering; perhaps slight discoloration on major discontinuity surfaces.
Slightly Weathered	W2	Discoloration indicates weathering of rock material and discontinuity surfaces. All the rock material may be discolored by weathering and may be somewhat weaker than its fresh condition.
Moderately Weathered	W3	Less than half of the rock material is decomposed and/or disintegrated to a soil. Fresh or discolored rock is present either as a discontinuous framework or as corestones.
Highly Weathered	W4	More than half of the rock material is decomposed and/or disintegrated to a soil. Fresh or discolored rock is present either as a discontinuous framework or as corestones.
Completely Weathered	W5	All rock material is decomposed and/or disintegrated to soil. The original mass structure is still largely intact.
Residual Soil	W6	All rock material is converted to soil. The mass structure and material fabric are destroyed. There is a large change in volume, but the soil has not been significantly transported.

After ISRM (1981)

Differences Between Soil and Rock Behavior

Rock mass descriptions are even more complex than soils. Aside from the obvious greater strength of the intact rock, the major difference in the behavior of rock from that of soils is the dominant effect of the anisotropy or other discontinuities in the rock mass. These discontinuities range from a texture or fabric in the rock, such as schistose foliation, up to major faults in the rock mass. Attention to the details of jointing is extremely important for rock tunnels. Classifications for joint spacing and for joint roughness are given in Table 4-12. The joints and the interaction between the various joint sets play a dominant role in the kinematic behavior of rock masses. Movement of rock blocks usually only takes place along joints, and thus movement can only be along the plane defining that joint. Further, the presence of joints drastically changes the rock mass engineering properties. Joints determine the kinematic freedom rock wedges have, and they dramatically affect the rock mass modulus. Further, they play a major role in the design and effectiveness of support of an underground opening say, by, rock bolts.

The significance of the discontinuities in the rock mass is that they have an enormous effect on the engineering properties of the rock mass, and therein lies the great difference in approach necessary when comparing soil behavior with rock behavior. An undisturbed sample of soil taken from a core boring can often (but not always) be considered representative of the soil mass and strength test values (or any other engineering property). Values obtained on such samples are frequently attributed to the soil mass. In rock, even more than in soil, a clear distinction must be made between the engineering properties of the "intact rock" and the engineering properties of the "rock mass," which include the effects of the discontinuities.

Size effects are important to rock tunnel projects. An intact rock sample from a core boring suitable for testing most likely does not have the discontinuities that govern the behavior of the whole rock mass. Accordingly, the intact rock properties are generally only an upper bound of the likely behavior of the rock mass. This same relationship of the size of any tested zone to the scale of the discontinuities must be understood during site characterization to assure that test results are not used incorrectly. For instance, a 3-ft-diameter plate load test may give representative results if the average spacing of discontinuities is a few inches, but not if the average spacing of discontinuities is greater than 1.5 ft. Similarly, a 10-ft-diameter tunnel may behave well in a certain jointed rock, while a 30-ft-diameter tunnel in the same rock may experience difficulties. A greater number of joints intersect the larger opening with a greater chance for unfavorable joint orientations relative to the tunnel boundary and greater kinematic freedom.

Table 4-12. Classification of Discontinuities

Joint Spacing	
Spacing of Discontinuity	Descriptive Term
Less than 60mm	very close
60mm to 200mm	close
200mm to 600mm	moderatel
0.6m to 2 m	wide
Greater than 2m	very wide
Fracture Roughness	
Term	Description
Rough	Large, angular asperities can be seen
Moderately rough	Asperities are clearly visible and fracture surface feels abrasive
Slightly rough	Small asperities on the fracture surface are visible and can be felt
Smooth	No asperities, smooth to the touch
Polished, slickensided	Extremely smooth and shiny

Rock characterization must address the inherent anisotropy of rock and the consequent strong directionality of any engineering properties. This is true not only in the small scale, where the fabric in the intact rock samples such as foliation makes the rock anisotropic, but also in the larger scale, where the joints and shear zones or faults make the rock mass behave anisotropically or directionally.

The joints in the rock make the rock mass behave like a jigsaw puzzle. Interlocking and the interaction of rock blocks place certain constraints on the kinematics of movement of any mass of rock in response to any external forces, including those caused by the excavation of a tunnel. Shear zones or persistent, through-going discontinuities interrupt any interlocking effects, dramatically changing the behavior of the rock (see Figure 4-5).

In situ stresses (both high and low) play an important role in the behavior of the jointed rock system. Many tunnels have experienced considerable difficulties, particularly at portals or topographic lows, where stresses may be unusually low as a result of stress relaxation during valley erosion. Such cases result in open joints that carry water and are difficult to support in a tunnel heading.

Finally, the permeability of a rock mass is also dramatically affected by the discontinuities. Sometimes a discontinuity such as a shear zone may be permeable and result in considerable water inflow. However, sometimes the shear zone may be impermeable and act as a dam holding back abundant water on the other side at high heads that will be encountered suddenly upon tunneling through the zone. Even the concept of using an "effective" permeability coefficient is challenged in rock projects. The effective permeability of the rock mass is governed by the discontinuities or other defects in the rock rather than the permeability of the intact rock. Not only is the permeability of a rock mass anisotropic (as it is in soil) but groundwater may pass easily through some joints (or even some portions of the same joint) but not through others.

Rock Mass Classification Systems

In order to provide a degree of organization and standardization to seemingly complex and confusing masses of field data, the concept of rock mass classifications evolved. These classification systems give the geotechnical engineer or geologist a method of assessing the relative importance of each rock parameter or geologic condition contained in the

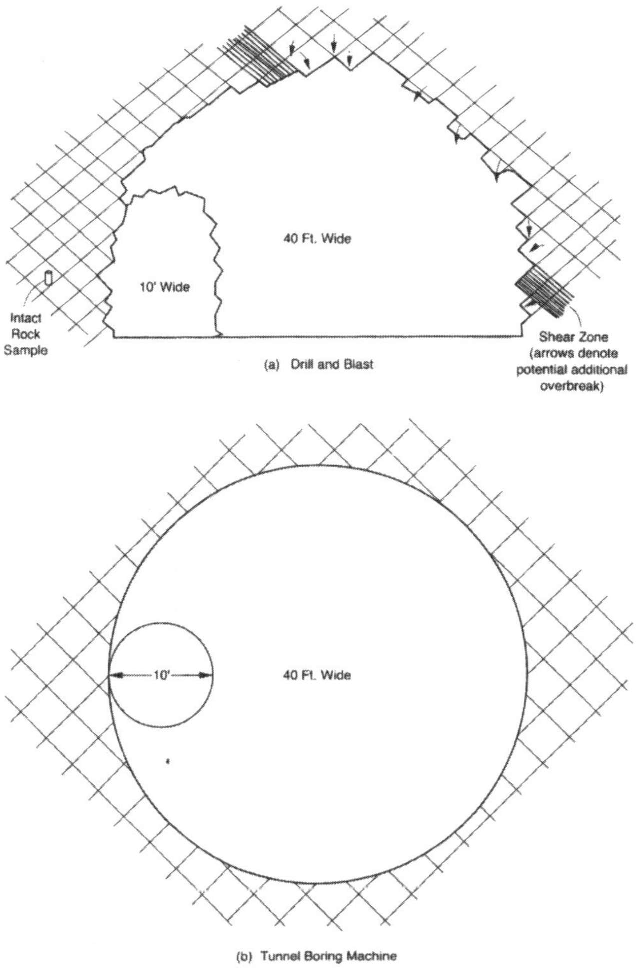

Fig. 4-5. Significance of jointing to tunneling.

given classification system that is being used. They are, in a sense, an extension of the same concept used in soil mechanics in the Unified Classification System and a forerunner of what are now called "expert systems." Barton (1988) states, "The Q System is essentially a weighing process, in which the positive and negative aspects of a rock mass are assessed." This is a good description of what all rock mass classification systems attempt.

In practice, rock classification systems are generally used to predict behavior by applying the concept to data obtained by logging rock core and by logging rock outcrops. None of the systems is perfect, and it may be necessary to adjust or fine-tune the particular rock classification system to fit the type of rock encountered and to account more properly for a special behavior of the rock condition being characterized. Further, during the construction of the tunnel, the exposed rock can be logged to obtain input data, which can be used to "calibrate" a rock mass classification system for use on future tunnels in the same geologic environment.

History of Development. Terzaghi (1946) developed a method of classifying rock mass types that could be used to describe and evaluate rock load on steel supports. Although this classification is decades old, the rock mass terms and descriptions coined by Terzaghi are still useful in communicating about the behavior of rock masses. The basic descriptions are given in Table 4-13.

In practice there are no sharp boundaries between these rock categories, and the properties of the rocks indicated by each one of these terms can vary between wide limits.

Lauffer (1958) developed a Stand-up Time Classification method that correlated stand-up time of unsupported spans with rock mass quality. Many others throughout the world have contributed to the development of rock classifications.

Deere's Rock Quality Designation (RQD). In the mid-1960s, Dr. Don U. Deere, of the University of Illinois, developed an improved method of logging rock core to calculate a modified core recovery percentage, called the Rock Quality Designation (RQD). RQD is essentially a simple measurement of the percentage of "good" rock in the core run (intact pieces 4 in. or more in length) and has been found to have a much better correlation to the actual behavior of the rock than the standard percent core recovery. It correctly reflects the fact that a rock mass that is so fractured or weathered that there are no pieces of sound rock longer than 4 in. will also behave poorly during construction. It was the first time that the engineering behavior of the rock mass could be estimated from a simple parameter obtained from core logging.

Subsequently, rock mass classification systems were developed that provided quantitative numerical values to conditions such as groundwater inflow; joint spacing, orientation, and condition; and intact rock strength, rock stress, etc., whose combined effect were previously described in more qualitative terms. Wickham, Tiedemann, and Skinner (1972) developed a numerical system that rates and weights rock mass parameters to correlate rock mass quality and excavation dimensions to ground support requirements, (Skinner, 1988).

RQD gained immediate acceptance and is now in common use worldwide. In fact, RQD is the basic parameter

Table 4-13. Terzaghi Rock Mass Descriptions

Intact rock contains neither joints nor hair cracks. Hence if it breaks it breaks across sound rock. On account of the injury to the rock due to blasting, spalls may drop off the roof several hours or days after blasting. This is known as *spalling* condition. Hard, intact rocks may also be encountered in the *popping* condition involving the spontaneous and violent detachment of rock slabs from sides or roofs.
Stratified rock consists of individual strata with little or no resistance against separation along the boundaries between strata. The strata may or may not be weakened by traverse joints. In such rock, the spalling condition is quite common.
Moderately *jointed* rock contains joints and hair cracks, but the blocks between joints are locally grown together or so intimately interlocked that vertical walls do not require lateral support. In rocks of this type both the spalling and the popping condition may be encountered.
Blocky and seamy rock consists of chemically intact or almost intact rock fragments which are entirely separated from each other and imperfectly interlocked. In such rock vertical walls may require support.
Crushed but chemically intact rock has the character of a crusher run. If most or all of the fragments are as small as fine sand grains and no recommendation has taken place, crushed rock below the water table exhibits the properties of a water-bearing sand.
Squeezing rock slowly advances into the tunnel without perceptible volume increase. Prerequisite for squeeze is high percentage of microscopic and sub-microscopic particles of micaceous minerals or of clay minerals with a low swelling capacity.
Swelling rock advances into the tunnel chiefly on account of expansion. The capacity to swell seems to be limited to those rocks which contain clay minerals such as montmorillonite, with a high swelling capacity.

used in the two most widely used comprehensive rock classification systems that have developed in the last couple of decades. Actually, RQD did not replace the traditional core recovery percentage; both are usually reported for each core run. The two percentages tell a lot about the quality of the rock and its likely behavior during construction. As simple as RQD is, it still requires a full understanding of how to drill and how to measure and count the pieces in the core run. The minimum standards for RQD are

- Good drilling techniques are essential.
- Minimum NX-size core.
- Drilled with double-tube core barrel, generally no greater than 5 ft long.
- Count only pieces of core that are at least 4 in. long.
- Count only pieces of core that are "hard and sound."
- Count only natural joints and fractures; ignore core breaks from drilling.
- Log RQD in the field immediately after recovery before any deterioration.

Core diameters larger and slightly smaller in diameter than NX (2.155 in.) can be used, but the smaller BX and BQ sizes are not recommended because of a greater potential for core breakage and loss. The widely used NX and NQ size (1.875 in.) wire-line cores are used extensively for RQD determinations and are the optimal core size.

There are strict rules and procedures for the determination of RQD. An example of an RQD core logging procedure is illustrated in Figure 4-6. Core length should be measured along the centerline of the core. Core breakage caused by drilling or handling (as evidenced by fresh rough surfaces) should be ignored, with the pieces fitted together and counted as one piece. If in doubt, Deere recommends considering the break as natural. With respect to soundness of the core, ISRM Weathering Grades I and II are counted. Grade III (Moderately Weathered) is counted but given an

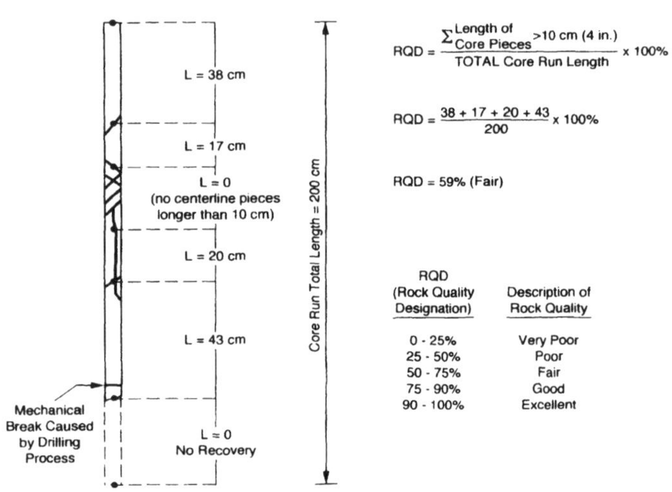

Fig. 4-6. Example RQD core logging procedure (Deere and Deere 1988).

asterisk to indicate that it is not quite sound. Grades IV and V (Highly and Completely Weathered, respectively), are not counted. This is an important aspect that is often overlooked. Details of RQD's history and of its correct use are given by Deere and Deere (1988).

There are many useful correlations between RQD and other parameters or descriptions of behavior. The basic classification comparing RQD with a qualitative rock quality and tunneler's description of the rock is given in Table 4-14.

RQD is very useful, but like all index properties, RQD has limitations. In particular, it does not take into account joint orientation, joint continuity, or the nature of the joint surfaces or filling. Some of its advantages and disadvantages are given in Table 4-15.

Deere and Deere (1988) describe the use of a graphical procedure on the core log, which highlights the "red flag" zones of concern, defined as rock with an RQD of less than 50%. Both RQD and the unmodified core recovery percentage are plotted with depth on the same graphical column on the boring log. Whenever RDQ is less than 50%, the area included between a line representing an RQD of 50% and the actual RQD is shaded or colored red (hence the name red flag), so that the reader easily gets a visual "red flag" of those zones which may present problems to construction. Moreover, the worst rock zones stand out more because they have a thicker band of red.

In the development of rock mechanics, several authors developed classification systems that take advantage of the strengths of RQD and use other parameters to make up for the limitations of RQD; some are described in the following sections.

Bieniawski's Classification System (RMR). Bieniawski (1973) introduced the Rock Mass Rating (RMR) Geomechanics Classification System, which provides a valuable weighted parameter system that accounts for the relative effects of six rock mass parameters. Based on a careful analysis of the behavior of numerous rock tunnels in different kinds of rock, Bieniawski has given each parameter a relative importance in the RMR weighted point system as shown in Table 4-16. The entire table of weighted points for the RMR Geomechanical System is given in Table 4-17 (Bieniawski, 1989).

Each of the five basic parameters is evaluated and given a numerical value of rating in accordance with Table 4-17A. The total points derived from the sum of the five basic parameters are then adjusted according to Table 4-17B to account for the effects of the strike and dip of joint discontinuities on

Table 4-14. RQD Classification System

Rock Quality	RQD (percent)	Approximate General Tunneler's Description
Excellent	90–100	Intact rock
Good	75–90	Massive, moderately jointed
Fair	50–75	Blocky and seamy
Poor	25–50	Shattered, very blocky and seamy
Very Poor	0–25	Crushed

Table 4-15. Advantages and Limitations of RQD

Advantages	Limitations
Simple	Does not account for joint orientation
Repeatable	Does not account for joint continuity or persistence
Inexpensive	Ignores interlocking of joint blocks
International acceptance	Ignores block size
Nondimensional	Ignores external forces like groundwater
Quantitative	Ignores nature of joint surfaces and filling
Correlated to modulus Accounts for weathering	Does not account for geology
Red flags potentially troublesome zones	Ignores in situ stress

the type of structure under consideration. The resulting total adjusted rating is the RMR of the rock mass under consideration and the rock mass is classified into the Rock Mass Classes according to Table 4-18.

Correlations from numerous projects have enabled the above generic classifications to be extended to predict broad ranges of rock mass behavior such as ground behavior in unsupported openings, and support requirements such as rock bolt spacing and shotcrete, as well as estimates of rock mass modulus and other parameters. The application of this classification system to tunnel design is discussed in Chapter 7.

Bieniawski (1978) proposed the following formula for the relationship between RQD and the rock mass modulus in GPa.:

$$E(\text{Rock Mass}) = 2\,\text{RMR} - 10 \qquad (4\text{-}1)$$

Since an index property is used in making the calculation, the resulting value of the rock mass modulus should be considered to be no better than an index of the rock mass modulus and thus only a rough estimate. Generally, projects that require a rock mass modulus in design will warrant the actual determination by bore-hole jack or large-scale in situ tests. However, once several of these large-scale tests are conducted, a site-specific correlation between RMR and the modulus of a certain rock formation can be developed. Then an estimate of the variation of rock mass modulus throughout the site can be made inexpensively by using this site-specific correlation and the RMR obtained from borings, exposures of rock in outcrops, or exploration adits.

Barton's Q System. The *Q* System was introduced in 1974 (Barton, Lien, and Lunde 1974). Its background, basis, and implementation are well described in Barton (1988).

Table 4-16. Weighted RMR Parameters

Parameter	Maximum Number of RMR Points for Most Favorable Condition
Uniaxial Compressive Strength of Intact Rock	15
Rock Quality Designation (RQD)	20
Spacing of Discontinuities	20
Condition of Discontinuities	30
Groundwater Conditions	15
Total RMR	100
Adjustment for Unfavorable Orientation of Discontinuities	Reduce RMR from 2 to 12 points

The *Q* System rates six parameters with the weights given in Table 4-19.

Like all classification systems, the *Q* System provides tables that enable the user to assign a numerical value for each of the *Q* System rating parameters depending on the condition of the rock mass with respect to each parameter. Barton's methodology for assigning the numerical ratings to the six parameters is given in Table 4-20.

To obtain the overall Rock Mass Quality (*Q*), these six parameters are combined in the following equation:

$$Q = (\text{RQD}/J_n) \times (J_r/J_a) \times (J_w/\text{SRF}) \qquad (4\text{-}2)$$

Despite this seeming abstract equation, the physical significance of the three basic components of *Q* is given in Table 4-21.

It can be seen that the numerical range of these six parameters is large, and the combined effect of their use in the (4-2) results in the overall rock mass rating (*Q*) having an extremely large range, approximately from 0.001 to 1,000. Accordingly, the *Q* System can accommodate the entire spectrum of rock mass conditions ranging from sound, unjointed rock to heavy squeezing ground; the relationship between *Q* and Rock Quality is given in Table 4-22.

Implementation of the *Q* System is accomplished through calculating a *Q* and an effective span. The effective span is Barton's method of applying a risk/safety factor to the application. It is obtained by dividing the actual span by the Excavation Support Ratio (ESR). Guidelines for determining ESR are given in Table 4-24 on page 00. A lower ESR provides a higher degree of safety. An ESR of 1.0 is used for civil works projects such as power stations, major road and railway tunnels, civil defense chambers, portals, and intersections. For this case, the actual span is the effective span. An ESR of 0.8 is used for underground facilities where sensitive machinery or people will be located. On the other hand, ESR for temporary mine openings could be as high as 3 to 5.

Figures and tables are then provided to determine the anticipated support requirement (see Figure 4-7). Like any other rock mass classification system, the *Q* System should be used by qualified geotechnical and tunnel design engineers with rock tunnel experience. Guidelines for using the *Q* System in design are discussed in Chapter 7.

DESCRIPTION OF INVESTIGATION TECHNIQUES

Rock Borings, Sampling, and Testing

The folllowing section on rock exploration, written by Ken Dodds (Dodds, 1982) for the previous edition of this handbook, has been retained with minor editing.

Borings are the most common method for detailed exploration of civil works. As a result, the mere fact that they are made can provide a false sense of security to the designer. Even a good boring program will not provide all the answers about materials and their properties, but it will

Table 4-17. Rock Mass Rating Matrix

A. CLASSIFICATION PARAMETERS AND THEIR RATINGS

	Parameter		Ranges of Values						
1	Strength of intact rock material	Point-load strength index (MPa)	>10	4–10	2–4	1–2	For this low range, uniaxial compressive test is preferred		
		Uniaxial compressive strength (MPa)	>250	100–250	50–100	25–50	5–25	1–5	<1
	Rating		15	12	7	4	2	1	0
2	Drill core quality RQD (%)		90–100	75–90	50–75	25–50	<25		
	Rating		20	17	13	8	3		
3	Spacing of discontinuities		>2m	0.6–2m	200–600mm	60–200mm	<60mm		
	Rating		20	15	10	8	5		
4	Condition of discontinuities (See Section E of Table)		Very rough surfaces Not continuous No separation Unweathered wall rock	Slightly rough surfaces Separation <1mm Slightly weathered walls	Slightly rough surfaces Separation <1mm Highly weathered wall	Slickensided surfaces or Gouge <5mm thick or Separation 1–5mm Continuous	Soft gouge >5mm thick or Separation >5mm Continuous		
	Rating		30	25	20	10	0		
5	Groundwater	Inflow per 10 m tunnel length (L/min)	None	<10	10–25	25–125	>125		
			or	or	or	or	or		
		Ratio: Joint water pressure / Major principal stress	0	<0.1	0.1–0.2	0.2–0.5	>0.5		
			or	or	or	or	or		
		General conditions	Completely dry	Damp	Wet	Dripping	Flowing		
	Rating		15	10	7	4	0		

B. RATING ADJUSTMENT FOR DISCONTINUITY ORIENTATIONS

Strike and Dip Orientations of Discontinuities		Very Favorable	Favorable	Fair	Unfavorable	Very Unfavorable
Ratings	Tunnels and mines	0	−2	−5	−10	−12
	Foundations	0	−2	−7	−15	−25
	Slopes	0	−5	−25	−50	−60

C. ROCK MASS CLASSES DETERMINED FROM TOTAL RATINGS

Rating	100←81	80←61	60←41	40←21	<20
Class no.	I	II	III	IV	V
Description	Very good rock	Good rock	Fair rock	Poor rock	Very poor rock

D. MEANING OF ROCK MASS CLASSES

Class no.	I	II	III	IV	V
Average stand-up time	20 yr for 15-m span	1 yr for 10-m span	1 wk for 5-m span	10 h for 2.5-m span	30 min for 1-m span
Cohesion of the rock mass (kPa)	>400	300–400	200–300	100–200	<100
Friction angle of the rock mass (deg)	>45	35–45	25–35	15–25	<15

E. GUIDELINES FOR CLASSIFICATION OF DISCONTINUITY CONDITIONS[a] (see this table, Part A.4)

Parameter	Ratings				
Discontinuity length (persistence/continuity)	<1 m 6	1–3 m 4	3–10 m 2	10–20 m 1	>20 m 0
Separation (aperture)	None 6	<0.1 mm 5	0.1–1.0 mm 4	1–5 mm 1	>5 mm 0
Roughness	Very rough 6	Rough 5	Slightly rough 3	Smooth 1	Slickensided 0
Parameter	Ratings				
Infilling (gouge)	None 6	Hard filling <5 mm 4	Hard filling >5 mm 2	Soft filling <5 mm 2	Soft filling >5 mm 0
Weathering	Unweathered 6	Slightly weathered 5	Moderately weathered 3	Highly weathered 1	Decomposed 0

[a] Some conditions are mutually exclusive. For example, if infilling is present, it is irrelevant what the roughness may be, since its effect will be overshadowed by the influence of the gouge. In such cases, use Part A.4 directly.

F. GUIDELINES FOR EFFECT OF DISCONTINUITY STRIKE AND DIP (see this table Part B)

Strike Perpendicular to Tunnel Axis			
Drive with Dip		Drive against Dip	
Dip 45–90	Dip 20–45	Dip 45–90	Dip 20–45
Very favorable	Favorable	Fair	Unfavorable
Strike Parallel to Tunnel Axis			Irrespective of Strike
Dip 20–45	Dip 45–90		Dip 0–20
Fair	Very unfavorable		Fair

After Bieniawski (1989).

Table 4-18. Rock Mass Classifications Based on Total RMR

Rating	Class	Description
81–100	I	Very Good Rock
61–81	II	Good Rock
41–60	III	Fair Rock
21–40	IV	Poor Rock
Less than 20	V	Very Poor Rock

provide enough answers so that the designer can be prepared for the significant variations in geological conditions to be encountered. The usefulness of borings per dollar spent becomes less as the tunnel is placed farther from the surface.

Borings are usually made to provide more specific knowledge of rock units, the variations in material, and their physical properties. They provide small, local samples for detailed study. Only where there are insufficient surface rock exposures should they be relied on to investigate general rock types, as this knowledge should be available from the geologic reconnaissance.

Because tunnels are linear features, boring programs for them should concentrate in the areas of greatest potential difficulty. Except in special cases, borings should not be arbitrarily spaced at some given distance from each other along the alignment. Some of the areas requiring more detailed exploration are:

1. Portals—usually need more investigation than commonly given
2. Topographic lows above the tunnel, as these often represent structurally weak rock
3. Rock types with deep weathering potential
4. Water-bearing horizons
5. Shear zones

In deep rock tunnels, borings are made to obtain specific knowledge of the rock along the tunnel alignment. This usually requires sampling the rock above the tunnel for the proper preparation of geologic cross sections. Figure 4-8 shows a group of borings along a tunnel alignment that are poorly located because they leave large gaps in the rock units involved. The layout in Figure 4-9, on the other hand, produces a much more complete picture of the geologic situation and thus is to be preferred, even though the borings do not reach the tunnel grade.

Table 4-19. *Q* System Parameters

Rock Mass Parameter/Condition	*Q* System Rating Parameter	Range
RQD	RQD	0–100
Number of joint sets	J_n	0.5–20
Joint Roughness	J_r	0.5–4
Joint Alteration	J_a	0.75–20
Joint Water	J_w	0.05–1.0
Stress Reduction Factor	SRF	0.5–20

Alignments of drill holes often deviate appreciably with depth. This deviation is usually between 1 and 3 ft per 100 ft, but in special cases, it can be much more. Bore-hole deviation surveys are required if the precise locations of samples from deep holes are necessary.

The number of borings required is a function of the geologic complexity of the area. Projects through competent, nearly horizontal rock may require very few borings, those in geologically complex areas may require many more. The success of even a carefully planned boring program depends greatly on the quality of on-site inspection during the drilling operation. It should be assured that trained, experienced persons are in charge of the technical aspects of a boring program.

The recommended depth depends upon the likelihood that there may be strata at some depth that might permit cheaper or safer tunneling. Often, several borings are extended much deeper than the first-cut depth of the tunnel only to find a more suitable depth. Borings should go deep enough to characterize the hydrogeologic regime. Once a depth is selected, exploration depths of 1 or 2 diameters below the invert are common.

Soft ground can be encountered at any depth of cover in a tunnel due to large shear zones, hydrothermal alteration, poor cementation in sedimentary beds, and so on. However, exploratory techniques especially adapted to soft ground conditions are primarily used for tunnels with shallow soil cover; these are, principally, drive sampling and cable tool churn drilling.

Field Inspection. Field logs should include the most precise description possible of the material, and they should be prepared as the hole advances. The field log is the most important record of the work. It, and not the hole in the ground, is the real product of the boring program that is purchased by the project owner.

A good field drill log is prepared with the idea that the information will be used for a variety of reasons, only some of which are known at the time of its preparation. It, therefore, must record all available information. The scale of the field log should be no less than 1 in. to 1 ft. The log must also be designed so that it can be used with a minimum of searching for needed data. This requires a compromise between a graphic and a narrative log.

Boring logs are most efficiently prepared by persons experienced in the interpretation of engineering geology and in evaluating the behavior of drilling equipment in relation to subsurface conditions. The field boring log should at least contain the information given in Table 4-23.

Water Pressure Tests in Borings. Water pressure tests are part of most rock exploration programs. Carefully done, they produce a great deal of useful information about subsurface conditions.

Table 4-20. *Q* System Ratings Matrix

1. ROCK QUALITY DESIGNATION (RQD)

Description	RQD	Notes
A. Very poor	0–25	(i) Where RQD is reported or measured as ≤10 (including 0) a nominal value of 10 is used to evaluate Q in Q = $(RQD/J_n) \times J_r/J_a) \times (J_r/SRF)$[a]
B. Poor	25–50	(ii) RQD intervals of 5 (e.g., 100, 95, 90, etc.) are sufficiently accurate.
C. Fair	50–75	
D. Good	75–90	
E. Excellent	90–100	

2. NUMBER OF JOINT SETS (J_n)

Description	J_n Rating	Notes
A. Massive, no or few joints	0.5 – 1.0	(i) For intersections use $(3.0 \times J_n)$
B. One joint set	2	(ii) For portals use $(2.0 \times J_n)$
C. One joint set plus random	3	
D. Two joint sets	4	
E. Two joint sets plus random	6	
F. Three joint sets	9	
G. Three joint sets plus random	12	
H. Four or more joint sets, random, heavily jointed, "sugar cube," etc.	15	
I. Crushed rock, earthlike	20	

3. JOINT ROUGHNESS NUMBER (J_r)

Description	J_r Rating	Notes
(a) Rock wall contact and		(i) Descriptions A through G refer to small-scale features and intermediate-scale features, in that order.
(b) Rock wall contact before 10-cm shear:		
A. Discontinuous joints	4	
B. Rough or irregular, undulating	3	(ii) Add 1.0 if the mean spacing of the relevant joint set is >3m.
C. Smooth, undulating	2	
D. Slickensided, undulating	1.5	(iii) J_r = 0.5 can be used for planar slickensided joints having lineations, provided the lineations are oriented for minimum strength.
E. Rough or irregular, planar	1.5	
F. Smooth, planar	1.0	
G. Slickensided, planar	0.5	
(c) No rock wall contact when sheared:		
H. Zone containing clay minerals thick enough to prevent rock wall contact	1.0[b]	
I. Sandy, gravelly or crushed zone thick enough to prevent rock wall contact	1.0[b]	

[a] J_a = joint alteration number, SRF = stress reduction factor
[b] Nominal

4. JOINT ALTERATION NUMBER (J_a)

Description	J_a Rating	ϕ_r value[a] (approximate)
(a) Rock wall contact:		
A. Tightly healed, hard, nonsoftening, impermeable filling (i.e., quartz or epidote)	0.75	—
B. Unaltered joint walls, surface staining only	1.0	25°–35°
C. Slightly altered joint walls, nonsoftening mineral coatings, sandy particles, clay-free disintegrated rock, etc.	2.0	25°–30°
D. Silty- or sandy-clay coatings, small clay-fraction (nonsoftening)	3.0	20°–25°
E. Softening or low-friction clay mineral coatings (i.e., kaolinite, mica); also chlorite, talc, gypsum, and small quantities of swelling clays (discontinuous coatings, ≤1–2 mm thick)	4.0	8°–16°
(b) Rock wall contact before 10-cm shear:		
F. Sandy particles, clay-free disintegrated rock, etc.	4.0	25°–30°
G. Strongly over-consolidated, nonsoftening clay mineral fillings (continuous, <5 mm thick)	6.0	16°–24°
H. Medium or low over-consolidated, softening clay mineral fillings (continuous, <5 mm thick)	8.0	12°–16°
J. Swelling clay fillings (i.e., montmorillonite) (continuous, <5 mm thick); value of J_a depends on percent of swelling clay-size particles and access to water, etc.	8.0–12.0	6°–12°
(c) No rock wall contact when sheared:		
K, L, M. Zones or bands of disintegrated or crushed rock and clay (see G, H, and J for description of clay condition)	6.0, 8.0 or 8.0–12.0	6°–24°
N. Zones or bands of silty or sandy clay, small clay fraction (nonsoftening)	5.0	—
O, P, R. Thick, continuous zones or bands of clay (see G, H, and J for description of clay condition)	10.0, 13.0 or 13.0–20.0	6°–24°

[a] Values of ϕ_r are intended as an approximate guide to the mineralogical properties of the alteration products, if present.

5. JOINT WATER REDUCTION FACTOR (J_w)

Description	J_w Rating	Approximate Water Pressure (kg/cm^2)	Notes
A. Dry excavations or minor inflow (i.e., <5 L/min locally)	1.0	<1	(i) Factors C to F are crude estimates. Increase J_w if drainage measures are installed.
B. Medium inflow or pressure occasional outwash of joint filling	0.66	1.0–2.5	(ii) Special problems caused by ice formation are not considered.
C. Large inflow or high pressure in competent rock with unfilled joints	0.5	2.5–10.0	
D. Large inflow or high pressure, considerable outwash of joint fillings	0.33	2.5–10.0	
E. Exceptionally high inflow or water pressure at blasting, decaying with time	0.2–0.1	>10.0	
F. Exceptionally high inflow or water pressure continuing without noticeable decay	0.1–0.05	>10.0	

6. STRESS REDUCTION FACTOR (SRF)

Description			SRF Rating	Notes
(a) Weakness zones intersecting excavation, which may cause loosening of rock mass when tunnel is excavated:				(i) Reduce these values of SRF by 25 to 50 % if the relevant shear zones only influence but do not intersect the excavation.
A. Multiple occurrences of weakness zones containing clay or chemically disintegrated rock, very loose surrounding rock (any depth)			10.0	
B. Single weakness zones containing clay or chemically disintegrated rock (depth of excavation ≤50 m)			5.0	
C. Single weakness zones containing clay or chemically disintegrated rock (depth of excavation >50 m)			2.5	
D. Multiple shear zones in competent rock (clay free), loose surrounding rock (any depth)			7.5	
E. Single shear zones in competent rock (clay free) (depth of excavation ≤50 m)			5.0	
F. Single shear zones in competent rock (clay free) (depth of excavation >50 m)			2.5	
G. Loose open joints, heavily jointed, or "sugar cube," etc. (any depth)			5.0	
Competent rock, rock stress problems:	σ_c/σ_1	σ_t/σ_3		(ii) For strongly anisotropic virgin stress field (if measured): when $5 \le \sigma_1/\sigma_3 \le 10$, reduce σ_c and σ_t to 0.8_c and $0.8\sigma_t$. When $\sigma_1/\sigma_3 > 10$, reduce σ_c and σ_t to $0.6\sigma_c$ and $0.6\sigma_t$, where: σ_c = unconfined compression strength, and σ_t = tensile strength (point load), and σ_1 and σ_3 are the major and minor principal stresses.
H. Low stress, near-surface	>200	>13	2.5	
J. Medium stress	200–10	13–0.66	1.0	
K. High-stress, very tight structure (usually favorable to stability, may be unfavorable to wall stability)	10–5	0.66–0.33	0.5–2.0	
L. Mild rock burst (massive rock)	5–2.5	0.33–0.16	5–10	(iii) Few case records available where depth of crown below surface is less than span width. Suggest SRF increase from 2.5 to 5 for such cases (see H).
M. Heavy rock burst (massive rock)	<2.5	<0.16	10–20	
Squeezing rock; plastic flow of incompetent rock under the influence of high rock pressures:				
N. Mild squeezing rock pressure			5–10	
O. Heavy squeezing rock pressure			10–20	

7. EXCAVATION SUPPORT RATIO (ESR)

Description	ESR	Notes
A. Temporary mine openings, etc.	ca. 3–5?	(i) Actual span/ESR gives "effective span" to be used in Q-System support determination t ables and figures.
B. Permanent mine openings, water tunnels for hydro power (excluding high pressure penstocks), pilot tunnels, drifts and headings for large openings.	1.6	(ii) ESR is used to incorporate user requirements for different degrees of safety.
C. Storage caverns, water treatment plants, minor road and railway tunnels, surge chambers, access tunnels, etc.	1.3	(iii) Increased safety can be achieved by reducing ESR value.
D. Power stations, major road and railway tunnels, civil defense chambers, portals, intersections	1.0	
E. Underground nuclear power stations, railway stations, sports and public facilities, factories	ca. 0.8?	

Table 4-21. Physical Significance of Basic Components of Q System Equation

Basic Component	Approximate Physical Significance
(RQD/J_n)	Block Size
(J_r/J_a)	Effect of minimum inter-block shear strength
(J_w/SRF)	Effect of Active Stress (i.e., Effect of external forces on rock mass, i.e., groundwater flow, in situ stresses)

Table 4-22. Q Versus Rock Quality

Q	Rock Quality
400–1,000	Exceptionally Good
100–400	Extremely Good
40–100	Very Good
10–40	Good
4.0–10.0	Fair
1.0–4.0	Poor
0.1–1.0	Very Poor
0.01–0.1	Extremely Poor
0.001–0.01	Exceptionally Poor

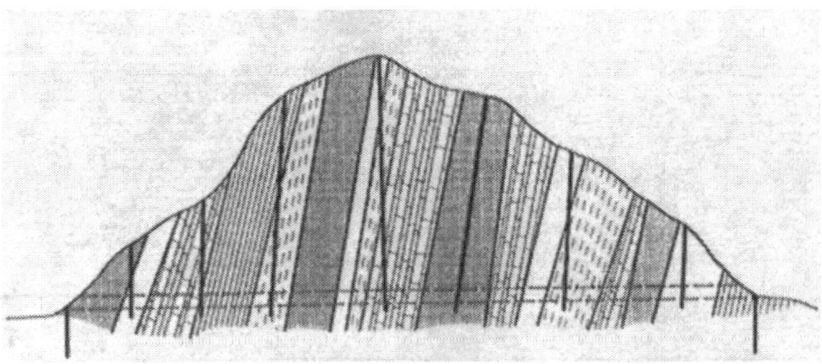

Fig. 4-8. A group of poorly laid out borings.

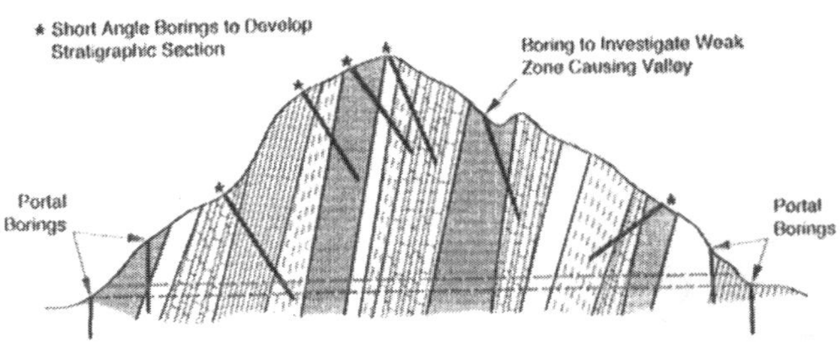

Fig. 4-9. Preferred boring layout, producing a more complete picture.

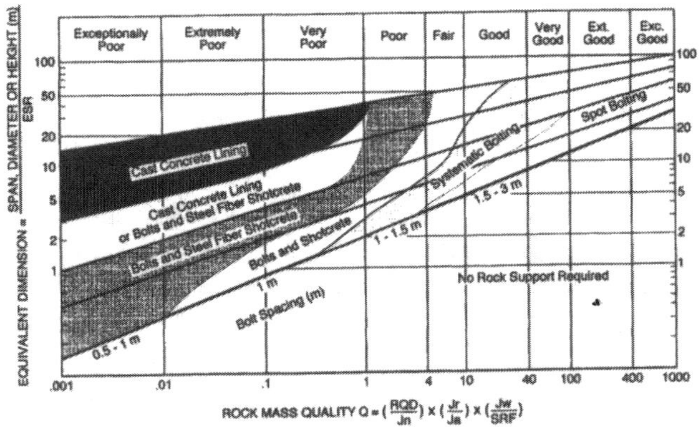

Fig. 4-7. Approximate correlation between Q and ground support requirements (Barton, 1991a, 1991b).

Table 4-23. Minimum Requirements for Field Boring Log

1.	Geologic descriptions of materials encountered.
2.	Descriptions of rock fabric elements.
3.	Records of field tests. a. Water permeability tests. b. Point load or rebound hammer tests. c. Pressuremeter or bore-hole jacking tests. d. Core caliper tests. e. Other special tests.
4.	Graphic logs of lithology, structure and core recovery (including likely tunnel location) for rapid scanning of data.
5.	Basic data on each core run. a. Length of run. b. Percent core recovery. c. RQD d. Longest piece of core. e. Location of core loss zones.
6.	Drilling rate.
7.	Return drill water condition and amount.
8.	Depth in hole where drill behaved abnormally.
9.	Opinion of logger on location and reason for core losses.
10.	Other conditions which could affect results of the work, such as the type and condition of equipment, bit and water pressure, and RPM of bit.
11.	Groundwater conditions. a. When groundwater encountered. Elevation at beginning of each day. b. Results of packer or other permeability tests. c. Details and graphic log of piezometer installation.
12.	Casing or cementation requirements.
13.	Location of core in box, for easy future reference.
14.	General administrative details. Small site sketch showing boring location Location or Coordinates Elevation Datum Record of photograph made of samples Dates and time of drilling and coring Project name and job number

After Dodds (1986).

Water Test Evaluation. In few rock types is the permeability more or less uniform throughout a rock mass. Generally, the value of k measured is an average of a wide range of values over the test section. Taken by itself, the test water loss often can give good approximations of the amount of water that will pass through a given stratum, providing, of course, that the test hole intersects rock and fractures which are typical of the rock mass overall. For other applications, such as grouting, the test water loss can be very misleading. For example, a test may give a permeability of 50 md (5×10^{-5} cm/sec), suggesting a fairly tight rock, while in fact most of the water may have been lost through one fracture with a permeability of 3,000 or 4,000 md (3×10^{-3} cm/sec). Consequently, when establishing grouting criteria, one must be careful about arbitrarily picking a value for k below which no grouting will be done. Furthermore, in some cases overall rock mass permeability is affected by the temperature, and so this must be measured and evaluated.

Interpreting Core Loss. It has often been said that, in borings, it is not so much the core that is important, but the places where no core is recovered. Some of the value of good inspection is found in interpreting core loss. When core loss occurs, the geologist must refer to his drilling notes to locate the actual point of the loss. Such things as a change in drill water color, a short period of rapid penetration, or a drill water loss may help to indicate the core loss zone and its extent. Where a weak clay filling of a joint has been washed out by drill water, some clay staining will usually remain on the joint surface, although it must be looked for carefully. Of course, such obvious causes as grinding of the core in the barrel or dropping core because of a weak core spring must not be overlooked, for these are core losses which do not necessarily indicate a weakness in the rock.

Some apparent core losses are not actual losses at all, and these must be identified if errors in interpretation are to be avoided. The most important examples of these apparent losses are

1. Voids in the rock due to solution, open fractures, lava tubes, etc.
2. Core left in the hole because the core did not break at the bottom of the hole when the core barrel was removed
3. Core not placed in the box in the proper order

Once a boring is completed and the core is in the box, the largest expense of the exploration is past, and other uses for the hole should be considered. Depending on the project requirements, some of the uses can be

1. Groundwater head or flow measurements, permeability tests
2. Groundwater temperature and salinity measurements
3. Rock structure studies using bore-hole or television cameras
4. Rock physical property measurements using down-hole seismic velocity probes or bore-hole deformation pressure cells
5. Geophysical logging to locate permeable strata, changes of material in core loss zones, voids, etc., using electrical resistivity, or gamma or neutron density logs
6. Hole diameter calipering used in conjunction with core calipering
7. Bore-hole directional surveys

Each of these methods is designed add insight into the composition and/or physical condition of the rock. None produces data that should be used without interpretation and reference to other exploration methods in the solution of an engineering problem.

Some recently developed methods that have merit for tunnel investigations are listed in Table 4-24.

Soil Borings, Sampling, and Testing

Drilling equipment for soil borings usually consists of rotary-wash or hollow-stem drilling equipment. Hollow-stem auger equipment is usually limited to depths less than 75 to 100 ft. In urban areas, the rigs are usually truck-mounted, which facilitates and speeds mobilization and boring setup. Elsewhere, they may be mounted on a skid-rig for mobilization in rough terrain or for sites requiring difficult access.

Table 4-24. Selected Additional Techniques for Geotechnical Investigations

DOCUMENTATION/MANAGEMENT	
Video	• Documentation of outcrops, portal locations. • Documentation of condition of core or soil sample when removed from sampler. Much better than still photos. • Documentation of drilling and sampling procedures. • Documentation of environmental conditions. • Excellent for briefing staff and/or management who could not visit site or conditions at site.
Global Positioning System (GPS)	• Invaluable for locating borings, test pits (etc.) at remote sites; helpful even in urban areas. Currently (1995) not expensive.
Geographic Information System (GIS)	• Extremely helpful in managing and presenting geologic data. • Invaluable for managing exploration data from entire city or major project.
ROCK INVESTIGATION METHODS	
Inclined Bore-holes	• Almost always desirable to drill some inclined borings to improve characterization of discontinuities and to improve understanding of geology (see Figures 4-8 and 4-9).
Horizontal Borings	• Ideal orientation for tunnel exploration but currently (1995) seldom done because special equipment and techniques required not readily available for long holes.
Oriented Core	• Invaluable for determining joint orientations for very deep tunnels or chambers. • Not easy and rather expensive.
Integral Sampling Method	• Excellent method to obtain 100 percent recovery of oriented core even in poorest of rock. Oriented pipe grouted into small pilot hole, also grouting up the surrounding rock. • Grouted core retrieved by overcoring. • Cannot conduct strength tests on core. • Time consuming and expensive.
Water Pressure (packer) Tests	• Relatively inexpensive method of determining rock mass permeability. Ideally, water pressure is measured by a piezometer in the test zone not at pump.
Hydraulic Jacking Test	• Packer test at pressures expected in pressure tunnels to estimate potentia for water loss and need for steel-lined penstock. Measures water pressure needed to open joints (Doe and Korbin, 1987).
Hydrofracture Test	• Relatively simple method to determine in situ stress in a bore-hole using water packer test methods (Haimson, 1978).
Impression Packer	• Special packer with inflatable sleeve covered with a soft moldable material. Inflation of packer against bore-hole wall leaves an impression of fracture in moldable material. Gives directions of maximum and minimum stress after hydraulic fracture test.
Borehole Camera or Video	• Excellent method to obtain additional information on rock quality and joint orientation to confirm core logging.
Triple-tube Core Barrel	• Somewhat more time consuming but currently (1995) is best method for recovery of poor rock short of integral core method. Should be specified to be on hand at drilling site to be used whenever double-tube core barrel results in inadequate recovery.
GENERAL INVESTIGATIVE METHODS	
Methane Tests and Other Hazardous Gases in Boreholes	• Important to measure in sufficient borings, perhaps all borings, to determine likelihood and nature of problem. Incidence of methane is probably more prevalent in both soil and rock than generally thought.
Air Photo and Satellite Imagery	• Now readily available and invaluable for reconnaissance.
SELECTED GEOPHYSICAL METHODS	
Seismic Refraction	• Most frequently used geophysical method. Should be calibrated together with borings. • Good for broad mapping of subsurface especially weathered profile, soil/rock interface, faults or extensive shear zones, dikes, etc. • Results useful for optimizing boring locations and for interpreting between borings. • Misses any layer with lower velocity than above.
Seismic Reflection	• Limited for shallow exploration except in favorable geology.
Resistivity/Conductivity	• Useful together with seismic refraction. Especially useful to determine soil/water interface. Useful to characterize contaminated groundwater plumes.
Ground-Penetrating Radar	• Useful only in favorable geology. Suitable for very shallow penetration (10 to 20 feet). Very limited penetration in wet clay or salty groundwater. • Generally oversold as an exploration method.
Bore-Hole Geophysical Logging	• Wide variety of in-hole methods adapted from petroleum industry are useful in special situations. Include sonic, electrical resistivity, density (gamma-neutron, neutron), etc. • Good for determining properties at depth but must be very selected to be cost-effective.
Vertical Seismic Profiling (VSP) in Bore-Holes	• Promising technique adapted from petroleum industry. • Methodology is improving, especially for defining permeable zones.
Cross Hole Seismic Techniques	• Improved geophysical definition of geology at depth. Excellent for detailed characterization for underground caverns. Can be expensive because need bore-holes close together.

When drilling beneath the water table, both rotary-wash and hollow-stem auger methods can result in significant sample disturbance and errors in blow count data, especially in sandy soils, when sufficient care is not exercised. It is difficult but essential that the drilling fluid level in the drill casing be maintained at or above the pressure head in the strata being drilled and sampled.

There are a variety of methods to sample soils, each offering different advantages and disadvantages for the type of soil being sampled and the conditions under which the soil is sampled. The Standard Penetration Test (SPT; ASTM, 1994) is still a commonly used method since the blow count (N) gives at least a relative measure of density or stiffness (ASTM D 1586). However, recent findings show that there are many ways for major errors to affect the blow count (N), by as much as a factor of 2 or more (Kovacs, 1977). Even though the SPT blow count (N) is frequently used in geotechnical investigations, special precautions should be taken in conducting the test and in interpreting the results, particularly if there appear to be inconsistencies in the data. This is a problem with all blow count data, not just the SPT.

Generally, the sampling method that creates the least disturbance to the soil should be used, so that a realistic representation of the soil in situ can be obtained. There are larger-diameter (2-1/2 in. I.D. or greater) drive-samples, some with soil retention rings, that retrieve samples that are less disturbed than the SPT, which is only 1-3/8 in. I.D. For soft to medium soils, thin-wall Shelby tube samples are generally appropriate, while piston samplers may be required to retrieve soft to very soft soils. These generally are pushed smoothly into the soil, eliminating a potential for disturbance from driving the sampler. An excellent sampler for harder soils is the rotary pitcher sampler.

Soil permeability can be inferred from grain size and from laboratory tests. However, since soil mass permeability is so important to tunnel behavior, geotechnical investigations should always emphasize the determination of permeability in situ if possible. Rough permeability tests can be conducted in any boring by either constant-head or falling-head test methods, but these give only rough indications of permeability. On major projects, special pump tests are often conducted to provide permeability data. These tests usually consist of a test well, which is pumped in increasing increments of pumping rates. The test also includes several observation wells that sense the effects of each increment of pumping in each observation well, which are at increasing distances from the test well, as shown in Figure 4-10 (Guertin and McTigue, 1982).

DEVELOPING THE INVESTIGATION PROGRAM

There is no fixed cookbook or check list to determine how to do geotechnical investigations. There is a wide latitude for determining what should be done. Guidelines are given here for selecting the scope of geotechnical investigations, for typical projects and for special situations. The required scope depends upon many issues, including

- Purpose of tunnel
- Consideration of alternative alignments
- Geology
- Groundwater conditions
- Contemplated excavation support and lining
- Sophistication of design methods required
- Time available for design and construction
- Project attitude toward risk
- Legal constraints, especially likely contract type
- Real and perceived budget constraints
- Likely protection needed for adjacent facilities
- Familiarity and tunnel experience of geotechnical engineer, designer, and owner
- Special safety issues
- Likelihood of contaminated ground/groundwater

The relationships among these features are so complex that no handbook could cover all eventualities, but this section will give examples and general trends to guide the reader. A list of tunnel types or purposes of tunnels is given in Table 4-25, together with special investigation considerations that might be applicable to that type of tunnel.

Certain types of geology create special demands for the scope of geotechnical investigations, as illustrated in Table 4-26.

Certain construction methods, pose special demands for the scope of geotechnical investigations. This is because the selection of the construction itself requires special data that

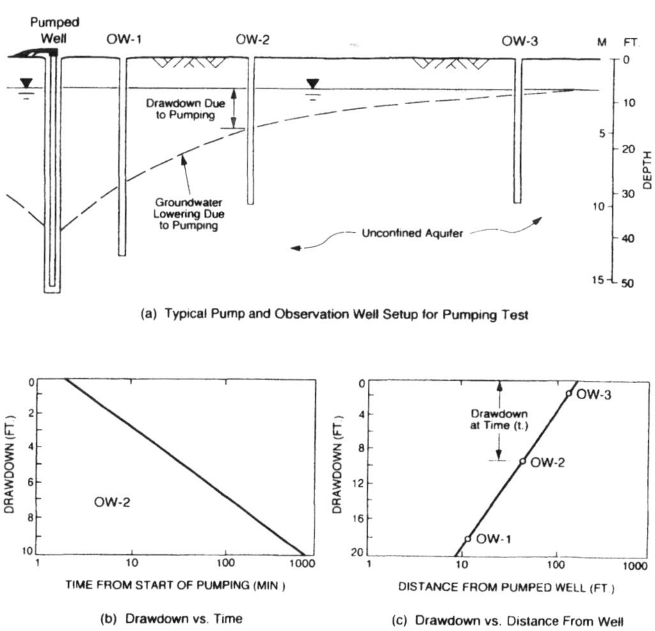

Fig. 4-10. Typical arrangement for pump test in soil (Guertin and McTigue, 1982).

Table 4-25. Geotechnical Needs Dictated by Tunnel Purpose

TRANSPORTATION	
Highways	• Generally large-size tunnels; clear span width affects general behavior. Usual flexibility in alignment and grade. Investigation should support optimization of tunnel location, length, grade, etc. Portals usually need special exploration to determine most economical location to place portal.
Railroads	• Less flexible in choice of horizontal and vertical alignment; may not be able to easily avoid adverse geological features or difficult portal locations by moving alignment. Need to explore for all potential adverse features.
Rapid Transit	• Need data to design protection of adjacent structures and to estimate potential settlement and groundwater inflow. • Need to investigate one or more corridors as early in program as possible to avoid expensive adverse geologic and/or environmental conditions. • Can benefit strongly from phased exploration approach.
ENERGY FACILITIES	

Hydroelectric Schemes	• Diversion - Some flexibility for optimizing location on basis of geology
	• Intake - Permeability for water inflow and discharge
	• Tailrace - Permeability for water inflow and discharge
	• Unlined or concrete-lined penstock
	  • Special emphasis on loading and unloading modulus of rock
	  • In situ stress
	  • Special attention to areas of low stress near surface
	  • Hydraulic jacking test
	• Steel-lined penstock
	  • Considerable geotechnical data needed to determine length of steel
	  • Special emphasis on loading and unloading modulus of rock. Need rock and groundwater data to predict external loads when dewatered
	• Underground powerhouse and valve chambers
	  • Need joint orientation data to orient caverns
	  • Often considerable flexibility to optimize location and orientation on basis of geotechnical data such as joint orientation and in situ stress data
	• Bifurcation chambers or header complexes
	• Access tunnels and shafts
	• Surge chambers—Requires loading and unloading moduli
	• Pumped storage schemes—Requires loading and unloading moduli
	• Compressed-air peak-shaving schemes
	  • Consideration of thermal-mechanical properties; need mass permeability data to predict potential air loss

Nuclear Waste Repositories	• Requires costly comprehensive site characterization including thermal-mechanical properties of the host rock at elevated temperatures. Also considerable emphasis on long-term permeability extrapolated over thousands of years. Extraordinary quality control (NQA-1) required in all aspects of investigation.
District Heating or Refrigeration Storage	• Consideration of thermal-mechanical properties
BULK STORAGE	
Underground Warehouses, Living, and Industrial Facilities	• Generally large spans needing geotechnical data on competency of roof rock • Potential for groundwater inflow must be predicted
Crude Oil	• Rock mass permeability to predict inflow of groundwater and/or product loss
Liquid Hydrocarbons	• Permeability to predict inflow of groundwater and/or product loss
Liquified Gas	• Consideration of thermal-mechanical properties
Compressed Air	• Consideration of thermal-mechanical properties; need rock mass permeability data to predict potential air loss
Wine or Other Foodstuffs	• Permeability to predict inflow of groundwater and/or any bulk liquid product loss
WATER RESOURCES	• Drainage—Need data to predict inflow into tunnel • Irrigation—Need data to predict inflow into tunnel • Transmountain diversion—Need data to predict inflow out into tunnel • Groundwater recharge • Fresh water supply • Underground water treatment plants
ENVIRONMENTAL AND SANITARY	• Sanitary sewer • Storm sewer • Pump station • Interceptor sewer • Wastewater storage (in-line) • Permeability for water inflow and discharge • Underground wastewater treatment plants Traditional geotechnical requirements
COMMUNICATION AND PIPELINE TUNNELS	• Traditional geotechnical requirements
HARDENED DEFENSE FACILITIES	• Missile Silos • Command Facilities
REHABILITATION PROJECTS	• Confirm presence and distribution of voids behind lining • Ground penetrating radar sometimes helpful • Search for weak/deteriorated materials—especially wood lagging, etc. • Careful structural evaluation of detailed crack survey may reveal causes of stress • Measure and evaluate existing water seepage during different rainfall or seasons • Borings/probes from inside the tunnel may be required

relate to the feasibility and risks associated with the method and provide input for cost estimates. Table 4-27 provides selected data on special investigation needs that are related to the construction method.

Table 4-26. Geotechnical Investigation Needs Dictated by Geology

Hard or Abrasive Rock	• Difficult and expensive for TBM or roadheader. Investigate, obtain samples, and conduct lab tests to provide parameters needed to predict rate of advance and cutter costs.
Mixed Face	• Especially difficult for wheel type TBM • Particularly difficult tunneling condition in soil and in rock. Should be characterized carefully to determine nature and behavior of mixed-face and approximately length of tunnel likely to be affected for each mixed-face condition.
Karst	• Potentially large cavities along joints, especially at intersection of master joint systems; small but sometimes very large and very long caves capable of undesirably large inflows of groundwater.
Gypsum	• Potential for soluble gypsum to be missing or to be removed because of change of groundwater conditions during and after construction.
Salt or Potash	• Creep characteristics and, in some cases, thermal-mechanical characteristics are very important
Saprolite	• Investigate for relict structure that might affect behavior • Depth and degree of weathering; important to characterize especially if tunneling near rock-soil boundary • Different rock types exhibit vastly differing weathering profiles
High Stress	• Could strongly affect stand-up time and deformation patterns both in soil and rock tunnels. Should evaluate for rock bursts or popping rock in particularly deep tunnels.
Low Stress	• Investigate for open joints that dramatically reduce rock mass strength and modulus and increase permeability. Often potential problem for portals in downcut valleys and particularly in topographic "noses" where considerable relief of strain could occur. • Conduct hydraulic jacking and hydrofracture tests.
Hard Fissured or Slickensided Soil	• Lab tests often overestimate mass physical strength of soil. Large-scale testing and/or exploratory shafts/adits may be appropriate.
Gassy Ground—always test for hazardous gases	• Methane (common) • H_2S

Adverse Geological Features	• Faults
	  • Known or suspected active faults. Investigate to determine location and estimate likely ground motion.
	  • Inactive faults but still sources of difficult tunneling conditions.
	    • Faults sometimes act as dams and other times as drainage paths for groundwater.
	      • Fault gouge sometimes a problem for strength and modulus.
	  • High temperature groundwater
	  • Groundwater
	    • Groundwater is one of most difficult and costly problems to control. Must investigate to predict groundwater as reliably as possible.
	    • Site characterization should investigate for signs of and nature of:
	      • Groundwater pressure
	      • Groundwater flow
	      • Artesian pressure
	      • Multiple aquifers
	      • Higher pressure in deeper aquifer
	      • Groundwater perched on top of impermeable layer in mixed face condition
	      • Ananalous or abrupt
	    • Aggressive groundwater
	      • Soluble sulfates that attack concrete and shotcrete
	      • Pyrites
	      • Acidic
	      • Always collect samples for chemistry tests

	• Sedimentary Formations
	  • Frequently highly jointed
	  • Concretions could be problem for TBM
	• Lava or Volcanic Formation
	  • Flow tops and flow bottoms frequently are very permeable and difficult tunneling ground
	  • Lava tubes
	  • Vertical borings do not disclose the nature of columnar jointing. Need inclined borings.
	  • Potential for significant groundwater flows from columnar jointing
	• Boulders (sometimes nests of boulders) frequently rest at base of strata
	  • Cobbles and boulders not always encountered in borings which could be misleading
	  • Should predict size, number, and distribution of boulders on basis of outcrops and geology
	• Beach and Fine Sugar Sands
	  • Very little cohesion. Need to evaluate stand-up time.
	• Glacial deposits
	  • Boulders frequently associated with glacial deposits. Must actively investigate for size, number, and distribution of boulders.
	  • Some glacial deposits are so hard and brittle that they are jointed and ground behavior is affected by the jointing as well as properties of the matrix of the deposit
	• Permafrost and frozen soils
	  • Special soil sampling techniques required
	  • Thermal-mechanical properties required

Manmade Features	• Contaminated groundwater/soil
	  • Check for movement of contaminated plume caused by changes in groundwater regime as a result of construction
	• Existing Obstructions
	  • Piles
	  • Previously constructed tunnels
	  • Tiebacks extending out into street
	• Existing Utilities
	  • Age and condition of overlying or adjacent utilities within zone of influence

Table 4-27. Special Investigation Needs Related to Construction Method

Construction Method	Special Requirements
Drill and Blast	Data needed to predict stand-up time for the size and orientation of tunnel.
Rock Tunnel Boring Machine	Data required to determine cutter costs and penetration rate is essential. Need data to predict stand-up time to determine if open-type machine will be ok or if full shield is necessary. Also, water inflow is very important.
Conventional Soil Tunnel Boring Machine Shield	Stand-up time is important to face stability and the need for breasting at the face as well as to determine the requirements for filling tail void. Need to fully characterize all potential mixed-face conditions.
Pressurized-Face Tunnel Boring Machine	Need reliable estimate of groundwater pressures and of strength and permeability of soil to be tunnelled. Essential to predict size, distribution and amount of boulders. Mixed-face conditions must be fully characterized.
Roadheader	Need data on jointing to evaluate if roadheader will be plucking out small joint blocks or must grind rock away. Data on hardness of rock is essential to predict cutter/pick costs.
Immersed Tube	Need soil data to reliably design dredged slopes and to predict rebound of the dredged trench and settlement of the completed immersed tube structure. Testing should emphasize rebound modulus (elastic and consolidation) and unloading strength parameters. Usual softness of soil challenges determination of strength of soil for slope and bearing evaluations. Also need exploration to assure that all potential obstructions and/or rock ledges are identified, characterized, and located. Any contaminated ground should be fully characterized.
Cut and Cover	Plan exploration to specifically define conditions closely enough to reliably determine best and most cost-effective location to change from cut-and-cover to true tunnel mining construction.
Construction Shafts	Should be at least one boring at every proposed shaft location.
Access, Ventilation, or Other Permanent Shafts	Need data to design the permanent support and groundwater control measures. Each shaft deserves at least one boring.
Solution-Mining	Need chemistry to estimate rate of leaching and undisturbed core in order to conduct long-term creep tests for cavern stability analyses.
Pipe Jacking and Microtunneling	Need data to predict soil skin friction and to determine the method of excavation and support needed at the heading.
Compressed Air	Borings must not be drilled right on the alignment and must be well grouted so that compressed air will not be lost up old bore hole in case tunnel encounters old boring.
Portal Construction	Need reliable data to determine most cost-effective location of portal and to design temporary and final portal structure. Portals are usually in weathered rock/soil and sometimes in strain-relief zone.
NATM	Generally requires more comprehensive geotechnical data and analysis to predict behavior and to classify the ground conditions and ground support systems into four or five categories based on the behavior.

Special Needs Related to Size of Excavation

As the dimensions of the underground excavation increase, so do the demands on the geotechnical exploration. For instance, a small tunnel may not experience the degree of construction difficulties or the loss of ground that a larger tunnel might experience. Similarly, the degree of sophistication and demands on the geotechnical team are much greater for a cavern that is 75 ft or more in size than for a 35-ft cavern or tunnel. Short caverns, such as hydroelectric powerhouses, can be oriented to take advantage of rock joint orientations. This increased demand on the geotechnical investigation is much more than just more borings or longer borings. Frequently, the method of construction changes with increased size and the risks are greater, so that different geotechnical tests may be required to support increasingly difficult analyses. Accordingly, the investigation requirements are more stringent.

Integration of Geotechnical Requirements into the Overall Scope of Project

The geotechnical exploration requirements must fit naturally into the overall needs of the project. If the project is small and the geology is not complex, the geotechnical needs may be only moderate. One way of visualizing the interaction of geotechnical investigations with a wide variety of other project elements is given in Figure 4-11, which can help you visualize how the geotechnical exploration requirements fit within the overall project. As illustrated in the figure, the geotechnical scope of work (and therefore the cost of geotechnical investigations) should increase as the geology becomes more complex. Generally, more complex geology demands additional exploration to determine regional

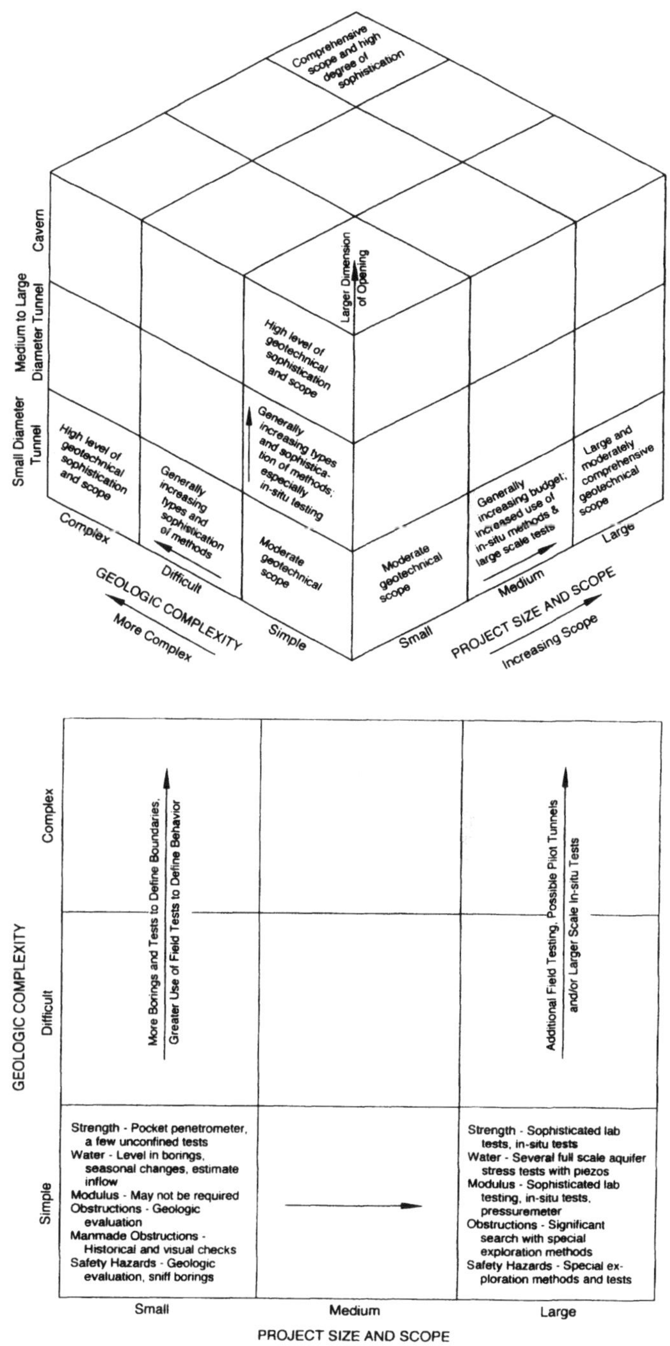

Fig. 4-11. (a) Interrelationships of geology and other project elements. (b) Illustration of variation of geotechnical requirements.

geology, and special methods of exploration are needed to characterize the more complex geology. As the size of the openings increases to cavern size and as the width or free span increases, there is an increased demand for detail in characterization and in prediction of behavior. This is usually accomplished by closer spacing of borings as well as sophisticated exploration such as oriented core, in situ tests, and in special cases, pilot tunnels or adits.

Naturally, the amount of work required also increases with the size and scope of the overall project. Generally, larger projects also can benefit from doing work that reduces the construction cost of the project, usually by increased use of in situ tests and large-scale tests.

Special Needs of Tunnel Boring Machine (TBM) Projects

Hard rock TBM projects present several special demands on geotechnical investigations. In particular, the determination of the feasibility of the use of a TBM and the selection of the type of TBM, cutters, and backup equipment require data beyond that required for conventional tunneling construction methods. For instance, knowledge of the compressive strength is not sufficient for a rock TBM investigation since data on the abrasiveness and other factors are also needed to predict penetration rates and cutter costs. Several machine manufacturers as well as universities and consultants have developed specialized tests to predict the overall effect of the many factors affecting rock TBM performance (McFeat-Smith and Tarkoy, 1980; Kaiser and McCreath, 1994; Trondheim, 1988). TBM tunnels in soil also have some special geotechnical data needs. Some of the special requirements for TBM tunnel projects are outlined in Table 4-28.

Table 4-28. Special Data Requirements for TBM Projects

BOTH ROCK AND SOIL TUNNELS	
Details of All Significant Mixed-Face Conditions	• Special need for rock or soil mass strength data and groundwater conditions in each strata to predict ground/TBM interactions in mixed-face condition. Soil/rock mixed-face needs special characterization.
In Situ Stress	• Special characterization required if particularly high or low.
Overload Ratio of Stress to Strength	• Prediction of control of shield or locking of machine.
Groundwater Inflow	• Prediction of inflow more important on TBM products.
Anisotropy	• Texture and fabric of rock or soil more important to TBM products.
Hazardous Gas (Methane, H_2S)	• For design of automatic shutdown system.
ROCK TUNNELS	
Unconfined Strength	• Use in combination with abrasion data
Abrasivity of Rock	• Cutter cost predictions
Mineralogy, Petrology, and Crystal Size	• Supplemental data to support engineering data
Rock Hardness	• Universities and TBM manufacturers use their special test methods with correlations to penetration rate and cutter costs.
Details of Jointing	• Details of joint spacing and orientation determine cutter/rock interaction. Intense jointing affects. Defines nature of muck. Look for rock wedges that might jam TBM.
Fault and Shear Zones	• Strength and deformability of fault on shear zone affect gripper behavior.
SOIL TUNNELS	
Cobbles and Boulders	• Important to predict size, distribution, and abundance of boulders and cobbles. Maximum size important for specifying type of machine.
Overload Ratio of Stress to Strength	• Generally more important in soil than rock. Factor in determining if pressurized-face TBM needed. Also factor in steering control of shield.
Stickiness of Soil	• Potential problem in EPB mode.
Groundwater Conditions and Likely Inflow to the Face	• May determine need for pressurized-face TBM.
Sensitive Structures within Zone of Influence	• May determine need for pressurized-face TBM.

Special Needs of Projects Requiring Sophisticated Analyses

Sometimes a project is so important, technically difficult, and politically sensitive that very sophisticated methods of analysis and design must be employed. These often take the form of complex computerized analyses such as finite element, discrete element, or finite difference analyses. Since the results of these analyses are no better than the input (garbage in-garbage out, or GIGO) the use of sophisticated analyses puts extra demands on the geotechnical investigation. More geotechnical data must be obtained to provide a statistical basis for each parameter over a broader range of conditions. In addition, such analyses often call for data on parameters that otherwise would be estimated or not even considered. The in situ stress ratio, K_0 is such an example. This parameter is usually ignored for most traditional tunnels except for large caverns or complex configurations under high stress conditions. Yet, once taken into consideration, this parameter has a dominant influence on the calculations but is difficult and costly to measure or even estimate.

Another example is when the sophistication of analysis takes into account the precise nature and sequence of loading or unloading. For most tunnel projects, simple unconfined or triaxial strength test results will suffice. However, when the sequence of construction is taken into account, triaxial extension tests may be desired. Long-term creep is not a major issue on most soil tunnels, even in soft soils. Yet once the need for a reliable quantitative estimate of creep movements or loads is established, time-consuming and expensive creep tests may be needed.

Finally, sophisticated analyses may require a determination of the magnitude and range of a given parameter with in situ tests, large-scale in situ tests, or even by conducting demonstration tests or pilot tunnels to indirectly measure the parameter at full scale. Some of the demands that special sophisticated analyses make on geoinvestigations are given in Table 4-29.

Special Needs for NATM. The New Austrian Tunneling Method (NATM) of construction involves a special approach to both the construction and contracting method. It generally requires more comprehensive geotechnical data and analysis to predict behavior and to classify the ground conditions and ground support systems into categories based on the anticipated behavior. These ground conditions become part of the construction contract for pay items and for selection of the ground support to be used. The investigation

Table 4-29. Special Demands on Geoinvestigations for Sophisticated Analytical Studies

• Generally require a very strong statistical base for the engineering parameters • This implies more exploration and lab tests
• Generally requires more sophisticated lab tests • Triaxial extension or other special strength test • Tests require stress-strain measurements • Creep tests
• Evaluation of engineering parameters requires considerable analysis
• Frequently require in situ tests for strength, permeability, and modulus

program should be directed toward developing clear, unambiguous descriptions of the stratigraphy, groundwater conditions, and behavior of each category of ground specified in the contract documents.

Parameters for Investigations and Tests. Since all projects are site-specific, no single list can be adequate. The geotechnical team and design study team should decide what types of construction and lining can be considered. Then the geotechnical engineer can formulate a program to obtain all the necessary data needed to size, design, specify, cost, and then inspect each particular element of the program.

TUNNEL MONITORING AND INSTRUMENTATION

Geotechnical instrumentation techniques to monitor and control tunnel construction have developed considerably in the last couple of decades. Some of the reasons for this development go to the value or purposes of instrumentation, which are

- To observe the behavior of the ground and groundwater to confirm design assumptions
- To confirm that specified influences have been achieved by the contractors
- To provide documented safety of construction methods and early warning of potentially adverse behavior
- To provide data that point to likely cause(s) of adverse behavior, so that remedial measures may be implemented
- To provide data to assure adjacent property owners and the general public of satisfactory construction behavior
- To confirm safety of innovative construction methods
- To provide facts on which future designs can be based to obtain greater safety and cost-effectiveness
- To provide factual data for planning future phases or extensions of product or to provide direction for research
- To provide factual data for legal claims management
- To control construction, such as for NATM

There are many sources of information on geotechnical instrumentation; the most recent comprehensive and useful reference is by Dunnicliff and Green (1988). Other valuable references on instrumentation for tunnels include Schmidt (1986), Thompson et al. (1983), Cording et al. (1975), and Dunnicliff et al. (1981).

Tunnel monitoring ranges from simple surveys of settlement points over the tunnel to sophisticated instrumentation that comprehensively documents the behavior of the tunnel and its surroundings. Traditionally, contract documents for soft ground tunnels specified a maximum settlement measured at the ground surface. In the last decade, where ground movements are critical, a few projects have specified a maximum settlement of an instrumentation point that is installed about 3 to 5 ft directly above the tunnel (Robinson et al., 1987). A sketch example of a highly complex and comprehensively instrumented tunnel, the 64-ft-diameter Mt. Baker Ridge Tunnel, is illustrated in Figure 4-12. More conventional use of instrumentation on an urban transit tunnel, the Downtown Seattle Transit Project, is illustrated in Figure 4-13 (Robinson et al., 1991).

Data collection is frequently computerized and the time needed to reduce, plot, and evaluate instrumentation data is measured in hours. Accordingly, instrumentation can truly provide an early warning system for control of tunnel construction.

GUIDELINES FOR LEVEL OF GEOTECHNICAL EFFORT

One of the most difficult aspects of any geotechnical investigation is deciding how much exploration to do for design and when the design geotechnical investigation should be stopped. The design geotechnical investigation should be planned and budgeted in consonance with design phases as discussed above. Each phase should have a finite life or at least have milestones where the results are carefully reviewed and a conscious decision made as to what further work needs to be done.

Controversy over how much exploration should be done results from geotechnical engineering being mostly an art or, at best, an inexact science. There is no guarantee that any given geotechnical task or procedure will provide sufficient information, even if properly planned and conducted. In fact, one of the purposes of exploration is to discover whether any conditions exist that may warrant further investigation, and the phased exploration procedure is based on this premise. This uncertainty must be accepted by the owner and designer, and risk management procedures developed to manage these eventualities.

Admittedly, many tunnels have been constructed and put into service with little or no geotechnical investigation. However, the case for not cutting corners on geotechnical services is substantial. It is well documented that insufficient investigation may result in misleading information and substantially increase the risk of not finding hazards and

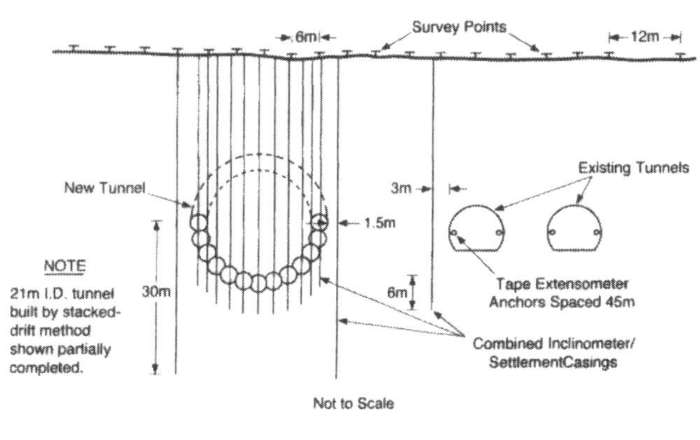

Fig. 4-12. Comprehensively instrumented tunnel, Mt. Baker Ridge.

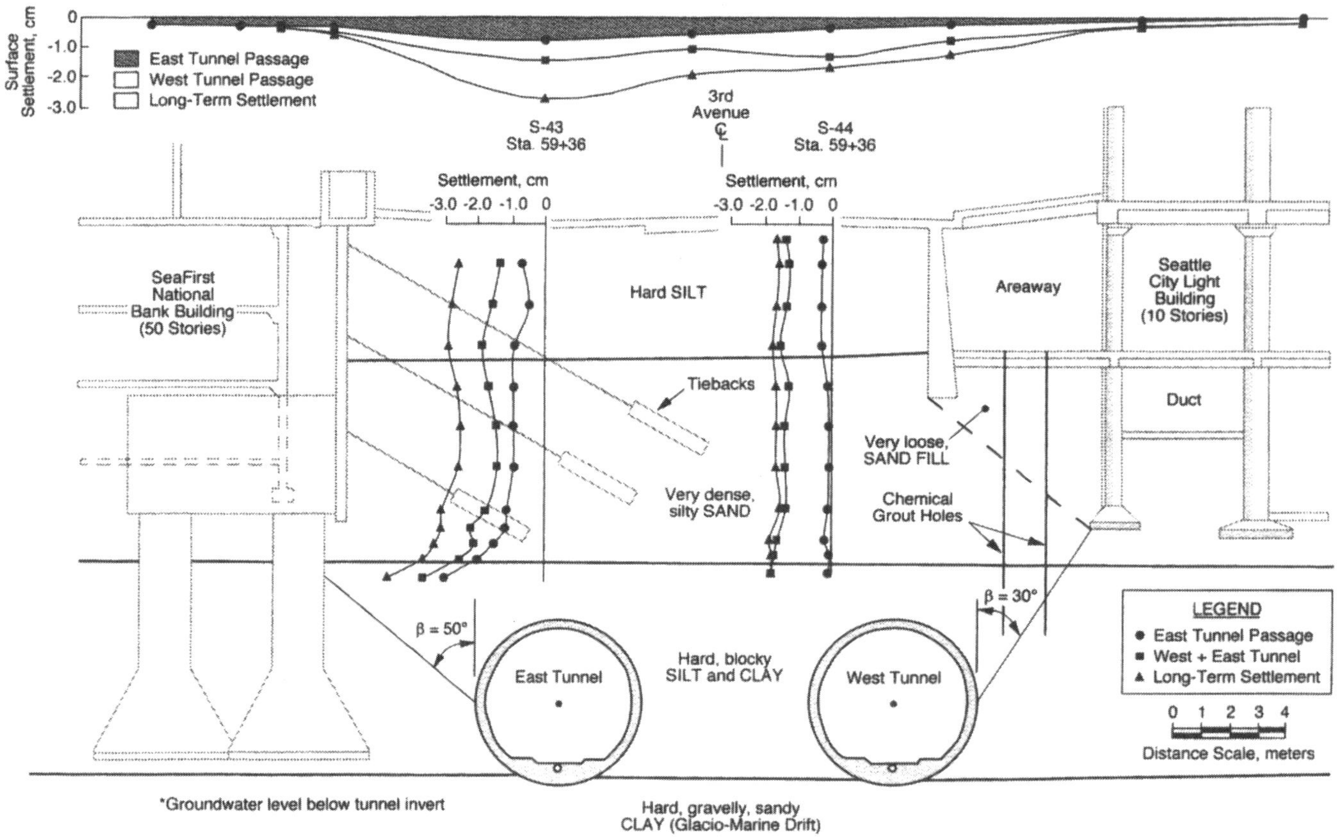

Fig. 4-13. Typical instrumentation on an urban transit tunnel (Robinson et al., 1991).

unknown conditions that seriously delay or stop construction, with costly consequences.

The cost of a geotechnical investigation is frequently related to a percentage of total cost of the project. This provides a helpful guideline: however, it is only one of several criteria that should be followed for planning geotechnical investigations for tunnels. Conventional projects in uniform geology might require less investigation than suggested by a cost guideline for a typical project. Complex projects in adverse geology might require much more investigation than the average. Sufficient funds should be budgeted to conduct a thorough investigation to reduce the possibility of finding unexpected ground conditions. Such unexpected conditions can make the final cost of the completed project many times higher than it would be if the right resources were applied to the investigation in order to disclose the condition. Guidelines for selecting the scope of the investigation are given in the following sections.

Past Practice and Recommendations

Currently, there is no accepted standard for the number of borings, their spacing, depths, etc. Each project must be evaluated on its own merits. With respect to depth, final design borings generally extend for a few tunnel diameters beneath the invert. However, during alignment selection and preliminary design, borings should extend to the greatest depth that the tunnel could conceivably be located. Frequently, owners and designers find a way to consider seriously different alignments, including deeper depths and dramatically different horizontal alignments, when the anticipated cost at their initially selected alignment is large for geological reasons. At these times, the deeper borings and borings off alignment become invaluable.

With respect to number of borings and their spacing, etc., recommendations for major projects in urban areas should be different from those of smaller or remote or difficult to access sites. Generally, however, geotechnical investigations are reported in the literature to cost from 0.5 to 3% of the total cost of the project.

To evaluate the proper level of effort of geotechnical investigations, the U.S. National Committee on Tunneling Technology (USNC/TT) created a distinguished blue ribbon subcommittee, headed by Eugene Waggoner, which made a comprehensive study of exploration practices as of 1984 (USNC/TT, 1984). The subcommittee found that more exploration reduces the uncertainties facing contractors to the extent that their bid will be significantly less, especially in the litigious contracting environment in the United States.

USNC/TT obtained and carefully evaluated data from 84 projects and presented the data and their evaluations in a two-volume report (USNC/TT, 1984). They found that claims for unexpected subsurface conditions were a significant part of the total tunnel cost. Of the 84 mined tunnels studied by the subcommittee, 49 projects reported claims related to subsurface conditions and nearly 95% of these claims were for large amounts. In fact, claims payments amounted to nearly 12% of the original basic construction cost.

In their study, data were plotted and evaluated in many ways to develop correlations between the scope as well as cost of geotechnical investigations and the cost of the completed tunnel. It was found that the as-completed cost (including claims) can differ from the engineer's estimate by up to 50% when the level of effort and funds devoted to geotechnical investigations are low. They also found that the average boring spacing (through final design) on nonmountainous projects was about 260 ft; mountainous projects had boring spacings on the order of thousands of feet. To evaluate the influence of both depth and number of borings, the data for nonmountainous projects were plotted and evaluated in terms of linear feet of bore holes per route foot of tunnel alignment. The definition of route feet refers to project stationing.

Of the many plots and correlations made, two example plots will suffice to illustrate the study results. Bore-hole footage data plotted against "changes requested" or claims by the contractor is illustrated in Figure 4-14.

The data showed that, in common practice for nonmountainous projects in the United States, the total boring footage was a median of 0.42 ft of bore hole drilled per route ft of tunnel. The magnitude of "changes requested" began to decrease at 0.6 linear ft per route ft of tunnel and dropped dramatically with increased boring footage to less than 10% at 1.5 linear ft of borehole per route ft of tunnel.

To evaluate the effect that the cost or scope of the geotechnical investigation may have on the cost of construction, the data were also plotted as shown in Figure 4-15. The median exploration cost of the geotechnical exploration was about 0.75% of the construction cost. Beyond an exploration cost of 3% of the "As Completed Cost," deviations between the Engineer's Estimate and the As Completed Cost were less than 20%.

USNC/TT developed numerous conclusions, all of which are still relevant (1995) and which are reproduced in Table 4-30. Their recommendations, also still valid, are reproduced in Table 4-31.

The top two items in Table 4-31 give USNC/TT's recommended level of effort for geotechnical site exploration as follows:

1. Site exploration budgets should average 3.0% of the estimated project cost.
2. Boring footage should average 1.5 linear ft of borehole per route ft of tunnel.

The recommended 3.0% budget is put into perspective by USNC/TT's feeling that the typical 1% expended for exploration was obviously too low when compared with the average 12% of project costs that were for payment of claims, usually for unexpected subsurface conditions. The levels of effort recommended by USNC/TT are guidelines for the scope and total cost of the geotechnical program. It should be noted that these recommendations are very aggressive; and it is usually difficult to convince clients that the recommended amount (and cost) of exploration is warranted. However, the study was carefully and comprehensively carried out by a distinguished blue ribbon panel and their recommendations deserve serious consideration, particularly on major projects. On the other hand, it is not appropriate for a geotechnical engineer to request that the full budgets recommended by USNC/TT be approved without considerable prior planning and justification. Guidelines for applying the spirit of these

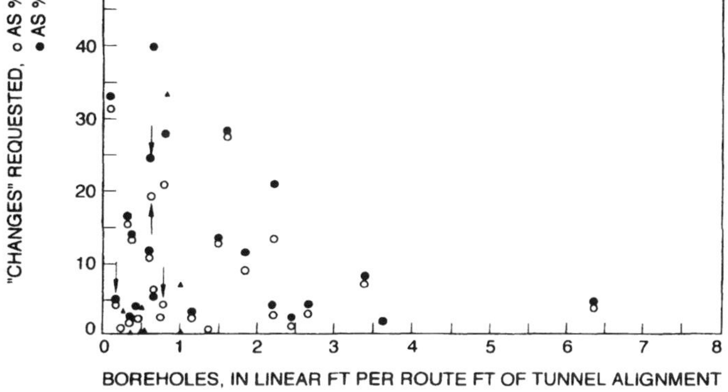

Fig. 4-14. USNC/TT bore-hole footage analysis.

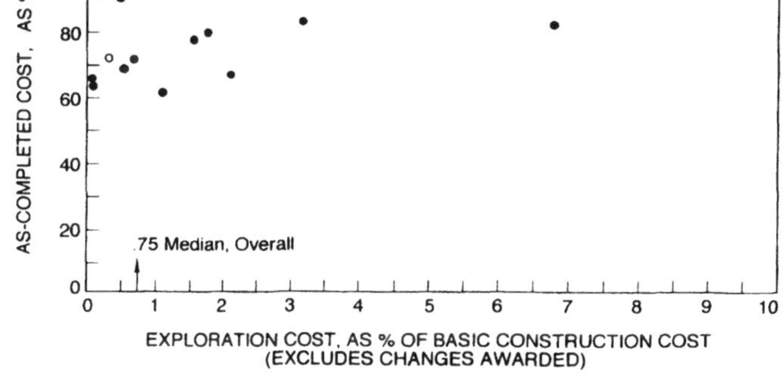

Fig. 4-15. USNC/TT geotechnical cost analysis.

Table 4-30. USNC/TT Conclusions on Site Investigations

1)	It is in the owner's best interests to conduct an effective and thorough site investigation and then to make a complete disclosure of it to bidders.
2)	Disclaimers in contract documents are generally ineffective as a matter of law, as well as being inequitable and inexcusable in most circumstances.
3)	Contracting documents and procedures can provide for resolution of uncertain or unknowable geological processes or conditions before and during construction, rather than afterwards.
4)	On major projects especially, it is important that: a) the owner employ a multi-disciplined team including engineering geologists, engineers, and a construction specialist to develop subsurface data and evaluate their impact on design and construction; b) designers and geologists possess a thorough working knowledge of construction method and equipment so that the proper geotechnical data are secured and design is consistent with construction systems; and c) contractors employ geologists experienced in underground work to evaluate and interpret the data provided at the time of bidding, thus ensuring that all the information obtained is fully considered in preparing bids.
5)	Site investigations have to proceed through, but should not always end with, completion of the feasibility/alignment setting/final design programs.
6)	Procedures for logging, documenting, and preserving samples from borehole require improvement.
7)	Geophysical methods can be use to advantage, especially in coordination with boreholes.
8)	Groundwater and its effects on the subsurface materials merit greater attention in exploration program.
9)	Laboratory testing of the subsurface materials generally needs to be increased.
10)	Exploratory adits and shafts are generally justified only when absolutely essential to obtain critical design data or when a substantial benefit to construction is indicated.
11)	Maintenance of technical data obtained during design and construction of underground projects often is not pursued by owners or demanded of their consultants and contractors.

recommendations in practice are given in subsequent sections; USNC/TT (1984) dealt with this issue as follows:

> The geotechnical site investigation cannot predict every problem that may be encountered, and attempts to do so generally result in programs that are disproportionately expensive for the value received. For every underground project, cost-benefit is a key element. Increasing the level of effort and funds for exploration is demonstrably beneficial and cost-effective.

This USNC/TT recommended program is more complete, aggressive, and costly because of the extreme need to provide the contractor with sufficient information that will reduce uncertainties, drastically reduce the contractor's contingency in the bid, and reduce claims. It is well documented that special programs developed specifically to reduce such contingencies have been instrumental in reducing the bid costs by more than 10 times the cost of the additional exploration. Even greater savings accrue to the owner because such exploration can lead to minimization of construction delays and of potential conflicts and claims from contractors. Both the contractor's and the owner's project managers' available time on the project is safeguarded, and claims leading to legal battles are avoided. Investigators showed that savings in the bid price have been on the order of 5 to 15 times the cost of exploration. That is to say, that if a million dollars is spent doing the right type of exploration, the savings in bid price could be 5 to 15 million dollars.

One such example was the special exploratory shaft and adits constructed as an integral part of the geotechnical investigation for the 64-ft-diameter Mt. Baker Ridge Interstate Highway tunnel project in Seattle, Washington (Parker, 1990). Prospective bidders were invited to visit the site when these excavations were being made so they could observe the ground behavior in essentially full-scale adits at the crown, springline and invert of the proposed tunnel. The cost of the shaft and adits was on the order of $1.5 million but the successful contractor estimated that his construction bid was at least $15 million less as a result of that exploration. This, combined with the special contract provisions regarding sharing of risk, disputes review board, etc., recommended by USNC/TT both in 1974 and 1984 and which were being adopted by the tunneling industry, resulted in a project that had no contested claims or legal proceedings and a final cost of several million dollars less than the contractor bid cost some three or four years before.

Table 4-31. USNC/TT Recommendations for Site Investigations

1)	Expenditures for geotechnical site exploration should be increased to an average of 3.0 percent of estimated project cost, for better overall results.
2)	The level of exploratory borings should be increased to an average of 1.5 linear feet of borehole per route foot of tunnel alignment, for better overall results.
3)	The owner should make all his geotechnical information available to bidders, while at the same time eliminating disclaimers regarding the accuracy of the data or the interpretations.
4)	All geologic reports should be incorporated as part of the contract documents.
5)	Designers of mined tunnels should compile a Geotechnical Design Report, which should be bound into the specification and be available for use by bidders, the eventual contractor, and the resident engineer.
6)	Monitoring of ambient conditions prior to construction should be undertaken to establish a baseline of information for comparison during and after construction.
7)	Pre-bid conferences and site tours should be conducted to ensure that all bidders have access to the maximum amount of project information.
8)	Geologic information from preconstruction explorations and as-built tunnel mapping and construction procedures should be compiled in a report detailing project completion.
9)	Investigation methods and predictions should be improved for three specific conditions: in situ stress, stand-up time, and groundwater.
10)	Improved horizontal drilling techniques should be developed that can recover rock core and penetrate long distances without wandering from line and grade.
11)	Research and development should be conducted to expand the capabilities of geophysical or other remote sensing methods for obtaining geotechnical data between boreholes and from the surface down to depths too great for boreholes.

How Much to Do, When to Stop

The amount of exploration is usually determined by experience and budgetary concerns. Because each project situation is unique, there are no standards and no "handbook solutions" to the amount of investigation that should be done. Accordingly, broad guidelines will have to suffice. An approach and recommended course of action to follow is given in the following sections. Since major or complex projects demand a greater level of geotechnical effort, the first step is to determine whether your project is either major or complex, or a smaller or conventional project, using the interrelationships shown in Figure 4-11 and the guidelines given in Table 4-32. This will determine whether the project will likely require a higher or lower level of geotechnical effort. General guidelines independent of project size are given in Table 4-33.

Ordinary projects include small-diameter, relatively short tunnels (as opposed to sewer systems), pipe jacking under a highway or other facility, and routine projects to be constructed at a reasonable depth through generally uniform geologic conditions and with no unusual groundwater conditions. Guidelines for the geotechnical effort needed for ordinary projects are given in Table 4-34.

Major projects include subway systems, large sewer systems, long aqueducts, large-diameter tunnels, caverns, etc. Projects of 12 ft or more in excavated diameter or tunnels longer than 1,000 ft are considered major projects. Guidelines for determining the additional level of geotechnical effort for major or complex projects are given in Table 4-35.

Table 4-32. Guidelines for Determining Project Magnitude

Major or Complex Projects

• Involve relatively large diameter (greater than 12 feet) or length (1,000 feet) of tunnel, or
• Involve identification and selection of optimum depth and horizontal alignment in an alternatives analysis, or
• Involve future phases or extensions of the project, or
• Require or involve unusual, complex, or unproven construction procedures medium to large diameter microtunnels would fall into this category, or
• Involve difficult or adverse geology, or
• Requires considerable sharing of risk, or
• Involve considerable risk because of factors such as:
  • Urban setting requiring special protection for utilities and buildings
  • Unusual soil or groundwater conditions
  • Relatively shallow depth for diameter of tunnel
• Demand a high level of geotechnical effort

Smaller or Ordinary Projects

• Generally involve tunnels 12 feet in diameter or less and generally less than 1,000 feet long
• Have relatively low risk because of factors such as:
  • Relatively uniform geologic conditions with few adverse geological features and groundwater problems
  • Traditional or relatively non-critical construction methods with adequate cover
  • Previous satisfactory experience with tunnels in same strata
• Can be done with a relatively moderate geotechnical effort

Table 4-33. Guidelines for Level of Geotechnical Effort for All Projects

• Determine all general and specific needs for geotechnical exploration, analysis, and design;   • Determine geotechnical parameters needed; prioritize them
• Use geologic expertise to the maximum extent possible
• Conduct exploration in at least two phases
• The initial exploration phase should be well funded so that sufficient geologic data is developed to confidently select the alignment, and to provide an initial estimate the likely construction methods, lining and cost
• Plan on using nontraditional techniques such as geophysics if they can be used cost-effectively
• Have a fixed budget for each exploration phase as well as a contingency fund   • Have contingency borings and other exploration techniques readily funded and ready to be approved to predetermined criteria in a timely manner to answer technical questions resulting from initial boring program;
• Do not use any more contingency more than necessary
• Get more information than needed for design! Even during the "Design Exploration Phase". Get enough data to be able estimate how the ground will behave under the contemplated construction methods (i.e., if excavated by TBM or by drill-and-blast techniques); and get enough data to minimize uncertainty
• Don't do any exploration unless it specifically fills a genuine need;   • Sometimes, reduction of uncertainty (just to be sure) is a genuine need but one must be very careful to be realistic about the likely cost and benefits

• Conduct a "Supplementary Cost Exploration Phase" after the alignment is fixed to obtain detailed data specifically to:
  • Confirm the design, and
  • Get enough information to minimize the uncertainties and to reduce contingencies. The contractor should be able to feel confident that the contractor can reliably select construction techniques and confidently estimate construction costs.
  • You have done enough exploration for this Cost Phase if, when you write the GDSR, you can be professionally comfortable writing that a certain parameter or factor *"will"* be (the stated value(s) or where you said it will be) rather than having to write that the factor *"may be"* or *"might be."*
  • The budget for this cost phase is likely to be covered by the overall budget.

Table 4-34. Guidelines for Level of Geotechnical Effort for Smaller or Ordinary Projects

• For all phases of design, budget, and fund between 1/3 to 1/2 of the USNC/TT guidelines (i.e., boring footage ranging from 0.5 to 0.8 times route footage and geotechnical costs ranging from 1.0 to 2.0 percent of construction cost).
• Have an overall budget including contingencies of 3.0 percent of overall cumulative cost.

Table 4-35. Guidelines for Level of Geotechnical Effort Major or Complex Projects

• Develop multi-phased program to fill actual needs.
• Plan on using non-traditional techniques such as geophysics, shafts, adits, pilot tunnels, pump tests, etc., as appropriate provided they can be shown to add significantly to the database and will reduce uncertainty.
• For all phases of design, budget and fund between 1/2 to 3/4 of the USNC/TT guidelines (i.e., boring footage ranging from 0.75 to 1.2 times route footage and geotechnical costs ranging from 1.5 to 2.25 percent of construction cost).
• Have a contingency budget up to full USNC/TT guidelines of 3.0 percent of construction cost. • Have contingency borings and other exploration techniques readily funded and ready to be approved in a timely manner to answer technical questions resulting from initial boring program.

When to Do Less Investigation

Sometimes there are reasons that less exploration should be conducted. Sometimes, a prominent river embankment, highway cut, or other exposure provides a rock exposure that is far better than any boring program could ever provide. At other times, the cost of borings is so expensive (because of difficult terrain that might require helicopter mobilization, etc.) that borings are not practical nor cost-effective and might result in misleading information (a little knowledge is a dangerous thing).

Table 4-36 provides a list of selected conditions when less exploration might be considered for a tunnel project.

When to Do More Investigation

It should be noted that there also are cases where there are compelling reasons for a comprehensive investigation even though there is abundant nearby data. Clearly, each tunnel must be taken on a case-by-case basis (see Figure 4-11). Some issues tending toward increased geotechnical budgets are outlined in Table 4-37. Note that some very complex projects in complex geology have required a geotechnical scope equal to or greater than 8% of the construction cost.

Table 4-36. Selected Conditions When Scope Might Be Reduced

• When ground and groundwater conditions are disclosed by existing outcrop exposures that can be reliably used to predict tunneling conditions at depth.
  • Natural valley exposures
  • Significant manmade cuts
  • Shallow soil cover with exposures
  • Lack of weathering
    • Arid regions with low soil cover
  • Exposures along river, or exposed in chutes or waterfalls
• Nearby soil and/or rock exposures may give as good or better information than borings
• Substantial quality database already existing in vicinity
  • i.e., city, investigator's, or owner's files
• When abundant recent good-quality borings suitable for tunneling exist nearby
• When nearby excavations or foundations have been constructed and sufficient borings and construction experience has been documented
• When a tunnel has been constructed nearby or parallel
• So complex a site that additional borings would only confirm scatter and not provide additional reliable, correlatable results
• When impossible to gain access to the site
  • Rugged mountainous site
  • No access because of obstructions

Table 4-37. Conditions Warranting Consideration of Increased Scope

• When project involves optimization of depth or alignment or orientation of caverns
  • Note some complex projects in difficult geology have required a geotechnical scope equal to or more than 8 percent of construction cost
• When there is reason to doubt the usefulness of existing data for a tunnel project
  • Borings/exploration not on exact alignment
  • Borings or other explorations or geo-data are suspect
  • Old borings taken for standard foundation investigation for pile or footing design
  • Borings did not go deep enough for new alignment
  • Exploration did not disclose geological stratigraphy in 3D
  • Previous geotechnical exploration/data did not emphasize groundwater inflow
  • Exploration did not provide enough data to estimate cost or to evaluate feasibility of construction method
  • Substantial uncertainty exists
    • Must have reasonable chance that more investigation will reduce uncertainty
  • When geological or historical evidence indicates that unusual geotechnical problems may be encountered
    • High in situ stress (K_0)
    • Gassy ground
    • Chemically unstable geology
      • Gypsum
    • Contaminated ground
      • Note: Contaminated ground does not fall within traditional cost guidelines
    • Archaeological finds
    • When project complexity and uncertainty warrants considerable sophisticated analysis
    • When construction methods or type of construction have never been used in area

How to Select the Additional Exploration Needed to Reduce Cost and Uncertainties

First, determine those parameters or constructibility issues that are big cost or big risk items; then, focus the additional exploration on those issues in a prioritized manner. Generally, these are parameters affecting the rate of advance. Selected big cost drivers for tunnels that are influenced or dominated by geotechnical issues as identified by USNC/TT (1984) are listed in Table 4-38. These can be judged by whether it is possible to confidently predict the behavior or response of the ground and groundwater to the various excavation and ground support methods that may be feasible.

Reliable Prediction of Behavior Needed to Predict Costs

There is a fundamental difference between the way geotechnical investigations are conducted for an underground opening and for any other project. This is because the tunnel project must be estimated on the *behavior* of the tunnel under several anticipated excavation and lining scenarios. These differences are, however, fundamentally important as to why more investigation must be done for a tunnel project and why even what would be regarded as a comprehensive investigation for another project may not provide the proper data.

One example will suffice. Borings for buildings or even for building excavations usually do not emphasize the prediction of groundwater. Instead they emphasize the level of the groundwater with sufficient information to grossly estimate the flow into an open excavation which, if estimated in error, can easily be corrected by additional wells. However, on an underground project, the effects of groundwater are so significant that much more information is needed to predict not only how much water will flow into the tunnel during excavation but how this inflow will affect the behavior of the ground at the face and behind the shield. This is a far more demanding requirement, which needs careful assessment of the exact effects;—not just one effect! Also, different estimates of flow volume and rates of groundwater and its effect on the ground must be assessed considering if the flow is in the crown, springline or invert.

Table 4-38. Selected Big Cost Drivers

Conditions	Percentage of Jobs With Problem	Percentage of Tunnels With Claims	Impact Rating[a]
Blocky/slabby rock, overbreak, cave-ins	38	16	4.2
Groundwater inflow	33	6	1.8
Running ground	27	9	3.3
Squeezing ground	19	8	4.2
Obstructions (boulders, piles, high rock, cemented sand)	12	11	9.2
Face instability, soil	11	5	4.5
Surface subsidence	9	2	2.2
Methane gas	7	2	2.8
Noxious fluids	6	4	6.6
Spalling, rock bursts	6	4	6.6
Hard, abrasive rock, TBMs	5	2	4.0
Face instability, rock	5	1	2.0
Flowing ground	5	4	8.0
Mucking	5	2	4.0
Pressure binding,-equipment	4	4	10.0
Roof slabbing	4	1	2.5
Soft zones in rock	4	2	5.0
Steering problems	4	0	0
Soft bottom in rock	2	2	10.0
Air slaking	1	0	0
Existing utilities	1	0	0

[a] Impact Rating refers to the likely cost consequences of the problem on a scale of 10.

GEOTECHNICAL REPORTS

Geotechnical reports range from short memos conveying geotechnical data to major multivolume interpretive reports. They are prepared and submitted throughout the design phase. The final report that supports the contract documents has received considerable attention because of its legal significance to the construction contract. However, this document is the culmination of all the previous geotechnical reports, all of which must be prepared with the requirements of this final report in mind.

One of the earliest and most important tasks is to plan the purpose and scope of all the reports, including those in future phases of work, so that even the earliest memos and reports will provide useful input to the final report. In fact, with proper planning, each report builds on the previous one, resulting in a strong base of documentation supporting the final report. This requires that the format of boring logs, lab test data, and names and descriptions of strata, and numerous other decisions be made as an early geotechnical task in such a way that they will support the final report, which may be prepared perhaps years later during a subsequent phase. This does not mean that changes cannot be made in the future if an improved presentation is developed; but it does mean that such changes will be even more effective since the entire relationship among all the reports has been planned.

As with most geotechnical tasks, there are no strict guides or specifications for these reports, since they must be project- and site-specific. More likely, these reports are the result of many compromises.

Fortunately, outright disclaimers of geotechnical information are not common any more. Reports still tend to emphasize the uncertainties and, to the extent that this gives the user a better understanding of what can and cannot be predicted, such emphasis is good. However, it should not be used as a pseudo-disclaimer. In fact, there is a trend in the opposite direction, as described in the section below on Geotechnical Baseline Reports (GBR). In this new concept, the geotechnical team makes their best prediction, which is presented as the geotechnical basis for the bid. All requests for changed conditions are compared with this datum.

It is crucial that the geotechnical reports distinguish clearly between fact, opinion, and interpretation. Unfortunately, there are several levels or degrees of hard data or facts as well as several levels of interpretation. Also, it is important to differentiate between interpolation and extrapolation. If the borings are closely spaced and the geology not too complex, geologic data can be interpolated between borings. Often, however the geologist must extrapolate beyond the field data or interpolate between such widely spaced borings that it is almost like extrapolation. Geotechnical reports should be prepared so that it is clear to the reader whether any given data point is factual or is an interpretation by either interpolation or extrapolation.

Finally, the geotechnical report, and especially the GDSR or GBR described below, frequently make predictions of the behavior of the ground and groundwater. This is very difficult to do and it requires the best tunneling experience be made available for these predictions. The most important aspect is that these predictions be accompanied by a clear description of the assumptions made regarding the dimensions of the excavation such as size and depth, the method of excavation, groundwater control, the speed and timing of the construction, the speed and timing of ground support, and so on.

Logs of Borings and Field Tests

The logs of all field tests include borings, test pits, groundwater measurements, in situ tests, geophysics, and geologic observations. The most used is the final boring log, which is the refined field log after all the field data have been digested in a total context. The final log is the basis for all subsequent sections and profiles. Using the geologic profile as a tool, the geotechnical team puts together the final boring logs so that all of the interpretations and descriptions of data are consistent from log to log. This may require several iterations and numerous inspections of the soil or rock samples. Special attention should be given to those items in the logs that may have particular significance to the behavior of the materials or to the cost of tunnel construction.

There is a wide variation in how the tunneling industry presents logs for the contract. Sometimes they are placed on drawings, sometimes they are left in 8-1/2 × 11 format as part of a geotechnical report or a GDSR. Again, there is a wide range in how the data are presented, often depending on personal preference. A good sample log of an outcrop is reproduced as Figure 4-16, and a good boring log is reproduced as Figure 4-17.

Geotechnical Design Summary Report (GDSR)/Geotechnical Baseline Report (GBR)

Past practice in the United States and elsewhere in the world involved the general disclaimer of all geotechnical data. This resulted in costly claims, delayed project completions, and considerable litigation. As a result, an entirely new strategy was developed for improving contractor/owner relationships, particularly in contracting practices (USNC/TT, 1964). This new risk-sharing strategy calls for sharing of risk of unknown conditions fairly, complete disclosure of all factual geotechnical information to all bidders, and the preparation of a special report that documents the designer's reasoning behind the selection of construction methods, lining types, anticipated ground behavior, and so on.

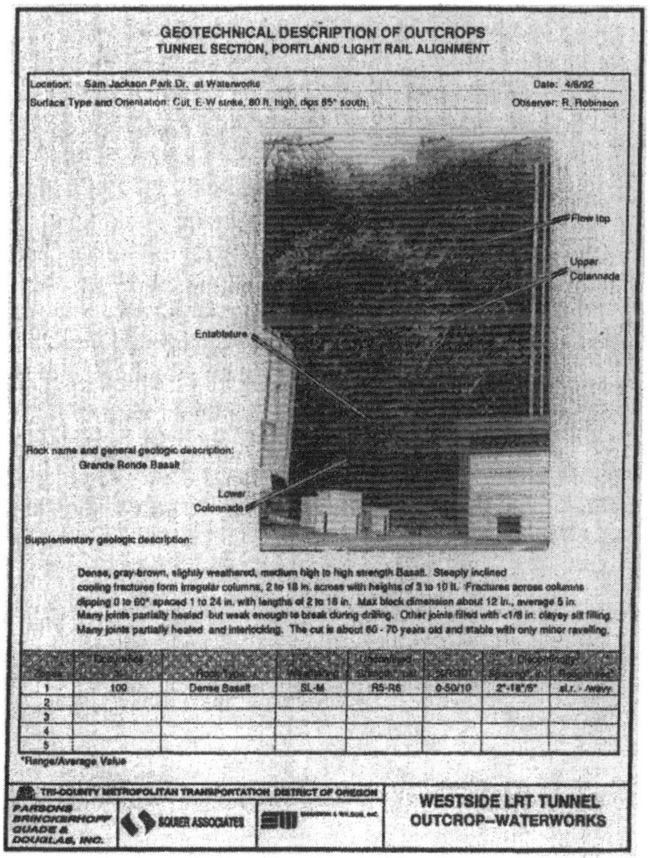

Fig. 4-16. Sample log of an outcrop.

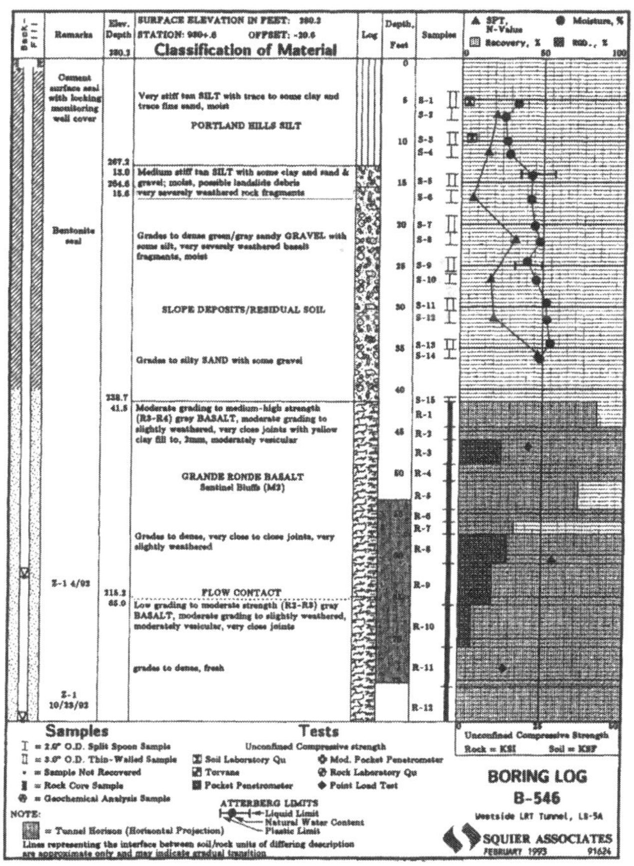

Fig. 4-17. Sample final boring log.

Key to this new contracting strategy is a new special report known as the Geotechnical Design Summary Report (GDSR), which is a powerful tool in presenting geotechnical data for tunnels of all types and sizes. Since one of the main purposes of this report is to define a specific set of geotechnical conditions that will be used as a baseline in contract administration, this report has recently been renamed the "Geotechnical Baseline Report (GBR)."

Geotechnical Design Summary Reports were recommended by the USNC/TT Better Contracting Practices Committee (USNC/TT, 1974) and International Tunnelling Association ITA. The most complete description of the GDSR is given in the booklet *Avoiding and Resolving Disputes During Construction*, published by the American Society of Civil Engineers (ASCE, 1991).

The GDSR is a formal means of communication of the essential geotechnical issues surrounding the project between the contractor and the owner, as well as between the designer and both the contractor and the owner's construction manager. Frequently in U.S. practice, an entirely new organization is brought on to be the construction manager, which has had no previous contact with the project until they arrive. In addition, the GDSR serves as a benchmark or baseline in the documentation of the project and the conditions anticipated to be dealt with during construction.

The factual portions of the GDSR should be prepared by the geotechnical engineer. The interpretive portions ideally should result from a partnership between the geotechnical and tunnel engineer. However, should differing interpretations become irreconcilable, the decision depends on the contractual relationship between the two or on who signs the contract drawings and thus has final authority. Since the current use of the GDSR is to set a baseline for geotechnical conditions when such a baseline cannot reliably be predicted, the actual effect is to establish how much risk the owner will share with the contractor. This places a great demand on clear and meaningful communications with the owner regarding the cost implications that might result from setting the baseline.

This also demands that a level of trust be established between the owner and the geotechnical engineer, tunnel engineer, and designers who prepare this risk-sharing document.

The ability to communicate not only facts but opinions and design philosophy to the future contractor and the future construction manager is powerful, both for the geotechnical engineer/designer and for the contractor and construction manager. The fact that the report conveys, among other things, design philosophy is powerful because the reasons that certain construction or lining methods were considered but not adopted can be conveyed to the bidders and to the construction managers in ways that were heretofore impossible. This falls in line with the full-disclosure philosophy; the disclosure being, in this case, the designer's thinking.

The GDSR also permits the geotechnical designer and tunnel designer to collaborate in describing the project, giving particular attention to unusual features or features that require special attention, or even features that might be unusual and should be brought to the attention of those constructing the tunnel. Heretofore, such descriptions in a form that was readable and understandable by the contractor and the construction manager were not possible because of the terse legalistic language commonly used to write all specifications.

The GDSR is frequently not a stand-alone document, and many variations exist for the backup documents. On very large jobs, the owner puts together a project library in which major reports, even progress reports, are archived and made available to bidders and, of course, the construction team once they begin. Usually the GDSR is preceded by a geotechnical data report, which documents the results of the geotechnical investigation in detail. This would include all boring logs (unless the decision is made to put the boring log only as part of the plans or GDSR), cross sections, detailed laboratory test results, geotechnical evaluations, etc., at a level of detail and complexity suitable for reading and evaluation by a geotechnical engineer. The GDSR, on the other hand, is written at a far more universally understandable level; the results of these detailed geotechnical data are interpreted, evaluated, and presented in a manner that can be easily understood by the bidders, the construction manager, the Disputes Review Board, and, if necessary, a legal entity such as a judge or jury. Thus, the GDSR is presented in commonsense language with simple descriptions conveying facts and interpretations.

This poses extraordinary demands on the writers of the GDSR. First, there must be a clear distinction between what is fact and what is opinion or interpretation. Second, in view of the great uncertainty inherent in all geotechnical issues, it is difficult to establish and describe the baseline for each issue or item discussed in the GDSR. In the past, geotechnical reports have not been precise in their description of ground conditions nor in their interpretation of its behavior. Instead they have tended to describe a range of behavior that includes almost every possibility. The GBR concept, however, demands a single, unambiguous description of every important issue. Language for GBRs is currently evolving that will allow both the description of the range and the establishment of a single, unambiguous value or description of behavior. This alerts the contractor that, although a certain parameter may vary widely, a specific value for that parameter should be assumed, as illustrated in the following hypothetical example:

> Accordingly, depending on the season and the number of joints encountered, inflow could range from 1,000 gpm to 3,000 gpm. For purposes of bidding, the Contractor shall assume a sustained total inflow of 1,500 gpm from the entire stretch of the tunnel.

or

> Though not encountered in any of the borings, boulders and cobbles should be expected in these glacial soils. For purposes

of bidding, the Contractor shall assume that no more than five boulders, with a maximum dimension of 3.0 feet, and no more than 200 cobbles will be encountered throughout the entire stretch of the tunnel.

Table 4-39 is the outline given by ASCE (1991) of a table of contents for a GDSR.

Construction Reports

There are numerous times when geotechnical services are required during construction for both the owner and the contractor. These services might include preparation and review of submittals, inspection of work, documentation of ground and groundwater conditions actually encountered and comparison with those predicted, comparison and evaluation of predicted to actual behavior, monitoring and evaluation of instrumentation, documentation of geotechnical conditions to determine cause of and solutions to construction problems, and to provide geotechnical input to evaluation of differing claims of site conditions.

In keeping with the nature of construction, reports for such services are usually less formal than reports made during the design period. They are, however, extremely important because they document conditions that might become part of a claim at some future time. Accordingly, they should be prepared by properly qualified and briefed staff. Statements and conclusions should be made carefully even when written at the end of a very long day.

As-Built Reports

The preparation of as-built reports was strongly recommended by USNC/TT (1984). These are particularly helpful to large projects, such as subways, in which subsequent phases or extensions of the system could benefit from a carefully prepared as-built report. Also, as-built reports are valuable to understand and improve future maintenance of a system or to have available when, inevitably, some third party will want to build some facility nearby or over the tunnel. Although this may seem a remote possibility to an owner at the time, as-built reports take on a different degree of importance than aboveground as-built reports because the geotechnical conditions reported on are buried and inaccessible to future inspections.

Table 4-39. Suggested Outline for Geotechnical Design Summary Reports

Chapter	Subject
1	Title
2	Introduction
3	Project Description
4	Sources of Information
5	Geological Setting
6	Geologic Features of Engineering and Construction Significance
7	Man-made Features of Engineering and Construction Significance
8	Anticipated Ground Behavior and Construction Difficulties
9	Excavation Method
10	Ground Support
11	Design of Ground Support (if applicable)
12	Construction Specifications
13	Anticipated Quantities

Note: Adapted from ASCE (1991)

REFERENCES

AASHTO. (1988) American Association of State Highway and Transportation Officials *Manual on subsurface investigations,* Washington D. C., 391 p.

American Society for Testing and Materials (1994) Test method for penetration test and split–barrel sampling of soils, 1994 Annual Book of ASTM Standards, D 1586–84 (1992) v. 04.08, p. 129–133. Philadelphia, Pennsylvania.

ASCE (1991) Avoiding and resolving disputes during construction—successful practices and guidelines, Underground Technology Research Council, Technical Committee on Contracting Practices, 1991, American Society of Civil Engineers, New York.

ASCE (1974) Engineering Foundation Conference on Subsurface Exploration for Underground Excavation and Heavy Construction *Subsurface exploration for underground excavation and heavy construction:* American Society of Civil Engineers, New York.

ASH, J. L., RUSSELL, B. E., and ROMMEL, R. R., (1974), *Improved subsurface investigation for highway tunnel design and construction—subsurface investigation system planning,* v. 1, , Federal Highway Administration, FHWA–RD–74–29, Washington D.C.

BARTON, N., GRIMSTAD, E., AAS, G., OPSAHL, O.A., BAKKEN, A., JOHANESEN, E.D. (1992) Norwegian method of tunnelling, *World Tunnelling,* v. 5, no. 5, p. 231–238.

BARTON, N., TUNBRIDGE, L., LOSET, F., WESTERDAHL, H., KRISTIANSEN, J., VIK, G., CHRYSSANTHAKIS, P., (1991) Norwegian olympic ice hockey cavern of 60 m span, *7th International Congress on Rock Mechanics,* Aachen, Germany, Balkema, Rotterdam, v. 2, p. 1073–1081.

BARTON, N., (1988) Rock mass classification and tunnel reinforcement selection using the Q– System, *Rock classification systems for engineering purposes,* Louis Kirkaldie, ed., American Society for Testing and Materials (ASTM), Philadelphia, Pennsylvania.

BARTON, N., LIEN, R., LUNDE, J. (1975) Estimation of support requirements for underground excavations, *Symposium on Design Methods in Rock Mechanics,* 14th, Minneapolis, Minnesota, American Society of Civil Engineers, New York, p. 163–177; p. 234–241.

BARTON, N., LIEN, R., LUNDE, J. (1974) Engineering classification of rock masses for the design of tunnel support, *Rock Mechanics,* v. 6, no. 4, p. 189–236.

BIENIAWSKI, Z. T. (1989) *Engineering rock mass classifications,* John Wiley & Sons, New York.

BIENIAWSKI, Z. T. (1979) The geomechanics classification in rock engineering applications, *4th International Congress on Rock Mechanics,* Montreux, Switzerland, Balkema, Rotterdam, v. 2, p. 41–48.

BIENIAWSKI, Z. T. (1973) Engineering classification of jointed rock masses, *Transactions of the South African Institution of Civil Engineers,* v. 15, p. 335–344.

Cording, E. J., Hendron, A. J., Hansmire, W. H., Mahar, J. W., MacPherson, H. H., Jones, R. A., and O'ROURKE, T. D. (1975) Methods for geotechnical observations and instrumentation in tunneling, Department of Civil Engineering, University of Illinois, Report VILV–E 75 2022, Urbana, Illinois.

DEERE, D. U., and DEERE, D. W. (1988) The rock quality designation (RQD) index in practice *Rock classification systems for engineering purposes,* Louis, Kirkaldie, ed., American Society for Testing and Materials (ASTM), ASTM Special Technical Publication 984, p. 92–101, Philadelphia, Pennsylvania,

DEERE, D. U., PECK, R. B., MONSEES, J. E., and SCHMIDT, B. (1969) Design of tunnel liners and support systems, Washington, D.C. [Available from National Technical Information Service, Springfield, VA 22161 as NTIS Report PB–183799.]

DEERE, D. U., HENDRON, A. J., PATTON, F. D., and CORDING, E. J. (1967) Design of surface and near–surface construction in rock, *Failure and Breakage of Rock—Proceedings of the 8th Symposium on Rock Mechanics,* Charles Fairhurst, ed., American Institute of Mining, Metallurgical, and Petroleum Engineers, p. 237–302, New York.

DEERE, D. U. and MILLER, R. P. (1966) Engineering Classification and Index Properties of Intact Rock, Air Force Laboratory Technical Report No. AFNL–TR–65–116, Albuquerque, NM.

DODDS, R. K. (1982) Preliminary investigations, *Tunnel Engineering Handbook* J.O. Bickel, and T. R. Kuesel, eds., Van Nostrand, Reinhold Company, Inc., New York.

DUNNICLIFF, J. and GREEN, G. E. (1988) Geotechnical instrumentation for monitoring field performance, John Wiley & Sons, New York.

DUNNICLIFF, J., HAMPTON, D., and SELIG, E. T. (1981) Tunnel instrumentation: why and how, *Rapid Excavation and Tunneling Conference,* San Francisco, California, Society of Mining Engineers, New York, v. 2, p. 1455–1472.

GUERTIN, J. D. and MCTIGUE, W. H. (1982) Groundwater control in tunneling–recommended practice, FHWA/RD–81–075, v. 3, Federal Highway Administration, Washington, D.C.

HEUER, R. E., VIRGENS, D. L. (1987) Anticipated behavior of silty sands in tunneling, *Rapid Excavation and Tunneling Conference,* Louisiana, New Orleans, Society of Mining Engineers, Inc., Colorado, Littleton, v. 1, p. 221–237.

HEUER, R. E., (1974) Important ground parameters in soft ground tunneling, *Subsurface exploration for underground excavation and heavy construction,* New England College, Henniker, New Hampshire, American Society of Civil Engineers, New York, p. 41–55.

HUNT, R. E. (1984) *Geotechnical engineering investigation manual,* McGraw-Hill, New York.

ISRM (1981) *Rock characterization, testing, and monitoring—ISRM suggested methods,* International Society for Rock Mechanics, Pergamon Press, London.

KAISER, P. K. and MCCREATH, D. R. (1994) Rock mechanics considerations for drilled or bored excavations in hard rock, *Tunnelling and Underground Space Technology,* v. 9, no. 4, p. 425–437.

KOVACS, W.D., EVANS, J.C., and GRIFFITH, A.H. (1977) *9th International Conference on Soil Mechanics and Foundation Engineering,* 1977, Japanese Society of Soil Mechanics and Foundation Engineering, Tokyo, v. 2, p. 269–276

LAUFFER, H. (1958) Gebirgsklassifizierung fur den stollenbau, *Geologie und Bauwesen,* v. 24, p. 46–51.

MCFEAT-SMITH, I. and TARKOY, P. J. (1980) Site investigations for machine tunnelling contracts, *Tunnels and Tunnelling,* v. 12, no. 2, p. 36–39.

NAVFAC (1986) *Soil Mechanics,* Design Manual 7.01, Naval Facilities Engineering Command, U.S. Navy, Alexandria, Virginia, Change 1, September 1986.

PARKER, H. W. (1990) Concept and construction of the Mt. Baker Ridge Tunnel, *Tunnels and Tunnelling,* v. 22, no. 3, p. 57–58.

PECK, RALPH B., HANSON, WALTER E., THORNBURN, THOMAS H. (1974) *Foundation Engineering,* John Wiley & Sons, Inc., New York.

PECK, R. B. (1969) Deep excavation and tunneling in soft ground *7th International Conference on Soil Mechanics and Foundation Engineering,* Sociedad Mexicana de Mecánica de Suelos, Mexico City, v. State–of–the–Art, p. 225–290.

PROCTOR, R. V., and WHITE, T. L. (1977) *Earth tunneling with steel supports,* Commercial Shearing, Inc., Youngstown, Ohio.

ROBINSON, R. A., KUCKER, M. S., and PARKER, H. W. (1991) Ground behavior in glacial soils for the Seattle transit tunnels, *Rapid Excavation and Tunneling Conference,* Seattle, Washington, Society for Mining, Metallurgy, and Exploration, Inc., Littleton, Colorado, p. 93–117.

ROBINSON, R. A., KUCKER, M. S., FELDMAN, A. I., and PARKER, H. W. (1987) Ground liner behavior during construction of the Mt. Baker Ridge Tunnel, *Rapid Excavation and Tunneling Conference,* New Orleans, Louisiana, Society of Mining Engineers, Inc., Littleton, Colorado, v. 1, p. 309–328.

SCHMIDT, B. (1976) Monitoring soft ground tunnel construction, *a handbook of rational practices for planners and designers,* UMTA–MA–06–0025–76–6, U.S. Department of Transportation, Washington, D.C.

SCHMIDT B., MATARAZZI, B., DUNNICLIFF, C. J., and ALSUP, S. A. (1976) Subsurface exploration methodologies for soft ground rapid transit tunnels—a definition of needs and potential developments, report prepared for UMTA, Rail Program Branch under contract DOT/TSC–654.

SCHMIDT, B. (1974) Exploration for soft ground tunnels—a new approach, *Subsurface exploration and heavy construction,* New England College, Henniker, New Hampshire, American Society of Civil Engineers, New York.

SKINNER, E. H. (1988) A ground support prediction concept: the rock structure rating (RSR) model, Rock classification systems for engineering purposes, Louis, Kirkaldie, ed., American Society for Testing and Materials (ASTM), ASTM Special Technical Publication 984, Philadelphia, Pennsylvania.

TERZAGHI, K. (1950) Geologic aspects of soft ground tunneling, *Applied sedimentation,* P.D. Trask, ed., John Wiley, New York.

TERZAGHI, K. (1946) Rock defects and loads on tunnel supports, *Rock tunneling with steel supports,* Proctor R.V., White, T. L., and Terzaghi, Karl, eds, Commercial Shearing & Stamping Co., Youngstown, Ohio.

THOMPSON, D. E., EDGERS, L., MOONEY, J. S., YOUNG, JR., L. W., and WALL, C. F. (1983) Field evaluation of advanced methods of geotechnical instrumentation for transit tunneling: Washington, D.C., U.S. Department of Transportation, Final Report UMTA–MA–06–0100–83–2.

THOMPSON, D. E., HUMPHREY, J. T., and YOUNG, L. W. (1980) Field evaluation of advanced methods of subsurface exploration for transit tunneling, Contract No. DOT–TSC–UMTA–80–1. Report prepared by Bechtel, Inc., for U.S. Department of Transportation, Washington, D.C., Contract No. DOT–TSC–UMTA–80–1.

U.S. National Committee on Tunneling Technology (USNC/TT) (1984) *Geotechnical site investigations for underground projects,* National Research Council, Washington, D.C.

U.S. National Committee on Tunneling Technology (USNC/TT) (1974) Better contracting for underground construction, National Academy of Sciences, Washington, D.C. [Available from National Technical Information Service, Springfield, VA 22161 as NTIS Report PB 236973.]

University of Trondheim, Norwegian Institute of Technology (1988) Hard rock tunnel boring: Trondheim, Norway, Project Report, 1988.

WICKHAM, G. E., TIEDEMANN, H. R., SKINNER, E. H. (1972) Support determinations based on geologic predictions, North American Rapid Excavation and Tunneling Conference, Chicago, Illinois, Society of Mining Engineers, New York, v. 1, p. 43–64.

5 | Tunnel Stabilization and Lining

Thomas R. Kuesel
Consulting Engineer, Chairman Emeritus,
Parsons Brinckerhoff Quade & Douglas, Inc.

This chapter covers the behavior of tunnels during excavation, and how this behavior may be modified by installation of a variety of reinforcing or supporting systems. *Stabilization* designates systems installed before, during, or immediately after excavation to provide initial support and to permit safe, rapid, and economical excavation. *Lining* designates systems installed either shortly or considerably after excavation to provide permanent support and durable, maintainable long-term finishes. The type of system chosen depends primarily on the ground conditions and on the end use of the tunnel. Frequently, stabilization and lining are provided in two separate operations; this is called a "two-pass" system. In some situations, a "one-pass" system will combine the functions of stabilization and lining.

The objective of this chapter is to establish a practical basis for selection and design of tunnel stabilization and lining systems. Construction of these systems is covered in Chapters 6, 7, 8, 11, and 13.

CLASSIFICATIONS

It is useful to start with a catalog of types of systems, with brief comments on the conditions for which they may be suitable.

Unlined Rock

Many old mountain railroad tunnels have served for years without any lining at all or with linings limited to portals and weak rock zones. The Wawona Highway Tunnel (Figure 5-1) at the entrance to Yosemite National Park has served for more than 60 years with large sections of unlined rock. This system is generally limited to massive, stable rock formations.

Rock Reinforcement Systems

Where rock is generally sound but includes structural defects (primarily rock joints), it may be stabilized with rock reinforcement, whose purpose is to knit the rock mass together so that it is self-supporting. Metal straps and mine ties, secured with short bolts (Figure 5-2), may be sufficient to bridge surface and shallow defects. Deeper joints may be reinforced with untensioned steel dowels or tensioned steel bolts. (Fiberglass bolts have been used for temporary stabilization of rock faces of partial face excavations that are to be removed in subsequent construction stages.) Dowels or bolts may provide only temporary stabilization until the permanent lining is installed, or they may be designed (with appropriate long-term corrosion protection) to support part or all of the long-term loads. An example of this is the Peachtree Center Station of the MARTA transit system in Atlanta (Figure 5-3).

To protect against surface spalling and fallout of small rock blocks between dowels or bolts, a surface skin may be provided. Depending on the tunnel usage, this may range from chain link mesh (Figure 5-4 shows the NORAD Tunnel in Colorado Springs) through shotcrete to a thin poured concrete lining.

Shotcrete

Shotcrete is widely used for stabilization of rock tunnels excavated by drill-and-blast methods or boom type excavators. Its great attraction is that it can provide early construction support in rock with limited "stand-up" time (see Figure 5-5). It is difficult to use shotcrete with full-face circular tunnel boring machines, because the machine and its trailing gear (see Chapter 11) occupy so much space that it is difficult to apply the shotcrete, and shotcrete rebounding from the tunnel wall fouls the equipment.

Shotcrete is sometimes used as a permanent lining. Figure 5-6 shows the Positron–Electron Project, a physics research tunnel at the Stanford Linear Accelerator Tunnel in California. Permanent shotcrete linings are usually built up in layers. The surface layer may contain wire mesh to provide long-term ductility. Alternatively, reinforced shotcrete containing randomly oriented steel or synthetic fibers may be used where toughness and ductility are desirable.

Fig. 5-1. Wawona Tunnel, Yosemite National Park—unlined rock.

Fig. 5-2. WMATA Rock Tunnel—Mine ties.

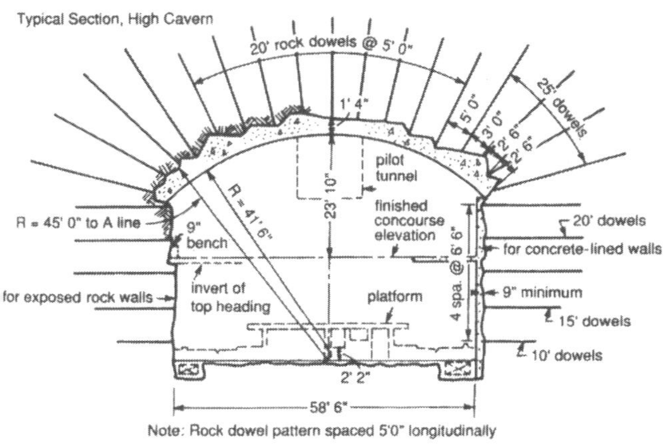

Fig. 5-3. Peachtree Center Station, MARTA, Atlanta—reinforced rock.

Fig. 5-4. NORAD Tunnel, Colorado Springs—Rock bolts with wire mesh.

Fig. 5-5. Shotcrete applied in tunnel heading.

Fig. 5-6. SLAC/PEP Tunnel, California—shotcrete finish lining.

Pre-excavation Stabilization

It is sometimes advantageous to pretreat the ground before excavation to improve its stability. Grouting of weak rock zones or loose granular soils is used to improve stand-up time and reduce the need for initial support, as well as to reduce groundwater inflow into the excavation. This may be done from the surface in advance of tunnel excavation, or from the tunnel face. Ground freezing may also be used for the same purpose.

Locally, spiling (rock dowels drilled ahead of the excavation face, generally inclined upward into the ground above the tunnel roof) may be used to create a reinforced rock umbrella over the next round of excavation. In very poor ground, interlocking steel channel spiling has been used to create a continuous canopy over the crown of the tunnel ahead of the excavation face. Similar umbrellas may be created by grouting or freezing, in suitable ground.

In underground chambers or tunnels of large cross section, it is frequently advisable to excavate in a series of smaller headings or drifts of limited cross section. The initial heading (which may be a preconstruction exploratory tunnel or an initial drift of a multidrift excavation) can be used as an access gallery for pretreating the ground over and around the remainder of the tunnel cross section, by any of the preceding methods.

Ribbed Systems

The traditional support for drill-and-blast rock tunnels is a two-pass system, in which stabilization is provided by ribs (once timber, later steel H-sections, now frequently precast concrete) and a second-stage lining of poured concrete is added subsequently. The system is peculiarly hostile to mathematical treatment but has considerable appeal to practical tunnel constructors, particularly in poor rock conditions. Figure 5-7 shows the Berkeley Hills Tunnel of the BART System, which was successfully driven through some of the worst tunneling ground imaginable (hundreds of feet of rock flour gauge in the Hayward fault earthquake shear zone), by drill-and-blast methods, and using steel ribs and a poured concrete lining.

Where shotcrete is used in soft ground for larger openings, it is sometimes supplemented with ribs formed of prefabricated reinforcing bar cages, commonly known as lattice girders. These may be incorporated into a permanent lining by encasing them with subsequent layers of shotcrete.

Ribbed systems may be used for soft ground tunnels, with the ground between ribs stabilized by "barrel stave" wood lagging or by pressed metal segmental liner plates. Figure 5-8 shows a barrel stave lining on the Los Angeles Metro.

Segmental Linings

Segmental linings are usually associated with soft ground tunnels, and they are erected within the protection of a cylindrical tail shield. In these conditions they can provide a one-pass system, providing both stabilization of the tunnel

Fig. 5-7. Berkeley Hills Tunnel, BART Project—steel ribs with wood blocking, second -stage pumped concrete lining.

Fig. 5-8. Los Angeles Metro—steel ribs with barrel stave lagging.

opening during construction and a permanent service lining. Figure 5-9 shows a metallic segmental lining in the Washington Metro System, and Figure 5-10 a precast concrete segmental lining in the Baltimore Metro System. These segments are manufactured and installed to close tolerances.

Segmental linings may also be used in two-pass systems, with rough cast segments providing only construction stabilization, and a second-pass poured concrete lining added for permanent service. The increased tolerances on manufacture and installation and the simplified details promote more rapid construction, which balances the cost of the poured-in-place inner lining. Figure 5-11 shows a tunnel of the Los Angeles Metro System with a rough precast concrete segmental stabilization system. A poured concrete service lining provides a smooth interior finish.

Segmental linings are usually smaller in diameter than the excavated tunnel, because they are erected inside a cylindrical shield that is part of the excavating equipment. The resulting annular void space is usually filled with grout. In wet ground, segmental linings are usually bolted, to compress gaskets to seal against water leakage. In dry ground, unbolted segmental linings may be used. These are structural mechanisms that derive their stability wholly from the support provided by the surrounding ground.

In ground with appreciable stand-up time, segmental linings may be expanded by jacking them after they have been cleared by the advancing shield. This eliminates the need for grouting the annular tail void. Figure 5-12 shows the system used for expanding segmental precast concrete rings on the Tacubaya Tunnel of the Mexico City Metro.

Poured Concrete

Except for one-pass segmental linings, it is common to follow initial stabilization with a poured-in-place concrete second-stage lining, either plain or reinforced. In wet ground, particularly where end usage may make water leakage objectionable, it may be advisable to install a waterproofing membrane layer between the initial stabilization system and the inner lining. For the Los Angeles Metro, concern for natural gas percolating through the ground prompted provision of a special gas-proof membrane between the two stages (see Figure 5-13).

Fig. 5-9. Washington Metro—fabricated steel segmental lining.

Fig. 5-10. Baltimore Transit System—precast concrete segmented lining.

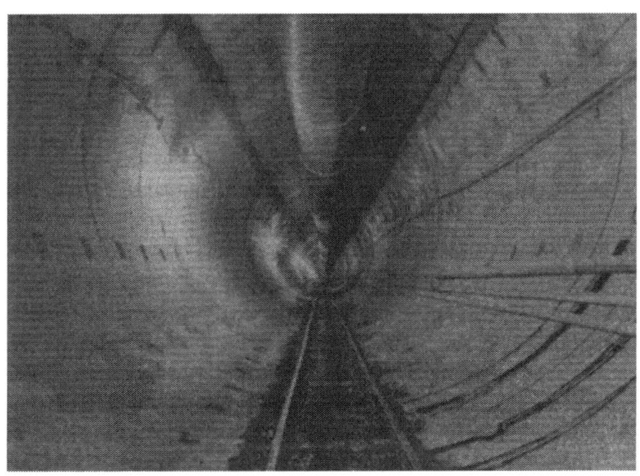

Fig. 5-11. Los Angeles Metro—segmented concrete construction lining.

Fig. 5-12. Tacubaya Tunnel, Mexico City Metro—expanded precast concrete rings.

Fig. 5-13. Los Angeles Metro—gas-proof membrane application in tunnel.

PRINCIPLES OF GROUND–STRUCTURE INTERACTION

Characteristics of Lining Behavior

The preceding discussion demonstrates that the term *tunnel lining* encompasses a broad range of concepts, materials, construction methods, and details. Several common characteristics pervade all systems:

- The processes of ground pretreatment, excavation, and ground stabilization alter the preexisting state of stress in the ground, before the lining comes into contact with the ground.
- A tunnel lining is not an independent structure acted upon by well-defined loads, and its deformation is not governed by its own internal elastic resistance. The loads acting on a tunnel are ill defined, and its behavior is governed by the properties of the surrounding ground. Design of a tunnel lining is not a structural problem, but a *ground*–structure interaction problem, with the emphasis on the *ground.*
- Tunnel lining behavior is a four-dimensional problem. During construction, ground conditions at the tunnel heading involve both transverse arching and longitudinal arching or cantilevering from the unexcavated face. All ground properties are time-dependent, particularly in the short term, which leads to the commonly observed phenomenon of stand-up time, without which most practical tunnel construction methods would be impossible. The timing of lining installation is an important variable.
- The most serious structural problems encountered with actual lining behavior are related to absence of support—inadvertent voids left behind the lining—rather than to intensity and distribution of load.
- In virtually all cases, the bending strength and stiffness of structural linings are small compared with those of the surrounding ground. The properties of the ground control the deformation of the lining, and changing the properties of the lining will not significantly change this deformation. The proper criterion for judging lining behavior is therefore not adequate strength to resist bending stresses, but adequate ductility to conform to imposed deformations. In short, the lining is a confined flexible ring.

Behavior of Flexible Rings

Consider first an unconfined circular elastic ring, of sufficient thickness that buckling is not a problem. Under uniform radial load (Figure 5-14), it will remain circular and will shrink slightly from the axial shortening induced by the uniform ring compression stress.

If this ring is subjected to concentrated loads at crown and invert (Figure 5-15), it will bulge inward at those points and outward at the spring lines. These deformations will be large compared with those of the uniformly loaded ring.

Consider now a partially confined ring, in which passive pressure is developed only at the springline (Figure 5-16). Compared with the previous case, the outward bulges are greatly reduced. Because of the continuity of the ring geometry, the inward bulges are equally reduced.

Figure 5-17 shows a fully confined ring subjected to local active pressure at the crown and invert, and developing passive pressure around the full balance of its perimeter. Both inward and outward bulging are further reduced from the previous cases.

Figure 5-18 shows the general case of a fully confined ring subjected to randomly distributed active pressure loading, in which the passive pressure is distributed relatively uniformly, and the deformations and curvature changes of the ring are slight and smoothly distributed.

It will be observed that the behavior of a flexible tunnel lining is much closer to that of a membrane than to that of a freestanding structural arch. Bending stiffness is generally undesirable: it induces parasitic stresses in the lining. Axial stiffness is of primary importance, as it facilitates the redistribution of unequally distributed active pressures and mobilizes passive pressures.

The most serious source of ring bending in segmental linings is the pressure of grout injected to fill the annular void left by advancement of the tunnel shield. Figure 5-19 shows the general conditions. The lining tends to sink to the bottom of the excavated space under its own weight plus that of the tunnel construction equipment. This leaves an eccentric an-

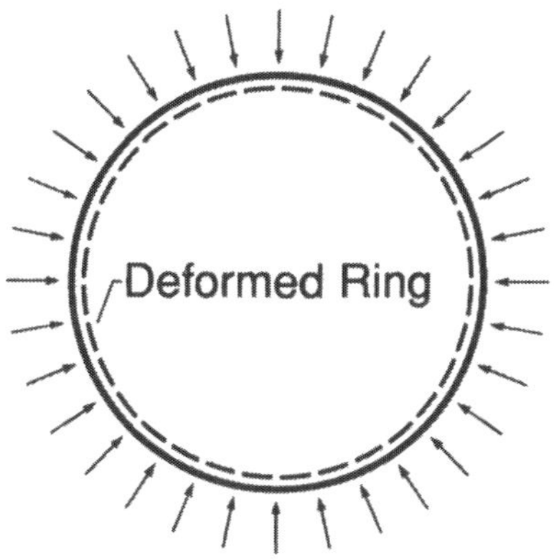

Fig. 5-14. Unconfined ring, uniform load.

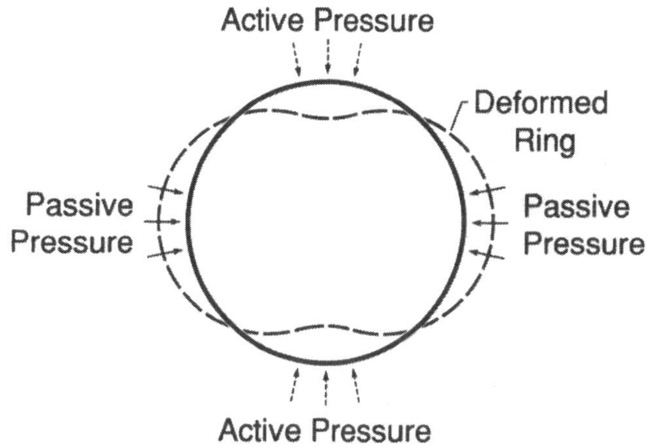

Fig. 5-16. Partially confined ring, concentrated load.

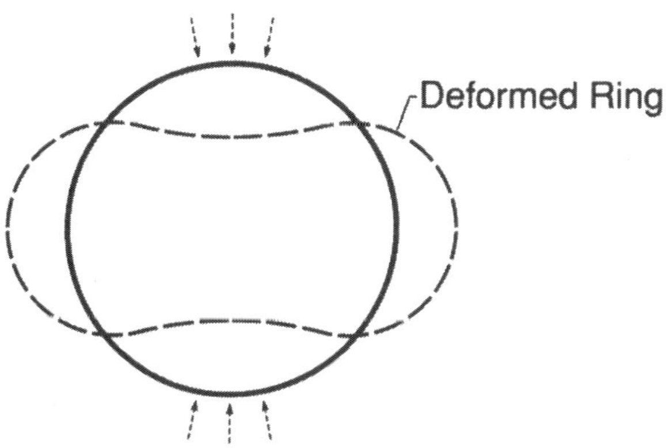

Fig. 5-15. Unconfined ring, concentrated load.

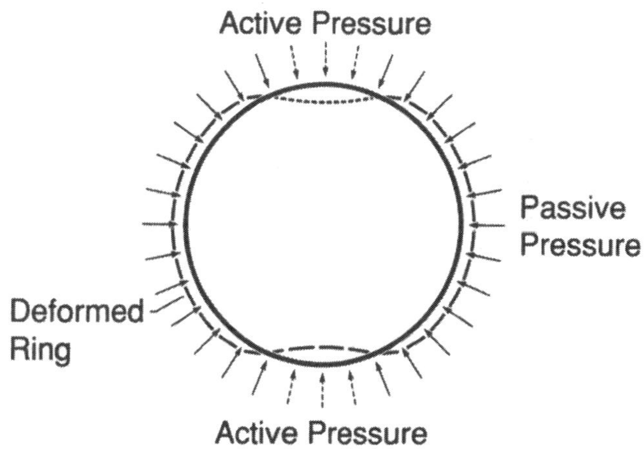

Fig. 5-17. Fully confined ring concentrated load.

nular void, larger at the crown than at the invert. Frequently, the ground collapses rapidly onto the crown of the lining but leaves substantial voids at the springline, which may remain open until the grouting is started. Grout injection is started at ports below or near the springline and proceeds upward toward the crown. Grouting pressure is a maximum at the injection ports, and it is reduced by friction at locations remote from the ports. The resulting pressures on the lining are frequently higher, and more irregular, than the ground pressures. Soil–structure interaction analyses generally neglect this action.

A lining cannot be loaded by ground deformations that occur prior to placement of the lining. It is still frequently contended that steel ribs must be designed to carry the full ground load (whatever that is), and then the concrete lining must be designed to carry the same load "because the steel ribs may rust out." But the ground will not rust out, and it is the shearing resistance of the ground (which has been mobilized by the deformation of the ribs) that carries the load, not the ribs themselves. A secondary concrete lining that is placed after the ground has deformed and stabilized sees

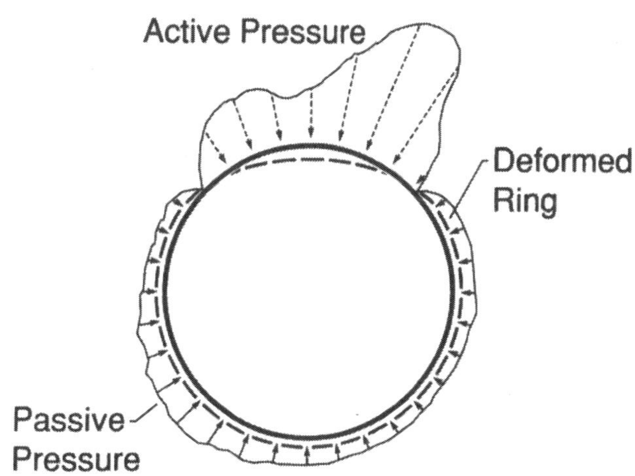

Fig. 5-18. Fully confined ring, random load.

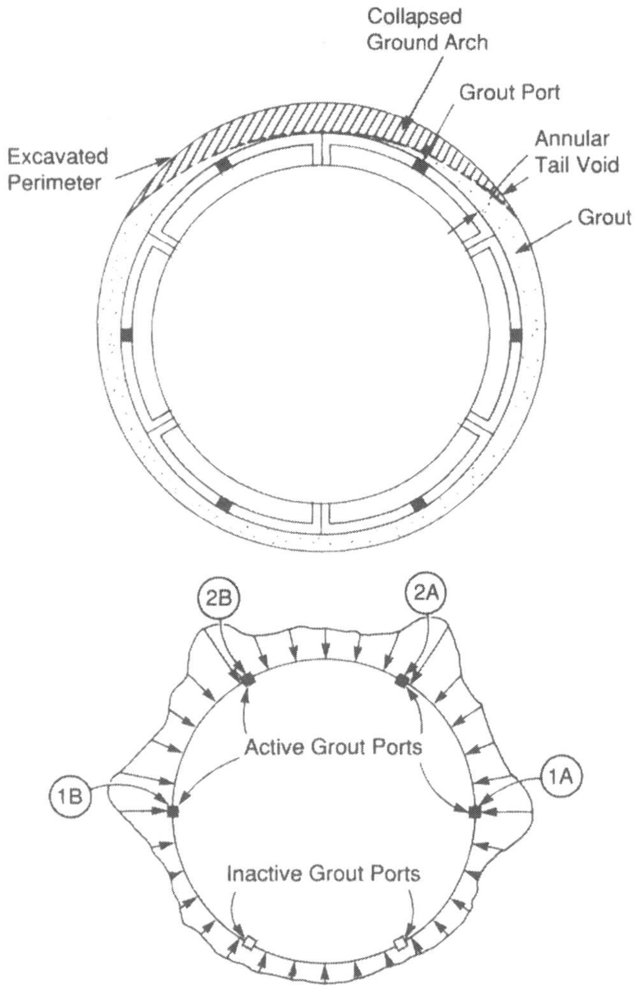

Fig. 5-19. Grout pressures on segmented linings.

only subsequent loads—groundwater pressures, long-term ground creep, and the effects of subsequent construction, such as excavation of parallel or intersecting tunnels.

DESIGN CONSIDERATIONS

Water

The dimensions and details of a tunnel lining are rarely governed by stress considerations. The first consideration is water. If the lining must resist hydrostatic pressure (external or internal), this probably governs the design. Note, however, that hydrostatic pressure produces essentially no flexure in a fully supported circular lining; also, eliminating voids behind the lining is much more important than increasing the lining strength.

Permanent groundwater pressure in soft ground tunnels is dealt with in two ways. The first system is a one-pass lining formed of segmented rings, with watertight gasketed joints. The second system is a two-pass lining, with an initial construction support layer backed by a waterproof membrane, and an internal concrete lining.

Design of tunnel linings to resist internal water pressure in excess of the external pressure is a specialized art, further discussed in Chapter 15.

Constructibility

The second consideration in lining design is constructibility—compatibility of construction processes with the expected ground conditions. The control is ground quality, generally measured by stand-up time. Unstable ground that requires immediate support will generally call for shield tunnel construction. Shields vary from open-faced cylindrical shells that permit manual access to the tunnel face, through simple tunneling machines with mechanical excavators (rotary or pivoted boom type), to elaborate pressurized-face tunnel boring machines. All provide a protected space within which a prefabricated circular lining may be assembled.

The choice of a type of lining system depends primarily on the stand-up time of the ground. If the ground is very soft and wet, support must be provided immediately behind the shield, and this support must be watertight. This generally calls for a gasketed segmental one-pass lining.

If the ground has an appreciable stand-up time, other types of lining become feasible, particularly if the ground is dry or can be dewatered. The initial lining does not then need to be watertight. A traditional form of construction lining was steel ribs with timber lagging. More recently, rough cast, ungasketed concrete segments have been used. Generally the quality of surface finish and joint alignment in these segments is not satisfactory for permanent service, and an interior second-pass concrete lining is added. If the ground is dry or the tunnel is subject only to limited percolation of groundwater, no further treatment is needed, but if the permanent groundwater level is above the tunnel invert, generally a watertight membrane is added between the construction support and the interior lining. It must be assumed that hydrostatic pressure will build up outside the membrane, and so the interior lining must be designed to withstand this pressure.

If the stand-up time is substantial, the initial support may be furnished by shotcrete. This is the basis for so-called New Austrian Tunneling Method (NATM) soft ground tunneling. The system cannot practically be applied if the ground is wet and invades the tunnel rapidly before the shotcrete can be applied, and it is not prudent to apply it to rapidly raveling ground, where unexpected sudden cave-ins and chimneys may lead to large surface settlements. Pretreatment of raveling ground by grouting may extend the range of shotcrete systems.

Shield Systems

In all shield tunneling systems, the diameter of the shield is larger than that of the prefabricated lining, first because the shield tail skin plate must overlap the lining to permit assembly of the lining rings, and second because clearance must be provided between the outside of the lining and the inside of the tail plate to permit steering the shield around curves and to correct misalignments.

This creates an annular "tail void" around the outside of the lining as the shield is jacked forward. If the ground does not immediately fill this void, owing to its three-dimensional, time-dependent arching capability, it may be filled by injecting grout through holes provided in the segments. Alternatively, if the ground has good stand-up time, the first stage ribs or rings of a two-pass lining may be expanded after they have come clear of the shield. This, of course, precludes any longitudinal connection between adjacent rings at this stage.

Lining Dimensions

Constructibility considerations are likely to govern the dimensions of the lining. The length and width of precast concrete or metal segments are governed by shipping and erection limitations, as well as by the design of the tunneling shield. The thickness of segmental linings generally ranges from 6 to 12 in. and is frequently governed by joint configuration, shield jacking loads, and handling stresses. Shotcrete linings may be constructed with thicknesses of as little as 4 in. and are usually built up of multiple 2-in.-thick layers, which are about as thick as can be made to adhere to overhead surfaces. "Poured" concrete linings are usually placed by pumping through a "slick line" pipe located in the crown. The minimum lining thickness is that required to permit the concrete to flow from the slick line down the arch and wall forms without excessive segregation and arching. A nominal clearance of 8 in. between the inside of the primary lining (ribs or liner rings) and the interior reinforcing steel is a commonly used criterion. This may be increased locally at the crown by providing a flat spot or dimple to allow for construction tolerances in placing the slick line (generally 6 to 8 inches in diameter).

LINING BEHAVIOR UNDER GROUND LOADS

Having determined the dimensions and details of the lining by a host of criteria that have nothing to do with ground loads, we may now consider how the lining behaves under ground loads. It behaves as the ground does. If the ground does not move, the lining cannot fail. On rare occasions a lining may fail by crushing and shearing (axial plus flexural compression). Unless there is an unfilled void, or an exceedingly soft surrounding medium, behind the lining, it cannot fail in flexure independent of ground deformation. The criterion for acceptable lining behavior is not a stress limit but compliance to imposed deformation.

Nonetheless, the tunnel designer can have some influence on the ground behavior. Increasing the lining's axial stiffness can reduce ground movement (that is, the ground movement occurring after lining installation). Reducing axial stiffness or delaying installation can permit ground movement to mobilize ground strength, thereby reducing the ground load on the lining. This is a four-dimensional ground–structure interaction problem, in which time is the most important dimension. Subdividing the excavation into multiple drifts limits the length and time during which the ground is required to be self-supporting, before initial support is installed. Ground pretreatment in advance of excavation improves natural stand-up time.

Increasing the lining's flexural stiffness cannot significantly affect ground movement, but it will increase parasitic lining stresses resulting from imposed ground movement. Therefore, if stress analysis indicates that the lining is overstressed, the lining stiffness (and strength) should be reduced. More properly, the assumptions used for the stress analysis should be reappraised and probably modified.

PERFORMANCE CRITERIA FOR FLEXIBLE RING DESIGN

Distortion Limits

In the early 1960s, Ralph Peck (1969; Peck et al., 1972) formulated the recommendation (which has become a widely accepted criterion) that flexible circular tunnel linings should be designed for a uniform ring compression corresponding to the overburden pressure at springline, plus an arbitrarily imposed distortion measured as a percentage change in radius. This criterion was based on observation and field measurement of the actual performance of many soft ground tunnels in a variety of soil conditions.

The arbitrary distortion limit is not just a measure of past performance, but may be specified (at least for segmental linings) as a construction requirement—that the ring distortion be controlled, if necessary, by temporary internal tie rods or struts, until grouting has been completed and the ground has stabilized.

Subsequent observations of soft ground tunnels during and after construction have added to the database. Birger Schmidt (1984) has recommended ranges of distortion ratios to be used for verification of the design of flexible ring linings in a variety of ground conditions. These are presented in Table 5-1.

In their 1972 RETC paper, Peck et al. recommended as a criterion for appraising the relative flexibility of the lining and the ground the following relation:

> Linings in clay soils may be deemed fully flexible if the ratio EI/R^3 is less than five times the unconfined compressive strength of the soil.

Table 5-1. Recommended Distortion Ratios for Soft Ground Tunnels[1]

Soil Type	R/R Range
Stiff to hard clays, overload factor < 2.5–3	0.15–0.40%
Soft clays or silts, overload factor < 2.5–3	0.25–0.75%
Dense or cohesive sands, most residual soils	0.05–0.25%
Loose sands	0.10–0.35%

Notes:
Add 0.10–0.30% for tunnels in compressed air.
Add appropriate distortion for external effects, such as passing neighbor tunnel.
Values assume reasonable care in construction, and standard excavation and lining methods.
[1] Schmidt, 1984.

In this relation:

E = effective modulus of elasticity of the lining
I = moment of inertia of lining per unit tunnel length
R = mean radius of the lining

This criterion is based on the approximate equivalence of the preceding relation and a "flexibility ratio" of 10, as demonstrated in the RETC paper.

Ring Flexibility

The implications of this criterion are shown in Figure 5-20, which is based on reasonable approximations that

- Plastic flow relieves stress concentrations in green (fresh) poured-in-place concrete.
- The stiffness (or effective modulus of elasticity) of a segmented ring is about half that of a monolithic ring.
- The moment of inertia of practical coffered precast segments ranges from 60 to 80% of that of solid sections of the same thickness.

The thickness-to-radius ratio for precast segmental linings generally ranges from 6 to 10%. It will be seen that at the lower end of this range segmental concrete linings are flexible with respect to even soft clays, while at the upper end there may be some interaction with soft soils. Sands are generally stiffer than clays, and so virtually all concrete segmental linings are flexible with respect to sandy soils. Metal pan or steel rib sections are even more flexible.

Cast-in-place concrete secondary linings, whose thickness may range up to 20% of the tunnel radius, are relatively rigid, but as will be seen in a later section, they are not subject to significant imposed distortions.

It is to be noted that the lining flexibility varies inversely with the cube of the thickness-to-radius ratio.

Range of Behavior of Practical Linings

To place a scale on the behavior of flexible concrete rings, Figures 5-21 and 5-22 are presented. Figure 5-21 shows bending stresses induced in an elastic ring (with allowances for plastic deformation of monolithic rings and joint flexibility of segmental rings), for a range of ratios of lining thickness to radius. It is noteworthy that the induced stress is directly proportional to the lining thickness, and that thin (or flexible) linings conform to imposed distortion more easily than thick (or rigid) ones.

Figure 5-22 shows ring compression stresses in circular concrete rings. The stress varies in direct proportion to the overburden depth, and in inverse proportion to the lining thickness.

It will be seen that thin linings are preferable at shallow depth, where inequality of ground pressures is likely to be pronounced, and that thicker linings are required at greater

$\frac{EI}{R^3} < 5q_u$*

$q_u > \frac{EI}{5R^3}$

$q_u > \frac{E}{60}\left(\frac{d}{R}\right)^3$

$q_u > \frac{1}{30}\left(\frac{d}{R}\right)^3 \times 10^6$ #/in²

$q_u > 4.8\left(\frac{d}{R}\right)^3 \times 10^6$ #/ft²

$I = \frac{d^3}{12}$ for unit tunnel length

Say $E_{eff} = 2 \times 10^6$ #/in² for monolithic concrete, considering plastic flow, or for concrete segments, considering joint deformations.

q_u = unconfined compression strength of soil

* From Peck, Hendron & Mohraz 1972 RETC Proceedings, p. 285

$\left(\frac{d}{R}\right)$	$\left(\frac{d}{R}\right)^3$	q^u (min) #/ft² = $4.8\left(\frac{d}{R}\right)^3 \times 10^6$
4%	64 X 10⁻⁶	307
6%	216 X 10⁻⁶	1,037
8%	512 X 10⁻⁶	2,458
10%	1,000 X 10⁻⁶	4,800
12%	1,728 X 10⁻⁶	8,256

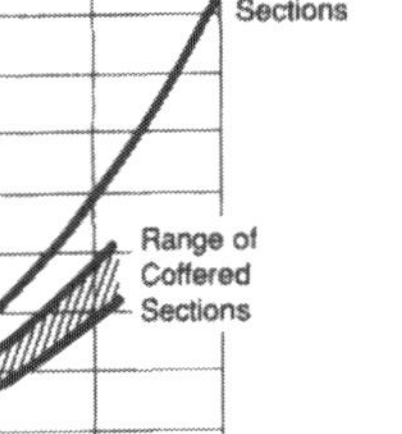

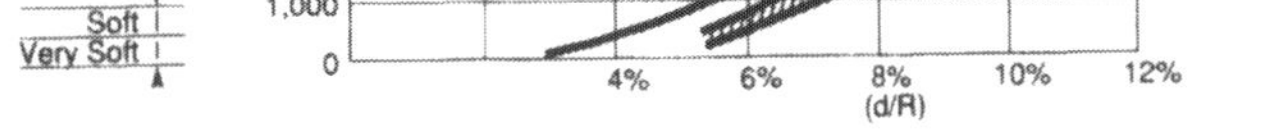

Fig. 5-20. Minimum unconfined compression strength for flexible concrete linings in clay.

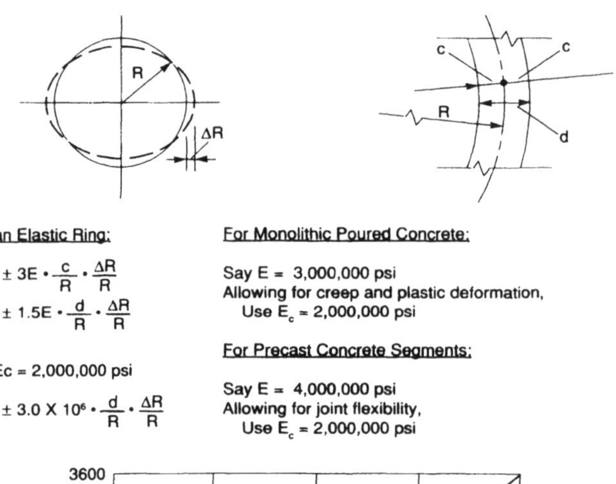

For an Elastic Ring:

$\sigma_b = \pm 3E \cdot \frac{c}{R} \cdot \frac{\Delta R}{R}$

$\sigma_b = \pm 1.5E \cdot \frac{d}{R} \cdot \frac{\Delta R}{R}$

For Ec = 2,000,000 psi

$\sigma_b = \pm 3.0 \times 10^6 \cdot \frac{d}{R} \cdot \frac{\Delta R}{R}$

For Monolithic Poured Concrete:

Say E = 3,000,000 psi
Allowing for creep and plastic deformation,
Use E_c = 2,000,000 psi

For Precast Concrete Segments:

Say E = 4,000,000 psi
Allowing for joint flexibility,
Use E_c = 2,000,000 psi

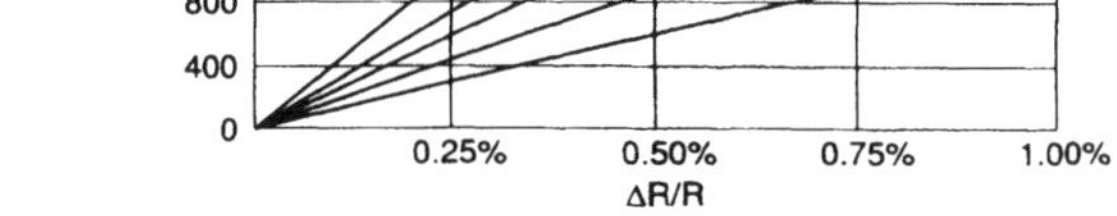

Fig. 5-21. Bending stresses in flexible concrete rings.

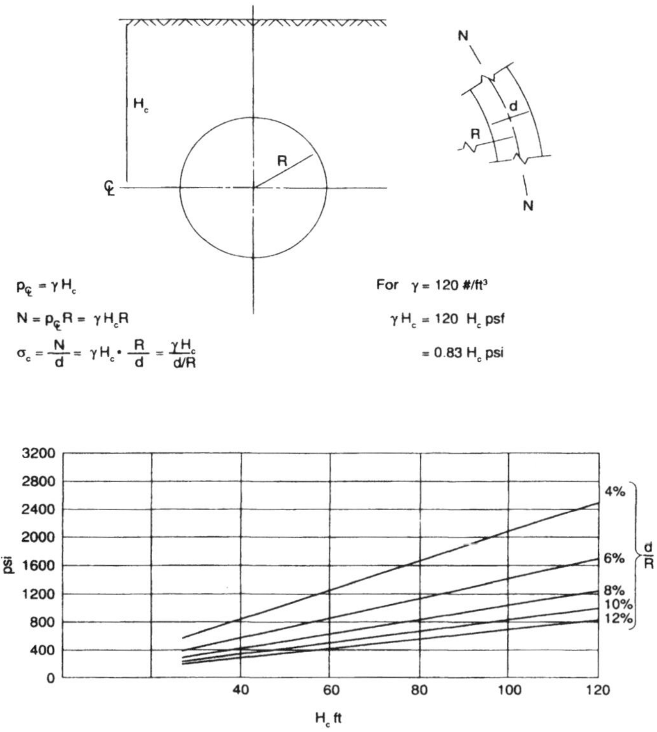

Fig. 5-22. Compressive thrust stresses in flexible concrete rings.

depths to resist axial thrust loads. Fortunately, at great depths the distribution of ground (and ground water) pressures tends to become more uniform, and so the imposed distortion becomes smaller.

It was noted that absence of support—voids behind the lining—is a significant hazard to the structural behavior of flexible linings. Since the presence of such voids (e.g., from incomplete grouting) is difficult to detect, it is advisable that linings have at least a minimum local flexural strength. Figure 5-23 gives a basis for appraising the eccentricity of thrust related to the size of voids.

BEHAVIOR OF TWO-STAGE LININGS

The behavior of a two-stage lining consisting of steel rib initial support and a poured-in-place inner concrete lining is shown in Figure 5-24. The behavior of two-stage soft ground linings, where the initial support is provided by rough concrete segments instead of ribs, is essentially the same as described here. The initial support is flexible and generally ductile—it is therefore capable of deforming easily to mobilize passive pressure and relieve active pressure. The concrete inner lining is installed only after any original imbalance of ground loads has been redistributed and smoothed out by the deformation of the flexible ribs. Loads originating after the inner lining is installed—e.g., groundwater pressure, long-term ground pressure adjustments, rib shortening effects—are relatively uniformly distributed and produce primarily axial loading in the concrete. Voids remaining behind the lining from incomplete concrete filling of trapped air pockets, and pressures resulting from grouting such pockets, are generally located in the crown and result in inward bending, if any.

There is no rational justification for placing reinforcing steel in the outer face of secondary poured concrete linings. On the inner face, a good case can be made for longitudinal steel to resist shrinkage cracking, and for sufficient circumferential steel to hold the longitudinal steel in place against the pressure of pumped concrete sliding down the forms. Many unreinforced concrete linings have behaved entirely satisfactorily for years, however, and it is not clear that the investment in inner ring reinforcing steel is cost-effective.

Where waterproof membranes are installed between lining stages, reinforcement in the inner lining is particularly objectionable, since it is undesirable to puncture the membrane with temporary supports to hold the reinforcing steel in place against the pressure of pumped concrete and difficult to stabilize a freestanding single-plane reinforcing steel mat. It is also difficult to preclude accidental puncture of the membrane during placement of the reinforcing steel.

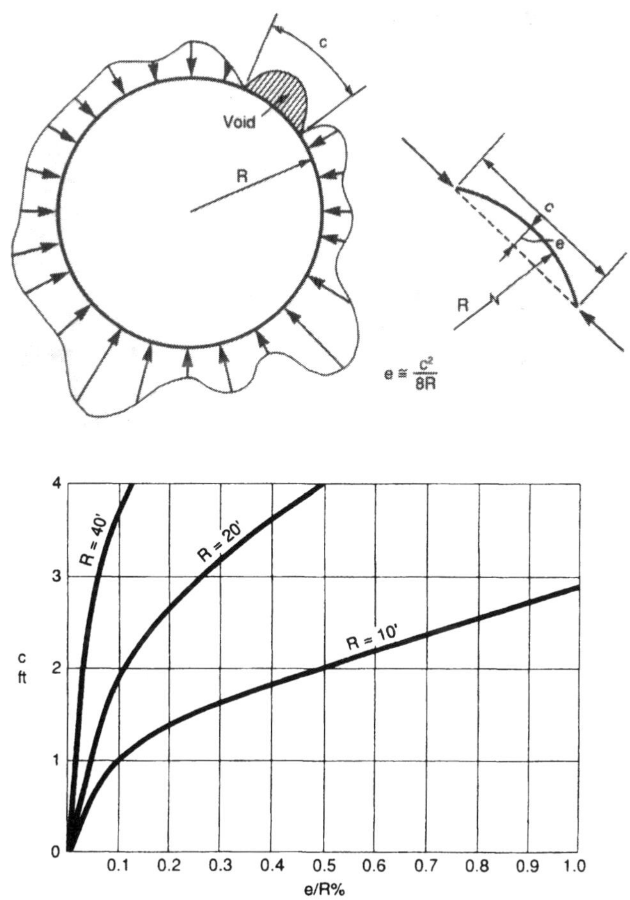

Fig. 5-23. Eccentricity due to lack of support.

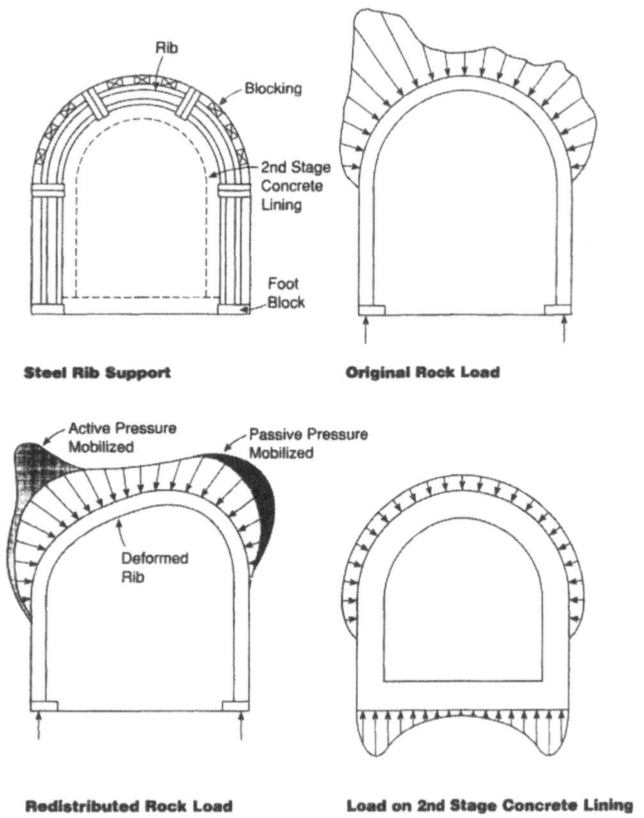

Fig. 5-24. Loads on two-stage linings.

LINING ANALYSIS

A principal problem in applying stress analysis to tunnel lining design is that the analyses are ill conditioned. The calculation of stresses is sensitive to assumptions of governing ground parameters and ground support distribution that are subject to wide uncertainty. Several years ago, Professor Duddeck of Germany undertook (Duddeck and Erdman, 1982) a comparison of the German, Swiss, Norwegian, British, U.S., and Japanese methods for analyzing stresses in a circular tunnel liner ring. One conclusion from this comparison was that the stress in the lining depends on the country in which the calculations are made. However, the same stress can be obtained by any of the proposed formulas simply by varying the "modulus of subgrade reaction" by 20%. It is doubtful that this value has ever been determined within 20% for any actual tunnel, or that whatever its average value may be, it remains within 20% of this value over the length of the tunnel. A thoughtful appraisal of closed form tunnel lining ring analyses is given by Schmidt (1984).

This indicates that the proper application of stress analysis to tunnel linings is in parametric studies of lining behavior, and in the evaluation of laboratory tests made under controlled conditions. Even here, it is important to simulate the ground support conditions as realistically as possible, and to evaluate the sensitivity of the analysis to variations in ground support parameters.

In recent years, much effort has been expended in attempts to subjugate tunnel lining behavior to mathematics. Although some of the mathematics has been impressive, the correlation between behavior predicted by the analyses and that observed in the field has been poor. The principal problem has been the development of a mathematical model that realistically simulates the performance of the ground.

Early models of tunnel linings represented the lining as a string of interconnected pin-ended structural beams, and the ground as a series of radial (and somewhat tangential) springs. Beam–spring models are discussed extensively by Paul et al. (1981, 1993). The original beam–spring model was developed to simulate conditions in a shield-driven, segmentally lined circular tunnel in noncohesive soil, where the ground over the tunnel collapsed rapidly onto the crown of the lining as it emerged from the shield, and this active loading deformed the lining ring against the passive resistance of the ground below the crown. The ground was represented as a series of independent, linear elastic springs, acting at discrete nodes located at the ends of the lining "beam" sections. There was no representation of the cohesion or internal friction of the ground.

While this model gave reasonably good correlation with laboratory tests of lining rings surrounded by loose, clean sands, it did not provide a reasonable prediction of the behavior of tunnels driven through cohesive ground. The next generation of mathematical models attempted to simulate the ground by representing it with a finite element mesh, with internal friction and cohesion properties and linear elastic axial and shearing stiffnesses. Some models used elasto-plastic stiffnesses and permitted varying ground properties in different layers. Such models are discussed in Gnilsen (1989).

Finite element and other simulations of ground mass media as geotechnical continua provide an improved representation of the ground, and they bring recognition of the shearing stiffness of the ground as the dominant element in the lining/ground system. However, virtually all mathematical analyses of the lining/ground interaction system assume that, in Ralph Peck's memorable description, "gravity is suspended while the lining is wished into place, in intimate contact with the undisturbed surrounding ground, after which gravity is turned on again." These analyses do not simulate the effects of construction dewatering, ground movement, and stress relief ahead of the shield and into the annular tail void, nor the effects of lining distortion during erection, ring expansion, grouting, and unfilled voids around the perimeter of the ring. All of these can have major effects on the behavior of, and stresses in, the lining. For two-stage linings, in which a cast-in-place concrete lining is placed after a flexible initial construction support system has permitted the ground to deform and the internal ground stresses to readjust and develop a ground arch, theoretical analyses of single linings in direct contact with undisturbed ground bear no relation to reality.

Cases of closely parallel, diverging, and intersecting tunnels are encountered with some frequency. The latter two categories almost certainly involve two-stage construction. 2D and 3D finite element models can give some qualitative insight into the ground behavior in such complex geometries, but the influences of unmodeled construction procedures cast considerable doubt on the validity of quantitative results of analyses of such situations.

From all this, it appears that the most reliable guide to proportioning soft ground tunnel linings is the criterion proposed by Peck and discussed above—that the lining be designed to resist an axial thrust based on the overburden and groundwater pressure at springline, plus bending stresses resulting from an arbitrary percentage distortion of the diameter of the ring. It should be noted that the usual limits of concrete stresses specified in standard *building* codes do not apply to tunnels, where outward bulging of the lining is resisted by passive pressure of the surrounding ground, and inward bulging is inhibited (at least for circular or curved lining sections) by the curvature of the lining. This is in contrast to a structural member in a rectangular building frame, where an excursion beyond the yield strength leads to unrestrained bulging and collapse.

A principal virtue of the Peck criterion is that it is amenable to field verification by simple monitoring of tunnel diameter changes. It is important to recognize that circular tunnel rings are rarely erected in a truly circular configuration. Lining distortions (particularly in segmental linings) that occur during erection engender little resistance or stress. It is only the distortions that occur after the lining has come into contact with the ground that are structurally significant. It should also be recognized that all tunnel lining sections are underreinforced—i.e., the compression capacity in bending is greater that the tension capacity. Excessive ring distortion that causes tension cracking does not imply potential structural failure, but merely the formation of additional structural hinges and the relief of bending stresses in the ring.

Special cautions are in order for the design of linings for tunnels subject to internal water pressure in excess of the confining pressure of the surrounding ground and groundwater. The axial ring load is tension, which tends to open cracks rather than to close them. Exfiltration tends to erode water channels and to increase leakage and external voids, whereas infiltration drags in fine sediments that decrease leakage and clog cracks. Internal pressure tunnels are particularly sensitive to ungrouted voids and air pockets around the perimeter of the ring (most commonly in the crown). Most lining sections have little shear capacity to bridge across voids on the convex side of the ring. In regions of low ground cover, the internal pressure may be sufficient to lift the ground and cause major ground failure, if the lining is permeable or has structural or shrinkage cracks. In such cases, supplementary steel penstock liners may be considered. Further discussion of internal pressure water tunnels is given in Chapter 15.

BEHAVIOR OF ROCK REINFORCEMENT SYSTEMS

The concept of rock reinforcement is fundamentally different from that of an internal lining. The objective is to use the rock as a structural material to support itself. Most rocks are considerably stronger than concrete, and so the problem conceptually is merely to knit together a rock arch of a capacity comparable with that of a traditional poured concrete lining, and to cure rock mass defects such as joints and shear zones that intersect the excavated surface at unfavorable angles.

Rock Arch Stresses

Figure 5-25 shows a relatively flat reinforced rock arch, with an interior rise-to-span ratio of 1:4 (higher arches are more favorable). With representative assumptions of rock bolt spacing and length, the effective thickness of the reinforced rock arch may be estimated at 1/8 of the span, and its radius at 3/4 of the span.

To place a scale on the stresses involved in global rock arch action, the construction shown in Figure 5-26 is adequate. The resulting loads on the reinforced rock arch are shown in the table in Figure 5-27. Although the analysis is relatively crude, it is sufficient to demonstrate that the compressive stresses in reinforced rock arches are quite small and of little practical concern for rock masses of any reasonable quality.

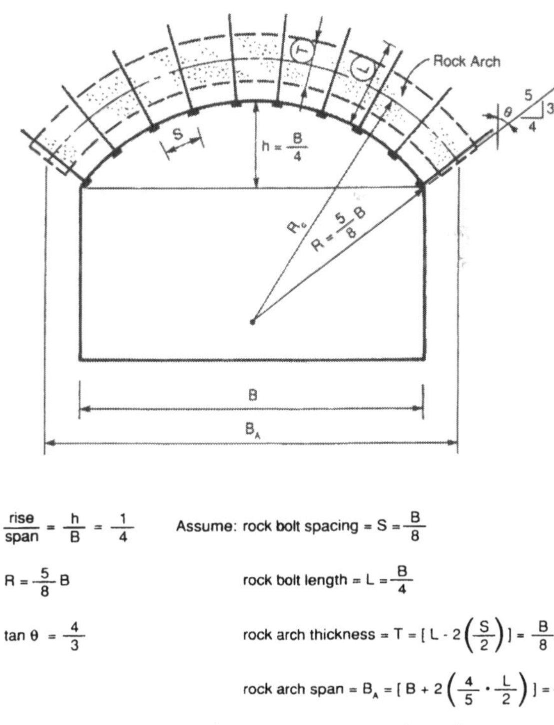

For arch $\frac{\text{rise}}{\text{span}} = \frac{h}{B} = \frac{1}{4}$

$R = \frac{5}{8} B$

$\tan\theta = \frac{4}{3}$

Assume: rock bolt spacing $= S = \frac{B}{8}$

rock bolt length $= L = \frac{B}{4}$

rock arch thickness $= T = [L - 2(\frac{S}{2})] = \frac{B}{8}$

rock arch span $= B_A = [B + 2(\frac{4}{5} \cdot \frac{L}{2})] = \frac{6}{5} B$

rock arch radius $Rc = (R + \frac{L}{2}) = \frac{6}{5} B$

Fig. 5-25. Reinforced rock arch geometry.

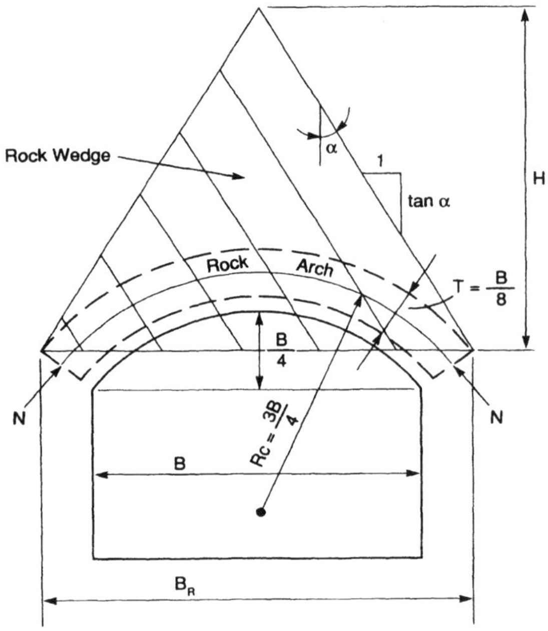

Width of rock wedge ≈ $B_R = [B + 2(\frac{4}{5} \cdot \frac{B}{4})] = \frac{7}{5}B$

Height of rock wedge = $H = \frac{B_R}{2} \cdot \tan\alpha = \frac{7}{10} B \tan\alpha$

Assume average rock arch load = $p = \frac{\gamma H}{2} = \frac{7}{20}\gamma B \tan\alpha$

Thrust/ft = $N = pR_c = \frac{21}{80}\gamma B^2 \tan\alpha \cong \frac{\gamma B^2 \tan\alpha}{4}$

Arch compression $\sigma c = \frac{N}{T} \cong 2\gamma B \tan\alpha$

Fig. 5-26. Reinforced rock arch loading.

Rock Arch Compressive Stress – psi

tan α		0.5	1	2	4
Avg. Rock Height Ratio = $\frac{0.5H}{B_R}$		0.35	0.7	1.4	2.8
B-ft	20	22	45	90	180
	40	45	90	180	360
	60	67	135	270	540

$\gamma = 160\ \#/ft^3$

$\sigma_c \cong 2\gamma B \tan\alpha$

$\cong 320 B \tan\alpha$ (psf)

$\cong 2.2 B \tan\alpha$ (psi)

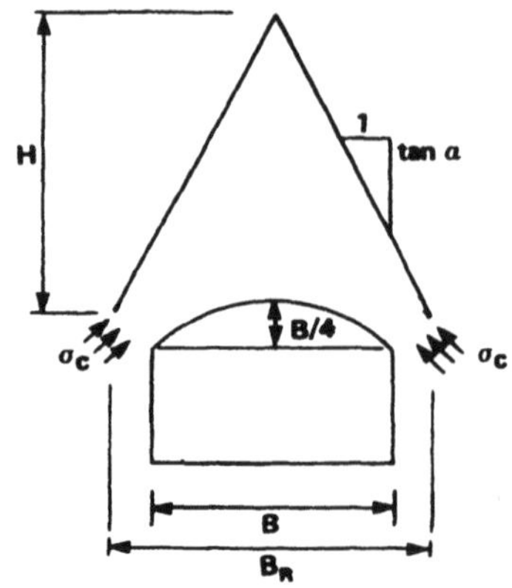

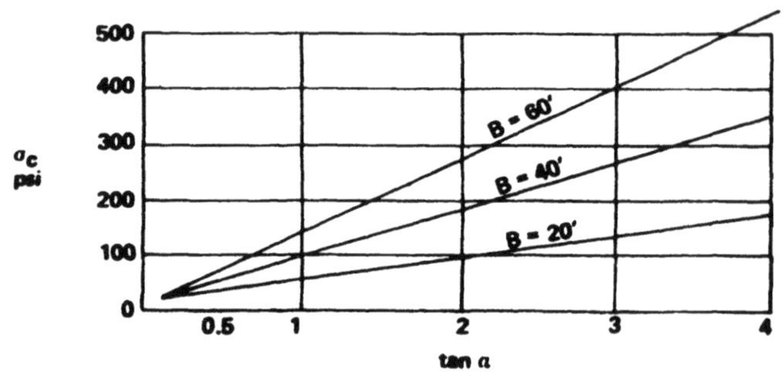

Fig. 5-27. Reinforced rock arch compressive stress.

Figure 5-28 offers a further perspective and may be used to compare the rock arch stresses with the free field stresses in the rock mass. Unless the tunnel is at great depth or of extraordinary span, or there are unusual locked-in tectonic stresses (high K_0), these stresses are also small compared with the compressive capacity of reasonable quality rock masses.

Rock Joints

The real questions related to reinforced rock systems concern the shear capacity along rock joints and shear zones. Figure 5-29 shows typical conditions in which intersections of rock joints with the tunnel perimeter create potentially unstable rock blocks. The geometry of such blocks (which are three-dimensional) can be estimated by projecting rock joint patterns determined from geological investigations to the most unfavorable configurations. Rock block analysis, including allowances for normal stress and friction coefficients across the boundary rock joints, can establish a basis for estimating the size, spacing, and length of bolts needed to tie potentially loose blocks back into the stable rock mass.

Unless a more massive concrete inner lining is provided for functional purposes, rock reinforcement systems usually include a thin surface layer of shotcrete to contain surface spalls and small rock blocks that could form between bolts. As shown in Figure 5-29, it is prudent to design this layer on the basis that any one bolt might be lost or become ineffective (e.g., through faulty installation).

Rock joints that intersect the tunnel perimeter at a shallow angle may pose a special problem, as illustrated in Figure 5-30. Tangential compressive stresses tend to force the acute side of such joints out into the tunnel and thus may call for special cross-bolting to assure stability.

Proportioning of Rock Reinforcement Systems

A thorough geological investigation is a prerequisite to the proper layout and sizing of a rock reinforcement system. For important tunnel projects, an exploratory or pilot tunnel is advisable. This permits collecting site-specific data on rock quality and on the location, frequency, orientation, and characteristics of rock mass defects such as joints and shear zones. These form the basis for rock mechanics analysis, as discussed in Goodman and Shi (1985).

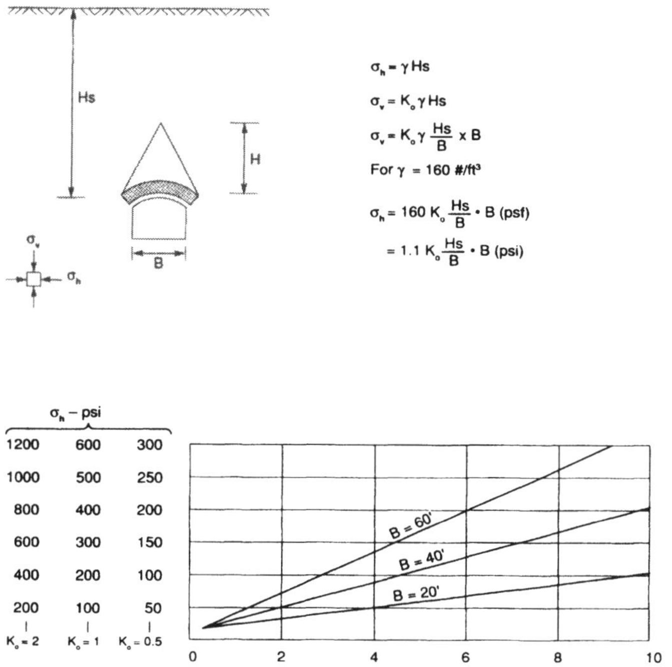

Fig. 5-28. Free-field rock mass stress.

For preliminary proportioning and estimating, guidelines are available that are based on observation of the performance of many reinforced rock tunnel and cavern excavations worldwide. Among the most comprehensive are those of Cording et al. (1971), Bieniawski (1979), Barton et al. (1975), and Heck and Brown (1980), and Wickham et al. (1974). One of the more elusive characteristics is the friction coefficient on rock joints. Cording and Mahar (1974) have recorded measurements made during the construction of tunnels in Washington, D.C. , and New York. These are indicative, but they should be supplemented with site-specific measurements for designs under consideration. Cording and Mahar (1974) also give guidance on selection and analysis of rock bolts. A further paper by Cording and Mahar (1978) gives a more comprehensive discussion of the design of reinforced rock tunnels and chambers. The design of rock tunnels is treated in more detail in Chapter 7.

An important element of rock reinforcement design is verification of performance by field instrumentation. Individual rock bolts may be tested by pullout jacks, and the overall reinforced rock system may be monitored by convergence measurements and rock extensometers. Critical zones for such monitoring may be established from observations of rock conditions and behavior disclosed during excavation. If any trend toward continuing rock movement is observed, additional bolts (or rock anchor cables) may be installed, until all movement ceases and the rock mass is stabilized.

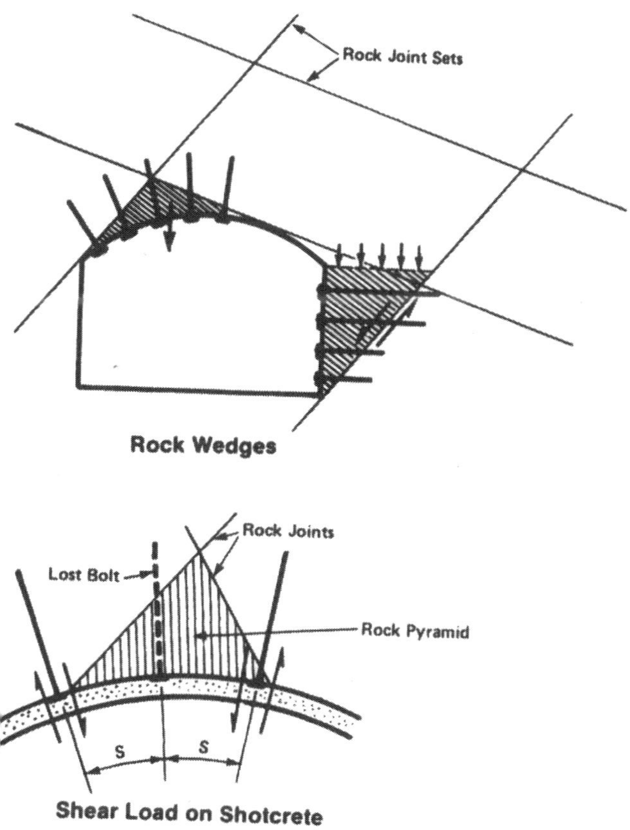

Fig. 5-29. Rock joint problems.

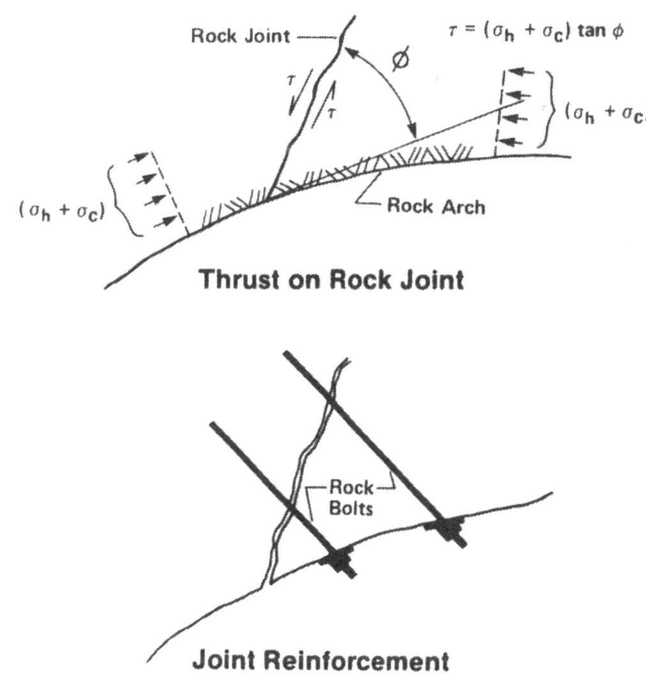

Fig. 5-30. Special reinforcement for critical rock joints.

TANDEM LININGS

In traditional American practice, a clear line was drawn between design and construction, and engineers worked for the owners. Initial stabilization of a tunnel during excavation was the contractor's responsibility, and the engineer designed a "permanent" lining to carry his perception of long-term ground loads. This design generally ignored the effects of construction processes on these ground loads, and it gave no credit for any long-term value of the contractor's initial stabilization system. The result was a dual system, in which the contractor designed "initial support" to carry the full ground load, and the engineer designed "permanent support" to carry the same load. The dual system was inefficient and uneconomic.

In contrast, in European practice the engineers worked for the contractors, and the lines between design and construction were blurred. The contractor's primary interest was safe, economical construction. As a result, engineers concentrated on initial stabilization systems, which were designed with considerably greater care and theoretical underpinning than were accorded to corresponding American contractors' designs. To secure economy, European engineers developed ground–structure interaction theory. Stabilization systems became increasingly sophisticated, culminating in the New Austrian Tunneling Method (NATM), in which rock bolts and shotcrete are applied in carefully controlled and monitored sequences to provide both construction stabilization and permanent lining. The original discussion of NATM by Prof. L. Rabcewicz (1964) is illuminating. NATM designs generally produce a rough interior surface, and a separate interior finish lining is provided. This may be designed to carry groundwater pressure, if a waterproof membrane is included between the initial support shotcrete and the inner lining, but its contribution to support of ground loads is neglected.

The result of these developments was that American design practice ignored the value of first-stage stabilization systems, and European practice ignored the value of second-stage inner linings. Belated American recognition of ground–structure interaction has led to increased attention to design of stabilization systems, particularly for rock tunnels. Permanent rock reinforcement systems are increasingly designed by engineers and included in the design drawings, with extensive specifications to control and verify their installation and testing, and to assure long-term durability and corrosion resistance. These systems approach their European counterparts in sophistication and capability, and they are frequently designed to carry both short-term construction stage loads and long-term service loads. In such cases an inner poured concrete lining may be added as a substrate for interior finishes, but it is accorded no structural value in supporting ground loads.

A further refinement of tunnel design is the concept of tandem linings. In many tunnels, functional requirements dictate a smooth inner surface for the inner lining—e.g., hydraulic conductivity for water tunnels, compatibility with durable and maintainable surface finishes for highway and pedestrian tunnels, reduced frictional drag for ventilation ducts. This generally requires a poured-in-place inner concrete lining. This lining has a minimum thickness controlled by practical construction details. Further, as noted previously, ground movements are almost universally stabilized before the inner lining is placed, and so the lining is subject primarily to axial loading, with very little flexure. It therefore has a substantial capacity to carry long-term ground loads.

Tandem linings use the stabilization system to carry short-term loadings, plus the excess of long-term loadings over the capacity of the inner lining. Where the inner lining capacity exceeds the expected long-term ground loads, the rock bolts or dowels and shotcrete of the stabilization may be designed for short-term performance only—they must support the rock only until the inner lining has been completed. Fully encapsulated, grouted, solid shank bolts are not required, and temporary thin-wall, hollow, friction rock stabilizers may be used instead. Specifications for materials, installation, and testing, devised to assure long-term performance of the elements of the stabilization system, may be relaxed or deleted.

Where long-term ground loads exceed the capacity of the minimum practical inner lining, a hybrid stabilization system may be designed to provide long-term supplemental support. Engineer-designed and -specified rock reinforcement is provided to cover the additional long-term capacity requirements. A minimum short-term rock reinforcement system may be stipulated by the engineer (to appropriately reduced specifications), but this is to be supplemented at the contractor's discretion to assure construction safety.

The tandem system uses both the initial stabilization system and the permanent lining system to their full capacities, and it optimizes the performance and cost of both systems. Three cautions are in order:

- Prediction of ground loads is an inexact art. Prediction is improved by good geology, thorough geotechnical investigations, and experienced interpretation, but there will always remain a range of uncertainty that can be resolved only in the field when the tunnel is excavated. The design must therefore be flexible, and this flexibility is most conveniently and appropriately provided in the initial stabilization system.
- Interaction of the two systems can be secured only if the inner lining is in intimate contact with the stabilization system, so that deterioration of the latter does not generate a loosening of the rock mass. If a waterproof membrane, backed by a compressible drainage fabric, is introduced between the two systems, the performance is somewhat degraded. This can be accommodated by designing the inner lining for an imposed distortion based on the compressibility of the drainage layer. This is generally a fraction of the distortion ratios recommended in Table 5-1 for soft ground tunnel linings, and so the degradation need not be significant.
- As in all tunnel designs, the avoidance of voids behind the lining is critical. If the inner lining is designed to carry long-term loads, contact grouting must be conscientiously carried out to fill air pockets trapped in the crown of the inner lining.

If a waterproofing membrane is included, care must be taken not to damage it during grouting.

The tandem lining system was first used on the 9-mi Mt. MacDonald Tunnel beneath Rogers Pass in Canada (Kuesel and Hansmire, 1987) with up to 5,000 ft of rock cover. The great majority of the tunnel length was stabilized with random location temporary thin-wall rock dowels, supplemented with wire mesh to contain rock spalls, and limited shotcrete. In locations where field observation indicated high rock stresses combined with low rock strength, permanent solid shank grouted rock bolts were installed. This system successfully stabilized the tunnel excavation for up to two years, until the poured-in-place concrete inner lining was added. The lining was required primarily to reduce frictional drag on ventilation air, to purge diesel exhaust fumes from the tunnel.

The tandem lining system design was subsequently used for the Trans-Koolau Tunnel in Hawaii and the Cumberland Gap Tunnel between Kentucky and Tennessee.

RELATION OF DESIGN AND ANALYSIS

This discussion has deliberately avoided the illusion of mathematical precision in order to direct attention to the behavior of various classes of tunnel linings, and to put a general scale on the effects of varying significant parameters within realistic ranges. This may serve to establish preliminary configurations and dimensions as a basis for more detailed design.

Much benefit may be derived from parametric mathematical analysis of various aspects of tunnel behavior to increase understanding of these phenomena. It should be recognized, however, that the precision of evaluation of the controlling properties of the ground is much less than the precision of the analytical methods, and that in any real tunnel the ground properties vary considerably along the length of the tunnel. Therefore, the applicability of elaborate analysis to the design of linings for specific tunnel projects is questionable.

Note that the practical design of tunnel linings is dependent on several matters whose importance is generally equal to or greater than that of structural behavior under ground loads. These include sealing against water leakage, loads imposed by construction procedures (e.g., shield jacking forces), and joint details. More detailed discussions are given in the Underground Technical Research Council's "Guidelines for Tunnel Lining Design" (1984).

PRINCIPLES OF TUNNEL STABILIZATION AND LINING DESIGN

1. The most important part of the tunnel lining is the ground that surrounds it.
2. The most important component of the ground is the groundwater.
3. The most important element of lining construction is to secure full, continuous contact between the lining and the ground.
4. The objective is to stabilize ground movement, not to carry ground loads.
5. The most efficient tunnel stabilization and lining system is one that mobilizes the strength of the ground by permitting controlled ground deformation.
6. Axial stiffness of the lining permits it to distribute nonuniform ground loads by mobilizing passive pressure from the surrounding ground, and it can thereby modify ground deformations.
7. Flexural stiffness of the lining is inefficient (and usually ineffective) in modifying ground deformations.
8. For multistage linings, the initial construction support is very flexible compared with the ground, and it can absorb large flexural deformations associated with redistribution of ground stresses.
9. If the installation of secondary linings is deferred until the ground has stabilized, they will not be subject to significant flexural deformations.
10. Single-stage linings (generally segmental) are flexible with respect to the ground, except for very soft clay. Such linings should preferably be thin, to minimize parasitic flexural stresses resulting from ground deformations.
11. Selection of the type of lining depends on excavation methods that are suited to the ground characteristics, of which stand-up time is usually most significant. Timing of lining installation can substantially affect the magnitudes of ground deformation and lining loads.
12. Dimensions of the lining are controlled by considerations of water sealing, constructibility, and facility usage, rather than by ground loads.
13. Estimates of ground loads and passive pressures are subject to wide uncertainty owing to redistribution of in situ stresses related to ground deformations before and after lining installation, and construction procedures such as contact grouting. Loads and pressures vary along the length of the tunnel owing to variations in geology and in construction proficiency.
14. The largest loads on the lining may come from construction processes such as shield jacking loads and contact grout pressures.
15. The precision of mathematical analyses of stresses in structural rings vastly exceeds the precision of estimation of the loads and support conditions on which the analyses are based.
16. A tunnel liner ring confined by the surrounding ground cannot deform in flexure independently from the ground. Independent structural failure in flexure is impossible unless there are unfilled voids behind the lining.
17. The structural performance of lining elements installed before ground deformations have been stabilized can be appraised by analyzing them for axial thrust plus an imposed deformation measured by an arbitrary change in lining diameter. Appropriate design values may be based on prior tunneling experience in similar ground, may be specified as construction requirements, and may be verified by instrumentation monitoring.

18. Lining elements installed after the ground deformations have been stabilized can be analyzed for axial loads only, plus allowances for anticipated effects of future construction and long-term ground squeezing effects, if appropriate.

CONCLUSION

The design of tunnel linings must account for large uncertainties and variations in ground conditions, construction procedures, and ground behavior during construction. These inherent uncertainties are incompatible with the apparent precision of mathematical analyses of stresses in elastic structural rings. Nonetheless, reliable guidance for economical lining design may be derived from experience with construction of tunnels in similar ground. To repeat, the most important part of the lining is the ground.

REFERENCES

BARTON, N., LIEN, R., and LUNDE, J. (1975) "Estimation of Support Requirements for Underground Excavations," Norwegian Geotechnical Institute, reprinted in *Design Methods in Rock Mechanics* Proceedings of the 16th Rock Mechanics Symposium, Minneapolis, 1975 (published 1977), ASCE.

BIENIAWSKI, Z.T. (1979) *Tunnel Design by Rock Mass Classification* Office of Chief of Engineers, U.S. Army (NTIS AD-A076 540).

CORDING, E.J., and MAHAR, J.W. (1974) "The Effect of Natural Geologic Discontinuities on Behavior of Rock in Tunnels, *Proceedings, Rapid Excavation and Tunnel Conference, San Francisco,* AIME/ASCE.

CORDING, E.J., and MAHAR, J.W. (1978) "Index Properties and Observations for Design of Chambers in Rock," in W.R. Judd (editor), *Near Surface Underground Opening Design,* Engineering Geology 12. Elsevier Scientific Publishing Company, New York.

CORDING, E.J., HENDRON, A.J. JR., and DEERE, D.U. (1971) "Rock Engineering for Underground Caverns," *Symposium on Underground Chambers, Phoenix, Arizona,* ASCE.

DUDDECK, H., and ERDMAN, J. (1982) "Structural Design Models for Tunnels," - *Tunneling '82, Proceedings of the 3rd International Symposium,* Institution of Mining and Metallurgy, London, pp. 83–91.

GNILSEN, R. (1989), "Numerical Methods", in R.S. Sinha (editor), *Underground Structures, Design and Instrumentation.*Elsevier, New York.

GOODMAN, R.E. and SHI, G. H. (1985) *Block Theory and Its Application to Rock Engineering.* Prentice-Hall, Englewood Cliffs, N.J.

HOEK, E., and BROWN, E.T. (1980) "Empirical Strength Criterion for Rock Masses", *Journal of the Geotechnical Engineering Division, ASCE, September.*

KUESEL , T.R., and HANSMIRE, W.H. (1987) "Rogers Pass Tunnel, Bore and Blast," *Tunnelling and Underground Space Technology,* Vol. 2, No. 4, pp. 385–389.

PAUL, S.L., HENDRON, A. J., CORDING, E. J., SGOUROS, G.E. and SAHA, P.K. (1981) *Design Recommendations for Concrete Tunnel Linings. Volume I—Results of Model Tests and Analytical Parameter Studies.* U.S. Department of Transportation, Urban Mass Transportation Administration, Washington, D.C.

PAUL, S.L., HENDRON, A. J., CORDING, E. J., SGOUROS, G. E. and SAHA P.K. (1993) *Design Recommendations for Concrete Tunnel Linings. Volume II—Summary of Research and Proposed Recommendations.* U.S. Department of Transportation, Urban Mass Transportation Administration, Washington, D.C.

Peck, R.B. (1969) "Deep Excavations and Tunneling in Soft Ground," *Proceedings, Seventh International Conference on Soil Mechanics and Foundation Engineering, Mexico City, State of the Art Volume,* pp. 225–290.

PECK, R.B., HENDRON, A.J. JR., and MOHRAZ, B. (1972) "State of the Art in Soft Ground Tunneling," *Proceedings, First Rapid Excavation and Tunneling Conference, Chicago,* ASCE/AIME.

RABCEWICZ, L. (1964) "The New Austrian Tunneling Method," *Water Power,* November/December 1964 and January 1965. Also "Stability of Tunnels Under Rock Load," *Water Power,* June/July 1969.

SCHMIDT, B. (1984) "Tunnel Lining Design—Do the Theories Work?" *Proceedings, Australia/New Zealand Geomechanics Conference, Perth, Western Australia.*

Underground Technology Research Council (1984) - *Guidelines for Tunnel Lining Design.* ASCE.

WICKHAM, G.E., TIEDEMANN, H.R., and SKINNER, E.H. (1974)—"Ground Support Prediction Model—RSR Concept," *Proceedings, Rapid Excavation and Tunneling Conference, San Francisco,* AIME/ASCE.

6 | Soft Ground Tunneling

James E. Monsees
Vice President, Principal Professional Associate,
Parsons, Brinckerhoff, Quade & Douglas, Inc.

As the population of urban areas increases, so will the congestion at the earth's surface. To provide the services (e.g., transportation, water and wastewater, utilities) required by the populace, more and more of those services must be provided by going underground, simply because economical space to provide those services does not exist at the surface. Since tunneling is less disruptive and destructive than cut and cover and since tunneling depth in most large cities lies within the soft ground zone, it is clear that the need for soft ground tunnels will increase. This chapter covers the major considerations that influence the design and construction of soft ground tunnels, which are defined as tunnels that could be excavated using hand tools and methods, although they seldom are in today's mechanized world.

The objective of this chapter is to establish an awareness that tunnels are complex structures, but that basic engineering principles and approaches can be applied to obtain acceptable and economical completed facilities. Because of this complexity, it is not intended that this chapter represent an exhaustive cookbook for design of tunnels in soft ground. Rather, in parallel with Chapter 5, the major considerations and engineering principles are discussed and appropriate references for the remaining details are given.

GEOTECHNICAL INVESTIGATIONS

Geotechnical investigations for tunnels have been discussed in Chapter 4. These investigations are usually conducted in steps, with each step (reconnaissance, preliminary program, and final program) being more site-specific and designed to augment and complete the data found in the earlier exploration(s). Chapter 4 describes the general information obtained in these investigations.

For soft ground tunnels, the properties or conditions of special interest are those associated with groundwater conditions along the alignment and the stability (or strength) characteristics of the soil to be encountered in the tunnel face. Often, especially in granular soils, these two considerations are intimately interrelated.

ANTICIPATED GROUND BEHAVIOR

Anticipated ground behavior in soft ground tunnels was first defined by Terzaghi (1950) by means of the Tunnelman's Ground Classification (see Table 4-6). It can be expanded by considering behavior above and below the water table as summarized in the following paragraphs.

Clays to Silty Sands

Following Peck's (1969) lead for cohesive (clay) materials or materials with sufficient cohesion or cementation to sample and test for unconfined compression strength, an estimate of ground behavior in tunneling can be obtained from the equation

$$N_t = \frac{P_z - P_a}{S_u} \qquad (6\text{-}1)$$

where N_t is the stability factor, P_z is the overburden pressure to the tunnel centerline, P_a is the equivalent uniform interior pressure applied to the face (as by breasting or compressed air), and S_u is the undrained shear strength (defined for this purpose as one-half of the unconfined compressive strength). The ground behavior can then be estimated as follows.

Cohesive Soils

Cohesive (clayey) soils behave as a ductile plastic material that moves into the tunnel in a theoretically uniform manner. This is the type of material for which Broms and Bennermark (1967) initially developed the above stability equation. For this material, tunnel stability may be estimated approximately as shown in Table 6-1 (modified from Peck, 1969, and Phienwaja, 1987).

Table 6-1. Tunnel Stability: Cohesive Soils

Stability Factor, N_t	Tunnel Behavior
1	Stable
2–3	Small creep
4–5	Creeping, usually slow enough to permit tunneling
6	May produce general shear failure. Clay likely to invade tail space too quickly to handle

(After Peck, 1969, and Phienwaja, 1987)

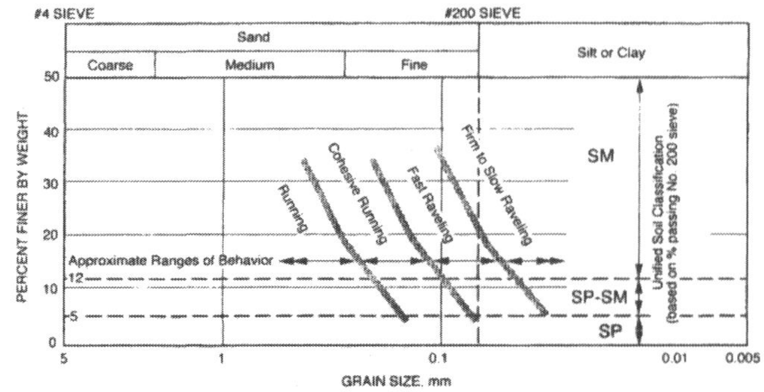

NOTES

1. Based on D_{10} size shown for dense soil, N>30, above water table developed from Terzaghi (Proctor and White, 1977).
2. Very loose soils (N<10) or rounded particles may behave 1 or 2 classes poorer.
3. Very angular sands, bonds, or cementation may behave 1 or 2 classes better.
4. Behavior below water table may be flowing and is a function of water head and permeability and other factors.

Fig. 6-1. Anticipated ground behavior based on D_{10} size.

Silty Sands Above the Water Table

These materials may have some (apparent) cohesion, but they typically behave in a brittle manner adjacent to the tunnel opening. Predicting their behavior by the above equation is more subjective but may be attempted as shown in Table 6-2 (Heuer, 1994).

Silty Sands to Gravels

When the material lacks sufficient cohesion or cementation to sample and test, the behavior is more subjective. Heuer (1987) describes these soils as having "undefined strength" and suggests with "some judgment and audacity" that Figure 6-1 be used as a means of estimating ground behavior based on the D_{10} size of the soil. He notes that this figure is for dense soils ($N > 30$) above the water table, relatively uniformly graded (uniformity coefficient $C_u < 6$), and typical of grain shape and packing for soils transported and worked by water. Thus, loose or very rounded soils might have a behavior one or two classes poorer, and angular or cemented soils might have a behavior one or two classes better. The writer applied Figure 6-1 to the first portions of the Los Angeles Metro and found the comparison of predicted versus observed behavior to be quite good for the alluvial deposits on that project.

Silty Sands Below the Water Table

These soils will be flowing if the uniformity coefficient C_u is not less than 3, and flowing to cohesive running if C_u is less than 6 (Terzaghi, 1977).

Table 6-2. Tunnel Stability: Silty Sands Above the Water Table

Stability Factor, N_t	Tunnel Behavior
1/4–1/3	Firm
1/3–1/2	Slow Raveling
1/2–1	Raveling

(Heuer, 1994)

Sands and Gravels

The behavior of sands and gravels in tunneling was summarized by Terzaghi (1977), and that summary still applies (Table 6-3).

SOIL STABILIZATION AND GROUNDWATER CONTROL

Forty years ago, Dr. Karl Terzaghi wrote the following when discussing ground conditions in soft ground tunnels:

> All the serious difficulties that may be encountered during the construction of an earth tunnel are directly or indirectly due to the percolation of water toward the tunnel. Therefore most of the techniques for improving the ground conditions are directed toward stopping the seepage.

Table 6-3. Tunnel Behavior: Sands and Gravels

Designation	Degree of Compactness	Tunnel Behavior	
		Above Water Table	Below Water Table
Very Fine Clean Sand	Loose, $N \le 10$	Cohesive Running	Flowing
	Dense, $N > 30$	Fast Raveling	Flowing
Fine Sand with Clay Binder	Loose, $N \le 10$	Rapid Raveling	Flowing
	Dense, $N > 30$	Firm or Slowly Raveling	Slowly Raveling
Sand or Sandy Gravel with Clay Binder	Loose, $N < 10$	Rapid Raveling	Rapidly Raveling or Flowing
	Dense, $N > 30$	Firm	Firm or Slow Raveling
Sandy Gravel and Medium to Coarse Sand		Running ground. Uniform ($C_u < 3$) and loose ($N < 10$) materials with round grains run much more freely than well graded ($C_u > 6$) and dense ($N > 30$) ones with angular grains.	Flowing conditions combined with extremely heavy discharge of water.

(Terzaghi, 1977)

The control of groundwater is of utmost importance in soft ground tunneling. The presence of a small amount of water in granular soils above the water table may be beneficial in providing an increase in stand-up time because of apparent cohesion brought about by negative capillary forces (until they dissipate), but below the water table the presence of water serves to reduce the effective strength drastically, and seepage pressures cause rapid and complete failure in noncohesive soils. The presence of water in clays is of primary importance in determining the strength, sensitivity, and swelling properties of the material. The type of control to be exercised, whether in construction or in design of the final lining, is directly dependent on these properties.

To control groundwater, the engineer has four principal methods: dewatering, compressed air, grouting, and freezing. Figure 6-2 shows the applicability of these methods for grain sizes of the soil. In the following section each of these methods is discussed in somewhat greater detail. However, for the reader who is faced with designing a soft ground tunnel in conditions where one or more ot these methods may be needed, it is recommended that a specialist be consulted or (at a minimum) that the references cited be studied in detail.

Dewatering

Sumps and pumps typically are the most economical means of dewatering, but they most commonly are more applicable in large open excavations. Thus, in a tunneling operation they likely would be found only at the portal excavation. In addition, they may be used in a tunnel in sand where the head is low and/or it is feasible to permit the groundwater to drain into the face of the tunnel.

Wellpoint systems (Figure 6-3) are the most versatile dewatering tool and can be used to draw the water table down in soils ranging from sandy silts to coarse sands and gravels. The usual practice is to install two lines of wellpoints, one line on either side of the tunnel. Spacing of wellpoints in each line typically ranges from a few feet (3 to 6 ft in fine-grained soils) up to approximately 15 ft. Water discharge per well typically ranges from 5 to 60 gpm.

The effectiveness of wellpoints is limited to shallow tunnels because the suction lift limitation restricts the drawdown to 15 to 22 ft, the latter limit requiring special techniques. In contrast to open cuts, for tunneling it is usually not possible to extend the effective depth of wellpoints by installing them in stages. The effectiveness of wellpoints in fine-grained soils sometimes may be extended by special techniques such as applying a vacuum or electo-osmosis. Eductors or ejectors can be used to lower the groundwater table significantly more than 15 ft, but they have a lower efficiency and more complex design requirements than well point systems (Abramson, 1994). They may have greatest advantage in enhancing or augmenting a wellpoint system.

Deep wells have a larger diameter than wellpoints. Thus, under applicable geologic conditions, each deep well generally has a larger drawdown area than does each wellpoint. In addition, because the pumps are submersible, each deep well typically attains a greater drawdown; wells can be

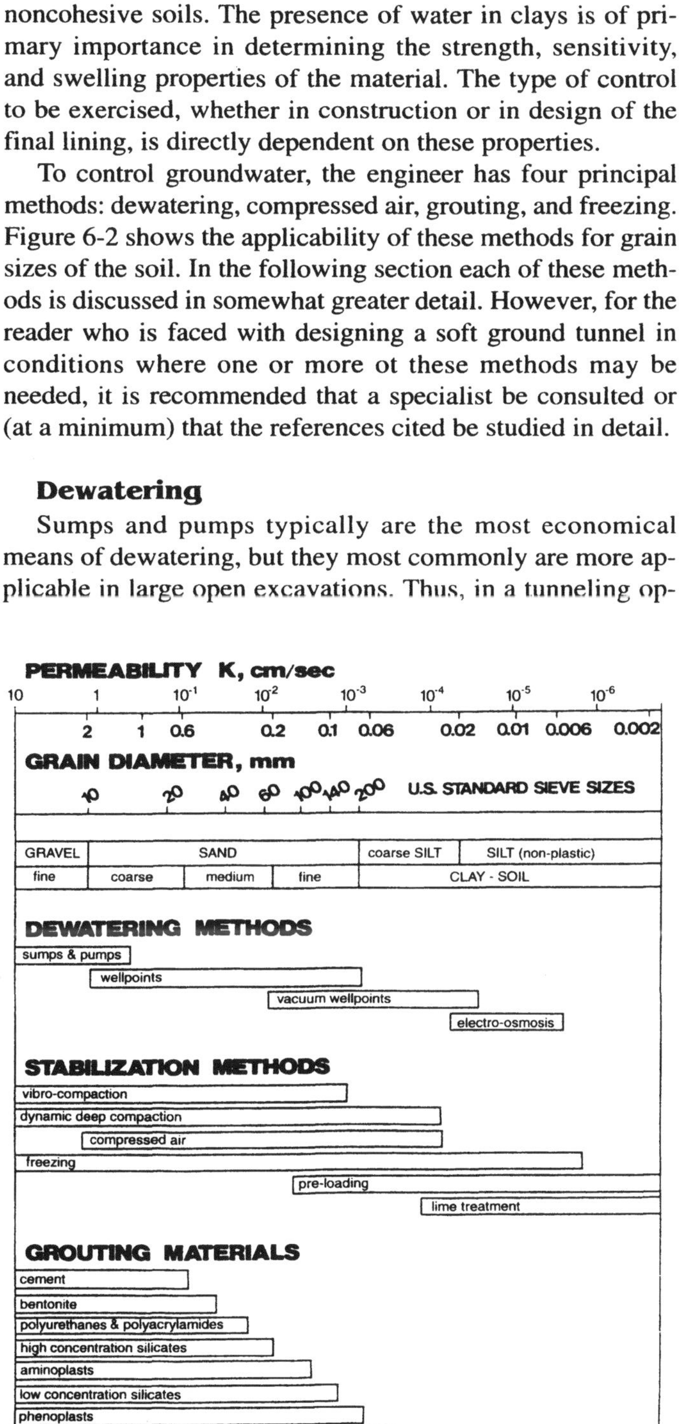

Fig. 6-2. Methods of controlling groundwater (Karol, 1990).

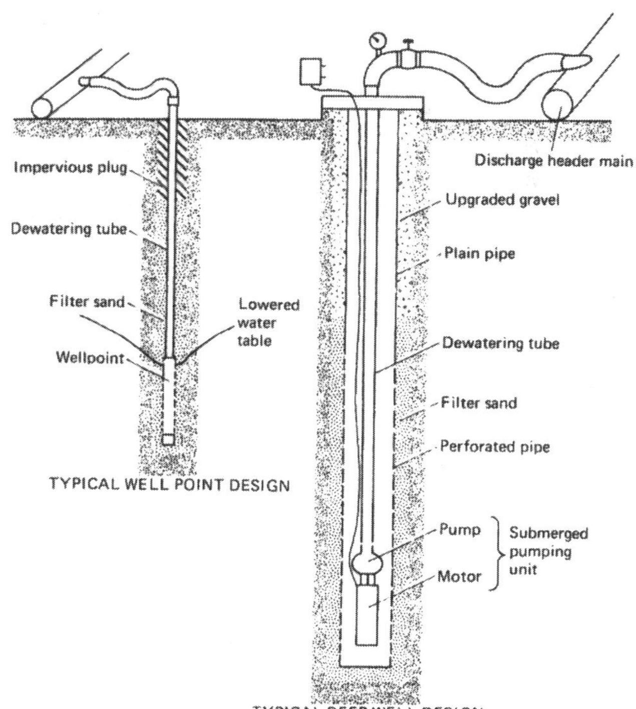

Fig. 6-3. Schematic design of wellpoint and deep well.

installed at depths up to 100 ft, with drawdown typically three-quarters of that. Details of installation (depth, spacing, diameter, screening, filtering, and the like) require special expertise and experience (Powers, 1981, 1992).

Deep well installations typically range from 4 to 12 in. in diameter with the corresponding range in capacity being 70 to 1,000 gpm. Compared with wellpoints, deep wells have a high unit cost for both installation and operation. Thus, the designer and contractor must continue to observe and test continuously as the construction proceeds, making changes (even during installation) if actual conditions prove to be different from those assumed, (Powers, 1981, 1992; Xanthakos, 1994).

In any dewatering operation, the drawdown of the groundwater level causes a corresponding and often significant increase in the effective stresses in the soil within the drawdown zone. This increase in effective stresses results in a settlement that can be roughly estimated by consolidation theory in cohesive soils and elastic compression in granular soils. Particularly in soft cohesive soils, this settlement is often large enough to produce worrisome differential settlements at overlying structures. Each case must be evaluated independently, but a few approaches to mitigating this situation are summarized below:

- Allow the settlement to occur and then repair the building. This may be acceptable where the settlements are small, the building is forgiving of differential settlement, and/or the building is low-level.
- Install cut-off walls (typically slurry walls) along both sides of the alignment, restricting the drawdown (hence the settlement) to the immediate area of the alignment.
- Recharge the groundwater immediately outside the alignment. This also has the effect of restricting the drawdown to the immediate area of the alignment.

In recent years, dewatering applications have encountered increased challenges arising from contaminated groundwater. This has resulted from an increased awareness of contamination concerns and from increased demands for underground construction in congested areas that have previously been processing, storage, or commercial sites for chemicals, petroleum products, and the like. In urban areas, especially the older parts of town, old gasoline station sites can be a serious problem.

Often, these old sites are not reflected on existing maps or in land-use records. In addition, even if the locations of old stations are found, exploration may not reflect the amount of leakage into the ground nor the direction, size, or concentration of the resulting plume of contaminated groundwater. Whether identified early or found during construction, these plumes can present a number of problems, including

- A dangerous working atmosphere (fumes or explosive concentrations) for workers and equipment not properly prepared
- Contaminated muck for disposal
- The necessity for seals or membranes to prevent seepage back into the tunnel

An example of this occurred on the Baltimore subway, where gasoline caused all of these problems and resulted in more than a year's delay to modify the design and construction to adjust for the gasoline.

Even "natural" contamination can be a problem. During construction on the first portion of the Los Angeles Metro, groundwater containing natural hydrogen sulfide was encountered. In the "old days," this water might have been returned to the same aquifer outside the construction area. However, in this case it was necessary to build a multimillion dollar treatment plant to clean up the contamination before the effluent could be discharged into the drainage system.

Compressed Air

With the advent of modern soft ground tunneling equipment, the use of compressed air in tunnels has undergone a corresponding decrease. However, compressed air can be an effective and productive means of stabilizing the soil and controlling groundwater, especially in granular (sandy) soils below the water table or in squeezing soft cohesive (clay) soils.

In granular soils, compressed air is used to offset the water pressure at the tunnel face, thereby preventing the flow of the water (and accompanying fines) into the face. However the engineer is faced with a dilemma. Because the water pressure increases at 0.43 psi per ft of depth, it is impossible to strike a perfect balance:

- If the air pressure is balanced at the invert, the water at the crown will be driven away from the tunnel. This can dry the sand, leading to possible running conditions and even the risk of an air "blow."
- If the air pressure is balanced at the crown, the water at the invert will be under a pressure of $0.43D$, enough for troublesome flows of water into the tunnel and possibly for flowing conditions to exist.

The usual approach to this dilemma is to adjust the air pressure so that it balances the water pressure at the tunnel springline. This compromise generally is the best solution available, but the engineer and contractor must recognize that both of the problems enumerated above remain a possibility, although each should be minimized.

When used in cohesive ground, the goal is to provide enough air pressure so that, in combination with the soil's natural strength, the tunnel will be stable for the tunnel excavation and support operations. To obtain a measure of tunnel stability, Peck (1969) has proposed Equation (6-1). For this application, P_a in that equation is the air pressure above atmospheric. Tables 6-1 and 6-2 summarize the approximate relationship between the stability factor (N_t) and the behavior of a tunnel.

The use of compressed air requires design and operation of the system required to provide and control the requisite air. Design of such a system is beyond the scope of this book, but the reader should be aware that the requirements are numerous:

- Compressors (with spares)
- Material and muck lock(s)
- Man lock(s)
- Medical lock
- Medical facility and staff
- Standby (trained) hospital staff and facility

The quantity of air required can only be found by experience on a site-by-site basis, but for initial planning purposes it may be estimated according to

$$Q = KD^2 \tag{6-2}$$

where Q is the quantity (CFM) of compressed air at the design pressure, K is a constant (approximately 12 in fine sands and 24 in gravel), and D is the tunnel diameter in feet. Finally, exposure of workers to the air pressure and their depressurization must be very carefully monitored and controlled. Failure to do so can induce the bends, just as experienced by divers. This condition can be debilitating, even fatal, if not properly controlled. Control includes physical examinations, close monitoring of working conditions, and rigid adherence to all decompression schedules. To illustrate, Table 6-4 shows the total time required to decompress after working various hours and pressures. Note that the actual decompression schedule is stepwise and governed by the time of exposure and the working air pressure.

Table 6-4. Decompression Table

Work Pressue, psig	Hours Under Pressure										
	1/2	1	1-1/2	2	3	4	5	6	7	8	Over 8
0–12	3	3	3	3	3	3	3	3	3	3	3
14	6	5	6	6	5	6	6	6	16	16	32
16	7	7	7	7	11	17	48	63	63	73	87
18	7	7	7	8	11	17	48	63	63	73	87
20	7	7	8	15	15	43	63	73	83	103	113
22	9	9	15	24	38	58	98	108	118	128	133
24	11	12	23	27	52	92	117	122	127	137	151
26	13	14	29	34	69	104	126	141	142	142	163
28	15	23	31	41	98	127	143	153	153	155	183
30	17	28	38	62	105	143	165	168	178	188	204
32	19	35	43	85	126	163	178	193	203	213	226
34	21	39	58	98	151	178	195	218	223	233	248
36	24	44	63	113	170	198	223	233	243	253	273
38	28	49	73	128	178	203	223	238	253	263	278
40	31	49	84	143	183	213	233	248	258	278	288
42	37	56	102	144	189	215	245	260	263	268	293
44	43	64	118	154	199	234	254	264	269	269	293
46	44	74	139	171	214	244	269	274	289	299	318
48	51	89	144	189	229	269	299	309	319	319	—
50	58	94	164	209	249	279	309	329	—	—	—

Notes: Working chamber pressures. Total decompression time, minutes; California Code of Regulations August 1985.

1. 14–22 psi require a two-step decompression;
   24–38 require a three-step decompression; and 40–50 require a four-step decompression.
2. When decompression exceeds 75 minutes, a special decompression chamber must be provided.
3. At least one physician, licensed in the state, must be available at all times to provide medical supervision. The physician must meet all requirements of the compressed air workers and be willing to enter the pressurized environment as needed.
4. See applicable regulations for additional details.

Because the use of compressed air has decreased so dramatically with the advent of modern soft ground tunneling machines, the recent literature generally lacks articles on compressed air. As a start, readers who wish to pursue this subject should consult Richardson and Mayo (1975), Deix (1987), and Farjeat (1991).

GROUTING

Several types of grouting are used to modify and/or stabilize soils in situ in preparation for soft ground tunneling. Recent improvements in grouting have made it a valuable tool in both groundwater control and soil stabilization for tunneling projects. It is a very effective method for improving tunneling under a number of situations such as the following (Baker, 1982; Gularte, 1989): (1) to strengthen loose or weak soil and prevent cave-ins due to disturbance of loose, sensitive, or weak soils by the tunneling operation, (2) to decrease permeability and hence groundwater flow, (3) to reduce the subsidence effects of dewatering or to prevent the loss of fines from the soil, (4) to stabilize sandy soils that have a tendency to run in a dry state or to flow when below the water table. Gularte breaks grouting into three general applications: *permeation grouting* that fills the voids in the soil with either chemical or cement binders, *jet grouting* that uses high-pressure jets to break up the soils and replace them with a mixture of excavated soil and cement, and *compaction grouting* that densifies the soil during tunneling by injection of a stiff grout. These applications are illustrated by Figure 6-4, and the types of soils in which each grouting method may be effective are indicated approximately by Figure 6-5.

Groutability is primarily determined by the permeability of the soil. For initial planning purposes, this may be expressed in broad terms such as shown on Figure 6-6 where grouting materials are shown and matched to the general soil descriptions. In more detailed terms, groutability may be expressed directly in terms of soil permeability or of the percentage of fines (percent passing a No. 200 sieve) in sand, as shown in Table 6-5. Finally, in greater detail, one can compare the grain size curve for a given soil with a plot of groutability, as shown on Figure 6-7.

Permeation grouting may be by either cement- or chemical-based grout, with the latter being necessary for satisfactory penetration of finer soils. As shown by Figure 6-6, (Baker, 1982), cement grout is feasible only in gravels and some sands, whereas chemical grouts are feasible in soils containing more fines (Mitchell, 1981). Historically, the rule of thumb has been that soil having less than 10% of fines passing the No. 200 sieve could be successfully grouted and those with more than 20% of fines could not. Recent experience in Los Angeles (Gularte, 1991) indicates that current methods and materials have raised these limits approximately 5% each.

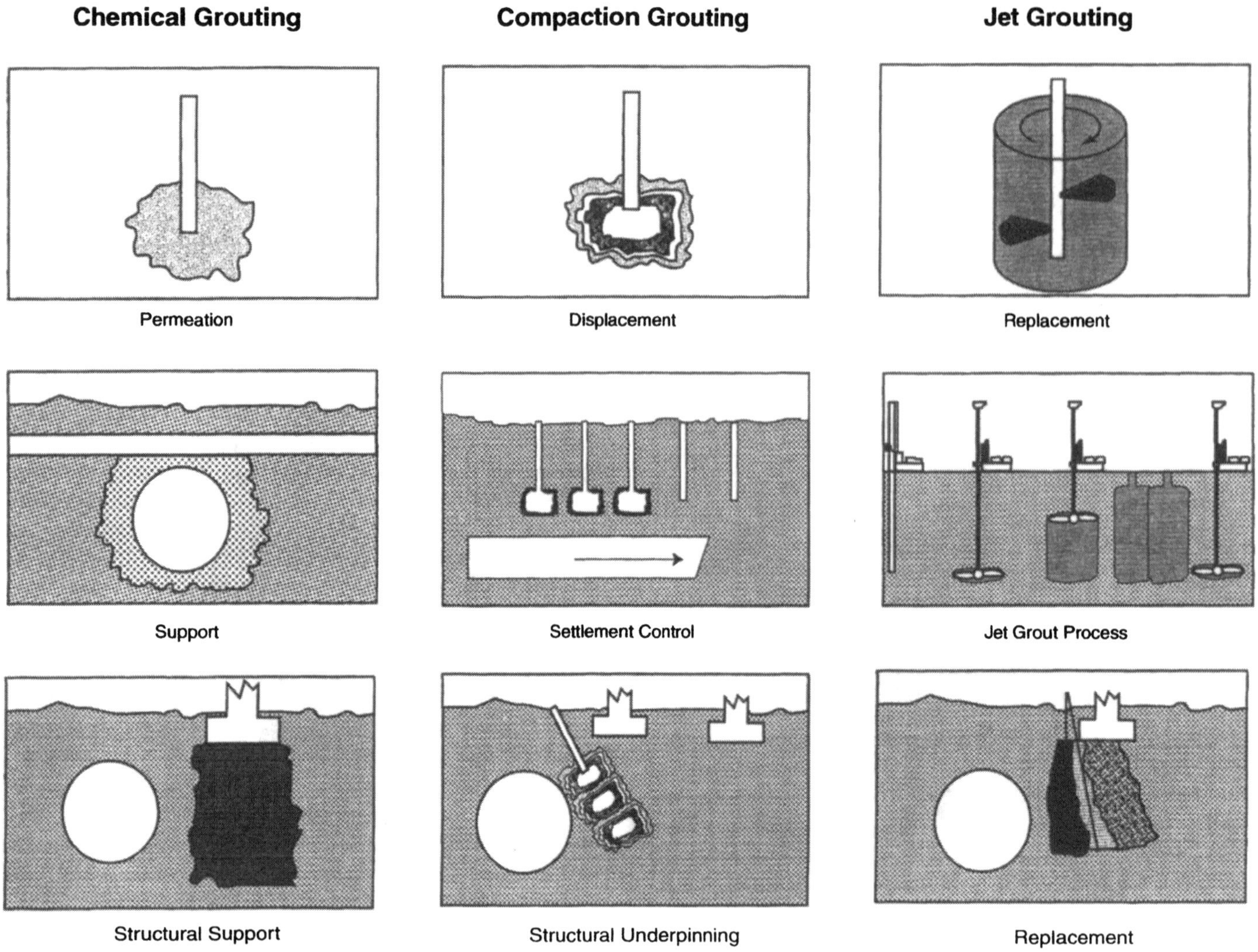

Fig. 6-4. Applications of grouting for tunneling (after Gularte, 1989).

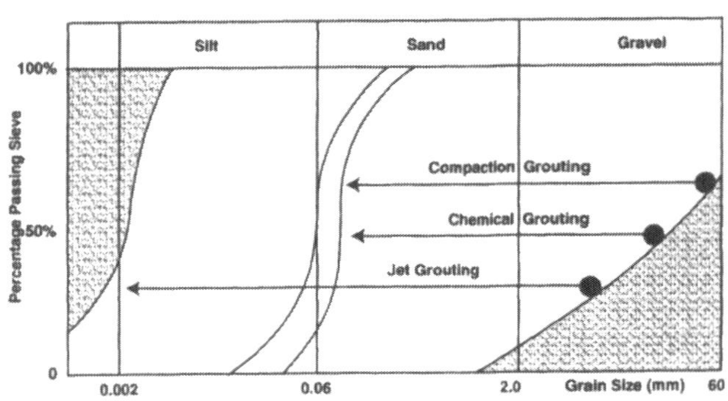

Fig. 6-5. Grouting method and soil type (after Gularte, 1989).

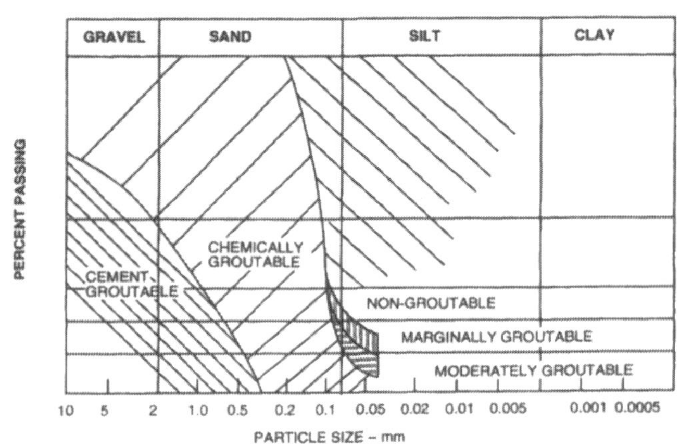

Fig. 6-6. Grain-size ranges groutable soils (after Baker, 1982).

Table 6-5. Groutability Related to General Soil Permeability and to Percentage of Fines in a Sand (from Baker, 1982).

Groutability	General Soil Permeability, cm/sec	Percentage of Fines in Sandy Soil
Easy	10^{-1} to 10^{-3}	<12
Moderate	10^{-3} to 10^{-4}	12 to 20
Marginal	10^{-4} to 10^{-5}	20 to 25
Ungroutable	$<10^{-5}$	>25

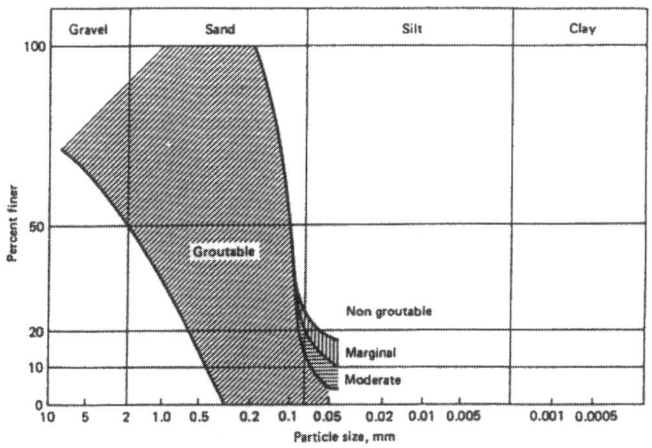

Fig. 6-7. Groutable soils (Karol, 1990).

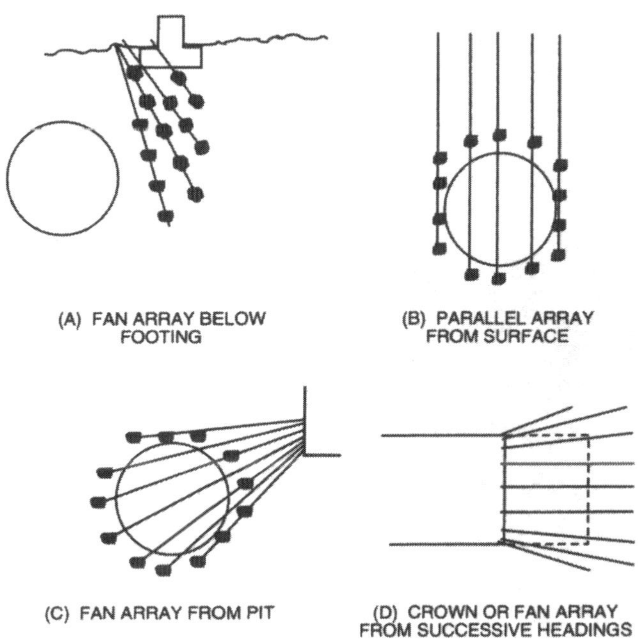

Fig. 6-8. Grout pipe layout plans (after Baker, 1982).

In recent years environmental and worker safety regulations have increasingly restricted the use of toxic chemicals, and they have effectively retired some kinds of chemical grouts formerly in common use. New grouts are also being developed. Current regulations and specialists should be consulted.

Techniques and equipment are too varied to discuss in great detail here; the reader is referred to Baker (1982) for a discussion of such details. It should be noted, however, that the final grouted soil mass is produced by the logical geometrical arrangement of small, contiguous grouted masses and that secondary and tertiary grouting patterns are often required to fill ungrouted zones left by the original patterns. Thus, pipe spacing may vary from 18 in. to 5 ft.

Preferably, the pipes should be placed in an array in which all pipes are parallel, with the engineer determining the grout zone pattern required for the problem at hand. However, the geometry of the problem, soil permeability, and the type of grout very often dictate other patterns around a tunnel, as illustrated in Figure 6-9.

In the past, four grouting methods were used: stage, series, circuit, and packers. Current practice, however, consistently uses packer grouting because it has been shown to provide (economically) the best control of the grouting operation by

- Controlling the location of the injection along the grout pipe
- Controlling the formation of unexpected flow channels along the grout pipe
- Providing access for second- or third-stage grouting

The goal of the grout injection is to fill the defined volume of soil with overlapping grout bulbs so that, under ideal conditions, the total volume is grouted. In reality the bulbs do not take the assumed shape, and the total volume is not grouted. However, when properly executed, a grouting program involving primary and secondary grouting should impregnate 90% or more of the required volume, and that is generally enough to provide the stability required for the tunneling operation. Schematically, these operations have been illustrated by Baker (1982); see Figure 6-9.

Tertiary grouting may be desirable in some cases to try to fill most or all of the ungrouted zones left by the primary and secondary grouting. The most obvious need for tertiary grouting is where the prime goal is to shut off groundwater. Especially where moving groundwater may carry fines with it, any ungrouted channels left from the primary and secondary grouting might concentrate the flow and potentially compound the transport problem. In these circumstances, tertiary grouting may be critical. For further information on grouting theory and practice, interested readers are encouraged to study a number of recent references on this rapidly advancing technical field (Baker, 1982a, b; Karol 1990).

The equipment required for grouting typically consists of storage tanks, mixing tanks, pumps, meters, gages, packers, and hoses. The pumping plant is of primary importance: its basic function is to proportion, mix, and pump the grout. The grout pumping plant's versatility is increased if the flow rate can be varied at any time, allowing the flow rate to be maximized for a given soil and the design limit pressure. The dimensional capability can be again increased if the relative proportions of the grout components can be varied easily and quickly at any time, allowing changes in gel times while pumping. This is important when temperature changes

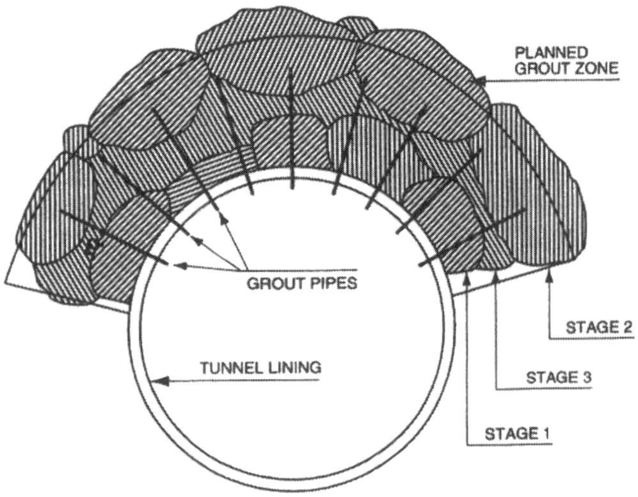

Fig. 6-9. Crown grouting (after Baker, 1982).

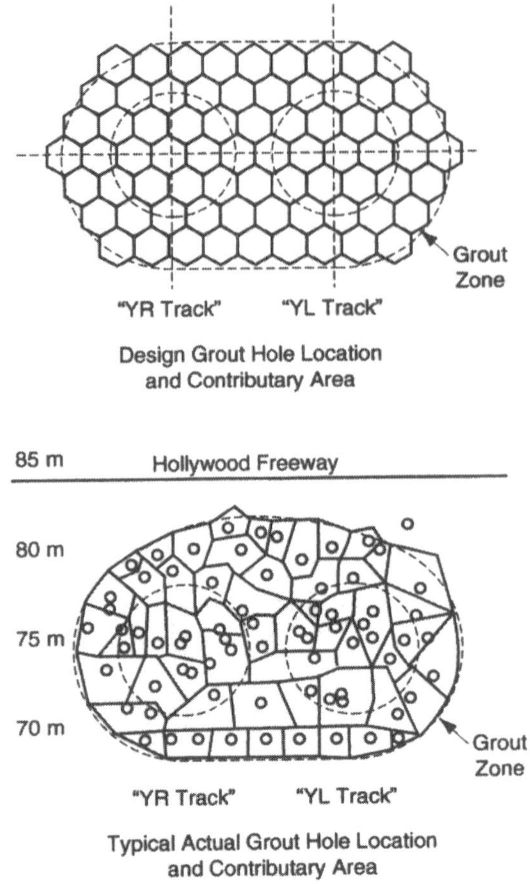

Fig. 6-11. Design and actual grout hole locations (Monsees, 1993).

during pumping affect the gel time, or when trying to develop resistance to flow in given situations. An accurate and appropriate means of monitoring flow rates and pressures involved is essential. Positive displacement meters should be placed on the inlet or outlet of the pumps, and pressure gauges should be placed at the pumps and the top of the grout pipe.

On the Los Angeles subway project, it was necessary to perform the grouting from pits alongside a major (eight-lane) freeway (see Figures 6-10 and 6-11). From the pits, the grout holes were drilled horizontally under the freeway, in that way avoiding the traffic disruption that vertical drilling would cause. The contractor elected to drill all the way under the freeway from one side. This made the holes approximately 275 ft long, a length that challenged the contractor's ability to control the long horizontal holes and keep them aligned as planned. Although there was some horizontal and vertical deviation of the holes, the contractor was able to bring them back to acceptable limits (and established a record for horizontal grout holes). Measurements of the grout hole locations and of the grout injected indicated that the grout program was successful. This was proven later when the tunnel was driven successfully and still later was proven again when a fire in the tunnel burned away all the timber lagging of the initial sup-

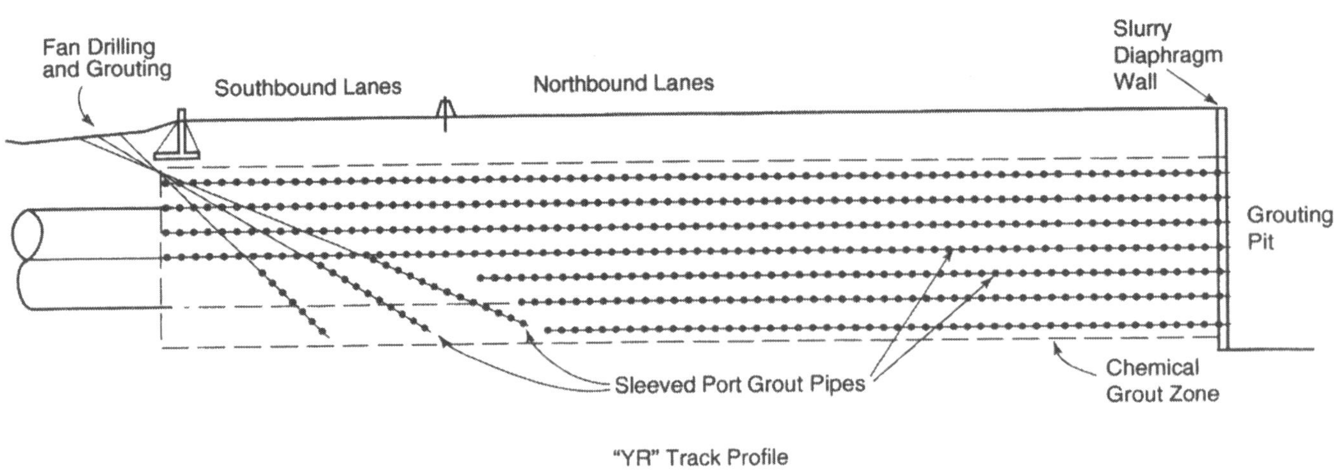

Fig. 6-10. Horizontal and fan chemical grouting program for the Hollywood Freeway, Los Angeles Metro (Gularte et al., 1991).

port system, but the grouted ground remained stable even with freeway traffic moving only 6 to 12 ft overhead.

Grout Pipe Installation. The function of the grout pipe is to allow a designated grout zone to be fully permeated. This may require the introduction of the grout at one or more depths and in multiple injection stages (see Figure 6-12). A number of types of grout pipes are used to meet the goal of fully permeating a designated grout zone. A driven or jetted pipe is best used for grout injection at a single depth as long as soil conditions will permit installation. A slotted grout pipe can be used for grouting a relatively pervious, captive soil layer.

However, experience has shown that the sleeve port pipe (Figure 6-13) provides the most versatile grout pipe installation. By locating the internal grout packer at a particular sleeve port, grout can be injected at any specified depth, and regrouting can be performed at that depth in any number of stages. Four distinct advantages occur with the use of the sleeve port: (1) grout can be injected at specific depths in each bore hole, increasing the likelihood of penetrating the desired area; (2) the elastic sleeve at each port prevents the grout from returning into the grout pipe and gelling up after grouting, allowing the port to be reused; (3) the sleeve ports allow the pipe to be completely sealed in the bore hole, reducing grout leakage away from the desired zone, along either the pipe or bore-hole interface; (4) chemical and cement grouting can be performed in the same grout pipe at any particular depth to grout more permeable soils (Gularte, 1989).

The installation of the grout pipe must be done with care to assure that the sealing grout surrounds the entire pipe and fills the annulus. The pipe must be positioned below ground surface and a protective cap provided where pedestrian or vehicle traffic, and/or vandalism is possible. The grout used to fill the annulus is usually made up of cement, bentonite, and fly ash. The grout should be thick enough to prevent infiltration into the soil; ideally, it should be low strength and brittle.

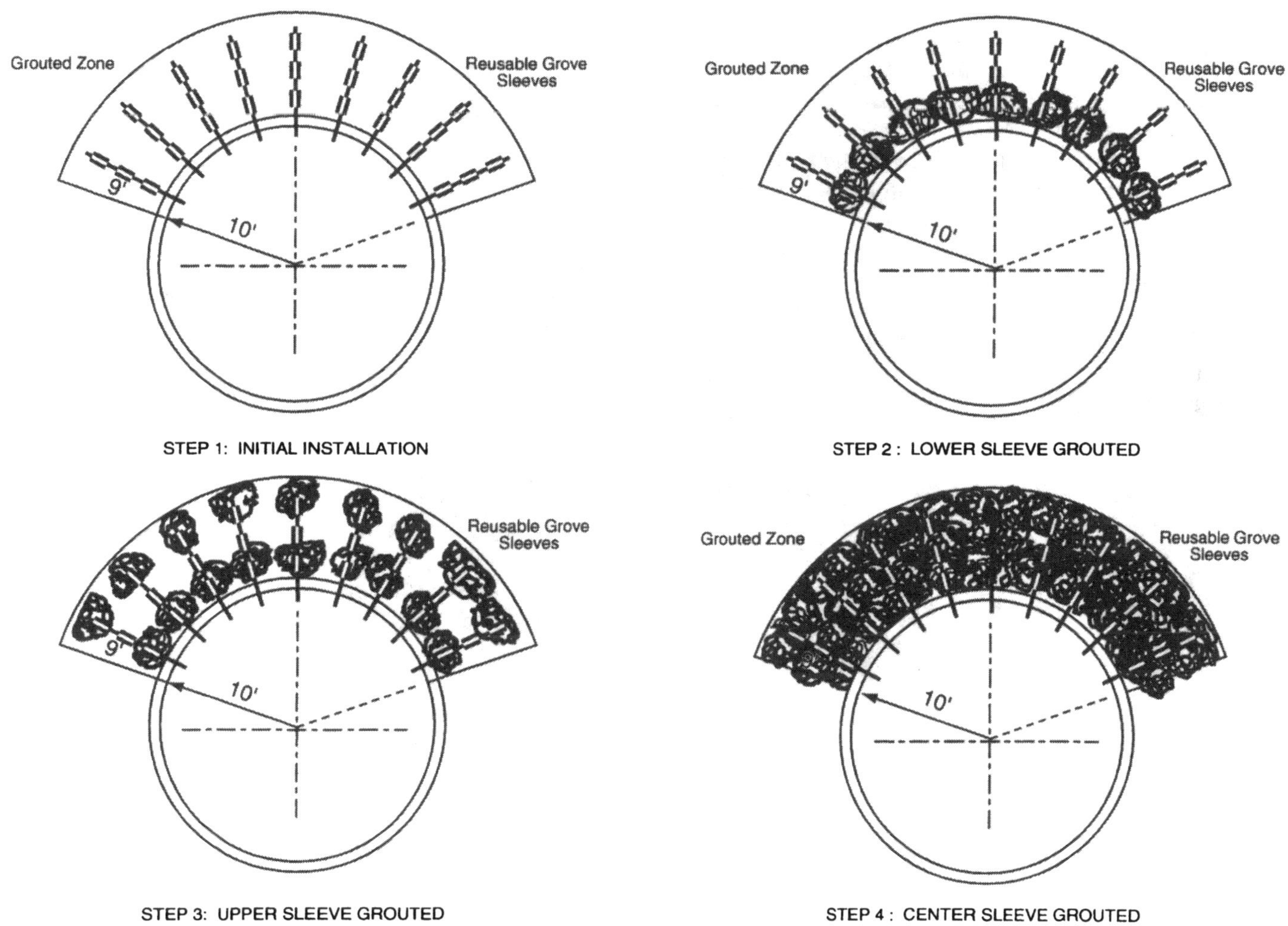

Fig. 6-12. Multiple injection for crown grouting (after Gularte, 1989)

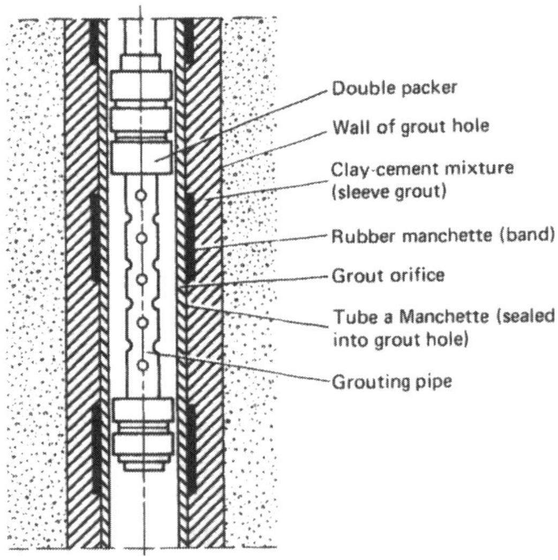

Fig. 6-13. Tube á Manchette.

Freezing

Basic Conditions. Freezing is more often used in shaft sinking than in tunneling, but the method is useful where nothing else will serve, provided that surface access over the tunnel alignment is available. It should be noted at the outset that, for the method to be successful, sufficient water must be present: freezing will not change the stability characteristics of dry soils. For freezing to be successful, the in situ porewater must be converted to ice by extraction of the latent heat. The ice then acts like a cement to bind the soil grains together, thereby raising the strength and lowering the permeability of the soil mass. It should also be noted that the presence of organic material (common in silts and clays) or salt water will result in greater difficulty in freezing, since the groundwater freezing point will be depressed.

Another major deterrent to successful freezing is moving groundwater. To convert the porewater to ice, the freezing system must remove heat so that the water temperature is lowered below freezing. However, moving groundwater can easily bring heat into the area faster than the freezing system removes it. Even with modern freezing techniques, it may not be possible to obtain a freeze if the groundwater is moving more than approximately 3 in. per hour. At lower velocities, the flowing water may create an elongated, rather than circular, bulb of frozen ground. Such an unpredictable shape may make difficult the closing of a freeze wall even though the volume of ground frozen is theoretically correct. Finally, note that in many soils, the expansion of water by freezing will first cause heaving of the ground surface, but subsequently, upon thawing, will cause subsidence. The net effect is generally a compaction and permanent settlement of the ground.

Refrigerated Brine. The typical freezing installation consists of a refrigeration plant that cools a brine solution, which is then pumped down the center of an annular freeze pipe to the bottom of the hole, returning via the outer annulus in contact with the soil. The warmed brine is returned to the refrigeration plant, and the cycle continues. In practice, a number of freeze pipes are connected to a pair of headers for the flow and return lines. The expanding frozen zones outside each pipe overlap. Thereafter, during construction, the circulation need be sustained only to the extent necessary to maintain the freeze wall.

For tunnel construction, it is only necessary to stabilize the soil; strength is usually secondary. There is, therefore, no need to maintain continuous freezing to achieve lower temperatures at which the ice is stronger. In fact, it may be a disadvantage in tunneling to have ground any stronger than necessary.

Instead of calcium chloride brine (the most commonly used), other brines with lower freezing points are available, but at greater expense. In addition, liquid nitrogen or liquid carbon dioxide can be used, but their cost is high and special precautions must be taken when they are used.

Liquid Nitrogen. For special purposes, especially for projects of limited extent and duration, boiling of liquid nitrogen in the freezing elements may be appropriate. It is important to provide close control, since there is a risk that the liquid nitrogen will be forcefully ejected from the open pipe by bubbles formed during the boiling process. In addition, considerable nitrogen is supplied to a freezing element in the same manner as the brine in a normal system; but with the supply pipe nearly as large as the freezing element, the flow can be regulated to match the rate of boiling.

Positive exhaust must be provided for the vapor, which is heavier than air and will cause suffocation if it accumulates in sufficient quantity. Also, steel pipes will be embrittled by the low temperature: thus nonferrous metals or stainless steel are usually required.

Design. It is necessary to determine the thermal characteristics of the soils to be frozen and the freezing point of the groundwater. The effect of variations in thermal conductivity is illustrated in Figure 6-14. The analysis will be governed by the conditions giving rise to the thinner frozen zones.

The computation of thermal energy and time to complete freezing is usually based on simplified assumptions as to uniformity of heat transfer within the soil. Computer analyses can be made for more complex or critical projects. Figure 6-15 shows the relationship between freezing time and spacing of freezing elements for brines and for forced conversion and boiling of liquid nitrogen.

Operation. It is essential to include instrumentation as part of any freezing installation. Thermocouple strings can be used to monitor the ground temperatures between freezing elements and to monitor refrigerant temperature. A simple method for determining closure of a frozen ring in free-draining soils is to monitor a centrally located piezometer. The advancing freeze front pushes water out of the pores as it ex-

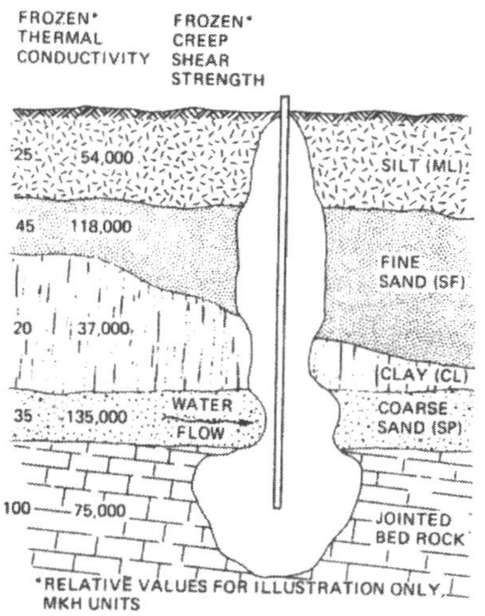

Fig. 6-14. Typical effect of thermal properties on frozen ground.

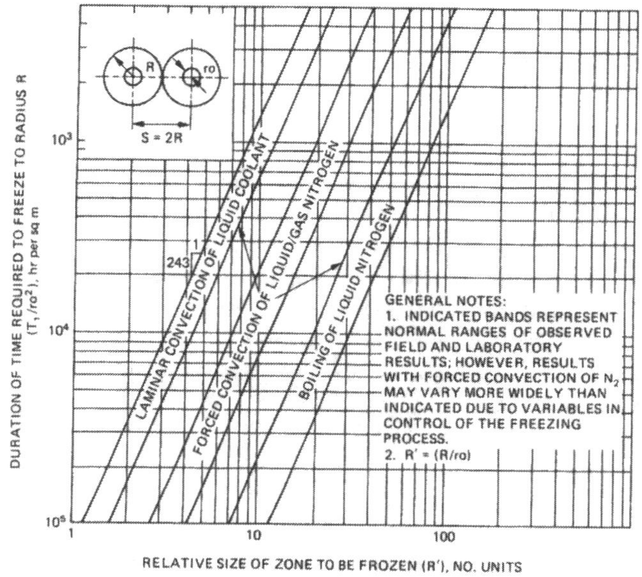

Fig. 6-15. Determination of desired freezing time.

pands on cooling, and once the wall is closed, there is no means of exit from the area and water rises in the piezometer.

During installation, it is important to ensure that the outer casing of the freezing element retains its integrity. Loss of brine into the ground will render completion of the freeze wall difficult, and it may be necessary to resort to a more expensive circulating fluid.

Any utilities in the area, such as water, sewer, or steam lines, subject to being affected by freezing must be protected by insulation or isolation if they are within the freeze zone. They should be monitored to ensure that this protection is effective.

The freezing method is likely to be most appropriate in silty soils, not subject to stabilization by dewatering or grouting. These soils also are the most likely to be affected by frost heave, caused by water migrating toward the freezing zone under capillary action and freezing into lenses of ice. The possibility of damage from this phenomenon must be kept in mind.

Excavation of the frozen ground may be done by drilling and blasting, although it is sometimes difficult to keep the drill holes open. Conventional mining equipment has been used successfully in permafrost, and roadheaders appear to be effective in the conditions created by freezing.

Short tunnels may be frozen by driving the freezing elements into the ground horizontally around the perimeter of the excavation. If the surface is inaccessible, it may be practical to drive a working tunnel at a level higher than that of the tunnel that required frozen ground stabilization, and to then install the freeze system from the higher tunnel.

SOFT GROUND TUNNELING MACHINES

Tunneling machines, including those for soft ground, are discussed in detail in Chapter 11; only a few points will be made here. Because nearly all soft ground tunneling machines are circular, it is that cross section that will be described, although it must be noted that other cross sections are used.

Marc Isambard Brunel is generally credited with the invention of the first soft ground tunneling shield, the forerunner of modern soft ground tunneling machines. Brunel's invention became a viable possibility because of the combination of two factors: (1) the observation of the capability of the shipworm to penetrate hard woods by a screwlike action, and (2) the recognition that small openings or drifts could be made in soft ground where larger openings would be unstable. Thus was born the basic concept of advancing the tunnel by successively working smaller cells or compartments within the overall protection of the larger outer shield. All soft ground tunneling machines, including shields, have five basic components.

The *cutting edge* trims the outside perimeter of the tunnel. In a digger shield, a rotating "cheese grater" mounted on the face of a drum, the cutting edge cuts only those portions of the perimeter not excavated by the digger mechanism. With a fully mechanized tunneling machine the front edge occasionally may not cut the ground depending upon the configuration of the cutterhead, specifically when gauge or perimeter cutters are used.

The *body* of the machine is usually a steel cylinder, stiffened by generally vertical and horizontal bracing members, by the housings and attachments for the face support and propulsion jacking systems in a simple shield, and by the equipment in a tunneling machine. In a shield the bracing members also divide the face into the number (and size) of working pockets or zones that are necessary to provide safe and stable working cells for mining and mucking. As discussed in Chapter 11, stability at the face of a mechanized shield is usually provided by control of the number and size of these working pockets or openings in the face of the machine.

The *tail* of the machine extends rearward from the body and provides protective cover for the workers and the outline for the erection of the tunnel support system. To assure the support system is always within the tail (or, conversely, that unsupported ground is never exposed) the tail length must be such that at least one-half of the length of the last fully erected supporting element stays within the tail shield when the shove jacks are fully extended.

The *shove jack system* provides the forward propulsion for the machine. These jacks are located around the perimeter of the shield and are housed within the body. They usually obtain reaction for their thrust from the tunnel support system to the rear.

At the front of a classical shield, the *hood* projects ahead of the body thereby providing protection for workers at the face. Working within this protection, the breasting or face support system is advanced stepwise to complete the excavation required for each shove of the shield. With a full-face or closed machine, there often is no hood, and face stability is provided by control of the openings in the machine face or cutting head.

SELECTION OF SOFT GROUND TUNNELING MACHINE

Selection of a soft ground tunneling machine requires consideration of soil conditions, water conditions, tunnel size, support system, excavation conditions, and the excavation environment. The number of variables that must be considered in selecting a machine is quite large, and the author is unaware of any attempt in the United States to systematize that procedure. The Japanese seem to have an edge in systemazing the selection of a soft ground machine; perhaps they have encountered a greater number of difficult soft ground tunnels, or perhaps they have simply collected the data. In either case, the following summary has been adapted from information furnished by Hitachi Zosen (1981, 1984).

Figure 6-16 shows the applicability of various soft ground machines in relation to the grain size distribution curve. Tables 6-6 and 6-7 contain a summary of the types of machines and their application to various ground conditions.

In today's world of production-driven tunneling, most soft ground tunnels are excavated by circular tunnel boring machines (TBM's, Chapter 11). However, special circumstances and configurations sometimes dictate that other tunnel shapes or types of tunnel construction be considered. For the most part, these other types often might be considered as a return to earlier days of soft ground tunneling.

Simple Shields

Simple shields have the basic components described earlier with excavation and ground support done by hand labor or basic mechanical excavators. The most simple of this group essentially is a Brunel shield. Excavation is by hand with face support by breasting jacks and tables with breast boards (Mayo 1945; Terzaghi, 1952). By selectively removing breast boards (or groups thereof), excavating a pocket or drift, and then rebreasting, the tunnel is moved forward incrementally. Ground support is by ribs and boards or perhaps liner plates, erected by simple erector arms.

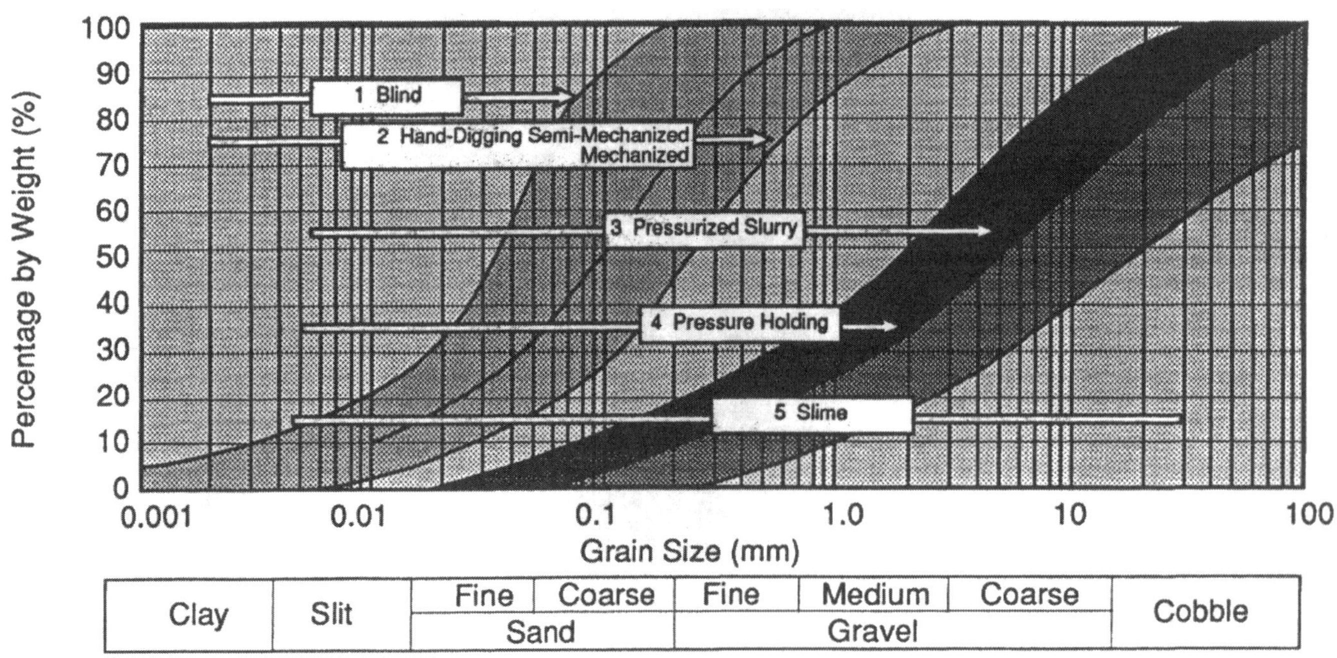

Fig. 6-16. Applicability of soft ground machines vs. grain size.

Table 6-6. Conventional Machines

Type	Description	Notes	Sketch
Blind shield	A closed face (or blind) simple shield used in very soft clays and silts. Muck discharge is controlled by adjusting the aperture opening and the advance rate.	Used in harbor and river crossings in very soft soils. Often results in a wave ror mound of soil over the machine.	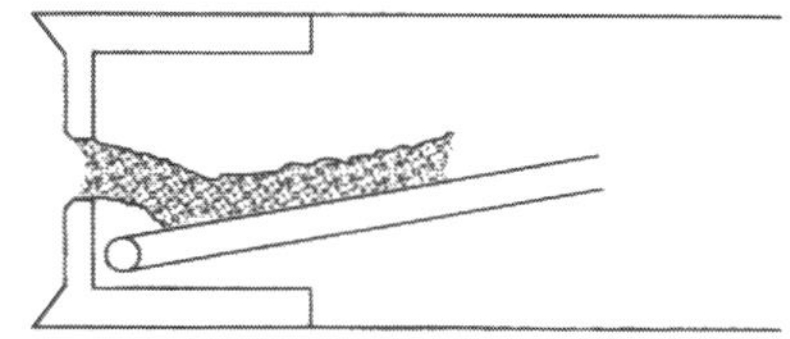
Open face, hand-dug shield	Good for short, small tunnels in hard, non-collapsing soils. Usually equipped with face jacks to hold breasting at the face. If soil conditions require it, this machine may have movable hood and/or deck.	A direct descendent of the Brunel shield. Now largely replaced by more mechanized equipment. Sometimes used at the head of large cross-section jacked tunnels.	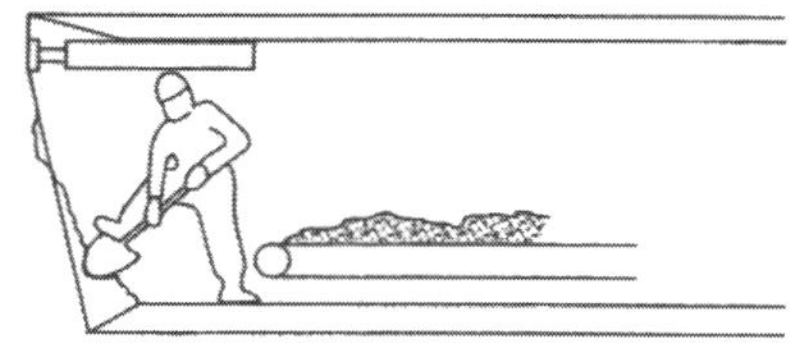
Semi-mechanized	Similar to the open face, but with a back hoe, boom cutter (roadheader) or the like.	Until very recently, the most common shield. Often equipped with "pie plate" breasting and one or more tables. Can have trouble in soft, loose, or running ground. Compressed air may be used for face stability in poor ground.	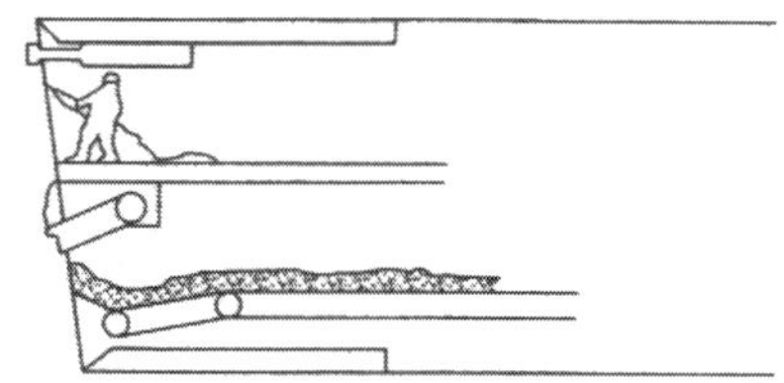
Mechanized	A fully mechanized machine. Excavates with a full face cutter wheel and pick or disc cutters.	Manufactured with a wide variety of cutting tools for various soils. Face openings (doors, guillotine, and the like) can be adjusted to control the muck taken in versus the advance of the machine. May also be used with compressed air for face stability in poor ground.	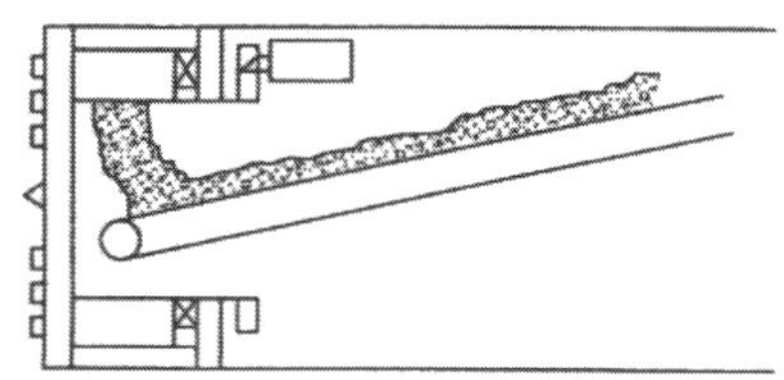

Table 6-7. Special Machines

Type	Description	Notes	Sketch
Slurry face machine	This machine uses pressurized slurry to balance the groundwater and soil pressure at the face. It has a bulkhead (closed face) to maintain the slurry pressure on the face; that slurry must be piped down and recycled from the surface. It may also be equipped with a stone crusher for occasional cobbles. This machine is good for water bearing silts and sands with fine gravels.	Best for sandy soils; tends to gum up in clay soils; with coarse soils, face may collapse into the slurry. Coarse soils are defined as • Gravel content >60% • Clay and silt content <10% • Water content <18% • Coefficient of permeability ≥ 10^{-2} cm/s • Cobbles greater than 8 in.	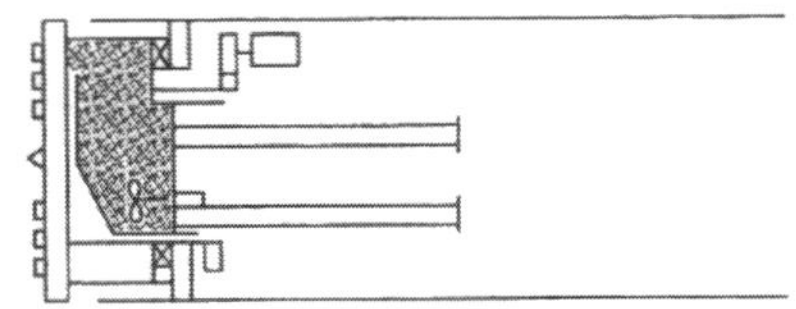
Earth pressure balance (EPB) machine	This machine has a closed chamber (bulkhead) face that uses trapped water and soil material to balance the groundwater and/or collapsing soil pressure at the face. It uses a screw discharger with a cone valve or other means to form a sand plug to control muck removal from the face and thereby maintain face pressure to "balance" the earth pressure. It is good for clay and clayey and silty sand soils, generally below the water table.	Also best for sandy soils, with acceptable conditions defined as • Clay and silt content >7% • Gravel content <70% • Cohesive soils (not less than 40% clay and silt) have *N*-value <15. • Water content >18% in sandy soils and >25% in cohesive soils	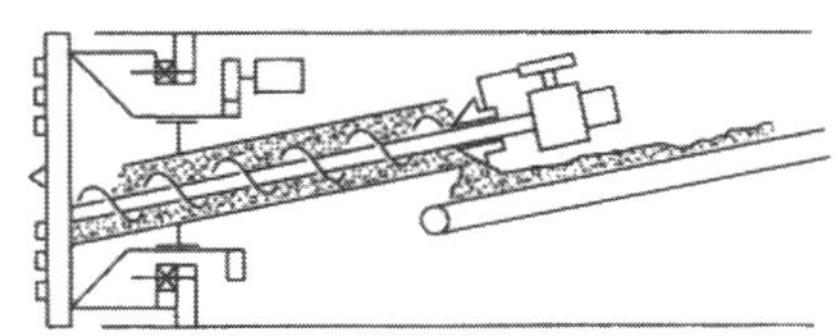
EPB high-density slurry machine	A hybrid machine that injects a denser slurry (sometimes called slime) into the cutting chamber. Developed for use where soil is complex, lacks fines or water for an EPB machine, or is too coarse for a slurry machine.	Has worked in soil with 85% gravel content and cobbles and boulders up to 20 in. ×10 in. × 7 in. Has worked in sandy gravel soil with $N = 30$ to 50 and sandy or silty soil with $N = 5$ to 35.	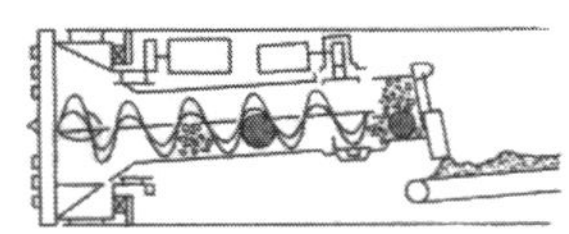 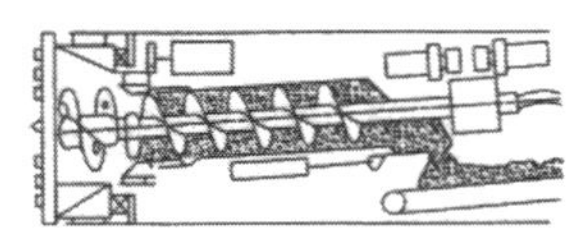

The next step up in complexity uses an articulated digger (backhoe type) or roadheader cutting head to excavate within the shield. Face support is by hydraulically activated breasting tables and/or breasting jacks. Additional face support may be provided by allowing the muck to stand at its angle of repose on the breasting tables

Multiple Headings

Some tunnels are still excavated by multiple, manual headings (see Terzaghi, 1977; Mayo, 1945; McCusker, 1982). This may be practical where the tunnel is short and mechanization is not practical, where obstacles would preclude use of a machine, or where special size or shape requirements exist. A typical example of the latter is a twin track subway in a shallow crossing under a freeway. In this instance a wide horseshoe tunnel might be necessary, and it would be driven in multiple drifts.

Multiple heading (or drift) tunnels may be supported by ribs and boards or by liner plates. McCusker (1982) discusses various approaches for driving these types of tunnels. In recent years the use of shotcrete to support multiple drift tunnels has become more frequent. This technique relies on the ground to have sufficient stand-up time that the drifts can be excavated sequentially and with limited advance. It also requires instrumentation of the tunnel and the ability to make field changes in the excavation and/or support procedure quickly.

SOFT GROUND TUNNEL SUPPORT AND LINING

Just as most soft ground tunnels are now excavated by one or another type of TBM (see Chapter 11), it is also true that most initial support systems consist of one of three schemes:

1. The first scheme consists of *ribs* and *lagging* (also called ribs and boards) in which steel wide flange sections (ribs) are rolled to the radius of the tunnel and erected in the tail of the machine on 3- 5-feet centers. Spanning between the steel ribs, and completely around the perimeter of the tunnel, are timber lagging members. As the steel ribs sequentially clear the tail shield, they are expanded into place to contact the ground, with props (called dutchmen) put in the expansion gap(s).
2. The second scheme consists of *unbolted precast concrete segments* in which the completed ring usually consists of approximately four precast concrete pieces or segments. These segments are erected within the tail shield and expanded into place as they clear that tail shield. To "complete the ring" the expansion gap (or gaps) is stabilized by means of steel props and then filled with fast-setting concrete.
3. The third scheme consists of *bolted* (or *pinned*) *precast concrete segments* made up into rings usually consisting of approximately six pieces. In contrast to the above segments, however, these members are bolted (or sometimes pinned) together at all circumferential and longitudinal joints. The variations for bolted segments are almost endless: the rings all may be tapered, requiring a 180° rotation in alternating rings for a straight drive; the segments may be identical except for special tapered ones for curves; a key segment may or may not be used; various dowels or pins may be used in place of or in combination with bolts; bolt pockets, curved bolts, and special locking devices have all been used to make up the connections. Obviously these segments are more complicated than those in the second scheme.

As discussed in Chapter 5, the successful performance of any tunnel support system is based on the interaction between support system and surrounding ground. In turn, this interaction can be effective only when the contact between support system and ground is installed early and is both uniform and continuous. For nearly all soft ground tunnels (the one exception being those supported by shotcrete), such contact can be obtained only by expansion of the support system, contact grouting between the excavated tunnel surface and the support system, or a combination of the two. Whatever the method, the contact must be obtained completely and early so that the ground does not have the opportunity to move (e.g., ravel, run, shear, squeeze) and render ineffective, if not destroy, the inherent strength of the ground and consequently the interaction between ground and support.

At this point a dichotomy exists: proper functioning of the tunnel depends on obtaining contact quickly, but all actions to do so require commitment, initiative, and concentrated effort by the contractor. The latter directly conflicts with the contractor's desire to drive tunnel quickly. Similarly, the design should leave means and methods to the contractor, but it must include enough detail to obtain the desired end product and provide an incentive for the contractor to assure that full contact is obtained even if it is necessary to forgo some advance rate to do so.

Schemes 1 and 2 typically have a secondary or final lining of cast-in-place concrete placed as a later operation. Thus, these are called two-pass lining schemes. The concrete lining is used to provide the design life support for the "temporary" scheme 1, to sandwich drainage fabric or waterproof membrane in both scheme 1 and 2, and/or to provide the requisite inner surface of the tunnel for user requirements such as fluid flow. Scheme 3 is called a one-pass lining and usually does not have a final lining unless a nominal one is dictated by user requirements.

It should be obvious that the combinations of support and lining variations are legion: hence, the author has not tried to develop a cookbook for soft ground tunnel lining design. Chapter 5 should be consulted for general principles, and other references (e.g., guidelines presented in the first edition of this book) should be consulted for more detailed information.

SURFACE EFFECTS OF TUNNEL CONSTRUCTION

When a tunnel is driven, changes are produced in the soil and groundwater regimes. These changes will manifest themselves primarily as settlement at the ground surface or at structures overlying the tunnel. It is the goal of the tunnel engineer to minimize the extent and impact of such settlements, recognizing that the contractor has most of the control of the subsidence through his selection of tunneling methods and equipment.

Subsidence Due to Water-table Depression

Water-table depression will occur because of external dewatering, as discussed earlier, or because the tunnel itself functions as a groundwater drain. In either case, the water-table depression increases the effective stresses on the geologic strata; the magnitude of this increase (and resulting settlement) can be estimated by the usual theories of soil mechanics.

For tunnels in sands and gravels, the settlements are approximated by elastic theory and are generally small, unless the sands are very loose. For tunnels in clays, silts, or peats, the settlements are approximated by consolidation theory and may be of greater magnitude, depending on the thickness and properties of these compressible strata.

Subsidence Due to Lost Ground

Until the 1970s, not much data concerning ground movements around tunnels had been obtained. Since then, the available information has increased, primarily because of measurements and observations on the Baltimore and Washington, D.C., subway systems. These studies have defined the following terms (see Figure 6-17):

- *Volume Change* (ΔV). Increase or decrease in soil volume caused by the tunneling.

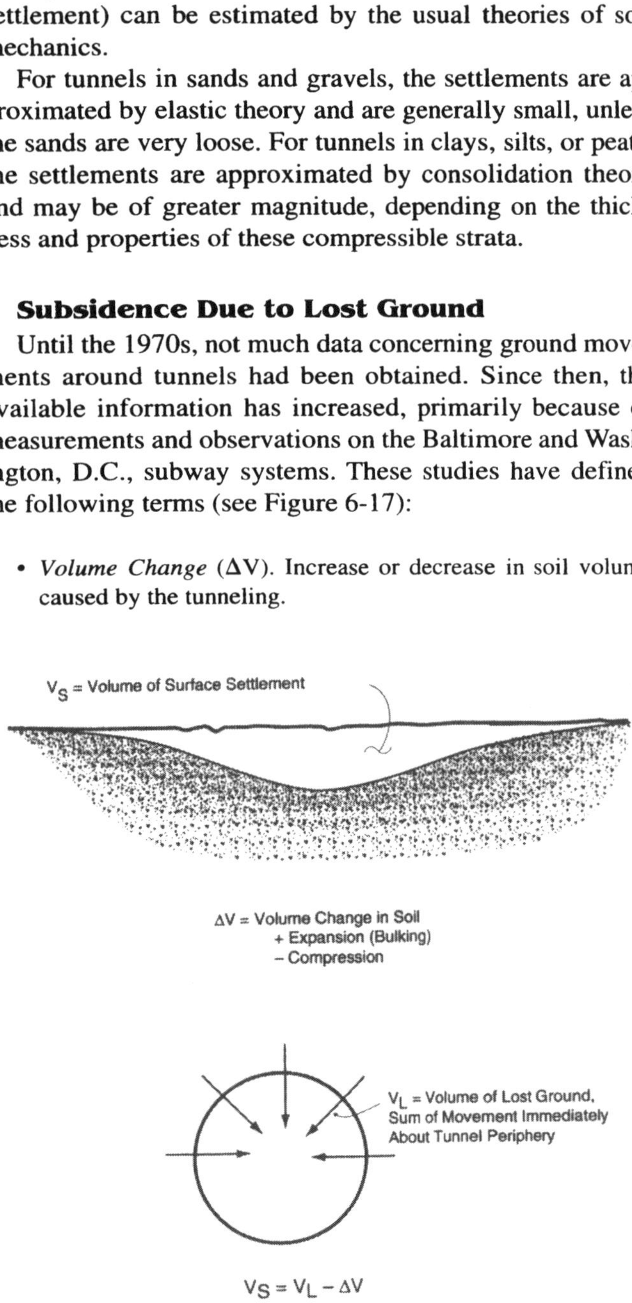

Fig. 6-17. Terminology for tunneling settlement and lost ground.

- *Volume of Surface Settlement* (V_S). The volume of the settlement trough at the ground surface.
- *Volume of Lost Ground* (V_L). The volume of all ground movements taking place about the tunnel.

The relationship among these three quantities is complex and incompletely defined. However, for most purposes it is usually possible to assume that the volume of surface settlement is equal to the volume of lost ground. This assumption is generally workable except in soils exhibiting significant increases in soil volume (bulking) or decreases in soil volume (consolidation).

Estimation of Lost Ground

Lost ground in a soft ground tunneling operation may occur in at least nine ways, which can be summarized in three forms, as follows:

- *Face Losses.* Soil movement out in front of the shield and into the shield by means of raveling, caving, flowing, running, or squeezing.
- *Shield Losses.* Soil movement toward the shield body between the cutting edge and the tail of the shield. This movement results from shield actions such as plowing, pitching, or yawing and from the void created by overcutters.
- *Tail Losses.* Soil movement toward the support system as it leaves the shield tail. This results from soil moving to fill the tail void that is created by the volume of the tail skinplate plus incomplete support expansion, failure to grout, delay in grouting, or a combination of these factors.

Two means are used to estimate the volume of lost ground that may be anticipated at a tunnel project. The first relies on the accumulation of data from a number of projects in a given area. Thus, for example, it can be estimated that tunnels constructed with good techniques in the Washington, D.C., and Baltimore, Maryland, area might be expected to achieve approximately 1% by volume of lost ground. It must be noted, however, that the usual project (in other areas) more typically achieves 1-1/2 to 3%.

The second means of estimating the percentage of lost ground is, in reality, a judgment call by the engineer. This judgment call includes an evaluation of the applicable combination of ground conditions, tunneling methods, and workmanship. Table 6-8 illustrates such estimates for several representative cases.

Both of these estimates assume a machine of typical dimensions on a straight drive. However, machines seldom run perfectly straight, and geometry considerations can add to the lost ground approximations. For a representative machine, diving or rising only 1% adds a corresponding 1% to the volume of the excavated hole. Similarly, yawing or traversing a horizontal curve can add 1 to 2% to the volume of the hole. If a larger than usual overcutter has been installed, it is easy to increase the hole another 1%. With care and good luck (and good ground), most of these extra volumes will be filled by support expansion or grouting. However, because they are abnormal, some or all of the extra volume(s) may be manifest as an addition to the percentages in Table 6-8.

Other factors have a major impact on surface settlements, but they are even more difficult to quantify than those discussed above. These latter factors include the following.

Start of Work. A successful tunnel is the result of close melding of geology, people, machines, techniques, and operations. Every tunnel project is a unique creation: while some elements may be likened to a previous project, there will always be enough differences to make the current project unique in some ways. Thus, there is always a "learning curve" during which methods, coordination, control, and production must be developed and fine-tuned, as must the machine itself. While in the learning curve, it is prudent to assume that settlements will be greater than those that will be experienced during the majority of the drive.

Stoppage and Restart of Work. Stoppage of work can occur for any of a host of reasons: strike, equipment failure, accident, loss of ground, support failure, interruption of funding, unexpected conditions, and more. Whenever work is stopped, the tunnel must be stabilized and secured, often under extreme time limitations. Such unexpected and unplanned activities are outside the established regimen in the tunnel and typically result in a (localized) area of greater than usual settlement. Then, when the work is restarted it is necessary to recover the tunnel and reestablish the orderly flow of operations, often not unlike the initial start of work.

Obstructions. By their nature, soft ground tunnels have a risk of encountering natural and manmade obstructions, such as boulders, foundations, abandoned wells, contamination, tie-backs, explosive or poisonous gases, and old piling or walls. When encountered, especially if unexpected, these obstructions trigger a stoppage and restart as discussed above. In the more usual case, they usually also trigger an increase in settlement because of the difficulty and extra (unusual) work required to clear and tunnel through the obstruction.

Disturbed Areas. Because many soft ground tunnels are built in urban areas, they frequently are built under disturbed zones existing because of other construction, such as in street intersections with crossing utility lines. Where the tunnel is near the surface, say less than two diameters, the residual disturbance from this earlier construction can interact with the new tunnel to produce larger displacements than might otherwise be expected. It is also possible that the settlements from the new tunnel may aggravate or initiate a problem; for example, the settlements may open a water or drainage line overhead, wetting the ground and causing consolidation or loss of strength (loss of apparent cohesion). These phenomena increase the settlement even further.

Table 6-8. Volume of Lost Ground and Quality of Tunneling Practice

Case	V_L, %
Good practice in firm ground • Applies to better soils and excellent ground control	0.5
Good practice in slowly raveling ground • Considered good ground	1.5
Fair practice in fast raveling ground • More shield and tail loss	2.5
Poor practice in cohesive running ground • Yet more shield loss • Tail void mostly unfilled by grouting and/or support expansion of the initial supports	4.0 or more

It is apparent that there are a number of factors (the above is not an all-inclusive list) that will affect the total settlement over a tunnel. Even for those that have been articulated above, the variables are too numerous and undefined to attempt to give guidelines for their full assessment. As with the discussion on multiple tunnels in the next section, the engineer may take a pass at these influences by applying judgment in raising the severity class of lost ground estimates for the offending tunnel(s).

Even greater ground losses can occur, especially if large losses are allowed at the face as the result of such behavior as

- Boulders caught in a rotating cutter
- A run or flow of sands or silts
- Squeezing of clays or silts
- Unequal excavation of a soft material in a mixed-face situation

Thus, predicting the amount of lost ground in advance of tunneling is mostly art. It is complicated even further by the engineer's lack of knowledge or control over the details of the machine or construction methods that the contractor will choose.

Distribution of Settlement

For a single tunnel, the volume of surface settlement for the individual tunnel is assumed equal to the volume of lost ground, estimated as above. Generally, the shape of the resultant settlement trough at the ground surface resembles that of the bell-shaped probability curve. This concept was used by Peck (1969) and others (Schmidt, 1969) to correlate field measurements of trough width for several cases. Geometry of the settlement trough is illustrated in Figure 6-18. In all cases in the calculations, the ground surface is assumed at the bottom of the building footing and the influence of building footing and building stiffness is ignored.

For geometrics of situations other than a simple (single) tunnel in reasonably good ground, the settlements depend even more on construction methods and are correspondingly more difficult to predict. For these situations, experience and judgment play an even greater role. As a start, the engineer may make assumptions along the following lines:

- For *parallel tunnels* three or more diameters apart (center to center), surface settlements are usually reasonably well

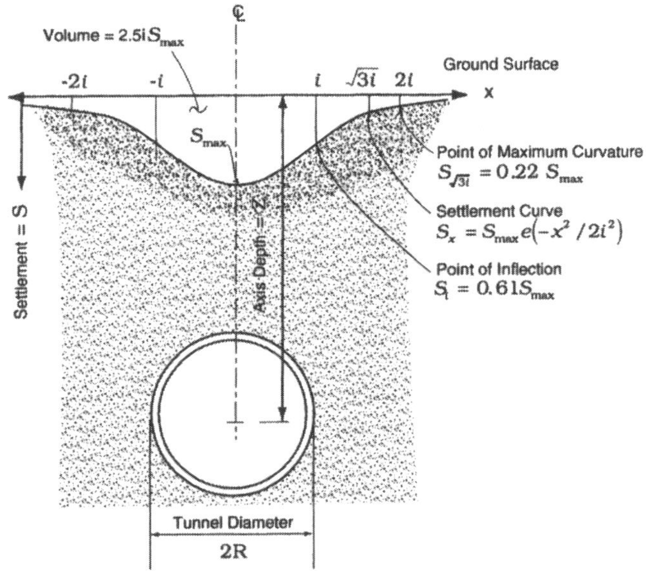

Fig 6-18. Properties of probability curve as used to represent cross section of settlement trough above tunnel (Adapted from Peck, 1969).

predicted by adding the individual bell curves of the two tunnels. In good ground and with good practice, this will often give workable approximations up to the point where the tunnels are two diameters apart. On the other extreme, when the tunnels are less than one and one-half diameters apart, the volume of lost ground assumed for the second tunnel should be increased approximately one level in severity in Table 6-8 before the bell curves are added. Intermediate conditions may be estimated by interpolation.

- For *over-and-under tunnels*, it is usually recommended that the lower tunnel be driven first so that it does not undermine the upper tunnel. However, driving the lower tunnel will disturb the ground conditions for the upper. This effect may be approximated by increasing the lost ground severity of the second (upper) tunnel by approximately one level in Table 6-8 before adding the resulting two settlement estimates to approximate the total at the surface.

The width of the settlement trough is measured by an i value, which is theoretically the horizontal distance from the location of maximum settlement to the point of inflection of the settlement curve (see Figure 6-19). The maximum value of the surface settlement is theoretically equal to the volume of surface settlement divided by $2.5i$.

Detailed field measurements have illustrated the complex nature of tunnel settlements and the difficulty in trying to reduce them to a simple correlation. For example, especially with very shallow tunnels, settlements may vary in a nonlinear way with the absolute magnitude of settlement (such as an ultimate case of forming a sinkhole at the ground surface). Also, geologic details such as stiff or loose layers of soil will strongly affect how ground loss about a given tunnel becomes surface settlement. Detailed examples of actual measurements of ground movement about tunnels in sand are given by Cording and Hansmire (1975), and for clay by Palmer and Belshaw (1979, 1980). However, despite the inherent complexity, the simplicity of the correlations in Figure 6-19 makes possible a preliminary estimation of the settlement resulting from tunneling.

Influence of Settlement—Potential for Structure Damage

Having estimated the settlements (and recognizing that the estimates are often quite crude), the engineer should then evaluate the possible impact of that settlement on overlying significant structures. For reference in discussion, the following definitions from Boscardin and Cording (1989) of categories of building damage are made:

- *Architectural Damage.* Damage affecting the appearance but not the function of structures, usually related to cracks or separations in panel walls, floors, and finishes. Cracks in plaster walls greater than 1/64-in. wide and cracks in masonry or rough concrete walls greater than 1/32-in. wide are representative of a threshold where damage is noticed and reported by building occupants.
- *Functional Damage.* Damage affecting the use of the structure, usually related to jammed doors and windows, cracking and falling plaster, tilting of walls and floors, and other damage that would require nonstructural repair to return the building to its full service capacity.
-

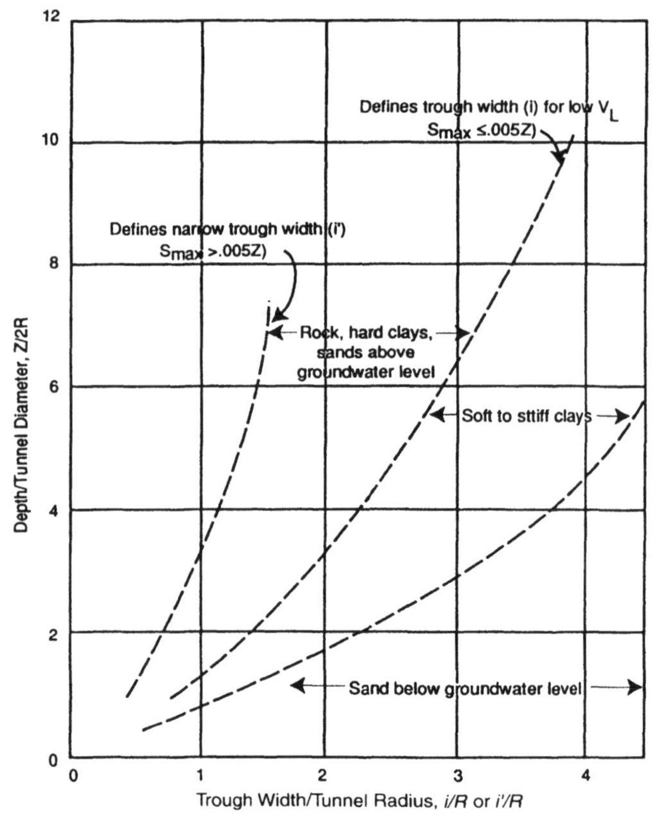

Fig. 6-19. Assumptions for width of settlement trough (adapted from Peck, 1969).

•*Structural Damage.* Damage affecting the stability of the structure, usually related to cracks or distortions in primary support elements such as beams, columns, and load-bearing walls.

Alternative methodologies exist for evaluating the influence of tunneling settlement on buildings. Many works exist on the subject, but a widely used one in civil works is Skempton and MacDonald (1956), who studied data from 98 buildings, 40 of which were damaged to some extent. The study included both framed and load-bearing wall structures. Grant, Christian, and Vanmarked (1974) extended the study to include 95 additional buildings, 56 of which had some damage. These results, along with the work of other investigators, were combined in a study by Wahls (1981). The results are astonishingly close in all cases, and permit evaluation of the influence of settlement on buildings by means of empirical correlations of building distortion (relative settlement between columns) to building damage. Such a correlation is given in Table 6-9, taken from Wahls.

It is generally agreed by all investigators, including the Russians (Polshin and Tokar, 1957), that bearing wall structures and plaster partitions are first cracked at an angular distortion of approximately 1/300. Considering all these factors, and to be conservative, it is recommended that the tolerable angular distortion be taken as one-half this amount, or 1/600. In addition it is recommended that absolute settlement of a column or footing be limited to 1 in.

The first line of defense in limiting possible damage from settlement induced by tunneling is to require and use tunneling equipment and methods that reduce the lost ground, including

- Full and proper face control at all times, especially while shoving the shield
- Limiting the length-to-diameter ratio for the shield, making directional control easier and reducing the effects of pitch and yaw (long shields may be subdivided with articulation joints in the shield skin plate)
- Rapid installation of ground support
- Rapid expansion, pea-graveling, and/or contact grouting of ground support

Table 6-9 Limiting Angular Distortion

Category of Potential Damage	Angular Distortion
Danger to machinery sensitive to settlement	1/750
Danger to frames with diagonals	1/600
Safe limit for no cracking of building[a]	1/500
First cracking of panel walls	1/300
Difficulties with overhead cranes	1/300
Tilting of high rigid building becomes visible	1/250
Considerable cracking of panel and brick walls	1/150
Danger of structural damage to general building	1/150
Safe limit for flexible brick walls[a]	1/150

[a] Safe limit includes a factor of safety.

Wahls, 1981)

In special cases, other steps also should be considered, including

- Use of compressed air
- Consolidation grouting of the ground before tunneling
- Consolidation grouting from the tunnel face
- Freezing the ground before tunneling
- Compaction grouting between tunnel and foundations
- Underpinning structures by any of a group of methods

BUILDING PROTECTION METHODS

Historically, when it was determined that potential settlements at a building would be damaging, only two measures were possible: underpinning the building before the tunnel was driven and/or repairing the building. Underpinning involves "picking up" the foundation loads by one of a number of methods and carrying those loads to a deeper bearing stratum that is not affected by the tunneling operation. Underpinning is typically performed before the tunnel is driven and, properly done, can provide effective support for the building. However, it must be recognized that underpinning itself can sometimes be damaging to the building. In addition, it must also be noted that underpinning can be very expensive and, in some cases, disruptive to operations in the building. Thus, it often has been decided that the preferred alternative is to tunnel past the structure using good tunneling techniques and tight control and then to repair any building damage that was caused. This alternative is especially attractive when the repair cost for the estimated damage to the building is less than the cost of underpinning.

Where access exists, and where the soils are coarse-grained enough to accept it, permeation grouting may be used to develop an improved mass through which to drive the tunnel. This technique was described earlier and may be applied from the face, from the surface or buildings on the surface, or (as was the case in Los Angeles) from specially constructed grouting pits. In those cases where it is appropriate, permeation grouting can be a very effective technique in controlling settlements about tunnels; as a general rule, it is also very expensive.

A new technique that is being used to mitigate possible damage to structures from nearby tunneling is compaction grouting. This technique was first used in Baltimore and contributed to reducing settlements from more than 3/4 in. in ungrouted test sections to less than 1/4 in. in grouted sections. More recently, compaction grouting has been used in Los Angeles, where twin subway tunnels were successfully driven through sandy soils under 16 downtown buildings.

Compaction grouting uses a stiff grout mix injected above the tunnel (and below the structure). Injection is done through pipes installed in a pattern. The pipes are installed, grout lines are in place and hooked up, and the grout and pumps are all ready as the tunnel approaches.

The concept behind compaction grouting is to inject the grout into the lost or loosened ground zone (developing settlement trough) that is tending to migrate from the tunnel to the structure. By controlling the time, pressure, and amount of grouting, the loosened zone is "cut off" or recompacted by the grouting before the settlement trough reaches the structure foundations. To assist in making the decisions of when and how to do the grouting, engineers use deformation readings from settlement instrumentation. These readings include settlement indicators on the surface and on the structure, and multiple position bore-hole extensometers anchored at various depths in the soil.

Another possibility for protecting structures from settlement due to tunneling is the use of protective walls. These walls may be slurry walls or soil–cement structural walls, which are now finding their way into U.S. practice from Europe and Japan, respectively. Usually thought of as possible protective schemes for surface buildings adjacent to cut-and-cover excavations, these walls also can be effective in tunnel applications. Due to their stiffness these walls serve as cut-off structures to stop the lateral spread of tunnel-induced settlement. Their effective use typically requires embedment below the tunnel and may require that tie-backs be installed near the surface. For a more thorough discussion of these types of walls and especially of equipment and procedures for their installation, refer to Xanthakos (1994).

PRACTICALITIES OF TUNNEL ENGINEERING

The writer would like to emphasize that tunneling, among all types of civil engineering projects, must always be approached with a great deal of practicality, which typically can only be learned from experience. The following two sections are taken nearly verbatim from a private report by Kuesel for a rapid transit rail tunnel project in Los Angeles. These two sections discuss the practicalities of tunnel engineering (and construction) and apply equally in principle to Chapters 5 and 7 as well.

Tunnel Construction

Tunnel construction has little in common with Swiss watch manufacture. Public works contracts are awarded on the basis of the lowest responsive bid. This means that the owner places a premium on rapid (and therefore economical) construction, and implicitly agrees to accept some degradation in the precision of the tunnel alignment and of the dimensions of the tunnel lining.

The design cannot assume perfection in alignment and concrete form placement, particularly on curves, where precise control of the direction of the tunnel boring machine (TBM) is difficult, and where the actual alignment in any case must be a series of chords that approximates the theoretical curve. The tunnel alignment, and the shape and dimensions of the tunnel lining, as shown on the design drawings, should therefore be regarded as an ideal to be approached, but rarely obtained precisely. To insist on literal enforcement of the theoretical alignment and dimensions would greatly increase the time required for construction, and the cost of the work, with no corresponding benefit in the safety or usefulness of the completed tunnel facility.

To allow for the inevitable tolerances of good-quality tunnel construction, it is customary to specify a "bull's-eye," concentric with the theoretical tunnel centerline, within which the actual centerline of the tunnel cross section is required to fall. The size of the bull's-eye is selected by the engineer, to represent a reasonable target for the prevailing ground conditions, which experienced and careful tunnel contractors can meet as a prevailing standard of workmanship, with a reasonably rapid rate of progress. The bull's-eye is deliberately made tight to discourage sloppy work, which would result if a large tolerance were specified, thereby tempting the contractor to reduce accuracy for increased speed of construction.

Nonetheless, a prudent design must recognize that in the best of circumstances a small percentage of actual tunnel lengths will fail outside the bull's-eye. This may occur as a result of difficult or unexpected ground conditions, equipment malfunctions, or human error. If the deviation is great, it is necessary to stop tunnel excavation, remove some of the tunnel lining, resupport the ground, mine around the TBM by hand methods, reset it on the correct alignment, and reconstruct the removed tunnel lining sections. Although this procedure is specified in the tunnel construction contract, it is rarely enforced, not only because it results in large delays and costs, but more particularly because it generally endangers overlying and adjacent structures and facilities.

A good tunnel design succeeds in avoiding this unpleasantness by allowing for adjustment of the track or roadway alignment to a "best fit" with the actual tunnel lining alignment as constructed. If corrections to deviations from the bull's-eye are made gradually, it is usually possible to work out a compound curve alignment that brings the actual alignment within the bull's-eye and does not deviate from the theoretical simple curve alignment sufficiently to cause any perceptible variation in the smooth ride of trains or highway vehicles. As one example of this procedure, on the Trans-Bay Tube of the San Francisco Bay Area Rapid Transit System, a deviation of over 2 ft from the theoretical alignment was successfully mitigated by this method, with no cost other than the minor cost of a computer run of the compound alignment, and no one other than the engineer and the contractor was ever aware that it was done. Similar "best fit" track alignment adjustments have been made at several locations in the Los Angeles Metro.

After the "best fit" adjustment, there frequently remain a few isolated locations where, to maintain the operating clearances inside the tunnel, it is necessary to move the tunnel lining forms out and reduce the lining thickness. In such situations there are three courses of action:

1. Remine the tunnel in that location and reconstruct the lining to the specified thickness.
2. Reinforce the lining locally, either by adding reinforcing steel before the concrete is poured, or by adding steel plates or other reinforcing to the inside surface after it is poured.
3. Accept the lining of reduced thickness, as constructed. This may be acceptable if the lining can be shown to have ample capacity for the local conditions at that spot (see next section).

To make a proper selection among these courses, it is necessary to consider the basis of tunnel lining design.

Lining Analysis and Design

Fortunately, the thickness of a tunnel lining is rarely controlled by considerations of loading or stress—virtually all linings are considerably stronger than they need be to support the design loads. This comes about through consideration of practical construction requirements. For instance, for precast concrete segmental liners the thickness is usually governed by the space required for bolts to connect adjacent segments. For the Los Angeles Metro standard tunneling design, the precast segments are 8 in. thick. Considerations of design loadings alone would lead to a thickness of 6 in., which has been used in other subway systems.

For cast-in-place circular concrete liners, the Los Angeles standard design specifies a concrete thickness of 12 in. The basis for this is that the concrete must be placed by pumping it into the annular space between the initial tunnel support (circular steel H-section ribs with timber lagging, covered with a high-density polyethylene (HDPE) gas-proof membrane) and the inner concrete forms. Experience has shown that if the nominal design thickness is less than 12 in., it will be difficult to get the concrete to flow through the restricted space, particularly where internal reinforcing steel is required (as it is for seismic considerations in Los Angeles), and excessive segregation and honeycombing of the concrete (inadequate coating of cement paste over the stone aggregate) will occur. It is to be noted that even with a 12-in. thickness, small amounts of honeycombed concrete were discovered during the regular inspections of portions of the concrete lining, and they were dug out and repaired.

Specifying a nominal design thickness of 12 in. also recognizes that there will be isolated locations where the actual lining thickness might be thinner, and this would further impede proper concrete placement. The thickness of lining required for strength considerations alone would not exceed 8 in. It would be impractical to construct such a thin cast-in-place concrete lining in the tunnel, but if it were practical it would be amply strong to carry all design loads.

Although experience teaches that stresses in tunnel linings are usually well below allowable values, the design generally includes an analysis to demonstrate this. Tunnel lining analysis is an imprecise art compared, say, to analysis of the stresses in a steel bridge or building frame carrying well-defined loads. The ground and groundwater loads acting on a tunnel are difficult to assess, and they vary widely along the length of a tunnel as ground conditions vary. They are also strongly affected by the processes of excavation, advancing the TBM, and installing the initial tunnel support. Assumptions regarding how the ground loads the lining, and at the same time supports it, are therefore arbitrary. A number of mathematical procedures for the analysis of stresses in tunnel linings have been developed and propounded by academic researchers. In general, the precision of the analysis greatly exceeds the precision, and accuracy, of the arbitrary assumptions made for the loading and support conditions. To say that analysis "requires" a lining thickness of "*X*" inches indicates an unjustified faith in the power of mathematics to describe complex natural conditions and how they are affected by construction operations.

Generally, lining analysis develops numerical values for the axial thrust and bending stresses in the lining ring. The basis for thrust calculations is fairly straightforward and reliable. The basis for bending calculations is much more uncertain and arbitrary. Empirical experience indicates that bending stresses in tunnel linings are usually greatly overestimated.

In the case of two-stage linings (initial support plus a cast-in-place inner lining), such as used in Los Angeles, the bending stresses computed by the conventional analysis act wholly on the initial support, causing the steel ribs to deform and mobilize internal shearing stresses in the ground surrounding the tunnel. These ground shearing stresses stabilize the opening, in the same natural action that stabilizes cave openings around the world. Once the ground deformation has stopped, the tunnel opening is self-supporting and stable. In this condition the inner concrete lining is placed. Initially, it carries no load except its own weight. Over a long time period, relaxation of internal ground stresses may transfer some thrust to the lining, but bending stresses must necessarily be minor, a small fraction of those calculated in the conventional tunnel lining analyses. This is why the thickness of a cast-in-place concrete tunnel lining required to support the actual loads applied to it is markedly less than that indicated by the design analyses, and much less than that required for practical construction considerations.

One special consideration applies where earthquakes are a concern (e.g., Los Angeles). In the absence of active faults or other major ground instabilities (such as landslides or liquefaction of loose soils), the effect of the earthquake is to distort the ground temporarily, generally returning it to its undisturbed condition after the earthquake subsides. This action distorts a rectangular section of ground into a slightly rhomboid or diamond shape, and distorts a circular section (or a circular tunnel lining) into an oval. The effect is to induce additional bending stresses into the lining. It is important to recognize that the tunnel is very small compared with the ground mass deformed by the earthquake. The strength or stiffness of the tunnel structure (the lining) cannot modify the earthquake distortion of the ground—the lining simply "goes along for the ride," conforming to the ground distortion. Making the lining thicker or thinner does not change the distortion. In fact, thinner linings are more flexible and offer less resistance to the imposed distortion; hence, they are stressed less strongly.

The engineering program for design of the Los Angeles Metro extended the state of the art for earthquake analysis of tunnel linings. Particular attention was given to predicting the intensity of the ground distortion in various types of ground, and to evaluating the response of the tunnel lining to this distortion. This analysis demonstrated that the distortions imposed on the linings by the design earthquake, in the ground conditions encountered to date in Los Angeles, would not overstress the 12-in. linings. The bending stresses in thinner linings for the same distortion would be reduced by the greater flexibility of the thinner linings. The next section discusses the seismic analysis method developed for use in competent soils in Los Angeles.

SEISMIC DESIGN OF SOFT GROUND TUNNELS

Although underground structures are much less vulnerable to earthquakes than surface structures, there still is a potential for damage to buried structures in strong-motion earthquakes. The actual risk must be assessed on the basis of both seismological and geotechnical evaluation of the site. For this assessment, the required seismological information includes

- Historical data on earthquake recurrence intervals, magnitudes, and associated parameters of ground shaking
- Proximity of faults
- Historical evidence of slippage on the faults and magnitude of actual offsets with their recurrence intervals

Similarly, the required geotechnical information includes

- Depth to and nature of underlying bedrock
- Stratigraphic section and properties of the individual components of soil/rock in the overburden
- The location of the water table, presence of perched water, and degree of saturation of the soil
- Geophysical data, especially seismic shear wave velocity in each major segment of the soil/rock horizon

These data must be carefully evaluated, and an appropriate philosophy of design established. In most cases, it is impractical to design a structure to survive the conditions expected from the most severe earthquake that statistically might occur during the useful life of the project. If the service life is in the range of 50 years, for example, it is usually inconsistent to design to survive an earthquake with a 500-year recurrence interval. This is not to say that the 500-year earthquake may not occur during the service life of a project with a 50-year life. There certainly is a finite probability that it will, but the costs normally associated with creating a structure to survive unscathed are prohibitive and often not justified. On the other hand, it may be mandatory that operations within the structure be safely shut down or diverted. For example, if the structure is occupied, the structure may be damaged but must not collapse, and it must be possible for the people to be quickly and safely evacuated. Thus, the operational philosophy and associated command and control system for the system must be selected and designed to allow for a completely safe shutdown and evacuation in case the design earthquake occurs or is exceeded (Monsees, 1991).

Seismologists generally refer to the maximum credible earthquake, which is the largest earthquake that a given fault is believed capable of ever creating. By comparison, engineers generally refer to the maximum design earthquake (MDE) and the operating design earthquake (ODE). These typically are defined as follows:

- The MDE is the earthquake event that has a return period of several thousand years. It has a small probability of exceedance, approximately 5% or less, during the 100-year facility life. The MDE defines the level at which critical elements continue to function to maintain public safety, preventing catastrophic failure or collapse and loss of life. However, some elements will experience inelastic deformation, and the structure(s) may require major repair before being returned to full service.
- The ODE is the earthquake event that has a return period of several hundred years. It can reasonably be expected to occur during a 100-year facility design life; the probability of exceedance of this event is approximately 40% during the facility life. In the ODE, critical elements of the facility maintain function and the overall system continues to operate normally. Any needed repairs are cosmetic in nature and can be done as part of maintenance operations.

Major Effects Of Earthquakes

The major effects of earthquakes on tunnels may be broadly grouped into two general classes: faulting and shaking.

Faulting. Faulting includes direct primary displacements of bedrock, which may or may not carry through the overburden to the ground surface. Such physical shearing of the rock or soil is generally limited to relatively narrow, seismically active fault zones, which may be identified by geological and seismological surveys. In general, it is not feasible to design tunnels to restrain such major ground faulting. Useful design measures are generally limited to identifying and avoiding fault crossings, or if this is not possible, to accepting the displacement, localizing and minimizing damage, and providing means to facilitate repairs. Although designs for fault displacement, for conservatism, normally assume that the displacement occurs on a single plane, the actual ruptures very often occur over a zone and thus tend to mitigate the distress imposed on the buried structure.

Location of fault crossings and design for mitigation of their effects are especially challenging. Nyman (1983) indicates that faults should always be crossed, if possible, such that the structure is placed in tension. However, unlike pipelines running across the country (the principal subject of Nyman), many structures cannot be freely located relative to the fault. For example, seismic analysis of the fault crossing by a recent sewer tunnel in Los Angeles showed that the structure would be in compression if the fault moved. Similarly, for the Los Angeles Metro it is expected that other considerations will govern the alignment, and the designers will not be able to change the alignment based solely on the angle of fault crossing.

For the sewer tunnel, the needed length of special design was quite limited, and it was practical to consider a small conduit in a larger excavation for this limited reach. Thus, it was decided to design an articulated concrete lining surrounded by cast-in-place cellular concrete, which could literally accommodate the design value of 8 in. of lateral displacement and limit the compressive load on the conduit to about 20 psi. This proved practical and cost-effective.

For the original alignment of the Los Angeles Metro, a ductile steel lining was planned should the tunnel cross suspected faults. Segmented concrete did not appear to be appropriate due to the large displacements, the relative brittleness of concrete, and the need to prevent leakage of methane gas into the tunnel from surrounding deposits. However, for the portions of subway constructed to date, no fault crossings have been identified, and thus special linings have not been used.

Whenever an underground structure crosses a fault, the potential for slippage to occur on the fault as a result of an earthquake must be considered. Both the magnitude and recurrence interval are important. In cities and suburban areas, development often has destroyed all surface evidence of the exact location of faults. Additionally, development often precludes the use of trenches to uncover the faults and define the history of motion on them.

Finding the fault early and defining potential displacement on it are crucial. Practical schemes for fault location may require use of special, often expensive, exploration procedures. However, being able to minimize the need for special construction procedures will often repay, by several times, the cost of the effort to find the fault and gather definitive data on the magnitude and recurrence interval of the fault displacement. This was true for the Los Angeles sewer discussed above, where the final design required only 400 ft of special lining system using special backpacking instead of the one mile originally believed necessary.

Prediction of fault displacement is done using procedures such as those developed by Bonilla (1984). These procedures relate fault displacement to the length of the rupture created by the earthquake in the epicentral area. Typical historical values on major faults vary from a few centimeters to meters. In the last case, the potential impact on design is obvious.

Shaking. In response to earthquake motion (shaking), the soil transmits energy by waves. Seismologists identify various types of earthquake waves, but underground engineers are generally interested in the effects of transverse shear waves, which produce a displacement of the ground transverse to the axis of wave propagation. The orientation of propagation is generally random with respect to any specific structure. Waves propagated parallel to the long axis of a tunnel will tend to produce a corresponding transverse distortion of the structure. Waves traveling at right angles to the tunnel will tend to move it back and forth longitudinally and may tend to pull it loose at zones of abrupt transitions in structure type or in soil conditions, where wave properties may vary.

The major contribution to deformations and corresponding stresses in long linear structures (tunnels) is traveling wave effects. These generally can be accounted for by assuming that the tunnel and surrounding soil move together as the wave passes. As a result of these soil movements, two primary types of deformation, "snaking" and ovaling (or racking), are imposed on the tunnel structure.

It should be recognized that although the absolute amplitude of earthquake displacement may be large, this displacement is spread over a long length. Thus the gradient of earthquake distortion is generally small, often within the deformation capacity of the structure. If it can be established that the maximum deformation imposed by the earthquake will not strain the structure beyond the elastic range, no further provisions to resist the deformation are required. If certain parts of the structure are strained into the plastic range, the ductility of such parts must be investigated. If continuity of the structure has been assumed in the design for static loads, the residual effects of plastic distortions induced by earthquake motions may require special consideration.

Strains From Snaking Motion. The maximum longitudinal strains due to snaking in the lining can be conservatively approximated as (Kuesel, 1969; Metro Rail Transit Consultants, 1984)

$$\epsilon_L = \frac{V_{max}}{2C_{se}} \pm 0.7R\frac{A_{max}g}{C_{se}^2} \tag{6-3}$$

where

ϵ_L = longitudinal strain in lining
V_{max} = maximum particle velocity
A_{max} = maximum acceleration in gravity units
g = acceleration of gravity
C_{se} = effective shear wave velocity
R = centerline radius of tunnel lining

The first term is the free-field compressive/tensile strain parallel to the tunnel axis (assuming no slippage between lining and surrounding ground). The second is the longitudinal bending from the snaking of the tunnel. Usually the strains ϵ_L will be found to be small enough that they can be accommodated easily and will not impair structural behavior. This is especially true if, as is often the case, there is slippage between the lining and ground that typically reduces the strains from the first term by approximately an order of magnitude.

Strains from Racking Motion. Circumferential strains induced by shear waves propagating perpendicular to the tunnel axis change the shape of the lining by causing an ovaling or racking motion. It is the strains from this motion that are usually most critical to the design and behavior of a tunnel.

Often it will be found that these strains can also be accommodated by the structure. However, in some cases it may be found that these strains become critical to the design and behavior of the structure. In this latter case, it may be necessary to perform a detailed (usually 2D) finite element solution using such programs as ABACUS. However, it is recommended that in all cases the following approximate and conservative calculation be made first, because if the results are acceptable no further analysis is needed. For a circular tunnel:

$$\epsilon_{\text{rack}} = \frac{V_{\max}}{C_{se}}\left[2\frac{t}{R} + \frac{3E_m R}{16E_L t}\right] \quad \text{in compression} \qquad (6\text{-}4)$$

$$\epsilon_{\text{rack}} = 2\frac{V_{\max} t}{C_{se} R} \quad \text{in tension} \qquad (6\text{-}5)$$

where

ϵ_{rack} = strain induced by racking
t = lining thickness
E_m = modulus of elasticity of soil
E_L = modulus of elasticity of lining

$V_{\max}$, C_{se}, R are as defined earlier.

These strains are combined (statically) with the results of the basic tunnel design calculations. Preferably, the resulting acceptability analysis should be based on the capacity of the lining to withstand the deformations (strains) imposed by the earthquake rather than on factors of safety expressed in terms of load or stress ratios. The latter do not properly reflect the reserve deformation capacity of the lining. That reserve capacity is quantified by the ratio of maximum seismic induced lining strain to the strain at which failure of the concrete takes place. However, if comparisons of section moment-thrust demand with available strength according to ACl-recommended procedures are favorable, the designer can be assured the section has ample capacity, and the analysis need go no further.

In using Equations (6-4) and (6-5), it must be recognized that the following assumptions are being made:

- The flexibility ratio of the tunnel structure is approximately 20 or greater.
- The wave length of peak velocities is at least eight times the width of the opening.
- Tensile cracking of the section under seismic distortions is acceptable provided the remaining compression section is not strained beyond 0.003 in./in.

It should also be noted that for the large deformations associated with design earthquakes, the shear wave propogation velocity and the "modulus of elasticity" of the soil are substantially reduced from the values determined from small vibrations, such as those produced by seismic exploration techniques. For soft soils, the velocity ratio can be 2 to 3, and since $E \propto C^2$, the "modulus" ratio 4 to 9.

REFERENCES

ABRAMSON, L. W. (1994). Groundwater lowering and drainage techniques. In *Ground Control and Improvement,* P.P Xanthahos, L. W. Abramson, and D.A. Bruce, eds. New York: John Wiley and Sons.

BAKER, W. H., ed. (1992). *Grouting in Geotechnical Engineering.* New York: American Society of Civil Engineers.

BONILLA, M.G., R. K. MARK, and J. J. LIENKAEMPER. (1984). Statistical relations among fault displacement. *Bulletin of the Seismological Society of America, 74.*

BROMS, B.B. and H. BENNERMARK. (1967). Stability of clay at vertical openings. *Journal of the Geotechnical Division 3.*

CORDING, E. J., and W. H. HANSMIRE. (1975). Displacements around soft ground tunnels. *Proceedings, Fifth Panamerican Congress on Soil Mechanics and Foundation Engineering,* Buenos Aires.

DEIX, F., and B. BRAUN. (1987). The use of NATM in combination with compressed air and ground freezing during Vienna subway construction. *Proceedings, Rapid Excavation and Tunneling Conference,* Littleton, CO: Society of Mining Engineers Inc.

GRANT, R., J. T. CHRISTIAN, and E. H. VANMARKE. (1981). Differential settlement of buildings, *Journal of Geotechnical Engineering, 107,* American Society of Civil Engineers.

GULARTE, F. B., G. E. TAYLOR, J. E. MONSEES, and J. P. WHYTE. (1991). "Tunneling performance of chemically grouted alluvium and fill - Los Angeles Metro Rail Contract A-130, *Proceedings, Rapid Excavation and Tunneling Conference,* Littleton, CO: Society of Mining Engineers, Inc.

GULARTE, F. B. (1982). Grouting Practice for Shafts, Tunnels, and Underground Excavations", Short Course Notes, University of Wisconsin-Milwaukee.

GULARTE, F. B. (1989). Grouting Practice for Shafts, Tunnels, and Underground Excavations", Short Course Notes, University of Wisconsin-Milwaukee.

HEUER, R.E. (1995). Personal communication.

HEUER, R.E., and D. L. VIRGENS. (1987). Anticipated behavior of silty sands in tunneling. *Proceedings, Rapid Excavation and Tunneling Conference,* Littleton, CO: Society of Mining Engineers, Inc.

HEUER, R.E. (1974). Important ground parameters in soft ground tunneling. *Proceedings of Specialty Conference on Subsurface*

Exploration for Underground Excavation and Heavy Construction, New York: ASCE.

HITACHI ZOSEN (1984). *Hitachi Zosen's Shield Tunneling Machines,* Company Brochure, Tokyo.

HITACHI ZOSEN (1981). "Shield Tunneling Machines", Company Brochure, Tokyo.

KAROL, R. H. (1990). *Chemical Grouting,* second edition, New York: Marcel Dekker, Inc.

KUESEL, T. R. (1969). Earthquake design criteria for subways. *Journal of the Structural Division, 95,* ST6.

MCCUSKER, T.G. (1982). Soft ground tunneling. Chapter 5 in *Tunnel Engineering Handbook,* J.O. Bickel and T. R. Kuesel, eds., New York: Van Hostrand Reinhold.

Metro Rail Transit Consultants (1984). Supplemental Criteria for Seismic Design of Underground Structures, SCRTD, Los Angeles.

MITCHELL, J. K. (1981). State-of-the-art report on soil improvement, *Tenth International Conference on Soil Mechanics and Foundation Engineering,* Stockholm.

MONSEES, J. E., and J. L. MERRITT (1991). Earthquake considerations in design of the Los Angeles Metro, in Lifeline Earthquake Engineering, M.A. Cassuro, ed., New York: ASCE.

NYMAN, D. J., et al. (1984). *Guidelines for the seismic design of oil and gas pipeline systems.* ASCE Technical Council on Lifeline Earthquake Engineering, New York: ASCE.

PALMER, J.H.L., and D. J. BELSHAW, (1980). Deformations and pore pressures in the vicinity of a precast, segmented, concrete-lined tunnel in clay. *Canadian Geotechnical Journal, 17.*

PALMER, J.H.L., and D. J. BELSHAW (1979). Long-term performance of a machine-bored tunnel with use of an unreinforced, precast segmented concrete lining in soft clay, *Proceedings Tunnel '79,* London: Institution of Mining and Metallurgy.

PECK, R.B. (1969). Deep excavations and tunneling in soft ground. *State-of-the Art Volume, Seventh International Conference on Soil Mechanics and Foundations,* Mexico City.

PHIENWAJA, N. (1987). Ground response and support performance in a sheared shale, Stillwater Tunnel, Utah, Ph.D. Thesis, University of Illinois at Urbana-Champaign.

POLSHIN, D.E., and TOKAR, R.A. (1957). Maximum allowable non-uniform settlement of structures. *Proceedings 4th International Conference on Soil Mechanics and Foundation Engineering, Vol. I,* London.

POWERS, J. P. (1992). *Construction Dewatering—New Methods and Applications,* Second Edition, New York: John Wiley and Sons.

POWERS, J. P., Construction Dewatering—A Guide to Theory and Practice, John Wiley and Sons, New York, 1981.

PROCTOR, R.V. and T. L. WHITE (1977). *Earth Tunneling with Steel Supports,* Commercial Shearing, Inc., Youngstown, OH.

RARJEAT, P. E. (1991). Evolution of tunneling methods in Mexico during last 20 years. *Proceedings, Rapid Excavation and Tunneling Conference,* Littleton, CO: Society of Mining Engineers, Inc.

RICHARDSON, H. W., and R.S. MAYO. (1941) *Practical Tunnel Driving.* New York: McGraw Hill.

SCHMIDT, B. (1979). Settlements and ground movements associated with tunneling in soil. Ph.D. Thesis, University of Illinois.

SKEMPTON, A.W., and D. H. MACDONALD. (1956) The allowable settlements of buildings. *Proceedings, Institute of Civil Engineers, Part III,* London.

STACK, B. (1982). *Handbook of Mining and Tunneling Machinery,* New York: Wiley.

TERZAGHI, K. (1977). Earth tunneling with steel supports. Commercial Shearing and Stamping Co., Youngstown, OH.

TERZAGHI, K. (1950). Geologic aspects of soft ground tunneling. Chapter 11 in *Applied Sedimentation,* R. Task and D. Parker, eds. New York: John Wiley & Sons.

WAHLS, H. E. (1981). Tolerable settlement of buildings. *Journal of Geotechnical Engineering, 107,* ASCE.

XANTHAKOS, P. P., L.W. ABRAMSON, and D.A. BRUCE (1994). *Ground Control and Improvement,* New York: John Wiley and Sons, Inc.

7 | Rock Tunnels

Elwyn H. King
Principal Professional Associate (retired),
Parsons Brinckerhoff Quade & Douglas, Inc.

As late as World War II and for a number of years thereafter, rock tunnels were "designed" by assigning the responsibility (and all risks) to the contractor for excavating and stabilizing the ground and then constructing a concrete lining of contractual dimension and detail. Lining thickness was based on judgment and prior experience. This frequently boiled down to roughly an inch of thickness per foot of span, with some adjustment for quality of rock anticipated. The lining was conceived of as a conventional freestanding arch, such as a bridge arch, carrying a uniform load of (dead) rock of arbitrary thickness, or alternatively, a half-span loading of similar thickness in order to "quantify" moments by which to design the reinforcing steel.

In 1941, Terzaghi set forth the first rational basis for estimating the loads to which the lining might be subject (see Table 4-13). These varied from no load for hard and intact rock to 250 ft of rock depth for swelling rock. This was a significant improvement but, in hindsight, failed to recognize adequately the innate capacity of the rock, which often has a unit compressive strength that is greater than that of the concrete that is relied on to support it. In some situations, this approach is still valid; it may be termed the conventional or classical approach.

Deere et al. (1967) showed the Terzaghi loads were sometimes overly conservative and modified the load values. He also developed the RQD, or rock quality concept, detailed in Chapter 4. Concurrently, he and others were actively pursuing other design concepts, especially those recognizing that even highly fractured rock masses possess inherent strength that, if judiciously improved by simple reinforcing or restraint, can reduce significantly the manmade structures previously considered necessary.

The "modern" approach prevalent today requires an inextricable interweaving of rock technology, engineering design, and construction practice and techniques. The rock mass characteristics, particularly the details of the discontinuities, place constraints on or permit freedom in construction methods. The construction method selected—tunnel boring machine (TBM) or drill-and-blast—may result in different rock stabilization techniques being appropriate to each method for identical rock masses. Overall economy may require stabilization choices that seem to be uneconomical. Preconceived ideas of what is necessary, permissible, or acceptable need continuous reexamination. While the basic concepts and procedures in each specialty area are now generally understood and much written about, the relationships among them and effects of one on another are not. Both basics and relationships are considered in this chapter.

Construction aspects of rock tunnels must be among the designer's earliest considerations. Needless constraints on tunneling operations inevitably result in increased cost. Prohibition of blasting for reasons of noise and vibration rather than establishment of acceptable limiting levels of these irritants and arbitrary limits on working hours are two examples.

Rather than considering the technical aspects of using explosives, discussion in this chapter is limited to setting forth what needs to be known about explosives so that urban dwellers and high-level decision makers can be reassured about the actual effects of using a supposedly—but not necessarily—dangerous and annoying procedure.

Although computer analysis can be a valuable tool for parameter studies, tunnel design still requires a thorough understanding of the physical characteristics of rock masses and their action and interaction.

CLASSICAL CONCEPTS

Once basic principles are set forth correctly, as Terzaghi did in 1941, they remain valid and useful through time. When additional fundamental and useful concepts are developed and when experience proves the direct application of the original work may result in unnecessarily conservative designs, the original is modified accordingly. Cases in point are the development of the RQD rock classification system and the experience-derived modification of Terzaghi's load coefficients (Deere et al., 1967).

The description of the classical approach contained in the first edition of this book is as valid today as it was then and is reproduced below (with minor condensing). It was written by Lyman D. Wilbur, retired vice president of Morrison Knudsen Company Inc. and esteemed Honorary Member of the American Society of Civil Engineers. Table 4-13 provides necessary definitions of the rock conditions. In practice there are no sharp boundaries between these rock categories, and the properties of the rocks indicated by each of these terms can vary within wide limits.

Rock Loads

When a tunnel or cavern is excavated in intact rock, an adjustment takes place in the stresses that previously existed. Prior to excavation, the vertical load is equal to the weight of the rock mass above (see Table 7-1). Horizontal unit pressures may vary within wide limits. Circumferential stress at the wall of a tunnel in intact rock is approximately twice that existing prior to excavating, whereas radial stress equals zero. The effect of the excavation rapidly decreases with increasing distance from the tunnel walls. In intact rock, at a distance from the wall equal to about twice the diameter of the tunnel, the stress approaches the condition prior to excavating. Because of the high crushing strength of rocks, the walls will not fail in compression except at great depths—more than 2,000 ft for softer sandstones and more than 19,000 ft for the strongest rocks.

However, slabbing of the wall may occur at lesser depths, particularly if high horizontal in situ stresses occur. Because of arch action, a tunnel or cavern excavation may require no support or very heavy support, depending on the strength of the rock and its condition. Rocks with few fractures require little or no support, while badly fractured rocks, regardless of how hard they are, require support. Uncontrolled blasting, the normal practice until recently, tends to fracture the rock and open existing joints for several feet from the tunnel periphery. For this reason, tunnels excavated with the use of explosives, except in the most intact rocks, developed heavier loads on supports than those excavated without blasting.

Deere et al. (1967) refer to many theoretical analyses of rock loads but conclude that none can be applied to the usual hard rock situation without also considering the effect of rock fractures and faults that are usually the governing factors. Because of the complex nature of fractures, joints, and faulting, it has been necessary for tunnel support designers to rely on empirical formulas.

Table 7-1. Weights of Various Rocks (lb/cu ft)

ROCK TYPE	RANGE	AVERAGE
Sandstone	132–170	154
Limestone	145–172	162.5
Granite and gneiss	163–171	167.5
Slate	174–175	175
Trap rock	178–187	185

The loads that must be supported in hard rock tunnels are due to the tendency of rocks to drop out of the roof, largely due to fractures. In some hard and brittle rocks, thin slabs of rock are suddenly detached from the walls after the rock has been exposed. This is known as *popping*. In some regions popping appears to be due to horizontal pressures leading to the formation of geologically young mountain chains; in other cases, the causes are unknown. Pressures required to prevent popping are small. Tunnel lining in areas of popping rock should be able to withstand 400 lb/ft^2, and footings of ribs should be able to resist horizontal displacement. Support should be wedged tightly against the tunnel wall.

Terzaghi (1964) was the first to develop a rational approach to loading for design of tunnel supports. This method was developed for use in tunnels excavated by conventional drill-and-blast techniques. His approach is still used by many designers, and the loads he developed are given in the following paragraphs.

Tunnels in Unweathered Stratified Rocks and Schists. The load on the supports depends on the dip and strike of the strata, the spacing between the joints, the shattering effects of blasting, the distance between the working face and the roof support, and the length of time between removal of natural support and installation of artificial support. The vertical load should not exceed the weight of rock in an area with a height of $0.5B$ in horizontally stratified rocks and $0.25B$ in vertically stratified rocks, where B represents the width of the tunnel. In tunnels in steeply inclined stratified rocks, the lateral force on tunnel walls can be estimated as shown in Figure 7-1.

If the bedding or cleavage planes rise at a steep angle to the horizontal, a wedge-shaped body of rock, **aed** in Figure 7-1, tends to slide into the tunnel and subjects the post at **ac** to bending. The lateral force, *P*, per unit of length of the tunnel, which acts on the post, can be estimated as indicated in Figure 7-1. The estimate is based on the assumption that the rock indicated by the shaded area to the right of **ce** in Figure 7-1a has dropped out of the roof and that there is no stress transfer along **de.** On these assumptions, the wedge-shaped

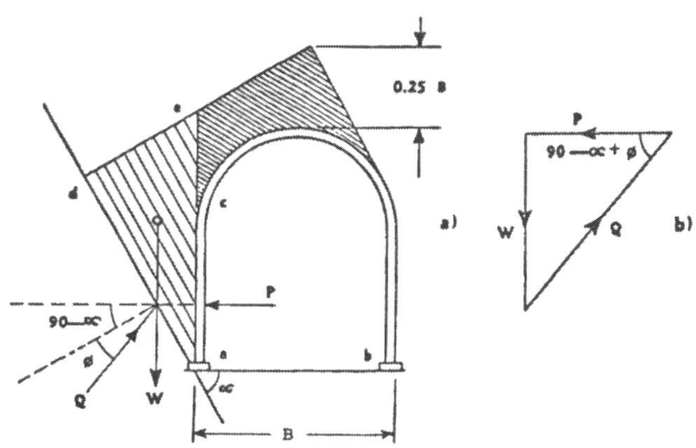

Fig. 7-1. Forces acting on tunnel; support in inclined strata (Terzaghi, 1964).

body of rock **ade** is acted upon by its weight, W, and the reaction, Q, on the surface of sliding **ad.** To prevent a downward movement of the wedge, the vertical post **ac** must be able to resist a horizontal force, P. The reaction, Q, acts at an angle ϕ to the normal on the surface of sliding **ad.** The angle ϕ is the angle of friction between the wedge and its base. The weight, W, is known. The intensity of the forces Q and P can be determined by means of the polygon of forces shown in Figure 7-1b.

The angle of friction, ϕ, depends not only on the nature of the surfaces of contact at **ad** but also on the hydrostatic pressure in the water that percolates into the space between the two surfaces. Experience with slides in open cuts in stratified rocks indicates that the value of ϕ for stratified rocks with clay or shale partings may be as low as 15 degrees. If no such partings are present, 25° seems to be a safe value.

The highest value for the unit pressure on the roof depends on the slope of the strata. For steep strata, it will hardly exceed $0.25B$, whereas for gently inclined strata it may approach $0.5B$.

Tunnels Through Moderately Jointed Massive Rocks. The ultimate load on the roof should not exceed $0.25B$. Horizontal loads are negligible except where the rock is in a state of elastic deformation, due to tectonic stresses or other causes, in which case support may be required as stated for popping rock.

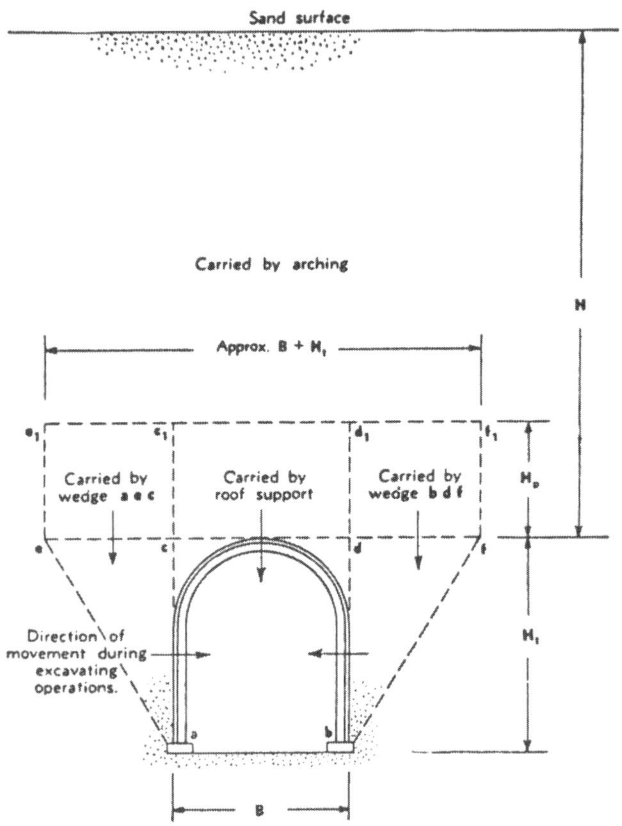

Fig. 7-2. Loading on tunnel support in sand (Terzaghi, 1964).

The hard rock tunneler must be prepared to handle crushed zones near faults that behave similarly to tunnels in cohesionless sand. Above the water table, the loads, as determined by model tests with dry sand, are affected by arch action. After the tunnel support is in place and backpacked, the load will increase about 15%. (Refer to Figure 7-2 and Table 7-2 for a summary of these loads.)

The load $H_{p\ min}$ is produced by a very small downward movement of the arch, which satisfies the deformation condition for arching. If the arch is allowed to settle more, the load increases to $H_{p\ max}$. It is important, therefore, to install and backpack the support as quickly as possible after a section of the tunnel is excavated, to keep the load as low as possible. Experience shows that the loadings on most tunnel supports are nearer the minimum than the maximum.

The average unit pressure in crushed rock zones on the sides P_h may be determined by

$$P_h = 0.30W(0.5H_t + H_p)$$

in which W is the weight per cubic foot of the rock. In crushed rock below the water table, tests on sand with percolating water indicate that H_p is about double that in the dry.

Blocky and Seamy Rock

When a new round opens up new ground, the rock temporarily supports itself as a half-dome. Immediately after blasting a round, some rock will fall from the roof between the face and the last rib of tunnel support, leaving a gap in the half-dome. If the newly opened ground is left unsupported, additional blocks will fall, and the rock in the roof will loosen at the joints with more and more rock falling out until a general breakdown of the roof occurs. The time interval between blasting and the general breakdown of the rock half-dome is called the bridge-action period, or *stand-up time.* Support must be installed before the bridge-action period expires. The earlier it is installed, the less will be the ultimate load on the support.

Lauffer (1958) has emphasized the importance of the relationship of active span (width of tunnel or distance between face and supported tunnel, whichever is less) to stand-up time (Figure 7-3; Brekke and Howard, 1972).

Table 7-2. Rock Loads in Dry Crushed Rock

Type of rock	H_p	Yield
Dense sand	$H_{p\ min} = 0.27C$	$0.01C$
	$H_{p\ max} = 0.60C$	$0.15C$ or more
Loose sand	$H_{p\ min} = 0.47C$	$0.02C$
	$H_{p\ max} = 0.60C$	$0.15C$ or more
Notes: C = width of tunnel plus height of tunnel ($B + H_t$, Figure 7-2)		

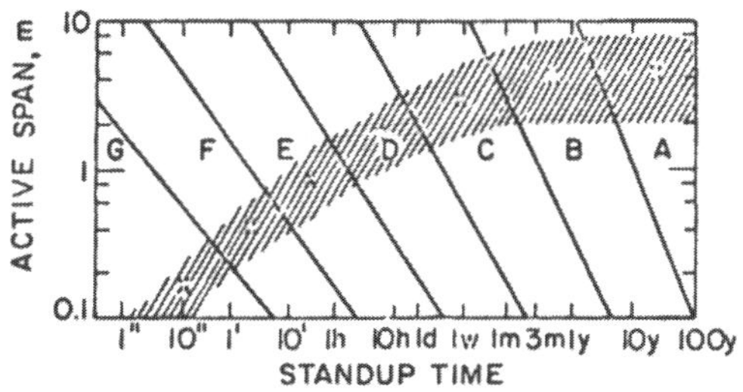

Fig. 7-3. Active span vs. Stand-up time (Brekke and Howard, 1972).

According to Terzgahi (1964), the load on the roof of a tunnel in blocky and seamy rock will increase with time and distance from the working face. Later tests made in the pilot bore of the Straight Creek Tunnel (now known as the Eisenhower Tunnel) at Loveland Pass, Colorado, showed that loads build up sharply near the working face and then taper off to lower stable loads about 200 ft from the heading (*ENR*, 1964).

As determined by tests on various railroad tunnels in the Alps, loads on the roof will be in the range shown in Table 7-3.

In squeezing ground at depths of about 300 ft, the load to be carried by roof supports can be estimated as $H_p = 2.1C$, and at depths of more than 1,000 ft, $H_p = 3.5C$, where C = width + height of tunnel in feet.

In the squeezing ground of the Eisenhower Tunnel, Zone III loads were estimated using an elastoplastic analysis that was developed in soil mechanics (Hopper, 1972). This section of the tunnel was 56 ft, 2 in. wide by 64 ft, 8 in. high. Rock cover was 700 to 1,000 ft. The analysis indicated loads of 47 kips/sq ft (ksf). The final lining was designed for 75 ksf. The maximum load measured at the time work was nearing completion was 40 ksf. This compares with Terzaghi's estimate of 42 ksf at a depth of 300 ft and 70 ksf at depths of more than 1,000 ft. The loads on the sides and floor of the tunnel are, respectively, about 1/3 and 1/2 of that on the roof. In swelling rock, pressures as high as 10 tons/ft^2 are not uncommon in deep tunnels and, in some cases, have been as high as 20 tons/ft^2 (Ledbetter, 1938).

A summary of Terzaghi's recommendations is shown in Table 7-4. In swelling rock, pressures as high as 10 tons/ft^2 are not uncommon in deep tunnels and, in some cases, have been as high as 20 tons/ft^2. Deere et al. (1967) have shown that in many cases Terzaghi's rock loads are ultraconservative. With more advanced methods of smooth wall drilling and blasting that limit the shattering of the rock beyond the tunnel periphery, and the complete elimination of blasting through the use of tunnel boring machines, the loads on supports are appreciably reduced.

A more recent engineering classification of rock for use in determining the behavior of rock in tunnels is known as *rock quality designation*, or RQD (see Chapter 4). Rock with an RQD of 25 or less is in a class with soft ground tunnels insofar as tunneling methods are concerned. It has been shown that a qualitative relationship exists between the RQD and the support required for tunnels in rock. Tentative recommendations (Deere et al., 1967) for load to be carried in tunnels 20 to 40 ft in diameter are shown in Table 7-5.

These recommendations for tunnels excavated by blasting, which are about 20% below the average estimated by Terzaghi's method, will probably be satisfactory in most cases. Rock load factors for machine-driven tunnels are about 25% below those for tunnels driven by drilling and blasting. Additional experience is needed before final values can be adopted.

Table 7-3. Rock Loads (in feet) in Blocky and Seamy Rock

Type of Rock	Initial Value	Ultimate Value
Moderately blocky rock	$H_p = 0$	$H_{p\,ult} = 0.25B$ to $0.35C$
Very blocky and shattered rock	$H_p = 0$ to $0.60C$	$H_{p\,ult} = 0.35C$ to $1.10C$

H_p = height of rock to be supported (ft)
B = width of tunnel (ft)
C = width + height of tunnel (ft)

CHANGING CONCEPTS

Review of the Terzaghi method shows clearly the basic concepts of increasing "dead" loads with increasing tunnel span (B), of greater "heights" of affected rock as formation "looseness" increases (rock condition coefficients selected), the further load increase as the rock arch abutments deteriorate ($C = B + H$), and the substantial effect of presence or absence of water.

The next major concept, RQD, was published some 25 years after Terzaghi's pioneering work (see Chapter 4). This

Table 7-4. Rock Load H_p in Feet of Rock on Roof of Support in Tunnel with Width B (ft) and Height H_t (ft) at depth of More than $1.5C$, Where $C = B + H_t$

Rock Condition	Rock Load H_p in feet	Remarks
1. Hard and intact[a]	zero	Light lining, required only if spalling or popping occurs.
2. Hard stratified or schistose[b]	0 to $0.5B$	Light support.
3. Massive, moderately jointed	0 to $0.25B$	Load may change erratically from point to point.
4. Moderately blocky and seamy	$0.25B$ to $0.35C$	No side pressure.
5. Very blocky and seamy	$(0.35$ to $1.10)C$	Little or no side pressures.
6. Completely crushed but chemically intact	$1.10C$	Considerable side pressure. Softening effect of seepage toward bottom of tunnel requires either continuous support for lower end of ribs or circular ribs.
7. Squeezing rock, moderate depth	$(1.10$ to $2.10)C$	Heavy side pressure, invert struts required.
8. Squeezing rock, great depth	$(2.10$ to $4.50)C$	Circular ribs are recommended.
9. Swelling rock	Up to 250 feet irrespective of value of C	Circular ribs required. In extreme cases, use yielding support.

[a] The roof of the tunnel is assumed to be located below the water table. If it is located permanently above the water table, the values given for types 4 to 6 can be reduced by 50%.

[b] Some of the most common rock formations contain layers of shale. In an unweathered state, real shales are no worse than other stratified rocks. However, the term shale is often applied to firmly compacted clay sediments which have not yet acquired the properties of rock. Such so-called shale may behave in the tunnel like squeezing or even swelling rock. If a rock formation consists of a sequence of horizontal layers of sandstone or limestone and of immature shale, the excavation of the tunnel is commonly associated with a gradual compression of the rock on both sides of the tunnel, involving a downward movement of the roof. Furthermore, the relatively low resistance against slippage at the boundaries between the so-called shale and rock is likely to reduce very considerably the capacity of the rock located above the roof of bridge. Hence, in such rock formations, the roof pressure may be as heavy as in a very blocky and seamy rock.

Table 7-5. Support Recommendations for Tunnels in Rock (20 to 40 ft in Diameter)

Rock Quality	Tunneling Method	Support Systems: Steel Sets[c]	Rocks Bolt[d]	Shotcrete
Excellent[a] RQD>90	A. Boring machine	None to occasional light set. Rock load 0.0 to 0.2B[c]	None to occasional.	None to occasional local application.
	B. Conventional	None to occasional light set. Rock load 0.0 to 0.3*B*.	None to occasional.	None to occasional local application 2 to 3 inches.
Good[a] 75<RQD<90	A. Boring machine	Occasional light sets to pattern on 5 to 6 foot center. Rock load 0.0 to 4*B*.	Occasional to pattern on 5 to 6 foot center.	None to occasional local application 2 to 3 inches.
	B. Conventional	Light sets, 5 to 6 foot center. Rock load (0.3 to 0.6)*B*.	Pattern 5 to 6 foot center.	4 inches or more on crown and sides.
Fair 50<RQD<75	A. Boring machine	Light to medium sets, 5 to 6 foot center. Rock load (0.4 to 1.0)*B*.	Pattern, 4 to 6 foot center.	2 to 4 inches on crown.
	B. Conventional	Light to medium sets, 4 to 5 foot center. Rock load (0.6 to 1.3)*B*.	Pattern, 3 to 5 foot center.	4 inches or more on crown and sides.
Poor[b] 25<RQD<50	A. Boring machine	Medium circular sets on 3 to 4 foot center. Rock load (1.0 to 1.6)*B*.	Pattern, 3 to 5 foot center.	4 to 6 inches on crown and sides. Combine with bolts.
	B. Conventional	Medium to heavy sets on 2 to 4 foot center. Rock load (1.3 to 2.0)*B*.	Pattern, 2 to 4 foot center.	6 inches or more on crown and sides. Combine with bolts.
Very poor RQD<25 (Excluding squeezing or swelling ground)	A. Boring machine	Medium to heavy circular sets on 2 foot center. Rock load 1.6 to 2.2*B*.	Pattern, 2 to 4 foot center.	6 inches or more on whole section. Combine with medium to heavy sets.
	B. Conventional	Heavy circular sets on 2 foot center. Rock load (2.0 to 2.8)*B*.	Pattern, 3 foot center.	6 inches or more on whole section. Combine with medium sets.
Very poor (Squeezing or swelling)	A. Boring machine	Very heavy circular sets on 2 foot center. Rock load up to 250 feet.	Pattern, 2 to 3 foot center.	6 inches or more on whole section. Combine with heavy sets.
	B. Conventional	Very heavy circular sets on 2 foot center. Rock load up to 250 feet.	Pattern, 2 to 3 foot center.	6 inches or more on whole section. Combine with heavy sets.

Note: From Deere et al. (1967)

[a] In good and excellent quality rock, the support requirement will in general be minimal but will be dependent upon joint geometry, tunnel diameter, and relative orientations of joints and tunnel.

[b] Lagging requirements will usually be zero in excellent rock and will range from up to 25% in good rock to 100% in very poor rock.

[c] *B* = tunnel width

[d] Mesh requirements will usually be zero in excellent rock and will range from occasional mesh (or straps) in good rock to 100% mesh in very poor rock.

concept accelerated the already developing concept of improving the ability of rock to support itself reliably, i.e., rock reinforcement. It has taken another 25 years to develop fully and assimilate the theory and practice of rock reinforcement to its current state. Realization that it is only actual internal movement in the rock mass resulting from excavating the tunnel that produces subsequent loading has increased during the last period. Unfortunately, the concept has yet to be fully and universally assimilated into the overall design process. The latest concept is that of "tandem" linings wherein the frequently excess capacity of a cast-in-place concrete tunnel lining can be used retroactively to reduce initial rock reinforcement requirements (see Chapter 5).

Because only actual movements create loads, it follows in most cases that reduction of internal movement reduces the resistance man must supply to achieve stability of the rock mass. That reduction can be accomplished by increasing internal friction in the mass or by reducing the amount of space permitting movement. Rock dowels—either friction or encapsulated—accomplish the former. Shotcrete accomplishes the latter. Initially, filling the interstices between rock fragments on the perimeter reduces space otherwise created by raveling; then membrane action as the shotcrete cures adds strength and composite action. (A third method—using multiple drifts to provide incremental support that changes the reaction of the mass—has been used for scores of years.)

The procedure of rock reinforcement changes the design characteristics of the rock mass from passive to active and results in significant tunneling economies.

To take advantage of rock reinforcement, considerably more must be known about the characteristics of the rock mass than was needed with the Terzaghi method. It should be noted that in 1941, rock mechanics was not even in its infancy—it was only gestating. Overall, geotechnology and tunnel construction know-how are now inextricably linked.

ROCK DISCONTINUITIES

Advances in heading mechanization have voided the old definition of rock (i.e., "can be excavated readily only by drilling and blasting") and made the equally old "cannot be excavated readily only by pick and shovel" more appropriate. The supposedly precise "ground having a strength of more than 150 psi" is also invalid, as rock excavation methods have been used in ground with specimen tests of less than 50 psi. For present purposes, a definition such as "ground whose reaction to mined excavation is controlled primarily by distinct discontinuities," while insufficient, covers the broad range considered here.

Types

Principal discontinuities are fractures and bedding planes. Fractures result from cooling of magma, tectonic action such as seismic events, less dramatic natural phenomena such as formation of synclines and anticlines, or other geologic stresses. An example of the last is the Balcones fault zone in Texas, where massive formations of Austin chalk were broken as a result of bending created by centuries-long regional consolidation of softer underlying materials. Bedding planes frequently control the reaction of otherwise strong material such as limestone. These bedding planes are layers of weaker materials, sometimes only a fraction of an inch thick, creating a definite discontinuity, interspersed between thicker layers of more competent sedimentary material.

Faults

To a geologist, faults are any fracture showing relative displacement. Typically, they are the result of ancient or geologically recent seismic activity at great depth. They are categorized in several ways. Strike slip faults have essentially horizontal movement between the two sides, while "dip slip" indicates vertical or near vertical primary movement. Thrust faults are somewhat similar to dip slip, with the upper mass moving over the lower at a relatively shallow angle to the horizontal. If the upper mass drops, it is a *normal* fault; *reverse* indicates the opposite movement. Its surface and near-surface movements are especially disruptive.

Only those faults showing major displacement are of concern to the tunnel engineer. These must be considered individually and necessary stabilization provided on each side of the fault. Frequently, it is the damaged rock ("gouge") on each side that is the primary concern. This "fault zone" varies from a few inches to many feet. Additionally, the active fault zone—the width within which the actual rupture may occur—may be narrow, as for the San Andreas fault, or very wide, as for the nearby Hayward fault zone (less than 15 mi away), which may rupture anywhere within a zone that is more than 500 ft wide in some locations.

A special hazard of faults is that movements may have juxtaposed a dry and tight formation against a heavily water-bearing formation. (The fault gouge on one side of the fault is also frequently heavily water bearing.) If the water condition has not been anticipated, the resulting inflow may destroy the heading.

Joints

Fractures along which there is no evident displacement are commonly termed joints. These joints may be continuous or discontinuous and occur randomly or at regular intervals. Repetitive parallel joints are called sets. Typically, sedimentary rocks have three sets, one parallel to the bedding plane, the others intersecting approximately at right angles. Joints are defined spatially by their strike and dip. Strike is the direction of a contour along its surface and referenced by its azimuth; dip is its angle of inclination referenced to horizontal, measured normal to the strike. Strike and dip are important because they determine both stability (design) and overbreak (construction).

Joints have several characteristics that are important, especially in rock reinforcement design. Continuous and discontinuous have been mentioned; discontinuity causes only a very localized problem, if any. Another characteristic is general joint shape. Planar joints may cause problems; curved generally do not because movement along the joint is inhibited by the weight of overburden that must be overcome before movement can occur. Joint roughness is another characteristic. Roughened surface joints must shear through the rock asperities before significant movement occurs. Roughness depends on size of the asperities and can by measured by its initial and residual angles of friction. Smooth joints permit movement easily. Slickensides are the result of movement polishing the surface. In addition to creating a geologist's fault, slickensides permit movement most easily of all. Interlocked joints (i.e., joints so rough that jagged offsets are frequent) resist movement strongly.

Joint alteration is another important characteristic. Alteration may be caused by weathering over time or may be of chemical origin. In some cases, deposits of minerals or chemical action may result in healed joints as strong as the original rock. In others, excessive weathering or joint infiltration may result in clayey deposits that markedly reduce or eliminate resistance to movement.

Bedding planes are the result of change in the material deposited. The thickness may be only a small fraction of an inch but still be sufficient to reduce or eliminate the transfer of shear stress between successive competent layers. Bedding planes may be feet apart in limestones and sandstones, but less than an inch apart in shales.

ROCK MOVEMENT

In structural theory for surface steel design, practitioners work easily with the elastic portion of the stress–strain "curve" and worry only when necessary about how much, if any, of the plastic portion can be used. Underground, however, in the design of the principal structural element (the rock arch or rock shell as appropriate), the elastic movements are taken for granted and are insignificant, except for a few special cases. (See the discussion of squeezing in Chapter 8.) Design then consists of evaluating the erratic and irregular movements of joints and blocks and how they can be controlled. "First crack" is essentially meaningless except for the initial shear value changing to the residual value, and collapse does not occur until stressed blocks are crushed, which is rare, or moved substantially out of position in the matrix, which is not as rare.

Types of Movement

Movements can be divided into frictional, or those with high resistance where shearing of rock asperities is involved, and sliding, where resistance is low. Friction is supplied by the shape and surface of the joint; the amount of jointing is also a factor. Joint shapes can be, from highest to lowest resistance, discontinuous, curved (or undulating), or planar. Similarly, joint surfaces can be rough, smooth, or slickensided.

Sliding movements are those where intrusive material separates rock fragments or where the rock itself has been altered to lower resistance. Examples of joint-altering effects are healed (as if no joint), unaltered (clean or sharp), and altered (nonsoftening coatings or particles to softening degradations or intrusions). The amount of separation between the intact rock walls is a major factor here.

Other Factors

In addition to the above, presence of water, in situ stress conditions, and special zones of weakness must be considered in evaluating the rock mass. At the minimum, free water in the discontinuities acts as a lubricant and frequently, whether by pressure or quantity, is a significant tunneling deterrent. Disintegrated rock or clay seams or low confining (near-surface) stress or high in situ stresses detract from the rock capacity and must be considered when evaluating conditions. Finally, the purpose of the mined opening affects the amount of movement that can be tolerated, i.e., size of the safety factor; obviously, a temporary tunnel does not need the same reinforcing as a major transportation artery.

WATER

Water is found in almost all rocks in the upper limits of the earth's crust, where tunneling takes place. Joints in the rock, even if tight, permit water to penetrate by capillary action. The larger the openings of the joints, the more water that can enter a cavity (such as a tunnel). When a tunnel is mined, friction in the joints helps to hold the individual blocks in place. When the joints are dry, the rock may stand, whereas if they later become wet during a rainy season, the roof and sides of an unsupported tunnel might fail. Except in well-keyed formations, water, regardless of quantity, causes extra problems.

Because joints in rocks are usually more open the closer they are to the surface, less water is likely to be found in deeper tunnels. Below 100 or 150 ft, joints tend to be tight and produce little water. This is not always true, however. The San Jacinto Tunnel in Southern California produced flows of 40,000 gpm at depths of more than 800 ft, 15,800 gpm near a single heading and pressures up to 600 psi (Ledbetter, 1938).

The excessive water flows in the San Jacinto were stopped by building a concrete bulkhead in the tunnel and grouting behind it. This method has been used successfully in several cases, but it is costly because of delay to tunnel progress. In limestone formations, water channels are sometimes formed by the solution of the limestone, creating large caverns such as the Mammoth Cave of Kentucky. Large streams can flow in such caverns. Sometimes caverns or faults in rock formations are filled with sand saturated with water. These cause some of the most difficult tunneling conditions. If an otherwise hard rock tunnel intersects such a sand formation, it may stall progress for many months unless anticipated and pretreated. Many feet of the tunnel previously driven may be filled with sand, as in the Litani Tunnel in Lebanon (*ENR*, 1960). When that tunnel advanced into a loosely cemented sandstone, a flow up to 95,000 gpm filled 9,800 ft of the tunnel with 130,000 yd^3 of sand before finally receding to 3,000 gpm. The work was stopped for two years.

The worst kind of water in a tunnel is hot water. The Tecolote Tunnel in the Coastal Range in southern California encountered water temperatures of 112 degrees F. Personnel were transported through the hot water in muck cars filled with cold water (*ENR*, 1954). The Graton Tunnels in Peru had high water flows (134,000 gpm), very hot water (156°F), and hot rock temperatures. Two tunnels, each 10 ft wide by 13 ft high, were driven 60 ft on centers. Both were used for ventilation; additionally, one was used for haulage, and the other, 6 ft lower, was used for drainage. It was only by transferring water from one tunnel to another that progress was possible. An elaborate refrigerated cooling system was installed in a station near the hot water area to make it possible for personnel to work. As soon as possible after excavation, the hot water and hot rock areas were lined with concrete with arrangements to drain the water from the haulage tunnel into the drainage one (*ENR*, 1964, 1970; Kincaid, 1970).

Construction Groundwater Control

Fortunately, the examples cited above are rarities and illustrate extreme conditions. Water, however, is one of the most troublesome obstacles to tunnel advance. Most tunnels have some water inflow. Even a small amount in a tunnel driven downgrade collects at the face and is a nuisance. Appreciable amounts at the face interfere with progress regardless of grade.

An appreciable volume of flow is not required for some problems to become threatening. Among these are ground susceptible to air or water slaking, swelling ground, invert softening, and very flat grades. The first two can be handled by requiring that a protective coating be applied to such ground soon after exposure. A thin layer of shotcrete will suffice and can be included as part of the initial ground support. Invert softening is a problem only in the softest rocks and lenses of heavily altered rock. Water supply tunnels sometimes and wastewater tunnels generally are driven on very flat grades, i.e., less than 0.2% (Terzaghi, 1964). This can create a housekeeping problem, primarily for rail transport. Spillage from muck cars can create small dams, long sags can result from lack of profile control, or the flow quantity can require excessive water depth. The resultant ponding may obscure the trackbed or even the rails, making proper maintenance difficult and causing time-consuming derailments.

Designs frequently stipulate a granular drainage layer beneath and across the entire concrete invert. This can lead to construction gravity drainage problems (above). An alternative of two side channels, each less than a quarter of the overall invert width, would suffice for the long term, would reduce the haulage-way water problem during construction, and should be evaluated for cost and benefit.

The most obvious problem is inflow in excess of reasonable capacity for gravity drainage or economic pumping cost. This also indicates ground conditions that will create long-term problems. Therefore, the designer should require corrective action that will solve the construction problem and reduce the long-term pressure problem.

Construction Aspects

Normally, handling water during construction is considered a routine contractor's problem. This is acceptable when driving upgrade until the water flowing downgrade begins to erode or otherwise deteriorate the invert, or where the combination of gradient and quantity interferes with servicing the heading. When driving downgrade, pumping is required if work is to be done efficiently. The limit of acceptable inflow here depends on costs of pumping and decreased crew efficiency.

When heavily water-bearing strata are known or suspected to exist at tunnel elevation, consideration should be given to specifying a pregrouting program in the contract documents, as grouting will undoubtedly be required for long-term protection as well. A contractor needs to reduce flows only enough to get through the zone, whereas long-term control probably requires much more complete shutoff. Unfortunately, the location, extent, and other necessary data are rarely known with any exactness. Then, an exploratory program performed at the heading is a prerequisite. The pregrouting program must be presented in such a way that it will be executed only if needed, can be changed to suit the conditions actually disclosed, and necessary work will be performed with minimum delay to the project.

The exploratory program should begin before the suspected zone is reached. At least one probe or "feeler" hole should be required to be kept well ahead of the face. Drilling the holes may result in delay time for tunnel boring machines (TBMs), but not necessarily for drill-and-blast operations. When drilled, the probe should extend at least 100 ft ahead of the face. When its tip is only 30 to 50 ft ahead of the face or the probe becomes ineffective, a new hole must be drilled. When a probe hole shows a significant increase in flow, the relative seriousness is immediately apparent. It may be advisable to determine its beginning location and the approximate quantity under gravity flow and hydrostatic head as well.

Additional probe holes may be necessary to determine lateral and vertical extent, among other things. If conditions are severe enough, the advance must be halted and the ground ahead grouted to avoid a sudden inrush that would threaten both safety and ground stability. If it is known or determined that the condition will persist for a considerable distance, it is advisable to require periodic drill/grout niches in the tunnel sidewalls so that tunnel advance will not be halted completely.

Additional discussion of water problems will be found in the later sections on concrete lining and leakage.

During a tunnel's service life, infiltration may be a problem. The maximum probable hydrostatic head must be determined. For an unlined tunnel, the reduction in head and inflow quantities can be estimated with flow net procedures. Where inflow must be restricted for whatever reasons, the problem becomes more complicated. The worst-case scenario calls for resisting the unreduced hydrostatic head. In tight formations, however, the unlined inflow from even substantial heads may be within the established infiltration limits, and water pressure is not a design factor. As the formation becomes more pervious or the head increases, the flow must be restricted. A revised flow net allowing the limiting inflow will establish the actual hydrostatic head to be resisted. A hydrostatic head of 50 ft above springline should pose no problem for a 4-in. continuous shotcrete lining in a 20-ft rock tunnel. Either thicker stabilization lining or leaving drainage windows in the shotcrete and adding a secondary liner later will be necessary at some point.

FORMATION GROUTING

Two basic types of grouting, contact and formation, are used in rock tunnels. Contact grouting is low-pressure grouting, less than 30 psi, performed after the final concrete lining has cured. Filling the voids behind the lining serves both groundwater control and structural functions. See discussions later in this chapter under "Cast-in-Place" Linings and "Leakage."

Formation, or consolidation, grouting also provides groundwater control and structural functions. Contact grouting improves the performance of the lining. Formation grouting does the same for the rock mass. When the problem is insufficient stand-up time, even with a multidrift approach, grouting can improve the mass stability sufficiently that support can be erected. If the problem is water inflow in quantities that essentially prevent heading work, grouting can reduce the inflow sufficiently to permit work. Unless ground conditions permit local sealing of fractures and the like, wherever they are exposed, to prevent excessive escape of grout into the excavation, the grouting must have been completed before mining begins.

Where practical, the grouting should be done from the surface well ahead of the advancing face. If the tunnel is deep or the surface too congested for access, the operation must be carried out from within the tunnel. Minimizing production mining downtime then becomes important. Use of niches or local enlargements for grouting stations on long reaches permits maintaining at least partial production. Driving the first drift of a multidrift tunnel early to permit its use for advance grouting is also cost-effective.

Chapter 6 discusses permeation grouting, grouting plant, and the like for soft ground. Much of that discussion is applicable to rock tunnels as well. The principal difference between soft ground and rock for water control grouting is that the former generally requires treating a porous water-bearing medium, whereas in the latter the groundwater is confined to discrete channels (fractures, contacts, etc.) between impervious blocks. Fissures less than 0.01 in. wide can require grouting if subject to significant hydrostatic head. Interconnected channels increase the amount of drilling and grouting required.

The grout zone will extend a considerable, but not readily predeterminable, distance from the tunnel. A variety of grout components, admixtures, set and gel times, viscosities, and pumping pressures are employed. Cement grouts, the most common type, range from thick, sanded grouts pumped at moderate pressures for large fractures and extensive voids to thin, less viscous, neat cement grouts pumped at the higher pressures necessary to achieve penetration in more heavily fissured but less open formations. While cementitious grouts are generally adequate to improve rock mass stability, chemical grouts also are necessary when water control is the primary objective. Viscosities only slightly greater than that of water, and gel times accurate to within a few seconds can be obtained.

A second stage of grouting is frequently necessary because of gaps left by first-stage work. The thicker, more viscous grouts will not penetrate the smaller voids and, upon setting, may leave additional small, continuous voids. Depending on conditions, grouts may be required to penetrate into, around, and beyond the first-stage plugs. On occasion, a third stage may also be required.

Grouting is a schedule-consuming and expensive process. Accordingly, contractors will use only the minimum amount necessary to facilitate mining through the troublesome zone. However, inflow sufficient to disrupt lining placement locally may remain. When this condition can be anticipated during design, a mandatory and specific advance grouting program should be prepared.

ROCK REINFORCEMENT

The two types of rock reinforcement are rock bolts and shotcrete.

Rock Bolts

Rock bolt is the general term that includes rock dowels and cable tendons. Specifically, bolts are pretensioned, while dowels are initially unstressed. Originally, the prestress was considered necessary to increase the internal friction in the rock mass. However, it was soon realized that any movement in the mass would stress the untensioned dowel as well as be inhibited by the dowel. As a result, the economy and simplicity of the dowels reduced the use of bolts to special situations, such as very narrow pillars where the additional confinement provided by the pretension force is considered necessary.

The use of tendons is limited to long distances between anchorage and excavated surface; for example, an insufficient thickness of sound rock overlain by a substantial thickness of incompetent rock can be supported and used by anchoring it to a second, overlying layer of competent rock. Similarly, incompetent rock at roof level can be supported by adding a layer of shotcrete to the tendon support. At times it is advantageous to install the tendons from a higher-level auxiliary drift before opening the main excavation.

The earliest bolts were merely steel bars with a split end and a steel wedge for anchorage, and end hardware consisting of a bearing plate and nuts at the rock surface end. The bolt was anchored by driving the bar onto the wedge until it was held firmly in place.

The capacity of such bolts was highly indeterminate and therefore unsuitable for the design of civil engineering structures. The expansion bolt was developed to provide a reliable anchorage. The anchorage hardware consists of an expansion cone and mating shell. Turning the bolt expands the shell until the rock capacity is reached. Longer or multiple cone and shell anchors can be used. Standard end hardware supports the rock surface.

It is necessary to fill the annulus between bolt and drill hole to avoid the possibility of corrosion if the bolt is to be considered permanent and a part of permanent structure. The hollow, groutable expansion bolt was developed to fill this requirement. The hole in the faceplate was changed from circular to keyhole shape to permit entry of a short plastic tube. As grout was pumped in, any entrapped air escaped, ensuring complete encapsulation. When not grouted, this type of bolt is removable and useful where drifts are to be enlarged later, because it eliminates steel from the muck.

The search for economical installation procedures resulted in the development of epoxy resin cartridges to eliminate both the anchorage hardware and the extra grouting process. The cartridges have an internal membrane separating the two components. When the components are mixed, they harden in well-defined time increments. This permits use with either dowels or bolts. The installation process is simple: drill the hole; insert the requisite number of cartridges; insert the steel bar with a spinning action and continue spinning for about 60 seconds to ensure complete mixing. Setting of the mixed material occurs shortly thereafter. It should be noted that after years of usage, the long-term performance of the resin has been called into question. As a result, there is movement toward returning to cement grouting. Cement emulsion cartridges have been available for some time but have not been widely used to date. Both types suffer somewhat from the plastic membrane of the cartridge remaining in the hole.

Permanent bolts are not always required. Economical alternatives have therefore been developed. The generic name of friction bolts (or dowels) has been adopted. The most common types are trade-named Swellex and Split-set. Both are made of thin metal. Swellex comes undersized (relative to drill hole) with one half folded into the other. The bolt is expanded to full size and tight to the rough hole surface by applying hydraulic pressure on the order of 5,000 psi. The Split-set is furnished oversized and slotted. It is forced into the drill hole, shrinking it to proper size and ensuring a tight fit. Friction bolts are considered temporary because they are not grouted and their thin walls are vulnerable to corrosion. However, the economy inherent in both material and labor highly recommends them for use where appropriate.

Size/length combinations are limited for both friction types. There are some differences in action between the two. Split-sets act entirely in friction. This may provide a desired limited yielding as the anchor yields gradually at high loads. The pressure required to set the Swellex bolt molds it to the unevenness of the hole circumference; this permits less slippage and essentially fixes the bolt.

Rock Bolting Details. The structural generality that failure occurs more often because of inadequate details than of errors in major design concepts applies to rock reinforcement as well.

Rock arches develop naturally as the rock deforms, moves, ravels, etc., after an opening is excavated and the rock attempts to reestablish stability. Resulting openings are controlled and the amount of excavated materials minimized by tying the rock mass together at regular intervals with rock dowels. Such a design is applicable to either TBM or drill-and-blast excavation.

If the dowel capacity is divided by the tributary profile area, a uniform internal "confining" pressure on the order of 5 to 10 psi results. This generally would be insufficient to stabilize the ground according to the classical concept. The answer of course is that the constraint has reduced internal movements, thereby reducing the load. Additionally, the constraint helps the rock resist more of its self-load.

Rock arches can also be formed or augmented with a shotcrete membrane. The smaller the typical fragment is, the more effective shotcrete is relative to bolting. Membrane action of the shotcrete is not the only resisting element mobilized. Shotcrete fills interstices between fragments and creates a larger element that resists movements outside its own dimension. As movement beyond the second element is constrained, additional fragments are held in place and are thereby enabled to take load. Thus a small element, shotcrete, exerts an active resistance and increases stability far beyond what its thin dimension would indicate.

The most effective arch reinforcement occurs when rock bolts and shotcrete are combined. The rock bolts act as described in Chapter 12; the shotcrete supports and activates the lower "dead" areas. The total result is a significantly thicker arch than either system alone produces.

Rock beams similar to laminated timber beams can and should be created in flat or shallowly dipping sedimentary formations. Attempts to create an arch of nominal span in limestone frequently result in small falls of thin wedges and a resulting corbeled appearance. This is as true for TBM excavations as for drill-and-blast. Where individual beds are sufficiently thick, the arch may be formed; however, stresses within the rock will be from beam, not arch, action.

Figure 7-4 is an example of a reinforced rock beam. Shear along a parting is generally the dominant consideration

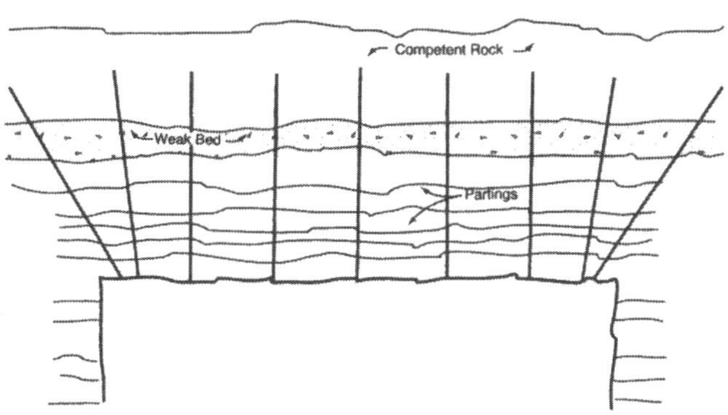

Fig. 7-4. Reinforced rock beam.

because shear capacity of the rock is only a fraction of compressive strength and shear strength along a parting is normally less than in the intact rock. Moment and shear capacities are both increased by the clamping action of the dowels. The end dowel should extend well into the abutment, and the adjacent dowel should split the angle change. Key areas are at the wall and in the block over the wall rock. While the shear varies across the span, a uniform spacing such that total beam shear is developed is more practical.

Wall wedges are frequent problem areas. A simple analysis of the mechanics provides necessary design data. When the peak friction along the failure joint is less than the paralleling component of the weighted block, dowels must be used to tie the wedge onto the support block and to supplement the residual joint friction.

While only a short stub of bolt need extend across the joint to develop shear resistance of the bolt, the tension supplement requires full anchorage. Cement grout anchorage requires on the order of 3 ft; resin anchorage will be shorter. However, both the skew of the joint strike to the tunnel axis and the dip must be considered when determining additional anchorage length and length of pattern bolting required.

Shotcrete

Shotcrete is discussed in detail in Chapter 12. It functions as rock reinforcement in two distinct ways. Being applied to the rock at high pressures, it forces its way into the spaces between intact rock pieces. With final set occurring within a very few minutes (and initial set sometimes in seconds), it effectively prevents raveling or falling of individual pieces, thereby eliminating the nil confining pressure at the surface and constraining the movements within the mass. It gains strength rapidly (typically 150 psi in 30 min, 700 psi in 8 hours), permitting it to begin functioning quickly as a membrane and thereafter gaining strength as the newly confined rock struggles to obtain a new equilibrium condition.

CURRENT CONCEPTS

Terzaghi's 1941 work on qualitative evaluation of geologic conditions and associated rock loads was discussed previously. Deere's 1967 quantitative assessment of rock quality (RQD) and the modification of Terzaghi's loads for both TBM and blasting excavation were also described under "Classical Concepts." RQD has since become a cornerstone for geotechnical analysis of rock tunnel conditions.

Three works in 1972–1974 provided further refinements and quantitative procedures for assessing rock conditions for proposed tunnels. The Rock Structure Rating (RSR), the geomechanics classification (Rock Mass Rating, or RMR), and the Q System all consider essentially the same geologic factors, but some are explicit in their handling while others are implicit or hidden within other factor determinations. All three expressly consider joint frequency, joint condition, and groundwater regime. Two of the three consider also consider

RQD, intact rock strength, and joint orientation. Only one considers differences in excavation method, and then by use of an auxiliary rating multiplier. Only one considers the relative importance of projects for various types of facilities. All produce auxiliary information useful in other calculations.

Superficial use of any one just because quantitative numbers are assigned can make obtaining a rating easy but dangerous. The texts describing all three are filled with special notes and conditions. Careful study of the full description of each, not just the summaries provided herein, is necessary if a reliable rating and consequent design is to be obtained.

Rock Structure Rating

A study of data for 190 sections from 53 tunnel projects, together with various practical and empirical applications relating to tunnel construction, has resulted in a correlation of Rock Structure Rating and rib ratio to determine support requirements (Wickham et al., 1972, 1974). RSR is a number based on geologic conditions usually known during the design stage and prior to construction. Six conditions are combined into three parameters and given maximum values (Table 7-6).

Parameter A (Table 7-7) is a general appraisal of rock structure through which the tunnel is to be driven. Parameter B (Table 7-8) is related to the joint pattern and direction of drive. Parameter C (Table 7-9) takes into account (1) the overall quality of rock as indicated by the numerical sum of values assigned to parameters A and B; (2) the condition of the joint surfaces; and (3) the anticipated amount of water inflow.

The RSR value is the numerical sum of parameters A, B, and C and will range from a minimum of 19 to a maximum of 100.

The rib ratio (RR) is the ratio of actual support required to that developed from Terzaghi's formula for determining roof loads for loose sand below the water table, expressed as a percentage. Terzaghi's empiricial formula for maximum roof load for loose, cohesionless sand below the water table (datum condition) is

$$P = 1.38(B + H)\gamma$$

where:

P = load per square foot (lb)
B = width of tunnel (ft)
H = height of tunnel (ft)
γ = unit weight of sand (assumed 120 pcf)

The rib ratio developed from the studied tunnels is shown in Figure 7-5.

Spacing of ribs for circular tunnels or tunnels with approximately the same height as width, using the RSR method, would be determined by dividing the spacing shown in Table 7-10 by the rib ratio taken from the RSR–RR curve in Figure 7-5 and multiplying by 10.

The rib ratio shown in Figure 7-5 reflects the condition for tunnels excavated by drilling and blasting. Information available for tunnels excavated by boring machines was insufficient to determine a correlation of support requirements between the two methods. It is well known that support re-

Table 7-6. Rock Structure Rating Concept Parameters

Parameter	Conditions	RSR
A	Rock type and folding or discontuniuities	30
B	Joint pattern and joint orientation	45
C	Water inflow and joint condition	25
	RSR value, maximum	100

Table 7-7. Rock Structure Rating, Parameter A, General Area Geology*

MAX. VALUE 30

BASIC ROCK TYPE	HARD	MED.	SOFT	DECOMP
Igneous	1	2	3	4
Metamorphic	1	2	3	4
Sedimentary	2	3	4	4

	GEOLOGICAL STRUCTURE			
	MASSIVE	SLIGHTLY FAULTED OR FOLDED	MODERATELY FAULTED OR FOLDED	INTENSELY FAULTED OR FOLDED
Type 1	30	22	15	9
Type 2	27	20	13	8
Type 3	24	18	12	7
Type 4	19	15	10	6

*From "Ground Support Prediction Model: RSR Concept," *Second North American Rapid Excavation and Tunneling Conference* 1:691–707, AIME, 1974.

Table 7-8. Rock Structure Rating, Parameter B, Joint Pattern, Direction of Drive*

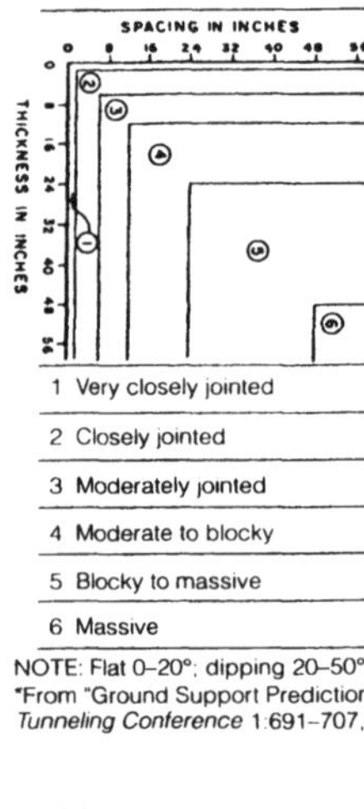

MAX. VALUE 45

	STRIKE PERPENDICULAR TO AXIS					STRIKE PARALLEL TO AXIS		
	DIRECTION OF DRIVE					DIRECTION OF DRIVE		
	BOTH	WITH DIP		AGAINST DIP		BOTH		
	DIP OF PROMINENT JOINTS					DIP OF PROMINENT JOINTS		
	FLAT	DIPPING	VERTICAL	DIPPING	VERTICAL	FLAT	DIPPING	VERTICAL
1 Very closely jointed	9	11	13	10	12	9	9	7
2 Closely jointed	13	16	19	15	17	14	14	11
3 Moderately jointed	23	24	28	19	22	23	23	19
4 Moderate to blocky	30	32	36	25	28	30	28	24
5 Blocky to massive	36	38	40	33	35	36	34	28
6 Massive	40	43	45	37	40	40	38	34

NOTE: Flat 0–20°; dipping 20–50°; vertical 50–90°.
*From "Ground Support Prediction Model: RSR Concept," *Second North American Rapid Excavation and Tunneling Conference* 1:691–707, AIME, 1974.

Table 7-9. Rock Structure Rating, Parameter C, Groundwater, Joint Condition*

MAX. VALUE 25

	SUM OF PARAMETERS $A + B$					
	13–44			45–75		
ANTICIPATED WATER INFLOW (GPM/1000')	JOINT CONDITION					
	GOOD	FAIR	POOR	GOOD	FAIR	POOR
None	22	18	12	25	22	18
Slight (<200 gpm)	19	15	9	23	19	14
Moderate (200–1000 gpm)	15	11	7	21	16	12
Heavy (>1000 gpm)	10	8	6	18	14	10

NOTE: Joint condition—good = tight or cemented; fair = slightly weathered or altered; poor = severely weathered, altered or open.
*From "Ground Support Prediction Model: RSR Concept," *Second North American Rapid Excavation and Tunneling Conference* 1:691–707, AIME, 1974.

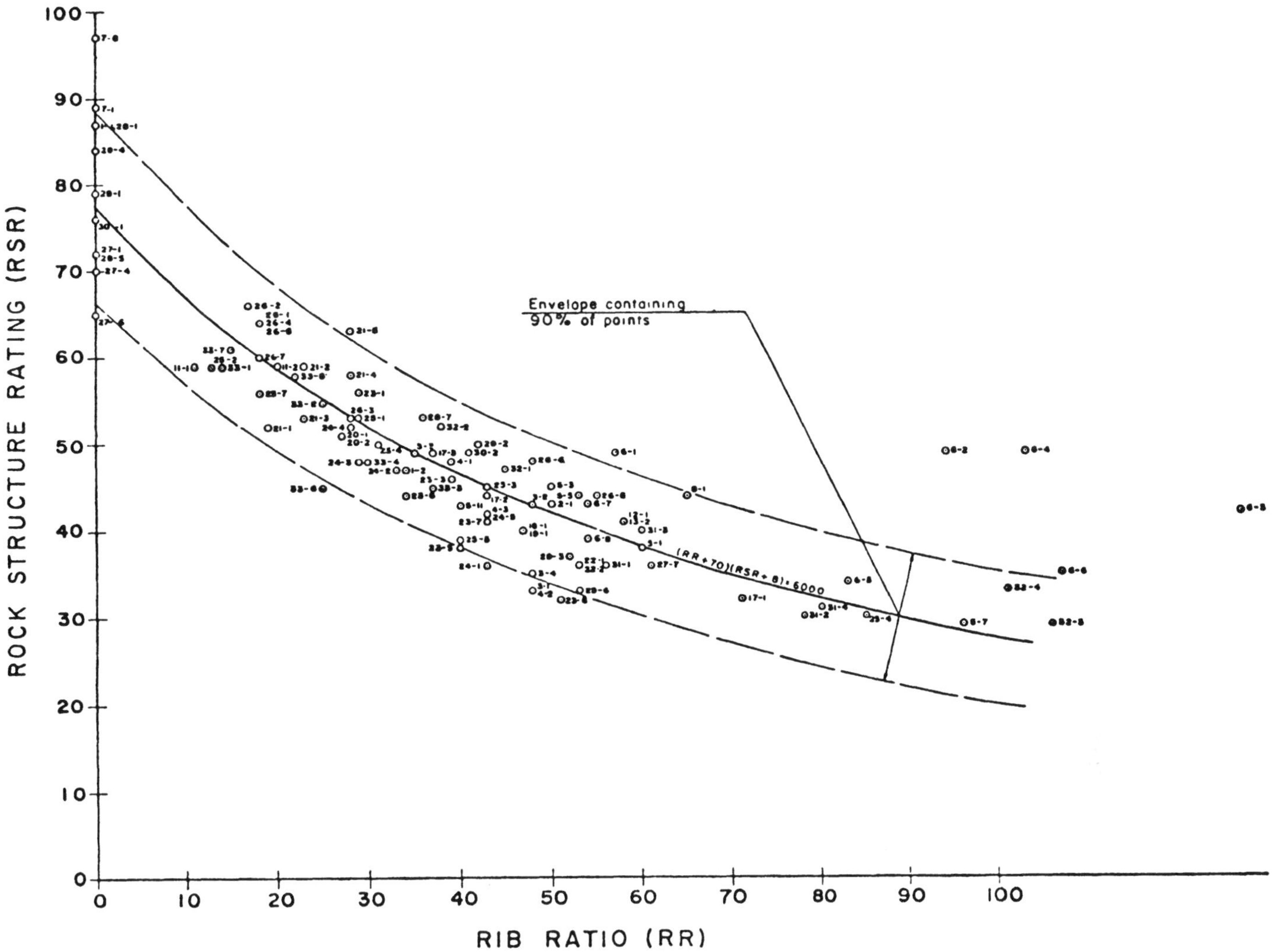

Fig. 7-5. Rock structure rating; correction of RSR and RR.

quirements for a TBM tunnel are less than for one excavated by drilling and blasting. From available data, it is suggested that, for TBM tunnels, the RSR value should be adjusted upward by multiplying by a factor of 1.2 for a 10-ft tunnel, 1.15 for a 25-ft tunnel, and 1.05 for a 30-ft tunnel.

RSR values can be expressed in terms of unit loads for various sized tunnels, as shown in Table 7-11. The correlation between RSR and rock loads can be extended to show the general relationship between the ribs, rock bolt, and shotcrete types of support (as shown in Figure 7-5).

Table 7-10. Theoretical Spacing (ft) of Typical Rib Sizes for Datum Condition

	TUNNEL DIAMETER										
Rib Size	10	12	14	16	18	20	22	24	26	28	30
417.7	1.16										
4H13.0	2.01	1.51	1.16	0.92							
6H15.5	3.19	2.37	1.81	1.42	1.14						
6H20		3.02	2.32	1.82	1.46	1.20					
6H25			2.86	2.25	1.81	1.48	1.23	1.04			
8WF31				3.24	2.61	2.14	1.78	1.51	1.29	1.11	
8WF40					3.37	2.76	2.30	1.95	1.67	1.44	1.25
8WF48						3.34	2.78	2.35	2.01	1.74	1.51
10WF49								2.59	2.22	1.91	1.67
12WF53										2.19	1.91
12WF65											2.35

Table 7-11. Correlation of Rock Structure Rating to Rock Load and Tunnel Diameter

Tunnel Diameter (D) (Feet)	(W_r -Rock Load On Tunnel Arch (K/sq ft)											
	0.5	1.0	1.5	2.0	3.0	4.0	5.0	6.0	7.0	8.0	9.0	10.0
	Corresponding Values of Rock Structure Ratings (RSR)											
10	62.5	49.9	40.2	40.2	32.7	21.6						
12	65.0	53.7	44.7	37.5	26.6	18.7	13.8					
14	66.9	56.6	48.3	41.4	30.8	22.9	16.8					
16	68.3	59.0	51.2	44.7	34.4	26.6	20.4	15.5				
18	69.5	61.0	53.7	47.6	37.6	29.9	23.8	18.8				
20	70.4	62.5	55.7	49.9	40.2	32.7	26.6	21.6	17.4			
22	71.3	63.9	57.5	51.9	42.7	35.3	29.3	24.3	20.1	16.4		
24	72.0	65.0	59.0	53.7	44.7	37.5	31.5	26.6	22.3	18.7		
26	72.6	66.1	60.3	55.3	46.7	39.6	33.8	28.8	24.6	20.9	17.7	
28	73.0	66.9	61.5	56.6	48.3	41.4	35.7	30.8	26.6	22.9	19.7	16.8
30	73.4	67.7	62.4	57.8	49.8	43.1	37.4	32.6	28.4	24.7	21.5	18.6

ROCK MASS RATING (RMR)

Chapter 4 discusses obtaining the basic rating and adjustment for relative joint orientation. The geomechanics classification also known as the rock mass rating (RMR) was developed by Bieniawski (1973, 1974, 1978). The tunnel is first divided into appropriate lengths of similar geologic conditions, and a basic rock mass rating then is established for each. See Part A of Table 4-20. The basic ratings are then adjusted for the relationships between direction of tunnel drive and the strike and dip of the discontinuities. Table 7-12 provides necessary guidance and descriptors. Returning to Figure 4-21, Part B sets forth the numerical reduction to the basic rating to obtain the actual RMR. (The rating has a maximum value of 100.) Parts C and D provide additional information.

Bieniawski's conclusions as to excavation method and permanent (not initial) support/reinforcement are shown in Table 7-13.

It is of passing interest to note that the RMR gives maximum importance, 30 possible points, to conditions (roughness, alteration, etc.) of the discontinuities. RQD and joint spacing together (possible 20 points each) rate a possible 40. This, in effect, increases the relative importance of RQD because higher RQDs are normally associated with fewer discontinuities. On the other hand, driving through flat dipping strata, through steeply dipping joints running parallel to the tunnel or against moderately dipping discontinuities across the tunnel, reduces the RMR by 10 or 12 points.

Q System

The *Q* System was developed by Barton, Lien, and Lunde, all of the Norwegian Geotechnical Institute (Barton et al., 1974). Methodology and necessary constants for developing the rock quality rating for this procedure are given in Chapter 4.

Originally, 38 categories of support were proposed. The current version is shown in Figure 4-7. The substantial difference in the two is primarily due to the development and widespread use of fiber-reinforced shotcrete after publication of the original. In either one, the appropriate category is determined by the *Q* value and an equivalent span dimension (see Tables 7-14 through 7-16). The development of the latter through an ESR is covered in Chapter 4.

Both major changes in support and chart usage occur in a small part of the chart (Figure 7-6).

Table 7-12. Relative Joint Orientation—Tunnels

Strike Direction	Drive Relative to Dip	Description
Perpendicular to tunnel axis	Drive with dip 45 to 90	Very favorable
	Drive with dip 20 to 45	Favorable
Same	Drive against dip 45 to 90	Fair
	Drive against dip 20 to 45	Unfavorable
Parallel to tunnel axis	Dip 45 to 90	Very unfavorable
	Dip 20 to 45	Fair
Irrespective of strike	Dip 0 to 20	Unfavorable

As in RMR, the author's recommendations are for permanent (total) support. When the support is temporary, e.g., tandem linings, the authors recommend increasing *Q* to 5*Q* or increasing ESR to 1.5 ESR.

Figure 7-6 is shown here primarily to demonstrate that support selection requires considerably more than just locating a *Q*/(equivalent span) point on a chart. For spans of 20 to 40 feet and $0.1 < Q < 10$, seven of the original support categories are involved as set forth in Table 7-14. Note that ratios RQD/J_n, J_n/J_a, and Span/ESR are also involved. In addition, but not shown, there is a P and Span/ESR relationship. *P*, from empirical studies by others, is a uniform roof pressure in psi required to provide the same support as the appropriate category. In addition, various comments (not shown) provide constraints.

Wedge Failures

Empirical methods such as RMR and *Q* provide, in effect, an intensity of support required, such as the *P* described above. The support is provided by pattern roof doweling and various thicknesses of shotcrete. This uniform arrangement generally, but not always, also protects against small or moderate wedges sliding out of the walls or falling from the arch. The simplest procedure for detecting potential failure wedges is the graphical stereonet procedure. While this method determines which combination of discontinuities can produce a failure wedge, it does not determine the all-important size or weight. Again, graphics is the simplest choice, and cross sections in three dimensions should be drawn. When the blocks are large, a more rigorous method such as found in Goodman and Shi (1985) should be used.

Reinforcement Details

The three parts of reinforcement design are size (diameter), length, and spacing. Dowels are normally standard (ACI) deformed reinforcing steel bars cut to length and threaded on one end to match the end hardware nuts; however, specially designed proprietary dowels are also quite popular. Standard practice uses ACI standard designations (i.e., #3–#9 for the 1/8-inch intervals between 3/8 and 1-1/8 in.). When special conditions warrant, the #11, #14, and the 4 in.^{2} #18S can be used. Undoubtedly the most common rock dowel size is #8 × 8 ft (i.e., a 1-in. diameter deformed bar 8 ft long). Dowels smaller than #6 (0.75 in.) should not be used for stress loads and are generally unsuitable long term. Some prefer a #9 bar (1.128-in. diameter) because of its 1.00 in.^{2} area.

Arch dowel lengths are functions of the spans and the rock conditions. Normally, the longer the span, the thicker the rock arch required. Also, there should be an interaction area between adjacent dowels to assure a reinforced rock arch and consequent participation in the rock above in the arching. Early on, dowel lengths were typically one-third or more of the span. Tests demonstrated the dowels were

Table 7-13. Geomechanics Classification Guide for Excavation and Support of Rock Tunnels (Tunnel Widths: 20–40ft, Construction: Drilling and Blasting

		Support		
Rock Mass Class	Excavation	Rockbolts* (Length 1/3 to 1/2 Tunnel Width)	Shotcrete	Steel Sets
Very good rock I RMR: 81-100	Full face. 10 ft advance.	Generally no support required except for occasional spot bolting.		
Good rock II RMR: 61-80	Full face. 3-5 ft advance. Complete support 60 ft from face.	Locally bolts in roof 10 ft long, spaced 8 ft with occasional wire mesh.	2 in. in roof where required.	None
Fair rock III RMR: 41-60	Top heading and bench 5–10 ft advance in top heading. Commence support after each blast. Complete support 20 ft from face.	Systematic bolts 12 ft long, spaced 5–6 ft in roof and walls with wire mesh in crown.	2 to 4 in. in roof and 1 in. on walls.	None
Poor rock IV RMR: 21-40	Top heading and bench 3–5 ft advance in top heading. Install support concurrently with excavation.	Systematic bolts 12–15 ft long, spaced 3–5 ft in roof and walls with wire mesh.	4 to 6 in. in roof and 4 in. on walls.	Light to medium ribs spaced 5 ft where required.
Very poor rock V RMR: <20	Multiple drifts. 1.5–3 ft advance in top heading. Install support concurrently with excavation. Shotcrete as soon as possible after blasting.	Systematic bolts 15–20 ft long, spaced 3–5 ft in roof and walls with wire mesh. Bolt invert.	6 to 8 in. in roof, 6 in. on walls and 2 in. on face.	Medium to heavy ribs spaced 2 ft 6 in. with steel lagging and forepoling if required. Close invert.

Notes:
*Length of bolts specified here is applicable to tunnels 30 ft wide.

1. When borehole core is unavailable, RQD can be estimated from the number of joints per unit volume, in which the number of joints per meter for each joint set are added. A simple relation can be used to convert this number to RQD for the case of clay-free rock masses:

   RQD = 115 – 3.3 J_v (approx.)
   where J_v = total number of joints per m^3
   (RQD) = 100 for $J_v < 4.5$)

2. The parameter J_n representing the number of joint sets will often be affected by foliation, schistosity, slaty cleavage or bedding, etc. If strongly developed, these parallel joints should obviously be counted as a complete joint set. However, if there are few joints visible, or only occasional breaks in the core due to these features, then it will be more appropriate to count them as "random joints" when evaluating J_n.

3. The parameters J_r and J_a (representing shear strength) should be relevant to the weakest significant joint set or clay filled discontinuity in the given zone. However, if the joint set or discontinuity with the minimum value of (J_r/J_a) is favorably oriented for stability, then a second, less favorably oriented joint set or discontinuity may sometimes be more significant, and a higher value of J_r/J_a should be used when evaluating Q. The value of J_r/J_a should in fact relate to the surface most likely to allow failure to initiate.

4. When a rock mass contains clay, the factor SRF appropriate to loosening loads should be evaluated. In such cases the strength of the intact rock is of little interest. However, when jointing is minimal and clay is completely absent the strength of the intact rock may become the weakest link, and the stability will then depend on the ratio rock-stress/rock-strength. A strongly anisotropic stress field is unfavorable for stability and is roughly accounted for as in Note II in the table for stress reduction factor evaluation.

5. The compressive and tensile strengths (σ_c and σ_t) of the intact rock should be evaluated in the saturated condition if this is appropriate to present or future in situ conditions. A very conservative estimate of strength should be made for those rocks that deteriorate when exposed to moist or saturated conditions.

most highly stressed in the 8 ft or so nearest the excavated surface, with the stress decreasing beyond. Currently, the length–span ratio is on the order of one-quarter to one-third. Table 7-17 is a generalized guideline for minimum lengths.

Precise spacing distances on drawings for reinforcement are impractical. Normally, 6 in. should be the minimum increment, and 3 in. an absolute minimum. There is neither need nor time in the heading for greater accuracy.

Table 7-14. Q System: Support for Rock Masses of "Fair" and "Poor" Quality (*Q* Range: 10–1)

Support Category	Q	Conditional Factors RQD/J_n	J_r/J_n	Span/ ESR (m)	P kg/cm^2 (approx.)	Span/ ESR (m)	Type of Support	Notes (Table 7-16)
17	10-4	> 30	—	—	1.0	3.5–9	sb (utg)	I
		≥ 10, ≤ 30	—	—			B (utg) 1–1.5 m	I
		< 10	—	> 6			B (utg) 1–1.5 m	I
		< 10	—	< 6			+S 2–3 cm	I
							S 2–3 cm	
18	10-4	> 5	—	≥ 10	1.0	7–15	B (tg) 1–1.5 m	I, III
		> 5	—	< 10			+clm	I
		≤ 5	—	≥ 10			B (utg) 1–1.5 m	I, III
		≤ 5	—	< 10			+clm	I
							B (tg) 1–1.5 m	
							+S 2–3 cm	
							B (utg) 1–1.5 m	
							+S 2–3 cm	
19*	10-4	—	—	≥ 20	1.0	12–29	B (tg) 1–2 m	I, II, IV
		—	—	< 10			+S (mr) 10–15 cm	I, II
							B (tg) 1–1.5 m	
							+S (mr) 5–10 cm	
20 N/A	—	—	—	—	—	—	—	—
		—	—	—				
21	4-1	≥ 12.5	≤ 0.75	—	1.5	2.1–6.5	B (utg) 1 m	I
		< 12.5	≤ 0.75	—			+S 2–3 cm	I
		—	> 0.75	—			S 2.5–5 cm	I
							B (utg) 1m	
22	4-1	> 10, < 30	> 1.0	—	1.5	4.5–11.5	B (utg) 1m + clm	I
		≤ 10	> 1.0	—			S 2.5–7.5 cm	I
		< 30	< 1.0	—			B (utg) 1 m	I
		≥ 30	—	—			+S (mr) 2.5–5 cm	I
							B (utg) 1m	
23	4-1	—	—	≥ 15	1.5	8–24	B (tg) 1–1.5 m	I, II, IV, VII
		—	—	< 15			+S (mr) 10–15 cm	I
							B (utg) 1–1.5 m	
							+S (mr) 5–10 m	
24 N/A	—	—	—	—	—	—	—	—
		—	—	—				—

* Authors' estimates of support. Insufficient case records available for reliable estimation of support requirements. Key to support Tables 7-14 and 7-15: sb = sport bolting; B = systematic bolting; (utg) = untensioned, grouted; (tg) = tensioned (expanded shell type for competent rock masses, grouted post-tensioned in very poor quality rock masses; see note XI); S = shotcrete; (mr) = mesh reinforced; clm = chain link mesh; CCA = cast concrete arch; (sr) = steel reinforced. Bolt spacings are given in meters (m). Shotcrete, or cast concrete arch thickness is given in centimeters (cm).

Dowels normally are installed radial to the arch and perpendicular to the tunnel axis. In very poor ground, spiling may be necessary for temporary stabilization. Angling the bars 30° off the tunnel axis and lengthening them provides a measure of pre-reinforcement and stability. Doubling the length and the longitudinal spacing requires no additional reinforcement but does increase safety.

Empirical Procedures

Use of the most appropriate of the three methods above will be sufficient to carry through the preliminary design phase. In all probability, the design will not change through completion of contract documents unless significant additional geotechnical information becomes available. A numerical model analysis may be required by the owner, but it is unlikely to demonstrate that the design is deficient.

The RST method is based on evaluation of 164 sections, including 147 using steel sets, 14 rock bolts, and 3 shotcrete. Therefore, it is prudent to use the method only when steel sets are planned.

The *Q* System requires the most detailed geotechnical knowledge of the three; when this is available and informed

Table 7-15. Q System: Support for Rock Masses of "Very Poor" Quality (Q Range: 1.0–0.1)

Support Category	Q	Conditional Factors		Span/ ESR (m)	P kg/cm^2 (approx.)	Span/ ESR (m)	Type of Support	Notes (Table 7-16)
		RQD/J_n	J_r/J_n					
25	1.0–0.4	> 10 ≤ 10 —	> 0.5 > 0.5 ≤ 0.5	— — —	2.25	1.5–4.2	B (utg) 1 m + mr or clm B (utg) 1 m + s (mr) 5 cm B (tg) 1 m + S (mr) 5 cm	I I I
26	1.0–0.4	— —	— —	— —	2.25	3.2–7.5	B (tg) 1m +S (mr) 5–7.5 cm B (tg) 1m + S 2.5–5 cm	VIII, X, XI I, IX
27	1.0–0.4	— — — —	— — — —	≥ 12 < 12 > 12 < 12	2.25	6–18	B (tg) 1 m +S (mr) 7.5–10 cm B (utg) 1m +S (mr) 5–7.5 cm CCA 20–40 cm +B (tg) 1m S (mr) 10–20 cm +B (tg) 1 m	I, IX I, IX VIII, X, XI VIII, X, XI
28* See note XII	1.0–0.4	— — — —	— — — —	≥ 30 ≥ 20, < 30 < 20 —	2.25	15–38	B (tg) 1 m +S (mr) 30–40 cm B (tg) 1 m +S (mr) 20–30 cm B (tg) 1 m +S (mr) 15–20 cm CCA (sr) 30–100 cm +B (tg) 1 m	I, IV, V, IX I, II, IV, IX I, II, IX IV, VIII, X, XI
29* N/A	—	— —	— — —	— — —	—	—	— — —	— — —
30	0.4–0.1	≥ 5 < 5 —	— — —	— — —	3.0	2.2–6	B (tg) 1 m + S 2.5–5 cm S (mr) 5–7.5 cm B (tg) 1m +S (mr) 5–7.5 cm	IX IX VIII, X, XI
31	0.4–0.1	> 4 ≤ 4, ≥ 1.5 < 1.5 —	— — — —	— — — —	3.0	4–14.5	B (tg) 1 m +S (mr) 5–12.5 cm S (mr) 7.5–25 cm CCA 20–40 cm +B (tg) 1 m CCA (sr) 30–50 cm +B (tg) 1 m	IX IX IX, XI VIII, X, XI
32 See note XII	0.4–0.1	— — —	— — —	≥ 20 m < 20 m —	3.0	11–34	B (tg) 1 m +S (mr) 40–60 cm B (tg) 1 m +S (mr) 20–40 cm CCA (sr) 40–120 cm +B (tg) 1 m	II, IV, IX, XI III, IV, IX, XI IV, VIII, X, XI

Table 7-16. Q System: Supplementary Notes for Support Tables

I. For cases of heavy rock bursting or "popping," tensioned bolts with enlarged bearing plates often used, with spacing of about 1 m (occasionally down to 0.8 m). Final support when "popping" activity ceases.
II. Several bolt lengths often used in same excavations, i.e., 3, 5, and 7 m.
III. Several bolt lengths often used in same excavations, i.e., 2, 3, and 4 m.
IV. Tensioned cable anchors often used to supplement bolt support pressures. Typical spacing 2–4 m.
V. Several bolt lengths often used in some excavations, i.e., 6, 8, and 10 m.
VI. Tensioned cable anchors often used to supplement bolt support pressures. Typical spacing 4–6 m.
VII. Several of the older generation power stations in this category employ systematic or spot bolting with areas of chain link mesh, and a free span concrete arch roof (25–40 cm) as permanent support.
VIII. Cases involving swelling, for instance montmorillonite clay (with access of water). Room for expansion behind the support is used in cases of heavy swelling. Drainage measures are used where possible.
IX. Cases not involving swelling clay or squeezing rock.
X. Cases involving squeezing rock. Heavy rigid support is generally used as permanent support.
XI. According to the authors' experience, in cases of swelling or squeezing, the temporary support required before concrete (or shotcrete) arches are formed may consist of bolting (tensioned shell-expansion type) if the value of RQD/J_n is sufficiently high (i.e. > 1.5), possibly combined with shotcrete. If the rock mass is very heavily jointed or crushed (i.e., $RQD/J_n < 1.5$, for example a "sugar cube" shear zone in quartzite), then the temporary support may consist of up to several applications of shotcrete. Systematic bolting (tensioned) may be added after casting the concrete (or shotcrete) arch to reduce the uneven loading on the concrete, but it may not be effective when $RQD/J_n < 1.5$, or when a lot of clay is present, unless the bolts are grouted before tensioning. A sufficient length of anchored bolt might also be obtained using quick setting resin anchors in these extremely poor quality rock-masses. Serious occurrences of swelling and/or squeezing rock may require that the concrete arches are taken right up to the face, possibly using a shield as temporary shuttering. Temporary support of the working face may also be required in these cases.
XII. For reasons of safety, the multiple drift method will often be needed during excavation and supporting of roof arch. Categories 16, 20, 24, 28, 32, 35 (SPAN/ESR > 15 m only).

choices can be made, its use is warranted. Although not evaluating the relative orientation of the joints is a weakness, detailed analysis of failure wedges such as those discussed earlier is necessary in any case and will result in appropriate modifications.

Bieniawski (1976) compared in detail the results of six rating procedures on a railroad tunnel 2.4 mi long and 18 ft wide in rock. The procedures include the four already discussed (Deere et al., RSR, RMR, and Q); one described in a 1974 paper (in German) by Rabcewicz, Pacher, and Golser, prime developers of what is now known by the acronym NATM; and a French procedure (Louis, 1974). He tabulated the results for five of the sections with rock varying from excellent to very poor. For the best rock (RQD $>$ 90), the first three required only spot bolting, the last three pattern bolting. Although the French procedure permitted more widely spaced bolts than NATM, it also required 2 in. of shotcrete. In the remaining sections, the NATM procedures required the heaviest support, only slightly more than the French procedure but substantially more than the others. RMR resulted in the next heaviest. The Q System consistently required the least support, with Deere's bolt and shotcrete recommendations only slightly greater. Results from Bieniawski's own method were in the middle; rock bolting requirements

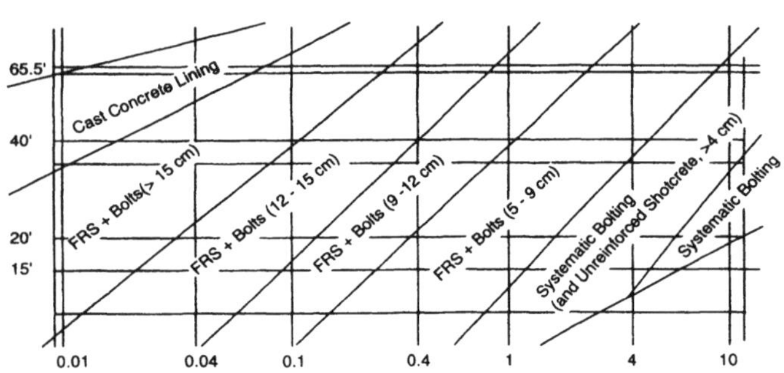

FRS = Fiber Reinforced Shotcrete

Fig. 7-6. Detail of Q System support requirements (see Figure 4-7)

Table 7-17. Generalized Minimum Dowel Lengths

Bolt Length (ft)	Span (ft)
6	to 12
8	13 to 30
10	31 to 40
12	41 to 50
14	51 to 60

[a] Length of dowels should be varied in 2 ft increments, depending on ground conditions.

agreed well with Deere et al. and Q, but more shotcrete was generally used, although the amount was significantly less than with NATM. Finally, in the two sections of poorest ground (RQD < 50), NATM procedures required concrete linings 12 and 20 in. thick in addition to the heavy ground support.

EXCAVATION METHODS

The three principal excavation procedures are

- TBM, for full-face, circular sections only
- Roadheader, for partial face advance, any cross section, or full-face for small sections
- Drill-and-blast, full or partial face advance, any cross section

These methods are generally used separately, but may be used in combination. For example, the 9-mi Mt. MacDonald Tunnel in British Columbia was driven from both portals. The eastern contractor elected a keyhole section and used a TBM for the top heading and drill-and-blast for the bench; the western contractor (for good geologic reasons) used drill-and-blast throughout.

TBM Excavation

Chapter 11 is devoted to TBMs and includes a discussion of factors making a TBM the proper choice. Very few comments, therefore, are necessary here. TBM tunnels have very high start-up (pre-excavation) costs and accompanying long lead time; the high rate of advance reduces the per-foot excavation cost. The total machine length (TBM proper plus necessary trailing gear) may approach 1,000 ft in length.

Roadheader Excavation

This method could also be called *partial face mechanization.* Whereas TBMs are generally purpose-built, roadheaders (also called *continuous miners*) are nearly "off-the-shelf" equipment requiring relatively little lead time. Each of a number of manufacturers builds a series of sizes (capacities).

Crawler-mounted and electric powered, they are particularly suited to noncircular sections; a proportionately small cutterhead rotating either transverse or parallel to the tunnel axis is mounted on long articulated arm and swept across the face. The width of tunnel excavated varies from slightly more than the width of the machine body plus treads to nearly twice that width. Much less heading equipment is required, and start-up costs are only a fraction of that for TBM excavation. The compactness, mobility, and relatively small size of the roadheader combined with simultaneous mucking makes it practical to install rock bolts and/or shotcrete quickly and easily as soon as and in any amount required.

The principal constraint on roadheaders is that they currently are usable only in rock of less than about 12,000 psi compressive strength. Somewhat stronger rock can be cut, or chipped away, if it is sufficiently fractured. Given favorable geology and a properly sized and equipped machine, they are capable of single pass advances of up to 100 ft per day.

The manner of cutting results in fairly small-sized muck fragments. Mechanically collected on the invert apron of the roadheader, they are delivered to the rear of the machine by an integral conveyor. A roadheader coupled with conveyor transport makes an efficient tunneling system.

One of the more spectacular uses of roadheaders was for the two large crossover caverns on the English Channel Tunnels. The British cavern was constructed from the 16-ft-diameter service tunnel, which was to one side and had been enlarged locally. A small roadheader first excavated an access tunnel to the future crossover crown. The largest roadheader available then excavated the chamber, which is 64 ft in clear span, 49 ft high, and 500 ft long. Lower sidewall drifts about 21 ft wide and 24 ft high were excavated first, followed by the 17-ft-high crown heading, which intersected the sidewall excavation crowns. Excavating the remaining central core and finish invert cuts completed the excavation. The French crossover was also excavated by roadheader. However, poorer geology and prior existence of the two running tunnels dictated somewhat different sequencing.

Drill-and-Blast Excavation

This excavation method has been used from time immemorial (or at least from the pre-explosives time when rock tunnels were excavated by building fires at the face and cracking the rock by throwing cold water on the hot surfaces). It is still the conventional method for noncircular cross sections and also for circular tunnels too short to amortize the high start-up costs of a TBM. Drill-and-blast also is preferable when encountering too great a variety of geologies or other specific conditions such as mixed face, squeezing ground, etc.

This method breaks down into a well-defined three-stage cycle of drilling the face; loading, shooting, and ventilating; and removing the spoil (mucking). Use of explosives is discussed later in this chapter.

Advancing the Face

Full-Face Advance. Excavating the complete tunnel section in one operation (full-face advance) is preferred to the maximum extent practical. That extent is determined by the geology, the size (especially span) of the opening, and the stand-up time. The Lauffer diagram (Figure 7-3) displays qualitatively the range of stand-up time for various geologies.

Stand-up time is usually assessed on a two-dimensional basis. Near the face, however, a three-dimensional half-dome stress condition exists. This is especially useful in the narrower partial face drifts required in the poorer mass rock conditions. It is also the reason behind frequently specifying that rock reinforcement be installed close to the face, thus keeping most of the next round length, when blasted, under the half-dome.

Heading and Bench. When stand-up time is not adequate to install support, the round length must be shortened or partial face advance used to reduce cycle time. The most common approach is heading and bench. A top heading is excavated first; this can extend the full length of the tunnel, or it may be as short as a single round length. The key to heading size is the time required to install necessary support or reinforcement for the arch.

Once the roof (or back) is secure, the bench can be excavated. With the stand-up time problem eliminated (unless there will be a problem with wall stability), longer increments of bench can be excavated. Also, the second free face created by the top heading permits a lower powder factor to be used in the bench. When the specifics, including a long and relatively high top heading, permit, the bench can be taken by more economical quarrying methods (vertical holes drilled from the heading invert) rather than continuing with face drilling. This is common practice in underground powerhouse excavation.

Installation of rock reinforcement in a top heading presents no special problems. However, when steel ribs are used, the necessity for a full-strength steel rib arch creates a problem at the abutments. The arch must have a temporary foundation while the bench beneath it is excavated. One solution is to increase the arch span and set footblocks outside the wall excavation line. For this solution, short lengths of rib posts must be welded to the arch to simplify the later arch–post connections.

More commonly, wall plates are installed longitudinally beneath the ribs at the top heading invert. A wall plate is a horizontal structural steel member placed under the arch to act as an abutment and spread the reaction while the bench is being taken. For smaller or lightly loaded ribs, the wall plate is a single wide flange beam with its web horizontal; arch and post segments fit inside its flanges. For heavier loads or larger spans, a pair of beams joined together, with webs vertical, is located directly beneath or over the arch/post flanges. Connecting short wall plates together is somewhat difficult; sometimes small wall plate drifts are excavated ahead of the face, permitting longer members to be installed.

Multidrift Advance. If the stand-up time is insufficient for heading and bench advance, because of either the geology or large spans, the top heading must be divided into two or more drifts.

This is advantageous because the reduced span increases stand-up time, the reduced volume decreases mucking time, and time required to install support or reinforcement is also reduced. When using steel sets, the appropriate final arch segment is used and supported temporarily on one or more steel posts. When the adjacent drift is excavated, the next arch segment is erected, connected, posted, and so on. Once the wall plates are in place and the full arch erected, it is swung by removing the temporary posts.

When stabilization is by rock reinforcement, there is special advantage in using a tripartite division for spans of about 50 ft or more. Driving the top center drift first permits installing rock bolts and creating a massive rock crown bar as the drift advances. This improves the rock mass characteristics and may reduce the amount of remaining reinforcement required. Driving the drift full length early on creates a pilot or exploratory tunnel from which detailed geologic and geotechnical information can be obtained, and the design can then be modified if necessary. In fact, this drift can be driven during the design phase, perhaps eliminating some change orders. When driven early for exploration, there is temptation to use a small cross section, say, less then 100 ft^2. If kept near the construction drift size, final reinforcement can be installed, and a more massive crown bar results.

Similarly, for spans of about 40 ft or more, it may be desirable or necessary to subdivide the bench also. The principal reasons are to permit advancing the heading drifts and bench concurrently and for wall stability. The former requires maintaining a ramp to service upper-level operations. The latter is to excavate and reinforce the walls within subcritical stand-up time. The bench could then be taken in layers related to the wall bolting pattern, or a vertical subdivision can be used. The latter permits either taking a large central cut, leaving only minimal side buttresses for later excavation, or initial narrow side drifts permitting timely installation of wall reinforcement, followed by completion of the excavation.

While tunnel or cavern excavation from top down is preferred, in exceptionally poor ground it may be necessary to work from the bottom up. Driving bottom sidewall drifts first permits concreting abutments and eliminates having to establish, undercut, and reestablish support as described earlier. It may also apply when lattice girders are used (see the section on NATM later in this chapter).

Hanging Lake Multidrift Tunnels. These 4,000-ft twin tunnels on I-70 in Glenwood Canyon, Colorado, consist of two pairs of two-lane tunnels about 2,500 and 1,200 ft in length separated by a cut-and-cover section at Cinnamon Creek, which also houses ancillary control facilities.

A 10-ft-wide by 12-ft-high exploratory tunnel was driven near the crown of the southern (eastbound) tunnels. Geologic reconnaisance and a drilling program had indicated highly fractured and metamorphosed granitics and pegmatites, exhibiting a wide range of weathering. Significant reaches defied reliable estimating of the lengths for various support types. The decision to drive the exploratory tunnels before completing the design was most wise because the RQD evaluations changed substantially from those based on the boring logs, and support requirements were reduced accordingly. Fully encapsulated permanent rock dowels were specified for initial support. Their contribution to the final lining permitted reducing the lining thickness by half.

The tunnels were excavated by drill-and-blast using multidrift advance for both top heading and bench. All tunnels were excavated from the Cinnamon Creek canyon construction portals and daylighted in the near-vertical south walls of the Colorado River canyon. Six equipment-sized construction cross passages were driven through the 100-ft (nominal) pillar to facilitate crew and other movements. One cross passage, near the west portals, was enlarged to a width of 50 ft to provide for emergency U-turns by fire vehicles.

Efficiency and Advance Rate. An additional consideration is the number and efficient use of crews. If two (or four) headings can be advanced concurrently, crew efficiency and overall advance rate increase. Alternating crews between active faces ("swinging headings") permits using efficient separate drilling and mucking crews and reduces idle time. For twin tunnels driven full-face, equipment-sized cross passageways should connect the two. When use of only one TBM is planned for twin tunnels, its advance rate overall can be matched by drill-and-blast methods driving from all accessible portals concurrently, and/or driving from either design-required or "construction-only" shafts.

NATM. It has been largely forgotten that the New Austrian Tunneling Method (NATM) has as its nucleus an observational procedure for verifying installed support adequacy. It was developed for mining Alpine tunnels at depths where cost of geotechnical exploration from the surface was prohibitive and ground conditions frequently difficult and variable. Best judgment and past experience were combined to select an initial drift size and accompanying stabilization system. Observations (measurements) were then made to determine if inward movements were decreasing or if additional stabilization was necessary. Theory was developed gradually and much later. As experience and confidence grew, more and more difficult ground conditions were attacked.

NATM is generally a multidrift approach, even when fewer openings could be used. The drift advances are generally short because of the early recognition of the importance of the time dimension to incremental installation of economical support in unstable or very weak rock, i.e., timing of "closing the ring." It is also an advantage because each top heading face can be worked with an articulated boom excavator from the bench invert. Shotcrete and rock bolts are generally used because they are easily available, inexpensive, quick and simple to install, and readily amenable to augmenting when the initial array must be reinforced. Shotcrete can also be used to stabilze the face of each advance, temporarily when necessary.

Lattice girders are a frequent component of NATM construction. These consist of three or four sizable concrete reinforcing bars arranged in triangular or trapezoidal section, prebent to the shape of the excavation periphery, and joined together into a prefabricated unit with continuous small-diameter diagonal lattice bars. After erection, the girder is filled and encased with shotcrete and becomes an integral part of the initial support membrane.

Like steel ribs, lattice girders frequently must be erected in segments as excavation proceeds and then bolted together. Because rock reinforcement results in most of the stress resulting from tunnel construction remaining in the ground rather than requiring direct support, it is more practical to begin excavating in bottom sidewall wall drifts and continue upward.

Rock reinforcement principles embodied in the NATM have been a distinct advance in the art of tunneling. The success of the method is due to what underlies the philosophy of Chapter 5 (Tunnel Stabilization and Lining), and Chapter 6 (Soft Ground Tunneling), the present chapter, and Chapter 8 (Tunneling in Difficult Ground) as well—logic, reason, understanding, technical proficiency, and willingness to innovate.

EFFECT OF EXCAVATION METHOD ON DESIGN

For a given rock mass, the choice of TBM or blasting affects the rock reinforcement design in at least three ways. Use of a TBM requires somewhat less rock reinforcement in the moderate range of rock quality.

On a TBM drive, maximum economy may result from using different stabilization patterns than for drill-and-blast excavation (e.g., circular steel ribs throughout, rather than only where required by bad ground).

Use of a TBM currently precludes use of shotcrete at the TBM or within the length of its trailing gear because the bulk of the machine inhibits access to the tunnel walls and shotcrete rebound would foul the TBM. This constraint precludes shotcreting within 500 ft of the cutterhead, perhaps even farther. Accordingly, rock reinforcement patterns requiring shotcrete within less than two to five days after rock exposure cannot be used. Mine straps may or may not compensate adequately. In raveling ground, chain-link mesh or welded wire fabric is generally required for safety and to preclude clogging the TBM equipment with rock fragments. If combined with rock bolts, this may be sufficient to support the tunnel until a final concrete lining is poured.

In rock of high mass strength (modulus), the TBM can be advanced by thrusting outward laterally with gripper jacks to provide necessary resistance to the thrust of the TBM ram jacks. In this case, pattern bolting undoubtedly is the most economical solution. However, when the mass is sufficiently poor, TBM longitudinal thrust jacks reacting against the initial rock support system will be required. Nevertheless, for both conditions the stabilization measures for both types of excavation will be essentially the same, except for the shotcrete factor.

When tunnels have both good and bad reaches, TBM tunnels must be excavated full length for the largest diameter required, i.e., the steel rib (or similar) system. The rib or similar system then must be used full length unless the TBM tailskin is slotted to permit rock bolting or is equipped with both thrust jack and gripper propulsion systems and time is taken to change over from one system to the other whenever necessary. The TBM also must be capable of handling both types of ground efficiently; this may require mounting alternative cutting systems on the face of the TBM. On the other hand, drill-and-blast excavation can cope with the changing condition relatively easily and take advantage of the lesser costs of rock bolting in the good reaches and using ribs and lagging only in the poor ground.

An inexpensive segmental concrete lining alternative to ribs and lagging has found favor with some contractors when there is to be a final cast-in-place lining. The segments are unbolted and reinforced only sufficiently for handling and for resisting shield or TBM thrust. They are easy and quick to erect within the tailskin. Once out of the tail, the segmental ring is expanded against the ground by hydraulic jacks at roughly the upper quarter points. The two slot spaces created are then stabilized temporarily with steel dutchmen plus necessary shims while the remainder of the slots are dry packed.

Unless the specifics of a tunnel clearly indicate the superiority of one excavation method over the other, the contract documents should leave the choice to the contractor. This is true even though two sets of stabilization documents are then necessary. The owner will realize all the savings if the documents recognize both methods and the contractor is required to choose during the bid period, but only half as much if only one method is shown and a less expensive alternative is eventually approved under "value engineering" after contract award.

SEISMIC EFFECTS

For a general discussion of seismic effects, see Chapter 6. The following material applies only to tunnels in rock.

Tunnels should not be located across known active faults if possible. Even faults thought to be inactive may come alive. Earthquakes will cause displacement—horizontal, vertical, or combined—at the causative fault, with the result that the use of the tunnel may be impaired or lost. A tunnel that crosses an active earthquake fault is almost certain to be completely ruptured along the fault plane and part of the fault zone for any major earthquake on the fault. In addition, any lining probably will be heavily damaged for many feet adjacent to the rupture. The design objectives in this area, therefore, must be to accommodate as much movement as practical to reduce the damaged area to the minimum practical length, and to keep the maximum practical damaged length from complete collapse even if offset somewhat from the earlier line and grade.

Except at the immediate fault area, tunnels in competent rock (with high RQD) and with strong lining should not be seriously damaged by earthquake shaking. In poor rock formations, close to a fault, shaking may loosen the rock so that loads increase to those shown for crushed zones in Table 7-4.

There is little record of earthquake damage to tunnels. Four concrete-lined 16 × 22 ft Southern Pacific Railroad tunnels were severely damaged or destroyed in the 1952 Bakersfield, California, earthquake by movement along a fault intersected by the railroad. Two of the tunnels were within the zone of faulting. The most distant damaged tunnel was within 1,000 ft of the faulting. This line in the Tehachapi Mountains of California was out of service for 21 days. Repairs were made by daylighting two tunnels and part of a third with cuts up to 150 ft deep. In the 1,170-ft-long fourth tunnel, a 372-ft section was relined with a 2-ft thickness of concrete and a 600-ft section with 8 in. of fine aggregate shotcrete. A 12-in. concrete invert was placed throughout the tunnel. The most amazing and graphic evidence of movement occurred in one tunnel where one wall ruptured and lifted at track level while the track buckled and disappeared into the wall void as the ground dropped back; the track reappeared out of the wall some distance beyond its disappearance.

The 18-ft-diameter San Fernando tunnel, 29,000 ft long, was under construction by the Metropolitan Water District of Southern California, and two-thirds excavated at the time of the February 9, 1971, earthquake. The primary support system consisted of four 8-in.-thick by 4-ft-wide precast concrete unbolted segments placed in abutting rings. Surveys indicated that during the earthquake the east portal of the tunnel rose 7.2 ft relative to the western terminus. There was no evidence of shear offsets resulting from the earthquake. The support system adjusted itself to the irregularities that developed throughout the excavated length so that damage to the tunnel was very minor, consisting principally of cracking and spalling of a few of the segments. Because it operates as a pressure conduit, the change in grade due to the earthquake has not had any serious effect on the operation of the tunnel.

The elastic and plastic ground deformations near a fault rupture must be understood in order to plan any tunnel damage mitigation. On a sharp-cleavage strike slip fault, such as in the generally narrow San Andreas fault zone, elastic and plastic deformation build up on both sides of the fault prior to an earthquake. Although there is no relative movement at the fault, a straight line drawn perpendicular to the fault is deformed into a shallow S-shaped curve, with the curvature and incremental deformation decreasing asymptotically as the distance from the fault increases. During this pre-earthquake period, frictional resistance along the fault builds up to a peak value. At the instant of ground rupture, the frictional resistance decreases to a residual value. This permits the ground to recover all of the elastic deformation and part of the plastic deformation, thereby decreasing both the relative deformation and the curvature that had built up on each side of the fault before the earthquake. The relative displacement across the fault is the sum of the partial releases of all accumulated strains on both sides. The S-shaped line now has a sharp Z-break at the fault, with shallower curves leading away from it. If a tunnel is driven perpendicular to a fault through warped (preearthquake) ground, after the rupture it will exhibit a sharp shearing deformation at the fault trace, plus an imposed curvature on both sides (from the partial release of strain in the adjacent ground) that is greatest at the fault and decreases as the distance from it increases. If the lining can adapt to these lateral deformations, the length of damaged area and the severity of damage will be reduced.

The second most active fault in the San Francisco Bay area is the Hayward fault, which lies subparallel to and only about 15 mi northeast from the San Andreas fault. It also is strike slip in nature. It exhibits the same basic seismic response, but with much smaller single event displacment—perhaps one-third. Collectively, the results along the Hayward fault do not have a well-defined repetitious fracture surface; instead, the surface trace for any given event may occur anywhere within a fault zone on the order of 500 ft wide or even wider. Accordingly, any line that had been perpendicular to the fault trace prior to the seismic event would be expected to exhibit more of an S than a Z alignment afterward.

An exploratory drift for the Berkeley Hills Tunnels of the BART system was driven 1,400 ft across the Hayward fault to determine geologic characteristics and fault zone location more exactly. In addition to heavily fractured and completely crushed rock, there was an abundance of heavily squeezing gouge. The low-mass modulus in the fault zone should result in a more gradual permanent deformation pattern on each side of the rupture. Advantage was taken of this in the tunnel design.

The twin single-track Berkeley Hills tunnels are 16,800 ft long. The Hayward fault zone begins within a few hundred feet of the west portal. The steep and unstable portal slope precluded open cutting through the fault zone. Accordingly, the tunnels were oversized by 1 ft of diameter, allowing for that much lateral offset movement at a future rupture line. Depending on the shape of the resulting inelastic deformation curve on either side, the total movement that can be accommodated may be twice that amount. Heavy steel ribs with curving invert struts are embedded in the concrete lining at 2-ft centers throughout the fault zone. To date (25 years after construction), seismic cracking, if any, cannot be distinguished from the shrinkage cracks.

Recognizing the objectives of tunnel design for crossing an active earthquake fault, the following conceptual design is suggested as a good beginning:

1. Oversize the tunnel, even if it is TBM-driven. Depending on relative lengths and economics, local drill-and-blast enlargement of the TBM excavation is an alternative if the enlargement is 3 ft or more.
2. Use definite and complete cold joints. Unless there is only minimal groundwater present (not likely), provide hydrophilic gaskets in joints (i.e., rubberlike material containing continuous bentonitic strips). Dumbell or other waterstops are not appropriate. Provide these joints at the assumed limits of permanent ground deformation and at appropriate intermediate limits.
3. Provide excess rock reinforcement or steel supports for initial stabilization and heavy circumferential reinforcing in the lining to aid in maintaining local stability.

Thrust faults are likely to create more tunnel damage than strike slip or the steeper dip slip faults because the rupture plane intersects a greater length of tunnel, and longitudinal compressive movements can crush and explode a concrete lining. In addition, most tunneling is at shallow to medium depths. The overburden weight thus is less, meaning lower confining pressures and, therefore, less resistance to movement along the rupture surface. The basic principles of oversizing and independent tunnel sections are still valid. However, oversizing is somewhat less effective unless significant increased invert depth below profile grade line is used and backfilled initially with granular material; more frequent cold joints with some inclination in the direction of thrust appear prudent.

USE OF EXPLOSIVES

The excavation method for rural and mountain tunnels generally can be selected based on suitability and relative economy. In urban areas, nontechnical factors now generally control the choice. Human sensitivity to levels of vibration far below the level of incipient damage to plaster in a building is of as much concern to the typical citizen as the actual structural damage level is to the engineer. An early, informative, and concerned public relations effort is essential if an urban tunnel is to be excavated efficiently with explosives.

Proper design and careful field control can limit the surficial effects of underground blasting to any degree reasonably desired without prohibiting the use of explosives. Blasting technology is sophisticated beyond the primary objectives of this text. The intent here is to provide only sufficient information to permit informed decisions to be made about the use of and limits on blasting, particularly in urban areas, so that the contractor is not constrained unnecessarily in his operations.

The surficial effects (vibration, air blast, noise, etc.) of an underground blast are not necessarily a function of the total pounds of explosives used; rather, they are a function of the pounds used per delay within the round that controls. For example, although a total charge of 500 lb may be required, the effect can be reduced to below that of a single 25-lb charge. Twenty such small charges fired in rapid succession with appropriate delays between charges will not increase the overall vibration effect.

Many factors must be considered in the design of a blasting round. Controlling surface effects is essential in an urban area; this is accomplished by specifying limits on the pounds of explosive per delay, the sound increase in decibels, the maximum particle velocity at adjacent building foundations and the permissible air overpressures. Further limiting of both vibrations and decibels during the nighttime hours is preferable to limiting working hours because it specifies what must be accomplished, not how to accomplish it. The tunnel designer's prime concerns are that the quality of the host rock and the shape of the tunnel, especially the arch, be maintained. Controlled blasting, including requiring closer spacing of the profile (perimeter) holes plus reduced charges in these holes, normally suffices for this. Limiting round length may be advisable, particularly in small tunnels, near portals, and when intersecting an existing tunnel.

The remainder of the blast design pattern is properly the contractor's responsibility. Firing sequence (delay pattern) to provide efficient breakage and muck pile concentration, proper rock fragmentation to facilitate mucking operations, drill-hole diameter, spacing, and burden (measured parallel and perpendicular, respectively, to the momentary free face) and explosive strengths are only a few of the construction concerns.

Controlled Blasting

The purpose of controlled blasting is to reduce overbreak and minimize damage to rock outside the neat or design line. The four principal techniques are line drilling, presplitting, cushion blasting, and smooth blasting. (Neat line or design line normally refer to the outline within which no rock is permitted.) The first three are predominantly surface blasting techniques; normally, only smooth blasting is used underground.

Smooth blasting involves drilling a line of holes along the walls and arch at closer spacing (i.e., 9 to 15 in.) than the production holes, loading them lightly with lower strength and well-distributed charges, and shooting them on the last delay or two in the production round.

Delays. Three delay series are in use—mine (generally inappropriate for civil projects), tunnel (based on time intervals of multiples of one-half second), and millisecond. The millisecond delays are generally based on multiples of 20 or 25 milliseconds, even though intervals as short as 8 or 9 milliseconds are sufficient to decouple vibration effects of successive charges. In addition to reducing vibration velocities, use of millisecond delays decreases total round blast time to a relatively few seconds, thereby decreasing audible perception time; it frequently also improves excavation efficiency.

Vibration. A vibration velocity limit of 2 in./sec adjacent to building foundations is commonly specified when there is concern over any form of damage. Plaster cracking does not ordinarily occur below 3 in./sec, and structural damage is unlikely below about 6 in./sec. The age, type of construction, and general building conditions affect the ability of a structure to withstand vibration; the poorer the overall condition, the more susceptible it is to damage.

Velocity is used for vibration control, rather than acceleration or displacement, because it relates most closely to structural damage. If the velocity constraint is satisfied, other criteria normally will be also. Various methods can be used to establish initial limits on amounts of explosive per delay so that the vibration limits will not be exceeded. The Scaled Distance equation is simple and gives reasonable results for the usual dynamites. In revised form, it is

$$W = (D/D_S)^2$$

where W is the maximum weight of charge per delay in pounds, D is the direct distance from shot to target location, and D^S is the scaled distance. Using a D^S of 50 generally keeps the vibrations below 2 in./sec. Similarly, a D^S of 60 would be appropriate for 1 in./sec.

The frequency of vibration also affects vibration response. Under nearly all rock tunneling conditions, the frequency will exceed 40 cycles/sec (Hertz, Hz) and the 2-in. limit is valid. As noted previously, lower vibration limits may be appropriate under special conditions involving public reaction. On occasion, the general constraint has been set at 1 in./sec. Generally this can be traced back to considerable published material on surface mining. The design of these charges is quite different from those for tunneling. Frequencies considerably below 40 Hz then result, and the allowable velocity near a surface excavation should be reduced.

The only reliable way to determine actual site vibrations is to install and monitor a seismograph registering movements in all three directions. This should be required during tunnel construction. This not only will verify the severity of actual vibrations, it will permit field adjustment of loads per delay to the maximum consistent with negligible damage.

Air Overpressure. Broken windows are the first indication of excessive air blast. Overpressures to 0.03 psi cause no damage. Window breakage should be expected before 1.0 psi, and structural damage may occur by 3.0 psi. Improperly set and poorly maintained windows will break well before the 1.0 psi level is reached. Very little window damage will occur if the overpressure is kept below 0.5 psi. Once well underground (more than a diameter of cover) and away from the portal or shaft, there is little likelihood of creating excessive surface overpressures.

A spectacular example of air blast and its vagaries is the exploratory tunnel for MARTA's Peachtree Center Station in Atlanta, driven through virtually unjointed gneiss. The tunnel was 800 ft long, with a cross section of 12 ft by 14 ft, and was accessed by a lateral drift near its beginning from an off-street shaft collared at basement level. In the early stages of excavation, the air blast traveled from the face down the tunnel to the drift, across to the bottom of the shaft, and up the shaft. It lifted the shaft cover, squirted out against a bulkhead, ricocheted from this to a brick wall of an adjacent building, and "caromed" up and across one street, where it still had enough strength to break windows in a two-story building. As the tunnel lengthened, it developed sufficient internal volume to absorb some energy, and the problem subsided. Contrariwise, it did no damage to a nine-story building across an intersecting street that was about the same distance from the shaft as the two-story building. It is always advisable to make local test blasts to calibrate the general transmission and attenuation estimates.

Noise. Audible shock waves are called *noise*; inaudible ones are *concussions* (if they are felt). Technically, the difference is whether the waves have frequencies greater or less than 20 Hz. Blasting noise is frequently the subject of complaints, but in itself rarely the source of physical damage. Reduction of noise really means the reduction of volume, or sound levels.

Only a block from the location of the aforementioned air blast example and for the same tunnel, there was excellent proof of what good blasting control can do. It was a local area of high-rise buildings (to 70 stories). A fashionable restaurant was located at basement level with its outer wall at the sidewalk line and less than 100 ft direct (diagonal) distance to the initial top center heading of the station cavern excavation. The sound of the blast was clearly heard by the observer. (It sounded much more like a stack of planks falling somewhat leisurely onto the floor above than the conventional public conception of a blast of explosives.) Nevertheless, it passed unnoticed in the completely filled room, as evidenced by no change in the conversational noise level. No vibration whatsoever was felt by the standing observer.

Premature Initiation Hazard

Lightning is by far the most dangerous source. A strike close to a blasting circuit will almost certainly cause detonation. Conductors such as fences, transmission lines, or conductive ground can greatly increase the strike's range. Manmade sources of energy include electrostatic or electromagnetic induction, stray ground currents, and radio transmission. Only rarely, however, is their energy sufficient to cause detonation.

Nonelectric initiation is generally the answer. Detonating cord was one of the earliest nonelectric initiation systems. The Nonel system, sometimes called the shock tube, probably is the best known of the later ones. It has a single hollow tube coated on the inside with a thin layer of reactive material, higher-strength caps than most electric blasting caps, and various accessories. The shock wave travels in the tube at 6,000 ft/sec, but it has insufficient force to initiate high explosives. Millisecond delays are available for the system.

However, when closely controlled blasting is required, the greater accuracy of electric initiation may warrant its use. Enlarging an existing underground powerhouse in the province of Salamanca, Spain, brought all the potential premature initiation hazards together with all the electric initiation, velocity, and acceleration vibration criteria and monitoring problems. Limiting air overpressures, and providing for the protection of operation machinery during the break through of the rock plugs between cavern extensions and existing operational chambers were extremely important. The paper by Oriard et al. (1990) provides an excellent description of the problems and solutions.

Overbreak

Overbreak in a drill-and-blast tunnel is unavoidable. It can be minimized by using controlled blasting. Such overbreak is due primarily to characteristics of the rock mass.

Limited overbreak beyond the neat line must be expected, even with controlled blasting, because the drill rod cannot be located closer than about 1 ft from the unexcavated rock surface. As a result, the hole must be started on the neat line and drilled with a "lookout" of about 1 ft to avoid a "tight" at the beginning of the next round. Overbreak also can be the result of a poor blast design, excessive hole load, poor workmanship, or payment by other than the linear foot of tunnel, or solid cubic yard based on neat or pay lines.

Amount of Explosives Required

Only experienced blasters can forecast the amount of explosives required with any accuracy. The powder factor (weight of explosives per cubic yard of in situ rock) will vary from about 2 lb/yd^3 for a large opening (i.e., more than 1,000 ft^2) in soft fractured rock to more than 9 lb/yd^3 for a 100 ft^2) drift in hard intact rock. Figures 7-7 and 7-8 provide more detailed information on quantities and required number of drill holes. These are for conventional face drilling where there is only one free face; when there are two free faces, as in slashing into an existing drift or in benching, the amount can be reduced.

CAST-IN-PLACE LININGS

Cast-in-place concrete linings, so frequently taken for granted as necessary, account for a significant percentage of tunnel costs. Close examination of actual need and relative costs of

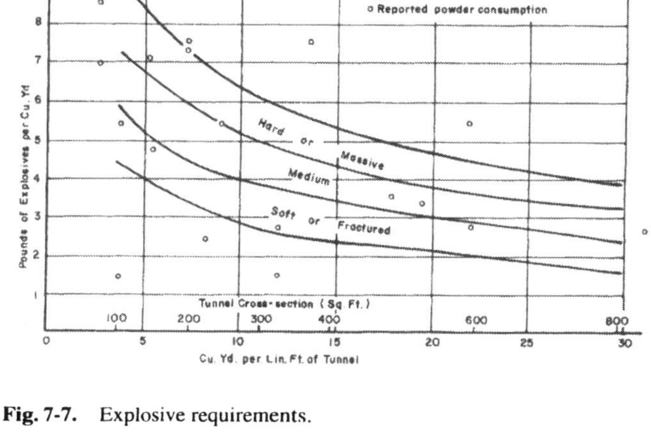

Fig. 7-7. Explosive requirements.

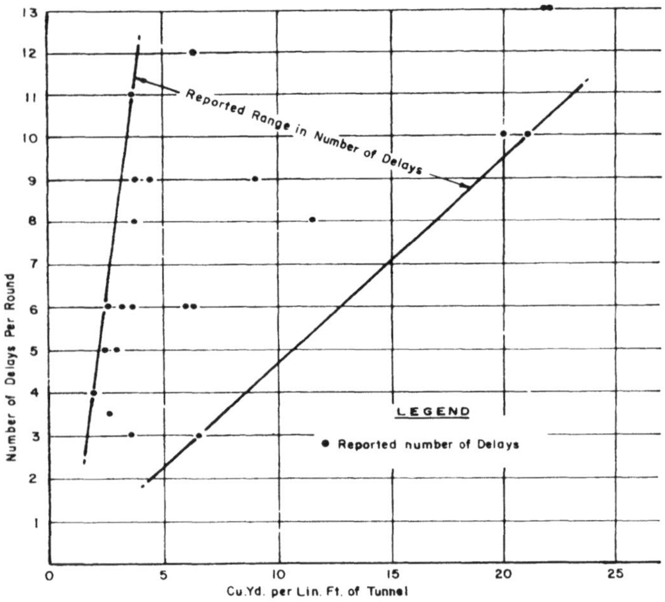

Fig. 7-8. Number of delays per round.

less expensive alternatives is necessary if avoidable cost is to be eliminated.

Proper tunnel linings include none, local, precast, shotcrete, and cast-in-place concrete. In competent and relatively massive rock, no lining may be required, even for freshwater tunnels. This would be true particularly if a TBM were used. In water tunnels and in other tunnels requiring continuous, purging, or emergency ventilation, economy sometimes can be achieved by omitting the lining except in localized areas requiring it and using the increased net opening area to compensate for the increased friction losses created by the rougher perimeter surface. If additional net area is required, a comparative cost study will decide the issue. The added excavation costs much less than the basic excavation because total equipment cost is essentially unchanged and additional labor costs, if any, are minor. If the reason for lining is only protection of the rock surface, a thin shotcrete lining will suffice, especially if the shotcrete is troweled shortly after applying. Presence of hydrostatic groundwater pressure to a level well above the crown is sometimes given as the reason for not using a shotcrete lining; an example in Chapter 12 refutes that argument.

Lining Design

Cast-in-place (CIP) concrete tunnel linings for rock tunnels are the primary focus for this discussion. While tunnel linings in any geology have much in common, such things as differences in interaction between lining and tunneling medium do result in uniqueness.

Before continuing, it would be well for the reader to review Chapter 5, particularly the sections on lining behavior under ground load, lining analysis, and tandem linings.

A UMTA-sponsored investigation (Paul et al., 1983) included interviews with several tunnel design firms. Nearly two-thirds of those experienced in rock tunnel design selected a CIP lining based on minimum constructible thickness, i.e., 8 to 12 in. A majority of those firms also consider water pressure. The remainder used rock loads in some form (e.g., wedges, single failed rock dowel, rock quality indicators, etc.) as well as full or partial water pressure.

It is reasonable to assume that current rock tunnel excavation includes stabilization using rock reinforcement wherever practical and that the design is for maximum and permanent effectiveness of that reinforcement. Under those conditions, a CIP lining in rock is cast in a completely stabilized excavation. Accordingly, it is difficult to imagine how additional ground load can be imposed on the lining. A closely adjacent paralleling or intersecting tunnel driven after concreting might do so. This would be very poor practice unless dictated by long-term project planning. Two other loads must be considered, however. One is pressure resulting from contact grouting. It equals the maximum grout pressure permitted by the specifications. The other, of course, is the hydrostatic pressure, about which there is considerable lack of consensus.

Many designers apply additional ("what if") loads to the lining, e.g., wall wedges that somehow have been destabilized, failure of a permanently encapsulated rock dowel, and subsequent triangular rock wedge becoming supported by the lining, an arbitrary roof load, etc. The first two are good practice because they attempt to define more closely a prudent safety factor. At the same time, unfortunately, conventional aboveground ultimate load design codes are sometimes applied together with these "what if" loads. It is submitted that this is both unrealistic redundancy and an incompatibility. It is redundant because both "load factor" and "what-if load" consider the solution to the same factor-of-safety problem, because one deals with stopping movement, the other with resisting loads.

ACI 318's minimum load factor is 1.4; the strength reduction factor ranges from 0.9 to 0.7. Taken together, these factors provide a safety factor of 1.5 to 2.0. This is excessive for underground structures where loads are not and cannot be codified; where good practice requires conservatism and determining maximum possible "loads," not just probable ones; and where conditions required for collapse are so different from those covered by a code that implicitly assumes freedom of movement to collapse. A tunnel lining cannot deflect or deform freely because it is continuous, and except for straight walls, curvature prevents buckling on the concave side and the ground itself resists movement on the convex side.

Temporary Stabilization

Initial stabilization needs summing up here because it can affect lining design. Two general systems and auxiliaries have been described previously in various locations in this chapter. One is ribs and lagging (or simply steel sets), the other rock reinforcement (rock bolts and/or shotcrete). When and where is each used? Simply put: bolts and shotcrete whenever possible, steel sets otherwise. The geology is the primary determinant. Very poor or unstable ground is the prime candidate for ribs and lagging; however, excellent results can be obtained with ribs or lattice girders and shotcrete when used by experienced personnel in multidrift construction. Steel sets are sometimes used on TBMs where bolts and shotcrete would be satisfactory to maintain a rapid advance by avoiding time-consuming changes of propulsion systems and accompanying disruption in crew activities. Shotcrete sometimes is used as lagging between steel sets with drill-and-blast excavation. Chain-link fabric and mine ties or straps have been mentioned to prevent raveling fragments from falling when using rock bolts in a TBM drive. Use of economical, temporary stabilization with unbolted segments also has been described; the system is used in TBM or shield drives.

Concrete linings in rock tunnels are frequently underused (low stresses). Although well described in Chapter 5 (see the discussion of tandem linings), it needs to be emphasized again that use of friction rock bolt temporary stabilization rather than the more expensive permanent bolts can effect an appreciable economy simply by using the concrete lining to support the ground loads as well as any moderate hydrostatic load that may occur.

Concrete Mix

The concrete mix is probably the most important part of lining design because it is the best way to increase lining impermeability. The elements are well known:

- Class 2 cement (unless groundwater or minerals require Type 5)
- Limiting cement content
- Use of finer cement grinds if there is a choice
- As low a water cement ratio as practical
- Use of fly ash as a cement substitute for up to 25% of cement content
- A low cement aggregate ratio

The reasons for most of the constraints is to minimize heat of hydration and thereby reduce shrinkage cracking. The finer grinds of cement and substitution of fly ash also improve both workability and impermeability because of their extreme fineness.

A slump of 4 to 5 in. is required for reasonable ease of placement in the forms by pumping. At the same time, the amount of cement and the water–cement ratio must be maintained at minimum levels. If all these are to be accomplished, a superplasticizer may need to be added to the mix to provide pumpability.

Several of the aforementioned mix elements conflict with one another; therefore, compromises are necessary between optimal theoretical design and construction practicality. Examples include a low cement factor, which is desirable for shrinkage control, but will decrease pumpability; the desirable low water–cement ratio creates a harsh mix that increases difficulties in filling the lining space completely and in pumping; a relatively high slump is necessary for placing in and filling the forms and for pumpability as well, but it has an adverse effect on strength; sharp (crushed) aggregate improves strength, but rounded is preferred for pumpability; a superplasticizer is desirable for pumping and placing in the form, but the delayed set is coupled with a greatly shortened time interval for the set, which can create placing problems. Obviously, designing the optimum concrete mix is not a simple matter.

Grouting

Contact grouting is required behind all CIP linings. It is impossible to eliminate all voids above the concreted crown during placing and at random locations elsewhere. The purpose of grouting is to provide direct contact between lining and ground to ensure the lining acts as designed. Cement grouts should be used. A thin grout is appropriate because most voids are small, except at the crown, when the concrete mix is placed and compacted properly. Depending on crown void sizes and placement efficiency, a thicker grout may be beneficial initially. Contact grouting should not begin until at least two weeks after concreting because the resulting lining load is considerable. (At a maximum of 30 psi, the loading is over 4,000 pounds lb/ft^2.)

Grouting of crown voids is essential. It is imprudent not to require that at least the central 120 of arch be grouted; there is no consensus about additional amounts. While there is lack of consensus, contact grouting preferably should extend at least to below springline.

An important secondary reason for additional contact grouting is to reduce water flows, both to reduce infiltration and to minimize any flow along the tunnel exterior.

Formation grouting may also be required. This was discussed earlier in the section titled "Water."

Construction Plant

Forms, mixing plant, and transport are all elements of the construction plant. Mixing plant and transport are also discussed in Chapter 13, Materials Handling and Construction Plant. Several types of concrete forms can be used; the construction methods for each are described in the following sections.

Steel Telescoping Forms. These are designed to collapse sufficiently to be transported as units through other sections of forms that are in place and ready to receive or already supporting concrete. This type is used for continuous placing of concrete with only a sloping joint at end of pour, and permits maximum production per shift. Sufficient forms must be provided, usually several hundred feet, to permit concreting at a rapid rate and to allow time for the concrete to set and cure to limited strength before stripping and moving the forms from the back end through the remaining forms to the front end of the operation. This type of form will be found most economical for long tunnels. If the thickness of concrete or presence of reinforcing steel does not allow room for a concrete placing pipe (called a *slick line*) on top of the arch form, it is necessary to provide a dimple or a thickened flat soffit in the crown of the form to house it.

Steel Collapsible Forms. These are designed to collapse only sufficiently to permit stripping of the forms. This type is used for short tunnels not justifying the capital expenditure for continuous pouring or where the mass of reinforcing steel does not permit continuous placing. The disadvantages of this form are the requirement of placing a vertical bulkhead and accompanying waterstop at the end of each pour, and lessened crew efficiency. Bulkhead forms are costly in both labor and materials, and sometimes the cold joint results in leakage.

Hinged windows in both types of forms are provided to permit internal vibration of concrete and, in larger tunnels, to place wall concrete at lower levels. They are closed as the concrete reaches them. Additional vibration is provided by mounting external vibrators on the forms.

Wood Forms. Wood forms are used only for very short tunnels, transitions, and intersections. Full-circle forms may require strutting to the rock at the top to prevent floating.

Placing Equipment

The basic problem in placing concrete in tunnel lining is to elevate the concrete to the top of the arch and pack it as tightly against the roof as possible. Generally, the concrete is pumped, sometimes for long distances, to the slick lines. Compressed air jets may be used to improve concrete movement and reduce plugging. The end of the placing pipe is kept well buried in the fresh concrete and retracted gradually as the concrete mass advances.

Sequence of Operations

The lining of a small circular tunnel is usually placed in one operation. Larger circular tunnels may be placed in the same manner, but more frequently, the invert is placed separately. If the invert is placed first, longitudinal forms on radial planes between 30° and 45° from the vertical centerline are placed first, concrete is placed on the unformed flatter part of the invert and then within the forms, and a heavy screed, shaped to the desired radius and riding on the forms, is dragged along them. Frequently, it is desirable to leave the haulage track in place on the rock bottom until the arch is placed. In this case, longitudinal forms are set to form curbs on both sides of the track with their tops shaped to the tunnel radius. The curbs will contain inserts into which tapered bolts supporting the arch forms are screwed. After the arch concrete is placed and the tunnel invert prepared for concrete, the curbs act as a guide for the heavy invert screed that is dragged along them as the concrete is placed. After the screed has passed, the green concrete may be troweled if a smoother surface is required.

For tunnels of horseshoe section, it is usually desirable to first place concrete curbs at the base of the walls on each side of the tunnel to support the arch forms plus a few inches of the invert to provide a shelf for the invert screed to ride on. The advantage of this sequence is that the haulage track or road for continuing excavation operations is retained and serves the concrete arch placing as well. After the concrete arch is in place, the track is removed, followed by placing of the invert concrete.

Form Stripping. Concrete lining in tunnels does not need much strength to support itself, generally less than 1,000 psi. The bond to the rock, the rough surface of the rock and the confined situation enhance the ability of tunnel concrete to stand by itself. It has been found in practice that arch forms can be stripped safely 12 hours or less after placing concrete.

Tunnel-cured concrete normally has higher early age strength than its laboratory-tested counterpart. A considerable economy can accrue from this fact, particularly on longer tunnels, by shortening the pour cycle time by only a few hours.

Conventionally, concrete strength is defined by breaking of concrete test cylinders, laboratory-cured, while freestanding at constant temperature and humidity. These three elements do not represent typical tunnel conditions. As a result, the laboratory tests generally understate the actual strength. Air temperature and humidity aside, the heat of hydration alone accounts for a significant difference in the strengths. In the laboratory, the internal temperature of the 6-in.-diameter cylinder remains close to the air temperature; in the tunnel, the lining is 8 to 12 in. thick and probably at least 40 ft long. As a result the tunnel concrete internal temperature is appreciably higher during and following hydration. This causes the tunnel concrete to gain strength more rapidly. Testing equipment, methodology, and analyses are available to determine the actual strength.

CAVERNS

At least 400 hydroelectric power stations have been built underground. Heights of the caverns for existing stations are up to 175 ft, with widths to 95 ft and lengths to 1,233 ft.

Exploration and Instrumentation

Because of the great widths and heights involved, it is necessary to do much more exploratory work, testing of the rock quality, and instrumentation than is necessary for the usual size tunnel. If support fails or rock falls from the roof or sides of a tunnel 20 to 30 ft in height during construction, there is no serious problem in reaching the point of trouble and correcting it. The falling rock will choke itself off before irreparable damage occurs. However, in a powerhouse cavern, a failure of the roof or wall support could make it necessary to abandon the site. It is necessary to determine the dip and strike of the rock, its strength, joint patterns, and particularly, the location of weak joints. A study of geological records and surface geology will determine whether further consideration of a site is warranted. The second step may be seismic surveys, but usually it is taking cores from a number of drilled holes. The drill holes may be used for water absorption tests, testing strength of the rock, and bore-hole photography. If the results appear favorable, the following steps may be necessary to prove the adequacy of the site.

1. An exploratory drift to and along the top of the proposed cavern, from which
   a. Observations of the geology can be made.
   b. In situ stress measurements can be taken.
   c. Plate load tests can be accomplished.
   d. Flat jack tests can be accomplished.
   e. Orientation of joints and their strength can be checked.
2. Checking of in situ stresses by the bore-hole deformation method developed by the USBM. This consists of drilling a 1.5-in.-diameter pilot hole and measuring the change in diameter when the pilot hole is stress-relieved by overcoring with a 6-in.-diameter core barrel (Wild et al., 1971).

During construction, it is important to instrument the rock to determine how it reacts to the excavation. Extensometer readings and seismitron measuring of noise will help locate any trouble spots in time to reinforce them before serious damage is done.

Arch Excavation

Because of the great heights and widths, it is necessary to excavate the arch first in such a manner that the openings are always under control. Of the great variety of excavation sequences possible, only a limited number are adaptable to any given site, where rock conditions will limit the choices.

At Kemano in British Columbia, in granite formation, the arch was taken out by excavating first a center drift, followed by successive rings of rock around the drift designed so that the broken rock from successive rings would approximately fill the space left by the previous excavations. To excavate the final ring, at the perimeter of the arch, slots were excavated at 200-ft centers, from which holes were drilled 100 ft in each direction along the excavation line. These provided for blasting with little overbreak (Libby, 1959; Wise, 1952).

Arch Support

As the arch is excavated, it must be supported to the extent required before proceeding with the balance of the excavation. Except in the most exceptional rock, some support will be required. Because spalling rock can damage equipment installations below, most underground power stations require the rock surface to be covered. Most stations built to date have a solid concrete lining over the full length of the powerhouse, even though rock bolts may be designed to support the load. Of 40 stations reviewed, 5% had no support, 12.5% had rock bolts without concrete, 10% had concrete ribs, and 72% had solid concrete lining, mostly designed to take the rock load. Most new stations are now being supported by rock bolts with wire mesh and shotcrete covering the rock surface. Loads may be estimated as for tunnels.

Rock Bolts in Caverns

The loads capable of being supported by rock bolt systems installed in large caverns, with RQDs in excess of 50 based on their yield strength, have usually been the equivalent of columns of rock between $0.10B$ and $0.25B$ in height. This is less than for steel ribs with relatively soft wood blocking that permits more adjustment of the rock and therefore greater loads on the support. Bolts should be installed as quickly as practicable after opening a section of the crown so as to limit rock adjustment to a minimum and thereby reduce loading accordingly. Joint orientation and quality of joints (clay-filled, slickensided, smooth surfaces, or tight, rough surfaces) will considerably affect crown stability. Rock with a high RQD of 90 to 100 might not behave as well (be considered "good–excellent") if the joints were smooth planes filled with a thin film of clay. The wedges of rock that must be supported by bolts depend on the angle of friction along joint surfaces and their orientation. It has been shown that, for caverns at large depth, the wedge that must be supported cannot be forced into the opening if the angle of friction is greater than half the angle of the wedge (Cording and Deere, 1972). This is illustrated in Figure 7-9. As the wedge displaces, the following relation applies:

$$P_i = P_n\left(1 - \frac{\tan\phi}{\tan\theta}\right) + \frac{B_r}{4\tan\theta}$$

where

θ = one-half of included angle of wedge
P_i = internal pressure required to hold the wedge in place
P_n = average normal pressure acting on the side of wedge
ϕ = angle of friction along the joint surface

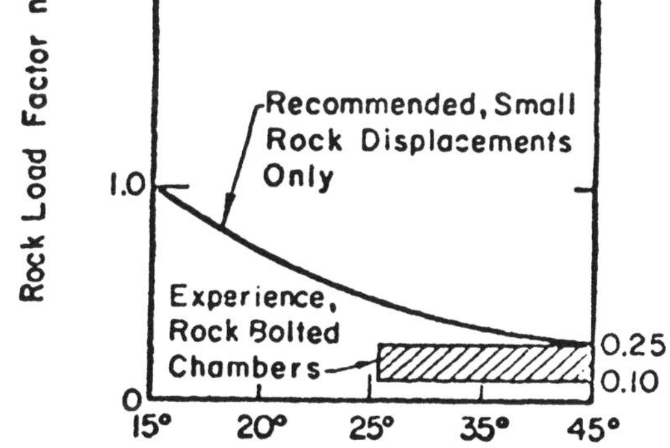

Fig. 7-9. Relation of rock and load factor to angle of friction, ϕ, along wedge boundary.

The loads occurring from the largest wedge that could move into the flat crown of a deep cavern are represented by Figure 7-10, providing the rock joints above are not permitted to open up. Joint orientations may reduce the possible load. It will be noted that recommended loads are conservative for fair to excellent rock. For shallow depth of cover, loads can be greater.

At Oroville Dam, the arch was supported by 20-ft rock bolts spaced at 4-ft centers. The walls were reinforced with rock bolts at 6-ft centers, and the time allowed for installation was increased from 3 to 48 hours after each blast. At Churchill Falls, the arch was supported by bolts spaced at 5-ft centers installed in two patterns, labeled I and II, at 7.1-ft centers each. Design load of bolts with an ultimate strength of 68 kips was 45 kips. Pattern I was installed within 10 ft of the working face, with lengths as shown on the plans, within 8 hours after blasting. Lengths of bolts in pattern II were varied from 15 to 25 ft and were required to be installed to within 60 ft of the working face within 3 days. In poor-quality rock, installation of both patterns was required to within 5 ft of the face. Wall rock bolts were installed on a 7-ft grid and had a design load of 30 kips with an ultimate strength of 45 kips. They were placed at 20° to the horizontal so as to reinforce the most joints and were designed to hold gravity wedges inclined at 50° to the horizontal. Recommendations and trends in the length of rock bolts are shown in Figure 7-11.

Excavation Below the Arch

Excavation below the arch is usually taken out in benches by down drilling, blasting, and loading with shovels or front-end loaders into trucks. While this excavation is less costly than arch excavation, many estimators frequently overlook the extra cost of line drilling and presplitting, as well as minor, but complex, excavations at the bottom of the cavern.

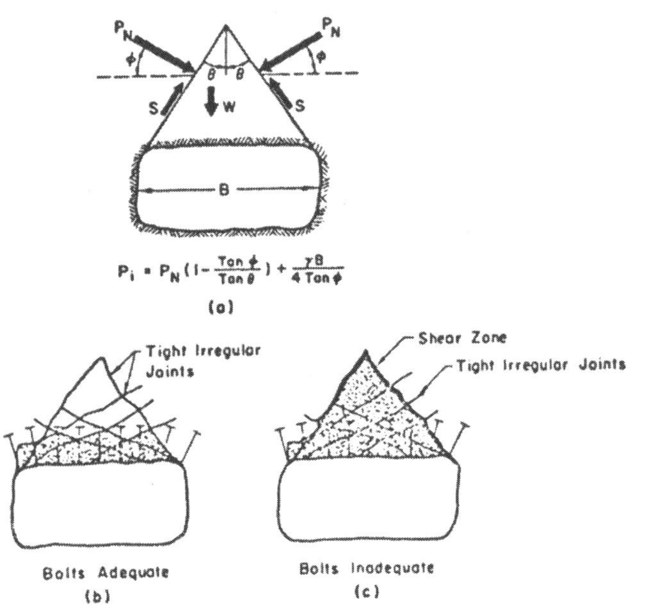

Fig. 7-10. Effect of joint orientation on crown stability.

Wall support in underground powerhouses is frequently limited to rock bolting.

Access

Because the excavation must be carried on from many different levels, access is always a concern. The first work in the powerhouse must be done near the arch, whereas the normal access tunnel comes in at the powerhouse floor and the draft tubes come in at the extreme bottom. Anything that the designer can do to make possible the use of design excavations for access to the powerhouse at the springline of the arch would tend to reduce costs of construction. Because of the congested working space and interconnections between penstock shafts, access tunnels, trailrace tunnels, and other openings into the powerhouse, it is necessary to coordinate all operations that may be going on at one time. No operation can be carried on by itself once a connection is made to the main cavern. If a "shot" is to be made, it may be necessary to clear all personnel from the cavern, thus interrupting all work, as well as that where the blast is to take place. Nowhere is proper scheduling and sequencing of the operations more important than in an underground powerhouse.

Underground Industrial and Storage Facilities

Because of danger from land or snow slides, war damage, or other causes, it may be desirable to place storage tanks,

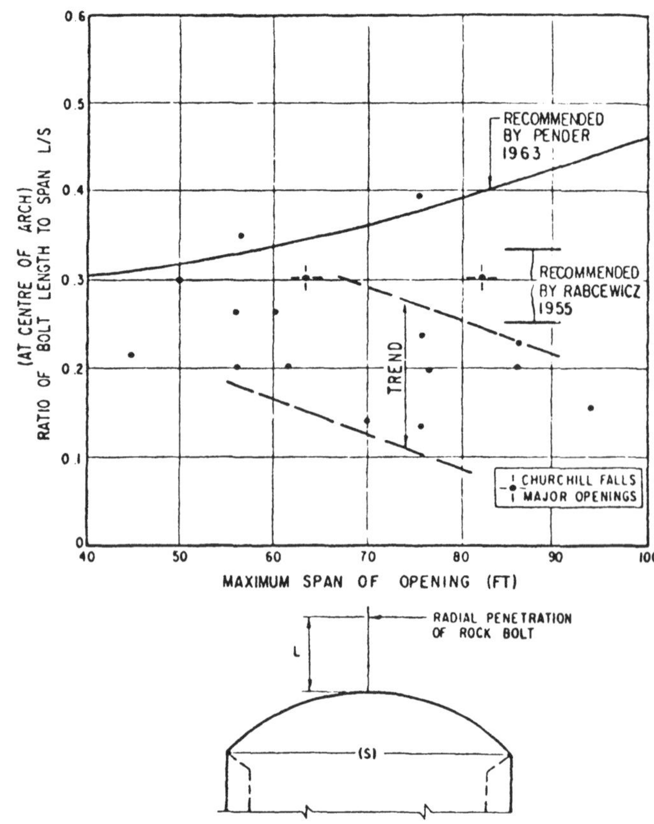

Fig. 7-11. Arch bolting for precedent openings.

manufacturing plants, and warehouses underground. In Chicago, large caverns are being excavated to provide temporary storage for rainwater run-off where sanitary sewers are not in a separate system. These caverns are built as very large tunnels. In Chile, because of snow avalanches, a copper ore deposit at the Rio Blanco mine site was not developed in the area until it was decided to place the mill underground. The large caverns required were excavated in a manner similar to that used for underground powerhouses.

Concrete Work

Because of limited access, concrete in underground power stations and facilities must be carefully scheduled. The saving in form cost by forming on one side only instead of two, as for outside sections, is more than offset by the cost of anchoring forms and overbreak backfill. The lack of storage and equipment maneuverability space within the cavern requires materials to be brought into the station only as they are needed and to be removed and returned between uses. It is usually necessary to install the temporary powerhouse gantry cranes before placing much of the concrete. This more difficult and expensive handling of materials results in costs much higher than for outside construction.

LEAKAGE

Early in the 20th century, subway stations were constructed by cut-and-cover methods. The preferred "waterproofing" was brick-in-mastic constructed outside the concrete structure. At mid-century nearly all stations still were cut-and-cover; however, the waterproofing preference was for an external multi-ply felt and bitumen membrane (see Chapter 17 for details on cut-and-cover tunnels). Near the end of the century, mined stations were increasing in number; a tough plastic membrane with welded seams was showing considerably improved protection.

Leakage through a concrete tunnel lining occurs primarily because even a good-quality mix with 28-day strength in the 4,000 to 5,000 psi range will have a drying shrinkage of 3/8 to 1/2 in. per 100 ft with or without longitudinal shrinkage reinforcing.

Secondary causes include the trend to ever-longer pumping distances, the difficulty of properly placing concrete in tunnel forms, excess heat buildup in the tunnel during early curing, deficient workmanship, etc.

Methods for minimizing leakage are both preventive and corrective. Preventive methods include extreme care in designing the concrete mix, minimizing rebar interference, formation drainage or grouting to provide an impermeable membrane and eliminating as many vertical (unstressed) construction joints as possible. Grouting programs are both preventive and corrective. An effective corrective measure is epoxy grouting of joints.

Waterstops have long been considered a necessary barrier to water penetrating joints. However, many of those protected joints have continued to leak. Horizontal joints with an upstanding water stop on the lower face become inadvertent trash collectors that are rarely adequately cleaned and prepared before pouring the next lift. Vertical joints with outstanding waterstops suffer from lack of compressive stresses. Proper placement of concrete is difficult in both types. Both types can be improved by elimination of the waterstop. The horizontal joint surface can be cleaned and roughened better, and a mortar layer applied minutes before the pour begins will improve bond between the two faces. A sloping joint similarly cleaned, scarified, and given a mortar coat also has improved bond plus a crack closing load across the joint.

The limit of acceptable infiltration is determined by tunnel purpose. Public use as in passageways and rapid transit stations make overhead drips and wet, slippery surfaces unacceptable. A limit of 0.2 gpm per 250 linear ft of tunnel was set for the Bay Area Rapid Transit around 1965. It has proven satisfactory over the years and has also been used with equal success by several other transit properties.

The effectiveness of formation grouting was demonstrated on the Buffalo (Niagara) transit system where leakages of 250 gpm in 300 ft were reduced to less than 1 gpm per 1,000 ft.

CONCLUSION

My objective has been to provide sufficient information on basics—which experienced practitioners are already well aware of—to provide some understanding and knowledge of principles. Enough detail to demonstrate the multitude and complexity of considerations involved has also been included.

A tunnel engineer must have a combination of skills. First and foremost is a thorough knowledge of geotechnical factors and construction know-how, followed by specifications writing ability, and finally, good design knowledge and familiarity with innumerable details. There is no substitute for experience, but marked ability, combined with capacity to reason, does run it a close second. The best overall design procedure is to build the tunnel in one's mind first and then to design it.

Use of empirical methods for selecting stabilization quantities remains the best detailed design approach, whether one of the methods described here or another is used. Owners sometimes require mathematical verification of a design. Rarely, if ever, will use of RSR, RMR, or *Q* not be verified by subsequent analysis using either a continuous finite element method, or preferably, a discrete solution such as the computer FLAC code.

The reader undoubtedly will have noted that major changes and significant improvements come slowly, but are inevitable. The future is unpredictable, but like the last 25 or 50 years, it will be exciting, and the advances a tribute to the innate abilities of tunnel engineers.

REFERENCES

BARTON, N., LIEN, R., LUNDE, J. (1974). Engineering classification of rock masses for the design of tunnel support, *Rock Mechanics,* v. 6, no. 4, p. 189–236.

BIENIAWSKI, Z.T. (1973). Engineering classification of jointed rock masses, *Transactions of the South African Institution of Civil Engineers,* v. 15, p. 335–344.

BIENIAWSKI, Z.T. (1979). The geomechanics classification in rock engineering applications, *4th International Congress on Rock Mechanics,* Montreux, Switzerland, Balkema, Rotterdam, v. 2, p. 41–48.

BREKKE, T.L., and HOWARD, T.R. (1972) "Stability Problems Caused by Seams and Faults," *Proceedings of First North American Rapid Excavation and Tunneling Conference,* AIME, Chicago, Illinois, June 5–7.

CORDING, E.J., and DEERE, D.U. (1972) "Rock Tunnel Supports and Field Measurement," *Proceedings of First North American Rapid Excavation and Tunneling Conference* AIME, Chicago, Illinois, June 5–7, Vol. 1, pp. 601–622.

CRANDALL, F.J. (1960) "Transmission Coefficient for Ground Vibrations Due to Blasting," *Journal of the Boston Society of Civil Engineers,* Vol. 47, No. 2 (April).

DEERE, D.U., et al. (1967) *Design of Tunnel Liners and Support Systems,* PB 183799. Customer Services Clearing-house, U.S. Department of Commerce, Springfield, Virginia.

E.I. DU PONT DE NEMOURS AND CO., *Blasters Handbook,* 15th ed. pp. 157, 139, 170–171.

Engineering News-Record (1953) "Heat Being Conquered at Tecolote Tunnel," Vol. 151, p. 25 (November 26).

Engineering News-Record (1954) "Heat Being Conquered at Tecolote Tunnel," Vols. 152, pp. 45–48 (June 17).

Engineering News-Record (1960) "Flowing Sand Plugs Two Miles of Tunnel," Vol. 165, pp. 38–40 (August 25).

Engineering News-Record (1964) "Pilot Bore Is Laboratory for Twin Road Tunnels," Vol. 173, pp. 38–39 (August 13).

Engineering News-Record (1965) "Sand-Filled Water Tunnel Holed Through," Vol. 174, p. 20 (May 6).

Engineering News-Record (1967) "Tunnel Will Store Storm Runoff," Vol. 179, p. 32 (November 30).

Engineering News-Record (1970) "Graton Tunnelers Nightmare Complete with Scalding Deluge," Vol. 185, pp. 22–25 (November 26).

GOODMAN, R.E., and SHI, G.H. (1985) *Block Theory and Its Application to Rock Engineering.* Prentice-Hall, Englewood Cliffs, NJ.

HOPPER, R.C., et al. (1972) "Construction of Straight Creek Tunnel, Colorado," *Proceedings of First North American Rapid Excavation and Tunneling Conference,* AIME, Chicago, June 5–7.

JAEGER, C., (1955) "Present Trends in the Design of Pressure Tunnels and Shafts," *Water Power,* p. 127, April.

KINCAID, C.G. (1970) "The Graton Tunnel Project in Peru," *Tunnels and Tunneling,* Vol. 2, No. 5, pp. 281–285.

LAUFFER, H. (1958) "Gebergsklassefezierung fur den Stollenbau," *Geologie und Bauwesen,* Vol. 24, No. 1, pp. 46–51 (in German).

LEDBETTER, B.C. (1938) "Driving an Extremely Difficult Tunnel," *Engineering News-Record,* Vol. 121, pp. 669-673 (November 24).

LIBBY, J.W. (1959) "Planning and Construction of Underground Facilities for Kemano, Kitimat Project," *Proceedings of Second Protective Construction Symposium,* Rand Corporation, Santa Monica, California, Vol. 2, p. 711.

MASON, E.E. (1968) "The Function of Shotcrete in Support and Lining the Vancouver Railroad Tunnel," *Proceedings of the Tunneling and Shaft Symposium,* Minneapolis, Minnesota.

MASON, R.E. *Instrumentation of the Shotcrete Lining in the Canadian National Railway Tunnel,* Vancouver, B.C., M.S. Thesis in the Department of Mineral Engineering, University of British Columbia.

Mine Gases and Methods for Detecting Them (1954) U.S. Bureau of Mines Circular 33, Washington, D.C.

TERZAGHI, K. (1964) "Rock Defects and Loads on Tunnel Supports," in J.E. Proctor and R.L. White (editors), *Rock Tunneling with Steel Supports,* Commercial Shearing, pp. 17–99.

WICKAM, G.E., TIEDMAN, H.R., and SKINNER, E.H. (1972) "Support Determinations Based on Geologic Predictions," *Proceedings of First North American Rapid Excavation and Tunneling Conference,* AIME, Chicago, Illinois, Chicago, Illinois, Vol. 1, pp. 43–46.

WICKAM, G.E., TIEDMAN, H.R., and SKINNER, E.H., (1974) "Support Determinations Based on Geologic Predictions," *Proceedings of Second North American Rapid Excavation and Tunneling Conference,* AIME, Chicago, Illinois, Vol. 1, pp. 691–707.

WISE, L.L. (1952) "Methods Spur Underground Powerhouse Tunnels," *Construction Methods and Equipment,* p. 72.

8

Tunneling in Difficult Ground

Terrence G. McCusker
Tunnel Consultant

INTRODUCTION

In this chapter, the emphasis is placed on creating and maintaining stable openings in ground that actively resists such efforts. As adjuncts to tunneling, mention is made of the problems of creating permanent portals in predetermined locations and some of the stability problems encountered in cavern excavation.

The factors that make tunneling difficult are generally related to instability, which prevents timely placement or maintenance of adequate support at or behind the working face; heavy loading from the ground, which creates problems of design as well as installation and maintenance of a suitable support system; natural and manmade obstacles or constraints; and physical conditions that make the work place untenable unless they can be modified.

Instability

Instability can arise from lack of stand-up time, as in noncohesive sands and gravels (especially below the water table) and weak cohesive soils with high water content or in blocky and seamy rock; adverse orientation of joint and fracture planes; or the effects of flowing water. The major problems with mixed-face tunneling can also be ascribed to the potential for instability, and this class of tunneling will be discussed under this heading. Similarly, the problems commonly associated with portal construction are stability problems.

Heavy Loading

When a tunnel is driven at depth in relatively weak rock, a range of effects may be encountered, from squeezing through popping to explosive failure of the rock mass. Heavy loading may also result from the effects of tunneling in swelling clays or chemically active materials such as anhydrite. Adverse orientation of weak zones such as joints and shears can also result in heavy loading, but this is dealt with as a problem of instability rather than loading. Combinations of parallel and intersecting tunnels are a special case in which loadings have to be evaluated carefully.

Obstacles and Constraints

Natural obstacles such as boulder beds in association with running silt and caverns in limestone are examples of natural obstacles that demand special consideration when tunneling is contemplated. In urban areas, abandoned foundations and piles present manmade obstructions to straightforward tunneling, while support systems for buildings in use and for future developments present constraints that may limit the tunnel builder's options. In mining districts, abandoned workings can create problems in both urban and rural settings. In urban settings, interference conflicts, public convenience, or the constraints imposed by the need or desire for gravity flow in water conveyance facilities will sometimes result in the need to construct shallow tunnels, which have a range of problems from working in confined spaces, avoiding subsidence, and uneven ground loading and support.

Physical Conditions

In areas affected by relatively recent tectonic activity or by continuing geothermal activity, both high temperatures and noxious gases may be encountered. Noxious gases are also commonly present in rock of organic origin; and elevated temperatures are commonly associated with tunneling at depth. In an urban setting, contaminated ground may be encountered, and it will be especially troublesome when found in association with other difficult conditions.

Where appropriate, some information is provided as to the reasons that the condition under discussion creates problems for construction. Some examples of each of the conditions referred to above are discussed briefly to yield insight into the problems and to define the range of solutions available to current technology. Both manual and machine technology are considered.

INSTABILITY

Noncohesive Sand and Gravel

Cohesion in sands is more than a matter of grain size distribution. For instance, beach-derived sands normally contain salt (unless it has been leached out), which aids in making sand somewhat cohesive regardless of grain size. The moisture content then becomes a determining factor.

The age and geologic history of the deposit is also important since compacted dune sands with "frosted" grain surfaces may develop a purely mechanical bond; and leaching of minerals from overlying strata may also provide weak to strong chemical bonding.

A very low water content amounting to less than complete saturation will provide temporary apparent cohesion as a fresh surface is exposed in tunnel excavation because of capillary forces or "negative pore pressure." This disappears as the sand dries and raveling begins. Nevertheless, some unlooked-for stand-up time may be available. In this case, it is important not to overrate the stability of the soil. As it dries out, the cohesion will disappear, and it cannot be restored by rewetting the ground.

If groundwater is actually flowing through the working face, any amount may be sufficient to permit the start of a run that can develop into total collapse. There is no such thing as a predictably safe rate of flow in clean sands. Uncontrolled water flows affect more than the face of the excavation. If the initial support system of the tunnel is pervious, water flowing behind the working face will carry fines into the tunnel and may create substantial cavities—sometimes large enough to imperil the integrity of the structural supports.

While factors such as compaction or chemical bonding may permit some flow without immediate loss of stability, this is not a reliable predictor. Soil deposits are hardly ever of a truly uniform nature. It has been observed in soft ground tunnels in recent deposits that all that is necessary to trigger collapse may be the presence of sufficient water to result in a film on the working face; i.e., there is no negative pore pressure to assist in stabilizing the working face. Of course, there is never a safety factor arising from surface tension (capillary action) in coarse sand or gravel.

The cleaner the sand, the more liable it is to run when exposed in an unsupported vertical face during tunnel construction. Single-sized, fine-grained sands (UCS classification SP) are the most troublesome, closely followed by SP-SM sands containing less than about 7% of silt and clay binder. Saturated sands in these classes have been observed to flow freely through the clutches of sheet piles and to settle into fans having an angle of repose of less than 5 degrees. Unconfined SP sands will run freely, as in an hourglass, whether wet or dry, having some stability only when damp but less than saturated (no piezometric head). The large proportion of the sand particles of the same size allows the sand to move almost as freely over one another as would glass marbles.

The effects of water flow on the stability of tunnel excavations will be examined in more detail later in this section.

Silt, intermediate in grain size between sand and clay, may behave as either a cohesive or noncohesive material. It is common around the Great Lakes to find thin seams of saturated fine sandy silt trapped between clay beds in glacial deposits. In general, unless the seams are thicker than about 9–12 in., when the silt layer is exposed in the wall of an excavation, the soil slumps out at intervals leaving a series of small shallow caves like entrances to burrows. The water appears to drain fast enough from the increased surface area exposed so that the remainder of the exposed material stabilizes.

Tunnels constructed in noncohesive sands are also at risk during their service life. Sewer tunnels constructed in the Detroit area through beds of SP sand suffered eventual collapse after several years in service. The sewers do not run full ,so that infiltration occurs when the tunnels are below the water table—especially through honeycombed areas around water stops at construction joints and through shrinkage cracks. Cavities are created slowly, but they eventually result in structural failure. In the Twin Cities, most tunnels are constructed in the St. Peter Sandstone. This sandstone is a variably lithified SP sand that loses its intergranular bonding readily in the presence of water. It has been found that shrinkage cracks as narrow as 0.02 in. will permit the passage of sand, whether below the water table or in ground that is perfectly dry around utility tunnels carrying steam heating pipes. It is now the practice in this area to require sealing of all such cracks as part of utility tunnel construction contracts.

During construction of twin bores of the Seattle Bus Tunnel, part of the new rapid transit system under development in the downtown area, tunneling was carried forward in ground composed of successive deposits of alluvium, colluvium, lacustrine, and glacial materials. The tunnel follows the city street above along the strike of a steep hill sloping down toward the Pacific. The depth of construction, as much as 90 ft, was such that the water table was as much as 60 ft above the tunnel.

The utilities and manholes in the street were protected by pre- and post-injection of compaction grout and consolidation grout in the near-surface region, while dewatering wells were used to control groundwater in the deeper pervious materials. Near the end of the project, the tunnels make a right-angled turn and pass beneath occupied structures. At the approach to this turn, very fine-grained loose silty sand was encountered in the lower part of the tunnel. The leading shield settled several inches before it could clear the zone affected. Examination of the material showed that it had unexpected properties. When recovered from borings made to install additional dewatering wells, it was collected in chunks of almost dry material. However, when a jar sample was disturbed by shaking, it deliquesced and gave up about 10% by volume of free water. In the tunnel, where it had been disturbed by tunneling operations, the same material appeared to the eye as a black, very fine sand with free water present. A successful attempt was made to control the behavior of this material by installing eductor wells. While the well yields were low, this was sufficient to control the behavior of the ground and to permit the safe resumption of tunneling.

SP sands were at the root of problems encountered in construction of the Johns Hopkins/Shot Tower section of the Baltimore Metro. The instability of thick continuous beds of dry sand caused initial problems in controlling lost ground. The shield used initially had extending poling plates. The theory of action of this design is that unstable ground will be restrained at its natural angle of repose between the tips of the poling plates and the bowl of the shield. Unfortunately, effective use of such a system requires skill and judgment, during both the shield advance and the excavation cycle. The ground did not permit time for the learning process to be effective. The net result was that sand would ravel into the crown of the excavation from outside the shield. At this point, effective control was no longer possible with the means at hand, in spite of efforts to stem the flow of material by stuffing salt hay into the voids outside the shield. After a collapse to the ground surface, which broke utilities in a street intersection, a program of consolidation grouting was instituted, and this proved effective in protecting critical structures and utilities.

Another problem appeared. The tunnels were initially driven in free air, but they were to be driven in large part using compressed air methods to limit the amount of dewatering required. In a location not far ahead of the collapse referred to above, gasoline fumes had been detected at tunnel elevation. This part of the work was to be under compressed air once the locks were installed. When it had been determined that the carcinogenic components of the gasoline were no longer present, compressed-air working began.

However, the tunnel lining was not airtight, and the fumes were now transported by the escaping compressed air through a continuous sand layer for a distance of several blocks, where their detection in an apartment building basement caused concern. Although the fumes were neither toxic nor explosive, they were certainly unpleasant, and not everyone was comforted by the assurances they received.

A unique problem of tunneling in noncohesive sands was encountered during construction of the Toronto subway tunnels along Bloor Street just west of Sherborne Street. In this area, tunneling was generally in varved clay laid down along the northerly margin of the glacial Lake Algonquin, the precursor of Lake Ontario. In one reach, a bed of pure fine perfectly dry compact sand was encountered. The tunnel that reached it first was being driven using a prototype soft ground rotary head tunnel boring machine (TBM) designed and manufactured by the contractor. It had been noted over a few days that when the shield was shoved, a low-frequency booming sound could be heard at the shaft, as though the tunnel were behaving like an organ pipe. A complaint was received from a building owner two blocks away from the site in a direction perpendicular to the line of drive that intense vibrations were damaging boiler foundations in the building. Since Metro construction was the only construction activity in the area, it was logical to suspect a connection. A brief investigation did tie timing of the foundation vibration to the timing of the shield advance. There were no effects elsewhere and—especially noteworthy—not in the intervening block of buildings. The problem was solved simply by wetting down the ground exposed in the excavation.

The usual problem encountered with running sand is settlement and cratering at the surface with damage to utilities in the area. If the ground is permeable, consolidation grouting of the entire sensitive area can be undertaken to stabilize the soil before tunneling. If dewatering is successful in depressing the water table below the tunnel invert, it may be found that the sand is just as unstable dry as wet. The alternative of using compressed air is attractive, provided the working pressure is very carefully controlled; but even so, the ground may be too dried out for stability.

If the face is a full face of sand and similarly weak materials, then a slurry shield or an earth pressure balance shield, with provision for reintroducing overexcavation into the plenum chamber to maintain pressurization, may suffice to permit the advantages of machine excavation. The alternative is to use a manual shield and abandon the advantages of mechanical excavation in favor of the capability of providing full-face breasting when this is required. The mixed face situation is discussed below.

In general, rotary head tunneling machines for soft ground tunnels require very similar physical properties over the entire working face and the entire job. If these conditions do not prevail, then weaker ground, and running sands in particular, cannot be prevented from entering the shield more rapidly than is proper for the rate of advance. Slurry shields have the best opportunity of controlling variable conditions where running sands are present; but they will prove difficult to keep on line and grade in mixed-face conditions if one of the beds present is even a strong clay. If the sand and clay beds are more or less evenly distributed (e.g., a varved clay), then this problem may not arise. Of the digger-type shields, neither extensible poling plates nor orange peel breasting have proved to be generally successful; except that the former have enjoyed a high degree of success in the difficult tunneling conditions found in Milwaukee when used by local contractors who assisted in their development. Digger shields necessarily rely more on a sloping face of muck and stand-up time up to 30 min for face support than on direct mechanical support when active tunnel excavation is in progress.

A problem with all shield construction is the necessary difference in diameter between the shield and the lining. If the soil has no stand-up capability by the time it is exposed in the upper part of the tunnel before expansion of a primary lining or introduction of pea gravel or grout into the annular space for nonexpanded linings, then there will be loss of ground. Even if only local raveling takes place, it may choke off the flow of grout before the void can be filled with a continuous supporting fill material. This loss of ground results in a contribution to settlement. It may also establish a drainage channel outside the lining since the raveled material will be much more pervious in general than the native soil. If such drainage is likely to create permanent problems, then it becomes necessary to establish cut-offs at intervals along the tunnel. Such cut-offs are usually in the form of grouted rings several feet thick and deep around the entire perimeter of the tunnel. When the ground is susceptible to erosion from flowing water, infiltration control must be as near perfect as practically possible.

Soft Clay

For the purposes of this discussion, soft clay includes any plastic material that will close around a tunnel excavation if free to do so. This will be the case if the overburden pressure at springline exceeds the shear strength of the clay by a factor of about three or more. However, if the clay is sensitive and loses strength when remolded, the remolded strength will govern some of the clay behavior during tunnel construction. The phenomenon of sensitivity is mediated by several factors that cannot be discussed here. In general, sensitivity may be suspected in clays with a high moisture content. Particularly at risk are marine clays from which the salt has been leached. The loss of strength may lie within a wide range, the ratio of undisturbed to remolded strength being from 2 to 1,000. Moderate sensitivity of 2 to 4 is quite common. During remolding, the void ratio in the clay is reduced and free water is released. When this free water has access to a drainage path such as a sand bed or the tunnel itself, there will be a volume change in the soil mass, which will have consequences in surface settlement.

If the soil is to be stabilized so that closure around the tunnel lining is minimized and stable control of line and grade are maintained, the critical number must be reduced below about 5; this will enable reasonable control of alignment and grade. The critical number is given by

$$N_{crit} = (P_z - P_a)/S_u \tag{8-1}$$

where N_{crit} is the critical number, P_z is the overburden pressure at tunnel springline, P_a is the working pressure in a compressed-air tunnel, and S_u is the undrained shear strength of the soil in compatible units. As an example, if N is to be maintained at a value of 5, the overburden pressure is 40 psi and the unconfined shear strength of the soil is 1,000 psf = 7 psi, then from

$$P_a = P_z - N_{crit} \times S_u \tag{8-2}$$

the required working pressure in the tunnel will be $(40 - 5 \times 7) = 5$ psig. From this same equation, it can be seen that if the shear strength of the soil is reduced by remolding caused by passage of the shield through the ground to a value of 250 psf, then the required air pressure for stability increases to more than 30 psig, transforming the project from a relatively straightforward one to a very difficult one.

In practice, it must be accepted that some closure and consequent settlement will occur. Referring again to the above example, if it is decided that the maximum working pressure allowable for health, safety, and economic reasons is 14 psig, then the critical number will be nearly 15, using the remolded strength for the value of S_u. In practice, it has been found that for moderately sensitive clays, reasonable results are obtained by using the undisturbed value for S_u. However, it must then be accepted that settlement will occur. In these circumstances, prudence dictates that provision be made for independent support of utilities and structures in the expectation that the preexisting ground support will not be available or will not be reliable. As mentioned in the discussion of tunneling in Seattle, local compaction grouting beneath manholes and foundations and consolidation grouting or hanging supports for utilities may be used. In parks or rural areas, settlement can often be allowed to proceed, with regrading and restoration as necessary at a later date.

Attempting to calculate the required volume of grout injection into an annular void between shield excavation and lining in clays is not a fruitful exercise. It will certainly be possible to inject the requisite volume of grout, but it will not be possible to make it flow around the tunnel perimeter in an even layer. Even with multiple simultaneous injection points in use, it may still be found that the grout accumulates in bulbs near the point of injection. If the ground has lost water as a result of the remolding effect of tunneling operations, then the actual volume required to restore even a semblance of the previous support condition will be substantially higher than the calculated neat volume. Injections of this type are best regarded as a form of compaction grout designed to ensure that the final loading on the tunnel lining is reasonably uniformly distributed.

The best results are obtained by establishing multiple simultaneous injection points permanently fixed within the shield tail and passing through the tail seals. Grout is injected throughout the time the shield is in motion. For this system to work, the lining must be a bolted segmented lining with built-in seals between segments, and the annular space must be on the order of 15 cm thick. Projects on which such systems have been used are still few in number and have been confined to subaqueous tunnels. The only one in North America as this is written is the St. Clair River tunnel under construction for Canadian National Railways between Sarnia and Port Huron. It must be expected that for simultaneous injection through multiple ports while the shield is in motion, there will be a substantial learning curve before all elements of the system are functioning smoothly to achieve the desired result.

It is generally difficult to use any mechanical excavation equipment in this type of ground except for a slurry shield or earth pressure balance shield (EPB). These days, the two types of machine are substantially interchangeable with the slurry formed from the native clay, using some chemical additives to control the flocculation and consistency of the clay in slurry mode. The EPB is preferred as being more flexible in varying conditions and somewhat less expensive than a slurry shield. To control pressure in the plenum chamber behind the cutterhead, a screw conveyor is required. The rotational speed of the screw is matched to the advance rate of the EPB, and pressure in the plenum is monitored using multiple sensors. If boulders are likely to be encountered, especially if they will be larger than can pass through the screw conveyor, the cutterhead must be fitted with disk cutters in addition to the drag bits normally associated with this type of machine. This topic is covered in more detail in the section dealing with boulders.

For short tunnels, the use of manual shields is still common practice, especially in urban settings when detailed ground control is an objective, and more especially when there is a possibility of encountering obstacles such as abandoned piles, tiebacks, and existing foundations.

An interesting alternative that gives good results in obstacle-free ground where surface effects are unimportant and there are no utilities to worry about is to use a fully bulkheaded shield fitted with a door of controllable aperture. Such a shield can be shoved blind, admitting only as much soil as is necessary to maintain control of grade and alignment. The ground will be remolded for a substantial distance around the shield. During construction of the Lower Market Street tunnels of the BART system in San Francisco, the portion of the offshore section clear of the piles associated with the Ferry Building piers was constructed by this method after bulkheading the manual shields used in the drives. The earth pillar between the tunnels was steadily reduced in width to less than 5 ft at the connection to the offshore ventilation structure, which formed the transition between conventional and immersed tube construction. The first bore was braced with horizontal 6 × 6 timbers in each ring before the second bore passed. Control of muck intake was sufficiently sensitive to permit correction of a small misalignment of the first bore.

The work was not without its problems. During one night shift, a check valve on the hydraulic system failed, with the result that when a certain shove was completed, the external earth pressure forced the shield backward. Instead of sending for the mechanic, the shift boss decided that he could maintain pressure on most of the jacks, retracting only those necessary for installation of one segment at a time. By the time that he was forced to give up on this idea, the shield was out of alignment 3 in. high. In the following days, the Resident Engineer naturally expressed his dissatisfaction with this outcome and instructed the Contractor to "do something about it." The response was to remove grout plugs throughout the area in the lowermost 120° of the tunnel. Naturally, no grouting had been done in this reach. The clay was encouraged to flow into the tunnel by vibrating it with pneumatic drills whenever there was anyone available to do this. The muck was cleaned up periodically and always flowed reluctantly. In this way, about half the misalignment was recovered. Doing more would obviously have been dangerous. It is interesting that, at the time of construction, this was the cleanest and driest part of the tunnel, even though it was 80 ft below a free water surface. Unfortunately, subsequent concreting of the ventilation structure caisson resulted in settlement, which opened up the joints. This put the end of the tunnel on a par with the rest. Although the joints between the fabricated steel segments were lead caulked, persistent small differential movements arising from reconsolidation of the disturbed clay made it impossible to make the tunnel completely watertight. Logs showed that the water was entering the tunnel at different points at different times, even after all grommet and grout plug replacement and recaulking efforts had ended. The driving of foundation piles for the new Hyatt Regency hotel and for an offshore platform above the tunnels also reactivated the leaks. In all, it was about three years before they were sealed off—largely due to the deposition of carbonate deposits derived from oyster shells in the clay.

Some additional discussion of tunneling in sensitive clays will be found in a later section dealing with the effects of water on tunneling.

Blocky Rock

Rock is a basically strong material that requires little or no structural support when intact; although it may require protection from exposure to air, water, or from fluids conveyed in the tunnel. However, when the rock joints and fractures are sufficiently open that the natural rugosity of the block surfaces will not prevent movement of rock blocks or substantial fragments, the rock is said to be "blocky." If the joints and fractures contain claylike material resulting from weathering or light shearing, then the rock is described as "blocky and seamy." In zones where the rock has small folds but is open along the direction of the folds, it may be free to move in only one direction. Such rock is still blocky.

When rock is subjected to the action of explosives, high-pressure gases flow into any fissures in the rock before they have finished their explosive and rock-fracturing expansion. Even in hard granite, a result of blasting is the creation of microfissures extending well outside the blasted perimeter. In blocky rock, the effect may well extend more than a tunnel diameter outside the desired finished surface; a good deal of overbreak is likely to result.

Another problem with this type of rock is that it is highly susceptible to the destabilizing effects of water flowing through the fracture system with sufficient energy to dislodge successively more rock. This action is dealt with more fully in a later section. Finally, it is quite likely when blocky and seamy rock is encountered in a tunnel excavation, especially in heavily folded strata, that there will be zones where the weathering has proceeded to a conclusion resulting in the presence of weak earthlike material with little capacity to sustain loads or to preserve the tunnel outline.

All of the rock conditions described require early and carefully placed primary support to preserve ground stability and to provide a safe workplace. Even before support installation, it is necessary to minimize surprises by scaling off any loose rock that will present a hazard to the crews installing the support system. Many still prefer to use steel ribs and wood lagging in this type of rock. It provides positive support and is quickly installed in tunnels less than about 5 m in diameter. Unfortunately, crews still have to work under the unsupported rock to install the ribs and lagging; the material costs are high; the presence of timber results in the possibility of future uneven loading on the permanent tunnel lining as wood rots out and steel corrodes; and it becomes relatively difficult to ensure good contact between the lining concrete and the rock even after contact grouting.

For these reasons the use of shotcrete and rock bolts has become popular. In rock known to be blocky and therefore to need support, an initial layer of shotcrete about 5 cm thick should be applied as soon as possible in the tunnel crown. This is followed by the installation of pattern rock bolts whose length and diameter are governed principally by the tunnel diameter. Additional shotcrete may be applied later at a time and location where it will not interfere directly with the work of advancing the heading. This delay in shotcrete placement is rarely a problem. Remember that however flimsy the rock structure is in an unsupported condition, the first thin layer of shotcrete begins to do its job of keeping the rock fragments in place as soon as any adhesion has developed between the shotcrete and the rock. This adhesion will always be weak when the rock substrate is weak, because it can never exceed the tensile strength of the rock perpendicular to the exposed surface. Also, the action of blasting leaves a great deal of microfine dust in the air, which is not precipitated by water spraying. Such dust is electrically charged and it clings by electrostatic attraction to the remaining rock surface. It takes a great deal of work to remove the dust and, in the case of blocky rock, there is always the risk that high-pressure air and water jets will dislodge more rock. It is therefore a good idea to use fiber-reinforced shotcrete to increase the value of shotcrete that has not adhered to the surface.

Thicker initial shotcrete serves little purpose. If the shotcrete is thicker, it will take longer to apply; more significantly, it will not provide any more useful support, because if any block of rock is heavy enough to break through the thinner shotcrete, it is likely to be heavy enough to peel off thicker shotcrete, especially if adhesion is low. The provision of structural support in a rock tunnel should always be examined carefully to see what loads are present once initial failure has been prevented. It is certainly prudent to allow for future damage and deterioration, but the necessary allowances are often overdone. Note that wet-mix shotcrete permits the application of thicker layers than the more common dry mix. It also results in somewhat less rebound. The addition of silica fume is considered helpful by many.

Alternatives to rebar or Dywidag rock bolts are to be found in Split sets (marketed by Ingersoll Rand) and Swellex bolts (marketed by Atlas Copco; see Chapter 7). Both have the advantage of much faster installation than conventional resin or cement cartridge designs, but they may have other disadvantages, depending on how they are used. Split sets are essentially steel tubes with a longitudinal split. They are driven into a slightly undersized drilled hole and gain their purchase by friction against the sides of the hole. Swellex bolts are closed cylinders collapsed into a C-shape. They are introduced into drilled holes and then expanded by forcing in compressed air through a fitting on the exposed end. The cylinder blows up like a balloon and fits itself to the rough interior shape of the drill hole, thereby providing anchorage along its entire length.

See Chapter 7 for further discussion of design philosophy and the technological details of the various support components.

Adverse Combinations of Joints and Shears

Jointing systems in rock arise from many causes, some of which are noted here. Sedimentary rocks, and particularly limestone, typically have three more or less orthogonal joint sets arising from the modes of deposition and induration that formed them. Not all joints are continuous, but those in any set are parallel. There may be many sets or, in weak, massive sandstone, for instance, only one or two.

Intrusive igneous rocks that have cooled slowly may be massive with widely separated open joints. Basalt often shrinks so as to exhibit a columnar structure reminiscent of crystal forms. The joints are then distinct and are usually somewhat open. When erosion has created a deep valley, stress relief in the rock in the valley sides commonly causes the formation of a joint set parallel to the valley axis. Rock that has been sheared or weathered along continuous planes may combine with the joint systems to create the adverse conditions discussed here.

Joints and fracture systems combine to break up the rock mass into interlocking fragments of varying sizes and degrees of stability. The situation where open spaces or weathering products are present in the joints and fractures has been discussed earlier in this chapter.

In the absence of direct evidence to the contrary, it should be assumed that shears and faults are continuous throughout their intersection with the tunnel excavation. In schistose materials, weathering usually follows a foliation plane to great depths, even in temperate climates when a weak zone has been formed by slippage along that plane. Other faulting may cause the development of extensive fracture systems in any direction. A section through the project area perpendicular to the strike of the exposed surfaces in schistose materials will generally reveal a sawtooth profile with one of the surfaces parallel to the foliation. Continuation of the plane thus defined to tunnel elevation will be a preliminary indicator of the presence of sheared and weathered rock in the excavation.

Continuous joints and shears can define large blocks with little or nothing to hold them in place once the tunnel excavation has been completed. It is important to identify the locations of blocks with the potential for falling out in order to provide support during cautious excavation. For large-diameter tunnels in particular, this requires an assessment of the potential before construction begins, mapping during construction, and control of drift size and round length to ensure against complete exposure of an unstable block in a single round. The processes required can be simplified by the use of computer programs such as have been developed by Richard Goodman at U.C. Berkeley.

The difficulty of controlling the correct placement of steel sets in multiple-drift headings works against the use of this kind of support. Initial rock bolting followed by reinforced shotcrete is a reasonable approach. In all cases where rock bolts have to be located to take direct and reasonably predictable loads, it is better that they be installed ahead of the shotcrete while the joint locations are still visible. If mechanical rock bolt installers cannot be used, then the crews must be protected by overhead cages.

During construction of a 14-ft-diameter section of the Manhattan West Side Interceptor tunnel beneath Riverside Drive just north of 96th Street, drill-and-blast tunneling was in progress ahead of a shield that had been installed to cross a reach of mixed-face tunnel. An adit had been driven previously to minimize blasting requirements during shield tunneling. The roof of the adit was several feet beneath the soil/rock interface. Nevertheless, the firing of a round, using only 15 lb of dynamite in all, caused a fallout extending 15 ft ahead of the end of the round just drilled. Before the roadway could be barricaded off, a crater developed to the surface below a car driving north. No one was injured, but it could have been a different story had a bus fallen into the crater. No utilities were involved, and the hole was immediately backfilled with tunnel muck and paved over. Nevertheless, in spite of a major consolidation grout program, it was necessary to repave the area on an almost daily basis until about 8 ft of asphalt were in place. Problems are most likely to arise in wide noncircular tunnels for highways.

Faults and Alteration Zones

Tectonic action, high pressure, and high temperatures may metamorphose rock into different structures with unpredictable joint patterns. The uplift and folding of rocks by tectonic action will cause fracturing perpendicular to the fold axis along with faulting where the rock cannot accommodate the displacements involved, so that shears develop parallel to the fold axis. Other types of faults arise as the earth accommodates itself to shifting tectonic forces. Faults or shears may be thin with no more significance than a continuous joint, or they may form shear zones over a kilometer wide in which the rock is completely pulverized with inclusions of native rock, sometimes of large size.

Intrusive dikes and sills of igneous rock may be injected along planes of weakness. These dikes and sills, if sufficiently hot and substantial in width, may induce contact metamorphosis along their margins. Depending on the mineralogy, the intrusive material may weather at a faster rate than the intruded formation.

The injection of magmatic fluids causes substantial alteration (and mineralization) of the host rock and may leave the altered rock in a condition mechanically similar to all stages of weathering. Water may accumulate preferentially in the altered material.

All of the conditions briefly described above may be additionally complicated by the presence of locked-in stress or high overburden loads.

Dealing with the conditions encountered in such fault zones and weathered intrusive zones depends on the excavation method in use, the depth below the ground surface, the strength of the fault gouge, the sheared material or the weathered or altered rock, and the water conditions. Water problems in general are discussed in the next section, including consideration of the difficult water conditions commonly found in association with faults; however, to the extent that they affect the selection of construction methods appropriate to fault crossings, they are referred to here.

Crossings of major transform faults at plate boundaries, such as the 1,750-km (1,110-mi) long San Andreas fault may attract more interest as an engineering challenge, but the vast majority of faults are not of this type. They are, indeed, ubiquitous, and most tunnels more than about 500 m (1,500 ft) long will cross at least one fault or shear. The difficulties caused by faulting are generally a result of the nature of the material in the sheared zone and the weakening of the adjacent rock. Without going deeply into the reasons underlying the change of form, suffice it to say that the fault gouge may take any combination of the following forms: completely claylike; clay with blocks and boulders of the original material (these blocks, while superficially intact are often very highly fractured); sandy gravel with varying amounts of clay; or sand. The material is formed in place initially by the energy provided by tectonic action; later, weathering as a result of seepage of free water within the fracture zone can result in additional alteration (weathering) to substantial depths as a result of weak chemical reactions of the various minerals present. Much more significant changes are wrought by the injection of hot, chemically active fluids from below, leading to substantial local alteration of the rock affected. All forms of alteration generally act both to weaken the rock and to make local forecast of conditions difficult. Some types of alteration are more predictable as a result of experience. For instance, foliation shears in the schists commonly found along the eastern seaboard usually result in clay seams on the order of 10 to 30 cm in thickness. The shear strength of the clay is reasonably high, and the shears only contribute to instability when they are part of an adverse jointing system.

In shallow tunnels, preconstruction geotechnical exploration can identify the presence and nature of significant faulting with reasonable accuracy, although minor shears may escape detection. Nevertheless, since their presence is probable and the jointing system and foliation directions are known, the design and construction methods can be designed to take them into account.

The topography of Manhattan Island is largely determined by faulting in the schistose bedrock, so that the locations of significant faults are easily determined in relation to project alignment. An interceptor sewer tunnel on the Upper West Side was largely constructed as two, 2-mi-long TBM tunnels, 8-ft, 6 in. and 11-ft, 0-in. diameter, in the early 1970s. The 11-ft, 0-in. tunnel was expected to cross a 600 ft long sheared and weathered zone beneath high-rise apartment buildings at a shallow depth. There are two construction problems inherent in this situation for TBM construction when ground control had to be maintained at all times. It was known at the time TBM construction was undertaken that the rock was generally very hard—in fact it was the hardest successfully penetrated by a TBM up to that time. The machine selected was therefore the strongest and stiffest design available. This meant a machine fitted with radial grippers to provide the thrust reaction for the rotary cutterhead operation. As the TBM entered a weak zone, there came the time when the grippers had to bear on the soft material. As long as the cutterhead was also in weak material, a balance could be found provided that no attempt was made to take too much advantage of the easy cutting conditions. However, once the grippers were bearing on soft rock or clay while the cutterhead was bearing against hard rock, the grippers could no longer provide sufficient reaction. In the meantime, the weak rock in the tunnel crown was not self-supporting for the time necessary to install a conventional temporary support system.

The solution to the gripper problem was found by placing baulks of heavy timber beneath the grippers and then extending the grippers to force the timbers into the clay. It proved possible to sustain sufficient frictional forces in this way to permit controlled progress of the machine. The solution to the temporary support system was found by installing unconnected partial rings of light liner plates in the tunnel crown, immediately behind the cutterhead, and holding them in position with short rock bolts. Once the grippers were past the liner plate, additional plates were installed, generally extending down to springline. There was sufficient support from the rock bolts to sustain the partial liner plate rings in place until the concrete lining was poured at a later date, without adversely affecting the overlying structures.

Current technology would provide other solutions, such as the use of precast concrete lining in the weak ground with supplementary jacking capability to enable the lining to provide the jacking reaction. However, the more primitive solution is an indication of what could be accomplished if such a situation were encountered unexpectedly.

In general, fault crossings offer conditions akin to those of mixed-face tunneling, and the same methods are available to deal with them. Different circumstances come into play with deeper tunnels, especially if these are of large diameter. Such tunnels are usually long, and logistics are important. The comparative lengths of fault zone and normal tunnel dictate that the construction method be efficient for the normal tunnel. Nevertheless, sufficient flexibility is required to permit safe and reasonably expeditious construction through the worst conditions likely to be encountered. Drill-and-blast excavation is still commonly used in such tunnels. Rock bolts and shotcrete then become the preferred support system, although steel ribs and lagging or steel ribs with shotcrete are also still used. TBM successes in these conditions have been few. There are two principal problems: the loose material in the fault runs into the buckets and around the cutters and stalls the cutterhead; and if the fault contains cohesive material, it squeezes and binds the cutterhead and shield with similar results.

Such conditions plagued the pilot tunnel of the Ping Ling highway tunnels in Taiwan. The material excavated in the portal zone is a terrain with characteristics similar to those of the Franciscan formation in California. The rock is heavily sheared in random directions with substantial faulting. The TBM excavation was routinely interrupted at fault crossings to permit the excavation of a bypass drift by manual methods to enable clearance of the raveling material from the front of the cutterhead.

The best physical solution to the problem of loose or loosened raveling and running material is to establish a grout curtain ahead of the TBM and then to maintain it by continuing a grout and excavation cycle throughout the fault-affected portion of the drive. Even if imperfect—as consolidation grouting tends to be, especially when placed from within the tunnel in conditions providing limited access—it is likely that a properly designed and executed program will add sufficient stability to the ground to permit progress. It should be noted that any such program will be expensive and time consuming. It is therefore unlikely that any contractor will willingly do the necessary work unless it has already been envisaged in the contract as a priced bid item. It is also important to recognize that if water is running into the tunnel through the working face, a bulkhead will be required to stop the flow while the initial grouting is in progress. Grouting into running water is a slow and expensive way to establish a grout seal.

Within limits, the squeezing problem can be dealt with in part in TBM tunneling by tapering the shield and making its diameter adjustable within limits; and by beveling the cutterhead itself to the extent that this is possible without interfering with the efficiency of the buckets. Expandable gauge cutters are also used, but this is not yet mature technology. One of the problems is that there is a tendency for local shearing of the cutter supports to result in an inability to withdraw the cutter once it has been extended. Also, since such cutters are acting well outside the radius of the buckets, muck that falls to the invert is not collected but provides an obstruction the cutters must pass through repeatedly. This grinds the debris finer and finer and abrades the cutter mounts as well as the cutter disk. This makes it necessary to provide means for eccentric cutterhead rotation so that the invert is properly swept. Unfortunately, squeezing is commonly, if not most often, manifested preferentially in the tunnel invert. This cannot yet (in 1994) be regarded as a solved problem.

The Hades and Rhodes Tunnels in Utah were successfully completed by TBM in faulted and somewhat karstic limestone. To cross the fault zones, the contractor stopped up above the cutterhead and hand mined and supported an enlarged roof cavity for the length of the loose rock zone. The TBM was then able to proceed, and minimal time was lost.

The Yacambú–Quibor water supply tunnel in Venezuela, 25 km long and 5 m in diameter, is being driven in the

Venezuelan Andes. The rock types are generally phyllites throughout the inlet half of the tunnel and interlayered sandstone and weak shale (lutita) with quartzitic veining in the outlet half. The two formations are separated by the Bócono fault. The Bócono is a transform fault similar to the San Andreas fault; its width is about 800 m. The tunnel passes through it at a depth of more than 600 m. In recognition of the sheer difficulty and the time it would take to excavate this section of the tunnel, a separate inclined adit was driven at a 15% gradient to intersect the tunnel alignment downstream of the fault zone. Substantial difficulties with squeezing ground made adit construction itself difficult for the first several hundred meters. Driving toward the fault, light squeezing was experienced in material whose original structure was difficult to determine by visual examination. After an interval of delay as funds were exhausted, tunneling work resumed using conventional drill-and-blast methods. In the fault proper, the fault gouge generally presented a claylike appearance, and it contained very large blocks of the native rock, so that the tunnel was at times in clay and at others in rock with numerous but tight fractures. Conditions were generally dry and the rate of squeezing was quite low, enabling support with shotcrete and mesh. Heading and bench tunneling was generally used. At one point, following the long Christmas and year-end break (about three weeks in all) it was found that more than 100 m of the heading had failed by invert heaving. The crown did not fail. The heading was recovered, and excavation continued using a mix of backhoe and drill-and-blast excavation. On the whole, conditions in the fault zone proved to be quite favorable for tunneling, although productivity was severely reduced by access limited to a long and narrow tunnel at 15% grade and the continuous, low-volume flow of water into the heading from the adit. The water ponded in the invert, and it was a long time before adequate pumping facilities were installed. The problem was exacerbated by the use of a rubber-tired haulage system. In terms of support requirements, more severe conditions were found in the zone contained between the Bócono and the associated Turbio fault. This is discussed in more detail in a later section dealing with squeezing ground.

The Crystal Springs lake tap in San Mateo County south of San Francisco was constructed in the Santa Clara formation within the trace of the San Andreas fault. The tunnel was driven from an on-shore shaft to a terminus beneath preinstalled risers in the lake bed. The material had the basic characteristics of clay with rock fragments and sand seams. Backhoe excavation was used in conjunction with specified grouting to seal off the prevalent small leaks. When allowed to flow for any substantial length of time, the leaks caused raveling; the raveling became visibly progressive. Near the shaft, steel ribs and lagging were used in the horseshoe tunnel, but as water control became a problem, shotcrete was first applied over the lagging. This effort was not very effective, and in the last reach, shotcrete was used with hit-and-miss lagging or without lagging. The grout pattern was not designed to create a fully grouted zone ahead of the working face, and it generally accomplished little while being a substantial impediment to tunneling productivity. Near the end of the tunnel, where cover was reduced to less than 30 ft, the flow rate of water through the face increased significantly in spite of repeated attempts to grout it off. It was decided that it would be unsafe to proceed further without a drastic change in working methods; and since the risers were accessible from the tunnel at its current station, the working face was bulkheaded off and all shotcreting was brought up to date. On the whole, the shotcrete did more than the grouting to control water inflow. While the flow rates were never large in absolute terms, they took on an added significance for the miners who could hear microseismic activity of the San Andreas amplified by the resonance of the tunnel cavity.

Water

It was Terzaghi's view that the worst problems of tunneling could be traced to the presence of water. Among other things, he considered that (except for circular tunnels) it was prudent to double the design rock load on the tunnel lining when the tunnel was below the water table. This in itself would not be a serious problem, since most tunnel linings are already limited as to their minimum dimensions by problems of placement rather than by design considerations. However, there are many other problems are associated with the presence of water. Several are discussed here, working in sequence from clay to rock and, within rock, from weak and fractured to strong and intact.

The routine methods of groundwater control for tunnel construction are dealt with elsewhere in this book. The discussion that follows here will therefore be limited to special cases not ordinarily contemplated and conditions not frequently encountered or not well understood.

Clay. Most clays are at least slightly sensitive. This arises from the microstructure of clay soils, which are composed largely of platy minerals. As with a heap of coins, the packing is not perfect, even though the clay is relatively impermeable. Each fragment is held in place by some combination of free-body equilibrium forces, ionic interaction, and chemical or mechanical bonds at the contact points. The pores of the clay are generally filled with water, which may contain salts in solution. Disturbance of the clay results in disruption of the bonding, migration of water, and at least temporary weakening of the clay structure. The free water will be released at any temporary boundaries formed by shearing. As the clay reconsolidates, it is likely to gain strength over the initial condition, but this will be a protracted process. The immediate effect, and the one that affects tunnel construction, is loss of shear strength throughout the disturbed mass. In organic silty clays, such as the Bay Mud found in the San Francisco Bay area and mentioned earlier, the sensitivity is commonly about 4, indicating a fourfold loss of strength upon remolding. This is associated with an initial water content of about 60%. In any one material, the sensitivity may vary greatly, depending on the water content. Sensitivities as high as 500 to 1,000 may be found in some clays, such as the Leda clay commonly encountered in previously glaciated areas and especially in Scandinavia and parts of Canada and Alaska. Marine clays such those found in Boston lose salt by diffusion when situated below the water table. Such clays are typically highly sensitive.

Tunneling is already sufficiently challenging in moderately sensitive clays, as the critical number (described previously in the discussion of the behavior of soft clays) suffers a local fourfold increase. For shielded tunneling, it is very important to avoid excessive efforts to correct line and grade, as it is easily possible to create a situation in which control is lost.

A further effect of disturbance of sensitive clays is directly dependent on the loss of porewater expressed from the clay. The volume change results directly in rapid subterranean and surface settlement. In addition, the clay closes rapidly on to the tunnel lining, resulting in even greater settlement unless adequate compensation grout can be injected promptly.

Finally, the effects in any natural clay are somewhat controlled by local differences in soil structure and amount of disturbance. As the soil reconsolidates, flexure of the tunnel lining will occur to an extent sufficient to open sealed joints. Observations in the Lower Market Street BART tunnels showed that an area leaking on one day would be dry on another without intervention. The converse was also true. Efforts to make the whole tunnel dry at one time proved to be wasted. The sealing method used was lead caulking in conjunction with polyethylene grommets. At a time when foundation piling was driven for an adjacent new hotel, the leakage rate increased. In all, it was more than three years before the tunnels were dry. It is tempting to think that calcite deposition from the abundant shells within the Bay Mud was the final sealing agent. However, the pilot tunnel for the first bore of the Dartford–Purfleet Tunnel beneath he Thames east of London was driven in chalk and used for pregrouting of the chalk and overlying gravel. Because of the interruption in construction imposed by World War II and its economic aftereffects, the tunnel remained untouched for 20 years after it had been completed. When construction recommenced in 1956, the pilot tunnel was leaking quite freely, even though stalactites and stalagmites were extensively developed in many areas.

A problem encountered in the Detroit area affected a sewer on Nine-Mile Road. The whole area is underlain by clays, which are typically weak near Lake St. Clair and which gain strength with distance from the lake. Tunnel lining design is specified as circular, unreinforced concrete 1-in.-thick per ft of tunnel diameter plus 3 in. additional thickness. Construction joints were vertical, with waterstops between full-circle pours. There was at first no obvious reason for the collapse, the visible clay being self-supporting. However, investigation revealed that the small clear water flows through honeycombed concrete at the construction joints was in fact carrying a substantial burden of fine-grained granular material. This led to the identification of a stratum of black SP sand within the clay below springline. Over the years, sufficient material had been eroded from this stratum by the small leaks (especially at times of low flow in the sewer) to remove the support required by the tunnel.

Weak Rock. Similar conditions are more easily recognized in the Twin Cities (Minneapolis–St. Paul), where extensive tunneling is done in the St. Peter Sandstone. This is a weak "rock" of variable strength, depending on degree of cementation. Otherwise, it is a fine-grained dune sand that gains its strength from intergranular mechanical bonding, with iron salts and calcite providing some chemical bonding. There are substantial natural and manmade caves, which have endured for long periods. However, flowing water easily breaks down the bonding, and it has been observed that the sand (which has a characteristic grain size of 0.3 to 0.5 mm) will flow quite freely through narrow cracks. In steam tunnels in the University of Minnesota campus area, the ambient temperature is sufficiently high to have dried out the surrounding sandstone, and places can be found where sand trickles in through similarly fine cracks in the concrete tunnel lining. To deal with the problem posed by mobility of the sand, current specifications require the sealing of all cracks wider than 0.05 in. in concrete tunnel linings.

The control of water in weak and unstable rock was a major concern during the initial segment of the Channel Tunnel drives offshore from the Sangatte shaft in France. For several kilometers, the tunnel drives had to penetrate a soft, fractured, and highly permeable chalk unit. The machine design chosen to perform this difficult task was put together by Robbins and Kawasaki, using a cutterhead capable of acting in the closed mode in the bad ground and designed to withstand external pressures of 10 bars; and which could be operated in open mode in the better ground for enhanced productivity. The combination design proved successful.

Blocky and Seamy Rock. Similar problems are often found on a smaller scale in blocky and seamy rock where continuous seepage is occurring. This seepage may occasionally become a high-volume flow or inrush. Such inrushes are usually self-limiting, so that when the reservoir of water has been discharged, the rate of flow will return to nuisance levels. Occasionally, however, a relatively inexhaustible reservoir is encountered. In deep tunnels, the flow may also be at high pressure. In such cases, the preferred approach is to contain the flow behind an impervious bulkhead as a preliminary step before consolidation grouting. A partial timber bulkhead is erected near the face and is fitted with an open pipe at low level; this pipe is fitted with a valve at the outer end, which is left open during preparatory work. Concrete is placed up to pipe elevation for the full width of the bulkhead wall. Corrugated sheeting may then be placed against the rock face, irregularities being filled with stone; this will permit flow behind the bulkhead to reach the discharge pipe. Depending on the rate of flow, it may also be possible to use geotextile membrane for this purpose. The main timber bulkhead is then completed and concreted to provide a reasonably strong and impervious barrier. Once the concrete has cured, the valve is closed so that the flow of water will not interfere with the placing of consolidation grout. Grouting of the rock in which the bulkhead is located is highly advisable.

Depending on conditions, cement grout with a 1:1 to 2:1 water/cement ratio will be used to seal off the flow to at least 15 ft behind the face. The use of higher water/cement ratios leads to early breakdown of the grout seal and is not recommended. Initial injection of water to test permeability and to flush out fines is acceptable. Other materials may be found necessary if the fissures are narrow or the rock is generally pervious (e.g., weathered limestone), such as microfine cement or expansive polyurethane grouts. The progress and success of the grouting program can be tested by cracking the bulkhead valve. Once the water is under control, tunnel advance can continue with successive rounds of consolidation grouting until better rock is encountered. It is, of course, necessary to angle the outermost holes along the surface of a cone so as to establish a barrier around as well as ahead of the tunnel. It may prove necessary to do extensive additional grouting to establish the integrity of the bulkhead.

It should be noted that unless the work is done in a thorough and workmanlike manner, the chances of success in establishing consolidation grout from the tunnel face are small. The use of silica gels pumped into flowing groundwater is generally a waste of time and money. It is essential to stop the flow of water first if grouting is to be successful in this application. Success has been reported in fissure-grouting of bare tunnel walls in dolomitic limestone using expansive polyurethane grouting agents, but grouting off inrushes through the working face presents a different problem.

Clay Overlay on Bedrock. In certain Norwegian cities, especially Oslo, the presence of thick deposits of soft marine clays overlay a bedrock consisting mainly of Cambro-Silurian Age shales and limestones. Reductions in pore pressure in the thick clay mantle lead to damaging settlements of surface structures. It has therefore been necessary to introduce pressure grouting to seal off fractures and consequent leaks ahead of tunnel construction even though natural water inflows would be small. This has been achieved, even in TBM tunnels, by grouting ahead outside the cutterhead from a point well behind the working face. Such grouting has been controlled by observing water pressures in drill holes and adjusting grout pressures accordingly on an experience basis to preserve porewater pressures monitored in the foundation soils. Postgrouting has been found necessary from time to time when the initial program has failed to provide adequate sealing. Since the grouting rounds are generally performed every 20 m, substantial interference with productivity results in increases in tunnel cost per meter from about 50% to 150%. The cost increase in drill-and-blast tunnels where more grout is needed has been reported to be as high as 330% (Garshol, 1983).

In the blocky sandstone-siltstone-shale reach of the Yacambú-Quibor tunnel, the excavation encountered a fault several meters wide. The fault was filled with broken rock and sand with a high content of water under hydrostatic head. After the initial inrush, water flow diminished but did not cease. Cement grouting further diminished the flow, but the next attempt to advance the heading using crown spiling resulted in another inrush. Probing indicated the presence of a void with sand collecting in the bottom and a rising water level in an open cavity. It was apparent that loose material was ravelling from the crown of the cavity while water continued to flow. It was also clear that the spiles had not been sufficiently long to span the fault. Rather than attempt to inject sufficient material to fill the cavity (which might still have remained unstable), cement grout was injected to stabilize the debris and 10-m long tube spiles were installed on close centers in the affected area. These spiles had slits to enable the injection of grout once in place. This method succeeded and the grout injection was also sufficient to dry up the water flow. Since a long reach of tunnel had already been excavated in very wet conditions, use of the spiles was continued to traverse the remainder of the reach of wet ground and was completely successful in preventing water flow into the tunnel. The drawback to this method was that installation of a complete spile canopy required about 24 hours every 8 or 9 m. The system is more fully described in the following section.

Mixed-Face Tunneling

Tunneling in mixed-face conditions is a perennial problem and fraught with the possibility of serious ground loss and consequent damage to utilities and structures as well as the prospect of hazard to traffic. The term *mixed face* usually refers to a situation in which the lower part of the working face is in rock while the upper part is in soil. The reverse is possible, as in basalt flows overlying alluvium encountered in construction of the Melbourne subway system. Also found are hard rock ledges in a generally soft matrix (e.g., basalt in limestone encountered in Montreal); beds of hard rock alternating with soft, decomposed, and weathered rock; and noncohesive granular soil above hard clay (as in Washington, D.C.) or above saprolite (as in Baltimore). The definition can also be extended to include boulders in a soft matrix (discussed elsewhere in this chapter) and hard, nodular inclusions distributed in soft rock (e.g., flint beds in chalk, or garnet in schist).

The primary problem situation is the presence of a weak stratum above a hard one. With modern tunneling equipment, the other cases are now trivial and will not be discussed here.

There will always be water at the interface, which will flow into the tunnel once the mixed condition is exposed. This increases the hazard because of the destabilization of material already having a short stand-up time. Stabilization therefore calls for groundwater control as well as adequate and continuous support of the weak material. Moreover, this support must be provided while energetic methods, such as drill-and-blast excavation, are required to remove the harder material.

It is not yet possible to recommend the use of rotary head tunneling machines in this application. It is not assuredly practical to control the inflow of the soft material so as to match the forward progress of the machine as it cuts the hard material. Efforts to do this have resulted in total loss of control of line and grade of the tunnel as well as surface settlements substantial enough to close traffic lanes in an urban situation in more than one tunnel where the attempt has been made.

Dewatering can reduce the head of water, but it cannot remove the groundwater completely; nor can it realistically be expected to offer control on an undulating interface with pockets and channels lower than the general elevations established by borehole exploration. Compressed air working will not deal with water in confined lenticular pockets, and it is usually inappropriate when the length of the mixed-face and soft ground conditions amount to only a few percent of what is otherwise a rock tunnel. Also, recent experience where extensive beds of clean (SP and SP-SM) sands have been major components of the weak ground shows that compressed air alone will not stabilize the ground, which becomes free-flowing as soon as it has dried out. Therefore, on the whole, consolidation grouting is to be preferred in this situation.

A commercial system developed in Europe uses tubular spiles, slotted for grout injection. The system has been very successful in controlling such conditions. The spiles are about 10 cm in diameter and 10 m long, and grouting pressures compatible with face stability are used. Packer grouting could be used to provide an initial seal in the first 2 or 3 m beyond the working face, with higher pressures in the remainder of the spile. Spacing depends on permeability of the soil, but an initial separation of about 30 cm should suffice to demonstrate the needs of the situation. If a shield is being used, it will be necessary to drive the spiles far enough ahead so that the inner ends will clear the shield skin. Removable extensions can be used to accomplish this end. It will normally take 12 to 24 hours, more or less, to complete installation and grouting of one set of spiles, depending on spacing and tunnel diameter. The system can also be used to deal with troublesome water in a rock tunnel. In this application, grouting pressures up to 30 bars may be appropriate.

Again, the best time to seal off groundwater is before it has started to flow into the tunnel. Once the water is flowing, it is extremely difficult to stop it from within the tunnel except by establishing a bulkhead.

Portal Construction

It is common experience that when the first round is shot into the rock face, a major slide occurs on adverse joints, on hidden shears, or in unstabilized blocky rock. Some such slides have caused major damage and loss of life, although more commonly, there is only the frustration of having to clean up the situation and start again. This situation can be controlled, as has been demonstrated in the construction of several recent highway tunnels (Peterson, et al., 1991).

Detailed mapping of joints should be a part of the initial geotechnical exploration program. As the portal is excavated, the mapping should be confirmed bench by bench to ensure that the proper initial rock bolt stabilization is installed. It is prudent to cover all newly exposed rock surfaces with wire mesh to control the raveling that will normally occur. At a convenient stage of the excavation, a ring of rock around the portal is heavily rock bolted, and the ends of the bolts are left protruding to provide a tie-in to a structural concrete portal poured against the rock face. The structural tie between the concrete and the rock may be additionally increased by installing tensioned rock bolts through the concrete if this is merited. The initial excavation of the tunnel may now begin with much greater assurance. Both portals should be treated similarly, and a minimum of 20 or 30 m of tunnel should be excavated from each portal, even if the tunnel is driven from one end, so as to ensure that the final connection will be made at a distance from the free rock face. Construction of the architectural portals follows completion of construction.

While it is highly desirable that the location selected for the portal be in fresh rock with cover of the same order as tunnel width and height, environmental constraints or other relevant considerations will sometimes dictate that the portal be located where there is low cover, weathered rock, or even soil. Where rock is exposed, the preconstruction of a reinforced concrete portal structure will still be of substantial assistance. As tunneling proceeds (probably in multiple drifts) it will be necessary to install considerable rock bolt and shotcrete support, using wire mesh tied to the bolts as reinforcement for the shotcrete. Multiple layers may be needed, depending on the ground loading that will be necessary to support to prevent failure.

Ultimately, in weathered rock or soil, it will be necessary to use either conventional spiling of sufficient length and short enough distance between successive rounds of spiles to ensure that there are at least two and preferably three layers of spiles at any location in addition to steel ribs and/or reinforced shotcrete to provide the necessary initial support. The grouted spile method referred to in the discussion of mixed-face tunneling will also be applicable. Substantial success has attended the use of preinstalled drilled-in pipe spiles 8 in. or more in diameter penetrating the full depth of the weak ground to form a roof over the tunnel crown and to extend down a part, at least, of the walls.

Where there appears to be little risk of developing a slide or collapse at the portal, it is still prudent to establish a false portal up to 8 to 10 m outside the rock portal to provide protection during construction.

HEAVY LOADING

Squeezing Rock

When a tunnel opening is formed in rock, the local stress regime is changed. The radial stress falls to zero and the tangential stresses increase to three times the in situ overburden load (neglecting the effects of any locked-in stress resulting from past tectonic action that has not been relieved). If the unconfined compressive strength of the rock mass is less than the increased tangential stress, a mode of failure will be initiated, which is described as "squeezing failure." As elastic failure occurs, with consequent reduced load-bearing capacity of the rock, the rock load is transferred by internal shear to adjacent rock until an equilibrium condition is reached. If the rock develops brittle failure and is shed from the tunnel walls, then there will be no residual strength of the failed rock to share in the load redistribution. If the rock is sufficiently weak or the overburden load too great, the unrestrained tunnel will close completely. In deep gold mines, this can happen explosively as rock bursts shatter the tunnel or drift walls. In milder form, although still dangerously, popping rock shows the same phenomenon with lesser force. The action of load redistribution still continues, and it may well be that reexcavation of a failed tunnel opening will encounter much less difficult conditions.

The Squeezing Process

The detailed mechanism of rock movement is complex and depends on the presence or absence of water and swelling minerals as well as on the physical properties of the rock. For the purposes of this discussion, however, the squeezing process may be described as follows.

Initial Elastic Movement. As the tunnel is excavated, stress relief allows elastic rebound of rock previously in compression to relieve stress. This stress relief occurs beyond the working face as well as around the tunnel excavation. In thinly laminated rocks such as schist and phyllite, the modulus of elasticity parallel to the foliation is likely to be much higher than that in the perpendicular direction. Therefore, the elastic movement immediately distorts the shape of the excavation as the rock moves a greater distance perpendicular to the foliation than parallel to it. Moreover, since the rock can move more easily along regular foliation planes than perpendicular to them, more than one factor is at work determining the actual distortion of the tunnel shape. The elastic rebound takes place in all tunnel excavations and is not properly a part of squeezing, which is associated with changes in the rock structure. However, the associated increase in tangential stress in the rock initiates the next phase of movement as the rock fails. As the rock moves toward the tunnel opening, the circumference of the tunnel shortens. There is a limit imposed by the modulus and strength of the rock on how far this process can continue before elastic failure is initiated. Consider a rock of compressive strength 35 Mpa and an elastic modulus of 35,000 Mpa. The circumferential strain per unit length at failure will be 35/35,000 cm/cm or 1 mm/m. For a tunnel of 2-m radius, therefore, a shortening of this radius by about 2 mm implies the initiation of elastic failure at the exposed rock surface. This does not mean that the rock suddenly loses all strength (unless it is brittle enough to flake off the wall), but rather that its residual strength is greatly reduced. As the tangential shear stress builds up, there will come a time when the differential stress is sufficient to cause internal shear failure. This is manifested by the development of new parting surfaces where the overstressed rock separates from the neighboring rock.

The development of the high-tangential-stress regime was clearly demonstrated during construction of the Pfaender Tunnel in Austria. A small-diameter pilot tunnel was excavated by TBM in the overstressed rock in which the tunnel was to be driven. This pilot tunnel was essentially unsupported. Drill-and-blast enlargement to final size was undertaken to complete the two-lane highway tunnel using conventional heading and bench and NATM (New Austrian Tunneling Method) support methods common to this type of construction in Europe. As the heading progressed, it revealed strongly defined circles resembling joints concentric with the pilot tunnel and spaced less than 10 cm apart over the whole of the exposed working face. The strong definition of these pseudojoints was such as to obscure the original jointing system. In addition, the working face itself assumed a domed shape.

Strength Reduction. When the rock remaining is insufficiently strong to carry the increased load passed to it as shearing progresses, it will fail in turn. In strong and brittle rocks, this failure can result in explosive release of rock fragments from the surface in a phenomenon known as "rock bursting." A somewhat gentler expression of the same phenomenon is known as "popping rock," which is still dan gerous. Because these occurrences actually remove rock from the surface, there is obviously no residual load-carrying capability of the failed rock. In weaker and less brittle rock, the failed material stays in place and enters a plastic or elasto-plastic regime. Its modulus of elasticity and its unconfined compressive strength (which represent its load-carrying capacity) may be reduced by two orders of magnitude, but it can still support some load. In the meantime, the load shed by the failed rock at the perimeter of the opening is transferred deeper into the rock mass, where the degree of confinement is higher and the ultimate load-bearing capacity is therefore also higher. The phenomenon may be modeled stepwise, but it is truly a continuous process and will cease only when the total load has been redistributed. Depending on the amount of excess load-carrying capacity available in the partially confined rock around the tunnel perimeter, the stress regime will be affected up to several tunnel diameters away from the opening.

Compounding the stress increase, which leads to failure, is the similar regime in the dome ahead of the working face. The abutment of this dome is the already overstressed rock behind the working face. The problem is therefore three-dimensional in the region affected. The initial movements associated with strength reduction take place quite fast, so that as much as 30% of the final loss of tunnel size may be completed within one to one-and-a-half tunnel diameters behind the working face.

Creep. As a consequence of the reduced elastic modulus and the reduced strength of the rock, additional radial movement of the tunnel walls occurs. In the zone outside the tunnel, the rock properties are substantially changed. In particular, both the elastic modulus and the unconfined compressive strength decrease continuously (but not in a linear fashion) from their original values still existing in undisturbed rock toward the tunnel wall. The tunnel decreases in diameter as the weakened material creeps toward the tunnel boundary. The rate of movement is roughly proportional to the applied load. The movement is therefore time-dependent (after the initial elastic stress relief, which may be regarded as essentially instantaneous). As the ground is allowed to strain, so the strength of the support required to restrain further movement is reduced. However, depending on the amount of squeezing, shear failures and dilatation accompanying failure may result in unstable conditions in the tunnel walls and crown. Since the timing, location, and amount of such failures are not subject to precise definition, support is usually introduced well before the full amount of potential movement has occurred.

Modeling Rock Behavior. Because of the nature of the failure mode, elasto-plastic and visco-elasto-plastic mathematical models have been developed to describe the resulting movements and to evaluate the stress regimes. These models are not exact but correspond sufficiently well with experience to be useful. Unfortunately, for any given tunnel they depend on the use of information that can only be derived from experience in the specific tunnel involved. This is the origin of the observational approach to tunnel support exemplified by the NATM.

It has been noted from experimental work that the net load appearing at the tunnel surface varies with the tunnel diameter as a power function. The loading is also dependent on the rate of tunnel advance. It is therefore clear that when such conditions are encountered, the smallest tunnel diameter adequate for the purpose should be selected. Experience also shows that circular tunnels are easier to support than any other shape.

Other Factors. If the rock contains porewater, negative pore pressures are set up as the rock moves toward the tunnel. This provides limited initial support until the negative pore pressure is dissipated. In addition, the new pressure gradient set up by the release of confining pressure results in seepage pressures toward the tunnel boundary. In regions of high hydrostatic head, significant increases in rock loading can occur. It is also thought that even small proportions of swelling clay minerals in the rock can contribute significantly to rock loads when water is present. This water need not be flowing—only present in the pores. When all factors contributing to rock mass behavior have been identified and quantified, it may be possible to develop more exact predictive models and to devise new means for controlling and improving ground behavior. In the meantime, we must make do with approximations based on experience.

Monitoring. Rate of squeeze and rock loads are somewhat dependent on tunnel size and rate of advance. It is essential in squeezing (or swelling) conditions—or even in blocky and seamy rock where joint closure may create problems—to establish a program of convergence point installations that will be routinely used to monitor the amount and rate of movement of the tunnel walls. This information collected over time and collated with the behavior of the tunnel support system will provide the information needed both to predict and to install the appropriate amount of support as tunneling progresses. This technique lies at the heart of NATM tunneling in rock.

It has been noted that a visible deformation of the tunnel boundary are most obvious in the portion below spring line. Shear failures commonly result in surface fracture and intrusion a meter or two above the invert, while the invert itself will also heave persistently, especially in noncircular tunnels. This has been clearly demonstrated in several European tunnels (Einstein, 1989).

DRILL-AND-BLAST TUNNELING

Early tunnels were constructed by drill-and-blast methods to variations on the horseshoe shape. This shape has always been popular because for transportation tunnels it combines the maximum of usable space with the minimum of excavated volume. Nevertheless, it was always clear that in squeezing ground a closer approximation to circularity offered improved stability and longer service life.

Techniques

Concrete-Filled Drifts. One of the most severe problems of excessive ground loading on a tunnel encountered in the United States in recent years was during construction of the Eisenhower Tunnels on I-70 in Colorado. After the problem had brought the first bore to a standstill, the shear zone causing the problem was penetrated using multiple small-diameter drifts around the periphery of the desired tunnel profile. Each drift was filled with concrete until a complete canopy permitted easy access to the remainder of the work. The same system was adopted by design during construction of the second bore and was equally successful.

Steel Supports. The replacement of timber supports by purpose-made steel supports was virtually complete 50 years ago, driven by the great reduction in labor involved in their use and the consequent increase in productivity. In squeezing ground, it soon became apparent that closure of the invert was essential to preserve stability of the supports and the size of the tunnel opening. Experience also showed several other things.

If light supports are used, especially without ties and collar braces, early failure and twisted ribs result. On the other hand, if heavy sections are installed, the ground will tend to squeeze between the ribs without unduly distorting them, although lagging or shotcrete will be broken. The remedial work necessary is reduced to trimming off excess material.

If the amount of support necessary has been underestimated, the installation of jump sets will usually cure the problem.

Yielding Supports

The Yacambú–Quibor water transport tunnel is about 25 km long and, for most of its length, has a nominal horseshoe section of 5 m by 5 m. About half the tunnel from the upstream portal is in a formation of weak graphitic, micaceous, and siliceous phyllite of Cretaceous Age that has been tilted and deformed by tectonic action. Squeezing of this rock has caused substantial problems ever since tunneling began because of the low rock strength and abundant faulting. From the downstream portal, the tunnel is in Tertiary sandstone, siltstone, and claystone (lutita). The claystone is an essantially weak material, which in many places has been infiltrated with veins and nodules of quartz. It is highly fractured, generally blocky and seamy rock with frequent small fault crossings and abundant water flows. It is under low cover compared with the phyllite reach and has suffered only minor squeezing. Separating the two formations described is the Bócono fault, a transform fault similar in extent and importance to the San Andreas fault. Additionally, between the Bócono fault and the phyllite reach, there lies the Turbio fault, which has not yet been penetrated, so that its condition can only be surmised the main part of the Bócono, the tunnel has penetrated about 1 km of fault gouge with some extremely large blocks of highly fractured rock belonging to the Tertiary formation. At the time this is written, the excavation from the outlet portal is in phyllite, indicating that the heading has passed the main shear. However, extremely difficult squeezing conditions have been encountered, possibly associated with higher overburden loads as well as drag folding of the phyllite. The tunnel overburden rises to a maximum of about 1,200 m in this reach.

In late 1994, 10 km of tunnel remained to be excavated in the phyllite and the Turbio fault zone. Since the beginning of the drive from the upstream portal under the present contract, close attention has been paid to rock convergence behavior, and a predictive model has been prepared relating expected convergence to ground cover and rock type (Senthivel, 1994). The rock has been classified into four main classes with subclasses for transitional conditions. Initial experimentation was performed to determine the type of support best suited to the conditions, using a drill-and-blast extension of the existng heading as the experimental chamber. Since that time the support system has been modified and developed to give improved service reliability.

Some changes were forced by the decision to use a roadheader rather than drill-and-blast excavation. This method was expected to cause less initial damage to the rock structure and therefore less total convergence. Unfortunately, perhaps, this also meant that rock bolt drilling in the numbers and lengths required could not be accommodated readily because of the volume occupied by the large machine. The rock bolts were still preferred, but the installation time was simply too long to allow any reasonable production rate to be sustained. Steel sets (roughly equivalent to 6WF25) replaced the rock bolts, and heavy shotcrete was placed between and over the sets. Later it proved necessary to provide for greater convergence for stress relief, and yielding sets were introduced and are still being used.

The design of the yielding sets is illustrated in Figure 8-1. The number of yielding joints can be modified to provide the needs of the rock currently being excavated since all components are manufactured on site. Each joint permits up to 22 cm of closure. Shotcrete is applied to space between the ribs and to cover them; and gaps are left to permit the yielding joints to move freely, impeded only by frictional forces applied by the sleeve bolts and the additional force necessary to allow the steel rib sections to slide through the sleeve, which is not exactly aligned to the rib curvature. It has been found essential to shotcrete the gaps once the closure nears the limit allowed without the steel sections actually butting together. Failures have been common when this has been allowed to happen. It has also been found that allowing the invert to heave freely for 20 to 30 days before making an invert closure allows the total support system to resist all remaining loads with some reserve capacity for long term load increases.

In summary, the support system provides a relatively low initial support pressure and permits almost uniform stress relief for the rock in a controlled manner around the entire circumference of the tunnel while preventing the rock from raveling. The shotcrete is not damaged by the convergence because of the yielding joints and so maintains its integrity, provided that timely closures are made. After allowing practically all the stress relief required by the elasto-plastic stage of rock deformation, the support system is made rigid when

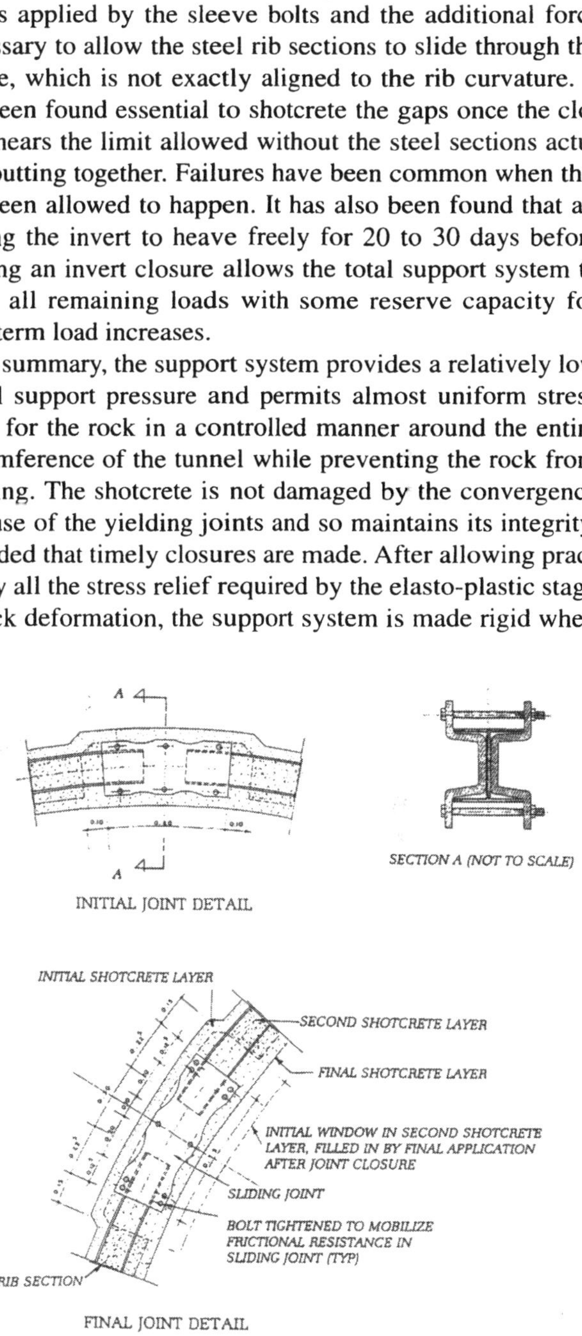

Fig. 8-1. Details of yielding sets used at Yacambú in squeezing phyllite under high cover; Original design (top) and modified design (bottom) with four more bolts to prevent premature closure of the gap in the rib section. (Courtesy of Sistema Hidraulica Yacambú-Quibor, C.A.)

it can support a pressure of 3.8 MPa, which is available to deal with long term creep pressure.

The long-term support resistance can be increased substantially by adding small amounts of shotcrete at the junction between the side-walls and the invert slab and in the roof arch as illustrated in Figure 8-2. This additional shotcrete can be applied at such a distance from the working face as will avoid interference with the main production operations.

Timber Wedges and Blocking. The use of blocking tightened against the ground by pairs of folding wedges introduces a structural element that can be allowed to fail by crushing. Observation of progressive failure coupled with experience provides warning that the behavior of the steel supports should be closely monitored in case a decrease in spacing or an increase in section becomes necessary.

Precast Invert. When squeezing is sufficiently severe to be troublesome, it will be seen that the prime tendency is for the lower sidewalls to move in and for the invert to heave. The loss of strength of the rock in the invert leads to the rapid development of muddy and unstable conditions under the tunnel haulage operations. At the Pacheco Tunnel No. 2, the contractor elected to use precast concrete invert slabs kept up close to the working face in place of invert struts. In locations where the worst squeezing occurred, the slabs were heaved up and required maintenance to keep the track on grade. However, they did provide a good and trouble-free surface otherwise.

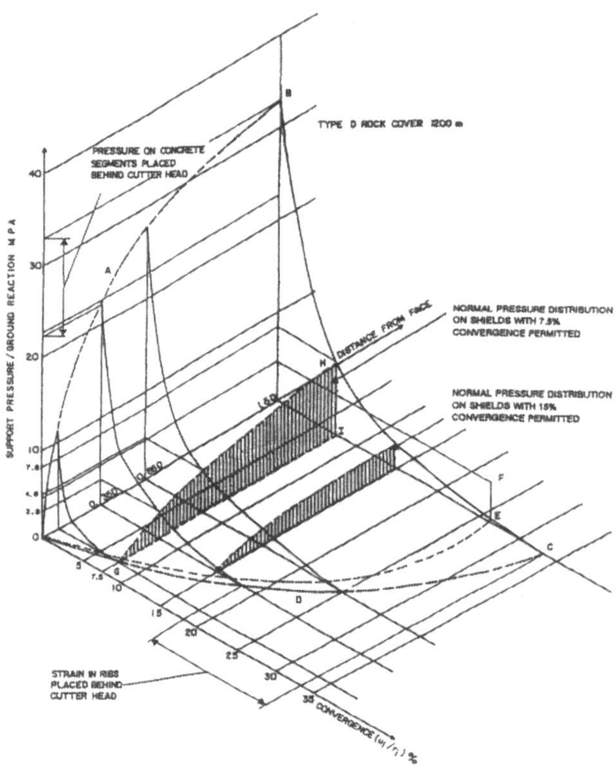

Fig. 8-2. 3D graph of relationship between convergence, distance behind working face and ground pressure for poor quality rock with 1,200 m of cover (Senthivel, 1994).

TBM TUNNELING

Because of the number of large tunnels now under consideration where the use of TBMs is contemplated and where squeezing conditions may become important, the following discussion is extended, even though not based on a great deal of current experience.

It is only with the advent of the modern tunnel boring machine that construction methods have been freely adapted to tunnels of circular cross section. Even so, it was not until after 1970 that TBM design had advanced to the point where more tunnels were completed by TBM than by drill-and-blast methods. In the mid-1970s, Japanese slurry shields for deep soft ground tunnels were being manufactured in sizes up 10 m in diameter. Machine tunneling in truly hard rock did not become generally successful until about this same time. Improvements in mechanical design and development of larger-diameter single-disk cutters improved the performance of TBMs exceeding 6 m in diameter before 1980. Even larger machines, 9-m diameter and greater were used in limestone with a compressive strength of 175 Mpa in Chicago. At the present time, 12-m machines are being constructed for major tunnels in Taiwan with difficult conditions anticipated near one portal and at several fault crossings.

The Stillwater tunnel second contract was the first project for which a TBM was specifically designed to penetrate squeezing ground. The machine was successful, although a second machine modified from a conventional design was more successful (Phien-wej & Cording, *1991*). Both designs recognized the need for adaptation to the squeezing conditions, but the actual amount of radial reduction in tunnel diameter was comparatively small. Had it been much greater, it is unlikely that the conventional machine could have worked.

The majority of examples of tunnels in squeezing ground are related to the crossing of faults. TBMs have been troubled in this situation by inrushes of water carrying sand and finely divided rock or by blocks of rock jamming between the cutters. The second of these problems has been dealt with in many tunnels in otherwise normal conditions. The primary solution is the use of a cutterhead design that allows only a limited projection of the cutters forward of the cutterhead using a face shield ahead of the structural support element. The second development is a design that permits worn cutters to be changed from within the tunnel, so that no access is required in front of the cutterhead. There has, as yet, been no easy solution for the problem of the cutterhead and its buckets being choked with sand and rock fragments while unrelenting water flows are in progress. It becomes a difficult and slow process of cleaning out and gaining progress slowly until the affected area has been cleared. Also in such conditions, the presence of a shield is important to protect the machine and to provide temporary support to material with no stand-up time. In some circumstances, if the condition is known to exist or to be likely to exist, probing ahead to identify the precise location can give an opportunity to stabilize the ground with grout injections, keeping a bulkhead thickness ahead of the excavation at all times. It is sometimes possible to allow most of the water to drain out of the ground, but this is not a reliable approach to prediction of construction methods. Shielded TBMs have been used successfully in such conditions, but unfortunately the use of a long shield militates against successful use in squeezing ground.

The other major problem, whether or not in a fault or shear zone, is the closure of the ground around the cutterhead shield and any protective shield behind the cutterhead. Many TBMs, including that used on the first contract at Stillwater, have been immobilized because the load on the shield system was too high to permit the machine to advance. An indication of the way to approach this problem was offered by the use of a short shrinkable shield on the machine used on the second contract from the inlet portal.

It is not anticipated that tunneling in squeezing ground or fault zones will ever become a simple, routine operation because of the erratic variability of conditions. However, current opinion is that virtually all tunnels can be attacked by TBM methods whenever there is an economic advantage in doing so.

As previously noted, the difficulty of predicting rock behavior in squeezing ground has played a major role in the development of observational methods for determination of rock support requirements. However, if tunneling by TBM is selected, some of the flexibility of the observational method is removed, and decisions must be made at the time the TBM is designed as to the amount of ground movement to be anticipated or permitted and the design of the support system to accept the loadings implied at different stages.

Since squeezing of soft rock does usually lead to immediate instability, it should be possible and practical to delay major support installation until a high percentage of the total strain has taken place and ground loading has been reduced. Sixty to seventy percent of the potential ground movement has usually taken place within about three diameters of the working face. If the total amount of squeezing is not great, it may not be necessary, or even desirable, to delay support installation so long.

Ideally, final support is not installed until convergence is less than one millimeter per month. The loading associated with a given amount of convergence depends on the parameters of the project. One must also consider any long-term requirement for the tunnel to carry water. Finally, it must be realized that if groundwater is to be totally excluded from the tunnel, the final lining must be designed to carry the full hydrostatic head unless the aquifer is fully sealed off by consolidation grouting. If groundwater is admitted, whether in a controlled manner or by allowing local cracking of the lining, then only seepage pressures need to be accounted for. In the case of weak squeezing ground or faulted rock with an unknown potential for swelling behavior, the latter alternative appears undesirable.

Steel Rib Support System

Steel ribs set close to the tunnel surface and blocked from it are normally used as the initial support system for rock tunnels, especially those constructed by conventional drill-and-blast methods. Wood, concrete, or steel lagging may be placed between the ribs to secure blocky or raveling ground, or welded wire fabric can also be used. The same system can also be used in TBM tunnels, but it is necessary to allow an initial small distance between rib and ground so that the last rib segment can be positioned conveniently. In normal tunneling, this space is later closed by expanding the rib against the ground. Especially in squeezing ground, the rib must be blocked to the rock all around its perimeter. As the ground movement occurs and continues, it will squeeze past the ribs, and stress relief will occur. In this type of installation, it is necessary for the ribs to be as stiff as possible to prevent displacement and buckling. The chief safeguard is to install steel ties and collar braces at intervals around the rib. The collar braces are typically steel pipe sections set between the ribs. The ties then pass through holes in the web of the steel section and through the pipe forming the collar brace. These members are also subject to deformation by the invading ground. If this creates any substantial problem, angle irons welded to the inner face of the ribs can be substituted.

It is significant that in tunnels where the ribs have buckled under squeezing load but have been left in place, they commonly retain enough structural strength to provide support. The problem is that the squeezing intrudes on the required final profile of the tunnel and remining involves removal of the distorted supports.

Concrete Segments

Segmental concrete linings take two quite different forms. The traditional bolted and gasketed lining is meant to be a final lining erected in one pass. The more common application these days is to use unbolted, ungasketed segments, with light reinforcement to allow handling, as a sacrificial primary lining. This latter type of lining is sacrificial only in the sense that it is allowed to sustain fractures resulting from jacking loads or redistribution of stress; it retains most of its initial load-bearing capacity. A final lining is always placed within this type of lining; it is usually an unreinforced concrete lining of nominal thickness. The combination lining is less expensive than the one-pass system and has the merit of flexibility. Problems arose with the precast concrete tunnel lining at Stillwater because there was insufficient erection space to allow for deviations normal to tunneling.

In electing to use a precast concrete lining, one must decide the amount of ground movement to be allowed and the backfill material to be used between the lining and the rock. From the experience to date on tunnels having large annular cavities, such as the Channel Tunnel, the Great Belt Tunnel, and the St. Clair Tunnel, it is possible to design a system that will permit at least 15 to 20 cm of radial distortion from the initial excavated cross section.

In allowing for a large amount of potential ground movement, certain problems of erection stability arise. The lining will require support clear of the invert and a horizontal tie or blocking to keep it in shape during and after erection. Also, since it is not in friction contact with the rock, it would be prudent to tie segments back to a preceding ring to insure against toppling. Time and skill is involved in executing the work but no significant difficulty. Segments cast with a lip on the circumferential flange will provide a guide for erection of rings after the first one. The leading edge of the invert segment or segments can be blocked and shimmed with removable material. A tie (known as a "hog-rod") at spring line or a ring reformer placed within the ring last erected will stabilize the shape and additional removable blocking can be used as well. An angle clip fitting over the leading edge of a segment and secured via a cable and hook at the other end to a grout hole in the preceding segment will serve to prevent toppling. After the ring is secured in place, an inflatable cuff can be inserted in the annulus between ring and rock to provide both secure temporary support and a seal to contain grout injected into the void. If the annulus is small, lengths of circular cross section polystyrene can be used for the same purpose. The inflatable cuff can be reused, but the polystyrene is usually be left in place. There is no reason why the same type of wire brush seals used in EPB machines cannot also be used in TBMs.

The completed precast concrete rings can be used as a reaction for a jacking system to propel the TBM as is done, for instance, in the Lovat-type earth pressure balance machines being used in Caracas Metro tunnels and the St. Clair tunnel. Otherwise, they can be erected independently of the TBM at a distance behind the face sufficient to ensure that most of the short-term ground movement has already occurred. If the rings are to be used to support propulsion loads, this loading may govern their design. If only ground loads are to be supported, a lighter section may be used. Since the rings will eventually be heavily loaded, it may be useful to introduce strips of bituminous felt (roofing felt) into the longitudinal joints to assist in uniform load transfer across the joints.

It is anticipated that such rings would be cast and cured in a controlled factory environment and that they would be of high-strength concrete (about 550 kg/cm^2) for high resistance and high elastic modulus. Such a ring 4 m in diameter and 20 cm thick could support an overburden load of more than 200 m. It is unlikely that there would be any interest in making the rings less than 15 cm thick; such a ring could support 150 m of overburden load.

It is important that the moving ground should not come into contact with the completed ring at any point. Distortion would necessarily result, with a possible consequence of reducing load-bearing capacity. It also seems unwise to use compressible backfill in the annular void, since such a material might not offer enough resistance to mobilize passive reactions sufficient to withstand distortion of the lining. At the least, careful consideration would be needed in specifying the strength and deformability of any compressible material to be used.

From the foregoing discussion it appears that two adequate conventional systems of tunnel support are available for use in conjunction with TBM excavation of a circular tunnel in squeezing ground.

TBM Tunneling System

The principal components of a TBM affected by the difference between tunneling in squeezing and nonsqueezing ground are discussed here.

Cutterhead. Many different cutterhead designs have been used over the years, from the earliest flat heads with multiple disk cutters through domed heads, rounded edge flat heads, and conical designs. These days the cutterhead geometry is selected on the basis of the ground it is expected to penetrate. It has been found preferable to arrange that at least the gauge cutters be designed to be changed from behind, and it is possible to arrange this system for all cutters. A spoke design allows ready access to the working face and simplifies design in some respects. However, such machines offer little support if weak ground is encountered, and it is generally considered prudent to use a closed-face machine. Also, to protect the cutters and cutter mounts, a lighter false face is provided so that the cutter disks protrude only a short distance.

In conventional designs, the cutterhead is provided with its own shield as part of the cutterhead bucket system. The conventional design creates a drum about 1.2 m long almost in contact with the ground. In squeezing ground this shield is vulnerable to the pressure exerted by rock movement. It is therefore better that the shield be smaller in diameter than the excavation and that it be tapered toward the rear. The gauge cutters should be arranged to protrude beyond the main body of the cutterhead. Figures 8-3 and 8-4 illustrate the designs of the machines used successfully at Stillwater. Figure 8-5 shows a design for extending gauge cutters devised by Wirth. Other methods are available. Figure 8-6 shows the design for a TBM used successfully in squeezing clay in Italy (Wallis, 1991). Note that the exposure to squeezing loads on the machine has been all but eliminated. It is probably not proctical to adopt such a design for a rock tunnel, but the principle is important.

If the cutterhead is not in close contact with the ground, provision must be made to provided stable support in its place. This will be the equivalent of a sole plate as used for overcutter compensation in earth pressure balance machines. However, to provide for varying amounts of overcut, the support will need to be hydraulically actuated. Since it will be subjected to substantial shear loading, the design will have to be very stiff.

Propulsion. A TBM requires a reaction against which to propel itself forward. This reaction can be obtained by shoving directly against the tunnel support system with

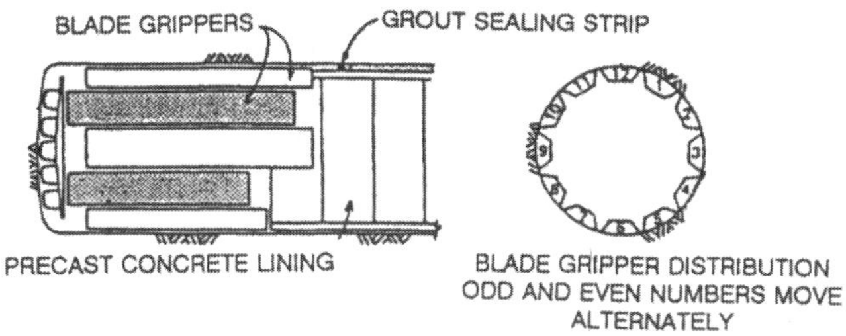

Fig. 8-3. Prototype TBM specially designed for the Stillwater tunnel and driven from the downstream portal.

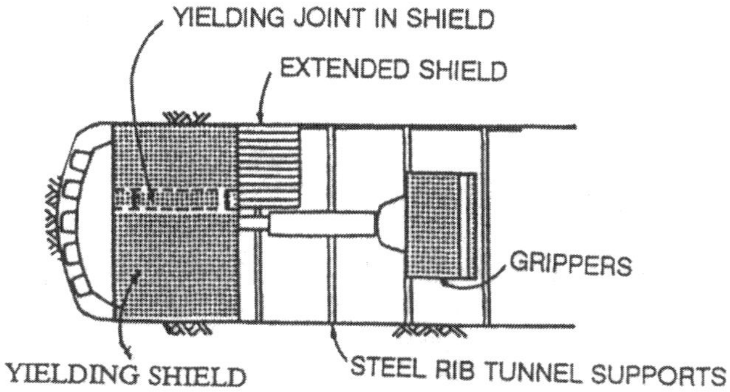

Fig. 8-4. Refurbished TBM designed with a full-circle adjustable front shield that can be retracted or enlarged depending on the amount of squeezing during tunnel excavation.

jacks spaced around the perimeter of the machine or by developing frictional resistance against the tunnel sidewalls.

The thrust needed to keep the cutterhead moving forward would be about 14,000 kg per cutter. Because the ground is weak, it would be desirable to limit the bearing pressure on the tunnel walls because the weak rock would fail under even light loads, especially perpendicular to the direction of foliation. This would accelerate the rate of squeezing and might increase the total strain. At the same time, it would be desirable to limit the length occupied by the grippers so as to minimize the necessary distance between the working face and any support system. This would probably require that there be multiple grippers covering most of the circumference but of limited length to minimize uneven bearing on the squeezing rock surface.

Shield. If any shield is felt to be desirable or necessary, it should be short and shrinkable, following the example of the inlet portal TBM at Stillwater. Many TBMs have been stuck because the ground has moved on to the shield and exerted sufficient load to stall the machine.

Erector. It is desirable to have complete flexibility in selecting the point at which ring erection is to take place. Therefore the erector should be free to move along the tunnel, mounted on the conveyor truss. A ring former should also be used to maintain the shape of the last erected ring until it has been grouted, if concrete segmental lining is used.

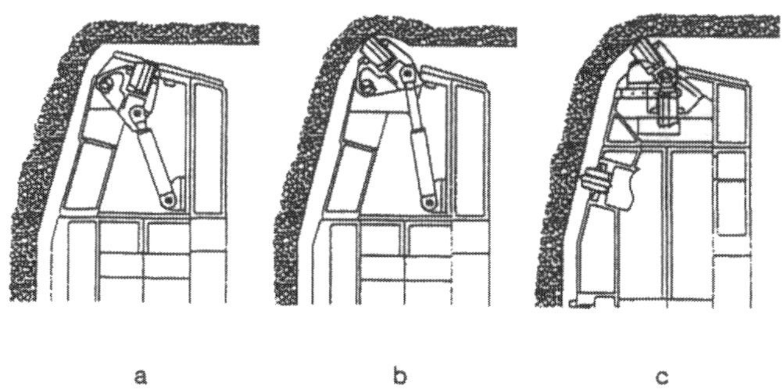

Fig. 8-5. Hydraulically extendible cutters: (a) extending cutter in its retracted position; (b) result of gradual extension of the cutter by extending the hydraulic cylinder as it cuts a conical profile; (c) continuing excavation with the cutter extended. (Courtesy of Wirth)

Spoil Removal. Conventional conveyor to rail car systems or single-conveyor systems designed for the tunnel size selected are appropriate.

Back-Up System. To keep the area between the grippers and the ring erection area as clear as possible, any ancillary equipment such as transformers, hydraulic pumps, etc., should be kept clear of this space at track level.

Operational Flexibility

It is envisaged that the system outlined above would be capable of handling either steel ribs or precast concrete supports. If shoving off the supports were to be selected for TBM propulsion, the degree of flexibility would be less than the use of a gripper system. It would also be more vulnerable to problems in any circumstance where the convergence rate was markedly higher than expected.

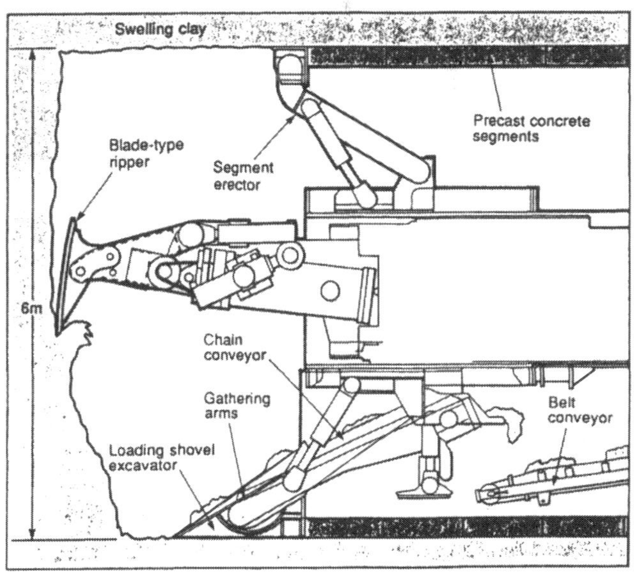

Fig. 8-6. Machine developed in Italy for use in heavily squeezing clay. (Reproduced by permission of Tunnels and Tunneling).

SWELLING

Swelling phenomena are generally associated with argillaceous soils or rocks derived from such soils. In the field, it is difficult to distinguish between squeezing and swelling ground, especially since both conditions are often present at the same time. However, except in extreme conditions, squeezing is almost always self-limiting and will not recur vigorously, or at all, once the intruding material has been removed; while swelling may continue as long as free water and swelling minerals are present, especially when the intruding material has been removed, thereby exposing fresh, unhydrated rock. Many European rail and highway tunnels are constructed in formations noted for their susceptibility to swelling. Most construction involves a more or less circular wall and roof section with an invert slab having a greater radius of curvature. Some of them are still being periodically repaired a century after construction. It has been noted in this connection that as the invert arches are excavated and replaced to more nearly circular configurations, the greater the time that elapses before the next repair is necessary.

Expansive clays are more common in younger argillaceous rocks, the proportions ranging from 65% in Pliocene and Miocene Age material to only 5% in Cambrian and Precambrian. Montmorillonite is found in rocks of all ages as thin partings or thicker beds. Sodium montmorillonite is much more expansive than calcium montmorillonite.

The Swelling Mechanism

Most swelling is due to the simultaneous presence of unhydrated swelling clay minerals and free water. Tunnel construction commonly creates these conditions. Minerals such as montmorillonite form layered platy crystals; water may be taken up in the crystal lattice with a resultant increase in volume of up to 10 times the volume of the unhydrated crystal. The displacements resulting from this increase in volume give rise to the observed swelling pressures, whether in soil or in rock. Swelling caused by anhydrite is a somewhat different phenomenon and is discussed separately below.

Tunneling. Water should, if possible, be kept away from rock or soil containing swelling clay minerals; however, it must be realized that water vapor from a humid atmosphere or porewater released from confinement within the rock will initiate the swelling process. Since the swelling will not passivate in the same way as squeezing generally will in rock, tunnel support must be designed to resist the swelling pressure, even if it proves possible to let some swelling take place without creating problems. The difficulties that the installation of support can cause are exemplified by experience on one of the Navajo tunnels of the Central Arizona Project. The tunnel was excavated by TBM at record rates in dry, soft rock containing momtmorillonite; but when the first concrete was poured, water from the concrete initiated swelling, and the concrete was severely damaged.

Other Rock Problems

Schists commonly contain clay minerals such as biotite, mica, and chlorite. All of these are platy minerals and are found aligned with the foliation. If present as continuous layers, they have to be considered planes of weakness when assessing questions of rock stability. Similarly, weathered material in shears and mylonite not yet weathered indicate planes of weakness.

Anhydrite converts to gypsum in the presence of water with a volume increase of up to 60%. However, beds of anhydrite are not affected in the same way as finely divided rock since the reaction does not penetrate below the surface. However, if the anhydrite is fractured, the conversion will proceed faster and faster as more fracturing is developed by the expansive reaction. The actual amount of expansion will depend upon the void ratio of the anhydrite. As with other water-sensitive minerals, every effort should be made to keep water away from anhydrite. This may be a particular problem when fluid transport tunnels are being constructed, since any leakage will result in major damage to the tunnel.

OBSTACLES AND CONSTRAINTS

Boulders

Practical experience of the value of cutterhead disks in such a situation was first developed in Warrington, England. A slurry shield was to be used for a tunnel originally expected to be in soils. A late decision to change the alignment because of local constraints forced the tunnel into an area where boulders and sandstone bedrock would be encountered in the invert. Since the equipment was already built, roller cutters were added to the head in the hopes of solving the unexpected problem. These hopes were fulfilled. More recent investigation in Japan has indicated from experimental models that even very soft clay will provide sufficient support to hold boulders in place so that they are broken up by the action of roller cutters. On the other hand, rotary head excavators of various general designs have failed to deal successfully with boulders when drag picks were relied on.

In general, if roller cutters are to be used on a tunnel in clay soils, extra protection should be provided so the cutters do not jam and so clay particles cannot penetrate the seals.

Small diameter shields developed by Herrenknecht for bouldery ground present a concave conical face to the ground. The cone is eccentric to the rotation axis and is hard-faced so it will behave like a cone crusher to break boulders into manageable pieces. Some slurry shields are fitted with crushers in the invert that collect and reduce boulders entering the plenum.

A particular difficulty sometimes occurs when boulder beds are encountered that have saturated fine silt in the void spaces between the boulders. This problem seems to be most often encountered in regions that have been subjected to glaciation. The loss of ground associated with flow of the saturated fines into the tunnel does not normally result in ground settlement, because the movement of any other material replacing the lost fines will generally be choked off. If this is not the case, or if it is felt undesirable to leave such voids unfilled, various courses of action are available. Compressed air working will drive water out of the silt and thereby stabilize it, provided that the boulder bed is not confined within impervious material. In such a case, compressed-air working will not be very effective.

The use of an EPB fitted with roller cutters will be effective provided that the pressure in the plenum chamber is kept at a level high enough to balance the hydrostatic head in the silt. Slurry shield operation with the same restrictions would be even more effective, but at a higher cost. As a last resort, consolidation or replacement grouting may be employed behind the shield. The choice of method will depend on economics, as is often the case when selecting a construction method. If the condition exists in only a small part of a long tunnel, less efficient means may be selected for dealing with the boulder bed—even including local cut-and-cover work, if the tunnel is not too deep or the water table too high. In any case, full breasting of the face is required if the boulders are not in intimate contact with one another. It is conceivable that grout could be injected into the working face at a distance behind it so as to force out the flowing material. For such a program to be effective, it would be necessary to grout multiple points simultaneously so as to avoid development of a preferred path for escaping fines. The grout would also have to extend outside the tunnel perimeter for a sufficient distance to establish a plug that could be excavated without developing problems behind the shield. However, it must be said that in small tunnels, access for implementation of such a program is unlikely to be available.

Karstic Limestone

Karstic limestone is riddled with solution cavities. Depending on the geologic history of the locale in which it is found, cavities ahead of the excavation may be filled with water, mud, gravel, or a combination of these. Flowing water may be present in large quantities. There may be an insufficient thickness of sound rock at tunnel elevation to provide safe support for tunneling equipment. All of these possibilities point out the need for thorough exploration before undertaking tunnel construction in limestone, particularly in an area where there is no prior history of underground construction or mining.

Abandoned Foundations

During construction of the Lower Market Street tunnels for the BART system in San Francisco, 898 piles were encountered in the excavation. This was more than double the highest estimate. These piles were mostly unrecorded relics of earlier construction abandoned after fires, which regularly ravaged the area during the late 19th century, as well as those left behind by successive reclamation operations, which moved the waterfront several hundred meters into the bay over a few decades.

All but two of the piles were timber; they were removed by cutting them into short lengths as they were exposed in the face of the shield using a hydraulically powered beavertail chain saw purpose-made for the job. The other two posed a different problem. One was concrete and the other steel. Since the tunnel was being constructed in compressed air, both burning the steel and breaking the concrete were nontrivial problems. At 1970 prices, each cost about $6,000 to remove; this was a measure of the extra time involved—about 10 times as long as the time needed for removal of timber piles. As an additional problem, the lengths of pile left in place above the tunnel eventually crept downward as they sought to carry the weight of soil adhering to them as well as the artificial fill above. In several places it was necessary to reinforce the skin of the fabricated steel liner plates, which were dimpled by the point loads exerted.

The Second Street drainage tunnel in Minneapolis was constructed to deliver storm drainage from new highway construction to the Mississippi. For a short distance at the downstream portal, the tunnel was in soil; because of its short length, it was driven without a shield. Since the soil was largely, if not entirely, fill, it proved difficult to maintain the tunnel shape until steel ribs were introduced between alternate rings of liner plate. It was known that old mill foundations lay ahead, but their location was uncertain. It was therefore deemed prudent to continue this tunneling method into the St. Peter Sandstone for at least a short distance. In the event, some of the foundations were found in the soil tunnel. Careful breasting to isolate the concrete was successful in controlling soil movement while the concrete was broken out. With the next advance of normal tunneling, the voids were promptly and completely filled and there was no encroachment on the tunnel profile.

It would be possible to multiply examples endlessly, but the key to all such problems is to gather the maximum available information, project the worst scenario, and be prepared to deal with it as an engineering rather than an economic problem.

Abandoned Mines

Coal mining has been a significant activity in Britain for well over a century—two centuries in places. Many mines were exhausted and abandoned in times before this was a regulated activity. In particular, many mine shafts were never closed; instead, a high brick wall was built around the shaft collar and no more was done. Of course, in time these walls crumbled or were knocked down by someone wanting to build on the land. Timber caps were placed over the shafts some distance below ground, and then the holes were backfilled. The dangers of such conditions are illustrated by an occasion when a main road was being resurfaced with tarmacadam and the steam roller used for compaction vanished into one such shaft. Many others are now identifiable by ponds with sloping banks, the mine workings and shafts being completely filled with water. There is some risk of encountering such conditions in any mining town of substantial age.

Shallow Tunnels

A tunnel for a 24-ft-wide private road beneath the eight-lane Pacific Coast Highway near Newport Beach, California, was constructed without incident only 7 ft below the road surface in material described as loose sand. The construction was protected by a canopy of 8-in.-diameter pipes directionally drilled between the tunnel crown and the road foundation. The pipes were spaced at 16-in. centers and were filled with concrete. Because of the highway authority's concerns about future settlement, the principal support was steel arches set on 48-in.-diameter jacked pipes as a foundation to distribute bearing loads. The spaces between the arches were filled with light liner plate, and the whole was shotcreted. Similar construction was used for a smaller flood control channel bypass tunnel beneath a freeway in Corona, with equal success.

The problem with shallow tunnels is that side support is not reliable, and loading on the support system is far from the usual comfortable assumption of uniform radial load. It is quite common in urban situations to be restricted by the presence of significant structures—whether on the surface or underground. Consolidation grouting has been used where ground conditions are favorable, and compaction grouting has also been used successfully to avoid the need for underpinning.

Jet grouting was used in Caracas to permit Metro tunneling beneath the foundations for a two-level interchange structure in Plaza Venezuela. A complex raking pile system was installed to isolate the foundations from the tunnels. Vertical piles would have intersected the tunnels. It is not practical to define the range of conditions leading to selection of any particular solution, since all such projects have unique features.

PHYSICAL CONDITIONS

Methane

Methane is commonly found where organic matter has been trapped below or within sedimentary deposits, whether or not they have yet been lithified. It is particularly common in the shaley limestones around the Great Lakes, and in hydrocarbon—whether coal or oil—deposits in Pennsylvania, West Virginia, Colorado, California, and in many other localities. Note that "methane" commonly denotes all of the ethane series that may be present, although methane is distinguished as the major component usually present. It is also the only member substantially lighter than air. Methane forms an explosive mixture when mixed with air and between about 5 and 9% of the total volume is methane. It is readily diluted and flushed from a tunnel by ventilation when encountered in the quantities that are normally expected.

Safety rules require that action be taken when methane is present in concentrations of 20% of the lower explosive limit (Table 8-1). For practical purposes, this means a concentration of 1% by volume. It is necessary to use routine testing to determine whether or not explosive gases are present. This testing is carried out in all tunnels identified as being gassy or potentially gassy. A positive rating of non-gasy is required to relieve the contractor of the duty to test, although in many localities it is deemed prudent to continue testing on a reduced schedule even though no gas has been identified in the tunnel excavation.

Hydrogen Sulfide

Hydrogen sulfide is present in association with methane often enough that its presence should always be suspected in gassy conditions. Its presence is easily identified in low concentrations by its typical rotten egg smell. It is a cumulative poison and deadly in low concentrations; a whiff at 100% percent concentration is generally instantly fatal. Note that, when it is present in low concentrations, the nose becomes desensitized to its presence. Apart from testing and maintenance of high-volume ventilation, signs of its presence follow a sequence of headaches, coughing, nausea, and unconsciousness (Table 8-2).

Meticulous attention to ventilation, especially in work areas, is required when hydrogen sulfide is present. Ventilation must be maintained at high volumes for dilution 24 hours a day, seven days a week, regardless of whether work is in progress or not. Even so, no shaft, pit, or tunnel should be entered without first testing the air. This is especially important if the purpose of entering is to repair a defective fan. Where possible, the gas should be extracted directly and discharged into the ventilation system without ever entering the tunnel atmosphere.

Radon

While radon is commonly associated with granitic rock, it can be found anywhere, depending on the availability of a transport mechanism capable of moving the gas either verti-

Table 8-1. Upper and Lower Explosive Limits for Hydrocarbons and Hydrogen Sulfide as Percentage of Gas Vapor in Air

Gas	Lower Explosive Limit (%)	Upper Explosive Limit (%)
Gas-Oil	6	13.5
Methane (CH_4)	5	15
Ethane(C_2H_6)	3.22	12.45
Propane (C_3H_8)	2.3	9.5
n-Butane (C_4H_{10})	1.86	8.41
n-Pentane (C_5H_{12})	1.4	7.8
Gasoline	1.4	7.6
n-Hexane (C_6H_{14})	1.25	6.9
n-Heptane (C_7H_{16})	1.0	6.0
n-Octane (C_8H_{18})	0.84	3.2
n-Nonane (C_9H_{20})	0.74	2.9
Fuel Oil No. 1	0.7	5.0
n-Decane ($C_{10}H_{22}$)	0.67	2.6
Hydrogen Sulfide (H_2S)	4.3	—

cally or horizontally (or both, of course) in natural rock fissures and soil voids. There are three isotopes of radon, with half-lives of about 3 seconds, 1 minute, and 4 days. The last is the most significant. Since radon decays by alpha radiation, very little shielding is necessary to prevent access to human tissue from a distinct radiation source. Also, in a tunnel, it is more than likely that normal ventilation will keep the radon below acceptable levels. However, once it has been inhaled, the damage to tissue caused by radon decay is highly localized because it is stopped within a short distance; it is therefore more significant than radiation of a higher-energy decay, which passes readily through the body. In addition, the decay products, such as lead, can also have toxic effects.

Radon is reasonably soluble in water and may be carried for great distances from its point of origin in fast-flowing water. For this reason, limestone series in the vicinity of granitic masses may be more heavily contaminated than the granite containing the uranium salts from which it is originally derived. Airborne contamination in the tunnel is usually carried by other gases originally trapped in the rock; this is particularly true for carbon dioxide, which can carry the radon to much greater distances from its source than would otherwise happen.

An intermediate decay product between uranium and radon is radium. Uranium is very mobile within the strata in which it is found, while radium tends to stay in place. All these and other relevant data need to be considered when assessing the possibility of the presence of harmful amounts of radon. Testing methods that determine the yield from rocks by dispersion to water in a sealed container over a measured period are available.

In general, it would appear that adequate ventilation will deal with the problem during construction. It seems prudent to take care with permanent caverns in which personnel will be continuously present in conditions in which ventilation will be at much lower rates than during construction. The length of time an individual will spend in traversing tunnels will not normally be significant in this respect, but thought should perhaps be given to operating crews of subway systems.

Table 8-2. Physiological Effects of Hydrogen Sulfide

Concentration (ppm)	Effect or Action Required
0.002–0.02	Odor perceptible
10	Limit for 8-hr exposure; begin continuous monitoring
15	Short-term limit of exposure (Cal-OSHA); headaches, coughing prevalent
20	Action required to reduce levels (OSHA)
20–100	Sense of smell lost; odor becomes undetectable
300	Health and/or life threatened
600–1000	Immediately lethal

High Temperatures

The geothermal gradient is different in different localities within a range of about 2:1. As a rule of thumb, one degree Celsius per 100 m of depth will be a reasonable guide. Where the tunnel is comparatively shallow—say less than about 150 m—there will be little effect. In fact, it will be found that the tunnel temperature is the average year-round temperature at that location.

Nevertheless, especially in areas of volcanism, geothermal activity, or tropical temperatures, the temperature in deep tunnels can rise to blood heat or higher. If hot water flows are present or if the tunnel is very humid (which is more common than not), conditions can be actively dangerous as sweating and evaporation are inhibited; heat stroke can be induced in such conditions. The only factor that can be directly controlled is the tunnel ventilation. By supplying air at a lower temperature, the local conditions can be kept bearable, especially if the incoming air is dry enough to accept evaporating moisture.

There have been cases (although not in recent years) of crews riding into the tunnel in water-filled muck cars to temper the heat during a long ride in. Chillers supplying cold ventilating air are to be preferred.

Contamination

There have been instances in recent years where remote-controlled microtunneling has been contemplated for installation of shallow utilities in contaminated ground as a cost and risk reduction method. Oil refinery owners in northern California, needing to change the flow of a stream on its property to make way for a highway improvement, decided to put the diversion in a shallow tunnel rather than doing cut-and-cover work. The costs of disposal of contaminated soil made the tunnel alternative financially attractive and avoided surface disruption. In any event, the municipality responsible for the highway found itself short of funds, and the work has not been accomplished to date.

OBSERVATIONS

The significant effects and the construction problems resulting from the various difficult tunneling conditions discussed here make it clear that all of the possibilities associated with the geology and occupational history of the region in which new tunneling is contemplated need to be borne in mind from the construction as well as the design standpoint when the preliminary and final geotechnical exploration and testing programs are designed.

Engineers designing a tunnel project must develop a full understanding of the nature of the ground conditions affecting the construction; so that not only the field investigation but also the design development, specifications, and geotechnical reports reflect a full understanding of the problems and the variety of potential approaches to their solution. In the end, the project owner's interests will be best served by thoughtful analysis and full disclosure of conditions, of the solutions foreseen, and of the underlying design approach rather than by avoiding the recognition of problems and their potential impact.

REFERENCES

EINSTEIN, H.H. (1989). Design and analysis of underground structures in swelling and squeezing rocks (Chapter 6). In *Underground Structures—Design and Instrumentation,* R.S. Sinha, ed. New York: Elsevier.

GARSHOL, K. (1983). Excavation support and pre-grouting of TBM-driven sewer tunnel, Norwegian Tunnelling Technology Publication No. 2, Tapir.

GOODMAN, R.E. (1995). Block theory and its applications—the 1995 Rankine Lecture. *Geotechnique.* In press.

PETERSON, D.L., C.R. NELSON, B.K. NELSON, and B.D. WAGENER. (1991). Design and construction of the Lafayette Bluff Tunnel, *Proceedings RETC,* Seattle.

PHIEN-WEJ, N., and E.J. CORDING. (1991). *Sheared Shale Response to Deep TBM Excavation.* New York: Elsevier.

SENTHIVEL, A. (1994). Report to the Board of Consultants for the Yacambú-Quibor Transfer Tunnel, November.

Technique fits tunnel to a tee. (1991). *Civil Engineering,* July 1991.

WALLIS, S. (1991). Non-shielded TBM holds squeezing clay in check, *Tunnels and Tunnelling,* Jan.

9 | Shafts

Robert J. Jenny
President, Jenny Engineering Corp.

There are generally two modes of access to tunnel construction: through a portal providing direct access at the surface or through a shaft providing vertical access to the level of tunnel operations. Since urban land is valuable and interference with existing services must be minimized, most tunnels built through urban areas require shafts to reach the working area and to provide for removal of tunnel muck.

Tunnel shafts can be temporary or permanent: temporary shafts are for the contractor's use during construction; permanent shafts may be used by the contractor during construction, but will become an integral part of the tunnel structure. Permanent shafts can be used for ventilation, pumping, utility lines or manholes, or they may be enlarged to house stations. Temporary shafts normally are backfilled at the end of construction.

The location of shafts is critical in planning efficient construction. Locating a shaft at the midpoint of a tunnel will permit tunnel driving in two directions; also, a single compressor plant, hoist, shop, and office can serve both headings. Locating a shaft near vacant land will facilitate the erection of temporary buildings. The proximity of muck disposal locations and routes should also be considered.

Once the shaft has been excavated to grade, a pump chamber and a sump may be excavated if required. The pump should provide sufficient capacity to handle the maximum anticipated flow. Information regarding the amount of water that will enter the shaft is obtained during the shaft excavation process, and an estimate of groundwater seepage can be based on previous experience in the same soil or rock medium. Unexpectedly large inflows may occur if water-bearing strata or seams are encountered during excavation (see Chapter 7).

SHAFT EXCAVATION IN SOFT GROUND

Shafts in soft ground are normally excavated with a crane using a clamshell bucket to hoist the muck from the shaft and drop it into a hopper or a stockpile or directly into a truck on the surface.

Temporary shafts in soft ground are often circular. A concrete collar, a ring of concrete usually 2 ft wide and 4 ft deep, with the top surface at least 12 in. above ground level, should be placed around the top of the shaft. The collar prevents distortion of the shaft's primary lining and prevents surface water and debris from falling into the shaft. Handrails or another type of safety protection around the top of the shaft must also be provided.

Primary shaft linings are normally installed at every 4 or 5 ft of advance. However, shafts have been sunk up to 30 ft without supports. The rate of installation depends on the type of lining and the nature of the soil medium.

A permanent shaft usually will have a final lining of concrete. Permanent shafts may be round, oval (NATM), or rectangular in shape, and the concrete for shaft lining may be cast either with forms on both sides or forms on the inside only with the ground support system on the outside.

Soft ground shaft sinking may disturb or damage neighboring buildings, utilities, pipelines, or streets. The problem is especially acute with soft plastic soils; when plastic soil is excavated, the load over the excavated area is reduced, and plastic yielding may result, causing ground yielding at the surface. Plastic yielding can occur below stresses that are associated with shear failure. A properly designed shaft support system can prevent plastic yielding.

Soil characteristics, shaft depth, diameter, and economic factors will dictate the choice among the many available sheeting and bracing systems.

Timber Sheet Piling

Timber piling is inexpensive and, once installed, very easy to work with (Figure 9-1). Timber sheet piling is normally used only in shallow shafts, since driving the thick timbers is difficult. The method can be economical to start excavating in soft material, not deeper than about 20 ft of soil overlying rock.

Three- or four-inch thick timbers are driven into the ground and excavation is usually performed simultaneously, thereby reducing friction in the pile driving and also preventing the soft timber tips from catching on obstructions

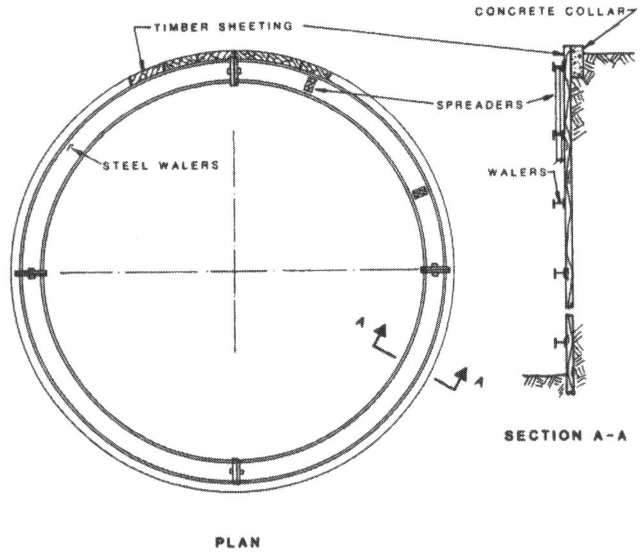

Fig. 9-1. Timber sheet piling.

and splintering. Horizontal, circular, or rectangular steel rib sets (wales) are installed against the interior of the sheeting. All the lateral earth pressure acting on the sheeting is transmitted to the steel ribs, which carry the stress in ring compression or bending, depending on whether the shaft is circular or rectangular.

Steel Sheet Piling

Interlocking steel sheet piles are commonly used to brace soft, water-bearing ground if the excavation exceeds about 20 ft.

Steel sheet piles must be driven carefully to ensure proper interlocking of the joints to cut off water seepage. Excavation usually begins after the pile-driving operation is completed, unless the shaft is unusually deep.

Horizontal steel rib sets (wales) can then be installed progressively at appropriate vertical depths as the excavation progresses downward. As with timber sheet piling with horizontal rib sets, the lateral earth pressure is transferred to the wales (Figure 9-2).

Soldier Piles and Lagging

To begin this type of shaft excavation, steel H piles called *soldier piles,* usually spaced from 6 to 10 ft apart, are driven to the required depth. Excavation begins after all the piles are driven.

As the excavation proceeds, horizontal timber lagging, usually 3- or 4-in. hardwood, is placed against the face of the excavation and wedged between the flanges of the H piles. In moist to wet soils, water usually drains between the lagging boards.

In extremely wet soil or in running sand, hay can be forced between the lagging boards to prevent the ground from flowing into the excavation. A small space is reserved between the lagging to allow for drainage. In very wet soil or in running sand, hay is forced into the spaces to prevent the ground from flowing into the excavation.

In deeper excavations, where larger earth pressures are encountered, horizontal steel rib sets (wales) are installed. The steel rib sets must be designed for either ring compression or bending, depending on whether the shaft is circular or rectangular (Figure 9-3). In large rectangular shafts, steel struts can be installed to span between wales.

Liner Plates

The main advantage of liner plates is that their small size permits ease of operation in limited working spaces, and it is not necessary to have special equipment to lift or place the liner plates (Figure 9-4).

Most liner plates are corrugated pressed steel pans with bolt holes in the flanges on the sides and ends to permit bolted erection of the ring.

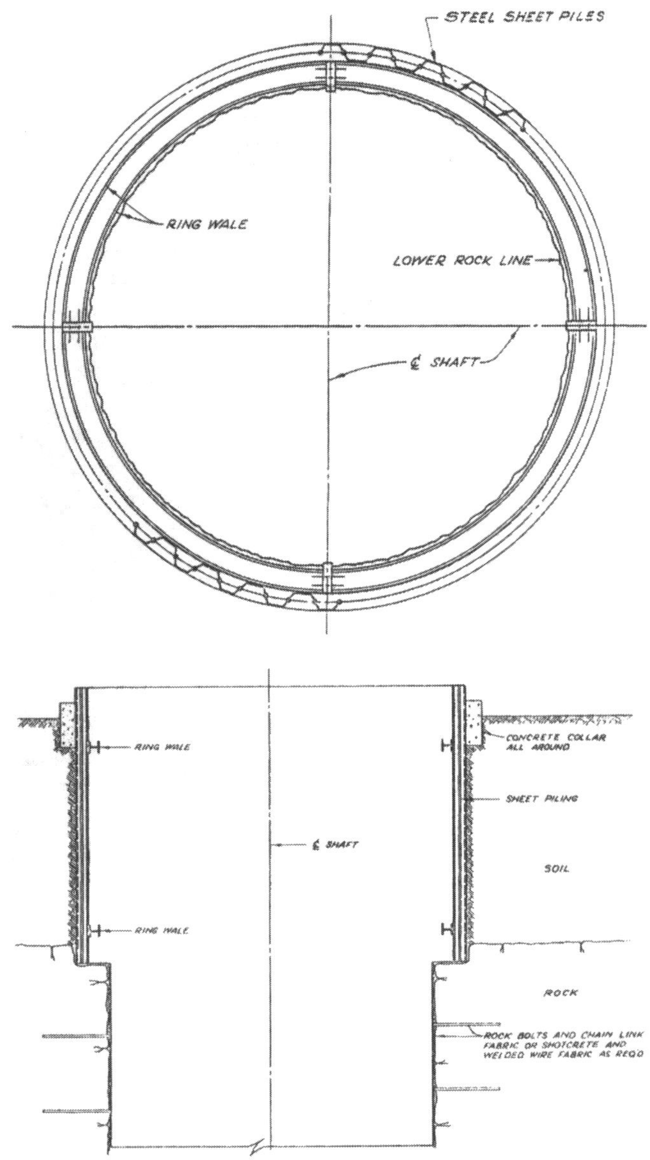

Fig. 9-2 Steel sheet piling: (a) cross section; (b) elevation.

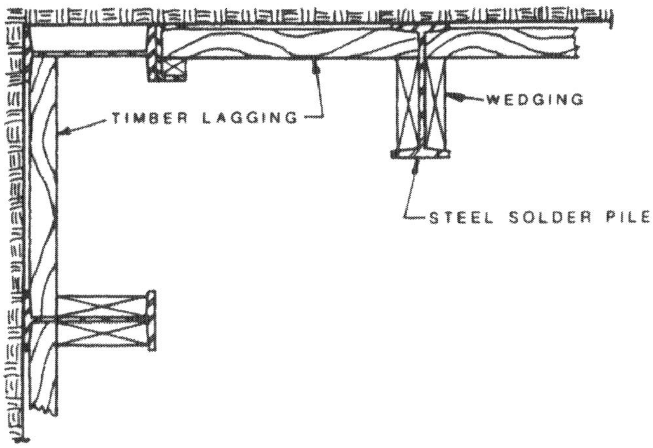

Fig. 9-3. Soldier piles and lagging.

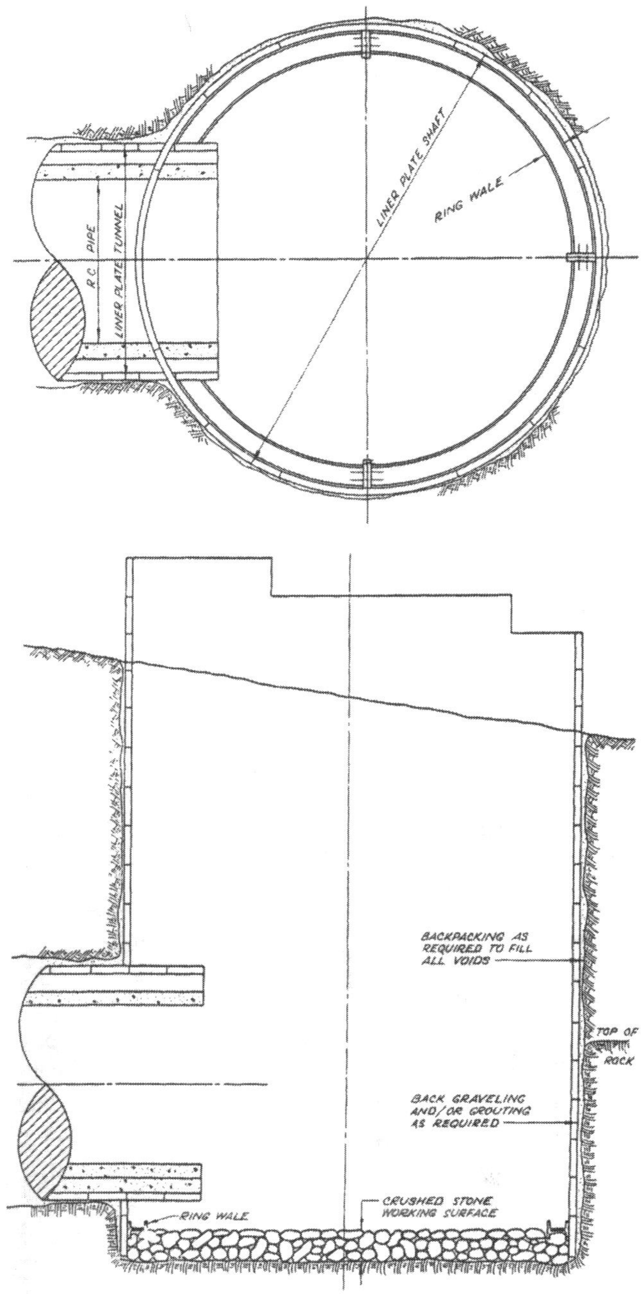

Fig. 9-4. Liner plate shaft: (a) cross section; (b) elevation.

Shaft excavation begins by precisely erecting the first ring of liner plates on the ground surface and placing a concrete or earth collar around it. The soil is then excavated within the ring, and as space becomes available a liner plate is bolted to the bottom flange of the first liner plate ring; assembly of the succeeding rings proceeds in the same manner. The joints of individual liner plates are staggered from the joint immediately above to increase strength.

The soil pressure is carried by the liner plates in ring compression. In larger-diameter shafts or in shafts where the lateral earth pressure is large, circular horizontal steel rib sets (ring wales) can be set at a predetermined vertical spacing inside the liner plate ring to increase strength.

Horizontal Ribs and Vertical Lagging

The horizontal ribs and vertical lagging method is somewhat similar to liner plate construction. Rings are made of structural steel members, cold-formed to required curvature, the sections butted at each end. Butt plates welded to the ends of the segments are provided with bolt holes. Six- to eight-foot lengths of timber are usually used for lagging (Figure 9-5).

This method requires excavation of the soil to a distance equal to the length of the lagging. Curved ring segments are bolted together and held in place by tie rods and spacers that are placed between the webs of the rings. Placement of the vertical lagging follows. The steel rib rings can be placed on varying centers to resist such lateral earth loads as may be encountered.

Since the soil must be initially somewhat self-supporting for the height of the lagging, this method is usually employed in cohesive soils, although it can be used when the ground has moderate stand-up time. During construction, the interval between excavation and lagging placement should be minimal to prevent ground loss.

Slurry Walls

Bentonite, a naturally occurring clay, has a large capacity for absorbing water. In suspension, it is a liquid when agitated, and a gel if left to stand. The liquid suspension can be injected into a permeable soil mass and allowed to congeal in the voids, thereby decreasing permeability. There are generally two methods of excavation with slurry.

The first method uses excavating equipment suitable for ground formations consisting of soft to medium hard, loose or cohesive soils without boulders or other obstructions. The excavation is performed by clamshell buckets that open the trench in a panel sequence, with the trench kept filled with slurry. The clamshell rigs are available in a wide variety of types, depending on the site conditions. Some are hydraulically or mechanically operated and are cable suspended,

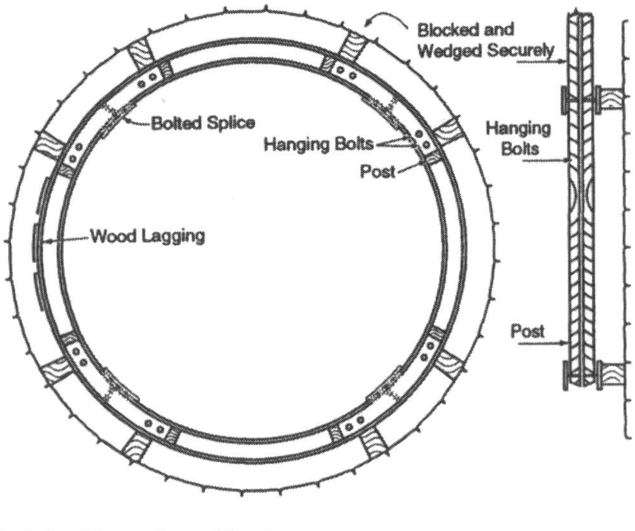

Fig. 9-5. Ring wales and lagging.

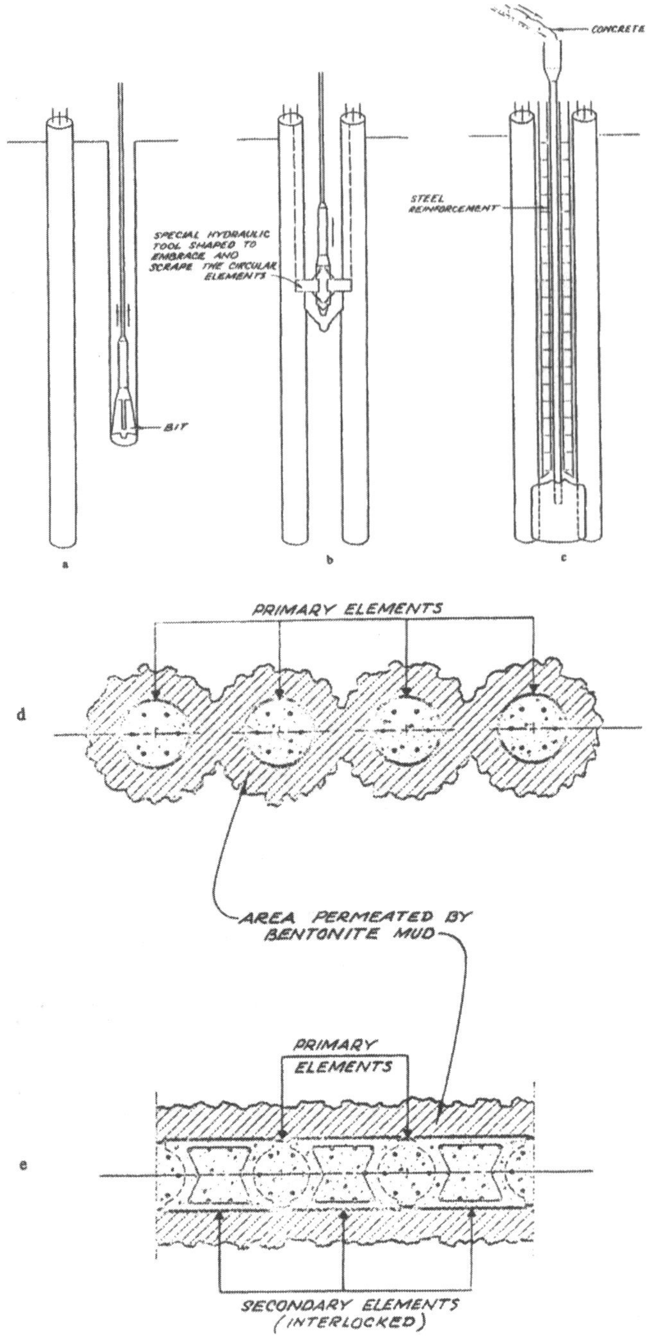

Fig. 9-6. Slurry wall construction by ICOS interlocked element method: (a) construction primary elements; (b) drilling secondary elements; (c) concreteing secondary elements; (d) completed primary elements (section); (e) completed wall section. (Adapted from Xanthakos, 1974)

while others are mechanical or hydraulic clamshells controlled with a "kelly bar." Concrete is placed in each panel by tremie with the bentonite being recirculated to the next panel excavated. A cofferdam can then be created by bracing the concrete with steel or reinforced concrete wales and struts.

The second method, known as the "interlocked" element type, is used when the soil is very hard and bouldery, or when the excavation must reach considerable depths. The technique used under these conditions is diagrammed in Figure 9-6. Since clamshells are no longer effective in these formations, the excavation must be performed solely with the aid of percussion tools. Primary holes are first drilled using percussion rigs with the assistance of a bentonite slurry. In most cases, the spacing of the primary holes (center to center) is twice the diameter, which means that the space to be filled by the secondary elements is equal to the diameter of the primary holes. This spacing is also convenient with respect to the permeability of alluvial deposits, since the penetration of the soil between primary holes by the bentonite slurry is almost complete, and therefore the excavation of the secondary elements involves a zone of soil stabilized through gelation in its pores.

Concrete is placed in the primary holes with the aid of tremie pipes. A hydraulically expandable chisel is then used to excavate the panels between the primary holes.

Structures built by these methods can be reinforced by placing the steel in the slurry-filled holes or trench before concrete placement begins. After the concrete walls are in place, the interior excavation for the shaft can begin.

NATM Shaft Support System

A recent development in the design of support for shafts in soft ground has been the adaption of the New Austrian Tunneling Method (NATM). NATM originally was conceived as a new way of looking at rock support in tunnels and shafts. However, particularly with improvements in and more general acceptance of shotcrete, it has been found that NATM concept support can be used in some soft ground shafts.

Key elements in the NATM support concept are, first, constant monitoring by convergence/divergence instrumentation of what is occurring within the exposed excavation; and second, immediate installation of the support components in contact with the exposed earth, as soon as excavation makes room for the component installation.

In general, the installation of the NATM support system follows somewhat the same sequence as the installation of liner plate. A lattice-type rib ring is a key element of the system. As soon as a section of shaft excavation is exposed, a section of the lattice ring is installed. A desirable feature of the lattice ring segments is that they are lightweight and can be easily installed by hand. Once the lattice ring segment is in place, shotcrete with wire mesh is applied to the exposed excavated surface. Usually about 6 in. of shotcrete is adequate, but if the instrumentation indicates movement of the earth or other undesirable characteristics, more shotcrete can be applied to obtain a thickness of 12 in. or more. See Figure 9-7 for a NATM system shaft. Note the oval shape, which replaces the more conventional rectangular shape normally used for subway ventilation shafts.

EXCAVATION IN SOFT, WET GROUND

Excavation in soft, wet ground can be accomplished in a number of ways. The most common method is to lower, by any of several means, the groundwater table in the working area. Other methods include freezing of the soil, the use of slurry, grouting, sinking a pneumatic caisson, and sinking a dredged drop caisson, with a tremie concrete seal.

The methods of open pumping, wellpoint systems, deep wells, and freezing will also be discussed. Typical pneumatic and dredged caissons are shown in Figures 9-8 and 9-9.

Lowering of Groundwater

Although dewatering operations in shafts can be time consuming, lowering the water table for shaft excavation will ensure dry, safe, and firm working conditions. To determine the proper type of groundwater control system to implement in advance of construction, geologic and soils information should be evaluated and a pumping test should be performed on the soil. The test should yield a distance-drawdown versus time curve, and the test should continue until the slope of the curve in relation to the pumping rate becomes constant. The pumping test should yield results such as water volume pumped, well yield, and time required to reach equilibrium.

It is advisable to perform a chemical analysis of the groundwater to check for dissolved salts or gases. Calcium salts and iron oxides in the water could corrode metal portions of the dewatering system. If well screens are to be used, the screens could become plugged by precipitating salts.

Dewatering operations can result in the lowering of the water table under adjacent areas, in some cases as far as 2,000 ft or more from the well. Therefore, extreme caution must be observed if large structures are in the vicinity of dewatering operations. Recharging of the groundwater or other

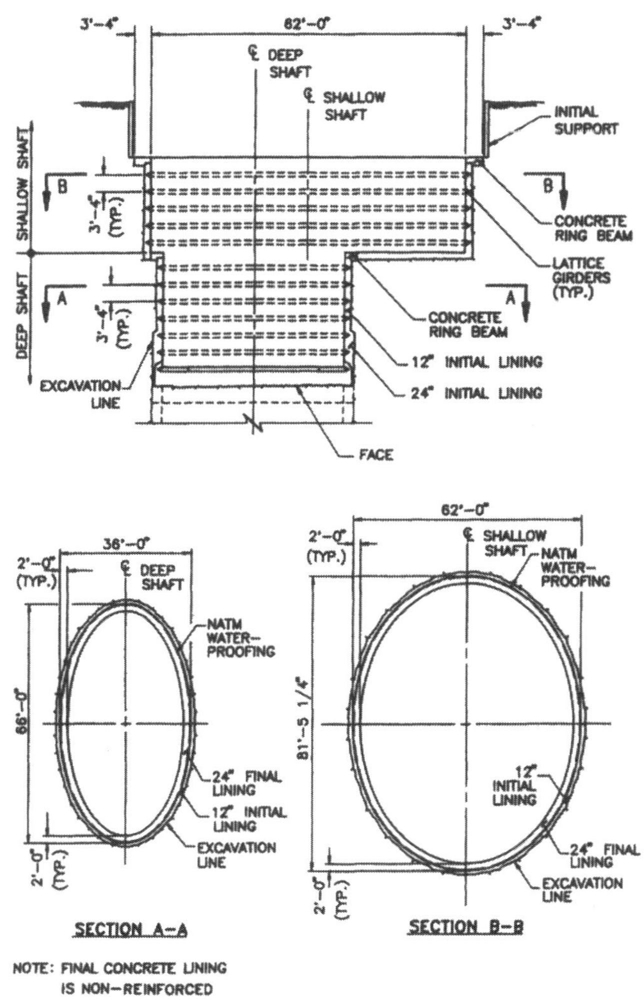

Fig. 9-7. NATM system shaft (final concrete lining is non-reinforced).

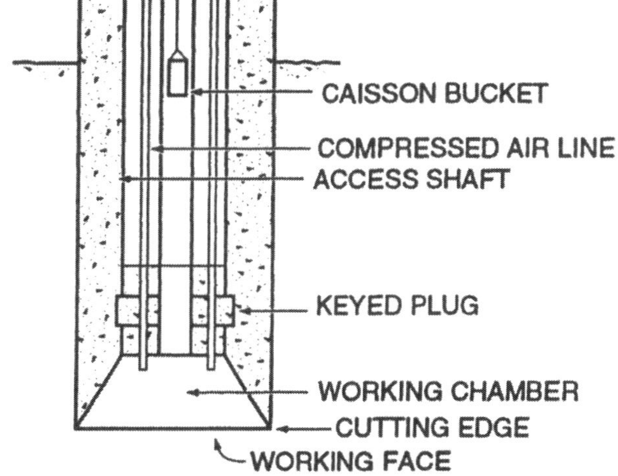

Fig. 9-8. Schematic representation of pneumatic caisson excavation.

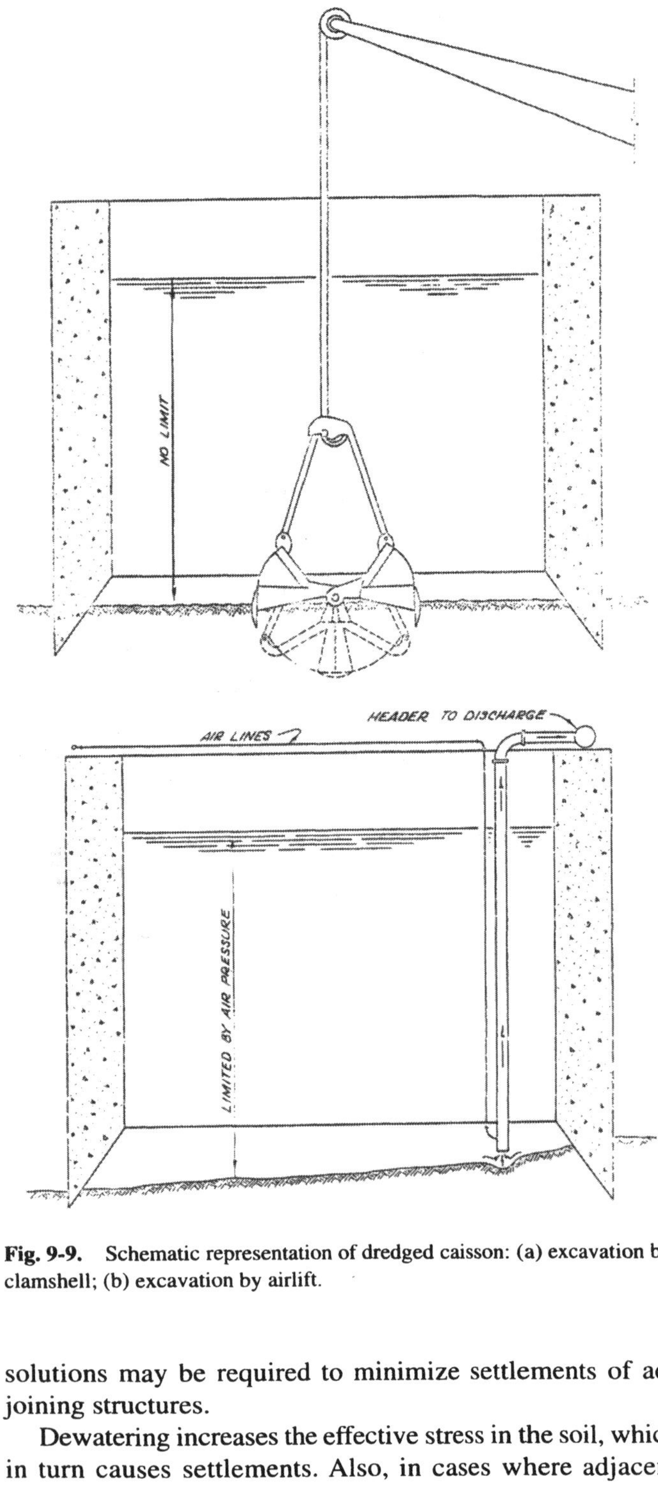

Fig. 9-9. Schematic representation of dredged caisson: (a) excavation by clamshell; (b) excavation by airlift.

solutions may be required to minimize settlements of adjoining structures.

Dewatering increases the effective stress in the soil, which in turn causes settlements. Also, in cases where adjacent structures are supported on piles, enough drawdrag can be developed on the pile foundation to cause settlements.

Open Pumping

The simplest shaft excavating method through water-bearing pervious soil is the sheeting and open pumping method. This consists of driving steel sheet piling, excavating, and pumping water from the bottom of the excavation.

However, the pumping operation may cause seepage of water (and loss of fines) around the toe of the sheeting. Furthermore, if the pressure due to the upward seepage of water becomes greater than the soil pressure at the bottom of the excavation, a quick or "boiling" condition in the soil can result. Also, if the seepage of water around the toe of the sheeting significantly dislodges soil particles, the sheeting can be undermined (Figure 9-10).

Wellpoint System

The wellpoint system is generally used for dewatering to a depth of about 15 ft. This is the most common method of dewatering shallow, open excavations in the United States. The technique is best suited for use in medium-to-fine sand for work of short duration.

The method consists of placing wellpoints on 3- to 12-ft centers around the area to be excavated. The wellpoints are attached to a common header pipe, which is connected to a pump. Wellpoints are well screens, which require suitable filter material around the screen to prevent the collection of soil particles with the water. Once the dewatering is accomplished, shaft excavation can begin. However, the wellpoint system must operate during excavation lest the water table return to its original level.

The main disadvantages of a wellpoint system for shaft excavation are that the system must be within the area of excavation, and that the depth of dewatering is limited to 15 ft (Figure 9-11).

Deep Wells

Deep wells can be used to dewater pervious materials to whatever depth the excavation requires, and they can be installed outside the zone of excavation.

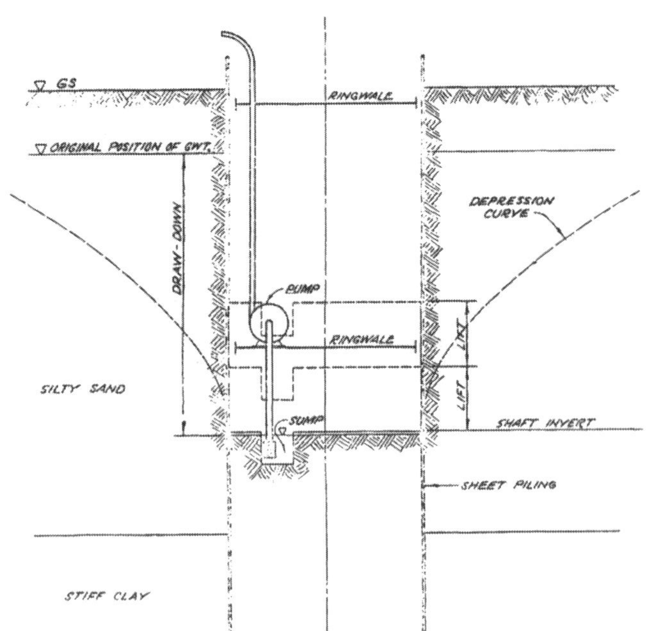

Fig. 9-10. Schematic representation of open pumping.

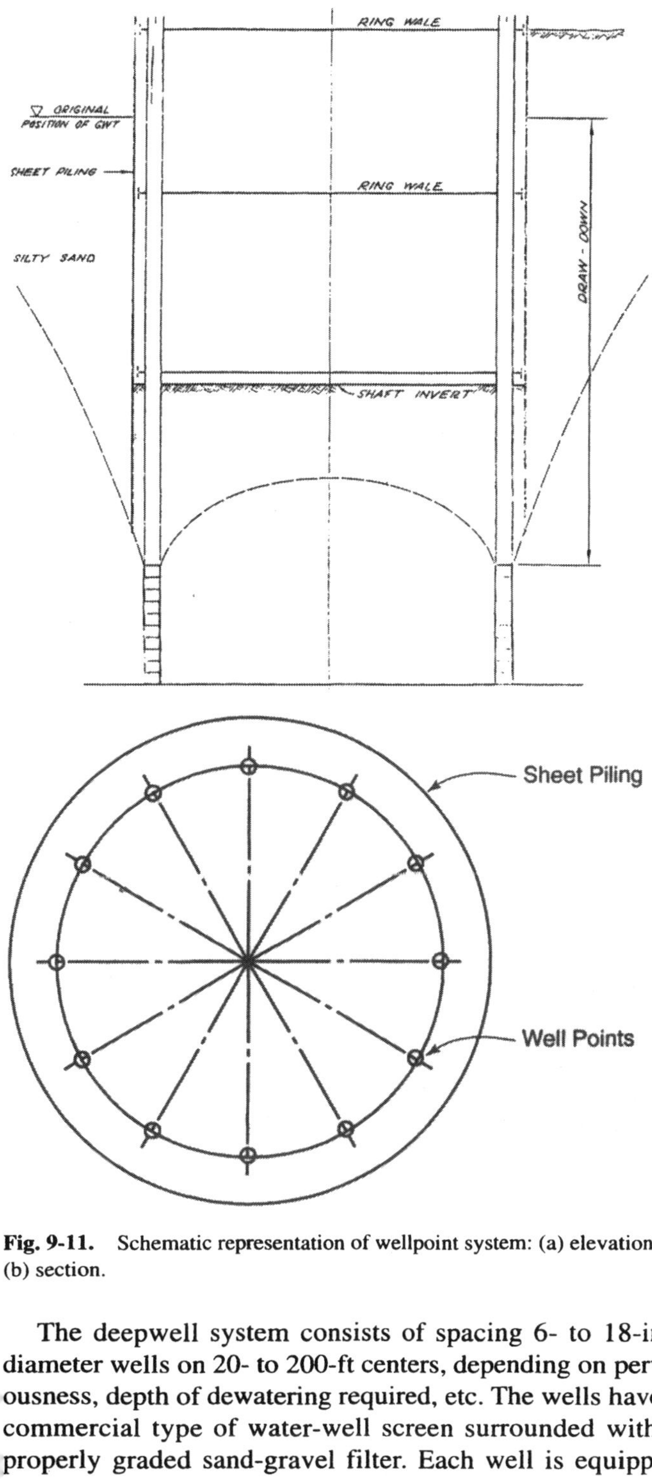

Fig. 9-11. Schematic representation of wellpoint system: (a) elevation; (b) section.

The deepwell system consists of spacing 6- to 18-in.-diameter wells on 20- to 200-ft centers, depending on perviousness, depth of dewatering required, etc. The wells have a commercial type of water-well screen surrounded with a properly graded sand-gravel filter. Each well is equipped with its own submersible pump. The excavation for the shaft can begin after water drawdown to the required elevation has been accomplished, or possibly somewhat earlier depending on the type of shaft supports to be installed and whether the rate of drawdown has been firmly established.

For shaft excavation, deep wells can provide dewatering to the depth desired; relatively few units have to be installed; and once installed properly, only maintenance is required. However, only pervious strata can be dewatered by this method (Figure 9-12).

Freezing

In water-bearing ground where even minimal surface subsidence cannot be tolerated, such as adjacent to large buildings, the most reliable method of handling the excavation is to freeze the soil and then excavate. The freezing eliminates both seepage and plastic soil flow. There is no limitation on the depth to which freezing may be used.

The procedure consists of sinking pipes around the area to be excavated and circulating a cold brine solution through the pipes, thereby freezing a wall of soil. Excavation can then begin. If a concrete shaft lining is to be installed, the concrete can be placed against the frozen soil.

The refrigerating circuit in each bore hole consists of a tube closed at one end, containing another smaller-diameter,

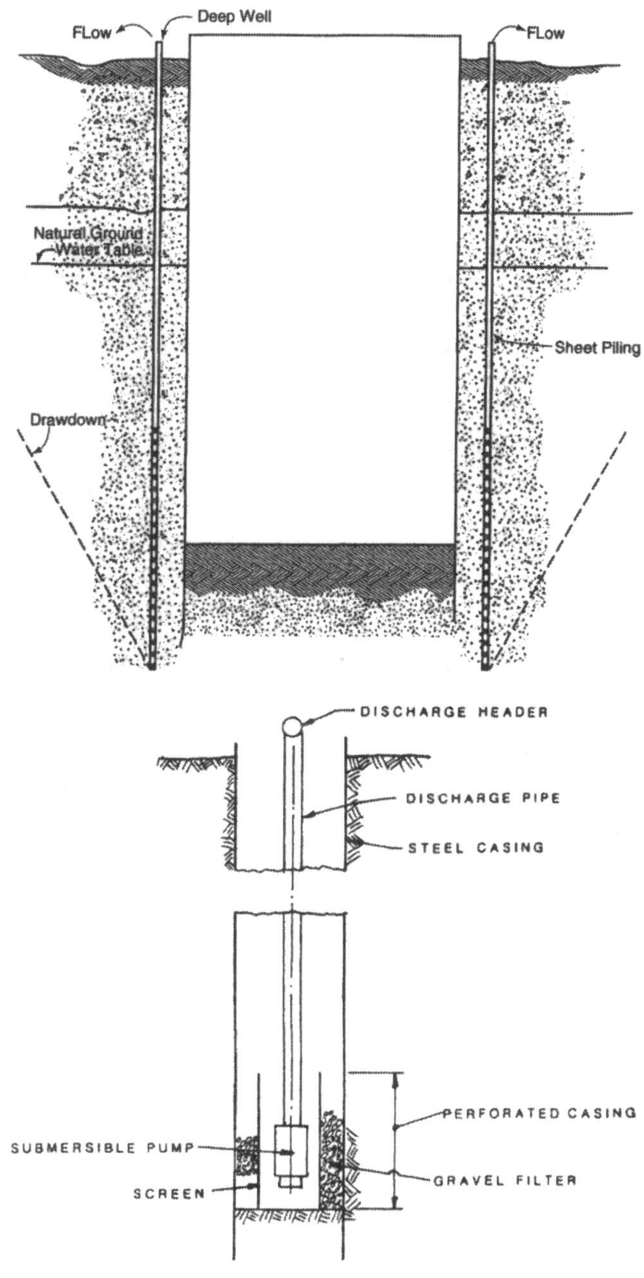

Fig. 9-12. Schematic representation of deep-well dewatering.

open-ended pipe. Brine is cooled at the surface refrigerating plant, then circulated down the inner pipe and up through the annular space between the pipes back into the refrigerating plant. The process is continued until sufficient soil is frozen. The time interval before shaft excavation can begin varies usually between two and five months, depending on depth, type of soil, and other factors.

The principal disadvantages to the freezing technique are the time required to freeze the soil and the cost of the equipment. The refrigerating plant can, however, be salvaged and reused (Figure 9-13). As an alternative to a brine refrigerant, liquid nitrogen is sometimes used to accelerate the freezing process.

SHAFT EXCAVATION IN ROCK

Shaft excavation in rock is usually performed by the drill-and-blast method. Shaft excavations for tunnels are normally less than 120 ft deep; therefore, the use of more-sophisticated shaft excavating equipment is not often economically feasible. When sinking shafts deeper than 120 ft, other methods can be used that are more within the realm of the mining engineer (see Chapter 10).

Prior to the start of rock excavation, it can be advisable to grout and seal the overburden if groundwater infiltration from the overburden can become a problem. If water infiltration is expected when excavating through poor rock, a grout curtain can be installed to achieve a water barrier. A ring of holes about 10 ft from the shaft is drilled for the depth of the shaft, then grouted to refusal.

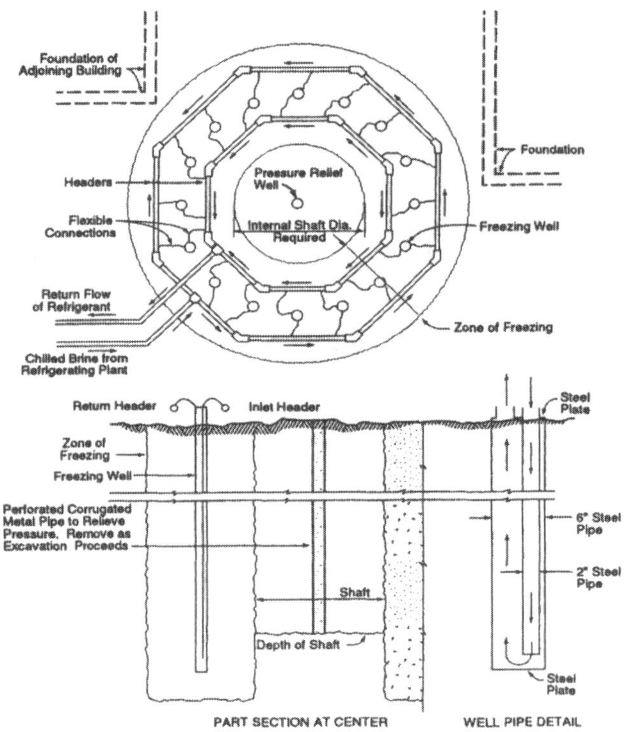

Fig. 9-13. Ground freezing method: (a) section; (b) elevation.

Drilling

The size and depth of the shaft and the type of rock expected to be encountered are the determining factors in choosing the type of drilling equipment to employ. Shaft drilling can be done by either handheld air drills, air-track drills, or collapsible shaft jumbos. The shaft jumbos can be set and leveled in the hole, the bull hoses attached to the boom-mounted drills, and drilling can commence. Usually, only the deeper and larger shafts can justify the use of a jumbo.

Blasting

The major considerations in designing a shaft round are ease of drilling and minimizing overbreak. Special arrangements can be made for creating sumps if there is a water problem.

The most suitable blasting arrangement for circular shafts is the pyramid cut. The pyramid cut consists of several holes drilled angularly to meet in a common apex near the center of the face. Peripheral relief holes and trim holes may also be incorporated (Figure 9-14).

Rectangular shafts are commonly drilled and blasted using a V cut (Figure 9-15). Each V consists of two holes drilled from points as far apart as possible on the face to meet at the bottom of each hole. A series of V cuts parallel to one another will control the width of the shaft.

The pyramid cut and the V cut are just two examples of blasting patterns. There are numerous other drill-and-blast patterns that can be used in shaft excavation. Competent blasters who have previous experience with the rock in the area of the shaft generally can suggest the type of round that will best accomplish the excavation without undue overbreak and that will produce muck that can be readily handled by the mucking equipment available.

Mucking

After blasting, mucking can usually be carried out by cranes with a clamshell bucket. Hand mucking is used in shafts up to 10 ft in diameter. When the shaft is less than about 60 ft deep, the muck may be hoisted in muck boxes that are lifted by a crane. Methods and equipment used for shafts over 100 ft deep are discussed in Chapter 10.

Raises

A raise is a vertical excavation proceeding from a lower elevation to a higher elevation, perhaps from one tunnel to another, or from a tunnel to the ground surface. The raised shafts may be used for manways, rock passes, or ventilation. The raise can be used to intersect another tunnel above an existing one. Shaft raising is sometimes used in urban tunnel construction to minimize surface disruption.

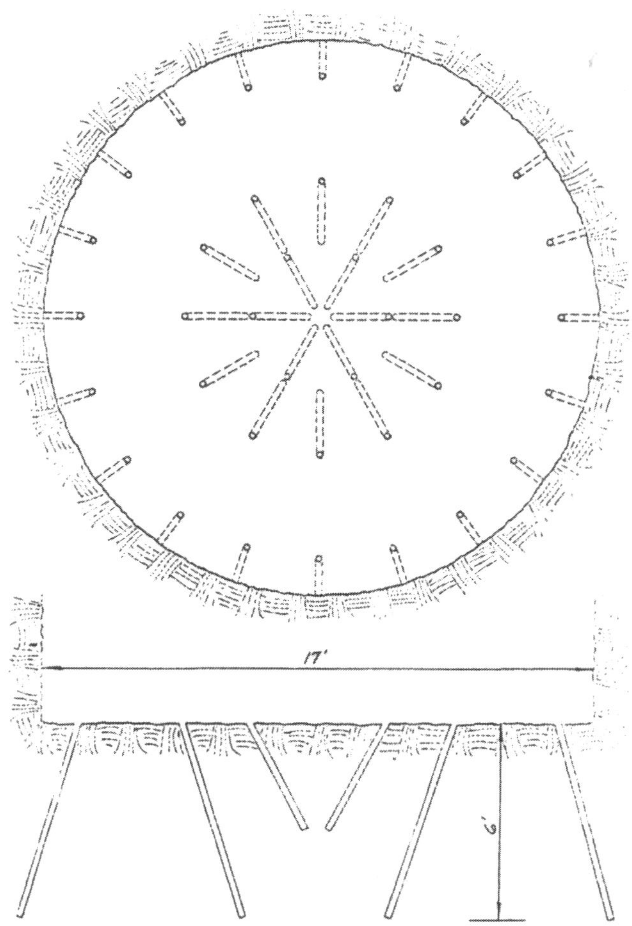

Fig. 9-14. Pyramid cut for shaft sinking.

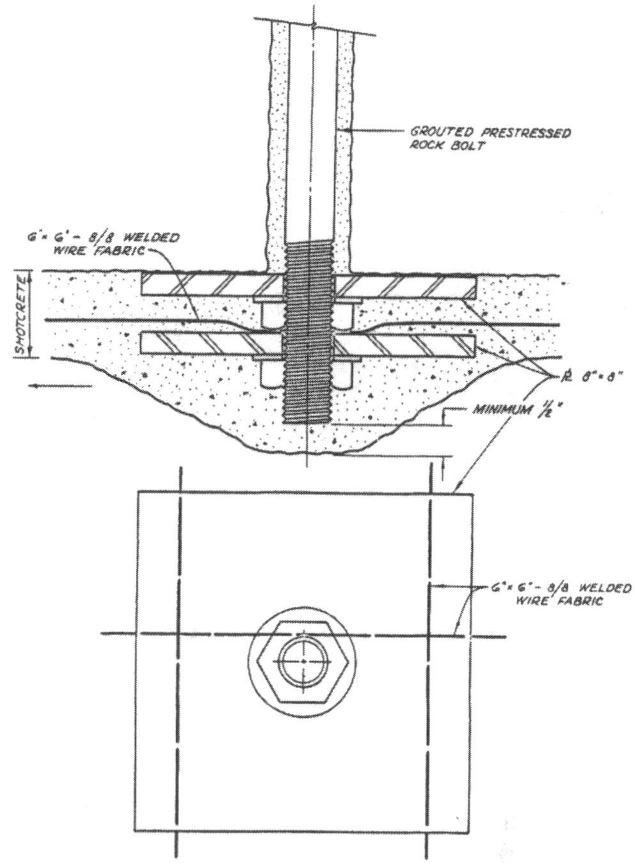

Fig. 9-15. Typical installation of rock bolts, wire mesh, and shotcrete, Washington Metro.

Raises are usually excavated by drilling pilot holes, usually of diameter about 9 to 12 in., then reaming the hole to the proper diameter. The deviation of the pilot hole can be limited to about 1% of the length, using modern techniques. Good crews have been known to achieve less than 0.5% deviation. The most common and successful system to date has been drilling the pilot hole down and reaming up the required raise. Sometimes the pilot hole is drilled up and reamed down. Only a few types of raise drills are widely used, and the nature of the rock through which the raise will pass should be studied carefully to assure the use of proper cutters for efficiency and economy.

Raise excavation can be accomplished by drilling and blasting but is often quite hazardous, especially the scaling down of loose rock after the blast. A comparison of drill-blast raise excavation versus raise boring in the Western Cordilleran Region of the United States revealed a much higher daily advance rate by the boring method.

Temporary Supports

In sedimentary, fractured, or blocky rock, the walls can be quite treacherous. When rock support is required, it should be placed quickly after excavation. The support can consist of steel ribs and liner plates, steel ribs with lagging, rock bolts with or without wire mesh, or pneumatically placed concrete (shotcrete). Generally, all the various types of supports described for support of soft ground shafts can be used, perhaps with some modifications, in rock shafts. The NATM shaft support system described earlier is particularly adaptable for rock shafts that require temporary supports. Prior to the placement of the support system, the loose rock should be scaled down.

If the rock is of reasonably good quality, it may be advantageous to merely install rock bolts into the wall and fasten wire mesh to the bolts to keep rock spalls from dropping on the workers.

It is sometimes good practice to apply shotcrete to the walls. Shotcrete linings are also becoming more popular as permanent linings (see Chapter 12).

A typical installation of rock bolts, wire mesh, and shotcrete is shown in Figure 9-16.

LINING OF SHAFTS

Permanent shafts are usually concrete lined; however, shotcrete is becoming popular and acceptable as a final lining. Naturally, the planned permanent usage of the shaft may determine the type of final lining.

Fig. 9-16. Forms for concrete shaft lining.

Concrete Lining in Soft Ground

Any shaft sunk through soft ground requires an initial, or "primary," lining for construction support. Prior to the construction of the final concrete lining, the primary lining should be true in shape, and direct continuous contact between the ground and the primary lining should be ensured.

The secondary lining can either be poured against the primary lining, or it can be formed from both the outside and inside. If the lining is formed on the outside, the annular space between the primary and secondary lining should be tightly backfilled or packed with pea gravel, well-graded sand, or other suitable material. If the shaft has been excavated through water-bearing ground, placing an impervious sheeting material on the face of the primary lining prior to placement of the final concrete lining has become recent common practice. Additionally, grouting may be required to prevent water seepage into the finished shaft.

Precast concrete segments can be used for the secondary lining where it can be placed from the bottom up. This creates an annular void, which must be filled subsequent to completion of the lining.

Rock Bolts and Wire Mesh

Rock bolts and wire mesh are usually used in relatively sound rock to maintain stability and prevent spalling. Sometimes shotcrete is applied in addition to the bolts as a primary, or final, lining. Since the rock is of fairly good to good quality, the construction of the secondary lining is rather straightforward, and the concrete can be poured directly against the rock. If water infiltration through seams in the rock is a problem, the seams should be well grouted before the construction of the secondary lining, and as noted above, an impervious sheeting material should be placed on the rock (or shotcrete) prior to placement of either a concrete or shotcrete final lining. Final contact grouting will also be required to fill any voids between the concrete secondary lining and the rock.

Shotcrete

Shotcrete is used for temporary support to prevent rock spalling. Sometimes the shotcrete is used with wire mesh for added strength. Currently, some work is under way using shotcrete as a permanent lining, since the quality and strength of the mixture has been greatly improved. The NATM shaft support system described earlier illustrates part of the broad spectrum of possible uses for shotcrete.

REFERENCES

Blaster's Handbook (1980) I.E. DuPont de Nemours & Co., Wilmington, Delaware.

BRADY, J.J., BHORE, J.S., GORDON, R.B., and HONEYCUTT, J.W. (1985) "Practical Methods of Shallow Shaft Sinking," *Proceedings of Rapid Excavation and Tunneling Conference, June 16–20,* New York.

HAMMOND, R. (1959) *Tunnel Engineering.* Macmillan, New York.

HAVERS, J.A., and STUBBS, F.W. JR. (1971) *Handbook of Heavy Construction.* McGraw-Hill, New York.

JESSBERGER, H.L. (1979) *Ground Freezing, Developments in Geotechnical Engineering,* Vol. 26. Elsevier Scientific Publishing Company, New York.

JUMIKIS, A.R. (1971) *Foundation Engineering.* Intext, Scranton, Pa.

LANE, K., and GARFIELD, L.A. (1972) *Proceedings of First North American Rapid Excavation and Tunneling Conference, Chicago, Illinois, June 5–7, 1972.* Port City Press, Baltimore.

LEONARDS, G. (1962) *Foundation Engineering.* McGraw-Hill, New York.

MCWILLIAMS, J.R., and ERICKSON, E.G. (1960) *Methods and Cost of Shaft Sinking in the Coeur D'Alene District, Shoshone County, Idaho.* U.S. Department of Interior.

MAYO, R.S., ADAIR, T., and JENNY, R.J. (1968) *Tunneling: The State of the Art.* U.S. Department of Housing and Urban Development.

MERRITT, F.S. (1968) *Standard Handbook for Civil Engineers.* McGraw-Hill, New York.

POWERS, J.P. (1972) "Groundwater Control in Tunnel Construction," *Proceedings of North American Rapid Excavation and Tunneling Conference,* ASCE-AIME, New York.

SHUSTER, J.A. (1985) "Ground Freezing for Soft Ground Shaft Sinking," *Proceedings of Rapid Excavation and Tunneling Conference, June 16–20,* New York.

STOSS, K., and BRAUN, B. (1985) "Sinking a Freeze Shaft with Installation of a Water-Tight, Flexible Lining," *Proceedings of Rapid Excavation and Tunneling Conference, June 12–16, 1983,* New York.

SZECHY, K. (1966) *The Art of Tunneling.* Akademiai Kiado, Budapest.

Use of Shotcrete for Underground Structural Support (1974) ACI/ASCE, Detroit.

XANTHAKOS, P.P. (1979) *Slurry Walls.* McGraw-Hill, New York.

XANTHAKOS, S.E. (1974) *Underground Construction in Fluid Trenches.* University of Illinois, Chicago.

10 Deep Shafts

Maurice Grieves
Consulting Engineer

Very deep shafts have traditionally been associated with the gold mining industry in South Africa, where depths of 6,000 to 8,000 ft are common. In good, hard rock with dry conditions, shafts with circular concrete linings of 25 to 30 ft internal diameter have been conventionally sunk and lined at rates of up to 1,250 ft in a month (e.g., Buffelsfontein Eastern Twin Shaft, 1962). In contrast to mine shafts worldwide, where 2,500 ft depths have been commonplace for over a century, the very deepest shafts for most civil engineering projects rarely exceed 2,000 ft depth. Applications include ventilation shafts more than 2,000 ft deep for long transportation tunnels through high mountain locations such as the trans-Alpine tunnels connecting population centers in Switzerland, Austria, and Italy, and high-pressure penstock shafts of 2,000 to 3,000 ft deep for hydroelectric pumped storage projects.

SHAFTS FOR TUNNELS AND CAVERNS

Deep shafts provide vertical or inclined access to existing or proposed underground openings, such as water distribution or drainage tunnels, road or rail transportation tunnels, service tunnels, power stations, storage caverns, or military and scientific facilities. Only circular shafts will be discussed here since they have replaced almost all rectangular and elliptical shapes, although some shafts such as partitioned ventilation shafts or equipment access shafts are often constructed by inserting flat extensions between semicircular end walls.

In most civil projects, deep shafts are shafts that are sunk to total depth from the floor of a previously completed foreshaft or presink. During the presink, a mobile crane or scotch derrick will have been used to hoist muck, materials, and possibly personnel, and temporary equipment will have been used in the excavation and lining process (Figures 10-1 and 10-2). The purpose of the presink is to provide sufficient depth and permit the installation and commissioning of the main sinking facilities. The main sinking facilities consist of a specialized shaft-sinking plant and machinery capable of constructing the shaft to final depth by performing and optimizing the repetitive processes of excavation, muck removal, temporary support, and installation of an interim or permanent lining. Shaft may be sunk to depths ranging from 300 to 2,500 ft or more.

Deep shafts for civil engineering projects are constructed to meet a range of special duties and requirements:

- *Site characterization.* Shafts may be used initially to investigate underground rock mass characteristics and response. An example would be the 2,200-ft-deep access and ventilation shafts at Burkesville, Kentucky, for Cominco American, which were blind-bored from the surface in 1977 and 1978 at diameters of 72 in. and 27 in., respectively.
- *Ventilation.* The 2,400-ft-deep, 25-ft-diameter concrete-lined Albana shaft for the Arlberg highway tunnel near Stuben, Austria, was slipformed with an internal partition wall. Watertight PVC sheathing was installed between an inner and outer lining. Another example is the 1,150-ft-deep, 28-ft-diameter mid-tunnel partitioned ventilation shaft and fan house for the 9-mi-long Mt. Macdonald/Rogers Pass Tunnel in British Columbia, which opened in 1988. Ground freezing was used to allow excavation of the shaft through unconsolidated water-bearing ground.
- *TBM access and muck removal.* Examples include 32-ft-diameter access shafts to 330-ft-deep tunnels with 1,000-tons-per-hour capacity vertical muck removal conveyors constructed to drive miles of 32-ft-diameter TBM-excavated tunnel for the Metropolitan Water Reclamation District of Greater Chicago. Another example is Boston's 34-ft-diameter, 400-ft-deep on-shore shaft with vertical muck removal conveyor, which provides access for construction of the 9.5-mile-long undersea Boston Effluent Outfall driven with a 26-ft, 6-in. diameter TBM.
- *Power generation.* Vertical and inclined penstock shafts deliver water from upper reservoir storage and surge facilities to drive the turbines of deep underground power stations. Constructed projects include the 1,780-ft combined depth of the surge/orifice/high-pressure shaft system at Dinorwic (Wales) and the 50° inclined, 320-ft-long penstocks of the HydroQuebec Outardes 4 powerhouse. Cable shafts also bring power cables from the underground transformers to the

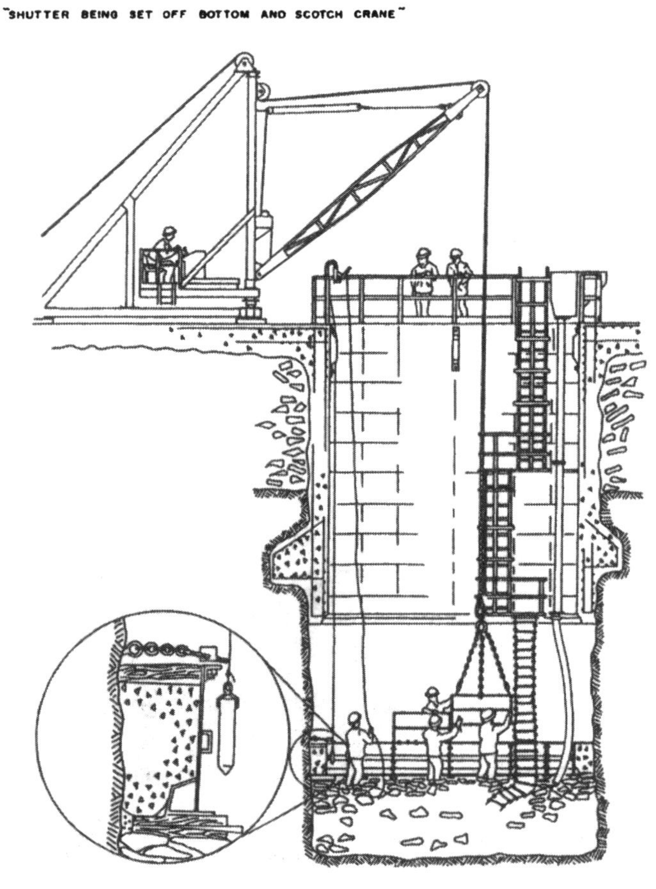

Fig. 10-1. Collar construction (Douglas AAB and Pfutzenreuter FRB Shaft Engineering IMM, 1989).

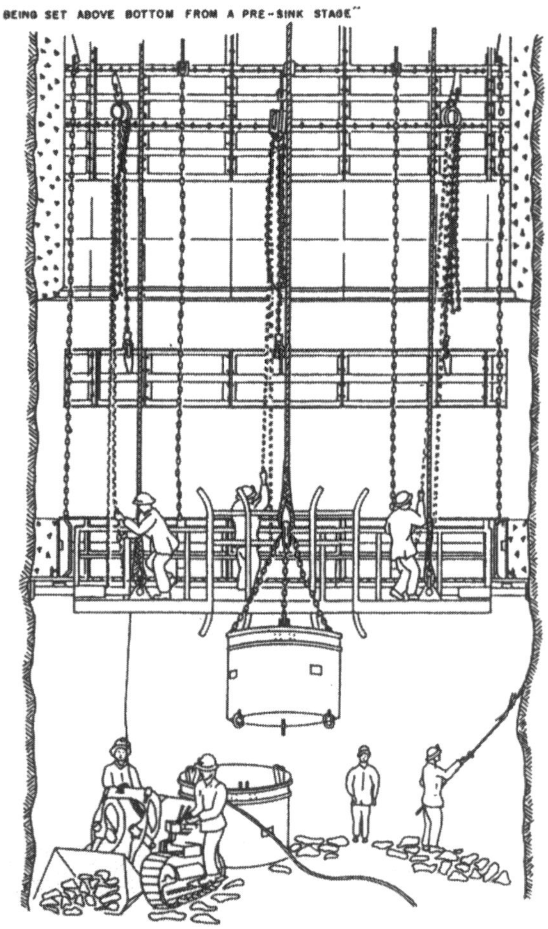

Fig. 10-2. Presink (Douglas AAB and Pfutzenreuter FRB Shaft Engineering IMM, 1989).

surface. The Canadian James Bay Hydroelectric power plant has 16 cable shafts, each 410 ft deep. Still awaiting construction is the Mt. Hope Hydro, with four 3,000-ft-deep, 28- to 35-ft-diameter shafts connecting to the 90 ft high by 60 ft wide, eight-and-a-half-million cu yd of caverns that comprise the underground reservoir.

ALTERNATIVES TO CONVENTIONAL DRILL-AND-BLAST METHODS

Traditionally, deep shafts in hard rock have been excavated by drilling and blasting, a method that changed little in more than 100 years until the introduction by the South Africans of turret-type lashing units and cactus grabs (Figure 10-3). In the last two decades, selective use of the shaft drill jumbo to replace handheld sinker drills has allowed deeper rounds to be more accurately drilled, with improvements in productivity. Other shaft construction systems in use today are blind drilling from the surface and shaft-sinking methods using pilot holes drilled to intersect existing underground openings. These include Alimak raises, raise bored shafts, and down-the-hole rodless borers such as the V mole, which is fast and very effective when alternative underground access is available at the shaft bottom for muck removal. A series of ventilation shafts were constructed by Wirth V mole for Jim Walters Mines in Alabama, which were 23-ft excavated diameter and approximately 2,000 ft deep, at a "best" rate of 1,600 ft of concrete-lined shaft in a single month (October 1982) and 1,600 ft of "unlined shaft" in the month that followed. Because of requirements for minimal to zero vibration only 144 ft away from the main penstock shaft, a similar V mole was used to construct the 23-ft-diameter, 550-ft-deep shaft for the Visitors Center at Hoover Dam in 1991.

CONSTRUCTION SEQUENCE

The construction sequence is dictated by shaft dimensions, in situ conditions (principally depth of overburden, rock quality, and groundwater), and the availability of an existing underground opening such as a pilot tunnel.

Each shaft has a collar at the surface that is designed to accommodate live loads during sinking, and the loads and

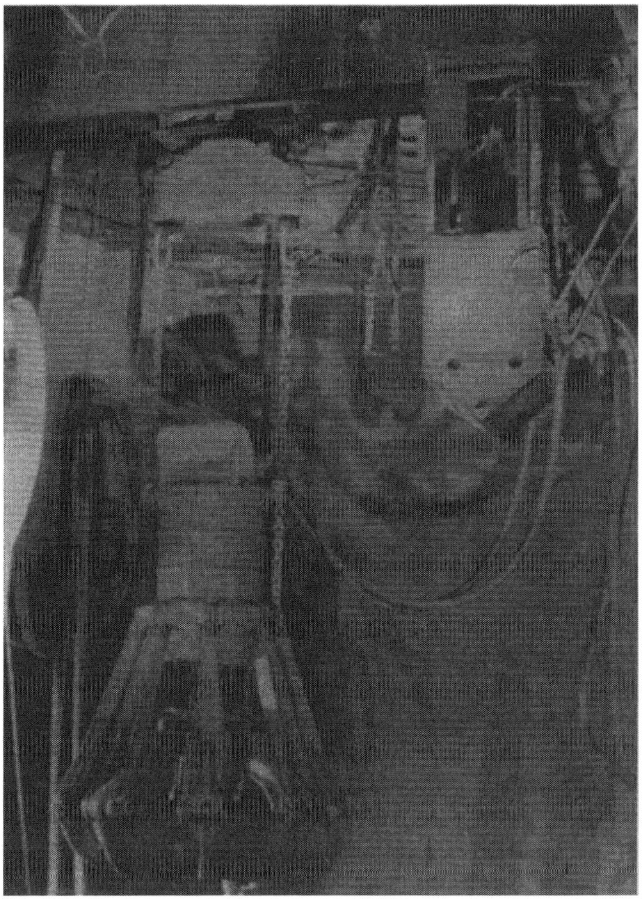

Fig. 10-3. Lashing unit and cactus grab.

duty of a permanent headframe or enclosed ventilation building. Initially, it provides a rigid, fenced structure that provides physical restraint and allows heavy cranes for muck, materials, and personnel hoisting to operate close to the shaft top.

Alternatively, if a hoist is used, it provides a stable foundation to support the shaft-sinking contractor's sinking headframe, muck dumping bins and chutes, and shaft-top cover doors. On completion of sinking and station construction, the temporary headframe can be used with a hoisting system (muck box or skip), or with a vertical belt conveyor for underground development. Later it may be modified for permanent use, or dismantled and removed to allow the permanent shaft-top structure to be constructed.

Below the collar, the shaft is deepened during the presink phase to allow sinking equipment to be installed. When the sinking equipment is commissioned the main sink starts, slowly at first, until equipment and crew problems are corrected, then gathering momentum as the learning curve progresses, and a cyclic rhythm is established. Shaft stations providing access to the tunnel or tunnels are partially or fully constructed as the shaft progresses from the upper to lower station limits. Sinking will normally continue for a short distance below the lowest station to provide space for conveyance over travel restraints, spillage handling, and a sump for water if present. Where water is predicted at specific horizons, it can be dealt with by pregrouting from the surface or by grouting from the shaft floor during sinking. Water inflows in excess of 10 gpm will impede the rate of sinking.

Shafts used for temporary access only may be lined by steel rings backed by wire mesh or corrugated sheet. Alternatively, good ground may be unlined, supported by random rock bolts as needed, or by a systematic pattern of rock bolts supplemented by wire mesh and/or shotcrete. A concrete lining is usually provided where guides and buntons are installed for high-speed hoisting and in shafts where a smooth lining is needed to optimize ventilation or fluid flow.

Normally, the concrete lining is installed during sinking in lifts of about 15 ft (or 50% of the guide length) following fairly closely behind the shaft floor. For specific applications, deep shafts may be slipformed from the bottom up to achieve a joint-free continuous smooth surface. An example is the shaft at the Dinorwic Pumped Storage Power Station in Wales. With a maximum lining thickness of 40 in., the 1,440-ft-deep, 31-ft-internal-diameter high-pressure shaft was slipformed from the bottom up at rates of up to 50 ft per day.

Collar Construction

The shaft collar is usually excavated by dragline or backhoe down to rock head (Figure 10-4). Excavation is then deepened into firm rock with handheld compressed air tools, hydraulic impactor, or cutter boom-type shaft excavator (roadheader). The first and second rounds are drilled and blasted in solid rock or excavated by roadheader, removing muck as necessary to allow the circular shaft formwork to be set on the muck pile at the selected foundation level. If blasting, the leveled-off muck from the muckpile left in the shaft bottom supports the first reinforced concrete lift, which is keyed into solid rock.

The collar is built up ring by ring from foundation level (Figure 10-5) to surface elevation by pouring reinforced concrete between the circular (internal) shaft form, and external formwork (Figure 10-6). The external formwork is

Fig. 10-4. Excavating shaft open cut.

Fig. 10-5. Constructing shaft collar.

Fig. 10-6. External form walls.

boxed out to provide openings for temporary sinking services (ventilation duct, water and air columns, plumb-bob, and laser brackets), and pockets are provided to accommodate the shaft-top doors and the final permanent structural steel. Holding-down bolts for the headframe steel are cast into the collar concrete (Figure 10-7). On completion of collar concreting, the external formwork is removed, and the remaining part of the open cut section is filled with engineered fill, in which ducts are constructed as appropriate to allow access for installation and maintenance of the temporary sinking services and permanent duty services (if any).

When the ground in which the collar is sited is not self-supporting, the unstable section can be presupported with closely spaced auger-drilled reinforced concrete piles founded at penetration refusal depth. The pile caps are connected into a reinforced concrete ring beam so that excavation can proceed within a structurally sound preformed cage. This was done effectively at a several of the 200-ft-deep access/ventilation shafts excavated during 1992 and 1993 for the 54-mi-long main collider tunnel ring for the Superconducting Super Collider at Waxahatchie, Texas. The shafts themselves were also excavated by a 16-ft-diameter auger to depths of 200 ft in the Austin chalk.

Fig. 10-7. Shaft collar concrete.

Foreshaft Construction

Where overburden thickness is extensive, a substantial, reinforced concrete structure must be extended down to and keyed into solid rock. This is known as the *foreshaft* (Figure 10-8). Muck hoisting in the foreshaft is usually by mobile crane or scotch derrick, with swiveling and slewing jib. A temporary movable work deck is used for lining installation and as protection for shaft-bottom workers. The muck digging system may be a clamshell, backhoe, or small crawler-mounted overshot loader.

Groundwater inflows are often met with during foreshaft construction and can be troublesome, especially in unstable ground. Water inflow may be handled by pumping, or pregrouting from the surface, or may be prevented from entering the excavation by dewatering to wellpoints around the

Fig. 10-8. Typical foreshaft with precast segmental concrete rings.

shaft site. In more severe cases, it may be necessary to isolate the shaft by drilling a ring of freeze holes and forming a frostwall around the ground prior to excavation. Every effort should be made to achieve a watertight lining through this section of the shaft; nevertheless, provision should also be made in the lining below the near surface water-bearing zone for installation of a water collection ring (Figure 10-9) with a small recessed sump and automatic pumping facilities to discharge residual inflows of water to the surface settling pond.

Presinking Requirements

Prior to commencing full-scale drill-and-blast shaft-sinking operations, the shaft bottom must be deepened sufficiently to permit the main sinking stage and mucking (lashing) unit to be installed in the shaft bottom, with enough clearance so that blasting and fly rock will not damage the shaft doors, the stage, or the lashing unit.

During presink, a temporary stage provides protection for workers and facilitates access to and handling of the circular shaft formwork and suspended curb ring. Stage suspension ropes provide guidance for the crosshead of the sinking bucket used as the men, muck, and materials conveyance.

The shaft is excavated to the required depth, say, 100 to 150 ft, or possibly deeper, using handheld-drill-and-blast methods. Muck is loaded into buckets by crawler-mounted rocker shovels, a wall-mounted backhoe, or a cryderman mucker. Alternatively, excavation can be carried out in suitable ground by roadheader with a transverse "ripper" cutterhead. A backhoe attachment to such a shaft-cutting machine effectively loads the muck box or hoist bucket, thus providing a compact excavate/load system. The main multideck

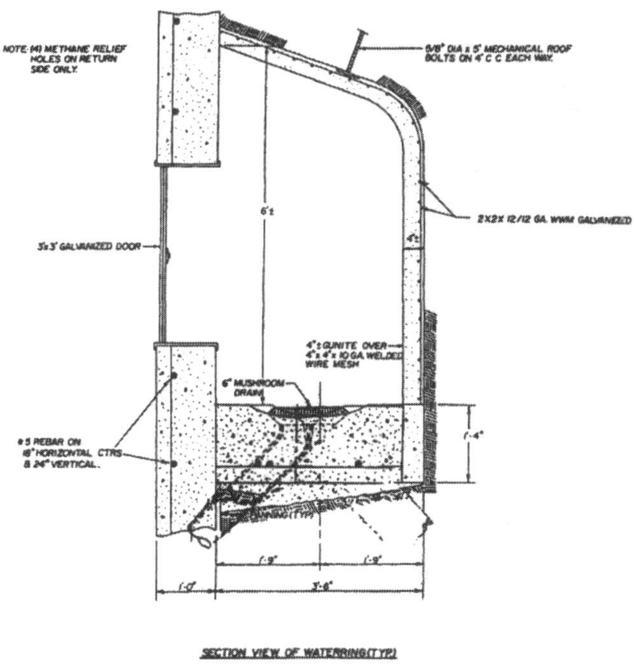

Fig. 10-9. Typical water ring.

sinking stage is installed on the shaft bottom (or may be preerected on the surface and lowered into the shaft by crane). After completion of erection of the headframe (which also may be preerected and rolled into position) the stage is roped up and raised, the lashing gear is installed, and all presinking services and equipment are removed from the shaft in readiness for sinking. Alternatively, a crane with limit switches and approved operational controls may be used to continue sinking to total depth (e.g., specially adapted cranes have been used at the Nevada Test Site to sink to depths in excess of 2,000 ft) provided that all cages, skips, and buckets are guided continuously by rope or rail guides throughout the entire length of the shaft.

Effect on Collar Design when Alternative Sinking Methods Are Used

For practical purposes, shaft collar construction to rock head is the same whether drill-and-blast methods or alternative construction methods using a pilot hole drilled by raise borer or raised by Alimak are planned. The depth of the presink, or the need for presink, will be determined by the method adopted. If raise boring by reaming in stages to full diameter is the preferred method, no presink is necessary except where the collar has to be extended to rockhead through deep, non-self-supporting overburden. If the V mole is to be used, only about 40 ft below the collar is needed to allow the partially assembled V mole to be introduced into the shaft. At that depth, it can start cutting the shaft bottom until the excavation is deep enough to allow the concreting work deck to be installed. The collar of a shaft blind-bored from the surface differs from conventional shaft collars in that it requires a large surface plant, an accurately installed, plumb starter casing, and a very strong concrete pad to carry the extremely heavy hook loads of the drill string and suspended in-hole equipment or steel liner.

Main Sink

During the time between the start of collar construction and commencement of the main sink, which can take as much as 4 to 6 months, power, water, and air services will be installed and ropes, hoists, and hoist houses for the main sinking bucket (kibble) and stage will have been installed and commissioned. The headframe, muck disposal chutes, and shaft-top doors will be erected, and facilities to lower or pipeline concrete (readymix or batch plant) down the shaft will be in place.

CONVENTIONAL SINKING EQUIPMENT

High-speed deep-shaft-sinking methods need safe, proven, up-to-date equipment of practical design. Brief specifications for sinking equipment for a conventionally sunk shaft, say, 2,000 to 3,000 ft deep, at 16 to 20 ft diameter, aiming for sinking rates of say 350 to 400 ft/month in reasonably good ground could typically include the following: headframe with decks at appropriate elevations for sheaves, deflector sheaves, overtravel protection, crash doors, lazy chain, and muck disposal chutes. The shaft-top safety doors and muck chutes are operated hydraulically or with compressed air. Both are controlled from bank level and are connected electrically to indicators in the hoist drivers' cabins. Each of the hoisting compartments of the shaft has its own muck disposal chutes and bins, into which the excavated material is dumped by the attachment of a lazy chain to the underside of the kibble or by an automatic tilting base into which the kibble is lowered. From the bin, muck is drawn off into trucks or onto a conveyor and transported for disposal on dumps adjacent to the site.

Concrete for the shaft wall is transported through the shaft by pipe column and distributed from a hopper on the top deck of the stage. Typically, the shaft is serviced by a 6-in.-diameter compressed-air column, a 2-in. water column, an 8-in. concrete column, and a 27-in. steel ventilation column. A 6-in.-diameter pumping column and four 1-1/2-in. high-pressure grouting pipes connected on the surface to the grout shed are needed to deal with water whenever it is present. The shaft (together with several hundred feet of tunnel) will be ventilated by six stage fans, combining to give 119 kW (160 hp) and producing 15,000 ft^3/min at 48.5 in. of water gauge.

The main kibble hoist should be on the order of 1,200 to 1,600 hp with 12- or 14-ft drums (both clutchable), maximum rope speed of 35 to 40 ft/sec, and high-strength 1.4-in.-diameter locked coil rope with 120-ton breaking strength. The stage hoist should be 200 hp with 12-ft drums (both clutchable) and high-strength 1.4-in.-diameter locked coil rope. (A spare rope of the same diameter should be available at the site for emergencies.)

Shaft formwork should be a 20-ft-high concrete shutter of appropriate diameter, consisting of one 2-ft-6-in.-high curb ring, six 2-ft, 6-in. sections, and one 2-ft, 6-in. matching ring. The shaft should be equipped with a 24-ton, 50-ft-long six-deck sinking stage. The stages should be equipped with four 1-ton capacity, air-driven winches for handling the shuttering. Standard fully rotational mechanical lashing units should be suspended from the stage. These units are fitted with 20-hp compressed-air-driven motors and equipped with extendable booms. Cactus grabs should have a capacity of at least 17 ft^3. Three to five compressors with combined output of 3,000 ft^3/min serve the site and shaft. The concrete batching plant should have an output of 40 yd^3/hour.

Key components and performance requirements are described in the following sections (Figure 10-10).

Surface Equipment.

- Headtower or headframe
  Adequate height for muck-tipping arrangements, overtravel, and jack-catch/crash door or other form of bucket catching protection
  Construction in steel or reinforced concrete if permanent, or pin-jointed or rapidly erected steel frame, if it is to be removed on completion
- Shaft-top doors and fence
  Adequate size and strength to accept loads from equipment being hoisted in and out of the shaft (e.g., drill jumbo)
  Pneumatic or hydraulic cylinder operation
- Crash beams and crash doors
  Adequate to shear hoist rope in overtravel emergency
  Doors released by trip wire
  Kibble suspension caught in jack catches
- Muck dump platform and muck chute and bins
  Automatic or semiautomatic emptying of muck kibble into chutes
  Pneumatic or hydraulic cylinder to tilt the muck of chutes to clear kibble path
  Adequate storage capacity in bins to allow the kibble to be discharged and stored until muck disposal vehicle returns
  Spillage proof and fail-safe design bin chutes
- Sheaves
  Easily maintained and of adequate diameter
  Deflector sheaves if needed to be positioned as appropriate
  Monorail lifting beam provided
- Kibble hoist(s)
  Modern construction
  Compliant with codes and recommendations of regulating authorities and industry practice
  Double-drum, double-clutched hoist
  Speed and load matched to planned rate of muck hoisting at depth

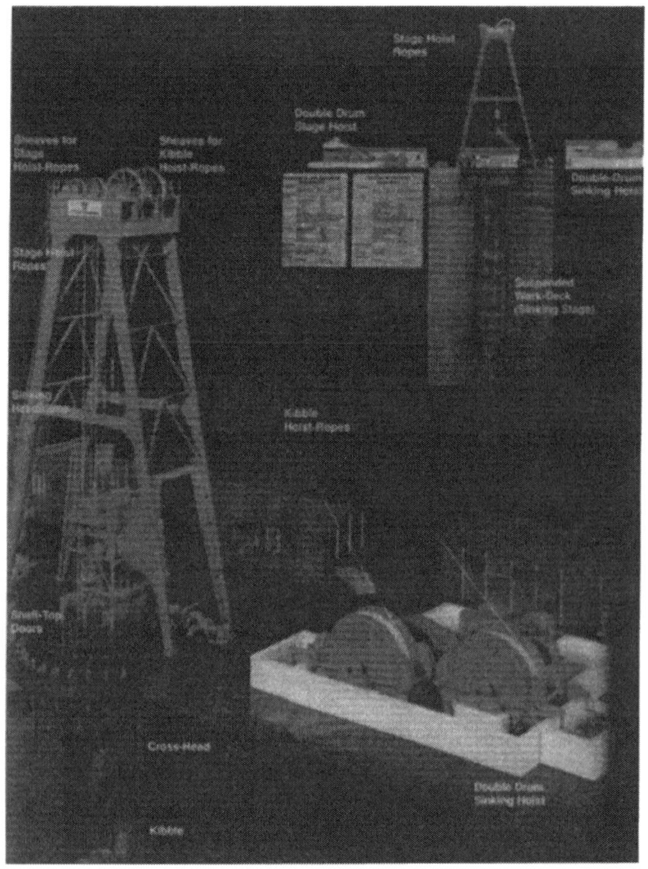

Fig. 10-10. Sinking headframe, multideck stage, and hoisting equipment.

- Stage hoist(s) (winches)
  Double-drum clutched (150 to 200 hp)
  Drums adequate to accept enough rope for depth of sinking in double purchase
  Alternative is three or four synchronized single-drum hoists of, say, 35-ton capacity each
- Emergency hoist
  50-hp, 15-ton capacity driven by diesel electric generators set to pull ladder-type emergency cage through the shaft unguided at a speed of 20 ft/min
- Compressed air
  System capacity of at least 3 compressors with total output of 3,000 ft^3/min

In-Shaft Equipment

- Sinking stage
  Multideck construction
  Adequate strength to house centrally mounted lashing unit (air or hydraulic operation)
- *Shaft Drill Jumbo.* Three-boom jumbo readily erected, easily hoisted in and out of the shaft and sized to pass through kibble openings in work deck.
- *Mucking Unit.* Cactus grab as large as possible for the size of the shaft (e.g., 28 ft^3 matched with 7-yd^3 kibbles for use in 26-ft-diameter shafts 3,000 ft deep). Grab is suspended from 360°-rotation turret boom of lashing unit, for rapid rates of shaft sinking. Alternative may be high-performance cryderman Brutus 3/4-yd^3 bucket, wall-mounted muckers.
- *Concrete Forms.* Collapsible steel form with curb ring suspended on 8 steel rods embedded in the liner (sacrificial) or from chains inside the liner suspended from anchor plates, embedded into the exposed wall of an earlier cured concrete lift. One complete lift of at least 15 ft to be poured with formwork lowered from winches on stage without need to move work deck.

THE SHAFT-SINKING CYCLE

Sinking by drill-and-blast is highly cyclic in practice, and the effectiveness of each cycle is critical to the whole operation and the productivity of the shaft-sinking crews. In deep shafts in good rock, with effective cycle times, shaft-sinking rates of 15 ft per day should readily be achieved when using a large diameter (15 in. or more) long hole burn cut. Experienced shaft-sinking crews will use a modern lashing unit and a large-capacity cactus grab or cryderman matched with muck kibbles and a hoisting system capable of removing muck at a rate to keep pace with the shaft-bottom equipment. The concreting cycle is integrated with lashing operations to achieve safe, semioverlapping activities. The work deck and the concrete form may be designed so that muck can be loaded and hoisted as concrete is being placed behind the form. Concrete must be delivered by pipeline if operations are to be overlapped. Shaft furnishings (buntons and guides) may be installed as sinking proceeds.

An average cycle for a modest 7-ft, 6-in. advance readily achievable in fair ground (friable sidewall with some spalling) is given in Table 10-1.

Table 10-1. Average Shaft-Sinking Cycle

Operation	Time Required
Raise stage, blast, re-entry and lower stage	1 hour
Lashing with grab	4-1/2 hours
Bolt and mesh sidewall	1-1/2 hours
Blow over	1 hour
Drill (60 holes with 8 machines or jumbo)	1-1/2 hours
Set curb ring (20 ft concrete lift) when required	2 hour
Charge up	1 hour
Total time	12 hours

Working 3 shifts a day, 6-1/2 days a week (1/2 day for maintenance) will give a sinking rate of over 400 ft of concrete lined shaft per month. In good ground with reasonably dry conditions, the lash, drill, charge, and blast cycle should ideally be completed within 8 hours.

Sinking

Normally, the shaft is reentered as soon as possible after blasting, and the stage and lashing crew travel down to the bottom of the shaft. The shaft wall is washed, and loose rock is barred down and made safe. The stage is then brought up to its working position about 40 ft above the muck pile and anchored with stage jacks. The muck pile is leveled with the grab, the first muck kibble is lowered onto the muck pile, and the three-legged suspension chain is released from the kibble and hoisted up to the surface. While the first kibble is being filled (Figure 10-11), an empty kibble is being lowered from the surface and is sited on the muck pile nearby. The suspension chains are then attached to the full kibble, which is hoisted to the surface. Using a double-drum, double-clutched hoist, an empty kibble is lowered as the full kibble goes to the surface to be dumped. This process is repeated until all the muck is hoisted. Water in the muck pile is pumped from a temporary sump into one of the kibbles, which is removed to the surface when sufficiently filled.

Near the end of the lashing cycle, solid rock is exposed and the effectiveness of the cactus grab is reduced. Large compressed-air hoses are used to blow the muck into small mounds, from where it is hand loaded by shovel into the tines of the grab. Sockets are searched for, and if located are blown clean and checked for misfires. Blowing over continues until the solid rock of the shaft floor is totally clean. At this stage, the drilling cycle is ready to begin again.

Sinking activities consist of lashing, drilling, and blasting.

Lashing

- Stage is lowered to lashing position.
- Lashing (blasted broken muck removal by cactus grab or cryderman).
- Temporary support (one or more of bolts/mesh/shotcrete or steel rings).
- Blow over to clean residual muck from shaft floor.

Drilling

- Drill jumbo is brought in (or drill machine kibble with manifolds and hoses if handheld).
- Compressed-air and water supply is connected.

Fig. 10-11. Muck kibble on shaft floor being filled by cactus grab.

- Plumb line is dropped from curb ring to transfer finished concrete face to shaft floor. The first blast hole is marked off with paint in the perimeter ring of the holes. The first vertical hole nearest to it is then measured and marked. This process is repeated around the shaft circumference for each perimeter hole (at approx. 18-in. centers).
- Short starter holes (12 in.) are drilled in sequence to follow each paint-marked blast hole while cut, easers, and burden rings of blast holes are painted.
- Blast holes are drilled to a depth of, say, 8 ft for a 7-ft, 6-in. round, then blown clean with a compressed-air pipe and plugged with a wooden peg or plastic pipe to keep the hole clean and mark its position.

Blasting

- Explosives kibble is brought to shaft floor just as the blast-hole drilling is finished.
- Bus wires and detonators are laid on the shaft floor to match the drill-hole pattern.
- Primers and sticks of explosives are installed in each hole and detonator leads are coupled parallel to the bus wires, which in turn are coupled to the suspended blasting cable.
- Personnel are cleared from the shaft bottom, the stage is raised to blasting level, and the cable is connected to the terminals of the parallel mains.
- All personnel come to the surface to allow blasting to take place.

Lining

Concrete is usually batched and mixed in a concreting plant near the shaft top, or located between shafts, if twin shafts are being sunk. If near enough, concrete may be transferred to the shaft by a chute, as a narrow ribbon of concrete on a conveyor belt, or by dumper, and delivered to either a bottom-dump concrete bucket or into a concrete slick line. Two pipelines are usually installed, one in use and one as a spare.

The slick line is a heavy-walled steel pipe manufactured and installed to precise tolerances to achieve minimum pipe wear. The pipe must be set perfectly plumb against dedicated plumb wires with precision alignment at the machined ends of each pipe. Concrete velocity is broken in a kettle bolted and secured by chains to the bottom of the pipeline. The kettle has an inverted Y discharge through which the remixed concrete "boils" into heavy duty hoses, delivering the concrete by gravity feed behind the formwork (Figure 10-12). The discharge ends of the concrete hoses are constantly moved circumferentially to prevent a buildup of concrete in any one location that could create an out-of-balance load to deform the shaft shutter. If delivered by bucket, concrete is discharged into a chute/hopper on one of the upper decks of the stage and then distributed by heavy-duty flexible hoses behind the shaft formwork.

Lining activities (which may take place simultaneously with lashing) consist of preparatory work, curb ring lowering, setting and concreting, main formwork rings lowering, setting, and concreting.

Preparatory work.

- Concrete pipes (if used) are extended and the stage is readied for concreting.
- The curb ring (containing bunton pockets, nut boxes, or other cast-ins) is readied for lowering.
- The flange and most of the key plate bolts are removed.
- Plumb wires and vertical measurement tapes are readied.

Curb Ring.

- Shutter winch ropes (or chains) are tightened to attachment points on curb ring. The remaining bolts are removed from the key plate.
- The curb ring (which is a heavy, rigid structure) has one ring of shaft formwork attached, which has been collapsed an inch or so circumferentially to pull it free of the concrete lining. The stage (carrying the weight of the curb ring) is lowered to the new elevation and jacked or wedged in position.
- Turnbuckles in the curb suspension adjust the curb to the correct elevation measured against shaft tapes.
- The curb is oriented against reference marks using turnbuckles and timber props as sprags against rock to maintain radius.

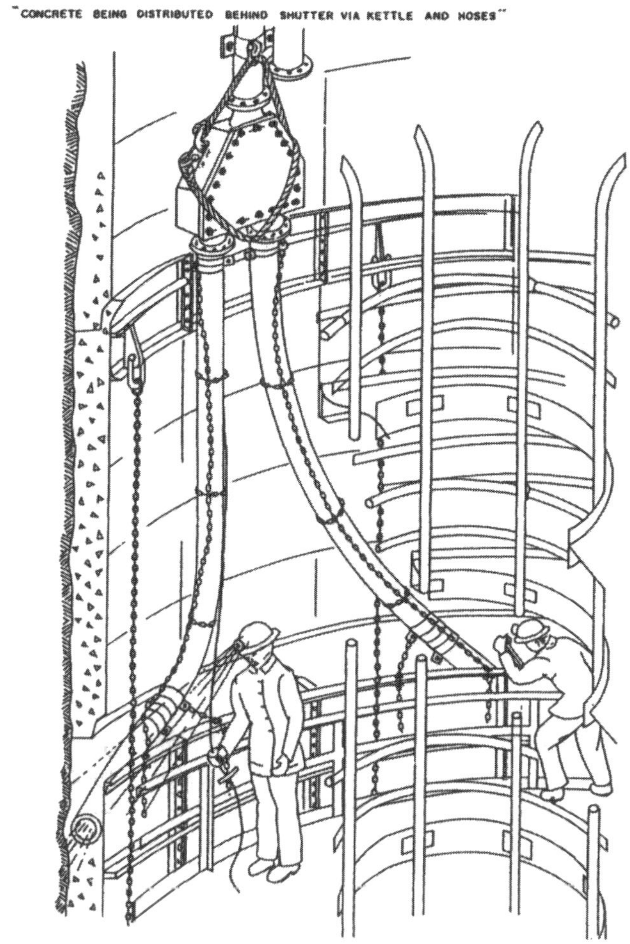

Fig. 10-12. Concrete lining (Douglas AAB and Pfutzenreuter FRB, Shaft Engineering IMM, 1989).

- Scribing boards are inserted, secured, and covered with paper or plastic to prevent escape of concrete fines. (Alternatively, the curb ring may be set directly on a broken round of muck that has been leveled off at the correct elevation.)
- Concrete with calcium chloride accelerator is poured to fill the curb ring and the first ring of shaft formwork

Main Formwork Rings.

- When the curb concrete takes its initial set (after 2 hours or longer), the remaining rings of shaft formwork are attached to the shutter winches (either individually or as the complete cylinder), key plates are removed, and ring(s) are collapsed (pulled free from the shaft wall).
- At this time, lashing can be resumed and muck removal and lowering, positioning, and concrete filling of the remaining length of formwork can proceed simultaneously.
- Shutter winches lower each shaft ring to allow it to be bolted to previously placed ring already filled with concrete.
- Each ring of formwork is filled with concrete to within an inch or so of the top, and the process repeated until the matching joint with the previous concreting lift is reached.
- The top ring (matching ring) has filler ports attached that extend above the elevation of the matching joint, and concrete poured through the filler ports flows to fill the matching joint.
- The rough surface and matching irregularities of the still-green concrete are scraped or plastered smooth when the concrete is exposed by removal of the matching ring.

Shaft deepening is continued, and the drilling, blasting, mucking, and lining, cycles are repeated continuously until the shaft is sunk to the elevation at which the station is to be excavated to provide entrance to the tunnel.

Station Construction

Initial excavation to form the station takes place when the shaft floor penetrates several feet below brow elevation. Drilling takes place from the top of the muck pile, to allow sufficient rock to be blasted to form an entry (or entries) to match the initial excavation profile of the station brow. The entry is extended as a top bench using crawler-mounted rock shovels to place muck where it can be rehandled by the shaft grab. In double-sided stations, drilling can be performed on one side while mucking takes place simultaneously on the other. Rock bolts, mesh (if needed), and/or shotcrete provide initial ground support.

The shaft bottom will then be deepened, say, two rounds below station level. As the shaft is deepened, successively lower levels of the station entry are excavated and extended. Below the station floor elevation, the circular shaft profile is extended to excavate openings for loading pockets or box-outs for station steelwork and equipment. At this point the shaft lining is extended down to brow elevation. In the station, rocker shovels and later load/haul/dump cycles (LHDs) of 1 to 2 yd^3 capacity are used to extend the entries away from the shaft, dumping muck into the deepened shaft, filling it until a convenient floor level is established from which the grab can rehandle the muck.

Station excavation is suspended when the tunnel entries reach a predetermined length sufficient to prevent blast damage to the shaft if driving the tunnels by drill-and-blast, or to allow erection and start-up of the TBM if the tunnel is being bored. The shaft sump is mucked out by the grab, and then one or two lifts of the concrete lining are poured below the station. Box-outs are formed in timber against the outside of the circular shaft formwork to provide space for permanent station steel decking and equipment, or to form loading pockets into which conveyors or LHDs will dump muck from the development of the tunnels.

Additional circular shaft formwork is extended upward to form a tower through the station. For a distance of 10 ft or so from the shaft shutter or at a convenient station profile change, the shape of the interpenetration between the horizontal station profile and the vertical shaft profile is formed in timber. The timber is braced internally and from the circular shaft shutter. Reinforcement, if specified, is installed. Stop-end formwork is positioned, and concrete is placed behind the formwork through the flexible concrete hoses (Figure 10-13). Sinking and lining is resumed to extend the bottom of the shaft sufficient to accommodate usual conveyance overtravel, arrestor columns, and sump and spillage arrangements.

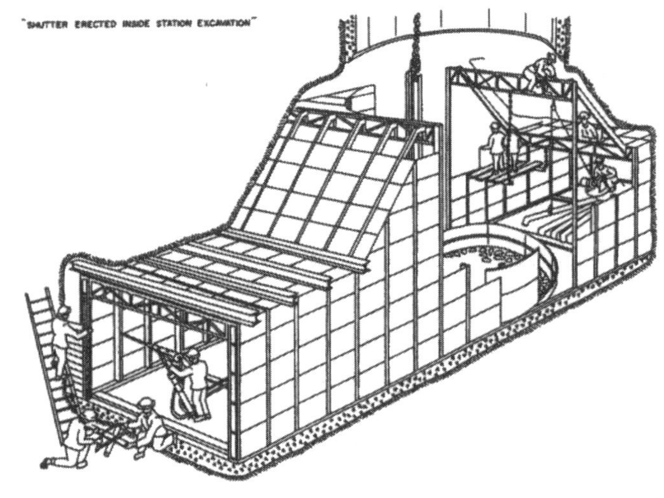

Fig. 10-13. Station construction (Douglas AAB and Pfutzenreuter FRB, Shaft Engineering IMM, 1989).

This concludes the shaft excavation and concreting operations. Construction operations now concentrate on the change over to tunneling, or on equipping the shaft for its final duty.

SHAFT EQUIPPING

Some shafts may have only minimal equipment installed to provide limited access for routine maintenance. At the other end of the scale, construction shafts may provide the only means of access to excavate and line the tunnel and must be equipped with high-speed hoisting facilities to handle personnel, materials, muck hoisting, and tunnel drivage support services.

Shafts with Minimal Equipment

Deep shafts where access is not needed for tunnel construction may be designed for ventilation duty only, requiring only limited access for inspection and possibly maintenance of walk-in water collection rings. Such shafts may also contain water drainage or pumping columns and power cables. A rigid personnel and light materials cage tall enough to support hinged-drawbridge-type cantilever decks and handrails will provide safe access to strategically placed services in the shaft.

To minimize air turbulence, service columns and cables can be attached to a recessed steel support bracket cast into a shotcrete or concrete liner during shaft sinking and lining operations. The same recessed bracket, if extended in length, will allow vertical steel guides for the conveyance to be attached directly to it. As an alternative, the conveyance may be designed to travel on rope guides or the shaft may be equipped with an Alimak-type rack-and-pinion raise climber, with the rack bolted directly into the liner. In any such configuration, the task of equipping is not a major effort and may be undertaken from the work deck as the shaft is being sunk, or alternatively, could be installed after sinking is complete when the sinking stage is being raised to the surface for removal from the shaft.

Other deep shafts, such as a slipform-lined high-pressure penstock shaft for an underground pumped storage power station, will contain no permanent in-shaft equipment whatsoever. If the deep shaft is to be used for limited construction, as for an exploratory facility or a short pilot tunnel, the hoisting facilities installed during sinking can be designed from inception to be modified to provide a personnel conveyance, and a muck skip can be substituted for the sinking kibble. The work deck will be left below station level to locate and maintain tension on the guide ropes for the conveyances. Again, shaft-equipping activities will be minimized, and needed facilities will be installed during sinking and lining.

Shafts Equipped to Service High-Speed Tunneling. It may be that the shaft is the only access for construction of a major tunnel or tunnels to be excavated at rapid rates of advance over a period of years, where high volumes of muck must be removed, large quantities of tunnel lining materials must be routinely handled, and large volumes of air are required to ventilate the tunnel drivage.

In such cases, equipping the shaft for its construction duty and/or final duty may require the installation of high-speed large capacity muck skips, conveyances for personnel and large cages to hoist materials, and heavy equipment. Other equipment such as a vertical muck conveyor (if used as an alternative to skips), large-diameter ventilation ducts, columns for pumping, compressed air, water supply, power and communications cables, and emergency hoisting must be considered. Shafts with a final ventilation duty may require installation of a concrete partition from bottom to top.

Planning, engineering, and quality control for a deep hoisting shaft outfitted with buntons (steel sets) and guides for high-volume muck hoisting and equipped with heavy equipment and personnel cages are major undertakings that must be initiated very early in the design phase of the shaft. Shaft furnishings (buntons and guides) more and more are being installed simultaneously with sinking, especially in cases where shafts are sunk using a large-diameter, long hole burn cut, which minimizes or eliminates flying rock damage.

The sequence adopted when shafts are equipped as a separate operation on completion of sinking is as follows. Temporary shaft-sinking equipment at the shaft bottom and station elevation together with the lashing unit must be removed and sent up to the surface. Main members of the steel decks at the station and loading pocket steelwork can be installed with the stage in position. The stage will then travel up through the shaft, stripping out all temporary pipes, cables, and steelwork that were used for sinking. When the stage reaches the collar, shaft-top doors and deck steelwork are removed, and permanent steelwork and fences outside the perimeter of the concrete liner are installed. The work deck is modified, and equipment conveyances are installed and commissioned. Permanent shaft-top steelwork is installed, and plumbing and measuring tape brackets are surveyed and fixed in position. The work deck and all facilities are now in place to install the permanent in-shaft equipment and services.

Shaft steelwork must be designed and procured well in advance of installation to allow time for on-site jigging, checking, and guide drilling, and matching. To avoid thermal expansion, this work should be performed in a covered area free from large temperature variations. All sets are marked to show matching ends and depth in the shaft. They are then stored in sequence near the shaft top, ready for installation from the shaft collar down. All steel is set to preinstalled plumb wires. To prevent oscillation, shaft plumb wires are locked in position by steady brackets at, say, 300-ft intervals. Templates are provided to locate the steelwork in the horizontal plane relative to the plumb wires and to check the tolerances after installation. Installation tolerances are generally ±1/8 in. horizontally and vertically. Mating ends of guides will be ground to fit during preinstallation jigging and matching at the surface. The shaft shutter will be designed so that tolerance from the center at any point in the shaft on radius, diameter, and circumference will not exceed 0.2% (say, 1/4 in. per 10 ft). Overall shaft verticality will be 1 in./1,000 ft and deviation over the entire shaft length will be not more than 1-1/2 in. (assuming a shaft of 2,000 ft or more total depth). Shaft steelwork installed in corrosive atmospheres should be hot-dip galvanized and or epoxy coated.

Shaft steel is installed from the top down, and the actual installation takes place from the top deck of the stage. Guides are brought down by the guide conveyances and lowered through prelocated openings in the stage. The guide is connected to the bottom of the previously installed guide. The conveyance then brings down the horizontal sets (buntons), which are also installed from the top deck of the stage. The buntons are set on packs in the pockets previously cast in the liner at the approximate needed elevation. The bunton framework is loosely assembled, lined up, surveyed, checked, leveled, and set to the guides. Buntons are adjusted laterally in the pockets to the final surveyed position by pairs of diametrically opposed pipe screw jacks, which also lock the ends of the buntons in position within the pocket. Grout is poured into each pocket behind formwork to permanently fix the position of the shaft steelwork.

When equipping reaches a station, main structural members are installed and additional personnel complete the station steel while shaft equipping continues to the bottom of the shaft. At that point, the debris in the shaft bottom is cleaned out, the stage is cut up and hoisted to the surface, and the stage ropes are removed. Permanent conveyances are installed, and the hoisting system is commissioned.

Design Considerations

An experienced shaft designer will develop a concept into a safe, cost-effective design that

- Is readily constructible by an experienced shaft-sinking contractor using up-to-date techniques and equipment
- Recognizes potential critical or problem areas and engineers the problem out of the job before the problem arises
- Maintains opening stability and lining integrity for the planned life of the project
- Conserves the inherent strength of the surrounding rock mass and mobilizes that strength as an essential component of ground support
- Meets design requirements and functional/operational criteria including environmental and aesthetic considerations

A design criteria document, usually prepared in close cooperation with the owner, formally outlines the design requirements (e.g., operational life and duties, ventilation quantities or hydraulic flow, hoisting duties, and residual water inflows, together with regulations, codes, standards, etc.). Design requirements are presented as functional requirements (what the system must accomplish), performance criteria (how well the system must perform), constraints that place limitations on the design, or assumptions made to identify features or characteristics.

A parallel design criteria document is the system database document, which contains specific site or other construction-related data. The database document provides a common reference for the design, bidding, and subsequent construction. It will provide information about geology, hydrology, and laboratory test data on in situ conditions, together with properties for materials such as concrete and grout formulations, steel liner materials, and so on.

The design criteria document is specifically targeted and contains only sufficient information, and no more, to fulfill the owners prime requirement that the project should be completed in a safe, timely, and economical manner.

The aim of the designer is to establish dimensional criteria (e.g., liner thickness, groundwater acceptance, length of unsupported ground from shaft floor, etc.) for temporary support (bolts, mesh, shotcrete, rings) and for the permanent shaft liner (shotcrete, concrete, segments, welded steel, tubing), and the method of groundwater treatment (pregrouting, grouting, freezing, depressurization, pumping, drainage behind liner). Geological criteria may be interpreted or extrapolated from outcrops or research into excavations in comparable ground and similar conditions, where strata are well defined over a large area and hydrogeological conditions are constant. Such an investigation, although a very useful exercise, will not usually produce sufficient hard data to allow a reliable estimate of cost to be made. As a consequence, criteria are normally obtained from a bore hole drilled on or very close to shaft center. A core from the bore hole is logged by an engineering geologist to provide a record of lithology and discontinuities (roughness, type, width, cementation, filling, etc), a fracture log, an assessment of rock strength, and so on. The engineering geologist selects representative samples for laboratory testing wherever significant changes occur and at regular intervals of 20 to 30 ft.

Core samples selected for testing should be wrapped carefully and protected. The testing program submitted to the laboratory should contain as much information as possible about the samples to enable the technician to decide where to cut the cores. The samples should be recored through the wrappings, and no attempt should be made to unwrap the cores. The laboratory testing program should include tests to determine physical properties and provide data on the mechanical properties (i.e., compressive, shear, and tensile strength, elasticity and the time-dependent properties of swelling, and creep), if any. Tests may also be conducted on samples obtained from aquifers where ground freezing is considered to determine the frozen strength and thermal properties of the strata. Porosity and permeability measurements on the core samples should be carried out using the connate fluid if possible. Test cores must be taken from the center of the drill cores to avoid the outer part of the core, which may be contaminated with drilling mud. Flushing reagents should not be used to remove drilling mud because naturally occurring interstitial clay minerals can also be flushed out, thus altering the test results.

A geophysical well-logging program can provide complementary information to derive a continuous well record of lithology, porosity, fluid content, permeability, clay content, hole size, and temperature.

The hydrogeological parameters of the most important aquifers will be assessed by in situ tests. At shallow depths, simple pumping tests will measure drawdown against time under a monitored rate of flow. Pumping water through a packer into selected sections of bore hole at various constant pressures and measured flow rate will provide estimates of in situ permeability.

At greater depths, drill stem tests in which a packer at the tip of the drill string is set above the zone of interest provide measurements on a pressure–time relationship of initial static pressure, flow pressure, and final static pressure of the formation as strata fluid is allowed to rise up the empty drill string. Samples of formation fluid should be taken at frequent intervals during testing and analyzed to obtain data on chemistry and hardness, chlorides, sulfates, total dissolved solids, conductivity, and freezing point. This data is essential for application of freezing and grouting techniques.

Much of the information produced from the near-shaft bore hole is vital to any potential shaft-sinking contractor. The shaft sinker needs to know the type and quality of the rock and joint orientations as they translate into rates of advance, temporary ground support, how much support, and how quickly it must be installed. He needs to know the condition of the core so he can see where zones of poor ground occur (high-quality photographs of every core box and clear labeling of elevations are a must). He needs to know where to expect water, at what hydraulic pressures, how much water, and water chemistry, especially as it affects the concrete lining. What is the permeability, porosity, and competence of the host rock? Can it be depressurized or grouted, or must it be frozen? If grouted, what grout formulation will be most effective, and how much time, drilling, and material will it take to grout the formation? If frozen, to which horizons, how much heat must be extracted from the ground, and at what rate can it be extracted?

GROUND STABILIZATION

Ground stabilization and control of water are necessary in poor ground conditions where the rock will not stand up without support, or where the quantity or hydrostatic pressure of the groundwater is likely to seriously interfere with the rate of shaft sinking or compromise the integrity and long-term performance of the shaft lining.

At shallow depths, soils may be stabilized by construction of a cut-off wall, such as a bentonite slurry wall, or by using a geotechnical improvement process in which the in situ soil is mixed with cement. This deep soil mixing system uses crane-supported auger/injection equipment. Below 100 to 150 ft, other forms of treatment are needed to stabilize the ground.

Ground treatment at greater depths to provide permanent improvement can be achieved by using cement or chemical grout injection to fill fissures, reduce permeability, and increase the strength of weak ground. Grouting may be accomplished by pregrouting from the surface or by drilling and grouting in stages from the shaft floor. Alternative forms of ground treatment, such as depressurization and freezing, provide temporary relief to facilitate sinking. Relief is provided by lowering or removing water pressure until excavation is complete and a permanent lining can be installed. The liner will be designed to withstand the hydrostatic pressure, or if nonhydrostatic, will have weep-holes and a drainage/collection/pumping system built in to prevent buildup of hydrostatic pressure.

A further alternative type of groundwater control (not applicable in noncohesive ground) is to continue sinking, controlling water inflows and installing temporary support as sinking proceeds. This is achieved in water-bearing ground by installing panning (overlapping corrugated steel sheets), liner plates, plastic-faced wire matting drains, or perforated pipes with anchor studs to cover fissures bleeding water from the shaft wall. A systematic application of shotcrete, bolts, and mesh and lattice girders is then possible to provide early protection in the form of an active ground support system.

Pregrouting from the Surface

Pregrouting is carried out prior to commencement of shaft sinking and can be very effective in cementing up interconnected fissures at depth. For example, the Jim Walters (Birmingham, Alabama) series of shafts were known to be in water-bearing strata that had seriously affected sinking rates in previously constructed shafts. Each shaft was pregrouted with a single hole, which injected cement slurry into every water-bearing horizon as encountered, reducing potential inflows of 100 gpm to provide a shaft considered dry for practical purposes.

Many of the deep South African shafts have been pregrouted, using two or three holes usually started equally spaced around the circumference of a circle approximately twice the excavation diameter of the shaft. An example is the St. Helena Gold Mines Ltd., No. 7 Ventilation Shaft (Table 10-2). The shaft intersected water on six occasions during sinking. Fissures were grouted up and sealed off during sinking with only 493 ft^3 of cement and with only 232 hours spent on cementing, proving the highly effective sealing achieved during the pregrout from the surface.

Dewatering and Depressurization

Dewatering and depressurization can be a useful follow-up to pregrouting to allow a shaft to be sunk through known water-bearing horizons in permeable to highly permeable ground.

Deep wells of 6-in. diameter or more are drilled outside the pregrouted zone into and through the aquifers. The wells will be fitted with a commercial screen and graded sand/gravel filter. A submersible pump delivers water from the aquifer, reducing the elevation of the water table in a localized zone around the shaft excavation, thus reducing the quantity of water entering the shaft and the hydrostatic pressure. This can significantly reduce the time needed for grouting to seal the aquifer.

Removing large quantities of water from the ground will, over time, cause settlement and damage to foundations, and the application of dewatering or depressurization must be restricted to such isolated locations where property and structures are not affected.

Stage Grouting

A suitable groundwater control system will generally use pregrouting from the surface, depressurization during sinking where appropriate, and drilling and grouting in stages during sinking to reduce potential inflows to an acceptable level. A drainage system will be installed behind a shotcrete lining placed as initial ground support. The drains collect residual water and channel water from above to a pumping platform, and water from below the platform to the shaft-floor sump.

Throughout the length of water-bearing ground, 30 to 50 m (maximum) grout covers using one, two, or three concentric rings (depending on ground conditions) of grout holes will be drilled and injected with cement grout. The outer rings of the holes are drilled with spin as well as dip to ensure that all near-vertical joints are intersected. At first, holes will be injected with thin cement grout, which will be thickened after prolonged acceptance. After setting time elapses, all holes are to be redrilled, and unless water flow is substantially reduced, all holes will be injected with cement again. Stabilized clay grouts will also be used, as well as chemicals compatible with the groundwater chemistry and the underground fracture/permeability regime. Grout for these cover rounds will be mixed at the surface and will be delivered by shaft columns to the grout cocks. Cocks will be attached to stand pipes grouted into the shaft floor in horizons where better-quality rock is located (Figure 10-14). All stand pipes will also be anchored to bolts drilled into the floor. The maximum pressure on the grout pumps at the surface will be controlled to avoid hydrofracturing.

Table 10-2. Pregrouting at St. Helena No. 7

Hole No.	Cementing Depth (ft)	Cased (ft)	Cement (cu ft)	Total Working Hours	Cementation Hours
1	5,733	1,230	9,935	7,698	1,752
2	3,982	1,220	3,664	4,618	783

Ground Freezing

The ground freezing method can be used to stabilize unstable water-bearing soils, to a depth of 3,000 ft and more during shaft sinking and installations of the permanent lining. An example is the freeze holes at Boulby Mine (North of England), which were drilled to a depth of 3,050 ft to provide a 12-ft-thick frostwall to isolate the shaft excavation from the 1,372-psi aquifer in the Bunter Sandstone (Figure 10-15).

The principle of ground freezing is the use of refrigeration to convert in situ porewater to a frostwall. The ice in the

Fig. 10-14. Grout cover from the shaft floor against 1,300-psi hydrostatic head.

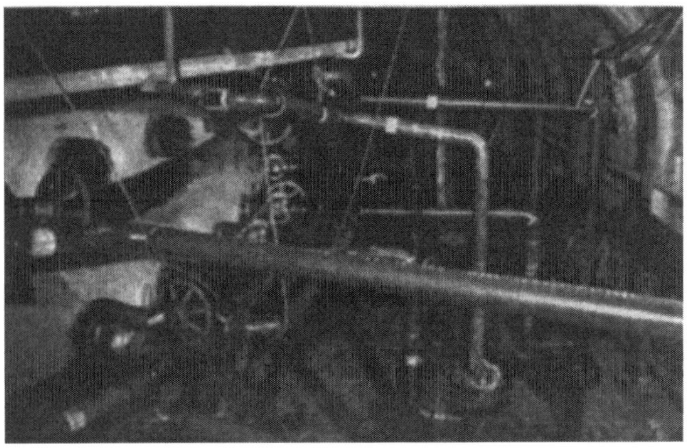

Fig. 10-15. Underground freeze chamber 3,000 ft below surface.

frostwall becomes a bonding agent, fusing adjacent particles of soil to increase their combined strength and make them impervious. A cylindrical freeze wall is formed around the periphery of the planned excavation. This provides a stable support and groundwater control system that enables safe sinking of the shaft. Ground freezing has been successfully used in shaft sinking for 100 years. The extensive experience available and the reliability of this method enables detailed planning of time and economic factors.

To create the freeze wall, refrigeration pipes are equally spaced (approximately 4 ft apart) on a circle that is greater than the excavation diameter. The refrigeration pipes extend as deep as necessary to tie the freeze wall safely into competent, impermeable rock.

To ensure a positive and more or less uniform closure of the freeze wall, it is essential to drill the refrigeration holes very accurately, minimizing any deviation. A target area for the bore holes is defined within two concentric circles around the center of the shaft. One is 3 ft larger than and the other 3 ft less than the diameter of the theoretical freeze pipe circle.

In addition, spacing between two adjacent bore holes should not exceed 5 ft. Continuous alignment controls during drilling operations and—if required—the use of directional drilling tools to correct excessive deviations guarantee that the holes remain within the specified limits.

Five-inch refrigeration pipes are installed in the bore holes. Three-inch plastic down-pipes will be lowered into them. The refrigeration pipes will be connected to the main manifold lines, and a coolant—usually calcium chloride brine—at a temperature of approximately −25°C will be pumped through the circuit. Circulation of the cold brine through the system causes continuous extraction of heat from the ground.

Initially, a cylinder of frozen ground is formed around each refrigeration pipe. The frozen cylinders grow with time, and when they merge a continuous freeze wall is formed. The freeze wall will then have to thicken with time at least to the required thickness according to the structural design. In general the freeze-wall thickness has to be increased with depth, as the water pressures are also growing.

During the prefreezing period—the time from the start of freezing until a continuous freeze wall is formed—the refrigeration plant operates under full load. Later on, to facilitate mucking operations, it will be necessary to control the freezing process so that as much of the area to be excavated remains unfrozen as possible.

Monitoring of the Freezing Process. Special measurement techniques are used to monitor the freezing process. The temperature of the brine in the main supply line to the shaft as well as the return temperature in each individual refrigeration pipe is regularly measured and recorded. Pressure and flow rates are also monitored in the main brine supply line. Furthermore, the brine flow in each refrigeration pipe will be checked once daily and adjusted if required. Ground temperatures at critical locations are continuously measured in temperature monitoring holes (Figure 10-16).

In addition, ultrasonic measurements are used to check the progress, extent, and continuity of the freeze wall. The principle underlying this method is the appreciable increase in the velocity of ultrasonic sound waves when passing through ice as opposed to water.

The data gained from all the measurements and their development with time provide information on the shape and condition of the freeze wall, such as

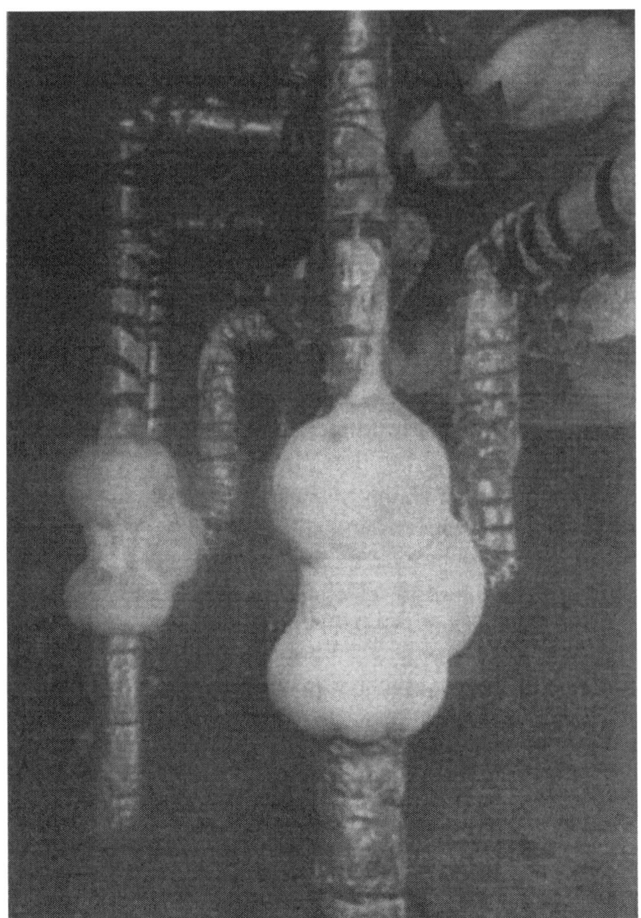

Fig. 10-16. Buildup of frost on freeze pipes.

- Wall thickness
- Wall temperatures that determine the strength of the frozen soil
- Possible irregularities

Causes of such irregularities may be previously unknown local changes in ground conditions or lateral groundwater movement affecting the uniform advance of the freeze wall. Precise evaluation and interpretation of the measured data will enable conclusions to be drawn as to possible causes of irregularities and to suggest the necessary remedial action that should be taken.

Thawing of the Freeze Wall. After the final watertight lining has been installed in the frozen section of the shaft, the freezing process can be terminated. Natural thawing, as opposed to the application of heat from external sources, is normally employed. When the thawing process has progressed sufficiently the down-pipes will be pulled and the refrigeration pipes, which will remain in the ground, will be backfilled with cement grout.

Typical Example of Freezing

A typical application for ground freezing is demonstrated at the access shafts for New York City's No. 3 Water Tunnel. The tunnel, traversing Manhattan, Queens, and Brooklyn, will distribute water from upstate reservoirs and replace an existing system in need of maintenance. It will lie 500 to 600 ft below the surface, in gneissic and granitic rocks with a soil overburden varying between 120 and 260 ft in thickness. The soils are, for the most part, below the water table and consist of Cretaceous sediments (clays, silts, sands) overlain by glacial outwash (sands, gravel, and often, boulder beds) and recent peat and fill. Adjacent to the sound rock, there may be layers of weathered or completely decomposed material with soil-like characteristics.

Ground freezing has been employed on all five shafts sunk in Brooklyn and Queens between 1987 and 1993 (see Table 10-3). Freezing is the preferred method for the following reasons:

- Ground settlement must not be allowed to occur in the urban environment. In addition to tight controls on soil movements, dewatering is not a possibility for ground control.
- A versatile ground control system is essential to deal with the large variation in soil conditions below the water table.
- The support system must be adequate for an excavation diameter in excess of 40 ft and at depths in excess of 250 ft below the water table.
- The support system must extend into sound granitic bedrock.

Table 10-3. Freezing Operation for Water Tunnel No. 3

Statistics	
Freezing circle	48 to 52 ft in diameter
Number of freeze holes	45 - 55 depending on soil conditions
Freeze depth	140 to 270 ft depending on geology
Toe into rock	15 to 50 ft depending upon rock conditions
Refrigerating horsepower	500 to 1,100 hp depending upon moving groundwater and schedule requirements
Schedule	
Install system	2 to 3 months
Form freeze	1-1/2 to 2-1/2 months
Maintain freeze	4 to 7 months (during construction)

BLAST DESIGN AND THE USE OF EXPLOSIVES

The drill/blast/muck-out cycle is highly critical, and blast design is a key component. A free face must be established by "cut" or sumper holes, and the remaining rock is blasted into this initial excavation.

V cuts and benching have typically been used in rectangular shafts with small cross-sectional areas and short rounds. Benching in circular shafts reduces blast damage because the broken rock is thrown toward the opposite side of the shaft floor. Another benefit is that it provides a sump that facilitates sinking in wet conditions. Benching is also used in circular shafts to eliminate flyrock in cases where shaft equipping follows closely behind the shaft floor and also in special circumstances in frozen shafts, where there is a need to reduce the shock impact of blasting in proximity to the frozen ground.

Another popular cut used in a circular shaft is the pyramid cut, which employs concentric circles of holes inclined toward the shaft center so as to bottom out close to each other. The cut normally consists of five to eight evenly spaced holes angled in at approximately 30° and somewhat deeper than the rest of the holes in the round, so as to provide a sump for the next mining cycle. Sometimes in massive rock, a couple of shallow buster holes are added to the center of the pyramid to help break up any boulders created by the initial cut holes of each blasting round. A drawback with the V cut and pyramid cut drilling patterns is the tendency for muck to be thrown up high into the shaft during blasting, requiring the sinking stage to be moved at least 200 ft from the shaft bottom prior to each blast.

Another blasting pattern popular in hard rock tunnels uses a burn cut, developed by Atlas Copco as the coramant cut, which can be drilled easily with lightweight handheld equipment. The success of the coramant cut requires drilling of holes that are truly parallel and to a precise pattern with two oversize (2-1/4 in.) holes, which must slightly overlap (figure eight shape) for the full depth of the cut (say, 10 ft). The necessary accuracy is achieved by use of a drill jig (template) weighing only 33 lb. It is fabricated from light metal plates spaced by tubes that serve as guide sleeves during drilling.

The greatly increased use of multiboom jumbos in shafts as well as tunnels allows very accurate parallel drilling, and it facilitates the drilling of a number of larger-diameter cut relief holes (say three at 3-1/2). This has been successfully used to pull 10.8-ft-deep rounds in a 15-ft-diameter shaft for Atomic Energy of Canada Limited's Underground Research Laboratories (URL).

At the Falconbridge Ltd. Craig Shaft, Ontario, a two-boom jumbo was used with an in-the-hole drill. One very large (8 in.) diameter cut hole replaced two overlapping 6-in. holes to enable a 16-ft-deep pull to be consistently achieved in the 20.5-ft-diameter shaft. In all cases, for effective blasting it is essential to expel water from the cut hole. A small charge (about 1/2 cartridge) placed in the bottom of each relief hole containing water initiated on zero delay before the first cut and helpers are fired, providing effective discharge of water. Larger-diameter deep cut holes of 12-in. diameter or more are now being used to increasing effect.

In each of the shafts referenced, Magnadet detonators were used to prime the charge. A mean powder factor of 3.1 kg/m^3 (5.2 lbs/yd^3) was recorded in the URL shaft.

RAISE DRILLING, BLIND DRILLING, AND OTHER ALTERNATIVES

The "conventional" method of sinking is a description applied to drill, blast, load out, and hoist the muck to the surface in a sinking bucket. Alternative methods fall into four major groups: shaft drilling, shaft boring, raise boring, and raise drilling.

Shaft Drilling

Shaft drilling is the process of excavating a shaft from the top down using a large surface drilling rig. The shaft may be excavated to its required diameter in one pass, or by successive passes with reaming bits. Muck removal is normally achieved by reverse-circulation mud flushing. The shaft is kept full of drilling mud to maintain stability and exclude groundwater prior to the installation of the lining.

Shaft Boring

Shaft boring is a process in which a shaft is excavated to its required diameter from the top downward using an in-shaft boring machine. The machine has on-board personnel for operational control. Often this type of machine follows a pilot shaft, so that muck may fall freely to an underground removal system. It may be used for blind boring when combined with systems for collecting and transporting drill cuttings to the surface. The excavating process may use either full-face or part-face cutting of the shaft sump.

Raise Boring

Raise boring is the excavation of a shaft by reaming from the bottom upward using a raise boring machine located at the shaft top. A pilot hole is first drilled into an underground chamber. A cutting head (Figure 10-17) is attached to the drill stem, which is then used to ream in a upward direction, with muck free-falling to the lower chamber. The process can be repeated to ream the shaft to larger diameters.

Raise Drilling

Raise drilling is the process of excavating a shaft from the bottom up using a raise (box hole) drilling machine located at the shaft bottom or using a self-contained drilling machine (similar to an unmanned, mini-TBM), which is launched from a crawler-mounted tube (Figure 10-18). This system, known as the *Borpak concept,* can operate from vertical to 30° inclination.

Fig. 10-17. Raise bore cutterhead.

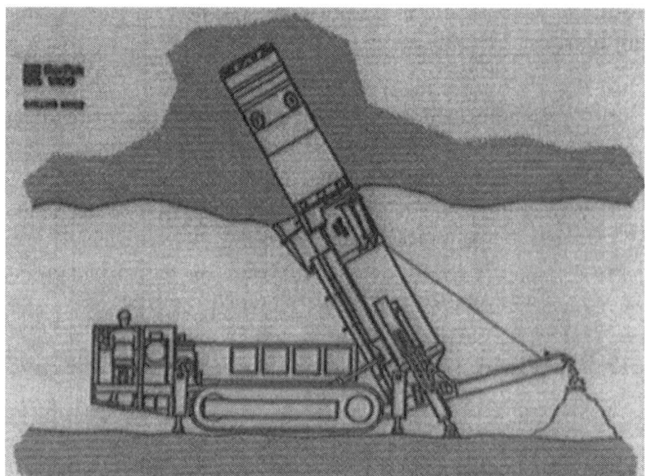

Fig. 10-18. Borpak concept.

11

Tunnel Boring Machines

Harry Sutcliffe
Consulting Engineer

Tunnel boring machines (TBMs) are used to excavate tunnels in virtually all types of ground and under widely different physical conditions. This chapter covers hard and soft rock TBMs with full-face rotating heads and their development from soft ground shields. Roadheaders (boom headers) are also discussed here because of their growing role in the softer rock tunnels and caverns. The innovative Ranging Mobile Miner, which combines some of the advantages of the roadheader and the TBM and can be used in harder rocks, is also covered.

The functions of a TBM are simple enough:

- To excavate the ground
- To remove the material excavated
- To maintain line and grade of the excavation
- To support the excavated tunnel temporarily until permanent support can be provided
- To handle adverse ground conditions

These five simple requirements become less simple when the necessary qualifying conditions are added. The functions must be performed:

- Safely
- Reliably
- Continuously for many months
- Through any and all ground conditions
- Quickly
- Economically

Throughout the development of TBMs, a term which implies a rotary action, designers have built on the successes and the failures of their predecessors. In recent years, there has been a great deal of healthy cross-fertilization between designs of machines for hard rock and those for soft ground. Today's TBM is likely to have been designed specifically for the anticipated ground conditions and is likely to use technology from both hard rock and soft ground machines.

A TBM is subjected to heavy stress and much abuse during its service life. It is worth reflecting that a typical hard rock TBM may be called upon to cut through a wall of rock considerably larger in area than the wall of an average room, the wall being made up of rock of up to 10 times the strength of normal concrete. It must do this steadily, day and night, for many months, perhaps under corrosively saline water inflows, and almost certainly, at some point, through uncooperative ground. Only the state of the art in metallurgy and mechanical design can create a machine that can accomplish this.

HISTORICAL DEVELOPMENT

Like many other technical developments, the TBM was designed in concept by men of genius long before the technologies of metallurgy and motive power were advanced sufficiently to meet the challenges the designs imposed. In the period starting in the early 19th century, numerous tunneling machines were built, largely in the United States and Great Britain, many of them with features that can be recognized in the modern TBM. Before that time, soft ground was excavated and supported by hand mining methods, then from the protection of a shield, then from increasingly mechanized shields. The rock tunnels were excavated by explosives set in holes drilled into the face. Subaqueous tunnels in both rock and soft ground were not attempted until the early 19th century, and the poor success rate led to the development of the shield, and its subsequent mechanization into a TBM.

The history of TBMs started with soft ground shields of the type developed by Marc I. Brunel (the senior Brunel) and J.H. Greathead in England. These shields progressed by breaking the excavation into small compartments excavated by hand. The first Brunel shield, patented in 1818, excavated these compartments and advanced the shield in a spiral pattern, with lining segments following in the same spiral. The shield did not rotate, but the spiral arrangement of the head meant that the miner excavated along a spiral path at right angles to the direction of the tunnel. Since the miner was the cutterhead, a case could be made that this was a rotating head "machine," but the verdict is best left to the reader. The first subaqueous tunnel under the Thames at Rotherhithe may have been started in 1827 using one of two Brunel designs patented in 1818, but it was completed nine years later by a newer Brunel-designed compartmented shield, and one with no hint of rotary action in its operation. The completed tunnel lay idle for many years until the dawn of the railway age put it to work, a function it still performs well 160 years later.

The coming of the railway age was a great stimulant to tunnelers. The next two shield tunnels were both successes, at least in execution, and were built concurrently in New York and London in 1869. In New York, A.S. Beach, editor of the *Scientific American,* drove a length of 8-ft diameter tunnel to promote a pneumatic railway. Beach used a circular shield that advanced in one piece, shoved forward by hydraulic jacks thrusting against a brick lining. The shield operation was a success, but the project did not go further. The hydraulic jacks, used here for the first time, proved their worth. In London, a second crossing under the Thames, in clay, was driven in 1869 by a shield patented by P.W. Barlow in 1864 and 1868; J.H. Greathead was the engineer in charge. This shield, like Beach's in New York, was circular and segmented to offer greater protection to the miners against face collapse. The shield was moved forward by screw jacks, but this time thrusting against cast iron liners erected behind the shield. A lime slurry was used to grout behind the liner segments, another first. These shields gradually became more mechanized and exchanged features with the developing rock TBMs. But the first machine to incorporate a full-face rotating head was not to appear until 1882 in either soft ground or rock machines, and then in a rock that is softer than most rocks, and softer than some soft ground.

The first, and almost the only successful rock machine of the period, a machine that embodied many of the features of a modern TBM, operated in the soft chalk marl on each side of the Straits of Dover and later in the relatively soft sandstones of the Mersey River in northern England. These soft rocks were not too demanding on the metallurgists of the day and permitted the developers the opportunity to concentrate on the other aspects of the equipment. The chalk excavation was the first start of construction of an English Channel tunnel in 1882. The machines were the Beaumont/English machines, designed by two officers of the Royal Engineers and put to work on each side of the Channel. Until political pressure stopped the work, the machines excavated 7-ft diameter tunnels at the very respectable rate of 50 ft per day for the final 53 days, with peak rates of over 80 ft per day. The unlined tunnels were still standing almost a century later, when they were crossed by the service tunnel bore of the new Channel Tunnel.

On this and the Mersey Tunnel projects, the soft rock meant that the demands on the cutting steels were not too heavy, and the tunnel diameters were small enough to permit the use of compressed air motors, the exhaust from which supplied relatively fresh, cold air to the working face.

In the United States, several machines had been built earlier. One was patented by John Wilson as early as 1851. This machine cut an annular slot a foot wide and of unspecified depth around the perimeter of the Hoosac Tunnel East Portal in Massachusetts before being withdrawn. A second machine, by Haupt, was tried at the West Portal in 1857, this time attempting to drive an 8-ft diameter pilot heading. Both these attempts in the hard igneous rocks of New York and New England were technical and commercial failures. They were in honorable company, however; several contractors using the conventional methods of the day were bankrupted before the tunnel was completed. These early endeavors remain as testimonies to men of vision and to the entrepreneurial spirit, which in later generations produced the successful machines of today.

Features now found in the modern rock TBM were being developed in soft ground TBMs. The demand for more and more underground rail lines in London, and the presence of the London Clays under the city, led to rapid development of mechanized shields. In 1893, J.J. Robins patented a shield with a full rotating head equipped with drag picks, hydraulic jacks thrusting against cast iron rings, and a mucking system employing an endless chain belt to remove the spoil. The patent drawing even appears to provide for circumferential seals between tailskin and segments, such as are necessary to operate under external groundwater head. There is no record that the machine was built. It had a weakness in that the cutterhead drive was located close to the axis of the cutter wheel, where it would have been subject to heavy torque loading. This weakness was shared by a successor, the Price Shield patented by J. Price in 1896. After modification of the drive to turn the head closer to the periphery, the machine was a success. The machine was modified and developed further, and Markham and Co. Ltd. began commercial production of the unit in 1901. The soft ground TBM had arrived.

A new era in rock tunneling began in the 1950s, when James S. Robbins, then of Chicago, designed and built a machine that operated successfully on the Oahe Dam project in South Dakota in 1952. It drove a 26-ft-diameter tunnel through faulted and jointed weak shale of 200 to 400 psi compressive strength. Cutterhead horsepower was 400, and thrust 50 tons. This and a similar machine supplied in 1955 drove 22,500 ft of tunnel, reaching maximum rates of 140 ft a day and 635 ft in one week. The machine drew considerable attention, but it had performed only in soft rock. Robbins then achieved notable success when one of his machines drove a 10-ft, 9-in. diameter tunnel in Toronto through limestone, sandstone, and shale with strengths of 8,000 to 27,000 psi. This machine was the first to be equipped with rotary disk cutters. It had 24 cutters, a cutterhead horsepower of 340, and a thrust of 157 tons. The performance of this machine drew worldwide attention and initiated a period of intense development, in which many new manufacturers entered the field. The industry has now matured, and the number of manufacturers has fallen to a level that the demand can support. A thriving market exists in refurbishing used machines, which sometimes complete a project in better working order than when they began. An extensive and readable history of TBMs may be found in Barbara Stack's *Handbook of Mining and Tunnelling Machinery* (Stack 1982).

EXCAVATION UNDER EXTERNAL WATER PRESSURE

Until recently, TBM designs were categorized by the hardness of the ground to be encountered or, more precisely, expected to be encountered. While this is clearly an important design consideration, the quantity and pressure of groundwater is as important an influence in the design or selection of the TBM. In a hard rock tunnel, the water inflows are likely to be localized, volcanic lava formations being a notable exception. Water inflows are unlikely to erode or weaken rock in the area of the inflow, whereas water from soft rock or soft ground is likely to carry some of the ground with it. In water-bearing ground, this usually calls for a full watertight lining to be erected within the tail shield of the TBM, the assembled lining ring being extruded from the tail shield through a pressure resisting circumferential seal. Owing to cost or schedule constraints, this is usually the permanent tunnel lining.

The idea of compressing the air inside a tunnel to reduce water inflows, and thereby support the ground, was first tried under the Hudson River in 1879. The technique was successful in a second attempt in 1889 and was used on many subsequent tunnels. Working under compressed air and decompressing after such work was found to be hard on the human body. Limits on time worked under compressed air developed such that other means had to be found to make such tunneling economical. In 1964, the Robbins Etoile TBM used air pressurization of the face ahead of a bulkhead in driving subway tunnels in Paris. But air is compressible, and carries almost explosive potential when compressed to several atmospheres. Preferable are the incompressible semiliquid slurries, which, like water, can transmit, but do not store pressure when under pressure.

Greathead had patented a shield in 1874 that had a face through which slurry could be circulated under pressure to support the face and remove the spoil. The ground was dislodged by being poked by excavating tools through stuffing boxes in the bulkhead. There is no record that it was built. In 1960, E.C. Gardner successfully built a tunnel in Houston, Texas, using a shield with a rotary arm excavator in front of a bulkhead. Water was pumped into the face, and water and spoil slurry pumped out of the invert. The slurry may have played some role in supporting the ground. In 1964, J.V. Bartlett patented a Bentonite Slurry TBM, which worked well in London, and in 1974 Wayss and Freytag A.G. used their Hydroshield, which used a slurry in combination with an air chamber pressure regulation system. These machines, together with a series built by the Tekken Kenetsu Co., have driven tunnels in very difficult ground and groundwater conditions. The main disadvantage of the slurry TBM is that the spoil removal systems are at the surface, and the inlet and discharge lines must be extended as the tunnel advances. During this extension of the pipes, care must be taken that pressure is not lost at the face. A small leak can produce a large pressure drop and loss of ground support. The circumferential seals between tail shield and the exterior of segment ring are difficult to keep fully watertight under pressure.

In 1963, the Japanese company Sato Kogyo designed an earth pressure balance TBM, which eliminated the need for slurry and, with improved tail seals, made it possible to maintain pressure at the face. Removal of spoil was by a screw conveyor inside a pipe, in which the pressure drop to atmospheric level took place. These systems are discussed in the section on spoil removal later in this chapter. Earth pressure systems are today the preferred system for TBMs operating below the groundwater table, and the TBM is designed so that it can operate with or without the system. Spoil removal from high- to low-pressure areas is key to the success of the system.

As soft ground and rock TBMs built on each other's experience, they shared more and more features. By the 1980s, of the 11 machines used in the soft chalk of the Channel Tunnel, most were rock machines, but at least 2 were adaptations of soft ground TBMs, and 3 had earth pressure balance (EPB) features. The latter were the French undersea TBMs, designed to excavate and erect lining under 10 bar external water pressure in fractured and faulted rock.

COMPONENTS OF A MODERN TBM

A modern TBM is a complex system of interdependent parts. A fully equipped TBM can occupy as much as 1,000 ft of tunnel, and be made up of mechanisms for cutting, shoving, steering, gripping, exploratory drilling, ground control and support, lining erection, spoil removal, ventilation, and power supply. All of these items must advance with the tunnel heading, and items such as trackwork, power supply, and ventilation ducting must be extended behind the TBM as it advances.

The key word in the preceding description is *interdependent,* since all mechanisms must be able to function at a rate consistent with the advance of the cutterhead. Deficiency in performance of any part handicaps the heading advance rate and affects project schedule and cost.

Cutting the Ground

A TBM excavates rock or soft ground by the rotation of an assembly of teeth or cutting wheels under pressure against the rock face. The teeth, known as drag bits and made of very tough alloy steel tipped with tungsten carbide, excavate softer rocks and soils by a ripping action. As a rough guide, they are suitable for rocks with strength of up to 13,000 psi (90 MPa). The earliest TBMs used drag bits. Cutter wheels, again made of tough alloy steel, excavate rock by a rolling crushing action under pressure against the rock sufficient to fracture the material. Cutter wheels are suitable for use in rocks up to 40,000 psi (275 MPa) at rates of advance competitive with or exceeding those attainable with drill-and-blast methods of excavation. For even harder rocks the rolling cutters are of different design, embodying carbide buttons in a single- or multiple-disk-shaped wheel. The buttons pulverize the rock. The degree to which the rock is broken is a measure of the energy needed to excavate a unit volume of tunnel. Much work has been done in the past two decades on the mechanisms of rock cutting by drag bits and cutter wheels. The references contain several reports on this continuing work.

The magnitude of the thrust against the face and the torque at the cutterhead are varied to suit the conditions anticipated or encountered. Rotational speeds are slow, typically around 5 rpm. The speed at the periphery of the cutterhead is held to 400–500 ft per min. The thrust of a modern machine can exceed 2,000,000 lb, and the load on each cutter 64,000 lb. The torque capability can be as high as 13,000,000 ft-lb. More than 150 Robbins TBMs ago, the machine that began the modern development of the TBM at Oahe Dam, had a thrust of 100,000 lb and a torque of 281,000 ft-lb.

Soft ground machines can require even higher torque, earth pressure balance machines requiring the highest torque. A rule of thumb for torque on such machines in ton meters is 2 to 2.5 times the diameter in meters cubed.

The cutter wheels move at different speeds across the rock face, depending upon their radial distance from the center of the head. The segment of the wheel in contact with the rock is in a straight line, but it is moving and cutting in a circular path. This subjects the cutter wheel to a sideways twisting action that demands great lateral strength and bearing capacity. The torsional loading on wheels and bearings is reduced by splaying the cutters slightly outward from the line of advance. The outer cutters and the gauge cutters are splayed outward. It is the gage cutters that control the diameter excavated and that give the tunnel its smooth finish. The loading is particularly severe on the cutters nearest the center of the head, where the cutters have a substantial component of rotation about a diametric axis as well as the conventional axial rotation. The effect increases on larger-diameter wheels. Some machines use smaller center wheels for this reason.

The cutter wheels or drag picks are arranged in what at first may appear to be a haphazard pattern over the cutterhead. Closer inspection will show that the wheels are arranged radially to cut grooves separated by a few inches, and circumferentially located to give as balanced a thrust as is practicable to the cutterhead. The pattern of the cutters on the head is an important design feature, since balanced loading on the cutters and the power train is a major factor in component durability.

Removing the Spoil

As the spoil, or *muck,* falls from the excavated face, it is picked up by scoops rotating with the head. In a typical arrangement, the scoops, or buckets, empty into a hopper, which feeds a conveyor belt, which in turn carries it to dump cars at the rear of the train for hauling out of the tunnel. In some tunnels a conveyer belt or trucks are used instead of a train for spoil removal. When the TBM is designed to operate under external water pressure, however, the first part of this handling sequence is not practicable because of the pressure differential between the groundwater and the inside of the tunnel. Means must be found to pass indeterminable mixtures of rock, soil, mud, and water through the watertight bulkhead of the machine through some form of pressure lock. This is not an easy task.

One spoil removal approach involves the use of two end-to-end encased screw conveyors rotating at slightly different speeds. The conveyors are set at an angle to each other, and the second rotates slower than the first, creating a moving blockage at the angle, in which the pressure drop is achieved. A second technique uses a single screw conveyor with a breech discharge mechanism to discharge onto a conveyer belt. The arrangement operates like a breech-loading rifle mechanism in reverse, with the spoil taking the place of the bullet.

Main Bearing, Motors, and Gearboxes

Behind the cutting head is the main bearing, which must absorb the full thrust of the machine. Because of its size and location, failure of the main bearing in operation is a serious occurrence, and as a result the maintenance and treatment of this key item is critical. In this same area the cutterhead main drive motors and gearboxes are located. Machines designed to operate under external water pressure incorporate a watertight bulkhead at this point.

Steering Shoes, Grippers, and Shove Jacks

Arranged around the body of the TBM just behind the cutterhead are the steering shoes, hydraulically operated pads that bear against the ground selectively to keep the TBM on line and grade. These shoes and the grippers also serve to prevent rotation of the body in reaction to the head rotation. If the ground is suitable, the TBM will be designed with gripper pads—large circumferential pads that are thrust hydraulically against the ground to provide a firm base against which the cutting head can be thrust forward. If the ground is soft, the pads will be unable to develop the requisite bearing, and this, together with the need for immediate ground support, leads to the adoption of precast linings. The machine is then designed to develop the thrust necessary for cutting from thrust or shove jacks bearing on the ends of the ring of segments. The erected segment rings are grouted in place, and the liner acts as a long hollow horizontal friction pile against which to shove. In the absence of grippers, this friction pile must also provide torsional reaction resistance for the rotating cutterhead. The grouting also serves the important purpose of providing early support for the heavy loads imposed by the segment erectors and other TBM equipment, which follow behind the cutterhead. Thus the TBM and the lining design must be designed to work together. It is common that the highest loadings the lining will experience will be the combination of grouting loads and the longitudinal forces caused by the TBM shove jacks.

Ground Support Equipment

The ground support equipment is installed behind the main bearings as close to the excavation as is practicable. This can consist of complex precast liner erectors, or relatively simple rock bolting drills, or steel rib erectors, or shotcrete machines, depending on the ground being designed for. In this area will be located any forward ground exploration drills or ground modification grouting rigs. If the machine is designed to use precast linings, it might be built in two parts, front and rear, the one telescoping over the other. This permits the cutting head to move forward a limited amount independently of the lining erector section, which must be stationary during the erection sequence. By this means the excavation, which is a continuous process, can proceed during the lining erection process, which is a cyclical process. This has not yet been achieved in a machine required to operate under external waterhead because of the need for annular seals where the two outer shells slide over each other, an arrangement that has not been made to work entirely satisfactorily.

Backup Facilities

At the rear of the train are located the backup facilities, which must be extended behind the TBM as it advances. These include the high-voltage electrical cable reel, ventilation ducting, track-laying equipment and catenary, if used, and water and drainage lines. At various points along the train, grouting equipment is located to fill the annular space between the segment and the ground, and to redrill the grouted areas to ensure adequate filling of voids by proof grouting. The support train for a fully equipped TBM can extend as much as 1,000 ft. If precast linings are used, a large part of the remaining TBM train will be taken up by conveyer belts, segment hoists and gantries, which move the segments from the cars that brought them in and orient them correctly for installation. Through the same area, conveyer belts will carry spoil to the muck cars. Fitted into this restricted space will be the transformers and switchgear, the TBM operator's cabin, and the worker amenities. The difficulty of replacing major items of equipment at the front end of this train during the work will be readily appreciated.

OPERATION OF THE TBM

Steering

If allowed to move forward unattended, a TBM would not advance in a straight line. Variations in ground conditions would deflect it; gravity would tend to move it downward; and buoyancy might tend to move it upward. In addition, the reaction from the rotation of the cutting head and the cutting forces themselves would tend to move the machine from a straight path. Steering a TBM is an art, and one not easy to practice off-site.

Steering is achieved by selective use of grippers and shove rams, with some machines being fitted additionally with steering shoes at the front end, these being smaller versions of gripper pads. Guidance on a modern machine is almost invariably by laser and targets. Front and rear targets can be adjusted by the surveyor such that when the laser beam passes through the center of both targets the TBM is on course (see Chapter 3).

Why is precise steering so important? In a drill-and-shoot excavation, the tunnel is of necessity oversized, and the forms used for final lining will straighten out short-term irregularities, converting them to lining thickness irregularities. As a result, linings in drill-and-shoot tunnels are usually more than adequate. A TBM bore is also oversize, but only by a few inches to permit steering of the machine, and possibly for tail-shield clearance to segmented liners. If the tunnel is to have a precast lining and the lining is to be expanded against the ground, the finished tunnel will follow the alignment and irregularities of the excavation. Such irregularities, and even some wandering off course, can be tolerated in, say, a water supply tunnel, but in a subway rail tunnel, and even more so in a high-speed rail tunnel, accuracy of alignment and grade is critical. In rail tunnels, the owner is buying a track laid to a precisely specified line and grade, and the tunnel is merely an inconvenience en route. The track geometry rules, and the tunnel must be set out to accommodate the track, not vice versa. Thus for every inch of cross section out of alignment allowed in the specification as a construction tolerance, the tunnel must be built two inches larger in diameter to maintain the clearances. Experience teaches that it is the most prudent and economical course to make the construction tolerance a generous one.

Accuracy of grade can be as important as alignment. For high-speed rail tunnels, both rate of change of grade and deviations of track alignment are very gradual indeed, so much so as to create virtual straight line alignment and grade. The change of grade can be deceptive because engineers are accustomed to studying geologic sections, which for clarity and convenience of presentation have a greatly expanded vertical scale. An example of this is the published profile of the English Channel Tunnel, a difficult thing to present pictorially. The impression given is of a vertical tunnel alignment that rises and falls to follow the good tunneling ground, with a change in elevation of about one eighth of the undersea length of tunnel. In fact, the change is about 100 m in the undersea length of 38 km, and to natural scale the entire profile would be contained within a straight pencil line 19 cm long and half a millimeter thick. There is little room for deviation from either azimuth or grade in a long high-speed rail tunnel.

Advancing the TBM

While a TBM operation is considered a continuous process, which it would be if excavation were the only operation, it is in fact a cyclical operation, particularly if a lining is being installed behind the machine as it advances. The excavation is continuous until the limit of extension of the thrust rams is reached. At that point the excavation stops, the rear of the machine is supported (either by a second set of grippers or jack legs), and the main grippers move forward as the thrust rams are retracted. At this point, temporary support or permanent lining is placed, and the cycle begins again. To advance, the TBM must develop thrust to enable the cutters or bits to penetrate the ground. In its simplest form, found in the hard rock TBM for an unlined tunnel, a TBM consists of the cutting head, motors, main bearing, and front grippers, connected by hydraulic cylinders to a rear frame and a second set of grippers. In some machines the hydraulic rams are skewed to resist rotational forces. Double-shielded TBMs, which are intended to make the cycle truly continuous, are discussed in the section on ground support equipment earlier in this chapter.

The operator controls the thrust on the cutterhead and also the speed of rotation. Too much thrust and too slow a rotation will tend to stall the machine or burn out the clutches. Too little thrust and too fast a rotation will not move the machine forward and may result in a churning action where the cutterhead turns in place and material falls from the roof, creating an overhead chimney, which can be dangerous, particularly in undersea tunnels.

The operator has no direct view or feel of the ground conditions or the parts of the TBM and thus must deduce the conditions from the array of instruments and video screens in the compartment. Ground is seldom uniform for long, and changes can be sudden or deceptively subtle. A great deal rides on the operator's skill in advancing and steering the TBM.

Where the TBM is designed to work under external water pressure, the designer and the operator face yet another challenge. In all advancing and steering operations, the operator must be constantly aware of changing ground conditions. A good operator can make for good advances. A poor operator can destroy the TBM.

Ground Support Erection

If the project schedule and ground conditions permit, a contractor will usually prefer to install the permanent lining in the tunnel after excavation is complete. This makes for a simpler TBM, and he can first concentrate on excavation and temporary support, with greatly simplified logistics. A tunnel can be lined faster, and the equipment used more efficiently, using movable forms when the support lines carrying air, power, communications, and water to and from the working face have been removed. However, project schedules and ground conditions are not always compatible. At the other end of the difficulty scale, the schedule is tight and the ground requires immediate support. The final lining in such a case is then likely to be precast and installed as close as practicable behind the cutterhead. If the ground is highly variable, the project will probably be equipped to handle the poorest ground condition.

In rock tunnels, temporary support erection equipment can be installed well forward on the machine, and this will consist of rock bolting equipment or a steel rib erector, usually the former. When precast segments are used, the temporary support is not needed, since the segments are erected as soon and as close behind the cutterhead as are practicable.

In all tunnels where they are used, the precast segments arrive a ring at a time into the heading. As there are usually several shapes of segment in a ring, they must be offered up in the correct order. They arrive at the face usually lying parallel to the line of the tunnel. They are picked up, rotated into position, and held in place by building bars cantilevered from the body of the machine, or by thrust jacks bearing on the front of the segment, until the final segment is in place. This last segment may be a wedge to expand the ring hard against the ground, in which case the rings may not be bolted but rely on ring compression and thorough grouting to hold them in place. Alternatively, the segments may be bolted and perhaps gasketed for watertightness. Where the TBM is operating under external water pressure, the completed ring must be built within the tail of the TBM and extruded from the tail of the TBM past a system of seals between the outside of the ring and the inside of the tail shell. The favored tail seal is a triple row of wire brush seals kept generously supplied with thick grease.

If the machine is designed to use precast linings, it might be built in two parts, front and rear, one telescoping over the other. This permits the cutting head to move forward a limited amount independently of the lining erector section, which must be stationary during the erection process, which is a cyclical process. This has not yet been achieved in a machine required to operate under external waterhead, because of the need for annular seals where the two outer shells slide over each other, an arrangement that has not been made to work entirely satisfactorily.

Once erected, the rings must be grouted into place as soon as is practicable, starting at the invert and working up to the crown. Proof grouting is repeated at various points behind the first-stage grouting to ensure complete filling of voids. The grouting serves several purposes. The TBM support equipment following behind the head transmits heavy loads from the erectors and gantries to the invert of the tunnel. A solid footing is necessary to support these loads. The horizontal tube formed by the segments is subjected to heavy axial loads during the shoving and needs to be grouted in place if the segments are not to move around during the shoves. The grout also serves to bind the segments and the surrounding ground into a coherent structural ring, which gives the lining its strength and stability. Lastly, the grout serves to cut off water flows outside the lining, both in radial and longitudinal directions.

THE TBM, TEMPORARY SUPPORT, AND PERMANENT LINING

A minority of rock tunnels require neither temporary support nor a permanent lining. The majority require some degree of structural support and lining for protection from rock falls or groundwater inflow, or for hydraulic smoothness. The tunnel use and the ground conditions will determine the need for a final lining. The amount and the extent of support and lining can only be estimated at the time the work is priced. The accuracy of such estimates, or in some cases, the lack of such estimates, is a common aspect of tunnel construction litigation.

TBM excavation is a continuous process, whereas installation of temporary support and permanent lining are cyclical processes. An early planning decision, therefore, is whether the two or three operations should be tied together. For example, if ground conditions permit, the contractor may elect to support the ground temporarily as he excavates and install the permanent lining later, after holing through. This gives him less logistic congestion, but it may extend the overall schedule. Alternatively, and depending on the requirements for the finished tunnel, he may elect to support and line the tunnel with a segmented precast concrete lining as he excavates, eliminating the need for temporary support. He may elect to go partway between these two approaches and lay a precast concrete invert behind the TBM, installing temporary support as necessary, and complete the lining of the arch later.

It should be recognized that tying the TBM and the lining operations together means that the work will progress at the speed of the slower operation, with the potential for inefficiencies in crew usage. Some alleviation of this disadvantage can be gained by designing a double-shielded TBM to permit limited relative movement of the excavation and lining systems as discussed in the previous section. However, this improvement will be minor if there is a fundamental mismatch between the rates of excavation and support installation.

It is considerations such as the preceding that make the question of what type of lining is best for the job in the conceptual stage a difficult one for the engineer to answer with assurance. Unless some imperative dictates the nature of the lining from the beginning, the best support and lining methods will not become clear until late in the design process. Even then, the decision will lack the important input of the contractor until the successful bidder is determined.

THE DECISION TO USE A TBM

The economics of tunnel construction are extremely site-sensitive and resist generalization. Most tunneling contracts are bid competitively. The decision whether or not to use a TBM is usually left to the contractor, although the bid specifications may lean, or appear to lean, toward or away from use of a machine. If the owner feels the case for use of a TBM is clear, he may require its use. In a privately financed project, the interest cost during construction and the urgency of the need for early revenue may take precedence over construction costs.

Listed and discussed here are some of the factors that are likely to influence a contractor in his decision of whether or not to use a TBM in a rock tunnel.

Factors Favoring Use of a TBM in Rock

1. Adequate geologic data.
2. Timely geologic data, which should be complete before issuance of invitation to bid, *not* just before bid date.
3. Geologic data that indicates reasonable uniformity of rock quality and behavior throughout the drive.
4. Low to moderate water inflow expectations, preferably based on pump tests with quantitative, rather than qualitative, estimates of flows.
5. Rock not excessively hard or abrasive.
6. Circular cross section of tunnel or crown. A noncircular section, such as a horseshoe section with a circular crown, can usually be driven circular and enlarged by a pass of an invert excavator or subsequent drill-and-blast benching excavation.
7. Single portal access. Unless the tunnel length justifies the use of two TBMs, access from both ends of a tunnel reduces the competitiveness of a TBM in terms of construction schedule. To complete a tunnel from one heading by TBM when both ends are available to drill-and-blast methods requires that the TBM advance at least twice as fast as a drill-and-blast heading. Access from shallow intermediate shafts or adits further increases the demands on the TBM advance rate.
8. Access by portal rather than shaft favors the TBM for large tunnel diameters, where heavy TBM sections must be lowered to tunnel elevation for assembly underground. A portal permits assembly of the TBM outside the tunnel, a distinct cost and schedule advantage.
9. Long tunnels. The drive must be long enough to warrant the heavy initial expenditure in purchasing the TBM. Equipment costs for drill-and-blast construction are usually lower.
10. Availability of the right type of and the right diameter TBM. During its working life a TBM must be well maintained, since its performance in the last quarter of the drive is as important as in the first three-quarters. Because many of the working traits of the machine are known, and because modifications improving the performance may have been made, availability of a suitable used TBM is a big advantage to a bidder, who also knows his competition may

have to allow many months of lead time for delivery of a new TBM.

11. Availability of skilled labor. TBMs require different skills and crewing from those for other tunneling techniques.
12. Ready sources of adequate electric power. A TBM can require as much as 5 MW of installed power.

Factors Not Favoring Use of a TBM in Rock

1. Inadequate or unconvincing geologic information.
2. Mixed-face excavation, or the likelihood of it.
3. Heavily faulted areas and/or wide fault zones. In this case and in item 4 below, a TBM may help support the poor ground, or it may be trapped by ground movement behind the face. Such situations should be considered separately on their merits and demerits.
4. Short stand-up time. Tunnels in rocks requiring support when excavated will stand unsupported for a period; this is known as *stand-up time.*
5. Very hard, abrasive, or swelling rock.
6. Geologic formations known to be unpredictable. For example, glaciated formations with deep valleys into the bedrock; karstic limestone or volcanic lava formations.
7. Short tunnels.
8. Variable geology requiring frequent changes in excavation and support methods.

None of the unfavorable or favorable factors should be taken as absolutes. Every project is different, and a careful weighing of all factors, with a good measure of judgment, will go into the final decision. A little luck has been known to help.

SELECTING A SOFT GROUND TBM

Selection of a soft ground TBM requires careful consideration of soil and groundwater conditions, tunnel size, and the excavation conditions and environment. This topic is covered in Chapter 6.

MEASURING TBM PERFORMANCE

The measurement of performance of a TBM must serve the needs of a variety of people: the contractor, the manufacturer, the owner of the work, and the tunnel engineer. Not all have the same interest. The contractor is interested in the percentage of time the TBM is cutting rock. The manufacturer is interested, in addition, in the percentage of time the TBM was available to operate since, for reasons beyond the manufacturer's control, the TBM may not have been used to its fullest capacity. The owner shares this interest.

The designer and the student of tunneling are also interested in the instantaneous rate at which the TBM advances through the ground: the rate at which the TBM can penetrate the ground for short periods. All are interested in forecasting future advance rates, in analyzing problems and correcting them, and in minimizing the often painful shakedown time during the early days. During this "learning curve," TBMs have problems, operators have problems, and crews have problems.

The objective during these days is to establish the optimum thrust, torque, and cutter wear for the ground, to minimize the cycle time, and to settle the work crew into a steady and safe routine. Performance is measured primarily for internal project reasons, with the objective of improving performance on the project. Because of this, what is included or excluded in measuring a particular activity is likely to be job-specific. However, when the performance of several projects is to be compared, we find there is little standardization of input.

Formulation

In measuring performance of the TBM, it is usual to separate the TBM proper from the extensive support system when defining what we mean by TBM, but since down time of the support systems affects the availability of the TBM, it may be taken into account in calculating availability. The usual and logical point for this separation is the tail end of the last moving unit in the train of equipment that advances with the TBM.

Six activities are of interest:

1. The time TBM is cutting rock
2. The time the TBM cutter head is idle because other parts of the TBM are functioning (e.g., regripping, segment erection, or ground support); design of the TBM to permit maximum overlap of these functions in steps 1 and 2 is desirable
3. The unplanned down time for TBM maintenance or repair
4. The unplanned down time for back-up equipment maintenance or repair
5. The planned down time for TBM maintenance
6. The planned down time for back-up equipment maintenance
7. The cycle time, which is the time taken to work through one cycle: excavation, muck removal, ground support and/or liner erection, regrip and restart excavation

It would seem to be a simple matter to calculate, from the project records, measures of performance, reliability, and durability, but such is not the case. Unless the detail of the derivation of the activities is known and standardized, the published figures on availability and utilization of a machine will mean little or nothing. As of today, it is not standardized. Even when it is standardized, the great variety of tunneling projects will make comparison difficult.

If a TBM is working two shifts a day, planned maintenance on both TBM and support equipment will be scheduled after working hours. Thus, planned maintenance does not affect the availability of the machine, which can be as high as 100%. However, on a project like the Channel Tunnel, where the TBMs worked 24 hours a day, there is no "after hours," and the planned maintenance would of necess sity reduce the achievable availability to below 100%. To rectify this, the planned downtime is deducted from the working hours, and a 720 hour month with 60 hours planned downtime would become a 660 hour month for availability calculations. The planned support downtime is assumed to be scheduled in this same period and is also eliminated from consideration. There may be some overlaps and inaccuracies in doing this, but they are not believed to be significant. The shorter the planned shift time, the more after-hours time is available for maintenance and repair, and the higher the availability is likely to be. Normal shifts last from 7 to 10 hours. The sums of items 1–6 is usually 24 hours. Item 7, cycle time, is separate measure of the overall efficiency of the operation.

If we exclude from the measurement the planned time for maintenance of TBM and support, we are left with four activities, which give us our parameters of performance (Figure 11-1):

1. Cutting rock
2. TBM activities other than cutting rock
3. Unplanned TBM downtime
4. Unplanned support downtime impacting TBM operation

We can now move to definition of terms.
Operating can be defined as A or $A + B$.

1. *TBM Availability.* This is defined as the percentage of the planned shift hours a TBM is available to excavate and load spoil into whatever conveyance system is used for mucking. Percentage availability can then be defined as:

$$\text{Percentage Availability} = \frac{A + D}{A + B + C + D} \times 100$$

or

$$= \frac{A + B + D}{A + B + C + D} \times 100$$

2. *TBM Utilization.* Utilization is a measure of how fully an available TBM is being used. Percentage utilization is defined as the percentage of time a TBM operates when it was available to operate. With two definitions of operate, it can be defined as:

$$\text{Percentage Utilization} = A \Big/ \frac{A + D}{A + B + C + D} \times 100$$

or

$$= (A + B) \Big/ \frac{A + B + D}{A + B + C + D} \times 100$$

There is nothing wrong with either definition if its derivation is clear. Unfortunately, it is generally used without definition, and it is then misleading or meaningless. When the terms are used, either standardization or clear definition is needed.

3. *Instantaneous Penetration Rate.* This is the rate, in units such as inches per minute, at which the cutting head of the machine moves forward through the rock. It is a function of rock strength and abrasiveness, design of cutting mechanism, thrust and speed of rotation of the cutter head. The rock characteristics must be extrapolated from samples and will vary; the cutting mechanism designer will use judgment and experience to provide for variations in ground and to optimize cutter-wear costs against advance footage. The operational speed and thrust are in the hands of the operator in a soft ground machine. In a hard rock machine, the operator usually controls only the thrust. There is usually good cooperation between manufacturer and contractor, but there are enough variables in nature that close agreement between laboratory cutter tests and end results in the field should not be expected on every project. Instantaneous penetration can be measured by dividing the advance of the cutter head of the machine in inches by the time the head is rotating. Alternatively, the penetration per TBM revolution is calculated. There is no reason why both should not be recorded. The rate is usually measured over a 5 min period of advance.
4. *Advance Rate per Month.* This end result measure of performance is the net result of driving time, down time, and unutilized time. It is a useful measure for charting progress, but gives no indication of reasons for good or poor progress. Multiplied by the cross-sectional area of the tunnel excavation, it becomes the neat-line volume of ground excavated per month, which may be preferred.
5. *Cycle Time.* A fifth and important measure of performance is the cycle time. Because tunnel projects are site specific, the cycle time is an internal measure of performance rather than one useful for comparing projects. Since good cycle time depends on good logistics, it is a measure of the

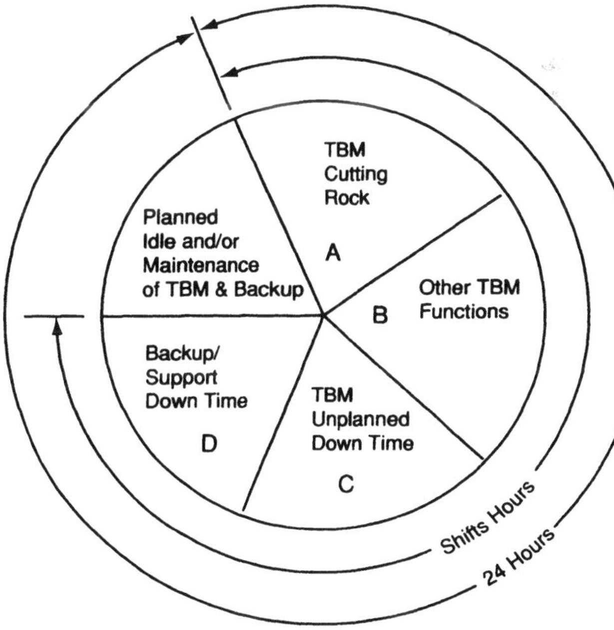

Fig. 11-1. TBM performance parameters.

efficiency of the whole operation. Cycle times vary with ground conditions, but in tunnels with precast liners erected directly behind the heading, twenty to forty minutes is common.

Other Performance Issues

The placement, scheduling, and type of the final lining will affect the cycle time and the monthly advance rate. The liner may be poured behind the machine, or it may be poured after hole-through, when the rest of the equipment has left and the tunnel is relatively clear of obstructions. An unlined rock tunnel requiring little or no temporary support will be driven by a simpler machine requiring less backup and maintenance than a machine that must erect its precast final lining as it excavates. The diameter of the tunnel also affects the advance rate, which is why some prefer the volume of excavation per unit time as a measure.

An important parameter of performance on a hard rock TBM is cutter wear. Advances in design and metallurgy have steadily improved the performance and cost of rock excavation. The TBM manufacturer's contract is likely to contain clauses guaranteeing cutter footage. In all but the harder rocks, downtime for other reasons is likely to be much greater than downtime for changing cutters.

The preceding measures of performance are of use in the planning and control of TBM drives. They are often written into the TBM manufacturer's contract, and because of this they must often be more closely defined than above. A contractor will, for instance, take the opportunity to reschedule planned maintenance on the TBM when the system is down for external reasons. He may even advance the date for a cutter change in order to accomplish it during a down period. While such operational inputs affect the measurement of the performance, a prudent owner will recognize them as contractor initiatives that are adding to the TBM effectiveness.

A review of the above may generate some sympathy for the TBM manufacturer. He is designing and manufacturing a machine that will be stressed to the limits, or beyond, of present-day bearing technology and metallurgy. He must design it to operate in an environment that is invariably hostile, sometimes extremely so. The TBM must work in conditions that cannot, by the nature of tunneling, be entirely foreseen. His machine will be operated by the party to whom he has guaranteed performance, and he knows that inept operation, overthrusting, or overspeeding of the machine can destroy the machine. Finally, the TBM manufacturer is his own worst competitor. A TBM, because of the economic imperative that it keep going in top shape through to the end of the job, is one of the few pieces of construction equipment likely to be in better shape, less some wear, at the end of a successful drive than at the beginning. It has proven its capabilities and has had most of its remediable faults corrected. It is likely to be available for a new project, after rehabilitation, at a lower cost, and more importantly, in a shorter delivery time, than a new machine.

THE LEARNING CURVE

Once the TBM has been assembled and checked, tunneling begins and, with it, the learning curve. The date is noted at which the TBM first cuts rock. The assembly period will extend beyond this date if the full backup train cannot be assembled before a considerable length of tunnel has been excavated to accommodate it. This will be the case, for example, if the TBM has to be assembled at the bottom of a shaft rather than outside a portal. In such a case it is usual to prepare an erection chamber at the bottom of the shaft by drill-and-shoot techniques to permit sufficient space for the essential part of the backup train. The complete backup train can approach 1,000 ft in length, a formidable erection challenge at the bottom of a shaft, and one that takes up a considerable amount of critical path time.

Once the TBM is assembled with enough backup to enable it to operate, it will begin cutting rock. The learning curve then begins. The term gets its name from the flat portion of the curve of excavation progress (*Y*) against time (*X*), which, one hopes, gets steeper as the work becomes routine. During this learning curve, crews are being assigned specific duties in the TBM advance cycle. The operator is getting the feel of the ground and the machine, and he is experimenting to find the best balance of the settings for thrust, rotational torque, and the resulting cutter wheel wear. A rhythm develops in the cycle and the work speeds up. Average progress rates often double during this period. The learning curve may start up again during the drive if changes in ground dictate changes in crew duties or machine operation. Because of the timing of the learning curve in the overall project, relatively late in the project after delivery and assembly of the TBM, any perceived delays tend to receive more than their share of management attention. Problems at the start of a project can be understood and accepted more readily than problems a year into the project.

On the Channel Tunnel Project, the undersea service tunnels were started first, ahead of the two running tunnels. The service tunnel drives were to serve as pilot tunnels to determine ground conditions ahead of the larger-diameter running tunnel machines. Any hole-through error in line or grade on the long "blind" survey traverses from either shore would also be evident in sufficient time to make any necessary corrections to the running tunnel line and grade. The service tunnel is used for ventilation, maintenance access, and emergency evacuation. Six TBMs were used on the undersea drives.

The learning curve for these machines was long and difficult on both sides of the Channel. There were reasons for this. First, the project was unique, and at relatively shallow depth under the ocean floor. Second, the headings were exceptionally long, up to 13.8 miles (22.3 km) and, in the case of the service tunnels, unsupported by any parallel headings. Third, on the French side, the TBM was to operate both in the dry and under external water pressures of up to 10 atm

(150 psi) in ground that was anticipated to be difficult for at least the first 3 mi (5 km). On the British side, there were few problems with the cutting of the rock, but many with segment erection, mucking, and electrical connections. As the headings lengthened, logistics took over as the controlling factor. The construction train operation was larger than many transit systems, employing 150 locomotives. The British service tunnel heading had to support the excavation of cross passages, a large crossover cavern, and pumping station excavations as well as its own advance. Each machine was erecting and grouting its segmented liner as it advanced.

Table 11-1 illustrates the learning curve on the six undersea machines. The length of the curve is somewhat subjective. It has been taken as the time needed to reach the average straight line advance rate between start and hole-through dates for each tunnel. TBMs B1 and T1 are the British and French Marine Service Tunnel Machines. B2 and B3 are identical British Marine Running Tunnel machines, as are the French T2 and T3. The two Running Tunnel machines from each side were parallel drives and traversed similar ground to each other and to the service tunnel.

Because of the heavily faulted and water-bearing ground on the French shore and the shallow cover near the British shore, the poorer ground and the learning curves overlap, distorting the figures and therefore the analysis. Even allowing for these distortions, the results are consistent.

Even so, the figures given in Table 11-1 are not typical for the industry. A review of underland tunnels in the United States shows learning curves of 5 to 10 weeks. These differences show that for the Channel Tunnel machines, other factors were at work. Logistic and support systems affect the learning curve because they, too, have their own learning curves.

An examination of availability and utilization of the machines will usually indicate the reasons for slow early progress. Low utilization and high availability, for example, can indicate logistic or backup support problems. On the British drives for the Channel Tunnel, only 7.0% of the construction workforce was employed at the TBMs; 11.2% were employed in other underground excavation, and the remaining 81.8% were rail operative, supply yard, and operation and maintenance personnel. Put another way, nine people supported every two employed in excavation and lining erection, but all made a vital contribution to the advance of the headings.

In spite of the difficult start-up, all tunneling for the Channel Tunnel was completed on June 28, 1991, three days ahead of the schedule set in 1985 at the start of the project, a remarkable achievement. The six undersea TBMs, made by four different manufacturers, Howden (B1), Robbins/Markham (B2 and B3), Robbins (T1), and Robbins/Kawasaki (T2 and T3), and varying in design concept from hard rock to soft ground pressure balance machines, averaged more than 500 ft of lined tunnel per machine per week. The highest weekly advance, by a Robbins/Markham machine, was 1398 ft of 27-ft-diameter lined tunnel. The highest monthly advance of 5,637 ft was made by the same machine. In the month of March 1991 the advance of the four running tunnel undersea machines totaled 3.48 mi.

VARIABILITY IN GROUND

Ground is the miner's term for anything from the hardest unjointed rock to water-bearing fine sand under pressure. A TBM can be built to handle any ground condition, but to build one to handle all potential conditions in a tunnel drive is to accept great expense and compromise in performance, without elimination of much risk.

The engineer cannot know in advance with certainty all the ground conditions that will be encountered along the line of the tunnel. This is particularly true of tunnels penetrating high mountain ranges, where the depth and cost of drilling bore holes to tunnel grade may be prohibitive, and the access to desirable drilling sites may be difficult. Such tunnels dictate that

Table 11-1. The Learning Curve—Channel Tunnel

TBM	Total drive (km)	Total time (weeks)	Average advance rate (m/wk)	Learning Period: Distance to reach average m/wk (km)	Learning Period: Time to reach average m/wk (wks)	Weeks rest of drive	Average rate during learning (m/wk)	Average rate after learning (m/wk)	Ratio during/after learning	Maximum weekly progress (m)
B1	22.3	151	147.7	2.6	42	109	61.9	180.7	2.91:1	293
B2	17.9	112	159.8	2.1	42	70	50.0	225.7	4.51:1	409
B3	19.0	101	188.1	3.0	40	61	75.0	262.3	3.50:1	426
T1	15.6	139	112.2	2.0	57	82	35.1	165.9	4.73:1	291
T2	20.0	128	156.2	2.8	45	83	62.2	207.2	3.33:1	295
T3	18.9	117	161.5	2.7	36	81	75.0	200.0	2.67:1	306

the geology be projected to grade from surface mapping, and even the mapping of high terrain may be impractical for a large part of the year. As a result, the core drilling on such tunnels may be limited to the shallow portal areas and to valleys crossing the alignment at acceptable depths.

Variability in ground is perhaps the biggest challenge faced by a modern TBM. The variability can range from a difference in conditions across the bore diameter to gross variations in geologic formations along the line of the tunnel. The mixed-face condition, where the face is in both rock and soft ground, is a particularly undesirable condition, which can make the steering of the machine difficult as the machine tends to steer away from the harder rock.

The human element is important in variable ground. A tunnel heading is a crowded and dangerous place. Routine crew duties make for efficiency and safety, and breaking the routine disrupts the smooth flow of the work for a considerable time. Different duties, skills, and crew sizes are needed for different ground, adding further to the disruption.

The design, cost, and construction schedules of tunnels are site-sensitive, and tend to resist generalization. The sensitivity flows through to the final lining type and design. It is in an owner's interest that the design permit as much flexibility of contractor planning, operation, and TBM selection as is consistent with the owner's requirements for the end product. Only then can the full skill and experience of the tunnel contracting industry, and the market forces of competitive and creative bidding, be brought to bear.

NONCIRCULAR TUNNELS

The TBM, as defined in this chapter, creates a cylindrical bore. While the cylinder is an efficient shape for resisting internal and external pressures, other shapes may be needed for other purposes. For example, railroads require tunnels about twice as high as they are wide to permit passage of special freight cars. Such tunnels have been driven by driving the arch of the tunnel by TBM and then excavating the lower half of the tunnel by drill-and-shoot techniques.

For great versatility in excavation cross section in softer rocks, the boom excavator, or roadheader, is proving its worth. In conjunction with NATM support techniques, it was used extensively on the Channel Tunnel, notably for the excavation and primary support of the large TBM erection caverns, and for the British crossover cavern one-third of the way across the Channel, the largest undersea excavation to date. The cavern is some 160 m long, 21 m in span, and 14 m high. Its location 12 km from the British shore meant that the excavation had to be supported logistically by the British service tunnel, which was at the same time being advanced beyond the crossover, and was, in addition, supporting the excavation of cross passages, electrical rooms, and pump stations. The rock was the excellent Chalk Marl, an essentially watertight clayey chalk with about one-third the strength of concrete, and which could be readily cut by the picks on the roadheader cutterhead.

The roadheader was developed from mining excavation equipment, and it is an excavation machine rather than a TBM, but ground support functions can be readily handled by follow-up equipment. In appearance, it resembles a piece of long-range track-mounted artillery, with the cutterhead revolving at the end of the barrel around the axis of the barrel or, in twin head designs, at right angles to the barrel. The cutting is done by tungsten carbide picks arranged in a spiral pattern on the cutterhead. The design of these picks, and the arrangement on the cutterhead, is critical to the efficient cutting action of the head. The current rock hardness limitation is governed by the cutting ability of the picks. Some models incorporate very high pressure water jets in the head to assist in the rock cutting. The excavated rock is collected by a scoop with crablike arms, another indication of its mining heritage, and deposited onto a conveyor belt for removal.

Another machine with both a mining and a TBM background is the Robbins Ranging Mobile Miner. This machine has a tracked chassis and a short maneuverable arm, on which is mounted a vertical cutting wheel carrying disk cutters around the circumference. The cutter wheel axis is at right angles to the tunnel axis, and the cutters face radially outward with axes parallel to that of the cutterhead. The cutters operate much as on a TBM, except that fewer are in contact with the rock at one time. The boom sweeps from side to side and up and down, and can be used to excavate a wide variety of tunnel or cavern sections in harder rocks than the roadheader. Like the roadheader, the Mobile Miner can cut a flat-bottomed tunnel section, a considerable advantage for logistic support behind the heading.

These two types of machine, which can also cut circular sections if required, have the advantage that, unlike the TBM, which is built for essentially one diameter of drive, they are adaptable to a range of diameters or cross sections without modification, and they are thus likely to be available for work sooner than a TBM. Both types of machine are likely to see intensive development in the future.

Fig. 11-2. Seven ft diameter trial tunnel driven in 1881 by Beaumont/English TBM at Abbotts Cliff. Unlined tunnel as it appears today.

Fig. 11-3. Robbins Hardrock open TBM, Svartisen Project, Norway. 19 in. diameter cutters; 14.1 ft in diameter, capable of enlargement to 16.4 ft in diameter.

Fig. 11-4. Robbins/Kawasaki 27 ft diameter TBM designed to operate under 10 bars external water pressure. Note both drag picks and wheel cutters I head; French Running Tunnel, Channel Tunnel.

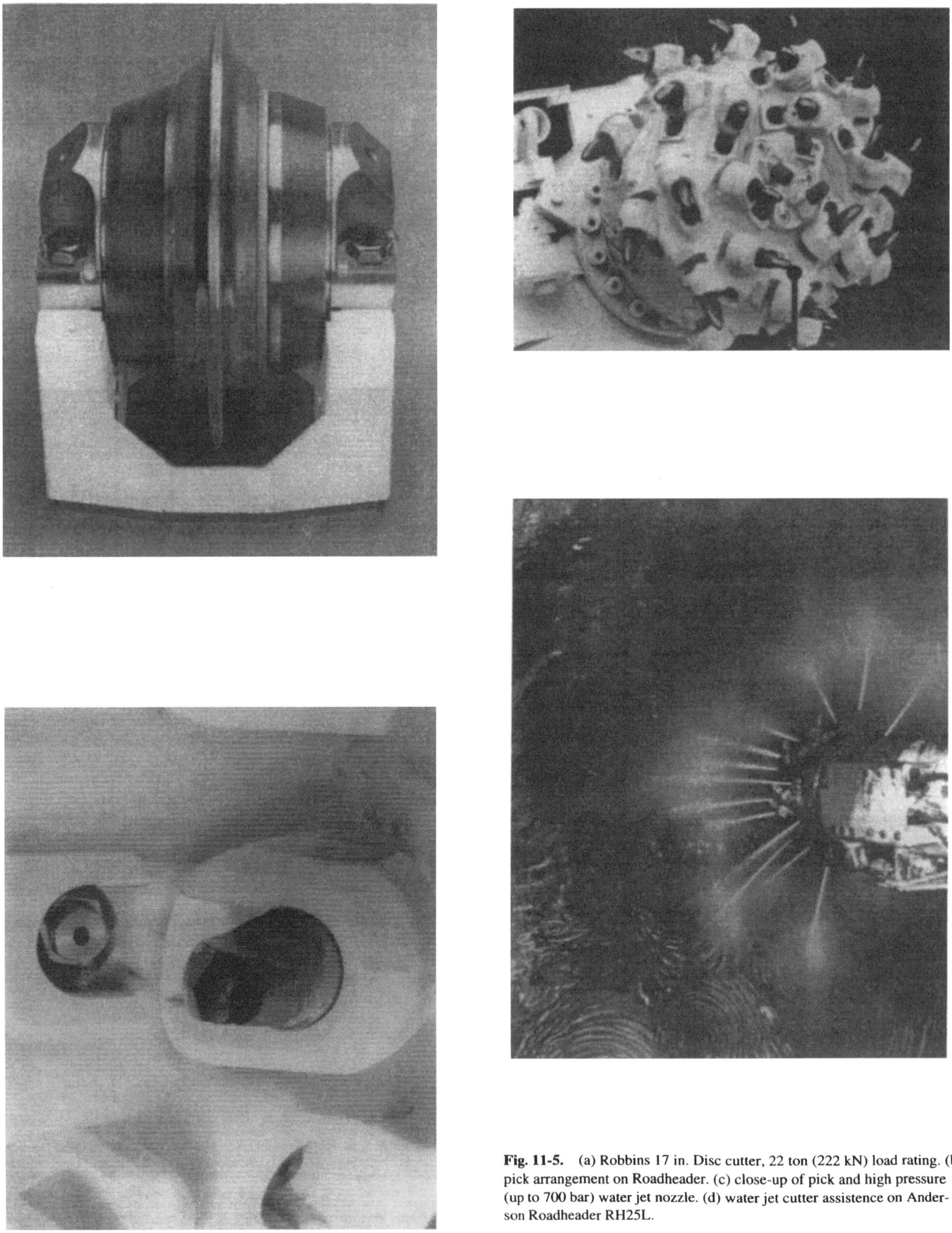

Fig. 11-5. (a) Robbins 17 in. Disc cutter, 22 ton (222 kN) load rating. (b) pick arrangement on Roadheader. (c) close-up of pick and high pressure (up to 700 bar) water jet nozzle. (d) water jet cutter assistence on Anderson Roadheader RH25L.

Fig. 11-6. Channel Tunnel. TBM Erection Chamber on UK shore. Excavated by Voest-Alpine AMT70 Roadheader in chalk.

Fig. 11-7. Channel Tunnel: Erection of Robbins/Markham UK Running Tunnel TBM in Erection Chamber.

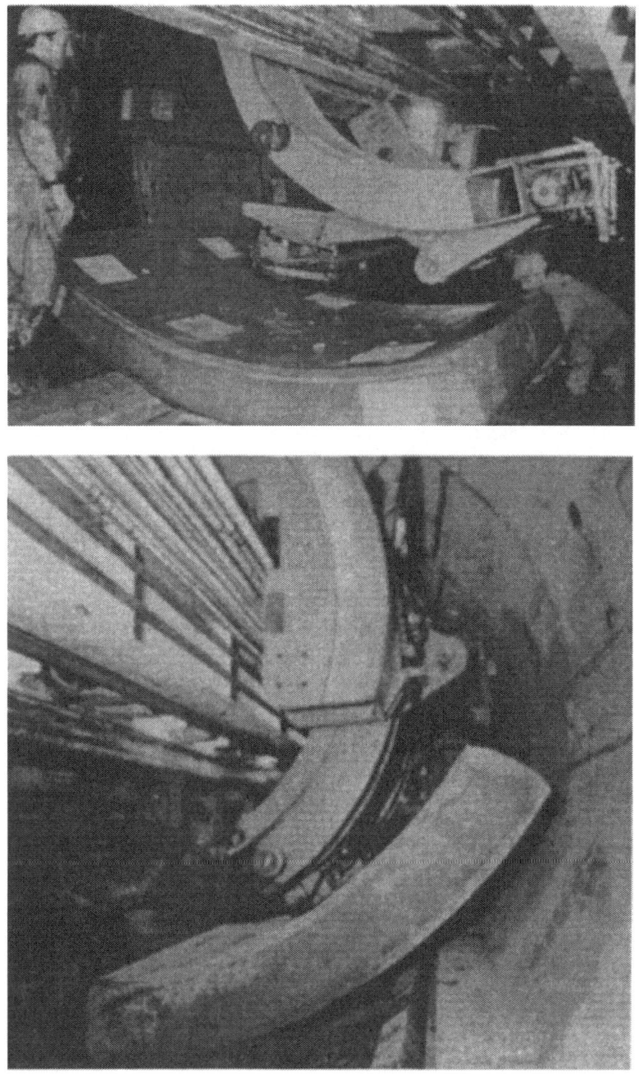

Fig. 11-8. A segment is (a) picked up and (b) moved into place.

Fig. 11-9. Voest-Alpine AMT70 Roadheader after completion of Channel Tunnel UK crossover excavation. Robbins/Markham UK Running Tunnel TBM pauses in cavern before resuming drive toward France.

Fig. 11-10. An unusual view of a TBM in mid-drive. A Robbins/ Markham TBM passes through the UK crossover cavern on its way toward France.

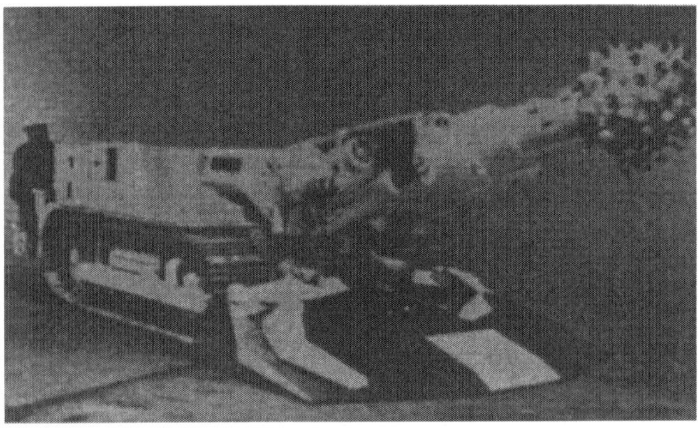

Fig. 11-11. Anderson RH25L roadheader.

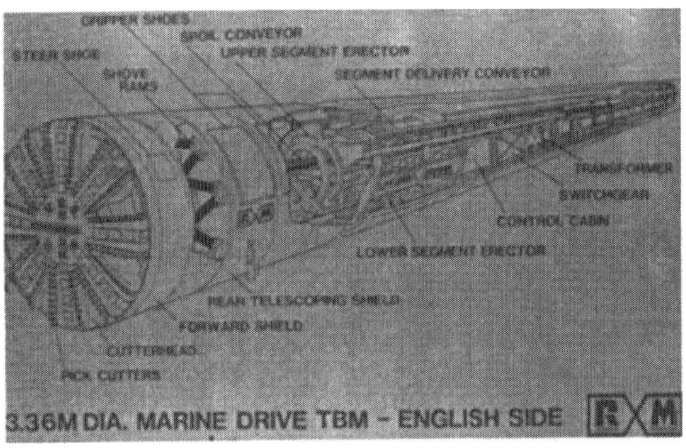

Fig. 11-12. m diameter Channel Tunnel, UK TBMs B3 and B2. These two Robbins/Markham TBMs drove 19,023 m and 17,921 m, and had best months of 1,911 m and 1,862 m, respectively. The nine TBMs on critical path drives all achieved a best month of over 1,000 m.

Fig. 11-13. Channel Tunnel UK crossover cavern under the channel, excavated by Roadheaders; volume 48,000 m^3, 164 m long, 2.10 m wide, 15.4 m high.

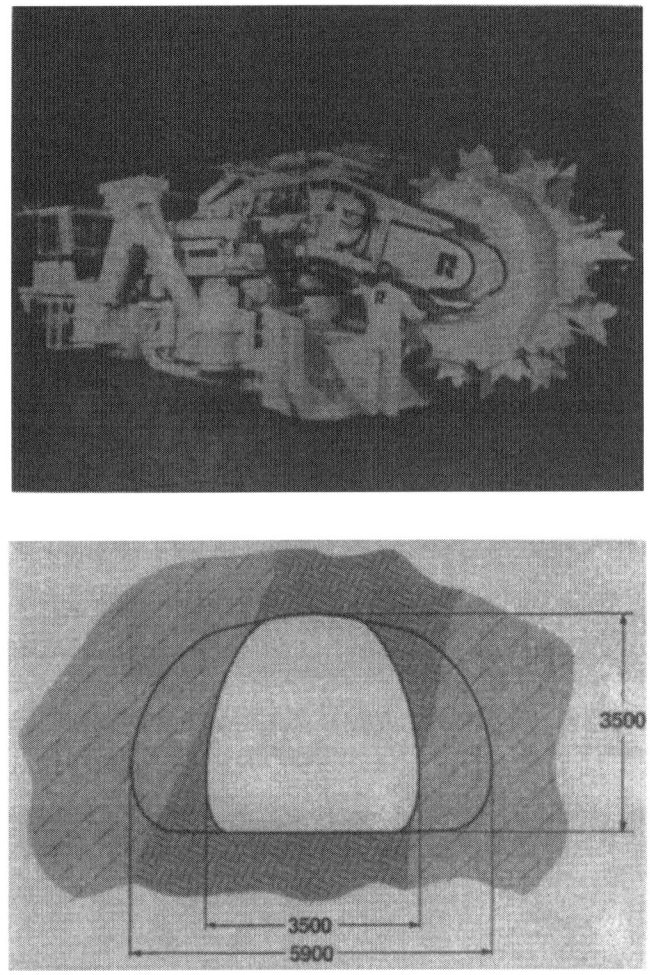

Fig. 11-14. (a)Robbins Ranging Mobile Miner and (b) 80 m^2 face profile.

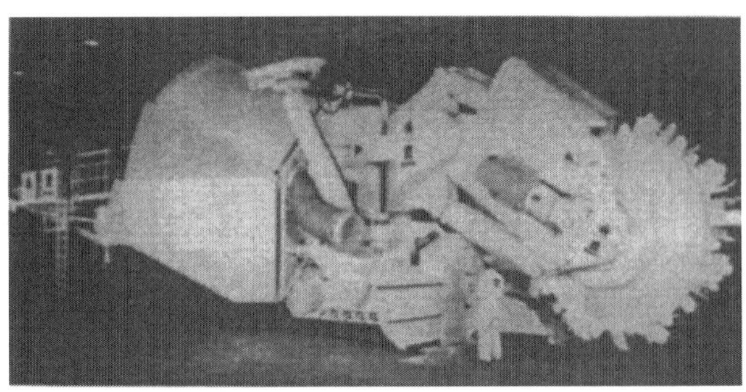

Fig. 11-15. Robbins hard rock ranging mobile miner built for the Taisei Corp. Of Japan. Capable of excavating a flat floor tunnel section 12 m wide by 8.1 m high.

BIBLIOGRAPHY

STACK, B. (1982) *Handbook of Mining and Tunnelling Machinery,* Wiley, New York.

Proceedings of the North American Rapid Excavation and Tunneling Conferences. (1972–1995) 19 volumes, SME/ASCE, New York.

Tunnels and Tunnelling, published monthly by Morgan-Grampian, London.

World Tunnelling, published 10 times a year by The Mining Journal, Ltd., London.

North American Tunneling, published quarterly by The Mining Journal, Ltd., London.

Trenchless Technology, published monthly, Peninsula, Ohio.

Tunnelling and Underground Space Technology, published quarterly by the International Tunnelling Association (ITA/AITES), Elsevier Science, Tarrytown, NY.

The Channel Tunnel, Part 1: Tunnels. (1992) *Proceedings of the Institution of Civil Engineers,* London.

12 | Shotcrete

Elwyn H. King
Principal Professional Associate (retired),
Parsons, Brinckerhoff Quade & Douglas, Inc.

Shotcrete is not just "pneumatically applied concrete." Although the basic materials (cement, aggregates, water) are the same and meet the same ASTM standards, additives (e.g., accelerators, microsilica, and steel fibers) change its character to make shotcrete unique and usable in quite a different fashion than concrete.

The shotcrete discussed in this chapter is specifically for use in underground civil engineering. Rapid initial set (1–6 min or less), rapid final set (typically about 12 min, 20 maximum), good adhesion to the ground surface, good bond between successive layers, high early strength (e.g., 700 psi in 8 hours), ductility and residual strength, reduced permeability, and thinner in dimension than possible with cast-in-place concrete are all requisites. However, all may not be required on the same project. Coarse-aggregate shotcrete is the norm for tunnels and caverns; the use of fine-aggregate shotcrete is limited to smoothing and other secondary applications.

Shotcrete references are scattered throughout the technical literature. An excellent, concentrated source is the series of conference proceedings of the Engineering Foundation (organized in 1916 by the United Engineering Trustees) on the general subject of shotcrete for underground construction. These conferences were held in 1973 (United States), 1976 (United States), 1978 (Austria, for NATM), 1982 (Colombia, for very difficult ground), 1990 (Sweden), 1993 (Canada), and 1995 (Austria). The overseas locations, all countries with widespread shotcrete usage, were selected especially to broaden the exposure of the U.S. contingents to practice elsewhere. Complete references are provided at the end of this chapter.

HISTORY

Principles of the cement gun (shotcrete nozzle) were developed in 1907 by Carl E. Akeley, naturalist, explorer, and sculptor, to build better trophy mounts. It was further developed and patented in 1910 by the Cement Gun Co. of Allentown, Pennsylvania, which also coined the word *gunite* for what is now formally designated as fine-aggregate shotcrete.

Introduction of sprayed mortar or gunite underground probably took place at the Brucetown Experimental Mine of the Pittsburgh Bureau of Mines in 1914. It failed to develop into an acceptable support material, however, due to a tendency to spall from minor rock pressures, need for applying in thin layers because of poor adhesion, and excessive shrinkage caused by the high cement content. Occasional successful applications were reported, such as its use in conjunction with rock bolts in a tunnel in the United States in 1952, and in a second tunnel reported by Keifer (1966).

European development largely paralleled that of this continent, although the European literature suggests more rigid quality control. Following World War II, emphasis was placed on underground development as an economic necessity in the multitude of hydroelectric and related civil engineering projects in the Alpine countries (Austria, Switzerland, and northern Italy) and Sweden. In 1952, sprayed mortar was used successfully as the sole support and lining of pressure and nonpressure tunnels in the Swiss Maggia hydroelectric development.

The next few years saw the development of sprayed concrete, or "shotcrete." Equipment capable of placing 1 in. aggregate was developed, facilitating the mixing of aggregates and cement without preprocessing and bringing the cement content into manageable proportions. A set-accelerating and hardening admixture also was developed, making it possible to place shotcrete in thicker layers, on wet surfaces, and against significant water flows.

In Austria, machines were developed for placing coarse-aggregate shotcrete, and shotcrete's function in limiting loosening in both chemically and structurally unstable rocks was demonstrated at the Prutz-Imst project (1953–54) and at Schwarzach (Los Birql, 1955–58) (Rotter, 1960). Its effectiveness was shown further in unconsolidated heterogeneous slide material and in soft, wet ground at Serra Ripoli and Monastero (Italy) (Zanon, ca. 1962). In 1960–62, shotcrete in one of twin highway tunnels at Planicia, Venezuela, halted or prevented loosening and kept the excavation stable for 12 months, while the conventionally supported twin suffered local failures from progressive loading (Rabcewicz, 1964). At the Kaunertal pumped storage project (Austria) in 1962–63, effectiveness of shotcrete in conjunction with grouted rock bolts was demonstrated in very heavy ground, wet plastic mylonitized sericite schists, where conventional steel supports combined with steel forepoling had failed (Rabcewicz, 1964).

In 1967, a section of tunnel was driven through unconsolidated gravels for the Milano Subway (Italy), incurring less surface settlement than had occurred in an adjacent shield-driven tunnel.

Meanwhile, parallel development of shotcrete practice was occurring in Scandinavia. The first major projects reported in Sweden were at the Holjes (1958–60) and the Lossens (1959–69) hydroelectric projects, and in Norway, at the Tokke (1963) hydroelectric project (Karlsson and Fryk, 1963). Development in Sweden emphasized the use of shotcrete without reinforcement, wire mesh, or other conventional tunnel support elements. By 1965–66, Japan also appears to have joined in the development, although little has been reported in English of their early experiences.

North America lagged, probably due to ample supplies of alternative economical support materials. Early experience with gunite underground also had left a general suspicion among engineers and contractors alike as to the integrity of the method, and some of this suspicion remains. Reported early examples of shotcrete usage in North America include the Canadian National Railways Tunnel (Mason, 1968), the Tehachapi No. 1 Tunnel (Cecil, ca. 1970), the Balboa Tunnel (Blanck, 1969), and the Lucky Friday Shaft (Miner and Hendricks, 1969).

In summary, the effectiveness of shotcrete for prevention of rock loosening has been demonstrated in a variety of geologic conditions.

Early shotcrete projects used the dry-mix process. Wet-mix shotcrete use began in the mid-1960s. Research into the use of steel fibers in shotcrete began in the late 1970s. Production use began in the early 1980s and increased slowly as various problems were overcome and acceptance increased. The impetus for this development came from the contractors and was due to problems with installation of welded wire fabric. Discovery of the benefits of adding silica fume to the shotcrete mix occurred about 1983. Within about three years, production use began and has increased steadily.

Case Histories

Case histories are almost invariably enlightening. Those which follow, selected from among the multitude available, represent milestones or demonstrate the versatility of shotcrete.

Vancouver Tunnel. The single-track Canadian National Railways tunnel is located beneath an industrial and residential area of Vancouver, British Columbia. Constructed in 1967–68, it was the first major tunnel in North America to use coarse-aggregate shotcrete for initial support and final lining. The "Boston-type section" (inclined leg horseshoe) was 29 ft high with a span of 20 ft. Total length was 10,760 ft, 9,340 ft of it in rock. The rocks were conglomerates (pebbles up to 4 in. in sand and clay), sandstones (coarse to fine grain, usually soft and saturated), and two types of shale (one massive, brittle, and fine-grained, the other coarse-grained and bedded). Work was carried on three shifts per day, five days per week, but no blasting was allowed between midnight and 7 A.M.

The rock heading began with steel sets (8WF28) in the soft weathered shales dipping gently toward the face and including a 12-in. coal seam overlying 18 in. of gouge and overlain by 3 ft of weathered, plastic shale. Forepoling was soon necessary. Voids above spiling were shotcreted; the face was also shotcreted. Short-wall plate drifts were driven and shotcreted; an arch ring cut excavated and shotcreted; and the steel arch erected. Finally, the bulk of the excavation was completed and shotcreted. The need for sectional excavation decreased with advance, and use of steel sets was abandoned after 25 ft.

A 200-ft test section followed. One water flow in the conglomerate resulted in a hail of pebbles followed by disruptive dripping. After 30 min of spraying, the shotcrete finally began to adhere to the ground. Following the experimental section, the work settled into a typical routine of drilling a 110-hole, 10-ft round; blasting; beginning application of a 2-in. layer of shotcrete on the arch within 45 min of the blast and from a flying deck extended over the muck pile, continuing shotcreting of that layer and bringing the prior 2-in. layer to full 6-in. thickness during the mucking cycle; and then spraying the walls to full 4-in. thickness during the drilling cycle.

Drakensberg Powerhouse Caverns. The Drakensberg Pumped Storage Scheme, located on the Drakensberg escarpment northeast of Lesotho, South Africa, includes an underground powerhouse mined in relatively poor ground (Sharp and Lawrence, 1982). The principle cavern, for the Machine Hall, has a span in excess of 55 ft, excavated height of 100 ft (excluding pits), and a length of 650 ft. The Transformer Hall and Valve Hall also are sizable. The geology was not well suited to such sizable excavations. Nevertheless, the caverns were stabilized by pretensioned rock bolts and permanently lined with shotcrete having a minimum thickness of 4 in. Lining placement was completed in 1979 (before reliable steel-fiber-reinforced shotcrete was available).

The geology consists of an interbedded sequence of mudstones, siltstones, and sandstones with dolerite dike and sill intrusions and occasional thin carbonaceous seams. Three near-vertical joint sets were also present. The dominant feature was the essentially horizontal bedding planes, generally spaced 0.2 to 20 in. apart. Slickensided compaction features, dipping at 40° to 60°, were found in the more argillaceous rocks, and rock deterioration by disaggregation when dried occurred in the finer ones. Preconstruction in situ stress determinations indicated vertical stress near the theoretical overburden value and horizontal stress normal to the high, vertical walls of 2.5 times the vertical at a depth of 500 ft. The geology thus dictated flat roofs with or without inclined haunches and early shotcrete application to avoid deterioration. It was recognized that the silty, muddy, and micaceous partings would result in poor adhesion of the shotcrete, and that the high horizontal stresses would create bedding plane slips, especially at or near the roof and invert levels.

The chamber excavations began with a top center heading having a flat roof span of 26 ft. This was followed by side slashes for and below the 45° haunches in the Machine and Transformer Halls. The remainder of the Machine Hall was excavated in a series of nine benches with a maximum lift of 12 ft. Fewer benches were needed for the other halls. One-inch-diameter bars were used for rock reinforcement; primary bolts were 20 ft long, secondary, 10 ft. The roof bolting pattern was generally 4 ft by 4 ft with both lengths used in the same pattern. Wall bolting pattern typically was 5 ft by 5 ft, again with both bolt lengths and with the bolts angled well down or well up (say, 30°) in alternating vertical lines.

Rock bolting generally preceded shotcreting. However, the initial shotcrete layer had to be kept within 25 ft of the face and applied within 48 hours of rock exposure. The initial 2-in. minimum layer probably averaged about 3.5 in. It was reinforced and contained 2.75% accelerator (by weight of cement). The layer would not adhere with a lesser percentage, and any significantly greater percentage degraded ultimate strength. Specified 28-day strength was 4,300 psi.

Face plates for the rock bolts had a special "spider" of four #3 mild steel bars 4 ft long attached by welding. Outstanding legs were downstanding to avoid shadowing in the initial shotcreting. Open welded wire fabric (8 in. × 8 in. @ 5.5 lb/yd^2) was installed between initial and final layers, and the spider legs bent to parallel the layer.

The final layer (2 in. minimum) proved more troublesome than the initial one because no accelerator was permitted. As a result, sublayers had to be limited to 1-in. thickness on the overhead because a greater initial thickness would not set fast enough to adhere. At times, up to four layers were required, all applied in the same overall operation.

Significant differential movement was expected horizontally between walls and roof and estimated at 4 to 8 in. It was expected to occur just below the flat roof soffit in the weakest bedding plane. Initial shotcreting was continuous across the roof–haunch junction. Stress buildup in the initial roof layer was sufficient to crush the shotcrete locally at the haunch. Damaged zones were removed and replaced where necessary before placing the final layer. The vault final lining was delayed until six of the nine bench stages were excavated to permit most of the differential movement to occur. A continuous slot was left in this layer, extending about 12 in. on each side of the junction. Infilling of the slot was delayed until wall lining was complete and all movement had ceased.

Du Toitskloof Highway Tunnel. The notable aspects of this tunnel are the use of freezing for initial stabilization for part of its length and the application of shotcrete against the freeze wall for the permanent lining. The tunnel is 2.5 miles long and located some 40 miles east of Capetown, South Africa.

All but the western 500 ft of the tunnel was driven uneventfully through good hard rock, nearly all of it granite; maximum overburden was 2,400 ft over the 42-ft (excavated) span tunnel. The western reach, which is the area of concern here, had a full face of saturated, completely decomposed granite for more than 400 ft, with the remainder a mixed face of decomposed, weathered, and fresh granite. The overburden was 25 ft at the portal, which increased relatively uniformly to more than 160 ft at the end of the difficult section. The natural water table rose from crown level about 40 ft from the portal to about 150 ft at section's end (Lawrence, 1982).

A 110-ft^2 exploratory tunnel was driven full length about 120 ft off the future tunnel centerline in 1979 to determine ground conditions in detail. Problems in driving through the decomposed granite below water table in the portal area were numerous, including sinkholes to the surface and inability to hold the ground with ribs and lagging. It was necessary to use localized ground freezing to stabilize this small tunnel.

A theoretical exercise carried out during the tender stage by the successful tenderer indicated that a 2-in. layer of shotcrete sprayed at 100°F onto a surface maintained at −4°F would be at 32°F in less than 5 hours. Similarly, a 4-in. layer would take 10 hours, and 8 in. 40 hours. A field test with a full-scale short tunnel section mock-up backpacked with disturbed decomposed granite maintained at −4°F with gaseous nitrogen in freeze lances also proved discouraging because the criterion of 4,000 psi compressive strength could not be obtained with either dry or wet mixes meeting specifications.

Fortunately, the anticipated temperature and strength problems were not realized in actual construction, and the required shotcrete strength was obtained routinely. A number of reasons probably account for this. The disturbed backpacking obviously held far more water than the in situ material, resulting in more total cold; brine rather than nitrogen was used as the freezing agent in the actual tunnel, and the ratio of freezing capacity to heating capacity of the shotcrete was quite different for the two. A 10% higher cement content (760 lb/yd^3) was used in the tunnel than in the test section, and the excavated surface began to warm up as soon as excavated. Finally, the insulating value of shotcrete is high so that the sacrificial 2-in. first layer allowed remaining layers to cure at nearly ambient tunnel temperature.

Enlargements for drilling niches 25 ft long were located at 100 ft. There the ice ring had a minimum width of 6.5 ft. The required lining thickness was 26 in. for 25 ft at the portal, 18 in. elsewhere. The portal section included steel

H-section ribs at 3-ft, 3-in. centers; the typical section contained smaller Alpine ribs (bell-shaped sections) at the same spacing. The typical section shotcrete was applied in four layers, the initial 2 in. containing a light mesh (welded wire fabric, WWF), a 4-in. layer with a heavy prebent mesh, an 8-in. layer containing the Alpine rib, and a final 4-in. layer, again with a heavy, prebent mesh. Considerable trouble was experienced in fixing the mesh.

English Channel Tunnel

Excavation of the large crossover chambers is described in Chapter 7 under "Excavation Methods." Shotcrete was used to stabilize the roadheader-excavated drifts and enlargements to final size.

Hanging Lake Tunnels

Also described in the same part of Chapter 7, rock bolts and shotcrete were used for initial support in these drill-and-blast tunnels. Other recent tunnels using shotcrete and rock bolts structurally for stabilization include the Trans-Koolau Highway Tunnels on Interstate H-3 in Hawaii and the Cumberland Gap Highway Tunnels for the National Park Service in Tennessee. All three used steel fibers and microsilica in the shotcrete and were designed by the same firm.

QUALITY ASSURANCE

A number of ASTM standards, when cited, will go far toward ensuring quality of materials, providing standardized testing methods, and determining strength characteristics of the finished product. They are listed below for convenience in later abbreviated referencing and will be discussed when and as appropriate. Several American Concrete Institute (ACI) publications are useful and are listed as well.

American Society for Testing and Materials Standards

The following standards help ensure quality of raw materials (six well-known standards for concrete and steel have been omitted):

- A820 Steel Fibers for Fiber Reinforced Concrete
- C311 Sampling and Testing Fly Ash or Natural Pozzolans for Use as a Mineral Admixture in Portland-Cement Concrete
- C1240 Silica Fume for Use in Cement Concrete and Mortar

The following test standards help ensure quality shotcrete (three well-known standards are omitted):

- C78 Flexural Strength of Shotcrete
- C642 Specific Gravity, Absorption, and Voids in Hardened Concrete
- C1018 Flexural Toughness and First-Crack Strength of Fiber-Reinforced Concrete (Using Beam with Third Point Loading)
- C1102 Time of Setting of Portland-Cement Pastes Containing Accelerating Admixtures for Shotcrete
- C1116 Standard Specification for Fiber-Reinforced Concrete and Shotcrete
- C1117 Time of Setting of Shotcrete Mixtures by Penetration Resistance
- C1140 Preparing and Testing Specimens from Shotcrete Test Panels

American Concrete Institute Publications

- 506R-90 Guide to Shotcrete
- 506.2 Specifications for Materials, Proportioning, and Application of Shotcrete
- 506.1R State-of-the-Art Report on Fiber-Reinforced Shotcrete
- 506.3R Guide to Certification of Shotcrete Nozzlemen
- 544.2R Measurement of Properties of Fiber-Reinforced Concrete

MATERIALS

The first three materials below are essentially the same as for concrete; the following three give shotcrete its necessary special properties.

Cement

- Type 1, 2, or 5. Typical amount about 6.5 to 8 sacks (94 lb each) per cubic yard.
- Use only one type on project to avoid delivery mix-ups.
- The finer grinds of Type 1 are preferable.
- Type 2 normally unnecessary in shotcrete for lower heat of hydration but acceptable when moderate sulfate resistance needed.
- Type 3 (for high early strength) not recommended in general. Accelerated initial and final sets frequently required on overhead and wet areas in excess of Type 3 capability alone. Many accelerators incompatible with Type 3.
- Type 4, developed for low heat of hydration (by slowing set) in mass concrete, not suitable for shotcrete.
- Type 5 acceptable when high sulfate resistance necessary.

Aggregates

Aggregates should be well graded and durable. ACI 506-2 Gradation No. 2 is the most popular for normal underground use. Gradation No. 1 is acceptable for the smoothing layer; Gradation No. 3 is also acceptable, particularly when thicker layers are necessary, or when it is used more like regular concrete.

The ACI 506-2 gradations are shown in Table 12-1 and plotted in Figure 12-1.

A simple descriptor for the fine aggregate is "concrete sand" with modified gradation. Note in Table 12-1 that 70 to 85% of total aggregate is normally fine aggregate. A sharp coarse aggregate is preferable, but rounded aggregate is acceptable. Reactive aggregate is unacceptable.

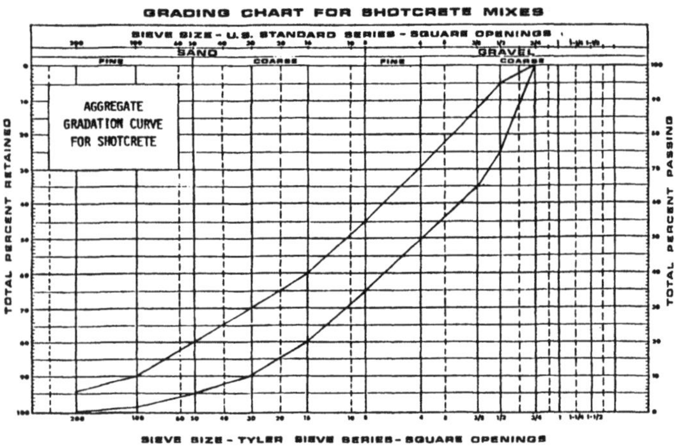

Fig. 12-1. Grading chart for shotcrete mixes.

Water

"Clean and potable; free from deleterious substances" is frequently all the specification required. Mixes with a water/cement (W/C) ratio of 0.35 or less and 0.50 or more will not produce satisfactory shotcrete. Typically, a W/C ratio between 0.40 and 0.45 is satisfactory.

Accelerators

The pneumatic delivery and the need to obtain adhesion to wet surfaces and to remain in place thereafter on vertical and overhead surfaces require the use of shotcrete accelerators to prevent slough and sag. The thicker the layer, the shorter must be the time to initial set. Accelerators come in solid, powder, and liquid forms. Use of the liquid form is usually most practical; quantities should be metered. For mixes without microsilica, 2% (by weight of cement) frequently is specified as a minimum for satisfactory performance; up to 5% may be needed on the arch; maximum should not exceed 8% even under very wet ground conditions. Additional layers will require less accelerator because of the better surface to be shot and lessened surface moisture. One frequent problem with high percentages of accelerators is a degrading of strength over the long term. When microsilica is an integral part of the mix, the aforementioned percentages will be reduced substantially; the accelerator should not be needed on near-vertical surfaces after the initial layer and may not be needed then, depending on the substrate. More than 2% is unlikely to be needed on the arch. Another accelerator problem has been causticity, which creates burns. This was particularly true in the early days, when powdered accelerator was measured by the heaping wooden paddleful! The unpleasant memory has lingered on in the specifications. A noncaustic liquid silica-based accelerator has been developed, and its use is increasing rapidly. Other noncaustic accelerators also are being developed.

Achieving a rapid strength gain for at least the first 12 hours is frequently of prime importance in satisfactorily stabilizing the ground. The frequently slower strength gain, compared with unaccelerated mixes, after about 8 hours and up to and even after 28 days is undesirable but acceptable if no better-performing accelerator is available. Given a higher than specified 28-day strength in place, a very minor long-term strength loss can be tolerated, provided there is no long-term loss to below the specified value.

Table 12-1. ACI 506-2 Gradations

Sieve size U.S. standard square mesh	Percentage by weight passing individual sieves		
	Gradation No. 1	Gradation No. 2	Gradation No. 3
3/4 in. (19 mm)	—	—	100
1/2 in. (12 mm)	—	100	80–95
3/8 in. (10 mm)	100	90–100	70–90
No. 4 (4.75 mm)	95–100	70–85	50–70
No. 8 (2.4 mm)	80–100	50–70	35–55
No. 16 (1.2 mm)	50–85	35–55	20–40
No. 30 (600 m)	25–60	20–35	10–30
No. 50 (300 m)	10–30	8–20	5–17
No. 100 (150 m)	2–10	2–10	2–10

Steel Fibers

Conventional reinforcing bars are not used with shotcrete except where ground conditions require the addition of lattice girders. Instead, the need for ductility, toughness, and a residual strength is generally met by incorporating short, thin pieces of wire or sheet steel into the mix. When the requirement is primarily for decreased permeability by limiting the number and opening of shrinkage cracks, polypropylene fibers may be used, when able to meet toughness and residual strength requirements.

Steel fibers are of four general types: cold drawn steel wire, slit sheet steel, melt-extracted, and other (ASTM-C820). Only the first two are considered satisfactory for underground shotcrete.

Two factors initially delayed the satisfactory use of steel fibers. "Bird's nesting" or "balling" of the fibers was frequent, and the fiber tended to pull out of the shotcrete without achieving the desired effect when cracking occurred. The first problem is solved by holding the aspect ratio (length divided by nominal effective diameter) to within well-defined limits of about 40 to 80; currently, fibers are from 3/4 to 1-1/2 in. long. Satisfactory pullout resistance is provided by deforming the fiber, either by upsets or bends at the ends of wire types or by corrugating or otherwise shaping the strips lengthwise.

Specifying steel fiber by the number of pounds per cubic yard is inappropriate because of major differences in engineering properties among the types. A performance specification stipulating ductility (toughness) and residual strength requirements should be used.

Silica Fume

Also called *microsilica,* this material replaces a small percentage of the cement content, not as much for its best-known attribute of improving concrete strength, as for its welcome side effects of increased adhesion, improved impermeability, reduction of required accelerator, and, for dry-mix shotcrete, considerably reduced rebound and dust when gunning. Both the latter are due in large part to the amount of material passing the No. 200 sieve. The replacement percentage should be kept within the 8 to 13% range, and preferably within the 9 to 11% range. Particular caution is necessary on the upper bound because of the tendency for silica to increase shrinkage and, therefore, cracking. Principal requirements for microsilica (ASTM C1240) are

1. SiO_2 content, 80% minimum (rather than 85% as in current specifications)
2. Amount retained on No. 325 sieve, 10% minimum
3. Maximum ignition loss of 2%
4. Maximum moisture content of 3%
5. Minimum 85% control, accelerated pozzolanic activity index with Portland cement at 7 days
6. Boiling absorption 6% maximum on shotcrete sample

Fly ash is used as cement replacement in concrete to good effect. It can also be used in aboveground shotcrete when fast setting is not a requirement. Underground, however, its variability and possible adverse effect on set warrant not recommending it.

The "stickiness" imparted to shotcrete by microsilica is of particular value for all overhead layers and for at least the initial layer on the walls when the host rock has inherent wetness of localized flows.

Wire Mesh

Welded wire fabric (WWF) is included here because of its former frequency of use. It was introduced into shotcrete usage to provide ductility; however, steel fiber now provides this characteristic more effectively. The reason for no longer recommending WWF is not theoretical; it is highly practical. Even properly spaced fabric (4 in. × 4 in. or 6 in. × 6 in.) is quite stiff, making installation time consuming, difficult, and therefore costly. When used in drill-and-blast tunnels, considerable excess shotcrete may be required to fill overbreak to which WWF cannot be properly formed. Furthermore, when spaced out from the existing surface, as required for maximum effectiveness, the shotcrete quality may deteriorate because of shadow effect due to both the solid wire and to the fabric's vibrating in the shot stream.

Woven wire mesh ("chain link") is frequently used in conjunction with rock bolts for safety when rock is reinforced. Its purpose is to catch ravel and small falls. However, shotcrete is more effective, provided it is applied early, because it prevents raveling as soon as it sets and prevents block falls as it gains strength by creating a limited arch that acts compositely with the rock. The smaller openings and larger wires exacerbate the shadow problem. Thus, mesh is also inappropriate with tunnel shotcrete.

Other Additives

When freeze–thaw cycling is anticipated, air entrainment should be added to wet mixes; it is ineffective with dry mix. Considerable air is lost during gunning; 10 to 12% at the pump may be necessary to produce 4 to 6% "on the wall." Water reducers and superplasticizers are needed in wet mix. Where construction environmental constraints (dust) are very severe, special dust suppressant additives can be added to dry mix, especially if microsilica is not used; however, special exhaust ducts at the gunning area should be sufficient. In early years, dry-mix shotcrete produced considerable rebound and high visual dust levels. However, this was due to several factors other than the mixing process. Remotely controlled nozzles were not available then, and the resulting improper nozzle position greatly increased both dust and rebound.

ENGINEERING PROPERTIES

Compressive strengths greater than 5,000 psi can be produced, but these are rarely needed. Thicker single or multiple layers are preferable because of difficulty in obtaining suitable mixes for the higher strengths. Unless microsilica is an integral part of the design mix, it is still preferable to increase the design thickness to obtain the required lining capacity. When it is obvious shotcrete will not be highly stressed, 4,000 psi will suffice. Except for special situations, only one strength of shotcrete should be used on a project.

Adhesion and shear strength are of greater importance than compressive strength. As with normal concrete, however, compressive strength is a good indicator of other properties; hence its general use as the principal test.

Necessary adhesion is most easily obtained by proper cleaning of the surfaces to be shot and by the addition of silica fume into the mix (see the section on surface preparation in this chapter). A well-designed mix should produce adhesion strengths on the order of 180 psi. The actual value required is quite indefinite because the surfaces shot are rarely smooth enough that adhesion is acting alone. The force of the shotcrete stream will result in all surface gaps and minor irregularities (not overbreak) being filled so that as soon as set occurs, other properties are brought into play. Even though it does not act alone for any significant time, good adhesion is an essential property of good shotcrete.

Normally accelerated mixes develop compressive strength of 40 psi immediately upon setting and 150 psi within 30 min. (These values are for information only.) Whenever early strength is necessary for initial tunnel stabilization, a requirement for 700 psi in 8 hours should be a specification item. Similarly, a 3-day strength, which will vary depending on required 28-day strength, should be specified. The conventional stipulation of 7-day concrete strength, however, is not normally significant during production. Hence, the cost of production testing of many cylinders of that age is not warranted.

Bond strength is difficult to state because of the substantial difference between wet and dry mixes. The bond of concern is not with rebar, as in concrete, but the bond between successive shotcrete layers to ensure that all layers act integrally strength-wise. Data is meager, but according to ACI 506R, when measured in shear it will vary from 8 to 12% of the compressive strength for dry mix, but only about half as much for a wet mix. When the new layer is placed within 24 hours of the old one, the bond normally is not a problem.

The flexural strength of plain shotcrete between 10 and 28 days is about 15 to 20% of the compressive strength. The fact that "green" (i.e., "young," "fresh," or partially cured) shotcrete is more ductile at early ages and will creep, thereby relieving stress, should be considered qualitatively.

Ductility (the ability to incur large deformation without rupture) is obtained by the use of fiber reinforcement. Accordingly, flexural strength and residual strength should be specification items. Currently, testing procedures are established but difficult to execute, and consensus on proper values is still forming (see discussion under "Testing").

Greater impermeability is obtained by avoiding excessively cement-rich mixes (to minimize shrinkage cracking), by using fiber (to minimize and distribute opening of shrinkage cracks), by using more finely ground cements, by adding silica fume, and by careful control of nozzle distance and attitude (so that the in-place shotcrete has maximum density).

Unlike structural steel or concrete, the final product "on the wall" will not be in exact conformance with specifications or with proportions developed in the laboratory. The in situ cement content will be higher; the amount of aggregate, especially the coarse fraction, will be reduced. Both are due to aggregate rebound. The total water is likely to be higher in wet-mix shotcrete (due to "retempering" for pumpability) and lower (or possibly higher) for the initial pass of dry-mix shotcrete because of adjustments required by the amount of water in/on the substrate. Testing during the field trials stage can give some indication of the changes.

Last of all, but far from least, the nozzleman is the final control on shotcrete quality. Most of the time he works under very difficult conditions, particularly in smaller tunnels; his gunning technique will determine density, adhesion, and quality. Nevertheless, the final in situ product generally will be within acceptable limits.

WET OR DRY?

Early on, all shotcrete was dry mix. Nozzles did not provide adequate mixing of solids and water, and remote control operation was nonexistent. As a result, both rebound and atmospheric dust were excessive. Correction was slow but is now effective. Nevertheless, some unwarranted criticism of the dry-mix procedure lingers. Similarly, wet-mix equipment in the early 1970s failed to produce shotcrete satisfactory for underground use, primarily for lack of proper proportioning. However, satisfactory shotcrete has now been produced by each method for many years. Advocacy of one over the other still produces lively discussions—and sometimes improper autocratic decisions. Central to the discussions are atmospheric dust, rebound, and production rates.

When worldwide usage is considered, a superficial overview gives the impression of favoritism by entire nations or regions. Closer examination discloses such preference is actually based quite properly on geology. At the 1991 conference of the shotcrete committee of the International Tunnelling Association (ITA), the proceedings covering both papers and discussions showed, for example, that Norway was then using wet mix almost exclusively and Sweden was using three times as much wet mix as dry. Both were typically using total thicknesses of 4 to 8 in. (10 to 20 cm). On the other hand, Germany was using dry mix almost exclusively, and Austria was heavily oriented toward dry mix. Both were also using lining thicknesses of 8 to 16 in. (20 to 40 cm), essentially the same dimensions as for concrete linings

Norway and Sweden are noted for their predominance of excellent rock. Thus thin linings and constant quantity accelerators are a natural result, especially since rapid and substantial support was not a major problem. Germany has used shotcrete predominately in the softer rocks and soft grounds. Austria, with its many Alpine tunnels, has had to contend with the whole range of rock and soft-ground conditions. In difficult ground, partial face advance is often appropriate and leads to small quantities of shotcrete and more frequent application. The short culottes (benches) typical of NATM require considerable temporary shotcrete as well as its removal. The shorter the time interval, the lower the shotcrete's attained strength and the easier its breakup and removal.

There are definite differences between, and advantages of, the wet-mix and dry-mix concretes, even though the end products can be nearly identical (see Table 12-2). The irrefutable conclusion overall is that both mix processes have their place underground. One type or the other, therefore, should not be arbitrarily excluded by the contract documents.

Wet-Mix Process

The wet-mix process consists of mixing measured quantities of aggregate, cement, and water, and introducing the resulting mix into a vessel for discharge pneumatically or mechanically through a hose to final delivery from a nozzle. It has the advantage of rigidly controlling the water/cement (W/C) ratio of the product. Existing equipment can handle maximum aggregate size of 3/4 in. Further, successful meth-

Table 12-2. Comparison of Dry- and Wet-Mix Processes

	Wet-mix process		Dry-mix process
1.	Mixing water is controlled at the delivery equipment and can be accurately measured.	1.	Instantaneous control over mixing water and consistency of mix at the nozzle to meet variable field conditions.
2.	Better assurance that the mixing water is thoroughly mixed with other ingredients.	2.	Better suited for placing mixes containing lightweight aggregates, refractory materials and shotcrete requiring early strength properties.
3.	Can use bulk ready mix.	3.	Less equipment investment.
4.	More accurate proportioning.	4.	Higher impact velocity; better adhesion.
5.	Normally has lower rebound resulting in less material waste.	5.	Start and stop placement characteristics are better with minimal waste and greater placement flexibility.
6.	Less dusting and cement loss accompanies the gunning operation.	6.	Capable of being transported longer distances.
7.	Can use air entrainment.	7.	Easier to use overhead.
8.	Capable of greater production.	8.	Capable of higher strengths.

ods have been devised to introduce quick-acting accelerators to the delivery hose. Pumping low-slump concrete is commonly a problem, and so a slightly higher than desirable water content is used. By use of accelerators, such concrete can be made to adhere overhead, but ultimate strength usually suffers. However, the method has been found convenient for use with less-skilled operators, particularly in the limited-size access workings of mines, most of which are generally dry.

Dry-Mix Process

Dry-mix shotcrete consists of a mixture of damp aggregate and cement fed into a placing machine, fed at a uniform rate into an airstream to travel through a hose to the nozzle. The water of hydration is added at the nozzle before discharge to the surface. Water is manually controlled, permitting adjusting to changing surface wetness. Powdered accelerators are added to the dry mix as it is fed into the placer. If liquid, the accelerator is mixed with the feed water before it goes to the nozzle.

PREPARATION, MIX, SHOOT, AND CURE

Surface Preparation

Good adhesion to the ground with the initial layer and good bond between successive layers are prerequisites of good shotcrete. The surface to be shot must be clean and moist, but not wet, immediately before shooting. This can best be accomplished with a high-pressure combination air–water jet applied with a long jet-pipe nozzle held relatively close to the surface. Merely washing the surface with water applied through the shotcrete nozzle is insufficient and unacceptable.

Similar cleaning is necessary when a considerable time (24 hours) has elapsed between applications or when work in the heading has resulted in deposits (including dust) on the shot surface. For shorter intervals, a definitely clean surface, and no surface pockets of inferior material (similar to laitance), a lower-pressure water wash/dampener can be used in the interest of saving time.

An exception to the above sometimes occurs in an argillaceous tunneling medium. In this case, an air-only jet may be necessary in order to avoid leaving a thin film of adhesion-resistant material. Air-only jets are also appropriate if the host material degrades rapidly upon exposure to water.

Dry-Mix Aggregate Preparation

When dry-mix aggregate storage is on-site, protection from the elements (rain, snow, ice) is necessary. Stockpiling by size groups should prevent subsize segregation. For best results, an aggregate dampness of 3 to 6% (essentially saturated) should be maintained. Less will absorb too much mix water; more will result in too high a W/C ratio. When prepackaged unit bags of dry materials are used, it is necessary to pass the materials through a premoisturizing unit immediately before delivery into the "pot" to achieve the 3 to 6% moisture condition.

The Mix

It is assumed here that both early strength for initial stabilization and maximum long-term impermeability are required and that steel fibers and microsilica are necessary. Elements of the mix were discussed earlier, under "Materials."

The best guide to initial proportioning is local experience; this is then supplemented by necessary routine and special testing. When such experience is not available, the initial trial dry mix might approximate the following:

Cement	670 pounds per cubic yard (pcy)
Silica Fume	89 pcy
Fine Aggregate	1,950 pcy
Water	270 pcy

Dry-mix water includes both aggregate moisture and mix water. Wet-mix proportioning should contain additional cement, less microsilica, less fine aggregate, and more water, with all changes initially on the order of 5 or 10%. In addition, a water reducer, superplasticizer, or air-entraining agent should be added as appropriate.

Product proportioning can be on either a weight or volume basis. If the latter, there should be a weight comparison check either daily or on a stipulated cumulative quantity analysis.

Inclusion of steel fibers will require little or no change from the plain shotcrete mix. The principal effect will be a somewhat harsher mix with less slump. The presence of microsilica, however, will increase workability.

Overall, the mix design should keep the W/C ratio and the cement factor as low as possible and the coarse aggregate fraction as large as practicable. It is always tempting to add cement for greater workability and to reduce maximum aggregate size, but this is counterproductive to minimum shrinkage.

Dry-Mix Nozzle

The nozzle for dry-mix shortcrete is of concern because the water is added here, and there is extremely limited time for interacting. A standard nozzle has a single water ring set back from the tip. Early on there was trouble obtaining complete hydration, which created very high dust and rebounds. Some nozzlemen solved this by adding a short length (say, 2 ft) of delivery hose to the nozzle tip, thereby confining the stream for longer mixing. Now, the water rings are set further back, the chamber interiors improved for better mixing, etc. Some nozzles have two separate water rings.

Nozzle Position

The best shotcrete "on the wall" is produced when the nozzle is kept within 3 to 5 ft of the surface being shot and perpendicular to the surface or within 15° of same. Deviation from this will result in less compaction (density) and more rebound (costly). Considering that particles in the shotcrete stream are traveling at 250 to 500 fps (170 to 340 mph), it is easy to understand why nozzlemen, even when wearing protective clothing and equipment, are rarely found manually holding the nozzle in proper position. The advantage of "robots" (more properly, remotely controlled nozzles) is obvious, and specifications requiring their use wherever possible are warranted.

Remotely Controlled Nozzles

Requirements for remote control are the mounting of the nozzle on the end of a long boom at the front of a weighted vehicle, the ability to move the boom end in three dimensions and to rotate the nozzle spherically, and provision of a set of controls to smoothly control the nozzle movement. In addition to the nozzle being properly positioned to provide maximum-density shotcrete at all times, it is possible to shoot over the muck pile, thereby more quickly stabilizing the tunnel.

In practice, the equipment ranges from simple job-built to complex complete shotcrete factories on wheels.

Other Shotcreting Factors

Good lighting during shotcreting is essential because appearance tells much about quality of the in situ product. In addition to earlier detection of incipient sloughing or sagging, appropriateness of water in the mix, onset of initial set, uniformity of surface texture, etc., can be better determined. Easily movable floodlighting, therefore, is necessary.

Depending on stringency of working environment constraints, auxiliary exhaust ducts from the main ventilation duct may be necessary.

Rebound material must be collected and wasted. This is particularly important as the wall shotcrete is being brought to invert level. Dry-mix rebound quantity is greater than for wet mix, probably twice as much as the 5 to 15% minimum possible with the latter.

Regular shotcrete equipment maintenance is essential if costly downtime is to be avoided. The gun (pot to nozzle), particularly the "innards," must be thoroughly cleaned after each shotcreting session to avoid plugging, etc. Regular replacement of parts subject to heavy wear is also necessary. Equipment downtime slows progress and increases costs overall, not just for the shotcrete crew. Maintenance is performed during noncritical periods while regular work is progressing.

Curing

In tunnels where relative humidity in the newly shotcreted areas is less than about 85% (higher humidity is more typical), shotcrete surfaces should be kept moist by spraying for up to 7 days. This is difficult in many tunnels because of space limitations. The actual time required varies and depends on how rapidly the shotcrete gains enough strength to function as required.

TESTING

Tests for Mix Acceptance

Testing is required of the raw materials, the operator (nozzleman), field applicability, and of the shotcrete product at various stages of mix development and age for all contract-specified shotcrete.

All shotcrete placed for the contractor's convenience is his sole responsibility and testing should not be required by the designer. However, "convenience excavations" that are not eventually backfilled and whose collapse might endanger the service operations should be permanently stabilized per specifications. The reduced sensitivity (external standard ratio in the system) should be appropriate.

Testing of the raw materials—cement, aggregate, steel fiber, and microsilica—is well established in methods and results required. Little more need be said other than the obvious, that changes in material source requires prior testing of the new material.

Testing of each nozzleman's ability to produce satisfactory shotcrete in place is an obvious necessity, as this could be the weak link in the materials–mix–application–product chain. ACI 506.3R provides for certification of nozzlemen. Holding a current certificate is sufficient evidence of ability, subject to verification by sample cores from early production work with approved production equipment. The written portion of the certificate test is desirable but not necessarily a prerequisite. If not certified, demonstrated ability with production equipment, approved mixes by field trials in conformance with field test procedures, and meeting required results should be mandatory before acceptance for production work.

Shotcrete testing is a three-part process. The first stage, compatibility checking, is required before the proposed materials and their sources are approved. Cement–accelerator compatibility is of prime importance. ASTM C1102 should be followed. Similarly, compatibility of the entire mix and proportions must be established by meeting the various requirements with proposed mixes prepared, cured, and tested in the laboratory.

The second stage, field trials, begins upon completion of the first part. Material from the approved sources should be combined per approved mixes by the production equipment to be used and then shot by a certified nozzleman into appropriate size boxes mounted in both vertical and overhead positions. The proposed remote control nozzle equipment should be used. (Later, the procedure should be used to certify nozzlemen.) After curing in the manner proposed for the production work, cores and beams should be taken and tested.

The third stage, production testing, has three parts. First, the field trial process should be repeated at the heading during production shotcreting upon demand by the engineer. Second, cores should be taken from the in-place shotcrete, at specified intervals. The primary purpose of these is to check thickness and adhesion; however, compressive strength should also be tested. The third part is the overall checking of the in-place concrete. In addition to a visual check for defects, the shotcrete should be sounded at frequent intervals (locations) by striking with a geologist's or similar hammer. Sound, adhering concrete will give a distinct ringing sound. Laminated shotcrete or voids behind the shotcrete will result in a drummy or hollow sound. If drummy, the area should be rechecked thoroughly and the approximate boundaries determined. Cores should then be taken and examined. Defective shotcrete should be removed and replaced with sound shotcrete.

Special Tests

When a high degree of impermeability is required, the mix design effectiveness should be tested according to ASTM C642 using a maximum boiling absorption value of 6%.

Toughness is an indication of the energy absorption capacity of shotcrete in flexure. Flexural toughness of fiber-impregnated shotcrete is determined from a plot of the load–deflection curve data obtained by ASTM Standard Test Method C1018-89 for a test beam. The toughness is represented by the area under the load–deflection curve from its origin to the selected deflection criterion.

Toughness indicies alone do not indicate how much shotcrete capacity remains after a given deflection. Accordingly, the addition of reserve strength requirements to the specifications is necessary to indicate how much reserve capacity is required for the project in hand. The residual strength factor ($R_{x,y}$) is calculated as the difference in the two toughness indices (I_x, I_y) multiplied by the appropriate constant.

Theoretically, a beam material having a perfectly elastic behavior up to first crack and a perfectly plastic behavior thereafter will have an I_5 value of 5.0, 10.0 for I_{10}, etc. The corresponding energies are 3.0 times that at first crack for I_5, 5.5 times for I_{10}, etc. Early on, only the I_5 and I_{10} values were specified. Currently, I_{30} values are also used and consideration is being given to using I_{50}. The theoretical residual strength factor for $R_{10,30}$ is 100. ASTM C1116-89 provides a guide to toughness indices for relative performance levels (see Table 12-3). Note that level IV cannot be achieved with current fiber.

DESIGN CONSIDERATIONS

Discussion to this point has focused on what shotcrete is and how to produce it. Overall design philosophies and procedures were discussed in Chapters 5, 6, and 7. Shotcrete thicknesses, based on empirical methods and case histories, were considered in Chapter 7.

In rock tunnels, shotcrete and rock bolts normally are used together to provide the rock reinforcement. Only in rare instances (e.g., impermeability the prime consideration, protection of the ground against degradation, and competent rock where good adhesion can be assured) may shotcrete be used alone. Conversely, shotcrete is frequently inappropriate in TBM drives (noted in Chapter 7). Elsewhere, shotcrete and bolts supplement one another. (In European nomenclature, bolts may be termed *anchors*, a more descriptive term insofar as the shotcrete is concerned.)

When the dowels do indeed provide anchorage and the shotcrete is primarily planar, the shotcrete may be designed as cantilevering from an anchor support, as a plate supported by four corner anchors, etc. However, adhesion to the rock and the composite rock–shotcrete beam action must also be considered, or too great a shotcrete thickness will result. The need for bolts as shotcrete anchors diminishes rapidly as the "plate" becomes curvilinear; the composite beam becomes even more effective, and the bolts serve more of their true function, i.e., constraining the rock so that it resists more of the ground deformation or convergence resulting from excavating the opening.

Thin shotcrete arches have substantial carrying capacity. A prime reason is the ground constraint, which essentially eliminates flexural stresses. Significant irregular roughness in the excavated perimeter also increases capacity, similar to the increase provided by corrugations to otherwise smooth thin steel plate arches.

Flat roofs are a condition where shotcrete should not be used alone. At least minimal rock bolting is warranted as insurance against an undetected weak bedding plane. In addition, adhesion may be inadequate locally. An initial sag has no protection against progressive delaminating or secondary bending unless anchored by the bolts.

Shotcrete can be used with confidence in many soft ground conditions. Firm clay is one. Although loaded to near its confined capacity, the reconfinement produced by an early ring of shotcrete should regain essentially all the original capacity.

The New Austrian Tunneling Method (NATM) and its observational technique have had good success in difficult ground. The individual drifts can be reduced in size until a reasonable amount of shotcrete provides stable opening.

The openings can then be enlarged or combined by applying additional shotcrete immediately after the larger opening is formed. This results in quite thick shotcrete arches and walls, but more importantly, it results in a completed tunnel.

Squeezing and swelling grounds are not amenable to shotcrete linings, at least not until long rock bolts and time have stabilized the ground. Some projects have used slots cut in the shotcrete to allow controlled deformation to continue.

Table 12-3. Performance Levels Defined in ASTM C 1116-89

Performance level	Toughness index, I_5		Toughness index, I_{10}	
	Specified value	Test result	Specified value	Test result
I	2.7	3.0	5.4	6.0
II	3.6	4.0	7.2	8.0
III	4.5	5.0	9.0	10.0
IV	5.4	6.0	10.8	12.0

Finally, the improvement in shotcrete technology over the past 25 years is nothing short of phenomenal. With its capability and versatility so well demonstrated, there can be little doubt that improvements will continue to a point now scarcely imaginable.

REFERENCES

BLANCK, J.A. (1969) *The Balboa Outlet Tunnel* Monorock Publication.

CECIL, O.S. (ca. 1970) "Site Inspection at Tehachapi Mountains Crossing of the California Aqueduct, Tunnel No. 1, North Portal."

KARLSSON, L., and FRYK, J.O. (1963) "Lining for the Intake Water Tunnel of the Holjes Hydroelectric Plant in Sweden," *Hoch- und Tiefbau,* Vol. 50.

KEIFER, O. (1966) *Multiple Layer Shotcrete Lining.* ACI Publications, SP-14.

LAWRENCE, D.L.J. (1982) "Shotcrete Lining for a Tunnel Requiring Ground Freezing for Initial Support," *Shotcrete IV,* pp. 155–170.

MASON, E.E. (1968) "The Function of Shotcrete in Support and Lining in the Vancouver R.R. Tunnel," *Tunnel and Shaft Conference,* University of Minnesota.

MINER, G.M., and HENDRICKS R.S. (1969) "Shotcreting at Hecla Mining," *Rapid Excavation Symposium,* Sacramento, California.

MORGAN, D.R. (1991) *Steel Fiber Reinforced Shotcrete for Support of Underground Openings in Canada,* Concrete International.

QUADRIO, C.P. (ca. 1964) "New Methods for the Construction of Tunnels," Il *Monitire Tecnico,* Vol. 1.

RABCEWICZ, L.V. (1964) "The New Austrian Tunneling Method," *Water Power.*

ROTTER, E. (1960) "Spiritzbeton und seine praktische Andwending, im Untertagebau," Symposium at Lesben, Austria.

SHARP, J.C. and LAWRENCE, D.L.J. (1982) "Final Shotcrete Linings for an Underground Power Station in Argillaceous Rocks," *Shotcrete IV,* pp. 115–154.

Shotcrete for Ground Support (1976) *Proceedings of the Engineering Foundation Conference,* Easton, Maryland (October), ACI Publication SP-54. American Society of Civil Engineers and American Concrete Institute.

Shotcrete for Rock Support (1992) *Guidelines and Recommendations—A Compilation,* Draft Document (B. Malmberg), February, for the ITA WG Shotcrete Use.

Shotcrete for Underground Support III (1978) Papers presented at the 1978 Engineering Foundation Conference, St. Anton am Arlberg, Austria (April). Engineering Foundation, New York.

Shotcrete for Underground Support IV (1982) Selected papers presented at the 1982 Engineering Foundation Conference, Paipa, Boyaca, Colombia (September). Engineering Foundation, New York.

Shotcrete for Underground Support V (1990) Proceedings of the Engineering Foundation Conference, Uppsala, Sweden (June). American Society of Civil Engineers, New York.

Shotcrete for Underground Support VI (1993) Proceedings of the Engineering Foundation Conference, Niagara-on-the-Lake, Canada (May). American Society of Civil Engineers, New York.

Shotcrete for Underground Support VII (1995) Proceedings of the Engineering Foundation Conference, Seefeld, Austria (June).

Use of Shotcrete for Underground Structural Support (1973) Proceedings of the Engineering Foundation Conference, South Berwick, Maine (July), ACI Publication SP-45. American Society of Civil Engineers and American Concrete Institute.

ZANON, A. (ca. 1962) "Excavation of Super Highway Tunnels in Very Difficult Formations," translated by B.E. Hartmann. Terrametrics, Golden, Colo.

13 | Materials Handling and Construction Plant

A. A. Mathews
President, Al Mathews Corp.

Materials handling is the key element in the tunneling process. To achieve their designed productivity, virtually all tunneling activities depend upon materials handling systems. Likewise, the facilities required to support the tunneling operation are largely oriented toward keeping the materials handling systems operating efficiently and at their planned rates of production. This requires systems for communication, power supply, and environmental control as well as facilities for storage, equipment maintenance, personnel needs, and administrative functions.

This chapter deals with the selection and design of the tunnel transportation and materials handling equipment and facilities, together with the plant required to support the entire tunneling operation.

BASIC TRANSPORTATION SYSTEMS

Every tunnel project requires a basic transportation system for moving personnel and materials in and out of the tunnel. This system is usually designed to transport the excavated material from the tunnel face as well.

Only two basic types of transportation are available for tunneling. One requires the use of conventional railroad track and various types of cars. The other uses a roadbed serving rubber-tired vehicles.

Rail

From the standpoint of energy consumption, rail haulage provides by far the most efficient handling of materials in the tunnel.

Advantages:

1. An easily maintained traffic way.
2. Minimum ventilation requirements.
3. Compatible with most excavating and loading methods.
4. Adaptable to nearly all sizes of tunnels.
5. Several power sources available.
6. Fixed guidance system permits relatively small clearance limits.
7. No limit to length of tunnel.

Disadvantages:

1. Requires constant extension at the heading.
2. Passing locations are either fixed or only semimovable.
3. In case of derailment or other accident, the entire system is generally shut down.
4. Unloading points are relatively fixed.
5. Inhibits work on the tunnel invert.

Limitations:

1. Three percent grade is a normal practical limit, although grades up to 6% have been used in special circumstances. Steeper grades require the use of auxiliary equipment, such as cables and hoists. Safety devices to prevent or mitigate runaway accidents should be considered for grades higher than about 1%.
2. With current construction and maintenance standards, speed is limited to about 15 mph. With superior track and roadbed, much higher speeds are feasible.

Propulsion

Rail-mounted vehicles are moved, singly or in trains, by locomotives powered by either internal combustion engines or electric motors. Compressed-air power is suitable only for light duty and short travel distances. For safety and environmental reasons, only diesel fuel is permitted for use underground. Diesel-powered locomotives provide maximum flexibility, with virtually unlimited range and power, as well as freedom from external power sources. The main disadvantage is the emission of toxic exhaust gases. This requires much greater tunnel ventilation capacity and may be an overriding factor in extremely long tunnels.

Environmental codes require that the exhaust of diesel engines be processed before discharge into the tunnel atmosphere. The main objective is to reduce the content of carbon monoxide and oxides of nitrogen to levels that can be satisfactorily diluted by the tunnel ventilation system. This is accomplished by means of catalytic- or washer-type scrubbers. The catalytic type requires less maintenance, but some regulatory agencies consider it to be less reliable and fireproof than the washer type.

Electric locomotives can be powered by rechargeable batteries or by an overhead trolley conductor. Battery-powered locomotives, within their operating range, have the flexibility of the diesel locomotive without the problem of the toxic exhaust gases. However, their range is limited by the amount of energy that can be stored in the batteries. They require spare batteries and a facility for charging, maintaining, and handling them. To assess their suitability for a particular application, manufacturers should be consulted for information on cost, weight, and operating characteristics.

Trolley-type electric locomotives are by far the most efficient of all, but they have the disadvantage of requiring an overhead trolley conductor and bonded rail. In the heading, this disadvantage usually dictates the use of auxiliary facilities. Separate battery- or diesel-powered units may be used beyond the end of the trolley conductor. When this distance is fairly short, a cable reel can extend the reach of the trolley locomotive. Combination battery–trolley locomotives are also available for special applications.

Performance Calculations. The critical performance factors for a locomotive are its weight and horsepower. Since the locomotive's drawbar pull is developed by means of the adhesion of its wheels to the rails, the maximum drawbar pull that can be developed is directly related to the weight on the driving wheels.

The maximum available drawbar pull in pounds, D, is found from the formula

$$D = L[(2{,}000)(A - G)]$$

where

L = weight of locomotive on drivers in tons
A = coefficient of adhesion (see Table 13-1)
G = grade resistance (plus or minus) in lb/ton (20 lb for each percent of grade)

The drawbar pull required to move the train is $D = TR$, where T = the total train weight (excluding the locomotive) in tons and R = the train's resistance in lb/ton.

R is equal to the sum of F, C, G, and P, where F = train rolling resistance. It is as low as 4–5 lb/ton under ideal conditions. For tunnel work using antifriction bearings, it will vary between 10 lb/ton for especially good track and 20 lb/ton for the average well-maintained tunnel track.

C is the resistance created by curves in the tunnel alignment. In the absence of substantial curvature in the tunnel alignment, it can usually be neglected. If desired, it can be computed by allowing 0.8 lb of drawbar pull/ton/degree of curvature. One degree of curvature is equal to 5,730/curve radius (ft). G is the resistance created by gravity as the train travels an incline. Within the grades that can be traveled by conventional locomotives (0–6%), the grade resistance is 1% of the train weight/percent of grade, or 20 lb/ton/percent of grade. It can be positive or negative.

P is the acceleration resistance. It is approximately 100 lb/ton for an acceleration of 1 mph/sec. For most tunneling applications, an acceleration of 0.1–0.2 mph/sec is satisfactory, corresponding to a resistance of 10–20 lb/ton.

Because of slack in the couplers, the drawbar pull required to start a train of cars moving does not depend on the static rolling resistance of the entire train. Sand can also be used to improve the adhesion of the driving wheels during starting. Consequently, the difference between static and dynamic rolling resistance is not considered when determining locomotive size and type. The usual determining factors in selecting a locomotive are its ability to move the loaded train and to maintain a satisfactory speed. If long, steep grades are involved, the determining factor may be the locomotive's braking ability or its ability to return upgrade with a train of empty cars.

For stopping a train on a grade, the velocity, the weight and rolling resistance of the train, and the weight and adhesion of the locomotive must be considered. In special cases, some of the cars may be equipped with air brakes.

From the foregoing formulas, the minimum weight of locomotive for any given application can be determined. Most manufacturers list the drawbar pull of their locomotives for various speeds and track conditions, making the selection of the proper locomotive quite simple.

The required horsepower can also be calculated from the following formula:

$$H_p = \frac{(T + L)RV}{375M_e(100\% - F_a)} + L_a$$

where

$T + L$ = weight of train plus locomotive, in tons
R = train's resistance, lb/ton
V = velocity, mph
M_e = mechanical efficiency of locomotive driveline, usually about 0.90 for direct drive and 0.80 for torque converter drive
F_a = altitude correction factor for diesel-powered locomotives; usually 3% for each 1,000 ft of elevation in excess of 1,000 ft above sea level for naturally aspirated diesel engines. For supercharged engines, there is usually no performance loss up to 5,000 with

Table 13-1. Coefficient of Adhesion, A

Rail Condition	Chilled Cast Iron Wheels		Rolled Steel Wheels	
	Unsanded	Sanded	Unsanded	Sanded
Clean, dry rail	0.20	0.28	0.25	0.34
Clean, wet rail	0.20	0.25	0.25	0.31
Slippery, most rail	0.15	0.25	0.15	0.25
Dry, snow-covered rail	0.10	0.15	0.10	0.15

1%/1,000 ft from 5,000 to 10,000 ft and 3%/1,000 ft thereafter. (See manufacturer's specifications.)

L_a = accessory losses (generator, fan, scrubber, compressor, etc.), in hp

Track and Roadbed

Common track gauges are 24, 30, 36, 42, and the railroad standard, 56-1/2 in. In selecting a track gauge, principal considerations are the tunnel width, requirements for rail-operated mucking equipment, gauge and size of rolling stock to be used, availability of equipment, and possible resale values. To provide passing clearances, the track gauge generally should be less than one-fourth the tunnel width, but for stability, should be at least one-half the maximum width of the rolling stock. Most rail-mounted mucking machines require a certain minimum track gauge for efficient operation.

The required weight of rail depends upon the maximum individual wheel loading of the equipment to be used and upon the spacing of the ties or other rail supports.

Figure 13-1 shows the minimum rail weight and maximum single-wheel load recommended for 1-1/2- and 3-ft tie spacings. Heavier rails than indicated for given loads will reduce track resistance and maintenance costs, while lighter rails should be used only for short-term jobs.

Track accessories include splice bars, bolts, ties, spikes, tie plates, and gage bars; the last two items are used only for high-speed track or special situations. The tie length should be at least 1 1/2 times the track gage, or 24 in. greater than the track gage, whichever is greatest. Ties may generally be of any good-quality, locally available wood. Steel ties have the advantage of strength, durability, and lower headroom requirements, but they are more expensive than wood. They are not adaptable to an unballasted circular invert without a supporting structure. Table 13-2 gives the approximate weights and dimensions of track and accessories per 1,000 ft of single track, all based on a 2-ft tie spacing.

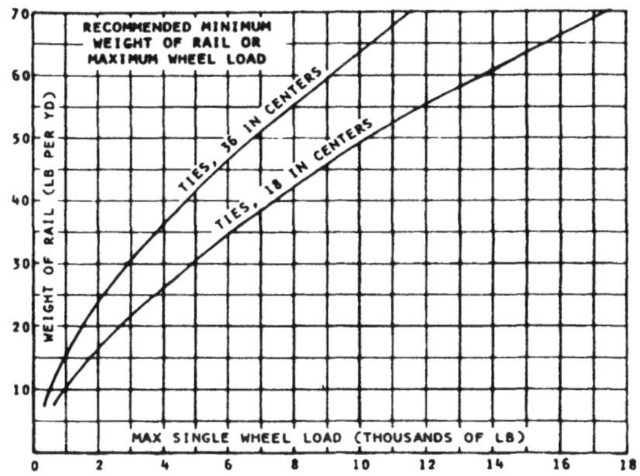

Fig. 13-1. Weight of rail and tie spacing.

Track ballast is usually derived from tunnel muck; however, good-quality ballast made from sound gravel or crushed rock, uniformly graded from 3/4 in. to No. 4 mesh, will pay for itself on many jobs.

In machine-bored tunnels in rock or where a segmented precast concrete tunnel lining is installed, ties may be set directly on the smooth, hard invert surface without ballast. Invert segments can be designed to support rails without ties, but they require accurate setting.

Table 13-2. Quantities Per 1,000 Feet of Single Track

Rail Weight (lb/yard)	Tons	Weight of Splice Plate, Including Bolts[a](lb)	Size of Spikes (in.)	Weight of Spikes (lb)	Number of Ties (2 ft center to center)	Sizes of Ties (in.)	MBM TIES 3 ft	4 ft	5 ft	6 ft	7 ft
12	4.00	245	3 × 3/8	315	500	3-1/2 × 4-1/2	2.1	2.8			
16	5.333	292	3-1/2 × 7/16	494	500	4 × 5	2.6	3.4	4.3		
20	6.667	334	3-1/2 × 1/2	627	500	4 × 5	2.6	3.4	4.3	5.1	
25	8.333	441	3-1/2 × 1/2	627	500	4 × 5	2.6	3.4	4.3	5.1	6.0
30	10.0	579/829	4 × 1/2	727	500	4-1/2 × 5-1/2	3.1	4.1	5.2	6.2	7.2
35	11.667	977	4-1/2 × 1/2	792	500	5 × 6-1/2	4.0	5.3	6.7	8.0	9.3
40	13.333	977/1,190	5 × 1/2	870	500	5-1/2 × 7	4.7	6.2	7.8	9.3	10.9
45	15.00	1,284	5 × 9/16	1,095	500	5-1/2 × 7	4.7	6.2	7.8	9.3	10.9
50	16.667	1,880	5 × 9/16	1,095	500	5-1/2 × 7	4.7	6.2	7.8	9.3	10.9
55	18.333	2,187	5 × 9/16	1,095	500	5-1/2 × 7	4.7	6.2	7.8	9.3	10.9
60	20.00	2,357/2,143	5-1/2 × 9/16	1,194	500	6 × 8	5.8	7.7	9.6	11.5	13.4
65	21.667	2,420	5-1/2 × 9/16	1,194	500	6 × 8	5.8	7.7	9.6	11.5	13.4
70	23.333	2,624	5-1/2 × 9/16	1,194	500	6 × 8	5.8	7.7	9.6	11.5	13.4
75	25.00	2,780	5-1/2 × 5/8	1,509	500	6 × 8	5.8	7.7	9.6	11.5	13.4
80	26.667	3,060	5-1/2 × 5/5	1,509	500	6 × 8	5.8	7.7	9.6	11.5	13.4
85	28.333	3,293	5-1/2 × 5/8	1,509	500	6 × 8	5.8	7.7	9.6	11.5	13.4
90	30.00	3,669	5-1/2 × 5/8	1,509	500	6 × 8	5.8	7.7	9.6	11.5	13.4
100	33.333	4,259	5-1/2 × 5/8	1,509	500	6 × 8	5.8	7.7	9.6	11.5	13.4

[a]30-foot rail up to 60 lb; 33-foot rail above 60 lb

Careful construction and adequate maintenance of the roadbed are essential. Their cost will be repaid many times over in increased traffic speeds, more efficient locomotive operation, fewer derailments, and longer life of rolling stock. This is not always appreciated.

A turnout is an arrangement by which rolling stock may be diverted from one track to another. Figure 13-2 illustrates the principal elements of a turnout.

Turnout and *switch* are not synonymous terms. A switch is part of a turnout that consists of the pair of switch points and appurtenant moving parts. Switches in general use in tunneling are the standard railroad split switches such as illustrated in Figure 13-2. A frog is the structure forming the intersection of two running rails, permitting flanged wheels on either rail to cross the other rail.

The general configuration of a turnout is determined by the frog number *N*, which is a measure of the frog angles. A larger frog number means a small frog angle and a more gradual turnout. Detailed dimensions of frogs, switches, and turnouts have been standardized among all major rail manufacturers.

Except for long car bodies binding one another or not clearing the tunnel ribs, the minimum possible radius of curve in the tunnel is limited by the wheelbase and wheel diameter on the locomotive and rolling stock, as given in Table 13-3.

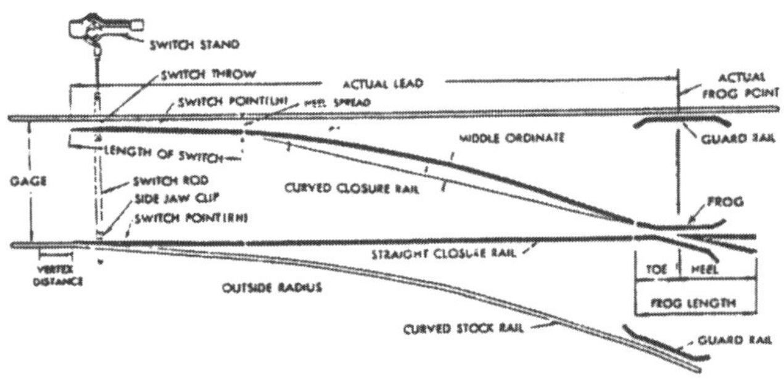

Fig. 13-2. Typical turnout with names of parts.

Table 13-3. Minimum Radius of Curve, in Feet, Over Which Rolling Stock Will Pass

Maximum Wheelbase (in.)	Diameter of Wheel (in.)										
	14	16	18	20	22	24	26	28	30	33	36
36	12	12	12	13	13	13	14				
38	12	12	12	14	14	14	15	15			
40	13	13	13	14	14	14	16	16	16		
42	14	14	14	15	15	15	16	16	16		
44	15	15	15	16	16	16	17	17	17	20	
48	16	16	16	17	17	17	19	19	19	22	22
54		18	18	19	19	19	21	21	21	25	25
60		18	18	19	19	19	21	21	21	25	25
66			22	23	23	23	26	26	26	31	31
72			25	26	26	26	28	28	28	34	34
84			29	30	30	30	33	33	33	39	39
96				34	34	34	37	37	37	45	45
108				39	39	39	43	43	43	51	51
144				52	52	52	56	56	56	68	68

Track gage on curves must be increased to prevent binding of wheel flanges. Gage is generally increased 1/16 in. for every 2-1/2° of curvature, to a maximum limited by the width of wheel tread.

Track Layout and Car Handling. Passing tracks in the tunnel and at shaft stations are provided by means of standard turnouts with sufficient passing track or tail track to satisfy the particular operational requirements. Switching cars at the heading, however, requires special facilities.

The portable or California switch comprises a section of double track with turnouts and ramps at each end, all of which slides on the main track. A typical example is shown in Figure 13-3.

Other car-changing methods include the lateral car passer and the overhead car passer or cherry picker.

A popular development for track layout in a drill-and-blast heading is the Jacobs sliding floor, illustrated in Figure 13-4. This comprises a structural steel floor some 200–250

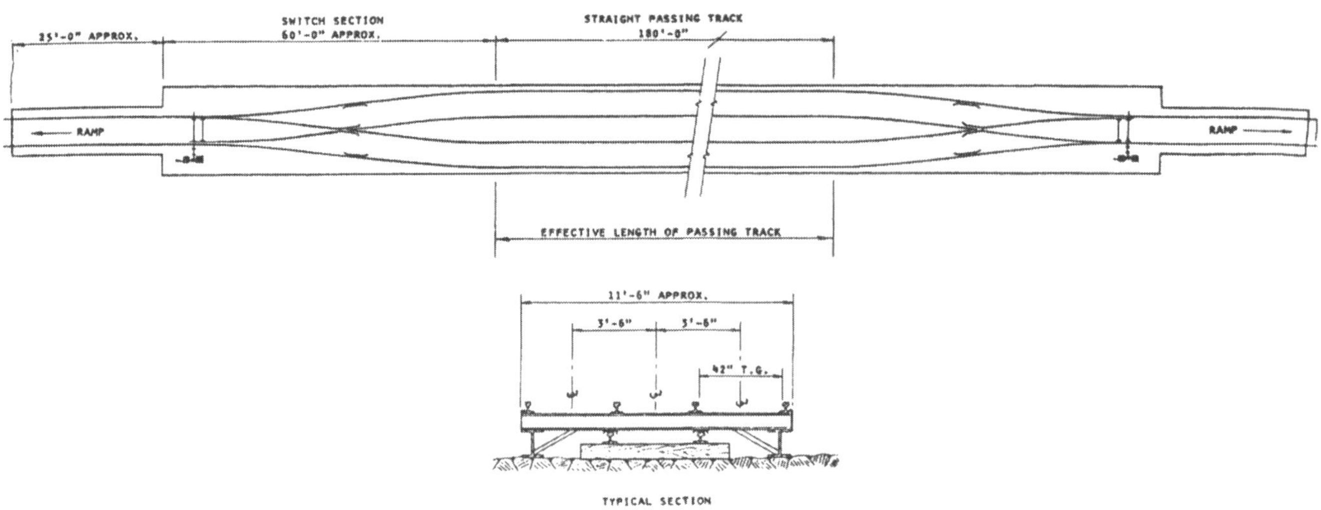

Fig. 13-3. California switch.

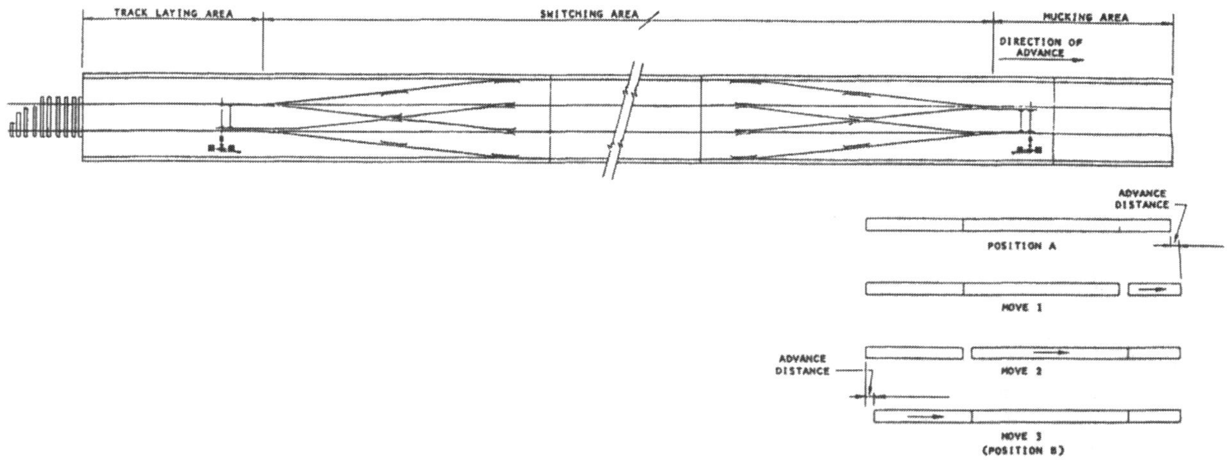

Fig. 13-4. Jacobs sliding tunnel floor.

ft long and occupying most of the invert width. It is built in three or more sections so that it can be moved along by means of integral hydraulic rams as the heading advances. The sliding floor has tracks for the jumbo, the mucker, and the loaded and empty trains.

The "Navajo blanket," shown in Figure 13-5, does not have all of the sliding floor features, but provides for extending the track in the heading in standard rail-length increments.

Rolling Stock

For any important tunnel job, all rolling stock should be equipped with roller bearings and springs. Automatic or manual couplers may be used. The selection of a coupler depends partly on the mucking system, the method of dumping, and the size of the car.

As a general rule, the muck cars should be as large as possible, considering the capability of the mucking system, clearance requirements in the tunnel, and the method of dumping. While the length of car that can be efficiently loaded with any given mucker is limited, this can be increased by means of accessory equipment such as belt conveyors or slushers.

In many tunnels, the dimensions of the car are limited by clearance requirements. The length of the muck car is also limited by structural design considerations and by the dumping method. The composition and gradation of muck may have some bearing on the choice of muck car. Sticky materials require special consideration. Table 13-4 lists the various types of muck cars in general use.

With bottom-dump cars, trap doors in the car bottom are unlatched, allowing the muck to drop out. With the end-dump and side-dump cars, the body is tipped up and the discharge side or end is automatically pivoted away, permitting the muck to slide out. Meanwhile, the chassis is held onto the rail so that the entire car does not tip. The following methods are used to tip the body:

- *Gravity.* Here the body is pivoted in such a manner that, when released by hand, it automatically dumps and swings back into position.
- *Power cylinder.* A portable compressed-air or hydraulic cylinder is used to raise the car body. With this type, the car

Table 13-4. Commonly Used Muck Cars

Car Type	Dump Method
Side dump	Gravity or power
End dump	Gravity or power
Bottom dump	Trap doors
Rigid body	Rotary dump

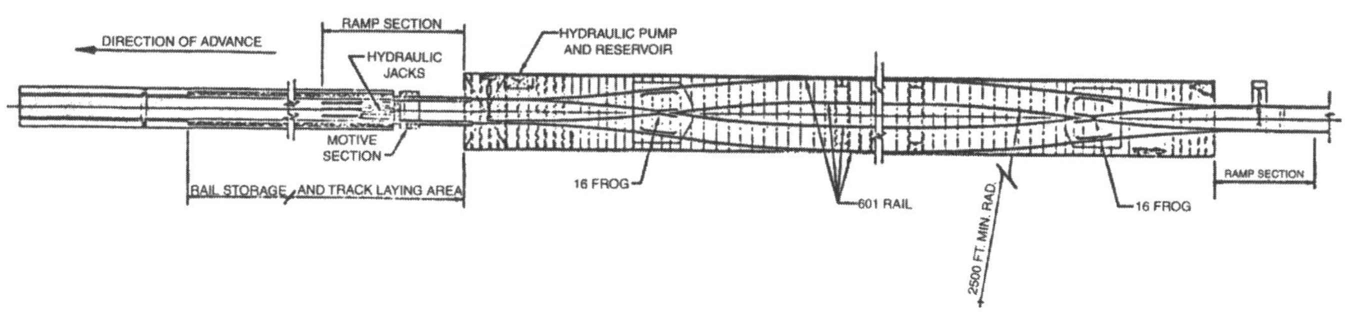

Fig. 13-5. Navajo Blanket with California switch (courtesy Ray D. Moran).

body can be mounted lower on the chassis and has less mechanism than the gravity type.

- *Traveling hoist* for dumping side-dump cars. A hoist, traveling on tracks independent from the muck car tracks, raises the car body to dumping position.
- *Camel back.* The Granby type of side-dump car has a roller attached to the side of the car body. At the dump, this roller travels up a ramp, thus causing the body to pivot and dump.
- *Rotary dumping.* For dumping rigid-body cars, the entire car is turned over in a rotating structure so that the muck is spilled out. The dumper may be designed to handle cars while they are still coupled in the train, using rotary couplers. Otherwise, single cars must be uncoupled from the train. Dumpers can also be designed to dump two or more cars at once.
- *Other cars.* Special cars or dumping systems have been devised for particular tunneling systems or for special conditions.

In addition to muck cars, various special cars must be provided to facilitate the work. These include man-haul, explosives, vent pipe, flat-, and concrete cars.

Rubber-Tired Vehicles

Transportation with rubber-tired vehicles offers great flexibility because its operation is not restricted to locations having fixed facilities, as is the case with railroad track. A wide range of vehicle sizes and configurations, all typically diesel-powered, with scrubbers on the exhaust, is available.

Excavated material can often be delivered directly to one or several disposal areas and at some distance from the portal. In multiple-heading operations, rubber-tired vehicles can move from heading to heading, in sequence with excavation cycles or variances in production rates.

Advantages:

1. Roadway extension not rigidly tied to heading progress.
2. In a wide tunnel, passing locations can usually be selected at will.
3. Disruption of entire system due to accident is minimized.
4. Maximum flexibility in operation.
5. Combination load-haul units are available.
6. Work on tunnel invert usually simplified.

Disadvantages:

1. Roadbed is difficult to maintain, particularly with soft invert or wet conditions.
2. Requires larger ventilation system.
3. Not compatible with all excavating and loading equipment.
4. In narrow tunnels, passing points are fixed.
5. Commercial-size vehicles are not usable in small tunnels. Productive capability, even with special equipment or passing niches, is reduced in small tunnels.
6. Diesel is the only readily available source of power.
7. Except in very large tunnels, clearances are generally a problem.

Limitations:

1. Ten percent grades are no problem; grades to 25% are possible.
2. With a good roadbed, 25 mph speeds are not impractical. Fifty miles per hour would be a likely practical limit for a superior type of roadway.
3. Because of ventilation and roadbed maintenance problems, truck transport is usually not economical for hauls in the tunnel of more than about 2 mi.

Roadbed. The tunnel muck remaining on the invert or produced at the face is sometimes suitable for the roadbed. If not, sound, well-graded aggregate from outside sources should be provided. It is good practice to stockpile roadbed materials near the portal and bring them into the tunnel during the drilling cycle.

Types of Vehicles:

- *Load-haul units.* Standard front-end loaders such as the Caterpillar 980 (Figure 13-6) and the low-profile Wagner MS-2 can be used for transport. Although primarily intended for loading, such units may be economically used for short haul distances. Shuttle or load-haul-dump units are designed in two configurations. In the first type, muck is carried in the bucket that does the digging, as shown in Figure 13-7. Bucket capacities range between 1 and 17 yd^3. In the second type, the loading bucket charges a hopper or bowl for hauling, as shown in Figure 13-8 (Joy Transloader), and capacities range between 2 and 12.5 yd^3. The economic limit of haul depends upon the size of the load carried and the operating speed. The larger sizes have been used on hauls up to 3,000 ft at speeds up to 15 mph. Usually, the speed is in the range of 6–8 mph.
- *Dump trucks.* Trucks designed especially for underground work are available in many different sizes. Usually, vehicles of this class do not require turning in the tunnel; they simply reverse direction, the gearing being the same for each direction. The driver either sits sideways or changes positions using dual controls. Figure 13-9 (Koehring 1860) is representative of this type, although much smaller units are available. Conventional rear-dump trucks such as the Euclid R22 are often used in large tunnels. The rocker-type rear dump, such as the Caterpillar-Athey PR 621, 631, and 651, shown in Figure 13-10, is particularly suitable for muck hauling. For their capacity, they have a short turning radius, which is further shortened when the body is in the raised position. A turntable or a tugger hoist pulling the front wheels on greased plates can be used for turning long-wheelbase units. Turning niches excavated in the tunnel sidewalls are also possible.
- *Special vehicles.* For man-haul, vent pipe erection, explosives delivery, and supply services, special vehicles are available or may be constructed on a truck chassis to suit the needs of the job.

Production Calculations. The production of the haul unit depends on its capacity; the time required to spot, load, pass, and dump; and the average travel speed of the vehicle. Speed of travel depends upon several factors. One is psychological. Speed in the confined space of a tunnel is magnified; 15 mph inside seems more like 30 mph outside.

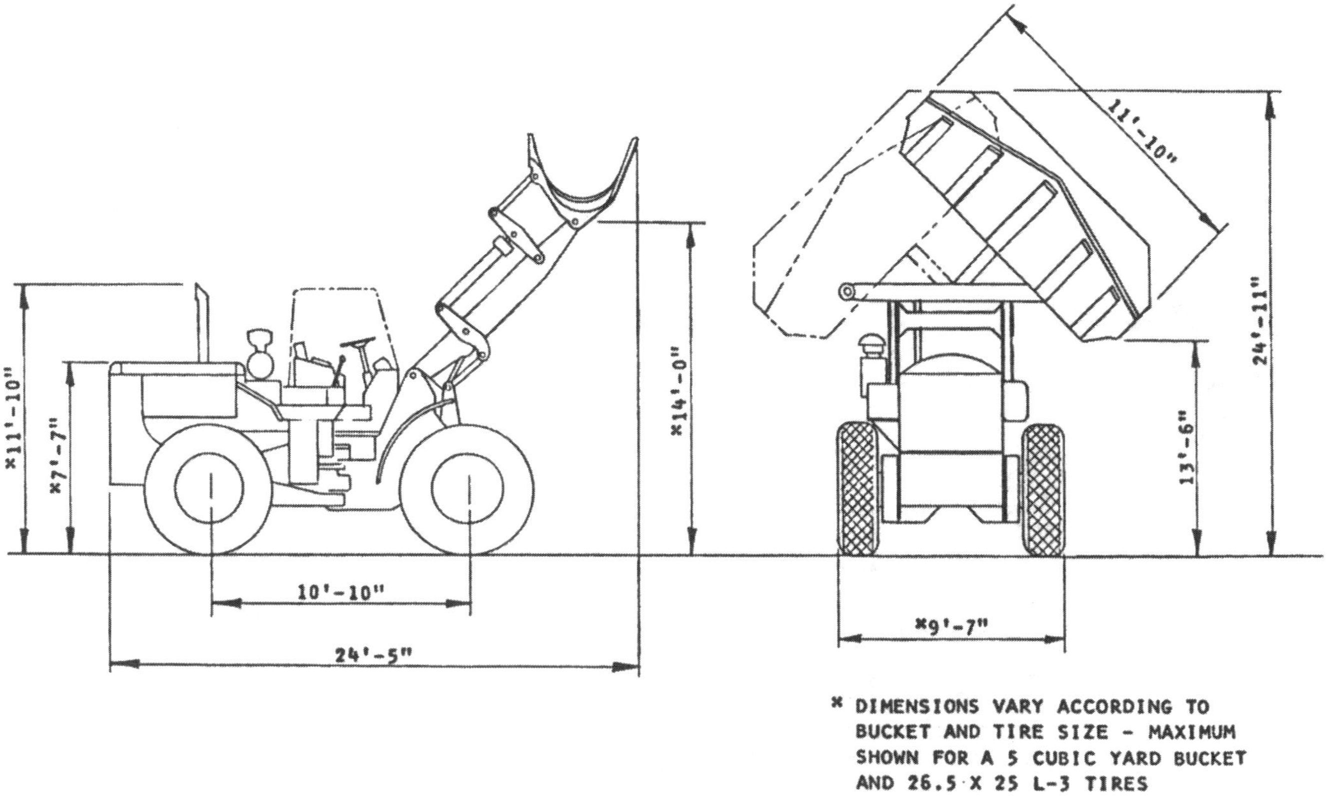

Fig. 13-6. Caterpillar Wheel Loader Model 980 W/ Libu Bucket (not to scale; courtesy Caterpillar Tractor Co. and Libu Shovel Co.).

Safety is another factor. Speed limits may be tightened due to other work in progress, passing requirements, or intersections. The other factors affecting speed are the condition and gradient of the roadbed.

For these reasons, sophisticated methods for calculating travel times are unnecessary for tunnel construction.

Grades and rolling resistances determine the maximum possible speed for any given unit. Actual average speed is determined according to judgment of the influence of the other above-described factors.

Performance specifications published by vehicle manufacturers give the maximum attainable speeds for various roadway and load conditions.

Allowance for loading time depends upon the mucking system being used.

Typical time allowances for other operations are given in Tables 13-5a and b. Unusually difficult spotting or turnaround problems will require allowances in excess of those stated.

In narrow tunnels, an additional allowance must be made for passing. To calculate the number of trucks required for a given situation, the total estimated travel time (travel loaded, turn and dump, travel empty, passing, and spot in heading) is divided by the estimated loading time. The result is the number of trucks required.

The number of standby units that should be provided depends upon the condition of the vehicles, the severity of service, and the extent of the job. Practice varies from one spare for each two units in service for unfavorable conditions to one for each eight units for favorable conditions. For average conditions, one spare vehicle for every four units should be provided.

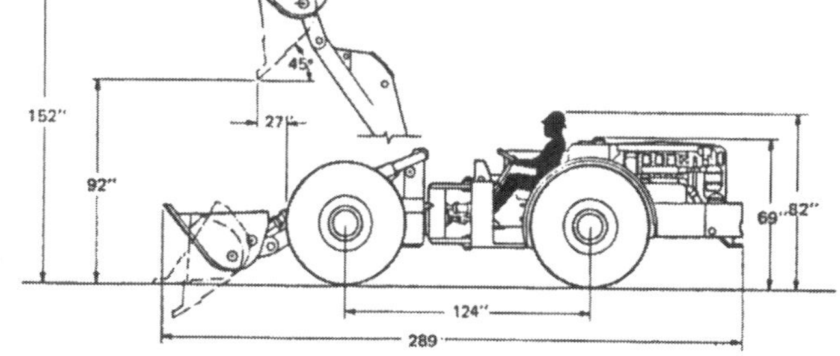

Fig. 13-7. Wagner Front End Loader Model MS-2 (2 cu yd; courtesy Wagner Mining Equipment Inc.).

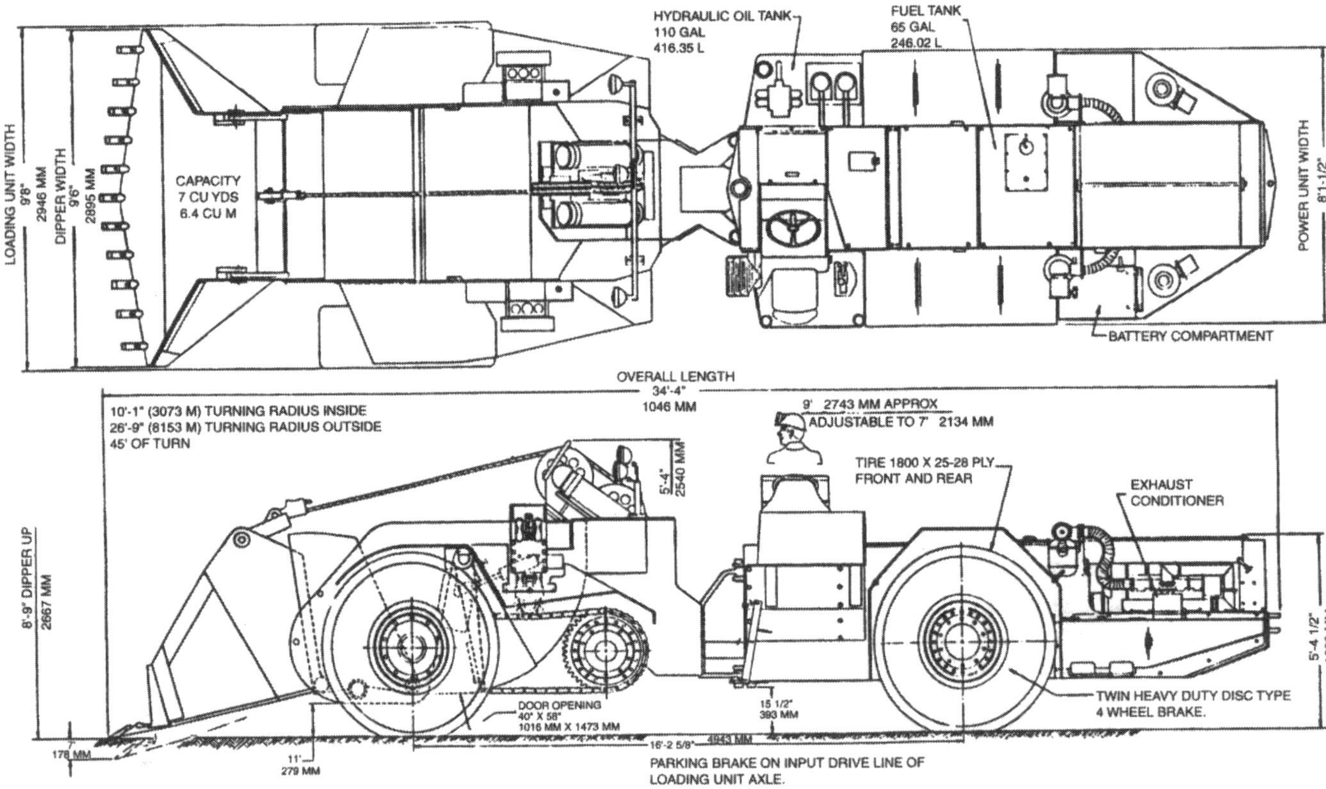

Fig. 13-8. Joy Transloader Model TL-72 (7 cy yd; courtesy Joy Manufacturing Inc.).

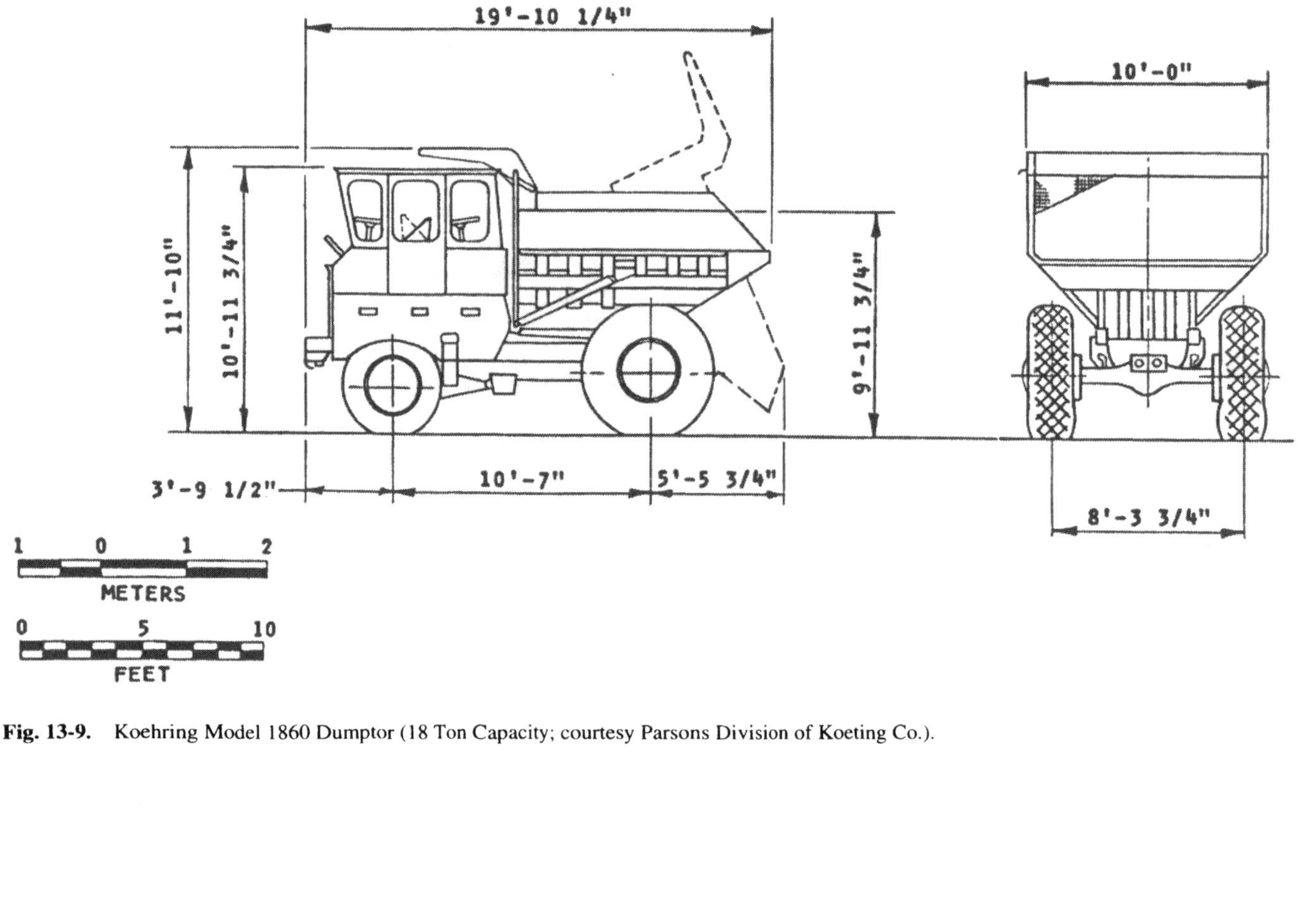

Fig. 13-9. Koehring Model 1860 Dumptor (18 Ton Capacity; courtesy Parsons Division of Koeting Co.).

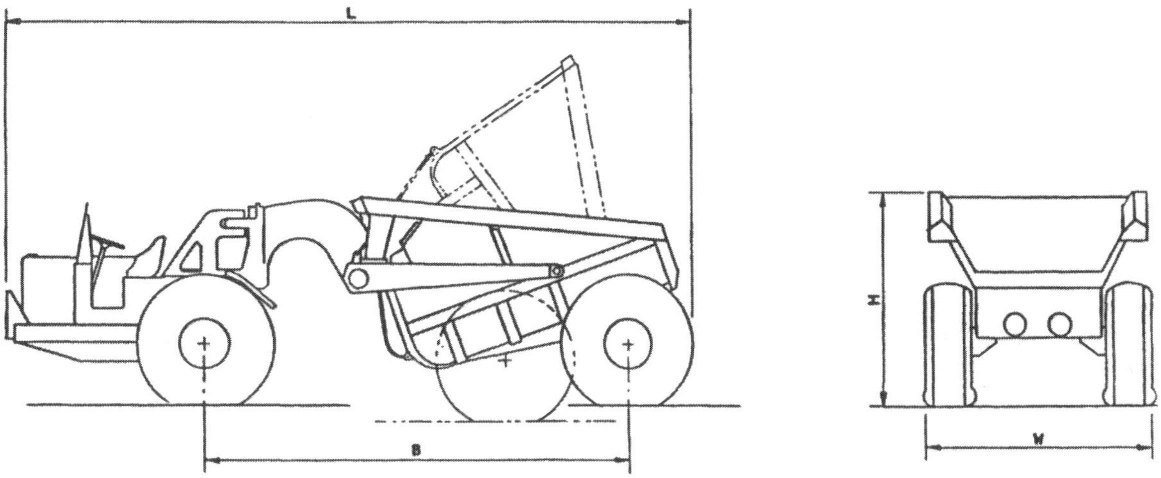

TYPICAL ATHEY PR 600 SERIES

ATHEY MODEL	HEAPED CAPACITY		DIMENSIONS			
	TONS	CU. YDS.	B	L	W	H
PR 621	30	24	20'-6"	32'-10"	11'-5"	9'-7"
PR 631	40	29	21'-10"	36'-1"	12'-1"	9'-9"
PR 651	60	47	25'-6"	40'-5"	14'-8"	11'-9"

Fig. 13-10. Rocker-Type Rear Dump (courtesy Caterpillar Tractor Co. and Athey Products Corp.).

SPECIAL MUCK TRANSPORTING SYSTEMS

To satisfy rigorous production demands or other special needs, a muck-conveying system completely independent from the tunnel's basic transportation system can be provided. The two presently available systems, belt conveyor and pipeline, are described here.

Belt Conveyors

The development of tunnel boring machines (TBMs) capable of very high sustained production rates has encouraged the use of belt conveyor systems. The conveyor can be sized to handle the maximum short-term TBM output and thus eliminate the delays associated with the logistic requirements of the basic transportation system. Although its attributes are particularly advantageous for a TBM operation, a belt conveyor could be used with nearly any mode of excavation as long as its operating requirements are met.

Advantages:

1. Capacity to handle excavated material for any conceivable rate of heading advance
2. Adaptable to nearly all sizes of tunnels
3. Relatively small clearance requirements
4. Excellent reliability and low maintenance
5. Continuous operation
6. Adaptable for conveying bulk materials into the tunnel
7. Suitable for large gradients

Disadvantages:

1. High capital cost.
2. Breakdown of one part shuts down entire system.
3. Maximum size of materials to be handled is limited, but this can be overcome by crushing.
4. Requires extensive structural support system.
5. Requires complicated system for extension in the heading.

Table 13-5(a). Turning and Dumping Time (minutes)

Operating Conditions	Rear-Dump	Shuttle Unit
Favorable	1.3	0.6
Average	1.5	1.0
Unfavorable	2.5	1.5

Table 13-5(b). Spot at Loading Machine (minutes)

Operating Conditions	Rear-Dump	Shuttle Unit
Favorable	0.6	0.3
Average	1.0	0.5
Unfavorable	1.5	1.0

6. Significant curvature of tunnel alignment requires special designs.
7. Large water inflows at the loading point may cause problems.

Limitations:

1. An 18–20° slope, either up or down, is practical limit.
2. Depending on conveyor width, maximum particle size is about 8–18 in.

Configurations. For light to moderate duty, wire-rope-supported conveyors may be used in place of rigid-frame-supported conveyors. The wire-rope conveyor may either be hung from the roof by chains or supported from the invert on light pipe standards as shown in Figure 13-11. It can also be supported on frames anchored to the tunnel sidewall.

Extension of the conveyor to keep pace with the heading advance may be accomplished through a variety of belt storage and take-up devices. When it becomes necessary to add belt length, a new section with a separate drive and material transfer station may be added, or a new length may be spliced into the belt. Both schemes involve potential delays, but they can usually be scheduled during planned maintenance periods.

The higher allowable tension in steel cord conveyor belting permits very long single flights, with lengths of several miles possible. Long flights should have prestressed steel cord belting to reduce required take-up travel. As excavation progresses, long flights of steel-cord wire-rope conveyors could replace a series of shorter conveyors, thus minimizing the problems associated with a large number of drives and transfer points.

Standard conveyors can be installed to negotiate fairly long radius curves. Also, the return side of a single conveyor flight can often be used to transport dry shotcrete or concrete material into the tunnel.

Flexible Conveyors. Conveyors having the flexibility for negotiating curves, climbing at steep angles, dumping to the side, and providing full load-carrying return capability are available. Figure 13-12 illustrates one of these nonlinear systems. A chain-driven device transmits the tension, thus allowing differential movement in the folds of the belt during twists or turns. Sectionalized construction permits fast extensions of the conveyor, and intermediate drive stations reduce chain tension.

Belt Conveyor Design. Belt width, speed, and idler spacing are dictated by the lump size, required capacity, and unit weight of the material (See Tables 13-6–13-8).

Generally, inclined conveyors should not exceed a slope of 18–20°, with flatter slopes necessary for material having a great tendency to roll. Skirt boards are necessary at transfer and loading points.

The power required to drive a conveyor can be obtained from Figures 13-13 and 13-14 and Table 13-9. The requirements for each element should be combined to evaluate the complete installation. Belt selection depends upon the material handled and the tension required. See Tables 13-10 and 13-11. Fabric belts are available for tension requirements not exceeding 1,000 lb/in. of belt width. Steel cord belts may be used for required tensions up to 8,000 lb/in. of belt width.

Typical pulley sizes and shaft diameter requirements are shown in Tables 13-12 and 13-13. Smaller drive components may be used for installations having lower belt tensions.

At charging or interchange points, the troughing idler spacing is reduced to protect the belt from the loading impact. For severe conditions, rubber or pneumatically cushioned troughing idlers should be used to absorb the loading shock. Each conveyor section must be equipped with a take-up device to compensate for stretch and shrinkage of the belt while still maintaining the required tension for proper operation. A large variety of drive arrangements is available to meet particular space, power, and reduction requirements. Direct coupled, fully enclosed, sealed, and lubricated drive systems are preferable for underground service.

Multisection conveyor systems should be interlocked electrically so that the last section to receive material is the first to start. Centrifugal switches provide one convenient interlocking system, which also serves to protect the drive system in case of failure of some component. Emergency

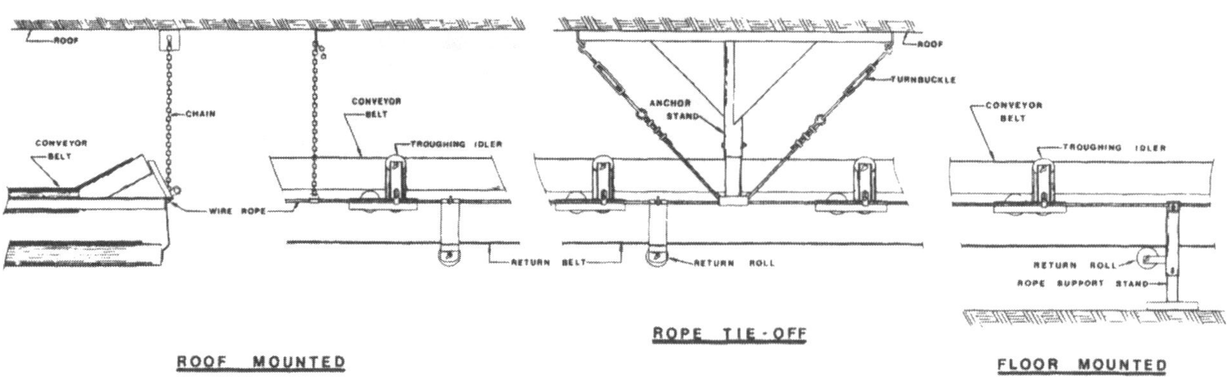

Fig. 13-11. Typical wire rope conveyor installation.

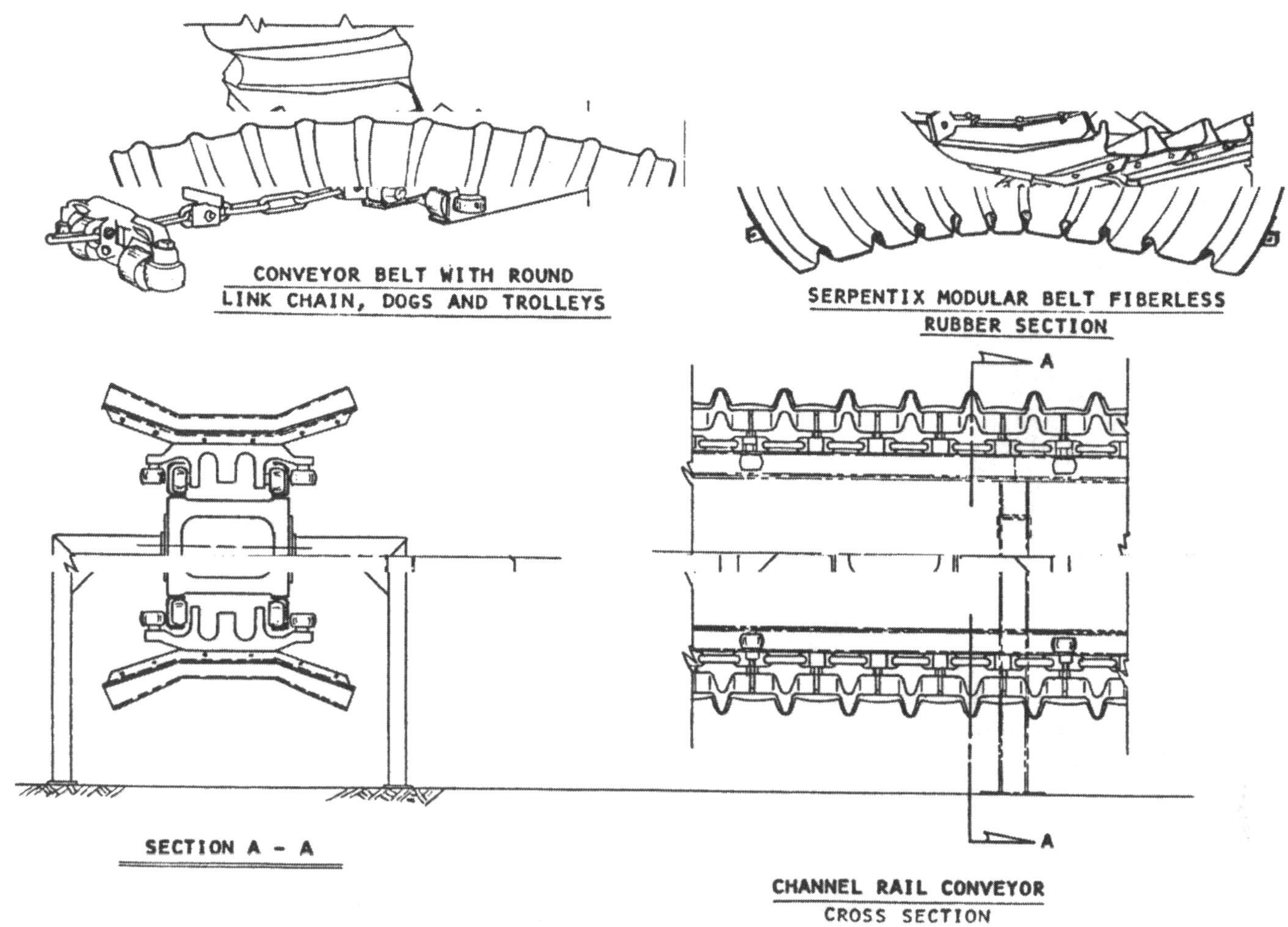

Fig. 13-12. Serpentix Conveyor (courtesy Serpentix Conveyor Corp.).

shutdown arrangements are necessary to prevent excessive damage to belts and other conveyor parts in case of accident. Braking devices may be employed to bring the system to a standstill quickly. A holdback device is required to prevent rollback of an inclined conveyor that is stopped under load. The belt speed of declining conveyors may be controlled by means of regenerative power systems.

Although for many TBM applications the overall economy of a belt conveyor system for transporting the spoil out of the tunnel, compared with railroad haulage, may be debatable, some cases unquestionably call for the belt conveyor. One is the 9-1/2-mi tunnel on the Boston Harbor Tunnel Outfall Project. The use of diesel locomotives for muck transport is ruled out because of the extremely high ventilation requirement. Battery locomotives would not be appropriate for such a long haul. Trolley-type locomotives would overcome the objections to diesel and battery locomotives, but problems generated by the live overhead conductor together with the problem of handling the gap between the end of the trolley and the heading also rule this out. The logical answer is to use a belt conveyor system.

The Tri-Met tunnels in Portland, Oregon, provide another example. There are about 2 mi of 5% grade in each of the two

Table 13-6. Maximum Lump Size and Idler Spacing for Typical Idlers (Courtesy Continental Conveyor and Equipment Co.)

Belt Width	Lump Size		Idler Spacing [a]			
			Bulk Weight of Material (per cubic foot)			
	Uniformly Sized	Mixed with 90% Fines	75 lb	100 lb	125 lb	Return Idlers
18	3	6	5 ft, 0 in.	5 ft, 0 in.	4 ft, 6 in.	10 ft, 0 in.
24	5	8	4 ft., 6 in.	4 ft, 0 in.	4 ft, 0 in.	10 ft, 0 in.
36	7	12	4 ft, 0 in.	4 ft, 0 in.	3 ft, 6 in.	10 ft, 0 in.
48	10	16	4 ft, 0 in.	4 ft, 0 in.	3 ft, 6 in.	10 ft, 0 in.

[a] Additional idlers at approximately 2 ft are recommended at loading points.

Table 13-7. Normal and Maximum Belt Speeds Recommended (Feet/Minute) (Courtesy Continental Conveyor and Equipment Co.)

Belt Width (Inches)	Lump or Moderately Abrasive Materials (Such as Well-Shot Gravel Sand)	Heavy-Sharp or Very Abrasive Materials (Such as Poorly Shot Rock)
18	300–400	250–350
24	500–600	400–500
36	500–600	400–500
48	700	600

Table 13-8. Speeds and Capacities for Typical Conveyor Loading on 35° Troughing Idlers in Tons/Hour for Various Weights of Materials and Widths of Belts. (Courtesy Continental Conveyor and Equipment Co.)

BW (in.)	Weight Per Cubic Foot of Material	Capacity (tons/hour) Belt Speed, ft/min													Cubic Yards Per Hour (at 100 ft/min)
		200	250	300	350	400	450	500	550	600	700	800	900	1000	
18	75	101	128	153	179	204	230	225							51
	100	136	170	204	238	272	306	340							51
	125	170	213	255	298	340	383	425							51
24	75	202	252	303	354	404	455	505	556	606					100
	100	269	337	404	472	539	606	674	741	809					100
	125	337	421	505	590	674	758	843	927	1,011					100
36	75	498	623	748	872	997	1,112	1,246	1,371	1,496	1,745	1,994	2,244	2,493	246
	100	664	831	997	1,163	1,329	1,496	1,662	1,828	1,994	2,327	2,659	2,992	3,324	246
	125	831	1,038	1,246	1,454	1,662	1,870	2,077	2,285	2,493	2,909	3,324	3,740	4,155	246
48	75	925	1,156	1,388	1,619	1,851	2,082	2,313	2,545	2,776	3,239	3,702	4,164	4,627	457
	100	1,234	1,542	1,851	2,159	2,468	2,776	3,085	3,393	3,702	4,319	4,936	5,553	6,170	457
	125	1,542	1,928	2,313	2,699	3,085	3,470	3,856	4,242	4,627	5,398	6,170	6,941	7,712	457

Note: Capacities given are for horizontal conveyors having a uniform feed and load. If there are peak loads, belts of sufficient capacity to handle material at maximum rate should be used.

parallel tunnels. Because of problems operating on such a grade together with the required size of locomotives compared with the capacity of the train, the use of a belt conveyor system for transporting the muck appears to be the best choice.

Even where the choice is not so obvious, belt conveyor systems for muck transport have been used successfully on several tunnel projects.

Pipeline

Bulk materials can be successfully transported through pipelines, using either air or fluids as the transporting medium.

Advantages:

1. High capacities available

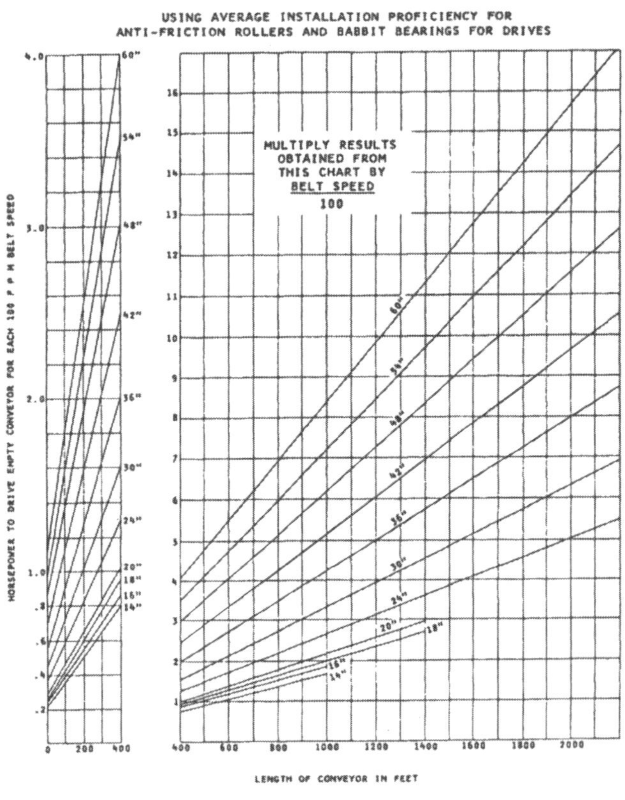

Fig. 13-13. Horsepower required to drive empty conveyor.

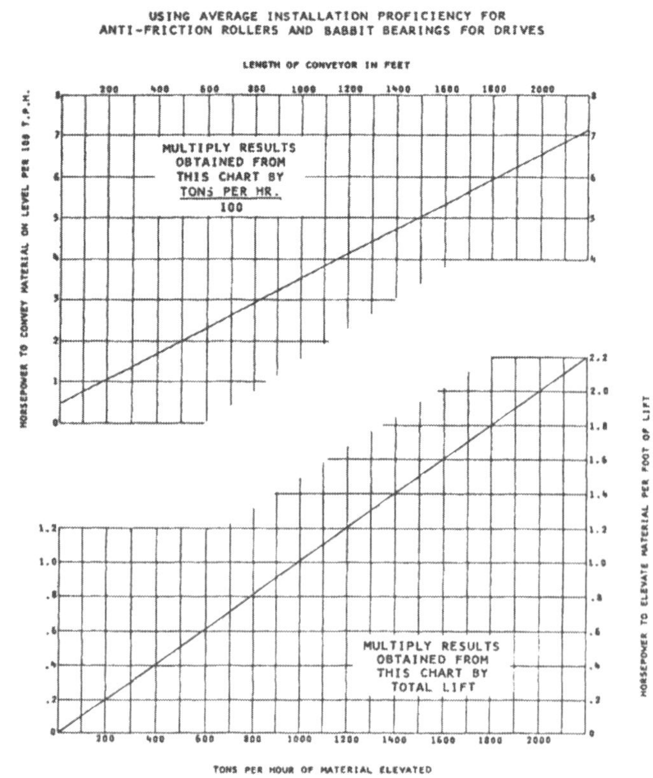

Fig. 13-14. Horsepower requirements to convey material horizontally and to elevate material.

Table 13-9. Additional Horsepower for Drive Losses (Courtesy Continental Conveyor and Equipment Co.)

Cast Tooth Gears	Roller Chain or "V" Belt Drives	Enclosed Speed Reducers Helical Type [a]
Add 6% for each reduction	Add 5% for each reduction	Add 2% for each reduction

[a] If a worm gear reducer is used, be sure to use manufacturer's efficiency ratings.

Table 13-10. Belt Tensions (Courtesy Continental Conveyor and Equipment Co.)

Arc of Contact	Type of Drive	Automatic Take-Up				Screw Take-Up			
		Tension (lb)				Tension (lb)			
		Tight Side		Slack Side		Tight Side		Slack Side	
		Bare Pulley	Lagged Pulley	Bare Pulley	Lagged Pulley	Bare Pulley	Lagged Pulley	Bare Pulley	Lagged Pulley
180	Plain	1.64 E	1.50 E	0.64 E	0.50 E	1.97 E	1.80 E	0.97 E	0.80 E
210	Snubbed	1.50 E	1.38 E	0.50 E	0.38 E	1.80 E	1.66 E	0.80 E	0.66 E
240	Snubbed	1.40 E	1.30 E	0.40 E	0.30 E	1.68 E	1.56 E	0.68 E	0.56 E
270	Snubbed	1.32 E	1.24 E	0.32 E	0.24 E	1.58 E	1.49 E	0.58 E	0.49 E
300	Tandem	1.26 E	1.19 E	0.26 E	0.19 E	1.51 E	1.43 E	0.51 E	0.43 E
330	Tandem	1.22 E	1.16 E	0.22 E	0.16 E	1.46 E	1.40 E	0.46 E	0.40 E
360	Tandem	1.18 E	1.13 E	0.18 E	0.13 E	1.42 E	1.36 E	0.42 E	0.36 E
390	Tandem	1.15 E	1.11 E	0.15 E	0.11 E	1.39 E	1.33 E	0.39 E	0.33 E
420	Tandem	1.13 E	1.09 E	0.13 E	0.09 E	1.36 E	1.31 E	0.36 E	0.31 E
450	Tandem	1.11 E	1.07 E	0.11 E	0.07 E	1.33 E	1.29 E	0.33 E	0.29 E
480	Tandem	1.09 E	1.09 E	0.09 E	0.06 E	1.31 E	1.27 E	0.31 E	0.27 E

$$E = \frac{(hp)(33{,}300)}{\text{Belt Speed, fpm}}$$

Maximum allowable belt tension, lb/ply/in. width rayon fabric belts; nylon or rayon filling						
Splice	MP 35	MP 43	MP 50	MP 60	MP 70	MP 90
Vulcanized	35	43	50	60	70	90
Fastener [a]	27	33	40	45	55	

[a] Best available metal fasteners.

Table 13-11. Conveyor Belt Construction[a]

Load	Belt Width	Belt Specification RMA Classification MP35	MP43	MP50	MP60	MP70	MP90	MP120	MP155	MP195	MP240
Minimum Number of Plies to Support the Load											
Material weighing not more than 75 lb/cu ft	18	3	3	3	3	3	3	3	3		
	24	4	4	4	4	4	3	3	3		
	36	5	5	4	4	4	4	4	4	4	4
	48	6	5	5	5	5	5	5	5	5	4
Material weighing not more than 100 lb/cu ft	18	4	4	3	3	3	3	3	3		
	24	5	5	4	4	4	4	4	3		
	36	6	6	5	5	5	5	5	5	4	4
	48	7	7	6	6	6	6	6	6	5	5
Material weighing not more than 125 lb/cu ft	18	4	4	4	4	4	4	4	3		
	24	6	6	5	5	5	5	5	4		
	36	7	6	6	6	6	6	6	5	5	5
	48	8	8	7	7	7	7	7	6	6	6
Maximum Number of Plies for Troughing											
Maximum number of plies which will trough on 35° idlers	18	4									
	24	5	4	4	4	4	4	4			
	36	8	8	7	7	6	6	5	5	5	4
	48	10	9	9	8	8	7	7	7	6	6

[a] Data contained in this table are based on present practices and ratings of equipment and belting; all of these data are subject to change with new developments.

Table 13-12. Recommended Minimum Pulley Diameters (Courtesy Continental Conveyor and Equipment Co.)

Plies	Rayon-Nylon Belt at Over 80% of Rated Tension MP35	MP43	MP50	MP60	MP70	MP90	MP120	MP155	MP195	MP240
4	20	20	24	24	24	24	24	30	36	36
5	24	24	24	30	30	30	30	36	42	42
6	30	30	30	36	36	36	36	42	48	48
7	30	30	36	42	42	42	42	48	54	60
8	36	36	42	48	48	48	48	54	60	66
9	42	42	48	54	54	54	54	60	66	78
10	48	48	54	60	60	60	60	66	66	84

Table 13-13. Typical Head Pulley and Shaft Diameter Requirements (Courtesy Continental Conveyor and Equipment Co.)

Belt Width	Head Pulley Diameter (in.)	Maximum Bearing Centers (in.)	Shaft Diameter (in.)	Maximum Tension (lb)
18	20	34	3-7/16	4,185
	24	36	4-7/16	7,709
24	20	40	3-15/16	6,166
	24	42	4-15/16	10,440
	30	44	5-7/16	14,481
36	24	56	5-7/16	12,072
	30	58	7	22,645
	36	60	7-1/2	25,096
	42	64	9	36,749
48	36	73	8	30,457
	42	77	9	36,749
	48	79	10	46,879
	54	79	10	45,219

2. Minimum space requirements in the tunnel
3. Adaptable to surface discharge through relatively inexpensive risers
4. Continuous operation

Disadvantages:

1. Maximum size of material to be handled is limited.
2. Requires complicated system for extension in the heading.
3. Reliability and maintenance costs are questionable for some materials.

Limitations:

1. Virtually no limitation as to slope, alignment, or capacity.
2. Limitations on particle size and material properties depend upon individual installations.

The Slurry System. The most straightforward application of a pipeline for muck transport is with a slurry-head or earth pressure balance soft ground tunneling machine. The excavated material is mixed with the slurry or earth and water at the face and then pumped to the surface for processing and disposal. Where a slurry is necessary, the reclaimed slurry, together with required makeup, is recirculated back into the tunnel. To meet the requirements of the particular job, the entire system should be designed in cooperation with the manufacturer of the tunneling machine.

A recent innovation involves the transport of tunnel muck through a pipeline by means of a conventional concrete pump. The muck must, of course, comply with certain grading requirements, and a calculated percentage of a specially designed slurry is added to provide sufficient lubrication and fluidity. Most of the muck from the French section of the Channel Tunnel was pumped from the Sangatte shaft to the dump at Fond Pignon.

The Hydraulic System. Although the use of water as a medium for transporting tunnel muck through a pipeline has been limited, it is certainly feasible.

When considering such a system, the following requirements should be kept in mind:

1. An adequate supply of water will be needed.
2. The design of the facility to extend the pipeline must be resolved so that uninterrupted operation can be assured.

3. The flow of muck from the tunnel excavation should be fairly continuous, without frequent shutdowns.
4. The raw muck from the excavation must be properly sized, and this may require crushing.
5. Where crushing is necessary, the effect of the water content of the muck on the crushing efficiency should be investigated.
6. The final muck characteristics should permit transport under heterogeneous or siltation flow conditions.
7. The pipe diameter should be at least three times the maximum particle size.
8. Dewatering of the material as well as treatment and disposal of the solids must be provided.
9. The possibility of serious wear problems should be considered.

There are two methods of introducing the solids into the pipeline. One is to feed the mixture of liquid and solids directly into the pump. The second is to inject the solids into the pipeline on the discharge side of the pump. The first method is simple and has a high capacity, but it is limited to low-pressure heads and has an inherently high pump maintenance cost. The second method is capable of handling higher heads, permitting long-distance movement. Pump maintenance costs are lower, but the design is more complex, with a complicated injection system.

Water is the usual liquid medium. Performance is affected by the nature and grading of the material. Pump manufacturers should be consulted for details.

The muck may be transported horizontally through a pipeline running the entire length of the tunnel, or the pipeline may be diverted vertically through shafts or drill holes.

Component specifications will be determined by operating parameters, such as hourly capacity, pipeline length, and static head, as well as by design criteria. Figure 13-15 presents a schematic layout of a hypothetical hydraulic transport system. It features a loop system where the head of the return water is used to partially balance the power demand on the transport line.

The Pneumatic System. Pneumatic transport has been used to backfill open areas of underground mines. Three-inch and smaller rock has been transported pneumatically at the rate of 300 tons/hour for a maximum horizontal distance of 1,000 ft. Transportation of 6-in. material for distances of 4,000 ft is the goal of current research.

A pneumatic system will operate most effectively with free-flowing granular material. Crushing may be required to limit maximum particle size. The system is not suitable for wet, sticky material. Vertical lifts should be limited to 1,000 ft and horizontal runs to 2,500 ft. Greater distances can be handled by installing systems in series.

One drawback to the pneumatic system is the large amount of power required. One 1,000-ft horizontal system conveying material at the rate of 300 tons/hour was reported to require 800 hp. This indicates a consumption of 14 hp/ton-mile/hour.

A high noise level and dust present serious problems. A high degree of wear can be expected, being much greater at bends. The wear problem can be reduced by rotating the pipe from time to time and by designing the bends for frequent replacement of wearing surfaces.

Because of environmental as well as economic concerns, the pneumatic transport of tunnel muck is not considered to be viable except in special circumstances.

SUPPLEMENTAL MATERIAL HANDLING SYSTEMS

Belt Conveyor

Belt conveyors are frequently used in the heading to facilitate the loading of the muck transport equipment. The Dixon conveyor, which was developed on the Colorado River Aqueduct, permits the loading of an entire train without changing cars. The conveyor is installed in a gantry-type frame, and the loading end can be raised to permit movement of the mucker and drill jumbo to and from the face. The discharge end is provided with a swinging chute to prevent spillage between cars.

The Dixon conveyor has also been adapted for loading the output of a TBM into trains of railroad cars. It is designed to remove the excavated material from the machine at its maximum penetration rate and to minimize delays in the heading progress. Originally, all TBMs operated on a cyclical basis, wherein the penetration phase is interrupted by repositioning of grippers or thrust jacks, by support installation, and by other requirements. However, machines are now being outfitted with gripper arrangements that do not interrupt the penetration phase. The interruption between cycles is minimal, and the conveyor system must be designed to facilitate the loading of successive trains with minimum interruption. Two typical systems are

- *Conveyor with single track.* Used in tunnels less than about 17 ft in diameter. The empty train is spotted under the conveyor and is pulled away by either a car mover or a locomotive as the cars are loaded. Mainline rail is installed between the TBM and the end of the train. The conveyor frame and the following California switch slide on the mainline track as the TBM advances. Trains are switched during the nonproductive part of the cycle.
- *Dead-end California switch with transverse discharge conveyor.* Used in tunnels greater than 17 ft in diameter. Incorporates a floor that slides on mainline track. Two parallel tracks are mounted on the floor, with a switch at the portal end. The conveyor is mounted on a frame above the muck cars and centered between the two tracks. A traveling transverse discharge conveyor diverts the muck from the main conveyor to the empty muck cars. While one train of muck cars is being loaded on one track, a locomotive is spotting an empty train on the other.

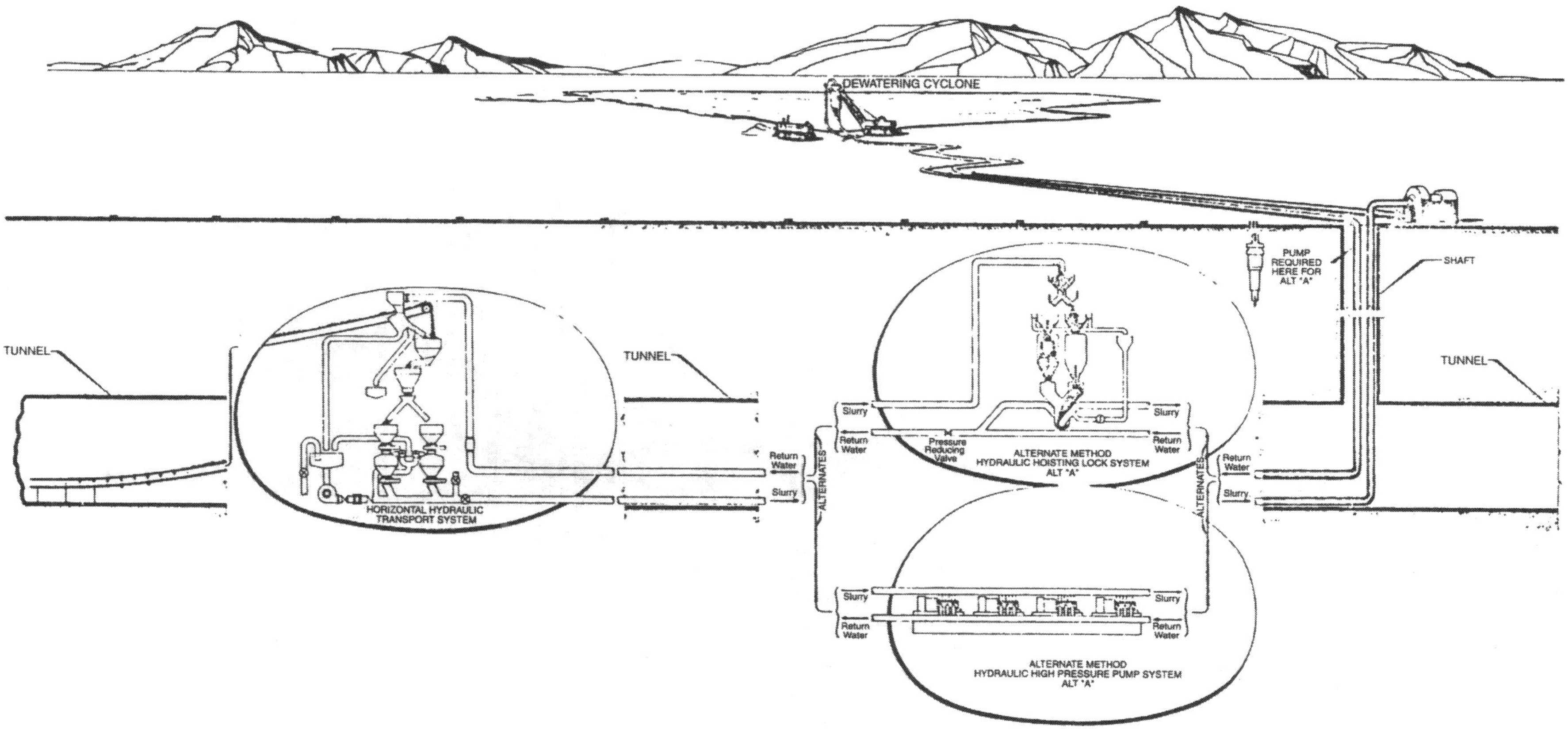

Fig. 13-15. Schematic layout of a hypothetical hydraulic pipe transport system.

These systems are pulled along behind the TBM and are moved either on a regrip cycle or during the boring stroke. If the pull required for this movement affects the action of the machine, it can be supplied by separate means. Mainline track laying must take place between the TBM and the trailing gear.

To negotiate curves, the trailing equipment must be articulated as necessary.

Any system should consider the character of the muck and the possibility that it may change. Means to transport support elements, track supplies, repair parts, and personnel should be incorporated into the overall system.

Belt conveyors also have many other uses in the tunnel. These include loading of vehicles during invert cleanup, transferring muck to a shaft hoisting system, and conveying concrete from transport vehicles into tunnel placing equipment.

Overhead Rail

This transportation method is generally used only for moving cumbersome items, such as support elements, track panels, and TBM cutters through the trailing gear and up to the vicinity of the heading. It consists of one or two trolley-mounted hoists traveling on overhead rails. The system is usually employed with mechanical excavators, but it may also be useful with other mining methods. It is also used for many other tasks in the tunnel, such as handling locomotive batteries at the charging stations and transferring concrete lining form elements. Except in special circumstances, it cannot compete with other systems for general transportation throughout the tunnel.

Slushers

A slusher installation moves broken rock or other bulk material from a point of origin, such as a muck pile, to a dumping point by means of a scraper pulled by a multiple-drum hoist. A pull rope and one or two tail ropes are attached to the scraper, providing forward and backward movement. The scraper itself may be either of the hoe or the box type.

Slushers are commonly used for short-distance movement of muck, such as from wall plate or footing drifts in a large tunnel. They can also be used for filling a train with muck without changing cars and, with a loading ramp, for filling other vehicles.

Advantages:

1. Requires no special roadbed
2. Performs both mucking and haulage functions
3. Can be set up by hand in restricted quarters

Disadvantages:

1. Low capacity, which decreases directly with travel distance
2. Restricts access to the heading during operation

Limitations:

1. With proper scraper design, slopes up to about 30° can be handled. No limit for downslopes.
2. Practical distance limitation is about 500 ft.
3. Straightline operation is simplest, although special rigging may provide some flexibility.

System Layout and Components. Normally, a slusher is used for special applications where the working area is restricted or congested, the face is on a steep slope, or when the prime objective is to move the material only a limited distance. Figure 13-16 shows several typical slusher layouts.

The three elements of the operating cycle are digging, pulling, and returning. Maximum power is needed during the digging period, and it may exceed the rated output of the motor. However, this is normally of short duration and within the overload capacity of the motor. The pulling period should not demand more than the rated capacity of the motor. After the scraper is emptied, the return pull requires from 1/3 to 1/2 of the pulling power.

Following are brief descriptions of the major components:

- *Slusher hoists.* Compressed-air hoist sizes are available up to 35 hp and electric hoists up to 150 hp in either double- or triple-drum styles. The movement of the scraper is controlled by the operator, or it may be automated. The range is limited only by the drum capacity, although normally it is kept within 100–150 ft. Slushing over longer distances is possible, and two installations may also be used in tandem. Rope speeds normally do not exceed 450 fpm.
- *Scrapers.* Scrapers are available in many sizes and types. They are usually made of an abrasion-resistant cast steel alloy and are built for rugged service. Wearing parts, such as teeth and blades, are replaceable. Figure 13-17 shows four typical styles.
- *Sheaves and anchor pins.* A tail sheaf must be mounted somewhat behind and above the material to be moved. Use of an anchor pin, as shown in Figure 13-17, is one convenient means of anchoring the tail sheaf. Diameter of sheaves depends on rope diameter, a ratio of 15:1 being acceptable.

System Design. The capacity of a slusher system depends on the volume of material transported during one cycle and the time required for the cycle. The former is influenced by the dimensions and shape of the scraper as well as the condition of the travel path. Maximum capacity for a given scraper is attained in a trough wherein side-spill is prevented. Capacity is also increased when scraping down slope and decreased when scraping up slope.

Productive cycles should allow for the fixed time of loading and dumping as well as any other lost time. Table 13-14 gives production rates for various slusher combinations, assuming a 50-min hour and a fixed time of 6 sec/cycle. Table 13-15 lists a number of typical air and electric slusher models.

Wire Rope Selection. The wire rope for a slusher application is subjected to very severe treatment. The effect of

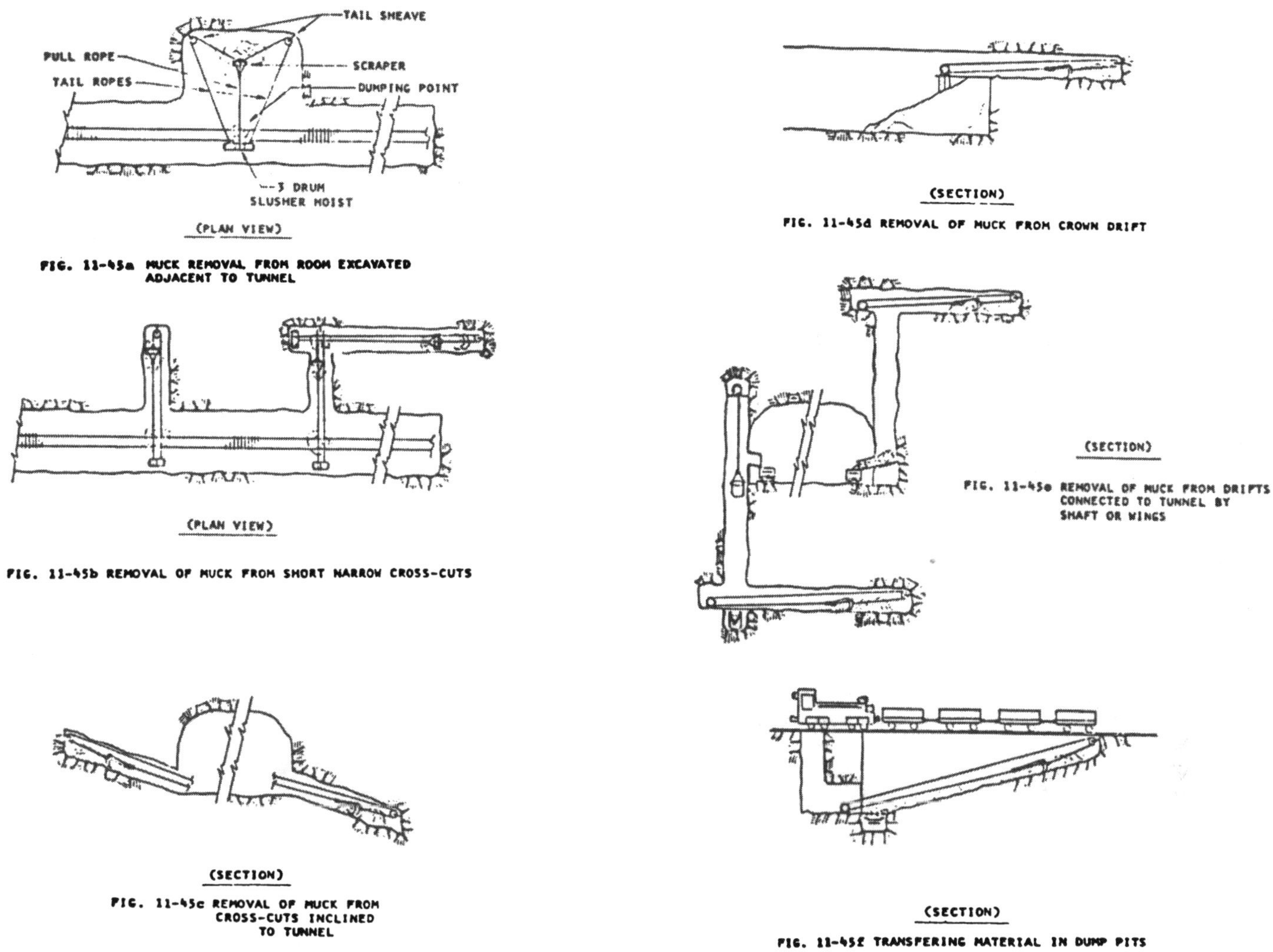

Fig. 13-16. Typical slusher layouts.

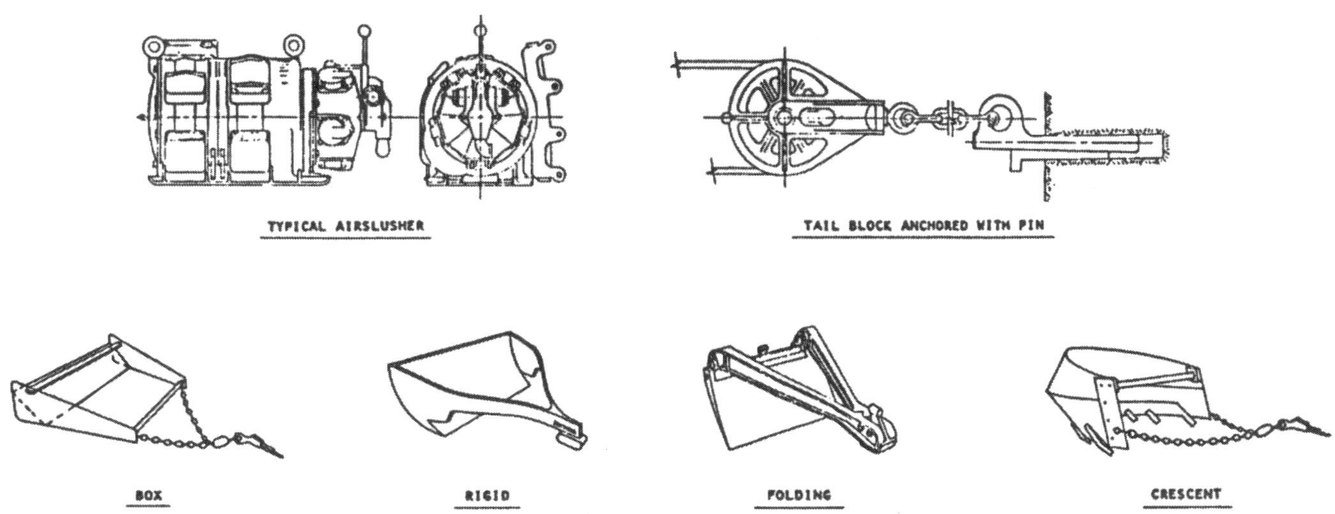

Fig. 13-17. Typical slusher styles, air slusher and tail block.

Table 13-14. Mucking Rates for Slushers

Scraper Size (in.)	Loose Cubic Foot Capacity	Scraper Weight (lb)	Muck Weight (lb)	Required Line Pull (lb)	Production-Loose Cubic Yards/Hour (50-Minute Hour, 6 Second Fixed Time) Haul Distance/Feet 50	100	150	200	500	Maximum Reach (Feet) and Cable Size (in.)
36	11	550	1,100	1,650						
	Air-130	fpm-2,000-lb	pull		23	12	8	6		300-3/8
	Air-215	fpm- 2,000-lb	pull		36	20	14	10	4	500-3/8
	25 hp-400	fpm-2,050-lb	pull		58	34	24	18	8	750-3/8
42	14	780	1,400	2,180						
	Air-170	fpm-2,500-lb	pull		37	20	14	11		230-7/16
	Air-250	fpm-2,620-lb	pull		52	29	20	15		250-7/16
	25 hp-350	fpm-2,350-lb	pull		67	39	27	21	9	950-7/16
48	19	1,290	1,900	3,190						
	Air-250	fpm-3,460-lb	pull		70	39	27	21		200-1/2
	25 hp-250	fpm-3,300-lb	pull		70	39	27	21	9	950-1/2
	35 hp-350	fpm-3,300-lb	pull		91	53	37	28	12	950-1/2
54	27	2,100	2,700	4,800						
	Air-200	fpm-4,950-lb	pull		83	45	31	24	10	635-5/8
	30 hp-200	fpm-4,950-lb	pull		83	45	31	24	10	635-5/8
	50 hp-300	fpm-5,500-lb	pull		116	65	45	35	15	635-5/8
60	39	3,280	3,900	7,180						
	75 hp-320	fpm-7,800-lb	pull		175	100	71	53		375-3/4
	100 hp -300	fpm-11,000 lb	pull		168	94	66	51		260-7/8
66	55	4,830	5,500	10,330						
	100 hp-300	fpm-11,000-lb	pull		235	133	93	72		260-7/8
	125 hp-300	fpm-13,750-lb	pull		241	134	94	72		202 1
72	75	6,800	7,500	14,300						
	100 hp-200	fpm-15,000-lb	pull		232	127	87	66		202 1
	125 hp-250	fpm-15,000-lb	pull		279	155	107	82		202 1

Factors for operating on a slope: Level = 1.0
-10 = 1.2 +10 = 0.8
- 20 = 1.5 +20 = 0.7
- 30 = 1.9 +30 = 0.5
- 40 = 3.4 +40 = 0.4

Allowance should be made for partially loaded trips and other adverse conditions.

abrasion can be minimized by selecting wire rope with large outside wires, such as the 6 × 19 Seale. Uneven and cross-winding on the drum, along with high pressures, tends to crush and distort the rope. This condition can be alleviated by using a 3 × 19 Seale class rope as it has no core, although flexibility will be sacrificed. Due to high impact loads, the pull rope for a slusher installation should have a safety factor of 7. Though the tail-rope diameter can be smaller, it is normal to size both ropes the same.

VERTICAL TRANSPORT

Access to tunnels located below the operating ground surface can be attained either by inclines or by vertical shafts. For temporary construction access, the tunnel builder may be free to choose either alternative. His selection will depend on the construction cost as well as compatibility with his chosen basic transportation system in the tunnel. The principles described in this section will assist in making a rational choice.

Slopes and Inclined Shafts

Following are practical grade limitations for various transportation systems:

- Belt conveyor for muck removal—18–20 maximum
- Rail for muck removal—locomotive—3%, with 6% maximum for short-haul or special situations
- Hoist—no limit
- Rubber-tired vehicles—10% for muck removal, 25% maximum for general access and special situations

Slopes can be excavated downgrade by various methods. For drill-and-blast with rubber-tired vehicles, 10% is practical limit with up to 25% possible with greater difficulty. For rail haulage, there is no limit for a drum hoisting operation. Locomotive haulage is limited to about 3%. Tunnel boring

Table 13-15. Medium-Capacity Heavy Duty Slushers (Courtesy Joy Manufacturing)

Air Slushers			
		Drum Hall Full 90 lb Air Pressure	
Number of Drums	Motor (HP)	Rope Pull (lb)	Rope Pull (ft/min)
2 or 3	20	3,200	210
	21	3,600	195
	23	4,300	180

Electric Slushers							
		Drum Half Full 60 Cycles, AC		Drum Half Full 25 and 50 Cycles, AC			
Number of Drums	Motor (HP)	Rope Pull (lb)	Rope Speed (ft/min)	Rope Pull (lb)	Rope Speed (ft/min)	Drum Capacity and Speed Relationship	Comments
2 or 3	15	2,475	200	3,000	165	Drum Capacity each Drum 640 ft of 3/8 in. 450 ft of 7/16 in. 360 ft of 1/2 in.	Standard tail rope speed is 20% faster than haul drum speed. If so specified, tail drum speed can be furnished the same as haul drum rope speed.
	15	1,980	250	2,360	210		
	15	1,650	300	1,980	250		
	20	3,300	200	4,00	165		
	20	2,640	250	3,140	210		
	20	2,200	300	2,640	250		
	25	4,125	200				
	25	3,300	250	3,930	210		
	25	2,750	300	3,000	250		
	30	4,950	200			Drum Capacity, each Drum 900 ft of 3/8 in. 650 ft of 7/16 in. 525 ft of 1/2 in. 350 ft of 3/8 in. 225 ft of 3/4 in.	Standard tail rope speed is 31.6% faster than haul drum speed. If so specified, tail drum can be furnished the same as haul drum rope speed.
	30	3,950	250	4,750	210		
	30	3,300	300	3,960	250		
	40	4,400	300	5,280	250		
	40	3,770	350	4,550	290		
	40	6,600	200	8,000	165	Drum Capacity, each Drum 1070 ft of 1/2 in. 685 ft of 5 /8 in. 475 ft of 3/4 in. 410 ft of 7/8 in. 270 ft of 1 in. 215 ft of 1-1/8 in.	Standard tail rope speed is 25% faster than haul drum speed. If so specified, rail drum rope speed can be furnished the same as haul drum rope speed.
	40	5,275	250	6,280	210		
	50	8,250	200	10,000	165		
	50	6,600	250	7,850	210		
	50	5,500	300	6,600	250		
	50	4,720	350	5,700	290		
	60	9,900	200	12,000	165		
	60	7,925	250	9,420	210		
	60	6,600	300	7,925	250		
	75	12,375	200				
	75	9,900	250	11,780	210		
	75	8,250	300	9,900	250		
	75	7,060	350	8,525	290		

machines can excavate downgrade up to 10% when properly equipped, but significant groundwater is an extremely severe handicap and may require grouting in advance.

It should be noted that grouting in advance of the TBM is a very difficult task. Once the machine has encountered a significant flow of water, it is extremely difficult to grout off the water without the risk of locking the TBM in place. Exploratory holes and grout holes should be drilled at least 100 ft ahead and it is difficult to orient the holes in the most favorable pattern.

Safety. Safety is a special consideration for material transport in all types of slopes. For muck transport by belt conveyor, the equipment should be provided with automatic braking devices to prevent rollback in case of power failure.

Trucks hauling muck up steep slopes should have tail gates or be otherwise configured to prevent material from rolling off the back of the body. For long downhill grades, emergency turnout ramps may be desirable. Good lighting and signal systems are essential.

Hoisting and lowering materials and personnel on railroad track on slopes requires particular attention to safety. Devices are available that automatically grip either the rail or an anchor cable when a certain speed is exceeded. If a number of different vehicles will be used, the safety braking equipment can be installed on a special car that is always connected to the hoist cable. At the bottom of the slope, a derail device should be installed to prevent runaway vehicles from traveling along the tunnel at high speed.

Vertical Shafts

Vertical shafts can vary from shallow openings with few features to deep shafts having many facilities together with equipment for high-capacity handling of the tunnel muck. For short-term work at shallow depths, a mobile crane may be all that is necessary for material handling. Even personnel can be lowered down the shaft with the mobile crane, although a protected ladderway may be required for emergency exit. For more important installations, a protected stairway may be indicated, or an electrically operated personnel elevator.

For shallow shafts of limited dimensions, the muck-hoisting area should be protected with some type of sheeting to prevent a bucket from hanging up during hoisting or lowering. Also, safety codes limit the depth for operating a shaft without guides to control the path of the load being hoisted or lowered. Generally, when shaft depths exceed 100 or 200 ft, a fixed installation is required to support the facilities at the surface and within the shaft.

Shaft Configurations. A circular configuration offers an advantage from the structural standpoint and is usually selected when any significant lateral ground pressures are expected. Rectangular sections provide more efficient use of the cross-sectional area and can be used where ground support problems are not significant. The shaft lining must resist expected ground pressures, preserve the excavated walls, and support the internal structure. Steel members, rock bolts, shotcrete, and concrete are all used, depending on circumstances (see Figure 13-18).

Shaft guides are needed for the skip, cage, and counterweight. Guides are made of wood, structural steel, or wire rope. The first two are fastened to the shaft lining, the shaft supports, or to buntons fastened to the lining. Steel guides are popular because they wear less than wood, maintain alignment better, are noncombustible, and are not affected by moisture. Wire-rope guides give a very smooth travel and are simple to install and maintain. Locked-coil rope is used and is tensioned by calibrated railroad car springs at the rate of 7–12 lb/ft of length. To prevent sway of the load due to the ropes having the same frequency of vibration, no two ropes should have exactly the same tension.

All shafts should be provided with ladderways to be used in case of hoist failure and for inspection and maintenance. In addition, provision must be made for compressed air, water supply, pump discharge, ventilation, and electrical lines. The shaft is usually divided into compartments to serve these needs as well as materials and personnel transport.

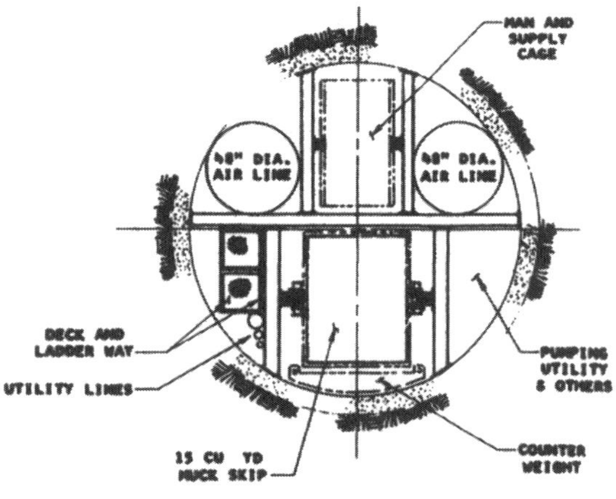

Fig. 13-18. Typical circular concrete lined shaft.

The configuration at the shaft–tunnel intersection will depend upon the tunnel driving method, muck transport system, and the muck-hoisting method. The shaft may be on the tunnel centerline, or it may be offset. Sumps are generally required for the collection and pumping of water. In the case of skip hoisting, the shaft depth below the tunnel will depend on the dimensions and configuration of the skip-loading system.

Material Transfer

When the tunnel muck is brought to the shaft by means of railroad trains or rubber-tired vehicles, it is often dumped into a temporary storage facility. This must be large enough to level out the peak production from the tunnel with the average production of the muck-hoisting system. Several typical shaft-bottom arrangements are available for transferring the muck from either the storage facility or the cars and into the skip for hoisting to the surface.

Measuring Pocket Fed by a Storage Bin. The amount of muck in the measuring pocket is controlled on either a weight or volume basis (Figure 13-19) .When the skip is in the proper location, an air cylinder moves the retractable chute into position and opens the gate of the measuring pocket. The proper amount of muck is discharged into the skip and is hoisted to the surface. While the skip is gone, the measuring pocket is automatically refilled. This system lends itself quite well to automation.

Muck Cars Dumping Directly into Skips. The muck car and skip size are equal, so that the skip is filled without spilling (Figure 13-20). This system can also be automated. In any case, it ties up a train of cars for a longer period of time compared with dumping the entire train into a storage facility.

Intermediate Conveyors. When the car dump is offset from the shaft, a belt conveyor can be used to transfer the muck from the car dump to the skip-loading area. Belt conveyors are often used in conjunction with the transfer methods described earlier.

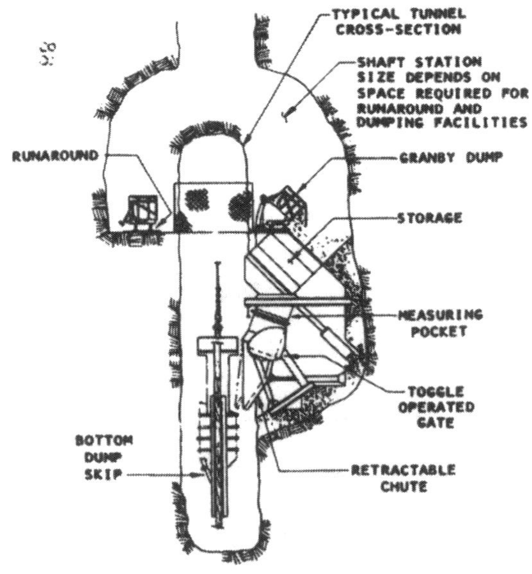

Fig. 13-19. Skip loaded by measuring pocket and retrackable chute.

Cages

Cages are used to convey personnel, material, and equipment. Loaded muck cars can also be hoisted in a cage. The cage can be attached to or part of the skip, or it can act as the counterweight in a separate compartment. It can also be handled by a separate hoist.

Cages and skips are usually of welded steel construction, although aluminum has been used. Replaceable wearing parts, such as guide shoes, are usually attached with bolts.

For shallow shafts, some contractors prefer to use a light-duty, self-service, passenger elevator for hoisting personnel. This relieves the muck hoist from that duty.

Broken rope and overspeed safety devices are required on cages used for hoisting or lowering personnel. One example is shown in Figure 13-21. Further details can be obtained from manufacturers.

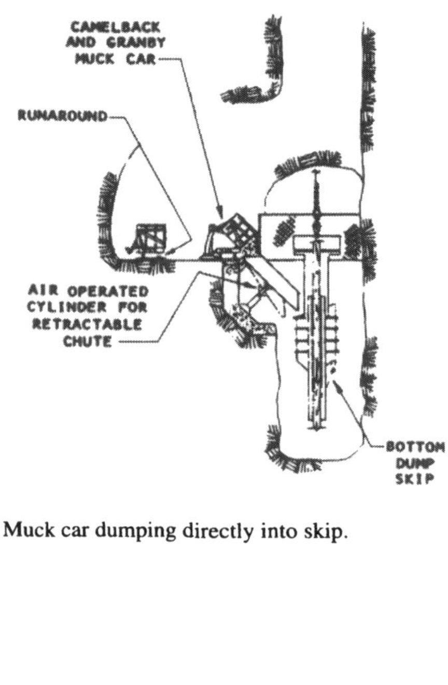

Fig. 13-20. Muck car dumping directly into skip.

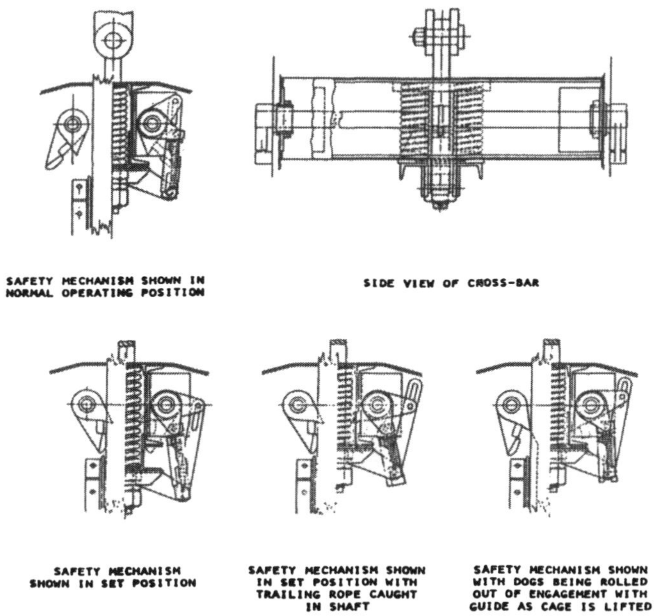

Fig. 13-21. Typical safety dog and anti-trailing rope device.

Hoisting Muck Cars

In addition to being hoisted in a conventional cage, muck cars can be hoisted in a special guide cage that provides for automatically dumping the car.

The guide cage consists of two frames. The outer frame runs in the shaft guides and is attached to the inner frame by a pivot at the bottom. A safety latch prevents the inner frame from accidentally pivoting at locations other than the dumping point. A guide wheel is attached to the inner frame so that the muck car will be emptied. The amount of rotation required of the inner frame will depend on the configuration of the muck car and the ability of the muck to flow. This cage can also be used to transport personnel if built with the proper safety features, such as cage enclosures, safety catches, and hinged bonnet.

It is also possible to hoist only the muck car body in a guide cage (Figure 13-22). In this case, the frames are designed to be lowered over the muck car to engage the body.

Skips

Compared with cage hoisting, hoisting in skips offers the advantages of less time loading and dumping, large capacity in shafts of small cross section, less labor for dumping at the top, and less dead weight. Empty weight is generally 40–60% of the payload weight. Two basic skip designs are available.

Bottom-Dump Skip. When a long slender skip is required, the bottom-dump type is preferable. Instead of rotating the entire skip for dumping, a guide roller engaging the dump scrolls opens a pivoting door on the side of the skip. A safety latch prevents the door from opening prematurely.

Kimberly-Type Skip. In this type, the skip body is pivoted on the bottom of the guide frame. It is dumped by means of guide rollers engaging the dump scrolls. The body should be relatively short. Sloping sides and a round bottom will facilitate dumping.

Headframes

The headframe is the structure at the top of the shaft for supporting components of the hoisting system. It can also contain muck bins and other facilities related to the tunneling operation. The headframe straddles the shaft and provides required working clearances at the shaft collar. It must be compatible with the skip and cage design. Headframes are fabricated from steel, reinforced concrete, or timber. For tunnel work, steel is generally used.

The conventional type of headframe has backlegs, and the hoist is mounted on the ground as illustrated in Figure 13-23.

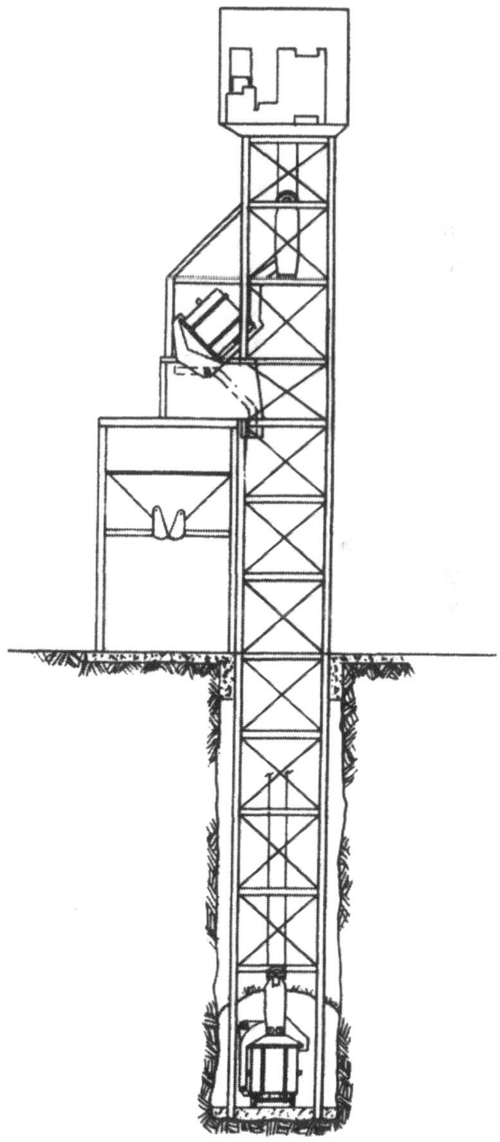

Fig. 13-22. Muck car body in guide cage.

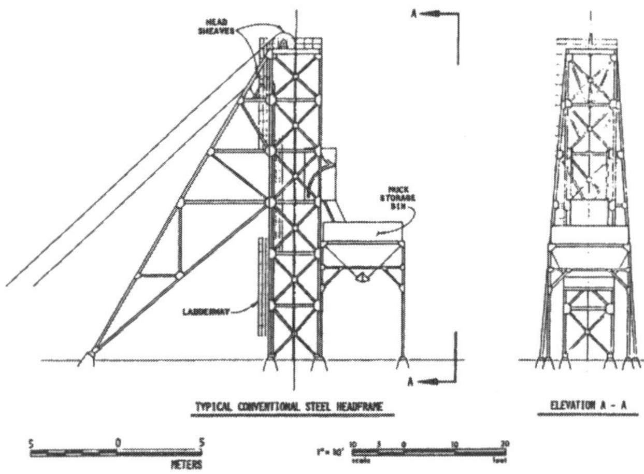

Fig. 13-23. Typical conventional steel headframe.

With a drum hoist, the distance from head sheave to drum must be great enough to limit the fleet angle to a maximum of 1° 30 sec (see Figure 13-24). The resultant of the rope forces on both sides of the head sheave should fall between the backlegs and the headframe. Otherwise, the headframe must be anchored down. Friction hoists are usually mounted in the headframe, eliminating need for backlegs (Figure 13-25).

The height of the headframe is determined by clearance requirements for equipment entering the cage, height of muck bins, overall height of skip and cage combinations, height required for dumping, and allowance for overwind-

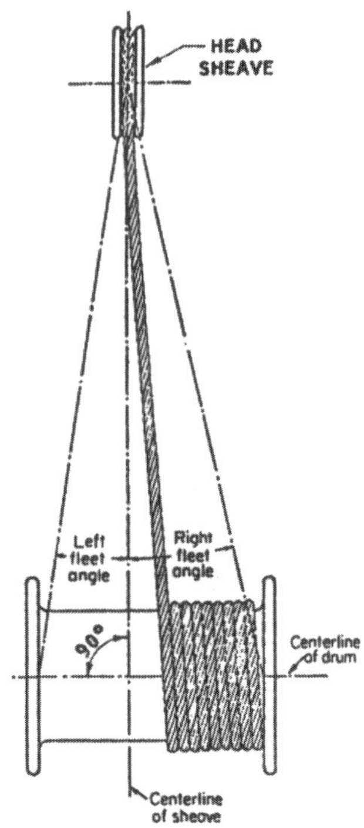

Fig. 13-24. Fleet angle.

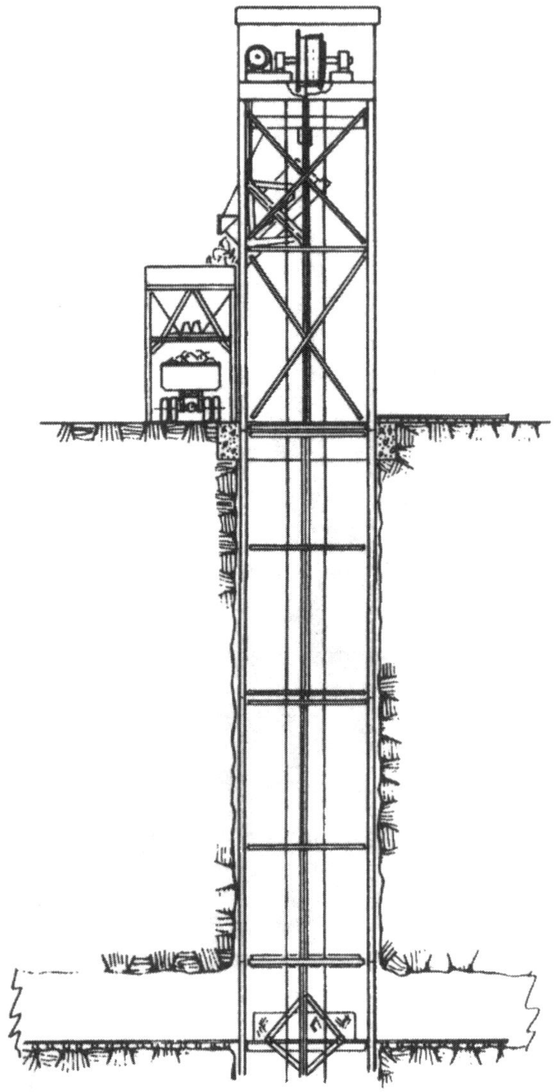

Fig. 13-25. Tower-mounted friction hoist.

ing. For 6 × 19 rope, the head sheave (and any other sheaves) should have a diameter of at least 48 times the rope diameter.

The headframe must be designed for stresses due to dead load, wind load, and live load. In some areas, seismic forces should also be considered and allowance must be made for shock and vibration. Rope forces applied to the tower should be assumed at one-half the rope-breaking strength.

HOISTING

The hoisting system must be capable of hoisting all the muck the tunnel produces and still have available time for handling supplies, equipment, and personnel. Alternatively, a second, separate material and personnel hoist or elevator may be added. The shaft should be large enough to implement this capability as well as provide space for utilities, emergency ladderways, and ventilation ducts.

The capacity to handle tunnel muck is a function of load, hoisting speed, and shaft depth. Since high speeds require greater acceleration and deceleration phases, low speeds are more appropriate for shafts of shallow depth. The shaft compartment size will frequently be determined by the size of equipment to be lowered.

System Design

The tunnel engineer is usually concerned with the development of the hoisting layout and selection of equipment for estimating purposes or for preliminary studies. For final system design, manufacturers should be consulted.

Hoist Types and Capacities

There are two main types of hoists, drum and friction. With the first, hoisting is done by wrapping the wire rope on one or more drums; with the second, hoisting is accomplished by friction between the wire rope and a powered friction wheel.

A drum hoist can operate as an unbalanced, a balanced, or a counterweighted system. The unbalanced system is primarily used for shaft-sinking operations, very shallow shafts, or small jobs. It is not as efficient as a balanced system, but the layout is simple and only one shaft compartment is required.

There are several drum configurations, but the most common is the single drum. It can operate as an unbalanced system or can be counterweighted, the counterweight being equal to the empty weight of the skip and cage plus one-half the payload. This greatly reduces the unbalanced line pull and the required horsepower. In a balanced system, the counterweight is replaced by another skip.

The friction, or Koepe, hoist has no drum and develops rope movement through friction between a friction wheel and the wire rope. The rope passes from the conveyance in one shaft compartment over the friction wheel and to the conveyance or counterweight in a second compartment. These hoists are commonly equipped with a tail rope to maintain the proper tension ratios of the hoist ropes. One end of the tail rope is fastened to the bottom of one conveyance, passes to the bottom of the shaft, and up the other compartment to the bottom of the other conveyance. Slippage between the rope and the friction wheel is prevented if the ratio of rope tension on the tight side to that on the slack side does not exceed a given value.

Friction hoists require a synchronizer that can be adjusted manually or automatically to correct the indicating device for creep and slip on the friction wheel. It is common practice to mount friction hoists in the headframe directly above the shaft.

Manufacturers should be consulted for details concerning motors, controls, brakes, emergency brakes, and clutches.

For a given productive capacity, the required size of the payload varies inversely with the hoisting speed. Larger payloads require heavier cable, and in deep shafts, the weight of the cable becomes an important factor in the hoisting requirement. Consequently, high hoisting speeds are necessary to reduce the required payload and minimize the size of cable. On the other hand, with high hoisting speeds, additional power is required for acceleration.

Generally, speeds from 500 to 1,000 fpm are common for shafts of moderate depth, up to 1,000 ft. For deep shafts, speeds as high as 5,000 fpm have been used. Accelerations of 2–4 ft/sec^2 are common, and extremes of 10–12 ft/sec^2 have been recorded.

In selecting a hoisting system, ample allowance must be made for nonproductive time. When the muck hoist must also be used for handling men and materials, productive time is not likely to be more than 75% of total time, and it may be as low as 50%.

Hoist Ropes

Many variations of hoist ropes can be found in manufacturers' literature, and each has its own advantages and disadvantages depending on the type of application. The final selection of a hoist rope will be based mainly on the strength and service requirements. Normally, a 6 × 19, regular lay, improved plow steel rope with independent wire rope center (IWRC) is used for hoisting purposes.

The safety factor of the hoist rope should vary with the depth of the shaft. Table 13-16 lists the factors of safety recommended by the U.S. Bureau of Mines depending on the length of rope. The minimum factor of safety decreases with rope length because the elasticity of a long rope compensates in part for the stresses due to loading and starting.

Table 13-17 lists the weight per foot and breaking strength of three types of 6 × 19 wire rope 3/4 in. and larger.

Design Calculations

The design of a high-capacity hoisting system is rather complicated and requires experience. However, for estimating purposes and for developing conceptual layouts, the following formulas can be used.

Rope Stress. Rope Stress S (estimated) is equal to $W(2.0 + 0.1a)$ where W = load (skip and/or cage plus payload) in lb, and a = acceleration in ft/sec^2. For example, assume live load of 7 tons, skip weight of 3 tons, and cage weight of 1.5 tons. W = 11.5 tons, or 23,000 lb. Assume acceleration = 4 ft/sec^2. S (est.) = 23,000(2.0 + 0.4) = 55,200. For a 1,000-ft-deep shaft, the safety factor should be 7, requiring a rope with breaking strength of 386,400 lb, or 194 tons. A 2-in rope of extra improved plow steel, with IWRC is indicated (see Table 13-17).

Table 13-16. Static Factors or Safety for Hoisting Ropes for Various Depths of Shafts When Men are Hoisted

Length of Rope (ft)	Minimum Safety Factor for New Rope	Minimum Safety Factor When Rope Must Be Discarded	Percentage Reduction
500 or less	8	6.4	20
500 to 1,000	7	5.8	17
1,000 to 2,000	6	5.0	16.5
3,000 and over	5	4.3	14
	4	3.6	10

Table 13-17. Breaking Strength (1 ton = 2,000 lb)

Approximate Weight (lb/ft)			Extra Improved P/S	Improved P/S	
Rope Diameter	Fiber Core	IWRC	IWRC	Fiber Core	IWRC
3/4	0.95	1.04	29.4	23.8	25.6
7/8	1.29	1.42	39.8	32.2	34.6
1	1.68	1.85	51.7	41.8	44.9
1-1/8	2.13	2.34	65.0	52.5	55.5
1-1/4	2.63	2.89	79.9	64.6	69.4
1-3/8	3.18	3.50	96.0	77.7	83.5
1-1/2	3.78	4.16	114.0	92.0	98.9
1-5/8	4.44	4.88	132.0	107.0	115.0
1-3/4	5.15	5.67	153.0	124.0	133.0
1-7/8	5.91	6.50	174.0	141.0	152.0
2	6.72	7.39	198.0	160.0	172.0
2-1/8	7.59	8.35	221.0	179.0	192.0
2-1/4	8.51	9.36	247.0	200.0	215.0
2-3/8	9.48	10.4	274.0	220.0	239.0
2-1/2	10.5	11.6	302.0	244.0	262.0
2-5/8	11.6	12.8	331.0	268.0	288.0
2-3/4	12.7	14.0	361.0	292.0	314.0

Total rope stresses can be expressed as follows:

$$P = S_r + \left[W + wl + a\frac{W + wl}{g}\right]$$

where

S_r = bending stress, lb
w = weight of rope, lb/ft
l = length of rope, ft

$= E_r \dfrac{dA}{D}$

where E_r can be taken as 12,000,000.
d = diameter of largest wire; for 6 × 19 rope, $d = d_r$ (rope diameter) ÷ 15.52.
A = total area of wires.
D = diameter of bend over sheave to centerline of rope, in.

For a 2-in. rope, $d = 0.1289$, $A = 1.4900$, and $dA = 0.192$. D should be at least 96 in. and $S_r = 24{,}000$ lb; $wl = 7.39 \times 1{,}000 = 7{,}390$ lb = weight of rope.

$$P = 24{,}000 + \left[23{,}000 + 7{,}390 + \frac{4 \times 30{,}390}{32.2}\right] = 58{,}165 \text{ lb}$$

For the factor of safety of 7, the rope should have a breaking strength of 407,000 lb, or 204 tons, indicating that the assumed 2-in. rope would be unsatisfactory. However, if the sheave diameter were increased to 120 in, S_r would be reduced to 19,200 lb, P to 53,365 lb, and the 2-in. rope would be acceptable.

Horsepower. Solving for the exact motor horsepower is a complicated process. The horsepower–time cycles have the general characteristics shown in Figure 13-26. When the acceleration period is short, the horsepower in excess of that required for full load speed can probably be handled by the overload capacity of the motor. The approximate required horsepower can be calculated from the following formula:

$$\text{HP} = \frac{LV}{33{,}000 \text{ eff}}$$

where

L = unbalanced load, lb
V = velocity, ft/min
eff = hoist efficiency, say, 0.85

In the previous example, L = load (W or 23,000 lb) plus weight of rope ($w1$ or 7,390 lb) minus weight of counterweight (16,000 lb) = 14,390 lb. If V = 800 fpm, HP would be 410.

VERTICAL CONVEYORS

Recent advances in the productive capability of large-diameter hard rock TBMs have encouraged the development and use of vertical transport systems having similar capacity. Conveyors are now available for lifting large volumes of tunnel muck, either vertically or on steep inclines, from the tunnel to the surface.

Advantages:

1. Continuous operation simplifies operations at both the loading and unloading ends.
2. More energy efficient than hoisting systems.
3. Excellent reliability.
4. Moderate maintenance costs.
5. Quick, simple installation.

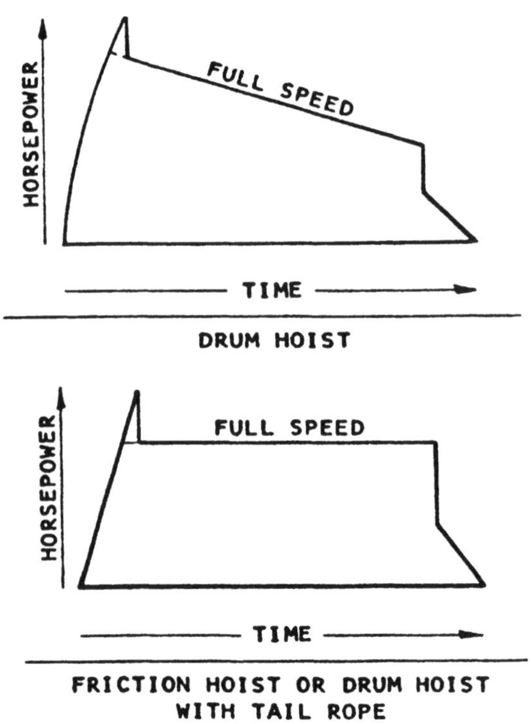

Fig. 13-26. Typical horsepower time cycles.

Disadvantages:

1. Particle size limits for muck may necessitate crushing in the tunnel.
2. Sticky materials might cause problems.
3. Height of lift is limited.
4. Requires auxiliary hoist for personnel and materials.

The most popular configuration consists of a specially reinforced belt having flexible sidewalls and reinforced cross-cleats, resembling a bucket elevator. The loading portion can be horizontal, the lifting portion either vertical or inclined, and the discharge portion horizontal, making either an "L" or an "S" configuration.

Vertical lifts of more than 100 m (328 ft) have been fully proven, and lifts up to 500 m are considered possible. Capacities up to 4,000 tons/hour are possible for moderate lifts. Typical installations include one with a capacity of 2,000 tons/hour, a 120-m vertical lift, and a 800-KW drive with speed of 3.6 m/sec. In Chicago and Milwaukee sewer tunnels, installations for 1,100 tons/hour with 82-m lift and 802 tons/hour with 104.5-m lift were highly successful. Figure 13-27 shows a typical installation.

UTILITIES

Power Supply

An electric power system for a tunnel project will have some or all of these major components:

- Power Sources: utility company; special generating plant
- Main Substations: transformers; switch gear (including circuit breakers)
- Main Feeder Lines
- Secondary Substations: transformers; switch gear (including circuit breakers)
- Secondary Feeder Lines
- Junction Boxes
- Service Lines: services for buildings, shops, and camps; light lines in tunnel; temporary power cable; trailing cables
- DC Conversion Units: rectifier; motor generator

The most economical source of electricity is usually the local power company. This will usually require building a high-voltage feeder from a trunk line to the job site. However, for remote sites, generators installed at the site may be more economical.

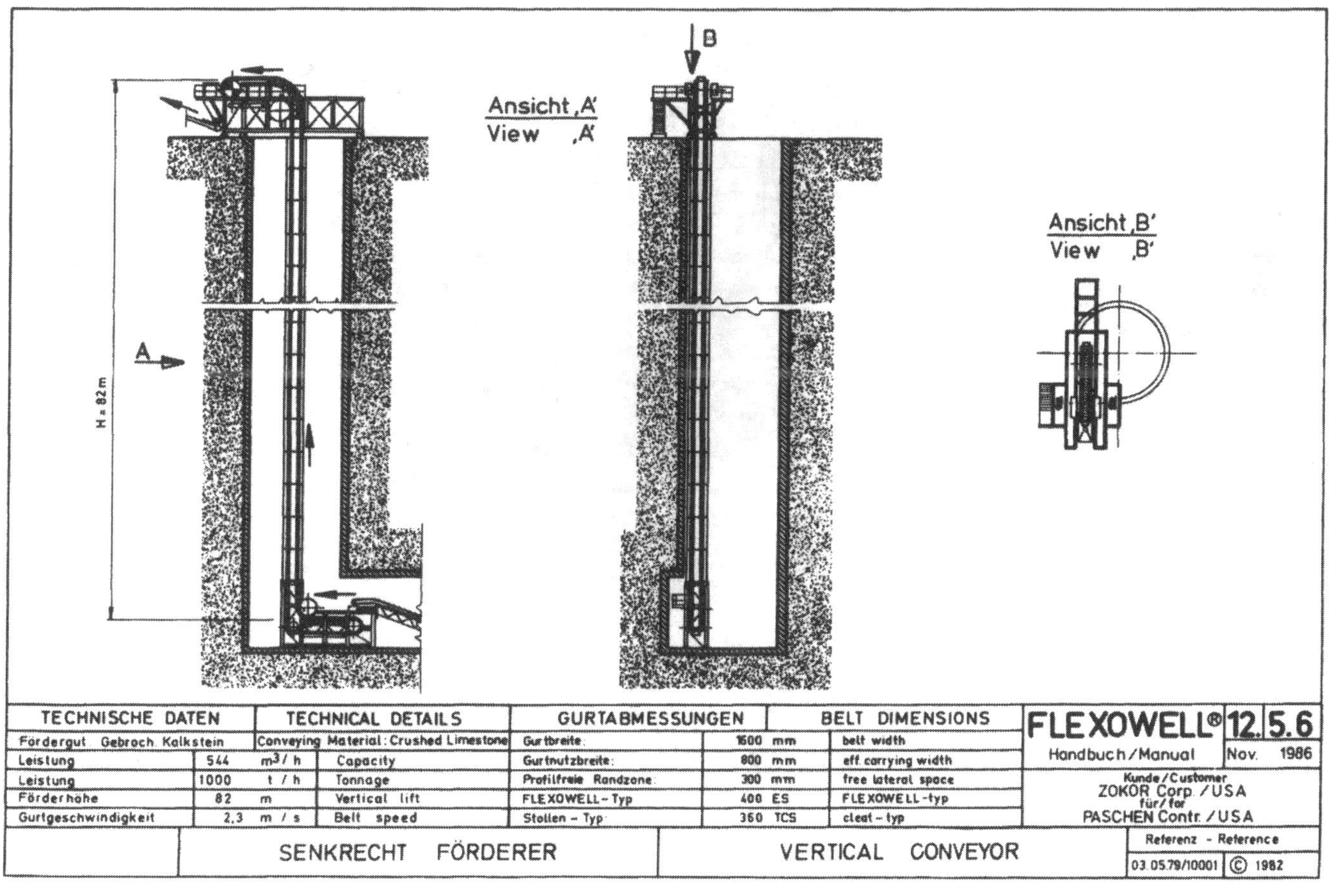

Fig. 13-27. Typical vertical lift (courtesy Flexowell Vertical Lift Systems).

A standby source of power must be available if the prime source is unreliable, the tunnel is potentially subject to disastrous flooding, or the tunnel is driven under compressed air.

The main substation reduces the primary voltage to some usable secondary transmission voltage, usually 4,160 or 2,300 volts. Transformers for this are generally provided and installed by the local power company. The voltage of the secondary side of the main substation must be compatible with the equipment and distribution system to be used in the tunnel.

Power for tunnel requirements is normally carried through a plastic-coated, armored, multiconductor cable (three conductors with one or more grounds).

Within the tunnel, common practice is to install the main power cable in 1,000-ft lengths, with junction boxes and lighting transformers at the end of each length. Transformers for fans, pumps, and other requirements are installed as needed.

Most power company rate schedules penalize customers with low power factors. Capacitors can be used to improve the overall job power factor.

A qualified electrical engineer should design the distribution system; select voltages, wire sizes, and transformers; and choose transformer locations, controls, and protective devices. It is the responsibility of the electrical engineer to design the most economical and practical system that will be compatible with the primary power source and comply with all applicable codes and standards.

The tunnel engineer, while not necessarily familiar with all of the details of a power supply system, should have a working knowledge of the fundamentals of electrical engineering so that power demands and equipment requirements can be anticipated, job layouts prepared, and adequate information made available to the electrical engineer.

Power demand can be estimated by listing all the anticipated load centers on the job, with a tabulation of the horsepower and lighting requirements at each. The combination of items, each modified by its respective duty factor, if any, that are likely to be consuming the most power at the same time determines the maximum overall demand. Energy consumption can be estimated from this same tabulation, using estimated operating hours and duty factors. Kilowatt-hours can be derived from hp hours by multiplying by 0.746 and dividing by the average power factor (usually about 0.85).

It is important to select the proper size of power cable for the tunnel. Ordinarily, the voltage drop should be limited to 5%. Figure 13-28 gives a graph for selecting 2,300-volt Parkway cable for a total load up to 700 hp. For other voltages, suitable correction factors can be applied and similar graphs can be made for other types of cable or higher horsepower ratings. In long tunnels, individual reaches can be calculated separately.

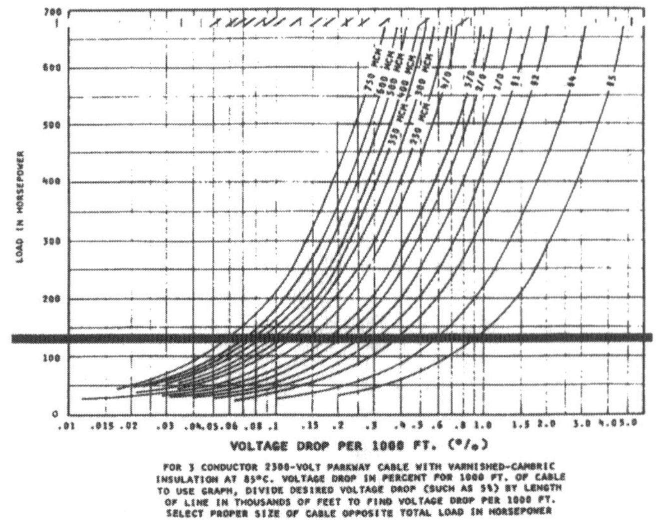

Fig. 13-28. Parkway cable selection graph,

Ventilation and Environmental Control

A tunnel may be ventilated either by blowing air in or exhausting it out through a duct. In the blowing mode, fresh air from the surface is forced through the vent line to a point near the face. The exhaust air travels back through the full length of the tunnel to the portal or shaft. This mode has the advantage of constantly supplying fresh air to the face where most of the work is done. It has the disadvantage of exposing the remainder of the tunnel to any contaminated air generated at the face.

In the exhaust mode, the foul air is exhausted through the vent line, and the fresh air enters at the portal or shaft and travels through the tunnel to the face. While this method creates a better environment along the tunnel, any heat, moisture, dust, and smoke generated along the tunnel is delivered to the vicinity of the face. This method is usually preferred for TBM tunnels driven in rock since any dust that escapes treatment by the dust collector is not driven out through the entire tunnel. On the other hand, in the blow mode, the fresh air is forced closer to the face where it is needed.

When the heading is advanced by means of drilling and blasting, the ventilation system may be operated in the exhaust configuration for 15–30 min after the blast to remove much of the smoke and dust and then changed to the blow configuration.

Auxiliary blowers are often used in any type of tunnel to properly distribute fresh air around the face.

Tunnel ventilation requirements are usually specified in the health and safety regulations applicable to the particular project. Usually, a minimum velocity of 50 ft/min in the tunnel is required together with a minimum volume of 200 ft^3/min for each worker in the tunnel. When diesel equipment is used, 100 ft^3/min of fresh air for each brake horsepower is often required. Normally, when more than one unit is being operated, they will not all be developing maximum rated horsepower at the same time, and some allowance can be made for that. Some regulations are based on the measured content of contaminants in the tunnel atmosphere.

Properties of various gases and diesel exhaust constituents, and their physiological effect on personnel, are listed in Table 13-18.

Table 13-18. Properties of Various Gases that May be Present in Tunnel

Gas	Density	Color	Odor	Source	Physiological Effect on Personnel
Oxygen (O_2)	1.11	None	None	Air is normally 20.93% O_2	At least 20% is required to sustain normal health. Personnel become dizzy if concentration drops to 15%. Some personnel may die at 12.5%; most will faint at a concentration of 9%; and death will occur at 6% or less.
Nitrogen (N_2)	0.97	Yellow	None	Air is normally 78.10% N_2	Nitrogen has no ill effect except to dilute air and decrease O_2%.
Carbon dioxide (CO_2)	1.50	None	None	Air is normally 0.03% CO_2. CO_2 is produced by decaying timber and fires, and is present in diesel exhaust.	CO_2 acts as a respiratory stimulant and may increase effects of other harmful contaminants. At 5% CO_2, breathing is laborious. A concentration of 10% can be endured for only a few minutes.
Carbon monoxide (CO)	0.97	None	None	Present in diesel exhaust and blast fumes.	CO is absorbed onto the blood rather than O_2. In time, very small concentrations will produce symptoms of poisoning. A concentration slightly greater than 0.01% will cause a headache or possibly nausea. A concentration of 0.2% is fatal.
Methane (CH_4)	0.55	None	None	Present in certain rock formations containing carbonaceous materials.	Has no ill effect except to dilute air and decrease O_2%. It is dangerous because of its explosive properties. Methane is explosive in the concentration range of 5.5 to 14.8%, being most explosive at a concentration of 9.5%.
Hydrogen sulfide (H_2S)	1.19	None	Rotten Eggs	Present in certain rock formations and sometimes in blast fumes.	Extremely poisonous—0.06% will cause serious problems in a few minutes.
Sulfur dioxide (SO_2)	2.26	None	Burning Sulfur	Present in diesel exhaust and blast fumes	Strongly irritating to mucous membranes at low concentrations. Can be kept below objectionable levels by limiting fuel sulfur content to 0.5%.
Oxides of nitrogen	Approx. 1.5	Yellow-brown	Stings Nose	Present in diesel exhaust and blast fumes.	NO_2 is most toxic. All oxides of nitrogen cause severe irritation of the respiratory tract at high concentrations. Acute effects may be followed by death in a few days to several weeks owing to permanent lung damage.

The use of TBMs at great depth can produce a real problem. The naturally high rock temperatures, augmented by the heat generated by the TBM, can produce very high temperatures in the heading. And since there is usually moisture present in the tunnel, the resulting high humidity further aggravates the problem. Then, as this hot, humid, dust-laden air is exhausted toward the portal, it becomes cooler, causing the moisture to condense and precipitate the dust on the bottom of the vent pipe.

Current health and safety regulations require that dust levels in the tunnel atmosphere be kept low. Wet drilling and wetting down the muck pile after blasting is required. With TBMs, the dust is somewhat confined to the face by a dust shield. From there it is conducted by an exhaust fan and plenum to a dust collector. The discharge air is then released into the exhaust fan line or the return air in the tunnel. Other dust-producing sources can be treated by water sprays.

The possibility of encountering toxic or explosive gas in the tunnel should always be kept in mind. Where this is likely, special precautions should be taken when planning the work and equipping the job. Permissible electrical equipment may be necessary, along with additional ventilation capacity. In conventional drill-and-blast tunnels, periodic testing should be done, while continuous monitoring is necessary with TBMs. Suitable instruments are available from safety equipment manufacturers. Attention should be given to the physical properties of the gases, since some tend to collect in high or low spots in the tunnel. Local, state, and federal regulations should be consulted when planning the work.

The ventilation equipment can be located outside the tunnel, using either an axial-flow fan, a centrifugal fan, or a positive-pressure blower. For certain conditions, OSHA regulations require the ventilation mode to be reversible. The latter two fans can be reversed by means of a piping and valving arrangement. Axial-flow fans can be equipped with a reversing switch. Greater pressure differential can be obtained by placing two or more fans in series, and greater capacity by placing them in parallel.

For very long tunnels, current practice is to install axial fans in the vent line as the tunnel progresses. With this system, the pressure extremes are reduced and capacity is added only as needed. The system can be operated in either the pressure or exhaust mode. Switches for reversing the flow can be controlled from the surface. Efficiency, and consequently capacity, of the fans is reduced when operated in reverse. In all installations, the vent line should be designed to resist expected external as well as internal pressure, the amount depending on the characteristics and the operating plan for the system.

In the past, collapsible vent tubing, also called bag line, was considered suitable only for short jobs, and for temporary and auxiliary use, such as supplemental ventilation to the working face. On long drill-and-blast tunnels, the bag line often became ripped or torn, and insufficient air was delivered to the face.

Recently, a good-quality bag line was used successfully in a long TBM tunnel. The system normally operates in the blow mode. A canister holding as much as 500 ft of collapsed bag is mounted on the TBM trailing gear, and the pressurized air is forced through the canister as the TBM moves forward and telescopes the bag. After traveling 500 ft, the empty canister is replaced. A second fan is located at the discharge of the canister and can be activated to operate the system in the exhaust mode.

Also in the past, steel vent pipe was considered the most satisfactory type for use in long tunnels. Spiral-weld pipe could be fabricated at the site from steel strip. Recently, vent pipe fabricated from sheet plastic and other synthetics has become competitive with steel in price and is becoming popular because of its light weight and ease of installation. Couplings should be strong enough for the particular job, and some provision, such as soft rubber strips, must be made for sealing the joints against leakage.

The two most important variables affecting the design of a system of given capacity are the vent pipe diameter and the power required. The greater the pipe diameter, the smaller the power requirement. But larger-diameter pipe costs much more than that of smaller diameter, being both larger and stronger, and in any tunnel there is a limit to the size that can be accommodated. Usually, several alternative combinations must be evaluated, considering both first cost and operating costs for the life of the job.

The required volume of air can be ascertained as previously noted. Lacking any specified criteria, a minimum velocity in the tunnel of 50 ft/min and a minimum of 200 cfm/workman plus 100 cfm/diesel brake horsepower should be provided.

From Figure 13-29, the total friction loss for various pipe diameters can be calculated. Fan manufacturers' catalogs can then be consulted for selecting a suitable fan for each likely pipe diameter. Most axial-flow fans develop static pressures of 5–15 in. water gauge (W.G.) and, generally, a fan spacing of 1,000–3,000 ft is appropriate.

For example, consider a 20-ft-diameter tunnel requiring 20,000 cfm of ventilation air. The chart shows a range of pipe diameters from 18 to 66 in. However, a 3,000-ft fan spacing with a 15-in.-W.G. fan would indicate a friction loss of 0.5 in. W.G./100 ft, requiring a 31-in.-diameter pipe. Similarly, a 1,000-ft fan spacing would indicate a 24-in.-diameter pipe. For a loss of 5 in. W.G., corresponding figures are 39-in. and 30-in. diameters. From this, it is apparent that vent pipe sizes in the 24–42 in. diameter range could be considered.

When making final calculations, the velocity head must be added to the total friction loss to find the total dynamic head required. The velocity head in inches W.G. is

$$H = \frac{V^2}{4{,}008}$$

where V = air velocity in the vent pipe in ft/min. Additional friction losses from filters, elbows, etc., if significant, can also be added.

Tables of friction loss and manufacturers' fan performance data are based on dry air at 70°F at sea level, having a density of 0.075 lb/ft^3. Since both the friction loss in the vent pipe and the pressure developed by the fan are proportional to density, the altitude of the installation can be ignored except for the determination of the required fan horsepower. The actual horsepower required is proportional to the density of the air and can be calculated as follows:

required hp = density ratio × manufacturer's specified hp

The density ratio for various altitudes and temperatures is given in Table 13-19.

Compressed Air

Most tunnel projects use tools and equipment powered by compressed air. However, with the advent of hydraulic drills for drilling blast holes in the heading, the demand for compressed air on most tunnel jobs has greatly diminished. In fact, many current tunnel jobs do not have a central compressed-air system and use small portable compressors, usually electrically driven, where needed in the tunnel or elsewhere.

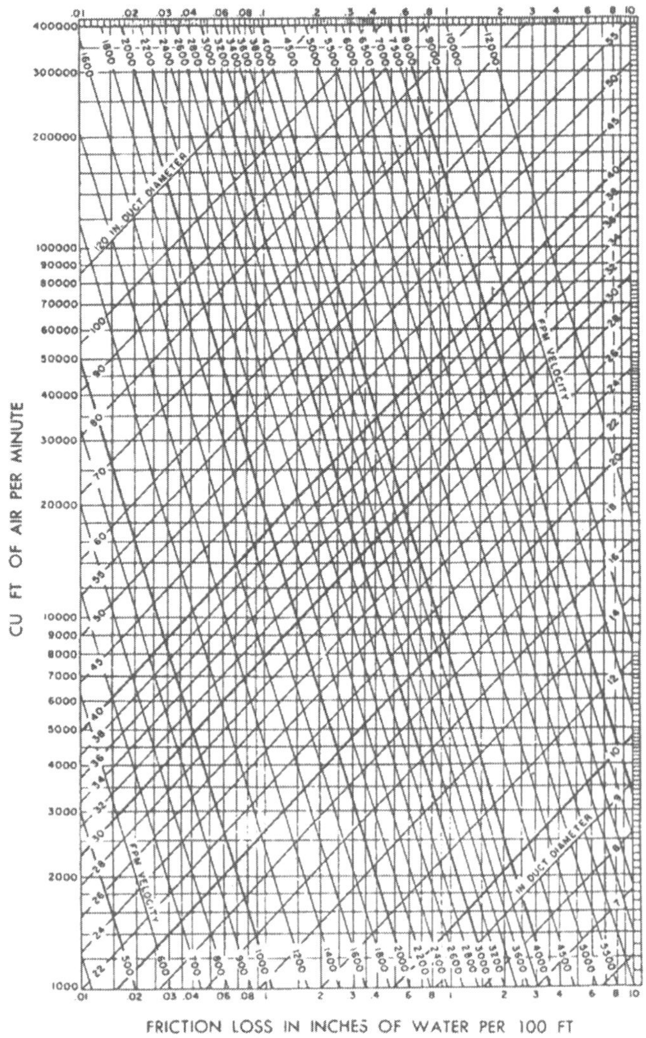

Fig. 13-29. Friction of air in straight ducts.

Table 13-19. Density Ratios (Based on Air Being Dry and Weighing 0.075 lb/foot^3 at Sea Level at 70°F)

Altitude above Sea Level (ft)	Standard Atmospheric Pressure		Air Density Ratios at Various Altitudes and Temperatures				
			Air Temperature (F)				
	psi	inches of mercury	50	60	70	80	90
0	14.69	20.92	1.039	1.019	1.000	0.981	0.964
1,000	14.16	28.86	1.003	0.984	0.965	0.947	0.913
2,000	13.66	27.82	0.966	0.948	0.930	0.913	0.880
3,000	13.17	26.82	0.931	0.913	0.896	0.879	0.847
4,000	12.69	25.84	0.898	0.880	0.864	0.848	0.817
5,000	12.23	24.90	0.865	0.848	0.832	0.817	0.787
6,000	11.78	23.99	0.833	0.817	0.802	0.787	0.758
7,000	11.34	23.10	0.802	0.787	0.772	0.758	0.730
8,000	10.91	22.23	0.772	0.757	0.743	0.729	0.702
9,000	10.50	21.39	0.743	0.729	0.715	0.702	0.676
10,000	10.10	20.58	0.715	0.701	0.688	0.675	0.650

For projects of short duration or in areas remote from electrical power sources, portable diesel-driven compressors may be used to supply a central system. These are available in various sizes up to 1,200-cfm capacity and can be quickly mobilized and put into operation.

Where electric power is available, electric-driven compressors are more economical for all but jobs of very short duration.

Slow-speed, double-acting compressors that are water cooled and have separate intercoolers and aftercoolers are the most efficient and offer the lowest ultimate maintenance cost. They are usually powered by synchronous motors providing a favorable power factor. However, the requirements for most current tunnel projects do not warrant the use of such equipment.

The volume of compressed air consumed by various items of pneumatic equipment is affected by

- Atmospheric pressure
- Compressed-air temperature
- Gauge pressure
- Equipment design
- Equipment condition
- Work conditions

Table 13-20 lists the approximate consumption rates of various rock drills and other items of pneumatic equipment, at sea level. To obtain the rated output at altitudes above sea level, the free air volume to be supplied must be increased above the volume required at sea level. Table 13-21 lists the multipliers to be used in calculating the minimum compressor plant capacity when a compressor plant is to be operated at various altitudes above sea level.

Compressors are usually designed for use over a specific range of altitudes. Two-stage reciprocating compressors are not significantly affected by changes in altitude below 3,300 ft. At greater elevations, the first-stage piston diameter must be increased to maintain the sea level production rating. If a compressor is to be used at an altitude other than that for which it was designed, the manufacturer should be consulted.

Compressor plant capacity should be increased 20% over theoretical to allow for miscellaneous losses such as leaks at couplings, valves, etc.

Table 13-20. Air Consumption Data (Based on 90–100 psi at Sea Level)

Drills											
	Cylinder Diameter of Drill (In.)										
Type	2-1/2	2-5/8	2-11/16	2-3/4	3	3-1/2	4	4-1/2	5	5-1/2	6
Air							400	575	650	750	900
Track							325	400	475	500	
Drifter				150	200	275					
Jackleg	125	150	175		210						

Miscellaneous Equipment		
Item	Specification	cfm/unit
Stopers	Light	155
	Heavy	180
Sinkers	Under 30 lb	70
	30 to 50 lb	100
	51 to 65 lb	125
Paving Breakers	30 to 40 lb	30
	41 to 65 lb	60
	66 to 85 lb	75
Spaders		40
Jumbo hydraulic pumps (1 required per 3 jibs)		125
Jumbo air lights		15
Muckers	20 hp	250
	37 hp	325
	50 hp	530
Diaphragm sump pumps; outlet hose		50
	2 in.	
	2-1/2 in.	85
	3 in.	100
Shotcrete machines:		
per CY per hour		50 to 100
Air saws		50

Friction losses in air lines create a loss of pressure. Table 13-22 lists the approximate losses due to friction in various diameter pipes and hoses over a range of flow rates. Pressure losses due to valves and fittings are tabulated in most compressed-air handbooks. The compressor plant discharge pressure should be adjusted so that the pressure at the point of use is kept at 90–100 psig.

Light-gauge, spiral-welded steel pipe is commonly used for the air line. Couplings should be light and simple, such as the "Victaulic" type. Outlet tees, with valves, should be installed periodically. The diameter of the main air line must be great enough so that sufficient pressure can be maintained where needed. Receivers should be provided where large intermittent demands occur.

Low-Air Plant

Before the advent of soft ground tunneling machines capable of handling subaqueous unstable soils, tunnels in such ground were often driven in compressed air. Now, the use of compressed-air tunneling is limited to special situations, and a comprehensive description of the required special facili-

Table 13-21. Air Consumption Multipliers for Altitude Operation (Courtesy Ingersoll-Rand Co.)

Altitude Above Sea Level (ft)	Multiplier
0	1.00
1,000	1.02
2,000	1.05
3,000	1.08
4,000	1.11
5,000	1.14
6,000	1.18
7,000	1.22
8,000	1.26
9,000	1.30
10,000	1.34
12,500	1.46
15,000	1.58

Table 13-22. Pressure Losses Due to Line Friction (Courtesy Ingersoll-Rand Co.)

cfm Free Air at 100 psig Line Pressure	Pressure Loss in psi per 100 ft of Straight Pipe							
	Nominal Pipe Diameter-Schedule 40 (in.)							
	2	3	4	5	6	8	10	12
150	0.15							
200	0.26							
300	0.57	0.07	0.05					
500	1.51	0.20	0.11					
750	3.36	0.44	0.19					
1,000	5.90	0.76	0.43	0.06				
1,500		1.68	0.75	0.13	0.05			
2,000		2.99	1.16	0.24	0.09			
2,500		4.67	1.64	0.36	0.14			
3,000		6.71	2.90	0.51	0.20	0.05		
4,000		11.90	4.50	0.90	0.35	0.09		
5,000			6.45	1.40	0.55	0.12		
6,000			11.50	2.00	0.73	0.19	0.06	
8,000				3.53	1.38	0.33	0.10	
10,000				5.47	2.13	0.52	0.16	0.06
12,500					3.30	0.80	0.25	0.08
15,000					4.75	1.14	0.36	0.12

cfm Free Air at 100 psig Line Pressure	Pressure Loss in Hose (psi) [a]						
	Hose Length and Inside Diameter						
	50 ft, 1 in.	50 ft, 1-1/4 in.	50 ft, 1-1/2 in.	50 ft, 2 in.	50 ft 2/12 in.	50 ft, 3 in.	25 ft, 4 in.
150	2.7						
200	4.8						
250	6.9	2.4					
300	9.9	3.4					
400		5.8	2.4				
500		8.9	3.7				
600		12.6	5.2				
700			7.0				
800			9.0	2.1			
1,000			13.6	3.2			
1,200				4.5	2.4		
1,500				6.9	4.2		
2,000				12.2	9.3	3.6	
3,000						6.3	
4,000						9.6	
5,000						13.6	1.4
6,000							1.7

For interpolation

To other pressure: $\Delta P_2 = \frac{P_1 + P_a}{P_2 + P_a} \times \Delta P_1$

To other volume: $\Delta P_2 = \frac{(Q_2)^2}{(Q_1)^2} \times \Delta P_1$

P_1 = Pressure loss from table for a certain volume and pipe diameter at 100 psig (table values are valid for a pressure range of 10% of the entrance pressure psia)
P_2 = Pressure loss in question for volume or pressure other than table value
P_1 = 100 psig
P_2 = Pressure other than 100 psig
P_a = Atmospheric pressures
Q_1 = Volume of free air from table
Q_2 = Volume other than table value.

[a] Assumes lubrication is at tool and air in hose does not have oil from line oiler.

ties is outside the scope of this handbook. In the following paragraphs, the principal items are noted.

The low-air compressors should be designed for the particular duty expected. Working pressures above 20 psi are seldom used because of the very high labor costs associated with the corresponding operating regulations. Some other means for building the tunnel is usually selected. In any case, the maximum capacity of the compressors should be about double the expected working pressure.

The total volume of required free air must provide for leakage through the ground, losses through the locks, and adequate ventilation. Leakage through the ground can be difficult to predict where the ground is pervious and lacks experience data from other projects. A rule of thumb formula for average conditions is

$$Q = 12D^2$$

where Q is the volume of free air in cfm and D is the diameter of the tunnel in ft. Actual requirements could vary from 50% under to 100% over this amount. Even more air may be required when an impervious lining is not installed in the shield tail, allowing air to escape throughout the tunnel.

In large tunnels, the bulkhead that isolates the pressurized tunnel should have three locks, one for muck and materials, one for personnel, and one for emergencies. In smaller tunnels, one lock may serve all three purposes; it consists of two bulkheads with suitable doors and other facilities. Separate personnel and emergency locks should be located near the tunnel crown to facilitate access in case of flooding. Safety screens and high-level runways are used in subaqueous tunnels subject to possible floods.

The size of the pipe required to convey a given volume of air can be estimated:

$$(P_1^2 - P_2^2)d^5 = 0.0005Q^2L$$

where

d = diameter of pipe, in.
Q = volume of free air, cfm
L = length of pipe, ft
P_1 = absolute initial air pressure, psi
P_2 = absolute terminal air pressure, psi

For a given condition, the values of P_2, Q, and L are fixed. Then, the required value of d depends on the value of P_1, which is controlled by the characteristics of the compressor. Note that the loss of pressure in the pipeline is greater when P_2, the terminal pressure, is lower. The reason is that a given volume of free air must flow through the pipe at a higher velocity when the pressure is lower. Although the pressure in the tunnel can be maintained with the air pipe terminated on the heading side of the bulkhead, it is good practice to carry the pipe up to the heading to assure an adequate supply of fresh air at that location.

A number of other facilities and appliances are needed for compressed-air tunneling. These include the following:

1. Water discharge pipe. The air pressure in the tunnel can blow out accumulated water.
2. Foul air pipe. When air leakage is very low, the foul air can be evacuated through a separate pipe. A relatively small diameter is sufficient.
3. Fire protection equipment and facilities.
4. Safety blow-off valves in the locks and bulkheads.
5. Air pressure gauges in the locks and working chambers.
6. Automatic valves for controlling the compression and decompression in the locks.
7. Recording monitors to continuously record the air pressures in the locks and working chamber and the quantity of air being used.
8. Constant availability of a medical lock in the vicinity of the job.
9. Other items dictated by the particular job conditions, local safety regulations, and the like.

Additional information can be found in Chapter 18.

Lighting

In the past, an evenly spaced string of light bulbs was the usual type of general lighting. Ordinary light bulbs no longer meet OSHA intensity and lifetime requirements. Recently, fluorescent and sodium vapor fixtures have become popular. On some projects, no general lighting is provided, and personnel are supplied with flashlights or cap lamps for emergencies and general use. Safety regulations may require specific lighting standards in tunnels and should be consulted when planning a particular job.

Floodlights are used for lighting work areas, and are permanently mounted in strategic locations on jumbos and other working structures.

Power is usually 110–120 V, single-phase, supplied by dry transformers spaced throughout the tunnel. Insulated wires separated about 6 in. are used for distribution. Power for additional lighting can be tapped off these lines by using "pigtails." The wires may be held to the rock, concrete, steel, or timber by various methods such as wooden dowels, expansion bolts, or concrete inserts.

Drill jumbos sometimes have compressed-air-driven floodlights since explosive handling should not be done with electric power on the jumbo. However, the use of non-electric detonators alleviates this problem.

Communications and Signal Systems

The efficiency and safety of any tunnel project is improved by the use of well-planned communications and signal systems.

Typical communications system terminals on a tunnel project are

- Tunnel headings
- Tunnel boring machine operator's console
- Intermediate points along tunnel alignment
- Shaft—top, bottom, and landings
- First aid station
- Superintendent's office
- Project engineer's office
- Project manager's office
- Mechanic's and electrician's shops
- Batch plant
- Compressor house
- Inspector's office

In addition, locations for signal system terminals might include the following:

- Materials transfer points
- Turnouts
- Passing sections
- Blind curves in alignment

One satisfactory means of controlling access to rail haulage passing sections uses a system of colored lights, visible to locomotive operators. Pull cords, hanging from the tunnel crown within reach of the locomotive operator, are used to regulate the lights on both ends of the passing section. Distance from the signal lights to the passing section must be greater than the stopping distance of the train.

A closed-circuit television system can be effectively used to monitor conditions and operations at many locations on a tunnel project, including materials transfer points, shaft landings, tunnel alignment blind spots, and the construction yard.

Telephones in the tunnel or at noisy locations should be equipped with signal devices that can be seen as well as heard when the telephone rings.

Water Supply

Waterline diameter is determined by the volume and pressure requirements and the length of line. Tables 13-23 and 13-24 give approximate friction losses for pipe and hose. Friction losses for pipe fittings are tabulated in most piping handbooks. Light-gauge, welded-steel pipe with bolted couplings is commonly used in tunnels. Tees and valves should be spaced at about 1,000-ft centers.

Drainage and Dewatering

A certain amount of groundwater entering a tunnel being driven upgrade will flow to the shaft or portal by gravity. It should be confined to a ditch located on one side, or it can flow on the invert of machine-excavated rock tunnels. To maintain their capacities, such waterways must be kept open and free from sediment.

When the water inflow exceeds the capacity of the open channels, when the tunnel is driven downgrade, or when the flowing water would have a deleterious effect on the tunnel, pumping through pipelines is necessary. The water is collected in sumps spaced as necessary along the tunnel and pumped through a discharge pipe by means of centrifugal pumps. The sumps should be large enough to handle expected surges and to cope with possible pump stoppages. They should be deep enough to reduce the need for frequent cleaning. If the water contains abrasive sediment, a settling compartment in the sump will save wear in the pumps and pipelines.

Water in the heading is collected in low spots or temporary sumps and pumped back to the first permanent sump by means of portable submersible pumps, powered by either compressed air or electricity. The discharge pipe is usually hung on the tunnel sidewall.

Since the quantity of groundwater flows cannot be predicted with any accuracy, and since the flows usually diminish with time, the dewatering system should have maximum flexibility. Automatic float switches should be provided at main sumps. The pump characteristics should be compatible with a broad range of pressures and flows.

Table 13-23. Water Friction in 100 ft. of Light-Gauge, Spiral-Welded Steel Pipe.

Discharge		4-inch I.D.		5-inch I.D.		6-inch I.D.		8-inch I.D.		10-inch I.D.		12-inch I.D.		
GPM	cu ft/sec	Velocity (ft/sec)	Loss (fl)	Velocity (ft/sec)	Loss (fl)	Velocity (ft/sec)	Loss (fl)	Velocity (ft/sec)	Loss (fl)	Velocity (ft/sec)	Loss (fl)	Velocity (ft/sec)	Loss (fl)	Discharge (GPM)
1	0.002228													1
5	0.01114													5
10	0.02228													10
15	0.03342	0.38281	0.2166											15
20	0.04456	0.51042	0.03519	0.01228										20
25	0.05570	0.63802	0.05176	0.40838	0.01801	0.28360	0.00759							25
30	0.06684	0.76563	0.07236	0.49002	0.02496	0.34032	0.01042	0.19146	0.00262					30
40	0.08912	1.0208	0.12135	0.65337	0.04135	0.45376	0.01739	0.25527	0.00435					40
50	0.11140	1.2760	0.18050	0.81671	0.06139	0.56720	0.02576	0.31910	0.00647	0.20425	0.00223			50
60	0.13368	1.5312	0.25118	0.98005	0.08518	0.68065	0.03576	0.38292	0.00901	0.24510	0.00311			60
70	0.15596	1.7864	0.33160	1.1434	0.11204	0.79409	0.04699	0.44674	0.01168	0.28595	0.00406			70
80	0.17824	2.0462	0.42713	1.3067	0.14315	0.90753	0.05959	0.51056	0.01493	0.32680	0.00515			80
90	0.20052	2.2969	0.52835	1.4700	0.17715	1.0209	0.07411	0.57438	0.01821	0.36765	0.00637	0.25530	0.00263	90
100	0.22280	2.5521	0.64019	1.6334	0.21374	1.1344	0.08871	0.63821	0.02219	0.40850	0.00768	0.28637	0.00318	100
125	0.27850	3.1901	0.97129	2.0417	0.32451	1.4180	0.13363	0.79776	0.03156	0.51063	0.01132	0.35459	0.00470	125
150	0.33420	3.8281	1.3448	2.4501	0.44738	1.7016	0.18613	0.95731	0.04567	0.61725	0.01611	0.42551	0.00655	150
175	0.38900	4.4662	1.7932	2.8585	0.59373	1.9852	0.24478	1.1168	0.06042	0.71488	0.02066	0.49643	0.00861	175
200	0.44560	5.1042	2.3422	3.2668	0.75949	2.2688	0.31330	1.2764	0.07664	0.81701	0.02624	0.56735	0.01099	200
225	0.50130	5.7422	2.8721	3.6752	0.94120	2.5524	0.38845	1.4359	0.09508	0.91913	0.03226	0.63827	0.01353	225
250	0.55700	6.3802	3.4890	4.0835	1.1494	2.8360	0.46956	1.5944	0.11502	1.0212	0.03886	0.70918	0.01624	250
275	0.61270	7.0183	4.1759	4.4919	1.3684	3.1196	0.56214	1.7550	0.13630	1.1233	0.04631	0.78010	0.01937	275
300	0.66840	7.6563	4.9150	4.9002	1.6105	3.4032	0.65821	1.9146	0.16051	1.2255	0.05428	0.85102	0.02249	300
325	0.72410	8.2943	5.7364	5.3086	1.8797	3.6868	0.76401	2.0741	0.18535	1.3726	0.06632	0.92194	0.02600	325
350	0.77980	8.9324	6.6157	5.7170	2.1556	3.9704	0.87631	2.2337	0.21266	1.4297	0.07122	0.99286	0.02969	350
375	0.83550	9.5704	7.5520	6.1253	2.4606	4.2540	0.99470	2.3932	0.24145	1.5318	0.08132	1.0637	0.03373	375
400	0.89120	10.208	8.5430	6.5337	2.7837	4.5376	1.1189	2.5528	0.27168	1.6340	0.09154	1.1347	0.03778	400
425	0.94690	10.846	9.6440	6.9420	3.1065	4.8212	1.2559	2.7123	0.30327	1.7361	0.10277	1.2056	0.04220	425
450	1.0026	11.484	10.689	7.3504	3.4627	5.1048	1.4000	2.8719	0.33809	1.8382	0.11333	1.2765	0.04680	450
475	1.0583	12.122	11.841	7.7587	3.8357	5.3884	1.5508	3.0315	0.37244	1.9404	0.12488	1.3474	0.05186	475
500	1.1140	12.760	13.044	8.1671	4.2253	5.6720	1.6984	3.1910	0.41028	2.0425	0.13681	1.4183	0.05684	500
550	1.2254	14.036	15.695	8.9838	5.0526	6.2392	2.0430	3.5010	0.49068	2.2467	0.16365	1.5602	0.06765	550
600	1.3368	15.312	18.566	9.8005	5.9413	6.8065	2.4026	3.8298	0.57713	2.5410	0.19253	1.7020	0.07916	600
650	1.4482	16.588	21.662	10.617	6.9304	7.3737	2.7860	4.1483	0.66933	2.6552	0.22199	1.8438	0.09184	650
700	1.5596	17.864	24.974	11.434	7.9892	7.9409	3.2114	4.4674	0.87877	2.8595	0.25444	1.9857	0.10469	700
750	1.6710	19.140	28.494	12.250	9.1142	8.5081	3.6867	4.7865	0.88045	3.0637	0.29033	2.1275	0.11877	750
800				13.067	10.307	9.0753	4.1436	5.1056	0.99572	3.2980	0.32833	2.2694	0.13434	800
850				13.884	11.564	9.6425	4.6198	5.4247	1.1103	3.4723	0.36617	2.4112	0.15076	850
900				14.700	12.883	10.209	5.1461	5.7438	1.2371	3.6765	0.40799	2.5530	0.16799	900
950				15.517	14.355	10.776	5.6977	6.0630	1.3699	3.8808	0.44899	2.6949	0.18494	950
1,000				16.334	15.807	11.344	6.2740	6.3821	1.5084	4.0850	0.49439	2.9367	0.202253	1,000
1,100						12.478	7.5432	7.0203	1.8022	4.4935	0.59068	3.1204	0.24190	1,100
1,200						13.613	8.9200	7.6585	2.1311	4.9020	0.69401	3.4041	0.28427	1,200
1,300						14.747	10.400	8.2967	2.4851	5.3105	0.80399	3.6877	0.32940	1,300
1,400						15.881	11.983	8.9349	2.8448	5.7190	0.92634	3.9714	0.37957	1,400
1,500						17.016	13.757	9.5731	3.2444	6.1275	1.0564	4.2551	0.43012	1,500
1,600								10.201	3.6668	6.5360	1.1940	4.5388	0.48300	1,600
1,800								11.487	4.6099	7.3531	1.5011	5.1061	0.50321	1,800
2,000								12.764	5.5777	8.1701	1.8283	5.6735	0.73742	2,000
2,500								15.955	8.6566	10.212	2.7980	7.0918	1.1245	2,500
3,000								19.146	12.294	12.255	3.9737	8.5102	1.5885	3,000
3,500										14.297	5.3321	9.9286	2.1276	3,500
4,000										16.340	6.8655	11.347	2.7588	4,000
4,500										18.382	8.6257	12.765	3.4662	4,500
5,000										20.425	10.572	14.183	4.2165	5,000
6,000										24.501	15.112	17.020	6.6974	6,000

Note: 1 ft of water = 0.4335 psi

Consideration must be given to the disposal of the tunnel drainage. Local environmental regulations may dictate the use of settling ponds or other forms of treatment.

SURFACE PLANT

The surface plant can be classified into the following categories:

- Access and yards
- Operational facilities
- Shops
- Warehousing and storage
- Sanitary and medical facilities
- Offices
- Camp facilities
- Utilities
- Environmental

The extent to which these items will be required depends on the job and its location. Large projects in remote locations require the most extensive facilities. The type of con-

Table 13-24. Water Friction in 100 feet of Smooth Bore Hose.[a] (Courtesy Gorman Rupp Co.)

Flow (GPM)	1/2 Inch		3/4 Inches		1 Inch		1-1/4 Inch		1-1/2 Inches		2 Inches		2-1/2 Inches		3 Inches	
	Velocity (ft/sec)	Friction Head (ft)	Velocity (ft/sec)	Friction Head (ft)	Velocity (ft/sec)	Friction Head (ft)	Velocity (ft/sec)	Friction Head (ft)	Velocity (ft/sec)	Friction Head (ft)	Velocity (ft/sec)	Friction Head (ft)	Velocity (ft/sec)	Friction Head (ft)	Velocity (ft/sec)	Friction Head (ft)
1.5	1.6	2.3	1.1	0.97												
2.5	2.6	6.0	1.8	2.5												
0.5	5.2	21.4	3.6	8.9	2.0	2.2	1.3	0.74	0.9	0.3						
10	10.5	76.8	7.3	31.8	4.1	7.8	2.6	2.64	1.8	1.0	1.0	0.2				
15			10.9	68.5	6.1	16.8	3.9	5.7	2.7	2.3	1.5	0.5				
20					8.2	28.8	5.2	9.6	3.6	3.9	2.0	0.9	1.3	0.32		
25					10.2	43.2	6.5	14.7	4.5	6.0	2.5	1.4	1.6	0.51		
30					12.2	61.2	7.8	20.7	5.4	8.5	3.1	2.0	2.0	0.70	1.4	0.3
35					14.3	80.5	9.1	27.6	6.4	11.2	3.6	2.7	2.3	0.93	1.6	0.4
40							10.4	35.0	7.3	14.3	4.1	3.5	2.6	1.2	1.8	0.5
45							11.7	43.0	8.2	17.7	4.6	4.3	2.9	1.5	2.0	0.6
50							13.1	52.7	9.1	21.8	5.1	5.2	3.3	1.8	2.3	0.7
60							15.7	73.5	10.9	30.2	6.1	7.3	3.9	2.5	2.7	1.0
70									12.7	40.4	7.1	9.8	4.6	3.3	3.2	1.3
80									14.5	52.0	8.2	12.6	5.2	4.3	3.6	1.7
90									16.3	64.2	9.2	15.7	5.9	5.3	4.1	2.1
100									18.1	77.4	10.2	18.9	6.5	6.5	4.5	2.6
125											12.8	28.6	8.2	9.8	5.7	4.0
150											15.3	40.7	9.8	13.8	6.8	5.6
175											17.9	53.4	11.4	18.1	7.9	7.4
200											20.4	68.5	13.1	23.4	9.1	9.6
225													14.7	29.0	10.2	11.9
250													16.3	35.0	11.3	14.8
275													18.0	42.0	12.5	17.2
300													19.6	49.0	13.6	20.3
325															14.7	23.5

Note: 1 ft of water = 0.4335 psi.

[a] For various flows and hose sizes, table gives velocity of water and feet of head lost in friction in 100 ft of smooth bore hose. Sizes of hose shown are actual inside diameters.

struction, weatherproofing, etc. will be influenced by the climate and other local conditions. For many applications, buildings can be composed of single or multiple trailers.

Access and Yards

Access to the work may vary from a simple driveway from a city street to miles of roadway through mountainous terrain. The quality of the access must be adequate to assure that personnel, supplies, and equipment can expeditiously reach the site. For large projects involving much traffic, roads should be paved. Railroad access may be indicated in some cases for equipment and material delivery. Access by waterway and by air are sometimes necessary.

When access to the work site is difficult, time can usually be gained if the owner constructs access roads, air landing strips, or wharfs during the design and bidding period for the tunnel.

Yards should be arranged for good communication between the various facilities. Special attention should be given to arrangement and location of storage and parking areas. If rail haulage is to be used in the tunnel, the railroad system must be extended from the tunnel or shaft to the yard for servicing of warehouses, powder magazines, repair shops, and storage yards.

In urban areas and rugged mountainous terrain, there may not be sufficient space for an ideal yard layout. In such cases, it is necessary to make the best use of whatever is available. In congested urban areas, the owner may avoid delays and claims by undertaking right-of-way negotiations for contractor work areas in advance of advertising the construction contract.

Operational Facilities

These include batch-mix plants, air-compressor stations, power substations, etc. Their layout should take local conditions into account. For example, the mode of delivery of cement and concrete aggregates will influence the location and layout of the batch-mix plant and associated storage facilities. In cold climates, a compressor house located adjacent to the repair shop will provide supplemental heating, but in a hot climate this would be undesirable.

Shops

Every job needs a repair shop for the repair and maintenance of equipment. Required facilities depend on the job, but might include metal-working machinery, welders, bit grinders, drill-steel reconditioning equipment, battery shop, electrical shop, tire shop, etc. Depending on the type of equipment to be serviced, pits, hoists, and cranes may be necessary.

Other shops may be needed for specialized job requirements. These might include a woodworking shop, a reinforcing-steel cutting and bending shop, a steel fabrication shop, etc.

Warehousing and Storage

Enclosed storage is needed for most repair parts, job supplies, and various materials and small tools. The warehouses should be designed for easy handling and good identification and retrieval of stored items. A computerized system for inventory control should be considered.

Large or heavy items can frequently be stored in the open. Again, provision for storing, handling, and retrieval must be provided.

If the job requires explosives, a magazine for high explosive and a separate magazine for detonating devices are necessary. The type of construction and location is dictated by safety regulations. Where safety regulations do not prohibit, a make-up house for preparing primers is also provided.

Sanitary and Medical Facilities

In addition to the usual sanitary facilities, a change house must be provided for personnel. This will include lockers for street clothes and racks or suspended hooks for storing and drying work clothes between shifts. Showers, wash basins, and toilet facilities must be provided, as well as adequate heat and ventilation. The change house should be convenient to the tunnel access.

Minimum medical facilities would comprise at least one first aid station with supplies for treating minor cuts, bruises, and burns and for handling emergency cases. In remote areas, or where other circumstances warrant, a job hospital may be indicated. Compressed-air tunnel jobs require access to a medical lock together with adequate examination facilities. Consult OSHA regulations for additional requirements.

Offices

Office space should be provided for project administrative, engineering, and clerical staffs. The extent and type of facilities will depend mainly on the size and nature of the job. Even in urban areas, combinations of trailers are frequently found most satisfactory.

Camp Facilities

In remote areas, a construction camp may be necessary. Its extent will vary with the duration and size of the job as well as its location. Quarters may include houses, cottages, and dormitories. A suitable mess hall with adequate kitchen facilities will be needed, as well as a commissary. Entertainment and recreational facilities are a must, their nature depending on the particular situation. The need for other services, such as laundry, post office, banking, and automobile service, will depend on the proximity to established communities.

Utilities

The yard and camp area must, of course, be provided with necessary utilities. Electric power, water supply, and sewage disposal are practically always needed. Other possibilities include natural or bottled gas, refuse pickup, telephone service, radio communication, etc.

Environmental Requirements

In remote areas, the location and final landscaping of muck disposal areas must be given adequate consideration. Water discharged from the tunnel may require treatment before disposal. Cleanup and landscaping of yard and camp areas is usually required, and replacement or protection of trees is demanded in National Forests.

In urban areas, many more restrictions are found. Yard areas may have to be guarded from public view by suitable fencing. Compressors, fans, and other equipment must be muffled and muck bins, batch plants, etc., insulated to minimize noise. Dust and other atmospheric pollutants must be controlled. Consideration must be given to traffic regulations and the flow of traffic near the project site.

CONCRETE PLANT

The concrete plant consists of storage facilities, batch plant, mixer, transport equipment, placing equipment, forms, and ancillary equipment. In urban areas, it is frequently possible to purchase transit-mixed concrete delivered to the job site, thus eliminating the need for storage and batch-mix facilities. However, the dependability of the supply should be confirmed. Some specifications have concrete temperature restrictions that demand ice-making or other cooling facilities.

Batching and Mixing

The batch plant should be located so that ample storage of aggregates can be provided, depending on the nature and dependability of local deliveries. Stockpiling and reclaiming systems will depend on the volumes to be stored and available storage areas. Protection against winter cold and summer heat may be necessary.

Cement is invariably handled in bulk, and so the batch plant will include cement storage facilities of adequate capacity. Bins for handling the required sizes of aggregates are also needed. Batching of cement and aggregate is by weight and that of water by either weight or volume. See Figure 13-30 for an example of a typical mobile batch-mix plant. Automatic batching is available and sometimes required.

Mixing is usually done at the batch plant unless the transport time would be excessive. Tilting mixers are most commonly used, although horizontal axis and turbine mixers are also satisfactory. When mixing in the tunnel, the horizontal axis type is usually indicated due to space limitations. The mixing plant should have ample capacity for the concrete placing schedule.

Transport

Mixed concrete is transported into the tunnel in transit mixers, or agitators. These can be mounted on either pneumatic-tired truck chassis or railroad cars. Since mixing in the tunnel is usually done only when the haul is extremely long, dry batches are normally transported in compartmented rail cars, each compartment holding one batch, in which the cement is isolated from the aggregates. Special agitators for use with rail cars can discharge successively through each other, so that switching in the tunnel is unnecessary. They are powered by compressed-air or electric motors that must be connected at the placing site.

Concrete can also be transported through the tunnel by pipeline, using a positive displacement pump. With modern concrete pumps, distances in the range of 2,000 ft can be achieved. Greater distances are possible when using two or more individual setups in tandem.

Mixed concrete can be dropped through vertical pipes into the tunnel, either at a working shaft or at other locations. It is not necessary to keep the pipe full, and there seems to be no practical limit to the distance dropped, friction and the displaced air apparently limiting the velocity.

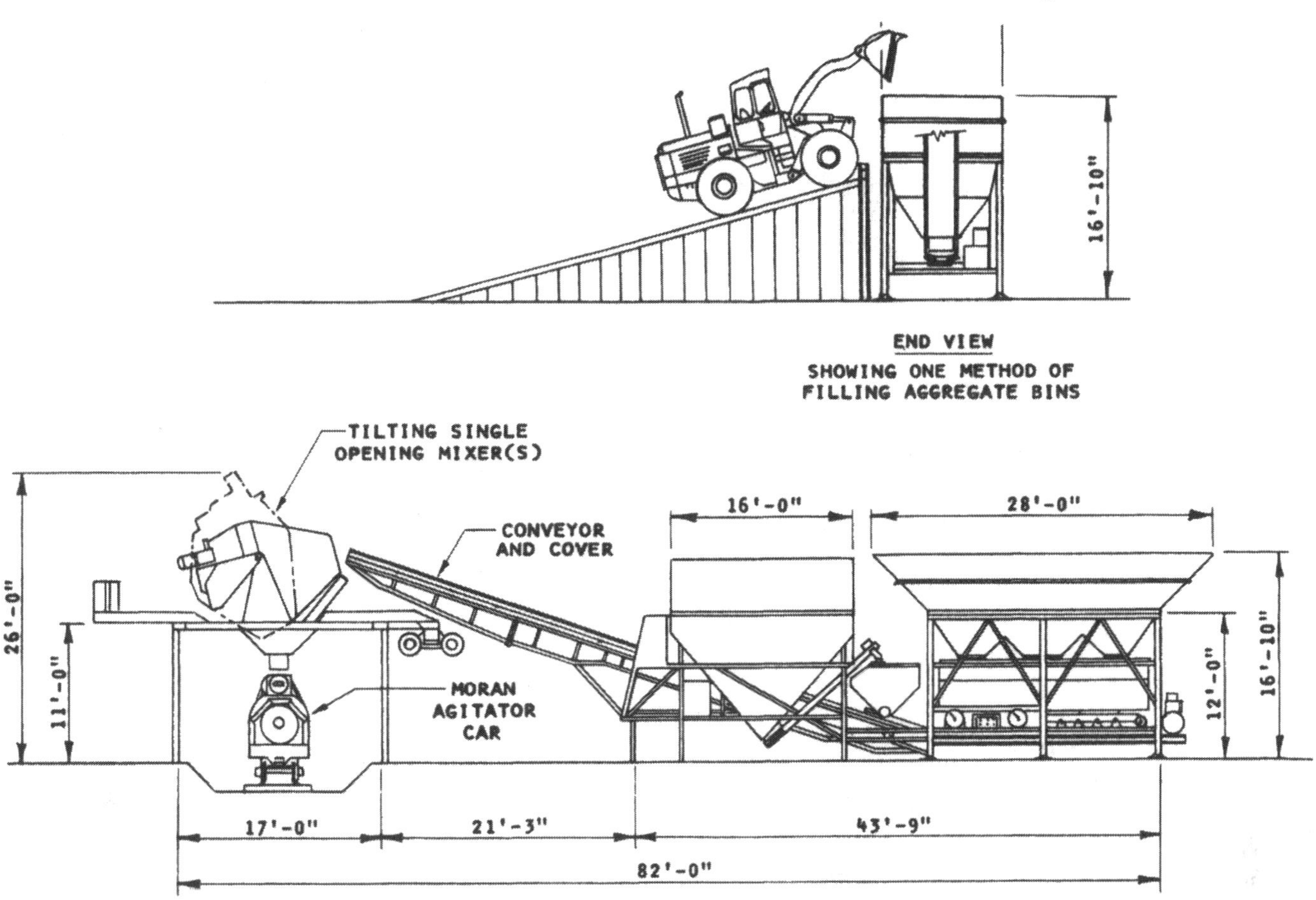

Fig. 13-30. Mobile batch plant.

The pipe must, of course, discharge into a terminal that can absorb the energy of the falling material and deliver it without segregation into the conveyance in the tunnel. Pumping concrete downward through vertical or steeply inclined pipes requires a device at the discharge end of the pipe to keep the pipe full and thus avoid segregation with subsequent plugging of the pipe. Investigate experience on other projects.

Placing

Various methods are used for placing the concrete in its final position in the tunnel. Simple low pours, such as a sidewall footing, can usually be made by merely dumping the concrete out of its transport vehicle, using a chute if necessary. For higher pours into open forms, the concrete can be elevated by means of a belt conveyor. Invert concrete must usually be placed on a prepared or cleaned invert and consequently must be conveyed from the end of the track or roadway to the point of deposit. This can be accomplished with a belt conveyor or a concrete pump. For fast-moving systems, an invert bridge may be used. This is a long movable platform supported on concrete curbs, on dowels in the invert, or on the invert itself. It provides space underneath for roadbed removal, invert cleanup, and placing and finishing the invert. It also supports the concrete conveyance, which may be a belt conveyor or the transport vehicles for bringing the concrete into the tunnel, as well as screeding and finishing equipment.

Concrete behind closed forms, such as the arch, or the entire circumference of a circular tunnel placed monolithically, is placed through a pipeline. See Figure 13-31 for typical arrangement of arch concrete placing equipment. Mechanically operated positive displacement pumps are now used almost exclusively, having displaced the pneumatic systems that were popular for many years.

Several makes of positive displacement concrete pumps are available. Basically, they are piston pumps designed to accept concrete from an integral agitator or remixer and force it through a pipeline. Various sizes and capacities are available, usually requiring discharge pipe of from 4 to 8 in. in diameter. Horizontal distances of 2,000 ft and vertical lifts of 300 ft or more can be handled. For maximum limits, a smooth, workable mix is necessary. Maximum aggregate size should be less than 1/4 the pipe diameter, and the coarse aggregate should preferably consist of well-graded, rounded particles. Slump should be at least 4–6 in., and the sand and

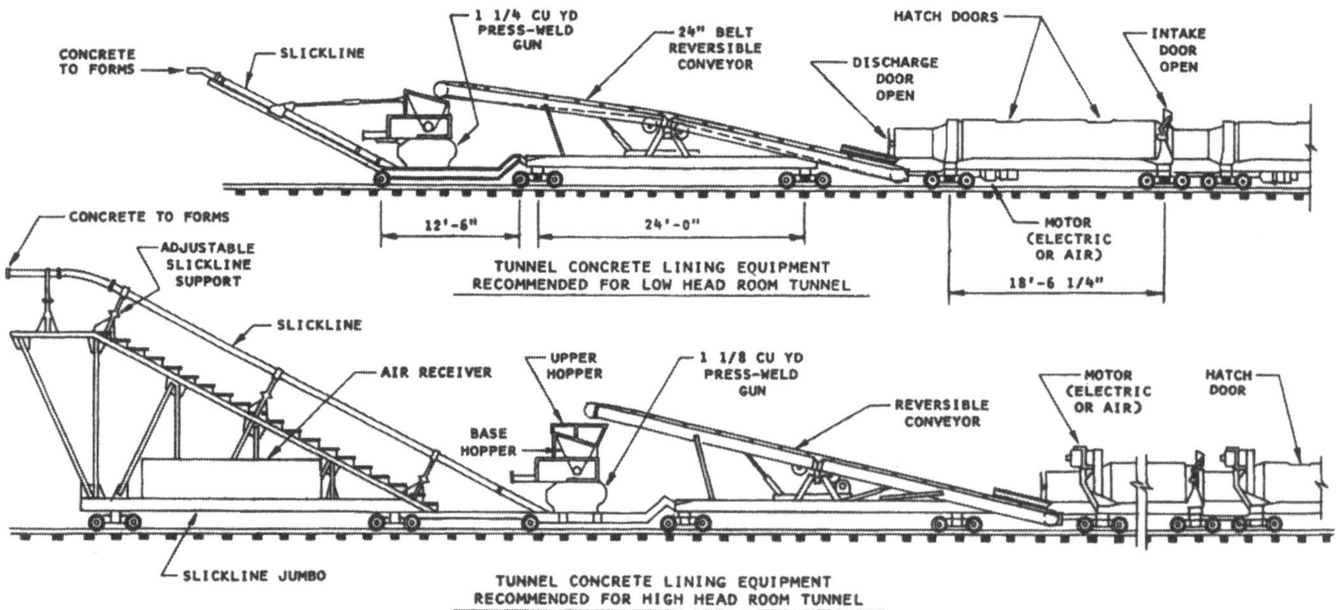

Fig. 13-31. General arrangement of arch concrete placing equipment.

cement content should be relatively high. Air-entraining admixtures are helpful in preventing segregation and consequent blockages at bends or other constructions. Water-reducing additives allow greater workability while maintaining the prescribed water/cement ratio. The concrete can be pumped through a horizontal slick line on top of the form, or vertically through ports in the crown, but care should be taken to avoid collapsing the form from too much pressure.

The low discharge velocity of the concrete pump permits its use as a conveyance for concrete to all other areas of the tunnel if desired. Figure 13-32 shows the relationship between pumping rate, pumping distance, and line pressure for positive displacement pumps.

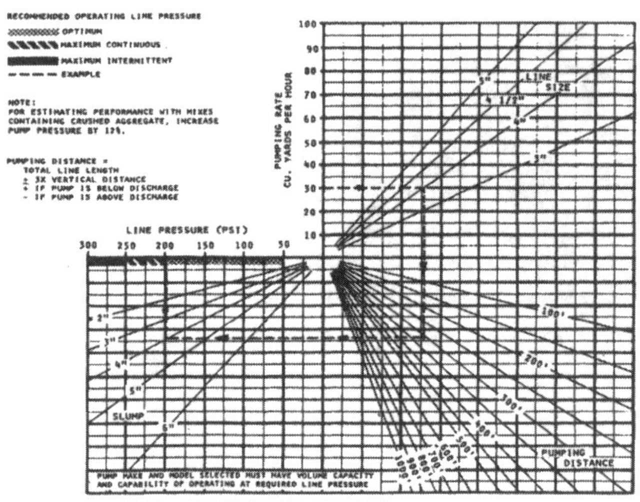

Fig. 13-32. Concrete pump performance estimator.

Pneumatic placers are rarely used any more, and many specifications forbid their use.

Forms. Most of the concrete placed in a tunnel must be confined and shaped with forms. While the design of concrete forms is beyond the scope of this chapter, moving the forms is a material handling problem, and the forms themselves may frequently be used in conjunction with other material handling activities in the tunnel. For example, invert forms may be used as support for an arch form traveler.

Because of the many reuses that can usually be made, most tunnel forms are fabricated from steel. In circular tunnels, the forms can be designed so that the full circle is concreted monolithically, or the invert may be placed separately. In either case, the forms are usually stripped and moved by means of a traveler. For continuous placing, the forms may be telescopic: the last form is retracted, moved through the entire line of forms, and reset at the front. See Figure 13-33 for examples of telescopic and nontelescopic forms.

A well-designed set of forms will include provisions for bulkheading, access doors for observation and vibration, provisions for mounting and operating form vibrators, provisions for bracing against uplift and horizontal movement, provisions for alignment, and provisions for utilities such as electric power.

SHOTCRETE PLANT

Depending on the contractor's preference, and sometimes on the requirements of the specifications, the shotcrete may be either the wet-mix or the dry-mix type. However, the

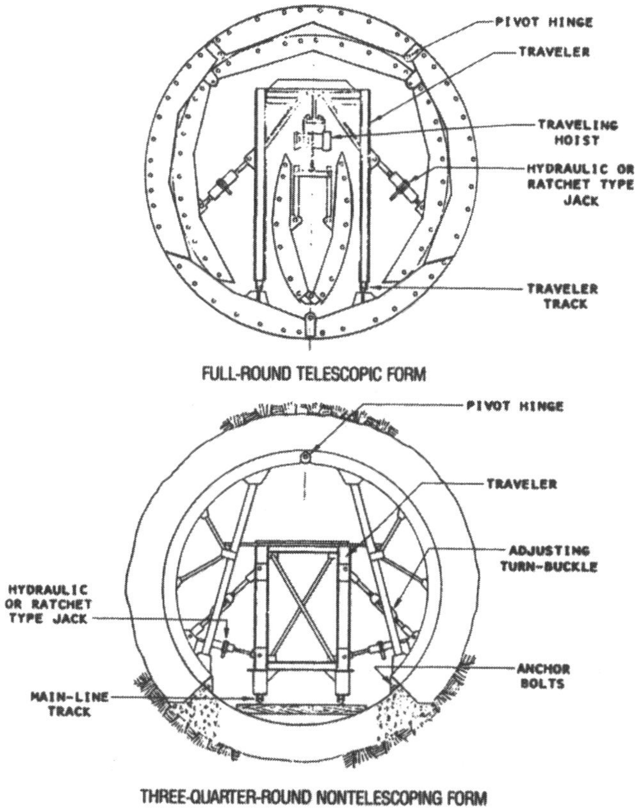

Fig. 13-33. Forms for placing concrete lining.

wet-mix type is now frequently specified because it reduces dust and rebound. In either case, it is often possible to have the ingredients mixed and delivered to the site by a local transit-mix company. However, whenever the work may require shotcrete to be available on short notice, it is best to have the materials and plant available on the job.

For dry-mix shotcrete, the moisture content of the mix is important, and so the aggregate storage should be protected from the weather. For either type of shotcrete, the aggregates and cement are usually batched on the surface and delivered into the tunnel with transit mixers mounted on trucks or rail cars. Compact, portable, batching-mixing units are also available for batching and mixing in the tunnel. Materials are transferred in the tunnel by means of belt conveyors.

The shotcrete machine is mounted on a truck or rail car, and compressed air is supplied from the central supply system or a portable compressor. The handling of the water and accelerator depends on the configuration of the overall system.

Access to the area being shotcreted must be provided. When the nozzle is manipulated by hand, a platform on a crane or boom truck can be used. For access to dangerous areas, high locations, and over the muck pile, the use of a robot is advantageous. It permits the application of shotcrete to commence immediately after the blast, safeguards the operator from rock falls, and lessens exposure to rebound. Generally, the use of a robot improves the quality of the shotcrete application. All shotcrete applications require good ventilation and lighting. Because of timing requirements for the application and the limited amount of time for the mixed ingredients to be stored, strict attention to logistics is essential. However, chemicals that allow wet-mixed shotcrete to be stored for many hours and then reactivated and applied with no significant loss of quality have recently been developed.

A great variety of plants and equipment are available for shotcrete work and the state of the art is advancing rapidly. Manufacturers should be consulted for details.

14 | Immersed Tube Tunnels

Ahmet Gursoy
Vice President, Principal Professional Associate, Parsons Brinckerhoff Quade & Douglas, Inc.

Immersed tube tunnels are composed of prefabricated sections placed in trenches that have been dredged in river or sea bottoms. The sections are usually constructed at some distance from the tunnel location and made watertight with temporary bulkheads. They are then floated into position over the trench, lowered into place, and joined together underwater. The temporary bulkheads are removed, and the trench is backfilled with earth to protect the tubes. Immersed tubes have been widely used for highway and rail crossings of soft-bottomed, shallow estuaries and tidal rivers or canals in which trenches may be excavated with floating equipment.

GENERAL DESCRIPTION

Under favorable conditions, the immersed tube method is the most economical construction for any type of underwater tunnel crossing. Tunnel sections in convenient lengths, usually 300–450 ft (Denmark's Guldborgsund Tunnel has the longest element length, 690 ft), are placed into a pre-dredged trench, joined, connected, and protected by backfilling the excavation. The sections may be fabricated in shipyards on shipways, in dry docks, or in casting basins depending on the type of construction and available facilities. A prerequisite for this method is a soil with adequate cohesion, which permits dredging of the trench with reasonable side slopes that will remain stable for a sufficient length of time to place the tubes and backfill.

The top of the tunnel should be preferably at least 5 ft below the original bottom to allow for an adequate protective backfill. Where grade limitations and bottom configuration make this impractical, the tunnel may project partly above the bottom and be protected by a backfill extending about 100 ft on each side of the structure and confined within dikes. The fill must be protected against erosion by currents with a rock blanket, protective rock dike, or other means. There have been cases where ships in confined channels have used anchors as turning pivots in wharfing operations. This practice, although contrary to navigation rules, may require deeper backfill over a tunnel in such a location or special protection by rock cover, concrete slabs, or other means.

Tides and currents must be evaluated to establish conditions to be met during dredging and tube-sinking operations. Nearby shellfish areas must be identified so precautions can be taken during construction to prevent damage from silting due to dredging or backfill spillage.

Dredging and backfilling operations should be executed so as to minimize disturbance to the natural ecological balance at the construction site. Permits for construction and jurisdictional conflicts with other governmental agencies over environmental protection, natural resources, and local conditions must be evaluated and resolved. Approval of these agencies should be obtained during the preliminary design stage.

Historical Perspective

The first use in the United States of immersed tube tunnel construction methods was for a water tunnel crossing the Shirley Gut in Boston Harbor in 1896. The first transportation tunnel constructed by immersed tube methods in the United States was the Michigan Central Railroad Tunnel under the Detroit River, completed in 1910 under the direction of William Wilgus. Another early immersed tube was the Harlem River crossing of the New York Subway, completed in 1914, presently part of the IRT Lexington Avenue line.

The first highway immersed tube in the United States was the Posey Tube, a concrete tube section, between Oakland and Alameda, California, completed in 1928. In 1930, the Detroit–Windsor Tunnel, a highway link between Detroit, Michigan, and Windsor, Ontario (Canada), was completed. This tunnel's octagonal double steel shell cross section became the model for widespread use of similar construction methods for steel tube tunnels, including

- Three two-lane tunnels under the Elizabeth River between Norfolk and Portsmouth, Virginia (1952, 1962, 1988)

- The Baytown Tunnel under the Houston Ship Channel in Texas (1953)
- Two parallel tunnels under Hampton Roads in Virginia (1957, 1976)
- Two tunnels in tandem on the Outer Chesapeake Bay Crossing in Virginia (1964)
- The two-track transit tubes for Boston's Orange Line Charles River crossing (1971)
- The two-track Trans-Bay Tube of the BART system in San Francisco (1970)
- The four-track 63rd Street transit and rail tunnel in New York City (1973)
- The Washington Channel crossing on the WMATA Yellow Line in Washington, D.C. (1979)
- The Fort McHenry Tunnel in Maryland (1986)
- The I-664 Tunnel under Hampton Roads, between Newport News and Craney Island, Virginia (1992)
- The I-90 Third Harbor Tunnel in Boston (1994)

After much discussion of concrete tubes in the United States over the years, Boston's Central Artery project will include a concrete subaqueous tunnel under Fort Point Channel. The tunnel crossing includes two 460-ft-long tube elements to be placed side by side on a caisson foundation. This unique structure will be the first concrete tube realized in the United States since the Posey Tube. The Deas Island Tunnel near Vancouver and the Lafontaine Tunnel at Boucherville, near Montreal, provide two Canadian examples of concrete tube construction in North America.

The first immersed tube tunnel built in Europe was the Maas Tunnel in Rotterdam, The Netherlands. In 1929 the municipal council of the city of Rotterdam commissioned three engineers, Van Dijk, Van Dunne, and Von Bruggen, to visit the United States to study new tunnel construction methods as part of the planning for the Maas Tunnel project. A contract was awarded in February 1937 to an alternative bid for a concrete tunnel with rectangular cross section, replacing the original design of two separate, two-lane octagonal double-shell steel tunnels very similar to the cross section of the Detroit-Windsor Tunnel. The Maas Tunnel was completed in 1942.

By 1986, at least 67 immersed tube road and rail tunnels had been completed worldwide. Of these, 32 are steel shells and the remainder, concrete (Culverwell, 1989). The International Tunneling Association's 1993 state-of-the-art report provides a technical inventory of 91 immersed tube tunnels completed since 1910 (ITA, 1993).

CONCEPTUAL CONSIDERATIONS

Alternative Tunnel Concepts for Subaqueous Crossings

All tunnel projects involve their own special problems. There is a wide range of conditions for which immersed tubes have been found suitable. However, immersed tubes are not invariably the best alternative for all underwater tunnels. Concept selection for underwater transportation tunnels generally requires performance of "alternative concept analysis" for comparative evaluation. Although their range of applications overlap, there are three proven and practical alternatives available:

- Bored or mined tunnels (generally in rock)
- Shield tunnels (generally in soft ground)
- Immersed tube tunnels

The relative suitability of each of these methods will depend primarily on the hydrographic and geotechnical conditions of the project site.

Recently, a number of floating tunnel alternatives have been studied for a fixed-link crossing of the Straits of Messina in Italy and for Hosfjord Crossing in Norway. These studies have advanced to a point where technical feasibility has been established. Realization of a floating tunnel project awaits economic and political decisions.

Several topics and parameters influence the selection of tunnel type among the alternatives listed above. These include, but are not limited to, required navigable depth and traffic volume of waterway, shoreline infrastructure, site topography, geology, and geotechnical considerations; special risks created by the (inexhaustible) water supply overhead during construction and service life; approach gradients and lengths; acceptable grades; ventilation requirements, and power usage by tunnel facilities and by vehicles using the tunnel. Some of these considerations have been discussed in Chapter 2. Ventilation is covered in detail in Chapter 20.

The required navigable depth establishes the absolute minimum tunnel depth below the water surface. The waterborne traffic volume affects the possible use of intermediate ventilation islands and the type and use of floating construction equipment.

The immersed tube tunnel invariably can be located the least distance below the water surface, generally minimizing the length of project affected by the water barrier. As discussed elsewhere in this chapter, immersed tubes need only a nominal allowance for protective backfill below the sea or river bed; where the water depth is substantially more then navigation requires, part of the tube cross section may actually be above the original bottom. Shoreline infrastructure affects terminal locations of tube-type tunnels and locations of ventilation structures for all alternatives.

Site topography affects approach gradients and may negate the otherwise pronounced advantages of a tunnel at minimum depth below the water surface. For example, a fjordlike shore topography will require long lengths of land approaches, thus allowing subaqueous (floating) tunnels an economic advantage.

Geology plays an important role in choice of alternative. In general, the poorer the subsurface geology, the more favorable a tube becomes. Rock depth below the water surface also affects the choice; occasional rock "peaks" within tube trench depth are only a minor problem, but long lengths or rock requiring blasting within trench depth increase trenching costs substantially. Problems due to "hard points" may be avoidable by revising the alignment.

The overlying body of water represents a special risk to all alternative types of subaqueous tunnels. However, some land-based rock tunnels have incurred nearly equal risk when unanticipated inflows of water have inundated and destroyed the tunnel. Even a remote possibility of such an occurrence makes the immersed tube tunnel the only logical choice because each tube section is a structurally complete unit protected against flooding by temporary end bulkheads before it is placed in the trench. Shield-driven, subaqueous, compressed-air tunnels have been used successfully in several instances (most before the use of tube type tunnels was fully accepted). Compressed-air tunnels are limited to depths of 130 ft or less; at such a depth only about an hour of productive work is possible in a full work shift, increasing costs to uneconomical levels. At lesser depths, the risk of a blowout and subsequent loss of life is always present. Use of modern pressurized-face tunnel boring machines (TBMs) has alleviated the risk to lives, but the depths to which these machines can be used without excessive breakdowns is not known.

Soft ground or rock tunnels must be located much deeper below sea bed than immersed tubes because there must be a sufficient thickness of impermeable materials over the tunnel to make free air construction possible without undue risk of flooding (and losing) the tunnel. The amount of extra depth required is site- and material-specific and also depends on the thoroughness and accuracy of the geotechnical investigation (see Chapter 4).

The problem of water infiltration during service life of successfully constructed subaqueous tunnels does not differ greatly among the three alternatives. However, the steel shell tube is considered superior to all others because the tube is constructed and tested in the dry under the best of conditions. In contrast, maximum attention to quality control is needed in the fabrication of concrete tunnels.

Advantages and Disadvantages

For underwater tunnels, the immersed tube concept offers several advantages compared with mined or shield-driven tunnels:

1. The tunnel has the minimum possible depth. For an approach gradient fixed by operating criteria, this usually means minimum tunnel length.
2. Almost all construction is accomplished from above the ground or water surface, in normal working conditions. This promotes better quality of construction, particularly for control of water seepage.
3. Most of the construction is related to standard materials and operation using readily available labor skills.
4. There are no major time-dependent construction constraints such as
   - Provisions for control of water inflow to permit the work to proceed under atmospheric pressure, such as in mined tunnels
   - Measures required to stabilize the working face, as in shield-driven tunnels
5. Repetition of construction activities offers opportunities for higher efficiencies.

Major disadvantages of immersed tunnels compared with mined or shield-driven tunnels are

1. Potential for disruption of existing facilities if the trench must be extended past the shorelines.
2. The need for special equipment to construct foundations.
3. The need for special equipment to place and join the tube sections.
4. Selection of acceptable dredging methods and construction of underwater dikes if needed in environmentally sensitive areas.
5. Environmental objections to disturbances caused by trench dredging and backfill placement.
6. Selection of suitable sites for a casting yard for concrete tube tunnels.
7. Selection of a disposal site for the dredged spoil material excavated from the trench. This can create serious environmental problems if not resolved in early planning. In the Fort McHenry Tunnel project, as a result of the early planning in preliminary engineering, the 3.5 million yd^3 (2.7 million m^3) of dredged material from the trench excavation was used in construction of a new port facility. The Canton Seagirt Facility, opened to operation in 1989, is now the largest in Baltimore Harbor.

Other limitations on immersed tube construction include

- There must be sufficient duration of slack tidal current to permit lowering the tube—preferably at less than 3 ft/sec (1 m/sec) over a duration of 2 hours.
- The bottom must not be so soft and unstable that the trench cannot be kept open.
- The site must be reasonably free from rapid deposition of fluid silts, which can alter the density of water in the trench and affect the balance of buoyancy at tube placement.

Water depth over the trench at terminal (end) tubes preferably should be sufficient to allow cross girders of placement barges to span over the top of the terminal tube during placement. Special equipment may be required to place shallower tubes.

Immersed tubes occasionally have been adapted for sites with irregular hard rock bottoms, and to sites exposed to ocean wave and storm conditions. In a few cases, they have been used for water supply and other utilities. The deepest trench for an existing tunnel extends 135 ft (41 m) below sea level, for the San Francisco Trans-Bay Tube (Figure 14-1).

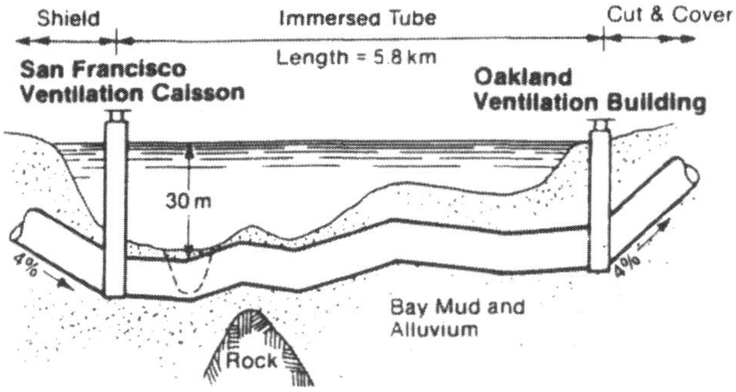

Fig. 14-1. San Francisco Trans-Bay Tube.

Tender design conducted in 1988 for an immersed tube crossing of Denmark's Great Belt proposed a trench extending to 150 ft (50 m) below sea level, and studies for the Tokyo Bay Bridge–Tunnel projected a trench depth of 138 ft (45 m) below sea level.

A hard rock bottom, particularly if it has an irregular or steep profile, will increase the cost of trenching substantially, but short sections of rock trench may be accommodated readily. The Boston Third Harbor Tunnel involved rock trench excavation over 3,000 ft, or 77% of the total length of immersed tubes. Immersed tubes are usually buried beneath the level of the existing sea bed, although short projections above the bottom across deep natural trenches may be mounded over for protection.

Settlement of Immersed Tube Tunnels

Under normal conditions, immersed tube tunnels are relatively insensitive to soil settlement, because their buoyant weight differs only slightly from that of the original submerged soil or the adjacent backfill. However, when the soil types and the loading pattern vary significantly along the tunnel alignment, differential settlements will take place. The redistribution of loads may create significant bending moments, which although secondary in nature, can be critical. The assessment of anticipated settlement is particularly important in the design of concrete tunnels. Concrete is brittle and susceptible to cracking compared with steel, which is ductile. In recent European practice, dilatation joints have been placed every 60 ft to minimize the bending stresses due to settlement (see "Tube Placement," later in this chapter).

Two examples given below illustrate the results of two major causes of settlement: varying soil conditions or an abrupt change in backfill loading. Both cases demonstrate how the ductile properties of steel tubes can be used to accommodate anticipated large settlements without damage to the structure.

The Second Hampton Roads Tunnel (1976) was constructed only 250 ft west of the first crossing (1957). Unusually variable soil conditions within the tunnel profile, ranging from coarse sand to silt and clayey silt, necessitated an extensive monitoring program during construction. For each element, displacements were recorded during critical stages of tube placement. Findings of this survey (Schmidt and Grantz, 1979) indicate an average settlement of 6 in. with an average change in slope of 0.033% for the total of 21 elements. A maximum displacement of 10 in. was recorded for one of the elements located near the manmade island. These changes have resulted in no structural or operational difficulties. The 1979 study also provides methods to correlate the field data with the theoretical basis for the settlement estimate.

In the Fort McHenry Tunnel in Baltimore Harbor (1985), three steel tube elements with binocular cross sections were placed beyond the western shoreline of the shipping channel to reduce the length of approach structures. These elements retained 50 ft of backfill to provide an operating platform for berthing facilities for 50,000-ton vessels, creating an immediate increase in loading. Maximum settlement of the land tubes during construction was 6 in. The remaining 16 pairs of (binocular) tunnel elements settled less than an average of 2 in. These changes have resulted in no structural or operational difficulties. No further settlement has been observed since completion of either project.

In a survey of eight European concrete tunnel projects, Rasmussen and Romhild (1990) report an average settlement of 2 in. A maximum settlement of 6-1/2 in. was recorded at one site. The number of elements for each project examined varied from five to six.

Immersed Tube Tunnel Types

Over the years, two distinct types of immersed tube tunnel construction have emerged on both sides of the Atlantic Ocean.

In 1930, the cross section adopted for the Detroit–Windsor Tunnel in North America captured the imagination of tunnel designers for generations to come. A number of variations have been used ranging from the twin-track, single-shell configuration of the BART Trans-Bay Tube in San Francisco (1970), the two-level, four-track, single-shell cross section of the 63rd Street Tunnel in New York City (1973), to the Fort McHenry Tunnel's twin double-bore double-shell cross section for eight lanes of traffic in Baltimore (1986).

Since the completion of the Maas Tunnel up to the end of 1986, a total of 24 concrete immersed tube tunnels have been constructed in Europe. In The Netherlands alone, 12 immersed tunnels have been built since the completion of the Maas Tunnel (Culverwell, 1989).

Steel Shell Immersed Tunnels

In the last 50 years, 20 steel shell tunnels have been constructed in the United States, each reflecting the needs of its time and location. The process evolved, resulting in two distinct variations in steel immersed tube tunnel construction: single-shell and double-shell construction.

In single-shell construction, an outer steel shell plate stiffened internally acts as a permanent watertight membrane, serves as an exterior form for the concrete lining, and acts as the structural element to carry flexural forces along the exterior face of the tube before and after the placement of the concrete lining. After the completion of interior concrete, the steel shell tube will behave as a composite steel–concrete structure. A single shell tube is generally protected against corrosion by a cathodic protection system or, on occasion, by pneumatically applied concrete protection. A convenient ballast box is placed at the top of the tube section, permitting the use of economical stone ballast. Attention needs to be given to stability of the section during towing and placing. On the Cove Point LNG Tunnel in Maryland, one tube flipped over when a support cable broke during placement.

In double-shell construction, an outer secondary form plate, usually octagonal in shape, is added. The outer shell plate, often called a *form plate,* provides needed space for placement of tremie concrete as ballast. The outside ballast concrete protects the inner shell plate from corrosion, leaving the outer shell plate as a sacrificial form plate. The inner shell plate for double-shell construction provides design features similar to the single-shell construction. However, unlike the single-shell construction, the stiffening elements of the inner shell plate (the transverse diaphragm and longitudinal stiffeners) are placed outside the inner shell plate. Figure 14-2 shows conceptual configurations of single- and double-shell construction. For single- or double-shell construction, compression and shear forces are carried by the concrete lining, and reinforcing steel carries the interior flexural tension forces. The Hong Kong Cross Harbour Road Tunnel used cylindrical inner steel shell plates and wooden exterior forms.

Concrete Immersed Tunnels

The advancement in immersed concrete tube tunnels in Europe was heavily indebted to the successful improvements of three major design and construction features: crack control, watertightness, and foundation and placement systems.

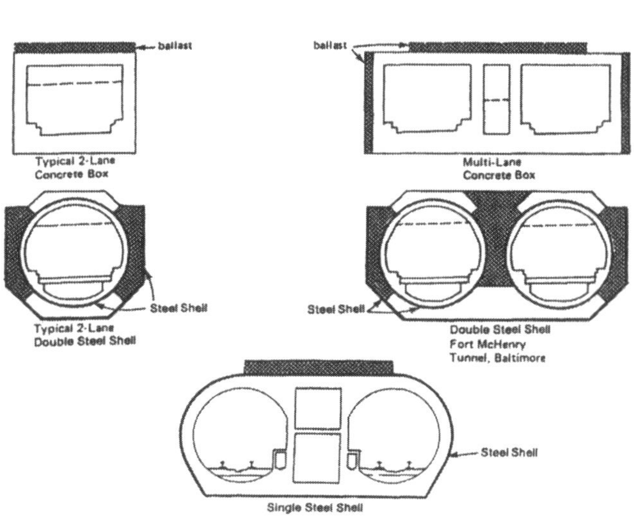

Fig. 14-2. Representative immersed tube cross-section configurations.

Crack Control. Uneven shrinkage is the major cause of cracking in concrete structures. For concrete immersed tunnels, a number of remedies have been used for crack control:

- Increasing impermeability of concrete by selecting suitable cement content and aggregate size and by other means.
- Reducing differences in temperature between the core and outer layers of the walls during concrete placement, including the bottom slab, roof slab, and between opposite sides of construction joints.
- Introducing dilatation (expansion) joints every 60 ft along the element. These joints reduce the high tensile stress due to shrinkage, and the stresses induced from possible settlement of the subsoil or foundation course.

Watertightness. Watertightness in concrete tunnels is an important part of the design. Major areas of improvements include

- The increased impermeability of concrete
- Improvements to construction joints between walls and top and bottom slabs
- Improvements in dilatation joints
- Improvements in element joints

In Holland, following the completion of the Maas Tunnel and up to the construction of the Vlake Tunnel in 1975, extensive waterproofing systems were used. In general, the system used contained a steel membrane for the bottom slab, copper stripping for the joints, and three exterior layers of impregnated fiberglass with asphaltic bitumen on sidewalls and the roof slab, followed by a layer of polyester foil with additional concrete cover for the roof slab. Recent European practice for shallow concrete tubes has favored reliance on crack control, impermeable concrete, and improved joints. For deep concrete tubes, exterior waterproofing is still favored.

Foundation and Placement Systems. After the introduction of the sand jetting method for the Maas Tunnel in 1942, tunnel designers in Holland developed the sand flow method. The new method eliminated the use of a gantry system riding on the roof of the tunnel element and provided better control for the sand flow pattern in locations where high velocity of currents, up to 6.5 ft/sec, are common.

Improvements in placement techniques include the introduction of three-point temporary support systems for the placement of elements during construction of the foundation system. The three-point support system used since 1969 makes use of a single pivotal support at the adjoining (outboard) end of the element, ensuring easier adjustments for alignment and reducing the number of corrections required in making the joint.

Steel and Concrete Tube Tunnel Comparison

Selection of immersed tube tunnel type is closely related to site conditions, which also affect the best usage of the steel and concrete. For this reason, selection of a tunnel configuration and cross section very much depend on the construction stages that are appropriate to the given site conditions and the existing local facilities to be used.

For a generic comparison, a steel tube tunnel offers the following advantages compared with conventional concrete tube tunnel construction:

1. Construction time is shorter, which may be a decisive advantage for privately financed projects. The time to mobilize a construction yard necessary for concrete tunnels is not required for steel tube tunnels, as the steel fabrication uses existing facilities. The steel tube elements can be completed more rapidly than the concrete elements. Early completion of the first tube elements permits an early start on placement of elements in the trench.
2. Use of existing ship fabrication and launching facilities eliminates the need to acquire a waterfront site for a casting basin. Difficulties, time, and cost of permits and environmental approvals for a concrete tube casting basin may be decisive in the selection of construction method.
3. Use of the steel shell as a permanent waterproof membrane permits eliminating the special temperature control provisions and dilatation (expansion) joints normally used to eliminate cracking in the concrete. This saves both direct cost and time.
4. Depending on the specific site of a project, steel shell construction is adaptable to use of labor-efficient processes. The reinforcing steel can be preassembled into very large units for efficient installation. This contrasts with labor-intensive reinforced concrete construction in a casting basin, and even more so with in situ concrete construction in deep trenches.
5. Using the steel tube as an exterior structural member eliminates half the reinforcing steel in a conventional concrete tunnel. The combined quantity of structural and reinforcing steel is comparable with the quantity of reinforcing steel in the concrete tunnel.
6. In general, the direct costs of launching and transporting the steel shells are offset by the elimination of the costs for opening and closing, dewatering, and equipping the casting yard for the concrete tubes. It can be assumed that costs of towing and placing tube elements are essentially equal. Concrete placement within the steel shell is more difficult than in the concrete element construction basin, but this is offset by the elimination of special procedures and details for crack control and others.

Steel shell construction has the following disadvantages:

1. Fabricated structural steel costs significantly more than an equivalent amount of reinforcing steel, and the quality of steel required for immersed tubes is comparable with that used in ship or tanker construction. As a result, the labor cost per ton of steel can be considerably higher if the steel fabricating facilities are distant and costly. These factors can make the concrete tunnel option more economical in many cases. The availability of graving docks or adaptable facilities near the project site can make the concrete option even more attractive, particularly when construction time is not a driving factor. In short crossings in Holland, excavations made for approach tunnels are commonly used as construction basins for the construction of concrete tunnel elements, which results in savings.
2. Since the single steel shell is the permanent watertight membrane, its durability against corrosion must be assured. Depending on local site and project conditions, this will require either a cathodic protection system or a sprayed concrete protection course.
3. Wide tunnel sections are more readily adapted to rectangular box elements, which are usual in concrete tube structures, than to groups of circular elements, which are customary for steel shell tube structures. However, both rectangular steel shells and circular concrete tube sections have been constructed.

It is usual practice to terminate immersed tube construction at a point where the profile of the top of the tunnel penetrates the normal high water level. Steel shell construction offers the opportunity to extend immersed tubes further inland by leaving out part of the concrete lining in one end of the terminal tube so that it floats higher. This requires special placing equipment, but it permits substituting incremental immersed tube construction for deep cofferdam cut-and-cover tunnel construction, which may reduce both cost and duration of construction.

STEEL SHELL TUBES

Double-Shell Configurations

The basic element of this type of tube is a steel shell that forms a watertight membrane and, in combination with a reinforced concrete interior lining, provides the necessary structural strength for the finished tunnel. The shell may have a circular cross section or any other configuration to suit the purpose of the tunnel. Figure 14-3 shows the cross section of a two-lane tunnel on an interstate highway. Its circular steel shell has a diameter of 36-ft, 2-in. and is made of 5/16-in. welded steel plate. It is stiffened by external diaphragms spaced 14-ft, 10-in. apart and external longitudinal stiffening ribs. The interior is lined with a minimum thickness of reinforced concrete. An exterior concrete envelope of 2-ft minimum thickness, confined by 1/4-in. steel form plates attached to the shell, protects the shell against corrosion and acts as ballast against buoyancy. The space below the roadway slab forms a fresh-air supply duct; the segment above the ceiling is an exhaust duct.

Single-Shell Configuration

Figure 14-4 shows a twin circular shell tube for a four-lane highway tunnel. The two shells are connected with transverse steel diaphragms spaced 17-ft, 6-in. on centers with horizontal stiffening trusses between diaphragms. The interior reinforced concrete lining has a minimum thickness of 1-ft, 6-in. Exterior keel concrete, 2 ft thick (minimum),

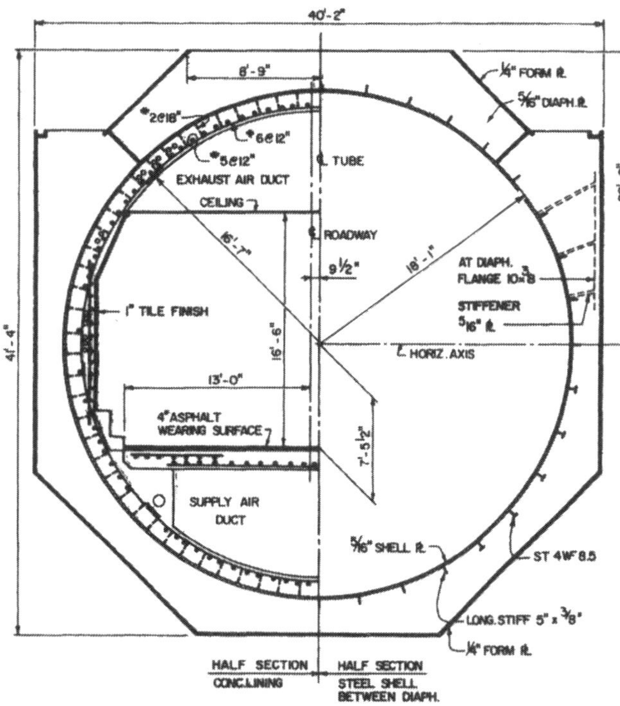

Fig. 14-3. Second Hampton Roads Tunnel.

and concrete filling the space between the two shells provides weight to overcome buoyancy and, in addition, to protect the shell plate against corrosion. Where not covered by the exterior concrete, the shells are protected against corrosion by 2-1/2 in. of pneumatic mortar reinforced with wire mesh.

Figure 14-5 shows a single-shell tube for two rapid transit tracks, separated by a gallery providing service access and an emergency ventilation exhaust duct for fire protection. The shell plate is 3/8 in. thick. It is stiffened by interior transverse steel ribs spaced 6 ft on centers and two longitudinal vertical interior trusses, encased in the reinforced concrete walls of the gallery. The interior lining of reinforced concrete has a minimum thickness of 2-ft, 3-in. The exterior

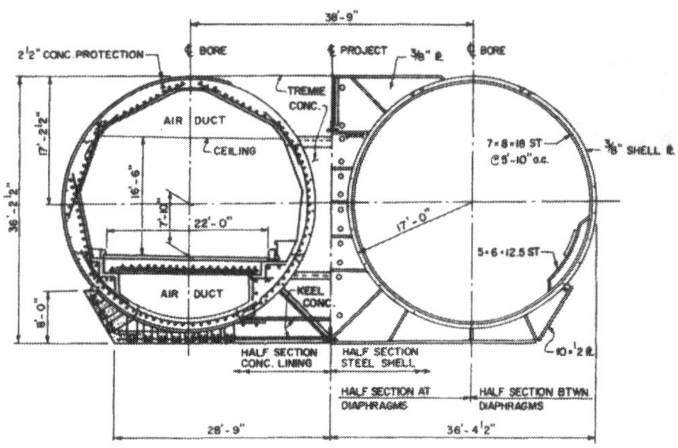

Fig. 14-4. Hong Kong Tunnel.

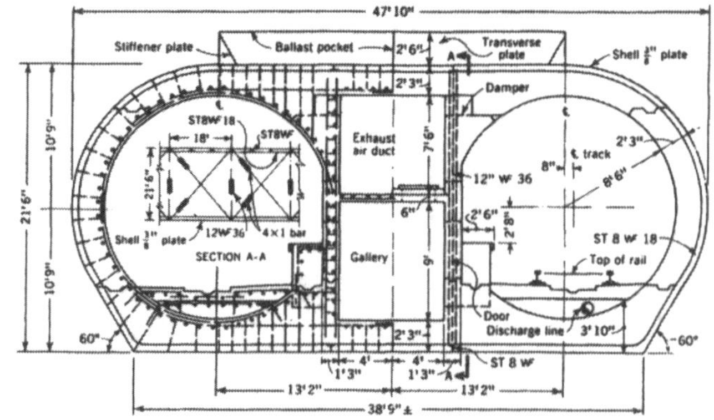

Fig. 14-5. Trans-Bay Tube (San Francisco Bay Area Rapid Transit).

shell is protected against corrosion by an impressed current cathodic protection system. Ballast pockets 2-ft, 6-in. deep on top of the tube are filled with gravel to provide adequate weight to overcome buoyancy.

Figure 14-6 shows a single-shell tube with four bores arranged in pairs vertically for rapid transit. The 3/8-in. shell plate is supported by interior transverse stiffeners spaced at a maximum distance of 4-ft, 4-1/2-in. longitudinally. The shell is stiffened by a vertical longitudinal truss between bores and horizontal pipe struts. The interior lining is reinforced concrete, and the shell is cathodically protected against corrosion. Ballast pockets are filled with concrete and rock.

Figure 14-7 shows the twin double-bore double-shell cross section of the Fort McHenry Tunnel. The Fort Mc-

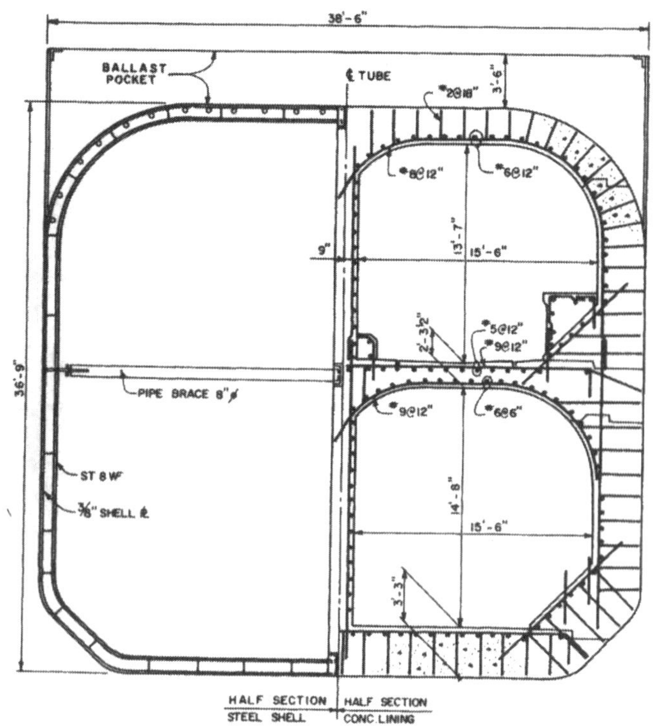

Fig. 14-6. Sixty-Third Street Tunnel.

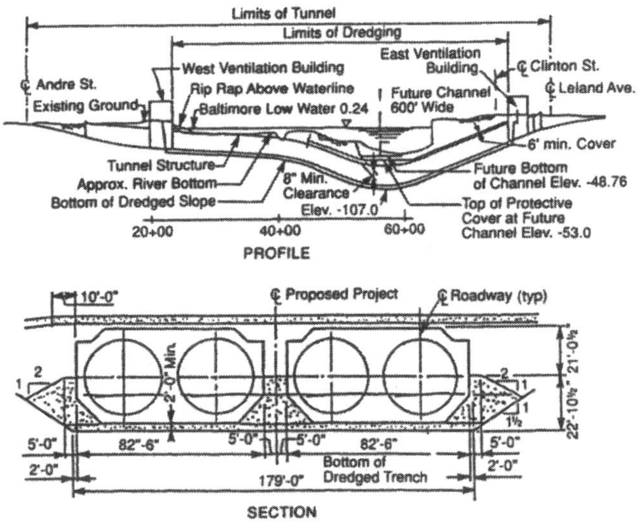

Fig. 14-7. Fort McHenry Tunnel.

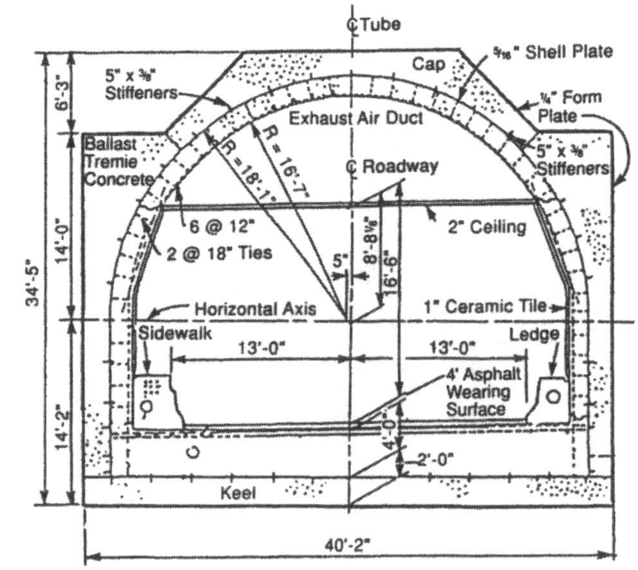

Fig. 14-8. Second Downtown Elizabeth River Tunnel.

Henry Tunnel is the largest highway immersed tube tunnel project ever undertaken in the United States. The tunnel carries eight lanes of I-95 beneath Baltimore Harbor. A typical double-bore tube section is 82 ft wide, 42 ft high by 350 ft long and displaced 35,000 tons at the time of placement. Sixteen pairs of these tubes make up the immersed tube portion of the tunnel. Elements are placed in a common trench 179 ft wide. Because a shipping channel is close to the eastern shoreline, the project required dredging a trench 1,300 ft long, 300 ft wide, and up to 70 ft deep into the eastern shore. Each bore is made up of a 5/16-in. shell plate supported by longitudinal stiffeners at 10° intervals and by transverse diaphragms spaced 14.8 ft apart. Temporary bulkheads made up of a steel plate supported by bulkhead beams close the end of the tube element. Two weldments called *annular rings* protrude from the dam plate in line with the shell plate. Together with two sets of rubber gaskets, the annular ring facilitates the underwater closing of the elements. An innovative design feature of the dam plates is a system of wedges, rigid on one end of the tube and movable on the other end. Unlike shims, the hydraulically controlled wedges provide a convenient and predictable way to compensate for horizontal misalignment during placement of tubes. A fully transverse ventilation system is used with 12 supply and 12 exhaust fans located at each ventilation building near the portals.

In Figure 14-8, the cross section of a horseshoe-type immersed tube tunnel is shown. The Second Downtown Tunnel is a two-lane vehicular tunnel, crossing the southern branch of the Elizabeth River between Portsmouth and Berkley, Virginia. The immersed tube tunnel portion consists of eight prefabricated steel tube elements. The horseshoe cross section and the use of a semitransverse exhaust ventilation system reduced the depth of construction and resulted in savings in construction cost. The double-shell cross section is composed of a 5/16-in.-thick inner shell plate, an inner concrete lining, roadway slab, outer form plate, structural cap and keel concrete, and ballast tremie concrete. The shell plate is stiffened longitudinally by flat bars and transversely by diaphragms, all welded on the outside of the shell plate. A 1/4-in. form plate envelops the diaphragms to contain the ballast concrete. The interior concrete lining is cast-in-place with shear connectors to provide composite action with the shell plate.

In 1988, tender documents for the Great Belt rail crossing in Denmark specified a bored tunnel and two types of immersed tube tunnels, steel and concrete. Although the immersed tube tunnel option was not selected for construction, bids received demonstrate that immersed tube tunnels are competitive with bored tunnels and that steel can be an economical alternative to the concrete. In Figure 14-9, the cross sections of the steel and concrete alternatives are shown. The concrete tunnel was specified to comprise 38 tunnel elements with a length of 488 ft each. The steel tunnel was specified to have 39 tunnel elements with a length of 475 ft each.

For the steel alternative, a single-shell cross section was adopted using an 0.4-in. steel shell plate with transverse stiffeners spaced every 5 ft. A longitudinal steel center bracing system was provided to stiffen the top and bottom shell plates against launching and towing stresses. The interior concrete includes a 40-in.-thick base slab, 32-in.-thick sidewalls and roof slab, and a 44-in.-thick separation wall between the tracks, providing space for electrical and other equipment. A typical steel tube is 54.5 ft wide, 29.4 ft high, and 475 ft long. A 3-ft-high ballast box was added for use of gravel ballast.

The concrete alternative was designed as reinforced concrete sections without prestressing. A continuous 1/4-in. steel plate was used as membrane waterproofing. Concrete lining includes a 42-in. bottom slab, 44-in. sidewalls, and

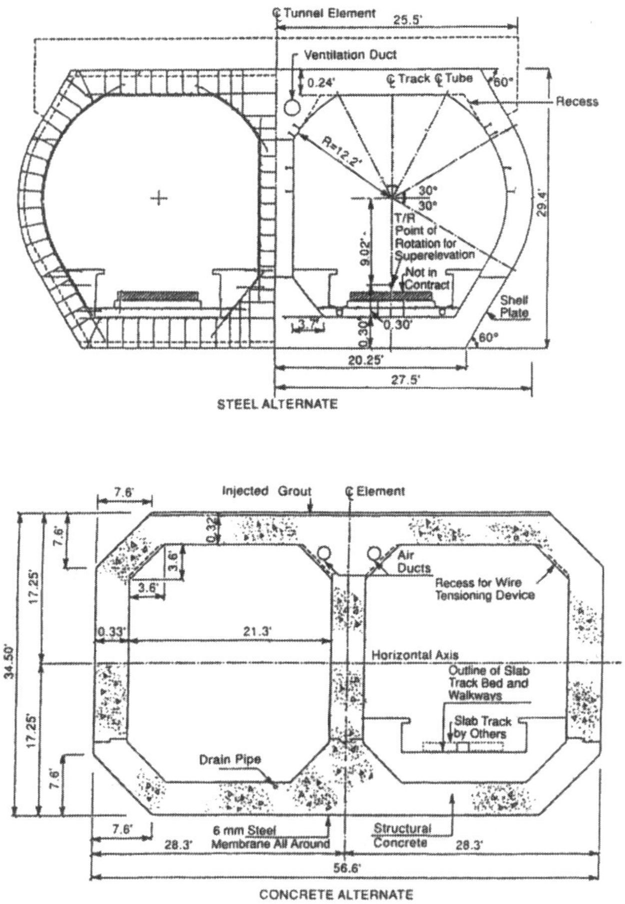

Fig. 14-9. Great Belt crossing (concrete and steel).

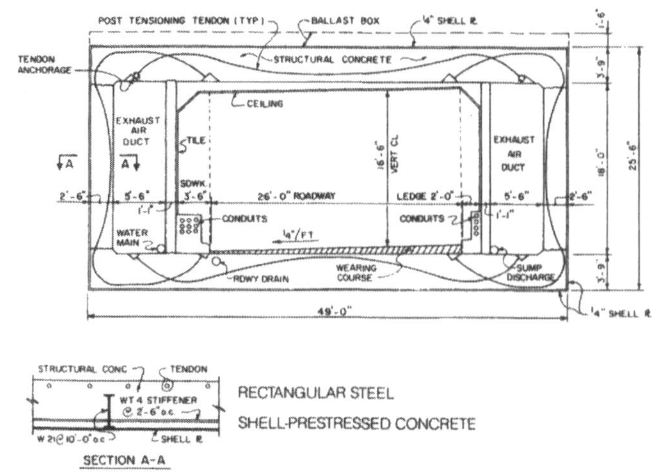

Fig. 14-10. Rectangular steel shell—prestressed.

41-in. top slab, with a 44-in. partition wall at the center used for electrical and other equipment. The thickness of the concrete lining was increased by 16 in. at the ends of each element to accommodate the construction of joints between the elements. Joints are made by the use of Gina and Omega gaskets. The Gina gasket is supported by 22-in. I-beams and a plate assembly located at the enlarged portion of the lining, covering the entire circumference of the cross section. The Omega seal is bolted to the steel assembly after the initial seal is made by the Gina gasket. The entire steel joint assembly is covered with a steel plate prior to the placement of cast-in-place concrete, which also serves as shear block for the joint within the designated shear key areas.

Figure 14-10 shows a rectangular two-lane vehicular tunnel cross section with side air ducts. The section consists of an exterior steel shell plate supported by transverse structural steel stiffeners spaced at 2-ft, 6-in. on centers with an interior prestressed concrete lining. The steel shell is designed to permit off-site fabrication of the steel shell and, after launching, to place the interior concrete lining while afloat. After concreting, posttensioning tendons made up of high-strength strands are pulled through the previously placed ducts. The tendons are then partially tensioned. The section is towed into position over the trench and lowered into position by placing ballast on top of the tube section. After the section is placed and joined to the previously placed unit, the posttensioning force is increased in stages as the tunnel is backfilled. The tendon anchorages are inside the cross section and can therefore be tensioned from the interior of the tunnel.

The technique of incrementally applied prestressing force to counteract successive backfilling is defined as staged posttensioning and has been successfully applied in North America for high-rise buildings and foundations. It may be necessary to use staged posttensioning in cases where the dead load at the time of initial tensioning is not sufficient to counteract the force of the final prestress force.

An alternative technique to staged posttensioning would be to add temporary tendons inside the tunnel cross section. This system was used on the La Fontaine (Boucherville) Tunnel in Montreal. The tendons spanned between the roof and invert slabs and were tensioned simultaneously with the other tendons. The temporary tendons were released as the tunnel was backfilled.

This type of cross section permits the use of a steel shell, which provides a watertight membrane and permits outfitting while afloat, with the advantages of prestressing, which can be more suitable for the wide rectangular configuration.

Materials and Fabrication of Steel Shells

Steel tubes can be constructed on shipbuilding ways of adequate size or in casting basins or dry docks. These may be remote from the project site; because the prefabricated steel tubes draw minimum draft, they can be towed considerable distances.

Material for the shell plates, stiffening members, and end bulkheads are usually ASTM designation A36 steel. Tubes are fabricated in convenient subassemblies and fitted together on the shipways in a casting basin or in dry dock. The maximum tolerance of the assembled shell measured from the theoretical longitudinal axis is ±1/2 in. in radius and ±1-1/4 in. in overall length.

All joints are welded by the metal arc-welding process. Joints in shell plates are full-penetration butt welds. Where practical, welds are made by the automatic machine process. Welds are lightly stressed in permanent conditions and are primarily for watertightness. Welds (Figure 14-11) are visually inspected throughout fabrication, and a certain percentage are tested by radiographic, ultrasonic, dye-penetration, magnetic particle, or other methods. Usually, 10% of butt welds in the end bulkheads and in the end sections of the steel shells, and 1% of the remaining butt welds in the shell plates, are tested radiographically. About 1% of all fillet welds can be tested by any of the other (most suitable) methods.

All shell plate welds are tested for watertightness before launching by coating the outside of the shell plate with soap solution and jetting compressed air at 40 psi against the inside, the nozzle being held not more than 3 in. from the surface, or by using a vacuum box. Welds in end bulkheads are similarly tested. All leaks are repaired and retested.

During fabrication at a shipyard, tubes are supported on the ways by blocking under diaphragms or stiffener rings. Before launching, the ends of the tubes are closed with watertight bulkheads. It is advantageous to place the interior reinforcing steel before the bulkheads are attached.

Launching of Tubes

Structurally, the tubes are particularly suitable for side launching, but they may also be end launched.

Side Launching. The tubes are supported by launching sleds on a series of parallel ways at right angles to the tube and located under diaphragms, which are designed to transfer the weight of the tube to the sliding ways. The actual launching may be free or restrained, depending on the declivity of the ways, and the depth and width of the waterway at the launching site. After completion of fabrication of the tube, its weight is transferred from the fixed building ways to the launching cradles or sliding ways that rest on a coating of grease applied to the groundways. Either the groundways or the sliding ways should be equipped with guides to prevent side slipping during launching. The tube is held in position until launch time by a number of triggers, usually four, which are cut simultaneously at launching. Jacks are provided to start the tube, if necessary.

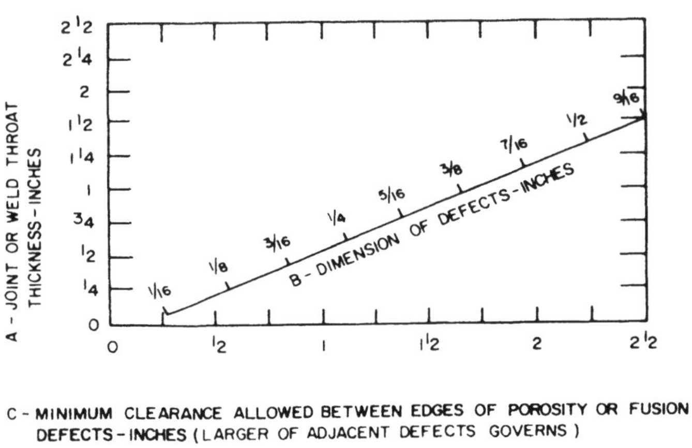

Fig. 14-11. Modified weld quality requirements (limitation of porosity or fusion).

Declivity of the launching ways ranges from 0.5 to 1.8 in./ft to ensure free sliding.

Restrained Launching. The speed during launching is controlled by chains or other suitable drag devices. These control the runout distance of the tube in the water.

Controlled Side Launching. The tube supported on the sled is gradually allowed to slide down the ways at a suitable speed, controlled by synchronized winches until it floats off the sleds.

End Launching. The tube slides on two ways parallel to its longitudinal axis. A single trigger holds the tube, after its weight is transferred to the launching ways, until released for the launch. At the inboard end of the sliding ways, the forepoppet is of special construction to support approximately one-half the weight of the tube when the outboard end of the tube is lifted off the ways by buoyancy.

Launching Stresses. These stresses have to be analyzed. Those for side launching are nominal, consisting of hydrostatic pressure plus impact on the shell and form plates when the tube enters the water. End launching causes bending moments when the outboard end of the tube floats while the inboard end is still supported on the ways. Buckling stresses in the top of the shell may be critical. In some cases, temporary strengthening by longitudinal steel beams welded on the top of the shell may be required.

In a dry dock, the steel shell is fabricated and may be filled with concrete lining. The completed tube element may be floated for transport and immersing. This procedure minimizes temporary stresses during construction.

Concrete Lining

After launching, the tube is towed to an outfitting site for placing the interior concrete lining. Structural concrete generally has a 3,000–4,000 psi compressive strength. This includes the interior lining, roadway slab, keel, and cap concrete, as shown in Figure 14-12.

The keel concrete is usually placed prior to launching in order to provide stability and to reinforce the shell and form plates. If the keel concrete is placed after launching, the shell and form plates probably will require additional stiffening. Exterior concrete acting as ballast and protection (Figure 14-12 and later Figure 14-14) has 2,200 psi compressive strength. If tremie concrete is used in joints between tubes, it should have a 3,000 psi strength. Before starting concreting, trial mixes are made to determine unit weight of concrete to conform to the design weight used in buoyancy calculations.

Reinforcing Steel. The interior concrete lining is reinforced with a layer of steel at the inside face, as required by

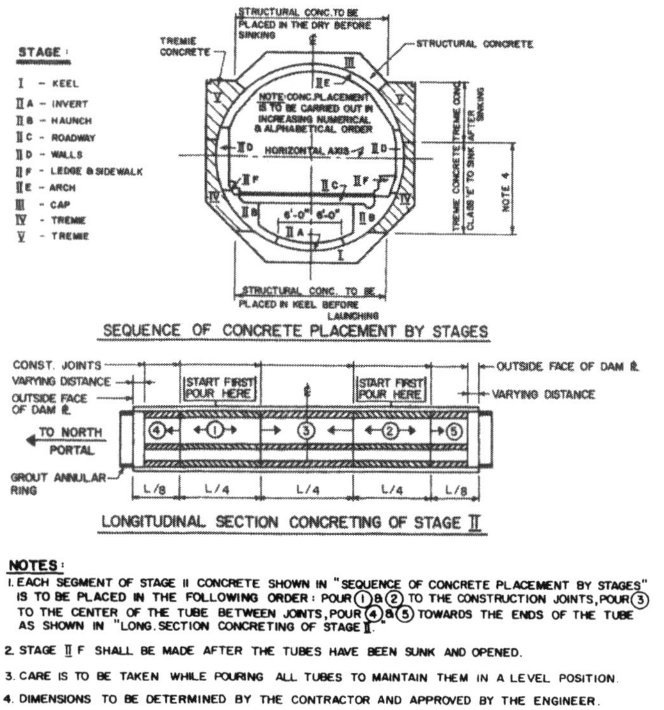

Fig. 14-12. Concrete pouring sequence.

design loads and temperature for minimum code requirements. Steel reinforcing bars are 60,000 psi yield point ASTM Designation A615. Bars spliced or welded to structural steel should not exceed a 0.30% carbon content. Ring and longitudinal bars are spaced 12 in. maximum. Ties, 18 in. on centers maximum, are not less than 1/4-in. diameter and welded directly to the steel shell, or upset bars are threaded into Nelson stud sockets welded to the shell. Additional reinforcing is provided at niches and openings in the concrete lining.

Placing of Concrete

Concrete is usually supplied by a plant at the outfitting site for large projects, or may be supplied by transit mix for smaller projects. Concrete is placed into movable steel forms by pump lines extended through access hatches in the top of the shell. For concrete placement while the steel shell is floating, minimum hatch openings of about 6 –8 ft by 4 ft are provided at approximately 75-ft centers for form handling. Small openings (12 in. by 12 in.) may be provided in the top of the steel shell to facilitate placing of the arch concrete. The pouring sequence is controlled to prevent excessive hydrostatic pressure on the shell at successive increases in immersion and to maintain the tube in level condition for trim and list. Placing of concrete starts at the invert and proceeds in increments on the haunch, sidewall, and arch pours. The roadway slab in highway tunnels is also placed at this stage. Length of pours is determined by the above considerations and convenience in the length of forms. Placing should start approximately at quarter points of the tube and proceed in both directions simultaneously. Construction joints are provided with keys. Figure 14-12 shows a typical pouring schedule and sequence.

Buoyancy and stability of tube sections for progressive stages of concreting are checked from the time of launching to towing to the tunnel site. Keel concrete (Figure 14-12) is usually placed before launching, and the structural cap concrete is placed in the dry at the outfitting pier. Shell plates above the keel concrete pour are sounded for voids, which must be grouted at pressures not to exceed 10 psi. Grout plugs are provided and sealed by welding. After all interior concreting is completed, all access hatches and concreting openings in the shell are closed by steel plates welded watertight. Voids between the shell plate and the interior concrete lining are similarly grouted.

Construction Hatches

The end tubes are equipped with construction hatches and steel stacks extending above the water to provide access to the interior for finishing and other work prior to completion of the adjacent cut-and-cover sections. Hatches are usually circular, with a diameter of 10–12 ft. They are closed with welded steel covers after the end bulkheads have been removed, and access is available from the cut-and-cover section. The stacks are cut off at least 3 ft below the finish line of the backfill placed over the tube. During launching and towing, the top of the stack is closed with a temporary cover. Figure 14-13 shows a cross section of a construction hatch.

Tremie Concrete

Exterior protective and ballast concrete (Figure 14-12) is placed by tremie while the tube is floating at the outfitting pier. Pockets are filled in stages to reduce pressure on form plates. The concreting sequence is controlled to keep the tube in longitudinal and transverse balance. Sufficient con-

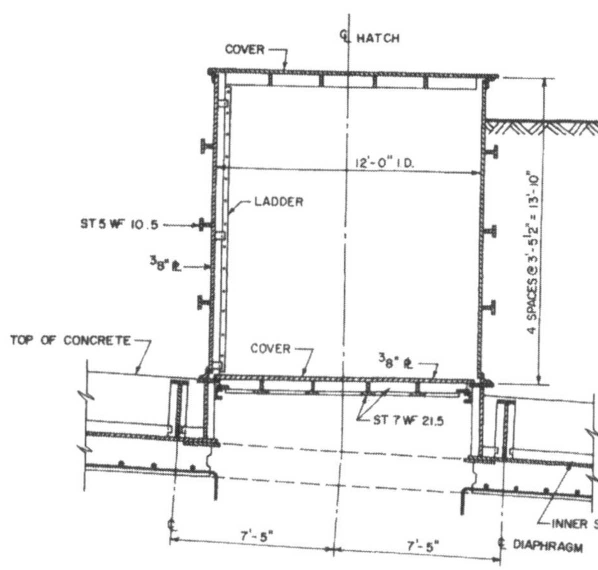

Fig. 14-13. Construction hatch (Second Hampton Roads Tunnel).

crete is placed at the outfitting pier to reduce freeboard of the tube to about 2 ft for towing. The remainder is placed after the tube is suspended from the placing barge.

Ballast

To provide weight for lowering the tubes into place, they are ballasted. In a section such as shown in Figure 14-3, the exterior concrete is sufficient for sinking. In sections such as Figure 14-5, gravel ballast is placed into pockets on top of the section. Water or sand temporarily placed in the interior of the tubes has also been used. Water has to be confined in tanks or in bulkhead compartments to prevent unbalancing in case the tube takes a longitudinal tilt, either accidentally or to conform to final grade. Exterior watertight ballast pockets, filled with water, on top of the tube have also been used, as have pontoons. Interior ballast is removed only after sufficient backfill and/or interior finish concrete or other permanent weight has been added to overcome buoyancy.

CONCRETE TUBES

Configuration

As discussed earlier, rectangular reinforced concrete sections are generally used for tunnels with four or more traffic lanes, particularly where concrete is more economical than steel. Figure 14-14 shows a four-lane highway tunnel with two two-lane compartments and ventilation ducts on both sides. A design for a highway tunnel with two three-lane roadways (with central upper compartments for cable ducts and lower compartments for emergency exit) is shown in Figure 14-15. It is the cross section of Zeeburger Tunnel, which completes Amsterdam's peripheral ring road system under the heavily used navigation channel of Buiten. The 2,900-ft-long crossing contains three 370-ft-long immersed tube elements, all supported on pile foundations. The six-lane tunnel has two 34.5-ft-wide roadway sections with a total width of 87.6 ft and a height of 23.7 ft. The major features of the tunnel include the elimination of external waterproofing and special connection devices for pile connections.

Construction of Tubes

The sections are constructed in dry docks or in temporarily dewatered basins located near the site, which are flooded

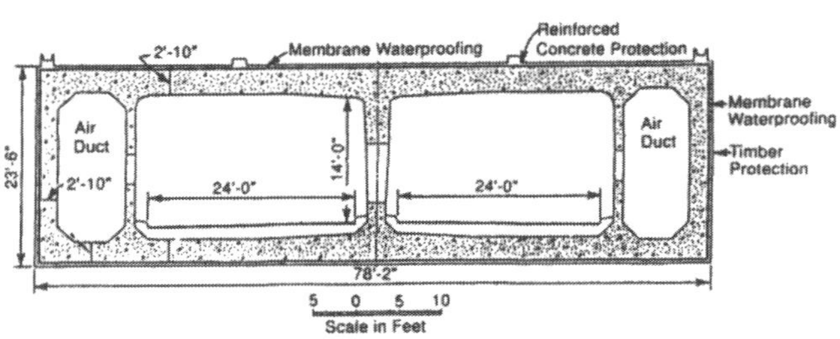

Fig. 14-14. Deas Island Tunnel (four-lane prestressed box section).

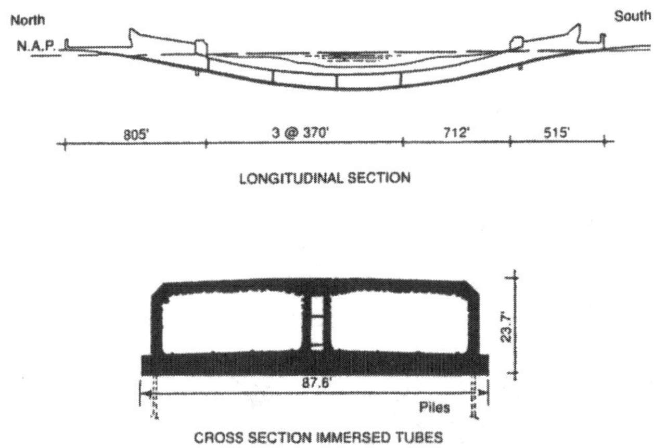

Fig. 14-15. Zeeburger Tunnel in Holland.

for floating the tubes. Unless a dry dock or outfitting basin is available for construction in the dry, these facilities must be constructed for the project, with resultant increased costs and time of construction. The Hong Kong Cross Harbour Road Tunnel was originally designed as a four-lane, rectangular, prestressed section, but an alternative tender was made for the twin steel single-shell section shown in Figure 14-4. The steel section was constructed, at a saving of several million dollars and 1-1/2 years of construction time. The base and roof slabs and sidewalls of rectangular concrete tubes are of massive reinforced concrete as determined by structural and functional requirements. In general, concrete has a compressive strength of 4,000 psi, and reinforcing steel a 60,000 psi yield point corresponding to ASTM Designation A-615. Since the tubes are solidly supported in the construction basin, the concreting sequence is not critical; however, concrete placement requires stringent quality control to assure crack control. Usually, alternate longitudinal sections varying from 20 to 60 ft are poured, and then the intermediate sections are placed, to minimize shrinkage. The base slab is poured over a waterproofing membrane, and walls and top slab are placed by means of movable steel forms to provide smooth and accurate interior surfaces.

The concrete tube may be heavy enough for sinking, in which case it has to be supported by pontoons for floating into position. If the completed section is buoyant, it has to be ballasted for sinking. Temporary ballast may be placed internally, using water, sand, or ballast blocks, or permanent ballast consisting of gravel or lean concrete may be placed on top of the section.

End bulkheads may be steel, timber with waterproofing, or reinforced concrete. Selection of end bulkhead materials is a matter of individual preference and economy.

Prestressing

Since concrete thickness is determined largely by the weight required to prevent uplift, prestressing is economical only for very wide sections. Obtaining watertightness by prestressing alone, without membrane, is controversial. Figure 14-16 shows a four-lane tunnel, prestressed to a residual

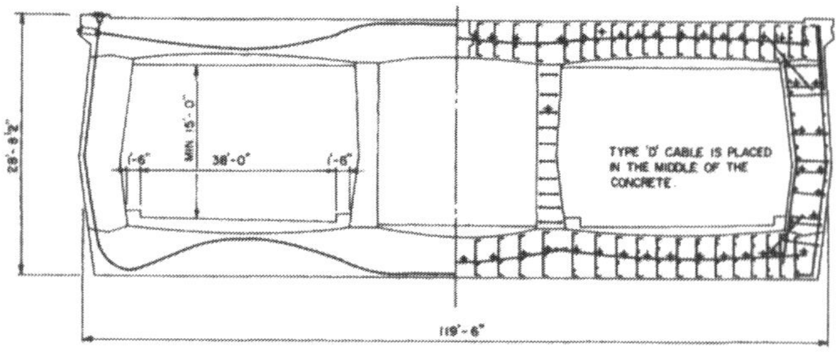

Fig. 14-16. Four-lane prestressed concrete section of LaFontaine (Boucherville) Tunnel.

compressive stress of about 200–300 psi, which is assumed to be watertight.

Waterproofing

As indicated earlier, the improvement in achieving watertightness was one of the factors for the advancement of concrete immersed tube tunnels in Europe. According to recent Dutch practice, a concrete immersed tunnel can be watertight if one of two design parameters are used effectively: crack control to achieve impermeability of the concrete, or an independent waterproofing system. The measures necessary for watertightness and crack control were discussed previously. In the following paragraphs, the waterproofing types are examined.

Steel membranes made of 1/4-in. structural steel plate have been applied to several concrete box tunnels as a waterproof enclosure. All joints are welded, and anchors are welded to the skin to bond it to the concrete. With temporary stiffening, the plates may serve as a form for the concrete. For the tender design of the Great Belt project in Denmark, a continuous steel plate was used along the circumference of the cross section. In most cases, a steel plate has been used to waterproof the base slab and at least a portion of the sidewall. The recently completed Conwy Tunnel in England used a 1/4-in. steel membrane for bottom slab and sidewalls. The top slab is protected by two layers of bitumen covered by concrete. Special attention needs to be given to the joint between the steel plate and the bitumen membrane. The steel membrane is protected against corrosion by cathodic protection.

Multi-ply membranes of fabric and coal-tar layers can be applied to the top of the box sections, at least four plies being used. To protect the membrane against damage during handling and placing of tubes, it is covered with a thin layer of poured concrete, bricks, or planks of concrete, asphalt, or wood. Due to the difficulty of attaching such a membrane, and in protecting it later, it is not practical to use it on the sidewalls or the bottom.

Plastic membranes made of synthetic neoprene or vinyl-type rubbers have been developed for waterproofing concrete structures. Sheets of this material about 1/8-in. thick can be attached to the roof and sidewalls with special adhesives. Due to the difficulty of protecting the membrane on vertical sides against damage, use of this method has been limited.

Epoxy coatings, mostly of coal-tar epoxy to waterproof the sides and tops of tunnel sections, have also had a few applications. Great care is needed in preparing an epoxy mixture in order to retain adequate elasticity and adhesion and in its application to the concrete surface. The coating may be applied by brush or sprayed on a completely dry surface. Protecting such a coating or membrane on vertical sides against mechanical damage is extremely difficult.

Historical development of waterproofing applications in Holland was discussed earlier. Since the late 1980s, however, primary emphasis has been on using concrete as an impermeable barrier with improvements in crack control and in expansion joints, details, and element joints.

WEIGHT CONTROL OF TUBES

Final Weight

Final weight of the completed tube, including main structure, interior finish, and backfill, has to be controlled to counteract buoyancy forces. Thickness of concrete is more often determined by the weight required than by structural strength.

Weight of Tubes at Various Stages

This has to be controlled in accordance with construction procedures. Tubes built in dry docks are solidly supported by keel blocks, and the only criteria are the loading on the supports and on the dock floor. These do not impose any serious restrictions on the construction sequence. When concrete is placed while the tube is floating, as in the steel shell type, the sequence of pouring concrete is governed by several factors: keeping the tube on an even keel; increments of immersion producing hydrostatic pressure on the steel plate; and longitudinal bending moments from unequal loading, as discussed in this chapter.

Factors of Safety Against Buoyancy

The fundamental difference between recent European and U.S. practices for the selection of safety factors against buoyancy arises from the difference in construction methods between concrete box and steel shell design. There are two contributing factors for the difference; the use of distinctly different materials for watertightness (structural steel versus structural concrete) and the use of final ballast. In U.S. practice, steel shell plate design generally is not controlled by stress considerations. It is primarily a watertight membrane, and the maximum stress occurs during construction. Also, the dimensions of the concrete lining of a steel tube tunnel are not controlled by stress, but by buoyancy. The steel shell's ductility provides a variety of options in placing the final ballast and the structural concrete. The ballast is generally placed during sinking, allowing the steel shell to be used as a vessel for long-distance towing with the minimum freeboard available.

In contrast to the steel shell tunnel, concrete tube tunnel design is controlled by the quality of concrete for crack control, temperature control, shrinkage, and creep control. The concrete tube usually cannot be used as a vessel for transporting elements to the site because the entire ballast concrete (usually at the bottom slab) is placed in the casting yard. This allows better quality control. As a result, a concrete tube generally will have a negative buoyancy, requiring pontoon-type equipment for transport. European immersed tubes are usually constructed relatively close to their permanent locations, as the large concrete sections are difficult to transport, By contrast, U.S. steel tubes are frequently fabricated at a considerable distance from the tunnel site. The outer Chesapeake Bay Tunnel and the Second Downtown Elizabeth River Tunnel in Norfolk were fabricated in Texas and floated to Virginia, a distance of 2,000 miles.

In European practice, the required factor of safety for buoyancy at the element's permanent state is 1.05, based on minimum weight and maximum water density, excluding the friction forces from backfill and excluding the weight of operational installations, finish work, pavement, safety walks, and backfill. Generally, no design guidance is given for the factor of safety to be observed during concreting, transporting, or sinking. The required additional temporary ballast during the dewatering of joints between tube elements is left to the contractor. The final factor of safety against buoyancy including total weight of the element and backfill, excluding friction, is generally greater than 1.15. It must be noted that Dutch practice is not to specify a factor of safety against uplift, but rather to specify instead a minimum residual bearing stress, to be maintained under maximum-uplift, minimum-weight conditions.

For U.S. practice, the following is recommended: After placing the tube in the trench and before backfill is placed or interior finish, including sidewalks, ledges, and roadway pavement, are completed, a minimum factor of safety of 1.10 is required. This may be achieved by adding temporary or permanent ballast. The specific gravity of the water in the bottom of the trench and the possible accumulation of fluid silt from adjacent dredging operations or sludge from other sources must be checked.

After dewatering of joints and removal of the end bulkhead, and prior to completing the interior concrete lining and backfill at the joint, the structure should have a minimum factor of safety of 1.02. The completed structure should have a minimum factor of safety of about 1.2, not including backfill. With backfill in place, the factor of safety against uplift of the tunnel is usually at least 1.5 or more.

Checking of weight is done by computation of weight of structural and reinforcing steel and accessories and continuous checking of unit weight of concrete samples. In determining the volume of the concrete, the volumes of embedded structural steel, reinforcing steel, conduits and pipes more than 2 in. in diameter, and all boxed-out niches should be deducted. The weights of available aggregates are determined, and the mix is designed for weight as well as for strength. If local aggregates have insufficient weight, it is usually more economical to use more distant sources of supply of heavier material instead of increasing the concrete volume of the tubes.

PREPARATION OF TRENCH

Dredging Sections

The trench must be deep enough to allow for a foundation course below the tunnel, the height of the tubes, and preferably a minimum of 5 ft of protective backfill. Under certain conditions, a shallower trench may be used for short distances, with the tunnel projecting partly above the natural bottom of the channel. Figures 14-17 and 14-18 show typical sections of the dredged trench. Figure 14-19 shows a configuration of a dredged trench with the tube projecting above the natural bottom with a protective dike. The side slopes depend upon the soil characteristics and may vary from near vertical in rock to 1:4 in soft material. Generally, a slope of 1:1.5 is feasible. In unusually deep trench sections, an intermediate berm may be required.

Payment for dredging may be based on fixed quantities determined by a side slope selected on the basis of borings and slope stability studies, say 1:1.5 in reasonably firm soil, giving the contractor the option to steepen the slope to, say,

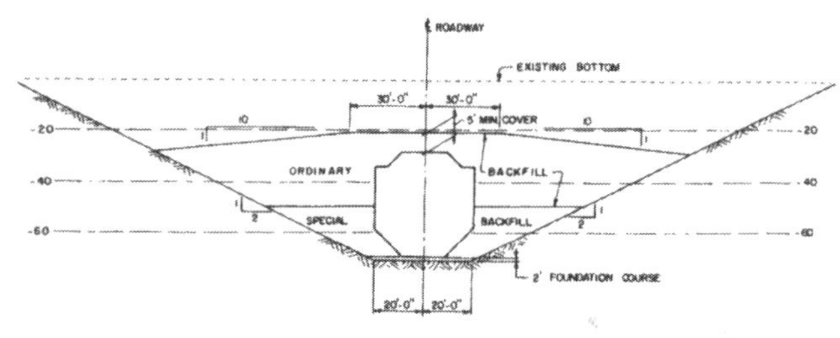

Fig. 14-17. Typical section, dredged trench.

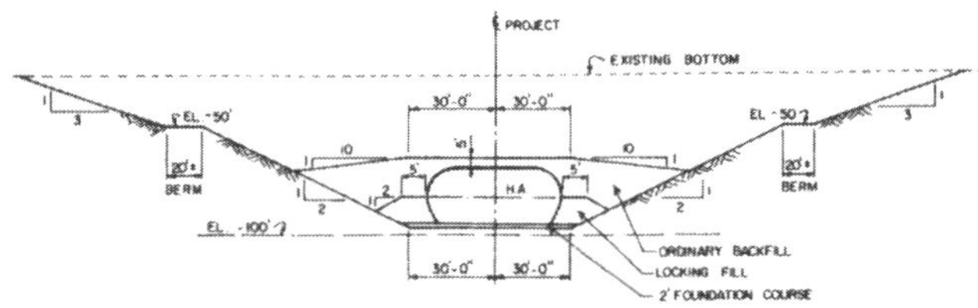

Fig. 14-18. Cross section, dredged trench with berms.

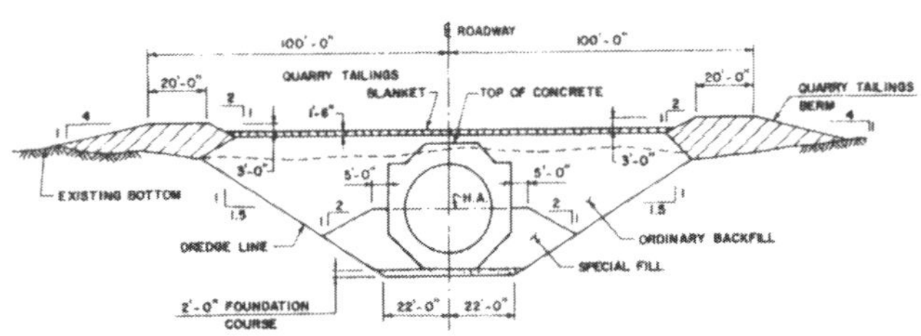

Fig. 14-19. Cross section, dredged trench, protective berms.

1:1 at his own risk or flatten it to 1:2 with a unit price payment on overdredging if actual soil conditions require this.

Rough Dredging

This is usually done, at the contractor's choice, as a continuous operation. Hydraulic dredges are used where suitable for depth and material and where disposal areas are within economical pumping distance. For greater depths or barge transportation of spoil, clamshell dredges are used. For harder materials, dipper dredges are used.

Fine Dredging

Fine dredging to final dimensions is done two or three tube lengths ahead, to keep the time interval between it and the placing of the tube to a minimum.

Removal of Silt

Just prior to the placing of the foundation course, the trench is checked for accumulation of silt or sloughing of the dredged slopes. Any such accumulation is removed by clamshell, suction dredge or air lift.

TUBE FOUNDATIONS

Several methods have been used to prepare the foundation supporting the tubes. Since the weight of the tubes including backfill is not much greater than that of the displaced soil, accuracy of alignment is the most important function, bearing capacity being a second consideration.

Screeded Foundation Course

The finished dredging is carried to a minimum of 2–3 ft below the bottom of the tube. A cover of coarse sand or well-graded gravel is placed in the trench and leveled to accurate grade. The gradation of the material may vary from 1/4 to 1-1/2 in. in light currents, or may go up to 6-in. stones where heavy currents occur. The leveling is done by dragging a heavy screed, made of a grid of steel beams, over the surface in successive passes, adding material as needed. The screed is suspended from winches on a carriage rolling on tracks supported on an assembly of two steel barges yoked together. The tracks are adjustable to parallel the grade. The rig is anchored in correct position. Adjustments are made in the screed suspension to compensate for tide level changes. The foundation for a 300-ft tube can be screeded in two or three setups. For the 58 tubes of the BART tunnels in San Francisco, a special screed rig was built consisting of a steel truss of 240 ft in length, supported by flotation tanks that could be partly ballasted with water to reduce buoyancy, permitting anchoring the assembly against tide level changes, thus eliminating adjustments during screeding.

Accuracy of screeding usually allows a tolerance of ±1-1/2 in. in elevation for a 300-ft tube for a highway tunnel. The tubes for the San Francisco Trans-Bay tunnel (Figure 14-20) were set within a tolerance of ±1-1/2 in. of theoretical grade.

Jetted Sand and Sand Flow Foundations

The first jetted sand foundation was used for the Maas Tunnel in Holland in 1946. Since then, a number of tunnels have been built using this method, where a sand–water mixture is jetted under the tunnel element while the element is positioned to line and grade on temporary supports. The jetting is accomplished through a horizontal pipe within the opening between the tunnel and the dredged trench. A sand jetting gantry with a tower is constructed over the top of the tunnel, moving along the tunnel to facilitate the jetting operation. The reach of the jetted sand is adjusted by a nozzle on the jetting pipe to change the exit velocity. On the average, the sand constitutes about 10–15% of the sand/water volume. Packing of the sand is usually loose, with a void ratio of approximately 40%. The minimum thickness of the sand foundation was determined to be 3 ft to ease the jetting operation. In early tunnels, including the Maas Tunnel, the establishment of a regular flow pattern was not easy to achieve because of the fixed position of the supply pipe (perpendicular to tunnel axis). In later tunnels, this was changed, where the supply pipe revolved around the vertical pipe located at one side of the tunnel and connected to a gantry that rode along the temporary bridge on top of the tube. The primary disadvantage of the jetting method is that the major part of the operation takes place at the surface, which can be a handicap for a busy channel. Because of this disadvantage, the design of the Western Scheldt Tunnel in Holland in 1969 proposed a sand flow method.

The sand flow method, as it was named later (in the Vlake Tunnel construction in 1975), was an improved version of the method proposed for the Western Scheldt Tunnel (which was not constructed). In the Vlake Tunnel project, a sand barge and pumping units were moored along the shoreline allowing a clear channel for navigation. Two sand flow lines were provided within each tube, connecting to two sets of discharge openings located in the bottom slab with an assumed radius of deposit of 36 ft for each. For a 412-ft element, a total of 26 discharge points were used, 13 for each half of the bottom slab. Discharge valve openings contained rubber-coated steel balls on the underside of the slab. During the pumping, sand and water mix will push the ball down, allowing inflow to pass. When pumping is stopped, the ball will be forced back to its seat by hydrostatic pres-

Fig. 14-20. Screeding rig—Trans-Bay Tube.

sure. The sand flow method not only provided a clear channel for navigation, it also secured better control for a uniform sand flow pattern, which was very desirable for conditions with high currents.

Pile Foundations

In unusual circumstances, where the soil under the tunnel is too weak to support the tunnel and backfill, and cannot be economically excavated and replaced with firm material, the tubes may have to be supported directly on pile bents. These are driven and cut off underwater to grade. Various cushioning systems are used between the tops of the piles and the bottom of the slab to mitigate the tolerance in practical pile cut-off elevation.

A weak layer of soil may be strengthened by driving a series of compaction piles. These are driven so that their tops are 1 or 2 ft below grade and are covered with a screeded foundation course.

Recently in Holland, several immersed concrete tunnels with interesting features of pile foundations have been constructed. In 1961 the IJ Tunnel in Amsterdam was constructed on pile foundations consisting of eight 3.5-ft diameter cast-in-place concrete piles with pile caps at 100-ft intervals. The boring and concrete placement for the piles were accomplished with the help of fixed pontoons with four legs. Only the top 30 ft of the piles were reinforced. The pile caps placed over the piles were constructed by diving bells suspended from floating double pontoons. The connections of tunnel elements to the pile caps were achieved by connecting 1-in.-thick rubber bags to the bottom slab of the tunnel elements, set at the preestablished pile cap locations to receive the bags, and then filling the retained volume by pressure grout from inside each tunnel element. The elements were held in theoretical position by temporary support during the connection operation.

In 1990, Amsterdam's Zeeburger Tunnel was constructed on a pile foundation using 1.65-ft-diameter steel piles driven to a depth of about 140 ft along the outer walls of the tunnel elements, spaced at 9-ft centers. The accuracy required for connecting the piles to the tunnel elements was achieved by using vertically adjustable serrated locking pieces. Initially, tunnel elements were supported temporarily by two piles while the serrated connection piece was lowered to engage the pile head. Final connection was established by engaging the upper piece of the serrated connection to the lower piece by applying pressure grout to the connection devices.

Mortar Injection as Foundation Course

Over the past decade in Japan, the mortar injection method has been frequently used as a foundation course. The system includes placement of mortar while the tube is supported by temporary support jacks. The base slate is designed to include 4 m × 4 m spaced injection orifices through which a low-viscosity bentonite mortar is injected at pressures 0.1–0.2 kg/m^2 higher than the local water pressure to form a continuous foundation.

Injection of the mortar is monitored by ultrasonic gap meter to check the clearance between the bottom slab and the top of the grout. Generally, low-viscosity, low-segregation bentonite mortar is used with balanced mortar pressure to obtain required strength in the foundation course. Generally a 500-mm-thick grout is used. To retain the mortar within the foundation limits, a stopper bag and crushed stone combination is used along the edges of the foundation. Before placing the tubes on the temporary supports, stopper bags are placed at the inner end of the temporary supports and the crushed stone along the outer edge of the foundation. The bags are inflated with water before injection of the bentonite. Bags used in combination with crushed stone prevent the grout from moving out of the foundation limits. The system was first used on the Tokyo Port Tunnel in 1976, and its use has since become common practice in Japan.

Tube Placement

The tubes are moved from the construction basin or outfitting site with a freeboard of about 12–18 in. Heavy concrete box sections may have less freeboard, and to ensure their buoyancy they may be supported by pontoons or barges, which then form part of the lay barge. The methods of placing vary somewhat according to the type of foundation support and type of joints between tubes. In tidal waters, the actual placing is scheduled during slack tide. In rivers with constant flow or where slack tides are extremely short, the effect of the current on the tube while it is being lowered must be analyzed. This is particularly important for wide rectangular sections, where water pressure may upset the equilibrium of the tube while it is suspended on wire ropes with a relatively small positive weight. The specific gravity of the water in the bottom of the trench is checked to gauge the adequacy of the ballast.

The *lay barge* consists of two or more steel barges. These are placed parallel, with sufficient space between them to clear the tube, and connected with transverse girder bridges that carry for lowering winches. The assembly is further stabilized with diagonal wire ropes. The lay barge is held in position by wire ropes extended from winches to heavy anchor blocks placed on the bottom at distances of several hundred feet. The anchoring devices must have an adequate factor of safety to resist the maximum current pressure and wind forces on the barge and tube. The capacity of the lowering winches and wire ropes is determined by the net positive weight of the tube during placing. This may vary between 100 and 400 tons, depending on the size of the tube and the contractor's choice of sinking equipment. Load-limiting devices may have to be installed on the lowering winches to prevent overload on the wire ropes, if there is danger of sudden surges in water level to which the tube cannot respond quickly enough. By imposing speed restrictions on passing vehicles during lowering, this danger can be minimized.

In some cases, two whirly cranes mounted on barges have been used to support the tubes during placing. This limits the lowering weight to about 100 tons, but it permits more precise adjustments of the position of the tube without moving the barges.

The tendering of Denmark's Great Belt Tunnel project in 1988 provided a range of interesting immersing techniques proposed by four construction consortia. The proposals varied from the classical lay barge design commonly used in the United States, to the variation in theme of a monolithic catamaran rig with rigid cross frames.

The saga of the construction of Virginia's Second Downtown Elizabeth River Tunnel started with its fabrication in Corpus Christi, Texas. Once steel fabrication was completed, pairs of tubes with keel concrete in place were skidded onto oceangoing barges and towed more than 2,000 mi via the Gulf of Mexico and Atlantic Ocean to Virginia. Special details were developed to make the tube elements and the barges act as a unit during the long journey. The tubes were floated off their supporting barges in a floating dry dock in Norfolk and then transferred to an outfitting yard in Portsmouth with 4-ft draft. After outfitting in Portsmouth, the tubes were taken to the project site with 18-in. freeboard and placed in the prepared trench with the help of ballast concrete. The keel concrete was placed in Texas for needed stability during transport to the outfitting yard near the project site.

Ballasting of the tube takes place after it is suspended from the lay barge and positioned over the trench. The type of ballast depends on the construction of the tube. For the Hampton Roads Tunnel (Figure 14-3), a number of the outside tremie pockets were filled with concrete to provide sufficient negative buoyancy to sink the tube. The remainder of the pockets were filled as soon as the tube had been lowered into position and connected to the previously placed tube, to provide sufficient weight to hold the tube in place during placement of backfill. The Trans-Bay Tube (Figure 14-5) was ballasted by filling the pockets on top of the tube with gravel. The Deas Island Tunnel (Figure 14-14) used water ballast in tanks inside the tubes.

Placing on Screeded Foundation. After ballasting, the tube is slowly lowered until it nearly touches the foundation. Its inboard end is kept clear of the outboard end of the previously placed tube. After a final alignment check, it is then moved against the other tube as required to make the connection at the joint and set down on the foundation. If the weight of the tube is light, as when handled by cranes, additional ballast must be placed as soon as the connection is made—either by filling all tremie pockets with concrete or by temporary heavy concrete blocks—to secure the tube against uplift.

Placing Tubes on Jetted or Sand Flow Foundation. In Europe the selection of temporary support for the jetted sand or sand flow foundation is usually left to the contractor. The temporary support may vary from concrete foundation blocks to inflatable bags, depending on the selected immersing system.

In early tunnels, temporary foundations were cast in dry dock and attached to the tunnels, usually four per tunnel unit. In recent construction, temporary blocks are placed on the bottom of the trench separately before the tunnel unit arrives. This reduces the depth requirements for the dry docks. Generally, temporary foundation blocks are placed on gravel foundations by a screeding gantry with hydraulic legs, or they are placed directly on the trench and have jacks with longer strokes. An average concrete block will have 15 ft × 15 ft × 3 ft dimensions to receive the rams from hydraulic jacks located inside the tube, in order to keep the tunnel element on alignment and at proper grade. Recently, the number of temporary foundations have been reduced to two supports located at the outboard end of the element being placed. Two seats on the unit in place provide the support for the inboard end. Unlike the screeded foundation, one important factor that must be recognized by the designer is the settlement characteristics of the jetted sand or sand flow foundations. For both systems, after the removal of temporary supports, the tunnel element can settle as much as 3 in. An average settlement of 3/8 in. can be expected under normal conditions. For this reason, in Holland, immersed concrete tunnels are designed with dilatation joints every 60–75 ft, coinciding with the ends of modular units. These joints are designed to transfer shear and therefore will prevent differential settlement between segments and, at the same time, will allow rotation. As a result, the concrete tube will act like a series of linked structures following the combined settlement of the sand foundation and the subsoil strata without inducing longitudinal bending moments. For a successful jetting or sand flow system, the sand particles have to be uniform, approximately 0.02 in. in average diameter, which is the cause of settlement of a foundation course (not a well-graded material).

One of the everlasting challenges of this type of foundation system is the quality assurance during sand flow or jetting. Over the years, a number of instrumentation systems have been developed to correlate the water pressure at discharge, sand packing of the mixture, the amount of sand supplied, jacking pressures at the temporary supports, and echo sounding of the outside walls in order to assess the uniformity of the sand foundation and to minimize the upward pressure against the base slab. During the Vlake Tunnel construction, 2.8 in. settlement took place at the last injection point. This was assumed to be related to silting during the sand flow operation. For two 412-ft-long and 90-ft-wide units of the Vlake Tunnel, a total of 13,000 yd^3 of sand was used within 200 working hours (30 days). In spite of all the instrumentation mentioned above, correlation between the mixture and pressure at the nozzle and the upward pressure at the base slab was not conclusive. Since then, the foundation placement time for sand flow foundations has been improved. The recently completed Liefkenshoektunnel in Antwerp required 60 working hours for the placement of foundations of an element 460 ft long and 106 ft wide.

Alignment Control

To check the horizontal and vertical position of the tube during placing operations, temporary survey towers (Figure 14-21) are mounted on the tube near the ends. These project above the water and carry survey targets. To control transverse level (roll) of tubes and to ensure coincidence of the target location with that of the tube below, a vertical plummet was installed in the survey towers of the Trans-Bay Tube in San Francisco in relatively deep water.

On the first tube placed at each end, two survey towers are mounted, one near each end to correctly place the tube. The tower at the outboard end is left in place until the next tube is placed. The position of the inboard end of each following tube is governed by the tube already in place, so that a survey tower on these tubes is needed only at the outboard end. After the towers have served their purpose, they are disconnected by divers and reused.

Instruments for checking the alignment are mounted on each shore. On long tunnels or where horizontal curves occur, additional instruments are set on pile-supported platforms in the water. Theodolites are used. Where more elaborate survey towers with vertical plummets are used, instruments can be mounted on them, sighting on targets located on the shores. A final check of the position is made after the tube has been placed and connected to the previously placed section.

Generally, in the United States, tunnel elements are placed by suspending them from catamaran barges, where all load lines and alignment control cables are anchored. In Europe, the horizontal control is made through fairleads from the elements connecting to freestanding anchors. This option permits positive control of the outboard end of the tunnel. However, the operation is more complex and time consuming. In the United States, before construction of the BART tunnel, the tunnel elements were joined by use of a tremie seal joint. Gasketed joints were used for the first time in the United States on the BART tunnel. In this type of joint, horizontal corrections for small errors cannot be accomplished by control cables from the barge. Corrections are made usually by the placement of shims, which can be frustrating at times. A method using an adjustable wedge system was first developed for construction of the Fort McHenry Tunnel. A set of wedges, about 16 ft long, 1.5 ft wide, with varying thicknesses up to 4-1/2 in., were placed on each side of the rubber gasket seal at the horizontal centerline of the section. Once the joint is dewatered and the wedges bare, a follow-up survey of the in-place tube will indicate how accurate the initial setting is. If further correction is needed, the joint is refilled, permitting the wedges to be reset for final desired positioning. The entire operation can be controlled from the surface by a hydraulic system that takes not more than 1/2 hour to operate on average. The time-saving feature of the wedge system can be an added advantage in constructing tunnels under busy harbors.

Fig. 14-21. Survey tower (Second Hampton Roads Tunnel).

JOINTS BETWEEN TUBES

Tremie Joints

Tremie joints (Figure 14-22) have been used in a number of steel shell tubes and are still applied when relatively few sections are required. A circular steel collar plate, of the same diameter as the tube shell, is welded to the outside of the end bulkhead and projects 4 ft. To the lower half of this collar on the outboard end of the tube, a 3-ft-wide hood plate is welded with a filler, half its width projecting beyond the end of the collars. A similar hood plate is welded to the upper half of the collar at the inboard end of the tube. The

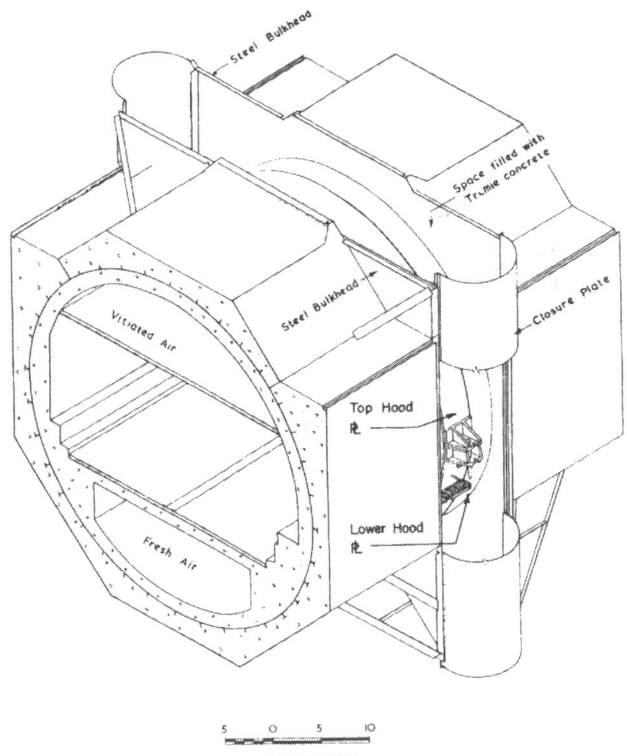

Fig. 14-22. Tremie joint.

fillers provide an annular clearance of 1 in. between hood plates and collar. When the following tube is lowered, its inboard collar plate fits into the lower hood of the previous tube. On the horizontal diameters, where upper and lower hood plates meet, each carries a heavy welded bracket with a hole for a 5-in.-diameter steel pin. The hole in the upper bracket is round; the one in the lower is elongated to provide about ±1.5 in. tolerance. After the tubes are pulled together by four steamboat ratchets, operated by divers, to match the brackets, the divers insert the tapered end steel pins, which are secured to the brackets by wedges driven through slots in the pins (Figure 14-23).

Curved closure plates are inserted in guides attached to the vertical edges of the rectangular end dam plates. Either sheet pile sections or steel angle guides are used to form the guides to connect the curved plates to the dam plates. Sometimes, a series of sheet piles in a curved configuration has been used instead of closure plates. The piles are driven into the bottom of the trench, the foundation course having been omitted at the joints, so that an enclosed space is provided by them and the end dam plates, which is filled with tremie concrete. To prevent concrete from flowing inside the joint, the annular space between hood plates and collars is caulked by divers. Care must be taken to ensure that the tremie concrete flows across the bottom of the joint, and that at the top it is level with the top of the dam plates. This provides an adequate seal of the joint with only minor leaks, if any. Any leakage must be sealed off by grouting, before the joint can be completed. To dewater the joint, valves are provided in the end bulkhead of the previous tube. The joint is then entered through the watertight doors mounted in the end bulkheads. Before opening the bulkhead doors, the air in the joint should be tested for explosive gases due to possible decomposition of organic matter in the water trapped in the joint.

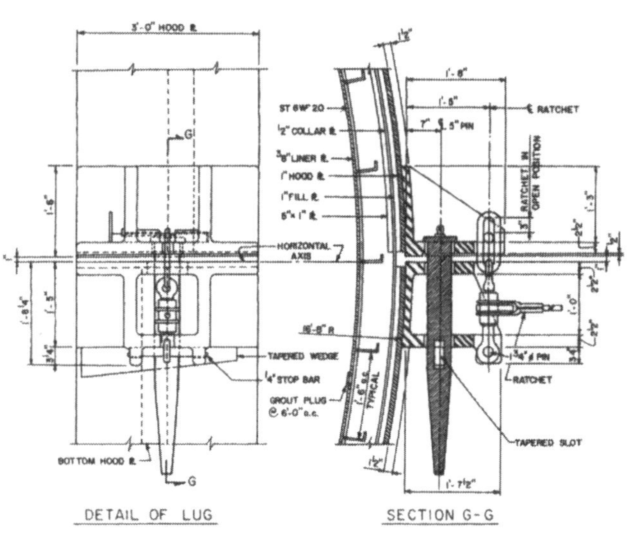

Fig. 14-23. Pin, wedge, and bracket detail.

Sectional linear plates are then welded to T-sections on the interior of the collar plates to form a continuous watertight steel membrane for the tunnel. The space between these liner plates and the collar plates is filled with grout under 10 psi pressure. Then the interior portions and the dam plates are removed, and the interior concrete lining is placed, thus completing the joint (Figure 14-24a). The same process is used for completion of regular gasketed joints. Liner plates are welded to the T-sections after successful completion of initial sealing and before concreting. Steel liner plates for tremie joints and regular gasketed joints are shown in Figure14-24a and later in Figure 14-27.

Although the figures show the joint between each element perfectly lined up, in reality the relative position of the two elements can vary. The interior dimensions as set for the cross section assume a number of tolerances in order to maintain the required clearances for operational purposes. These tolerances must be recognized from the beginning. They will depend on the fabrication practices exercised and the placement techniques employed. Nevertheless, the accumulated tolerances (from fabrication to placement) will finally have to be dealt with in the closure plates. In recent projects, a bent plate instead of T-sections proved to be more practical for adjusting and negotiating the closing conditions. The variations finally will show up in application of the tunnel sidewall and ceiling finish contract. Proper allowance must be made in setting the cross section to ease the installations of the element's finish contract. In Boston's Central Artery/Third Harbor Tunnel project, an Omega gas-

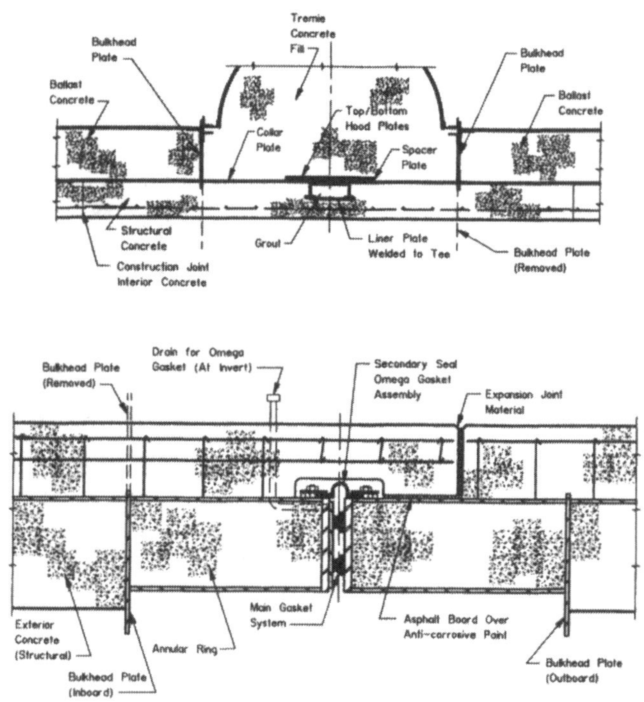

Fig. 14-24. (a) Tremie joint detail. (b) Proposed typical joint, Boston Third Harbor Tunnel.

ket assembly system that acts as a flexible joint between the interior concrete lining and exterior annular ring is used in lieu of liner plates, as shown in Figure 14-24b.

Rubber Gasket Joint

In this type of joint, the initial seal of the joint is provided by the compression of rubber gaskets attached to the face of one tube and bearing against a smooth surface on the adjoining tube. While various shapes of gaskets have been used, the principle is the same. The tube being placed carries a continuous gasket on the periphery of its inboard face. After lowering, the tube is moved close to the outboard smooth-faced end of the tube in place so that hydraulic coupling jacks, extending from one of the tubes, can be engaged into mating parts on the other tube. These jacks pull the tube into contact and give an initial compression to the gasket. This seals the joint sufficiently for it to be drained from the inside, which brings into action the entire hydrostatic pressure on the far end of the tube, compressing the gasket, temporarily sealing the joint. Positive stops are provided to limit the compression of the gasket to its design range. After dewatering, the joint then can be entered through the doors in the end bulkheads, and the permanent connection made.

A single gasket of large dimensions was used on the Deas Island Tunnel and on similar concrete box tunnels in Europe, as shown in Figure 14-25. The tip of the gasket provides the initial seal. Under final pressure, the body of the gasket is compressed to about half its height and carries the entire load.

A double-ring gasket of smaller cross section was used in the Trans-Bay Tube of BART (Figure 14-26). The gaskets are attached to a continuous bracket-type extension of the tube structure. The cantilever lip on the outer gasket is deflected by the pull of the jacks and provides the initial seal. Upon dewatering of the joint, the gaskets are compressed to one-half of their height. A steel bar welded to the face provides a definite (limit) stop to the movement that avoids any variation in the final spacing of the joint, as shown in Figure 14-27. Valved drain lines bleed the water between the gaskets to the interior during compression. The gaskets made

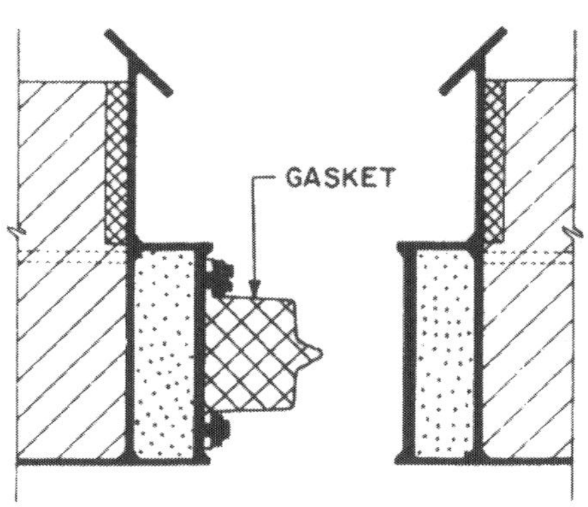

Fig. 14-25. Rubber joint with single gasket (Deas Island).

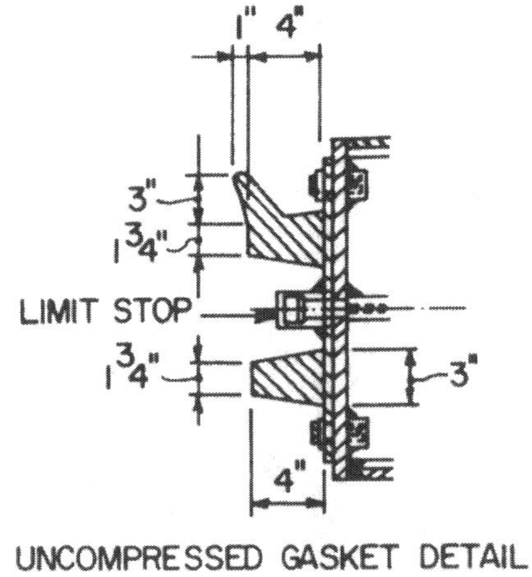

Fig. 14-26. Rubber joint with double gasket (Trans-Bay Tube).

the joint watertight (Figure 14-28), permitting the welding of the interior liner plates and completion of the concrete lining.

Coupling Jacks. Large single coupling jacks (Figure 14-29) mounted at the center of the bulkhead in each bore have been used in Deas Island and other similar tunnels. They must be controlled from inside the tube, and a diver has to enter the open joint to check that they are properly engaged. Once pulling has started, they are no longer accessible if trouble develops.

Multiple, externally mounted jacks—similar to large automatic railroad couplers—were used in the San Francisco Trans-Bay Tube. Two of these were mounted on each side of the tube: one near the top, the other near the bottom. They were at all times accessible to divers for checking. Should one have malfunctioned, the other three would have been

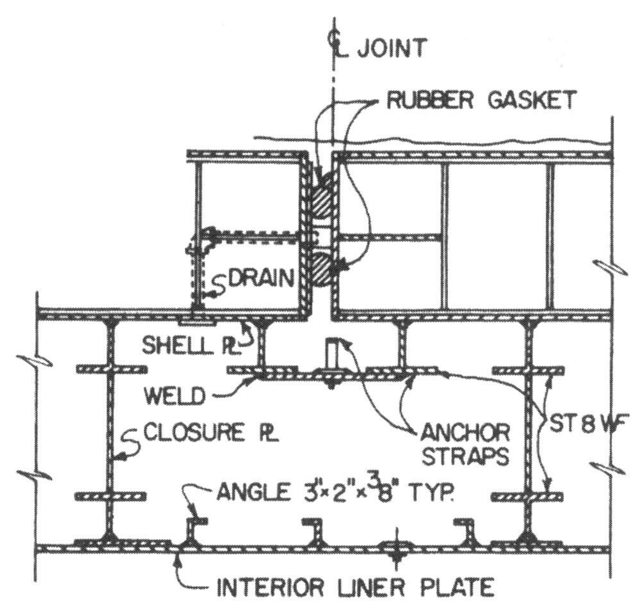

Fig. 14-27. Rubber joints with double gasket (Trans-Bay Tube).

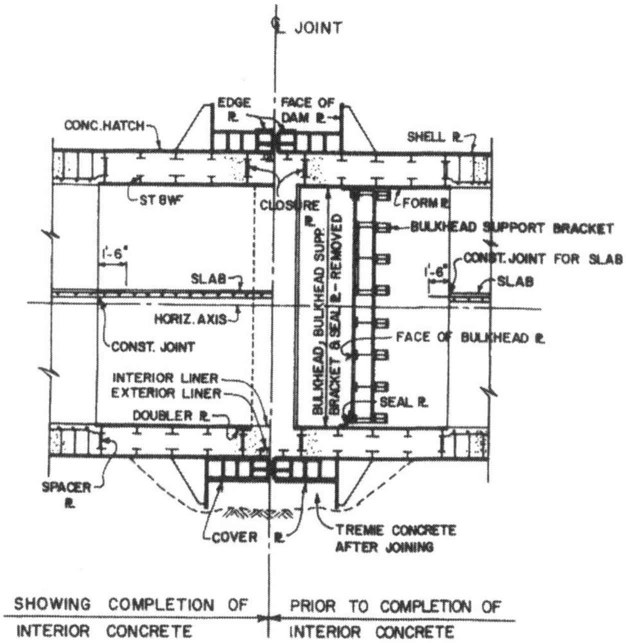

Fig. 14-28. Typical joint cross section (Trans-Bay Tube).

sufficient to close the joint. After the gasket was compressed, the coupler bars were wedged in position and the hydraulic jacks were removed for reuse. The hydraulic control hose lines were extended to a control panel on the lay barge. The system worked exceptionally well for all 58 tubes. Since completion of the BART tunnel, similar couplers have been used successfully in other projects.

Closure Joints

In long tunnels with many tubes, it is usually most expedient to start placing the tubes from both ends, with the closure joint (Figure 14-30) somewhere out in the waterway. In the 58-tube Trans-Bay Tunnel, Tube No. 37 was the last one placed. To provide for fabrication tolerances, which may be cumulative, and for adequate clearance in placing the last section into the space, the closure joint must allow for adjustment. Rubber gaskets cannot be used because no hydrostatic compression is available. A tremie joint with extra long hood plates was used in the Trans-Bay Tube, providing a 24-in. allowance for clearance. It was sealed in the same way as described for the regular tremie joint. Due to very tight fabrication control and limit stops in the joints, the actual tolerance required was less.

Special Joint for Seismic Movements

The San Francisco Trans-Bay Tube BART is located in a highly seismic area. Although it does not cross any active faults, the tunnel structure, including the joints, is capable of elastically absorbing deflections from seismic waves originating anywhere in the area's major faults. Where the ends of the submerged tube section join the massive ventilation buildings, particularly at the San Francisco side, differential movements in any direction may occur between tube and building, amounting to as much as 4 in. A patented composite telescoping and sliding joint allows for these motions (Figure 14-31). For transverse sliding or rotation displacement, two neoprene gasket rings are attached to the tube end, held under compression against a Teflon-coated steel plate on the building side by a series of short steel wire ropes. A telescoping joint of similar construction permits longitudinal motion. Provisions are made for retightening of the ropes by threaded sockets and nuts if needed. Flexible neoprene-impregnated fabric covers attached to the outside of the joints protect the outer gaskets and Teflon areas from the backfill.

Tolerances in fabrication and placement of tubes generate small misalignments between adjacent tubes, which require adjustments to the installation of interior finish elements such as ceiling panels, lighting fixtures, roadway pavements, curbs, and walkways. For this reason it is usually advisable not to mount attachments for finish elements on the interior of the tubes during fabrication, but rather to drill attachments into the concrete lining after the tubes have been placed and their final alignment and profile have been surveyed. This will ensure the necessary visual continuity within the entire tunnel.

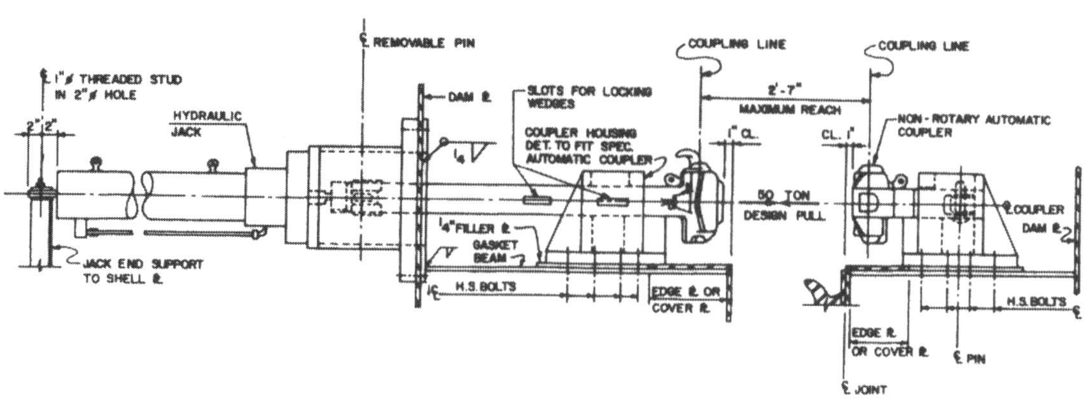

Fig. 14-29. Coupling jacks (Trans-Bay Tube).

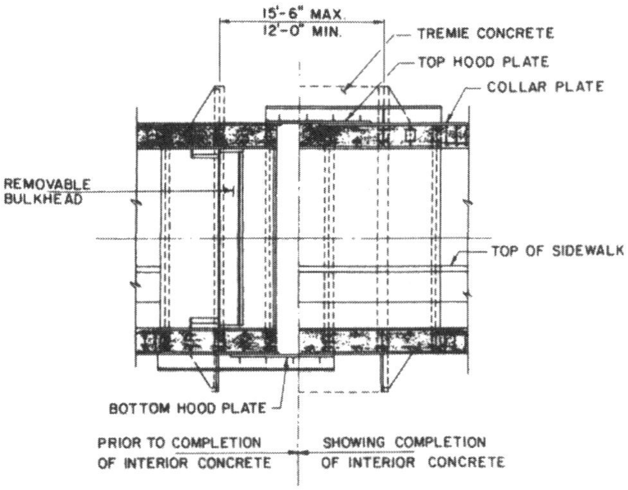

Fig. 14-30. Closure joint (Trans-Bay Tube).

BACKFILL

Locking Fill

A special fill is placed in the trench to about half the height of the tube to securely lock the tubes in position after they are connected. This is a well-graded material that compacts easily on placing. It may be sand or coarser material, depending on available sources. Material used has been well graded, usually ranging in size from 1-1/2 in. to No. 100 mesh size, but sizes up to 8 in. have been used in strong currents.

Sand may be placed through tremie pipes or by clamshell bucket, which is opened only when reaching the bottom. This is also used for coarser material.

Ordinary Backfill

Ordinary backfill to fill the trench to a depth of at least 5 ft above the tube may be any reasonably firm material available. In long tunnels, soil excavated from the trench may be used as backfill over tubes that have been placed in other parts. Backfill should be free from clay balls, chemically inert, and material passing the No. 200 mesh sieve should not exceed 20% by weight.

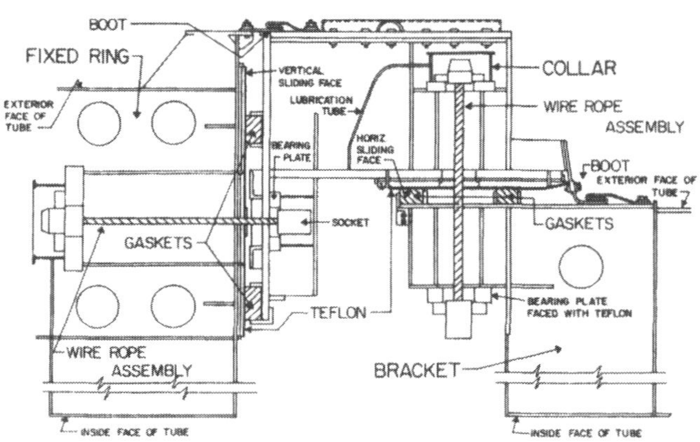

Fig. 14-31. Seismic joint (Trans-Bay Tube).

Protection Against Scouring

Where part of the tunnel projects above the original bottom, or where strong currents prevail, the backfill is protected with a rock blanket 2 or 3 ft thick. The size of the stone depends upon erosive action of current and usually varies between 1 in. and 10 in. The width of the blanket should be at least 100 ft on each side of centerline or confined within dikes.

DESIGN OF TUBES

Design Considerations

Although construction methods for steel tube and concrete tube tunnels are different, the design process is similar. For each tunnel type, evaluation of the following design considerations will be most beneficial in establishing the basic design parameters:

- Project limits and interfaces
- Design codes and standards
- Selection of materials (structural steel, structural concrete, and others)
- Applicable loads and surcharges
- Probability and magnitude of accidental loads such as sunken ships, anchor dropping, dragging ship anchors, and others
- Acceptable design flood levels
- Realistic tunnel fire scenarios and fire safety and other emergency requirements
- Determination of self-weight for applicable loads and variation in specific gravity of construction materials
- Assessment of water (sea) conditions such as wave heights, current velocities, current directions, swell and surf conditions, and density variations
- Yearly statistics of weather conditions and weather windows (pressure, wind, rain, temperature visibility, and others)
- Establishing design-basis earthquake if applicable
- Tidal effects
- Sedimentation
- Internal or external explosions due to terrorist attack or other causes
- Applicable operational system and safety requirements
- Navigational requirements
- Concrete cover, minimum reinforcement, shrinkage, and crack control limits
- Corrosion control system selection
- Selection of method of analysis, use of ultimate limit state, serviceability limit state, or both
- Selection of loading combination and load factors for each design stage or condition

The fundamental difference between recent European and U.S. practice for the design of immersed tube tunnels stems from the use and limitations of two distinctly different materials, concrete and steel. In general, for the U.S. practice, steel shell plate design is not controlled by stress consideration. It is primarily a watertight membrane. The shell plate thickness is chosen to limit distortion during fabrication, or to permit reasonable spacing of stiffeners to avoid excessive buckling and to ease placement of concrete for the liner. Maximum stress occurs during construction, launching, towing, and placement of interior concrete if the placement is performed while tubes are afloat.

Under final service conditions, stresses in steel tube tunnels are relatively uniform, resulting in smooth stress flow in a lightly stressed structure. This is highly desirable since the shell plate serves as the primary waterproofing membrane. Concrete lining of the steel tube tunnel is not controlled by stress, but by buoyancy. Since neither the steel shell nor the concrete lining are controlled by stress, the use of ultimate limit state analysis is not warranted. For this reason, design analysis for steel tube tunnel in the United States generally is controlled by serviceability limit state and the design is verified for accident loads using the ultimate limit state. Cracking of the concrete lining is of little concern, since the waterproofing is provided entirely by the ductile steel shell plate.

In contrast to the steel tube tunnel, concrete tunnel design is controlled by the quality of concrete for crack control, temperature control, shrinkage, and creep control. Therefore, the proper use of material control factors, durability, and safety class factors of ultimate limit states plays an important role in the design and analysis of concrete tube tunnels. This is a particularly important issue for the recent Dutch immersed concrete tube practice, where watertightness depends wholly on the concrete being crack-free, and exterior membrane waterproofing is eliminated.

Selection of Cross Section

For a railroad, transit, or vehicular tunnel, selection of cross section is dependent on vertical and horizontal clearances, number of lanes or tracks, type of ventilation system, and required air duct areas. Typical cross section configurations are circular, octagonal, arch, and rectangular. The number of bores depends on the number of tracks or lanes. Transit and railroad tunnels are usually ventilated by the piston action of the train, and except for fire protection they do not require air ducts. Vehicular tunnels more than 500 ft in length usually require ventilation. A full transverse ventilation system requires both exhaust and supply air ducts, and a semitransverse system requires either a supply or exhaust duct. Ideally, the exhaust duct is located above the roadway and the supply duct beneath it. The configuration for a two-lane single bore is best suited to a circular or octagonal shape tunnel, which structurally is the most efficient. The flat-bottomed arch shape is suited to single-bore tunnels with a semitransverse ventilation system. Rectangular shapes for multilane single or multiple roadways have center or side duct locations, which reduce the depth of the structure and dredging. For a single bore with a semitransverse supply ventilation system, location of the supply duct above the roadway has been used, but it requires reversible fans for emergency exhaust during a fire.

Setting the Interior Geometry

An optimized geometry satisfying the operational requirements of an immersed tube tunnel is the principal objective of the design process. Optimization of geometry applies not only to the dimensions of the cross section, but to the length of the tunnel element. For steel shell tubes, considerations include selection of fabrication methods for subassemblies; number of girth welds, number and location of mitered joints to accommodate vertical and horizontal curves, diaphragm spacing, use of stiffeners, constraints imposed by fabrication practices and tube-launching facilities, and the availability of material.

Optimization of the interior dimensions of a circular cross section can be accomplished by use of available software programs, where minimum radius for the shell plate can be obtained from coordinates of roadway or trackway, and the vehicle clearance envelope, including space requirements for signs, signals, and other installations.

Element geometry and layout can be a significant factor in cost optimization, particularly for tunnels with alignment curved in horizontal and vertical directions. For long tunnels, it may require a number of iterations before reaching an optimized element geometry that satisfies the alignment requirements and the local steel fabrication and production constraints. Before initiating an interaction of profile optimization and tube module length selection, discussions with local steel fabricators and shipyards or dry docks will be beneficial. In selecting the modular length, two items will control the process: the plate widths, and the lifting capacities. Separation of horizontal and vertical curves in the alignment is desirable. While miter locations need not relate to module length, significant cost savings can be achieved if the miter locations are close to the beginning or end of a module. Miters that actually *coincide* with the joint location would require complicated custom detailing of joints and are therefore not used. Precise alignment of the plate edges may involve repositioning of the module or the "strutting" of the plates. Optimization of these operations will result in significant cost savings.

Loading Conditions

For the design of the tubes, loads from the structure's weight, water pressure, earth pressure, and superimposed live loads must be considered. Other loads can consist of accidental loads such as a sunken ship, anchor dropping, vessel grounding, temporary load such as earthquake, or permanent loads due to special surcharge. The design loads are usually classified into two general categories: construction-phase loads and final in-place loads.

Construction-phase loads include those imposed during the fabrication, launching, towing, outfitting, and placing operations. The significance of these loads varies with the type of construction. For tubes fabricated in a dry dock or a graving basin, the stresses induced during fabrication, launching, and outfitting can be controlled easily. For steel shell immersed tube tunnels fabricated on shipways and outfitted in flotation, these stages can induce large stresses if not controlled by the proper sequence of concrete placement operation.

Final in-place loading conditions include normal and temporary loads. Normal loads include the dead load, water pressure, earth pressure, and superimposed live load that the structure is expected to encounter during normal operating conditions. Temporary loads include additional loads produced by unexpected events such as an earthquakes, floods, anchor dropping or vessel grounding, explosion loads, and others. For a quick preliminary analysis, basic components of the normal in-place loading conditions can be reduced to the elements shown in Figure 14-32 for a circular section. The loads shown in Figure 14-32 can be explained in the following manner:

1. *Uniform distributed load—top.* This is usually equal to the weight of the soil above the structure and the normal uniform surcharge load.
2. *Uniform distributed load—side.* This load results from the uniform load located above the structure and is equal in magnitude to the appropriate coefficient of lateral pressure times the magnitude of loading component in item 1.

Loading Condition	General Equation	
Uniformly Distributed Load – Top 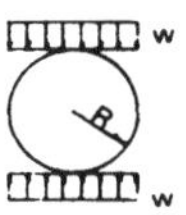 W = #/ft. R = ft. Variable	$M\theta = wR^2\left[0.25 - \frac{\text{Sin}^2\theta}{2}\right]$ $N\theta = -wR\ \text{Sin}^2\theta$ $V\theta = -wR\ \text{Sin}\,\theta\ \text{Cos}\,\theta$	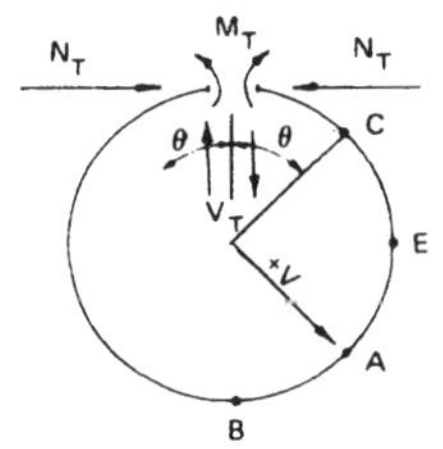 +M = Compression on outside of ring + Tension - Compression
Uniformly Distributed Load – Side 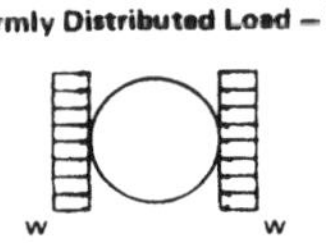 Variable	$M\theta = \frac{wR^2}{2}\left[\frac{1}{2} - \text{Cos}^2\theta\right]$ $N\theta = -wR\left[\text{Cos}^2\theta\right]$ $V\theta = +wR\ \text{Sin}\,\theta\ \text{Cos}\,\theta$	
Wedge Shape Horizontal Load 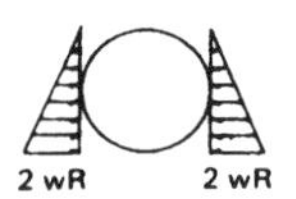	$M\theta = wR^3\left[0.250 - 0.125\ \text{Cos}\,\theta - \frac{\text{Cos}^2\theta}{2} + \frac{\text{Cos}^3\theta}{6}\right]$ $N\theta = -wR^2\ \text{Cos}\,\theta\left[0.125 - \frac{\text{Cos}^2\theta}{2} + \text{Cos}\,\theta\right]$ $V\theta = wR^2\ \text{Sin}\,\theta\left[0.125 + \text{Cos}\,\theta - \frac{\text{Cos}^2\theta}{2}\right]$	
Quadrant Loading 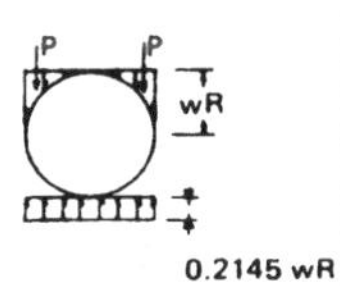	$0 \leqslant \theta \leqslant \pi/2$ "θ" in Radians $M\theta = -wR^3\left[0.327 - \theta\frac{\text{Sin}\,\theta}{2} + \frac{\text{Sin}^2\theta}{2} - 0.521\ \text{Cos}\,\theta + \frac{\text{Cos}^3\theta}{6}\right]$ $N\theta = -wR^2\left[\text{Sin}^2\theta - \theta\frac{\text{Sin}\,\theta}{2} - \frac{\text{Sin}^2\theta\ \text{Cos}\,\theta}{2} - \frac{\text{Cos}\,\theta}{48}\right]$ $V\theta = -wR^2\left[\text{Cos}\,\theta\ \text{Sin}\,\theta - \theta\frac{\text{Cos}\,\theta}{2} - \frac{\text{Sin}\,\theta\ \text{Cos}^2\theta}{2} + \frac{\text{Sin}\,\theta}{48}\right]$	
P = wR² (1 - π/4) P = 0.2145 wR²	$\pi/2 \leqslant \theta \leqslant \pi$ "θ" in Radians $M\theta = -wR^3\left[0.434 - 0.0208\ \text{Cos}\,\theta - \frac{\text{Cos}^2\theta}{2} - \frac{\pi}{8}\text{Sin}^2\theta\right]$ $N\theta = -wR^2\left[0.214\ \text{Sin}^2\theta - 0.0208\ \text{Cos}\,\theta\right]$ $V\theta = -wR^2\left[0.0208\ \text{Sin}\,\theta + 0.214\ \text{Sin}\,\theta\ \text{Cos}\,\theta\right]$	
Buoyancy Forces 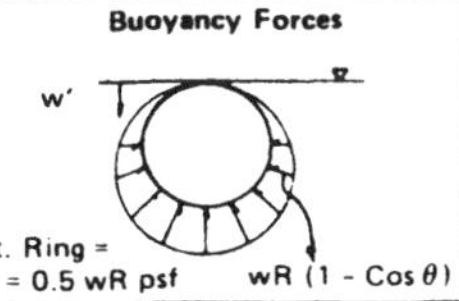 Wt. Ring = w' = 0.5 wR psf	$M\theta = 0$ $N\theta = -wR^2\left[1 - \frac{\text{Cos}\,\theta}{2}\right]$ $V\theta = 0$	
Water Surcharge 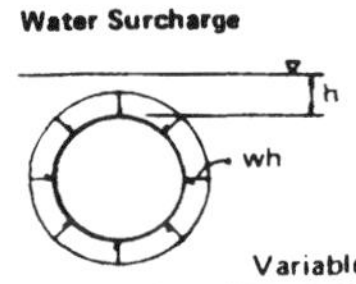 Variable	$M\theta = 0$ $N\theta = -whR$ $V\theta = 0$	

Fig. 14-32. Loading conditions and moments, thrusts, and shears.

3. *Wedge shape horizontal load—side.* This load results from lateral pressure of the soil located below the top of the structure.
4. *Quadrant loading.* This load results from the vertical soil load located below the top of the structure at the corners, which, for simplicity, is balanced by a uniform reaction along the bottom of the structure. For rectangular cross sections, quadrant loading does not exist unless the top of the tunnel section is sloped at its extremities.
5. *Buoyancy force.* This load consists of the buoyant force acting on the structure balanced by the weight of the structure. If the weight of the shell of the tube is uniformly distributed around the circumference and is sufficient to cause full submergence, there will be no bending in the shell. This can be proven by combining the moments and thrusts of Cases XVI and XVIII given by Paris (1921). For Case XVIII, the loads will be reversed from those shown by Paris. The assumption that the weight is uniformly distributed is not strictly correct since some of the weight is concentrated near the bottom, in the roadway slab, the sidewalks and ledge, etc.; also, part of the tremie concrete is not placed until after the tube rests on the bottom of the trench. Analysis has shown that these variations have a relatively slight effect on the stresses in the shell of the tube.
6. *Water surcharge.* This load results from the weight of water located above the top of the tube. When the tube is completely submerged, the additional uniform water pressure caused by the depth of water above the top of the tube causes a uniform compression around the ring, but no bending moments. The loading conditions are shown diagrammatically in Figure 14-33.

Design Loads

Design loads for stress analysis are based on boring information and the specific gravity of water at the site. Usual values are

Submerged earth	70 pcf
Moist earth in air	120 pcf
Water	64 pcf
Lateral coefficient of earth pressure for loosely placed backfill	0.27

For stability against uplift, assuming 3,000 psi concrete with 1-1/2 in. maximum size aggregate, the unit weight of concrete is usually 145 pcf; structural and reinforcing steel, 490 pcf; and water, 64 pcf.

For checking buoyancy during flotation, the usual weight of concrete is assumed to be 147 pcf, and steel and water weights are the same as for stability against uplift. Specific gravity of silt-laden water in the trench must be considered. During construction of the Washburn Tunnel in Texas, an overnight freshet washed silt into the trench before a recently placed tube had been backfilled, and the tube floated to the surface.

Factors of safety against uplift were discussed earlier in this chapter. Stability calculations should be based on the net volume of concrete. Weight of all conduit, pipe, reinforcing steel, and structural steel should be included. The calculations should be based on the full length of the tube. The critical condition for stability against flotation occurs after the tube is on the bottom and connected to the previously placed tube. After the joint has been dewatered and the end bulkhead has been removed, and prior to completing the interior concrete at the joint, with no backfill in place, the factor of safety should not be less than 1.02. If this factor of safety cannot be maintained, temporary ballast, such as sinking blocks, should be provided.

Water level is usually assumed to be mean sea level. Where the fill over the tube extends above water level, maximum moments occur in the tubes with minimum water level. Where the surface of the fill is below water level, the moments are not affected by the elevation of the water surface, but the direct thrusts in the shell of the tube will increase with higher water levels.

For a tunnel profile set below the existing sea bottom, the depth of submerged earth is usually calculated as the distance from the top of the structure to the existing sea bottom prior to construction. This accounts for any siltation into the trench that may occur in the future.

Live load surcharge depends on the future use of the area above the tube. Usually, 500 psf is assumed unless definite loads are known. In industrial areas, a surcharge of 1,000 psf is not unusual. On the Fort McHenry Tunnel in Baltimore, Maryland, a 2,000-psf surcharge was used on the east end of the project to accommodate a future cargo facility.

The seismicity of the project site must be investigated. The alignment should be located clear of any active faults. Particular attention should be directed to any abrupt changes in structure mass or in the stiffness of the subsurface materials. Seismic articulation joints may be provided where unacceptable movements or stress levels are anticipated.

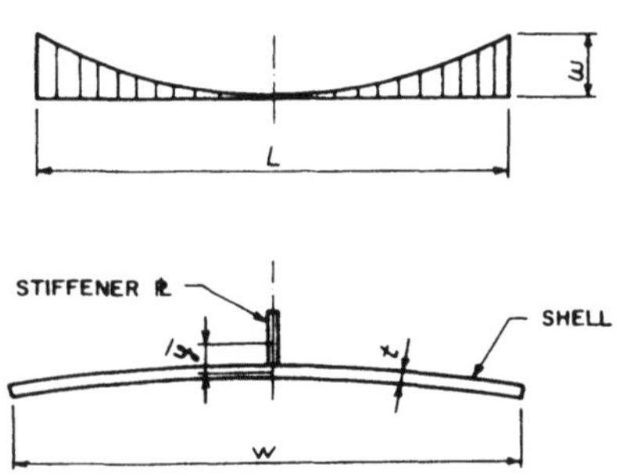

Fig. 14-33. Stiffener load.

Design Stresses and Codes

Design practice for tubes in the United States is usually governed by the requirements of the AISC Specification for Design, Fabrication and Erection of Structural Steel Buildings and the ACI Standard 318, Building Code Requirements for Reinforced Concrete Design. These codes are used in lieu of AASHTO Standard Specifications for Highway Bridges or AREA Specifications for Railway Bridges since the loads on a tube are essentially nonrepetitive. However, independent structural members subject to repetitive loads, such as a roadway slab, are designed according to the appropriate highway or railway design code.

Welding requirements are in accordance with AWS Structural Welding Code for Bridges.

Materials for the tube usually consist of structural steel meeting the requirements of ASTM Designation A 36; reinforcing steel of 60 ksi yield strength meeting the requirements of ASTM A 615; 3,000- or 4,000-psi structural concrete, 2,200-psi tremie concrete, and 5,000-psi prestressed concrete, all meeting the requirements of ACI 318.

Allowable stresses for serviceability limit state can be selected in accordance with the referenced codes. Standard design practice permits a 25% increase in allowable stress levels for temporary loading conditions.

Structural Systems

Structural systems for trench tunnels consist of a structural steel shell with a reinforced concrete lining acting compositely; reinforced concrete; or prestressed concrete.

Octagonal composite sections of double-shell systems have a circular steel shell backed by external longitudinal stiffeners and ring diaphragms, lined with reinforced concrete on the inside for structural strength. Outside the circular steel shell are octagonal-shaped form plates, attached to the shell by the diaphragms and radial angle struts. The space between the octagonal steel form plates and shell plate is filled with concrete, a portion placed in the dry to provide structural strength in the in situ condition, and the remainder placed under water to furnish corrosion protection to the shell and diaphragms, and weight to overcome buoyancy (see Figure 14-3).

Circular composite sections of single-shell systems have a circular steel shell stiffened by internal transverse structural steel tees. The shell and tees are designed to act as a ring girder during construction and, in the final in situ condition, to act compositely with the reinforced concrete lining. The steel shell usually requires protection against corrosion. On the Trans-Bay Tube (Figure 14-5) and 63rd Street (Figure 14-6) tunnels, a cathodic protection system was used, whereas for the Hong Kong Tunnel (Figure 14-4), a pneumatic mortar coating gunite was used to provide corrosion protection.

Arch and rectangular composite sections with steel shells and reinforced concrete linings have structural systems similar to the circular composite section.

Reinforced concrete and prestressed concrete sections usually have a rectangular rigid-frame structural system and require construction in the dry.

Design Analysis

As discussed previously, the two distinct phases that require analysis and design are the construction phase and the final in-place condition.

Construction Phase

The construction phase consists of the fabrication, launching, towing, outfitting, and placing operations. Fabrication, launching, towing, and outfitting stresses are a function of the construction method. For construction in a dry dock or casting basin, stresses produced by these operations are usually not significant. For fabrication on shipways and outfitting in flotation, these stresses can be very large. In the following paragraph, shipway construction and outfitting in flotation will be discussed.

Fabrication and launching stresses depend on several items, including the type of structural system (single- or double-shell), the method of fabrication, the length of the tube modules, the method of erection on the ways, the blocking between ways for support of the modules, way spacing, slope of the ways, and launching method. During these assembly stages, each module and tube section should be self-supporting or should be locally reinforced for local buckling, lifting, and launching stresses. In addition, transverse deflections of the modules should be checked, and if required, internal spider bracing frames should be installed to maintain the diameter of the shell plate, to permit proper fit-up for the circumferential butt welds between the tube modules.

Depending on the launching system, keel concrete may have to be placed prior to launching. If keel concrete is poured prior to launching, after the pouring the weight of the steel shell and keel concrete should be transferred from the temporary blocking to the launching sleds prior to launching. End launching may require reinforcement of the shell and diaphragms. Temporary longitudinal reinforcing of the top of the shell to preclude longitudinal buckling may be necessary. For side launching, the steel shell and diaphragms may require local strengthening over each sliding way. In evaluating these stresses, the keel concrete should be considered to be a longitudinal beam acting with the shell plate and diaphragms. Stresses in the shell and form plates should be investigated for water pressure at launching.

There have been occasions where steel immersed tubes have been towed in excess of 1,000 mi to an outfitting site. On several projects, the route from the fabrication site to the outfitting site involved towing on the open seas. Wave conditions for the selected route must be investigated and the tube analyzed to span between wave crests. For simplicity, the tube is modeled as a thin-shelled beam stiffened by the longitudinal stiffeners. The critical unit compressive stress for buckling is the same as discussed below for the outfitting operation. Fabrication, launching, and towing stresses generally are the responsibility of the contractor, since they depend on the method of fabrication and transport.

During outfitting while floating and except for the end bulkheads, the weight of the tube is uniformly distributed longitudinally and is supported by the buoyancy forces. The end bulkheads, however, impose concentrated loads, creating hogging moments along the tube element. Depending on the weight of the end bulkhead, the hogging moments will result in tensile stresses at the top and compression stress at the bottom of the tube element. This is highly desirable since the shell plate can sustain higher tensile stresses than the compression stresses at this stage of the construction. Placing the keel concrete prior to launching not only protects the bottom portion of the shell plate during launching, but it will also help to resist the compression stresses and buckling tendencies created by end bulkheads' hogging moments. As the placement of interior concrete proceeds, the added longitudinal moments produced by concrete placement will counteract the hogging moments, neutralizing the shell plate stresses along the tube element.

In addition to the longitudinal moments, the steel shell and stiffening diaphragms are subjected to circumferential bending moments resulting from the exterior water pressure. For double-shell construction the form plates are not made watertight, to minimize the pressure acting on them and to permit the shell plate to resist all the water pressure.

A typical sequence of concrete placement by stages is shown in Figure 14-12. The length of pour for each stage is a function of convenience in form-placing operations, the quantity of concrete that can be conveniently placed in a single shift, and the longitudinal moments and shears that the shell can withstand. Insofar as possible, pours should be symmetrical about the transverse and longitudinal centerlines of the tubes in order to maintain trim of the tube and reduce the water pressure on the shell plate. Experience has shown the stages and sequence indicated on Figure 14-12 usually result in the most favorable design and construction conditions.

Moments, shears, and head of water acting on the shell plate and diaphragms due to the unbalanced weight of the bulkheads and the fresh concrete are calculated for each stage and length of pour. Usually, the critical condition occurs during the placing of the haunch pour for sequence 5 (Figure 14-12). At this stage, the shell and diaphragms are fixed at the top of the keel concrete, and the head of water against the shell extends from the top of the keel to approximately the horizontal axis. The pressure of the fresh concrete against the inside of the shell is usually neglected. The resulting circumferential moments are resisted by the shell plate and the diaphragms, with longitudinal distribution of loads by the shell plate and longitudinal stiffeners.

The critical unit compressive stress of buckling of a curved panel under uniform compression is given in Table 35, Case 13, of Roark and Young (1975).

The moment capacity of the shell is determined by taking the critical buckling stress, dividing by a factor of safety—usually 2—and multiplying the result by the section modulus of the shell.

Within the elastic range, the local buckling strength of the shell, loading in torsion or transverse (beam) shear, can be computed by formulas set forth in Brockenbrough and Johnston (1974: p. 91, formula 4.27).

The water pressure on the shell plate is assumed to be uniformly distributed between longitudinal stiffeners, and therefore, since the load is acting radially against the plate, the stress induced will be axial compression only, with no bending moment. The axial stress should be checked against the minimum pressure that could cause buckling of the arch (see Roark and Young, 1975: Table 35, Case 21).

The longitudinal stiffeners maintain the shape of the shell plate and transfer by beam action between diaphragms the reactions of the plate elements considered above.

The shell plate and the longitudinal stiffeners should not be considered as separate structural elements, but rather as components of a cylindrical shell spanning between diaphragms. The shell plate alone is capable of resisting water pressure as both in ring structure and as a cylindrical beam. Therefore, as an approximation, the longitudinal stiffeners at the portion of the shell where the exterior pressure is applied is considered as a beam on an elastic foundation. The load on such a beam is proportional to its deflection; a parabolic deflection is assumed. Also, the shell plate is assumed to support the full load at the midpoint between diaphragms. Thus, the net load on the longitudinal stiffener beams is the uniform load minus the parabolic load supported by the shell plate. The stiffener beams (Figure 14-33) are considered as simply supported at the diaphragms.

The effective width, b, of the stiffener beam is = $600t/f_y^{1/2}$, as recommended in Brockenbrough and Johnston (1974: p. 102, Table 4.3).

The diaphragm is designed as a ring structure, fixed at the top of the keel concrete, resisting the external water pressure on the length of tube between diaphragms. Values for moments, shears, and thrusts at critical sections are tabulated in Figure 14-32 for rapid design. Calculation of bending moments is made per linear foot of tube.

The critical moments, thrusts, and shears are resisted by the diaphragm (see Figure 14-34), which consists of an exterior flange, a web, and a shell plate. The form plates, since they are not protected against corrosion, are neglected. In determining the unit stresses, section properties can be computed taking the effective width of the shell plate, $l = \sqrt{1.58\,Rt}$, acting with the web and flange, based on an article by Schorer (1933). Here, R = the centerline radius of the shell plate in inches, and t = the thickness of the shell plate.

Final Conditions

For structural analysis, an idealized model of the tunnel cross section can be devised to investigate the number of loading conditions by use of various software programs,

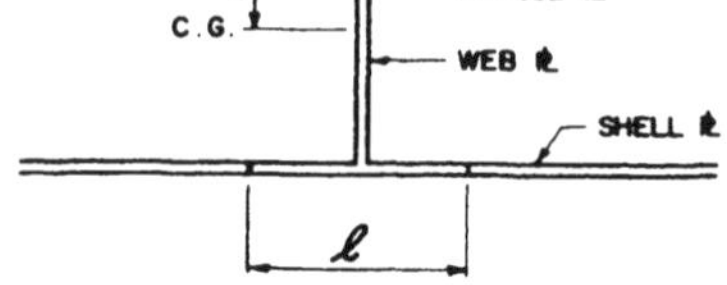

Fig. 14-34. Diaphragm cross section.

such as STRUDL, STAAD III, and others. From the output, the values of critical conditions can be plotted along the given geometry to observe the load deflection characteristics of the structural system. For stress analysis, available software for composite sections can be adopted or developed to investigate stress flow along the steel and the concrete sections.

Use of these software programs, with the advancement of personal computers, is relatively easy and convenient. However, a simplified method of analysis available for a rapid design verification or for a preliminary investigation can be a valuable tool for the designer. This analysis was presented in detail in the first edition of this handbook, Chapter 13, pp. 383–390.

Design of Rectangular Tubes

Discussions in "Design of Tubes" covered steel shell and concrete tube tunnels with customary reference to the concrete tunnels being rectangular box elements and the steel shell tubes being circular. However, both rectangular steel shells and circular concrete tube sections have been constructed.

Tender design conducted in 1988 for an immersed tube crossing of Denmark's Great Belt project proposed a steel shell and a concrete box section with striking similarities in their cross sections. (See Figure 14-9.) Two major differences are the configuration of sidewalls (for the steel shell tube being semicircular, and for the concrete tube having vertical walls with haunches), and the steel shell tube having a gravel ballast box above the top slab and the concrete tube having internal concrete ballast at the bottom slab.

Structural analysis of a rectangular section is rather straightforward, and there are a number of software programs available. An example of the application of such a program is given here for a double-shell steel tube. The same process is applicable for a rectangular cross section.

Design Analysis by Software Programs

Current tube design techniques use computer software programs to analyze the forces and stresses during fabrication and outfitting, and in the in situ structure.

A frame analysis program, such as STAAD III or STRUDL, can be used to analyze the circular or rectangular sections. For the structural analysis, the cross section of the structure is subdivided into a series of segments that are joined at nodes as shown in Figure 14-35. Tremie concrete is not considered in the analysis. For the composite structure, a more accurate analysis would locate the member nodes at the center of gravity of the composite steel and concrete. Usually, this refinement is not required.

Based on the gross section properties of the structural concrete, the area, moment of inertia, and shear area of each segment are computed. The section properties are those of the average section between nodes. This is a simplifying assumption, since the effects of the steel in the section are neglected in computing the section properties. A further refinement could be introduced by using the cracked section properties of the composite structure, but this would require many cycles of computation and would not greatly increase the accuracy of the results.

The frame is analyzed for the loading conditions shown in Figure 14-32. These loadings are computed for each section. The loadings are then analyzed and combined to produce output that gives maximum and minimum thrust, and shear and moment for each beam at each node. While this can be easily stated, it is, in fact, a complicated task that requires judgment to ensure that the proper combinations have been considered that will result in the maximum tension, compression, and horizontal shear in the tube.

While the above discussion refers to an octagonal tube, the same technique is valid for any cross section.

A further refinement is to consider the uniformly distributed base pressure to be replaced by a series of springs that join the frame at the beam nodes. These springs represent the elastic modulus of the foundation soil. This analysis considers the tube to act as a beam on an elastic foundation and results in a more accurate analysis of the forces in the tube. This is particularly true when analyzing a tube with a wide, flat base. It also has the advantage of simplifying some of the computer input since the output automatically gives the base pressure.

After completion of the analysis for the tube, the next step is to check the tube for stresses and to adjust the reinforcing steel, diaphragm plates, or possible concrete thickness to ensure that all stresses are at or below the permissible levels. This can be done with any software program, which, given the size and spacing of reinforcing steel and steel plates and the concrete thickness, as well as axial force and moment, will compute the stresses in the section.

For steel shell tunnels the program uses a transformed-area method of analysis. The steel in the section is transformed to an equivalent area of concrete. Since all tension concrete is neglected in computing the moment of inertia of the section, the location of the neutral axis is determined by trial and error. The section may or may not be cracked, depending upon the magnitude and sense of the applied axial force.

The axial force is assumed to act through the centroid of the gross section (i.e., tension concrete included), whereas the moment of inertia is computed about the centroid of the net section. Thus, in a cracked section, an eccentricity, e, results, inducing a moment equal to Ne. In the program, this moment is added to the original applied moment.

Following this analysis, the horizontal shear between the shell plate and concrete is computed using VQ/It. For the exterior portions of the concrete, this shear can be resisted by the longitudinal stiffeners, which act as shear lugs. For the interior portions of the concrete, the horizontal shear is resisted by the ties that support the reinforcing bars as well as by bond. The bond stress in some cases can be large. If extremely high bond stress is found, shear studs can be welded to the interior steel shell plate.

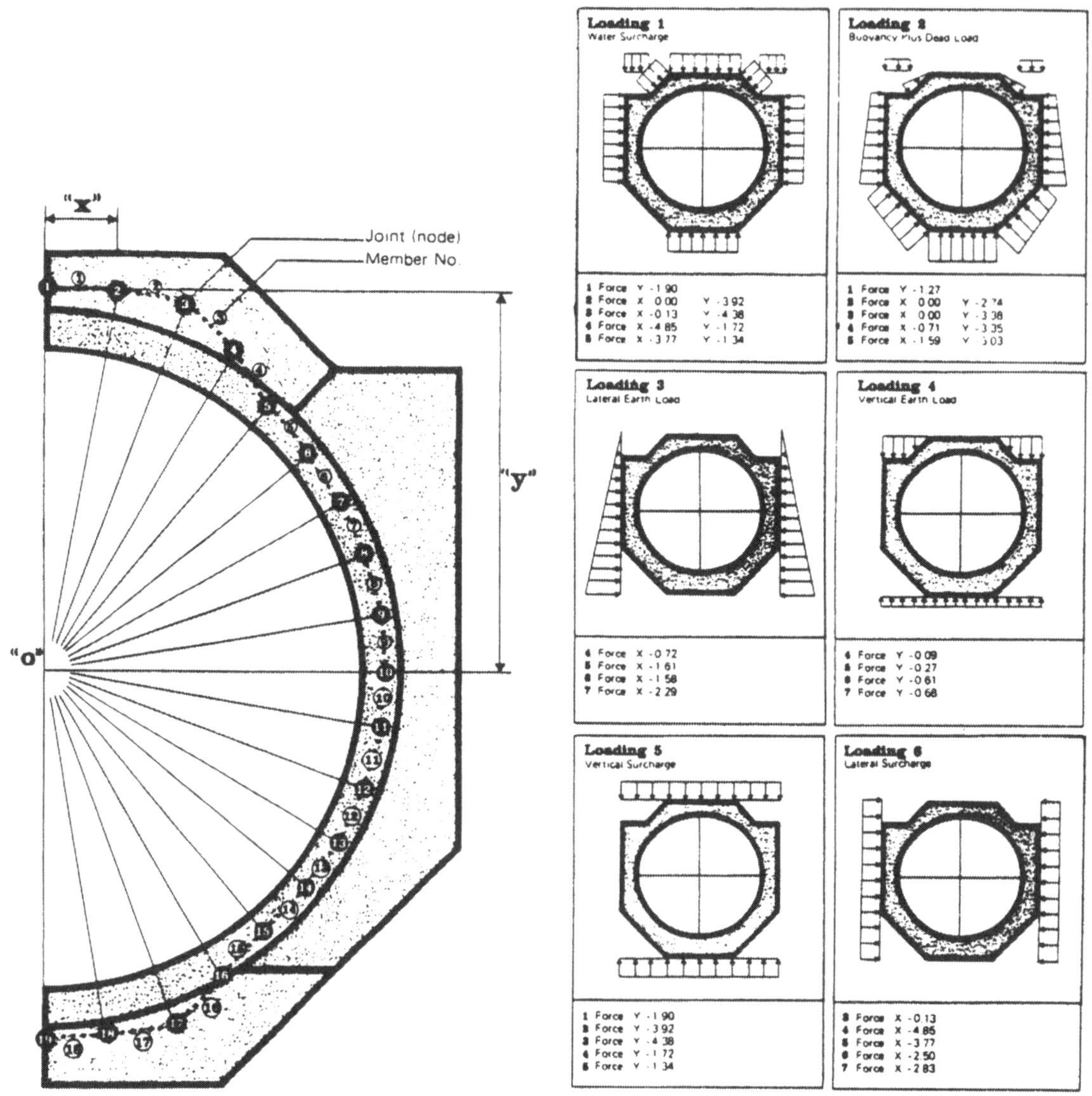

Fig. 14-35. Design analysis by computer.

Another software program can be developed to check the moments and shears in a tube during outfitting. This program can determine the depth of the tube section in the water after each concrete pour. The moments and shears in the tubes are used to check the shell and form plates for shear and bending stress, while the water depth is used to compute the hydrostatic pressure on the shell plate as well as the forces in the diaphragm.

An initial loading consisting of the uniform weight of the tube plus the concentrated weight of the end bulkheads is followed by any number of loadings, each consisting of one or more sets of uniform loads representing the concrete pours. It is not necessary to place the loadings symmetrically about the center of the tube, although the unsymmetrical loadings will cause the tube to list with an uneven draft. The uneven draft will increase the waterhead on one end of the tube, which may cause maximum pressure on the shell or form plate. The concrete unit weight for each set of pours may be different.

A shear and moment diagram is produced for each loading on the tube. The results are generated at each end of every pour and at all maximum and minimum points.

REFERENCES

BROCKENBROUGH, R.L., and JOHNSTON, B.G. (1974) *U.S.S. Steel Design Manual.*

CULVERWELL, D.R. (1989) "Comparative merits of steel and concrete forms of tunnel," Proceedings of the Immersed Tunnel Techniques Symposium, Manchester, UK.

International Tunneling Association (1993) Immersed and Floating Tunnels State of the Art Report, Working Group No. 11., Tunneling and Underground Space Technology, Pergamon Press, Vol. 8, No. 2 (April).

KUESEL, T. (1986) *"Immersed Tube Tunnels: Concept Design and Construction,"* Civil Engineering Practice, Boston Society of Civil Engineers, Spring.

National Public Works Department (1993) *"Underground Transport Connections (Rijkswaterstuat)."*

PARIS, J.M. (1921) "Stress coefficients for large horizontal pipes," *Engineering News Record, 87*(19) (November 10).

RASMUSSEN, N., and ROMHILD, C.J. (1990) "Settlement of immersed CN tunnels," *Proceedings of the Second Symposium on Strait Crossings,* Norway.

ROARK, R.J., and YOUNG, W.C. (1975) *Formulas for Stress and Strain,* Fifth Edition, McGraw-Hill, New York.

Royal Institution of Engineers and The Netherlands Tunnels (1987) Immersed Tunnels Proceedings of Delta Tunneling Symposium, Amsterdam, November.

SCHMIDT, B., and GRANTZ, W. (1979) "Settlement of immersed tunnels," *Journal of the Geotechnical Division,* ASCE, September.

SCHORER, H. (1933) "Design of large pipe lines," *Trans. ASCE 98.*

TIMOSHENKO, S. (1940) *Theory of Plates and Shells,* McGraw-Hill, New York, Article 3, p. 15, Figure 9.

U.S. Department of the Interior (1975) Special Report, California Undersea Aqueduct, January 1975.

15

Water Conveyance Tunnels

David E. Westfall
District Manager, Vice President, Principal Professional Associate, Parsons Brinckerhoff Quade & Douglas, Inc.

Water conveyance tunnels require special considerations regarding friction losses, drop shafts for vertical conveyance, air removal, control of infiltration and exfiltration, tunnel linings, lake taps and connections to live tunnels, and maintenance.

FRICTION LOSSES

The required finished inside dimensions of water conveyance tunnels can vary considerably due to the roughness of the proposed internal surface. As a tunnel or pipe becomes longer, the proportion of the total head loss attributed to boundary friction increases substantially. The measure of energy loss due to pipe roughness is developed from either the Darcy–Weisbach formula,

$$h_f = \frac{fLV^2}{D \times 2G} \tag{15.1}$$

or the Manning equation,

$$h_f = \frac{2.875V^2n^2L}{D^{413}} \tag{15.2}$$

in which

h_f = friction head loss (ft)
f = friction factor (dimensionless)
n = Manning's roughness factor (dimensionless)
L = length (ft)
D = diameter of tunnel (ft)
V = velocity (ft/sec)
g = acceleration of gravity (ft/sec^2)

The most common finished interior surfaces of water conveyance tunnels may be categorized as follows:

Case I. The tunnel is excavated by drill-and-blast methods and is left unlined.
Case II. The tunnel is excavated by the use of a full-face tunnel boring machine (TBM) and is left unlined.
Case III. The tunnel is lined with precast concrete segments.
Case IV. The tunnel is lined with cast-in-place concrete.

Various researchers have made field measurements of operating tunnels with each of these surface finishes and developed recommended values of roughness factors for use in design. These values expressed in terms of Manning's n values are as follows:

Case I, $n = 0.038$
Case II, $n = 0.018$
Case III, $n = 0.016$
Case IV, $n = 0.013$

Designers of water conveyance tunnels should consider the economic value of presenting alternative tunnel liners and, hence, tunnel diameters in their designs. Although a tunnel with a Case III liner may require a larger diameter than one with a Case IV liner, the actual construction costs may be higher for the latter. The marketplace as determined by the bidding contractors will give the most economical tunnel.

DROP SHAFTS FOR VERTICAL CONVEYANCE

Drop shafts are used in water conveyance tunnels to transfer flows from a higher elevation to a lower elevation. The drop shaft should be designed to dissipate the energy increase associated with the elevation drop, to remove any air that mixes or entrains with the water as it descends, and to minimize hydraulic head losses when the tunnels are surcharged.

Drop Shaft Components

Drop shafts have three essential elements: an inlet structure, a vertical shaft barrel, and a combination energy dissipater and air-separation chamber. The inlet structure's function is to provide a smooth transition from horizontal flows to the vertical drop shaft. The drop shaft barrel then transports the water to the lower elevation and, in the process, dissipates as much energy as possible. At the bottom of the drop shaft, a structure that will withstand the impact forces, remove any entrained air, and convey the water to the tunnel must be provided.

Basic Considerations in Drop Shaft Design

Several factors must be considered in the design of drop shafts:

- Variable discharge
- Impacts on the drop shaft floor
- Removal of entrained air
- Head loss associated with the drop shaft

The selection of an appropriate deep shaft for a particular use involves determining which of these factors are most important. When the difference in elevation between the upper level flows and the tunnel is minimal, impacts on the drop shaft floor may be alleviated with a simple plunge pool. When the difference in elevation increases, removal of entrained air is necessary and floor impact potential becomes more severe. In cases where the tunnel hydraulic gradient can rise all the way up to the hydraulic gradient of the upper-level flows, head loss also becomes a critical factor.

Variable Discharge. A drop shaft may be operated for steady-state flows, only during storm discharge periods, or as a combination of the two. The flow variability of a drop shaft has a considerable influence on the design. For instance, for steady-state flow, the water surface elevation in the tunnel may be below the base of the drop shaft. In that case, a plunge pool is required at the drop shaft floor to dissipate energy. A shaft that handles only storm flows will not normally require a plunge pool, because the water surface in the tunnel will submerge the drop shaft base and cushion the impacts.

Impact on the Drop Shaft Floor. The impact of the water on the floor of the drop shaft can be quite high, and steps should be taken to minimize it. This is accomplished by forcing a hydraulic jump within the shaft, by increasing the energy dissipation due to wall friction as the water descends, by entraining sufficient air to cushion the impact, or by providing a plunge pool at the bottom of the shaft.

The plunge pool may be formed by a depressed sump or by the use of a weir located in the chamber at the base of the shaft and downstream of the shaft barrel. The depth of the plunge pool can be determined by the Dyas formula:

$$\text{Depth} = \frac{h^{1/2} d_c^{1/3}}{2} \tag{15.3}$$

where

h = height of drop (ft)
d_c = critical depth in inlet (ft)

Removal of Entrained Air. As the water falls through the drop shaft, it entrains, or mixes, with air. There are several advantages and disadvantages associated with air entrainment. The advantages are

- The presence of air minimizes the possibility of subatmospheric pressures and thus negates the harmful effects of cavitation.
- The impact of the falling water on the drop shaft floor is reduced by the cushioning effect of the air entrained in the water.

Disadvantages of air entrainment are

- The flow volume is bulked up and requires a larger drop shaft.
- To prevent the formation of damaging high-pressure air buildups, entrained air must be removed before entering the tunnel.

Under certain conditions, the tunnel hydraulic gradient may rise to levels equal to those of the upper-level inflows. In these circumstances, the head losses become important because a large head loss may cause severe flooding in the upper-level flow delivery system. For example, if this upper-level delivery system is a sewer, large drop shaft head losses will result in flow backups into streets or basements.

Types of Drop Shafts

Various types of drop shafts have been designed and constructed based on hydraulic laboratory model studies. Drop shafts as deep as 350 ft have been constructed. The smaller structures, normally used for drops of less than 70 ft, consist of drop manholes in local sewer systems. The larger structures, for drops greater than 70 ft, are divided into several categories:

- Vortex
- Morning glory
- Subatmospheric
- Direct drop, air-entraining

Drop Manholes. Drop manholes are generally used in local sewer systems to transfer flows from a higher sewer to a lower sewer. These drop manholes are designed to minimize the turbulence that can release odorous gases and damage the manhole. A typical design has a vertical downcomer upstream of the manhole, which allows maintenance personnel to enter the lower sewer without climbing down the wet shaft.

Vortex Drop Shafts. Flow enters the vortex-flow drop shaft tangentially and remains in contact with the drop shaft wall, forming a central air core as it descends. Since the flows through the inlet are spun against the shaft wall, the entry conditions are relatively smooth. Vortex drop shafts are effective for a wide range of discharges. The air core helps to evacuate the entrained air and to prevent cavitation by maintaining near-atmospheric pressure throughout the shaft.

Vortex drop shafts generally entrain less air than other types of drop shafts for two reasons. First, the flows are highly stable due to the entry conditions. Secondly, a reverse flow of air occurs in the core of the vortex, which causes much of the air entrained in the flow to be released and recirculated in the zone above the hydaulic grade line. Below the hydraulic grade line, the helical flow has a pressure gradient that forces bubbles to move toward the center of the drop shaft, where they are able to rise against the relatively slower moving water. Therefore, much of the air that is entrained by the flow is allowed to dissipate before it enters the tunnel.

As the flows are spun against the walls of the drop shaft, a significant amount of energy dissipation is attained before the flow reaches the floor of the drop shaft. The dissipation is a consequence of the wall friction as the flows spiral down at high velocity. The remainder of the energy is dissipated in the air-separation chamber by either a plunge pool or by the formation of a hydraulic jump.

Several inlet configurations have been adopted to create vortex flows down drop shafts. A model study performed by the St. Anthony Falls Hydraulic Laboratory analyzed the votex drop shaft's hydraulic head loss. This model incorporated a helix inside of the drop shaft to generate the vortex flow. This study showed the impact forces to be minimal and the air removal to be high, but it also determined that for the tunnel-full condition, head loss through the system was excessively large.

Based on various model studies, a vortex drop shaft is highly efficient when the tunnel gradient does not approach the level of the upper incoming flow. It is a good energy dissipater and has a high air removal rate. However, if it is possible for the tunnel to surcharge, another type of drop shaft should be considered due to the inherently high head losses.

Morning Glory Drop Shafts. Morning glory drop shafts use a circular crested inlet structure. Generally, they are used as outflows from reservoirs. Model studies have determined that the flow characteristics are controlled by three conditions: weir control, orifice control, and differential head control. The capacity of the morning glory drop shaft is limited by the size of the circular crest.

No cavitation is expected in this type of drop shaft, but induced head losses could occur if the circular crest is inadequately designed. The U.S. Bureau of Reclamation recommends that the outlet tunnel be designed to flow 75% full to eliminate instability problems.

Subatmospheric Drop Shafts. St. Anthony Falls Hydraulic Laboratory and Bauer Engineering Inc. have modeled high-velocity, subatmospheric-pressure drop shafts. They analyzed velocities of approximately 50 ft/sec. Such high velocities would permit large discharges to be carried in relatively small shaft diameters.

The models attempted to limit the amount of air entering the system at the inlet. The Bauer Engineering shaft was a 12-in. pipe that discharged into a plunge pool. A float valve was used initially in the inlet structure. Lack of air in the high-velocity shaft caused the local pressures to drop below atmospheric pressure, inducing suction within the shaft. This produced a downpull through the shaft that increased the velocity. But the hydraulic downpull also affected the float valve and caused it to tilt. This produced operational problems and required a sophisticated design to overcome. The study recommended abandoning the float valve and adopting an open inlet basin. Vanes would be needed in the basin for vortex control, which is necessary for the high velocities.

The large negative pressures in the system would tend to cause cavitation damage to the shaft barrels. The model experienced no surface pipe damage, but the rock bottom of the plunge pool was eroded by the jet, causing extensive damage. Dissolved air comes out of solution in the areas of low pressure and does not completely return to solution before entering the tunnel, thus causing air-removal problems. The cavitation in the shaft produced vibrations and considerable noise problems that would restrict the use of the shaft to nonresidential areas.

Direct Drop, Air Entraining Drop Shafts. Flow enters these drop shafts radially and descends through the shaft. The shaft diameter is designed to flow full with air entrained in the water, bulking it up enough to fill the drop shaft. This helps to ensure the maximum possible energy dissipation by wall friction. The air entrained also provides a cushion for the water, reducing the floor impact. A large air-separation chamber at the base of the shaft is used to remove the entrained air. An air vent is necessary to allow the air to vent before entering the tunnel.

Model studies performed by the St. Anthony Falls Hydraulic Laboratory associated with the design of the Metropolitan Water Reclamation District of Greater Chicago's Tunnel and Reservoir Plan (TARP) concluded that this type structure is very effective in dissipating energy and removing entrained air. These experiments used a shaft inlet that was simply a horizontal sewer connected to the vertical shaft. The air-separation chamber is provided at the base of the shaft. This chamber is large enough to remove the insufflated air.

The St. Anthony Falls Hydraulic Laboratory analyzed several variations on this structure and selected two as the most efficient. The first, the E-15, consists of a sump chamber with a sloping top. The air vent is located inside of the drop shaft and separated from the drop shaft by a vertical slotted wall (Figure 15-1). The slots in the wall allow air to be recirculated into the falling water in the drop shaft, resulting in the reduction of large air slugs and providing a more homogeneous mixture of air and water.

At the bottom of the shaft is the sloped-roof air-separation chamber. As the air is released from the mixture, it follows the sloping wall of the air collector back up to the air vent side of the vertical shaft and rises to the surface, and

Fig. 15-1. E-15 drop shaft air-separation chamber, looking toward downcomer (note sloping roof).

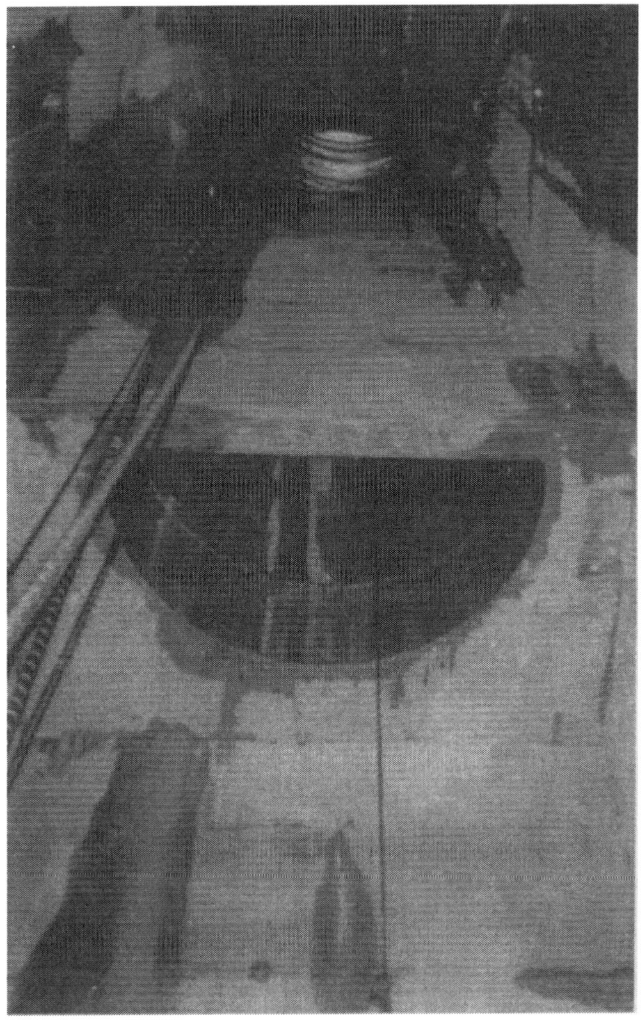

Fig. 15-2. E-15 drop shaft, looking up downcomer from air-separation chamber (note dividing wall).

some of it is recirculated through the slots into the drop shaft. If the drop shaft is to be used for steady state flows, a plunge pool is built directly beneath the shaft barrel to dissipate the energy.

This structure requires a rather large air-separation chamber, as shown in Figure 15-1. Larger drop shafts require a high chamber roof. During the design of the TARP system in rock, it was determined that this type of shaft was economical up to shaft diameters of 9 ft with a maximum discharge capacity of 600 cfs.

A second design was analyzed for drop shafts larger than 9 ft in diameter. The D-4 has a separate shaft for the air vent that is downstream from the downcomer, and connected to the downcomer, above the crown of the incoming sewer (Figures 15-2–15-6). The air-separation chamber has a horizontal roof (Figure 15-3). The air vent recycles air into the downcomer. The D-4 can be used in much larger drop shafts, up to 20 ft in diameter with a maximum discharge capacity of 4,500 cfs.

Both structures handle a wide range of discharges and have head losses that are 1/5 of those for the vortex-type shafts. These shafts are the only commonly used drop shafts that can adequately handle variable discharges, accept impacts on drop shaft floors, remove entrained air, and have minimum head losses to prevent backflow problems when tunnel gradients reach the levels of incoming flows. More than 200 of these shafts have been built in the Chicago area, and many have operated successfully for more than 20 years. The dimensions and maximum design discharges for these multipurpose shafts are given in Figure 15-3.

The large dimensions of the E-15 and D-4 drop shafts, particularly the air separation chambers, necessitate mining a major chamber in rock with attendant rock reinforcement and structural concrete linings. If founded in earth, these mined structures become more massive, and the design must be able to accommodate the forces and vibrations caused by the air–water impacts.

The larger versions of these drop shafts can be overexcavated and used as construction shafts. When used in this

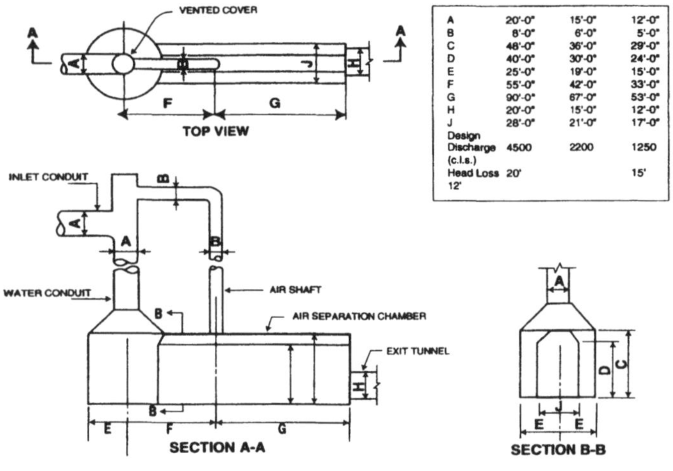

A	20'-0"	15'-0"	12'-0"
B	8'-0"	6'-0"	5'-0"
C	48'-0"	36'-0"	29'-0"
D	40'-0"	30'-0"	24'-0"
E	25'-0"	19'-0"	15'-0"
F	55'-0"	42'-0"	33'-0"
G	90'-0"	67'-0"	53'-0"
H	20'-0"	15'-0"	12'-0"
J	28'-0"	21'-0"	17'-0"
Design Discharge (c.l.s.)	4500	2200	1250
Head Loss	20'		15'
12'			

Fig. 15-3. D-4 drop shaft.

manner, the final lining of the shaft barrels must comply with the following equation or the drop shaft will fail to operate properly:

$$\frac{Q}{g^{1/2}D^{5/2}} = 0.2 \qquad (15.4)$$

where

Q = design discharge (cfs)
g = acceleration of gravity (ft/sec^2)
D = finished shaft diameter (ft)

St. Anthony Falls Hydraulic Laboratory also performed an analysis of low-height air-separation chambers for the City of Rochester, New York. The shaft would be constructed where thin rock layers prevented the use of the designs developed for the TARP system. The air vent is located inside the drop shaft and is separated from the drop shaft by a vertical slotted wall. The Rochester drop shaft modified the air-separation chamber of the D-4 shaft to decrease its height. To provide for this, a false crown with air passages through it was positioned close to the top of the air-separation chamber. The released air collects at the top of the chamber, passes through the air passages in the false crown to the chamber above, and returns up the air vent in the drop shaft.

AIR REMOVAL

High-velocity streams of water may entrap and contain large quantities of air. Air entrainment causes the flow to be a heterogeneous mixture that varies in bulk density throughout the flow cross section and exhibits pulsating density variations.

Potential Problems

Design engineers should eliminate the harmful effects brought of high-energy hydraulic jumps within the tunnel,

Fig. 15-4. D-4 drop shaft air-separation chamber, looking toward downcomer (note separate air vent in foreground).

transient phenomena induced by rapid filling of the downstream end of a tunnel without provisions for adequate surge shafts, the formation of air traps within the tunnel system, the introduction of entrained air into the tunnel from drop shafts, and the formation of vortices, which may enter the tunnel through shafts. In addition, the design should provide for the easy egress of air from a tunnel while it fills with water. Improper design can lead to one or more of the following phenomena, which may lead to structural damage:

- Blowbacks, high-pressure releases of air and water in the opposite direction of the flow
- Blowouts, high-pressure releases of air and water in the same direction of the flow
- Geysering, in which air or water vents above the ground surface through shafts located at points along the tunnel
- Transient and surging flows that cause rapid dynamic instability and possible tunnel collapse

As long as the depth downstream of a hydraulic jump does not reach the tunnel crown, jumps within tunnels are not a severe problem. When the downstream depth seals

Fig. 15-5. D-4 drop shaft air-separation chamber, looking downstream toward exit conduit (note separate air vent).

Fig. 15-6. D-4 drop shaft exit conduit from air-separation chamber into tunnel.

against the roof of the tunnel, the shock effects of air trapped downstream of the jump can create violent impacts and associated damage. High-energy hydraulic jumps have caused both blowouts and blowbacks. These rapidly escaping air pockets result in water rushing in to fill the voids, creating loud noises and pressure waves, which have stripped the lining from tunnels and shafts and caused partial tunnel collapse and severe erosion.

Even without the formation of hydraulic jumps, blowbacks, blowouts, and geysering, dynamic instability due to transients can take place whenever the downstream end of a tunnel is filling rapidly while air trapped within the system cannot escape at a reasonable rate. When the pressurization surge reaches an upstream end of the tunnel during the filling process, water will rise rapidly in the drop shafts near the upstream end. Water levels in other drop shafts will also rise as the surge reflected by the upstream end travels downstream.

In pressure tunnel flows, an air void can form at a bend that connects a vertical shaft to a horizontal tunnel. A sudden reduction in the flow rate can cause this void to vent back up the shaft and cause geysering.

Inlet No. 2 of the Oroville Dam Diversion Tunnels experienced the vortex development as the reservoir pool filled during a flood. The vortex grew in size and strength as the reservoir filled during the December 1964 flood. After the flood, the tunnel was dewatered and inspected. Although the observed damage was relatively minor, it did consist of many rough scoured surfaces throughout the entire tunnel length.

Solutions

By taking proper precautions during design, these problems can be avoided. The following steps should be taken:

- Check the tunnel slopes for the development of supercritical flow, and calculate whether a hydraulic jump can occur for any conceivable discharge. A hydraulic jump may not occur during the maximum design discharge, but it can occur for some lesser discharges. The tunnel slopes should be reduced if the check shows the potential for a hydraulic jump.
- Provide surge shafts of diameters at least equal to the diameter of the tunnel at both the upstream and downstream ends of the tunnel. A transient analysis should be made during the design phase to determine how high these surge shafts should be.

- Whenever branch tunnels or drop shaft exit conduits meet another tunnel and whenever a tunnel changes diameter, always match tunnel crowns rather than inverts, to prevent the formation of air pockets.
- Prevent entrained air from entering the tunnel from drop shafts.
- Provide a splitter wall to suppress the development of vortices in the inlet to tunnels whenever it is apparent that strong vortex development may occur.
- Provide some form of inlet control to regulate or completely shut off all flows into each inlet tributary to the tunnel. This is usually accomplished by the use of remote-controlled gates at each shaft inlet.

GAS BUILDUPS IN SEWER TUNNELS

Gas buildups are another consideration the designer must address in sewer tunnels. The buildup of hydrogen sulfide (H_2S) is characteristic of tunnels with low velocities and warm ambient temperatures, or tunnels that are relieving a number of existing sewers that have large sludge deposits on the inverts. Hydrogen sulfide results in odor problems and severe corrosion. It is also toxic.

To prevent the formation of hydrogen sulfide, aerobic conditions should be promoted by the provision of high wastewater flow velocities and the introduction of oxygen and/or the complete dewatering of the tunnel system within two days following complete filling.

CONTROL OF INFILTRATION AND EXFILTRATION

The phenomena of infiltration and exfiltration are of critical importance to water conveyance tunnels.

Infiltration during construction should be reduced to acceptable levels in all types of tunnels. Significant levels of infiltration after a water conveyance tunnel is completed are also unacceptable. Inflows can cause loss of ground into the tunnel and result in surface settlements and damage to neighboring structures. The inflows may cause the adjacent groundwater table to be seriously lowered, with resulting adverse impacts on water supply, trees, and vegetation. In flood control tunnels, groundwater infiltration can reduce the carrying capacity available to handle peak flows. Infiltration in water supply tunnels may lead to pollution of the supply. In sewer tunnels, infiltration contributes to increased water reclamation and pumping costs.

Exfiltration from water conveyance tunnels also has potential for undesirable effects. In flood control and sewer tunnels, exfiltration may cause pollution of the adjacent groundwater. Exfiltration from water supply and power tunnels can result in serious reductions in available drinking water and energy supplies as well as revenue loss.

Infiltration During Construction

Deep wells, wellpoints, grout curtains, slurry cut-off walls, freezing, and air pressure within tunnels and shafts are all used to control groundwater inflows during construction.

Monitoring Wells

Monitoring wells with casings open only at tunnel depth can be used to observe drawdown in groundwater levels caused by infiltration and can also be used to measure changes in groundwater quality that are indicative of tunnel exfiltration. When infiltration and/or exfiltration are considered to have significant potential, these wells should be installed at appropriate intervals prior to construction. The selected sites should be easily accessible for monitoring and located at least two tunnel diameters away from the tunnel neat line to obtain meaningful measurements.

Prevention of Infiltration and Exfiltration

The extent to which infiltration and exfiltration should be reduced must be determined before the design of the tunnel commences. It may be appropriate to apply different standards of watertightness to different sections of the tunnel. It is common practice to specify in the contract documents control of water inflows during construction and the permissible infiltration limits after the construction.

Infiltration and exfiltration can be prevented or minimized by the use of grouting techniques, waterproof membranes, and appropriate liner types.

Grouting. Four basic types of grouts are commonly used to alleviate infiltration and exfiltration:

- Portland-cement grout—a mixture of Portland cement, water, and frequently, chemical and mineral additives.
- Colloidal grouts—these can be subdivided into two classifications: natural soils found near the project site, and commercially processed clay such as bentonite.
- Asphalt grouts—more commonly used where major inflows are difficult to stop with cement, clay, or chemical grouts.
- Chemical grouts—particularly advantageous when low viscosity and fast setting times are desirable.

The grouting procedures most commonly used are

- *Curtain grouting.* The construction of a barrier of grout by drilling and grouting a linear sequence of holes to reduce permeability. In tunneling, this may be done from the surface prior to construction to prevent inflows from reaching the tunnel or shafts.
- *Consolidation grouting.* The grout is injected into short holes drilled on a pattern from within the tunnel or shaft to improve the bearing capacity and/or reduce the permeability of broken or leached rock.
- *Contact grouting.* The grout is injected through the tunnel or shaft lining to fill the voids between the liner and the surrounding excavation

- *Cavity filling.* Two- or three-phase grouting to control seepage into or out of the tunnel when filling large cavities encountered during the excavation of the tunnel or shaft.

Waterproof Membranes. Waterproofing membranes have been used extensively in transportation tunnels, but their use in water conveyance tunnels is relatively new. The chief difference is the need for full circumferential coverage in water conveyance tunnels. In addition, full circumferential cut-off rings should be provided at the junction points between tunnel sections whose lining is backed by waterproof membranes and sections that are not waterproofed.

Tunnel Linings

The geologic strata and lining work together, depending on the relative stiffness of the two elements, to resist internal pressure. External pressures, however, act upon the outside of the watertight element in the lining (the steel or plastic membrane in a sandwich liner, the steel internal facing in a steel-lined tunnel, the prestressed ring in a prestressed concrete liner, etc.).

The factors to be considered include the following.

Deformability of the Surrounding Strata. In strata with a low modulus of elasticity, or a potential for time-dependent creep or radial expansion due to internal water pressure, deformability will be relatively large and the lining will be severely stressed. A lining must then have sufficient strength to accommodate these large strains. For example, if a concrete lining cracks and watertightness is essential, it could be provided by a flexible, impervious membrane such as plastic.

The flexibility of a steel lining is reduced simply by increasing its stiffness (thickness). The steel then takes a greater share of the load, and the system deformations are limited. Because the steel is strong in tension, this is a permissible solution. However, if the steel liner is subject to external groundwater pressure, its thickness will have to be increased (sometimes significantly) or otherwise stiffened or supported to prevent buckling. Nonmetallic materials used for membranes have a limited tensile strength, generally exhibit creep, and can only be used for structural and watertightness purposes in appropriate circumstances.

The stiffer the lining, the greater the share of the load it carries. For example, concrete linings do have increased stiffness due to their relative thickness. However, the elastic modulus of concrete is only about 1/10 that of steel, which significantly reduces the load taken by it relative to the surrounding strata. On the other hand, concrete has a limited tensile strength and will crack in tension even under low internal pressures. Ability of the concrete to deform in extension can be developed by precompressing the concrete, which then always operates in compression. The tensile strain that can be tolerated is a function of the precompression that can be maintained. While concrete can only be assumed to provide a watertight barrier when backed by very stiff strata, prestressing of the concrete will provide watertightness where less stiff strata are encountered and larger deformations occur.

Where some leakage through the lining is acceptable, larger deformations can be tolerated and the concrete lining can be allowed to crack. Provided that considerations of durability are met for the supporting strata and reinforcement in the lining, the basic design requirement is to ensure that the lining will stay in place. In strata with a very high elastic modulus, where deformations will be small, a plain concrete lining may suffice. If deformations are likely to be large and thus cause large singular cracks, the lining must be reinforced to distribute those cracks and maintain the concrete as an integral ring.

Strata Durability. Where the surrounding strata is subject to deterioration and fluctuating internal water pressures may occur, a risk of eroding degraded strata into the waterway and developing cavities behind the lining exists. It is then necessary to control crack widths in the concrete to prevent the movement of fine degraded material through the lining, provide additional erosion barriers, or ensure that the lining is watertight.

Permeability of Strata. Where the strata are relatively impermeable, leakage from the tunnel will be controlled to a degree acceptable in terms of economic considerations. If leakage is considered environmentally acceptable, the lining does not necessarily have to be watertight. It must then be designed to be a stable hydraulic surface and to contain the surrounding strata, if necessary.

In relatively permeable strata, the lining must be designed as the water-retaining element in the lining/rock system. Depending on environmental sensitivity, some leakage may be acceptable, as typically occurs in some reinforced concrete water-retaining structures. If, however, absolute watertightness is essential, then the lining must be designed accordingly. In an environmentally sensitive situation, the risk of leakage from a less expensive design must be balanced against the consequences of leakage.

If the surrounding strata is relatively free-draining, there is limited potential for development of external hydrostatic pressures when the tunnel is dewatered.

Hydrofracture. Where the depth of cover and the hydraulic pressure in the tunnel are such that there is risk of tensile fracture (hydrofracture) within the strata, the lining must be designed as the water- (pressure-) retaining element.

A range of possible tunnel lining concepts includes concrete (plain, reinforced, and prestressed), steel linings (internal liners or membranes), and plastics (membranes). Less common materials, including fiber-reinforced concrete, fiberglass linings, and high-tensile-strength concrete formulations may also be considered.

The logic process that incorporates the functional requirements outlined above is summarized in Figure 15-7. From the top to the bottom, a series of questions (rectangular boxes) are posed regarding anticipated service conditions.

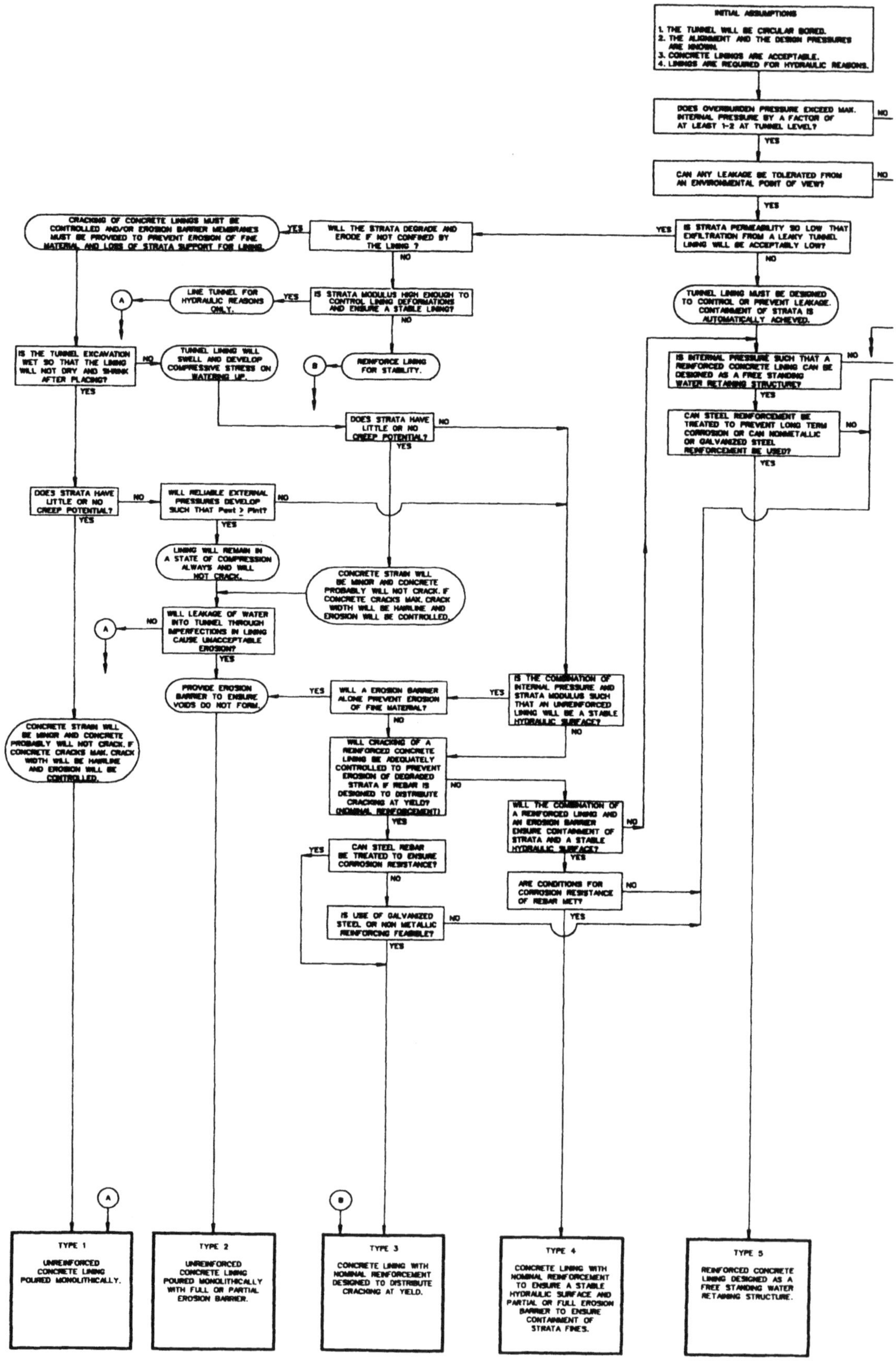

Fig. 15-7. Logic diagram for selecting lining type.

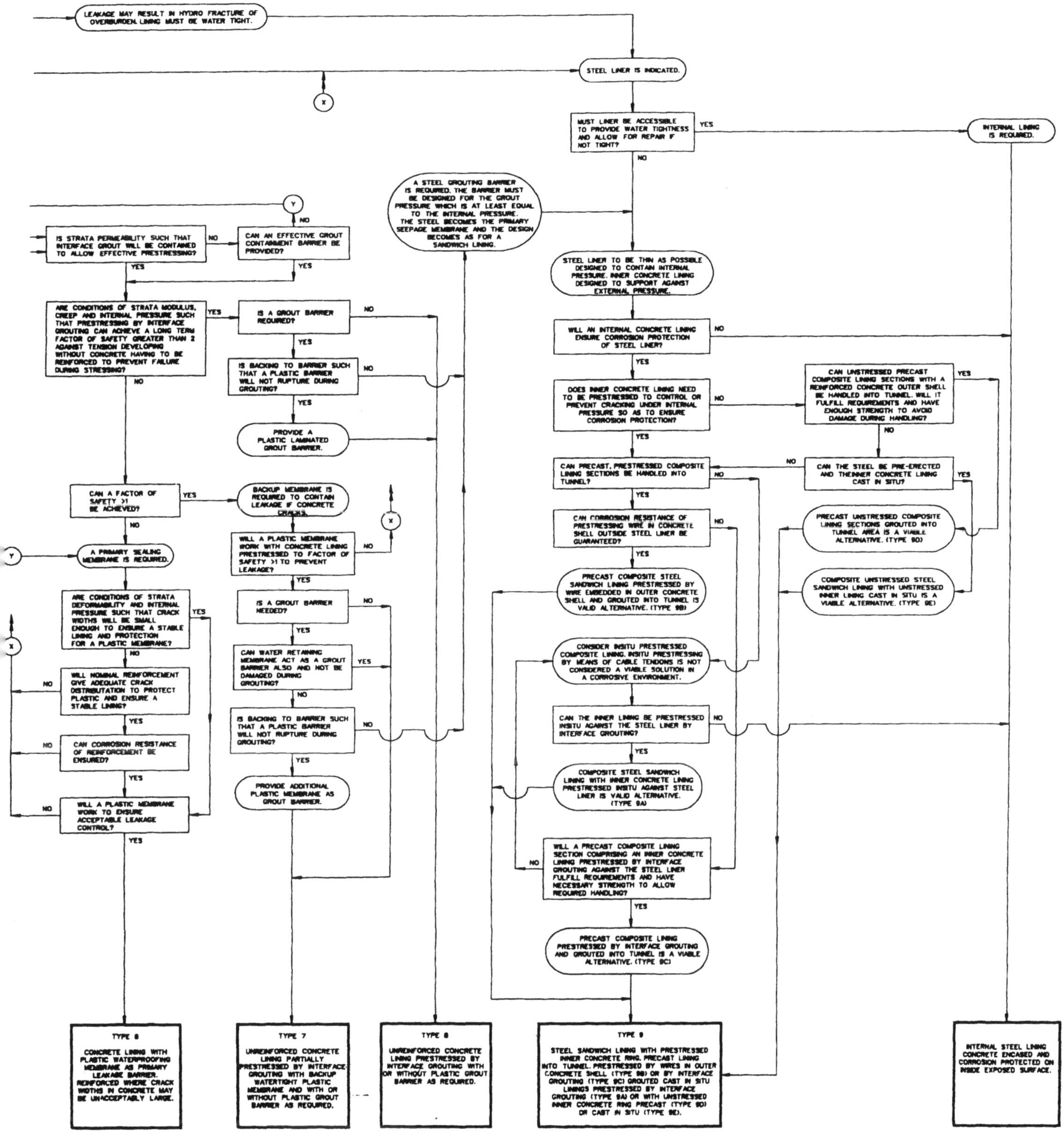

NOTE
FOR A GIVEN SET OF CONDITIONS, LINING TYPES TO THE RIGHT OF THAT INDICATED WILL GENERALLY OFFER GREATER SECURITY. THE CHOICE OF WHICH LINING TO USE MAY BE DECIDED ON ECONOMIC GROUNDS.

Moving through "yes" or "no" scenarios and a series of informational statements (elliptical boxes), a set of lining type concepts are reached at the bottom (rectangular boxes with heavy outlines). For a given set of service conditions, a path through the logic diagram will lead to a lining type that will perform the required function. The diagram is arranged so that all lining types to the right of the selected type will offer greater functional security. Thus, the final selection can be made based on the degree of security desired and economic considerations.

A brief description of each of the lining types shown in Figure 15-8 follows:

- *Type 1.* Unreinforced cast-in-place concrete lining.
- *Type 2.* Same as Type 1 except with partial- or full-circumference laminated plastic geomembrane.
- *Type 3.* Cast-in-place concrete lining with nominal reinforcement, fusion-bonded epoxy-coated or galvanized.
- *Type 4.* Same as Type 3 except with partial- or full-circumference laminated plastic geomembrane.
- *Type 5.* Cast-in-place concrete lining with fusion-bonded epoxy or galvanized reinforcing designed as a freestanding water containment structure.
- *Type 6A.* Unreinforced cast-in-place concrete lining with full-circumference laminated plastic waterproofing geomembrane as primary water containment barrier.
- *Type 6B.* Same as Type 6A except with reinforcing as in Type 3.
- *Type 7A.* Unreinforced cast-in-place concrete lining partially prestressed by interface grouting with a full-circumference laminated plastic watertight geomembrane and without a geomembrane grout barrier.
- *Type 7B.* Same as Type 7A except with a geomembrane grout barrier.
- *Type 8A.* Unreinforced cast-in-place concrete lining prestressed by interface grouting and without a full-circumference laminated plastic geomembrane (nonwaterproof) as a grout containment membrane.
- *Type 8B.* Same as Type 8A except with a full-circumference laminated plastic geomembrane (nonwaterproof) as a grout containment membrane.
- *Type 9A.* Unreinforced cast-in-place concrete lining, prestressed by grouting against inside of internal steel membranes. This is backed up by precast concrete segments with a combination of pea gravel and grout filling the interface between the internal steel membrane and the precast concrete segments.
- *Type 9B.* Precast concrete pipe with internal steel membrane prestressed by wires in outer concrete shell and grouted into place in the tunnel. The wire prestressing is used to compress the core of high-strength concrete containing the steel membrane. The concrete core behaves as an uncracked section under working loads.
- *Type 9C.* Same as Type 9A except precast outside of tunnel and with pea gravel grouted into place.
- *Type 9D.* Precast concrete pipe with internal steel membrane and conventional reinforcement, with pea gravel backing grouted into place. The internal steel membrane is supported against buckling from external groundwater pressure by the inside surrounding concrete of the precast pipe.
- *Type 9E.* Same as Type 9D except cast in situ within the tunnel.
- *Type 10.* Internal steel lining with corrosion protection coating, encased in unreinforced cast-in-place concrete.

All designs are based on the lining materials discussed above and have been predicated on the requirement that they should, as far as possible, fit within a uniform bored diameter.

A unique tunnel lining was developed for the Dez Dam penstock tunnels in Iran. A concrete lining was first cast against the rough rock surface to provide a smooth-walled cylindrical space. A steel interval liner was then constructed so as to leave a thin air annulus between the steel and concrete. Finally, the annulus was filled with high-pressure grout to prestress the inner steel lining, whose thickness was thereby cut in half.

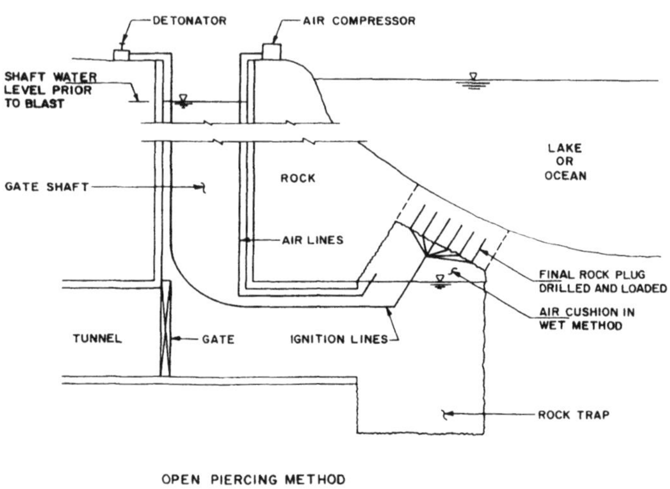

Fig. 15-8. Open-piercing method.

LAKE TAPS AND CONNECTIONS TO LIVE TUNNELS

Connecting a new water conveyance tunnel to an existing high-pressure water tunnel or tapping a lake or reservoir requires careful advanced planning. Obviously, such connections are best made in the dry, but in certain cases this is not economically possible. The following discussion highlights some alternatives.

Cofferdam

For tunnels that are to connect to a relatively shallow lake, a ring cofferdam can be constructed from tunnel level below the bottom of the lake to an appropriate elevation above the water surface. The enclosed area can then be dewatered to make the connection between the lake and the future tunnel in the dry.

In-Line Tunnel Diversion

To connect a new tunnel to a live high-pressure tunnel, an in-line diversion pipe or series of pipes can be installed within the existing tunnel after it has been temporarily dewatered. A flow cut-off must then be installed around the in-line diversion pipes on both sides of the proposed connection to prevent water from flowing along the backs of the pipes into the connection. With the in-line diversion in place, the new tunnel connection can be made in the dry while the existing tunnel is fully pressurized. When the connection is completed, the existing tunnel must be dewatered again, and the diversion pipes and cut-offs removed and the project completed.

Open-Piercing Method

The first known lake tap using the open-piercing method was constructed at Annex Lake near Juneau, Alaska, in 1915. This method is restricted to the construction of a connection in rock. The new tunnel is advanced below and then raised as close to the existing high-pressure source as possible, leaving a rock plug in place above the tunnel raise (Figure 15-8). The tunnel near the connection should be constructed so that when filled with water, a compressed-air cushion will be created below the plug. This air cushion should be maintained until the final connecting blast is made. A rock trap is provided in the invert of the raise below the plug. A shaft from ground surface to the new tunnel invert is also required as close as possible to the connection. A gate is provided on the side of this shaft furthest from the rock plug to seal off any water from entering the tunnel beyond the shaft–rock plug section. The rock plug is then drilled and prepared for blasting to make the final connection. Next, the gate is closed, and the tunnel (on the rock plug side of the shaft) and the shaft are filled with water to a depth slightly below the hydraulic grade line in the live tunnel or lake to be tapped. At this point, the air cushion below the plug should be checked for adequacy by remote monitoring and additional air pressure pumped in if necessary. The charge is then detonated, and the air cushion below the plug interrupts the water column to dampen the pressure shock and prevent damage to the new tunnel. Since the water pressure at the time of the blast is lower inside the newly constructed tunnel, most of the rock blasted in the connection will collect in the rock trap. In this procedure, the final connection is left unlined.

TUNNEL MAINTENANCE

The inadequate maintenance of water conveyance tunnels can best be summed up by the old phrase, "out of sight, out of mind." The reason that periodic maintenance of these tunnels is virtually nonexistent lies in the inherent dangers of entering them. Sudden collapse due to residual water pressure behind the lining and in the surrounding ground; the presence of suffocating, flammable, or toxic gases; the possibility of sudden unexpected inflows into the tunnel; the lack of adequate lighting for visibility; and the inherent difficulty of removing injured personnel from such tunnels all create a natural aversion to accomplishing the task.

Safety

Prior to entering any water conveyance tunnel, all inlets to the tunnel should be positively closed off by gates, bulkheads, stop logs, or other means. Temporary flow diversions around the inlets and away from the tunnel should be provided when necessary. After the potential inflow problem has been handled, the tunnel system (main tunnel, branches, and/or shafts) should be dewatered slowly to allow hydrostatic pressures to be relieved and prevent possible damage to the vertical shafts and tunnels.

Because the potential for the presence of gas is always high in these tunnels, temporary forced air ventilation should be provided in sufficient quantities to satisfy the OSHA requirements for the construction of tunnels in gassy conditions. The temporary ventilation system should be carefully engineered to allow for the size and spacing of the shafts, which were placed for hydraulic convenience and not necessarily for optimum ventilation. The ventilation should be operated continuously until all maintenance personnel have completed their work and exited from the tunnel system.

When the ventilation process has been in operation for at least 24 hours, a gas test should be made to detect the presence of any hazardous gases. No one should be allowed to enter the tunnel until this test is made and the air declared acceptable. Gas levels should also be continuously monitored while personnel are in the tunnel system. The four gases most commonly found in water conveyance tunnels are

- Carbon dioxide, which in high concentrations can cause suffocation. Its source in water conveyance tunnels is usually attributed to air releases from the groundwater surrounding the tunnel, which may have higher concentrations of this gas than that normally found in surface water.
- Carbon monoxide, which can also cause suffocation. It is commonly caused by exhaust from internal combustion engines.
- Hydrogen sulfide is a toxic gas formed from decaying organic matter and is commonly associated with stagnant water with warmer temperatures.
- Methane is an explosive gas whose source may be decaying organic matter, the ground surrounding the tunnel system, or both.

All repair, maintenance, and safety equipment used in the water conveyance tunnel system should all meet all the requirements for gassy tunnels.

Periodic Tunnel Cleaning

As water conveyance tunnels age, various materials that can reduce the effective cross-sectional area or increase hydraulic head losses build up around the tunnel periphery. Some typical examples include

- Calcium carbonate buildup of ridges, stalactites, and stalagmites caused by water seeping into a tunnel through minute shrinkage cracks. Can be found in all types of water conveyance tunnels.
- Bacterial slimes, primarily found in flood control and sewer tunnels.
- Mucilagenous material buildups formed by manganese absorption by algal polysaccharides, common in power tunnels.
- Sludge deposits along the tunnel invert in sewers.
- Sediment deposits in flood control, power, and raw water tunnels.

Various methods have been used to remove these materials including brushing, scraping, and high-pressure water jets. These methods are not particularly successful in removing calcium carbonate buildups. In the Clear Creek Tunnel and Colorado River aqueducts in California, a diesel-powered vehicle with hydraulically activated steel brushes was inserted into tunnels to clean away the obstructions.

REFERENCES

Bauer Engineering (1972) *Report on Drop Shaft Investigation for Crosstown Expressway (I-94) and Flood and Pollution Control Plan,* February, Project No. 1146.001.

CAMP, C.S. (1937) "Determination of shape of nappe and coefficient of discharge of a vertical sharp-crested weir, circular in plan with radially inward flow," State University of Iowa thesis.

COLEBROOK, C.F. (1958) "The flow of water in unlined, lined and partly lined rock tunnels," Proceedings I.C.E., London.

Construction Industry Research and Information Association (1979) *Tunnel Waterproofing.* Report 81.

DEWITT, M.J. (1983) "Constructing the Drakensberg pressure tunnels," *Water Power and Dam Construction,* June.

ELDER, R.J. (1956) "Friction measurements In Appalachia tunnel," *Journal of Hydraulics Division of ASCE, 82*(HY3).

GOLDRING, B.T. (1983) Air voids at downshaft—tunnel bends, *Journal of Hydraulic Engineering,* ASCE, *109*(2).

HOROWITZ, G., and LEE MAY, N.N. (1971) "Effects of mechanical cleaning operations on flow in large conduits," *Journal of Hydraulics Division of ASCE, 97*(HY5).

JEVJEVICH, V., and LEVIN, L. (1953) "Entrainment of air in flowing water and technical problems related with it," *Proc. Minnesota International Hydraulic Convention,* International Association of Hydraulic Research.

LANNING, C.C. (1967) "Oroville Dam diversion tunnels," *Journal of the Power Division of ASCE, 93*(PO 2).

LAZZARI, E. (1954) "Ricerca sperimentale sullo sfioratore a pianta circolare," *L'Energia Elettrica,* November, 838–840.

MIXON, E.O., ROSTUK, J.W., and SEHULSTER, J.P. (1989) "Crosstown 5/6 collector system soft ground tunneling with an EPB machine," *Proceedings of the Rapid Excavation and Tunneling Conference,* June.

PITT, J.D., and ACKERS, P. (1982) *Hydraulic Roughness of Segmentally Lined Tunnels.* CIRIA Report 96.

POLLARD, A.L., HOUSE, H.E. (1959) "An unusual deposit in a hydraulic tunnel," *Journal of the Power Division of ASCE, 85*(PO 6).

PRATT, H.K. (1970) "Diversion at Portage Mountain project," *Journal of the Power Division of ASCE, 96*(PO 3).

PROKOPOVICH, N.P. (1985) "Carbonate deposits in Clear Creek tunnel, California," *Bulletin of the Association of Engineering Geologists, XXII*(1).

RAMAGE, J., OLSON, D.J., and LUTZENBERGER, T.J. (1989) "Water control measures, north shore inline interceptor phase 1A tunnel, Milwaukee Water Pollution Abatement Program," *Proceedings of the Rapid Excavation and Tunneling Conference,* June.

RICH, G.R. (1951) *Hydraulic Transients,* McGraw-Hill Book Company, New York.

SCHLUTTER, A. (1988) "Keeping tunnels dry using plastic membranes," *Tunnels and Tunneling,* October.

SEEBER, G. (1976) "Prestressing a power tunnel," *Water Power and Dam Construction,* April.

STORMO, S.H. (1983) "Tunnel breakthrough beneath 85 m water at Ringedalsvatn," *Norwegian Tunneling Technology Publication No. 2,* Norwegian Soil and Rock Engineering Association.

Technical Manual No. 5-818-5 (1983) Dewatering and groundwater control, Department of the Army, U.S. Corps of Engineers.

Technical Manual No. 5-818-6 (1970) Grouting methods and equipment, Department of the Army, U.S. Corps of Engineers.

University of Iowa, Iowa Institute of Hydraulic Research, *Vortex-Flow Drop Structures for the Milwaukee Metropolitan Sewerage District Inline Storage System,* IIHR Report No. 264.

University of Minnesota (1968) St. Anthony Falls Hydraulic Laboratory, *Model Studies—Lawrence Avenue Sewer System, City of Chicago,* October, Project Report No. 1 00.

University of Minnesota (1971) St. Anthony Falls Hydraulic Laboratory, *Hydraulics of Long Vertical Conduits and Associated Cavitation,* June, Project 10034 FLU.

University of Minnesota (1973) St. Anthony Falls Hydraulic Laboratory, *Model Studies of Drop Shafts for the Chicago Underflow Plan, Dry Weather Flow Drop Shaft—Helix,* Project Report No. 150.

University of Minnesota (1975) St. Anthony Falls Hydraulic Laboratory, *Model Studies of Drop Shafts for the Tunnel and Reservoir Plan,* January, Project Report No. 154.

University of Minnesota (1982) St. Anthony Falls Hydraulic Laboratory, *Rochester Drop Shafts Model Studies,* April, Project Report No. 206.

University of Minnesota (1988) St. Anthony Falls Hydraulic Laboratory, *Hydraulic Transient Modelling of TARP Systems,* Project Report No. 270.

U.S. Bureau of Reclamation (1965) *Friction Factors for Large Conduits Flowing Full,* Water Resources Technical Publication, Engineering Monograph No. 7.

WAGNER, W.E. (1956) "Shaft spillways: determination of pressure-controlled profiles," *Transactions, ASCE, 121.*

WAGNER, W.E. (1967) "Glen Canyon dam diversion tunnel outlets," *Journal of the Hydraulics Division,* ASCE, *93*(HY 6).

WEMICKE, J.C. (1987) "Sewers get a breath of fresh air," *Civil Engineering,* ASCE, *57*(8).

WESTFALL, D.E. (1989) "Friction losses for use in the design of hydraulic conveyance tunnels," *Proceedings of the Second Intentional Symposium on Design of Hydraulic Structures,* Fort Collins, CO.

WILLIAMSON, J., and PARTNERS. (1952, 1964). Unpublished reports to North of Scotland Hydroelectric Board.

16 | Small-Diameter Tunnels

David E. Westfall
District Manager, Vice President, Principal Professional Associate, Persons Brinckerhoff Quade & Douglas, Inc.
Glenn M. Boyce
Senior Professional Associate, Parsons Brinckerhoff Quade & Douglas, Inc.

Conventional tunnels are driven by workers excavating, supporting, and advancing the tunnel at its face. As the tunnel diameter decreases, work becomes more difficult. There is less room for personnel. Efficiency and advance rates decrease. It becomes increasingly difficult to erect the tunnel lining system from within the tunnel. At diameters of 60 to 18 in., it becomes more efficient to push the lining into place from a shaft or portal. This lining is a string of pipe sections. This method of tunnel construction, commonly referred to as pipe jacking, was introduced in the United States during the 1890s. Pipe jacking is commonly used at highway and railroad crossings to install conduits or sewers through soft ground instead of trenching.

The early forms of pipe jacking had shields located at the leading edge of the pipe segments. Individuals working in the confined space would excavate through the soft ground with picks and shovels as the pipe was jacked forward (Figure 16-1). Over time, the individuals were replaced with a single-flight auger, which excavated the material at the tunnel face and pulled the cuttings/muck back through the pipe (Figure 16-2). The term *bore and jack* became more common, as it described the process. Horizontal auguring is commonly used for relatively short distances to install sewer connections, electrical lines, water connections, and oil and gas lines beneath existing highways, streets, railroads, or other structures where open-cutting a trench is impossible or undesirable. This procedure can be used to install pipes of from 4 to 60 in. in diameter.

The use of augers has also been surpassed by the miniaturization of the tunnel boring machine (TBM), such as those discussed in Chapter 11. In 1979, the first miniature TBM was developed by the Japanese, and a few years later one was developed by the Germans. This machine development led to the term *microtunneling*. In its simplest form, microtunneling is the use of a remote-controlled, computer-assisted, miniature, earth pressure or slurry pressure balance machine, advanced by jacking pipe. Unlike pipe jacking, the machines can be equipped with rock cutters, and small-diameter tunnels can be advanced through not only soft ground, but rock and mixed conditions. Figure 16-3 shows a present-day microtunneling machine.

What follows is a description of the basic procedures of pipe jacking. The importance of site investigations, the front end component of the pipe jacking operation, the pipe, and special ways small-diameter tunnels can be used are discussed.

BASIC PROCEDURE

Microtunneling, or pipe jacking, advances pipe from a main jacking pit. The pit is constructed in an appropriately supported shaft excavated to tunnel depth. The jacking pit is usually excavated so that its length (or diameter, if circular) is about 8–20 ft, with consideration given to the length of the individual pipes to be jacked. The width of the pit is normally 2 to 3 times the outside pipe diameter, with 6 ft as a minimum. These interior dimensions are needed to provide working room and adequate space for the thrust wall, pressure plate, hydraulic jacks, spacer blocks, thrust ring, and a length of pipe. This pit also requires a concrete base and guide rails to properly align the pipe. The shaft/pit can be converted to a manhole in the case of gravity sewer lines. Receiving pits are required at the far end of the tunnel. These pits may be smaller, as they only receive the shield or machine.

The forward end of the jacking operation involves a tunneling machine or shield with a cutting edge, which serves to guide and protect the first pipe and to support the excavation at the face. The leading edge of the shield or machine can adjust the horizontal or vertical alignment of the tunnel by use of steering jacks housed in the articulated machine.

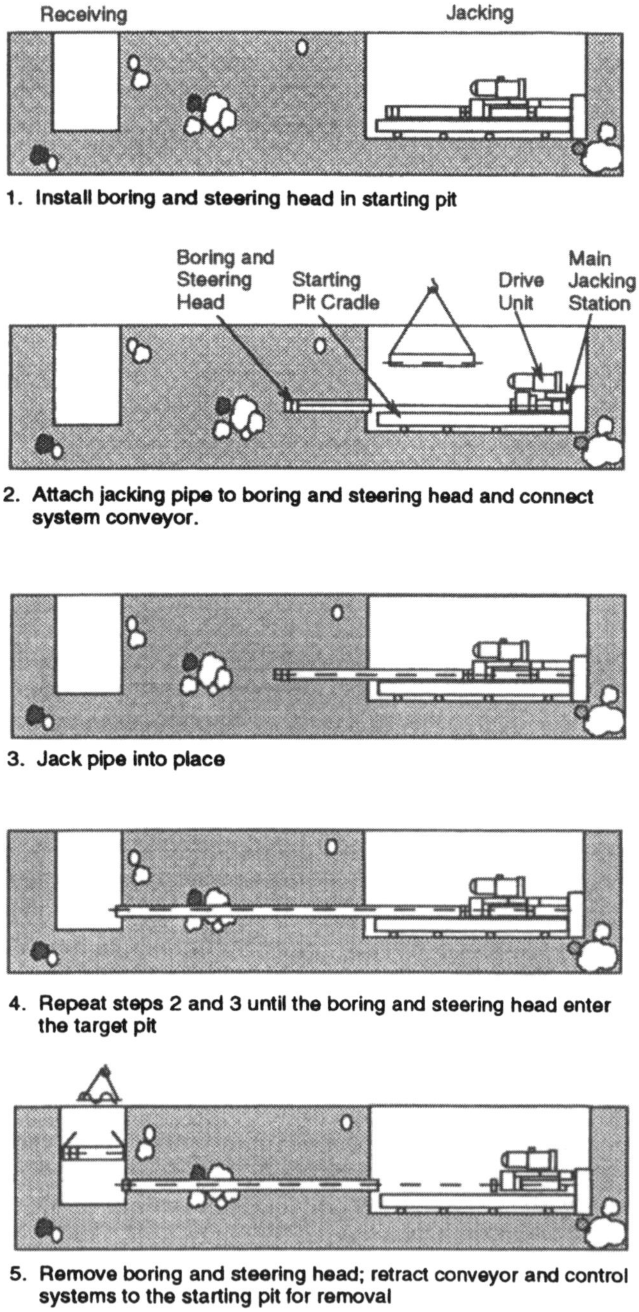

Fig. 16-1. Pipe jacking procedure.

Fig. 16-2. Jacking pit.

Fig. 16-3. The 45-in. diameter cutter head of a present-day microtunneling machine.

For microtunnels, the tunneling machine steering jacks are controlled remotely from the surface. The equipment and control house for the remote control microtunneling equipment are usually set up on ground level either directly over or adjacent to the starter shaft. All work is carried out from the container house and shaft, with a crew of only two or three people required for the entire operation.

The first length of pipe has its forward end protected by the shield or machine. It is installed in the tail of the shield at the bottom of the main jacking pit, with the jacks reacting against a thrust block. Excavation occurs at the tunnel face, with the pipe advanced at a rate compatible with the excavation rate. When the jacks reach their full extension, the jacks are retracted. A new length of pipe is lowered into the jacking pit and positioned on the guide rails. The main jacks are used to shove this pipe against the rear of preceding pipe, and the whole process is repeated. In many cases, the pipe used in the jacking process is the product pipe, producing a finished installation once the tunnel drive is complete. In other cases, a casing is jacked into place and the final product pipe sliplined into the casing.

The length of conventional tunnels is unlimited because the tunnel support is erected behind the tunnel shield and remains stationary. For the pipe jacking operations, the lining is moving along the entire tunnel length. To overcome the friction that builds up along the interface between the pipe and the ground, a mixture of bentonite and water is often introduced through injection holes along the tunnel or the tunnel is slightly oversized. Intermediate jacking stations also can be introduced within the pipe string to reduce the frictional distance over which the main jacks must push. With these intermediate stations in place, the order of advance of the pipe becomes caterpillar-like.

Tunnel spoil is normally removed by slurry or augers, depending on the size and type of tunneling machine. All electrical conduits, pressure hoses, water service, communication lines, and other utilities must be disconnected, extended, and reconnected as the pipe string advances.

When the leading edge reaches the receiving pit, the excavation equipment is removed. The jacked pipe is connected to the final shaft/pit/manhole lining. Contact cement grouting can be performed to fill voids if required along the tunnel length.

The maximum jacking distance from one starter shaft has been approximately 1,500 ft. Future developments should enable longer distances to be excavated from a single shaft. Jacking is usually maintained nonstop from the main jacking pit to the receiving pit to reduce the buildup of frictional resistance. Rates of progress vary, but a well-organized contractor can easily obtain 60–80 ft per 10-hour shift.

SITE INVESTIGATIONS

Tunnel diameter and obstructions are all a matter of comparable size. For large-diameter tunnels, such as transit, highway, and outfalls, the size of the obstructions such as boulders, existing utilities, and geological anomalies are small compared with the tunnel diameter. Their impact is minimal. The larger-diameter tunnels are unaffected by these obstructions. If the obstructions do prevent advance, the face of the tunnel can be accessed and the obstructions removed from the machine's path.

For pipe jacking and microtunneling, the size and scale of the operation is reduced to a point where the size of the obstruction does matter. The smaller microtunnels also cannot provide the access to the obstructions. Hitting an obstruction with a microtunnel can shut a project down. This is why a thorough geotechnical investigation is needed. Potential problems that must be identified include boulders, existing utilities, cemented zones, mixed-face conditions, loose sand lens, fat clay seams, and zones disturbed by past trenching. These conditions must be identified so the proper alignment, type of machine head, amount of overcut, use of pipe lubrication, and jacking forces can be established.

PITS AND SHAFTS

A starter pit (also known as a thrust pit) and a target pit (also known as a reception or receiving pit) are constructed to install a pipeline using pipe jacking or microtunneling techniques. These shafts/pits can be quite costly and are generally positioned for maximum usefulness. Two microtunnel drives from one starter pit is typical. Pits are usually located at manhole positions. The dimensions and construction method of the pit vary according to the specific requirements of the tunnel drive.

Pits are supported in a number of different ways depending on the size, depth, ground conditions, and groundwater conditions. For dry conditions in cohesive and cohesionless soils, the following support methods are used:

- Concrete segments
- Steel liner plate
- Sheet piling
- Secant piles
- Ring beam and lagging
- Trench sheets
- Precast caisson
- Casing
- Ground anchors with shotcrete
- Sloped excavation

For wet conditions in cohesive and cohesionless soils, the following support methods are used, sometimes in conjunction with dewatering:

- Concrete segments
- Steel liner plates
- Sheet piling
- Secant piles
- Grouting
- Precast caisson
- Casing

Caisson shaft construction with precast concrete ring segments is used extensively in Europe. It will be the method used in the United States in the coming years.

Microtunneling can be carried out from small shafts to meet special site circumstances. Eight-foot-diameter starter pits and 6-ft-diameter receiving pits are the minimum sizes. These size shafts are for machines that are 24 in. or less in diameter. Most jacking pits are typically 10–20 ft in diameter. The size is governed by site constraints and the length of the pipe to be jacked.

LEADING EDGE

A steel cutting edge, shield, or TBM must always precede the pipe string being advanced by jacking. The outside diameter of these shields or machines is usually a few inches greater than the outside diameter of the pipe. This overcut reduces the friction along the pipe string and allows the head of the tunnel room to maneuver. For short tunnels, such as railroad or street crossings in competent ground, the cutting edge is a short steel shield, and excavation is done by hand-held power tools. For longer tunnel lengths and in difficult ground, various types of excavators or microtunnel machines may be used. The selection of the appropriate equipment involves the same considerations as for any other type of tunneling operation.

Machines

In the United States, five manufacturers make or furnish microtunneling machines: American Augers, Herrenknecht, Iseki, McLaughlin Markham, and Soltau. Microtunneling machines and operation are supported by six systems. Each is described here.

Excavation System. The microtunneling machines are designed with a cutter face that is rotated by a drive motor. The face panel is configured to rotate clockwise or counterclockwise. The face panel is similar to the larger TBMs. Picks and slots are included to excavate clays, silts, and sands. Disk cutters or "strawberry" cutters can be included if rock or cemented zones are anticipated. In most cases, both face elements are included to allow the machine to cut through almost anything encountered.

Once excavated, the material is pushed into a chamber behind the face panel. The material is compacted and forms a plug. This plug provides an earth pressure that balances the material ahead of the machine. As the machine and pipe are advanced, the plug material is removed. The rate of this removal controls the earth pressure at the face. If the face pressure is less than the active earth pressure of the plug ($P_A > F$; i.e., material iflows into the chamber), settlement can occur. If the face pressure is greater than the active earth pressure and less than the passive pressure ($P_A > F > P_P$), settlement and heave can be controlled. And lastly, if the face pressure is greater than the passive earth pressure ($P_P > F$), surface heave can occur.

Two methods are used to control how fast the plug material is removed from the chamber. An auger system can be used. The system is ideal in dry conditions and for drive lengths less than 500 ft. The second method is a slurry system. Slurry is pumped to the chamber and mixed with the excavated material. The slurry/soil mix is pumped back to the shaft and up to the surface for separation. Slurry systems work well in wet conditions and for drive lengths greater than 500 ft. A decision tree is provided in Figure 16-4 to help determine the type of machine to use.

Some machines are equipped with a cone-shaped crusher that allows boulders 30% of the outside diameter of the machine's size to be excavated.

Alignment Control System. The alignment of small-diameter tunnels is controlled by either a steerable or a non-steerable system. Pipe jacking that uses a straight nonarticulated shield or the augered bore and jack setup are examples of nonsteerable systems. The alignment and grade for these systems are established by the initial placement of the cradle in the jacking pit.

For the steerable systems, the pipe jacking shield or microtunneling machine is articulated. Four jacks are placed

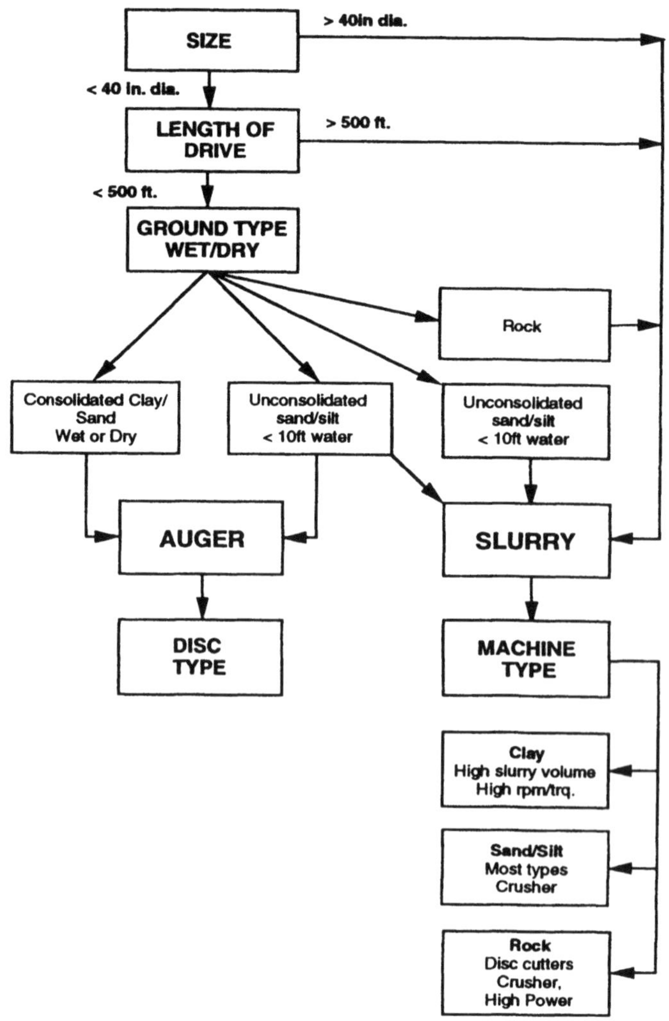

Fig. 16-4. Microtunneling machine selection process.

around the articulated joint. Controlling the four jacks allows the head of the machine to be directed. The steering capability is used to make minor course corrections to allow the tunnels to be driven at a precise line and grade. Actual curves in alignment are not readily possible, because curves impose large frictional forces on the pipe sections, which limit jacking distance. Curved alignments also prevent the pipe joints from being flush, which can be a disadvantage.

Line and grade control is monitored by a laser system using an illuminated target with graduated rings or squares located at the back end of the machine. The panel is observed by the use of television camera and/or computer screens. Other bubble indicators can be located on the back of the machine. The operator can use these indicators to monitor machine functions such as tilt-and-roll angle of the articulated head, lead angle variance of the pipeline from intended line, steering cylinder extensions, and water and bentonite flows.

Guidance System. An active method of controlling line and grade of the microtunnel during installation includes lasers in combination with closed-circuit televisions. Standard pipe lasers are designed to check line and grade of pipelines installed in sections during open-cut installation. The standard lasers do not have the optics or power to maintain their accuracy beyond a few hundred feet. They may be used for short drive lengths of less than 300 ft or so.

The most desirable type of laser for use in microtunneling is a long-range self-leveling laser. Its accuracy is high for distances up to 2,000 ft. Even for short drive lengths, the long-range lasers are preferred.

Guidance of the microtunnel depends on the laser setup and operation. A number of factors can adversely affect the laser:

- Temperature
- Dust
- Shaft construction
- Laser placement and setup

The laser should not be placed on the reaction back wall. Reaction of the propulsion jacks can cause deformation (1/4 in. is possible) of the back wall, causing the laser to deflect off line and grade.

Propulsion System. The objective of the whole process is to jack or push the pipe through the ground. To accomplish this, a jacking system is placed in the jacking pit. Jacking systems vary, but they typically consist of four jacks that react against the back wall of the pit. The wall of the pit must be designed to withstand the jacking forces generated by the face-pressure resistance of the machine and the skin friction along the pipe.

These force components depend on the ground type. In particular, the grounds' properties and the microtunneling system design contribute. These elements are

- Arching characteristics of ground
- Depth of the overburden
- Presences of surcharge loads
- Drive length
- Pipe diameter
- Advance rates and time off
- Overcut of the machine relative to the outside diameter of the pipe
- Use of pipe lubrication
- Use of intermediate jacking stations
- Consistency of pipe diameter and roundness
- Pipe material smoothness
- Alignment straightness

The main jacks are supplied with pressure fluid from a pump delivering pressures of from 1,000 to 7,500 psi. The jacks are usually in the 150- to 250-ton capacity range with a stroke of up to 11 ft (slightly greater than normal pipe length). The jack selection is based on judgment since the frictional resistance between the pipes and the ground is highly indeterminate. Some practitioners use an assumed frictional coefficient (e.g., 0.2) to apply to the vertical dead weight of soil plus pipe, mucking equipment, and other dead loads to obtain a first approximation of total jacking force and then multiply by a factor of 4 to 6 to obtain the total required jacking thrust. The resulting forces, if allowed to reach their maximums, will exceed the pipe compressive strength and are a constant concern during construction.

The number of jacks in the main jacking pit varies from a minimum of two to a maximum of six depending upon the size of pipe to be jacked and the anticipated force required. If spoil removal is by muck car or auger skip, it is desirable to arrange the jacks below the springline to permit easy access. This, however, introduces the possibility of eccentric pipe loadings; this may be somewhat overcome by the use of the thrust ring, which enables the jacking force to be distributed evenly around the periphery of the pipe. In addition, the ram head and foot are normally designed as ball-and-socket joints to withstand eccentric stresses.

The thrust from the jacks must ultimately be carried by the thrust wall, which is designed as a footing. Since this wall must be perpendicular to the direction of pipe movement, it must be installed carefully. Jacking forces may be distributed uniformly around the concrete face of this wall by a steel pressure plate.

There are two main factors to consider when deciding to use intermediate jacking stations. Insert intermediate jacking stations into the pipe string when the anticipated jacking load is greater than what the pipe has been designed to handle. Intermediate jacks are inserted at that point in the drive where the total expected push will exceed the available thrust of the main jacking frame. Other aspects to consider:

- Geotechnical conditions
- Use of bentonite and additives

- Accuracy of the pipe (trueness of the joints as they go together)
- Type of pipe
- Keeping the pipe on line and grade and straight

Intermediate jacking stations are required for long-distance runs from a single-run jacking pit. The intermediate station is composed of a steel can containing 6–34 equally spaced jacks around the circumference of the tunnel (see Figure 16-5). The capacity of these jacks is usually in the 10- to 100-ton range. Their stroke is on the order of 1–3 ft. These jacks use the main jacking station or the pipe section behind them for an abutment as they push the pipe between them and the shield or machine forward. When they are fully extended, the main jacks then push the rearward pipe forward while the intermediate jacks are retracted.

When the tunnel is completed, the intermediate jacks are removed and the steel can is left in place. If possible, the intermediate jacking station is then filled with concrete or grout to match the finished inside dimensions of the pipe.

There are no real restrictions on the total length of tunnel that can be constructed by the jacking method. On sewer projects, manhole spacings are usually sufficiently close that jacking can proceed from manhole shaft to manhole shaft without difficulty. For distances greater than say 1,200 ft, a trade-off study is needed to determine whether it is more economical to jack two shorter drives with a shaft or use intermediate jacking stations, higher forces, and lubrication to jack the longer distance.

Fig. 16-5. Intermediate jacking stations.

The microtunneling operation is usually maintained nonstop from the starter pit to the receiving pit. The friction from the ground is reduced by keeping the operations advancing. Jacking can be stopped for a shift or more, but the microtunnel machine should be rotated periodically in the counterclockwise direction to prevent the pipe string from freezing up.

Spoil Removal System. Auger systems are used in cohesive or cohesionless soils. In cohesionless soils, the auger system can effectively work in dry conditions to approximately 10 ft below the groundwater table. At depths below 10 ft, the cohesionless soil and water apply too much pressure, and the auger system has no way to counterbalance the forces. Augers can be used to install pipe as deep as 40 ft in clay. The auger system is typically restricted by length, travel along the auger, torque limitations, and the presence of groundwater.

Below the depths noted, the slurry systems work in more difficult ground types. The slurry system has the mechanize to counterbalance the earth pressures at the face and along the pipe. Larger-diameter slurry machines do need more settlement tanks for the slurry.

Pipe Lubrication System. The pipe jacking shields and microtunneling machines are designed to produce a hole diameter slightly larger than the trailing pipe diameter. This overcut is typically 3/4 in. To help reduce the jacking forces along the pipe, a lubricant can be injected into the annular space around the pipe. The lubricant can be clay-based or a polymer. Use of a lubricant reduces the jacking forces and allows for longer jacking lengths. Fluid losses can occur in the surrounding ground, depending on the soil type. This loss can cause ground collapse and squeezing onto the pipe. Some of the benefits of using a lubricant are eliminated. Use of a slurry-type machine will automatically lubricate the pipe.

JACKING PIPES

The pipe or casing pushed to form the lining falls into one of three categories. The pipe can be jacked as a single-pass lining where the pipe becomes the permanent lining. Pipe placement in the form of a double-pass lining uses a temporary casing, which is removed and replaced by a permanent pipe. The third approach involves using a casing. The initial tunnel lining is jacked into place and then the final pipe is sliplined inside the initial casing with the annular space filled. The double-pass lining system is rarely used, because the quality of the pipe used in the single-pass system is high enough. The casing system is mainly used for crossings under railroads. Installing casing is a requirement still imposed by railroad companies.

Pipes used in the jacking procedure are made of steel, ductile cast iron, reinforced concrete (RCP), reinforced glass-fibers (GRP), asbestos cement, vitrified clay (VCP), or polyvinyl chloride (PVC). The choice of material is influenced by diameter, drive length, ground conditions, and final use. The pipe must be rugged to resist the high stresses imposed by the jacking process, and their exteriors must be relatively smooth to minimize frictional resistance. Pipe with an outside diameter of 18 in. (up to 144 in.) has been successfully jacked. In the United States, however, pipe sizes manufactured off-site exceeding 108 in. in diameter can be difficult to transport with typical bridge clearances of 14 ft and highway lane widths of 12 ft. Asbestos cement pipe, while installed in the past, is not commonly used today.

The pipe used for jacking should be the highest strength available, with extra reinforcing steel placed in the shoulders of the pipe. In the United States, it is common practice to use bell-and-spigot pipe with joints sealed by O-ring gaskets. This type of pipe makes curved alignments of pipe jacked tunnels virtually impossible to build. European practice uses butt joints and flexible connections, which enables jacking to proceed around bends with a radii of as little as 650 ft.

It is common practice throughout the world to provide a layer of cushioning material between each pipe. This is usually composed of plywood or particle board, which is installed over the entire joint to assist in load distribution.

A discussion of the various types of pipe available for jacking applications follows.

Steel. The type of steel pipe installed by pipe jacking or microtunneling will vary depending on the applications. Steel pipe may serve as encasement for other carrier lines, or it may serve as a direct carrier pipe for water, gas, sanitary sewer, or other fluid products. A third application is as a load-bearing soil support system.

All steel pipe materials should be new. Standards developed by the American Society for Testing and Materials (ASTM), the American Water Works Association (AWWA), or the American Petroleum Institute (API) should be followed. Straight seam pipe is preferred, as seams can be points of weakness when jacked. Roundness, circumference, straightness, and pipe end treatment must meet the specified dimensional tolerances.

Pipe ends must be perpendicular to the longitudinal axis for transfer of uniform jacking forces. Steel pipe segments can be jointed by field butt welding, internal weld sleeves, or integral press-fit connections.

Exterior coatings provide corrosion protection and can help minimize skin friction buildup during installation. Coatings include epoxy-based polymer concrete, fusion bond epoxy, liquid epoxy, or other products that provide a hard, smooth surface. If steel pipe is field welded, a procedure must be established to repair the coating. Internal linings are shop-applied coatings such as liquid epoxy, polyurethane, cement mortar, or other materials.

Ductile Cast Iron. Centrifugally cast, ductile iron pipe is manufactured up to 64 in. in diameter for gravity and pressure applications. Ductile iron is used mostly in pressure applications. Applications include conveyance of sewage, wastewater, stormwater, treated water, and raw water. ASTM and AWWA standards should be followed.

Characteristics of pipe ends, segments, and coatings are the same as discussed for steel pipes.

Reinforced Concrete. Concrete is the most common material used as a primary lining for microtunnels. Problems with the reinforced concrete pipe (RCP) include the lack of standards governing pipe manufactured specifically for jacking. ASTM standards do exist for RCP. Many manufacturers exist throughout the United States. It is important to find a manufacturer that can make pipe meeting the specified dimensional tolerances for roundness, circumference, straightness, internal diameter, and pipe end treatment.

Reinforced concrete pipe is manufactured to meet specifications. It is important to use concrete with the appropriate axial compressive strength, require wall thicknesses sufficient to prevent cracking and spalling, and use and place reinforcement to allow the pipe to have the desired strength and ductility.

Pipe ends must be perpendicular to the longitudinal axis for uniform transfer of jacking forces. Concrete pipe for jacking applications uses two types of joints, concrete or concrete and steel. All concrete joints are typically good only for low jacking forces. Factors influencing the selection of the joint include anticipated jacking forces, joint deflection characteristics, and joint shear strength. Rubber gaskets or mastic can be provided to form a watertight seal.

Reinforced Glass Fiber. Glass-fiber reinforced pipe (GRP) is a very popular type of jacking pipe. GRP is produced in a centrifuge and is manufactured with a thick wall. The pipe is designed and manufactured with high strength, smooth inside and outside wall finish, and good corrosion resistance. Since it is manufactured in a centrifuge, the pipe is rounder, straighter, and more uniform in composition than is concrete pipe. The GRP weighs less than reinforced concrete pipe but costs about three times more. The joints are double spigot with a flush loose GRP sleeve to form the gasketed seal. Diameters ranging from 12 to 108 in. are available. Pipe lengths are a standard 20 ft. Shorter lengths are available but are machined from the 20-ft sections. Hobas USA in Houston, Texas, is the only manufacturer in the United States.

Vitrified Clay. Vitrified clay pipe (VCP) is a standard pipe material used for sewers. It can be manufactured with sufficient thickness to be jacked. The pipe is manufactured from fire clay, shale, and surface clay. The material is formed into pipe and fired to suitable temperatures to bake the clay. The resulting pipe has a high compressive strength. ASTM standards do exist for VCP. Pipe diameters ranging from 4 to 42 in. are available. Two manufacturers of VCP are located in the United States. The benefits of VCP are its resistance to absorption and chemical deterioration due to its smooth vitreous surface, its compressive strength for withstanding high jacking forces, and its economical production.

Polyvinyl Chloride. Polyvinyl chloride (PVC) pipe was successfully installed using an innovative microtunneling system. The system uses an internal sleeve to carry much of jacking force. The PVC pipe is generally weak and commonly carries low compressive loads. ASTM standards do exist. Roundness, circumference, and straightness must meet specified dimensional tolerances. Pipe diameters ranging from 21 to 48 in. are available.

APPLICATIONS

Most case histories involving pipe jacking are associated with crossing highways, railroads, runways, and other structures. Microtunneling, on the other hand, typically involves the installation of gravity-feed sewer lines.

Microtunneling is not limited to utility installation, however. New applications include crossings under active taxiways and runways, interstate highways, or gas plumes; driving intake structures from the middle of lakes; or establishing initial support for larger-scale tunnels. More inventive applications included underpinning buildings, placing pipes for ground freezing, installing monitoring instrumentation under landfills, crossing under wetlands, and installing an impermeable layer under a containment structure. The applications are endless. It is simply a matter of engineers beginning to think about how they can use pipe jacking and microtunneling to help solve some of their more challenging problems.

Water Conveyance Systems

Microtunneling is becoming the preferred method of designing and installing gravity water supply and sewer systems. Pipe jacking can be accomplished (i.e., graded) to allow the force of gravity to aid in maintaining water flow. In the event of a power outage, water can still be supplied by opening a valve along the pipeline accessible from the surface.

Seismic Lifeline

Tunnel design using microtunneling technology is one method available to minimize the effects of earthquakes. Fail-safe systems can be designed to remain intact even if a fault rupture occurs. Designs made to withstand earthquakes can realize great savings, especially in seismically active areas.

Environmentally Sensitive Areas

Projects that involve tunnels in environmentally sensitive areas require special design considerations. Microtunneling can help mitigate the environmental impacts of tunnel and pipeline construction, especially when compared with the adverse effects of trenching. Damage to valuable geological and archaeological deposits, as well as to wetlands, public parks, and the like, can be avoided. Fully automated, remote-controlled tunneling can also prevent personnel exposure to and handling of hazardous materials.

Initial Support for Larger Underground Openings

Pipe jacking methods have been used to excavate multiple elements under railways, roadways, and airport runways to form a "stacked drift" lining for a soft ground tunnel. In this application, the final tunnel excavation is performed under the protection of the previously installed "stacked drift" lining. This construction method was used on the MARTA system in Atlanta to permit passing twin single-track tunnels under I-285 freeway (see Figure 16-6).

Pipe jacked elements have been used in subway station construction in Tokyo to form a pipe roof over two parallel subway tubes in order to allow the removal of the soil between the tunnels and the construction of a subway platform.

A massive concrete-arch underpass for the Don Valley Parkway was continuously jacked through a Canadian Pacific Railway embankment in 12 days. The arch was cast in two segments separated by an intermediate jacking station on the thrust base. The forward portion was 40 ft long, and the rear section 65 ft long. Both portions were 36 ft wide by 27 ft high. The hydraulic jacking system used included 35 150-ton jacks at the intermediate jacking station and 40 similar jacks at the 3-ft-thick thrust base. Following each stroke of the jacks, workers excavated within the structure beneath the protection of the shield. This massive structure was jacked a total of 92 ft from its initial position without interruption to overhead rail traffic. The alignment was controlled within 3 in. by laser beam.

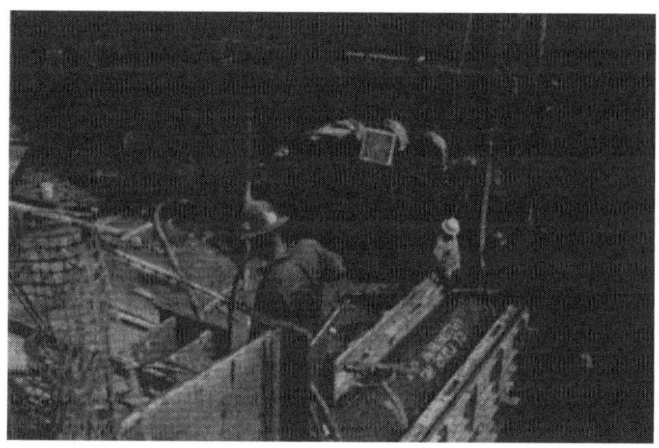

Fig. 16-6. A "stacked drift" lining.

REFERENCES

BOYCE, G.M. (1995) *Microtunneling Design Guidelines,* William Barclay Parsons Fellowship, Parsons Brinckerhoff, Monograph 9.

CASEY, E.F. (1974) "Objectives of UCRC Commission on mini-tunnels," North American Rapid Excavation and Tunneling Conference.

Concrete Pipe News (1958) "Large diameter pipe jacked into place at Memphis," November.

DRENNON, C.B. (1979) "Pipe jacking: State of the art," *Journal of the Construction Division:* ASCE September.

Engineering News Record (1985) "Jacked pipes form subway roof," April 18.

FANTON, D.L. (1973) "Mining and jacking for sewer tunnel construction," *Concrete Pipe News.*

GREEN, P. (1989) "Jacks push tunnel into place," *Engineering News Record,* November 16.

HULL, J.H. (1969) "Jacked-in-place tunnel triumphs," *Concrete Pipe News,* April.

KAWATA, H., VENO, T., and YAMAZAKI, J. (1979) "Effective new tunneling method under railways," North American Rapid Excavation and Tunneling Conference.

The Metropolitan Sanitary District of Greater Chicago (1978) Tunnel Construction Reports 1968 to 1976, Chicago.

MIYOSHI, M., MIYAZAKI, H., and FUJITA, K. (1979) "Shield-driven tunnel enlargement by pipe-roof method for subway station construction," North American Rapid Excavation and Tunneling Conference.

National Utility Contractors Association (1981) "Horizontal earth boring and pipe jacking manual with related job safety."

NICHOLAS, P., and ROBINSON, A. (1988) "Microtunneling in Yorkshire," *World Tunneling,* June.

RHODES, J., and HAGAN, R. (1969) "Jacking sewer pipe in tunnel," *Concrete Pipe News,* April.

THOMSON, J. (1993) *Pipejacking and Microtunneling,* Chapman and Hall, New York.

SULINSKI, S.J. (1972) "The jacking method in tunnel construction," *Proceedings North American Rapid Excavation and Tunneling Conference,* June 5–7.

WASHBOURN, J. (1982) "Sliding resistance of jacked tunnel lining," *Tunnels and Tunneling,* November.

17 | Cut-and-Cover Tunnel Structures

James L. Wilton
Engineering Consultant, Chairman (Retired),
Jacobs Associates

INTRODUCTION

Shallow-depth tunnels, such as large sewer tunnels, vehicular tunnels, and rapid transit tunnels, are frequently designed as structures to be constructed using the cut-and-cover method. Tunnel construction is characterized as "cut-and-cover" construction when the tunnel structure is constructed in a braced, trench-type excavation ("cut") and is subsequently backfilled ("covered"). For depths up to 35–45 ft this method is often cheaper and more practical than underground tunneling, and depths of 60 ft or more are quite common for rapid transit cuts. This chapter discusses the design and construction of the larger cast-in-place concrete structures used as sewer tunnels or transportation tunnels for pedestrian, vehicular, or rapid transit traffic. The tunnel is typically designed as a box-shaped frame, and due to the limited space available in urban areas, it is usually constructed within a braced excavation. Where adequate space is available, such as in open areas beyond urban development, it is often more economical to use open-cut construction. Where the tunnel alignment is beneath a city street, cut-and-cover construction interferes with traffic and other activities. This disruption is lessened through the use of decking over the excavation, placed immediately following removal of the first lift of excavation. The deck is left in place with construction proceeding below it until the stage is reached for final backfilling and surface restoration. Figures 17-1–17-4 show cross sections of the more common types of cut-and-cover tunnel structures.

In this chapter, discussions or characterizations of usual practice in the design and construction of cut-and-cover tunnel structures refer generally to practice in the United States. The design and construction of these structures in Canada, Mexico, Europe, Asia, and elsewhere abroad is similar in many respects, but it can differ in many respects as well.

Subway Line Structures

In cut-and-cover construction between stations, the subway tracks are usually enclosed in a reinforced concrete double box structure with a supporting center wall or beam with columns. These tunnel structures are commonly referred to as "line" structures. The track centers are normally located as close together as possible. In a typical double box section, each trainway will have a clear width of about 14–15 ft, depending upon width of vehicle and clearances to be provided for equipment and manways (see Figure 17-1). The configuration of subway line structures can depart from the typical section to accommodate atypical track alignment or grade. When, for example, system standards mandate that stations be designed with a center platform (see below), the track centers will need to be widened through an appropriate transition length upon approaching the station. Occasionally, it will be advantageous or necessary to gradually change the alignment and grade to the "over and under" position, in which one track lies above and in line with the other. In general, the configuration of the subway line structure must be subordinate to system requirements for track alignment and grade.

Subway Stations

Station structures include the trainway for trains, boarding and off-boarding platforms, stairs, escalators, concourse areas for fare collection, and service rooms. If the line structure is two circular tunnels constructed by tunneling methods, the station may be designed with a single center platform. If the line structure is a cut-and-cover double box subway line structure, the station is usually of the side platform type, except at terminals where center platforms are normally used to comply with system standards.

The usual perception of a cut-and-cover subway station is that of a two- or three-story reinforced concrete structure constructed in a rectangular excavation 50–65 ft wide, 500–800 ft long, and 50–65 ft deep. Typically, the station is a two-story structure with two tracks, as well as the center or side platforms, supported on the invert. A mezzanine floor typically lies between the roof slab and the invert/platform level. Figure 17-2 shows a basic cross section at the 7th/Flower Sta-

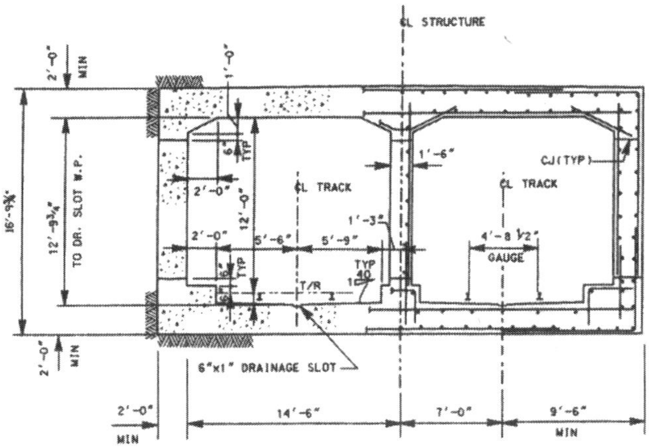

Fig. 17-1. Standard double box section, WMATA.

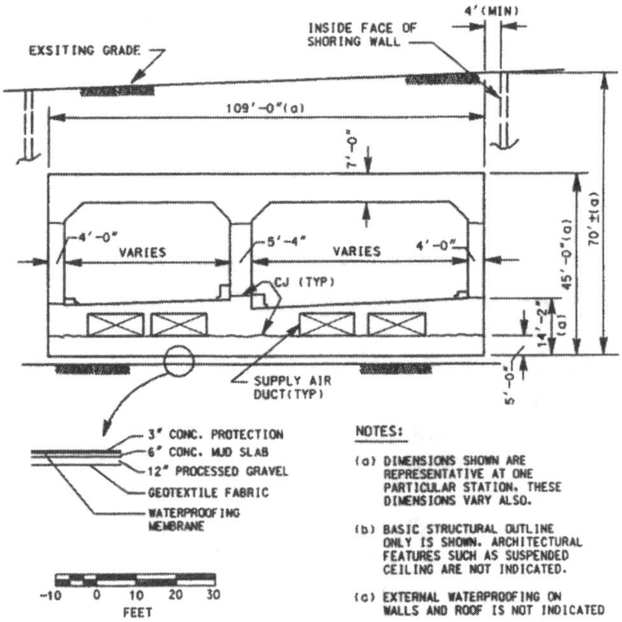

Fig. 17-3. Cross section (varies), double box configuration, Central Artery, Boston.

tion, constructed for the Los Angeles County Metropolitan Transportation Authority (MTA).

The complete subway station will be much more complex than is indicated by Figure 17-2. Internal configuration will be significantly affected by the need to provide escalators, stairs, ventilation requirements, rooms for mechanical and electrical equipment and other maintenance, and safety and service facilities. Architectural treatment of the station will also affect internal configuration and may have an important effect on the design of the basic structure as well. The external configuration of the station will ordinarily be irregular at and above the mezzanine level, where station entrances must be provided.

Although the two-story station may be considered conventional, more complex station structures are not unusual, particularly in and near the center of urban areas. At these locations the subway stations may be constructed at the intersections of principal system trainway lines, for example.

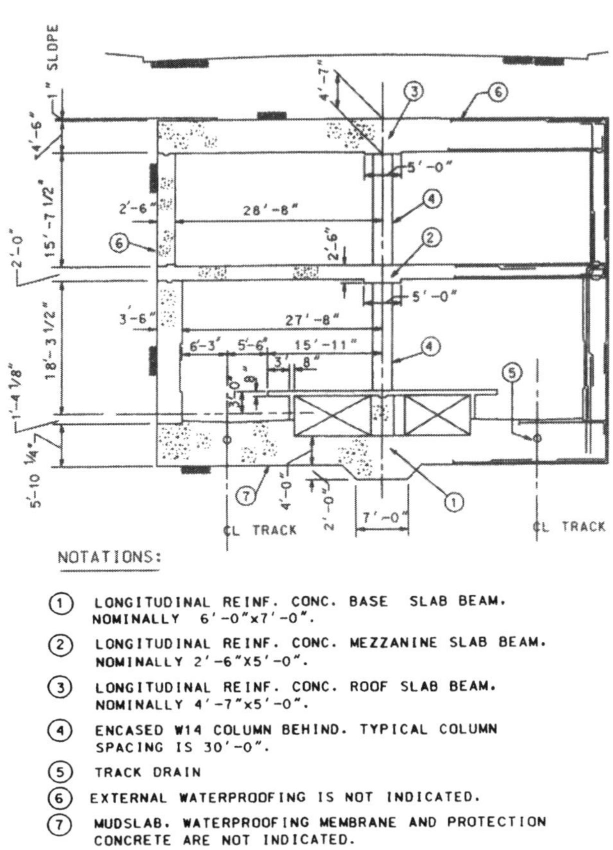

Fig. 17-2. Basic cross section, 7th/Flower Station, Los Angeles Co. MTA.

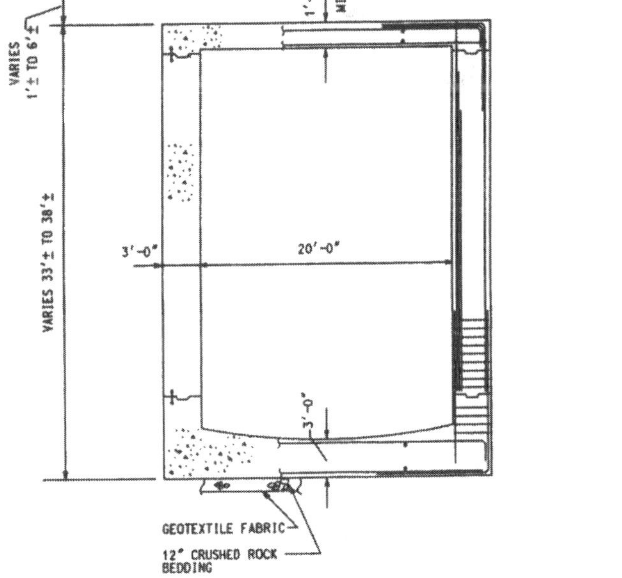

Fig. 17-4. Typical cross section, 20 ft transport/storage sewer tunnel, San Francisco Clean Water Program.

In such cases boarding and off-boarding platforms in two directions are normally required, so that in plan view the station structure is constructed in the shape of a cross, with, usually, more than two stories. Whenever there is a requirement that the subway station accommodate more than two tracks, the station structure will ordinarily be significantly more complex than the conventional subway station.

Subsurface Highway Structures

Subsurface highway structures of significant length, and constructed by the cut-and-cover method, are not common. Because of the high cost of the construction of these vehicular tunnels, the initial and operating cost of ventilation, and cost and practical problems of access and egress, surface and elevated highways are usually much more feasible. There are, however, occasions when a cut-and-cover vehicular tunnel is the most feasible or desirable alternative for a proposed section of a highway.

Cut-and-cover vehicular tunnels are probably most commonly constructed at the approaches to subaqueous vehicular tubes, due to the depth required for the highway structure to serve in this capacity. In this application they are usually under the groundwater table and, due as well to their width, they are typically very massive reinforced concrete structures. Figure 17-3 shows a typical cross section of the Central Artery (I-90) Tunnel as it approaches the I-90 Immersed Tube Tunnel in Boston.

Cut-and-cover vehicular tunnels can be appropriate as well where there is a compelling reason to restore the highway right of way to its original condition or to provide for other productive use of the surface property. Cut-and-cover vehicular tunnels have been constructed under major urban airport runways, for example. A significant portion of the I-90 freeway on Mercer Island in Washington State was constructed as a cut-and-cover vehicular tunnel.

Sewer Structures

The cut-and-cover sewer tunnels referred to or discussed in this chapter are large reinforced concrete box structures used for the transport and/or storage of wastewater. These structures are larger in area than the largest reinforced concrete pipe or precast concrete cylinder pipe commercially available. The larger of the transport/storage tunnels constructed for the San Francisco Clean Water Program are internally 20 ft wide and up to 33 ft or more high. (See Figure 17-4.)

TUNNEL DESIGN—STRUCTURAL

Loading—General

The cut-and-cover structure must be designed to have structural capacity sufficient to resist safely all loads and influences that may be expected over the life of the structure. The principal loads to be resisted are ordinarily the long-term development of water and earth pressures, dead load including the weight of earth cover, surface surcharge load, and live load.

All loads or potential loads should be categorized similarly to the load categories specified by the American Association of State Highway and Transportation Officials (AASHTO). The load categories should represent the requirements of the particular cut-and-cover structure under consideration. For example, the Washington Metropolitan Area Transit Authority (WMATA) specifies that rapid transit structures must be proportioned for the following loads and forces when they exist:

- Dead Load (DL)
- Live Load (LL)
- Impact (I)
- Centrifugal Force (CF)
- Rolling Force (RF)
- Longitudinal Braking and Tractive Force (LF)
- Horizontal Earth Pressure (E)
- Buoyancy (B)
- Flood (FL)
- Shrinkage Force (S)
- Thermal Force (T)

In locations where there is a potential for significant seismic activity, earthquake forces (EQ) must be added.

Loads and forces that must be considered in formulating design criteria for vehicular cut-and-cover tunnels typically are similar. The loads and forces identified and tabulated in AASHTO Standard Specifications are ordinarily appropriate for this purpose.

Loads and forces applicable to cut-and-cover sewer tunnels should be similarly identified. In terms of dead load, forces due to horizontal earth pressure and other applicable loads and forces are normally similar to those of rapid transit tunnels, except that the internal live load consists of the loads and pressures imparted to the tunnel structure by the contents of the sewer. In some cases, the hydraulic grade line may be above the roof of the cut-and-cover sewer, and the corresponding internal pressures must be properly evaluated. In the latter case, transient forces may also need to be considered.

Dead Load. The dead load to be considered for the design of cut-and-cover structures normally consists of the weight of the basic structure, the weight of secondary elements permanently supported by the structure, and the weight of the earth cover supported by the roof of the structure and acting as a simple gravity load. The design unit weight for the earth cover should not be taken as less than 120 pcf for dry fill or less than 130 pcf for moist fill. Some authorities specify a minimum design unit weight of 130 pcf for the earth cover both above and below the groundwater table.

Shallow structures of the size and width of subways or vehicular tunnels may impose undue restrictions upon future loadings over the tunnel if the structure is designed only for the actual earth cover. For example, structures that pass under city streets must be designed to permit special vehicle loadings in excess of normal axle loads, such as trucks moving transformers, boilers, vaults, or other large structural loads. It is often desirable in developing air rights in off-street areas that they be capable of directly supporting moderate building loads without distress. For these reasons it is common practice that the structure be designed for a minimum vertical load equivalent to 8 ft of earth cover, regardless of actual cover. Dead load should be considered as applied in stages representative of all conditions likely to be encountered during the life of the structure. For example, where removal of earth cover is a foreseeable possibility, and where such removal could affect design requirements, a separate load case addressing this condition should be analyzed.

Live Load, Impact, and Other Dynamic Forces. Subway vehicle loads are dependent on the passenger and maintenance vehicles operating within the system. Standard vehicle loadings (LL) used in San Francisco, Los Angeles, Toronto, and Washington, D.C., are shown in Figure 17-5. In addition, loads due to impact, centrifugal force, rolling force, and longitudinal braking and tractive force must be accounted for. Impact loads may be defined as statically equivalent dynamic loads resulting from vertical acceleration of live loads. For subway tunnels the impact allowance (I) is ordinarily formulated by appropriate modification of the expressions specified by either AASHTO or the American Railway Engineering Association (AREA), or both. For preliminary design purposes this allowance may be taken as 30% of the static vehicle loading. Centrifugal force (CF) can be calculated theoretically as a function of static subway vehicle loading, design speed, and degree of track curvature. A resulting expression for centrifugal force is given by both AASHTO and AREA and is used for subway tunnel design. For subway design CF is assumed applied horizontally at a distance representative of the location of the center of gravity above the top of rail (usually about 5 ft). Rolling force

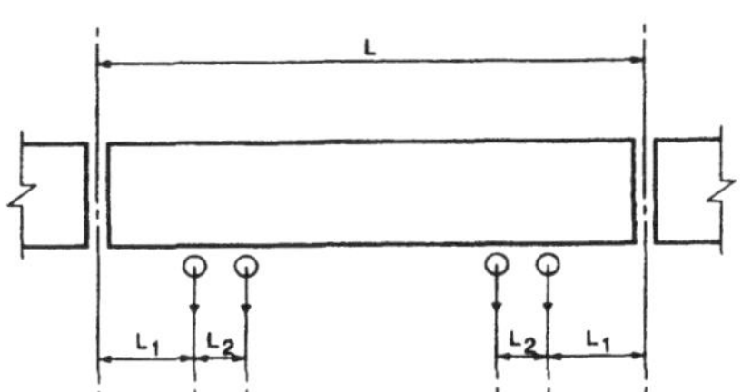

	San Francisco	Washington	Toronto	Los Angeles
L	70'-0	75'-0	75'-0	75'-0
L_1	5'-9	8'-3	8'-0	7'-1
L_2	8'-0	6'-6	5'-0	6'-10
Total Vehicle Weight	50 TONS	60 TONS	70 TONS	64.5 TONS
Axle Load	25 KIPS	30 KIPS	35 KIPS	35.3 KIPS

Fig. 17-5. Subway vehicle loadings.

(RF) is usually considered to be 10% of static subway vehicle loading applied downward on one rail and upward on the other. Longitudinal breaking and tractive force (LF) is usually considered to be a longitudinal force equal to 15% of static vehicle loading and applied at the center of gravity of the subway vehicle. For double-track structures, LL, I, CF, RF, and LF forces must be considered as applicable to either one or both tracks, and all foreseeable combinations of these forces, in terms of application and direction, must be analyzed.

Live loads may include pedestrian loading as well as subway vehicle loads. However, in a tunnel one-story high, pedestrian live load, subway vehicle live load, impact load, and other dynamic loads (CF, RF, LF) are ordinarily transmitted through the invert slab directly to the supporting ground and, as a consequence, have little or no effect on the proportioning of the structural elements of the tunnel. These loads will normally affect the basic structural design only if the tunnel is two or more stories in height.

Cut-and-cover subway structures must also be designed to support surface traffic loading or other live loading. Usual practice is to base roadway live loads on AASHTO HS 20-44 loading. For structures below a depth of 8 ft, a uniform live load of 300 psf is commonly used. For structures having less than 8 ft of earth cover, common practice is to design the roof of the subway structure for the more severe of the following two conditions:

1. Actual depth of cover (DL) plus superimposed HS20-44 wheel load distribution (LL) is accordance with AASHTO requirements
2. An assumed future cover of 8 ft (DL) plus a uniform live load of 300 psf

AASHTO requirements for impact due to roadway live loads were interpreted by WMATA as follows:

Up to 1-ft, 0-in. earth cover	I = 30% LL
1-ft, 1-in. to 2-ft., 0-in. earth cover	I = 20% LL
2-ft, 1-in. to 3-ft., 0-in. earth cover	I = 10% LL
Greater than 3-ft, 0-in. earth cover	I = 0% LL

For the most part, live load, impact, and other dynamic forces imparted to cut-and-cover vehicular tunnels are similar to the corresponding loads and forces imparted to cut-and-cover subway structures. These loads and forces, and their application, typically conform to or exceed the requirements contained in the AASHTO specifications for HS20-44 loading.

The roof, walls, invert or base slab, and other elements of cut-and-cover sewer structures must be designed for all foreseeable net internal pressures that can be imparted to these elements by the fluid contents of the structure. Net internal pressure must be considered to be internal pressure at any particular operating condition less minimum reliable external pressure due to retained or supported earth over the life of the structure. Maximum internal pressure must include transient pressure, if any. Net internal pressure is then treated as internal live load. (Internal live load is often zero on the wall and roof elements of the structure.) In addition to internal live load, live load and impact due to surface traffic loading must also be considered. Loading criteria for the combination of earth cover plus live surface traffic is commonly considered less for shallow cut-and-cover sewer structures than it is for otherwise similar subway or vehicular structures. It is recommended, however, that the external roof load on shallow-roof cut-and-cover sewer tunnels be not less than the weight of the earth cover (DL) plus AASHTO HS20-44 live load plus impact.

Horizontal Earth Pressure. Horizontal earth pressure (E) may be considered in this chapter to be lateral pressure due to both retained soil and retained water in soil when water is present. Horizontal earth pressure may include the effect of surcharge loading resulting from adjacent building foundation loading, surface traffic loading, or other surface live loading. All of these components of E must be evaluated both in terms of present conditions and future conditions. This admonition applies particularly to groundwater levels. Where future changes could adversely affect the subsurface structure, needed protective measures to mitigate the adverse effects might not be foreseen and might be extremely costly to add to an existing structure.

The soil component of E depends on the physical properties of the original ground through which the cut is being taken. For design and construction purposes, these physical properties are typically determined by an experienced geotechnical consultant retained to perform a comprehensive geotechnical investigation of subsurface conditions (see Chapter 4). These data, as well as recommended diagrams showing long-term and short-term horizontal earth pressures, are normally contained in a site-specific geotechnical report.

Table 17-1 is a guide showing how different soils may affect horizontal earth pressure on the structure. However, any attempt to categorize the behavior of retained soils must be viewed with caution. Stiff, overconsolidated fissured clays, for example, can impart high lateral pressures to the walls of the subsurface structure. Many subsurface structures will be in mixed or stratified soils, which cannot be represented in a guide such as Table 17-1. Also, many of these structures will be under the groundwater table. Below the groundwater table, lateral pressure due to retained soil is or may be considered to be a function of vertical effective stress in the soil. As a result, the soil component of E may be small compared with the total horizontal pressure due to both retained soil and retained water. Finally, in the modern practice of formulating lateral pressure diagrams for use as criteria for analysis and design, recommended maximum horizontal pressure is often equal to or approaching full vertical stress (i.e., the vertical effective stress plus water pressure), even for soils normally regarded as competent.

The short-term and long-term changes in horizontal earth pressure must be considered. During the life of the tunnel there may be substantial changes to this loading. Immediately following construction, the actual short-term earth pressure may be considerably less than long-term design pressure. In the event of a future excavation parallel and adjacent to the tunnel, unbalanced lateral pressures may occur with a pressure equal to long-term pressure applied to one side and a lesser pressure applied to the other. For these reasons, cut-and-cover tunnels should be designed for both short-term and long-term loading. There are differing opinions on whether or not the structure should be considered restrained against horizontal translation or proportioned for stresses resulting from side sway caused by unbalanced horizontal pressures. The approach taken may also depend on local requirements. A common recommendation for subway structures is that the tunnel be proportioned for side sway if it is a single story structure, and in the case of two or more stories, side sway should be considered in the upper story only, with the lower stories assumed to be restrained against horizontal translation. These assumptions should provide a competent factor of safety against a future mishap resulting from adjacent construction, and it will normally be found in proportioning the structural elements of a cast-in-place concrete box tunnel that they are not unduly increased in size through consideration of this loading. Where construction bulkhead walls are incorporated into the permanent structure, the design of the connections between walls and roof or invert slabs is governed by the unbalanced lateral load condition, and a more detailed geotechnical evaluation is appropriate.

The assumed magnitude of long-term and short-term horizontal earth pressure on subsurface tunnel structures has varied considerably with public authorities. These values have depended in part on the type of tunnel, the location of the tunnel, and other local factors, as well as the physical properties of the soil. However, maximum horizontal earth pressure (soil component) should never be taken as less than vertical effective stress multiplied by an appropriate at-rest (K_0) coefficient.

Similarly, the criteria to be used to account for unbalanced horizontal earth pressure have varied. In the design of the Toronto Subway, where groundwater lay below the tunnel structures, short-term reduced loading resulting from future adjacent construction was determined by multiplying the design value of horizontal earth pressure by a reduction

Table 17-1 Effect of Soil Type on Lateral Pressure (Generalized)

Soil Type	N Value blows/foot	Characteristics	Lateral Pressure
Dense sand	Greater than 30	Difficult to drive a 2 × 4 stake with a sledge hammer	Low
Loose-to-medium sand	Less than 30		Moderate
Very stiff clay or silt	Greater than 16	Can be indented by a thumb nail	Moderate
Medium-to-stiff clay	Less than 16	Can be indented by a thumb	Moderate-to-high
Soft clay	Less than 4		High

factor of 0.5. For the design of the WMATA subway, unbalanced horizontal earth pressure was defined as long-term horizontal earth pressure applied to one side of the tunnel structure opposed by short-term horizontal earth pressure applied to the other.

Diagrams from which maximum and minimum earth and water pressures were computed, for use as loading criteria in the design of cut-and-cover subway structures by WMATA, are presented in Figure 17-6. Both long-term and short-term loadings are shown. These generalized diagrams were applicable to a wide range of noncohesive and cohesive soil types to be found in the greater Washington, D.C., area. The soil component of maximum horizontal earth pressure was taken as equal to effective vertical earth pressure, to account for the effect of continuing vibration of the subway and the outward deflection of the structure.

Experience in or with the design, construction, and performance of vehicular cut-and-cover tunnels is not comparable with that for rapid transit systems. However, engineering practice with respect to horizontal earth pressure loading on vehicular cut-and-cover tunnels should clearly be appropriately similar to that for rapid transit tunnels.

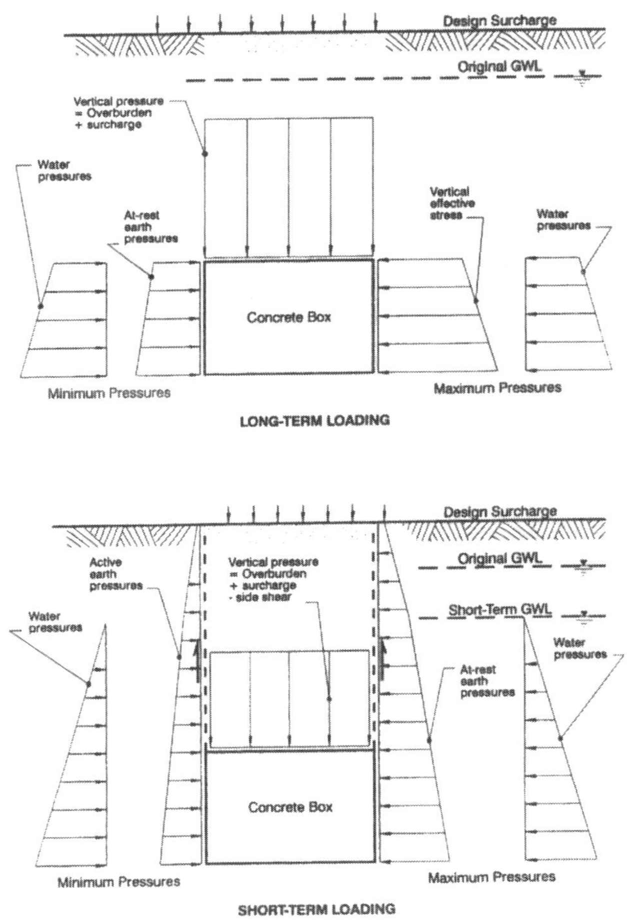

Fig. 17-6. Loading diagrams for the design of concrete box structure, WMATA.

A somewhat less conservative approach is normally taken in evaluating long-term horizontal earth pressure applicable in the design of cut-and-cover sewer structures. The reinforced concrete box structures constructed as part of the San Francisco Clean Water Program, for example, were designed for a long-term horizontal earth pressure equal to at-rest pressure plus hydrostatic pressure plus lateral pressure due to surface traffic surcharge. For noncohesive soils and other strong soils, at-rest pressure was computed with the coefficient, K_0, equal to $1 - \sin \phi$, where ϕ equals the effective stress friction angle. For saturated, soft to medium clays and silty clays (bay mud), K_0 was determined by laboratory test and analysis to be on the order of 0.55 (when applied to effective vertical stress). Additional horizontal pressure due to surface traffic surcharge was taken as a nominal, uniformly applied pressure of 150 psf.

Buoyancy. When the groundwater table lies above the bottom of the invert or base slab of a subsurface structure, an upward pressure on the bottom of the base slab, equal to the piezometric head at that level, must be accounted for. For a rectangular box, this upward pressure multiplied by the width of the base slab is the buoyant force (B) per lineal ft of structure. When the reliable minimum weight of the structure plus the fill above the structure (DL min.) exceeds B by an adequate factor of safety (FS), the structure is considered stable against uplift due to B. Opinions differ somewhat on an appropriate value for FS. For the design of the San Francisco Bay Area Rapid Transit System (BART), an FS $\geq$ 1.1 was considered satisfactory when side friction was neglected. Side friction between the walls of the structure and the retained soil offers additional resistance to B, but this resistance, except for its influence on the selection of FS, is normally neglected.

When B exceeds DL min., other resisting features must be incorporated into the design. Some of the features that may be provided for this purpose are the following:

- The weight of the structure may be increased by thickening the walls, roof, or base slab. Also, the base slab may be widened to increase the weight of earth resistance.
- Tension piles designed to provide a tensile force on the base slab can be provided. Both steel piles and prestressed concrete piles have been used in this application.
- Tie-down anchors, resembling permanent tie-back anchors, can be provided. Drilling for the anchors is accomplished at some convenient time after the base slab is placed. The anchor heads are located in formed recesses in the base slab. After completion of the tie-down installation, the recess is filled with concrete. The type of anchor used will depend in part on whether the anchor can be founded in bedrock beneath the structure or in competent soil.

Flood (FL). Where there is a potential for river floods or other flooding that could add loads to subsurface structures, the design of the structures should allow for this loading as required by the particular type of structure and the conditions affecting each location.

Shrinkage and Thermal Forces. Between transverse joints in cut-and-cover tunnel structures constructed of reinforced concrete, shrinkage forces (S) and thermal forces (T) are accounted for by the longitudinal reinforcement in the walls, roof, and invert slab. The stresses produced by these forces are typically normal to the principal stresses caused by DL, LL, and I and, therefore, do not enter into frame analysis of the structure. (See further discussions below.)

Earthquake Forces. Major codes that address the seismic design of surface structures in the United States contain no provisions for underground structures. The general view is that underground structures are much less affected by seismic motion than are surface structures. Although this view is substantiated by limited observations of the performance of underground structures during seismic events, some severe damage has been reported. In areas identified as subject to significant seismic activity, it is therefore necessary to determine the extent to which earthquake forces (EQ) should be considered in the design. In making this determination, the importance of the structure, the consequences of damage due EQ, the type of soil in which the structure is founded and the potential for liquefaction, and the location of potentially active faults at or in the vicinity of the site should all be considered.

The geotechnical data needed to assess earthquake forces and risk should be furnished by a geotechnical consultant experienced in this field. In addition to these data, the geotechnical consultant should, where appropriate, develop design values for the magnitude and distribution of EQ. Most such evaluations of EQ have resulted in an assumed lateral pressure to be superimposed on the horizontal earth pressure, E, on one side of the structure. Thus, a representation of both the increase in lateral pressure due to retained soil and the racking effect on the structure during earthquakes is provided for design purposes. Where evaluations of EQ have been made, it has to date been common to find that the resulting loading condition is less severe than other load cases without EQ. The effect of EQ on the proportioning of the structure can also often be considered negligible because of the stiffness of the soils in which the structure is founded, and because of increased allowable stresses in the elements making up the structure when EQ is added to other loads. The analysis of underground structures for earthqukes is discussed in Chapter 6.

Load Combinations and Unit Stresses. Except for earthquake forces, permitting allowable stresses to exceed basic unit stresses for particular combinations or groupings of the loadings identified in this chapter does not apply to the design of cut-and-cover tunnel structures. When EQ is omitted from consideration, any combination of these loads that can exist simultaneously should not produce stresses exceeding basic unit stress. Basic unit stress is considered to be allowable unit stress in each of the structural materials comprising the structure, as specified in applicable codes.

When earthquake forces are being considered, an increase in allowable stress is permitted. Usual practice for reinforced concrete design, using service-load methodology, is to permit 133% of basic unit stress for the load combination of DL + E + B + FL + EQ. Using load factor methodology, the comparable increase in stress is represented by the reduced design strength permitted in applicable codes.

Loading Cases

Cut-and-cover tunnel structures retain earth but are not free to yield significantly. Nevertheless, there is a need to consider unbalanced and other atypical loading on these structures. The particular load cases to be analyzed will depend on the type of structure, its location, the type of ground in which the structure is founded, the location of the groundwater table, and other local factors. All reasonably foreseeable temporary and permanent loading cases that would affect the design of the structure should be investigated. The system design criteria developed by WMATA for reinforced box and station sections specify that, as a minimum, the following three basic loading cases be investigated:

Case I: Full vertical and long-term horizontal load, as recommended by the General Soils Consultant.

Case II: Full vertical load, long-term horizontal load on one side, and short-term horizontal load on the other side, as recommended by the General Soils Consultant. In underground box structures potentially subject to unequal lateral pressures, the structural analysis shall consider the top slab as both restrained and unrestrained against horizontal translation in arriving at maximum shears, thrusts, and moments.

Case III: Full vertical load with short-term horizontal load neglecting groundwater pressure on both sides, as recommended by the General Soils Consultant.

The system design criteria adopted by the Los Angeles MTA is the same as above except that one additional case is added:

Case IV: Only dead vertical load with long-term horizontal load including hydrostatic pressures.

Both WMATA and MTA specify that for stress analysis, variations in the elastic support of the subgrade shall be considered for the different loading cases as appropriate.

A similar approach to the establishment of loading cases as criteria for design should be taken in designing cut-and-cover vehicular tunnels, and, ordinarily to a less severe extent, sewer tunnels. For the design of large reinforced concrete box structures constructed as part of the San Francisco Clean Water Program, the following loading cases, with the box empty, were established:

Case I:	Vertical	DL + LL
	Horizontal, both sides	E with traffic surcharge
Case II:	Vertical	DL + LL
	Horizontal	E without traffic surcharge
Case III:	Vertical	DL only
	Horizontal	E with traffic surcharge
Case IV:	Vertical	DL + LL
	Horizontal	No load

In the above sewer tunnel case, LL was assumed to be AASHTO HS20-44 traffic loading, DL was assumed to be the weight of the fill above the structure plus the weight of the structure as applicable, and horizontal forces were as previously given herein. Also, with the box full of wastewater, the hydraulic grade line lay below the permanent groundwater level, so that internal live load essentially did not affect the structural design.

Frame Analysis

In the analysis of the structural frame of the tunnel, loads and pressures representing each loading case are applied, and the shears, thrusts, and bending moments for each element of the frame are (in most cases) determined through rigid frame analysis using commonly accepted methodology. In modern practice this methodology is contained in specific computer programs for structural analysis. Except for particularly wide invert spans, usual practice is to assume that vertical reactions are uniformly distributed over the bottom of the invert slab. This assumption results in maximum slab bending moments and is therefore conservative. Figure 17-7 contains illustrations of the three cited WMATA loading cases applied to a typical WMATA reinforced concrete subway line structure and the configuration of each resulting bending moment diagram.

Frame analysis of reinforced concrete is complicated by its nonlinear nature, and more importantly by the fact that

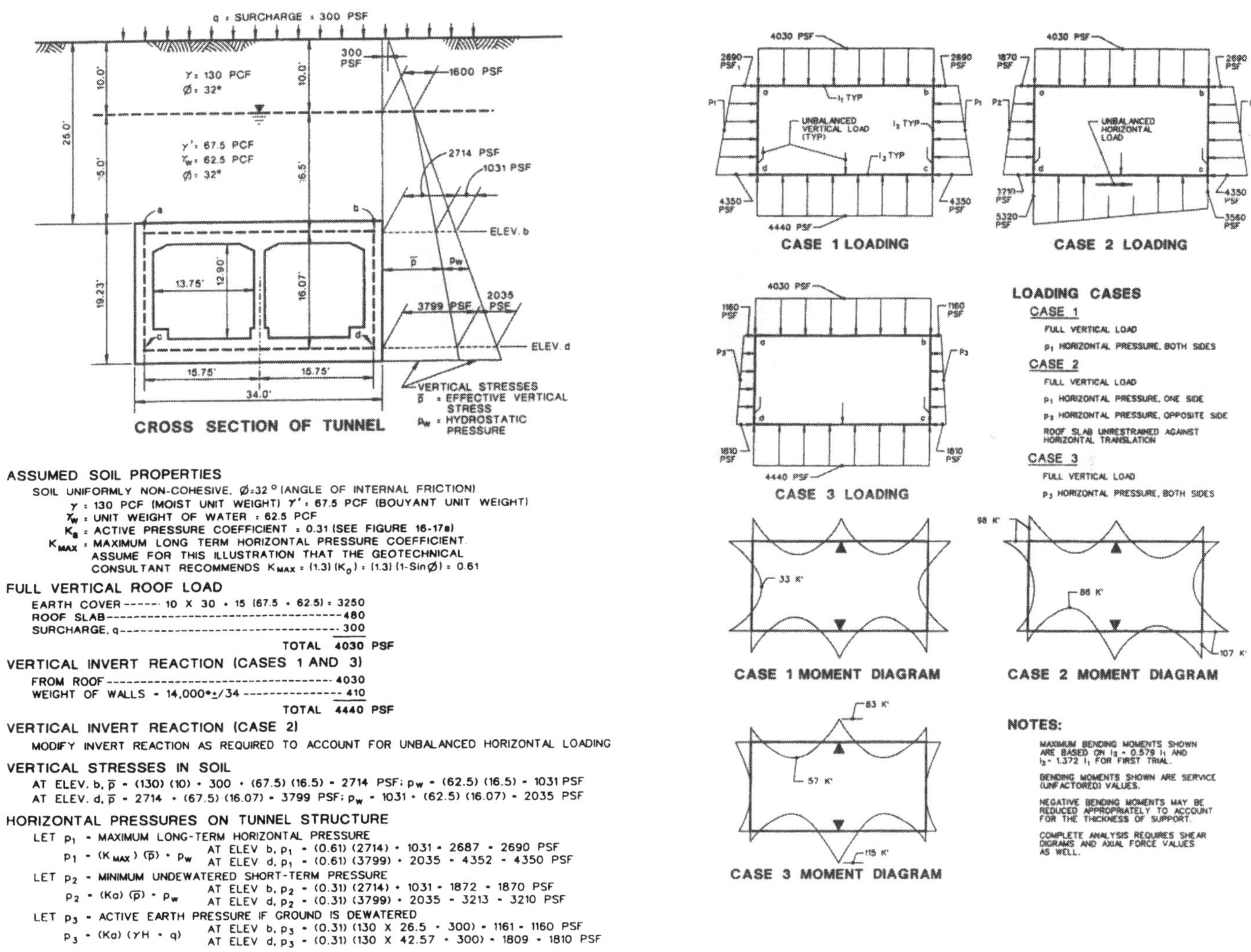

Fig. 17-7. Illustrative design calculations for a cut-and-cover box structure.

the cracked moment of inertia (I_c) is typically much less than the gross moment of inertia (I_g). If strains are not a concern, it is common to use gross values of El, since only relative internal reactions (forces and moments) are desired.

Foundations

For almost all cut-and-cover tunnel structures, the vertical pressure imparted to the subgrade beneath the structure is less than the in situ total vertical stress in the soil at that level before construction. Further, most of these structures are founded in competent soils with safe bearing capacity considerably exceeding applied vertical pressure. For these reasons, pile support for cut-and-cover tunnels is an uncommon requirement.

There are, however, occasions when the geotechnical consultant will recommend that the subsurface structure be supported on piles. Such cases can occur when the structure is founded in weak soils such as soft clay or silt and it is determined that the behavior of the supporting soil cannot be reliably predicted, that predicted differential settlement would adversely affect the structure, or that there is not adequate bearing capacity. The requirement for pile support under any of these conditions is more common for cut-and-cover sewer tunnels.

Reinforced Concrete Design

In the modern practice of reinforced concrete design, the predominant code defining design methodology is the *ACI Manual of Concrete Practice,* issued by the American Concrete Institute (ACI). Other major codes or specifications for design are similar in format and often identical to the ACI code. There are, however, significant differences between ACI requirements and AASHTO requirements, for example, with respect to the level of stress actually permitted in reinforced concrete.

The principal design methodology addressed in the ACI code is the Strength Design Method or Load Factor Design Method. Although the Strength Design Method is the predominant design method in use today for reinforced concrete structures, this methodology is not universally employed in the design of cut-and-cover tunnel structures. The much older Service Load Design Method (commonly referred to as the Working Stress Method) is still employed by many public authorities.

Modern practice in the design of reinforced concrete for cut-and-cover rapid transit structures is probably best characterized by the system design criteria specified by Los Angeles MTA. MTA directs that for all underground rapid transit cast-in-place reinforced concrete structures, including box lines and stations, the design shall be by the Strength Design Method of ACI-318, or Service Load Design Method of AASHTO Specifications, as applicable. For cut-and-cover vehicular tunnels, current practice may be characterized similarly, except that both the Strength Design Method and the Service Load Design Method are carried out according to AASHTO Specifications.

Most cut-and-cover sewer tunnels are combined sewers carrying both sewage and stormwater, and they may carry some industrial wastes as well. Such structures are normally designated Environmental Engineering Concrete Structures. Reinforced concrete for these structures should conform the requirements of ACI 350 R as well as the applicable requirements of ACI 318. Both the Strength Design Method and the Service Load Design Method (ACI Alternate Design Method) are used in the current practice of the design of these structures.

Reinforced concrete design for cut-and-cover tunnel structures must also conform to all local and other mandated codes, except when particular provisions of such codes are clearly, or can be shown to be, not applicable. In general, the codes and specifications cited above will contain the same or more severe provisions than other, mandated codes. Where earthquake forces are a factor, the structure should be designed for a desired degree of ductility and toughness as well. To incorporate these provisions into the reinforced concrete design, authoritative and pertinent literature on the subject of seismic design should be consulted.

The cited codes and specifications may be regarded as definitive for the proportioning of the subsurface structure for resistance to principal shears, thrusts, and bending moments. Engineering practice with respect to providing shrinkage and temperature reinforcement, however, is less well defined. Cut-and-cover structures are normally subjected to a much smaller range of ambient temperature than are surface structures. Also, these structures are commonly more massive than are the surface structures for which the cited codes are applicable. In the opposite vein, and particularly when these structures lie below the water table, it is almost always desirable that they be designed so that only minimal cracking is to be expected. These factors are typically taken into account in formulating criteria for shrinkage and temperature reinforcement. For the design of rapid transit walls and roof slabs, with transverse joints about 50 ft apart (see below), it is common to provide temperature and shrinkage reinforcement, on both faces of the wall or slab, in the amount of 80–100% of normal ACI 318 (7.12) requirements, up to a specified maximum. (No. 7 at 12 in. at each face is a common specified maximum.) Treatment of invert slabs has been similar to that for walls and roof slabs. In some cases subgrade drag may need to be investigated.

Minimum requirements for shrinkage and temperature reinforcement, as specified by AASHTO, have not been considered applicable for cut-and-cover highway tunnels. For these structures, engineering practice in this regard has been similar to that for rapid transit tunnels.

In the design of cut-and-cover sewer tunnels, both infiltration (usually) and exfiltration are concerns. ACI 350R offers definitive criteria and discussion applicable to the design of shrinkage and temperature reinforcement for these structures. The text of ACI 350 R includes an evaluation of the effectiveness of shrinkage-compensating concrete in reducing the required amount of shrinkage and temperature reinforcement.

Joints

The amount of shrinkage and temperature reinforcement that should be provided is also considered to be a function of the distance between transverse joints that will dissipate shrinkage and temperature stresses in the direction of the reinforcement. Engineering practice has differed with respect to both the spacing and type of transverse joint that may be used successfully (Figure 17-8). In the United States, typical spacing of transverse joints has ranged from 30 to 65 ft, but more commonly is in the range of 35–50 ft. The WMATA cut-and-cover structures contain transverse contraction joints spaced 50 ft (maximum) apart, with occasional expansion joints as well. For many other cut-and-cover tunnels there has been more recognition of the relatively small range of ambient temperature to which these structures are subjected by providing transverse construction joints in the design. (A construction joint is generally defined here as a joint through which reinforcement passes.) Cut-and-cover tunnels (line structures and stations) on the BART system were provided with transverse construction joints spaced typically 35 ft apart, with occasional expansion joints. This or similar practice has been common for more recently constructed cut-and-cover tunnels. Some examples are the following:

- Los Angeles MTA. Transverse construction joint spacing was typically 45 ft.
- Central Artery, Boston. Transverse construction joint spacing was typically 40 ft.
- San Francisco Clean Water Program. Transverse construction joints were specified to be 30 ft apart. Increased spacing was allowed up to 60 ft with a specified increase in longitudinal reinforcement. Longitudinal reinforcement specified was 0.3% for 30-ft joint spacing. For each additional 10-ft increment of spacing of joints, an additional 0.1% longitudinal reinforcement was specified. Actual spacing was usually 50 ft.

Horizontal construction joints are normally specified to be located at the underside of roof slabs, floors, beams, or girders and at the top of invert slabs. Waterstops are usually required at all transverse joints and sometimes at horizontal joints as well. (See discussions below on waterstops, joint treatment, and watertightness.)

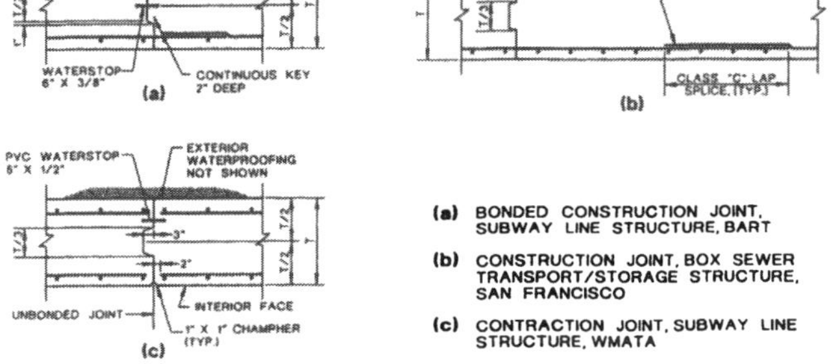

Fig. 17-8. Common transverse joints.

SHORING SYSTEMS

Neat-line excavation in soils requires that temporary walls (commonly referred to as "sheeting" or "shoring walls") be in place before significant cut-and-cover excavation commences. As the excavation progresses, the shoring walls must be supported so that they retain the vertical faces of the excavation. The function of the supported shoring walls (shoring system) is usually to prevent detrimental settlement of the ground, utilities, and buildings at the side of the cut as well. The design of the support system will depend on many factors, including the following:

- The physical properties of the soil throughout and beneath the cut.
- The position of the groundwater table during construction.
- The width and depth of the excavation.
- The configuration of the subsurface structure to be constructed within the cut.
- The size, foundation design, and proximity of adjacent buildings.
- The number, size, and type of utilities crossing the proposed excavation. Also, the presence of utilities adjacent to the excavation.
- Requirements for street decking across the excavation.
- Traffic and construction equipment surcharge adjacent to the excavation.
- Noise restrictions in urban areas.

The depth of excavation required for the construction of subsurface structures discussed in this chapter is rarely less than 25 ft. For the sake of clarity and reference in this chapter, it is arbitrarily assumed that "cut-and-cover excavation" exceeds 25 ft in depth. Additionally, cases where bedrock lies above the bottom of the excavation are not addressed in this chapter.

COMMON TYPES OF SHORING WALLS

Cross sections through each of the common types of shoring walls discussed in this chapter are illustrated on Figure 17-9. Soldier pile and lagging walls and steel sheet piling walls are often classified as "flexible" walls. In the same context, continuous concrete diaphragm walls, which are ordinarily much stiffer, are classified as "rigid" or "semirigid" walls, depending upon actual stiffness.

Soldier Piles and Lagging

A very common type of shoring wall is soldier piles and lagging. The soldier piles are usually steel wide flange (WF) or bearing pile (HP) shapes installed prior to the excavation by driving or drilling. Spacing of soldier piles can range from 3 ft or less to 10 ft or more; 5–8 ft spacing is most common. The soil between the soldier piles is retained with horizontal wood lagging placed between the soldier piles as

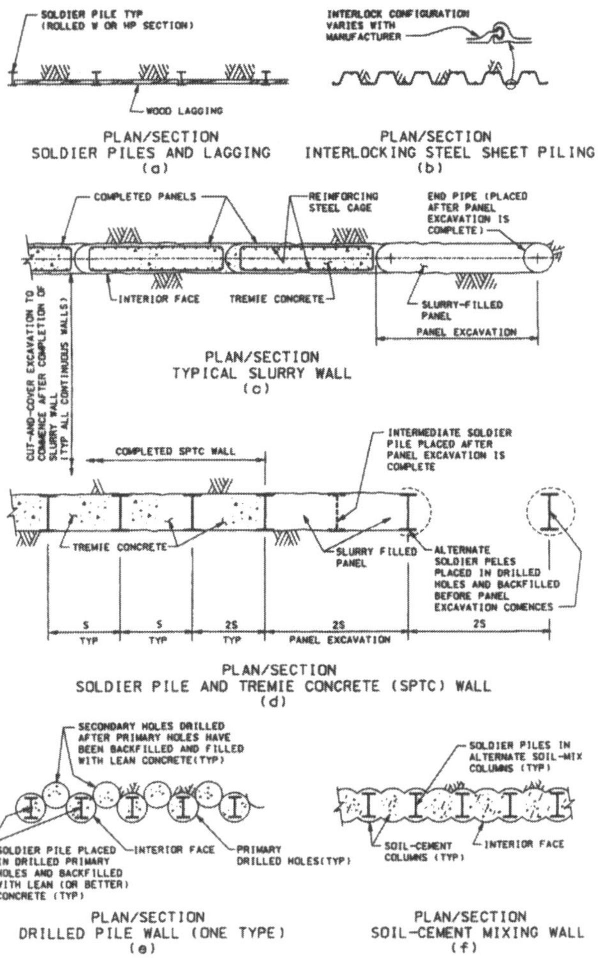

Fig. 17-9. Common types of shoring walls.

the excavation progresses (Figures 17-9a and 17-10). The wood lagging is usually placed behind the inside flanges of the soldier piles but is occasionally placed in front of the soldier piles ("contact lagging") using proprietary hardware or welded, threaded studs with retaining plates to secure the lagging to the soldier piles. During excavation, the depth of exposure below the last-placed lagging may be as little as 1 ft, as in the case of say saturated silt (see below), or as much as 4 or 5 ft in competent, cohesive or semicohesive soils. The restriction in depth of unsupported cut is the height of the cut that will remain stable until the lagging can be placed, but in soils it should not be more than 5 ft.

When soldier piles are placed in drilled holes, the portion of the drilled hole below excavated subgrade is sometimes backfilled with 3,000 psi (or more) concrete, either to provide for needed bearing capacity when the soldier pile is axially loaded or to improve available passive resistance below subgrade. More commonly, however, this portion of the drilled hole is backfilled with a lean concrete. Above excavated subgrade, good construction practice can range from filling the hole with lean concrete to filling it with a soil–cement mix, depending on the importance of avoiding

Fig. 17-10. Internally braced soldier pile and lagging walls, San Francisco Clean Water Program.

loss of ground when excavating the soil face to fit the lagging boards in place.

Soldier pile and lagging walls are particularly advantageous when many underground utilities will cross over the excavation. The utilities are exposed prior to installation of the soldier piles, and in most cases the soldier piles can be placed so that they straddle the utilities. After the soldier piles are installed, and during the first step in the cut-and-cover excavation, the utilities are either independently supported across the excavation, or they are suspended from street decking that spans the excavation.

When the cut is in saturated, pervious, or semipervious soils; or when there are zones of saturated soils within the cut, soldier piles and lagging are rarely feasible unless the groundwater table is lowered (see later discussion). Soldier piles and lagging have been used in some cases where the groundwater table lies a few feet above the excavated subgrade in semipervious soils, but this practice is not recommended, except in some otherwise very difficult circumstances.

When the groundwater table lies at or has been lowered to a level below excavated subgrade, soldier piles and lagging can be used more or less routinely in a wide variety of soils. Soldier piles and lagging are also used when the cut lies in weak or decomposed rock not capable of supporting itself or of being supported with modern shotcrete and rock bolt techniques. Soldier piles and lagging are not ordinarily considered suitable for cuts in very soft to soft clays, saturated silts, loose silty sands, or in any soil that is potentially unstable during excavation.

Steel Sheet Piling

Continuous steel sheet pile walls are composed of rolled Z-shaped or arch-shaped interlocking steel sections. Because of their greater stiffness and resistance to bending, Z-shaped sections are almost exclusively used as steel sheet piling for cut-and-cover construction (Figure 17-9b).

Interlocking steel sheet piling is typically used in saturated pervious or semipervious soils and other loose or weak soils that do not permit the easy placing of lagging. When the groundwater table cannot be lowered, interlocking steel sheet piling is normally adequately effective in cutting off concentrated flow through pervious ground both above and below the excavated subgrade.

Interlocking steel sheet piling is used as well in competent sandy soils when groundwater is not a concern, if there are sufficiently few utility crossings and other subsurface obstacles and it is found economically advantageous to do so. The sheet piles are driven with either impact-type or vibratory-type hammers, depending in part on the type of soil. Vibratory hammers are more commonly used in sandy soils. In modern pile driving practice, with the assistance of preauguring techniques, sheet piles can be driven with a vibratory hammer in dense sand to a depth of 65 ft or more with relatively good driving production. For the San Francisco Clean Water Program, in soft to medium clays, sheet piles up to 90 ft long were driven with a vibratory hammer.

Diaphragm Walls

In shoring system design and construction, the term *diaphragm wall* refers to continuous shoring walls that are reinforced concrete, a combination of concrete and structural steel, or similar systems. The walls are constructed from the ground surface and are ordinarily designed so that they are, for construction purposes, watertight. The more common diaphragm walls, and their applications, are discussed briefly below.

Slurry Walls. Although the term *slurry wall* sometimes has a broader meaning, it usually refers to a reinforced concrete wall placed in a deep trench usually 2–3 ft wide. The wall is constructed in increments, or panels. Panel lengths have ranged from 7 to 20 ft, but in cut-and-cover construction they are usually 10–15 ft long. The panel is excavated with a special clamshell-type bucket. The sides of the panel excavation are stabilized by filling the panel with a bentonite slurry and maintaining the level of the slurry at or near the ground surface throughout the excavation. Upon completion of the panel excavation, a preassembled steel reinforcing "cage" is lowered into the slurry-filled panel. Concrete is then placed in the panel by tremie techniques, displacing the slurry.

It is important (almost always) in slurry wall construction that the joints between panels be watertight. The most common type of joint is formed with a circular end pipe. Figure 17-9c illustrates the joint configuration formed by this method. The end pipe is a steel tube inserted at one end of the excavated panel as a stop for tremie concrete. Some time after the start of the tremie concrete pour, the end pipe is rotated to break bond; it is subsequently slowly extracted to produce a formed, semicircular joint, which can be cleaned when the next panel is excavated. The procedure shown in Figure 17-9c indicates a continuous operation in which one panel at a time is constructed, setting the end pipe at the leading edge. Another procedure is to construct alternate "primary" panels, setting end pipes at both ends. The resulting "secondary" panels between primary panels are then constructed.

There are many variations to the more common procedures described. Polymer drilling fluid has been used in lieu of bentonite slurry on several recently constructed slurry wall projects. More sophisticated types of slurry wall joints have been used to improve watertightness at the joints.

Slurry walls can be constructed in soil to depths exceeding 180 ft. For cut-and-cover construction, slurry walls deeper than about 100 ft are not common, however. Occasionally it is necessary to key the bottom of the slurry wall into rock. Shallow keys in soft to medium-hard rock have been successfully excavated using percussion tools developed for this purpose. Keys into medium-hard to hard rock have been difficult to achieve.

Soldier Pile and Tremie Concrete Walls. Soldier pile and tremie concrete (SPTC) walls are composed of soldier piles spaced at relatively close centers with a good-quality concrete placed between the soldier piles, thus forming a very stiff continuous wall. Soldier piles are typically 24–36 in. deep, rolled beams or deeper, built-up sections. Figure 17-9d illustrates the principal steps in the construction of an SPTC wall:

1. Alternate soldier piles are placed in predrilled holes. Bentonite slurry is used for hole stabilization if required. Each hole is then filled with sand or other weak backfill.
2. The slot between alternate soldier piles is then excavated with a specially designed clamshell tool. Bentonite slurry is used to stabilize the sides of the excavated slot in a similar manner to that for slurry walls.
3. The intermediate soldier pile is lowered into the slot. Unreinforced concrete is then placed in the slurry-filled slot on both sides of the intermediate pile simultaneously, using the tremie method.

Soldier pile spacing for SPTC walls typically ranges from 4 to 6 ft, with the upper limit being held to about twice the nominal wall thickness. Apart from the great strength that can be achieved using these walls, there is the added advantage that the soldier pile element of the wall can be extended deeper than the tremie concrete element. Thus, the soldier piles can be extended below the tremie concrete into very strong soil or into bedrock when there is a structural reason to do so.

Drilled Pile Walls. In this chapter, "drilled pile walls" refers to walls formed by abutting cast-in-place concrete or reinforced concrete piles, concrete and soldier beams placed in drilled holes, or combinations of these concepts. Drilled holes range in diameter from about 2 to 4 ft. Depending on soil and groundwater conditions, the excavation can be made with or without casing, either in dry or slurry-stabilized holes. Cast-in-place concrete or reinforced concrete piles placed in a single line or row, tangent, nearly tangent, or slightly overlapping with each other, have been called "contiguous," "secant," or (sometimes) "tangent" pile walls. However, in recent years the configuration shown on Figure 17-9e has been utilized on several cut-and-cover projects and is probably more common. With this configuration, considerably more strength can be built into the drilled pile wall. Drilled pile walls are especially suitable when the soil–rock contact lies near or above the cut-and-cover subgrade and a diaphragm wall is either required or most suitable above the rock line. This type of wall (or a wall of similar configuration) can be keyed into rock or constructed as a wall to support weak rock, using commercially available drilling equipment.

Soil-Cement Mixing Walls. Soil–cement mixing walls are composed of soldier piles placed in overlapping columns of in situ soil that has been mixed with a measured amount of cement or cement and bentonite. The soil–cement columns are constructed with special proprietary equipment, which employs a bank (usually three) of hollow-shaft augers. The augers penetrate the soil to the depth of the wall and then are withdrawn, mixing the soil with cement (and bentonite if needed) in the process, thus creating the overlapping circular columns of soil–cement. Before the soil–cement has hardened, soldier piles are pushed into the alternate (usually) soil–cement columns. The augers are normally about 24–36 in. in diameter. Soldier piles range from W18 to W30 spaced approximately 3 ft on center (with 24-in. columns) to 4 ft on center (with 36-in. columns). The unconfined compressive strength of the soil–cement ranges from about 75 psi in soft clays and saturated silts to about 200 psi in stronger sandy soils. Soil–cement mixing walls may be constructed to any reasonable depth required for cut-and-cover construction. Wall depths up to 180 ft have been reported in other applications (see Figure 17-9).

Applications for Diaphragm Walls. Diaphragm walls are used generally where it is required that the shoring wall be watertight and where, at the same time, more wall stiffness or resistance to bending is needed than can be provided by heavy steel sheet pile sections. Diaphragm walls have been constructed in virtually all soil types, but usually in very soft to medium clays, saturated silts, or saturated, loose silty or clayey sand. They are usually constructed where surface settlement adjacent to the cut must be minimized. The stiffer of the diaphragm walls have been specified in many urban cut-and-cover projects to obviate the need for underpinning adjacent buildings.

The construction of diaphragm walls causes much less noise and vibration than does the driving of sheet piles, and a diaphragm wall may be chosen over a sheet pile wall in some cases on this account alone. Where a diaphragm wall is to be used, the choice of the type will depend primarily on the required stiffness and resistance to bending and shear, actual subsurface conditions, and cost.

Occasionally, slurry walls or SPTC walls are utilized both as shoring walls and as the wall element of the permanent structure to be constructed within the cut. SPTC walls were utilized in this capacity in the design of several BART subway stations.

COMMON TYPES OF SHORING WALL SUPPORT

Internal Bracing

Most cut-and-cover excavations are relatively narrow and, as a result, internal bracing composed of multiple tiers of horizontal, structural steel framing is the most common type of shoring wall support used. In a typical excavation, the principal components of each internal bracing tier are longitudinal beams, or "wales," and transverse compression members, "or struts," arranged generally as shown on Figures 17-10 and 17-11.

The bracing tiers must be positioned so that they support the shoring wall and permit efficient construction of the permanent structure. Figure 17-12 shows the general sequence of construction operations typically employed during the construction of a subway line structure. The shoring walls are shown as constructed at the structure neat line (except for wall tolerance allowed), as is usually the case. During

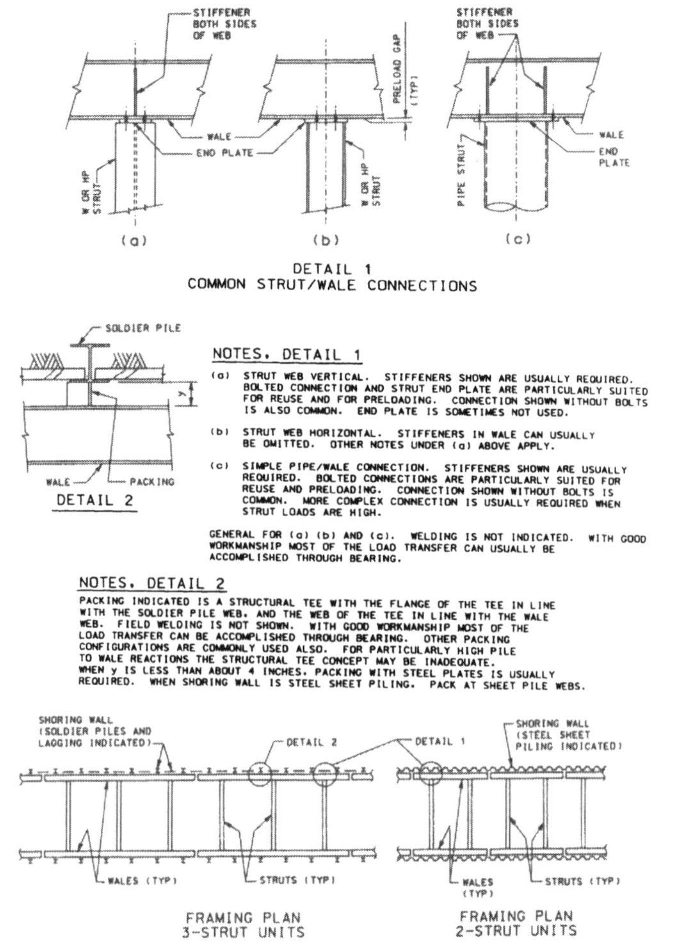

Fig. 17-11. Internal bracing—commonly seen framing plans and details.

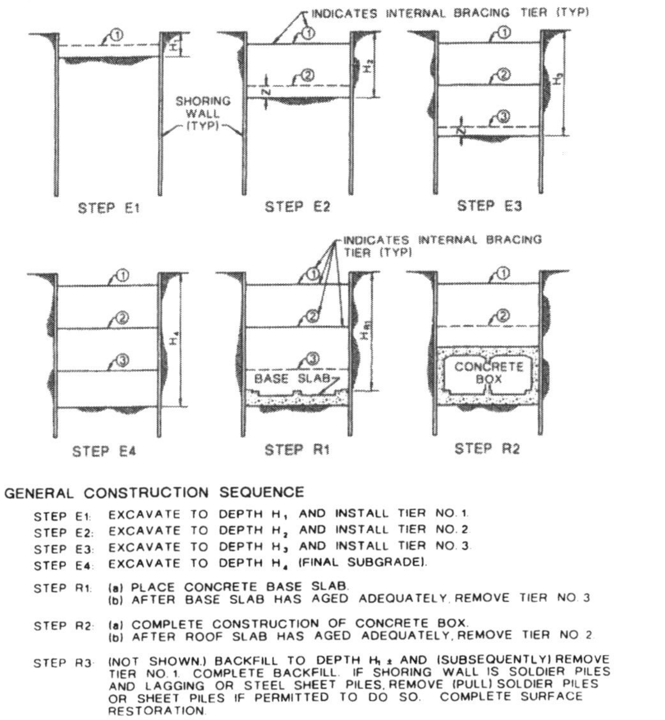

Fig. 17-12. General construction sequence typically employed during the construction of a subway line structure.

the excavation stage, vertical spacing of bracing tiers is often specified to be a maximum of 15–16 ft, sometimes 12 ft when it is crucial to minimize adjacent ground settlement. The maximum depth of cut in any excavation step is usually specified to be 3 ft below the centerline of the next bracing tier to be installed (dimension Z, Figure 17-12). The amount of settlement of the ground adjacent to the cut is generally considered to be largely a function of vertical spacing of bracing tiers and shoring wall stiffness. It is sometimes better to increase wall stiffness to permit larger vertical spacing of bracing tiers, when the larger spacing is needed to avoid interference of the bracing with the reinforced concrete construction. Occasionally, it is not feasible to avoid such interference, and different removal techniques or supplementary bracing methods are required. During the bracing removal stage, the shoring wall does not depend upon the soil below subgrade for support, and larger vertical shoring wall spans can ordinarily be permitted.

Struts in the internal bracing framing need to be spaced far enough apart so that excavating equipment can operate efficiently. Strut spacing is usually in the range of 10–15 ft, but larger spacing (up to 25 ft) is sometimes used to permit more clearance for construction activities. However, such large spacing is often very costly because of the much heavier wales that result, and it can be undesirable as well because of the inward wall movement that accompanies the increased wale deflection. Where there is no axial load in wales it is usually more economical to make the wales two to three strut spaces long and discontinuous at the wale ends. Figure 17-11 shows two framing plans representative of common internal bracing framing.

Internal bracing is compatible with any of the shoring wall types already discussed. Horizontal force from the shoring wall is transferred to the wale at each soldier pile, at each sheet pile web, or in the case of slurry walls, at heavy steel bearing plates embedded in the slurry wall at its inside face. (The bearing plates are fastened to the reinforcing cage when the slurry wall is constructed.) The shoring walls cannot be placed with sufficient accuracy to permit the wales to bear directly on the soldier piles, sheet piles, or slurry wall bearing plates. The gap between these wall elements and the wale is typically filled with a structural "packing" (see Figure 17-11). The wales are supported by structural steel brackets ("lookouts") mounted on the soldier piles, sheet piles, or slurry wall bearing plates.

Many framing concepts different from the typical framing shown in Figure 17-11 are employed. Irregular framing is usually required for irregularly shaped cut-and-cover excavation. Secondary framing may be required to brace the weak axis of struts in the wider excavations. When slurry wall panels can be constructed so that they are installed symmetrically about the longitudinal centerline of the cut-and-cover excavation, slurry wall panels are sometimes braced directly by struts, thus eliminating the need for wales.

Tie-Backs

A tie-back is a form of support in which the horizontal earth pressure (E) acting on the shoring wall is resisted by an anchor assembly, which in turn deposits its load into soil or rock far enough behind the wall to have no significant effect on the wall. Tie-backs anchored in soil are commonly referred to as "soil anchors"; tie-backs anchored in rock are "rock anchors." A tie-back consists of three principal elements (see Figure 17-13):

1. An anchor zone, which acts as a reaction to horizontal earth pressure (E) on the shoring wall.
2. A tie element, which transfers the load from the wall to the anchor zone.
3. A wall reaction assembly at the point of wall support.

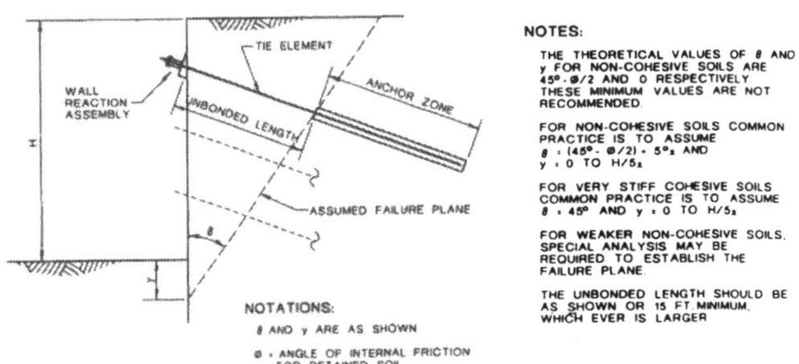

Fig. 17-13. Principal tie-back components. (For detailed discussions of tie-back theory and practice see Goldberg et al. (1976) and Schnabel (1982)).

To locate the anchor zone safely behind the shoring wall, the anchor zone is typically placed behind an "assumed failure plane." Theoretical establishment of this plane so that it is representative of a safe installation requires complex analysis. In most cases, the assumed failure plane is established by an experienced geotechnical consultant, who in turn relies on both experience from past performance of tie-back installations and theoretical considerations. Figure 17-13 shows the range in which the assumed failure plane usually falls.

There are many types of soil and rock anchors. The tie element for all of the more commonly used tie-backs is either a single threadbar 1–1-3/8 in. in diameter (usually high strength, as manufactured by Dywidag), or multiple high-strength strands, each (in most cases) 0.6 in. in diameter. The tie elements are installed in holes drilled from inside the excavation through the shoring wall at an inclination usually in the range of 15–30° from horizontal for soil anchors, and up to (commonly) 45° for rock anchors. The anchor zone is created by filling the drilled hole throughout the anchor zone with sand–cement grout or neat cement grout, depending upon the type of anchor. The zone between the anchor zone and the shoring wall is commonly referred to as the "unbonded zone" or "unbonded length." Throughout this length the tie element is covered with a plastic tube so that none of the tie-back load is transferred into the ground. (Other techniques are also used to create the unbonded length.)

Drilled holes for the most common tie-backs range from 8 to 18 in. in diameter; about 12 in. is typical. Grout is placed in the drilled hole by gravity or at modest pressure (about 50 psi) when hollow-stem augers are used to drill the hole. The anchor resistance is developed through grout-to-soil friction. For these "conventional" tie-backs, working capacities in competent soils ranging from 70 to 140 kips are common.

More sophisticated tie-backs of much higher capacity are also utilized on cut-and-cover projects. The type commonly referred to as "regroutable" employs equipment, hardware, other materials and methods that permit the injection of grout within the anchor zone at very high pressure (up to 300 psi or more) and in multiple applications. These regroutable tie-backs require a relatively small drilled hole (usually about 5 in.). In competent soils, tie-back capacities up to 400 kips or more have been installed. Regroutable tie-backs have also been installed successfully in medium clays (S_u = 500–1000 psf) with a working capacity up to 150 kips.

Tie-backs are also compatible with any of the shoring wall types discussed. When the shoring wall is soldier and lagging, sheet piles, or any of the diaphragm walls that use soldier piles, usual practice is to mount the anchor head assembly on double-channel wales, which in turn are mounted on and react against the soldier piles or sheet piles. When the shoring wall is a slurry wall, the anchor head assembly typically reacts directly on the slurry wall concrete (see Figures 17-14 and 17-15).

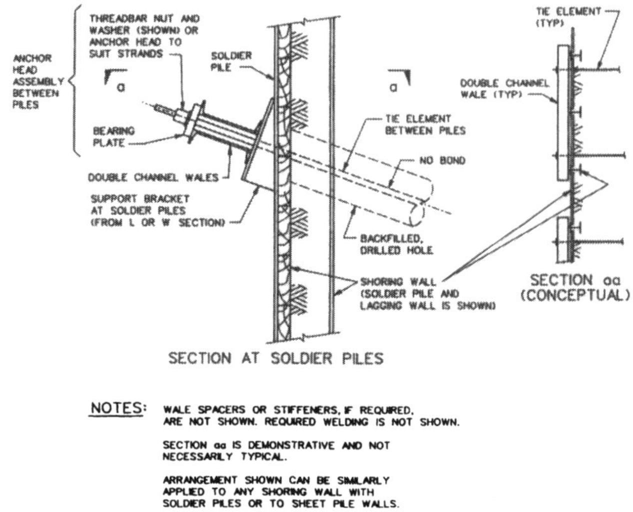

Fig. 17-14. Section showing tie-back wall reaction assembly.

When tie-backs are utilized as support for shoring walls, excavation proceeds in lifts that correspond to the vertical spacing of the tie-backs. Each succeeding increment of excavation cannot commence until the tie-back at that lift has been successfully posttensioned to its design working load at the wall reaction assembly. Normally, the tie-backs cannot be posttensioned until 5 days after completion of the grouting of the anchor zone.

Tie-backs can be considered an alternative to internal bracing when the following conditions exist:

- There is ample width within the excavation for tie-back installation.
- Permission is granted by the property owner to install tie-backs in the ground adjacent to the cut.
- There is no significant piezometric head behind the shoring wall at the level of the tie-back installation. (To date, attempts to install tie-backs when there is piezometric head exceeding a few feet behind the shoring wall have had limited success.)

Fig. 17-15. Tied-back slurry wall (foreground) and steel sheet piling wall (background), Central Artery construction, Boston.

- The soil behind the shoring wall is sufficiently competent to permit successful tie-back installation.
- There are no subsurface obstacles such as deep basements beneath adjacent buildings.

Although all these conditions are often present, it is uncommon to find that tie-backs are an economical alternative when the excavation is less than about 65 ft wide.

Soil Nailing

In some of the shallower cuts, installation of a shoring wall prior to the excavation can be avoided by utilizing the soil nailing method. Soil nailing is an in situ reinforcing technique that consists of installing passive inclusions (nails) into the undisturbed natural soil mass to retain excavation. The inclusions are usually steel reinforcement bars that either are placed in drilled bore holes and grouted along their entire length or (sometimes) are driven into place. To provide local stability between the nails, an outside facing is provided. The facing generally consists of 3–6 in. of reinforced shotcrete. Nailing differs from tie-back support: the nails are passive elements that are not posttensioned. In addition, the nail density in the retained soil is much larger and the nails are shorter than for tie-back systems. Figure 17-16 illustrates the process of constructing soil nail support.

The soil nailing technique is best suited to dry or moist noncohesive or semicohesive soils that will stand vertically for short durations during construction. In heterogeneous soils with cobbles, boulders, and weathered rock zones, this method offers the advantage of small-diameter, shorter drill holes. In cohesive soils subject to creep, however, soil nailing is usually either not feasible or is uneconomical, even at relatively low stress levels.

In North America, the retained depth of excavation using this system has generally been less than 30 ft. However, excavations as deep as 60 ft have been supported.

The mobilization of soil reinforcement interaction requires a relative displacement of soil and reinforcement. Therefore, in urban sites, the use of this technique can be limited by requirements for minimal settlement and movement of the ground adjacent to the excavation.

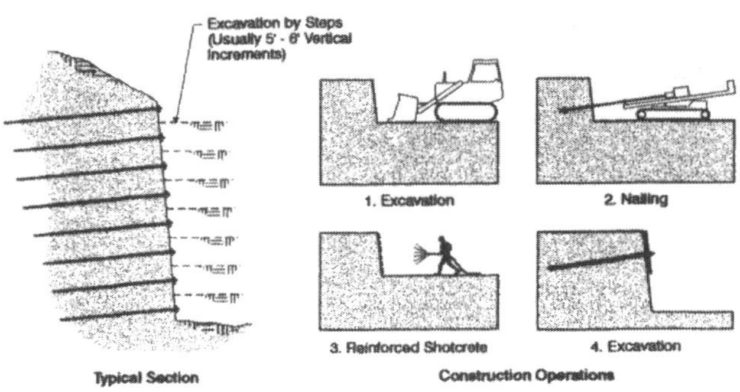

Fig. 17-16. Soil nailing construction operations.

DESIGN OF SHORING SYSTEMS

Lateral Pressure on Shoring Walls

In formulating the criteria for analysis and design of shoring systems, the following components of lateral pressure imparted to shoring walls must be evaluated when they exist:

- Lateral pressure due to retained soil
- Hydrostatic pressure
- Lateral pressure due to surcharge loads adjacent to the cut-and-cover excavation

The total lateral pressure on shoring walls is normally considered to be significantly less than the comparable lateral pressure on the walls of the permanent structure, because shoring walls are normally more flexible than permanent reinforced concrete walls. This flexibility permits small inward movements that relieve and redistribute the existing "at rest" lateral pressure within the soil mass. As discussed, lateral pressure due to retained soil, retained soil and water (when water is present), or adjacent surcharge loads depends on the physical properties of the retained soil. These physical properties should be determined from subsurface investigations and laboratory tests performed by an experienced geotechnical consultant. Further, on cut-and-cover construction projects, the geotechnical consultant usually provides in his geotechnical report lateral pressure diagrams suitable or useful as criteria for analysis and design of shoring systems. (This practice or its equivalent is recommended as a general rule for all cut-and-cover projects of significant size, and for any cut-and-cover project where the consequences of a shoring system failure would affect public safety.)

Earthquake forces are normally neglected as a component of lateral pressure imparted to shoring walls. This practice is based primarily on the temporary nature of the shoring wall and the prevalent view that shoring systems subjected to seismic activity have performed very well. On a few cut-and-cover rapid transit projects, a component of lateral pressure to account for earthquake forces has been specified. However, with the increased allowable stresses in the structural elements of the shoring system when lateral pressure due to earthquake is accounted for, there is usually no significant change in the design of the shoring system itself due to earthquake forces.

Lateral Pressure Due to Retained Soil. In the following discussion, the term *retained soil* refers to soil that has been predrained (dewatered), soil that lies above the groundwater table, or impervious soil. If the retained soil is supported in such a way that there is enough inward movement of the wall to permit all internal resistance to inward movement within the soil mass to be mobilized (as in the case of an ideal cantilevered retaining wall), the resulting pressure on the wall is referred to as *active pressure.* Common expressions for active pressure, based on Rankine formulas, are shown in Figure 17-17. These expressions apply (17-17a) to

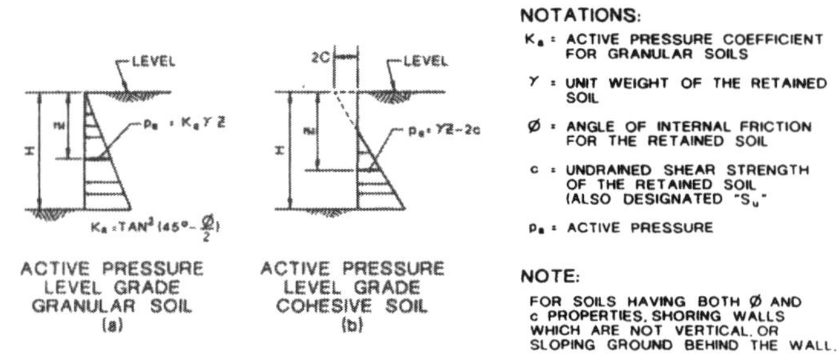

Fig. 17-17. Commonly used Rankine formulas for active pressure (Peck, 1969; Peck et al., 1974; Teng, 1962; Terzaghi and Peck 1967).

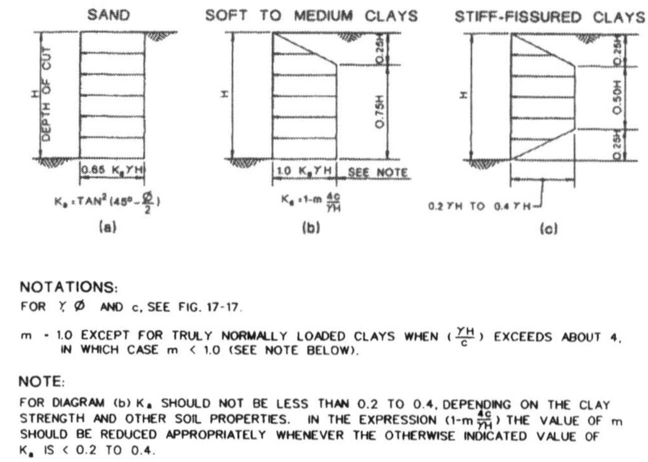

Fig. 17-18. Apparent pressure diagrams suggested (Terzaghi and Peck).

purely granular soils that derive all their resistance to movement from their angle of internal friction (ϕ) and (17-17b) to purely cohesive soils that derive all of their resistance to movement from their shear strength (*c*). Many soils are neither purely granular nor purely cohesive, but they are usually assumed to be one or the other, depending on whether they are predominantly granular or predominantly cohesive soils. The expressions in Figure 17-17 are also based on the case where the shoring wall is vertical and the surface of the ground behind the wall is horizontal. Using more-complex Rankine formulas, or other methods of analysis used in soil mechanics, active pressure diagrams can be drawn that take into account soils having both *c* and ϕ properties, shoring walls that are not vertical, and sloping ground behind the wall.

Braced shoring walls restrict the movement of the soil behind the upper portion of the wall (particularly). As a result, the pressure distribution behind the shoring wall is almost always considerably different from that represented by active pressure, and it cannot be predicted accurately based only on theoretical soil mechanics. Earth pressure diagrams have, however, been developed for the design of shoring systems from data made available from many measurements of strut loads in braced excavations. These diagrams are commonly referred to as "apparent pressure diagrams."

Apparent pressure diagrams for the purpose of estimating strut loads are commonly based on the empirical rules first developed by Terzaghi and Peck (1948, 1967) and later modified by Peck (1969) and Peck, Hanson, and Thornburn (1974). (See Figure 17-18.) Similar apparent pressure diagrams are recommended in other authoritative texts as well. Apparent pressure diagrams do not necessarily indicate actual pressure distribution on the shoring wall; rather, they represent an envelope inside which actual lateral pressure is ordinarily expected to fall. They are most suitable for the design of shoring systems with flexible walls that retain fairly homogeneous soils. For excavations in clay soils, however, additional analyses may be required (see later discussion).

When the apparent pressure diagrams in Figure 17-18 are applied to diaphragm walls, usual practice is to consider that the ordinate of lateral pressure should be increased to account for increased wall stiffness. Increases in the ordinate of lateral pressure in the range of 10–25% are common, depending upon actual wall stiffness and the consequences of wall movement.

For excavations penetrating through more than one stratum, straightforward application of the cited empirical rules (Figure 17-18) is not possible. Figure 17-19 illustrates a relatively simple comparable methodology, which is commonly used to construct apparent pressure diagrams for stratified soils. The basic horizontal earth pressures are computed as active pressure values. The total active resultant is then multiplied by a factor (MF) that takes into account the stiffness of the shoring wall, and the resulting load is distributed on the shoring wall in a trapezoidal pressure diagram similar to those shown in Figure 17-18. The multiplying factors used by WMATA in formulating specified minimum apparent pressure diagrams are representative of this practice. The WMATA values for MF are

- For sheet pile shoring walls, MF = 1.3 for general use. MF = 1.1 was allowed where resulting horizontal movement of retained earth was considered tolerable.
- For soldier piles and lagging shoring walls, MF = 1.3 for general use. MF = 1.2 where resulting horizontal movement of retained earth was considered tolerable.

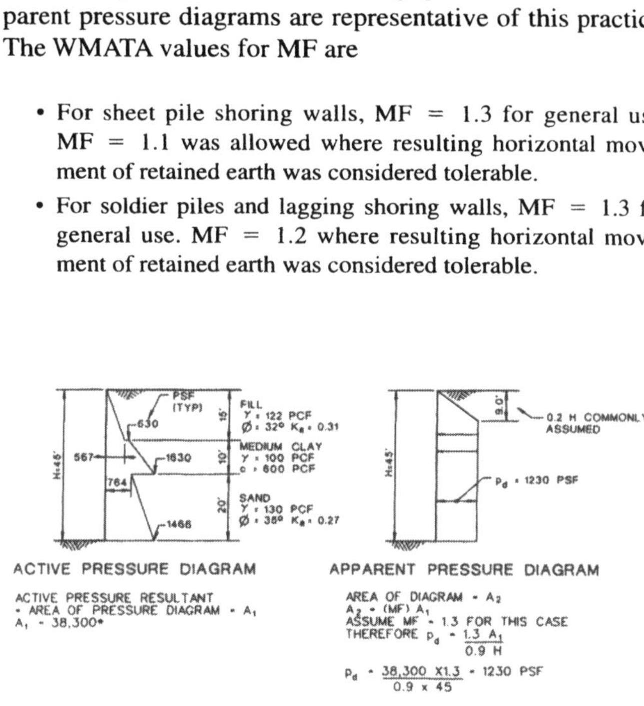

Fig. 17-19. Construction of an apparent pressure diagram for retained, stratified soils—illustrative case.

- For diaphragm walls, where movement is to be prevented to the extent practicable, MF = 1.4.

Lateral Pressure Due to Retained Soil and Water. When both pervious or semipervious soils and groundwater are being retained by a shoring wall, one should construct an apparent pressure diagram that represents the soil component of lateral pressure on the shoring wall. Triangular hydrostatic pressure is then added to complete the lateral pressure diagram above excavated subgrade. Figure 17-20 illustrates how a lateral pressure diagram for these conditions may be constructed using the same methodology to construct apparent soil pressure as was used in the construction of Figure 17-19. However, when both soil and water are being retained, additional investigation of the pressures acting on the shoring system is usually required.

Lateral Pressure Due to Surcharge Loads. Shoring systems for almost all cut-and-cover excavations should be designed for traffic and construction equipment surcharge adjacent to the excavation. Figure 17-21 shows the lateral pressure allowance recommended by BART, WMATA, and other public authorities for traffic and construction equipment surcharge applicable to cut-and-cover construction for subway structures. The recommended lateral pressure is approximately equivalent to a 600-psf (reasonably limited area) surface surcharge adjacent to the shoring walls, as determined by the Boussinesq equations, and has been found to be adequate for normal traffic and construction activity.

Most buildings not supported on pile foundations are in reasonably competent soils and are far enough away from the excavation, and/or have foundations deep enough, to have a minimal effect on the lateral pressure imparted to cut-and-cover shoring walls. Lateral pressure due to building surcharge can usually be neglected when $a/b \geq 0.6$ for dense sandy soils and $a/b \geq 1.0$ for stiff clay soils, where

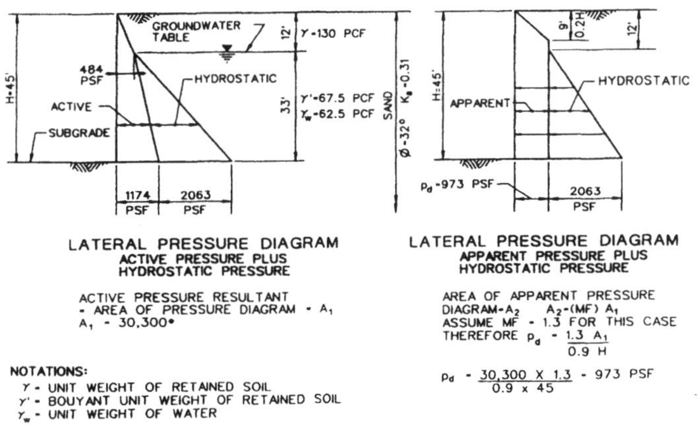

Fig. 17-20. Construction of an apparent pressure diagram when the groundwater table lies above subgrade—illustrative case.

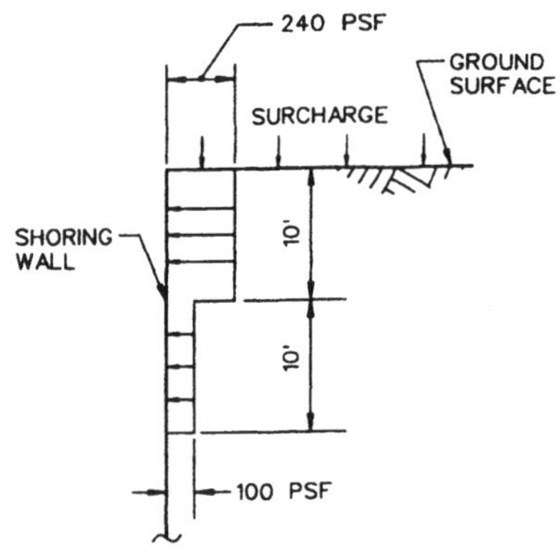

Fig. 17-21. Recommended lateral pressure due to traffic and construction equipment surcharge.

a = the horizontal distance from the building foundation to the shoring wall

b = the vertical distance from the level of the bottom of the building foundation to the level of the excavated cut-and-cover subgrade.

The limiting ratios of a/b should be confirmed by a competent geotechnical consultant for the particular cut-and-cover site. When the ratio of a/b is less than these limiting values, the geotechnical consultant should develop recommended criteria for estimating safe values of potential lateral pressure on shoring walls due to building surcharge. Potential lateral pressure due to building surcharge should then be computed and added to the other lateral pressures applicable to the design of the shoring walls.

Depth of Shoring Walls

The required penetration of shoring walls below excavated subgrade usually depends on the capacity of the soil below the excavated subgrade to resist the inward force imparted to the soil by the shoring wall. This capacity is known in soil mechanics as "passive resistance." Commonly accepted expressions for the available passive resistance of (a) purely granular soils and (b) purely cohesive soils are shown in Figure 17-22. In the case of granular soils, available passive resistance is a function of the coefficient of passive pressure, K_p. K_p in turn is a function of ϕ, the angle of internal friction, and δ, the angle of wall friction. Wall friction, which modifies both the direction and magnitude of the passive resistance, can be fully understood by referring to any authoritative text on soil mechanics. In formulating passive pressure diagrams for criteria for analysis and design of shoring walls, some public authorities have in the past chosen to neglect wall friction (i.e., $\delta = 0$). However, this practice is very conservative. It is more common practice to assume that $\delta = 1/2\phi$ is an amply conservative estimate in formulating passive pressure diagrams applicable to shoring walls.

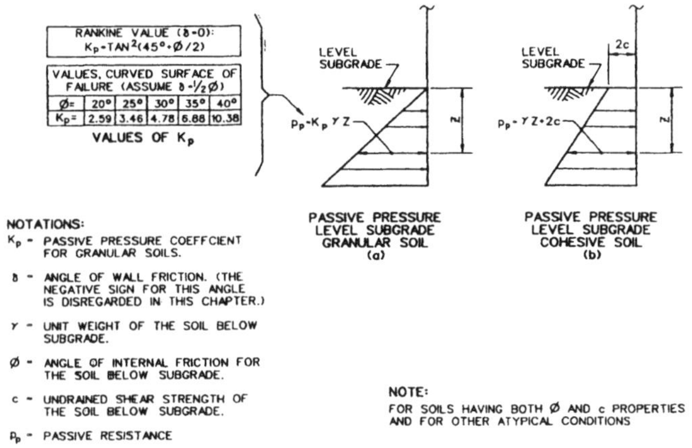

Fig. 17-22. Commonly used expressions for passive resistance.

As in the case of active pressure diagrams, passive pressure diagrams can be drawn for soils having both ϕ and c properties using more-complex theoretical expressions. However, it remains more common to consider the soil (or soil stratum) as either granular soil or clay soil, depending on the predominant physical properties of the soil in question.

Penetration of Shoring Walls in Competent Soils. When use is made of apparent pressure diagrams in soils capable of developing adequate passive resistance below subgrade, an equivalent bracing tier reaction is assumed to exist at the base of the cut. Figure 17-23 illustrates an empirical method for determining the depth of shoring wall penetration required to satisfy this assumption for continuous walls and for soldier pile and lagging walls. This empirical procedure may be applied to granular soils where the groundwater table is below subgrade, and to stiff-to-hard clay soils.

Figure 17-23b also illustrates the treatment of soldier piles below subgrade. Soldier piles are not continuous walls, and therefore both the passive earth pressure diagram and the active earth pressure diagram below subgrade must be constructed to reflect noncontinuous conditions. For single soldier piles, it is generally considered that to account for three-dimensional effects in the resisting soil, the effective width of the soldier pile may be taken as three times the actual flange width of the pile when computing passive resistance. When the pile is encased in structural concrete below subgrade, the effective width is usually taken as 2.25–2.5 times the diameter of the concrete encasement. Opinions vary on the effective flange width to assume when evaluating active pressure below subgrade. The value of 2.0 times actual flange width, used to construct Figure 17-23b, is recommended as sufficiently conservative for average conditions.

If it is necessary or desirable to reduce the depth of shoring wall penetration from that computed from Figure 17-23, the depth of penetration may be recomputed using "free-end" principles. The procedure is similar to that shown in Figure 17-25 or 17-26, to follow. However, this procedure results in an increase in the load imparted to the lowest brac-

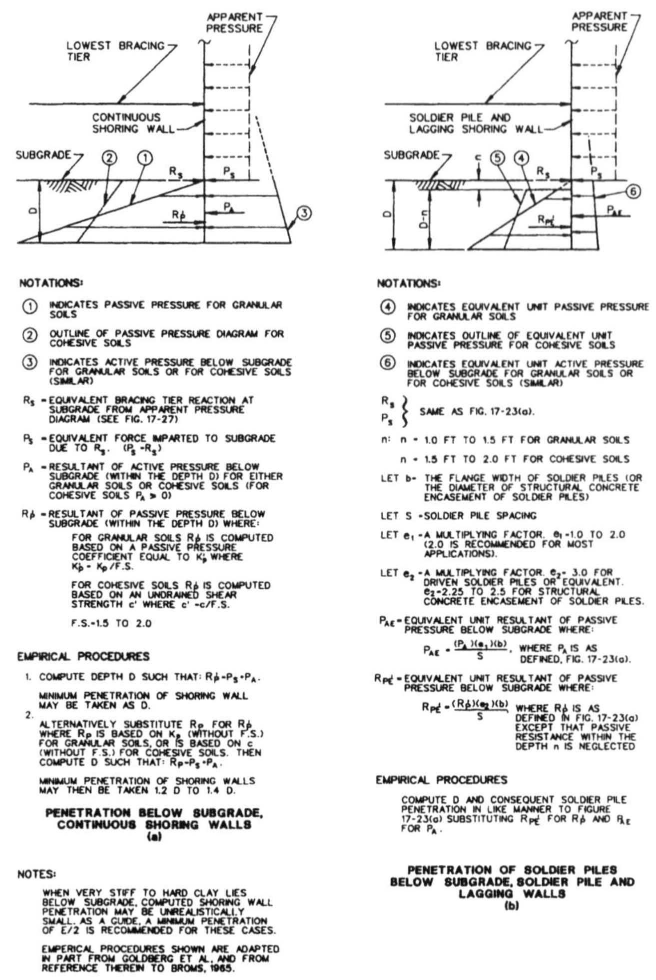

Fig. 17-23. Empirical determination of shoring wall penetration below subtgrade in competent soils.

ing tier, and an increase in the shoring wall bending moment at the final excavation step as well.

Penetration of Shoring Walls in Soft to Medium Clays. For relatively shallow cut-and-cover excavations in medium

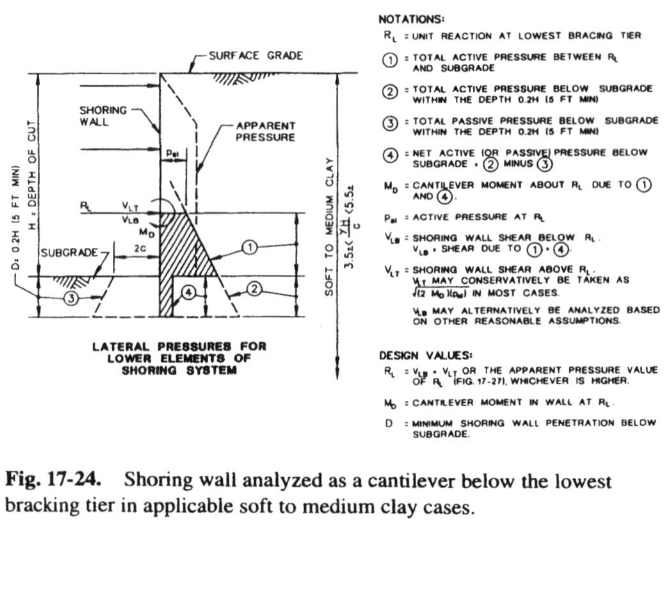

Fig. 17-24. Shoring wall analyzed as a cantilever below the lowest bracking tier in applicable soft to medium clay cases.

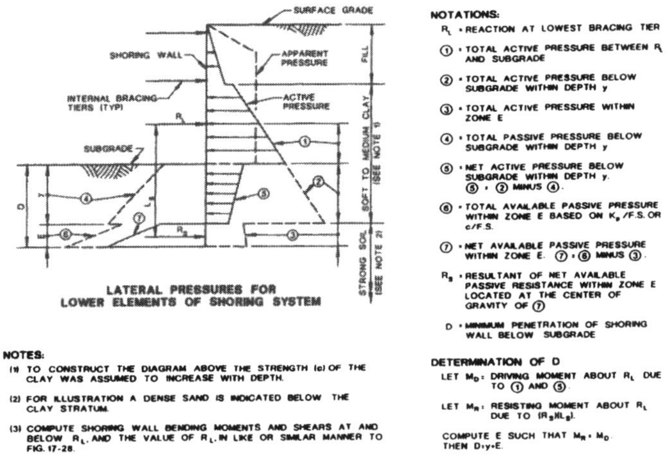

Fig. 17-25. Determination of shoring wall penetration below subgrade in soft to medium clay when free-end soil support is available as practicable depth below subgrade.

clays, there will ordinarily be enough passive resistance below subgrade to permit evaluation of the shoring wall penetration using the empirical method illustrated in Figure 17-23. When the "stability number" exceeds about 3.5–4, enough theoretical passive resistance is not available below the bottom of the cut to develop an adequate horizontal reaction. The stability number is defined as $\gamma H/C$, where

γH = the total overburden stress at a depth below the ground level corresponding to the level of excavated subgrade

c = the undrained shear strength of the clay at subgrade (also denoted as S_u).

Under these conditions, cut-and-cover excavation can be successfully prosecuted with modest shoring wall penetration below subgrade if the following conditions are satisfied:

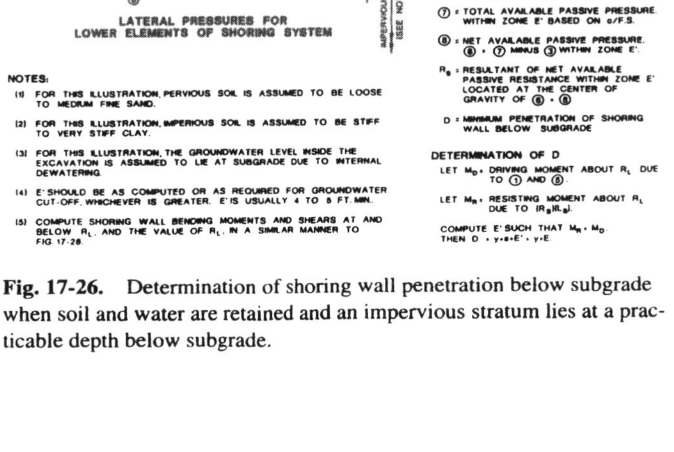

Fig. 17-26. Determination of shoring wall penetration below subgrade when soil and water are retained and an impervious stratum lies at a practicable depth below subgrade.

- The excavated subgrade must be stable against bottom heave. A minimum factor of safety (F.S.) of 1.25 against bottom heave is recommended. When the stability number, $\gamma H/c$, is less than about 5.5, this condition is usually satisfied. However, the F.S. against bottom heave should be evaluated for specific conditions by a competent geotechnical engineer using established methodology.
- Wall penetration should be adequate to provide an acceptable F.S. for the shoring system. Minimum wall penetration of 5 ft or $0.2H$, whichever is greater, is recommended, where H is the depth of the cut.
- The shoring wall must be designed as a cantilever below the lowest bracing tier, and the load imparted to the lowest tier must be calculated accordingly. (See Figure 17-24.)
- The consequent settlement and inward movement of the adjacent ground must be tolerable. (In urban areas the consequent ground movements determined from Figure 17-24 are often not tolerable.)

When the conditions necessary for the safe or tolerable application of the methodology shown on Figure 17-24 are not present, it is usually both possible and best engineering practice to extend the shoring walls into more competent soils that can provide required passive resistance. Stiff diaphragm walls are often ideal for this purpose, and commercially available stiff sheet steel piling is often feasible as well. For analysis and design of the shoring system in this case, the following general procedure is recommended:

1. Compute wall bending moments and shears, and bracing tier reactions, based on active and passive pressures applied to the shoring wall as shown in Figure 17-25.
2. Compute wall bending moments and shears, and bracing tier reactions, based on an appropriate apparent pressure diagram (Figure 17-18b or 17-19, as applicable).
3. Use for the design of the shoring system elements the larger bending moments, shears, and bracing tier reactions determined from steps 1 and 2.

Penetration of Shoring Walls when Soil and Water Are Retained. When the shoring system retains both pervious or semipervious soils and groundwater, active, passive, and hydrostatic pressures below subgrade must be analyzed correctly, and the shoring system must be designed so that all pressures on the shoring wall are safely resisted, and so that the excavated subgrade remains stable. In many cases the shoring wall can be extended into an impervious stratum, and the shoring system can be analyzed in a straightforward manner, as is illustrated in Figure 17-26. In some cases, however, the stratigraphy below subgrade may be such that the formulation criteria for a feasible shoring system design requires far more complex analysis. And in some cases the piezometric head below subgrade cannot be tolerated and must be reduced with relief wells, regardless of the consequences, if a shoring system basically similar to Figure 17-26 is to be used.

Factor of Safety. When the soil below subgrade is depended on as a resisting element in a shoring system, a factor of safety (F.S.) must be applied to passive pressure calculations. F.S. = 1.5 is usually recommended. For granular soils this requirement is typically accomplished by dividing the passive pressure coefficient, K_p, by F.S. In some empirical procedures, the calculated minimum wall penetration may simply be increased by a multiplying factor to accomplish the same result (as in Figure 17-23).

When soft to medium clays are depended on for passive resistance, it is ordinarily not realistic or workable to divide the shear strength, c, by F.S. in passive pressure calculations. In these cases, extending the shoring wall to a depth where the resisting moment about the lowest bracing tier will be F.S. times the driving moment about the lowest bracing level is acceptable practice when active and passive pressures are being assumed. Where empirical or semiempirical procedures can be used to compute wall penetration, the empirical procedure itself provides F.S. (as in Figures 17-23 and 17-24). Where stiff to hard clays are used for passive resistance, the shear strength c should be divided by F.S. in passive pressure calculations, unless empirical procedures that provide the same result are used.

Penetration of Tied-Back Shoring Walls. The preceding discussions of shoring wall penetration below subgrade apply to both internally braced shoring walls and tied-back shoring walls. The lowest row or level of tie-backs may be considered equivalent to the "lowest bracing tier." However, tie-back support of shoring walls is usually limited to cases where an equivalent bracing tier reaction may be assumed at subgrade.

Vertical Loads on Shoring Walls. Vertical load is commonly imparted to shoring walls from street decking that the walls support (see later discussion). In addition, the weight of the shoring system must be supported, and in the case of tied-back shoring walls, the vertical component of the tie-back loads must be supported. For internally braced shoring walls, the calculated or otherwise established shoring wall penetration is usually sufficient to provide the required resistance to vertical loading. For tied-back shoring walls, the requirement that the vertical component of tie-back load be resisted is often the determining factor in establishing satisfactory wall penetration. In both cases, however, the penetration of the soldier piles or continuous wall needs to be checked for adequacy in this regard, using appropriate engineering principles taken from authoritative texts in soil mechanics. Wall friction or adhesion above subgrade may be taken into account in analyzing resistance to vertical load. However, it is recommended that for internally braced walls a high factor of safety (2.0–3.0), and for tied-back walls a factor of safety of at least 1.5, be applied to calculated wall friction or adhesion.

Supplementary Methods for Providing Wall Support. In relatively deep or deep cut-and-cover excavations in soft to medium clays, or in similarly deep cut-and-cover excavations where both soil and groundwater must be retained, it may not be feasible or desirable to design the shoring system using the methodologies discussed above (Figures 17-25 and 17-26). These are cases where adequate passive resistance at an economically feasible depth are not available. Construction techniques that have been used to overcome this problem are discussed briefly here.

The Tremie Slab Method. A *tremie slab* or *tremie seal* is a plain concrete slab placed underwater at the bottom of the cut-and-cover excavation. The tremie slab must be continuous across the excavation and must be in intimate contact with both shoring walls so that it can (subsequently) act as a strut. The maximum elevation of the top of the tremie slab is located at or slightly below the bottom of the base slab of the permanent structure to be constructed (depending on the design of the base slab/tremie slab horizontal interface). Thus, the cut-and-cover excavation must be deeper than would otherwise have been required by at least (usually) the thickness of the tremie slab. To maintain subgrade and shoring wall stability during excavation, the excavation itself must be carried out under water, with the level of the water between the shoring walls (i.e., within the "cofferdam") remaining high enough to provide the required stability. The excavation and placement of internal bracing is prosecuted in a manner similar to that shown in Figure 17-12 except that the next step after final subgrade is reached is to place the tremie slab. After the tremie slab has aged, the cofferdam is dewatered. At this step, a lateral pressure diagram corresponding to Figure 17-18b or 17-20 (as appropriate) may be used to analyze the shoring system elements, with "subgrade" assumed at the top of the tremie slab.

It is usually not economical and is sometimes impractical to provide a tremie slab thick enough or heavy enough to resist upward hydrostatic pressure. In such cases, tension piles are commonly incorporated into the tremie slab design. On one WMATA cut-and-cover contract, a subway line structure was constructed in a soft to medium organic silty clay stratum up to 100 ft deep, and the tremie slab method with tension piles was specified and used. Excavated subgrade (bottom of tremie slab) was up to 67 ft deep. The tremie slab was nominally 6 ft thick and was placed in 200–300-ft-long increments where the cut was deepest. The tremie slab method permitted a shoring design based on a lateral pressure diagram corresponding to Figure 17-18b, which in turn made steel sheet piling a feasible selection for the shoring walls.

Jet Grouting. Jet grouting is a displacement grouting technique in which soil–cement cylinders are created by high-pressure (6,000 psi) injection from the ground surface. The cylinders can be installed as overlapping columns in a predetermined subsurface zone. Thus, by this technique a stratum of soil–cement can be created at or below subgrade before cut-and-cover excavation proceeds. In soft to medium clays, an average unconfined compressive strength of up to

500 psi or more has been attained for the soil–cement. In predominantly granular soils, an average unconfined compressive strength of up to 2,000 psi or more can be achieved.

The jet grouting technique is relatively new to the United States, and it has been used in only a few braced excavations. Successful application of the method, in creating a preinstalled soil–cement strut at or below subgrade, depends on existing subsurface conditions. The soil–cement stratum must remain stable against any uplift pressures beneath it when excavation to subgrade is carried out. In the right application, the jet grouting method can offer many of the same benefits offered by the tremie slab method.

Design of Shoring Walls

Apparent Pressure Methodology. When an equivalent bracing tier reaction (soil reaction) can be assumed to exist at or near subgrade, as in the cases illustrated in Figures 17-18a–c and 17-19, empirical rules may be employed to determine bracing tier reactions and wall bending moments and shears. These values are almost always expressed dimensionally as kips per ft for bracing tier reactions and wall shears, and kip-ft per ft for wall bending moments, meaning kips per ft or kip-ft per ft, per ft of wall width, respectively. Figures 17-27a and b illustrate two similar procedures, designated "Method A" and "Method B" respectively, commonly employed to determine bracing tier reactions and shoring wall bending moments and shears. Method A is based on Terzaghi and Peck's (1967) recommendations. Method B is a representation methodology commonly specified by rapid transit authorities. The unit bracing tier reactions (or unit tie-back tier reactions) are then utilized to design shoring wall support system.

In determining wall-bending-moment values for design based on apparent pressure, usual practice is to compute first the cantilever bending moment above the uppermost support and the simple bending moment between supports (i.e., between bracing tiers or tie-back rows). For flexible shoring walls, some authorities have recommended that before wall continuity is accounted for, the calculated simple bending moments may be multiplied by a factor ranging from 0.67 to 0.80 to account for arching of the soil between supports. However, with the common use of high-strength steels (F_y = 50 ksi), the reversal of stresses that ordinarily occur during the life of the shoring wall, and the common practice of permitting an increase in allowable bending stress in the structural steel elements of shoring walls (see later discussion), most public authorities now either discount the effect of soil arching or specify design procedures that accomplish the same result. This latter treatment of soil arching is the general recommendation in this chapter. Thus, the simple bending moment (M_S) may be computed in the same straightforward manner for both flexible shoring walls and rigid shoring walls. The design bending moment for the shoring wall below the uppermost bracing tier is then determined for any span by simply multiplying the simple bending moment (M_S) by a factor of 0.8. This empirical rule is commonly accepted as good practice for all shoring walls except slurry walls. The maximum wall bending moments are actually (almost always) negative moments at the bracing tiers.

For the design of slurry walls, using apparent pressure methodology, the wall should be analyzed as a continuous beam in order to provide properly placed steel reinforcement. However, neither positive nor negative bending moment should be taken as less than 0.8 M_S. Similarly wall shears should not be taken as less than simple span values. For analysis of reinforcement below the lowest bracing tier (in the absence of more rigorous analysis), it is usually simplest to provide A_{sn} steel reinforcement in both wall faces from the lowest bracing tier to the bottom of the slurry wall, where A_{sn} is the amount of negative reinforcement required at the lowest bracing tier when the wall is assumed pinned at subgrade.

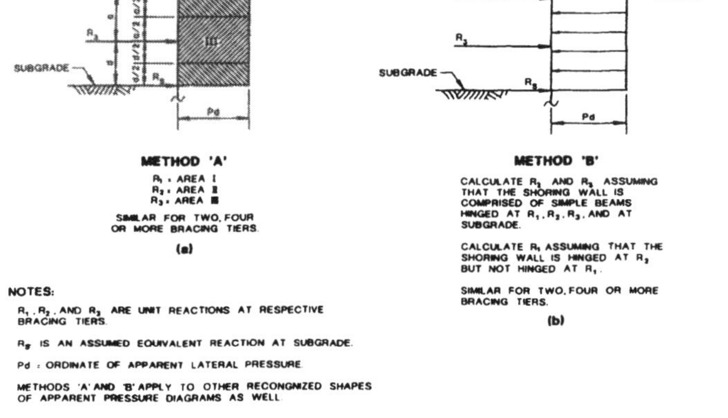

Fig. 17-27. Procedures commonly employed to determine bracing tier reactions from apparent pressure diagrams.

Methodology When Soil Reaction Is Below Subgrade. Figures 17-25 and 17-26 are examples of cases where an immediate soil reaction (R_S) cannot be assumed to exist at subgrade. The simplest method of analysis of conditions below subgrade is to assume free-end conditions for the lower portion of the wall. The shoring wall is extended deep enough into the soil below subgrade (penetration) so that adequate passive resistance is provided for shoring wall stability. Wall stability is reached when the resisting moment (M_R) provided by the soil reaction R_S about the lowest bracing tier is equal to the driving moment (M_D) about the lowest bracing tier due to net applied soil or soil and water pressure. At this step, the location of R_S is defined and the shoring wall may be analyzed as a continuous beam to determine bracing tier reactions and wall bending moments and shears. Figure 17-28 illustrates load, shear, and moment diagrams that result when the conditions shown in Figure 17-25 are used to construct an example.

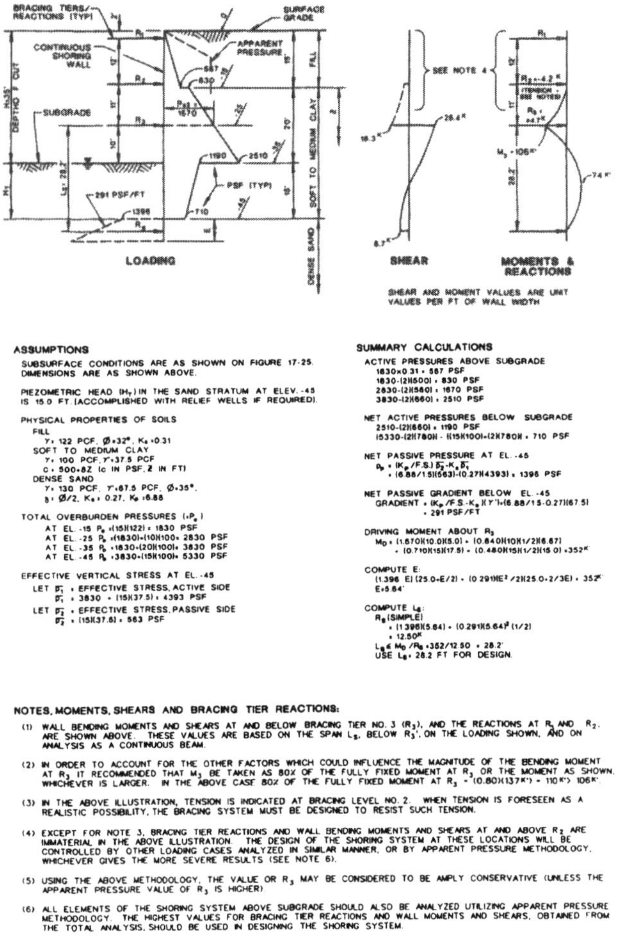

Fig. 17-28. Analysis of the lower elements of a shoring system when an equivalent wall reaction cannot be assumed to exist at subgrade—illustrative case.

The methodology described above should be refined so that it includes the effect of the resistance to wall bending, at the lowest bracing tier, offered by the shoring wall itself, when the ratio of wall stiffness to driving moment M_D is significant. Recognize that continuous beam theory must be considered approximate because soil pressures being applied do not necessarily correspond to the regular shaped diagrams assumed, and wall rotation at bracing tier supports does not necessarily correspond to theoretical values. These considerations should be taken into account, when they could be of significance, in finalizing design values for bracing tier reactions and wall moments and shears.

Loading Cases on Shoring Walls. Internally braced shoring walls must be analyzed for all conditions (loading cases) that might occur during the various stages or steps of cut-and-cover construction. The analysis of each loading case should be carried out to the extent necessary to determine whether or not bracing tier reactions or wall bendin-moments or shears, at the particular loading case being analyzed, will represent maximum reactions, moments, or shears. The design of the shoring wall should be such that the maximum positive and negative bending moments and shears at any horizontal plane along the wall, when all loading cases are considered, have been accounted for. Similarly, the design value of each bracing tier reaction should represent the maximum value determined from analysis of all loading cases.

Figure 17-12 illustrates the loading cases that might occur in a typical three-tier internally braced shoring system. Step E1 is analyzed as a cantilever and step E2 is analyzed as an anchored bulkhead, using commonly employed principles contained in authoritative texts. Step E2 (alternatively) and steps E3 and E4 are analyzed as braced excavations using lateral pressure diagrams appropriate for H_2, H_3, or H_4, respectively. At step R1, a reaction is assumed at the top of the base slab and the shoring system above the base slab is analyzed using the portion of the step E4 lateral pressure diagram remaining above the top of the base slab. Similarly, at step R2 a reaction is assumed at the top of the roof slab, and the shoring system above the roof slab is analyzed using the remaining portion of the same lateral pressure diagram. When apparent pressure diagrams are applicable to the design, the practice described immediately above for steps R1 and R2 is conservative because of the redistribution of applied lateral pressure that occurs when bracing tiers are removed. In this case, it is ordinarily considered adequately conservative to construct a new apparent pressure diagram at step R1 based on the depth H_{R1}, and to utilize this diagram for the remainder of the analysis.

Other loading cases may need to be considered, For example, when the lateral loading conditions on opposite sides of an internally braced excavation are not equal, the soil behind one of the shoring walls will be in a passive state, and that wall must be designed accordingly.

Design of Soldier Piles and Lagging. In a soldier pile and lagging shoring wall system, the soldier piles are designed as beams with bending moments and reactions at bracing tier or tie-back support locations as determined from the appropriate lateral pressure diagram. Unit values are simply multiplied by the soldier pile spacing to determine design values. When vertical or axial load is also imparted to the soldier pile, the combined bending and axial stresses are computed and appropriately accounted for. The soldier pile is considered fully braced against buckling in the plane of the lagging. In the plane perpendicular to the lagging, the column length is taken as the distance between supports. At the bottom of the excavation (subgrade), the soldier pile is considered fully braced in both directions. The methodology prescribed in *Manual of Steel Construction, Allowable Stress Design,* by the American Institute of Steel Construction (AISC or AISC Specifications), is used (predominantly) to determine the size of the soldier piles. Almost all public transit authorities and similar authorities permit design stresses in soldier piles equal to 120% (occasionally 125%) of the basic unit stresses permitted by AISC. In some cases, however, this practice may be overridden by local codes or other applicable codes.

The use of high-strength steel soldier piles (usually grade 50) has become common because of the economy offered by high-strength steel. However, on some cut-and-cover projects where wall stiffness is a concern, ASTM A36 structural steel (which remains commonly used also) is specified.

Lagging thickness design is necessarily based upon experience and/or empirical rules. Excavations in most soils retained with soldier piles and lagging will experience horizontal arching action transferring lateral pressure from the lagging to the soldier piles. The magnitude of the arching effect, in terms of reducing applied pressure on the lagging, depends on the type of soil and the soldier spacing and is not easily predicted by empirical rules or expressions. Some transit authorities have specified that wood lagging shall be designed to resist a lateral pressure equal to 50% of the lateral pressure indicated by the lateral pressure diagram on which the shoring wall design is based. This practice tends to be conservative, particularly where competent soils are being retained. Table 17-2 is a good guide to the selection of wood lagging thickness for different soil types and soldier pile spacing.

Design of Steel Sheet Piling. The procedure for determining the size of interlocking steel sheet piling in a cut-and-cover application is essentially the same as that for soldier piles, except that the analysis may be performed on a per-ft-of-wall basis. The data in literature published by the manufacturers of steel sheet piling include corresponding unit values for the design properties of each steel sheet pile section sold commercially. These properties may be used in all applicable AISC formulas. Basic unit bending unit stresses permitted may be taken as 0.65 times the yield stress of the steel ($0.65F_y$), which conforms to recommendations by U.S. manufacturers. Allowable stresses are then typically increased beyond basic unit stresses in the same proportion allowed for soldier piles.

Table 17-2 Recommended Thickness of Wood Lagging

Soil Type	Soil Description	Unified Classification	Depth	Recommended Thicknesses of Lagging (roughcut) for Clear Spans of:					
				5'	6'	7'	8'	9'	10'
Competent Soils	Silts or fine sand and silt above water table.	ML SM ML							
	Sands and gravels (medium dense to dense).	GW, GP,GM, GC, SW, SP, SM	0' to 25'	2"	3"	3"	3"	4"	4"
	Clays (stiff to very stiff); nonfissured.	CL, CH	25' to 60'	3"	3"	3"	4"	4"	5"
	Clays, medium consistency and $\frac{\gamma H}{S_u} < 5$	CL, CH							
Difficult Soils	Sands and silty sands (loose)	SW, SP, SM							
	Clayey sands (medium dense to dense) below water table.	SC	0' to 25'	3"	3"	3"	4"	4"	5"
	Clays, heavily overconsolidated fissured.	CL, CH	25' to 60'	3"	3"	4"	4"	5"	5"
	Cohesionless silt or fine sand and silt below water table.	ML, SM-ML							
Potentially Dangerous Soils	Soft clays $\frac{\gamma H}{S_u} > 5$	CL, CH							
			0' to 15'	3"	3"	4"	5"	—	—
	Slightly plastic silts below water table.	ML	15' to 25'	3"	4"	5"	6"	—	—
	Clayey sands (loose) below water table.	SC	25' to 35'	4"	5"	6"	—	—	—

Notes:

(1) In the category of "potentially dangerous soils," use of lagging is questionable.

(2) Lagging recommendations are based on douglas fir, construction grade (925f to 1000f) or equivalent.

(3) Where long-term settlement of the ground adjacent to the cut is a concern, common practice is to specify that treated wood be used for lagging that remains in place and is above the permanent groundwater table.

(4) Table adapted from Goldberg et al.

Design of Slurry Walls. The structural design of a slurry wall is carried out in essentially the same manner employed in designing any continuous reinforced concrete beam, using applied loading determined from lateral pressure diagrams appropriate for each loading case. The Strength Design Method as prescribed in the ACI code is most commonly used for analysis, but the Service Load Design Method, also as prescribed in the ACI code, remains in use also. In the modern practice of the design of slurry walls, construction methods and techniques that do not compromise the integrity of the finished wall are expected. Thus, neither load factors for Strength Design nor working stresses for Service Load Design are adjusted from corresponding ACI values to account for the method of construction used to construct slurry walls. Further, where cut-and-cover slurry walls are temporary structures, as is usually the case, permissible stresses in both concrete and steel reinforcement are typically taken as 120% of the corresponding basic unit stresses, where basic unit stresses are as hereinbefore defined for reinforced concrete.

The design of slurry walls, and the accompanying specifications, should be prepared by an engineer or engineering firm thoroughly familiar with slurry wall details and construction methods. Tremie concrete should have a minimum compressive strength of at least 3,500 psi for good results. A minimum compressive strength of 4,000 psi is commonly specified. The concrete should be a free-flowing mix that will displace the bentonite (or other stabilizing fluid) and bond to the reinforcing. An 8-in. maximum slump is usually permitted.

Design of Other Diaphragm Walls. Each of the other more commonly employed diaphragm walls described above (SPTC walls, drilled pile walls, soil–cement mixing walls) utilizes soldier piles as the principal structural element. The soldier piles are analyzed in the same manner as in a soldier pile and lagging wall. However, since the wall may be treated as a continuous wall below subgrade as well as above subgrade, the effective width of the soldier pile below subgrade may be assumed to be equal to the soldier pile spacing.

The compressive strength of the tremie concrete between soldier piles in an SPTC wall is normally at least 3,500 psi. Prior experience with SPTC walls should be studied when the quality of the tremie concrete is specified, since no realistic theoretical solution to the design of tremie concrete for SPTC walls is available. Similarly, prior experience with drilled pile walls should be studied when specifying concrete for drilled pile walls. In general, it is desirable that concrete for drilled pile walls be only as strong as is needed for safety and effective retention of soil and water, and for continuous passive resistance below subgrade. Unnecessary concrete strength will simply make the job of exposing the soldier piles at bracing tiers more difficult.

Design of Shoring Wall Support

In the following discussion, the wall support systems are considered temporary structures subjected to short-term maximum loads and stresses. This caveat applies particularly to tie-backs, which for cut-and-cover construction are not subject to the more severe design requirements and constraints applicable to permanent tie-backs.

Design of Internal Bracing. Unit reactions at each bracing tier are developed from the analysis and design of the shoring wall. If the shoring wall is soldier piles and lagging or a diaphragm wall where soldier piles are used as the principal structural member, the unit reaction is translated into horizontal point loads equal to the unit reaction multiplied by the soldier pile spacing. Similarly, if the shoring wall is a slurry wall, the unit reactions are translated into horizontal point loads equal to the unit reaction multiplied by the spacing of embedded plates. If the shoring wall is steel sheet piling, the unit reaction is usually treated as a horizontal, uniform load. The appropriate horizontal loading is then applied to the internal bracing framing for analysis and design. Figure 17-29 illustrates typical loading on internal bracing.

Ideally, soldier pile spacing and strut spacing are modular, so that the design of wales can be based on a regular pattern with allowances made for some departure of the soldier piles from their specified location. However, for long cut-and-cover projects, maintaining modular spacing of soldier piles is often unrealistic, and on many projects modular spacing is not used in any case. In designing wales, care must be taken into account for the varying relationship between soldier pile spacing (or slurry wall support point spacing) and strut spacing. In general, wale bending moment should not be taken as less than 0.8 times simple bending moment for spans between struts, nor less than the realistically foreseeable cantilever moment where wales are discontinuous between struts.

When the internal bracing framing consists of a systematic wale and strut configuration (as in Figures 17-11 and 17-29), it is usually satisfactory to calculate strut loads by simply multiplying the unit reaction by the strut spacing (or average strut spacing where applicable).

Internal bracing is typically designed using the methodology prescribed in the AISC Specifications. The practice of permitting design stresses in excess of basic unit stresses permitted by AISC varies with different public authorities, and it depends in part on the location of the cut-and-cover

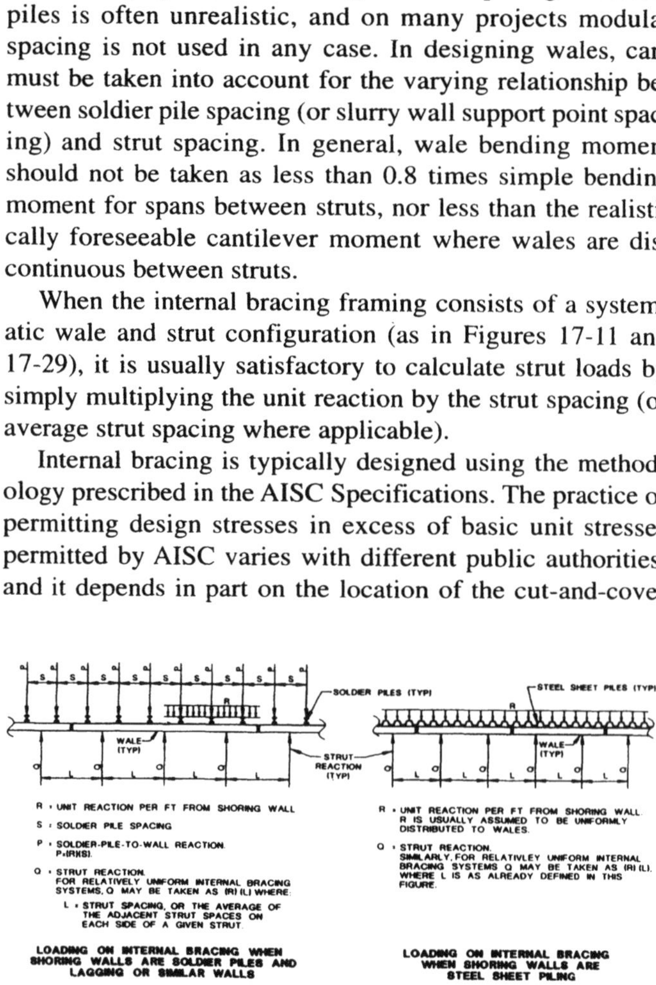

Fig. 17-29. Typical loading on internal bracing.

project, on the reliability of lateral pressure diagrams on which the design is based, and on the function of the structural elements of the framing. Design limitations on struts are often specified, for example, to account for the effects of ambient temperature changes, and for exposure of struts to impact loads during excavation and tunnel structure construction. Table 17-3 is recommended as a representation of modern practice with respect to allowable stresses and design constraints in the design of internal bracing.

Structural sections particularly suited for internal bracing should be selected. Relatively square wale sections are recommended (flange width greater than 0.45 times beam depth), for resistance to "rolling" and to facilitate the design of strut/wale connection details. Bearing pile (HP) shapes are often best as struts in relatively narrow cut-and-cover applications because of their thick webs. Where high strut loads preclude the use of bearing piles, column sections are usually preferred if the excavation is not too wide. Pipe sections are most commonly used in wide excavations. When pipe struts are used, the pipe walls should be sufficiently thick so that unnecessarily complex end details are avoided.

Preloading of the internal bracing system is almost always specified by public authorities on internally braced cut-and-cover projects. Preloading may be defined as the procedure by which stress is introduced into the internal bracing system before stresses due to applied lateral pressure are operative. Preloading is generally viewed as a desirable procedure that prevents initial inward movement of retained soil, minimizes deformation of the soil mass behind the sheeting, and (consequently) promotes better performance of the shoring system in terms of adjacent ground settlement. Usual preloading values range from 25 to 50% of the design load at each strut. Preloading is typically accomplished with hydraulic jacks operating between the wale and the strut. The opening created by the preloading operation is filled with steel shims.

Table 17-3 Recommended Allowable Unit Stresses for Internal Bracing

Bracing Elements	Design Loads		Loading Combinations and Allowable Stresses
	Vertical	Horizontal	
Wales	Weight of framing members (DL)	Wall reactions due to lateral earth and hydrostatic pressures (E)	DL + LL + E at 120% of basic unit stress for ASTM A 36 structural steel
	Weight of construction materials and other additional loads where applicable (LL)	Axial loads from end walls where applicable (E)	DL + LL + E at 100% of basic unit stress for grade 50 structural steels
Struts	Weight of strut plus connected framing (DL)	Axial loads from wales or directly from walls (E)	DL + LL + E at 100% of basic unit stress with the following additional constraints: • Slenderness ratio, l/r, not to exceed 120 • Axial stress in struts to be limited to 12,000 psi
Secondary Bracing	Own weight (DL)	Axial load equal to 3% of the design load in the more heavily loaded adjacent braced main member (LL)	DL + LL + E at 120% of basic unit stress. Slenderness ratio not the exceed 200
Welding, Bolts	As applicable		DL + LL + E at 100% of basic unit stress

Design of Tie-Backs. In a tied-back wall system, the horizontal component of load to be resisted by a given tie-back is determined by multiplying the unit reaction at the tie-back row or tier by the horizontal spacing of the tie-backs in that row or tier. The design value of tie-back load is then taken as the calculated horizontal component of force multiplied by the secant of the angle of inclination of the tie-back. The tie element is then sized to provide a capacity in tension equal to or exceeding the tie-back load. Where uniform tie elements and wall reaction assemblies are desired, this process is more or less reversed, so that the vertical spacing of tie-back rows is such that the tie-back loads are equal to or less than the capacity of the tie elements and wall reaction assemblies selected.

For conventional tie-backs, the length of the anchor zone is first evaluated by computing the area of grout-to-soil contact needed to provide within the anchor zone frictional resistance or adhesion corresponding to the design value of the tie-back load multiplied by an adequate factor of safety. Unit values for available grout-to-soil friction (for granular soils) or adhesion (for cohesive soils) are largely based on prior experience with particular soils, but they depend also on the depth of the tie-back (for granular soils) and the grout placing technique. For regroutable tie-backs, the length of the anchor zone and design of the special material and hardware within the anchor zone are established initially based on prior experience with the particular type of soil being retained. For both conventional tie-backs and regroutable tie-backs, a tentative design of the anchor zone is produced.

The design of anchor zone can be considered final only when the tie-back has been successfully tested in place. Every tie-back should be individually tested. Normal tests or "proof tests" consist of subjecting the tie element of the tie-back to a proof load ranging usually from 133% of design load at less critical locations to 150% of design load at more critical locations (i.e., highly urbanized areas). Typical specifications for proof testing require that the proof load be applied in increments and that after the proof load has been reached, it shall be held for a specified time (ranging from 10 to 30 min), during which the tie-back must conform to specified performance criteria. The tensile force in the tie element is then relaxed to 100% of the design load, at which time the anchor head is fixed in place. In critical applications, more rigorous "performance tests," requiring testing procedures similar to proof testing procedures for up to 24 hours, are usually specified also for a certain percentage (say 1–5%) of tie-backs.

Threadbars used as the tie elements in tie-back design usually have a guaranteed ultimate unit strength (f_{pu}) of 150 ksi. For strands used as the tie elements or tendon, f_{pu} is typically 270 ksi. The yield strength of these steels is very nearly equal to f_{pu}, and common practice is to specify allowable stresses in these steels as a percentage of f_{pu}. For tie-backs used in cut-and-cover construction, proof test stress is normally specified to be $0.80\,f_{pu}$ or less. Working stress is normally specified to be $0.60\,f_{pu}$ or less. The minimum size of the element is determined by the more demanding of these two requirements. For the components of the wall reaction assembly, stresses exceeding basic unit stress by 133% can be allowed during proof testing.

Tie-backs for short-term use have seldom needed any special corrosion protection. When, however, there is a potential for rapid corrosion, as may be the case when the tie element could be exposed to salt water, for example, the manufacturer of the tie element should be consulted.

Design of Soil Nailing Systems. There is no universally accepted analytical method for the design of soil nailing systems. Most methods in use in the United States today are limit equilibrium methods contained in specific computer programs. Typically the physical properties of the soil are entered, as are the design properties of the nails (such as length, diameter, spacing, tensile capacity, inclination, friction bond with the soil and other data, depending on the particular program). In each computer method, a number of failure planes, each passing through the base of the excavation, are generated, and a factor of safety is computed for each. The lowest factor of safety computed identifies the critical failure plane. Until engineering practice in the design of soil nailing systems is more codified, it is recommended that the design of these systems be undertaken only by individuals or firms thoroughly familiar with design methods now in use and with the performance of soil nailing systems under conditions similar to any proposed system to be designed.

PERFORMANCE OF SHORING SYSTEMS

The design and construction of shoring systems for cut-and-cover tunnel construction depends in part on the importance of preventing adjacent surface settlement. To evaluate the need for underpinning of adjacent installations or other precautionary measures, the engineer must be able to predict the probable amount of settlement and its effect on the installations. The amount of settlement will depend upon the soil type, the size of the excavation, the construction methods and quality of workmanship, and the design of the shoring system itself. Prediction of the amount of surface settlement based on soil tests and theory is very uncertain, and experience and observational data are needed as guides to good judgment.

The magnitude of ground settlement may also be considered as related to the vertical and horizontal displacement of the ground outside the support system. The causes of such displacement include the following:

- Horizontal movement of the shoring wall (translation, rotation, elastic deformation).
- In soldier pile and lagging systems, loss of ground due to failure to fill or pack any voids behind wood lagging with sufficiently cohesive soil.
- Base instability or near instability of cohesive soils, and consequent upward movement of the soil into the base of the excavation. Also, similar effects due to stress relief of the soil below subgrade due to excavation.
- Densification of loose granular soils due to the action of vibratory pile driving equipment.
- Loss of ground due to pulling soldier piles or sheet piles.

The curves shown on Figure 17-30 define zones in which many recorded cases of settlement behavior have fallen as reported by Peck (1969). The cases represented on Figure 17-30 may be used as a general guide in evaluating the potential for ground settlement adjacent to internally braced or tied-back flexible walls.

Settlement adjacent to the excavation can also occur when the site is dewatered and when, as a result, increased effective stress causes consolidation of compressible silts and clays or loose sands. In this discussion of the performance of shoring systems, any settlement due to dewatering is, for the sake of clarity, considered unrelated to shoring system performance.

Cuts in Granular Soils

Dewatered granular soils or granular soils above the groundwater table usually possess enough cohesion to facilitate excavation when the shoring walls are soldier piles and lagging. With careful lagging placement, wall performance very nearly comparable with that for continuous walls, all other considerations being equal, can often be achieved. If the cut is properly braced, the maximum settlement is not likely to exceed 0.5% of the excavation depth, and in competent granular soils it is often held to 0.2% or less. Significant ground settlement does not ordinarily extend beyond the potential plane of failure as established by Culmann graphical construction. If the adjacent ground carries no surcharge load, this criterion translates into insignificant ground

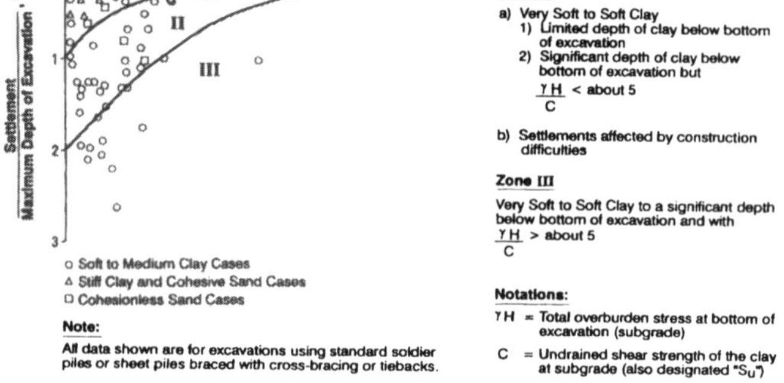

Fig. 17-30. Summary of settlements adjacent to cuts in various soils, as a function of the distance from the edge of the cut, as reported by Peck (adapted).

settlement at a distance behind the cut equal to 0.5–0.6 times the depth of the cut.

The experience reported above applies to both internally braced and tied-back walls. Also, the shoring system performance is not significantly affected by the wall type. One probable reason for little apparent difference between wall types or support methods is the fact that measured displacements are small, and many construction factors can contribute to displacement variation, thus masking such differences.

Cuts in Stiff to Hard Clays

The magnitude of maximum ground settlement adjacent to cuts in stiff to hard clays is normally in the same range as that for competent granular soils, although evidence of ground settlement may extend farther back from the cut. In soldier pile and lagging applications, these soils are usually not sensitive to ground loss, because of seepage pressures. In addition, the excavated face is stable, so that there should be no appreciable ground loss prior to placement of the lagging.

The exception to this comparison can be at cuts in highly overconsolidated, stiff-fissured clays. When these soils are supported by flexible walls, ground settlement behavior approaching that depicted by Zone II in Figure 17-30 can occur.

Cuts in Soft to Medium Clays

In this discussion, *soft clay* generally refers to very soft clay ($S_u \leq 250$ psf), clay normally classified as soft (S_u = 250–500 psf), and medium clay (S_u = 500–1000 psf). The weaker of the clays normally classified as stiff may in deep excavations behave similiarly to soft clays in a shallower excavation. In these discussions also, the shoring walls are assumed to be internally braced, since tie-backs are seldom used in clay applications, particularly where adjacent ground settlement is a concern.

Conventionally Supported Cuts. When a cut is excavated in soft clay, the clay located at the sides of the cut acts like a surcharge on the horizontal plane at the elevation of the excavated subgate. If the clay is unrestrained or partially restrained, the clay yields laterally toward the cut above and below subgrade, and the subgrade itself rises. When the shoring system is a more or less conventional system designed to conform to the empirical rules illustrated in Fgures 17-18b and 17-23 or 17-24, the inward movement of the shoring wall below the lowest bracing tier tends to correspond to the inward clay–soil movement that would occur if the shoring wall were only as deep as the excavated subgrade. The stiffness of the shoring wall affects the extent to which this behavior takes place, but ordinarily not very much. The total soil-mass movement greatly exceeds what occurs when granular or stiff-clay soils are retained. Thus, for cuts in soft clays both the magnitude and the extent of ground settlement adjacent to the cut are substantially greater than for the cuts in granular or stiff clay soils, as is shown by the recorded cases plotted on Figure 17-30.

Prediction of ground settlement behavior within a much narrower range than is indicated by Figure 17-30 can usually be made for cuts in soft clay in specific cases when all the principal factors affecting settlement are either known or can be estimated within reasonable limits. The most important of these factors are the following:

1. The depth of the excavation, the consequent value of γH as defined earlier, the undrained shear strength of the clay at subgrade, and the consequent stability number as also defined earlier.
2. The width of the excavation and the factor of safety (F.S.) against bottom heave. F.S. can be estimated knowing step 1 and the width of the excavation.
3. The stiffness of the shoring walls (particularly) and internal bracing, the spacing of level tiers, and the values H and Z at each excavation step on Figure 17-12.

Using realistic values or evaluations of steps 1, 2, and 3, theoretical criteria, and ground movement models, experienced geotechnical engineers can construct surface settlement profiles indicative of actual ground settlement to be expected adjacent to cuts in soft clay. However, deviation of 50% either way from such estimates of settlement should not be considered unusual.

Cuts with Deep Shoring Walls. In an excavation in soft clay where stiff shoring walls are taken down into competent soils below subgrade (as illustrated in Figure 17-25), bottom heave that might otherwise occur is typically prevented by the presence of the shoring wall. When the soil in which the bottom of the shoring wall is founded is sufficiently strong, there will be no significant movement of the bottom of the shoring wall. Under these conditions, the factors affecting inward movement of the shoring wall are largely limited to the stiffness of the shoring wall, the stiffness and spacing of internal bracing tiers, and the horizontal loading on the shoring wall during each of the loading cases to which the shoring system is subjected. In such cases, the amount of ground settlement adjacent to the cut can be held well below the limits of Zone I on Figure 17-30 when sufficient stiffness is built into the shoring system. Underpinning of buildings adjacent to cuts in soft clay can often be avoided when stiff diaphragm walls are utilized in this application.

Factors Influencing Allowable Movements

When a cut-and-cover tunnel structure is to be constructed in an urban area, the most important factor influencing permissible ground movement adjacent to the cut is usually the presence of adjacent buildings. Those portions of a building located inside the zone of movement will suffer vertical displacement. The difference in vertical displacement between any two settlement points (such as two footings) is referred to as differential settlement. The ratio of differential settlement to the distance between settlement points is referred to as the angular distortion. (See Figure 17-31.) The tolerable angular distortion that various struc-

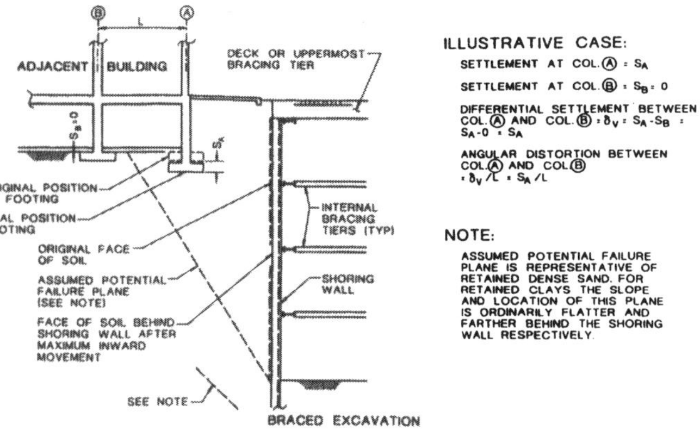

Fig. 17-31. Differential settlement and angular distortion.

tures may experience depends on the type of structure, the age and condition of the structure, and other variables. Figure 17-32 may be used as a guide to tolerable angular distortion to which buildings may be subjected. When conservatively predicted angular distortion exceeds the limits that may be established for a building, precautionary measures such as underpinning are required.

In general, horizontal soil movement corresponding to vertical ground settlement is accounted for in Figure 17-32. However, when portions of buildings on pile foundations are located inside the zone of movement, the horizontal drag effect on the piles should be analyzed by an experienced geotechnical engineer.

Differential ground movement that would damage an important utility also cannot be permitted unless appropriate precautionary measures are taken. In many such cases, this concern is obviated by requirements for building protection. However, in some cases it may be necessary to move the utility prior to proceeding with the excavation. In other cases it may be more feasible to underpin the utility.

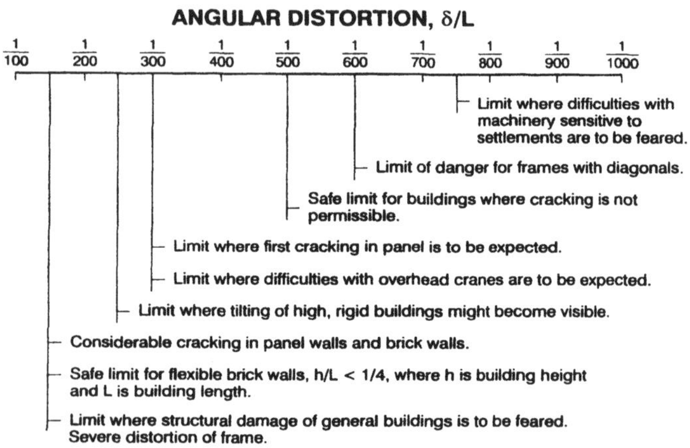

Fig. 17-32. Damage criteria for settlement of the buildings under their own dead weight.

DECKING

Decking is used in cut-and-cover construction to minimize surface disruption to normal activities during the construction period. Through the use of decking, the duration of complete or severe surface interference can be limited to a short time at the beginning of the construction when the initial lift of excavation is made and the desk itself is placed; and again for a short duration near the end of the construction when the deck is removed and final backfilling and surface restoration are completed. Decking is ordinarily required when vehicular traffic over and in the direction of the cut beneath must be maintained, or where the cut crosses a road or street on which vehicular traffic must be maintained.

Design

"Decking" or "street decking" may be considered to consist of deck framing and roadway decking. Figures 17-33a and b illustrate a typical general arrangement for street decking over a cut-and-cover excavation. The deck framing is composed of cap beams, deck beams, and secondary framing. Roadway live, impact, and dead loads are first imparted to the deck beams, which carry the loads to the cap beams. The cap beams in turn distribute these loads to the supporting shoring wall. The secondary framing serves to limit the unsupported length of the compression flange of the deck beams as well as to direct dynamic longitudinal loads to the cap beams. The deck beams are typically structural steel wide flange sections, and the cap beams are usually steel bearing pile (HP) sections. SFBART, WMATA, Los Angeles MTA, and other authorities specify that the deck framing shall be designed for AASHTO HS 20-44 loading or for loading due to the construction equipment that will operate on the deck, whichever is greater. Allowable stresses in the deck framing are limited to basic unit stresses as prescribed by AASHTO. For deck beams, however, maximum deflection due to service live load and impact equal to 1/600 of the span is often permitted. Some public authorities have specified that impact for construction equipment operating on the deck shall be considered to be 50% of the weight of the construction equipment.

Since the deck is a temporary structure, it is reasonable to recognize this fact when the deck is not carrying public traffic. For example, permissible deflection due to service live load and impact can be increased to, say, 1/500 of the deck beam span when the live loads are construction equipment. Similarly, the allowable shear stress in the webs of cap beams may be increased by 20% when the shear stress is due to construction equipment. Normally, the vertical load imparted to shoring walls need not include impact either for AASHTO loading or for construction equipment loading.

The structural steel deck beams are sometimes utilized also as struts, so that the deck structure becomes the uppermost bracing tier. Figure 17-33c illustrates a common application of this concept. Where soldier pile spacing is such that deck beams can rest directly on the soldier piles, cap

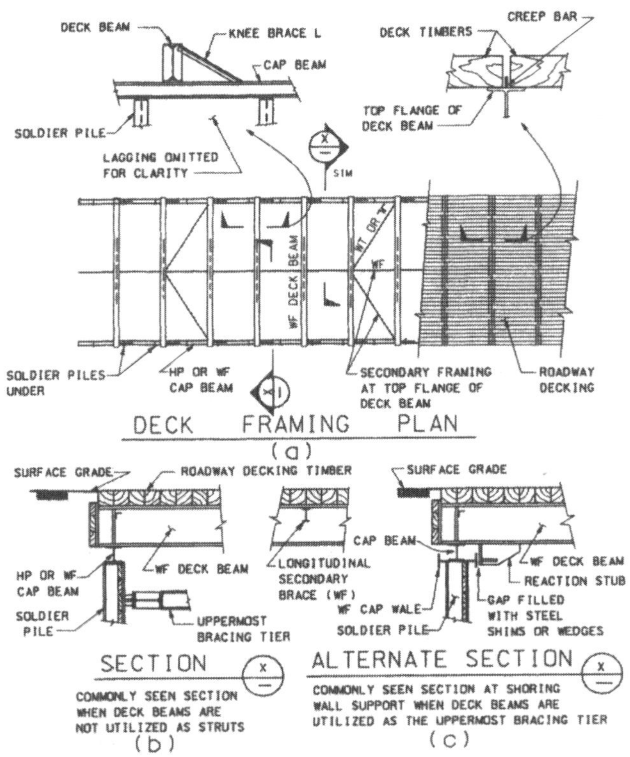

Fig. 17-33. Street decking—commonly seen framing plan and sections.

beams can be eliminated. However, this arrangement usually results in roadway decking of irregular length (see below). With properly designed deck-beam-to-shoring-wall connections, cap beams can more satisfactorily be eliminated when the shoring wall is steel piling or a slurry wall.

In the United States, the most common material used for roadway decking is timber. The timber is usually made into mats 4–6 ft wide and nominally equal in length to one or two spaces between deck beams. The roadway decking is frequently not fixed to the deck beams, although a positive means must be provided to prevent it from moving off the support due to vibration or other dynamic forces from the traffic (see Figure 17-33). Deck beam spacing in the range of 10–15 ft is normal, with 12-ft spacing probably the most common. Timber mats comprised of 12 in. × 12 in. timbers (1,350 f or better) are usually satisfactory for cut-and-cover construction when the deck beams spacing is 12 ft. In Europe and Asia it is more common to use precast reinforced concrete slabs as roadway decking. Concrete roadway decking could become more common in the United States as well. Partially hollow, precast reinforced concrete slabs were used on some Los Angeles MTA subway construction contracts.

Installation

Prior to deck installation, when the shoring walls are installed, it is usually possible to occupy only one or sometimes two traffic lanes at a time when the shoring walls are being constructed. Following installation of the shoring walls, when the initial lift of excavation is made and the deck is installed, it will be necessary to divert vehicular traffic to the side (or sides) of the cut, or to close the city street in increments to permit deck installation. When it is necessary to provide some public traffic lanes during the installation of a wide deck, split decking may be used. With this technique, pin piles are usually first driven in a row approximately midway between the shoring walls. A center cap beam is then placed over the pin piles. The deck is subsequently placed one-half width at a time. Each half-width deck beam is then supported at the center of the cut on the center cap beam. Several variations of this concept have been used. However, it is usually preferable to close the city street temporarily and construct the full-width deck.

EXCAVATION AND GROUNDWATER CONTROL

Excavation Methods and Equipment

Internally Braced Excavations. The two most commonly used pieces of equipment for excavating braced cuts are the backhoe and the clamshell bucket. Both are typically mounted on crawler units and operate from the surface either alongside the excavation or on the deck structure (when a deck is provided), depending on surface conditions. Both are particularly suited to excavate in the space between horizontal struts. Both may dump excavated material into trucks, a holding bin, or a stockpile alongside the excavation. The clamshell bucket is particularly efficient for cuts 20 ft or more in depth. However, modern backhoe equipment, when it can be properly positioned, can also excavate efficiently at a depth exceeding 30 ft. At depths exceeding 20–30 ft, excavating with a clamshell bucket, although often perceived as slow and inefficient, has nevertheless proved to be the most satisfactory construction method in most cases. As the excavation proceeds downward, low-profile tractor-dozers or tractor-loaders typically excavate the soil and doze or tram the material to a position where it can be handled by the backhoe or clamshell bucket. When the cut is sufficiently wide, the center of the excavation may be carried at a lower elevation than the sides of the excavation to facilitate dozing or tramming. However, the resulting berms should be massive enough to provide adequate passive resistance at the shoring walls. The required size of the berms at any stage of the excavation is easily evaluated. Nevertheless, the depressed center concept is often abused, causing unforeseen inward movement of the shoring walls and consequent, otherwise preventable, surface settlement. With care, it is quite feasible to excavate an internally braced cut in the steps shown on Figure 17-12, using the general excavation methods described.

Other methods of excavation are sometimes used in cut-and-cover construction when the shoring walls are internally braced. Extendible, vertical or inclined belt conveyors have been employed, for example, to raise the excavated material from the "hole" and deposit it into a truck-loading bin.

When this type of approach is used, low-profile load–haul–dump units or similar equipment operates in the cut and transports the excavated material to the conveyor feed point. In a few cases, where the soil to be excavated is competent and dry, and the cut is sufficiently wide, it has been possible to carry out the excavation using trucks with haul ramps from the surface to the bottom of the excavation. Ordinarily, however, it is not feasible to maintain the required vertical distance between bracing tier levels and the excavated subgrade using the haul-ramp method.

Tied-Back Excavations. When shoring walls are supported with tie-backs, the excavation is carried out in lifts equal in depth to the vertical spacing of the tie-back rows. Almost any excavation method that will limit the vertical distance between a tie-back row and bottom of the excavation to the prescribed amount, at any step in the construction sequence, can be considered. If the cut-and-cover excavation is sufficiently long, utilities or decking crossing the excavation are not a problem, and the soil to be excavated is dry and competent enough to act as a haul road, the most suitable excavation method that employs a haul ramp out of the cut will usually be the most efficient. At the end of the excavation, however, when the haul ramp itself must be removed, the more common methods described will need to be employed to complete the excavation.

Groundwater Control. At many cut-and-cover tunnel sites, the natural groundwater table will lie well above the level of the bottom of the excavation. When it is feasible to do so, it is almost always desirable and economical to lower the groundwater table to a level below the planned elevation of excavated subgrade before the excavation itself commences. This lowering of the groundwater table is often referred to as either "predraining" or "dewatering." However, the term "dewatering" has also had a broader meaning. When saturated pervious soils are to be retained, the principal advantages offered by predraining are the following:

- Cut-and-cover construction is performed "in the dry" (which is ordinarily necessary in any case inside the excavation).
- The lateral pressure imparted to shoring walls is reduced, and the shoring system, as a result, is simpler and less costly.
- Soldier piles and lagging becomes a feasible option for shoring walls.
- In many cases, an unstable condition at subgrade is prevented.

PREDRAINING WITH DEEP WELLS. The most common method used for predrainage of a cut-and-cover site is to remove groundwater with deep wells. Since the excavation is relatively deep, deep wells are ideal for this purpose when feasible. Although the range (in grain size) of soils in which deep wells can be effective is clearly limited, modern deep well design and construction techniques and methods of aquifer analysis have made it practical to use deep wells on many projects where they were once unsuitable. As a result, in the cases where predraining cannot be accomplished with deep wells, the cut-and-cover construction is usually prosecuted without predraining, using, where required, adequately stiff and deep, continuous shoring walls to avoid unstable excavating conditions. Figure 17-34 may be utilized as a guide in determining when a soil is sufficiently coarse for the effective use of deep wells.

When a site is to be predrained with deep wells, the geotechnical data needed to design the predrainage system should first be obtained. This data should include the results of representative mechanical (sieve) analyses of soil samples recovered from soil borings, and the results of pump tests conducted at test wells at selected locations. On larger projects these data are usually provided in the geotechnical report. The design of the wells themselves, including well spacing, should be developed with the assistance of a geotechnical consultant experienced in the field of construction dewatering, or by a well-drilling contractor with comparable geotechnical experience.

A typical deep well on cut-and-cover projects consists primarily of a well casing and screen placed in a drilled hole, a "gravel pack" or filter envelope in the annulus between the well screen and the drilled hole, a submersible pump at the bottom of the well casing/screen, and a discharge pipe (see also Chapter 6). The principal components and other features of the well, the well depth, and the well spacing are all designed to suit subsurface conditions and predraining requirements. Although great precision in analyzing well performance is ordinarily not attainable, the results of hydrologic analyses are usually either acceptable or are correctable with additional wells or revised well spacing. On cut-and-cover projects where the deep wells are normally located outside the shoring walls, well spacing can vary from 20 to 200 ft or more. When it is practical to do so, it is usually desirable to lower the groundwater table (both inside and outside the shoring walls) to at least 5–10 ft below the planned bottom of the excavation.

When it is planned or mandated that the groundwater table outside continuous watertight shoring walls is to remain at or near its natural elevation, deep wells may be installed inside the shoring walls to predrain the excavation (when it is safe, feasible, and otherwise desirable to do so).

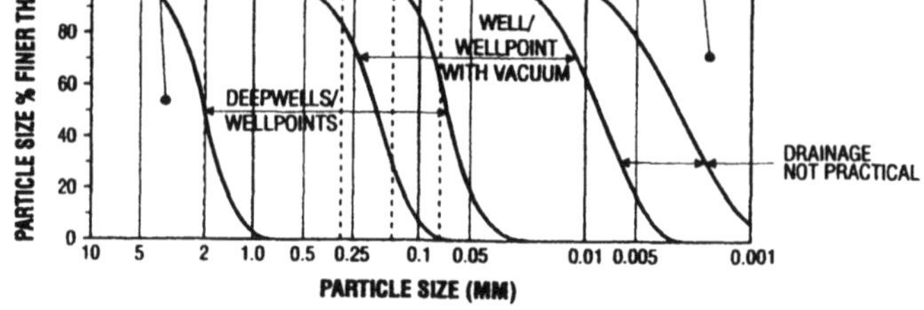

Fig. 17-34. Range of particle size for selection of a dewatering method (adapted from Harris).

Other Predraining Methods. Occasionally, particular groundwater lowering requirements can be best accomplished using wellpoint systems (Chapter 6). In some stratified soils (and in other cases) it may, for example, be difficult to effect an acceptably dry subgrade with deep wells. This problem can often be solved by installing conventional wellpoints inside the excavation when the excavation is partially or nearly complete. The wellpoint headers are typically placed adjacent to the shoring walls at an elevation low enough to permit internal predraining to a level below subgrade. Multistage wellpoint systems can also be used in this or similar applications. In some special cases, related systems employing wellpoints, such as vacuum wellpoints or eductor wells, have been utilized in predrainage applications.

Pressure Relief Wells. In this chapter a "pressure relief well" may be considered to be any well capable of physically pumping water from a stratum of pervious soil (aquifer) lying beneath impervious soil or soil of relatively low permeability. Most pressure relief wells are deep wells designed for this purpose. In a braced excavation there can be a dangerous potential for "blow-in" at subgrade when hydrostatic uplift pressure at the top of such an aquifer ($\gamma_w H_T$) exceeds the unit weight of the soil and water ($\gamma_s H_2$) in the "plug" below subgrade and above the top of the aquifer (see Figure 17-35) (The consequences of blow-in can be similar to those resulting from piping failure, which is discussed below.) When these subsurface conditions exist, pressure relief wells, which will reduce the piezometric head in the aquifer to an acceptable level, should be installed. Pressure relief wells are ordinarily not required when the factor of safety against uplift/blow-in (FS_B) exceeds 1.1, where $FS_B = \gamma_s H_2 / \gamma_w H_T$ and frictional effects (at the sides of the "plug") are neglected. The value H_T, which is the piezometric head at the top of the aquifer, should be determined from piezometers, since H_T may not correspond to the value that would be indicated by the elevation of the natural groundwater table.

The case illustrated in Figure 17-35 is representative of conditions that are not uncommon when a cut-and-cover site is not predrained or when a lower aquifer is confined and is unaffected by predraining efforts. The potential for uplift and consequent blow-in may also be present under other conditions of subsurface stratification. Whenever there is a potential for high piezometric head in pervious strata below subgrade, care should be taken to ensure that subgrade stability has been appropriately analyzed, and that any requirement for pressure relief wells is provided.

Stability Against Piping. When a cut-and-cover site is not predrained, and more or less homogeneous pervious soil lies at and below excavated subgrade, piping or "sand boiling" can occur if the continuous shoring walls are either not founded in a lower impervious stratum, or are not otherwise sufficiently deep. Under these conditions the resulting unbalanced hydrostatic head causes large upward seepage pressures in the soil below subgrade. At (piping) failure, a "quick" condition is produced, and the sand "boils" in the bottom of the excavation. Unless immediate remedial steps are taken, water and soil will flow upward into the excavation, thus undermining the shoring walls and disturbing the soil at and below subgrade, so that it may be unsuitable as a foundation for the tunnel structure. Loss of ground attendant to a piping failure will cause additional surface settlement adjacent to the cut including, in many cases, large "sinkholes" immediately behind the shoring walls. In extreme cases, a catastrophic failure of the shoring system may occur.

Where piping is a concern, the factor of safety against piping failure (FS_P) may be considered to be a function of the depth of the penetration of the shoring wall below subgrade (D), the unbalanced hydrostatic head (H_u), and the width of the excavation (B). (See Figure 17-36.) In general, FS_P will increase when D is increased, when H_u is decreased, and when B is decreased. To evaluate FS_P in homogeneous, pervious soils, a flow net may be constructed in ac-

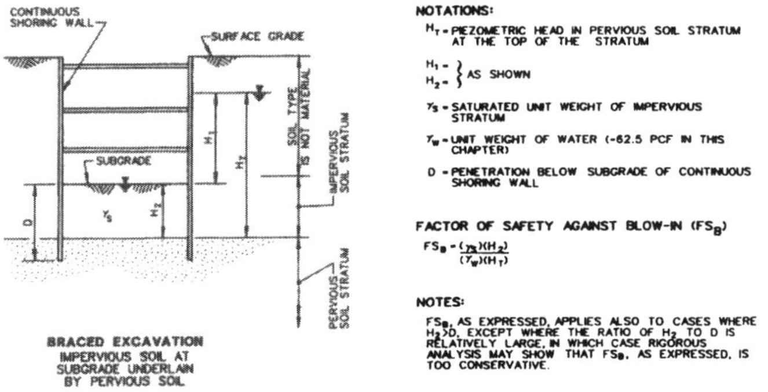

Fig. 17-35. Potential for blow-in at subgrade due to the presence of a pervious stratum at depth H_2 below subgrade (NAVFAC 1974).

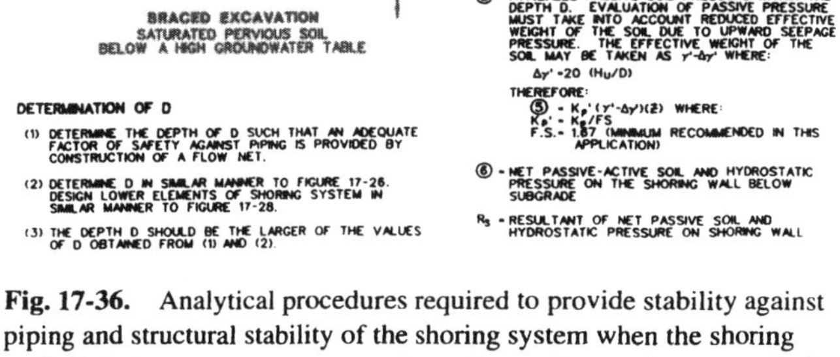

Fig. 17-36. Analytical procedures required to provide stability against piping and structural stability of the shoring system when the shoring walls lie in homogeneous, saturated, pervious soils, or when an equivalent condition exists below subgrade (see NAVFAC (1974) and USS (1975) for alternative methods of determining D).

cordance with the technique prescribed in standard textbooks on soil mechanics. In general FS_P should not be less than 1.5 and should usually be much more (2.0–4.0 or more) depending on the severity of the consequences of piping failure and on the reliability of assumed subsurface conditions. In the usual case, the determination of FS_P is complicated by the nonhomogeneous or stratified nature of the soils, and it should be evaluated with the assistance of a competent geotechnical engineer familiar with local conditions and experience.

As a general rule the conditions represented by Figure 17-36 should be avoided wherever possible, by predrainage or, in certain cases, by installing pressure relief wells. When, as in Figure 17-36, there is an otherwise manageable flow into the excavation, the flow must be directed into sumps from which it can be pumped. Also, the design of the shoring walls must take into account the reduced available passive resistance below subgrade, which results from upward seepage pressures, as well as properly accounting for hydrostatic pressure on the shoring wall (Figure 17-36).

When pervious soil outside of the shoring walls is not predrained, a "blow" can occur through a split in the sheet piles (out of interlock) or through an open joint in a slurry wall or other diaphragm wall. The consequences of such a "blow" can be similar to a piping failure.

INTERNAL CONTROL OF WATER. At some cut-and-cover sites, predominantly competent soils that cannot be effectively predrained will nevertheless contain lenses of pervious soils. When the shoring walls are soldier piles and lagging, some groundwater may flow into the cut and may carry soil fines with it, causing loss of ground and settlement. To reduce this loss of fines, straw is sometimes packed behind the lagging. Straw can be reasonably effective as a filter, but the generous use of straw can itself create voids, leading to subsidence. Burlap can be used as an alternative to straw, with generally better results.

When water seeps into the excavation through a soldier pile and lagging wall, there are some seepage pressures applied to the wall. In the design of the wall, however, these seepage pressures are usually considered both minor and temporary, and they are ordinarily neglected. Some water will accumulate in many excavations regardless of the wall type and the construction dewatering methods. This water should be directed into sumps and pumped to the surface more or less continuously as the excavation progresses.

SETTLEMENT DUE TO CONSTRUCTION DEWATERING. Dewatering can cause settlement adjacent to a cut in several ways. These include (primarily) the following:

1. Fines can be removed from the soil through improperly constructed wells or wellpoints.
2. Pumping that is required from within an excavation when the site is not predrained can cause blow-in, piping, or a blow through splits or open joints in shoring walls, as has been described.
3. Increased effective stress causes consolidation of compressible silts and clays or loose sands. ("Consolidation" in this chapter refers to any applicable mode of deformation of the soil).

The first two causes can be controlled or prevented with appropriate design procedures and construction techniques. Dewatering removes buoyancy from the soil, however, and therefore increases the effective vertical stress in the soil. This increased loading causes consolidation of the soil. The magnitude of the consolidation is ordinarily small in dense sands and silts, stiff clays, and stronger soils. But when the subsurface geologic profile contains compressible silts, clays, peats, loose sands, or other weak soils, the magnitude of consolidation and consequent ground settlement can be significant. The amount of consolidation to be expected in each compressible stratum, and the resulting surface settlement, can be estimated based on consolidation theory or, when appropriate, on other methodology. These analytical solutions (settlement analysis) must in turn be based on particular soil properties determined from laboratory analyses of soil samples. When settlement adjacent to a cut is an important concern, the settlement analysis should be performed by an experienced geotechnical engineer who will recognize realistically such considerations as the effect of overconsolidation of clay soils, time-related settlement behavior, and the significance of predicted differential settlement.

The ground settlement described is typically caused by predraining pervious strata that overlie or underlie compressible silts or clays. When the construction dewatering is limited to relief wells, similar consolidation of a compressible soil stratum can occur when the compressible soil stratum overlies the pervious stratum from which groundwater is being pumped. Settlement analyses may show that predraining outside of the shoring walls cannot be tolerated because of consequent unacceptable surface settlement. These analyses may show that there should be constraints on the use of relief wells as well. On important cut-and-cover projects, it is almost always preferable that these determinations be made by the geotechnical consultant, who is responsible for geotechnical input to the design of the tunnel structure.

PERMANENT SHORING WALLS AND SUPPORT

Occasionally, slurry walls or SPTC walls are used both as shoring walls for the cut-and-cover construction and as the permanent walls of the tunnel structure. When this concept is adopted, it is usual practice to place a reinforced concrete curtain wall inside the shoring wall for (largely) architectural purposes. When the permanent structure lies below the groundwater table, care must be taken to be certain that there is complete bond between the slurry wall or SPTC wall and the interior curtain wall.

Internal bracing tiers may also be designed to serve both as internal support during construction and as permanent support. In this application the permanent internal bracing tier (or tiers) typically also serves as the structural steel framing for an intermediate floor (floors) as well.

The concepts described above were successfully incorporated into the design and construction of several SFBART subway stations.

REINFORCED CONCRETE MATERIALS AND CONSTRUCTION

Subgrade Preparation and Protection

The procedures necessary for satisfactory preparation of the foundation on which the base slab for a tunnel structure is to be placed are governed both by design requirements and by the type and condition of the foundation material. In general the base slab should be placed on a foundation essentially equivalent in competence to the in situ soil, as perceived when the base slab was designed. When the stratum at the bottom of the excavation is a competent granular or clayey soil, the base slab can often be placed directly on the subgrade. Disturbed material should, however, be removed and replaced with concrete or other satisfactory material adequately compacted. Similarly, unsuitable in situ soil should also be removed and replaced (to the extent practicable without redesign of the base slab). The subgrade should be moist but not wet when the concrete is placed.

In many cases, due to construction activity, otherwise competent clayey or silty soils (usually saturated) will deteriorate to the extent that preparing an adequate natural subgrade is very difficult. In these cases it is often preferable to place a plain concrete leveling slab, usually 4–6 in. thick, on which the base slab is subsequently placed. When these conditions are foreseen during the design of the tunnel, leveling (or "working") slabs are often specified.

When the tunnel structure is to be founded on a soft to medium clay, usual practice is to specify that a geotextile fabric be placed on the excavated surfaces followed by a 12-in.-thick (usually) processed gravel bedding. The subgrade is thus able to support light spreading and compacting equipment, and a suitable foundation of uniform competence is created. The processed gravel usually contains enough moisture for subsequent placement of the base slab, but it may be moistened prior to concrete placement if the surface is not already sufficiently moist.

Concrete Mix Design

The objectives of good concrete mix design for cut-and-cover tunnel construction are (1) production of a concrete that conforms to realistically achievable design requirements and expectations and (2) production of a concrete mix that can be placed efficiently using modern placing methods and techniques. In many respects these two objectives are interrelated. In formulating the mix design, concrete ingredients should be selected and proportioned so that they will produce specified or desired results with respect to the following (principal) properties:

- Strength
- Impermeability
- Durability
- Workability

The strength, impermeability, and durability of the concrete depend greatly on the quality of the cement paste. The quality of the cement paste depends on the ratio of water to cement (water–cement ratio). In general, strength, impermeability, and durability are decreased as the water–cement ratio increases. For cut-and-cover tunnel construction, current practice is to specify a maximum water–cement ratio (by weight) ranging from 0.40 to 0.45. These low water–cement ratios are made practical by the use of admixtures (see below) in the concrete mix. Admixtures are used to promote other characteristics of the concrete mix as well, but they must be used judiciously if their effect on drying shrinkage (and consequent impermeability) is to be minimized. Drying shrinkage and other properties of concrete are affected by the type of cement in the mix as well. Type II cement (which can be very nearly regarded as a commercial standard) is normally specified.

The strength, impermeability, and durability of concrete can be affected by the quality of fine aggregate (sand) and coarse aggregate (gravel) as well. Commonly specified requirements for gradation, specific gravity, abrasion resistance (coarse gravel), soundness, deleterious substances, and organic impurities (fine aggregate), as determined from tests prescribed by the American Society of Testing Materials (ASTM) or other accepted testing practice, are often sufficient to ensure that suitable aggregates will be supplied. However, for cut-and-cover tunnel projects it is important to determine from local experience whether or not additional requirements for aggregates should be specified as well. For example, some local sources of aggregate may have undesirable properties that affect drying shrinkage. By specifying achievable maximum shrinkage of concrete specimens (ASTM C157 and ASTM C490), the better source(s) of aggregates, in terms of drying shrinkage of concrete, can in fact be specified.

"Workability" may be defined as the ease with which the ingredients in concrete can be mixed and subsequently handled, transported, and placed with minimum loss of homogeneity. "Consistency," or fluidity, of concrete, which is measured by means of a "slump" test, is an important component of workability. "Minimum loss of homogeneity" can refer to the degree to which the concrete is consolidated when placed, or to resistance to segregation and bleeding. In general the use of larger aggregate and lower cement content causes a decrease in workability.

To minimize the water–cement ratio in the mix design, air-entraining admixtures and modern water-reducing agents are almost always included. All else being equal, workability is improved dramatically with the addition of these admixtures. At the same time, on most cut-and-cover projects, the maximum size of coarse aggregate in the cement mix is relatively small, usually 1 in. or 3/4 in., thus further improving workability. The nominal, maximum slump specified varies with different public authorities, from about 4 to 7 in., generally corresponding to a concrete consistency compatible with routine placement by pumping (see discussion below). When the base slab, walls, and roof of a tunnel structure are particularly massive, as in the case of the Central Artery (I-90) Tunnel (Figure 17-3), the maximum size of the coarse aggregate is more commonly specified to be 1-1/2 in.

Many concrete mixes for cut-and-cover tunnel structures also contain pozzolan (commonly fly ash, but other pozzolans are used as well). Pozzolan is used as a replacement for a portion of Portland cement per cubic yard. This reduction in Portland cement will reduce the total heat of hydration, causing a reduction in thermal stresses and the consequent potential for surface cracking as the concrete cools. Concrete mixes containing pozzolan are thus particularly appropriate for cut-and-cover tunnel structures with, typically, relatively thick slabs and walls. The addition of pozzolan improves workability of the concrete mix as well. As a percent by weight of the Portland cement content, the upper limit of pozzolan content specified by public authorities ranges from as little as 5% to as much as 20% or more. Specified water–cement ratios usually apply to the ratio of water to Portland cement plus pozzolan.

With the advances that have occurred in concrete mix technology in the past three decades, there has been an increase in the 28-day compressive strength (F'_c) typically specified for principal concrete elements of cut-and-cover tunnel structures. In the late 1960s $F'_c = 3{,}000$ psi was specified by SFBART. In the 1970's, $F'_c = 3{,}500$ psi was specified by WMATA. The comparable value of F'c now specified for cut-and-cover tunnels is typically 4,000 psi.

Formwork

Concrete should be placed in forms that are sufficiently rigid and are designed so that they will not shift or bulge under the weight or horizontal pressure applied by freshly placed, fluid or plastic concrete, or due to construction loads imposed on them. In general, the design of all formwork should, as a minimum, conform to the requirements and methodology prescribed in ACI-347R.

In cut-and-cover tunnel construction, the shoring walls usually act as the outside forms for the tunnel concrete. When it is planned that soldier piles or sheet piles are to be pulled when the construction is complete, a bond-breaking material should be placed over the soldier piles or sheet piles. Plywood, masonite, or plastic sheets are commonly used for this purpose. Pulling elements of shoring walls will generally leave voids and increase both lateral and vertical movement of adjacent soils. This should usually not be attempted if sensitive structures or installations lie over or adjacent to the tunnel.

Since the cross section of most cut-and-cover tunnels can remain relatively uniform for long reaches, the opportunity for many reuses of formwork, and for repetitive construction operations, is often present (Figure 17-37). For overall economy of the tunnel structure, these opportunities should be promoted in the design of the tunnel.

For efficient use of reusable forming systems, it is almost always best to provide shoring walls that are stiff enough to allow removal of the bracing tier(s) that would otherwise interfere with wall forms (see Figure 17-12). This practice sometimes adds to the cost of the shoring walls and (sometimes also) to the cost of some of the internal bracing. But in the usual case, overall economy is achieved. The forming procedures likely to be used should also be recognized in the design of the tunnel structure itself. Tunnel structure roofs, and the intermediate floors of two- or three-story tunnel structures, should be designed to act as struts, not only for permanent loads, but also for the (occasionally larger) temporary loads imposed on them during construction.

The efficient use of reusable forming systems and overall tunnel construction economy are also affected by permissi-

Fig. 17-37. Ganged wall forms and carrier framework, San Francisco Clean Water Program.

ble stripping time for forms. When forms are stripped, the elapsed time following the placement of the concrete (stripping time) must be sufficient to ensure that there will be no extraordinary deflection or distortion of the concrete and no evidence of damage to the concrete, either from removal of support or from the stripping operation. These conditions are satisfied when the concrete has achieved sufficient compressive strength (i.e., a sufficient percent of specified design strength), which depends on the actual stresses to which the reinforced concrete is subjected after the forms have been stripped. For roof and floor slabs, sufficient compressive strength is usually evaluated by determining the ratio of the weight of the concrete slab to the total live and dead load for which the slab is designed (herein, "stripping stress ratio"). The time required for sufficient compressive strength to develop is affected by the type and proportion of cement in the concrete mix, the effect of admixtures in the concrete mix, the ambient temperature during the cure period, and other factors. Because of the many variables that can affect stripping time, engineering practice in specifying constraints on stripping time has varied considerably and is often conservative. ACI 347R contains recommended minimum stripping times under ordinary conditions when specifications for a construction project do not address the subject directly. However, most specifications for cut-and-cover concrete do address stripping time. Specified requirements for stripping time usually fall within the following (approximate) guidelines:

- *Sides or ends of invert slabs, beams, etc.* Minimum stripping time ranges from 12 h to 24 hours. Stripping time is usually not a practical constraint.
- *Freestanding walls.* Required concrete strength at time of stripping ranges from 25% (i.e., 1,000 psi) to 70% of design strength of the concrete (F_c'). Stripping time is usually not less than 24 hours. Stripping time up to 5 days is sometimes mandated.
- *Intermediate slabs.* Required concrete strength at the time of stripping soffit forms ranges from 70 to 90% of F_c' (usually 90% when the span exceeds 20 ft). Stripping time is usually not less than 5 days. Stripping time up to 14 days or more is sometimes mandated.
- *Roof slabs.* Required concrete strength ranges from 50% to 90% of F_c', depending largely on the stripping stress ratio. When the stripping stress ratio is high, 80% and 90% of F_c' are often specified for spans less than 20 ft and spans exceeding 20 ft, respectively. Stripping time is usually not less than 3 days. Stripping time up to 14 days or more is sometimes mandated.

The preferred method of determining stripping time is by the use of tests of job-cured specimens or of the concrete in place. High early strength cement (Type III) may be used in the concrete mix when low stripping time is desired, provided that there is not a drying shrinkage requirement that would be compromised.

Joint Treatment. At almost all horizontal construction joints in cut-and-cover construction, the reinforced concrete structure is assumed to be continuous across the joint. Accordingly, horizontal joints are typically specified to be "bonded" or its near equivalent. To achieve adequate bonding, typical specifications require that the surface of the set concrete should be roughened in a manner that will expose the aggregate uniformly and will not leave laitance, loosened particles of aggregate or damaged concrete at the surface of the joint. Wet sandblasting or cleaning with a high-pressure air and water jet (greencutting) are methods often specified and used for this purpose. Many specifications also permit other methods of joint preparation such as roughening and cleaning with a wire brush and water. Surface mortar retarders are sometimes applied to the concrete surface to facilitate cleaning. Just before placing fresh concrete on the joint, it is good practice to cover the joint with a layer of neat cement grout on which the new concrete is deposited before the grout has attained its initial set.

The structural function of most transverse joints differs from that for horizontal joints in that shear is normally not carried across transverse joints. Requirements for joint treatment are therefore largely limited to requirements related to watertightness. When a transverse joint is designated as a construction joint (Figure 17-8), and watertightness is an objective (as in most cases), specifications sometimes contain requirements for joint treatment similar to the described requirements at horizontal joints. Techniques used to avoid practical difficulties in cleaning transverse construction joints include the use of a surface mortar retarder and the use of proprietary metal bulkhead sheathing that remains in place.

Rebracing. Occasionally, it is either not practical or not feasible to remove an internal bracing tier that interferes with forming and placing the full height of a tunnel wall. In such cases, the wall is first placed to an elevation immediately below the interfering bracing tier. A supplementary bracing tier ("rebracing tier") is then placed near the top of the partially placed wall. When the partially placed wall has aged sufficiently, the rebracing tier is then preloaded, and the interfering bracing tier is subsequently removed. The preload in the rebracing tier should represent a realistic evaluation of the reaction to be expected at the rebracing tier when the interfering bracing tier is removed. The remainder of the wall and the floor or roof above are then formed and placed. When this latter concrete has aged, the rebracing level is removed. (See Figure 17-38.)

Where rebracing methodology has been used, it appears to have worked very well. Nevertheless, it is recommended that rebracing be avoided when feasible, because of the care required to avoid unforeseen stress and deflection in the wall that is rebraced. Rebracing can sometimes be avoided by designing the first pour of the wall as a temporary cantilever (as well a beam in its final configuration).

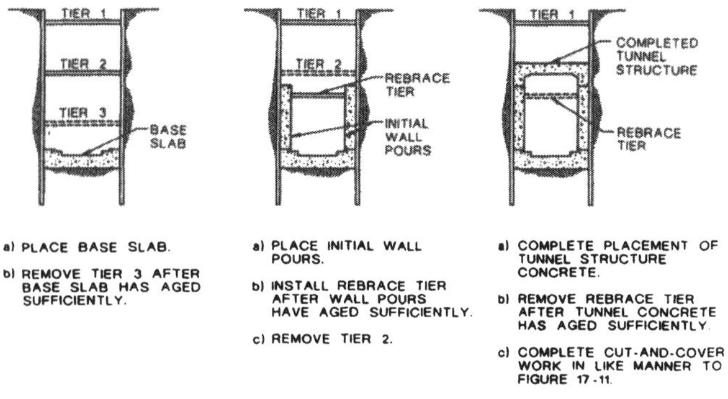

Fig. 17-38. Illustration of rebracing concept.

Rebracing is also sometimes avoided by locating shoring walls 3–5 ft or more outside the neat line of the tunnel structure. When this approach is taken, the outside wall of the tunnel structure is formed also. Internal bracing wales remain outside of the outside forms. "Blockouts" are formed in the walls, so that struts may pass through the walls without being embedded in concrete. The interfering bracing tier is removed after the tunnel structure is adequately complete, the blockout forms are then stripped, and the blockouts are filled with concrete.

Reinforcement. Most of the reinforcing bars used in cut-and-cover construction (for the past 20 years or more) are deformed bars having a guaranteed minimum yield strength of 60 ksi ("Grade 60"), and conforming to ACI requirements. To promote well-placed concrete, it is best to avoid spacing principal reinforcement in walls and slabs closer than about 6 in. To the extent practicable, reinforcement that protrudes from a horizontal construction joint should be designed so that it does not interfere with bracing tier wales. When such interference is unavoidable, the reinforcing bars may be spliced beneath the interfering wale. Where lapped splices are impractical or undesirable in this application, suitable welded splices or mechanical splices may be used. (Sleeve-type or coupler-type splice devices have gained acceptance on most cut-and-cover projects.) To avoid such splices altogether, holes may be cut in the web of the interfering wale.

Mixing, Placing, and Curing Concrete.

MIXING AND TRANSPORTING CONCRETE. Thorough mixing of the concrete ingredients is essential for uniformity of strength, durability, impermeability, and workability of concrete. Equipment and methods used for proportioning concrete ingredients and mixing concrete should be carefully checked for their ability to produce a consistent and uniform product. Two common methods for mixing and transporting concrete have been the following:

- *Central-Mixed Concrete.* Concrete is completely mixed in a stationary batching and mixing plant and is transported to the point of delivery either in a truck agitator or in a truck mixer operating at agitating speed.
- *Truck-Mixed Concrete.* The concrete ingredients are proportioned and fed into a truck mixer at a batching plant. The concrete is then completely mixed (and subsequently agitated) in a truck-mixer while en route to the job site.

Largely because of improvements in truck-mixing equipment (and related equipment), truck-mixed concrete has become the most common mixing and transporting method used on cut-and-cover tunnel construction in the United States. Mixing methods and equipment should conform to the requirements of ASTM C94.

PLACING CONCRETE. In transporting and placing concrete, methods and procedure must be followed that prevent segregation of the coarse aggregate from the cement paste or bleeding of water to the surface. Care must be taken at points of discharge, such as the ends of chutes, hoppers, elephant trunks, pump hoses, and conveyor belts, where a change in direction of the flowing concrete can cause segregation. The flow of the concrete should be controlled at all times using a placing technique that ensures that the concrete drops into the center of the formed wall or whatever container receives it. If the concrete is permitted to strike against the sides of forms or reinforcing bars as it drops, segregation will invariably occur. In general, rehandling of concrete should be minimized, and the discharge end of chutes, elephant trunks, or pump hoses should be located so that the concrete can be delivered to points within 5 ft (preferably less) vertically and horizontally of its final location. The concrete should be deposited in level layers of a thickness that can be properly consolidated.

The method most commonly used to convey concrete from the delivery truck to the concrete pour is pumping. Improvements in the efficiency, availability, and operating and maintenance costs of modern pumping equipment, along with improvements in concrete mix design, have made pumping preferred over alternative methods that were once more traditional. The concrete pump itself is a positive-displacement pump, usually truck- or trailer-mounted. Truck-mounted units are also fitted with a hydraulic boom, which carries the weight of the discharge hose and facilitates the positioning of the end of the hose at the desired horizontal and vertical location within the concrete pour. Discharge hoses are rubber and range in diameter from 4 to 6 in., depending on the maximum size of the concrete aggregate in the mix and other factors. In cut-and-cover construction, the preference for the pumping method is usually limited to concrete containing 1-1/2-in. coarse aggregate. As has been noted, concrete containing 1-in. or 3/4-in. coarse aggregate is more commonly used and is preferred. The pumping capacity of the concrete pumps used in cut-and-cover construction ranges from 10 to 150 or more yd^3 per hour. With the mobility of the truck-mounted units, it is usually not necessary to pump the concrete long distances. Routine pumping is usually limited to about 100–150 ft horizontally, but the concrete may be pumped great distances horizontally using steel pipe sections ("slick line") in place of the hose for most of the horizontal distance.

When the SFBART and WMATA subway systems were constructed, cut-and-cover concrete was only occasionally placed using pumping equipment. More commonly the concrete was conveyed from the delivery truck to the concrete pour using (1) chutes, (2) elephant trucks, (3) conveyor belts, (4) crane buckets, or (5) combinations of this equipment. Much of the cut-and-cover construction was decked so that chutes and elephant trucks (flexible drop chutes that are adjustable) could be positioned and supported conveniently and could be fed directly from truck mixers (or agitators). Portable conveyor belt units or sections, specifically designed to transport concrete, were often used as well to transport concrete horizontally to (typically) the feed points of elephant trunks or chutes. In specific applications, some concrete was placed with concrete buckets hoisted with cranes. Any of these methods of conveying concrete, properly executed, may be utilized on contemporary cut-and-cover construction, when it is expeditious or convenient to do so. Chutes, pipes, or other conveyances must be made of steel or rubber, as it has been found that aluminum can have a detrimental effect on concrete.

After being deposited in forms or elsewhere, concrete must be consolidated into a homogeneous mass. Almost all concrete in cut-and-cover construction is consolidated with internal vibrators. When used properly, vibrators transform low-slump concrete into a semiliquid, so that gravity causes it to flow into a compact mass. When using a vibrator, skilled workmanship is important. Insufficient vibration will cause honeycombing and underconsolidation. Excessive vibration will cause segregation.

CURING. Freshly placed concrete should be maintained without surface drying (cured) for the period of time necessary for the hydration of cement and the proper hardening of the concrete. In order that this requirement be satisfied (for practical purposes), most specifications require that all formed or exposed surfaces be kept continuously wet for a period for at least 7 days following placement of the concrete. This method is commonly referred to as "moist curing" or "water curing." In lieu of water curing, a liquid membrane–forming compound ("membrane curing") is sometimes applied to formed surfaces (after the forms are stripped) and to other exposed surfaces. Engineering practice with respect to permitting the use of membrane curing varies. On cut-and-cover subway and highway projects, membrane curing is often permitted only on surfaces where water curing is unfeasible or where the appearance of the concrete will not be jeopardized. Membrane curing is typically not permitted on surfaces that are to receive concrete, paint, tile, sealant, waterproofing, or other applications requiring a positive bond. On cut-and-cover sewer projects, membrane curing is usually considered acceptable for all exposed surfaces.

COLD WEATHER CONCRETE PLACEMENT. When the ambient temperature falls below 40°F, additional precautions should be taken to ensure a consistent concrete quality. The mixing water and aggregates should be heated so that the temperature of the concrete mix is sufficiently high at the time of placement. The minimum temperature of the concrete mix specified by public authorities is usually in the range of 55–60°F. The ambient temperature during the curing period must also be sufficiently high; 50°F is usually specified. Procedures and requirements for good practice in protecting concrete from cold weather are addressed in detail in ACI 306R.

HOT WEATHER CONCRETE PLACEMENT. When the ambient temperature exceeds about 85°F, additional precautions should also be taken. The temperature of the concrete mix at the time of placement should not exceed specified requirements (usually in the range of 85–90°F). In hot weather, aggregates should be kept as cool as is practicable as a first step to control the temperature of the concrete mix. The most effective next step is through the use of crushed ice as a substitute for mixing water. After the concrete is placed, curing methods that protect the concrete from the hot weather should be employed. Techniques and requirements for good practice in protecting concrete from hot weather are addressed in detail in ACI 305R.

WATERTIGHTNESS

Although the exercise of strict concrete controls will reduce the tendency of shrinkage cracks to form, their presence cannot be entirely eliminated. Structures located in permeable soils and below the level of the groundwater table will be subject to leakage (infiltration). Such infiltration tends to be concentrated at construction and contraction joints. To prevent or minimize leakage at the joints, waterstops are almost always incorporated into the joint design (see below). However, to prevent infiltration entirely (for practical purposes) the application of external waterproofing is usually required. In the United States, representative engineering practice with respect to watertightness, when there is a potential for infiltration, may be characterized as follows:

- *Subway stations and highway tunnel structures.* Infiltration is normally unacceptable since it would result in unsightly streaking of wall and ceiling finishes. Usual practice is to design for complete watertightness. Complete external waterproofing is typical for roofs and is usual for walls. External waterproofing of invert slabs is sometimes specified depending on the slab thickness, subsurface soil, and other factors.
- *Subway line structures.* Usual practice is to design for no free running water at any leak. External waterproofing is typical at transverse roof joints and usual at transverse wall joints. Exterior waterproofing not usual at transverse invert joints. External waterproofing is occasionally specified at horizontal construction joints. The extent of exterior waterproofing is

influenced by soil permeability. The accumulated influx of water into any 10,000 ft^2 of continuous interior surface is commonly limited to 0.2 gallons/min.

- *Sewer tunnel structures.* Usual practice is to specify that there shall be no free running water at any leak. External waterproofing is not usual. The accumulated influx of water into any 10,000 ft^2 of continuous interior surface below the water table is commonly limited to 1.0 gallon/min.

Waterstops

Where there is a potential for infiltration, waterstops are typically incorporated into the design of all exterior joints (Figure 17-8). Polyvinylchloride (PVC), neoprene rubber, and styrene butadiene rubber (SBR) waterstops are commercially available in many configurations. Manufacturers' representatives can be helpful in selecting the best waterstop for the type and configuration of joints under consideration, and they should be consulted. For construction joints, the most common nominal waterstop size is probably 6 in. wide × 3/8 in. (minimum) stem, either ribbed or plain with a standard bulb on each end. For maximum effectiveness in preventing joint leakage, the waterstops must be continuous and correctly installed (including splices), and the concrete should be carefully placed and vibrated around the waterstops. For complete watertightness, however, particularly at transverse joints, external waterproofing is usually provided as well. Waterstops in horizontal joints tend to collect debris during construction, and if this is not completely removed, it may form voids and promote leakage. Some authorities prefer to eliminate waterstops from horizontal joints, counting on the weight of overlying concrete and backfill to keep the joint watertight.

Common Types of External Waterproofing

Built-Up Membrane Waterproofing. This type of waterproofing is a multi-ply asphalt-saturated felt or asphalt-saturated fabric built-up membrane. Several suitable felts or fabrics are commercially available. Application of built-up membrane is similar as that for industrial roofing. Built-up membrane waterproofing is commonly used on the roofs of cut-and-cover structures when the complete roof is to be waterproofed, and it is sometimes used on formed vertical surfaces. Built-up membrane waterproofing may be used at the subgrade of a tunnel structure as well, to provide external waterproofing for the base slab. The exterior surface of built-up roofing (and other types as well) must be protected. Two inches of lean concrete protection is common for horizontal surfaces. Vertical surfaces can be protected with asphalt-impregnated glass-fiber rigid insulation boards or similar materials.

Bentonite Panels. Bentonite panels are board-type panels containing granular bentonite sealed between two layers of absorbent material. The panels are typically 3/16–1/4 in. thick and are furnished in convenient widths and lengths. When in place and subjected to water, the bentonite swells to form an impervious barrier, while the panel material ultimately dissolves or disintegrates. The advantage of bentonite panels is that the panels can be placed against the shoring walls (or against plywood or hardboard furring first placed on the wall). Thus, a practical solution to the problem of external waterproofing of the walls of the structure, when the shoring wall is at the neat line, is offered. For this reason, bentonite panels are commonly specified or used when exterior walls are to be waterproofed. When in this application only the wall joint is to be waterproofed, bentonite panels (usually 2 ft wide) are placed centered on the joint. Bentonite panel waterproofing is also used at roof joints when only the roof joint is to receive external waterproofing, and it is occasionally specified as an alternative to built-up membrane waterproofing for the complete roof. Bentonite panels, when used for external roof waterproofing, are usually protected with hardboard or similar material before backfilling above the roof of the structure. Bentonite is sensitive to salt water, and its use under such conditions may be undesirable or require special precautions. Care must also be taken to protect bentonite panels from rainfall before they are covered.

Other External Waterproofing

When external waterproofing is to be provided at the subgrade of a tunnel structure, the waterproofing is typically placed on a 4–6-in. lean concrete mudslab. Brick-in-mastic has been an excellent form of waterproofing when it is determined that waterproofing protection beyond that furnished by built-up membrane is required.

On two recent projects complete exterior membrane waterproofing was specified. In both cases the membrane at subgrade was placed on a lean concrete mudslab and was covered by a protective layer of concrete. The membrane waterproofing specified (for all surfaces) was as follows:

- *MTA (Los Angeles County Subway Construction).* High-density polyethylene (HDPE), 100 mils thick. (This material was selected for its impermeability and resistance to hydrocarbons.)
- *Central Artery (I-90) Tunnel.* "HPDE and bentonite sheet," which is composed of 20- or 40-mil HPDE sheet, bentonite, and spun polypropylene sheet (with the bentonite sandwiched between the two sheets).

Appropriate protection was required for these membranes at walls and roofs as well.

Internal Repair of Leaks

In determining the extent and type of external waterproofing to be specified for cut-and-cover tunnel projects, watertightness objectives, the potential for infiltration, the cost of external waterproofing, and the cost and effectiveness of modern methods for the internal repair of leaks must all be evaluated carefully. When there is a potential for infiltration and no external waterproofing is specified, or external waterproofing is specified at limited locations (such as transverse joints), some leaks must be expected. Normally, such leaks can be repaired by epoxy injection techniques. If the potential for infiltration has been evaluated reasonably, and if the construction of the structure has been carried out in accordance with good modern practice, the cost of repairing leaks should not be excessive, and commonly prescribed objectives for watertightness should be achieved.

REFERENCES

GOLDBERG, D.T., W.E. JAWORSKI, and M.P. GORDON (1976) *Lateral Support Systems and Underpinning,* U.S. Department of Transportation, Report No. FWHA-RD-75-128, Vo. I, April, 1976.

HARRIS, F.C. (1983) *Ground Engineering Equipment and Methods,* McGraw-Hill, New York.

MITCHELL, J.K., and C.B. VILLET (1987) *Reinforcement of Earth Slopes and Embankments,* Transportation Research Board, NCHRP Report 290.

NAVFAC (1974) *Design Manual—Soil Mechanics, Foundations, and Earth Structures,* NAVFAC DM-7, U.S. Naval Publications and Forms Center, March, Change 2.

PECK, R.B. (1969) *State-of-the-Art Volume,* Seventh International Conference on Soil Mechanics and Foundation Engineering, Mexico City.

PECK, R.B., W.E. HANSON, and T.H. THORNBURN (1974) *Foundation Engineering,* Second Edition, John Wiley and Sons, New York.

Schnabel Foundation Company (1982) *Tiebacks,* U.S. Department of Transportation, Report No. FWHA/FD-82/047, July.

SCHNABEL, H. (1982) *Tiebacks in Foundation Engineering and Construction,* McGraw-Hill, New York.

TENG, W.C. (1962) *Foundation Design,* Prentice-Hall, Inc., Englewood, NJ.

TERZAGHI, K. (1954) "Anchored Bulkheads," *Transactions,* ASCE, Vol. 119.

TERZAGHI, K., and R.B. PECK. (1967) *Soil Mechanics in Engineering Practice,* Second Edition, John Wiley and Sons, New York.

TSCHEBOTARIOFF, G.P. (1962) "Retaining Structures," *Foundation Engineering,* edited by G.A. Leonards, McGraw-Hill, New York.

USS (1975) *Steel Sheet Piling Design Manual,* United States Steel, Pittsburgh, PA.

18

Safety Provisions

Robert J. Jenny
President, Jenny Engineering Corp.

At one time, tunnel accidents claimed one life for every half mile of tunnel constructed. Increased concern over construction safety has led to improvements in the miner's working conditions, with a subsequent reduction in the frequency of deaths and disabling accidents. However, based on 1986 OSHA statistics, lost-time accidents still occur at more than twice the frequency for underground workers compared with other construction workers, and three times the rate for manufacturing workers. Approximately the same adverse ratio applies to fatalities. Obviously, there is a need for more improvements in safety for underground workers.

GENERAL SAFETY RULES

In discussing safety provisions, we must keep two facts in mind. First—and contrary to some beliefs—accidents are not inevitable. Second, accidents are extremely costly, and so accident prevention makes sound economic sense.

The major causes of accidents are

1. Uncontrolled contact between personnel and materials or equipment
2. Failure of temporary structures
3. Inherent constructional hazards such as the use of explosives
4. Unsafe practices or carelessness by individual workers

The employer must recognize these causes and establish programs, rules, regulations, guidelines, and whatever else might be necessary to reduce accidents. Reducing the number and severity of accidents reduces many costs in addition to insurance and other items obviously affected by accidents. When an accident occurs, there can be not only lost time of the injured employee, but also lost time of coworkers due to general work stoppage; lost time of supervisors attending to the injured person, investigating and preparing accident reports; costs associated with damaged equipment or property; as well as many other partially hidden costs. Adherence to safety regulations in the sensitive tunnel construction industry is the prudent economic course to follow.

Compliance with Official Regulations

The safety discussion presented herein is a merely an outline or guideline for safety procedures and accident prevention; it does not intend to present a comprehensive listing of criteria necessary to establish a proper safety program. The U.S. Bureau of Mines, state and local codes, as well as the Federal Safety and Health Regulations for Construction issued under the Construction Safety Act of 1969, and the Occupational Safety and Health Act of 1970, including all current amendments (Federal Regulations) must be complied with when applicable. Where there are differences between the regulations of different agencies, the more stringent will govern. On June 2, 1989, new and revised OSHA regulations regarding underground work were promulgated. At this writing, these regulations, which went into effect on August 1, 1989, represent the current thinking on the federal level concerning achieving better safety performance by underground contractors and workers.

Safety Engineer and Safety Program

On virtually every underground project, the contractor is required to employ at least one full-time, properly qualified safety engineer. On the same project, owners and engineers often have their own safety engineers who visit the project occasionally. In the early stages of construction planning, the contractor's safety engineer should design a safety program tailored to the project. All personnel should be obliged to comply with the provisions of the program. It is often desirable to offer incentives or rewards in the safety program to encourage active, enthusiastic participation.

All personnel should be instructed on the recognition of hazards, observance of precautions, and use of protective and emergency equipment in enclosed underground spaces. When personnel are underground, an accurate record of their location must be kept on the surface.

On tunnel operations with 25 or more workers underground, at least two rescue crews should be trained in rescue procedures and the use of oxygen breathing apparatus and fire-fighting equipment. At least one crew should be trained in routine safety operations. A safety inspection of the job site, including all materials and equipment, should be made at least once per shift.

Emergency Measures

Tunnel evacuation plans and procedures should be developed and made known. These plans should incorporate a separate communications system independent of the tunnel power supply as well as provisions for emergency hoisting or other egress in shafts.

Protective Clothing

Tunnel personnel should wear protective head gear, footwear, and any other special garments that applicable codes require. Safety glasses, rubber gloves, goggles, or face shields should be worn when handling corrosive, toxic, or injurious substances and should be made available for use when required. Working molten metal (e.g., welding) requires fire-resistant clothing and face shields. Moving machinery necessitates the wearing of snug-fitting clothes, and handling of sharp, rough, and splintery material requires protective gloves. In New York, the state's Industrial Code Rule governing tunneling operations (called the New York State Regulations) also requires waterproof clothing in wet areas and safety belts for shaft workers where the drop is 10 ft or more. Other jurisdictions may have other specific protective clothing requirements.

General Precautions

Specific working areas in underground construction can have their own unique hazards that personnel should be made aware of. These hazards should be addressed in the safety program.

Miscellaneous Areas. Safe means of access and egress should be available in all working areas, and all ladders and stairways should comply with applicable code requirements. Subsidence areas presenting safety hazards should be fenced and posted. In New Jersey, The Construction Safety Code requires the surface working site to be walled or fenced to a height of 8 ft. Again, other jurisdictions may have other requirements for overall construction sites and/or specific areas.

Tunnels. The crown, face, and walls of rock excavation areas should be tested frequently, and loose rock scaled down or supported. Rock bolts should be tested frequently for proper torque with a calibrated torque wrench. All steel or timber ground support sets should be properly installed and blocked to prevent movement of rock, and lateral bracing can be provided between sets to further stabilize the support. Damaged or dislodged tunnel supports must be repaired or replaced immediately. Properly designed shields, forepoling, or other devices should be provided as required for soft ground tunneling. Suitable provisions for breasting the face should also be incorporated (see Chapter 6).

Walls, ladders, timbers, blocking, wedges, and supports should be inspected for loosening following blasting operations. Corrections to unsafe areas must be made immediately.

Safety belts should be worn by crews on skips and platforms in shafts when the clear distance between the skip or cage and the sides of the shaft is greater than 1 ft, unless guard rails are provided. The New York State Regulations provide much more extensive and detailed precautions to be observed in shafts and hoisting operations.

Caissons. In caisson work in compressed air, a protective bulkhead should be erected in the working chamber when the chamber is less than 11 ft in height and the caissons are at any time suspended or hung so that the bottom of the excavation is more the 9 ft below the deck of the working chamber while work is in progress. Shafts must be made tight and hydrostatic, or air pressure tested. The test pressure must be stamped on the outside shell about 12 in. from each flange. Accurate and accessible gauges should be located in the lock and on either side of each bulkhead. Caissons greater than 10 ft in diameter or width should be provided with a manlock and shaft exclusively for personnel.

LOCALIZED OPERATIONAL HAZARDS

All equipment to be used during a shift should be inspected before use by either the prospective user or a supervisor. Unsafe equipment should be repaired immediately or removed from any location where it might accidentally be used.

Drilling

Before drilling, the roof must be scaled down by an experienced miner, and the area inspected and made safe from all potential hazards. All personnel should be warned of the possibility of residual explosives from previous blasting. Lifter holes should not be drilled through previously blasted rock. No one should be allowed on a drill boom while the drill is in operation, and no one except the driver should be allowed on a moving jumbo. Jumbos should have storage receptacles for drill steel, a mechanical heavy materials lifter, stair access to decks for at least two people, and removable guard rails on all sides, and at the back of platforms if the deck is more than 10 ft high. When a jumbo is being moved, equipment should be secured and booms should be in a safe position. Drills on columns should be anchored firmly before drilling, and retightened frequently. Scaling bars should always be sharp and in good condition. Water, air, or other utility lines in the area of the drilling should be clearly identified.

Blasting

Blasting operations must comply with Subpart U of the Federal Regulations, and with other federal, state, or local codes governing explosives and blasting procedures. Underground transportation and handling of explosives is described later in this chapter. The New Jersey Department of Labor and Industry Safety Regulations No. 23 governing the use of explosives is an excellent guide for engineers and contractors and is highly recommended for review whenever the use of explosives is contemplated.

Hauling

Powered mobile equipment should be provided with adequate brakes, audible warning devices, and lights at each end. Visible or audible warning should be given before equipment is started and moved. Cabs should have clean windows constructed of safety glass.

Adequate backstops or brakes should be installed on inclined conveyor drive units. No one should ride on power-driven chains, belts, or bucket conveyors, in dippers, shovel buckets, forks, or clamshells, in the beds of dump trucks, or on haulage equipment.

Electrically powered mobile equipment should not be left unattended unless the master switch is in the off position, all operating controls are in the neutral position, and the brakes are set. Parked railcars should be blocked securely. Means should be provided to prevent overtravel and overturning at dumping locations and at all track deadends. Rocker-bottom or bottom-dump cars should have positive-locking devices. Supplies, materials, and tools (other than hand tools) should not accompany personnel on man-trip cars. Equipment that is being hauled should be protected against sliding or spillage.

The most recent OSHA regulations as well as the regulations of other agencies address haulage specifics concerning braking requirements, whether personnel-carrying work trains can be pushed or must be pulled, the prevention of injuries when derailments occur, and many other items. These regulations must be consulted when formulating haulage plans.

Hoisting

Hoisting machines should be worm-geared or powered both ways, and if the power is stopped, the load should not move. Power hoist controls should have a nonlocking switch or control, and a device to deactivate the power should be installed ahead of the operating control. Hoist machines with cast metal parts should not be used. All anchorages of hoists should be inspected at the beginning of each shift, and every hoist should be annually tested to at least twice the maximum load.

Recently OSHA, other agencies, and even crane manufacturers have promulgated more stringent regulations regarding the hoisting of personnel. All of these sources must be consulted when equipment for hoisting personnel is being considered.

An enclosed covered metal cage designed with a safety factor of 4 should be used to raise and lower personnel in the shaft. The cage must be load tested prior to use, and the exterior should be free of projections or sharp corners. Only closed shackles should be used in the cage rigging, and a positive locking device should be installed on the cage to prevent the door from opening accidentally while in operation.

Maximum rates of speed for transporting persons should be established and adhered to in accordance with the applicable regulations, whether federal or local, and signal codes should be employed in the operation of the hoist. For instance, Table 18-1 shows the factors of safety for hoisting rope for passenger hoists provided by New York State Regulations.

FIRST AID STATION

Equipment

Weatherproof first aid kits should be provided at appropriate locations. These kits should contain materials recommended by the consulting physician and should conform to Red Cross standards with individual sealed packages for each item. The contents of the first aid kit should be checked by the safety engineer before being released for use and at least weekly to ensure that expended items are replaced. Equipment for prompt transportation of an injured person to a physician or hospital, and communications for ambulance service, should be provided. The New York State Regulations and most other jurisdictions also require that blankets and at least one stretcher per 100 workers underground be made available.

Attendant(s)

Sufficient competent personnel (with at least one person currently certified in first aid training by the U.S. Bureau of Mines or the American Red Cross) should be available either on or near the work site to perform first aid or any rescue work that may be required in the tunnel. Many agencies and/or project specifications may require more on-site first aid or other medical personnel. This usually depends on the size and complexity of the underground work.

Off-Site Medical Services

Provisions should be made to ensure the availability of medical personnel for continual consultation and of prompt medical attention in case of serious injury. The telephone numbers of the doctors, hospitals, and ambulances should be posted in conspicuous locations.

Table 18-1. Safety Factors for Hoisting Rope for Passenger Hoists (New York State Regulations)

Car Speed (ft/min)	Factor of Safety
50	7.5
100	7.9
200	8.6
300	9.2
400	9.8
500	10.3
600	10.7
700	11.1
800	11.3
900	11.4
1,000	11.66
1,200	11.8
1,500	11.9

FIRE HAZARDS

Limitations on Combustible Materials

Matches, lighters, or other flame-producing smoking materials must be prohibited in all underground operations where fire or explosion hazards exist. It is preferable that this restriction apply in all underground work, whether known fire hazards exist or not. Gasoline or liquefied petroleum gases should not be taken underground. Paper, combustible rubbish, and scrap wood should not be allowed to accumulate. Only the current day's supply of diesel fuel should be stored underground, and oil, grease, and fuel should be well sealed and kept a safe distance from sensitive areas. Only approved fire-resistant hydraulic fluids should be used in hydraulically operated equipment. Air that has passed through underground oil or fuel storage areas must not be used to ventilate working areas. When compressed-gas cylinders are being moved to a new location underground, the safety caps for protecting the cylinder valves should be secured in place.

Limitation on Burning and Welding

Noncombustible barriers should be installed below welding or burning operations in or over a shaft or raise. During, and for 30 min after welding or flame cutting underground, a person with a fire extinguisher should stand by.

Fire-Fighting Equipment

Fire extinguishers should be provided at the head and tail pulleys of underground belt conveyors, at 300-ft intervals along the belt line, and wherever combustible materials are stored. These extinguishers must be suitable for extinguishing fires of wood, oil, grease, and electrical equipment. Other fire-fighting equipment or fire barriers may be required on certain projects.

Smoke Mask and Inhalators

Bureau of Mines–approved self-rescuers (in good condition) should be available near the advancing heading, on the haulage equipment, and in any area where personnel could be trapped by smoke or gas.

Electrical Equipment

All electrical cables taken or used underground should be completely encased in an armored, noncombustible casing or jacket. Power lines should be well separated or insulated from water lines, telephone lines, and air lines. Oil-filled transformers should not be used underground unless they are in a fire-resistant enclosure.

Caution must be exercised when relying on local codes governing electrical equipment. For example, the New York State Regulations are far more comprehensive with regard to electrical equipment safety in tunnels than many other codes.

VENTILATION DURING CONSTRUCTION

Detection of Noxious and Explosive Gases

Major tunnel explosion disasters that occurred in the 1970s and 1980s provided much of the impetus for OSHA's updating of underground safety regulations. OSHA defines gassy, potentially gassy, and nongassy areas, and stipulates strict adherence to specific safety guidelines for each type of area. Other jurisdictions have followed suit in promulgating regulations designed to promote better understanding of the hazardous conditions of gassy atmospheres and the safety precautions that must be taken to prevent accidents. Top management as well as safety engineers must be aware of the possible gassy conditions on a project and must comply with whatever regulations govern in providing proper safety for the site and the personnel.

Testing for the presence of gases at principal work locations should be conducted continuously. Other locations should be spot-checked frequently. The allowable quantities given in Table 18-2 were taken from "Threshold Limit Values," Safety Regulation No. 3, by the State of New Jersey Department of Labor and Industry. Many similar tables are available. Some include quantities, and others include times of exposure and other more detailed information. The table applicable to your area or project is the one that governs.

The presence of flammable or toxic gases, dusts, mists, and fumes should be determined, and if 1.5% or higher concentrations are detected, personnel should be evacuated and power cut off to the affected area until concentrations are reduced to 1% or less. A record of all tests should be kept on file.

Gas detection equipment has been vastly improved and simplified in recent years. Easy-to-operate gas detectors that can be equipped to monitor virtually any gas that might be encountered in an underground project are on the market. All detectors have both visual and audio alarms, and most have digital readouts. Some detectors even record and retain the information from an entire shift of gas monitoring; this data can be plugged into a computer to generate a readout of the gases detected, concentrations, times, etc. Choosing the appropriate gas detection equipment for an individual project is an important task for the safety engineer and project manager.

Air Quality Maintenance

Tunnels should be provided with mechanically induced reversible-flow primary ventilation for all work areas. Venti-

Table 18-2. Allowable Quantities of Various Gases[a]

Gas	Maximum Allowable Quantity
Carbon monoxide	50 ppm
Carbon dioxide	5,000 ppm
Methane	1%
Nitrogen dioxide	5 ppm
Nitric oxide	25 ppm
Hydrogen sulfide	10 ppm

[a] From "Threshold Limit Values," Safety Regulation No. 3, New Jersey Department of Labor and Industry

lation doors should be self-closing and remain closed regardless of direction of flow. When primary ventilation has been discontinued for any reason, employees should be evacuated and qualified personnel should examine the tunnel for gas and other hazards before activating power or readmitting employees to the work areas.

The supply of fresh air should not be less than 200 ft^3/min per employee, and the velocity should be at least 30 ft/min where conditions can produce harmful dusts or gases. Respirators should not be used in place of environmental ventilation controls except in welding, blasting, and lead-burning operations. Internal combustion engines other than mobile diesel must not be used underground. After blasting, smoke and fumes should be exhausted through the vent line to the outside air.

HANDLING AND STORAGE OF EXPLOSIVES

Handling and Transportation Underground

The blaster should be fully qualified to handle and use explosives safely as required. All jurisdictions require the blaster to be licensed.

Only the explosives or blasting agents required for one blast should be taken underground. Explosives and blasting agents should be hoisted, lowered, or conveyed in a special powder car. The hoist operator should be notified before explosives are transported in a shaft, and personnel, material, supplies, detonators, or equipment should not be transported in the same conveyor with the explosives. Explosives should not be transported in unmarked conveyances or in a locomotive during man-haul trips. A physical separation of 24 in. should divide the compartments of detonators and explosives in a vehicle, and no one except the operator, his helper, and the powderman should be permitted to ride in a vehicle or train transporting explosives and blasting agents.

Trucks used for the transportation of explosives underground should have the electrical system checked for electrical hazards, and a written record of such inspections should be kept on file.

Storage of Explosives and Detonators

Explosives must be stored in the types of facilities required by the Internal Revenue Service Regulations, 26 CFR 181, Commerce in Explosives.

Permanent storage of explosives or blasting agents should not be permitted underground unless at least two modes of exit for personnel have been provided. However, most jurisdictions, New Jersey and New York State, for example, prohibit any storage of explosives underground where men are employed. Where permanent storage is allowed, magazines should be at least 300 ft from any shaft or active area.

Smoking and open flames should not be permitted within 50 ft of explosives and detonator storage magazines. Blasting caps, detonating primers, and primed cartridges should not be stored in the same magazine with other explosives or blasting agents. Nor should detonator magazines be located closer than 50 ft to a magazine containing explosives.

INACTIVE HEADINGS

Access to unattended underground work areas should be restricted by gates. Unused chutes, manways, or other openings should be tightly covered, bulkheaded, or fenced off and posted.

Short Duration

Should tunneling operations be interrupted, the heading should be securely supported. Hydraulic pressure or collapsible rams or struts should not be used for securing the faces of inactive headings. If a shield invert is below the water table, watchmen should be on duty at the heading at all times when excavation is suspended. However, closed-circuit television in lieu of maintaining the watchmen underground may be used in some cases to monitor the heading.

Long Duration

If a tunnel is inactive for a relatively long period of time, it is recommended that a bulkhead with a valve be installed at the face. The valve should be opened frequently to check for water pressure or noxious fumes and before the bulkhead is removed.

COMPRESSED-AIR WORK

Medical Regulations

A licensed physician, qualified and experienced in treating decompression illness and willing to enter a pressurized chamber, should be consulted prior to beginning work and be available whenever work is in progress. In addition, a fully equipped first aid station and a vehicle equipped with one litter should also be available.

Prospective compressed-air workers should be examined by the physician to determine if they are physically qualified for the work. If a worker is ill or injured, or has been absent for 10 days, he should be reexamined before returning to work under compressed air. In addition, compressed-air workers should be reexamined at least annually. The examination results should be kept on record along with the record of any decompression illnesses reported by the physician. The records should be available for inspection, and a copy should be sent to the Bureau of Labor Standards following a death, accident, injury, or decompression illness.

Under Federal Regulations, a medical lock must be maintained whenever air pressure in the working chamber is increased above the atmospheric pressure. The lock must conform to the specifications in the regulations.

Identification badges must be furnished to the compressed-air workers indicating the worker's name, the nature of his job, the address of the medical lock, the phone number of the physician on the project, and instructions that in case of unknown or doubtful cause of illness the wearer should be taken to the medical lock. The badge should be worn on and off the job at all times.

Records and Communication

There should be one person present who represents the employer and who is knowledgeable about and responsible for complying with compressed-air regulations. Every employee should be instructed in the rules concerning safety when working under compressed air. The time of decompression as shown in the applicable decompression table should be posted on each lock. Also, appropriate signal codes should be posted at workplace entrances. Communications should be maintained at all times among the following locations: working face, work chamber side of the manlock door, the manlock, lock attendant's station, compressor plant, first aid station, emergency lock, and special decompression chamber.

For each shift, a record of each employee's time under air pressure and his decompression time must be kept by an employee remaining outside the lock near the entrance, and a copy of the record should be submitted to the physician after each shift.

Compression and Decompression

All personnel going under air pressure for the first time should be instructed on how to avoid excessive discomfort.

Compression Procedure.

First minute: Up to 3 psig maximum, hold to determine if any discomfort is experienced.

Second minute: Raise uniformly at a maximum rate of 10 psi/min.

When personnel signal discomfort, hold the existing pressure for 5 min. If the discomfort does not cease after 5 min, reduce pressure gradually until the discomfort eases. If it persists, release the affected parties from the lock. No one should be subjected to a pressure greater than 50 psi. Decompression to normal atmospheric conditions must be in accordance with decompression tables. Table 18-3 shows the total decompression time for gauge pressures from 0 to 50 psi and working periods from 1/2 to 8 hours and over.

Decompression proceeds by two or more stages, with a maximum of four for a working chamber pressure of 40 psi or over. Stage 1 consists of a reduction in ambient pressure ranging from 10 to a maximum of 16 psi, but in no instance will the pressure be reduced below 4 psi at the end of stage 1. This reduction in pressure in stage 1 will always take place at a rate not greater than 5 psi per minute. Further reduction in pressure will take place during stage 2 and subsequent stages as required at a slower rate, but in no event at a rate greater than 1 psi per minute.

Table 18-3. Total Decompression Time (minutes)

Work Pressure PSIG	Working Period										
	1/2 hr	1 hr	1-1/2hrs	2 hrs	3 hrs	4 hrs	5 hrs	6 hrs	7 hrs	8 hrs	Over 8 hrs
0–12	3	3	3	3	3	3	3	3	3	3	3
14	6	6	6	6	6	6	6	6	16	16	33
16	7	7	7	7	7	7	17	33	48	48	62
18	7	7	7	8	11	17	48	63	63	73	87
20	7	7	8	15	15	43	63	73	83	103	113
22	9	9	16	24	38	68	93	103	113	128	133
24	11	12	23	27	52	92	117	122	127	137	151
26	13	14	29	34	69	104	126	141	142	142	163
28	15	23	31	41	98	127	143	153	153	165	183
30	17	28	38	62	105	143	165	168	178	188	204
32	19	35	43	85	126	163	178	193	203	213	226
34	21	39	58	98	151	178	195	218	223	233	248
36	24	44	63	113	170	198	223	233	243	253	273
38	28	49	73	128	178	203	223	238	253	263	278
40	31	49	84	143	183	213	233	248	258	278	288
42	37	56	102	144	189	215	245	260	263	268	293
44	43	64	118	154	199	234	254	264	269	269	293
46	44	74	139	171	214	244	269	274	289	299	318
48	51	89	144	189	229	269	299	309	319	319	
50	58	94	164	209	249	279	309	329			

If repetitive exposure to compressed air is required (more than once in 24 hours), the physician should establish and be responsible for compression and decompression procedures. The physician is also responsible for decanting methods, if these methods are required. In decanting, no more than 5 min should elapse in atmospheric pressure before recompression.

Manlocks and Muck Locks

Controlled decompression of employees from a compressed-air atmosphere must always take place, except in emergency. Except when the air pressure is below 12 psig and there is no danger of rapid flooding, each bulkhead in tunnels of 14 ft or more in diameter, or an equivalent area, should have at least two locks—one a manlock, the other a materials lock. If only a combination man and materials lock is required, the lock should be able to hold an entire heading shift. A lock attendant, responsible to the physician, should be at the controls of the manlock whenever men are in the working chamber or in the lock. If the air pressure is 12 psig or above, decompression must be regulated by automatic controls supplemented by manual controls to allow the lock attendant to override the automatic controls if required. Manual controls for an emergency must also be provided inside the manlock. The manlock must contain the following equipment: a clock and a continuous recording pressure gauge outside the lock; a pressure gauge, a clock, and a thermometer inside the lock. In addition, 4-in. minimum diameter observation ports should be installed so the lock occupants can be observed from the chamber and free air side. Ventilation should be provided, and the temperature should be at least 70°F in the lock. The lock must contain 30 ft^3 of air space per occupant and have 5 ft clear headroom minimum at the center. Also, each bulkhead should have a pressure gauge on both faces.

When locks are not in use and employees are in the working chamber, lock doors should be kept open to the working chamber. In an emergency, if the working force were to become disabled, provisions should provide for rescue parties to enter the tunnel quickly.

A special decompression chamber to accommodate the entire force of employees being decompressed at the end of a shift should be provided whenever the required time of decompression exceeds 75 min. This chamber is commonly known as the "Luxury Lock."

Special Decompression Chamber

The headroom in the special decompression chamber should be at least 7 ft. For each person there should be 50 ft^3 of air space, 4 ft^2 of walking area, and 3 ft^2 of seating space exclusive of lavatory space. The rated capacity of the chamber will be based on the stated minimum space per employee and should be posted. The capacity should not be exceeded except in case of emergency. Each special decompression chamber should be equipped with the following: clocks, pressure gauges, valves to control the supply and discharge of air, an oral communication system among the occupants, attendant and compressor plant, and an observation port at the entrance.

Seating space, at least 18 by 24 in. wide, should be provided per occupant, and normal sitting posture permitted. Proper and adequate toilet and washing facilities in a screened or enclosed recess should also be provided. Fresh, pure drinking water should be available. Community drinking vessels should be prohibited.

Unless the special decompression chamber is serving as the manlock to atmospheric pressure, the chamber should be adjacent to the manlock on the atmospheric pressure side of the bulkhead.

Compressor Plant

At all times a thoroughly experienced, competent, and reliable person should be on duty at the air control valves as a gauge tender, regulating the pressure in the working areas. During tunneling operations, one gauge tender only should regulate the pressure in two headings, provided the gauge and controls are all in one location.

The low air-compressor plant capacity should permit the work to be done safely and provide a margin to meet emergencies and repairs. Low air-compressor units should have at least two independent sources of power supply. The compressors should be of sufficient capacity to maintain the necessary pressure in the working chamber even during periods of breakdown, repair, or emergency.

Switching from one independent source of power supply to the other should be done periodically to ensure the workability of the apparatus in an emergency. Duplicate low-pressure air feedlines and regulating valves should be provided between the source of air supply and a point beyond the locks, with one of the lines extending to within 100 ft of the working face. All high- and low-pressure air supply lines should be equipped with check valves. Low-pressure air will be regulated automatically, but manual valves should be provided for emergency.

The air intakes should be located at a place where fumes, exhaust gases, and other air contaminants will be at a minimum.

Gauges indicating the pressure in the working chamber should be installed in the compressor building, the lock attendant's station, and the employer's field office.

Bulkheads and Safety Screens

Intermediate bulkheads with locks, or intermediate safety screens, or both, are required where there is a danger of rapid flooding. The New York Regulations limit the length between work face and bulkhead to 1,000 ft if the possibility of rapid flooding exists.

In tunnels 16 ft or more in diameter, where there is a danger of rapid flooding, hanging walkways should be provided from the face to the manlocks, as high in the tunnel as practicable, with at least 6 ft of headroom. Walkways should be constructed of noncombustible materials. Standard railing should be securely installed throughout the length of all walkways on open sides. Where walkways are ramped under safety screens, the walkway surface should be skidproofed by cleats or by equivalent means. Bulkheads used to restrain compressed air should be tested to prove their ability to resist the highest air pressure expected to be used.

Ventilation and Air Quality

The working chamber should be well ventilated. The air in the work areas should be analyzed at least once per shift, and a record of analyses kept. Test results must fall within the threshold limit values set forth by the applicable regulations; otherwise, immediate corrective action must be taken. During the entire decompression period, forced ventilation of fresh air must be provided.

Whenever heat-producing equipment is used, a positive means of removing the heat buildup at the heading should be provided. The temperature of all working chambers should be maintained at temperatures not in excess of 850°F.

Sanitation

Clean, heated, lighted, and well-ventilated dressing rooms and drying rooms should be provided for all employees engaged in compressed-air work. Such rooms should contain suitable benches and lockers. Bathing accommodations (showers at the ration of 1 to 10 employees per shift), equipped with running hot and cold water and with suitable toilet accommodations (1 toilet for each 15 employees per shift) should be provided. All parts of caissons and other working compartments should be kept in a sanitary condition.

Fire Prevention and Protection

Proper fire-fighting equipment must be available for use at all times. While welding or flame cutting is being done, a firewatcher with extinguisher should stand by. Shafts and caissons containing flammable material of any kind should be provided with a fire hose arranged so that all points of the shaft or caisson are within reach of the hose stream.

Tunnels should be provided with a 2-in. minimum diameter water line extending into the working chamber and to within 100 ft of the working face. The line should have hose outlets with 100 ft of fire hose attached and maintained as follows: one at the working face, and one immediately inside of the bulkhead. In addition, hose outlets should be provided at 200-ft intervals throughout the length of the tunnel, and 100 ft of fire hose should be attached to the outlet nearest to the location of flammable material or any area where flame is being used.

Fire hose should be at least 1-1/2 in. in nominal diameter, and water pressure and supply should at all times be adequate for efficient operation of the type of nozzle used. The powerhouse, compressor house, and all buildings housing ventilating equipment should have at least one hose connection in the water line. A fire hose should be maintained within reach of any wood structure over or near shafts. The compressor building should be constructed of noncombustible material.

In addition to the fire hose protection required on every floor of every building used in connection with compressed-air work, there should be at least one approved fire extinguisher of the proper type for the hazard involved. At least two approved fire extinguishers should be provided in the working chamber as follows: one at the working face, and one immediately inside the bulkhead (pressure side). Extinguishers in the working chamber must use water as the primary extinguishing agent and may not use any extinguishing agent that could be harmful to the employees in the working chamber. Highly combustible materials should not be used or stored in the working chamber.

Manlocks should be equipped with a manual fire extinguishing system that can be activated inside the manlock and also by the outside lock attendant. In addition, a fire hose and portable fire extinguisher should be provided inside and outside the manlock. The portable fire extinguisher should be the dry chemical type. Equipment, fixtures, and furniture in manlocks and special decompression chambers should be constructed of noncombustible materials. Bedding and like materials must be chemically treated to be fire-resistant. Headframes should be constructed of structural steel or open framework fireproofed timber. Temporary surface structures within 100 ft of any shaft, caisson, or tunnel opening should be built of fire-resistant materials.

Oil, gasoline, or other combustible material should not be stored within 100 ft of any shaft, caisson, or tunnel opening.

However, oil may be stored in suitable tanks in fireproof buildings if the buildings are at least 500 ft from any tunnel-connected building or opening. Positive means should be taken to prevent leaking flammable liquids from flowing into the tunnel-connected openings or buildings. The handling, storage, and use of explosives must comply with all applicable regulations in connection with compressed-air work.

Electricity

All lighting in compressed-air chambers should be by electric methods exclusively. Two independent electric lighting systems, with independent sources of supply, should be used. The minimum intensity of light on any walkway, ladder, stairway, or working level should not be less than 10 foot-candles, and in all work areas the lighting should at all times enable personnel to see clearly.

All electrical equipment and wiring for light and power circuits must comply with the requirements of the National Electrical Code for use in damp, hazardous, high-temperature, and compressed-air environments. External parts of lighting fixtures and all other electrical equipment, when within 8 ft of the floor, should be constructed of noncombustible, nonabsorptive, insulating materials, except that metal may be used if it is effectively grounded. Portable lamps should be equipped with noncombustible, insulating sockets, approved handles, basket guards, and approved cords. The use of worn or defective portable and pendant conductors should be prohibited.

DECOMPRESSION TABLE EXPLANATION

The Federal Regulations provide the following explanation concerning decompression times.

The decompression table is computed for working chamber pressures from 0 to 14 psi, and from 14 to 50 psi gauge inclusive by 2 psi increments, and for exposure times for each pressure extending from 1/2 to more than 8 hours (see Table 18-3). Decompressions will be conducted by two or more stages with a maximum of four stages, the latter for a working chamber pressure of 40 psi gauge or over.

Stage 1 consists of a reduction in ambient pressure ranging from 10 to a maximum of 16 psi, but in no instance will the pressure be reduced below 4 psi at the end of stage 1. This reduction in pressure in stage 1 will always take place at a rate not greater than 5 psi per minute.

The decompression table (Table 18-3) indicates the total decompression time in minutes for various combinations of working chamber pressure and exposure time.

Other decompression tables may indicate various combinations of working chamber pressures and exposure times for the following:

1. The number of decompression stages required
2. The reduction in pressure and the terminal pressure for each required stage
3. The time in minutes through which the reduction in pressure is to be accomplished for each required stage

4. The pressure reduction rate in minutes per pound for each required stage

The pressure reduction in each stage is accomplished at a uniform rate. Do not interpolate between values shown in the decompression tables. Use the next higher value of working chamber pressure or exposure time should the actual working chamber pressure or the actual exposure time fall between the calculated values shown in the body of the tables.

REFERENCES

Occupational Safety and Health Administration, Safety and Health Standards, Washington, DC.

New York State Industrial Code, Albany, NY.

New Jersey Department of Labor and Industry, Safety Regulations, Trenton, NJ.

Internal Revenue Service Regulations, 26 CFR 181 Commerce in Explosives, Washington, DC.

19

Fire Life Safety

Norman H. Danziger
Senior Vice President, Principal Professional Associate,
Parsons Brinckerhoff Quade & Douglas, Inc.

Major tunnel fires have been rare occurrences. However, the potential for entrapment and injury of large numbers of people who routinely use highway tunnels, underground mass transportation facilities, or mainline railroad tunnels warrants special considerations.

An effective program for fire life safety in tunnels depends on the coordinated interaction of several subsystems. These subsystems include detection, alarm, verification, incident location, communications, response plan, personnel evacuation, smoke control (ventilation), fire suppression, and electrical power supplies. Given the interdependence of the various subsystems, a deficiency in any one of them would reduce the attainable level of total system safety.

This chapter focuses on fire life safety applications to highway tunnels, rapid transit, and mainline railroad tunnels. How the interactive subsystems should be applied to a tunnel will vary as a function of the physical characteristics and planned use of the tunnel.

Details of the various subsystems—ventilation, lighting, power supply and distribution, water supply and drainage, and surveillance and control systems—are addressed in various other chapters of this book (Chapters 20–24).

BACKGROUND

Knowledge acquired in recent years through experiences with fires in tunnels, as well as from various research and development activities, including analytical studies and field measurement test programs, have contributed significantly to more effective design and performance of tunnel life safety programs.

Over the past 40 years, the most significant tunnel fires include the Holland Tunnel fire in 1949 (New York–New Jersey); the BART (Bay Area Rapid Transit) Trans-Bay Tube fire on January 17, 1979 (California); the Nihonzaka Tunnel fire in 1979 (Japan); the Caldecott Tunnel fire on April 7, 1982 (California) (Jackson, 1983; USDOT, 1984); and the Kings Cross Station fire in the London Underground in November 1987. From these experiences, it has been found that overall, tunnels do survive fires; however, people may not.

Several significant factors were learned from the world's major tunnel fires. For example, during the Holland Tunnel (a highway tunnel under the Hudson River connecting New Jersey and New York City) fire in 1949, enormous quantities of heat and smoke traveled along the ceiling of the tunnel for distances of more than 300 ft (100 m), subsequently causing secondary fires in vehicles trapped behind and uphill from the primary fire source. The Holland Tunnel has a full transverse system of ventilation. When the ceiling above the primary fire source collapsed, large quantities of air flowed longitudinally in the roadway toward this large opening in the ceiling exhaust duct. In effect, the ventilation system changed at that moment from a full transverse system to one of single-point exhaust. This served to reduce the spread of heat and smoke along the ceiling and eliminated further occurrences of secondary fires.

BART

The BART Trans-Bay Tube fire occurred in the inbound trainway January 17, 1979. The 3.6-mi-long subaqueous tunnel runs between ventilation structures at both sides of San Francisco Bay. There are no exitways, vent shafts, or other external openings throughout the tube length. One vent structure is on land on the Oakland side of the bay. The vent structure on the San Francisco side is located 400 ft offshore in the bay.

The two trainways are separated by a service passageway (gallery) between and parallel to the trainways. Cross-passageways at approximately every 320 ft (100 m) interconnect the two trainways via the gallery. Fire-rated doors isolate the trainways at each cross-passage.

An emergency ventilation duct in the upper part of the gallery, connected to fans in ventilation structures at both ends of the subaqueous crossing, provides smoke removal from selected locations via remote-controlled selective damper openings to each trainway, spaced at approximately

320-ft (100-m) intervals for the length of the tube. Beneath the ventilation duct, the maintenance passageway (gallery) provides access for fire-fighting personnel.

The fire began when a metal cover plate on a control box mounted on the undercarriage of a car on the train came loose, struck the third rail, and created a short circuit with the power rail. The fire began at about 6 P.M. on a normal work-day, and few passengers were on board the eight-car inbound train. Outbound traffic was heavy at this peak rush hour period. The train crew evacuated the passengers from the fire train to the tunnel walkway and through a cross-passage to the adjacent trainway, where they were picked up by a train traveling in the opposite direction. All passengers and crew escaped before the fire engulfed and consumed the train. The fire was hot enough to melt and eventually ignite the aluminum bodies of the transit vehicles.

The fire was declared out about 6 hours later. There was one fatality, a fire department lieutenant, who was overcome by smoke in the gallery due apparently to a fire door having been blocked open. The involved train was destroyed by the fire.

A report on the subsequent investigations of the fire noted that there were inappropriate responses by the various transit operating personnel and/or by the emergency response services agencies involved. The operating personnel had difficulty determining the location of the fire, and hence in deciding which dampers to operate. Rehabilitation included the installation of frequent location markers in the trainways.

Concrete spalling occurred in the arch of the structural lining for a length of about generally only to reinforcing steel depth, with a few isolated small spots extending 2–3 in. (5–7.6 cm) deeper.

Nihonzaka Tunnel

The following information, and that in the next section, was derived from a paper presented by Lawrence E. Jackson (1983), of the National Transportation Safety Board in January 1983.

The Nihonzaka Highway Tunnel is approximately 6,700 ft (2,042 m) long and consists of two dual-lane sections with three interconnecting passages located every 1,640 ft (500 m). The initial incident occurred in the westbound tube, which is on a downgrade, approximately 1,375 ft (420 m) from the exit portal. The fire spread upgrade to other groups of vehicles and most of the facilities (including the ceiling) within 3,750 ft (1,143 m) of the initial fire site were destroyed. Seven fatalities were reported and 189 vehicles were destroyed.

The fire started around 6:40 P.M. on July 11, 1979. It was brought under control on the afternoon of July 13th (almost 48 hours later), but it continued to burn until 10:00 A.M. on July 18th.

The tunnel was protected by a wet sprinkler system. The 6,000-ft^3 (170,000-liter) water reservoir feeding the sprinkler system was stated to be empty when fire erupted at the second location approximately 30 min after the fire had begun. It is not known why the water supply through the sprinklers could not be restored; that is to say, it is unknown whether there was an available water supply that could have been brought to the tunnel and pumped into the sprinkler system over the period of six days. At the same time, it is not known whether perhaps the sprinkler system pipes themselves were damaged by the heat of the fire so that further use of the system was not possible.

The system of ventilation was a reversible semitransverse type having a maximum capacity of approximately 1,120,000 ft^3/min (529 m^3/s) under normal conditions and 607,000 ft^3/min (286 m^3/s) under reverse operation for fire conditions.

When the tunnel was repaired and restored to service, what had been left of the original ceiling (which formed the ventilation air duct above) was removed, and the original semitransverse ventilation system was replaced with a jet fan longitudinal system throughout the tunnel.

Caldecott Tunnel

A third two-lane bore of the Caldecott Tunnel went into service in October 1964. It was in this bore that the accident occurred. The tunnel length is 3,370 ft (1,027 m). The tunnels were interconnected by periodic maintenance-only passageways (not accessible to passengers) protected by simple steel plate doors at each tunnel.

Traffic in the third tube was being operated unidirectionally from east to west. The tunnel is graded uphill from west to east, so that the traffic was going downhill. Prevailing winds are west to east, the direction of fire travel.

The ventilation system consists of an overhead full transverse system using a ceiling duct divided in two parts, with supply from ceiling outlets on one side of the tunnel and extraction through the adjacent ceiling duct.

The ventilation capacity for supply and exhaust was approximately 500,000 ft^3/min (236 m^3/s) for each system.

The tunnel traffic is monitored by the tunnel authority. In the control room, operators can observe the fire alarm annunciators, carbon monoxide recorders, television monitors, and fan operation. The operators were equipped with an emergency telephone connection to phones in the tunnels.

The fire began shortly after midnight in very light traffic conditions, following a collision involving an abandoned automobile, a transit bus, and a tank-truck trailer full of gasoline. The initial fire started in the tractor engine, and within several minutes the trailer tank exploded. The tanker explosion produced a fireball that filled the tunnel for several hundred feet uphill from the back of the tanker. The tanker was fully fire-involved with heavy black smoke moving uphill toward the eastern portal; no smoke traveled westward downhill.

From studies by the National Transportation Safety Board (NTSB), it was estimated that the maximum flame temperature during the fire did not exceed 1,950°F (1,066°C) (Jackson, 1983).

Fire-fighting personnel were able to approach the burning tanker (which was overturned) to within about 150 ft.

After 40 min from initiation of the fire, the firefighters were able to go in right up to the tanker itself, which at that time was still burning inside, with most of the metal already melted. The burning tank trailer was located at the approximate midpoint of the tunnel.

Near the rear of the tanker, the flames reached the ceiling and smoke extended from head height to the tunnel ceiling. Visibility was estimated by firefighters at about 30 ft (10 m) in all directions. As the firefighters moved east, visibility decreased. At a distance of approximately 550 ft (168 m) east of the tanker, visibility was reduced to about 3–6 ft (1–2 m) and smoke was within 1.5 ft (0.5 m) of the roadway.

Based on subsequent analysis by the NTSB, physical evidence indicated that temperatures had reached a maximum of 1,900°F (1,038°C) for at least 20 min in the areas of the involved vehicles and a minimum of 1,600°F (871°C) at the ceiling level from 1,600 ft (488 m) to about 900 ft (275 m) from the east end. In this same location, temperatures near the roadway exceeded 1,100°F (593°C). Between 25 ft (8 m) and 550 ft (168 m) from the east end, temperatures near the roadway ranged typically between 350 and 1,075°F (177 and 580°C) except in the areas of the burning vehicles, where temperatures reached 1,900°F (1,038°C).

Two elderly people trapped in their car only 115 ft (35 m) from the eastern end of the tunnel did not escape the smoke and fire, and died. The truck and bus drivers escaped through the west portal. Several people got out of their cars and tried to run back to the entering portal, but most were engulfed in the fire stream or overcome by smoke. Skeletons were found in some cars; others were on the roadway.

None of the motorists who tried to flee on foot used cross-passages. The cross-passage doors were not equipped with handles, signs, or lights.

The members of the NTSB were of the opinion that had the Caldecott Tunnel been equipped with video monitors or loop detectors, the number of fatalities would probably have been less than the seven that in fact occurred. The NTSB also concluded that due to the intensity of the gasoline fire, fire response or changes to ventilation probably would have had little effect in this accident (Jackson, 1983).

There was no ceiling collapse. Finish ceiling tiles had been removed at an earlier date. Concrete spalling was limited to the ceiling over the trailer and to the nearest sidewall. The depth of spalling was generally limited to mid-depth of the reinforcing steel with only a few longitudinal rears exposed.

Tiles on both walls were burned off and spread across the roadway for a considerable distance behind the tanker. Roadway debris was about 3 in. (7.5 cm) deep over this area.

The two fire hydrants' brass or bronze fittings nearest the trailer were badly damaged by the heat. A sizable jet of water was still spouting from one of them 36 hours after the fire.

Kings Cross

In November 1987, a fire in the Kings Cross Station of the London Underground resulted in 34 fatalities. Kings Cross is a multilevel underground station that includes three levels of trainways interconnected via stairs and escalators, and a ticketing hall (fare collection) concourse below the street level. The fire apparently began in an escalator between the uppermost trainway station platform level and the ticketing hall. It was concluded that the cause of the fire was probably a cigarette that had fallen through the risers of the escalator into the substructure below. The escalator was an old unit that included wooden treads. In addition, debris mixed with oil and/or grease from lubricants for the escalator mechanism in the escalator truss provided the initial source of fuel for the fire. Subsequent to the start of the fire in the truss structure of the escalator, the fire erupted through the moving stairs as a fireball and rapidly ascended to the ticket hall, which provided additional combustibles to feed the fire (ceiling construction, etc.). Most of the fatalities occurred in the ticketing hall or on the affected escalators.

The fire progressed upward from the original source. Thus, individuals who remained in the lower levels of the station were safe from the effects of the fire.

Smoking is now prohibited in the London Underground. Automatic sprinkler systems have been or are being installed in all London Underground escalator machine rooms and escalator trusses, and maintenance to preclude accumulations of combustible debris has been intensified. All of the older escalators in the London Underground that contain combustible components and are not sealed as effectively as the currently available escalator systems are being replaced with newer models.

Summary

The time of the fire occurrence in at least two of the five fires described above likely was a major factor in limiting the number of fatalities. In addition, several design elements for new tunnels or the retrofit of existing ones to improve fire life safety were learned. These include

- Improvements in signing on the walls of tunnels to guide people to an exit from the tunnel or to a place of refuge
- Low-height placement of illuminated exit signs
- Recessed fire protected enclosures for fire hydrants and related emergency equipment
- Duct banks or other fire rated enclosures for critical emergency power and communications circuits
- Illuminated electronic signs for communicating instructions to people in highway tunnels and at the portals during an emergency
- Radio rebroadcast systems to communicate with people via their car radios
- Escape from an emergency-involved tunnel via pedestrian cross-passages to a parallel bore

Research and Development

Over the past 25 years, several major research and development activities pertaining to fires in tunnels have been conducted. Many of these activities have been reported at the British Hydromechanics Research Association (BHRA) sponsored International Symposiums on the Aerodynamics and Ventilation of Vehicle Tunnels. Of particular relevance are various papers that have been presented or published, including the following:

- A paper by Heselden (1976) provided an excellent review of the available information at that time with regard to the behavior of fire and smoke in vehicular tunnels. Heselden provided data on fire statistics, information on past fires and fire tests, and drawing on experimental and theoretical work, he presented a method for estimating smoke quantities, temperatures, and rates of travel for fires of various magnitudes.
- Papers and reports describing fire tests in Japan (Kawamura et al., 1976; Tateishi et al., 1970; Mizutani, 1982), Austria (Fezlmayr 1976), and Switzerland (Offenegg, 1965), which provide full-scale test data and information on the behavior of tunnel fires.
- Work by the U.S. Bureau of Mines (Chaiken et al., 1979; Croce et al., 1978; Hwang, 1987), which has conducted a number of small-scale and some full-scale fire tests. These tests are primarily concerned with investigating the spread of fire in a fuel-lined passage (coal and wood lined), a common situation in mines. As a by-product of this work, the Bureau of Mines is also studying the effects of a mine fire on the ventilation system such as the throttling of the air flow, the conditions leading to the reverse flow of smoke (backlayering), and more importantly, they have developed a criterion for preventing reverse flow (Lee, et al., 1979a,b).

The above research, tests, and studies played an important role in the development of the design methodologies reported herein.

HIGHWAY TUNNELS

The Permanent International Association of Road Congresses (PIARC, 1987) data show that a fire occurs about once per six million vehicle miles. Most of these fires start without an external cause; thus, accident frequency is a minor influence on fire risk. With the high standards of equipment and control in road tunnels, death and/or injuries due to fire are extremely rare.

Nonetheless, while highway tunnel fires have been rare, they can be very serious due to the difficulties of fighting fires in an enclosed space, the possible concentration of poisonous gases, temperatures that can be in excess of 1,800°F, and the possibility of panic among the tunnel users. Serious fires resulting in loss of life have occurred in the past in the Holland and Caldecott Tunnels in the United States, the Velsen Tunnel in Holland and the Nihonzaka Tunnel on the Tokyo–Nagoya–Kobe motorway in Japan.

The experience of most operating tunnels indicates that the most common fires are small in size and attributable to electrical and mechanical faults in vehicles (brakes, tires, fuel systems, short circuits, etc.). The size of a fire is nearly always limited and can be easily extinguished, although rapid intervention is critical. Fire extinguishers are the most appropriate means for bringing this type of fire under control when it first begins. Hence, regularly spaced placement of easily accessible portable fire extinguishers is now common practice in most vehicular tunnels. It is also a fact that fires due to accidents (frontal or front–rear collisions, hitting of obstacles) are much less frequent inside than outside a tunnel. Drivers are more careful in a tunnel. They drive at lower speeds and are not exposed to risks due to weather and the condition of the roads that apply outside the tunnel.

The characteristics of heat and smoke propagation in the event of a vehicular fire in a tunnel are similar to those which occur in a fire of equal intensity on an open highway, in that buoyancy created by the heat at the roadway level causes smoke and heat to rise. On an open highway, the smoke and heat can dissipate and disperse vertically into the atmosphere. In a tunnel, however, the ceiling or roof precludes such dispersal, thereby creating a serious hazard to individuals in the tunnel. The smoke and heat that rises will move longitudinally at the ceiling or roof of a tunnel.

Experience has proven that toxic gases and particles in undiluted smoke are more life threatening than heat. Thus, people have a greater chance of escaping harm when the ventilation succeeds in keeping the lower portions of the traffic space free of smoke, and provides enough air and visibility to enable people to evacuate the tunnel. The ventilation system should be capable of removing smoke from the traffic space as soon as possible and preferably at a higher rate than it is being generated.

Basic factors that influence the determination of the safety equipment and systems to be installed in a tunnel include the tunnel's length and the amount of traffic. Other factors include the location of the tunnel (within or outside an urban area, under water), the number of traffic lanes, the amount of heavy-goods traffic and the regulations in force for the transit of dangerous materials through the tunnel.

The size of a fire is also a function of the type of vehicle involved, and if it is a truck, the nature of its cargo. In the event that vehicles are involved in a collision, there also may be a rupture of fuel tanks and consequently spillage of fuel, which may spread according to the grade of the roadway. If ignited, a spill can result in a rapid propagation and spread of the fire. The magnitude of a fire corresponding to various sources can be approximated by the values given in Table 19-1 (Heselden, 1976).

It is difficult to accurately predict the magnitude of a fire scenario. Even where hazardous cargoes are precluded from using tunnels, they cannot be detected. In fact, the Holland Tunnel fire of 1949 was a consequence of a prohibited hazardous cargo in the tunnel.

Documented reports of research on tunnel fires (Heselden, 1976; Kawamura et al., 1976; Tateishi et al., 1970; Mizutani, 1982; Fezlmayr, 1976; Offenegg, 1965) show the behavior of the fire and associated tunnel air flows to differ significantly from more familiar fire situations outside the confines of a tunnel. The most noteworthy distinction is the buoyant effect that tends to create a layer of hot smoke and gases flowing away from the fire near the crown of the tunnel, while air supporting combustion moves toward the fire beneath the smoke layer. For example, in a level, unventilated tunnel with the fire near the longitudinal midpoint, the buoyant effect will establish a symmetrical circulation pattern with the hot, smoky air leaving both ends of the tunnel and air outside the tunnel drawn in beneath it.

A longitudinal ventilation system forcing air to flow through the tunnel will shift the balance of heated air in the direction of the forced flow. If the ventilation is of sufficient capacity, it will cause all of the heated air to flow toward the downstream direction. If the ventilation capacity is more limited, the upper layer of heated air may flow in a direction contrary to the forced ventilation (a phenomenon called *backlayering*). Whether backlayering occurs depends upon several factors, including the intensity of the fire, the grade and geometry of the tunnel, and the velocity of the ventilating airstream.

Allowing for the uncertainties of the rate of fire growth and propagation in a tunnel, the objectives of a viable emergency ventilation system are to keep the smoke and heat of a fire away from people who may be in a tunnel.

In all of the above major tunnel fires, the common parameter that significantly influenced the magnitude of the fire, and the resultant impact on life safety, was time. The significance of time can best be identified by the key sequence of events in a fire situation as follows:

- Time to detect a fire
- Time to send an alarm
- Time to verify the source of fire
- Time to implement emergency response procedures

In recent years, technological advances have combined to minimize the time factors identified above. The subsystems appropriate to minimize total time, and therefore improve life safety capabilities in highway tunnels, are discussed in the following sections.

Detection

Thermal and/or smoke detectors come to mind when one thinks of fire detection. While some are in use in tunnels, they are limited in their capabilities. Smoke detectors of virtually any type are unsuitble for tunnel applications due to the products of combustion in the vehicle exhaust emissions. Most thermal detectors are also unsuitable.

First, detectors must be rugged and reliable in the harsh corrosive ambient atmospheric environment of a tunnel and be able to withstand the high-pressure water nozzles used on tunnel washing machines. Second, there is a possibility of frequent false alarms, particularly during congested operations, due to vertical exhaust stacks on large trucks or buses. Current state-of-the-art heat detectors suitable for tunnels include dual rate of rise detectors (spaced approximately 75 ft on centers for a two-lane tunnel), or linear thermal detectors (usually arranged in a serpentine pattern at the ceiling of the tunnel).

Given that most vehicular fires initiate under the hood or floors of the passenger compartment (or in the passenger compartment), there may be a substantial time delay before the heat is detected by thermal detectors at the ceiling. In fact, if and when such a situation occurs, it would be reasonable to conclude that the fire has had time to increase to some significant size. For the above reasons, thermal detectors are not often installed in tunnels, other than as a secondary detection system, if deemed necessary.

Subsystems facilities frequently will include traffic loop detectors embedded in the roadway pavement for monitoring traffic flow rates and speeds. They can detect stopped traffic and thereby usually will become the first means of detecting an incident and alerting the Operations Control Center staff of a problem.

Many modern highway tunnels include an Operations Control Center (OCC), which is continuously manned by trained operators. Operators in the OCC, using closed-circuit television (CCTV) cameras positioned to view the length of the tunnel roadway, can then assess the cause of the traffic being stopped and effect an appropriate response. In a fire situation, this means of early detection can be most effective toward initiating an appropriate and timely response.

Alarm

The detection of stopped traffic by a roadway loop detection system will be alarmed to the OCC. If a fire situation is verified, the OCC can transmit alarms as appropriate to the various emergency response agencies, so that suitable personnel and equipment may be dispatched to the scene. Thermal (heat) detectors at the ceiling will transmit alarms to the OCC and, if desired, directly to the emergency response authorities, including fire, police, and other emergency service units.

Manual fire alarm pull stations and/or emergency telephone installations in wall niches along the length of the tunnel roadway are also employed as supplementary alarm devices.

Incident Location

Knowledge of the location of a fire incident in a tunnel is critical to implementing an appropriate plan of emergency response. Loop detectors and heat detectors can be arranged in zones or, if a thermal linear detector is used, the location along the length of the detector (currently available up to 1,000 ft per power and control unit) can denote the specific location of the fire incident on a graphic display panel or CRT monitor. The OCC will also use the CCTV monitors, if possible, or dispatch personnel to the scene to effectively verify and establish an incident location.

Communications

Relatively recent advances in the state of the art in electronic communications have substantially improved incident management—including fires—in tunnels. Some of those which seem to be most effective include

- AM–FM rebroadcast antennas, which permit the OCC to convey messages, instructions, etc., to motorists via their car radios. Public address systems in tunnels are not suitable because of the high ambient noise level.
- Variable-message signs can provide information to notify and assist motorists in a fire situation, including guidance in evacuation to safety areas or exits by foot or vehicle.
- Traffic control signals can be set and warning notices can be flashed on an electronic sign at the tunnel entrance portals, activated simultaneously with a fire alarm, to preclude further entry to the tunnel by motorists. The OCC can also position lane traffic signal controls most appropriate to expeditious vehicular evacuation or to facilitate access by emergency response vehicles.
- Emergency telephones are available in tunnel wall niches at frequent intervals for motorists to contact the OCC as required. Other dedicated telephone and two-way radio support systems compatible with emergency response services communications requirements should also be provided in the tunnel installation.
- Locations of exit doors to safety areas are generally unknown to the average motorist in a tunnel. Illuminated fixed directional signs, flashing strobe lights at the exit doors, and emergency lighting systems are used to guide people in the event they are advised to abandon their vehicles and evacuate the tunnel on foot.

Planned Responses

Emergency response plans should be developed in conjunction with the fire department and other emergency response services as appropriate. The OCC should be equipped with computer hardware and software systems from which preprogrammed or manual emergency response plans applicable to all relevant subsystems can be implemented. Implementation may require a combination of automatic, semiautomatic, or manual actions as most appropriate to the situation. The emergency response plan(s) should be capable of addressing all types of incident management situations.

In the case of a fire situation, the emergency response plan should include the collection and verification of input data such as alarms and incident location, notifications to police, fire department and other response units as appropriate, dispatch of emergency tunnel personnel and vehicles, setting of traffic control signals, activation of smoke control ventilation and fire suppression systems, procedures and means for effective communications with motorists, and advice and guidance for their evacuation when necessary. The major elements and subsystems of the fire life safety program should function as follows:

1. The OCC should be manned 24 hours a day by trained operators. Operator proficiency in dealing with minor and major incidents and plant failure should be maintained by regular refresher training using incident simulation programs.
2. CCTV cameras should be located throughout the project roadways. Pictures from the cameras would be relayed to the OCC for display on monitors located in the OCC and available for selection by the operator.
3. Roadways should be provided with a system of loop detectors embedded in the pavement. The OCC computer should continuously monitor traffic passage time over successive loops within the tunnels to determine when a breakdown in traffic flow occurs, and alert the operator to a possible incident. An alarm would be displayed in the OCC. The appropriate CCTV output would be automatically directed to a dedicated monitor at the master control console. The operator should assess the situation and provide the appropriate response.
4. Traffic control systems should include lane-use signals mounted over each lane to control traffic flow in each lane and advisory message signs located on the tunnel approaches and throughout the tunnels to advise motorists.
5. The radio rebroadcast system, with override capability, should provide a means for the OCC to communicate directly with motorists via FM frequencies (variable-message signs should advise motorists to switch on radios)
6. Motorists' "lift to call" telephones located throughout the tunnels should provide communications between motorists and the OCC.
7. Zoned thermal detectors should provide additional data to the OCC to indicate and locate fire incidents, with simultaneous automatic notification of the fire departments.
8. Preprogrammed options for the operation of the mechanical ventilation system should be available for selection and implementation by the operator. Manual operation should also be available if required.
9. Cross-passageways between the roadways or emergency exits to the surface should provide easily accessible escape routes should evacuation be required.
10. Emergency response vehicles, manned by personnel trained to provide assistance ranging from fighting minor fires to clearing accident sites, should be provided. The personnel should respond to instructions issued by the OCC and should be able to locally assess the situation and render appropriate assistance. The vehicles should be equipped with two-way radios, public address facilities, emergency lights, fire-fighting equipment including foam extinguisher, metal-cutting equipment, and medical supplies.
11. The fire detection, alarm, and protection systems should include thermal detectors, thermal and smoke detectors, and sprinklers as appropriate in ancillary buildings, chemical suppression systems in OCCs, thermal and smoke detectors in switchrooms, fire indicator panels located in ancillary buildings, a main fire indicator panel located in the OCC, a continuous standpipe system with 2-1/2-in. fire department hose valves (or otherwise as appropriate) at approximately 150-ft centers, portable extinguishers, and a communications system using the radio rebroadcast system.
12. Emergency response plans should be developed in conjunction with fire departments and other local agencies as applicable. These plans should include the assembly of equipment operating modes, which may be selected during a fire incident and should be part of an integrated package. The

implementation of any emergency response plan option for fire fighting should be OCC operator-initiated at the direction of the senior fire department officer upon arrival on the scene. Manual operation of all systems should be available in case of control system failure, or to accommodate a particular situation.

In addition to the emergency response plan being documented and distributed to all necessary individuals of the tunnel operating authority, copies should also be provided to all emergency response units. Regular programs of training should be mandatory for all individuals who are likely to have to participate in a fire emergency situation. This training should be repeated at suitable regularly scheduled intervals if effective responses are to be expected. Coordinated simulation exercises involving all emergency response units and individuals must also be accomplished on a regular basis. Since the frequency of occurrence of a major emergency incident (especially one involving a major fire situation) is rare, only through all involved personnel participation in repetitive regularly scheduled training courses and field exercises will the emergency response team be effective when needed. Modern electronics and computers can minimize the critical response time, but the response personnel have to be ready to act. Additional details on surveillance and control are included in Chapter 24.

Personnel Evacuation

The need to evacuate personnel (i.e., motorists and passengers abandoning their vehicles in the tunnel) will be determined at the OCC. Evacuation pathways and procedures are presented as preprogrammed scenarios graphically displayed by the computer on monitors in the OCC. The display will reflect the incident input data and all the interdependent relevant information to support an evacuation including smoke control, fire suppression, and traffic control status. Evacuation is to be implemented when the OCC operator or fire commander at the scene verifies the applicability of all the indicated subsystems responses to the situation. When all required subsystems are activated at the OCC (by a single master control button or switch), the variable-message signs, appropriate exitways, and safety lighting will be energized to guide the personnel evacuation.

Smoke Control

The alternative ventilation concepts applied in tunnels and the details pertaining to the characteristics and design of such systems are addressed in Chapter 20.

As previously stated, the common parameter critical to all fires is time. In effect, therefore, it can be said that the objective of a good tunnel emergency ventilation system is to buy time. That is, time to allow people to escape from a tunnel involved with a fire incident. The longer the ventilation system can keep the concentrations of smoke and heat away from escaping motorists, the greater their chance of reaching a safety exit from the tunnel.

Today's technology is such that systems of detection, alarm, verification, and communication can effectively assist in guiding passengers to points of safety if they must evacuate a tunnel and abandon their cars. However, one must also design and operate an emergency ventilation system in a manner that will maintain a safe pathway of escape for the maximum amount of time.

While it is unlikely that an emergency ventilation system could be engineered to meet all possible conditions of fire and smoke propagation, it is possible to mathematize the effectiveness of certain emergency ventilation systems for a variety of fire scenarios.

Over the years, full transverse ventilation systems (a type usually associated with urban tunnels greater than 1/2 mi in length) have generally been regarded as the type of ventilation system that would provide the maximum safety in a fire situation. The popular theory associated with full transverse systems has been that the air flow in a tunnel is always transverse from the supply ports to the exhaust grilles and therefore no longitudinal flow will take place. Therefore, the theory contends that the smoke and heat can be contained local to the fire. Unfortunately, observation of actual tunnel fire scenarios demonstrates that this is not the case when the fire is large.

In the event of a fire involving a vehicle or vehicles disabled in a tunnel, the ventilation system should be able to control the direction of smoke movement to provide both a clear and safe path for evacuating people and to facilitate fire-fighting operations. The ability to prevent backlayering should therefore be a major objective in the design of the ventilation system and its operation during an emergency.

Given the advances in state-of-the-art analytical techniques, coupled with knowledge derived from the afore-described experience and research, a design tool to analyze the performance of certain concepts of mecahnical ventilation systems in a fire situation was developed in recent years.

This fire model simulation computer software program evolved from an original Subway Environment Simulation (SES) computer program capable of evaluating the environmental conditions under normal operations in tunnels and underground stations of a rapid transit system. That program, developed between 1970 and 1975, is a numerical simulation model that incorporates the results of theoretical research, scale model and field testing data, and has been verified through full-scale subsystem tests and by comparisons with measurements taken in operating transit systems.

The SES Program comprises four interdependent computation sequences developed for underground rapid transit systems simulations: a train performance subprogram, an aerodynamic subprogram, a temperature/humidity subprogram, and a heat sink/environmental control subprogram.

Sometime after 1975, the aerodynamic subroutine of the SES, which determines the piston action air flows (longitudinal air flow in a tunnel caused by vehicle travel), was extended to vehicular tunnel applications. While the piston effect is a function of traffic speed, density, and a number of other variables, in all cases, and for any type of ventilation system, it causes longitudinal air flow in the roadway in the direction of traffic movement.

By the late 1970s, with the additional knowledge acquired relevant to the effects of thermal buoyancy associated with fire propagation, it was recognized that the aerodynamic network balance that would occur at standard air conditions would not apply under the intense thermal effects of a fire. The U.S. Department of Transportaton subsequently commissioned the development of an SES fire model that would reflect the various aerodynamic and thermodynamic parameters associated with fires in tunnels. The work began in 1978, and in November 1980, a transit tunnel fire ventilation model had been developed. This was incorporated in an updated version of the original Subway Environment Simulation Program. Currently in use, it is known as the SES, Version 3 (Associated Engineers, 1980). It is this version that has been applied to vehicular tunnel fire evaluation.

Among the elements included in the fire model is the ability to simulate throttling effect. This pressure loss is caused by the rapid expansion of the air flowing past the fire site. Also, as a consequence of the law of conservation of mass, the viscous pressure losses in the section of the tunnel downstream of the fire increases, tending to reduce tunnel air flow. Density differences between the hot gases and the ambient air give rise to pressure differentials, which can either augment or retard the tunnel air flows, depending on the direction of ventilation (uphill or downhill). The effects of these density differences on fan performance have also been accounted for in the model.

Another element included in the fire model is that of wall surface temperature response. Accurate modeling of the transient heating of the wall surface at the fire site is an important factor in determining the conditions downstream of the fire, as it improves the accuracy of the predicted air temperatures, which are subsequently used to calculate the buoyant pressure differential.

The model treats the wall as a one-dimensional slab of infinite thickness with uniform thermal properties and an arbitrary time-dependent heat flux at the wall surface.

The heat conduction equation is solved by using an approximate integral method. This method was chosen because it requires relatively little computation time and provides good accuracy (results range within 3–9% of the theoretical value).

Heat is transferred to the wall by convection and radiation. Radiation will be the dominant mode of heat transfer at the fire site, while downwind of the fire, both modes will be nearly the same order of magnitude. At the site of the fire, heat is radiated from the flame directly to the tunnel wall at an "effective fire temperature." The effective fire temperature and a parameter called the equivalent fire area are input items. Downwind of the fire site, the hot smoke is assumed to be radiating to the tunnel wall at a temperature equivalent to the "bulk" air temperature at a given location. Only radiation effects in the transverse direction from smoke to tunnel wall are considered.

The changes in air density associated with elevated temperatures affect the performance characteristics (pressure versus volume flow curve) of the exhaust fans. These effects have been accounted for in the model.

The throttling and buoyancy effects, which are primarily caused by changes in density, are conveniently accounted for by noting that changes in density are inversely proportional to changes in the absolute temperature of the gas (air), a quantity that is computed by the program. Therefore, the effects of density changes have been accounted for in the computations without actually converting the program from an incompressible to a compressible flow model. As a result, the air flow quantities are "referenced" to the ambient air density. This notion of basing the computations on a reference air density has been used in mining ventilation computer programs (Greuer, 1973, 1977).

Examples of situations in which the program can provide important design information include the following engineering questions:

- What is the effect of emergency control procedures on the tunnel environment?
- What are the dynamic temperature and air flow conditions that prevail during a fire emergency?
- What ventilation system capacity is adequate to control the spread of smoke and heat during a fire emergency?

How the fire model has been developed limits its applicability to the analysis of tunnel ventilation systems that attempt to control smoke by creating a longitudinal air flow along the roadway. This would include longitudinal systems that use jet fans, intermediate shafts, or a Saccardo nozzle effect, and transverse or semitransverse systems that use a single-point exhaust (SPE) concept for smoke control.

The fire model is intended for use in a trial-and-error fashion to size the emergency ventilation system. The iterations are between the tunnel air velocity (approaching the fire site) predicted by the SES fire model and a required air velocity that precludes the backing of smoke against the ventilating air stream (i.e., backlayering). The required air velocity is a function of the fire heat rate, the tunnel dimensions, the average tunnel grade, and the temperature of the hot gases leaving the fire. A typical application of the fire model consists of the following steps:

1. Determine the required air velocity.
2. Perform an SES simulation to determine the predicted tunnel air velocity and hot air temperature.
3. If the predicted air velocity exceeds the required air velocity, the ventilation system is considered adequate. If the predicted air velocity is less than the required air velocity, change the system design and repeat steps 1 and 2.

The SES fire model has been designed with the ability to simulate the "overall" effects of a tunnel fire on the ventilation system. This level of detail is considered sufficient for evaluating the adequacy of an emergency ventilation system and is consistent with the state of the art in mining ventilation programs with the capability of simulating fires. However, the model does have its limitations. The SES is a one-dimensional model. Therefore, the results of a fire simulation will indicate whether or not the ventilation air flows are sufficient to prevent backlayering, but *not* the extent of backlayering (a two-dimensional phenomenon) if predicted.

The fire model that has been developed reflects a compromise between satisfying the basic needs of the ventilation system designer and using an analytical treatment of fires in a tunnel that is compatible with the basic structure of the SES program. In formulating the analytical treatment of the problem, it was concluded that a detailed simulation of the complex flow region near the fire was not required to be able to predict the occurrence of backlayering. Instead, it was concluded that only the bulk flows and temperature of the air moving toward and away from the fire, which depend on the coupled effects of the fire and the forced ventilation system, had to be simulated. Having done this, the occurrence of backlayering would then be determined by comparing the resulting velocity of the air moving toward the fire with a certain "critical velocity" above which backlayering is precluded.

The method selected for treating the problem uses the results of work performed by the U.S. Bureau of Mines (Lee et al., 1979a,b), which showed that backlayering will not occur if the velocity of the ventilating air moving toward the fire is equal to or exceds a certain critical velocity. For a level tunnel ($K_g = 1.0$), this critical velocity is determined from the following coupled equations:

$$V_c = K_g k \left(\frac{gHQ}{C_p A T_f} \right)^{1/3} \qquad (19\text{-}1)$$

$$T_f = \frac{Q}{C_p A V_c} + T \qquad (19\text{-}2)$$

where

V_c = critical velocity, m/s
g = acceleration of gravity, m/sec^2
H = tunnel height, m
Q = fire heat release rate, W
C_p = specific heat of air at constant pressure, $j/K_g - {}^\circ K$
A = net cross-sectional area of tunnel, m^2
T_f = hot gas temperature, °K
k = 0.61 (dimensionless)
K_g = grade correction factor (dimensionless)
T = ambient temperature, °K

For a tunnel in which the direction of ventilation is downgrade, the critical velocity is greater than that for a level tunnel. Although the effect of grade on the critical velocity has not been specifically studied in connection with tunnel fires, related studies on the control of methane layers in coal mines (methane, being lighter than air, tends to form layers along the crown of a mine gallery) have provided some useful data. Since the physical phenomena are similar in both cases, i.e., a low-density fluid flowing over a higher density fluid, the data presented for methane layers has been used to develop a grade correction factor.

The simultaneous solution of Equations (19-1) and (19-2) determines the critical velocity. This criterion determines the minimum steady state velocity of the ventilating air moving toward the fire that would be required to prevent backlayering. Note that this criterion determines the required air velocity during the fire and not the air velocity in the absence of the fire, which can be substantially different. The velocity of the ventilating air moving toward the fire must therefore be known to apply this criterion. This velocity is provided by the SES fire model.

The use of the fire model is illustrated by the following example, which considers a tunnel with two ventilation system alternatives.

The assumed tunnel is 3,940 ft long and level with two lanes of unidirectional traffic. The tunnel cross section is 800 ft^2. Tanker trucks are not prohibited passage through the tunnel; therefore, a 50-MW fire will be used as the basis for the calculations.

The performance of the following alternative ventilation systems will be evaluated during fire and nonfire conditions:

- *System No. 1.* A single-point exhaust system using a separate parallel tunnel as an air duct with large adits at 300-ft intervals connecting with the main tunnel. Each adit is provided with a remote-controlled motorized damper that permits the selective use of one or more exhaust points in the vicinity of a fire while closing the remaining openings. The total exhaust capacity of 1,500,000 cu/min. can be concentrated at a single point.
- *System No. 2.* A longitudinal system using jet fans. There are a total of 24 fan units arranged in pairs and located at 300 ft intervals. Each 4-ft diameter fan has a capacity of 82,000 cu/minute.

Smoke control is achieved by providing a longitudinal air flow with sufficient air velocity to force smoke to spread downwind of the fire site, thus protecting any motorists trapped upwind of the fire.

The magnitude of the air velocity required to achieve this control is calculated from Equations (19-1) and (19-2). For this example, a minimum air velocity of 570 ft/min approaching the fire site must be maintained during the fire. This value applies to both ventilation system alternatives.

The results of the calculations are shown in Table 19-1. In each case, the quantity presented is the approach velocity, which must exceed 570 ft/min to prevent the reverse flow of smoke.

Table 19-1. Fire Magnitude

Source	Thermal Output (MW)
Passenger cars	3
Van	10
Truck or bus	20
Gasoline tanker spill	50–100

Table 19-1 shows the approach velocity with System No. 1 during fire and nonfire conditions when the exhaust point is at the middle of the tunnel. Note the effect of the fire is to reduce the approach air velocity. The results indicate that smoke spread cannot be controlled unless there were no additional vehicles in the tunnel and there were no adverse wind.

With application of the SES fire model to vehicular tunnels, various ventilation systems can be simulated and a determination as to the most effective design for the ventilation system both in concept and magnitude can be determined for a variety of fire scenarios. When the emergency ventilation system design is integrated with the other subsystems that contribute to life safety in the tunnel, the ability to provide greater levels of protection for people trapped in a tunnel fire can be achieved.

Currently, plans are under way to conduct full-scale fire tests in an existing highway tunnel no longer in use, using the various alternative ventilation concepts applicable to highway tunnels. (The Memorial Tunnel Fire Ventilation Test Program is discussed in more detail in Chapter 20). The test program has been formulated by and is advancing under the guidance of the American Society of Heating, Air Conditioning and Refrigerating Engineers in conjunction with the State of Massachusetts Department of Public Works and the U.S. Federal Highway Administration, who are jointly providing the funding for the program. The results of the tests are expected to provide additional data, not currently available, which will be applied to the fire safety systems design and operations for one of the most extensive highway tunnel construction projects, located in the City of Boston, ever undertaken in the United States.

The test results will also be used to validate predictions of the SES fire model and to enable the model to be calibrated for greater accuracy than can be currently accomplished. Given the planned scope of the test program, and the extent of comprehensive data anticipated to be obtained, additional analytical design tools will be subsequently developed to evaluate various aspects of the interdependent fire size and alternative ventilation systems performance affecting fire life safety, with the physical and geometric parameters of a tunnel.

Fire Suppression

The design and application to tunnels of various fire suppression systems are collectively addressed for highway, rapid transit, and mainline railroad tunnels later in this chapter.

Power Supplies

The details of the design and application features of the electrical power supply systems for tunnels, including the means of power distribution and control, are addressed in Chapter 22. A review of that chapter will indicate the requirements for an electrical system that is highly reliable and therefore incorporates many redundant features and elements. In addition, power distribution to critical equipment (i.e., ventilation fans, sump pumps, and fire pumps, and communications and control wiring) essential to support of the various subsystems should be protected from physical damage or damage by fire in a tunnel roadway. This is frequently accomplished by embedding such circuits in duct banks within the tunnel construction.

Depending on the location and use of the tunnel, a number of emergency power backup systems are also applied. In general, these emergency systems range from a variety of uninterruptible power supplies (UPS) that operate from systems of battery power packs to diesel-powered emergency electric generators.

They are used for those subsystems where even a momentary loss of power could have serious consequences. Thus, they include central control computer systems and displays, traffic signals and controls, communications systems, detection and alarm systems, emergency lighting systems, illuminated emergency signs, including exit signs, and variable-message signs.

Emergency generator systems usually start up automatically. When available to carry load, they will provide power to those systems which have been operating on UPS systems as described above, and to other systems including fire pumps, tunnel drainage pumps, and other ancillary systems determined to be needed to support any emergency response plan. In general, however, where reliable dual power supplies and protected dual electric feeders are used to provide power to the ventilating fan motors, no additional emergency power source for the fans (from an emergency generator) is usually required.

RAPID TRANSIT TUNNELS

As considered in this handbook, rapid transit tunnels are defined as any combination of underground facilities (i.e., tunnels and stations) used solely for the transport of people by mass transportation vehicles on fixed guideways. Accordingly, fire life safety factors and considerations in such facilities would apply to electrified steel-wheel on steel rail trains, rubber-tired trains on concrete tracks, electrified buses, and fully automated driverless vehicles along an exclusive right of way.

As is the case with highway tunnel fires, most fires are small in size and attributable to electrical and mechanical faults in vehicles. However, several significant fires have occurred in fixed guideway systems during the past 25 years where, fortunately, the loss of life was limited.

In general, the subsystems essential to an effective fire life safety program for a rapid transit system are similar to those discussed previously in the section "Highway Tunnels." However, there are unique parameters and characteristics of rapid transit vehicles, and systems are compared with vehicular traffic in highway tunnels for which the features of the interactive subsystems must be uniquely designed.

For example, whereas the characteristics of heat and smoke propagation in a rapid transit tunnel fire may be similar to that observed in a vehicular highway tunnel, there are significant differences affecting fire life safety concerns. For example, electrical propulsion vehicles in rapid transit systems generally do not constitute the potential hazard attendant to on-board fuel supplies associated with cars, buses, or trucks propelled by internal combustion engines. However, the safety of 2,000 or more people on a single train during a peak-hour period in a fire situation presents some special problems:

- Time-consuming difficulty in evacuating people from a tunnel between stations if a train is immobile, due to constraints resulting from the physical and geometric characteristics of trains and trainways.
- Greater potential for panic, particularly due to people being in the environment of a facility that surrounds the train and is, for the most part, totally foreign and forbidding to them, and therefore an inherent danger.

The potential sources of fires in rapid transit systems can also be quite different from those for a highway tunnel. Most on-board train fires occur in the various components of the train's electrical/mechanical systems, which are usually concentrated below the floor of the passenger compartment.

The National Fire Protection Association (NFPA) has developed a fire safety standard for fixed guideway transit systems (NFPA, 1990). This standard includes minimal criteria requirements for the design of the components of a rapid transit system for fire safety considerations. For example, the floor system is required to have sufficient resistance to fire penetration to the interior of the vehicle by an external fire for a period consistent with the safe evacuation of a full load of passengers from the vehicle in the worst-case situation. However, many fires are started by arson in the passenger compartment.

The features of the interactive subsystems for an effective fire life safety program appropriate to rapid transit applications are as follows.

Detection

Most rapid transit tunnels are not equipped with either smoke detectors or fire detectors for reasons similar to those already discussed for highway tunnels.

In most instances, detection of an incident is reported by radio or phone to the OCC by a motorman or train conductor. The motorman or train conductor may have been made aware of the incident by a passenger using on-board communications facilities. In virtually all cases, the OCC is made aware of an incident by some on-board or wayside radio or phone communication. Transit workers in the tunnel may be the first to detect an incident.

Alarm

After an incident report is received, and verified to the satisfaction of the OCC, they will initiate alarms and dispatch emergency response units as appropriate. Verification of a reported incident is usually accomplished by sending designated transit employees to the scene of the incident.

Incident Location

A train's location within established electrical power blocks is graphically displayed in the OCC. Given lengths of trains (frequently) up to 800 ft in major urban systems and the usually longer lengths of electrical power blocks, a general location of a train in the system can be readily determined within the OCC. However, a more definitive location of the train and, in particular, the location of the fire in relation to the train, is usually necessary for the OCC to implement an effective emergency response plan. This information normally depends on voice communication by the motorman or conductor from the scene to the OCC. In the case of fully automated driverless vehicle systems, on-board passengers have facilities in each vehicle to enable direct voice communication with the OCC in an emergency.

Communications

Vehicle on-board radio or telephone systems for communication with the OCC are usually provided in the motorman's or train conductor's cab, or they will be provided for emergency use by passengers in each vehicle. In many transit systems, the OCC operator can communicate directly with on-board passengers via a public address system in each car, and provide appropriate instructions. Public address systems are required in all transit stations for communications from the OCC to passengers and employees (NFPA, 1990).

Wayside communication facilities are also provided in the tunnels. These facilities include radio and telephone capabilities. Radio capabilities and certain frequencies are reserved to enable the use by the fire department, or other designated emergency response personnel, of their own radio equipment.

Planned Response

The requirements for emergency response plans for a rapid transit system, in general, are similar to those for a highway tunnel. Notable differences of the major elements and subsystems of a rapid transit system in their functional requirements for an effective fire life safety program include the following:

1. CCTV cameras are normally not provided in rapid transit tunnels, although they are used for security and fire safety

in the station areas. Many of the rapid transit tunnels are not normally lighted, which precludes TV camera applications.

2. Provisions for selective emergency tripping of the sectionalized electrical traction power systems are essential for life safety. This capability should be available at the OCC and locally throughout the transit system.
3. Train operations and control systems are crucial elements in the support of an emergency response plan to assist in the evacuation of passengers from a tunnel or to transport emergency response personnel to the scene of a major incident.
4. Effective communications systems on-board, and wayside in tunnels (radio and/or telephone), having similar capabilities to those subsystems described for highway tunnels are mandatory in rapid transit systems as well .
5. Other than in very limited applications, heat detectors are not used in transit tunnels. Their effectiveness in a rapid transit system are minimal at best, given the nature of the potential fire hazard, and the likely places of a fire occurrence.

Personnel Evacuation

The physical constraints of a rapid transit tunnel and the physical relationships of a train to the trainway in which it operates present enormous problems in the event passengers must be evacuated from an immobilized train in a tunnel. Accordingly, the standard operating procedure in transit systems, if a fire is discovered on a train, is for the motorman to continue to drive the train into a station if at all possible, where he can disembark all passengers quickly and more safely. Experience of various rapid transit systems, where it has been necessary to remove passengers from an immobilized train in a tunnel during peak travel periods, without the presence of fire or other major emergency incident, is that it will take at least 30 min to complete an evacuation. In large urban systems, periods of 1-1/2–2 hours for effecting an evacuation are not uncommon, and in at least one case, 5 hours was reportedly required. Where a train is immobilized in a tunnel, consideration should be given to bringing a train to the scene on a parallel trainway to which passengers from the immobilized train can be transferred. This method has been employed and generally, in the event of a fire situation, will be most effective where a dividing wall and fire-protected cross-passageway doors separate the trainway, and safety walkways are provided. In lieu of this procedure, if not practicable in a given situation, evacuating passengers will have to be walked out in the tunnel to a vertical escape shaft or subway station. As with highway tunnels, emergency lighting systems and signing to direct people to safety exits are essential. In rapid transit systems, employees at or dispatched to the scene are required to implement an evacuation procedure and to guide and assist passengers in the process.

Smoke Control

Emergency ventilation systems are an essential component of a modern rapid transit system. Many older transit systems in recent years have been gradually retrofitted to incorporate appropriate systems of emergency ventilation fans and shafts.

Whereas the objectives of emergency ventilation are similar to those systems for highway tunnels, the concepts for the ventilation systems are usually quite different. In long highway tunnels, ducted ventilation systems are necessary for normal traffic operations to achieve effective air quality pollutant control, and therefore they are engineered to function in an emergency ventilation capacity also. Other than some unusual situations, ducted emergency ventilation systems are not used in rapid transit tunnels. Accordingly, the ventilation concepts in transit tunnels function to effect longitudinal flow in the trainways.

Two ventilation system concepts have been applied to transit systems built within the past 25 years. One involves the use of mid-tunnel shafts (one in the approximate center of a tunnel between two underground stations) equipped with exhaust fans (which may be reversible). On very long tunnel sections between two stations, more than one fan shaft might be used. When these fans operate in exhaust, they draw air from the outside via station entrances or via air intake vent shafts at the stations.

A second concept, which has had somewhat more extensive applications, uses fans in shafts at the ends of each station. These fans have a reversible air flow capacity of at least 90%. To ventilate a tunnel between two stations in a fire situation, fans at one end of the fire-involved tunnel will operate in a supply air mode while fans at the other end of the tunnel are operated in an exhaust mode. This concept is popularly known as a "push–pull" system. The reversible push–pull concept enables passengers to evacuate a train in the tunnel by walking toward a safety exit (or station) into a supply of fresh air while smoke is being exhausted from the site of the fire in the opposite direction.

Given the substantial time required to evacuate passengers from a train in a tunnel, the role of the emergency ventilation system in maintaining a clear evacuation pathway by controlling the direction and movement of smoke is critical to effective fire life safety.

In general, the types of fans most suited for rapid transit applications are axial fans. Their relative compactness and ease of establishing reversible air flows make them particularly appropriate for the push–pull concept. To achieve the high reversible air flow capacity, the required fan blade design inherently results in poor fan efficiency. Since these fans are generally used only for emergency ventilation, however, the energy consumption is not a significant factor. In addition, they are usually very noisy, and sound attenuation may be required as the fans should be tested periodically, and their noise level may well exceed acceptable ambient noise level criteria at the street level.

As with longitudinal ventilation systems for highway tunnels, the effectiveness of the transit tunnel system is a function of its ability to control the movement of smoke in a given direction (i.e., prevent backlayering). Because of the "openness" of most transit tunnels, and the relatively high aerodynamic resistance of an immobilized train (or trains) in a relatively tight tunnel section, the ventilation fans are of fairly large capacity, and sometimes more than one fan is required at a particular shaft location to develop the necessary air velocity past the fire site to prevent backlayering. Typically, fan capacities will be in the range of 100,000–175,000 cfm, with fans having approximately 6–8-ft-diameter fan rotors, and motors ranging from 75 to 200 hp. In addition, these fans and the related dampers necessary to the system must be of above average industrial applications–type equipment capable of withstanding transient pressure pulses and frequent reversing pressure forces as trains drive by the fan shafts.

The fire model of the Subway Environment Simulation program, Version 3, is used to determine the fan sizes and the number of fans serving a trainway that must be operated, both upstream and downstream of the fire incident location, to achieve the desired flow control. The SES is also used to develop the preprogrammed scenarios that are stored in the OCC computer.

Power Supplies

While similar requirements for reliable and redundant power supplies to serve all interactive subsystems of the fire life safety program apply for transit tunnels as for highway tunnels, some special considerations are necessary for rapid transit system applications.

All critical power supplies (and communications circuits) must be protected against physical damage and fire damage. Thus, such circuits that must be run through the tunnels should always be embedded in concrete-encased duct banks. Dual-feed power and control circuits to each emergency ventilation fan should be fed from two separate substations at different locations such that the dual feeds to each fan are brought from different directions.

MAINLINE RAILROAD TUNNELS

In North America, mainline railroad trains primarily use diesel-engine-powered locomotives for propulsion. Accordingly, they require different fire safety considerations than are applicable to electrically propelled vehicles.

In normal operations through most mainline railroad tunnels, the frequency of trains is limited such that the products of combustion of the locomotives are dissipated by natural ventilation. However, in very long railroad tunnels, normal train operations may result in overheating of the locomotives, particularly when multiple locomotive consists are used, as would be the case with long, heavy freight trains. To preclude overheating of the locomotives (and their resultant automatic shutdown), a system of ventilation is required. The ventilation system is also used to purge smoke from the tunnel, rather than depend on natural ventilation, in order that the tunnel can be cleared of smoke obscuring visibility sufficiently in advance of the next train entering the tunnel.

There are currently only four major railroad tunnels in North America (the Moffat Tunnel in Colorado, the Cascade Tunnel in the State of Washington, the Flathead Tunnel in Montana, and the Mount Macdonald Tunnel in British Columbia) that have major ventilation systems. These systems have been designed to furnish supplemental air for locomotive cooling and to purge the tunnels of the noxious and visibility-obscuring diesel exhaust fumes after the train has exited the tunnel. Each of the four tunnels has large gates (doors), which serve to control (increase) the air flow relative to the train for improved engine cooling capability, and also to enable the ventilation system to purge the tunnel more rapidly. These ventilation systems are then also available to be operated to control smoke resulting from a fire situation in the tunnel.

The nature of cargo transported by railroad freight trains includes volatile and toxic petrochemicals and other highly combustible products. Thus, the potential magnitude of a possible fire is extremely high, and generally beyond the reasonable capacity to address effectively with interactive subsystems as in the case of highway or rapid transit tunnels.

An illustration of the potential severity of railroad tunnel fires is provided by an incident on a 1,000-ft-long tunnel on a coal-hauling railroad in West Virginia. In the aftermath of a forest fire, suction following the rear of a train passing through the tunnel introduced burning embers, which ignited the timber lining. Before anyone became aware of it, the tunnel turned into a 1,000-ft torch. The fire was finally suppressed, after a week of strenuous effort, by bulldozing earth to seal both portals and pumping water in to flood the tunnel to the crown. After that, it took two further days for the tunnel to cool sufficiently so that it did not reignite when it was opened and oxygen readmitted to the hot interior. Fortunately, there was no loss of life, but the coal mine production was shut down for two weeks.

Fortunately, freight trains normally carry only a very limited number of train-operating personnel, who would be expected to have received specialized safety training and to know how to respond in an emergency situation. Therefore, with the exception of the major railroad tunnels that are ventilated as noted above, few of the fire safety–related subsystems described for highway and rapid transit tunnels are included in mainline railroad tunnels.

Due to a combination of factors, including the locations of mainline railroad tunnels in predominantly rural mountainous areas, frequently inaccessible except by rail, systems of conventional detection, alarms, emergency response, and fire suppression are normally inappropriate. Thus, mainline railroad tunnels are usually not provided with fire mains and standpipe systems. Given the materials of construction used in tunnels today, there are little or no combustibles within the tunnel itself. Train operating safety systems, including speed restrictions, minimize the risk of a potential accident and resultant fire on board a train while in a tunnel.

Train-operating personnel maintain radio communications with an OCC and are the primary means by which a fire situation would be reported. In a fire situation, the railroad might have to transport personnel and equipment by rail to the scene of a tunnel incident to address the situation as most appropriate. The ventilation system (for those tunnels so equipped) could be operated to control smoke so as to support fire suppression activities and/or personnel evacuation or rescue.

FIRE SUPPRESSION SYSTEMS

Systems of fire suppression in highway and rapid transit tunnels are usually provided with wet or dry pipe water fire mains feeding a standpipe system throughout the tunnel. The details relevant to the characteristics and design of such systems are addressed in Chapter 23. In addition, standards for application of various fire suppression systems for rapid transit tunnels are presented in the National Fire Protection Association standard, NFPA 130, Fixed Guideway Transit Systems (NFPA, 1990).

SPRINKLER SYSTEMS

In a limited number of tunnels, foam- or deluge-type sprinkler systems have been applied. While the fire safety effectiveness of automated sprinkler systems for most types of buildings is unquestioned, consideration of their possible use in tunnels requires evaluation of several factors that are significantly different from a building.

Much evidence suggests that sprinklers are not only ineffective in controlling a fuel fire, but can actually contribute to the spread and severity of the fire. Vehicular tunnel conditions cannot exploit sprinkler system strengths and turn most of them into disadvantages. This is reflected in various studies reported by PIARC and is also included in findings of the USDOT report "Prevention and Control of Highway Tunnel Fires" (USDOT, 1984).

Tests with sprinklers have been undertaken in Switzerland, Austria, and Japan. They all indicated that the water shower induces a strong vertical mixing that the cooled smoke cannot resist, thus eliminaing all chances for stratification; thus, smoke gets immediately spread over the entire cross section of a tunnel. There also is a risk of explosion of vapors generated by the sprinklers after the fire extinction. Depending on the fire intensity, the air vapor mixture may attain 212°F and more.

If consideration is given to the use of sprinklers in tunnels, the only type that has any chance at all of controlling large fires would be an AFFF (foam) discharge system. Experience dictates that it must be a deluge system and be manually activated. The complexity of a sprinkler installation with its attendant requirements for zoning valves, preaction valves, foam storage and dispensing facilities, control system, etc., in the corrosive atmospheric environment of highway and rapid transit tunnels would mandate a maintenance program of enormous proportions. Also, to ascertain a minimal level of service for such a system, periodic testing of the foam sprinklers would have to be performed, at great financial and operational expense (traffic limitations, etc.). Consequently, without regard to the massive capital cost incurred for a foam sprinkler system, its overall reliability would be very uncertain unless the required maintenance could be assured. With the anticipated years of service to be expected from highway and rapid transit tunnels and the probably infrequent use of a sprinkler installation, its potential deterioration with time would further reduce its reliability.

Where some sprinkler systems have been installed in tunnels, their use—in some cases—is no longer advocated by the local fire-fighting authorities, due mainly to inherent maintenance problems and resultant lack of reliability. Their preference is to rely on a fire standpipe and tunnel ventilation system in case of fire.

Consequently, it is generally the practice in almost all U.S. tunnels, and in most other countries, not to recommend the use of sprinklers in vehicular or rapid transit tunnels. This has been restated at the most recent PIARC conference on road tunnels. (Technical Committee Report No. 5, Road Tunnels, held in Brussels, September 13–19, 1987.)

In summary, therefore, sprinkler systems are generally not recommended for tunnels for the following reasons:

- Many of the perceived benefits of a sprinkler system that apply to building use do not apply to vehicular or rapid transit tunnels. Differences in combustible content and the physical relationships of the discharge from a sprinkler system in a tunnel upon a vehicle with an on-board fire, where the vehicle has been designed for exposure to water, limit a sprinkler's effectiveness.
- One of the key disadvantages (potential hazard) of a sprinkler system in a tunnel, which could adversely impact life safety, is the delamination of the stratified smoke layer, inducing turbulence and mixing of the air and smoke.
- Sprinkler systems for tunnels are not recommended by most tunnel operators and tunnel/fire protection related organizations, including the FHWA, NFPA, and PIARC.
- The cost of a proper foam sprinkler installation, with all its attendant special features and requirements necessary for a tunnel installation, would be extraordinarily high compared with that for a building of equivalent floor area.

Current state-of-the-art systems of detection, alarm, operation and surveillance, response, access and egress, and alternative suppression methods can be applied to constitute an effective integrated fire safety program for highway and rapid transit tunnels. A sprinkler system (if required) would have to be a preaction, manually remote-controlled zoned AFFF (foam) type. The complexity and maintenance needs of such a system would substantially reduce its reliability.

Halon Systems

These systems are most commonly applied to control and/or communication centers or critical electrical or electronic equipment areas. Subsequently the halon distribution systems are integrated within cabinet housings of the eqipment it is designed to protect. It is also frequently applied to the operating OCCs for most tunnel-type applications.

In all cases, due to the toxicity problems associated with the fumes from the discharge of halon gas, such systems must be installed with the appropriate safeguards, generally as described in NFPA (1990).

Fire Extinguishers

Fire extinguishers are the most appropriate means for bringing many small types of fire under control at initiation. Hence, interval placement of easily accessible portable fire extinguishers is now common practice in most vehicular tunnels. The extinguisher should be of the multipurpose, dry chemical type with an ABC rating. The minimum capacity should be one with a 4-A, 40-B:C rating. The maximum size considered convenient for this use is a 20-lb unit.

The extinguishers should be spaced so that the distance between units is not greater than 150 ft on each wall of the tunnel. It is preferable, where possible, to have fire extinguishers on both sides of a vehicular tunnel roadway. A staggered spacing arrangement would place the extinguisher at a maximum distance of 75 ft from any possible fire. For additional information, see NFPA 502.

The extinguisher should be mounted in a well-marked flush wall enclosure, preferably with a door. Often the fire extinguisher is located in the same enclosure with the fire hose valve.

The cabinet should be arranged so that when the door is opened or the extinguisher removed, an audible alarm is sounded in the OCC to alert the operator of the use of this equipment. This arrangement will discourage and signal unauthorized use or theft of fire-fighting equipment.

CONCLUSION

The appropriate subsystems and their details for achieving an effective integrated fire safety program for a tunnel can vary considerably with the physical characteristics, location, and planned use of the tunnel. In turn, careful attention to the design of each of the applicable subsystems will determine the overall level of fire safety.

REFERENCES

Associated Engineers, a Joint Venture of Parsons Brinckerhoff Quade & Douglas, Inc., De Leuw Cather and Company, and Kaiser Engineers (1980) *Subway Environmental Design Handbook, Volume II, Subway Environment Simulation (SES) Computer Program, Version 3.0, Part 1: User's Manual,* prepared for the Transportation Systems Center of U.S.D.O.T., November.

Associated Engineers, a Joint Venture of Parsons Brinckerhoff Quade & Douglas, Inc., De Leuw Cather and Company and Kaiser Engineers (1976) *Subway Environmental Design Handbook, Volume 1, Principles and Applications,* Second Edition, National Information Service No. PB-254-788.

CHAIKEN, R.F., SINGER, J.M., and LEE, C.K. (1979) *Model Coal Tunnel Fires in Ventilation Flow,* U. S. Bureau of Mines Report of Investigations, RI 8355.

CROCE, P.A., et al. (1978) *A Full-Scale Investigation of the Fuel-Load Hazard of Timber Sets in Mines,* Second Annual Report, No. RC 78-T-6, Factory Mutual Research Corporation, Norwood, MA.

FEZLMYR, A.H. (1976) "Research in Austria on tunnel fire," BHRA,Second International Symposium on the Aerodynamics and Ventilation of Vehicle Tunnels, March 23–25, Paper J2, Cambridge, England.

GREUER, R.E. (1973) *Influence of Mine Fires on the Ventilation of Underground Mines,* U.S. Bureau of Mines Open Fire Report 74-73, July (NTIS: PB-225-834).

GREUER, R.E. (1977) *Study of Mine Fires. Part 1. Computer Simulation of Ventilation Systems Under the Influence of Mine Fires,* U.S. Bureau of Mines Open File Report 115(1)-78, October (NTIS: PB 288231/AS).

HESELDEN, A.J.M. (1976) "Studies of fire and smoke behaviour relevant to tunnels," BHRA, Second International Symposium on the Aerodynamics and Ventilation of Vehicle Tunnels, March 23–25, Paper J1, Cambridge, England.

HWANG, C.C., and CHAIKEN, R.F. (1987) *Effect of Duct Fire on the Ventilation Air Velocity,* U.S. Bureau of Mines Report of Investigation No. 8311.

JACKSON, L.E. (1983) "Serious tunnel fires with emphasis on the Caldecott Tunnel fire, Oakland, California, April 7, 1982," pre sented to the Transportation Research Board Committee A2604, Tunnels and Underground Structures, January 29.

KAWAMURA, H., et al. (1976) "Study on a new ventilation system to effectively eliminate fire smoke in a tunnel," BHRA, Second International Symposium on the Aerodynamics and Ventilation of Vehicle Tunnels, March 23–25, Cambridge, England.

LEE, C.K., SINGER, J.M., and CHAIKEN, R.F. (1979a) "Influence of passageway fires on ventilation flows," Second International Mine Ventilation Congress, Reno, Nevada, November 4–8.

LEE, C.K., CHAIKEN, R.F., and SINGER, J.M. (1979b) "Interaction between duct fires and ventilation flow: An experimental study," *Combustion Science and Technology 20,* 59–72.

MIZUTANI, T. (1982) *Experiment of Vehicular Tunnel Fires,* Office of Highway Tunnel Research, Inst. of Civil Eng. Research, Department of Construction, Japan, February.

National Fire Protection Association (NFPA) (1990) Standard NFPA 130 Fixed Guideway Transit Systems.

[Offenegg Tunnel] (1965) *Final Report of the Tests on the Offenegg Tunnel from May 17–31, 1965,* Commission for Safety Measures in Road Tunnels. (Two Volumes—In German)

Permanent International Association of Road Congress (PIARC) (1987) PIARC report.

TATEISHI, S., et al. (1970) "Experiments on automobile fire in actual road tunnel," Annual Report of Roads, Japan Road Association, pp. 43–53.

U.S. Department of Transportation (USDOT) (1984) *Prevention and Control of Highway Tunnel Fires,* May.

20 | Tunnel Ventilation

Arthur G. Bendelius
Senior Vice President, Principal Professional Associate, Technical Director, Parsons Brinckerhoff Quade & Douglas, Inc.

Whereas tunnels themselves date back to early civilizations, the ventilation of tunnels has taken on greater significance only within the past 100 years, as increasing quantities of combustion products and heat have become more troublesome.

Exposure to the products of combustion generated by vehicles traveling through a highway tunnel can cause discomfort and illness to motorists and passengers. Ventilation is required to dilute the contaminants to provide a respirable environment. Visibility within the tunnel will also be aided by the dilution effect of the ventilation air. In recent years much effort has been made to improve the fire life safety within tunnels, thus focusing more attention on the emergency ventilation system (see Chapter 19).

A railroad tunnel presents a somewhat different problem. In addition to combustion products, the heat generated by the diesel-engined locomotives within the tunnel must also be removed to permit proper cooling of the engine.

An evaluation of the natural ventilation effects in a rapid transit (subway) tunnel must determine if enough of the heat emitted from the train is being removed from the tunnel during normal operations. Mechanical ventilation (or, possibly, cooling) is required if the natural ventilation does not adequately remove the heat. The thrust of current new subway design and of many rehabilitation projects is tied to the requirement for ventilation during fire emergencies.

HIGHWAY TUNNELS

The primary issues facing the tunnel ventilation engineer relate to controlling the level of vehicle-emission contaminants within the highway tunnel during normal tunnel operations and controlling smoke and heated gases during fire emergencies.

Vehicle Fuels

Most vehicles currently in the traffic population are powered by either spark-ignited or compression-ignited engines. Vehicles that use alternative fuels such as compressed natural gas (CNG) and liquefied petroleum gas (LPG) are being introduced into the vehicle population, but their percentage is still too low for their characteristics to be a driving force in the design of highway tunnel ventilation. However, growing concerns regarding the safety of vehicles fueled by some alternatives are beginning to have an impact on the fire life safety design of highway tunnels.

Alternative Fuels. It is evident that the use of vehicles powered by alternative fuels (i.e., fuels other than gasoline or diesel) will grow. LPG is currently the most widely used alternative fuel, although the use of CNG is growing. The American Gas Association estimates that by the year 2000 approximately 50% of the United States' 16 million fleet vehicles will be powered by alternative fuels. Under the Energy Policy Act of 1992 and the Clean Air Act Amendment of 1990, the following are considered potential alternative fuels for road vehicles:

- Methanol
- Ethanol
- Propane
- Natural gas
- Electricity
- Hydrogen
- Coal-derived liquids
- Biological materials
- Reformulated gasoline
- "Clean" diesel

The alternative fuels considered most viable in the near future are CNG, LPG, liquefied natural gas (LNG), and methanol.

COMPRESSED NATURAL GAS. CNG is the most attractive alternative fuel. Its physical and chemical properties make it clearly a safer automotive fuel than gasoline or LPG provided that well-designed carrier systems and operational procedures are followed. Although CNG has a relatively high flammability limit, its flammability range is relatively narrow compared with other fuels.

In air at ambient conditions, a CNG volume of at least 5% is required to support continuous flame propagation, compared with about 2% for LPG and 1% for gasoline vapor. Thus, considerable fuel leakage must occur to render the mixture combustible. Moreover, fires involving combustible mixtures of methane are relatively easy to contain and extinguish.

Because of the physical qualities of natural gas and the structural integrity of natural gas fueling systems, natural gas vehicles are reported to be safer than vehicles fueled by gasoline, diesel, or other alternative fuels

Since natural gas is lighter than air, in the event of a leak it normally dissipates harmlessly into the atmosphere. In tunnels, however, pockets of gas can collect in the overhead structures. Also, since natural gas can ignite only in a range of 5 to 15% volume of natural gas in air, leaks are not likely to ignite because of a lack of sufficient oxygen.

Additionally, CNG's fueling system is one of the safest in existence. The vigorous storage requirements and stronger features of CNG cylinders compared with gasoline contribute to the good safety record of CNG automobiles.

LIQUEFIED PETROLEUM GAS. LPG is an economically competitive vehicular fuel. It is normally delivered as a liquid and can be stored at 100.4°F (38°C) on board vehicles under a design pressure of 250–312.5 psi (1,624–2,154 kPa). It is costly to store because a pressure vessel is required. If engulfed in a fire, however, its heating may result in a rapid increase in pressure. This may be mitigated by venting the excessive buildup of pressure through appropriate relief valves.

LIQUEFIED NATURAL GAS. LNG has been proposed as an alternative fuel for vehicles. However, LNG is almost never considered as an alternative to petroleum-based fuels, primarily because it is thought to be more dangerous and costly. Also, it requires cryogenic storage at −323.6°F (−162° C) and does not appear to be suitable for extensive use in the transportation sector.

METHANOL (ALCOHOL-FUELED VEHICLES). Presently, methanol is being used as a substitute for lead-based octane enhancers in the form of methyl tertiary-butyl ether (MTBE) and as a viable method for vehicle emission control. MTBE is not consumed as a fuel substitute; it is used as a gasoline additive.

The hazards of methanol production, distribution, and use are comparable with gasoline. Unlike gasoline, however, methanol vapors in a fuel tank are explosive at normal ambient temperature. Saturated vapors above nondiluted methanol in an enclosed tank are explosive between 50 and 109.4°F (10 and 43°C). A methanol flame is invisible; a colorant or gasoline has to be added to enable detection.

Mitigation Measures. As the use of alternative fuels in highway vehicles has gradually increased, each tunnel operating agency has dealt with the question regarding whether to allow alternative fueled vehicles through their tunnels. Most tunnel agencies throughout the world do permit passage of alternative fueled vehicles.

The mitigation measures that can be taken by the tunnel designer relate primarily to the ventilation system, which in most circumstances can provide sufficient air to dilute the fuel below any troublesome levels. A minimum level of ventilation may be required to provide this dilution under all operational circumstances. The only other measure would be to reduce or eliminate any irregular surfaces of the tunnel ceiling where a pocket of gas could collect and not be diluted, thus posing a potential explosive hazard. A list of the key characteristics of principal vehicular fuels is shown in Table 20-1.

Future. Considerable technological development is under way for electrically powered vehicles, but their widespread use appears to lie many years ahead. It does not appear that alternative fueled vehicles will have a major impact on the size of tunnel ventilation systems in the next 10–15 years.

Table 20-1. Properties of Alternative Fuels

Property	CNG [a]	Gasoline [b]	Diesel [c]
Molecular weight	16.043	107	—
Storage state	Compressed gas	Liquid	Liquid
Storage pressure	(2,400-3,000 psig)	Ambient temperature and pressure	Ambient temperature and pressure
Density at NTP [d] (kg/L)	0.64×10^{-3}	0.74	0.82
Density of gas relative to air -1.00	0.555	3.4	>4.0 (est)
Boiling point (atmospheric pressure, K)	111.632	310-478	480-600
Diffusion coefficient in NTP air (cm^2/sed)	0.16	0.05	—
Buoyant velocity in NTP air (m/s)	0.16	Nonbuoyant	Nonbuoyant
Vapor pressure (atm)	1	0.508 @ 311 K	0.0005 @ 311 K
Heat of vaporization (KJ/kg)	509.88	309	—
Heat of combustion (low, MJ/kg)	50.02	44.5	—
Heat of combustion (high, MJ/kg)	55.53	48	—
Flammability limits, vol = in air	5.3–15.0	1.0–7.6	0.5–4.1
Auto ignition temp	813	501-744	533
Minimum ignition energy in air (MJ)	0.29	0.24	0.3 (est)
Flame temperature in air (K)	2148	2470	—
Flash point (K)	Gas	230	325 (min.)
Detonation limits vol % in air	6.0–13.5	1.1–3.3	—
Energy of explosion (g TNT)/(g fuel)	11	10	—
Threshold limiting value (TLV), ppm	Asphyxiant	500	500

[a] Properties are for pure methane gas. Natural gas, which varies in composition, will deviate slightly from the values listed.

[b] Unleaded gasoline

[c] Grade number 2 diesel fuel

[d] NTP = 1 atm and 93.15 K

Vehicle Emissions

The exhaust emissions from both types of internal combustion engines are of concern to the tunnel ventilation engineer. Each engine type generates exhaust gases with different characteristics.

Most passenger cars on the road in the United States today are powered by spark-ignited engines fueled by gasoline. Compression-ignited engines are more prevalent in trucks and large buses, but some small buses do have spark-ignited engines. In other countries, the ratio of spark-ignited to compression-ignited engines may vary significantly from that in the United States.

The spark-ignition, or Otto cycle, engine uses volatile liquid or gas fuels, most commonly gasoline. The major constituents of the exhaust are carbon monoxide, carbon dioxide, sulfur dioxide, oxides of nitrogen, and unburned hydrocarbons (Table 20-2).

The compression-ignited, or diesel cycle, engine uses a liquid fuel with a low volatility, ranging from kerosene to crude oil, but usually diesel oil. Nitrogen dioxide, carbon dioxide, and sulfur dioxide are the major components of diesel engine exhaust (Table 20-3).

Types of Vehicle Emissions

Carbon Monoxide. Carbon monoxide (CO) is an odorless, toxic gas present in the exhaust gas of both spark- and compression-ignited engines. In the human body, carbon monoxide combines with the blood hemoglobin much like oxygen to form carboxyhemoglobin (CoHb), thus preventing the formation of oxyhemoglobin, which is necessary for the transport of oxygen to the central nervous system and other tissues. Carbon monoxide has an affinity for blood hemoglobin 250 times that of oxygen; therefore, inhaled carbon monoxide interferes with the capacity of the blood to transport and release oxygen to the tissues, since carbon monoxide is absorbed in preference to oxygen. Figure 20-1 presents the relationships among the level of carboxyhemoglobin (CoHb) in blood, carbon monoxide content of inhaled air, exposure time level of activity, and the resultant human reaction. As noted in Figure 20-1, the fatal carboxyhemoglobin (CoHb) concentration is reached at approximately 65% CoHb in the blood. At about 10% CoHb, the first toxic effects become evident. Carbon monoxide poisoning is reversible; the carboxyhemoglobin level in the blood decreases as exposure to carbon monoxide–free air is continued (see Figure 20-2). Characteristics of carbon monoxide are given in Table 20-4.

Table 20-2. Typical Composition of Spark-Ignited Engine Exhaust[a]

Component	Percentage of Total Exhaust Gas Stream
Carbon monoxide	3.0000%
Carbon dioxide	13.2000%
Oxides of nitrogen	0.0600%
Sulfur dioxide	0.0060%
Aldehyde	0.0040%
Formaldehyde	0.0007%

[a] Adapted from Stahel et al. (1961).

Table 20-3. Typical Composition of Compression-Ignited Engine Exhaust

Component	Percentage of Total Exhaust Gas Stream
Carbon monoxide (maximum)	0.100%
Carbon monoxide (minimum)	0.020%
Carbon dioxide	9.000%
Oxides of nitrogen	0.040%
Sulfur dioxide	0.020%
Aldehyde	0.002%
Formaldehyde	0.001%

[a] Adapted from Stahel et al. (1961).

Carbon Dioxide. Carbon dioxide (CO_2) is a major product of combustion, but it is toxic only in concentration levels well above those found in highway tunnels (see Tables 20-5 and 20-6).

Oxides of Nitrogen (NO_x). Of the many oxides of nitrogen, only two are of concern as air contaminants produced by internal combustion engines: nitric oxide (NO) and nitrogen dioxide (NO_2). The toxic effects of these two are similar except that NO_2 is five times more toxic than NO.

Nitric Oxide. Nitric oxide is a colorless, odorless gas formed during high-temperature combustion. The amount of nitric oxide produced increases as flame temperature increases. It is slightly soluble in water and has a great affinity

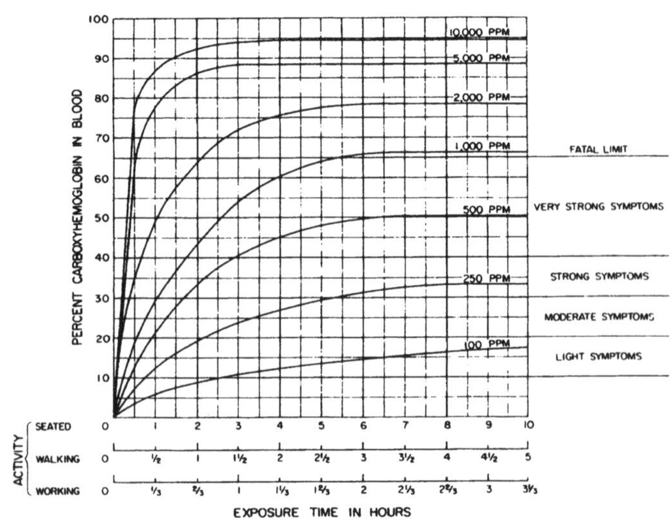

Fig. 20-1. Relation of carboxyhemoglobin in blood with carbon monoxide content of air, exposure time, and activity. (Adapted from Stahel et al., 1961).

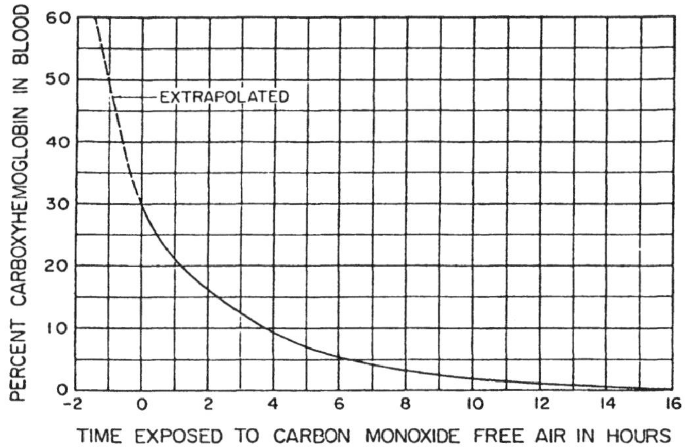

Fig. 20-2. Reduction of carboxyhemoglobin in blood when breathing carbon monoxide–free air and walking slowly.

for blood hemoglobin (forming methemoglobin). This then produces a shortage of blood oxygen. Nitric oxide is rapidly oxidized to nitrogen dioxide.

NITROGEN DIOXIDE. Nitrogen dioxide is a reddish-orange-brown gas that is almost insoluble. It has a characteristic pungent odor and is both irritating and toxic. Approximately 95% of the nitrogen dioxide inhaled remains in the body. Concentrations of 100–150 ppm are dangerous for exposures of 30–60 min. Nitrogen dioxide unites with water in the lungs to form nitrous acid and nitric acid. The nitric acid formed in the lungs destroys the alveoli and thus reduces the ability of the lungs to transport oxygen. Nitrous acid is a potent mutagen. Nitrogen dioxide also combines with hydrocarbons and sunlight to form smog, causing reduction of visibility by absorption of light.

Lead. Lead is a cumulative poison that affects the nervous, blood, and reproductive systems, ultimately leading to lead poisoning and death. Although lead is found in air, water, and food, most of the lead in the atmosphere is emitted by motor vehicles burning leaded fuels. Lead compounds such as tetra-ethyl lead were used, for many years, as a gasoline additive, beginning in 1924, to increase the octane rating of fuel. Of this total lead added to the gasoline, approximately 70% was emitted from the exhaust, of which 30% fell out, and 40% became airborne and, for the most part, remained in suspension for extended periods of time.

Table 20-4. Properties of Carbon Monoxide

Property	Value
Approximate molecular weight	28
Specific gravity relative to air	0.968
Density at 70°F (21.11°C)	0.0724 lb/cu ft (1.1597 kg/m^3)
Density at 32°F (0°C)	0.0780 lb/cu ft (1.2494 kg/m^3)
Gas constant, *R*	55.19 ft-lb/lb/∞R

Table 20-5. Dilution of Spark-Ignited Engine Exhaust Gas[a]

Component	Average Exhaust Gas Composition (ppm)	Level after Dilution[b] (ppm)	Threshold Limit Value Time-Weighted Average
Carbon monoxide	30,000	120.00	50
Carbon dioxide	132,000	528	5,000
Nitrogen dioxide and			5
nitric oxide	600	2.4	25
Sulfur dioxide	60	0.24	5
Aldehyde	40	0.16	NA
Formaldehyde	7	0.028	2

[a] Adapted from Stahel et al. (1961) and American Conference of Governmental Industrial Hygienists (1992).

[b] Diluted to maintain 120 ppm of carbon monoxide using a dilution ratio of 250 to 1.

The results of the long-term use of leaded fuels can be observed on the surfaces of the exhaust ducts of many of the older tunnels in the United States, where deposits of lead are frequently found.

In 1977 the U.S. government phased out the use of leaded fuels and introduced unleaded fuels. All vehicles produced after 1979 are restricted to the use of unleaded fuel.

Hydrocarbons. These are the most complex of all of the internal combustion engine emissions, although they comprise only a small portion of the total vehicle emissions. The major portion of the hydrocarbon emissions from vehicles are methane, ethane, propane, ethylene, acetylene, pentane, and hexane. Hydrocarbons are known to aid in the formation of photochemical smog, but by themselves are not present in large enough quantities in the tunnel environment to exceed the threshold limit value.

Sulfur Dioxide. Sulfur dioxide (SO_2) is the most prevalent oxide of sulfur found in vehicle exhaust gas. It is a nonflammable, nonexplosive, colorless gas, which oxidizes in the atmosphere to form sulfuric acid and then reacts with

Table 20-6. Dilution of Compression-Ignited Engine Exhaust Gas[a]

Component	Average Exhaust Gas Composition (ppm)	Level after Dilution[b] (ppm)	Threshold Limit Value Time-Weighted Average
Carbon monoxide	1,000	4.00	50
Carbon dioxide	90,000	360.00	5,000
Nitrogen oxide and	400	1.60	5
nitric oxide			25
Sulfur dioxide	200	0.80	5
Aldehyde	20	0.08	NA
Formaldehyde	11	0.44	2

[a]Adapted from *Industrial Ventilation* (1992) and Stahel et al. (1961)

[b]Diluted to maintain 120 ppm of carbon monoxide using a dilution ratio 250 to 1.

other pollutants to form toxic sulfates, which affect the respiratory system. At concentrations greater than 3 ppm, sulfur dioxide has a pungent irritating odor. Sulfur dioxide is not a major component of vehicular exhaust gas and, therefore, does not appear in the tunnel environment in harmful concentrations.

Aldehydes. These organic compounds are present in the internal combustion engine exhaust gas. All organic aldehydes are irritants either to the skin or mucous membranes or both. The irritant nature of these compound provides sufficient warning to preclude serious health problems.

Odor. The most offensive odor in vehicle emissions comes from the diesel engine. There is, however, little knowledge on this subject since the methods of odor identification are extremely subjective and thus prone to wide variation. Pollutants such as aldehydes, nitrogen dioxide, and sulfur dioxide have all been identified as possible contributors to odor.

Particulates. Most of the particulates in vehicle exhaust are produced by incomplete combustion of hydrocarbon fuels. They tend to stay in suspension indefinitely and are within respirable size. A further danger is the absorption of gases such as sulfur dioxide and oxides of nitrogen by these particulates, thus carrying these corrosive gases further into the lungs. The particulates are the cause of smoke and haze present in many tunnels. Smoke is usually most dense when a significant number of diesel-engined vehicles are present in the traffic stream.

Other Gases. Other gases are included in the vehicular exhaust stream, such as nitrogen, hydrogen, and water vapor, but they are not present in large enough quantities to be considered harmful.

Criteria

The primary criterion for the normal ventilation of highway tunnels is the permissible concentration level of contaminants within the tunnel roadway area. The primary standards have been established by the U.S. Environmental Protection Agency (US EPA) and Federal Highway Administration (US FHWA) in the United States, and by the Permanent International Association of Road Congresses (PIARC) on a worldwide basis.

Standards for the environment within manned tunnels in the United States are also generated by state departments of labor and public health agencies, along with the U.S. Department of Health. To fully understand these standards, manned and unmanned tunnels must be defined. Manned tunnels have attendants on foot patrol within the tunnel environment, usually on the walkway. Unmanned tunnels do not have any personnel working within the tunnel who are not inside vehicles, with the possible exception of an occasional maintenance crew.

Permissible Contaminant Limits. Permissible limits for the various contaminants present in the tunnel environment are required to protect the safety and health of the tunnel occupants. Carbon monoxide (CO) has been shown to be the key contaminant to health and safety within the tunnel. The primary limits have thus been set for CO, although with the continuing reduction of CO emissions, other contaminants such as oxides of nitrogen have become the critical criteria under certain circumstances. Consideration should also be given to contaminants that create visual obscurity within the tunnel and any other contaminant that might affect the occupants or the surrounding environment.

UNITED STATES. The first contaminant limits established in the United States were based on recommendations of the Bureau of Mines in a report regarding the ventilation of vehicular tunnels. (Fieldner et al., 1921) The recommendations were to provide sufficient air to dilute the CO concentration to 4 parts in 10,000 (400 ppm) at the vehicle driver position. It was felt that exposure to this level of carbon monoxide for not more than 45 min provided an adequate factor of safety.

Years later, although the maximum limit on CO concentration within the tunnel remained at 400 ppm, the nominal value used for most tunnel designs was 250 ppm. This meant that under most traffic conditions, the CO concentration was less than 250 ppm and, when an emergency situation occurred, a maximum of 400 ppm was permitted. When the diesel-engined truck and bus became prevalent it was found necessary to establish the design limit between 200 and 250 ppm to maintain adequate visibility within the tunnel (Stahel et al., 1961).

In 1975, the US EPA issued a supplement to their Guidelines for Review of Environmental Impact Statements for Highway Projects. In this supplement the EPA issued the following statement, "For the users of highway tunnels at or near sea level, it has been determined that an adequate margin of safety would exist if the concentration of CO does not exceed 125 (ppm) and the exposure time does not exceed one hour." Since the issuance of the EPA supplement, 125 ppm has been widely adopted in the United States as the design criterion for tunnels located at or below an altitude of 5,000 ft (1,500 m). This guideline was accepted by the US FHWA for all U.S. highway tunnels funded by the U.S. government.

This criterion remained in place until 1989, when US FHWA and US EPA jointly issued a set of revised guidelines based exclusively on time-dependent exposure to CO, limiting the exposure to 2% CoHb in the blood. The maximum CO level allowable under these new guidelines is 120 ppm for a maximum exposure time of 15 min. However, for exposures beyond 15 min, more stringent criteria are established as shown in Table 20-7. These new criteria do not impose a major hardship on U.S. tunnels since travel during normal operations through most tunnels in the United States is less than 15 min. If the travel time exceeds 15 min, it is due to an emergency, which would trigger other mitigation measures.

Table 20-7. Maximum allowable CO concentration for extended exposure[a]

Maximum Exposure Period (min)	Maximum CO Concentration (ppm)
15	120
30	65
45	45
60	35

[a] Adapted from US EPA Environmental Guidebook

PIARC. PIARC has in its periodic Road Tunnel Reports presented a set of criteria for the recommended CO levels for highway tunnels. The most current recommendations are summarized in Table 20-8. The upper limits of CO concentration shown in this table should be considered in countries where there are limited emission controls (PIARC, 1991).

There are other factors that must also be considered in the evaluation of tunnel environment, such as the visibility as it affects the traveling motorist. The visibility within a tunnel is difficult to evaluate, since the haze-producing contaminants are difficult to identify and quantify. PIARC has established limits of diesel smoke in tunnels in terms of the type of traffic; these are shown in Table 20-9.

HIGHER ELEVATIONS. If the tunnel is located at an elevation above 5,000 ft (1,500 m), the carbon monoxide limits must be adjusted accordingly (see Table 20-10).

MANNED TUNNELS. Manned tunnels fall under the jurisdiction of the U.S. Occupational Safety and Health Act (OSHA) of 1970 or other local agency agreements. OSHA has established the Threshold Limit Value (TLV) as adopted by the American Conference of Governmental and Industrial Hygienists (ACGIH) as the contaminant level for the working environment. Excursion beyond these limits can be tolerated if, as defined by ACGIH, these excursions are compensated by equivalent excursions below the level during the working day (ACGIH, 1994). These permissible excursions can then be applied to the tunnel environment. Table 20-11 illustrates the application of these time-weighted averages to the tunnel environment.

Table 20-8. PIARC Recommended CO Concentration Levels for Highway Tunnels[a]

Tunnel Type	Smooth Flow (ppm)	Congested Flow (ppm)
Urban Tunnels		
Congested daily	100–150	100–150
Seldom congested	100–150	150–250
Inter-Urban Tunnels (Highway or Mountain)	100–150	150–200

[a] Adapted from PIARC 1991

Table 20-9. Diesel Smoke Limiting Factor[a]

Type of Traffic	K_{lim} at Peak Traffic (1/m)
Fluid traffic:	
V_{max} = 60–80 kph	0.005–0.007
V_{max} = 100 kph	0.005
Congested traffic	0.007–0.009
Tunnel to be closed	0.0012
Maintenance work in the tunnel	≤0.003

[a] Adapted from PIARC 1991.

FUTURE. The continuing reduction of the contaminant emission of vehicle exhaust provides a question regarding the future design of tunnel ventilation systems. As the emission rates decline, the required air quantities should also decline. However, the trend toward reduced environmental limits could counteract that trend; thus, the required quantities for pollutant ventilation of a tunnel could remain unchanged.

Smoke Control

The control of smoke within a highway tunnel during an emergency incident is one of the critical aspects of a road tunnel ventilation system. For many years the air flow required to maintain a safe level of carbon monoxide was the dominant criterion. Today, after the effective implementation of emission controls on vehicles in many parts of the world, the air flow requirement to control CO is no longer as dominant as it was, and control or removal of smoke may be paramount.

The value most prevalent as a criterion for the removal or control of smoke has been 100 cfm per lane foot of tunnel (ASHRAE, 1991). However, this value does not have strong scientific support. In addition, for longitudinal ventilation systems, which generate a longitudinal air flow through the roadway area, a value of 100 cfm per lane foot has little meaning, since the same longitudinal flow rate would be required irrespective of tunnel length (provided all other factors are identical).

Table 20-10. Current US EPA CO Concentration Limit Guidelines

Exposure time (min)	Elevation: Concentration at 5,000 ft (1,500 m) (ppm)	Elevation: Concentration at 12,000 ft (3,660 m) (ppm)
15	125	140
30	65	70
45	45	50
60	35	35

Table 20-11. Permissable Excursions as Applied to the Tunnel Environment

Contaminant	Threshold Limit Value (ppm)	Permissable Excursion Limit (ppm)	Short-Term Exposure Limit
Carbon monoxide	50	75.0	4000.00
Nitric oxide	25	37.5	35.0
Nitrogen dioxide	5	5.0	5.0

Therefore, the critical velocity criterion concept became more viable. The critical velocity is defined as the longitudinal velocity needed to prevent backlayering of the smoke layer in the tunnel. The critical velocity depends on the tunnel cross section and grade, in addition to the fire size.

Traffic. The tunnel ventilation engineer must be aware of the traffic characteristics of the tunnel. The vehicle density, the traffic volume, and the traffic composition have a direct relationship to the amount of CO emitted. An evaluation is required to determine the proper design traffic volume. The maximum roadway capacity may not be the appropriate design value, due to factors such as tunnel location and lane width.

TRAFFIC VOLUME. The traffic engineer will furnish a traffic analysis including the tunnel design vehicle flow. There are times, however, when such an analysis is not available, such as during the early stages of a study, or when the traffic conditions to be studied occur at vehicle speeds other than the roadway design speed. In these cases, the ventilation engineer must be able to make a proper judgment and select an appropriate traffic volume to permit estimates of the vehicles' emissions.

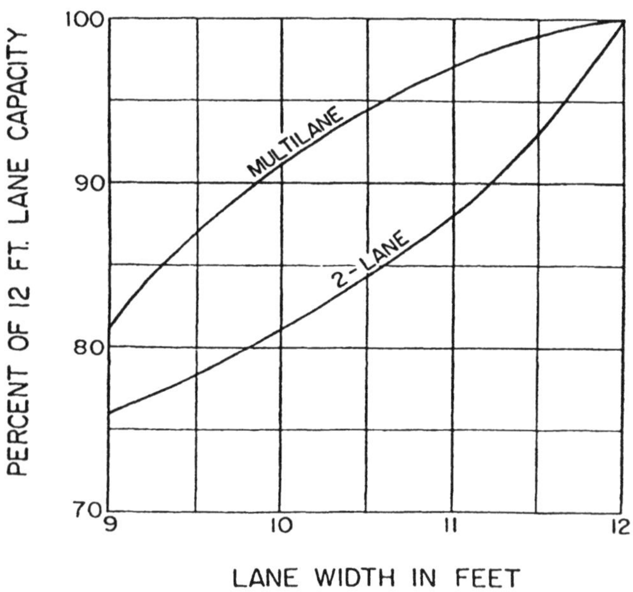

Fig. 20-3. Effect of lane width on lane capacity.

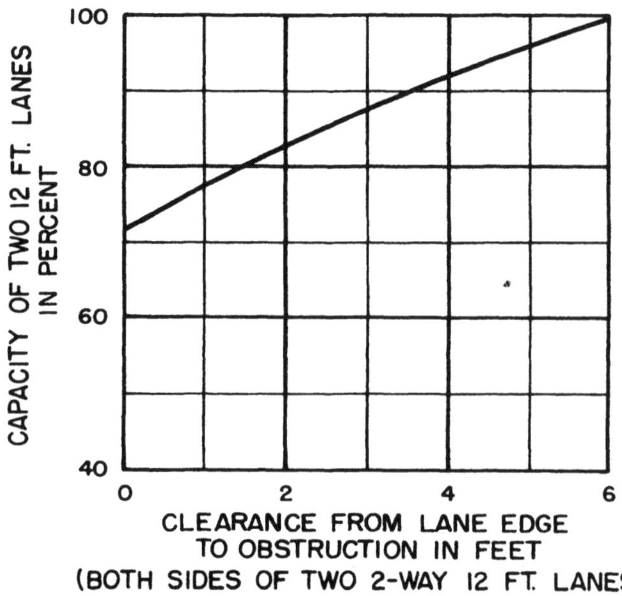

Fig. 20-4. Effect of lateral clearance on lane capacity.

A tunnel located in or near a highly populated urban area will undoubtedly use the maximum tunnel roadway capacity; however, a mountain tunnel located in a rural area will most likely not be used to full roadway capacity. The urban tunnel will be confronted with normal day-to-day congested traffic conditions, which also much be considered.

The maximum capacity of a standard 12 ft (3.7 m) lane in a well-lighted tunnel is approximately 2,000 vehicles per hour. This is based on two lanes moving in the same direction. This maximum volume will be reduced by opposing traffic, grades, reduced lane width, and vehicle mix. The effect of reduced lane width and lateral clearances (tunnel width) on free-flowing traffic can be noted in Figures 20-3 and 20-4.

An excellent set of guidelines to typical vehicle headways are contained within the *Highway Capacity Manual* (TRB, 1985). The average headway is rarely less than 1.0 sec and can go as high as 9.0 or 12.0 sec. Since these values relate to an open roadway and a tunnel is restricted, a typical minimum headway for a tunnel of 1.8 sec is appropriate.

The traffic volume in a tunnel can be estimated by using the average headway:

$$\mathrm{TRV} = \frac{3{,}600}{\mathrm{AHY}} \tag{20-1}$$

where

TRV = traffic volume (vehicles/hour)
AHY = average headway (sec)

The traffic volume can also be established by

$$\mathrm{TRV} = \frac{\mathrm{TRS} \times \mathrm{DCF}}{\mathrm{VES}} \tag{20-2}$$

where

$$TRS = \text{traffic (vehicle) speed (mph)}$$
$$DCF = \text{conversion factor (5,280 ft per mile)}$$
$$VES = \text{average vehicle spacing (ft) measured front to front}$$

In SI units,

$$TRS = \text{traffic (vehicle) speed (km/hour)}$$
$$DCF = \text{conversion factor (1,000 m per km)}$$
$$VES = \text{average vehicle spacing (m) measured front to front}$$

The average vehicle spacing can be estimated by

$$VES = AHY \times TRS \times SCF \qquad (20\text{-}3)$$

where SCF = conversion factor (1.4667 ft-hour/mile-sec); in SI units, SCF = conversion factor (0.2773 m-hour/km-sec). Vehicle spacing can also be estimated by

$$TRD = \frac{1}{TRV} \qquad (20\text{-}4)$$

where TRD = traffic density (vehicles per foot of roadway); in SI units, TRD = traffic density (vehicles per meter of roadway).

The preceding equations provide the necessary estimates of traffic flow data to combine with the methodology presented below to evaluate the vehicle-emitted contaminants within the tunnels.

Traffic Composition. The percentage of trucks and buses in any stream of traffic must be considered in determining the traffic capacity of a tunnel roadway. Trucks and/or buses will decrease the speed of the traffic stream. The percentage of trucks and buses also greatly affects the base emission factors, especially the oxides of nitrogen and smoke emitted by diesels.

Urban tunnels, such as those located in New York City, might experience a truck per bus percentage of 15%, whereas rural tunnels may see only 4 or 5% trucks per buses. The location of the tunnel should be the major factor in determining an appropriate truck per bus percentage in the traffic stream.

Analysis

Vehicle Emission Rates.

History. Early work was done in this area by the U.S. Bureau of Mines around 1920 (Fieldner et al., 1921). This early research was conducted, and a series of tests performed, to enable the governing commissions to establish design criteria for the Holland Tunnel. Results of these tests were used, for many years, as a basis for design of highway tunnels in the United States.

These tests were, of course, conducted with vintage 1920 cars, so that data are stale. Additional research for tunnels at elevations of 5,500–10,500 ft (1,680–3,200 m) was conducted by the Colorado Department of Highways in the mid-1960s, relating to the Straight Creek (Eisenhower Memorial) Tunnel.

A thorough evaluation was performed for tunnels located at or above 7,000 ft (2,134 m) by the Institute for Highway Construction of the Swiss Institute of Technology and published as Bulletin No. 10, "Report of Committee on Tunnel Ventilation to Swiss Department of Highways" (Stahel et al., 1961).

An emission rate for each of the exhaust pollutants mentioned above can be computed in varying degrees of accuracy for both the spark- and compression-ignited engines. However, it can be seen in Tables 20-5 and 20-6 that if the level of CO is maintained at or below 120 ppm (a 250:1 dilution ratio), all other constituents of vehicle exhausts will be well within the threshold limit value–time weighted averages TLV-TWA for each material. The TLV-TWA is defined by the American Government Industrial Hygienists as "the values for airborne toxic materials which are to be used as guides in the control of health hazards and represent time weighted concentrations to which nearly all workers may be exposed 8 hours per day over extended periods of time ('day after day') without adverse effects" (ACGIH, 1992, pp. 12-3–12-20).

United States. The method of computing the emission rate of CO is presented, but the same approach can be used for all vehicle-emitted pollutants.

During the 1970s the US EPA developed a mobile source emission factor model known as MOBILE. The MOBILE series of computer programs can be used to estimate hydrocarbon (HC), carbon monoxide (CO), and oxides of nitrogen (NO_x) emission factors for gasoline-fueled and diesel highway vehicles in the United States. The current version of this model is MOBILE5A (EPA, 1994). The program is based on the procedures outlined in the *Compilation of Air Pollution Emission Factors—Volume II, Highway Mobile Sources* (AP-42 Fourth Edition, September 1985, and Supplement A, January 1991).

The MOBILE model calculates emission factors for eight individual vehicle types at both low and high elevations. These estimates of emissions are based on input factors such as ambient temperature, vehicle speed, and vehicle age for any calendar year from 1960 through 2020. The 25 most recent model years vehicles are considered to be in the traffic stream in each calendar year.

The MOBILE5A model produces resulting emission rates in grams per mile per vehicle for carbon monoxide, oxides of nitrogen, and hydrocarbons. A sample output tabulation is shown in Table 20-12.

The basic emission rates used in the MOBILE models are continuously updated based on the analysis of emission factor program testing results in the United States.

The emission rates outlined above relate only to gasoline-fueled, spark-ignited engines. The emissions from compression-ignited engine are not as simple to determine. A

Table 20-12. Tabulation of Vehicle Emission Results from MOBILE5A

Vehicle Type:		A	B	C	D	E	F	G	H	MC	All Vehicles
Vehicle Speed:		10.0	10.0	10.0	10.0	10.0	10.0	10.0	10.0	10.0	10.0
Moisture Vapor Transmission Mix:		0.83	0.03	0.01		0.05	0.01	0.00	0.07	0.00	
Composite Emission Factors (grams per mile per vehicle)											
Volatile Organic Compounds	HC	2.50	2.88	3.57	3.05	5.56	0.99	1.38	3.35	5.01	2.72
Exhaust	HC	1.95	2.56	3.22	2.72	4.70	0.99	1.38	3.35	4.65	2.21
Evaporation	HC	0.12	0.05	0.05	0.05	0.18	—	—	—	0.18	0.11
Refuel	HC	0.07	0.09	0.09	0.09	0.14	—	—	—	—	0.07
Running	HC	0.34	0.16	0.18	0.16	0.51	—	—	—	—	0.31
Resting	HC	0.02	0.02	0.02	0.02	0.03	—	—	—	0.18	0.02
Exhaust	CO	24.83	30.93	36.81	32.40	52.42	2.69	0.00	20.23	0.00	25.97
Exhaust	NO_x	1.40	1.72	2.00	1.79	4.36	1.84	0.00	13.64	0.00	2.42

compression-ignited engine should be operated with 20–40% excess air to reduce the smoke produced to an acceptable level. This, however, reduces the maximum power capability of the engine. Some amount of smoke will have to be accepted to enable the diesel-engined vehicle to attain high power and good acceleration. This problem is intensified when these engines are not properly maintained, as they are often deliberately reset to gain power.

Because of the numerous unknowns inherent in the operation of compression-ignited engines, it has been the practice to assume all compression-ignited engines to be gasoline-fueled, spark-ignited engines for the purpose of computing the pollutant ventilation required. This smoke emitted from the diesel engine creates the haze found in many vehicular tunnels. It is composed mostly of carbon particles.

An acceptable level of diesel smoke must be established to ensure safety and comfort. From a safety standpoint, the stopping distance of a vehicle is an excellent guide.

PIARC. PIARC has established a set of passenger car emission values for both CO and NO_x. Table 20-13 includes values based on the type of emissions level in effect and the level of emission control in place in the specific country. All values in Table 20-13 are based on vehicles traveling at 37 mph (60 kmh). Correction factors are shown for elevation in Figure 20-5 and for vehicle grade and speed in Figure 20-6.

Table 20-13. PIARC Emissions from Passenger Cars

Current Emission Law	Control	Carbon Monoxide (CO) q^o (m^3/hr,veh)	Oxide of Nitrogen (NO_x) q^o (grams/hr,veh)
No law	no	1–1.5	120
EEC R 15/04	no	0.7	120
EEC R 15/04	yes	0.5	100
EEC 89/45B	yes	0.16	60
FTP 75	yes	0.14	40
Diesel engine	—	0.08	40

Based on vehicles traveling at 37 miles per hour (60 kilometers per hour); adapted from PIARC 1991.

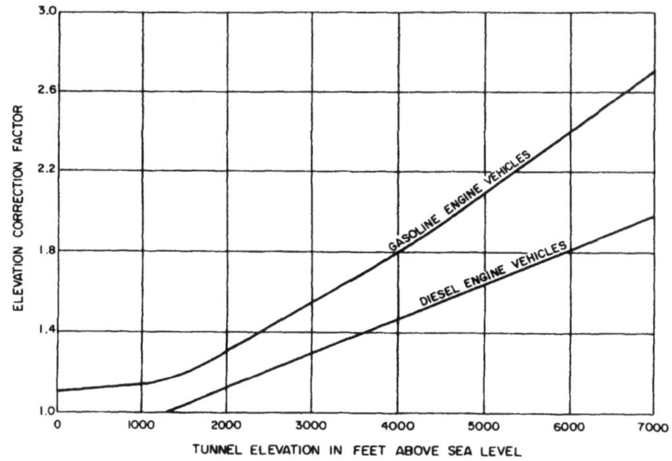

Fig. 20-5. Elevation correction factor for vehicle omissions.

In addition, PIARC has established a set of emission factors for diesel-engined trucks and buses, as shown in Table 20-14. The correction factor for tunnel elevation is shown in Figure 20-5 and for vehicle grade and speed in Figure 20-7.

Elevation Effect. At higher elevations, the air–fuel mixture in spark-ignited engines becomes richer due to the reduced air pressure. The resultant increase in gasoline consumption for both partial and full load generates a considerable increase in carbon monoxide production at these higher elevations. If the carburetor is adjusted for sea level and the car operated at a higher elevation, the emission rate is even further increased.

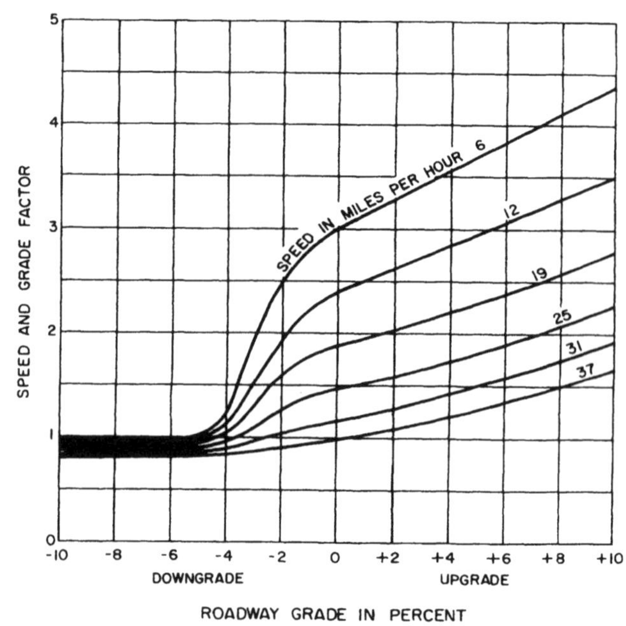

Fig. 20-6. Speed and grade factors for spark-ignited engine vehicles.

Table 20-14. PIARC Emissions from Trucks and Buses with Diesel Engines

Current Emission Law	Control	Smoke q^o (m^3/hr,veh)				Oxides of Nitrogen (NO_X) q^o (grams/hr,veh)		
		Truck weight (tons)				Truck weight (tons)		
		5	10	20	40	5	20	40
No law	no	80-130	160-25	300-40	400-600	500	1,400	1,900
EEC R 49 + 24	no	80	160	240	280	500	1,400	1,900
EEC R 49 + 24	yes	65	130	200	240	470	1,300	1,800
EEC 88/77	yes	50	100	160	200	360	1,000	1,400
US Transient 88	yes	50	100	160	200	330	900	1,200
US Transient 91	yes	30	60	100	140	270	750	1,000
US Transient 94	yes	2	40	70	110	220	600	800

Based on vehicles traveling at 37 miles per hour (60 kilometers per hour); adapted from PIARC 1991.

For the compression-ignition engine, the reduced air pressure at the elevated locations results in less excess combustion air and, therefore, creates an increase in smoke production. Smoke from diesel engines at higher elevations and steep grades can become intolerable within the tunnel environment, especially if the engines are not in proper adjustment and good operating condition.

Based on these facts, an adjustment to the vehicle emission rate is required at elevated locations. The MOBILE program includes an adjustment for elevation variation: low for elevation corrections from sea level to 4,000 ft (1,220 m) and high for all elevations "substantially" above 4,000 ft (1,220 m). In practice, the PIARC elevation correction factors shown in Figure 20-5 are applied to the MOBILE-generated emission values for elevations from sea level to 6,600 ft (2,000 m).

There is an extremely limited amount of data available on the emission rates of vehicles at elevations above 7,000 ft (2,100 m). The Colorado Department of Highways report prepared for the Straight Creek (Eisenhower Memorial) Tunnel (TAMS, 1965) (elevation, 11,090 ft [3,380 m] at center) contains test data for elevations from approximately 5,500 to 11,350 ft (1,675 to 3,460 m). A caution should be noted that the data available for these elevated conditions are limited. If possible, some elevated site testing should be considered prior to initiation of design for a tunnel in such a location.

The effect of increased elevation is also severely felt by the human body and its ability to react to a CO-laden environment.

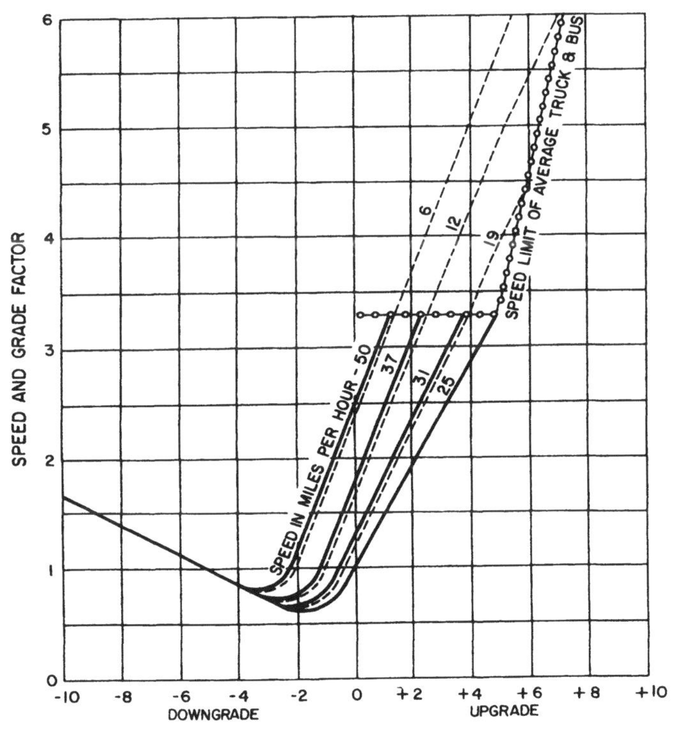

Fig. 20-7. Speed and grade factors for compression-ignited engine vehicles.

Grade and Speed Effect. The grade of the tunnel roadway and the vehicle speed have an impact on the emission rates.

On an upgrade, the increased fuel consumption results in a sizable increase in pollutant emission. The reduced fuel consumption on a downgrade produces a lower pollutant emission rate than on a level roadway. On extremely steep downgrades, however, the effect of a driver utilizing the vehicle's engine in lower gear ratios for braking will create a slightly increased pollution emission rate.

The MOBILE program does not include a prediction of the impact of grade and speed on the contaminant emissions; however, the PIARC method does include a set of charts shown in Figures 20-6 and 20-7, which present the impact of grade and speed on vehicle emissions. These PIARC-based factors have been applied to the MOBILE results specifically for tunnel applications.

For vehicle speeds beyond those shown in these two figures, special studies may be required to determine the effect on vehicle emissions.

Future. Vehicle emissions is another area that is presently being affected by legislation and will continue to be in the foreseeable future. The first vehicle emission values were established by the tests conducted prior to the Holland Tunnel design (Fieldner et al., 1921). These tests showed CO concentrations of from 5 to 9% by volume in the exhaust gas stream. These data formed the basis of design for many of the older highway tunnels in operation in the United States today. The CO concentrations in automobile exhaust gases have fallen off sharply through the years.

Within the United States there is a growing trend toward reduction of the emissions from motor vehicles. California state standards required reduction to 1.5% CO. Since then, US EPA has established standards for exhaust emissions.

The establishment of legal emission limits does not assure compliance of all vehicles on the road; therefore, some amount of judgment must be employed when using these legal limits for tunnel design.

Air Flow Rate Requirements

The air flow rate required to satisfy the dilution requirements of carbon monoxide and the other contaminants outlined earlier can be based on the average CO emission rate of the vehicles (see "Vehicle Emission Rates").

This will be true for an urban tunnel with stop-and-go traffic and a low percentage of diesel-engined vehicles. The air quantity for a tunnel with a high percentage of diesel vehicles will most likely be selected to satisfy the visibility criterion. The criterion for emergency ventilation will most likely be dominant in the case of a nonurban tunnel (without stop-and-go or idling and diesel traffic).

Before the total air flow rate is determined it is necessary to compute the contaminant or pollutant emitted per lane of tunnel. The basic emission rate per lane is given below.

The average carbon monoxide emitted by a lane of moving traffic can be computed using

$$\text{COV} = \text{COK} \times \text{TUL} \times \text{TRD} \times \text{FSP} \times \text{FGR} \times \text{FEL} \times \text{FSF} \tag{20-5}$$

where

COV = carbon monoxide emitted per lane of traffic (ft^3 per hour per lane)
COK = average carbon monoxide emitted for moving traffic (ft^3 per hour per vehicle)

For PIARC-based vehicle emission rates, see Tables 20-13 and 20-14:

$\text{COK} = q^0 \times 35.314$ (ft^3 per hour per vehicle)
$\text{NOK} = q^0 \times \text{TRS} \times 0.3045$ (ft^3 per hour per vehicle)

For MOBILE-based vehicle emission rates,

COK = (mobile composite emission factor) × TRS × 0.3045 (ft^3 per hour per vehicle)

TRS = vehicle speed (mph)
TUL = length of tunnel (ft)
TRD = traffic density (vehicles per foot per lane)
FSP = speed factor (Figures 20-6 and 20-7)
FGR = grade factor (Figures 20-6 and 20-7)
FEL = elevation factor (Figure 20-5)
FSF = reserve factor

In SI units,

COV = carbon monoxide emitted per lane of traffic (m^3 per hour per lane)
COK = average carbon monoxide emitted for moving traffic (m^3 per hour per vehicle)

For PIARC-based vehicle emission rates, see Tables 20-13 and 20-14:

$\text{COK} = q^0 \times 1.0$ (m^3 per hour per vehicle)
$\text{NOK} = q^0 \times \text{TRS} \times 0.3045$ (m^3 per hour per vehicle)

For MOBILE-based vehicle emissions rates,

COK = (MOBILE composite emission factor) × TRS × 720.6 (m^3 per hour per vehicle)

TRS = vehicle speed (kph)
TRD = traffic density (vehicles per meter per lane)
TUL = length of tunnel (m)

The average smoke produced by a lane of moving diesel-engined vehicles can be computed as follows:

$$\text{DSC}_i = \frac{\text{DSE} \times \text{TUL}}{\text{DTRV} \times \text{LCF}} \tag{20-6}$$

where

DSC_i = diesel smoke concentration per lane (ft^2 per hour)
DSE = average diesel smoke emitted from diesel-engined vehicle (ft^2 per hour per vehicle); see Table 20-14
DTRV = diesel traffic volume (vehicles per hour per lane)
LCF = length conversion factor (5,280 ft per mile)
FEL = elevation correction factor for diesel-engined vehicles (Figure 20-5)
FSF = reserve factor

In SI units,

DSC_i = average diesel smoke concentration per lane (m^2 per hour per lane)
DSE = average diesel smoke emitted from moving diesel-engined vehicle (m^2 per hour per vehicle); see Table 20-14
LCF = length conversion factor (1,000 m per km)

For idle traffic, the following can be used to compute the average carbon monoxide emitted:

$$\text{COV} = \text{COH} \times \text{TRD} \times \text{TUL} \times \text{FEL} \times \text{FSF} \times \text{LCF} \tag{20-7}$$

where

COH = average carbon monoxide emitted during idle (ft^3 per vehicle per hour)
TRD = Traffic density (vehicles per mile per lane).
LCF = length conversion factor (5,280 ft per mile)

In SI units,

COH = average carbon monoxide emitted during idle (m^3 per vehicle per hour)
TRD = traffic density (vehicles per km per lane).
LCF = length conversion factor (1,000 m per km)

Emission values must be computed for each type of vehicle in the traffic mixture; thus, the total carbon monoxide emission per lane is

$$COVL_i = COV_p + COV_t \qquad (20\text{-}8)$$

where

$COVL_i$ = carbon monoxide emitted in lane i (ft^3 per hour per lane)
COV_p = carbon monoxide emitted by passenger vehicles (ft^3 per hour per lane)
COV_t = carbon monoxide emitted by trucks and buses (ft^3 per hour per lane)

In SI units,

$COVL_i$ = carbon monoxide emitted in lane i (m^3 per hour per lane)
COV_p = carbon monoxide emitted by passenger vehicles (m^3 per hour per lane)
COV_t = carbon monoxide emitted by trucks and buses (m^3 per hour per lane)

Using this value of $COVL_i$, the total carbon monoxide emitted per bore can then be computed by adding the values for all traffic lanes.

At this point, the carbon monoxide values can be corrected for nonstandard atmospheric conditions, as noted earlier. The temperature experienced at the tunnel location has a minor effect on the amount of carbon monoxide emitted. This correction takes the following form:

$$TCOV_c = TCOV \times \frac{TEMA}{TEMS} \qquad (20\text{-}9)$$

where

$TCOV_c$ = corrected average carbon monoxide emission per bore (ft^3/hour)
TEMA = average maximum temperature at tunnel location (°R)
TEMS = standard temperature for emissions (°R)
°R = °F + 459.67

In SI units,

$TCOV_c$ = corrected average carbon monoxide emission per bore (m^3/hour)
TEMA = average maximum temperature at tunnel location (Kelvin)
TEMS = standard temperature at emission (Kelvin)
Kelvin = °C + 273.15

Using the emission rates computed in Equation (20-9), the total air flow rate requirements can be computed by

$$QAR_c = \frac{TCOV_c \times 10^6}{TCF \times (PPM - AMB)}$$

$$QAR_n = \frac{TNOX_c \times 10^6}{TCF(PPM - AMB)} \qquad (20\text{-}10a)$$

where

QAR_c = final air flow rate required to maintain allowable level of carbon monoxide (ft^3/min)
QAR_n = final air flow rate required to maintain allowable level of oxides of nitrogen (ft^3/min)
TCF = time conversion factor (60 min per hour)
PPM = allowable level of carbon monoxide (ppm)
AMB = ambient carbon monoxide concentrations at tunnel location (ppm)

In SI units,

QAR = final air flow rate required to maintain allowable level of carbon monoxide (m^3/sec)
TCF = time conversion factor (3,600 sec per hour)

The total air flow rate required to maintain visibility within the tunnel roadway area is given by

$$QARS = \frac{TDSC}{TCF \times DSL} \qquad (20\text{-}10b)$$

where

QARS = total air flow rate required to meet minimum standard of visibility (ft^3/min)
TDSC = total diesel smoke emitted (ft^2/hour)
TCF = time conversion factor (60 min per hour)
DSL = limit value of diesel smoke (1/foot)

In SI units,

QARS = total air flow rate required to meet minimum standard of visibility (m^3/sec)
TDSC = total diesel smoke emitted (m^2/hour)
TCF = time conversion factor (3,600 sec per hour)
DSL = limit value of diesel smoke (1/meter); see Table 20-9

An estimated value for the final air flow rate can be readily determined from the chart in Figure 20-8.

At this point, it should be determined if the air flow rate required for maintenance of a proper level of carbon monoxide and oxides of nitrogen at all traffic flow conditions in the tunnel will be sufficient to maintain a level of safe visibility. Since the smoke emitted from the diesel-engined vehicle will create a visibility problem in the tunnel, this is the factor used to evaluate visibility. The total diesel smoke emitted per bore, TDSC, is determined by adding the values per lane (Equation 20-6) for all lanes.

$$QAR_v = \frac{TDSC - 0.5885}{DSA} \qquad (20\text{-}11)$$

where

0.5885 = conversion factor (ft^3 hour per m^3 minute)
DSA = allowable concentration of diesel smoke (per m^3).

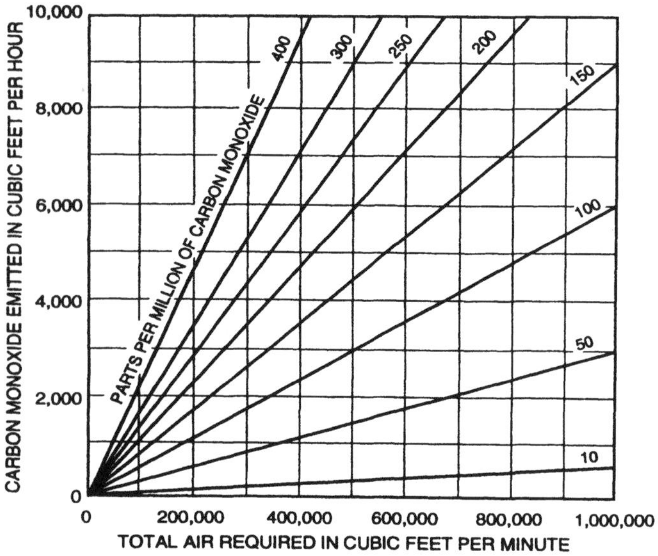

Fig. 20-8. Final air flow chart.

Emergency Air Flow Rate

The presence of smoke from a fire in a tunnel poses problems primarily for the motorist and the firefighters. In a highway tunnel there are two ways to deal with smoke from a fire: extract the smoke or control the smoke.

Smoke Extraction. There is limited understanding of the basis for smoke extraction in a tunnel. Currently, in the United States, the ASHRAE criterion of 100 cfm per lane foot is commonly used as a general guideline for transverse ventilation systems. The growing consideration of oversized extraction ports helps to find the proper solution to the extraction problem. The smoke should be extracted as close to the fire site as possible to permit evacuation and fire-fighting ingress.

Smoke Control. Smoke control includes the development of a sufficiently high longitudinal velocity (critical velocity) to prevent backlayering. Thus the smoke is contained in the portion of the tunnel beyond the fire incident and unless there is an extraction system present, it will exit at the downstream portal.

This critical velocity can be completed using the following coupled equations. The units for all variables in Equations (20-12)–(20-16) are in. of water. In the SI system, they are kilopascals.

$$V_c = K_g K \times \frac{(gH\mathring{Q})}{(\rho_0 C_p A T_f)} \times 1/3 \tag{20-12}$$

$$T_f = \frac{Q + T_0}{\rho_0 C_p A V_c} \tag{20-13}$$

where

V_c = critical velocity
g = acceleration of gravity
H = height of tunnel cross section
$\mathring{Q}$ = fire heat release rate
ρ_0 = ambient air density
C_p = specific heat of ambient air at constant pressure
A = net area of tunnel cross section
T_f = hot gas temperature
K=0.61 (dimensionless)
K_g = grade correction factor (dimensionless)
T_0 = ambient temperature

Adjustment of Air Flow Rate. Most tunnel ventilation systems are divided into ventilation zones to aid in emergency smoke-purge operations. In many cases, the zones are arranged based on the location of ventilation buildings. This is clearly shown on the diagram of the Holland Tunnel ventilation system in Figure 20-9.

One of the requirements established by the U.S. Federal Highway Administration (FHWA) regarding tunnel ventilation is that, in a single ventilation zone, a capacity equal to approximately 85% of the total capacity is required when one fan is out of operation. This can be achieved by providing a spare or standby fan or by installing a minimum of four fans in each ventilation zone (Figure 20-10). With this configuration, and if the fans are selected properly, the three fans when operating in parallel will provide approximately 85% of the total system capacity. This can be seen on the fan and system curves shown on Figure 20-11. The curves plotted in this figure assume that all system resistance varies directly with the total air flow rate. This may not be absolutely correct for some ventilation system components. As a result, the system resistance curve for three fans could shift upward from the four-fan curve, resulting in a lower air flow rate per fan. Therefore, in designing to 85% flow with three fans, the four-fan air flow will typically be greater than 100%, although the system flow with four fans operating may exceed the total flow requirements. Using this approach, an independent standby fan is eliminated.

The air volume requirements within the tunnel roadway will vary continuously throughout the day as the traffic volume varies. To minimize the power consumption, some

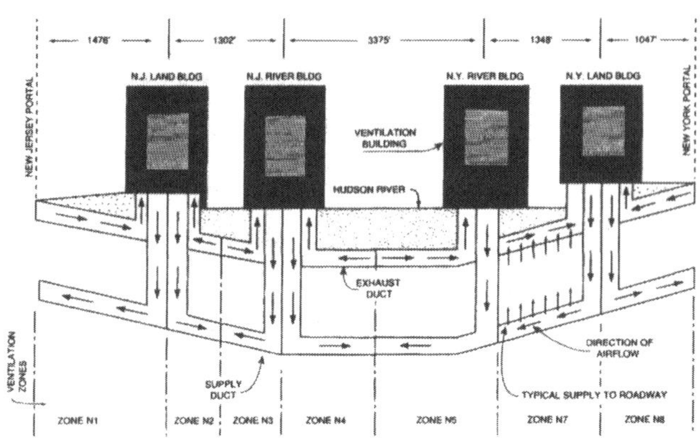

Fig. 20-9. Ventialtion diagram of the Holland Tunnel.

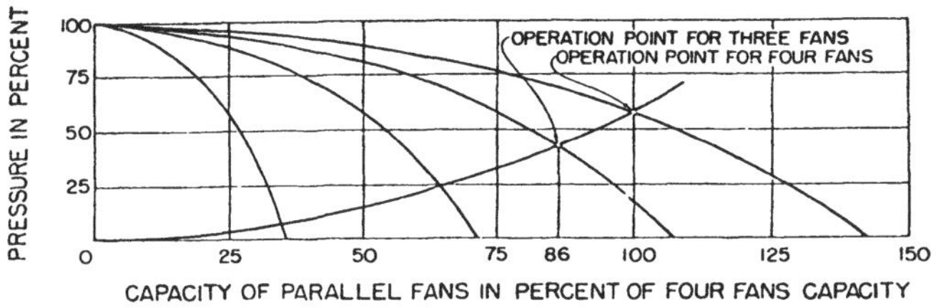

Fig. 20-10. Operation of four fans in parallel in one ventilation zone.

form of flow adjustment must be considered. This adjustment must be such as to reduce the fan horsepower along with the fan air flow rate. Numerous methods of accomplishing this are outlined under "Equipment and Facilities" in this chapter.

The fan air flow should be varied to maintain a preselected level of carbon monoxide within the tunnel. This variation can be achieved by varying the number of fans operating in the multiple-fan system, using multiple-speed motors, using controllable-pitch fans (axial), or using inlet vane control (centrifugal). The use of multiple fans and two-speed motors is a typical method of volume control (Table 20-15).

Pressure Evaluations. The air pressure losses in the tunnel duct system must be evaluated to compute the fan pressure requirements and the power required to drive the fan.

The fan selection should be based on fan total pressure instead of static pressure, since the fan total pressure is equal to the difference in total pressure across the fans, whereas the static pressure is not equal to the difference in static pressure across the fan. The fan total pressure (FTP) has been defined by the Air Moving Conditioning Association (AMCA, 1986; p. 4) as the algebraic difference between the total pressure at the fan outlet (TP_2) and the total pressure at

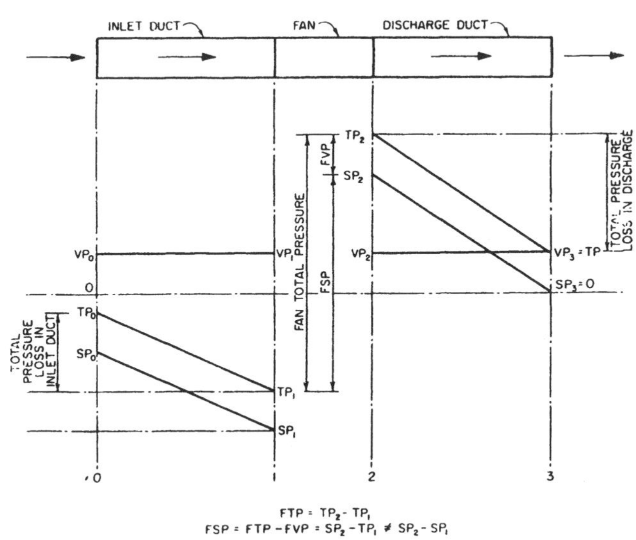

Fig. 20-11. Fan system pressure diagram.

Table 20-15. Typical Fan Operational Modes

Carbon Monoxide Level	Two Fans per Zone, Three-Speed Motors	Four Fans per Zone, Two-Speed Motors
1	1 fan, low speed	2 fans, low speed
2	2 fans, low speed	3 fans, low speed
3	2 fans, medium speed	4 fans, low speed
4	2 fans, high speed	2 fans, high speed
5		3 fans, high speed
6		4 fans, high speed

the fan inlet (TP_1), as shown in Figure 20-11 and expressed in

$$FTP = TP_2 - TP_1 \tag{20-14}$$

where

FTP = fan total pressure
TP_2 = total pressure at fan discharge
TP_1 = total pressure at fan inlet

The fan velocity pressure (FVP) is defined by AMCA to be the pressure corresponding to the air velocity and air density at the fan discharge as expressed in

$$FVP = VP_2 \tag{20-15}$$

where

FVP = fan velocity pressure
VP_2 = velocity pressure at fan discharge

The fan static pressure (FSP) is equal to the difference between the fan total pressure and the fan velocity pressure:

$$FSP = FTP - FVP \tag{20-16}$$

A fan must have the total pressure at the fan discharge (TP_2) equal to the total pressure losses ($\Delta TP_{2\text{-}3}$) in the discharge duct and the exit velocity pressure (VP_3):

$$TP_2 = |\Delta TP_{2-3}| + VP_3 \tag{20-17}$$

where

ΔTP_{2-3} = total pressure losses in the discharge duct
VP_3 = exit velocity pressure

Likewise, the total pressure at the fan inlet (TP_1) must be equal to the total pressure losses in the inlet duct system and the inlet velocity pressure:

$$TP_1 = |\Delta TP_{0-1}| + VP_0 \tag{20-18}$$

where

ΔTP_{0-1} = total pressure losses in the inlet duct
VP_0 = inlet velocity pressure

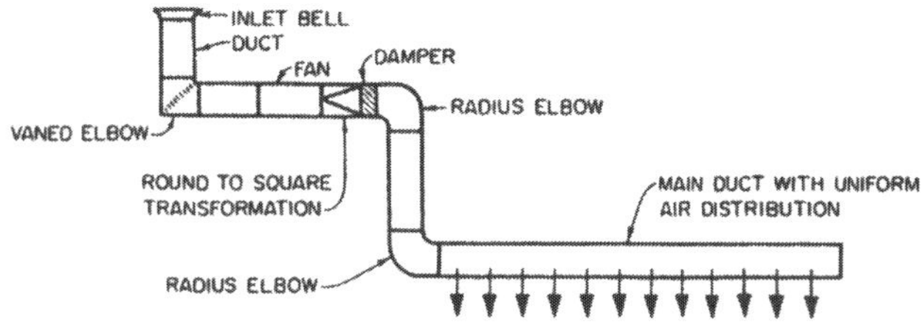

Fig. 20-12. Typical tunnel supply ventilation system elements.

However, a properly selected fan must deliver the total pressure required. The fan total pressure (FTP) is equal to the sum of the total pressure losses in the inlet system (ΔTP_{0-1}), the total pressure losses in the discharge system (ΔTP_{2-3}), and the exit velocity pressure (VP_3):

$$FTP = |\Delta TP_{0-1}| + |\Delta TP_{2-3}| + VP_3 \qquad (20\text{-}19)$$

The loss of pressure in an air duct system is due to the effect of friction and dynamic factors. These factors manifest themselves in the many components of a duct system, such as the main ducts, elbows, transformations, duct entrances and exits, obstructions, and equipment, as typified by Figure 20-12.

Straight ducts used in tunnel ventilation systems can be classified into two categories: those that merely transport air and thus have constant cross-sectional area and constant air velocity, and those that uniformly distribute (supply) or uniformly collect (exhaust) air and thus have constant cross-sectional area and diminishing or increasing air velocity.

TRANSPORT DUCTS. The pressure losses in a straight (without turns) transport duct having constant cross-sectional area and constant velocity are due to friction alone and can be computed using standard air flow equations (ASHRAE, 1991).

DISTRIBUTION/COLLECTION DUCT. The most significant method that predicts the pressure losses in a duct of constant cross-sectional area and uniform distribution or collection of air, presented in 1929, was developed for the Holland Tunnel design (Singstad, 1929) and has been in active use for many years.

The following relationship will give the total losses at any point in the distribution (supply) duct (as defined in Figure 20-13):

$$P = P_1 + \frac{12y}{D}\left\{\frac{V_0^2}{2g}\left[\frac{aLZ^3}{3m} - \frac{1}{2}(1-K)Z^2\right] + \frac{bLZ}{2gm^3}\right\} \qquad (20\text{-}20)$$

where

P = total pressure loss at any point in duct (in. of water)
P_1 = pressure at bulkhead (in. of water)
y = density of air (pounds per ft^3)
D = density of water (pounds per ft^3)
V_0 = velocity of air entering duct (ft/sec)
g = acceleration of gravity (ft/sec^2)
a = first numerical constant related to coefficient of friction for concrete (0.0035)
b = second numerical constant related to coefficient of friction for concrete (0.01433)
L = total length of duct (ft)
$Z = \dfrac{L - X}{L}$
X = distance from duct entrance to any selected location (ft)
m = hydraulic radius (ft)
K = numerical constant that takes turbulence into account (0.615).

The pressure P at any point in the collection (exhaust) duct (Figure 20-13) is given by

$$P = P_1 + \frac{12y}{D}\left\{\frac{V_0^2}{2g}\left[\frac{aL}{(3+C)m}Z^3 + \frac{3}{2+c}Z^2\right] + \frac{bL}{2gm^2(1+C)}Z\right\} \qquad (20\text{-}21)$$

where

P = total pressure loss at any point in duct (in. of water)
P_1 = pressure at bulkhead (in. of water)
Y = density of air (lb/ft^3)
D = density of water (lb/ft^3)
V_0 = velocity of air exiting duct (ft/sec)
g = acceleration of gravity (ft/sec^2)
a = first numerical constant related to coefficient of friction for concrete (0.0035)
b = second numerical constant related to coefficient of friction for concrete (0.01433)
c = numerical constant relating to turbulence at exhaust ports (0.25)
L = total length of duct (ft)
X = distance from duct exit to any selected location (ft)
$Z = (L - X)/L$
m = hydraulic radius (ft).

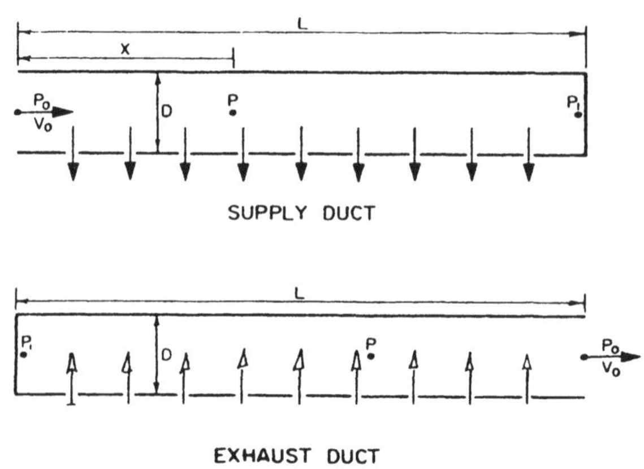

Fig. 20-13. Duct pressure analysis (Singstad, 1929).

Dynamic Losses. The dynamic losses created by eddying air flow are found in sudden changes in direction of air flow and changes in the magnitude of the air velocity. The general relationships necessary for the evaluation of pressure losses due to changes in flow direction or velocity are known (ASHRAE, 1991; Idelchik, 1994).

Supply Air Flues. The loss of pressure as the air passes through the supply air flue, if the flue area and cross section remain constant, can be computed as follows (Stahel et al., 1961) (a typical supply air flue can be seen in Figure 20-60, later in the chapter):

$$PLF = \frac{12}{D}\frac{Y}{2g}\left[\left(1 + C_{EF} + \lambda\frac{LF}{DF} + C_{BF} + C_{XF}\right)V_F^2 - (1 - C_E)V^2\right] \qquad (20\text{-}22)$$

where

PLF = pressure loss through supply air flue (in. of water)
Y = density of air (lb/ft^3)
D = density of water (lb/ft^3)
g = acceleration of gravity (ft/sec^2)
C_{EF} = loss coefficient due to change in direction of air flow at entrance to flue (Figures 20-14 and 20-15)
λ = flue friction coefficient
LF = length of flue (ft)
DF = equivalent diameter of flue (ft)
C_{BF} = dynamic loss coefficient relating to change of air flow direction within the flue (ASHRAE, 1989)
C_{XF} = dynamic loss coefficient relating to flue exit (if there is not exit deflection grille, then $C_{XF} = 0$)
V_F = air velocity in flue (ft/sec)
C_E = loss coefficient due to change in direction of air flow at entrance to flue (Figure 20-15)
V = air velocity in main duct (ft/sec)

Using the values of C_{EL} obtained from Figure 20-15, the following can be applied:

$$PLF = \frac{12}{D}\frac{Y}{2g}\left[(C_{EL})V_F^2 + \left(1 + \lambda\frac{LF}{DF} + C_{BF} + C_{XF}\right)V_F^2 - V^2\right] \qquad (20\text{-}23)$$

where C_{EL} = pressure loss coefficient at entrance to flue (Figure 20-15):

$$C_{EL} = C_{EF} + \frac{C_E}{(V_F/V)^2} \qquad (20\text{-}24)$$

The overpressure required in the main duct to maintain constant air distribution into the roadway must be at least equal to PLF; where these overpressures are larger than PLF, provisions must be made to adjust the size of the flue opening.

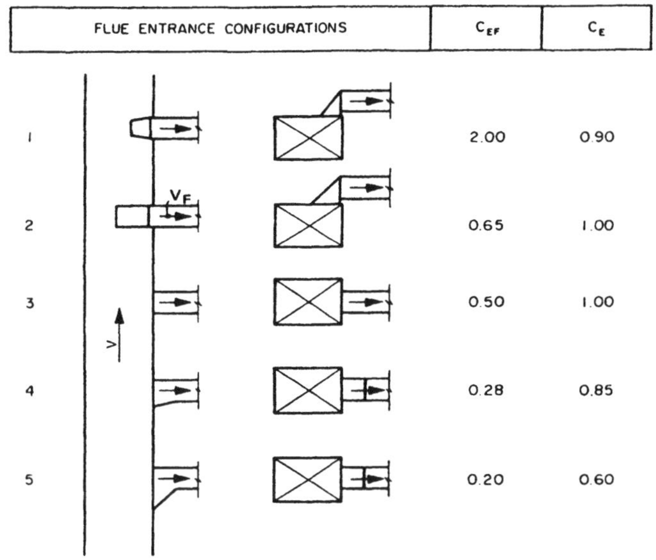

Fig. 20-14. Entrance pressure loss.

Exhaust Air Ports. Exhaust air ports, which are usually located in the tunnel's ceiling, can be evaluated as square-edged orifices, unless the exhaust port has a branch duct connection, in which case it should be evaluated in a fashion similar to the supply air flues. (A typical exhaust air port is shown in Figure 20-61, later in the chapter.)

Equipment Losses. All equipment located in the ventilation air stream will add resistance to the flow of air. These losses must also be added to the system pressure losses. Louvers and dampers are the most prominent pieces of equipment, other than the fan, to be found in a tunnel ventilation system.

Density Correction. The air density for a tunnel ventilation system should be selected based on the minimum temperature and the maximum barometric pressure antici-

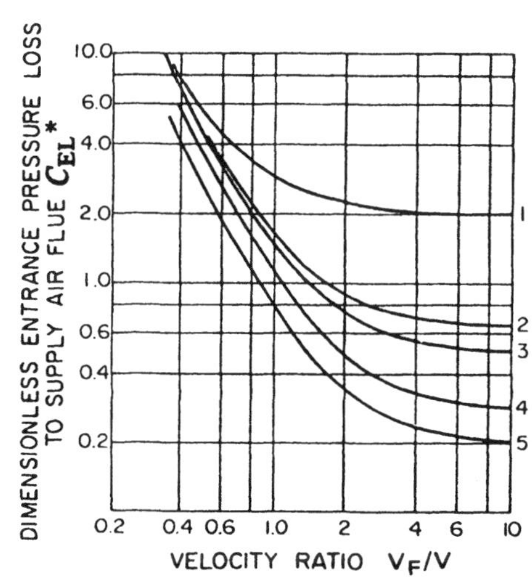

Fig. 20-15. Entrance pressure loss coefficient.

pated, thus assuring adequate fan power for all meteorological conditions. Figure 20-16 provides the necessary data for various tunnel elevations and air temperatures:

$$FTP_2 = FTP_1 \times \frac{p_1}{p_2} \qquad (20\text{-}25)$$

where

FTP_2 = fan total pressure at condition 2 (in. of water)
FTP_1 = fan total pressure at condition 1 (in. of water)
p_2 = density of air at condition 2 (lb/ft^3)
p_1 = density of air at condition 1 (lb/ft^3)

Fan Power. The fan power input required to deliver the specified air quantity at specified pressure (Jorgensen, 1983, p. 12–15) can be estimated from

$$POW = \frac{FAR \times FTP}{CFQ \times EFF} \qquad (20\text{-}26)$$

where

POW = fan input power (hp)
CFQ = fan air flow rate (ft^3/min)
FTP = fan total pressure (in. of water)
EFF = total efficiency (%)

Highway Tunnel Ventilation Systems

Highway tunnels, from a ventilation viewpoint, are defined as any enclosure through which highway vehicles travel. This definition includes not only those facilities that are built as tunnels, but those that result from other construction such as development of air rights over highways.

All highway tunnels require ventilation, which can be provided by natural means, traffic-induced piston effects, and mechanical ventilation equipment. Ventilation is required to limit the concentration of obnoxious or dangerous contaminants to acceptable levels during normal operation and to remove and control smoke and hot gases during fire emergencies. The ventilation system selected must meet the specified criteria for both normal and emergency operations and should be the most economical solution considering both construction and operating costs. For normal operations, naturally ventilated and traffic-induced ventilation systems are considered adequate for relatively short tunnels (less than 600 ft [180 m]) with low traffic volume.

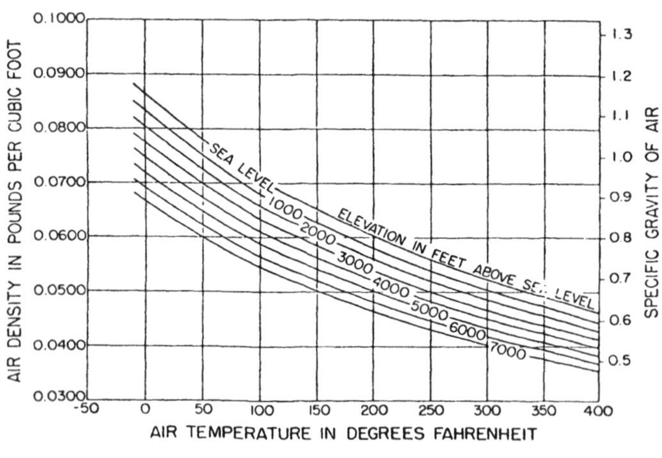

Fig. 20-16. Density correction.

Longer and more heavily traveled highway tunnels should be provided with mechanical ventilation systems. In addition many tunnels that do not require mechanical ventilation for normal operations may require a dedicated ventilation system for emergencies.

Natural Ventilation. Naturally ventilated tunnels rely primarily on meteorological conditions and the piston effect of the moving traffic to maintain satisfactory environmental conditions within the tunnel. The chief meteorological condition affecting the tunnel is the pressure differential between the two tunnel portals created by differences in elevation, ambient temperatures, or wind. Unfortunately, none of these factors can be relied on for consistent results. A sudden change in wind direction or velocity can rapidly negate all of these natural effects including, to some extent, the piston effect. The total of all pressures must be of sufficient magnitude to overcome the tunnel resistance, which is influenced by tunnel length, coefficient of friction, hydraulic diameter, and air density. None of the natural effects defined above can be considered when addressing emergency ventilation.

Air flow through a naturally ventilated tunnel can be portal to portal (Figure 20-17a) or portal to shaft (Figure 20-17b). The portal-to-portal flow-type system functions with unidirectional traffic, which produces a consistent, positive air flow. As can be seen in Figure 20-17a, the air velocity within the roadway is uniform and the contaminant concentration increases to a maximum at the exit portal. If adverse meteorological conditions occur, the velocity is reduced and the carbon monoxide concentration is increased, as shown by the dashed line in Figure 20-17a. If bidirectional traffic is introduced into such a tunnel, further reductions in air flow result.

The naturally ventilated tunnel with an intermediate shaft (Figure 20-17b) is best suited for bidirectional traffic. However, the air flow through such a shafted tunnel is also at the mercy of the elements. The added benefit of the "stack effect" of the shaft depends on air temperature differentials, rock temperatures, wind direction and velocity, and shaft height. The addition of more than one shaft to a naturally ventilated tunnel will most likely be more of a disadvantage than an advantage, since a pocket of contaminated air can be trapped between the shafts, causing high contaminant levels in the zone.

Because of the numerous uncertainties outlined above, historically it is rare that tunnels of greater than 1,000 ft (305 m) in length have been ventilated by natural means. There are exceptions to this, such as the Via Mala Tunnel in Switzerland and the Tenda Pass Tunnel between France and

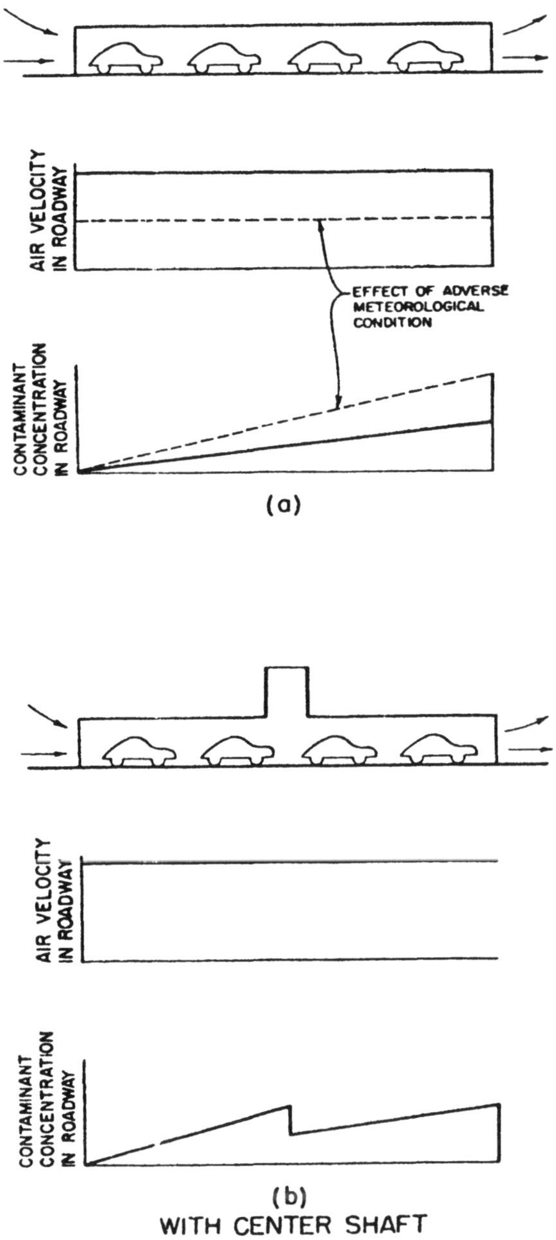

Fig. 20-17. Natural ventilation systems. (Adapted from Stahel et al., 1961).

Italy. Via Mala, which is 2,050 ft (625 m) long, is located in an area where the traffic flow is extremely low. The 9,000-ft (2,800-m) long Tenda Pass Tunnel has a large difference in portal elevations, which creates a large consistent pressure differential and thus adequate air flow. It also is located in an area having favorable wind conditions. Nonetheless, both of these tunnels have been outfitted with booster fans to supplement the natural ventilation if required.

An emergency mechanical ventilation system has been installed in most naturally ventilated urban tunnels more than 600 ft (180 m) in length. Such a system is required to purge the smoke and hot gases generated during a fire emergency and may be required to remove stagnant polluted gases or haze during severe adverse meteorological conditions.

The reliance on natural ventilation for tunnels more than 600 ft (180 m) long should be thoroughly evaluated, specifically the effect of adverse meteorological and operating conditions. This is particularly true for a tunnel where a heavy or congested traffic flow may require mechanically generated air flow. If the natural mode of ventilation is not adequate, a mechanical system with fans must be considered. There are several types of mechanical ventilation systems, outlined below.

Longitudinal Ventilation. A longitudinal ventilation system is any system where the air is introduced to or removed from the tunnel roadway at a limited number of points, thus creating a longitudinal air flow within the roadway (Figure 20-18). The injection-type longitudinal system (Figure 20-18a) has been used in railroad tunnels; however, it has also found application in highway tunnels. Air is injected into the tunnel roadway at one end of the tunnel, where it mixes with the air brought in by the piston effect of the incoming traffic (Figure 20-18a). This system is most effective in a tunnel with unidirectional traffic. The air velocity within the roadway is uniform throughout the tunnel, and the concentration of contaminants increases from ambient at the entering portal to maximum at the exiting portal. Adverse external atmospheric conditions can reduce the effectiveness of this system. The concentration of contaminants increases as the air flow decreases or the tunnel length increases.

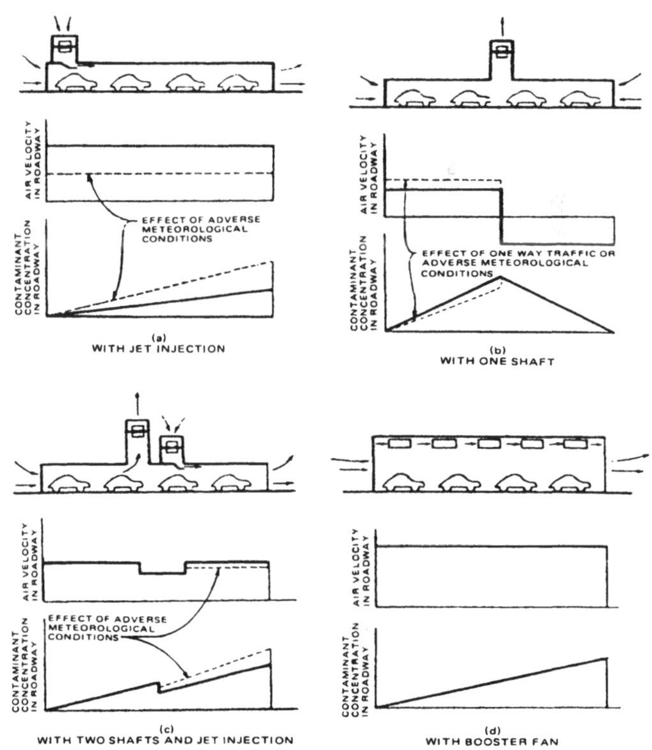

Fig. 20-18. Longitudinal ventilation systems. (Adapted from Stahel et al., 1961).

The longitudinal system with a shaft (Figure 20-18b) is similar to the naturally ventilated system with a shaft except that it provides a positive stack effect (fan induced). With bidirectional traffic, peak contaminant concentration occurs near the shaft location. This system is generally not used for unidirectional tunnels, because the contaminant levels become unbalanced and excessive amounts of air will therefore be required for ventilation.

An alternative longitudinal system uses two shafts located near the center of the tunnel, one for exhausting and one for supplying Figure 20-18c. This configuration will provide a reduction of contaminant concentration in the second half, because a portion of the tunnel air flow is exchanged with ambient air at the shaft. Adverse wind conditions can cause a reduction of air flow, a rise in contaminant concentration in the second half of the tunnel, and "short circuiting" of the fan air flows.

JET FANS. In a growing number of highway tunnels throughout the world, jet fans mounted at the tunnel ceiling (Figure 20-19) are being employed to generate the longitudinal air flow (Figure 20-18d). This system eliminates the construction required to house ventilation fans in a building, but it may require larger clearances within the tunnel to accommodate the jet fans and thus, a larger tunnel (Figure 20-19).

The concept of jet fans mounted at the tunnel ceiling relies on the high-velocity impulse jet of air produced by the jet fans to create an additional induced flow in the tunnel. These fans are usually mounted in groups of two or more and spaced longitudinally in the tunnel at about 300-ft intervals.

SACCARDO NOZZLE. The principal feature of a Saccardo-nozzle-based longitudinal ventilation system is the use of a high-velocity "nozzle" to inject air into the tunnel, usually in the direction of traffic flow. This concept uses a high-velocity jet flow that will induce additional longitudinal air flow. The Saccardo nozzle functions on the principle that a high-velocity air jet injected into the tunnel roadway at an extremely small angle to the tunnel axis can induce a high-volume longitudinal air flow in the tunnel roadway area. The level of this induced flow depends on the nozzle discharge velocity (V_n), the momentum exchange coefficient (K), and the nozzle angle (θ) (see Figure 20-20).

The induction effect of a Saccardo nozzle is proportional to the square of the discharge velocity. A reasonable value of the discharge velocity is 6,000 fpm (30 m/sec), which is about the maximum value that can be used without an adverse impact on the traffic stream.

The basic advantages of the use of a Saccardo-nozzle-based longitudinal system over a jet-fan-based system are

- Reduced tunnel height.
- Reduced number of moving parts to maintain.
- Maintenance can be accomplished without impeding traffic flow.
- Noise level in tunnel is decreased.
- High fan efficiency.

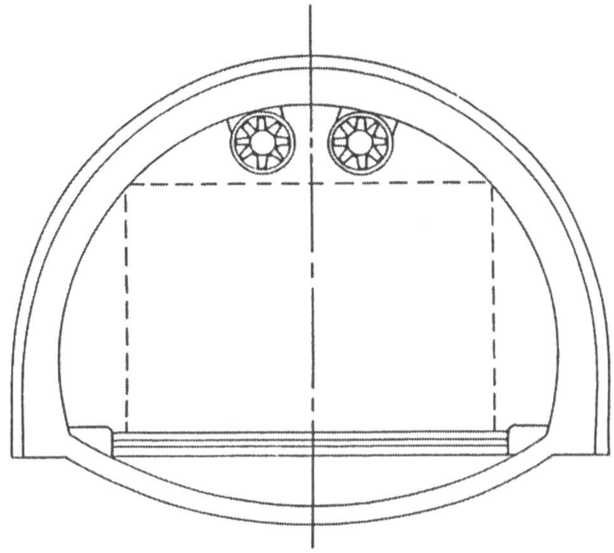

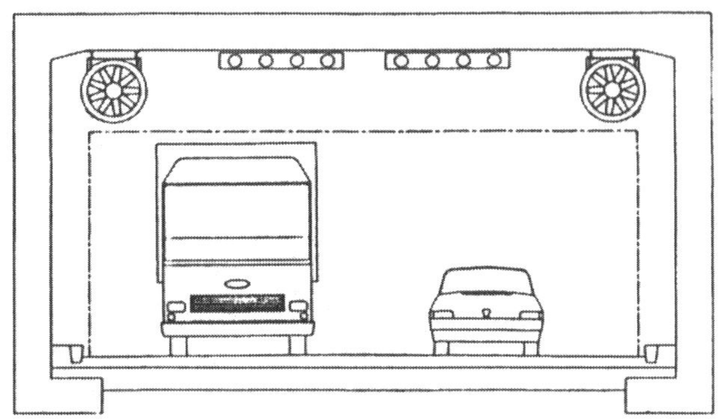

Fig. 20-19. Longitudinal ventilation systems with jet fans.

Longitudinal ventilation systems, excluding the jet fan system, that have either supply or exhaust at a limited number of locations within the tunnel are the most economical systems since they require the least number of fans, place the least operating burden on these fans, and do not require separate air ducts as part of the tunnel structure. However, as the length of the tunnel increases, the disadvantages of these systems become pronounced, such as the excessive air velocities in the roadway and the discharge of pollutants at the exiting portal. If the tunnel portal is located in an environmentally sensitive area or near a sensitive environmental re-

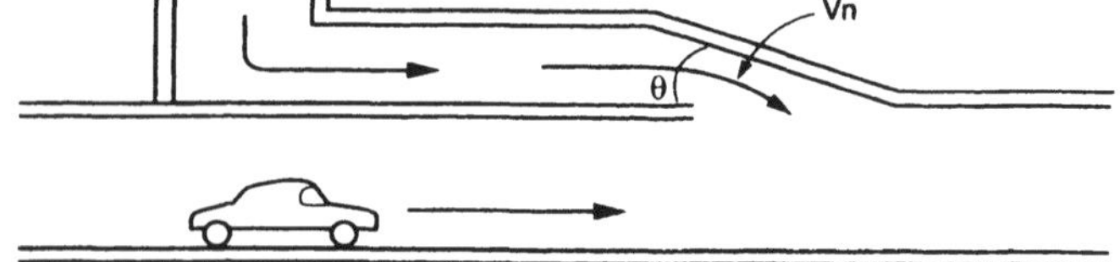

Fig. 20-20. Longitudinal ventilation systems with Saccardo nozzle.

ceptor, then the latter disadvantage can be severe, and a more complex system may be required. Longitudinal systems are not well adapted to long tunnels with bidirectional traffic, particularly for fire life safety.

Semitransverse Ventilation. Uniform distribution or collection of air throughout the length of a tunnel is the chief characteristic of a semitransverse system. A supply air semitransverse system (Figure 20-21a) produces a uniform concentration of contaminants throughout the tunnel because the air and the vehicle exhaust gases enter the roadway area at the same relative rate. In a tunnel with free-flowing unidirectional traffic, additional longitudinal air flow will be created by the moving traffic within the roadway area. Supply air is transported to the roadway in a duct and uniformly distributed through flues. The most suitable location for the introduction of air to the tunnel roadway is at the level of the vehicles' exhaust pipes to permit immediate dilution of the exhaust gases. To accomplish the air distribution described above, an adequate pressure differential must be generated between the duct and the roadway to counteract the vehicle piston effect and atmospheric winds.

During a fire within the tunnel, the air supplied from a semitransverse system will provide dilution of the smoke. However, to aid in fire-fighting efforts and in emergency egress, the fresh air should enter the tunnel through the portals to create a respirable environment for these activities. For these reasons, the fans in a supply semitransverse system should be reversible.

The exhaust semitransverse system in a unidirectional tunnel (Figure 20-21b) will produce a maximum contaminant concentration at the exiting portal. For low traffic speeds with unidirectional traffic or in a bidirectional tunnel, a zone of zero fresh air is located within the tunnel, which, of course, corresponds to the maximum concentration of contaminants.

A combination supply and exhaust system has been used (Figure 20-21c). Such a system is applicable only in a unidirectional tunnel where the air entering the traffic stream is exhausted in the first half, and air that is supplied in the second half is exhausted through the exit portal. This system, in effect, creates longitudinal air flow in the roadway area.

The supply system is the only semitransverse system not affected by adverse meteorological conditions or opposing traffic. Semitransverse systems are used in tunnels up to approximately 9,000 ft (2,750 m) in length (not altitude), at which point the tunnel air velocities near the portals become excessive.

Full Transverse Ventilation. For longer tunnels, a full transverse system is typically used. A full exhaust duct is added to a supply-type semitransverse system, which achieves a uniform distribution of supply air and a uniform collection of vitiated air (Figure 20-22). This system was developed for the Holland Tunnel in 1924. With this system, a uniform pressure will occur except that generated by the traffic piston effect, which will tend to reduce contaminant levels. A pressure differential between the ducts and the roadway is required to assure proper distribution of air under all ventilation operating conditions.

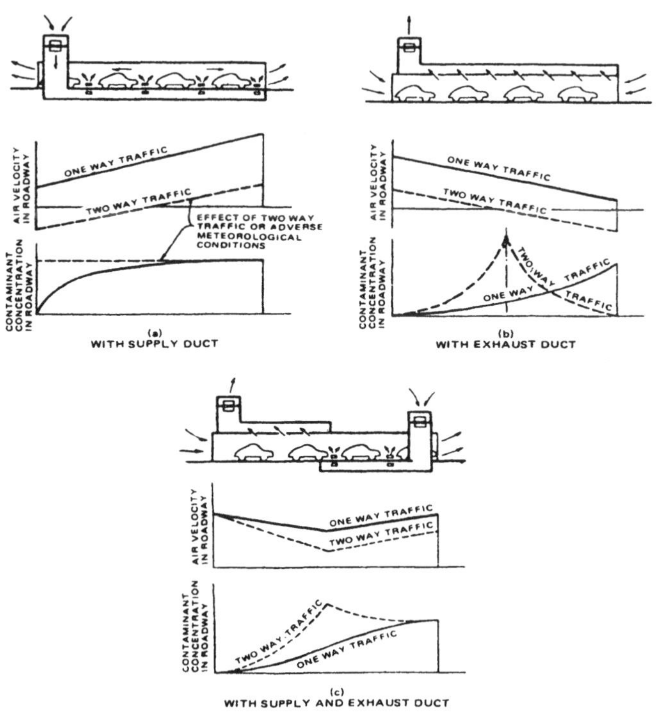

Fig. 20-21. Semitransverse ventilation systems. (Adapted from Stahel et al., 1961).

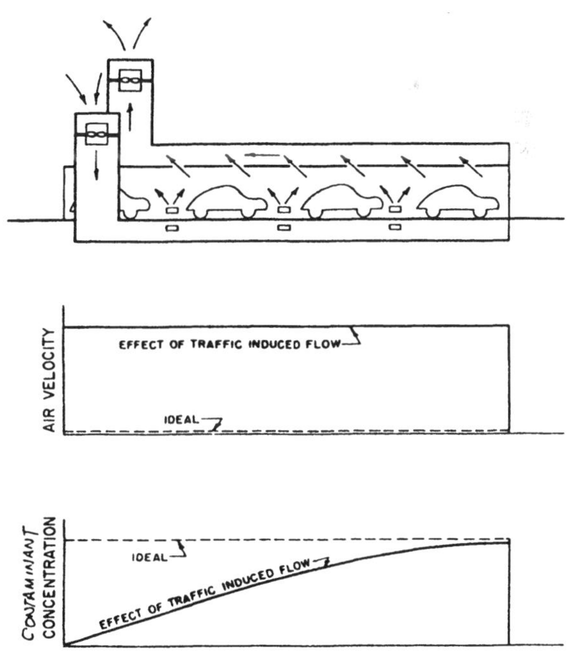

Fig. 20-22. Transverse ventilation systems. (Adapted from Stahel et al., 1961).

The desirable location of the supply air inlets from the standpoint of immediate dilution of exhaust gases is at the level of the vehicle emissions near the road surface, with the exhaust outlets located in the ceiling, thus creating transverse air flow. This arrangement originally was demonstrated by the full-scale tests conducted by the U.S. Bureau of Mines (Fieldner et al., 1921). The air distribution can be one-sided or two-sided depending on tunnel width and the number of traffic lanes.

Single-Point Extraction. Ever since the ceiling fell down during the Holland Tunnel Fire in 1949, ventilation engineers have been looking for ways to apply the concept of enlarged ceiling openings to process a larger rate of smoke extraction. What occurred at the Holland Tunnel fire after the demise of the ceiling was a sudden improvement in the extraction of the smoke from the roadway area. One could arrange for the tunnel ceiling to collapse, but a more meaningful approach is to provide a specialized enlargement of an existing exhaust port or a new large extraction opening. Both approaches have been tried, the enhanced exhaust port in a recent Holland Tunnel retrofit and in the Sydney Harbour Tunnel in Australia and several Swiss tunnels.

An expandable exhaust port is a typical exhaust port, probably 6 × 36 in. (152 × 914 mm), which is opened to some larger size in the event of a fire incident (Figure 20-23). This can be accomplished by using a meltable panel—used in Holland Tunnel—or an active damper, which can be opened from a central control system. The damper approach is unreasonable due to the high cost of both maintaining and operating the dampers. The meltable panel is a more cost-effective solution. However, finding a material having the proper melting temperature, nontoxicity, and structural strength has been extremely difficult.

An oversized extraction port is usually applied in a new tunnel when a large opening (approximately 8 ft × 8 ft) (2.4 m × 2.4 m) can be installed in the tunnel ceiling or wall to communicate with the fan system through the exhaust or extraction duct. An early application of the concept, although in a rail tunnel, was in the San Francisco Trans-Bay Tube. This concept has been exposed to a fire and subsequent tests in the 1970s as discussed by Chan (1990). The purpose of an oversized extraction port is to provide a larger opening close to the fire to provide a large extraction rate to remove the smoke gases. This concept was tested in the Memorial Tunnel Fire Ventilation Test Program in 1994.

HYBRID SYSTEMS. There are many variations and combinations of the systems described above. A combined hybrid system for a unidirectional tunnel approximately 1,300 ft (400 m) long is shown in Figure 20-24. Section 3 uses a full transverse system because of the upgrade roadway; section 2 uses a semitransverse supply system with a longitudinal exhaust; the remainder of the tunnel (section 1) is a semitransverse supply system.

Highway Tunnel Descriptions

Following are brief descriptions of the application of the above-mentioned ventilation systems to actual tunnels. These descriptions make the use of some hybrid systems developed to address specific project requirements.

Hampton Roads Tunnels. The first Hampton Roads Tunnel, connecting Hampton and Norfolk, Virginia, was opened to traffic in 1957. This first crossing of the Hampton Roads Ship Channel was joined by the parallel second crossing opening to traffic in 1976. This project includes two immersed tube tunnels between mid-channel portal islands, connected to shore by approach trestles.

Both tunnels have full transverse ventilation systems supported by four ventilation buildings, which house a total of 32 vertically mounted axial fans. Each fan has a capacity of 23,000,000 ft^3/min (10,855 m^3/sec). The ventilation system is controlled from a central control room located in the original north ventilation building.

Fort McHenry Tunnel. The Fort McHenry Tunnel on I-95 in Baltimore, Maryland, is constructed of two parallel tubes, each containing two bores with two traffic lanes each, for a total of eight traffic lanes.

The full transverse ventilation system employs 48 centrifugal fans housed in steel portal-located ventilation buildings. There are 12 supply fans and 12 exhaust fans in each ventilation building. The fans are driven by a single-speed 300-hp (223.8-kW) and two-speed 100/40-hp (74.6/29.84-kW) motors. Each fan delivers about 300,000 ft^3/min (141.58 m^3/sec) of air in supply or exhaust.

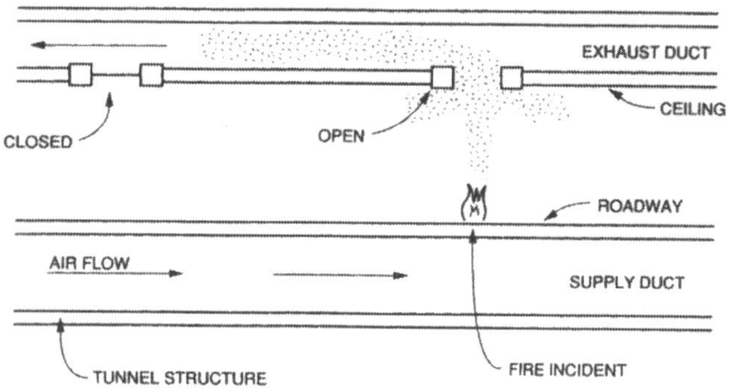

Fig. 20-23. Transverse ventilation with expandable exhaust ports.

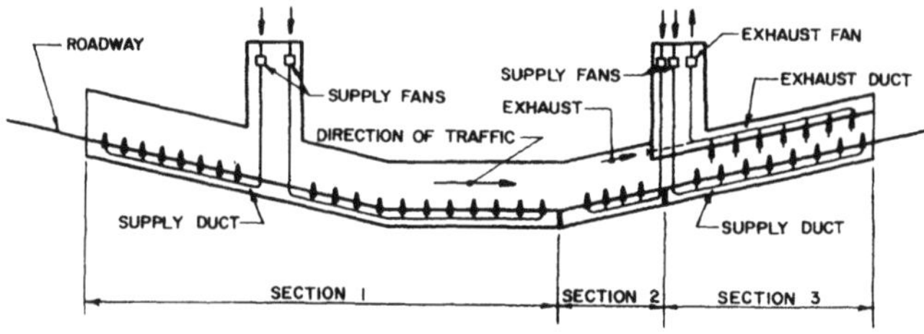

Fig. 20-24. Modified transverse ventilation system.

Sydney Harbour Tunnel. The Sydney Harbour Tunnel, which opened to traffic in 1992, incorporates three different tunnel construction methods (immersed tube, driven tunnels, and cut-and-cover tunnels), which result in the use of five different tunnel cross sections and air duct configurations (Figure 20-25).

The ventilation system is a modified semitransverse system with single-point extraction as shown in Figure 20-26. This system incorporates the positive features of a semitransverse system coupled with the advantages of a longitudinal system.

One of the unique features of this tunnel is the use of the Sydney Harbour Bridge pylon to house the total exhaust system (16 axial flow fans); see Figure 20-27. In addition, the supply fans are housed in an underground ventilation structure on the north shore.

The ventilation system uses the advantages offered by both of the ventilation concepts noted. The exhaust ventilation system can generate the required "critical" longitudinal air velocity by removing or supplying air via the central exhaust opening, located within each tunnel. The supply ventilation system, operating in reverse (exhaust) mode, can exhaust smoke locally through remotely activated emergency exhaust dampers located along the length of each bore. This system offers the motorists the required protection and also limits the length of tunnel exposed to the heated gases.

Elizabeth River Tunnel. The Second Downtown Elizabeth River Tunnel in Virginia, completed in 1988, is a two-lane tunnel under the southern branch of the Elizabeth River and connects Norfolk and Portsmouth, Virginia. It parallels the First Downtown Tunnel, completed in 1952, and incorporates an innovative ventilation scheme that combines the best of "classical" semitransverse and full longitudinal exhaust systems, allowing fresh air to enter the tunnel portals, mix with and dilute vehicle emissions, and be exhausted through air ducts over the roadway. This concept eliminated customary air ducts beneath the roadway.

Glenwood Canyon Tunnels. The Hanging Lake Tunnel has twin bores containing four lanes. It is ventilated by a reversible semitransverse system using a ceiling duct for supply air during normal operations and exhaust during smoke purge. There are four 300-hp (224-kW) vane axial fans for each bore.

Singapore Underground Roadway Tunnel System (SURS). The SURS complex is a unique tunnel project, a proposed urban ring road to be constructed totally below grade, which will result in a loop roughly 9.1 mi (14.66 km) long with 18.2 mi (29.3 km) of three-lane roadway underground and some 33 ramps with 27,600 ft (8,426 m) of these ramps also located below ground.

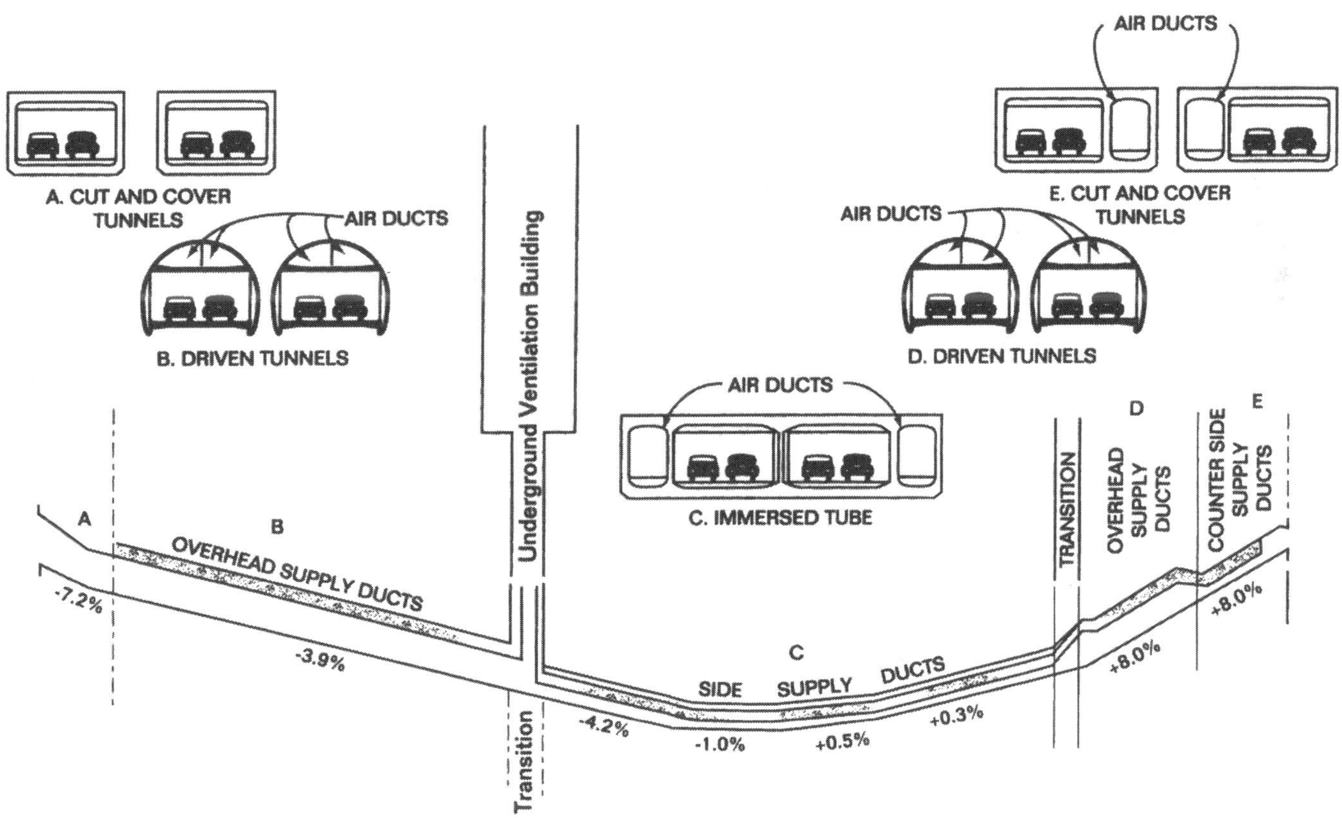

Fig. 20-25. Sydney Harbour Tunnel profile and cross section.

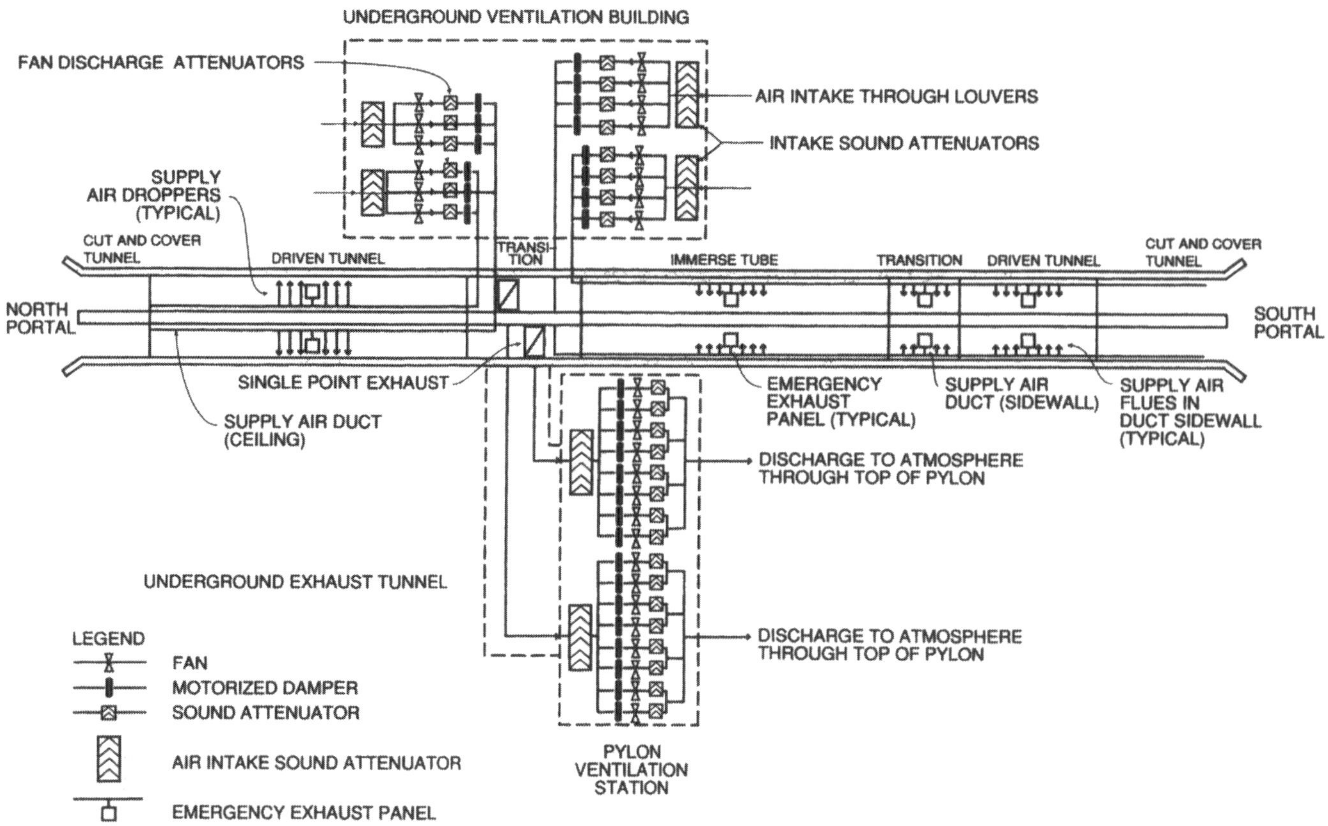

Fig. 20-26. Sydney Harbour Tunnel air flow diagram.

A longitudinal ventilation system is proposed for the main loop roadway. This system will be implemented by the use of Saccardo nozzles to inject air into the main loop roadway area and induce circular flow. The vitiated air will be extracted at the end of each ventilation segment. In addition, a jet fan longitudinal system will be employed in certain ramps where the injection system will not meet the required criteria.

RAILROAD TUNNELS

This section pertains to railroad tunnels; more specifically, tunnels through which rail vehicles travel propelled by diesel engine locomotive power. (Tunnels in which electrically propelled trains operate are covered later in the section on rapid transit tunnels.)

The most important environmental factors in diesel-powered rail tunnels are heat and the products of combustion. Excessive heat will raise the tunnel ambient air temperature to a point at which the locomotive engine can no longer function. The presence of excessive products of combustion will create a hazard to health and a haze within the tunnel. Any criteria established must therefore consider the health of the train occupants, both crew and passengers, and the ability of the locomotive units to function properly within the tunnel.

In a diesel locomotive, the diesel engine drives an electric generator supplying power to the electric traction motors. The throttle, which usually has eight positions, controls engine speed. The diesel engines, in 70% of the locomotives in the United States, are of the two-stroke cycle type. The capacity of these engines ranges from 700 to 4,000 hp (522 to 2,983 kW).

Diesel Exhaust Gases in Rail Tunnels

A general discussion of diesel engine exhaust gas composition was provided earlier. The composition of railway locomotive diesel exhaust is similar to that of vehicular diesel engines. However, in the railroad tunnel, the significant contaminant becomes the oxides of nitrogen instead of carbon monoxide. Thus the oxides of nitrogen must be studied in depth. It can be shown that, based on the composition of diesel exhaust gas, if the oxides of nitrogen are maintained within specified acceptable limits, all other exhaust gas contaminants will also be maintained within acceptable limits.

The establishment of criteria for oxides of nitrogen must consider the length of time personnel are exposed to the environment, since the nature of the effect on the human body is time-dependent. The American Conference of Governmental Industrial Hygienists, in the 1992 edition of *Indus-*

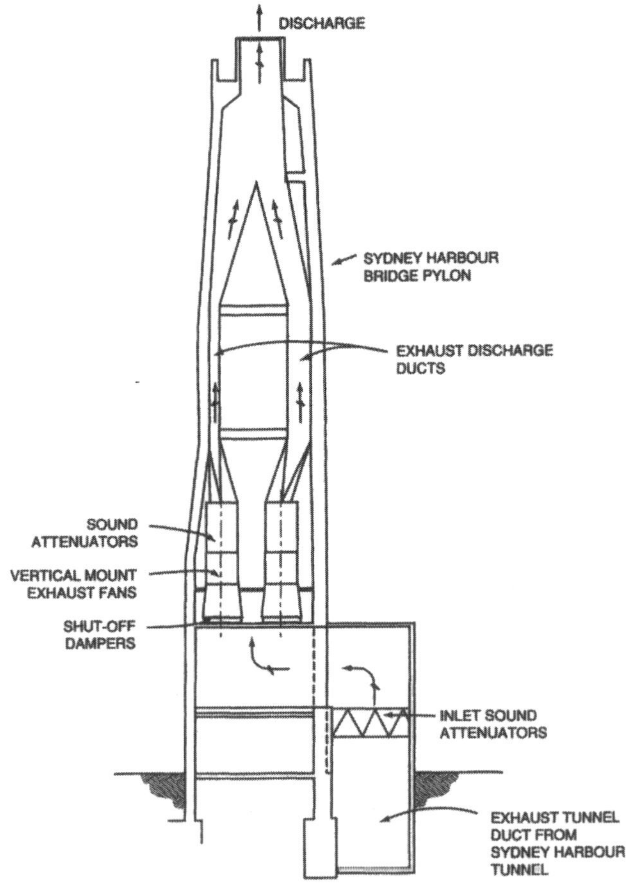

Fig. 20-27. Sydney Harbour Tunnel pylon ventilation structure.

trial Ventilation (ACGIH, 1992), presents a threshold limit value of 3 ppm for nitrogen dioxide and 25 ppm for nitric oxide. These threshold limit values are for exposures of 8 hours per day over a working lifetime. Therefore, since the travel time through a railroad tunnel is on the order of 1 hour, levels of 5 ppm for nitrogen dioxide and 37.5 ppm for nitric oxide can be tolerated for these short periods of exposure without adverse effects (ACGIH, 1992).

The exhaust gases are emitted from the top of most diesel electric locomotives (Figure 20-28). This creates a phenomenon that aids the ventilation of any railroad tunnel. The stratification effect created in the crown of the tunnel and in the annular space around the locomotive by the temperature gradients remains stable, and thus a percentage of the exhaust gas contaminants will remain in the crown of the tunnel and not interact with the train (Figure 20-28). Tests have shown that about only 45% of the emitted exhaust gases descend from the tunnel crown to interact with the air in the spaces at the sides of the locomotive.

The effect of air flow in the annular space has been determined from tests conducted at Cascade Tunnel (PBQD, 1966). These tests show that approximately 50% of the total air flow rate within the annular space surrounding the train is effective for cooling and combustion. A method for determining the oxides of nitrogen emitted from a diesel engine locomotive is described by Berger and McGuire (1946).

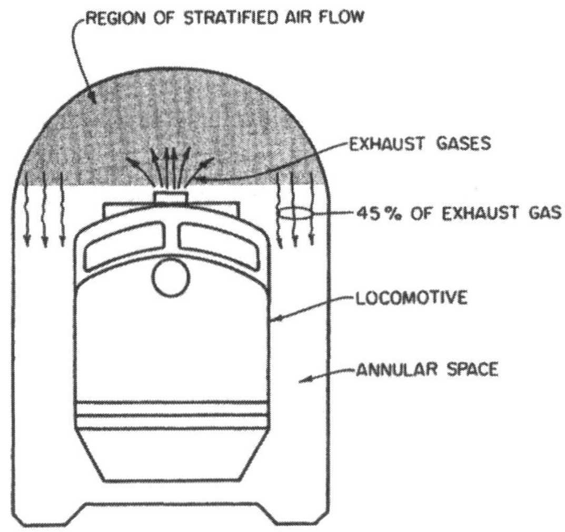

Fig. 20-28. Rail tunnel cross section showing stratified air flow.

Analysis

Heat. Heat from the engine combustion process is released by the locomotive as it travels through the tunnel. The total heat is composed of heat from the exhaust gas, jacket water, lubricating oil, braking resistors, and radiation. The heat from exhaust gases, jacket water, lube oil, and braking resistors is emitted from the top of the locomotive, whereas the radiated heat is emitted from all surfaces of the locomotive (Figure 20-29).

The final dissipation of this heat is important to the proper operation of the locomotive unit, since air is required for the engine cooling. If the inlet air temperature is raised by this heat to a point above the maximum operating temperature of the engine, the engine will shut down on a high-water-temperature condition. This usually occurs where a large number of locomotive units are operating in a single

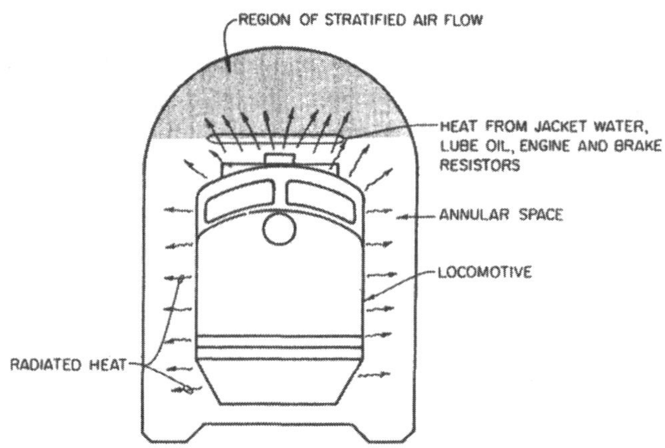

Fig. 20-29. Typical rail tunnel cross section showing stratified airflow and locomotive heat rejection.

consist. As in the case of the air contaminants, only a portion of this heat affects the locomotive inlet air conditions due to the stratified air flow in the crown of the tunnel.

The total amount of the radiated heat, along with the heat from the exhaust gas, jacket water, lube oil, and braking resistors, interacts with the inlet air. An expression for the temperature of the inlet air of a specific locomotive unit is

$$TIA = TAA \times \frac{HER + RHI(HEE + HJW + HLO + HBR) \times N}{DEN \times CP \times EAF \times QRT} \quad (20\text{-}27)$$

where

TIA = temperature of inlet air to $N + 1$ locomotive unit (°F)
TAA = temperature of tunnel ambient air ahead of consist (°F)
HER = heat radiated from one locomotive (BTU/min)
RHI = portion of heat from the exhaust gas, jacket water, lube oil and braking resistors that interacts with the inlet air
HEE = heat rejected with engine exhaust gas (BTU/min)
HJW = heat rejected from jacket water system (BTU/min)
HLO = heat rejected from lubricating oil system (BTU/min)
HBR = heat rejected from braking resistors (BTU/min)
N = number of locomotives in consist ahead of locomotive of interest
DEN = density of air (lb/ft^3)
CP = specific heat of air (BTU per min per °F)
EAF = portion of tunnel air flow that is effective for cooling and combustion
QRT = tunnel air flow relative to train (ft^3/min)

Values for engine heat rejection in the case of a specific design should be obtained from the engine manufacturer.

The effect that the inlet air temperature has on engine operation can be judged by the design air inlet temperature of most locomotive units of 115°F (46°C). This value applies to over-the-road service. However, most locomotive engines can operate at temperatures greater than 115°F (46°C) for short periods of time. Tests have shown that units have operated with inlet air temperatures exceeding 150°F (66°C). The allowable intake temperature for each locomotive type should be obtained from the engine manufacturer when a specific design is being contemplated.

Ventilation. Ventilation is required in all railroad tunnels to remove the heat generated by the locomotive units and to change the air within the tunnel, thus flushing the tunnel of air contaminants. Ventilation can take the form of natural ventilation, piston effect, or mechanical ventilation. While the train is in the tunnel, the heat is removed by an adequate flow of air with respect to the train, whereas the air contaminants are best removed when there is a positive air flow out of the tunnel (as shown also in Figure 20-30).

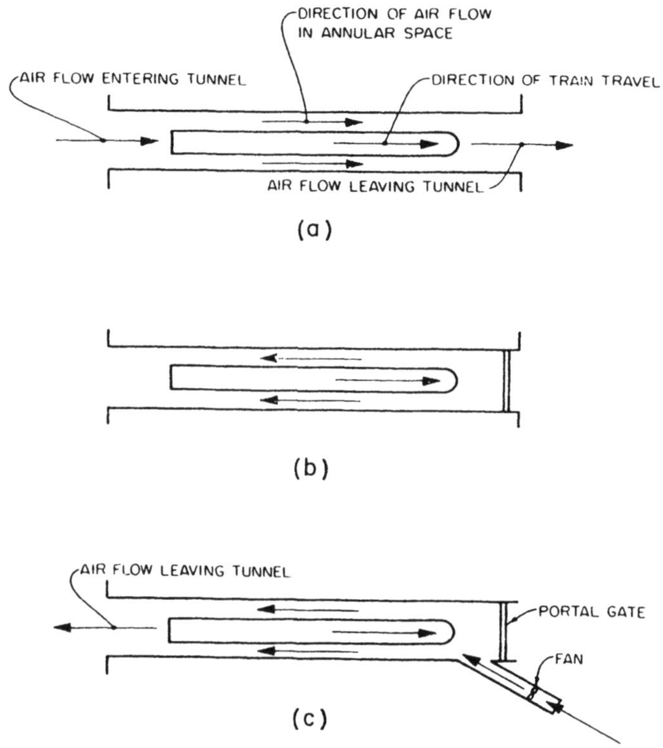

Fig. 20-30. Rail tunnel ventilation system configurations.

The three major forms of railroad tunnel ventilation are outlined below:

- *Piston effect with an open-ended tunnel.* In this type of tunnel (Figure 20-30a), the only means of ventilation is the piston effect of the train on the air. The air in the annular space flows in the same direction as the train, and there is a net flow of air through the tunnel in the direction of the train. This method is satisfactory for short tunnels since the flushing effect is good, and because the oxides of nitrogen levels and the heat have not had an opportunity to build up.
- *Piston effect with a portal gate.* The addition of an operable portal gate at the end of a tunnel (Figure 20-30b) will greatly enhance the ability of the ventilation system to remove the heat generated by the locomotive units. However, since there is no net flow from the tunnel, the tunnel under these conditions is not flushed of the air contaminants. This fact precludes the use of a portal gate on a railroad tunnel without mechanical ventilation.
- *Mechanical ventilation with a portal gate.* The addition of a mechanical ventilation system (Figure 20-30c) provides the ultimate in the ventilation of a railroad tunnel. The heat, along with the air contaminants, will be removed, since there will be adequate air flow with respect to the train and an adequate net flow through the tunnel.

Pressure. To properly design a railroad tunnel ventilation system, it is necessary to be able to predict the pressures generated in the tunnel for various train speeds as well as the air flow surrounding the train. These air flows must satisfy the cooling requirements of all locomotive units in the con-

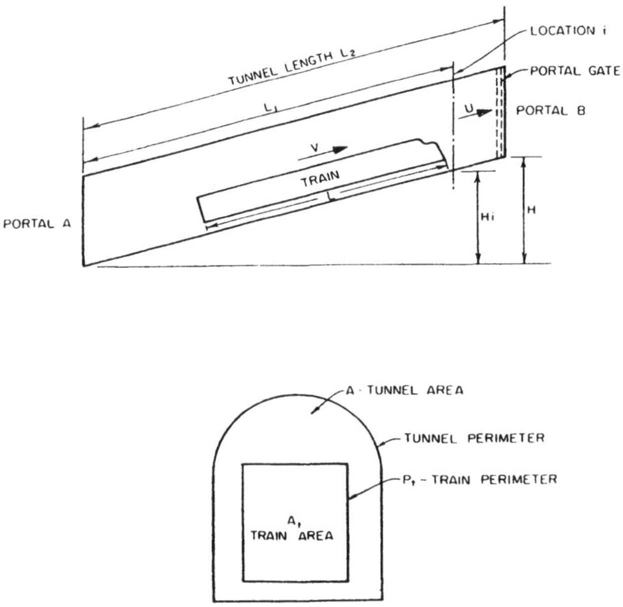

Fig. 20-31. Identification of variables in railroad tunnel.

sist. The ventilation system components must be designed to withstand the maximum pressures generated by the train and the fans.

The following relationships were developed (Aisiks and Danziger, 1969) to predict the piston effect of the train within the tunnel and the pressures in the tunnel. For a definition of the variables in these relationships, see Figure 20-31.

Tunnel Portal Gate Open.

$$U = \frac{\sqrt{\epsilon}}{\sqrt{\epsilon} - 1} V \tag{20-28}$$

where

U = air velocity in unobstructed tunnel (ft/sec)

$$\epsilon = \frac{\dfrac{\phi}{(1-\phi)^2}\left[\phi - \dfrac{Lf_t}{4R_t(1-\phi)}\right]}{\sum_1^n K_i + \dfrac{f_w(L_T - L)}{4R_W}} \tag{20-29}$$

where

V = train velocity (ft/sec)
ϕ = A_t/A
L = train length (ft)
f_t = friction factor of train surface
f_w = friction factor of tunnel wall
R_t = hydraulic radius of train (ft)
K_i = localized tunnel loss coefficient
L_T = total tunnel length (ft)
R_W = hydraulic radius of unobstructed tunnel (ft)
A = tunnel cross-sectional area (ft^2)
A_t = train cross-sectional area (ft^2)

Tunnel Portal Gate Closed, Fans Not Operating.

$$\frac{\Delta P}{\gamma} = \frac{P_A - P_B}{\gamma} - \Delta H + \frac{V^2}{2g}\left[\frac{P_t f_t L}{4A(1-\phi)^3} + \left(\frac{\phi}{1-\phi}\right)^2 + \frac{\phi^2}{(1-\phi)^3}\frac{Lf_w}{4R_W}\right] \tag{20-30}$$

where

P = pressure above portal B pressure at location 1 (psf)
P_A = barometric pressure at portal A (psf)
P_B = barometric pressure at portal B (psf)
H = difference in elevation between portals (ft)
V = train velocity (ft/sec)
g = acceleration of gravity (ft/sec^2)
P_t = train perimeter (ft)
γ = unit weight of air (lb/ft^3)

Tunnel Portal Gate Closed, Fans Operating.

$$\frac{\Delta P}{\gamma} = \frac{P_A - P_B}{\gamma} - \Delta H_1 \pm \frac{U^2}{2g}\left[\sum_1^n K_t + \frac{f_w(L_1 - L)}{4R_W}\right] + \frac{U \pm V)^2}{2g}\left[\frac{P_t f_t L}{4A(1-\phi)^3} + \left(\frac{\phi}{1-\phi}\right)^2\right] \pm \frac{(U \pm \phi V)^2}{2g}\left[\frac{Lf_w}{4R_W(1-\phi)^3}\right] \tag{20-31}$$

where

H_1 = difference in elevation between location 1, where pressure increment is wanted, and portal A (ft) (when location 1 is chosen at the gate, $H_1 = H$)
L_1 = total distance between location 1 and portal A (ft)

By continuity,

$$\text{QRT} = (U - V) \times A \tag{20-32}$$

Equations (20-28) and (20-29) give the explicit value of the air velocity in the tunnel, in front of and behind the train, for a level tunnel with the gate open. Only a straightforward substitution of numerical values is required. Equation (20-30) is used to determine the pressure change produced by piston effect in the portion of the tunnel between the train and the gate. That pressure change could be either above or below atmospheric pressure, depending on the direction of the train movement. However, that value must be determined by successive approximations, since the value of U, air velocity in the tunnel, depends on that pressure difference. The computational procedure starts by estimating a value for U and determining the corresponding P, which is also the pressure rise against which the fans are operating. With this computed P and the fan characteristic curve, the total fan discharge can be determined. That fan air flow must equal the tunnel airflow: i.e., U (assumed above) multiplied by the area of the tunnel; otherwise, the procedure must be repeated until an equality is attained.

To complete the above computations, the friction factors for both tunnel and train must be known. The tunnel friction factor, which corresponds to the coefficient f in the Darcy–Weisbach equation for friction losses in pipe flow, as developed from test data for a concrete-lined railroad tunnel with ballasted roadbed, is 0.0133. This value was verified with excellent correlation by computations taking into account relative roughness and the Reynolds number.

The air flow in the annular space between the train and the tunnel wall is influenced by the train friction factor. Friction factor values for both freight and passenger trains were developed based on tests conducted at Cascade Tunnel (PBQD, 1966). For freight trains, the friction factor is 0.143; for passenger trains, it is 0.065.

The makeup of the trains passing through the tunnel vary, as there are many sizes and shapes of railroad cars. A method was developed to determine the cross-sectional area, perimeter, and hydraulic radius of an equivalent train. A weighted average of the area and hydraulic radius can be computed by grouping the cars in classes on the basis of the Association of American Railroads' (AAR) mechanical designation list for each car type. This method permits rapid determination of the area and hydraulic radius of an equivalent train having the same length as the train being considered.

Flushing. There is a requirement in all diesel-powered railroad tunnels for a flushing cycle after the train has passed through the tunnel. The ventilation of the tunnel during the train passage is never 100% effective. Air contaminants will always remain in the tunnel. The natural ventilation effect created by the pressure differential due to a difference in elevation may be sufficient for a short tunnel to provide the necessary flushing. A fan system will be required to provide this flushing effect for longer tunnels or in cases where the train headway requires a more rapid flush cycle.

Railroad Tunnel Descriptions

Included in this section is a description of several railroad tunnel ventilation systems.

Flathead Tunnel. This is a 7-mi (11-km) long rail tunnel located in northwest Montana (USCOE, 1966). It was built as a part of the relocation of the Great Northern Railway (now the Burlington Northern) in 1971 as a part of the Libby Dam project. The tunnel is straight, with a uniform grade of +0.46% (with the exception of the 230 ft [70 m] at the eastern portal, which has a −0.20% grade; see Figure 20-32). The ventilation system is capable of flushing the tunnel of contaminants and providing cooling for the locomotive units of an eastbound (upgrade) train. The ventilation system (Figure 20-33) includes two axial flow fans, shut-off dampers, a relief damper and a portal gate, along with all auxiliary equipment. Two standby diesel engine-driven generators are provided to furnish power to the ventilation system and all auxiliaries during a power failure.

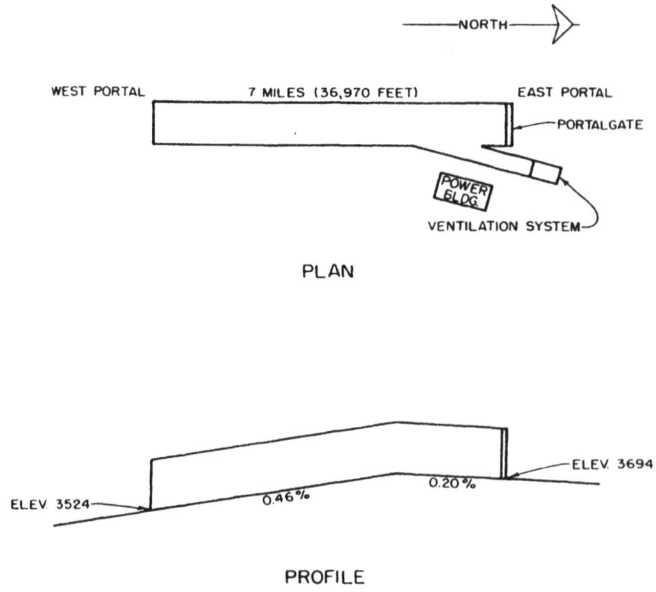

Fig. 20-32. Flathead Tunnel.

The fans are horizontal, two-stage reversible axial-flow fans direct-driven through floating shafts by AC induction motors. These fans are equipped with a blade pitch control system that permits automatic adjustment of the fan air flow to compensate for the wide ambient temperature variation experienced at this location. This system maintains a constant mass flow of air through the fan and limits the brake horsepower to 2,000 (1,491 kW) per fan.

The relief damper, provided to prevent an excessive buildup of pressure in the tunnel, is a four-module multibladed type damper, controlled by pressure regulators within the tunnel. The fan shut-off dampers are of the multibladed type designed to withstand a pressure of 60 in. of water gauge (15 kPa). The portal gate is of welded structural steel construc-

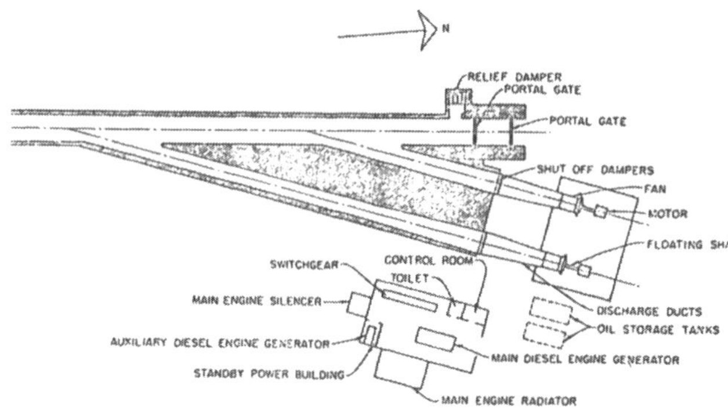

Fig. 20-33. Arrangement of ventilation system at Flathead Tunnel.

tion designed for uniform air pressure of 60 in. of water gauge (15 kPa). It is supported by rollers bearing against steel supports located in the walls. Rubber seals minimize leakage through the gate. The gate has a fail-safe mechanism that will open the gate upon loss of power. The center portion of the gate is built of frangible material that will allow a train to go through the gate, should any of the gate mechanisms fail, without causing major damage to the train or gate. Two gates are installed, but only one is used at any time (the second is a standby).

A 2,000-hp (1,491-kW) diesel engine–driven generator provides emergency power for one fan. A smaller unit provides power for the auxiliary loads.

One fan is used to provide cooling for the upgrade trains that require it, while both fans are used to flush the tunnel in less than 20 min after a train leaves the tunnel.

Mount Macdonald Tunnel. The Mount Macdonald Tunnel is a 9.1-mi (14.5-km) long railroad tunnel beneath Rogers Pass in Glacier National Park, British Columbia, Canada. The tunnel is a single-track tunnel and handles diesel-electric powered trains. It is the longest diesel railroad tunnel in North America. Previously, CP Rail's main line operated through the park via the 5-mi (8-km) Connaught Tunnel, a 0.95% grade single track serving bidirectional traffic. The westbound approach grades to this tunnel, however, are on the order of 2.2% and required the use of locomotive pushers to move heavy coal and freight trains through this portion of the line. With increased traffic forecast for the main line, the tunnel and its approach grades would have become a bottleneck, and so CP Rail decided to build a second tunnel. Westbound traffic now runs through the new 9-mi (14.5-km) long, 0.7% grade Macdonald Tunnel. Eastbound traffic runs through the existing tunnel. The new tunnel is also part of the railroad's overall grade reduction program. Trains are now able to run from Calgary to Vancouver over grades not higher than 1.0% and this, in turn, reduces the required hauling capacity to approximately 1.0 hp (0.75 kW) per trailing ton, considerably less than that required previously.

The railroad established a maximum time interval between trains of not more than 40–45 min. Due to the tunnel's length and the speed of the design trains, the standard portal-to-portal ventilation concept described earlier could not be applied without restricting the frequency of traffic. Accordingly, a unique system had to be developed.

The tunnel overburden permitted the location of an economically feasible ventilation shaft near the midpoint of the tunnel. The opportunity to use mid-tunnel ventilation provided the solution to ventilating such a long tunnel without unduly delaying entering trains.

Figure 20-34 shows the major components of the vent system, which includes a tunnel gate system at the east portal, a tunnel gate system at the mid-tunnel, a 1,200-ft-deep (366-m), 28-ft-diameter (8.5-m) vent shaft that is partitioned and connects to the tunnel on opposite sides of the mid-tunnel gate, a series of dampers, and a system of five fans. One exhaust fan is housed in a vent building at the east portal, and the remaining four fans are housed in a vent building atop the shaft. Two supply air to the east segment, and two exhaust air from the west segment. The combination of the partitioned shaft and the mid-tunnel gate serve to divide the tunnel into east and west segments, each with its own ventilation system. The system purges half of the tunnel while a train passes through the other half.

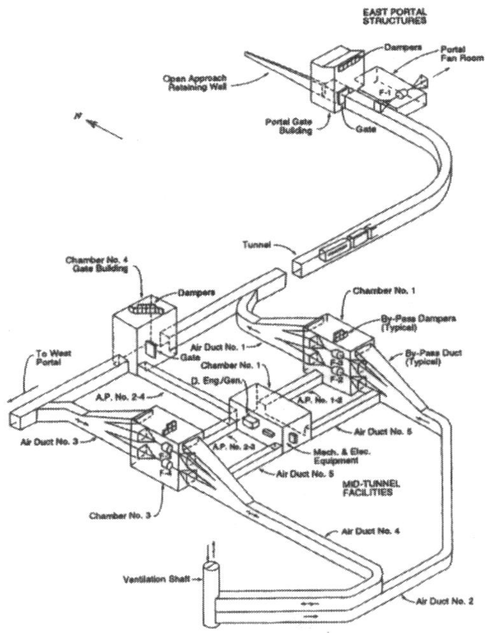

Fig. 20-34. Mount McDonald Tunnel.

As a train enters the east portal, the mid-tunnel gate is closed, the intake dampers at the top of the shaft serving the east tunnel close, and one of the two fans serving the east tunnel is operated in supply to cool the locomotives.

When the rear of the train enters the east portal, the east portal gate is closed, the fan at the top of the shaft is put into idle (a nondelivery mode), the bypass dampers at the top of the shaft are opened, and the fan at the east portal is operated in exhaust.

As the train nears the mid-tunnel area, the mid-tunnel gate is opened and the intake dampers at the top of the shaft are sequentially closed. During this period, the source of air for the fan at the east portal transitions from the top of the shaft to the west portal, thus providing continuous cooling as the train moves from the east to the west segment of the tunnel.

When the rear of the train passes the mid-tunnel gate, the gate closes behind it, the east portal fan is set to idle, the east portal gate is opened, and two east tunnel fans at the top of the shaft operate in parallel for 15 min to purge the east segment. While this occurs, one of the other shaft-top fans is operated in exhaust to provide supplemental ventilation while the train is in the west segment. Air is exhausted from the west portal across the train and up the shaft.

When the rear of the train leaves the tunnel, the two west tunnel shaft-top fans operate in parallel for 13 min to purge the west segment. Once the purge cycle in the east tunnel has ended, a train can enter the east portal while purge of the west continues.

Each tunnel gate system contains two vertical-lift, independently operating gates, but only one gate is operational at any given time. The inoperable, or standby, gate is deenergized and held in its up position by the force of its counterweights. The gate is of steel construction except for its center wooden frangible panel, which is designed to break away if hit by a train. Each gate is powered through a 5-hp (3.73-kW) motor per clutch per reducer arrangement located high above the gates. Upon loss of power, the counterweight system is designed to open a gate fully within 15 sec. When a train passes under a gate, the gates are deenergized and held open by their counterweights. During this period, the gates are maintained deenergized.

The prime movers of the ventilation system are five identical vane axial, controllable pitch in motion, fans. The fan is single stage, has a wheel diameter of approximately 9.5 ft (2.9 m), and is powered by a nominal 2,250-hp (1,678-kW), 1,200-rpm induction motor. The motor, floating shaft, and fan shaft are all outside the airstream. A unique feature of this fan is its antistall ring, which prevents it from going into surge. Surge is a dangerous operating condition and occurs when the fan operates in the unstable portion of its characteristics curve. The antistall ring virtually eliminates this portion of the curve.

During peak traffic periods, the fans are required to cycle (cooling-off, purge-off, cooling-purge). Because the size of the fan motors precludes frequent starting, the motors run continuously during these periods. The fan flow is varied by changing the pitch angle of the fan blades through a hydraulic blade pitch control mechanism. This system also includes a programmable controller that allows specific blade angles to be preset, such as the idle position where the blades are in the closed position and no air is delivered. The programmable controller also allows the rate at which the fan blades are opened and closed to be varied to minimize pressure transient effects.

The operating point of the fan when operating in cooling (one fan operating delivering 2 in. water gauge; 0.5 kPa) differs significantly from that when operating in purge (two fans operating in parallel, each delivering 11 in. water gauge; 2.75 kPa).

Each fan is equipped with an isolation damper. The damper is 14 ft (4.3 m) square, and the blades and connecting rods are made of stainless steel. The damper is hydraulically powered by a stand-alone unit. The damper actuator includes a spring return mechanism, which automatically closes the damper upon loss of power.

The ventilation system is controlled through a computer-based central control system. The system was designed to operate fully automatically with only minimal dispatcher interface. For a train approaching the tunnel, the dispatcher need only enter one of the 16 available fully automatic cooling modes. The system then executes the mode automatically through the monitoring of track circuits.

Mount Shaughnessy Tunnel. In 1988, the Canadian Pacific Railway (CP Rail) completed construction of the Mount Shaughnessy Tunnel in Glacier National Park, British Columbia, as part of the Rogers Pass Project to reduce grades and increase capacity on its main line. The 1.2-mi (1.86-km) long, single-track tunnel serves diesel locomotive operation.

The heaviest and slowest trains passing westbound through this 1.0% grade tunnel are 14,080 trailing ton (12,770 tonne) coal trains powered by five General Motors SD40-2 units, three at the head end and two in the remote position. Since the tunnel did not have a ventilation system originally, succeeding locomotives relied on the speed of the train to generate sufficient relative air flow to dissipate the heat emitted from lead units to preclude overheating in trailing units and subsequent train stall. This was found to be insufficient during the warmer months of the year, but a combination of generated air flow coupled with the short time the locomotives are in the tunnel (about 5 min) was found to be marginally sufficient for these trains to get through the tunnel.

To generate the required air flow, a system of jet fans was designed. The location of these fans along the length of the tunnel is shown in Figure 20-35. The location of the fan in the concrete-lined segments is shown in Figure 20-36. The physical size of the fan is based on the clearance requirement for a double-stack container (21.33 ft [6.5 m] above top of rail) and the minimum height of the crown.

Control of the jet fan system is either local manual, remote manual, semiautomatic, or fully automatic. Under manual control, fans can be activated individually. Under semiautomatic control, the dispatcher can automatically activate various levels of ventilation according to preprogrammed modes. Under fully automatic control, the system is designed to use existing track circuits to activate the jet fans in accordance with preprogrammed modes of operation. The fully automatic modes of operation are linked to the fully automatic modes of the adjacent Mount Macdonald Tunnel.

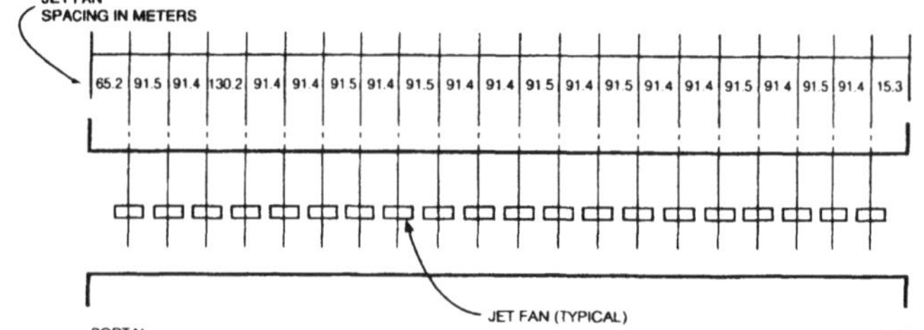

Fig. 20-35. Location of jet fans along length of Mount Shaughnessy Tunnel.

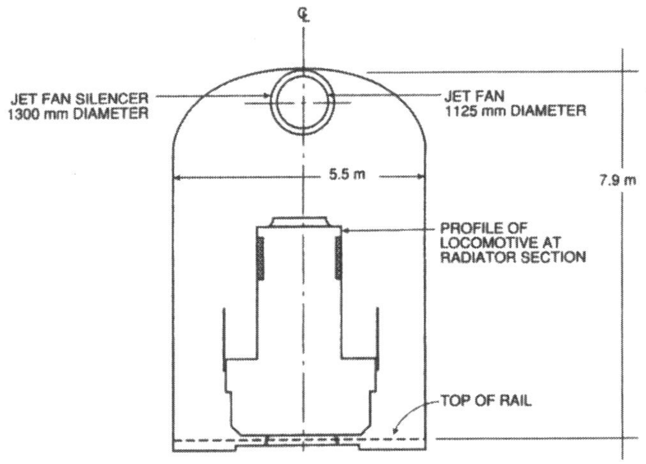

Fig. 20-36. Mount Shaughnessy Tunnel cross section with jet fan.

English Channel Tunnel. The English Channel Tunnel is made up of two running tunnels with one service tunnel situated between them as shown in Figure 20-37. Under normal operating conditions, the railway tunnels are connected directly by 194 pressure relief ducts equipped with dampers (one pressure relief duct every 820 ft [250 m]). During maintenance periods, the tunnels can be connected by opening doors that allow trains to cross from one tunnel to another (crossovers). There are two undersea crossovers (one on the U.K. side, and one on the French side) and one underland crossover on the U.K. side.

Each running tunnel is connected to the service tunnel by cross-passages. There are 135 cross-passages (spaced at approximately 1,230 ft [375 m]) per tunnel, which are usually closed off by doors. Some of the cross-passages (38 out of 146) are equipped with air distribution units, which allow the passage of air from the service tunnel to the running tunnels. Nonreturn valves prevent air movement in the other direction.

Tunnels are normally ventilated by air blown into the service tunnel by ventilation system plants, one located on each end of the tunnel. The air is distributed throughout the full length of the service tunnel, transversely to the two running tunnels by air distribution units, which are installed in the central part of the tunnel above the cross-passage doors. This air is moved by the train's piston effect and is returned to the atmosphere through the portals of the running tunnels. The normal ventilation system has an approximate air flow rate of 340,000 cfm (160 m^3/sec).

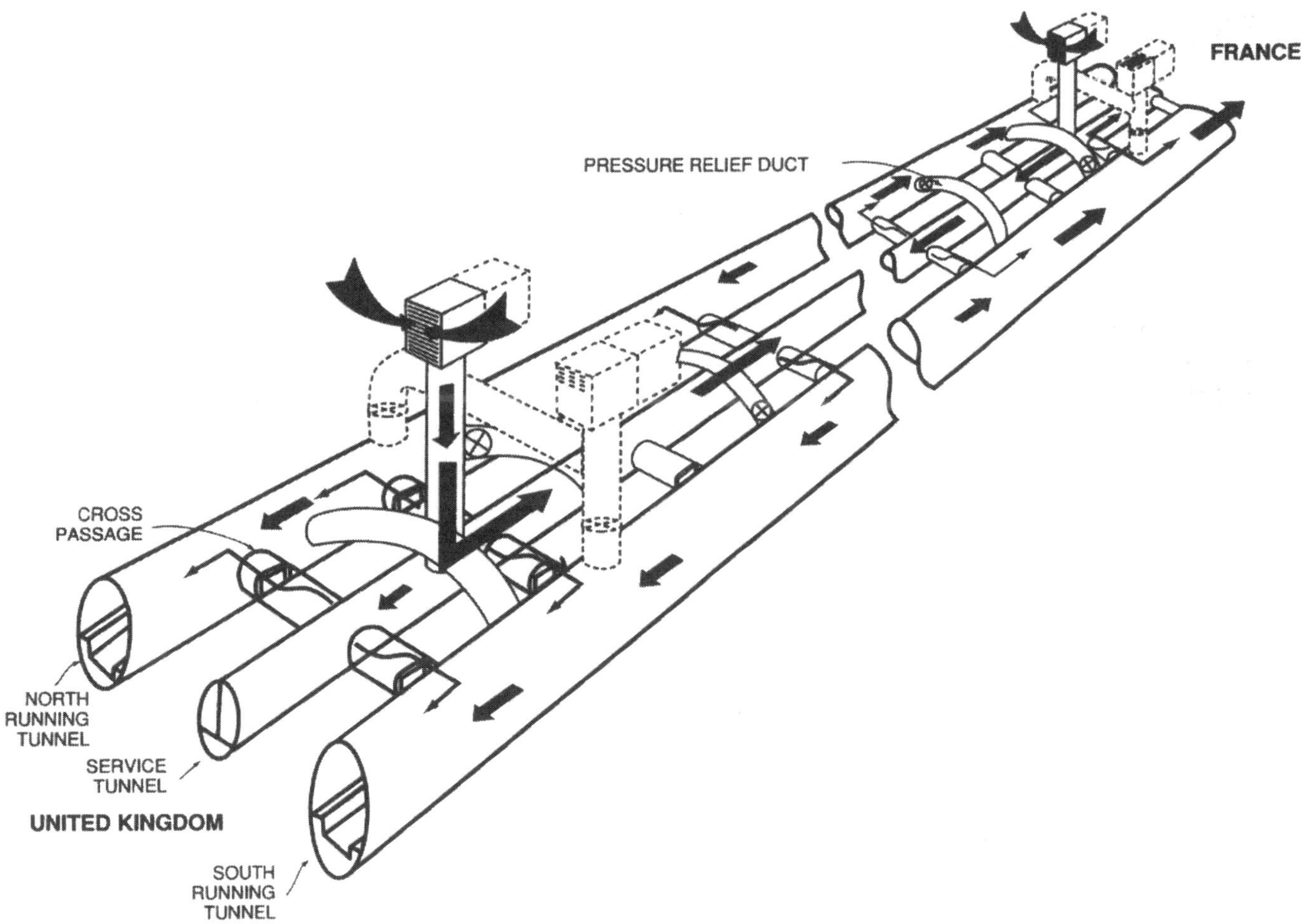

Fig. 20-37. English Channel Tunnel—air flow diagram.

In the event of an emergency in one of the running tunnels, the normal ventilation system maintains a pressurization of the service tunnel relative to the running tunnels to guarantee that the service tunnel remains free of smoke, even if some of the cross-passage doors are open. The air flows in tunnels and open cross-passages must not, however, exceed some admissible levels for passengers evacuation.

In case of fire in the service tunnel or any of the control rooms, the normal ventilation system ensures an air flow along the service tunnel, which directs smoke to one end or the other and finally removes it from the service tunnel.

Supplementary ventilation system plants, located at each end of the tunnel, supply or can extract air from either running tunnel or both simultaneously. The supplementary ventilation system can create a longitudinal air flow rate of approximately 200,000 cfm (100 m^3/sec) in the undersea part of either running tunnel (with no trains in the tunnel) by providing air to one end and extracting it from the other end.

Like the normal ventilation system, each supplementary ventilation system plant includes two fans, one for regular duty and one standby. It is also possible to use both fans in "dual mode" in both running tunnels simultaneously, especially when a crossover door is open. The supplementary ventilation system is used only in extreme cases of fire when it is necessary to increase air flow to get better smoke control in some desired direction. It is also used to evacuate smoke from the running tunnels after the fire is extinguished.

While the supplementary ventilation system is in operation, the normal ventilation system remains in duty to ventilate and pressurize the service tunnel and maintain it as a safe haven. Both the normal and supplementary ventilation systems are be adjusted to maintain a safe environment, especially for passenger evacuation.

RAPID TRANSIT SYSTEMS

The portions of rail rapid transit systems located below the surface in subway structures require control of the environment.

In rapid transit systems, there are two types of tunnels: the standard subway tunnel, usually located between stations and normally constructed beneath surface developments with numerous ventilation shafts and exits communicating with the surface, and the long tunnel, usually crossing under a body of water, such as the Trans-Bay Tube in San Francisco, or through a mountain, such as the Berkeley Hills Tunnel, also in California. The ventilation concepts for these two types will be different, since in the long tunnel there is usually limited ability to locate a shaft at any intermediate point, as can be done in the standard subway tunnel. The subway tunnel is considered in this section, while the long transit tunnel is considered in the section "Rail Tunnels."

The evaluation of the environment within the subway system has taken great strides in recent years. Since the BART system was the first new system to be designed (1965) in approximately 30 years, the available technology was limited. Efforts have progressed from the analytical tools that were developed for BART (Danziger and Aisiks, 1970) to the sophisticated analytical tools developed as a part of the Subway Environmental Research Project. The published handbook resulting from this project contains the most up-to-date technical data available for evaluation of the subway environment (*SEDH*, 1976).

Subway Environment

During normal train operation, the "piston effect" of the moving train provides air motion within the trainway and stations. In most of the systems built in early years, this air movement has been sufficient to maintain adequate environmental conditions within the subway facilities. However, new higher-performance transit vehicles, which are capable of attaining speeds of 80 mph (129 kph), often generated an amount of heat exceeding the ability of the piston effect ventilation to remove it from the subway. Coupled with this is the increasing use of air-conditioned transit trains and stations. Many of the newer systems being designed and constructed in the world today consider cooling of the station facilities. This occurs on systems in Caracas, Washington, Baltimore, Atlanta, Hong Kong, and Singapore.

The other aspect of subway ventilation is emergency operation. The major purpose of an emergency system is to remove heat and smoke generated during a fire to permit the safe egress of passengers and the entrance of fire-fighting personnel. When an emergency situation occurs, the trains are halted or slowed, and the "piston effect" ventilation ceases. This requires the incorporation of a mechanically driven ventilation system into the subway.

The objectives to be sought in controlling the environment within a subway system are, first, to provide a suitable environment for patrons as well as for operating and maintenance personnel; and second, to minimize the exposure of the equipment to high temperatures, thus prolonging its life. This must be considered for all public and nonpublic spaces within the stations and the normally occupied space in the trainways. Also to be considered is the control of haze, odor, and vapor, and the purging of smoke during a fire.

The Urban Mass Transit Administration (now the Federal Transit Administration) of the U.S. Department of Transportation funded a research project dealing with the subway environment. The rapid transit properties in the United States and Canada sponsored and partially funded this project through the Institute for Rapid Transit (now the American Public Transit Association) and the Transit Development Corporation. The Subway Environmental Research Project (SERP) had as its objective preparation of a handbook, along with the required methodologies to permit proper evaluation of the subway environment (*SEDH*, 1976).

The subject of tunnel ventilation in a rapid transit system has taken on a broader meaning over the last 20 years. It now incorporates the tunnel and station environment including both normal and emergency operations. The station environment is severely affected by events in the connecting tunnels. The following text addresses the factors to be considered in the evaluation and design of subway environment control. The *Subway Environmental Design Handbook* (*SEDH,* 1976) provides a detailed description of the evaluation and design process.

Consideration of the environment within the subway system involves three major components: criteria, analysis, and control.

Normal Criteria. Criteria for the subway environment must be based on the reaction of the human patron to the surrounding environment. The factors to be considered are temperature, humidity, air movement, noise, and vibration. The capacity of human beings to endure the environment is a function of age, occupation, health, acclimatization, and the natural variation in human beings. The most critical item is the thermal factor.

Criteria should also be established for maximum air velocity, rapid pressure change, and air quality, including gas particulate and odor contaminants. A thorough definition of these criteria and their method of determination is covered in the *Subway Environmental Design Handbook* (*SEDH,* 1976).

The primary normal criteria to be applied in the subway station are related to platform comfort and tunnel temperature. The criteria for the stations are usually based on the specific climatic conditions, the local comfort requirements (demands), and funding limitations set by the operating agency.

Thermal Comfort. Thermal comfort is the primary factor in setting the environmental criteria. This factor has the greatest influence on the comfort of the traveling patron. One of the basic underlying philosophies in setting criteria is that the patron in entering a station should not be exposed to a degradation in thermal environment upon entering the subway facilities from the street environment.

The establishment of a specific single value criterion is nearly impossible in this environment. The application of standard office environment criteria would be too costly. Therefore, a method that uses several thermal indices was developed (*SEDH,* 1976). The indices used are the Relative Warmth Index (RWI) and the Heat Deficit Rate (HDR), which were both developed (*SEDH,* 1976) for a transient or subway environment based on the Relative Strain Index derived by Lee and Henshel (1963) and comfort tests conducted by ASHRAE. The RWI was developed to be applied in warm environments, whereas the HDR was developed for cooler environments.

The RWI can be computed from the following:

$$\text{RWI} = \frac{M(I_{cw} + I_a) + 1.13(t - 95) + RI_a}{70(1.73 - P)} \tag{20-33}$$

where

RWI = relative warmth index
M = metabolic rate (BTU/hour ft^2)
I_{cw} = insulation (clo)
I_a = insulation affect of air boundary (clo)
t = dry bulb temperature (°F)
$t - 95$ = temperature difference between dry bulb temperature and average skin temperature just before a person feels uncomfortably warm (°F)
R = mean incident radiant heat from sources other than walls and room temperature (BTU/hour ft^2)
P = vapor pressure of water (in. of mercury)

The HDR in BTU/hour ft^2 can be computed from

$$\text{HDR} = \frac{D}{H} = -M - \frac{1.13(t - 87)}{I_{cw} + I_a} + 9 - \frac{RI_a}{I_{cw} + I_a} \tag{20-34}$$

where

HDR = heat deficit ratio (BTU/hour ft^2)
D = heat deficit (BTU/ft^2)
H = exposure time (hours)
$t - 87$ = difference between dry bulb temperature and average skin temperatures just before a person feels uncomfortably cool (°F)

Air Quality. The quality of air within a subway system can have an impact on the comfort and health of the traveling patron. The degradation of the air quality can be annoying (such as odor and haze), health related with an impact on the respiratory system, or disastrous as with methane. Most of these contaminants can be contained and the potential hazard mitigated by judicious use of ventilation.

Air Velocity. The movement of the train in the tunnels generates a piston effect within the tunnels and stations. Under many circumstances this air flow is useful to maintain a reasonable environment within the tunnels. However, there are circumstances when this air velocity can be objectionable, particularly to patrons on the station platform at the tunnel/station junction. The maximum air velocity for a station platform should not be greater than 1,000 fpm (5 m/sec). This is an intermittent flow.

The other location where air velocity can be a problem is on sidewalk gratings when the vertical velocity should not exceed 500 fpm (2.5 m/sec). However, gratings located in nonpedestrian areas could have velocity up to 1,000 fpm (5 m/sec).

Pressure Changes. The train movement that creates air flow in the subway also creates transient pressure changes within the subway. These pressure changes have varying physiological effects on subway occupants.

These pressure changes are usually only felt in the patron areas. Table 20-16 shows some of the reactions of the ear to various levels of pressure changes. Criteria must be set for this phenomenon.

Emergency Criteria. An emergency in a subway system, in general, is defined as any unusual situation or occurrence

Table 20-16. Ear Sensations at Various Levels of Pressure Change

Ear Sympton	Pressure difference between the middle ear and ambiant air pressure	
	(psi)	(kPa)
Perceptible feeling of fullness in the ear	0.06–0.10	0.414–0.689
Increased feeling of fullness in the ear	0.10–0.19	0.689–1.310
Distinct feeling of fullness in the ear, lessened sound intensity	0.19–0.29	1.310–1.999
Increased discomfort with ear ringing, hissing, roaring and cracking; may be pain and mild vertigo	0.29–0.58	1.999–3.999

that halts train movement and makes it mandatory that passengers leave the train and enter the tunnel environment, or that requires the evacuation of a subway station.

The only emergency criteria that have an impact on ventilation are those related to air temperature, air velocity, and air quality.

AIR TEMPERATURE. Exposure by subway patrons and employees to extraordinarily high temperatures during an emergency can be harmful to the individual. NFPA states in its Standard 130 for Fixed Guideway Systems that the "ventilation system shall be so designed that in an emergency situation the air temperature in existing pathways shall be controlled to a design goal of no more than 140°F (60°C)" (NFPA 130, 1990). This temperature criterion of 140°F (60°C) should be at shoulder or head level of the patron, not at ceiling level.

AIR VELOCITY. Significant air flow rates may be required to control or evacuate smoke in tunnels and stations during a fire emergency situation. In some circumstances, the evacuating patron may be exposed to this high velocity. This air flow in the faces of the evacuating patrons is helpful to the patrons, giving them a sense of direction (minimum air velocity of 500 fpm) and providing more respirable air. A maximum air velocity of 2,500 fpm in an emergency situation is acceptable.

The magnitude of the air velocity in the tunnel will depend on the size of the fire and thence on the velocity required to control the backlayering of smoke. If the backlayering is controlled, there is a clear evacuation path for patrons.

AIR QUALITY. The quality of air within a subway tunnel and/or station during a fire emergency deteriorates rapidly. The combustion can produce gases and aerosols, some of which may be toxic and all of which are, at the least, irritating to the patron.

While oxygen criteria are established by the American Congress of Governmental Industrial Hygienists (ACGIH) for most of the gases in the smoke, the type and volume of these substances will depend on the material being consumed by the fire. The best solution is to provide the maximum flow rate required to control the smoke and directing fresh air into the faces of the evacuating patrons.

ANALYSIS. After defining the criteria for a subway system, an analysis of where the heat is in the system and how the air flow created by piston effect of trains and/or fans serves to dissipate this heat is required. A series of sophisticated design tools that permit this form of evaluation have been developed as part of the Subway Environment Research Program (*SEDH,* 1976).

If the heat generated by the equipment and the people in a subway is greater than the capacity of the ground or of ventilation to dissipate it, the temperature within the system will rise. All forms of electrical energy input to the subway system eventually are dissipated as heat. The trains account for approximately 85–90% of all the heat generated within the system. Heat generated within line sections by trains from their traction, braking, and air conditioning systems will be at a substantially higher rate in subway systems now under construction, or being planned for the future, than that which exists in most systems today. This is due to the higher speed and acceleration requirements for vehicles, which necessitate significantly higher power input and resultant power losses. Of these inputs, the major portion is derived from braking and accelerating. There have been and continue to be many efforts to seek ways of reducing this heat source, such as by the use of speed restrictors, wayside dynamic braking resistors, regenerative braking, motor controllers, signaling, vehicle weight, rock profiles, fly wheels, etc.

Approximately 50%–60% of the total heat input attributable to train operation can be assigned to braking—or approximately 45%–50% of the total heat input of the subway system. Since braking occurs in the vicinity of the station platform, it is the basic cause of much of the environmental problem.

The heat released into the tunnels or stations is partly transmitted into the surrounding soil and partly carried forward by piston action, or mechanical ventilation (if such is used), and eventually exhausted to the atmosphere through the shafts and access openings in the stations. The soil acts as a heat sink when the air in the trainway is at a higher temperature than the soil's temperature, and as a heat source when the reverse takes place.

The heat sink, where it is effective, is a "natural" cooling mechanism, as is the piston effect of moving trains. Cooling is accomplished by the exchange of hot inside air with cooler outside air.

A reliable estimate of the piston effect ventilation, along with the heat sink, and the impact of both on the thermal conditions within the subway, is required prior to determining what measures must be taken to meet established environmental criteria.

One of the analytical tools developed as a part of the SERP is a public domain computer model called the Subway Environment Simulation (SES). For details on this model see the section "Simulation."

ENVIRONMENTAL CONTROL. Control relates to the methods used to maintain the desired environment within the subway facility. Many types of control systems can be employed in this regard.

Control of the temperature rise in the subway system involves balancing the heat from the system against losses of heat to the surrounding heat sink and to the air flowing in the system. When the heat gain from the system is greater than the losses to the sink and to the air, the temperature in the system will rise. Such a rise in temperature will increase the rate of heat flow to the sink. Additional air flow may be required to control the temperature rise. This increase in air flow would reduce the temperature rise, but it would also reduce the cooling effect of the heat sink. Thus, under extreme conditions of maximum heat gain and maximum outside air temperature, an imbalance of heat gain and heat losses may result in an overall temperature rise in the system. Under conditions of reduced heat gain and reduced outside air temperature, normal air flow rates resulting from train piston effects may provide the desired control over temperature rises. Frequently, the air flow rate required to provide this control will also be sufficient to replenish the system air with outside air at a desirable rate. If the rise in temperature is not desirable, a system of mechanical ventilation or cooling by refrigeration must be considered.

A knowledge of the available environmental control equipment and its application is required to effect a complete solution for the subway system. The environmental engineer must consider the variety of system concepts appropriate to subway environmental control and the applicability of these systems in the optimization of subway construction and operation. These systems include, in addition to station platform air conditioning, supply and exhaust ventilation systems, trainway or under-platform exhaust systems to remove heat, tunnel line section ventilation systems, and various other possibilities and combinations. Descriptions of standard environmental control equipment and systems for general cooling applications are found in the *ASHRAE Handbook* (ASHRAE, 1992).

The major contributor to the heat problem in the subway system is the transit vehicle and its propulsion system. Much greater attention must be addressed to this equipment, since optimization of the cost of the vehicle alone may not optimize the cost of building, owning, and operating the entire rapid transit facility. Consideration should be given in the future to propulsion systems that provide for rejection of the waste heat out of the enclosed portions of the system and possible reuse of this energy.

Rapid Transit System Descriptions

The ventilation and environmental control highlights of several of the systems designed within recent years are described below.

Mount Lebanon Tunnel. The Mount Lebanon Tunnel is a 2,913-ft (888-m) twin-tube rail tunnel located south of Pittsburgh, Pennsylvania. The tunnel carries the Allegheny County Light Rail Transit System and was opened in 1987. It has a unique ventilation system using impulse fans in portal ventilation buildings. The horizontally mounted, axial-flow fans are located one above the other because of the limited distance between track centers. The 5.9-ft (1.8-m) diameter fans each have a 177-hp (132-kW) motor and a nominal capacity of 90 ft^3/min (42.5 m^3/sec). However, it was decided during commissioning to adjust the fan blades to provide the maximum air flow possible without exceeding the continuous operating duty of the fan motors. The result was that two-fan operation provided a nozzle air flow of 591,000 ft^3/min (87.0 m^3/sec), discharging at 5,906 ft/min (30 m/sec).

Each fan has a 56-ft^2, 4.9-ft (5.2-m^2, 1.5-m) long sound attenuator to reduce the tunnel sound level to about 86 dBA. The purpose of this is to avoid patron discomfort during congested operations and interference with evacuation during emergency operations.

Each fan has a 63.5-ft^2 (5.9-m^2) shut-off damper so that single-fan operation is possible. The equipment layout allows the repair and removal of one fan while the other remains available for operation.

A 31-ft^2, 59-ft-long (2.9-m^2, 18-m) smooth-walled nozzle extends from the fan room to each tunnel, intersecting it at an angle of about 11°. Each nozzle has a 51-ft^2 (4.7-m^2) damper at its inlet to allow the supply of air to either/or both tubes. The floor of the nozzle is 8.5 ft (2.6 m) above the walkway, thus allowing the high velocity air to be discharged over the heads of patrons evacuating the tunnel. During detailed design it was thought possible that the air jet discharging from the nozzle would adhere to the tunnel wall rather than mix with the tunnel air. A portion of the jet energy would then be dissipated against the rough tunnel wall rather than increasing the air pressure in the tunnel. It was therefore decided that a lip would be added at the end of the nozzle so that its air flow would be carried out beyond the tunnel wall before expanding.

As shown in Figure 20-38, the tunnel has two tunnel-to-tunnel cross-passages at its third points. The purpose of the cross-passages is to provide (1) an additional path of emergency evacuation, (2) access by fire-fighting personnel, and (3) maintenance access. The south cross-passage is also the site of a tunnel low-point pumping station. The cross-passages have an area of 96 ft^2 (8.9 m^2); however, their doors provide a restriction to 25 ft^2 (2.3 m^2). The size of this restriction must be coordinated with the design of the ventilation system so that the short-circuiting of air flow, caused by an open cross-passage door, is allowed for.

Trans-Bay Tube. The Trans-Bay Tube Tunnel in San Francisco is a 19,000-ft (5,795-m) crossing of San Francisco Bay and is a part of the BART system (McCutchen, 1974). The tunnel is ventilated during normal operation by the piston effect generated by trains that travel at speeds up to 80 mph (129 kph). During an emergency, when purge of the tunnel is required, a unique mechanical ventilation system can be operated. This system consists of four large centrifugal

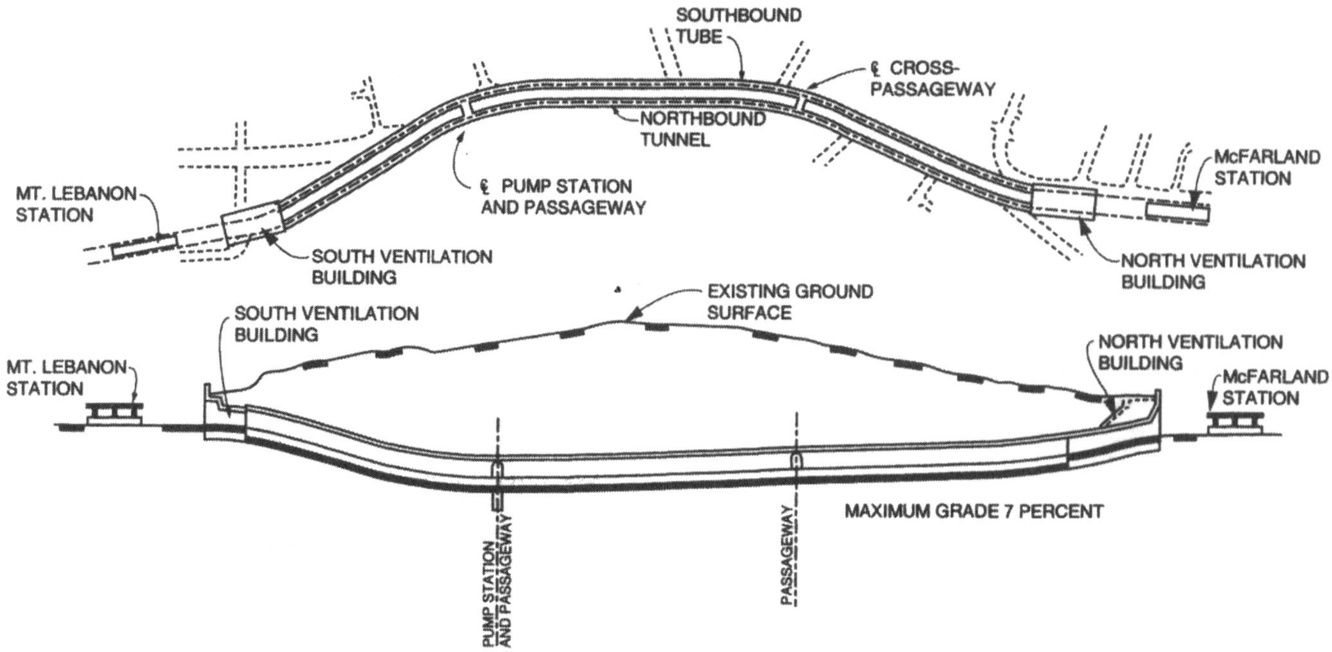

Fig. 20-38. Mount Lebanon Tunnel.

exhaust fans located in the two ventilation structures (Figure 20-39). These fans exhaust air through an exhaust duct located in the cross section, as shown in Figure 20-40.

A series of automatic dampers are set in the separating wall between the duct and the trainway. In operation, several of these dampers can be opened in either tunnel at or near the site of the heat or smoke generation. This type of operation will maintain a flow of air toward the emergency site and prevent movement of smoke to the remainder of the tunnel. This will permit safe egress of passengers and entrance of fire-fighting personnel toward the fire.

Bay Area Rapid Transit. The Bay Area Rapid Transit (BART) system, in the San Francisco area, was the first new system to be designed in the United States in a number of years. Of the total system length of 75 mi (121 km), approximately one-third (or 24 mi [39 km]) is aligned in subway. There are 20 underground stations. San Francisco has an extremely mild climate, and thus it was found that ventilation alone would be adequate to satisfy the patrons' comfort objectives in the subway station's public spaces. Mechanical ventilation at the rate of six air changes per hour will result in a maximum average ambient station air temperature of 4°F (2°C) above the maximum outdoor ambient. The piston effect ventilation is found to provide approximately three air changes per hour, thus achieving a station average temperature of 10–12°F (6–7°C) above outside ambient. In the tunnels, emergency ventilation fans were installed. These fans are reversible with approximately identical air flow generated in both directions. A maximum of 15 mph (24 kph) air velocity and a minimum of 6 mph (10 kph) air velocity in the annular space were selected as criteria. The purge time was set at a maximum of 30 min.

Fans were selected so that when operated in unison in supply or exhaust modes, these criteria will be met. The criteria established for air velocities were maximum exit velocities at surface gratings of 350 fpm (1.8 m/sec) in public sidewalks and 1,000 fpm (5.1 m/sec) in street areas other than crosswalks. Gratings were located to minimize disturbance to pedestrians.

Caracas Metro. The Metro in Caracas is a system approximately 12 mi (19 km) in length (Catia-Petare Line), which is the initial phase of a four-line system (Bendelius and Metsch, 1973).

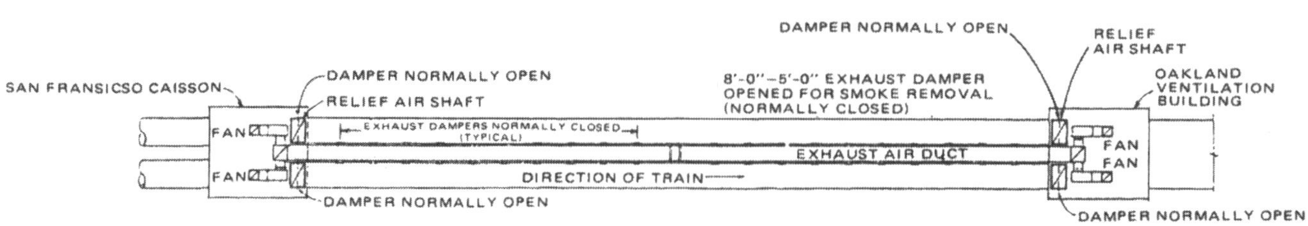

Fig. 20-39. Air flow diagram (Trans-Bay Tube).

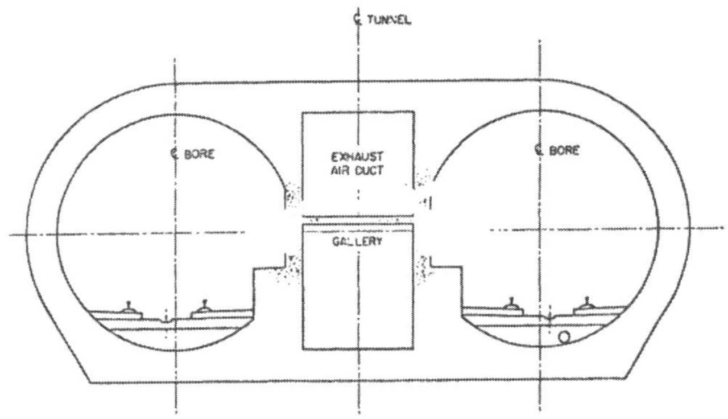

Fig. 20-40. Cross section (Trans-Bay Tube).

Approximately 90% of the Catia-Petare Line is underground, with 19 underground stations. The initial part of the system was designed based on system-wide environmental studies. It was found that the ground sink temperature is extremely high, 87°F (31°C). This meant that the use of mechanical cooling to remove the heat in the stations was necessary. The heat sink effect of the ground was negligible, and because of the limited temperature differential, the ventilation was inadequate to remove the heat.

A system of air cooling by mechanical refrigeration was developed along with under-platform exhaust and tunnel exhaust systems. This permitted consideration of the subway stations and the tunnel as separate environments with different criteria. A higher ambient temperature is to be maintained in the tunnels than in the stations. This required minimizing of the interaction of station environment with tunnel environment. This is achieved by a center exhaust fan in each tunnel segment located midway between stations, thus creating a flow of air from the stations toward the center of the tunnel. The environmental control equipment to be located in each station consists of two built-up supply air cooling systems of 80,000 cfm (38 m^3/sec) capacity each, one located at each end of the station mezzanine. These systems have a capability of 100% recirculation of air, which is used during the evening rush hour peak to take advantage of the 5°F (3°C) temperature differential between station and outdoor air temperatures. The air is distributed to the station through ducts located over the trainway (Figure 20-41). The return air is taken directly above the trainway, as shown in this figure. Chilled water is supplied to the station cooling systems from remote off-site refrigeration plants. The chillers and primary chilled water pumps are located in the plant, and secondary booster pumps are located in each station. The condenser water for these plants will be provided by cooling towers located on the roof of each plant.

Also included in each station is an under-platform exhaust system (Figure 20-41), which provides a means of removing some of the under-car heat from the train area while it is located in the station. These systems have a capacity of 70,000 cfm/trackway (33 m^3/sec/trackway) and are considered at this point to be 40% efficient.

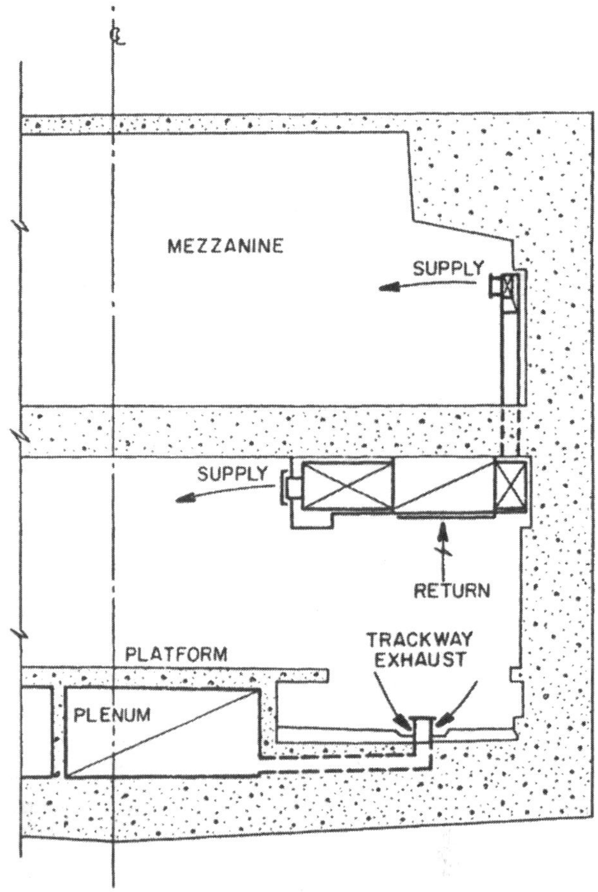

Fig. 20-41. Partial section through station air distribution system (Caracas).

Washington Metro. The metro in Washington, D.C., (WMATA) currently has 90 mi (145 km) of an ultimate 103 mi (166 km) in revenue service. There are 51 underground stations associated with 50 mi (80 km) of tunnel currently in expansion. The subway stations are air cooled, the vehicles are air conditioned, and the tunnels ventilated.

The basic tunnel ventilation system uses axial-flow fans with bypass damps in fan shafts located between stations. These are combined with vent shafts (no fans) at both ends of each station. The typical tunnel ventilation configuration is shown on Figure 20-42.

The subway stations are conditioned with cool air supplied through grilles in side platform stations, and through platform pylons in center platform stations (Figure 20-43). This is an envelope cooling concept (Solomon, 1973). In addition to the cooling system, each station is also equipped with a dome relief and under-platform ventilation system, which will further enhance station air movement. The station cooling systems are designed to maintain 85°F db when outdoor temperature is 91°F db and 74°F wb.

During a fire emergency the tunnel ventilation system is augmented (Bumanis, 1994) to create the needed flow past the incident train. As the regulations regarding transit fire life safety have become more stringent, WMATA has embarked

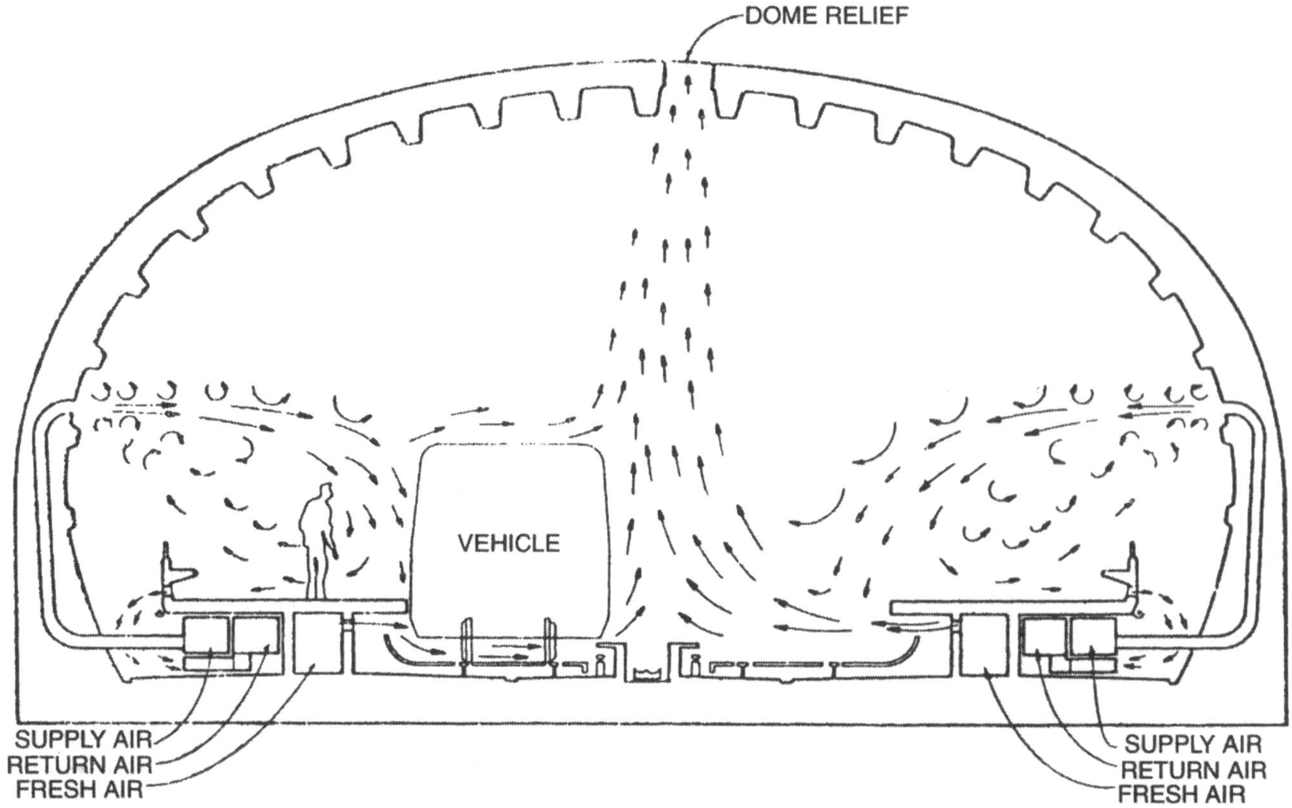

Fig. 20-42. Typical air distribution (WMATA).

on a program to evaluate the use of deployable tunnel barriers to create increased air flow losses in nonincident tunnels, thus increasing air flow past the incident trains.

MARTA. The Metropolitan Atlanta Transit Authority (MARTA) system will ultimately be 60 mi (84 km) of rail lines with 45 stations. As of mid-1994 there were 38 mi of line and 33 stations in operation. The system will include 10 mi (16 km) of subway (Bendelius, 1975). There are 12 distinct subway sections, six of which are contiguous to underground stations. The remainder are isolated, that is, not connected to stations.

The MARTA underground station environmental control system is designed to maintain average platform temperatures of 85°F (29°C) dry-bulb with 50% relative humidity during the peak evening rush period, thus providing a suitable transitional environment from the street to the transit vehicle. It was determined that a significant portion of the train heat can be kept out of the station by providing an effective tunnel heat relief exhaust system, which included the consideration of mid-tunnel exhaust fans and a 100-ft (31-m) break in the center dividing wall located at the station platform, as shown in Figure 20-43. The air pushed ahead of the train is relieved through the break in the dividing wall into the opposite trainway, thus minimizing the interaction of heated tunnel air with cool station air. The subway stations are provided with air cooled by refrigeration, supplied at the platform areas only. This air is transported by under-platform ducts to vertical risers, to horizontal headers, from which the air is distributed to the patron areas on the platforms (Figure 20-44). Mechanical equipment rooms are located at both ends of the platforms and house fans, filters, and chilled water coils. None of the supply air is recirculated, and it is permitted to exfiltrate through the station portals and serve as makeup air to the tunnel exhaust systems.

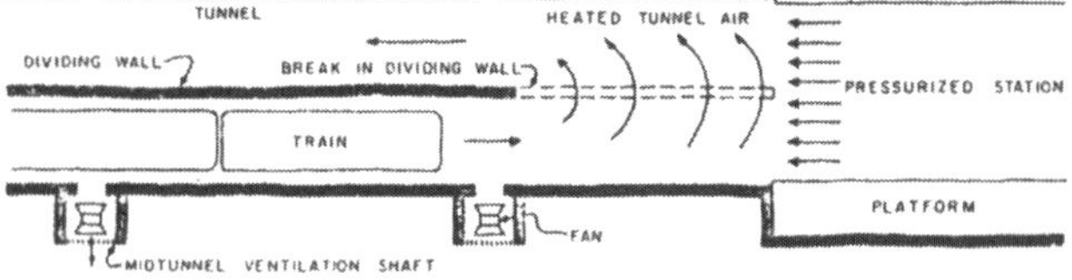

Fig. 20-43. Station/tunnel interface (MARTA).

The adjoining tunnels are provided with ventilation fans for both normal and emergency operations. During the peak traffic times of normal operation, mid-tunnel fans having a capacity of 140,000 cfm (66 m^3/sec) can be operated in each trackway near the midpoint of each tunnel to extract tunnel heat.

For emergency ventilation, specially designed, "100%" reversible fans are located in each trackway at the point where the break in the dividing wall begins (Figure 20-43). The mid-tunnel fans are also a part of the emergency venti-

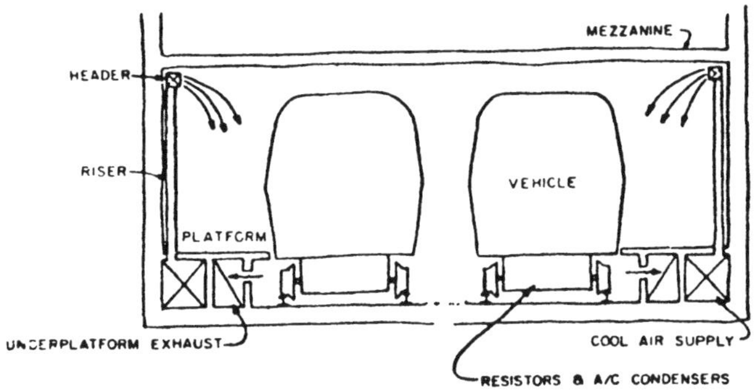

Fig. 20-44. Station distribution (MARTA).

lation system, and they are reversible so operation can be coordinated with the station-end emergency fans.

Chilled water for station air conditioning systems is generated by electrically driven chillers located in plants either local to the station or remote and serving several stations.

SIMULATION

Models

The Subway Environment Simulation (SES) computer program is a designer-oriented tool that provides estimates of air flows, temperatures, and humidity, as well as air conditioning requirements, for both existing and proposed multiple-track subway systems.

The SES model provides a dynamic simulation of the operation of multiple trains in a subway and permits continuous readings of the air velocity, temperature, and humidity throughout the stations, tunnels, and ventilation shafts. In addition, the program has been designed to provide readings of the maximum, minimum, and average values for air velocities, temperatures, and humidities during any present time interval. The program also will compute estimates of the station cooling and heating loads necessary to satisfy established environmental criteria, as well as the percentage of time that such environmental criteria are exceeded.

The SES program has the ability to model the effects of a tunnel fire. When the fire model is "turned on," the following aerodynamic and thermodynamic factors are considered by the program: A fire in a tunnel has the effect of throttling the ventilating air flow. This effect is caused by the rapid expansion of the air flowing past the fire site. Also, as a consequence of the law of conservation of mass, the velocity of the hot gases downstream of the fire increases inversely proportional to the density (or equivalently, directly proportional to the absolute temperature of the gases), hence increasing the viscous pressure losses in this section of the tunnel. These pressure changes will reduce the tunnel air flow. The density differences between the hot gases and the ambient air give rise to pressure differentials, which can either augment or retard the tunnel air flows, depending on the direction of ventilation (uphill or downhill). The elevated air temperatures produced by a fire cause the tunnel walls to heat up. The transient heating of the wall surface is an important factor in determining the conditions downwind of the fire. Allowing the wall surface temperature to respond properly improves the accuracy of the predicted air temperatures, which are subsequently used to calculate the buoyant pressure differential.

The model treats the wall as a one-dimensional slab of infinite thickness with uniform thermal properties and an arbitrary time-dependent heat flux at the wall surface. This approach is appropriate because temperature changes resulting from heating at the wall surface will be confined to within a short distance of the wall surface, and the wall surface temperature is of interest rather than the temperature at some depth below the surface.

The heat conduction equation is solved by using an approximate integral method. This method was chosen because it requires relatively little computation time and provides good accuracy (results range within 3 to 9% of the theoretical value). Heat is transferred to the wall by convection and radiation. Radiation will be the dominant mode of heat transfer at the fire site, while downwind of the fire, both modes will be nearly the same order of magnitude. At the site of the fire, heat is radiated uniformly from the interior of the burning vehicle through the windows and opened doors directly to the tunnel wall. The interior temperature of the vehicle is assumed to be at an "effective fire temperature." Both the effective fire temperature and the total area of the openings are input items. Downwind of the fire site, the hot smoke is assumed to be radiating to the tunnel wall as a "black body" at a temperature equivalent to the "bulk" subsegment air temperature. Only radiation effects in the transverse direction from smoke to tunnel wall are considered.

The changes in air density associated with elevated temperatures degrade the performance characteristics (pressure vs. volume flow curve) of the exhaust fans. These effects have been accounted for in the model.

Application of Models. The fire model is intended for use in a trial-and-error fashion to select the emergency ventilation system capacities. The interactions are between the tunnel air velocity (past the fire site) predicted by the SES fire model and a design air velocity criterion that precludes the backing of smoke against the ventilating air stream (backlayering). This "critical" air velocity criterion is a function of the fire heat release rate, the tunnel width, the average tunnel grade, and the temperature of the hot gases leaving the fire. A typical application of the fire model consists of the following steps:

- Perform an SES simulation to predict the tunnel air velocity and the hot air temperature.
- Determine the critical air velocity using the methodology given in Chapter 16 of the SES Users Manual.
- If the predicted air velocity exceeds the critical air velocity, the ventilation system is considered adequate.
- If the predicted air velocity air velocity is less than the critical air velocity, change the system and repeat the process.

The SES is essentially a one-dimensional, incompressible, turbulent, slug-flow model. The throttling and buoyancy effects, which are primarily caused by changes in density, are conveniently accounted for by noting that changes in density are inversely proportional to changes in the absolute temperature of the gas (air), a quantity that is computed by the program. Therefore, the effects of density changes have been accounted for in the computations without actually converting computations in the program from an incompressible to a compressible flow model. As a result, the air flow quantities printed out by the program are "referenced" to the ambient air density. This notion of basing the computations on a reference air density has been used in mining ventilation computer programs, most recently in a program prepared at Michigan Technology University for the U.S. Bureau of Mines.

The SES fire model has been designed with the ability to simulate the "overall" effects of a tunnel fire on the ventilation system. This level of detail is considered sufficient for evaluating the adequacy of an emergency ventilation system and is consistent with the state-of-the-art in mining ventilation programs with the capability of simulating fires. However, the model does have its limitations. As previously mentioned, the SES is a one-dimensional model. Therefore, the results of a fire simulation will indicate whether or not the ventilation air flows are sufficient to prevent backlayering, but not the extent of backlayering (a two-dimensional phenomenon) if it is predicted to occur. In addition, the early stages of a fire, before the ventilation system is activated, generally cannot be simulated since this period is dominated by buoyant recirculating two-dimensional air flows.

TEST PROGRAM

The issue of safety takes on many faces in a highway tunnel. One of the more critical is that of exposure to smoke—smoke generated by a fire incident within the tunnel. While fires have occurred in tunnels for decades, it has only been recently that the tunnel industry has begun to focus attention on the adequate control of smoke and hot gas emanating from a fire within the tunnel carriageway. Why do we want to control smoke in a highway tunnel? The primary purpose is to protect life—to permit safe evacuation of the tunnel. This involves creating a safe evacuation path for both motorists and any operating personnel located within the tunnel. The secondary purpose of smoke control ventilation is to assist fire-fighting personnel in accessing the fire site by again providing a clear path to the site if possible. The tunnel ventilation system in not designed to protect property; however, the effect of the ventilation in diluting the smoke and thus recovering some of the heat may reduce damage to the facilities and vehicles. The continued reduction of vehicle emissions has shifted the focus of the ventilation engineer from design based on dilution of emission contaminants to a design based on the control of smoke in a fire emergency. Despite the increasing focus on life safety and fire control in modern highway tunnels, no uniform standards for fire emergency ventilation or other fire control means within highway tunnels have been established in the United States.

The concepts that have been applied to these facilities were based on theoretical and empirical values, not on the results of full-scale tests. Accordingly, the design approach currently used to detect, control, and suppress fire smoke within highway tunnels has become a controversial issue among tunnel design engineers, owners, operators, and firefighters throughout many parts of the world. While most highway tunnels have ventilation systems with smoke control operating modes, there is limited scientific data to support opinions or code requirements regarding the capabilities of various types of ventilation systems to effectively control heat and smoke.

The Test Plan

One of the largest underground highway projects ever undertaken in the United States, the Central Artery/Tunnel (CA/T) project in Boston, Massachusetts, is estimated to cost more than $6 billion (U.S.). It is jointly funded by the Federal Highway Administration and by the Commonwealth of Massachusetts. As part of the CA/T project, engineering investigations of ventilation operating strategies and performance in full-scale fire situations were authorized to be performed in the Memorial Tunnel, a disused highway tunnel made available for these purposes by the State of West Virginia. The American Society of Heating, Refrigerating and Air-Conditioning Engineers Technical Committee 5.9 (ASHRAE TC 5.9), "Enclosed Vehicular Facilities," developed a Phase I Concept Report and work scope. This report outlined the objectives of the testing program, which included identification of appropriate means to account for the effects of fire size, tunnel grade, and cross section, direction of traffic flow (unidirectional of bidirectional), altitude, type of ventilation system, and any other parameters that may have a significant influence on determining the ventilation capacity and operational procedures needed for safety in a fire situation. Establishing specific approaches to permit effective reconfiguration for both new and existing tunnel facilities was deemed of equal importance.

The purpose of the Memorial Tunnel Fire Ventilation Test Program is to develop a database that will provide tunnel design engineers and operators with an experimentally proven means to determine the ventilation rate and system configuration that will provide effective smoke control during a tunnel fire. Of even greater importance is to establish specific operational strategies to permit effective reconfiguration of ventilation parameters for existing tunnel facilities. While the life safety issue is paramount, it should be recognized that significant cost differentials exist among the various types of ventilation systems. In the instance where more than one ventilation configuration offers an acceptable level of fire safety, the overall project life-cycle cost needs to be addressed to identify the option with the optimum cost benefit. In addition, the impact of ventilation systems that cause horizontal roadway-level air flow on the effectiveness of fire suppression systems (such as foam deluge sprinklers) can be better determined on the basis of full-scale test results.

Test Fire Heat Release. Fires with heat releases rates ranging from 20 MW, equivalent to a bus or truck fire, to 50 MW, equivalent to a flammable spill of approximately 105 gallons (400 L), to 100 MW, equivalent to a hazardous material fire or flammable spill of approximately 210 gallons (800 L), were produced. The fires have been generated in four floor-level steel pans in which a metered flow of No. 2 fuel oil up to 2 in. (5 cm) deep will be floated on top of a 6-in. (15-cm) layer of water.

Ventilation Systems. The ventilation systems to be configured and evaluated under varying flow rates and varying heat release rates, with one or two zones of ventilation, include

- Transverse ventilation
- Partial transverse ventilation
- Transverse ventilation with point extraction
- Transverse ventilation with oversized exhaust ports
- Natural ventilation
- Longitudinal ventilation with jet fans

After the first four series of tests were completed, the tunnel ceiling was removed to allow for the natural ventilation tests, followed by the installation of jet fans at the crown of the tunnel for the longitudinal ventilation tests. A fire suppression system is included, which can be used to suppress the fire in an emergency; however, it was also used during several tests to evaluate the effect of ventilation air flow on operation of a foam suppression system.

The Test Facility

The Memorial Tunnel is a two-lane, 2,800-ft (854-m) highway tunnel originally built in 1953 as part of the West Virginia Turnpike (I-77). The tunnel is located near Charleston, West Virginia, with a 3.2% upgrade from the south to the north tunnel portal. The original ventilation system was a transverse type, consisting of a supply fan chamber at the south portal and an exhaust fan chamber at the north portal. The tunnel has been out of service since it was bypassed by an open-cut section of a new six-lane interstate highway in 1987. The existing ventilation equipment was removed to allow installation of new variable-speed, reversible, axial-flow ventilation fans. The equipment rooms were modified to accept the required ventilation components needed to permit supply or exhaust operation from both ends of the tunnel. There are six fans, three each in the modified north and south portal fan rooms. Each of the fans has capacity to supply or exhaust 190,000 ft^3/min (94 m^3/sec). The existing overhead air duct, formed by a concrete ceiling above the roadway, is split into longitudinal sections, which can serve as either supply or exhaust ducts, and a mid-tunnel duct bulkhead has been installed to allow a two-zone ventilation operation. Openings in the duct dividing wall and duct bulkhead have been designed to obtain air flow patterns similar to that which would be observed if the dividing wall was not present. The widths of the ducts vary linearly along the length of the tunnel, to provide maximum areas at the point of connection to the fan rooms above the tunnel portals. High-temperature insulation was extensively applied to various structural elements, including the concrete ceiling and ceiling hangers, as well as all the utilities, instrumentation support systems, wiring, gas sampling lines, closed-circuit TV (CCTV) camera cabinets, and all other related items that are exposed to high tunnel fire temperatures.

Instrumentation

Data Collection. Specific instrumentation is provided to monitor and record the air temerature, air velocity, and gas concentration during the fire tests.

All of the measured values are provided as inputs to a data acquisition system (DAS), which monitors and records data from all field instruments for on-line and historical use. The DAS consists of five data acquisition units (DAUs), three located in the tunnel and two in the portal electrical equipment rooms. The DAS central processing units (CPUs), operator consoles, dataloggers, printer, and tape drives are located in the control trailer. The measurement of tunnel air temperature is accomplished through the use of thermocouples located at various cross sections throughout the length of the tunnel.

Instrument Trees. There are instrument trees located at 10 tunnel cross sections, which have been designed to measure air flow to a modified ASHRAE traverse method. At these locations, thermocouples are located at each air velocity sensor, and these measure air temperature from 32 to 2,500°F (0 to 1,370°C). Additional temperature measurements are taken at five other tunnel cross sections, and at two locations 15 m outside of the tunnel portals. The measurement of air velocity in the tunnel under test conditions is accomplished through the use of differential pressure instrumentation designed to measure very low pressure ranges from 0 to 61 Pa. Temperatures in the vicinity of the bidirectional pitot tubes and the ambient pressure are combined with the measured pressure to calculate the air velocity. A gas-sampling system draws sample gas from specific tunnel locations to analysis cabinets located in the electrical equipment rooms. Sample gases are analyzed within the analysis cabinets for two ranges of CO, CO_2, and total hydrocarbon context (THC). The analyzers are housed in climate-controlled cabinets. Methane gas can be detected at the test fire location through the use of individual in situ electrochemical cell detectors, located behind the tunnel wall. CO is monitored with electrochemical cell–type analyzers at the control trailer for personnel safety. In addition, portable detectors, capable of detecting carbon monoxide, total hydrocarbon, oxygen, and methane are provided for personnel safety when entering the tunnel after fire tests. Two meteorological towers located outside of the north and south tunnel portals include instrumentation to monitor and record ambient dry and wet bulb air temperatures, barometric pressure, and wind speed and direction.

Cameras. A CCTV system consists of six cameras, two located in close proximity to the fire area, two located outside of the tunnel (near the portals), and two located on the north and south meteorological towers. Instrumentation to monitor and record important parameters of the fire suppression system, the chilled water system used for equipment cooling, and the compressed-air and fuel oil systems are also included.

Conclusions

The Memorial Tunnel Fire Ventilation Test Program represents a unique opportunity to evaluate and develop design methods and operational strategies leading to safe underground transportation facilities. This comprehensive test program, which commenced in September 1993, will result in much-needed data acquired in a full-size facility under controlled conditions over a wide range of system parameters. The results of the program will be processed and made available to the professional community for the use in the development of tunnel ventilation design and emergency operational procedures. (See page 438a for additional conclusions.)

EQUIPMENT AND FACILITIES

Once a determination is made that mechanical ventilation is required and the air flow rate has been computed, the equipment needed to provide the air flow—fans, motors, dampers, and sound attenuators—and the facilities to house them must be selected and designed.

Fans

A fan is a rotary, bladed machine that maintains a continuous air flow created by aerodynamic action. A fan has a rotating impeller carrying a set of blades that exerts a force on the air, thereby maintaining the air flow and increasing the pressure. A fan is a constant-volume device: it delivers the same air volume regardless of the air density.

Two basic types of fans, axial and centrifugal, are used predominantly in tunnel ventilation systems. The type used is determined by the required air flow and pressure and the available space and its configuration.

Axial-Flow Fan. The flow of air through this fan is virtually parallel to the impeller shaft (Figure 20-45). The radial component of velocity is nearly zero. The axial fan impeller with airfoil blades rotates in a cylindrical housing. There are two types of axial fans, tube axial and vane axial. The tube axial fan is usually used in systems requiring de-

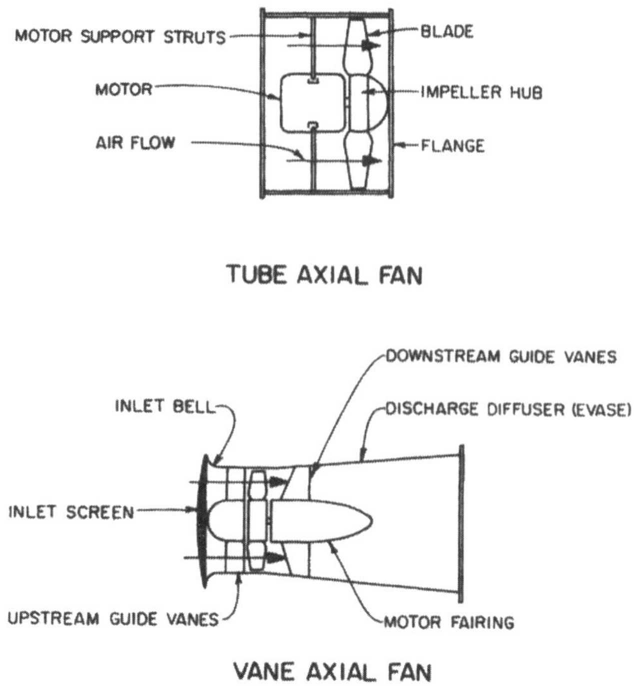

Fig. 20-45. Axial-flow fans.

velopment of pressures up to 2-1/2 in. water gauge (0.6 kPa). The vane axial fan is a tube axial fan with guide vanes on one or both sides of the impeller to correct the rotary motion imparted by the impeller. These vanes improve the pressure characteristics and operating efficiency, thus making it possible to use the vane axial fan for applications requiring up to 10 in. water gauge (2.49 kPa) of pressure, or higher for special design fans.

Axial fans exhibit performance characteristics, as represented by the curves shown in Figure 20-46. The characteristic dip in the axial pressure curve at 40 to 50% of wide open volume is caused by aerodynamic stall, during which the blades cease to function normally, although air flow continues. The maximum efficiency occurs nearer to free delivery than for the centrifugal fan. The optimum operating range is from 60 to 80% of wide open air volume. The shape

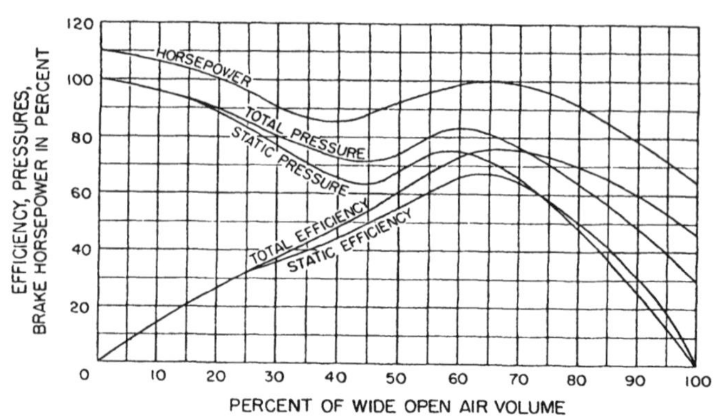

Fig. 20-46. Axial fan performance characteristics.

of the horsepower curve is significant, since it has a nonoverloading characteristic within the normal selection range. As can be seen in Figure 20-46, the horsepower characteristic falls off as flow increases above the maximum efficiency point. The total efficiencies of axial-flow fans can range between 70 and 80%. The level of sound generated by an axial fan is lowest near the maximum efficiency point.

The pressure developed by an axial fan is influenced by the hub diameter, tip clearance, and staging. These characteristics are established by the manufacturer during the design phase. A small hub-to-housing-diameter ratio results in low pressure, whereas a large hub-to-housing-diameter ratio results in high pressure. The maximum pressure is obtained with minimum tip clearance, which is usually between 0.1 and 0.2% of impeller diameter. To obtain high-pressure characteristics, strict manufacturing tolerance must be adhered to, to maintain minimum tip clearance. Staging involves the installation of two or more impellers in the same fan housing. This has the effect of fans operating in series, that is, doubling the developed pressure at the same flow volume. While both tube and vane axial fans can be staged, the tube axial must use contra-rotating impellers to eliminate air rotation and increase efficiency. Contra-rotating tube axial fans in series could produce 2-1/2 times the pressure developed by a single stage axial fan at the same flow rate (Daley, 1978, p. 116).

The volume delivered by a fan at a fixed speed can be varied by changing the blade pitch angle. Axial fans can be supplied with factory set, adjustable, or controllable blade pitch. The fan with fixed pitch is limited to one flow rate unless speed is changed, whereas the air volume delivered by the adjustable and controllable type can be varied by changing the blade pitch. The adjustable type requires that the fan be stopped to permit the blade pitch adjustment to be made. The ability to adjust the blade pitch while the fan is in operation is the feature of the controllable type. There are only a few rare cases where the controllable pitch feature can be justified in vehicular tunnel design; therefore, careful consideration should be given before controllable pitch fans are used for tunnel ventilation, since the controllable pitch mechanism is an added maintenance item. The necessary diversity of air flow in a tunnel can be achieved by the use of multiple two-speed fans, making the use of controllable pitch fans dubious. The cost of controllable pitch can only be justified on fans with large-horsepower motors.

The effect of blade pitch change can be noted in Figure 20-47, where at low blade angles the pressure increase is smooth with a drop in volume. However, at the higher blade angles, a point is reached where the fan can no longer develop the pressure required to deliver the air volume, as demonstrated by the accentuated stall points.

Inlet bells shown on the vane axial fan in Figure 20-45 should be added to an axial fan with a free air intake to reduce the inlet losses and obtain efficient operation. A discharge diffuser is added to an axial fan to conserve additional energy.

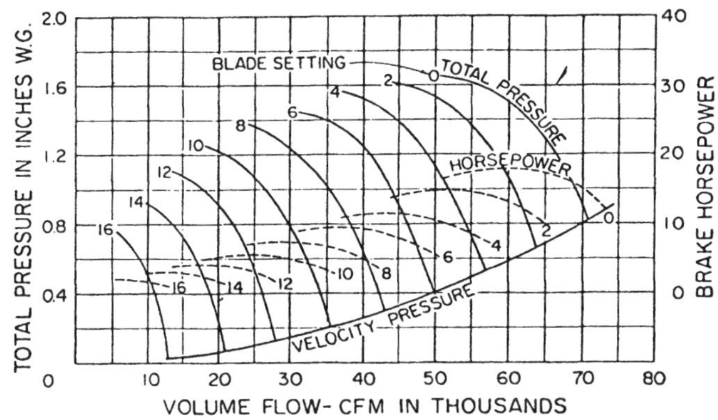

Fig. 20-47. Effect of blade pitch angle on axial fan performance.

Axial fans may be driven by any number of methods such as belt, chain, direct connected, or floating shaft (see "Fan Drives"). The most reliable method is the direct connected with the impeller mounted directly on the motor shaft. In this case, the motor bearings are the fan bearings; therefore, there are no separate bearings or drive mechanisms to maintain.

Propeller Fans. A simple special form of an axial-flow fan designed to operate without a housing is the propeller fan. The propeller fan has had a limited use in tunnels due to its severe pressure limitation. The maximum pressure for most propeller fans is from 1/2 to 1 in. water gauge (0.13 to 0.25 kPa). The singular advantage of propeller fans is the large volume of air they can handle at low operating costs along with low capital costs.

Propeller fans can be either direct-driven or belt-driven; however, if a direct drive is used, there is no way to adjust volume, unless the blade pitch angle is adjustable.

The horsepower of a propeller fan increases with increasing pressure. This can cause motor burnouts if the motor is not generously sized.

The air enters the propeller fan from all directions and leaves axially. As the pressure increases, recirculation of air occurs, thus reducing the net air flow. Therefore, steps must be taken to prevent this loss of air flow. In a free-flow condition, the discharge edge of the blades are mounted flush with the frame. When the propeller fan is installed in ductwork, the discharge edge of the blades is extended beyond the frame, thus permitting centrifugal discharge of air and reduction of recirculation. This will enable the fan to develop maximum pressure. The duct must be at least 25% larger than the impeller diameter to achieve this maximum.

Centrifugal Fans. These consist of a wheel rotating within a scroll-shaped housing or casing, as shown in Figure 20-48. The wheel has a series of blades on the periphery, and the casing has an inlet on the wheel axis with a discharge outlet at 90° to the inlet. The centrifugal force created by rotating the air trapped between the blades and the kinetic energy imparted to the air by its velocity when it leaves the wheel are the major producers of pressure in a centrifugal

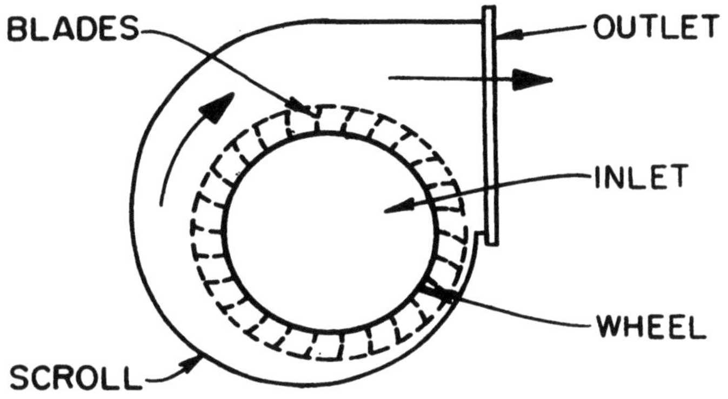

Fig. 20-48. Centrifugal fan.

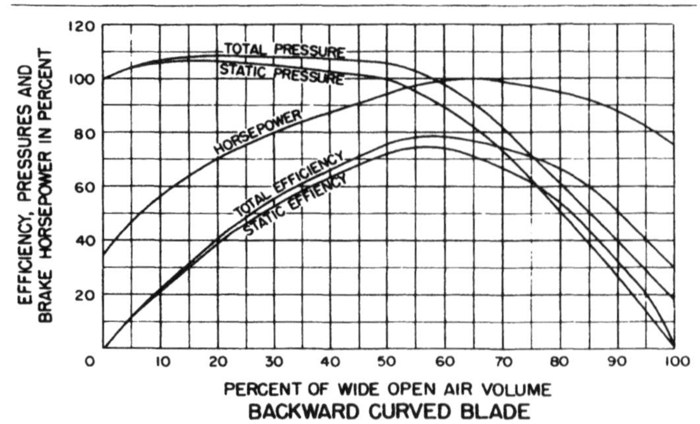

Fig. 20-49. Typical centrifugal fan performance characteristics.

fan. The scroll-shaped casing converts high-velocity pressure at the blade tip to static pressure.

The air enters the centrifugal fan parallel to the wheel shaft and is discharged at a 90° angle, tangential to the wheel. This turn reduces somewhat the efficiency of centrifugal fans due to the shock losses created. The efficiency of centrifugal fans will range from 45 to 85%.

Centrifugal fans can be obtained with either a single or double inlet. The single-inlet fan, called single-width, single-inlet (SWSI), has a single-width wheel with one casing inlet on the side of the fan opposite from the drive. The double-width, double-inlet fan (DWDI) has a double-width wheel with casing inlets on both sides. The DWDI fan will deliver twice the volume of air that a SWSI fan of the same wheel diameter and rotational speed can deliver.

Centrifugal fans are available with four basic blade configurations: radial, forward-curved, backward-curved, and airfoil. Only two are suitable for tunnel ventilation applications: backward-curved (BC) and airfoil-bladed.

The backward-curved (BC) blade provides the family of centrifugal fans with its highest efficiencies. By curving the blades backward, the air flow through the blades is improved by the reduction of shock and eddy losses. The BC blade develops more pressure by centrifugal force and less by velocity conversion. Higher tip speeds are used in BC fans, and so the wheels must be of sturdier construction than other types. The maximum horsepower occurs within the normal operating range, as seen in Figure 20-49.

The airfoil-bladed centrifugal fan is a refinement of the backward-curved type. The airfoil shape of the blade produces an increase in efficiency and a decrease in generated sound.

The backward-curved and airfoil-bladed fans are of greatest interest to the tunnel ventilation engineer because their high efficiency provides the least space requirements, along with the smallest horsepower requirements. These factors are extremely important when the size of tunnel ventilation systems is considered.

Fan Characteristics. Flow reversibility is frequently required in a tunnel ventilation system. An axial-flow fan or a propeller fan can be reversed electrically by reversing the rotation of the motor. Electrically reversing a standard vane axial fan will result in anywhere from 40 to 60% flow in reverse direction. There are axial fans available that approach 100% reversibility. These are being used on rapid transit systems as emergency ventilation fans. Some amount of efficiency is sacrificed in both flow directions to obtain this degree of reversibility. Care should be exercised in using a fan of this type for continuous operation because of the low efficiency.

The direction of air flow in a centrifugal fan cannot be reversed by reverse rotation. To use a centrifugal fan in a reversing situation, a duct runaround system must be constructed. This requires a number of additional dampers along with additional ductwork, which can be both space consuming and costly.

Parallel operation of two or more fans is possible and must be evaluated. Two identical fans operating in parallel will deliver twice the volume as one fan at the same pressure. Figure 20-50 illustrates the effect of parallel operation of three fans on the performance curve. When the fan curve has a more complicated shape, such as a dip, the relationships are not as simple. The dip creates an extremely complex combined curve. Care must be taken to select the fans so that none of the possible system resistance curves lie on the reclining figure eight. System curve E would not provide a stable operation, whereas D would. As seen on Figure 20-50, two fans will not deliver twice as much air as one fan alone when both are operating in the same system.

Two or more fans can be operated in series with the outlet of one fan connected to the inlet of the next fan. This arrangement for two fans will produce the same volume as one fan at an appropriate multiple of the total pressure. For an evaluation of a series operation, only the total pressures can be used, since the static pressures are not additive.

Volume Control. The volume of air delivered by a fan can be controlled in several ways, both mechanically and electrically. In tunnel ventilation applications, the primary concern is reduction of air flow and, in turn, reduction of horsepower during periods of light traffic to reduce energy costs.

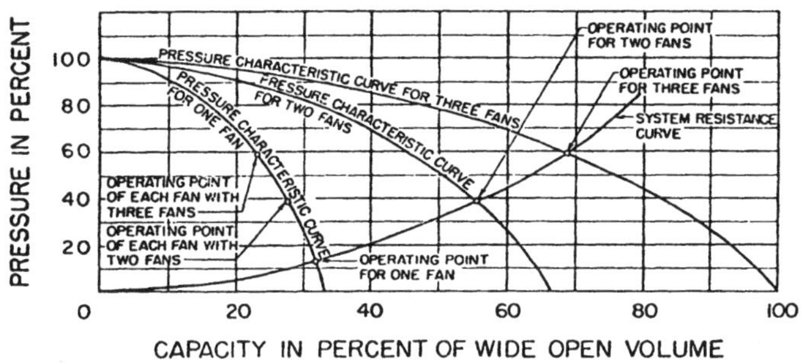

Fig. 20-50. Parallel fan operation.

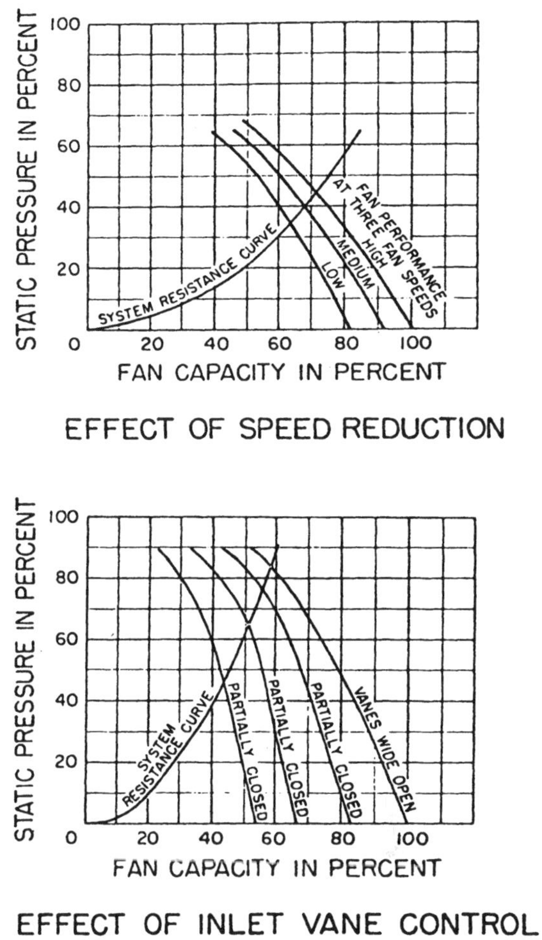

Fig. 20-51. Effect of volume control on fan performance.

Throttling dampers in the duct system at the fan discharge is the simplest, yet the most costly, method. Although this method does reduce the flow, it accomplishes this reduction by increasing the system resistance, but not necessarily reducing the horsepower; in fact, this may increase. This method also poses a problem if the pressure increase places the fan near the operational stall point of the fan. Throttling dampers should not be used in a tunnel ventilation system.

The bypass method uses relief of air to atmosphere, thus reducing the flow into the system, but the fan continues to handle the same volume. Some small pressure reduction is usually achieved due to the reduced flow in the system; however, any power reduction will be extremely small; therefore, the bypass method is not appropriate for tunnel ventilation systems.

Speed regulation is the ideal method of fan volume control. This can be accomplished by some form of variable-speed coupling, by gear box, by drive, or by electrical means, by a multiple-speed motor. The effect of speed reduction on fan performance is shown in Figure 20-51. This method is ideal, since as the flow is reduced, the horsepower is also reduced, thus cutting power consumption.

Inlet vane dampers can be used to control the air volume delivered by a centrifugal fan. As the vanes turn, the relative direction of the air flow entering the fan wheel is altered, thus varying the amount of work done by the blades. The effect of this is to modify the fan performance curve, as shown in Figure 20-51, instead of throttling the air flow. This results in a power reduction.

Blade pitch control can be used to change the volume of air passing through an axial fan. Such a system has the effect of changing the fan performance curves, as shown in Figure 20-51. A lower power consumption is achieved with this method, although the fan may actually be operating at a point of lower efficiency. A pitch control system of this type will permit pitch change while the fan is in operation. This approach has also been applied to multistage axial fans (see the section "Axial-Flow Fan" for limitations).

Air volume in a multistage axial fan can be controlled by proper operation of the stages. By shutting down one or more stages, a variety of flow rates can be achieved.

Fan Sound. The sound generated by a fan is a function of fan design, air flow, pressure, tip speed, and fan efficiency. The sound generated within a fan by air turbulence is transmitted partly through the fan outlet and partly through the fan inlet. Bearing, drives, and vibration harmonics are also sources of fan sound.

The only valid basis for evaluation and comparison of fan sound is the actual sound power levels generated by the fan. These data should be obtained from the fan manufacturer and, if possible, based on actual sound tests. The fan outlet velocity has no bearing on the fan-generated sound; thus, any selection based on outlet velocity alone is meaningless. To minimize the sound generated by a fan, the following items should be considered:

- Select the fan near its point of peak efficiency.
- Design the air system for smooth flow and low resistance.
- Provide good fan inlet conditions.

To properly evaluate the effect of fan selection on fan sound, it is necessary to consider the sound generated across the entire sound spectrum. Figure 20-52 shows the relative differences in sound generated by axial fans and centrifugal fans across the entire sound spectrum. The vane axial fan generates very little sound at low frequency but has a significant sound peak at the blade passing frequency (blade passing

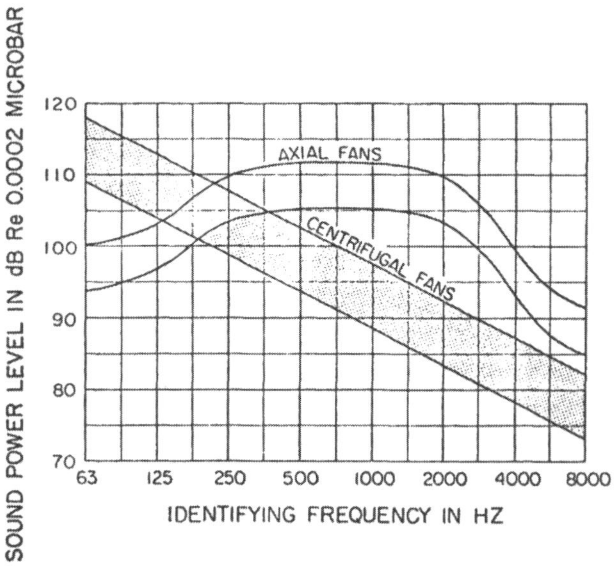

Fig. 20-52. Fan-generated sound.

frequency equals number of blades × rpm) and secondary peaks at multiples of this frequency. The centrifugal fan, on the other hand, exhibits high sound characteristics in the low-frequency ranges with less significant peaks at the blade passing frequencies.

Vibration. Vibration is induced in fans by unbalanced centrifugal forces and by aerodynamic forces. If a fan is rigidly mounted to the structure, the full force of this vibration is transmitted to the structure. When considering rigid mounting for a fan, the fan must be accurately balanced to minimize the unbalanced forces. To reduce the transmission of vibration to the structure, some form of antivibration mounting must be considered. These devices do not suppress the vibrations at the fan; in fact, they may permit an increase of vibration. They do, however, reduce transmission of these vibrations to the base structure. Transmissibility is the ratio of the force transmitted to the structure to the force generated by the vibration. The amplitude of any vibrations can be reduced by adding mass such as an inertia block under the fan.

The most commonly used materials for vibration isolation of fans are steel and rubber. Steel springs are used for all fans that have rotational speeds below 700 rpm, and they may be used for fans of all speeds. Rubber in shear isolators are used for fans with rotational speeds above 700 rpm. Steel springs exhibit nearly perfect elasticity and therefore require some form of restraint in a fan isolation application. The rubber used in vibration isolation devices is likely to decompose at temperatures above 150°F (66°C) or in the presence of grease and oil.

The amount of deflection of the isolator must be chosen carefully to give proper natural frequency. This depends on the disturbing frequency and allowable transmissibility.

Fan Selection. A fan can be selected only after the total air flow and pressure requirements have been determined. If at all possible, to reduce costs and possible manufacturing delays, standard manufactured fans should be selected for a tunnel ventilation system in lieu of those requiring special designs. Often several fans will meet the system performance requirements; therefore, capital cost, operating costs, and maintenance costs are the determining factors. Several factors must be considered before proceeding with the fan evaluation:

- Air density
- Possible combinations of fans to be operated
- Inlet and outlet conditions
- Type of drive
- Allowable sound levels

A system resistance curve can be plotted for each ventilation duct system. The majority of tunnel ventilation systems will be in the completely turbulent range, thus exhibiting a $P = CQ^2$ relationship. This results in a curve similar to that shown in Figure 20-53. When a fan performance curve is superimposed on such a system resistance curve, the point of operation can be determined readily, as noted on Figure 20-53. These curves form the best method of selecting a fan for a tunnel ventilation system, since the full fan curve is visible and the possibility of unstable operation can be detected and avoided. Although the use of total pressure in the fan selection process is the correct method, occasionally the fan diameter is not known and the static pressure is used instead of the total pressure. This approach is acceptable provided the curves and relationships are properly identified.

Some fan manufacturers present their fan performance data in what are called "multirating tables." These tables are relatively simple to use; however, they do not provide the overview of a fan's performance characteristics as does a performance curve. Therefore, before the final fan selection is made, a performance curve should be obtained for each

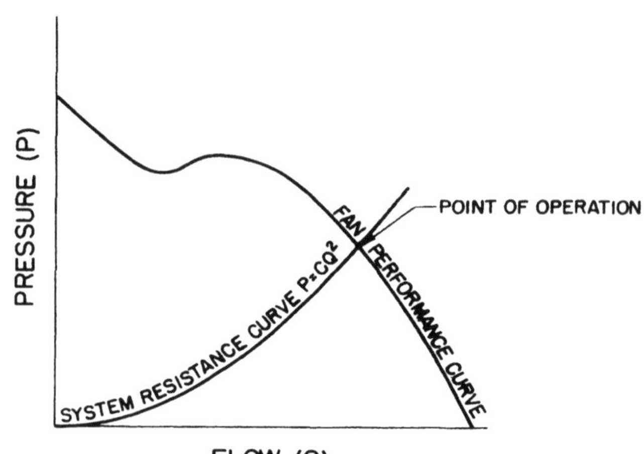

Fig. 20-53. Typical system resistance curve.

tunnel ventilation fan at each of its operating speeds and conditions.

In most tunnels, there will be more than one fan operating on a system, usually in parallel. All combinations of fans operating in parallel at all speeds must be evaluated to assure that all design operating points are well within the stable operating range of the fan (Figure 20-50 illustrates a set of fan performance curves prepared for three fans in parallel). The major problem with this type of fan operation is the possible instability created when the fan curve displays a noticeable dip. The fans must be selected near the maximum efficiency point and must remain within the stable portions of the curve for all points of operation. If there is a question regarding the accuracy of the fan duty calculations, a margin should be allowed for additional pressure rather than for additional air volumes.

Fan Laws. Fans exhibit certain properties, which are defined by the fan laws. These laws relate the performance variables for any homologous series of fans. The performance variables include fan size, rotational speed, air density, capacity, pressure, horsepower, and efficiency. To properly use the fan laws, the dimensions of the fans considered must be proportional to the dimensions of the rated fan.

Air Density. Fan performance curves and rating tables developed in the United States are based on standard air, which has a temperature of 70°F, a barometric pressure of 29.29 in. of mercury, and a density of 0.075 lb/ft^3. In SI units, these are 20°C, 101.325 kPa, and 1.204 kg/m^3, respectively.

When the density of the inlet air to a fan changes, there is a proportional change in power and pressure developed by the fan. The fan, being a constant-volume device, delivers the same total air volume regardless of the density. From the fan laws (Jorgensen, 1982, pp. 12-3–12-6):

- Pressure developed varies directly with density.
- Power consumed varies directly with density.

To correct the pressure and power for density variation, the values computed for standard air conditions must be multiplied by the ratio of the densities as follows:

$$\text{PRES} = \text{PRES}_o \times \frac{\text{DEN}}{\text{DEN}_o} \qquad (20\text{-}35)$$

where

PRES = pressure at modified density (in. of water gauge)
PRES_o = pressure at original density (in. of water gauge)
DEN = modified density (lb/ft^3)
DEN_o = original density (lb/ft^3)

$$\text{BHP} = \text{BHP}_o \times \frac{\text{DEN}}{\text{DEN}_o} \qquad (20\text{-}36)$$

where

BHP = brake horsepower at modified density
BHP_o = brake horsepower at original density

The density ratio can be computed as follows:

$$\frac{\text{DEN}}{\text{DEN}_o} = \frac{460 + \text{TEM}_o}{460 + \text{TEM}} \times \frac{\text{BAR}}{\text{BAR}_o} \qquad (20\text{-}37)$$

where

TEM_o = original air temperature (°F)
TEM = modified air temperature (°F)
BAR_o = original air barometric pressure (in. of water)
BAR = modified air barometric pressure (in. of water)

For U.S. standard conditions,

$$\frac{\text{DEN}}{\text{DEN}_o} = \frac{530}{460 + \text{TEM}} \times \frac{\text{BAR}}{29.92} = \frac{530}{29.92} \times \frac{\text{BAR}}{460 + \text{TEM}} \qquad (20\text{-}38a)$$

For SI standard conditions,

$$\frac{\text{DEN}}{\text{DEN}_o} = \frac{273.15 + 20}{273.15 + \text{TEM}} \times \frac{\text{BAR}}{101} = \frac{293.15}{10} \times \frac{\text{BAR}}{273.15 + \text{TEM}} \qquad (20\text{-}38b)$$

where

TEM_o = original air temperature (°C)
TEM = modified air temperature (°C)
BAR_o = original air barometric pressure (kPa)
BAR = modified air barometric pressure (kPa)

A tabulation of specific gravities of air is presented in Table 20-17.

$$\text{DEN} = \text{DEN}_o \times \text{SGB} \times \text{SGT} \qquad (20\text{-}39)$$

where

SBG = specific gravity based on barometer
SGT = specific gravity based on temperature

The system resistance should be computed and the fan selected assuming standard air conditions, since most available fan data are based on this standard. After the fan is selected, the necessary adjustments should be made. If the fan data are available and presented for the required nonstandard conditions, the system resistance must then be computed based on the actual density.

The effect of humidity on fan selection is usually neglected in tunnel applications. Saturated air, however, is slightly

Table 20-17. Specific Gravity of Air

Temperature (°F)	SGT [a]	Elevation (feet above sea level)	Barometer (inches of mercury)	SGB [b]
0	1.152	0	29.92	1.000
20	1.104	500	29.38	0.982
40	1.060	1,000	28.86	0.964
50	1.039	1,500	28.33	0.947
60	1.019	2,000	27.82	0.930
70	1.000	2,500	27.32	0.913
80	0.982	3,000	26.82	0.896
100	0.946	4,000	25.84	0.864
120	0.914	5,000	24.90	0.832
160	0.855	6,000	23.98	0.801
200	0.803	7,000	23.09	0.772
300	0.697	8,000	22.22	0.743
400	0.616	9,000	21.39	0.715
500	0.552	10,000	20.58	0.688

[a] SGT = specific gravity (temperature) of standard air at 29.92 inches of mercury.
[b] SGB = specific gravity (barometer) of standard air at 70° F.

lighter than dry air, the difference in density being only approximately 2.0% at 60°F (16°C) and 4.3% at 120°F (49°C).

Fan Construction. The fan selected for tunnel ventilation systems must be designed to withstand the environment encountered. This includes air quality, pressure, and temperature. Most tunnel fan housings are constructed of carbon steel with an applied protective coating such as paint. In certain environments the fans have been fabricated of stainless steel to minimize corrosion. The centrifugal fan wheel is usually also fabricated of steel, whereas aluminum alloys are being used for axial fan impellers.

The pressure developed by a fan will determine the amount of structural reinforcement required. The designations adopted by the Air Moving and Conditioning Association (AMCA), as shown in the *AMCA Standards Handbook* (AMCA,1985), are used to establish the class of fan construction required.

The supply air systems should be evaluated at the maximum density conditions, the lowest ambient temperature air to enter the fans, to assure sufficient motor capacity. In a tunnel, temperature consideration is also made for the exhaust portion of the ventilation system, since during an emergency such as a fire, high temperatures could be reached at the fan. The usual temperature requirement for these systems ranges from 250 to 300°F (121 to 149°C). It has been recommended to install a deluge water system on the inlet side of tunnel exhaust fans.

To facilitate installation and maintenance, the housings of most large tunnel fans must be split; that is, they must be of bolted, gasketed construction, thus permitting removal of a portion of the housing without disturbing the remainder of the fan. A typical example of split housing for an axial fan is shown in Figure 20-54. In the case of the horizontal axial fan, the removal of the top portion of the housing is usually required to gain access to the impeller and the motor of a direct-driven fan. Removal of the centrifugal fan wheel and shaft will require removal of the top portion of the centrifugal fan housing.

Motors

Most tunnel ventilation fans are driven by electric motors. The selection of a motor for a fan is based on the full-load horsepower requirements, the fan speed, and the starting characteristics. The type of drive to be used must also be considered.

The torque required to accelerate the air and the rotating mass and to overcome friction are considered retarding torques for the motor. During the starting phase, the motor torque must exceed these retarding torques if the fan is to be accelerated. Fan torque varies as the square of the speed; thus, the torque required is zero at a stop fan condition and increases with increasing speed. Therefore, all motor torque is available for fan starting, and a percentage is available for accelerating at other speeds.

Most tunnel fan motors are of the polyphase AC, either squirrel cage or wound rotor, nonsynchronous-speed induction type. Both of these types experience little slip at rated speed. The synchronous speed of a motor is a function of the alternating current frequency and of the number of poles in the motor.

Squirrel cage motors are constant-speed machines that, with the addition of special windings, can be made to operate at multiple speeds. Class B insulated motors are usually used for fans. They are suitable for continuous operation at slight overloads, and the starting current is normal (approximately six times the full-load current).

Wound rotor or slip ring motors of the continuous rated type are used where adjustable-speed motors are required. Wound rotor motors are not suitable for tunnel ventilation systems, as they require secondary resistance and their speed control is not positive under light load conditions.

Motor enclosures are either open or enclosed. Examples of open-type motors are dripproof and splashproof. The open motor should not be used if the ambient air at the motor contains any harmful material. The totally enclosed motor (TE) prevents free interchange of air between the in-

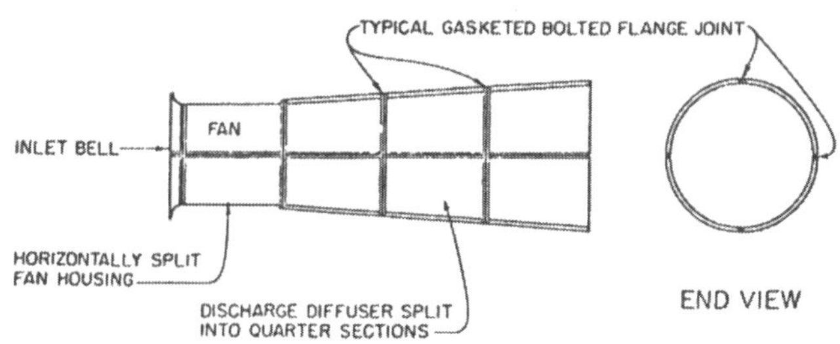

Fig. 20-54. Typical horizontal axial fan housing.

side and the outside of the motor. However, it is not to be considered airtight. The totally enclosed fan-cooled motor (TEFC) has a built-in fan which cools the motor by directing air over the motor. Totally enclosed nonventilated motors (TENV) have no internal fan. Therefore, by comparison, a larger frame size is required. If air is flowing over the motor, the frame size can be reduced. A totally enclosed air over motor (TEAO) is rerated to account for the cooling effect of the air flowing over the fan. This is the type of motor used for direct connected axial fans. The explosion-proof and the dust ignition–proof type motors are special designs not frequently used in tunnel applications.

Motor winding insulation is classified based on the maximum temperature for which it is designed. The maximum temperature may be considered to be the sum of the ambient temperature and the temperature rise above the ambient at the hottest accessible spot in the motor. The standard ambient condition for motor design is 40°C (104°F). If the ambient temperature is greater than 40°C (104°F), the temperature rise must be reduced. At elevations higher than sea level, the allowable temperature rise must be reduced to compensate for the reduction in air density, which reduces the cooling effect. In tunnel ventilation systems, the maximum gas temperature a motor will be exposed to must be evaluated.

The National Electric Manufacturers Association (NEMA) has adopted standards regarding all phases of motor design and performance and, therefore, should be consulted for in-depth information.

By NEMA standards, a motor should be capable of operating successfully under a $\pm 10\%$ voltage and frequency variation provided that the frequency variation does not exceed $\pm 5\%$.

Motor starters are either full voltage, reduced voltage, or split winding. Starters will also provide overload protection in addition to being able to energize and deenergize the motor circuits. The full voltage or across-the-line starter is the simplest and least costly type; however, if limitations are imposed on starting currents, then a reduced voltage starter will have to be used.

Fan Drives

Generally, fan drives may be the direct or indirect type. The direct connected drive means that the fan wheel or impeller is rotating at the same speed as the motor, and the torque is transmitted directly through a fixed shaft. This method is used mostly in axial and propeller fans where the impeller is mounted directly on the motor shaft. This arrangement places the motor bearings in the role of fan bearings. They must be designed to withstand the weight and thrust of the fan impeller. The use of a floating shaft is also considered to be a direct connected drive; however, in this circumstance, a separate set of bearings is required for the fan impeller. The indirect drives, which allow greater flexibility in motor location and fan speed, include both variable and nonvariable types. Belts, chains, and gears are forms of indirect, nonvariable-speed devices. This does not mean that these do not allow fan speed adjustment by modification of sheaves or gears. The V-belt drive is probably the most widely used type of fan drive; however, the chain-driven fan was frequently used in the past for tunnel ventilation systems. The chain drive produces less slip than the belt drive; however, it has more parts and requires lubrication and more maintenance in general. With the advent of efficient V-belt drive, the use of chain drive is minimal.

Variable-speed drives, either the magnetic or fluid type, are used where a fan speed variation is required, usually with a single-speed motor. These devices also eliminate transmission of mechanical vibrations through the connector. A variable-pitch sheave used with a V-belt drive will permit a wide range of speed variation.

Sound Attenuation

Prior to evaluation of the amount of sound attenuation required, a set of design goals must be established. The fan sound will be transmitted both through the discharge and the inlet. The portion of the sound that passes through to the tunnel roadway is usually not critical, since some degree of attenuation is achieved by the ductwork, and the ambient sound level within the tunnel due to the movement of traffic will be high.

The sound transmitted to the neighborhood surrounding the fan structure deserves the most critical look. An evaluation of the existing community sound surrounding the fan installation is necessary. The most accurate method of determining this is to survey the local conditions by having a set of sound readings taken. These tests must be related to the appropriate time of day when conditions are most severe, such as when the generated sound is highest and the surrounding sound is lowest. The effect of the fan sound on occupants of nearby buildings must be evaluated, especially during off-peak traffic hours. Before establishing the sound design goals, the local antinoise ordinances must be reviewed to assure that these goals will be within any limits that are established by these ordinances.

Once the fan sound level is determined and the design goals established, the necessary amount of noise must be removed by some form of attenuation. All natural and built-in attenuation should be explored before considering the installation of a sound attenuator device. Distance is an excellent natural form of attenuation. Barrier walls erected between the source and the receiver are also extremely good attenuators. Other factors, such as the sound absorption of the fan rooms, ducts, plenums, and other enclosures, are also to be considered.

If, after the evaluation of all of this natural and built-in attenuation, there remains an amount of noise yet to be removed, noise attenuation devices must be considered. These devices usually consist of a series of acoustical baffles designed for smooth air passage. The acoustical material is usually contained behind perforated metal walls. These can be weatherproofed to protect the acoustic fill. They are available in two shapes, cylindrical and rectangular. The cylindrical type is designed to be mounted directly on an axial fan inlet or outlet, while the rectangular model can be built in a modular manner, as shown in Figure 20-55, to be connected by duct to a fan or mounted architecturally.

Selection of these units should be based on the unit manufacturer's noise reduction data, usually presented as dynamic insertion loss (DIL).

Dampers

A damper is a device used to control the flow of air in a ventilating system duct. The control of air flow is accomplished by varying the resistance to flow created by the damper much as a valve does in a water system.

There are two general categories of dampers, those having sliding blades and those having rotating blades. The rotating type can be furnished either with a single or with multiple blades. The application of dampers in a tunnel ventilation system, in most instances, is a fan shut-off operation. This requires only two positions of operation, full open and full closed. There are some instances, however, where dampers are used to control air flow or pressure.

The sliding blade, or guillotine-type, damper is the simplest form appropriate for fan shut-off operation. There are a minimum number of moving parts to this damper, thus reducing the maintenance cost. There is also no damper resistance loss when the damper is in its full open position. One severe shortcoming, however, is the added space required to store the blade when the damper is in its full open position.

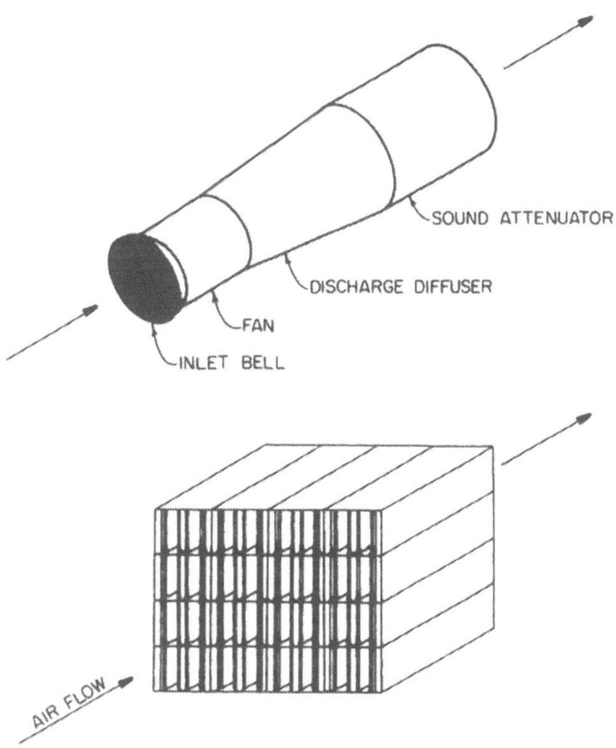

Fig. 20-55. Sound attenuation installations.

The simplest rotary blade damper is the single-blade type, which can either have a rectangular or a round shape. This damper is also highly suitable for a fan shut-off operation due to its low resistance when in the full open position. There are, however, restrictions on the size of the blade from a structural design standpoint. Also, the space required on both sides of the damper blade is often a problem.

The multiple-blade damper is the most widely used of the rotary blade types. The blades in the multiblade damper can be arranged to rotate either in parallel or in opposed action (Figure 20-56). The opposed blade action damper is most appropriate for air flow control because of its flow pressure characteristics, while the parallel blade action damper is used chiefly for a fan shut-off application due to its excellent low leakage characteristics.

The inlet vane dampers used for volume control on centrifugal fans contain a series of radial pie-shaped blades that rotate on a series of shafts set in a ring frame.

Dampers used in tunnel ventilation systems should be constructed to withstand the maximum pressure and temperatures anticipated. Fan shut-off dampers should be designed to withstand the maximum shut-off pressure of the fan, with a factor of safety added to allow for shock loadings due to sudden closure of the damper. The blades of all rotary-type dampers should have an aerodynamic shape to reduce resistance losses. The frames of all dampers should be of structural steel.

The damper operator should be of the type to provide a power drive in both the closing and the opening mode. A minimum allowance of 50% of required capacity should be added to allow for deterioration of the damper mechanism.

Leakage characteristics of a damper are most important in a fan shut-off application. The air leakage through a

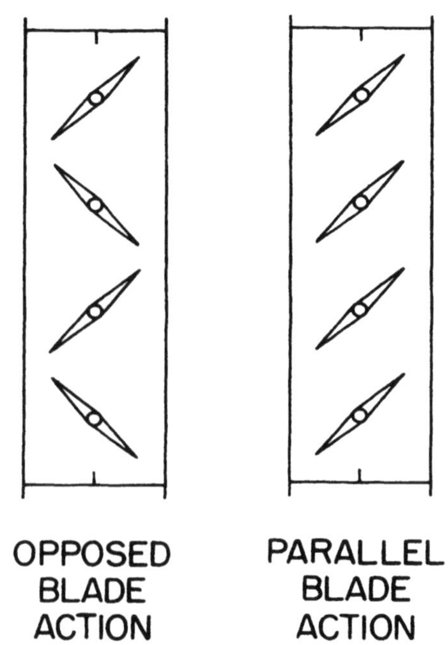

Fig. 20-56. Typical multiple-blade damper.

closed damper must be specified for a particular pressure drop across the damper. The allowable leakage must be that amount of air which will be acceptable to the ventilation system without severe deterioration of the ventilating effect. This leakage of air flow must not be of a magnitude to permit windmilling of the idle fan or fans.

The pressure drop characteristics across a full or partially opened damper should be included in the computation of the system resistance.

Turning Vanes

Turning vanes should be installed in all square duct elbows to reduce the resistance loss. The type of vane employed will depend on the duct velocity, as the loss is directly related to the velocity pressure, as shown in Figure 20-57. For example, the 25% velocity pressure saving shown for the double thickness vanes over the single thickness vanes at an air velocity of 4,000 ft/min (20 m/sec) becomes a 1/4-in. water gauge (0.0626 kPa) reduction in system loss. For the ranges of horsepowers considered in tunnel ventilation systems, this difference can be significant in the amount of power saved.

Turning vanes are designed to maintain a reasonable aspect ratio of the space between the vanes (Figure 20-58). Based on a curve ratio of 0.7 and an aspect ratio of 5:1 for the flow passages between the vanes, Equations (20-40)–(20-43) will provide the necessary configuration. For a clear picture, see Figures 20-52 and 20-59.

$$\mathrm{NV} = \frac{5W}{D} - 1 \qquad (20\text{-}40)$$

where

NV = number of turning vanes
W = width of duct (ft)
D = depth of duct (ft)

$$\mathrm{P} = \frac{W}{\mathrm{NV} + 1} \qquad (20\text{-}41)$$

where P = spacing between turning vanes (ft).

$$R = 3.33P \qquad (20\text{-}42)$$

where R = radius of inner turning vane surface (ft).

$$r = 2.33P \qquad (20\text{-}43)$$

where r = radius of outer turning vane surface (ft).

Turning vanes are usually constructed of corrosion-resistant metal.

Ducts

The ducts that transport the air within the tunnel are usually part of the tunnel construction. In most tunnels, these ducts are a constant cross-sectional area throughout. The surfaces of these ducts must be relatively smooth to assure minimal air friction losses. Streamlining of obstructions, such as hangers, especially in areas of high velocity, is recommended.

The maximum air velocity in these ducts has, in most instances, been held below 6,000 ft/min (30 m/sec). There are, however, cases where 7,000 ft/min (36 m/sec) velocities have been used.

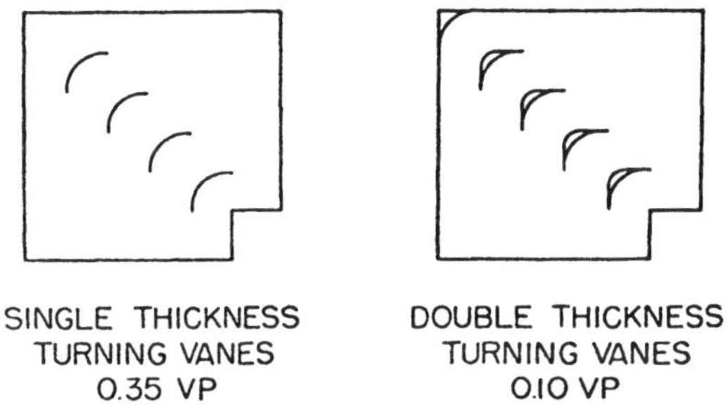

Fig. 20-57. Square duct elbow turns.

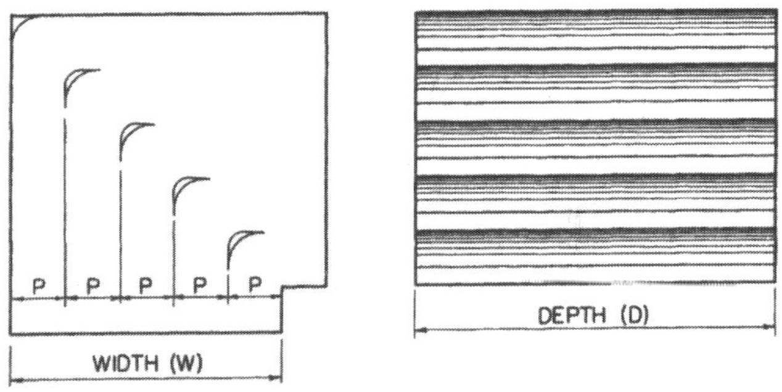

Fig. 20-58. Vaned elbow.

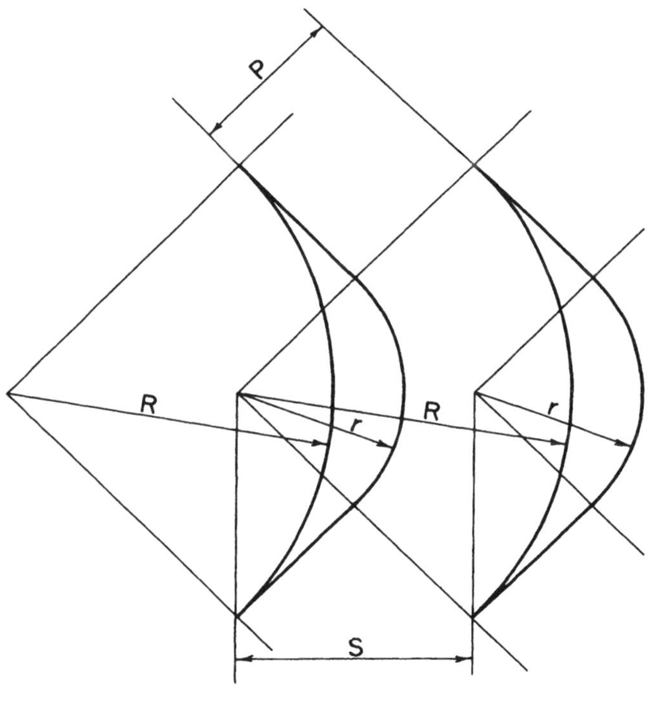

Fig. 20-59. Turning vanes.

The various duct connections from the fans to the tunnel ducts are usually either of concrete (as a part of the structure) or of sheet steel. The steel ducts must be reinforced against the maximum pressure conditions anticipated. The interior surface must be smooth, with all bracing located on the exterior. General practice has been to use 1/4-in. (6.4-mm) steel plate for most of these applications.

Access must be provided for all ducts and near all equipment.

Air measuring ports are required at strategic locations to permit balancing and testing of the tunnel ventilation system. These ports should be large enough to permit entrance of a proper sized Pitot tube. They should be fitted with a screw-type plug or cap to permit closure during system operation.

The ventilation air must be transported between the ducts and the roadway. Various types of flues and ports are used to accomplish this task (Figure 20-60). Supply air flues designed to carry the supply air from the duct to the roadway are usually constructed of steel. They may also be constructed of any material that will withstand forces, such as pouring of the concrete or other construction activities. The interior surfaces of these flues must be smooth to minimize air friction losses. All of the flues are usually the same size, and regulation of air flow through each flue is accomplished by use of a simple plate-type damper. A typical flue and damper arrangement is shown in Figure 20-60.

Ports are simply slots or openings in a wall or ceiling to permit passage of air. They are sometimes sloped in a direction of air flow, as shown in Figure 20-61. Ports can be used for either supply or exhaust air. To assure even air distribution along the length of the duct, a plate-type damper is often used or the spacing of the ports is varied.

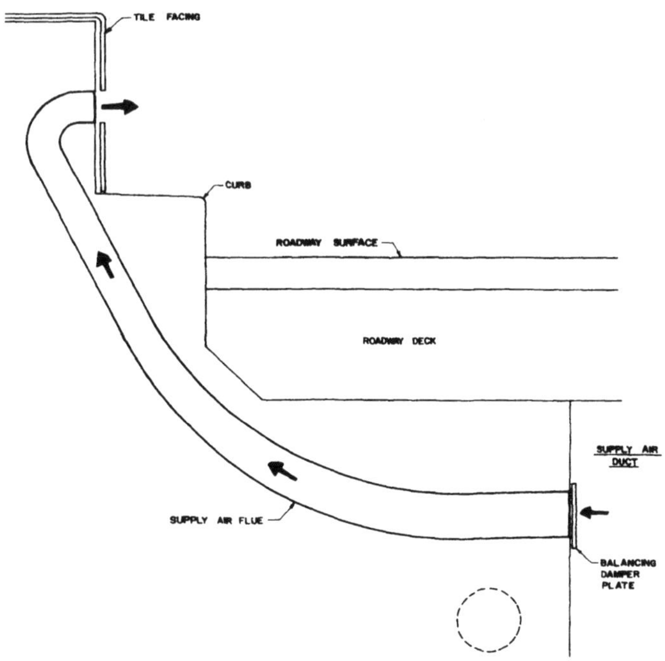

Fig. 20-60. Typical supply air flue.

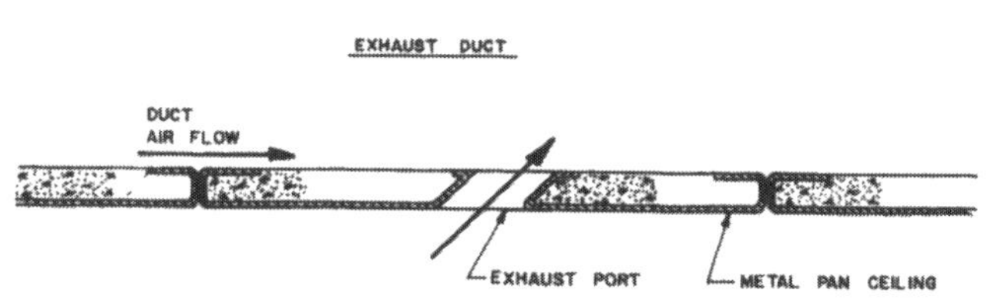

Fig. 20-61. Typical ceiling exhaust port.

Ventilation Buildings and Ventilation Shafts

Ventilation buildings are required to house the tunnel ventilation fans, electrical equipment, controls, and auxiliary equipment and facilities. Relatively short tunnels will most likely have only one ventilation building, whereas larger tunnels will have two, three, or more buildings. Each building will usually serve one or more ventilation zones.

The general arrangement of the ventilation building will depend to a great extent on the type of fans and ventilation system employed, and on the type and length of tunnel. A subaqueous tunnel will require a vertical building arrangement to take advantage of the space above the tunnel and below grade, as shown in Figure 20-62. A cut-and-cover tunnel in an urban area using an underground ventilation building could appear, as in Figure 20-63. In a tunnel such as this, often the space available for the ventilation structure is limited. A mountain tunnel using centrifugal fans is shown in Figure 20-64.

During the process of arranging the ventilation building layout, consideration must be given to the space required for the air flow to and from the fans and for the proper service and removal of equipment. When considering maintenance, both routine service and removal of equipment is important. Enough space must be allowed around each fan for service activities and for removal. Along with accessibility, provision must also be made to lift the equipment during service activities. On a centrifugal fan, the impeller, shaft, housing,

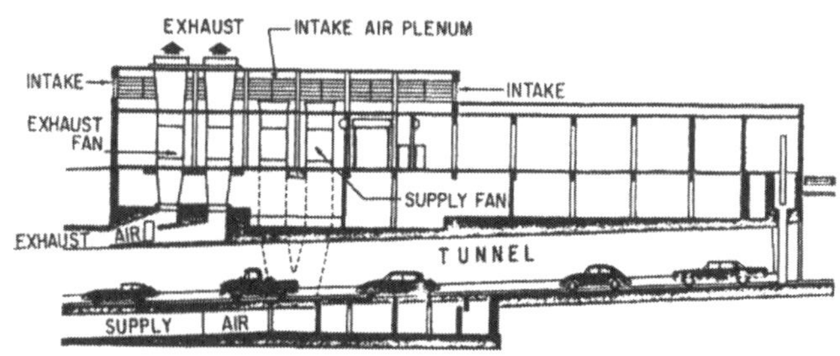

Fig. 20-62. Typical ventilation building for a subaqueous highway tunnel.

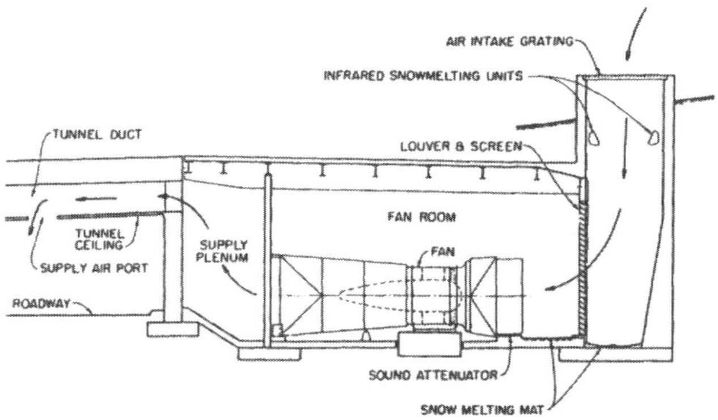

Fig. 20-63. Typical ventilation building for an urban highway tunnel.

and motor will be the items most likely requiring lifting. The impeller, motor, and housing on a horizontal axial fan will most likely be moved. On a vertically mounted axial fan, a removable rail system can be provided that will permit ready removal of the entire fan unit for service.

Testing

Factory Testing. Factory tests should be conducted on all specially designed and built major operating components of the tunnel ventilation system. These tests would normally cover fans, motors, and dampers to verify the predicted performance of this equipment. They will assure that there will be minimal delays when the equipment is permanently installed. It is also preferable to test the fan along with its associated motor and shut-off damper, if possible. The fan tests should be conducted in accordance with the standard AMCA test procedure. It may be necessary, however, under some circumstances, to adjust the setup outlined by AMCA if the fan is too large for the particular test facility. Sound tests, as outlined in the AMCA standards, should be conducted along with the air flow tests mentioned above.

Mechanical testing of the fan wheel or impeller should be considered, especially if the fan rotational speed is high, as in the axial fan. Overspeed testing in a vacuum pit is one method of mechanical testing the impeller.

Dampers should be tested for leakage, especially if the fan operating pressures are high. The dampers should be tested against the fan shut-off head, too. If the damper is being used to modulate pressure and flow, a more complicated testing would be required, where flow and pressure for all intermediate positions of the damper could be demonstrated.

Site Testing. Once the ventilation system has been installed, the air flow pressure and sound should be measured and recorded. This procedure will assure compliance with design and will aid in the balancing process. The shop test usually does not mimic the actual site conditions, while site testing will evaluate the actual flow conditions.

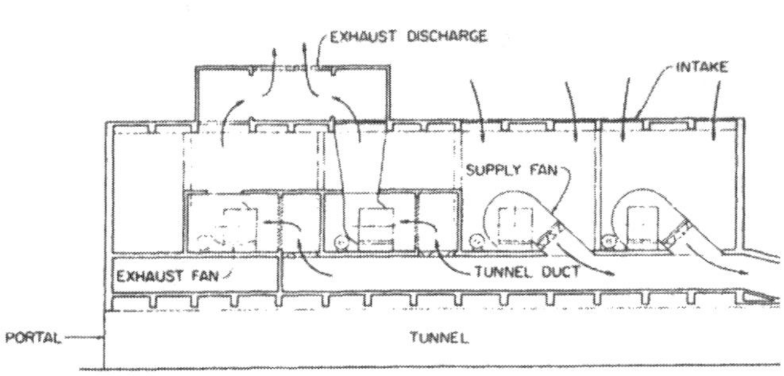

Fig. 20-64. Typical ventilation building for a mountain highway tunnel.

CONTROL AND MONITORING SYSTEMS

Tunnel ventilation systems can be controlled manually or automatically, and operated remotely or locally. A basic operational philosophy is involved regarding whether there will be a human operator continuously present at the tunnel control. Manually operated tunnels require an operator to be present at all times, whereas fully automatic systems can function without the attendance of an operator. However, fully automatic systems are not completely without human participation, in that a number of system operating conditions must be monitored to prevent serious equipment breakdown. Many tunnel designers still feel that, except for relatively short tunnels, the human element must be included in tunnel operation since only a human operator can exercise the required judgment in the event of an emergency. The type of control selected for the ventilation system should be consistent with those selected for all other equipment, such as pumps, lights, and power.

Manual Control

Manual control systems are usually operated from a room within the facility that provides a central location for all control indication and monitor functions. In many vehicular tunnels, the ventilation system control is combined with the traffic control at one location. All necessary control and monitoring systems must be provided to enable the operator to start, stop, or adjust the ventilation system to suit the tunnel traffic and tunnel air quality conditions. The operator responds to the output from some form of monitoring of these conditions.

The predominant environmental factors to be monitored for fan operation are carbon monoxide (CO) concentration, visibility, and traffic flow. Any one or combination of these factors can be used as a control factor. The use of a carbon monoxide analyzer system that produces an indication of the CO concentration in ppm and a visibility measuring system that produces an indication of the percent visibility, along with knowledge of the historical pattern of traffic flow, will allow the operator of a manual system to anticipate the air flow required to suit the tunnel conditions. For a full description of these monitoring systems, see Chapter 24.

Automatic Control

The operation of fans responding automatically to changes in CO concentrations, visibility, traffic flow, or the calendar is all possible with a fully automatic tunnel ventilation control system. Any one or a combination of the above factors can be used to control the fans.

A tunnel ventilation system operating automatically from either CO concentrations or visibility monitoring devices will have fan operation adjustments made whenever a change in the tunnel environment takes place. There is usually a damping or time delay built into the control system to prevent unnecessarily frequent fan speed adjustments. When the fans are controlled by a system based on traffic flow, it is not the specific tunnel environment that is monitored. The traffic flow information, along with the anticipated CO concentration and visibility generated by the traffic, is used to set the fan operation program. This is usually based on historical data gathered during tunnel operation. A calendar-based control system functions along the same lines using the anticipated contaminant levels at each hour of the day as a basis for a control program. This information is gathered during historical experience. The traffic- or calendar-based systems must also be provided with a manual or automatic override based on CO or visibility. The most elaborate control system would be one operating on a calendar basis with automatic overrides for CO, visibility, and traffic flow.

Local Control

All control systems, whether automatic or manual, must be provided with local control within sight of the equipment to facilitate safe maintenance and testing.

Indication

In any tunnel ventilation system, some operational conditions must be monitored to assure continued operation and prevent damage to the ventilation equipment. Provision must be made in the control room (or a remote location, in the case of a fully automatic system) to permit indication of the ventilation system operational status and announcement of the system operational conditions.

Indication of the operational status of the fans and associated equipment, such as fan on/off, damper open/close, and fan speed, can be accomplished by the use of indicating lights on the control board. The operational conditions being monitored can be set to trigger an alarm, which signals that a particular operational limit has been exceeded. The following are typical examples of these operational limits: motor winding overtemperature, bearing overtemperature, and high CO concentration.

A closed-circuit TV system installed within the tunnel can also be considered as part of the control system. It can be used to evaluate traffic flow and aid in selecting the proper action to be taken with regard to fan operation. For a full description, see Chapter 24.

Environmental Monitoring Systems

The systems used to monitor the environment within a highway tunnel and provide signals for fan operation and control include the carbon monoxide analyzing systems and the visibility monitoring systems. Other systems are used for the control of fans, such as traffic counters and closed-circuit TV; these are covered in Chapter 24.

VENTILATION DURING CONSTRUCTION

Ventilation is required during the construction of any tunnel. This is true whether the tunnel is constructed by blasting, boring, or placing prefabricated tubes in a trench. Temporary ventilation is necessary to provide a suitable, safe working environment for the construction workers. Since many flammable or airborne toxic gases, dust, mists, and fumes are released during the construction process, these contaminants can only be removed through ventilation. Temporary ventilation systems used during construction are usually separate from the permanent ventilation installed in the finished tunnel, since the purposes and the fan capacities are usually not compatible, although some portions of the permanent ventilation system, such as concrete rock ducts or shafts, may be used for both the temporary and the permanent system if properly designed.

There are a number of requirements for ventilation in construction activities. In general, fresh air must be supplied to all work areas in sufficient amounts to prevent any dangerous or harmful accumulation of dusts, fumes, mists, vapors, or gases. A minimum of 200 ft^3 of fresh air per min (0.1 m^3/sec) is to be supplied for each employee underground. Mechanical ventilation, with reversible air flow, is to be provided in all of these work areas, except where natural ventilation is demonstrably sufficient.

The requirements for ventilation during the construction phase of a tunnel project are covered by elements of the U.S. Occupational Safety and Health Act of 1970 contained in Title 29, subpart S, 1926.800 Underground Construction. The general subject of temporary tunnel ventilation considered appropriate to be used during the construction phase of a tunnel is addressed in depth in Chapter 13.

REFERENCES

Air Moving and Conditioning Association, Inc. (AMCA) (1985) *AMCA Standards Handbook,* Publication 99-86, Park Ridge, Illinois.

Air Moving and Conditioning Association, Inc. (AMCA) (1986) *AMCA Test Code for Air Moving Devices,* ANSI/AMCA Standards 210-85 (ANSI/ASHRAE Standards 51-1985), Park Ridge, Illinois.

Aisiks, E.G., and Danziger, N.H. (1969) *Ventilation Research Program at Cascade Tunnel,* Northern Railway Bulletin 622, American Railway Engineering Association (AREA).

American Conference of Governmental Industrial Hygienists

(ACGIH) (1992) *ACGIH Industrial Ventilation,* 21st edition, Lansing Michigan.

American Conference of Governmental Industrial Hygienists (ACGIH) (1994) *ACGIH TLVs: Threshold Limit Values and Biological Exposure Indices for 1993–1994,* Cincinnati, Ohio.

American Society of Heating, Refrigerating and Air Conditioning Engineers, Inc. (ASHRAE) (1991) *ASHRAE Handbook—HVAC Applications,* Atlanta, Georgia, Chapter 13.

American Society of Heating, Refrigerating and Air Conditioning Engineers, Inc. (ASHRAE) (1992) *ASHRAE Handbook—HVAC Systems and Equipment,* Atlanta, Georgia.

American Society of Heating, Refrigerating and Air Conditioning Engineers, Inc. (ASHRAE) (1992) *ASHRAE Handbook—Fundamentals,* Atlanta, Georgia.

BENDELIUS, A.G. (1975) "Subway Environmental Control for the MARTA System," Intersociety Conference on Transportation, Atlanta, Georgia.

BENDELIUS, A.G., and METSCH, W. W. (1973) "Environmental Control for the Caracas Metro," *ASHRAE Journal,* American Society of Heating, Refrigerating and Air Conditioning Engineers, Inc. (ASHRAE), Atlanta, Georgia.

BERGER, L.B., and MCGUIRE, L.M. (1946) *Observations on the Use of a Diesel Locomotive Through a Railway Tunnel,* U.S. Dept. of the Interior, Bureau of Mines, Washington, D.C.

Bureau of Mines (1968) *Mobile Diesel-Powered Equipment for Non-gassy, Non-coal Mines and Tunnels Approved by the Bureau of Mines 1951–1961,* IC8363, U.S. Dept. of the Interior, Washington, D.C.

Bureau of Public Roads (BPR) (1950) *Highway Capacity Manual,* U.S. Dept. of Commerce, Washington, D.C., p. 3.

CHEN, TI.C. (1973) "Air Quality Distribution in Highway Tunnel Ventilation," *Flow Studies in Air and Water Pollution,* America Society of Mechanical Engineers (ASME), pp. 9–11.

DALEY, B.B. (1985) *Woods Practical Guide to Fan Engineering,* 3rd edition, Woods of Colchester, Ltd., Colchester, England.

DANZIGER, N.H., and AISIKS, E.G. (1970) "Analysis and Control of Thermal Environment in Subway Systems," presented at the Rail Conference of the American Transit Association/Institute for Rapid Transit, Boston.

DONATO, G. (1972) "Montreal Metro Fire, December 9, 1971," International Metropolitan Railway Committee, International Union of Public Transport, New York.

Environmental Protection Agency (EPA) (1971) *Air Quality Criteria for Nitrogen Oxides,* AP-84, Washington, D.C., pp. 2–6.

FIELDNER, A.C., et al. (1921) *Ventilation of Vehicular Tunnels* (report of the U.S. Bureau of Mines to N.Y. State Bridge and Tunnel Commission, and N.J. Interstate Bridge and Tunnel Commission, and the American Society of Heating and Ventilating Engineers, Inc. [ASHVE]), New York.

HAERTER, A. (1973) "Fresh Air Requirements for Road Tunnels," Proceedings, International Symposium on the Aerodynamics and Ventilation of Vehicle Tunnels, BHRA Fluid Engineering, Canterbury, England.

Highway Research Board (HRB) (1965) *Highway Capacity Manual,* Special Report 87, Washington, D.C., p. 89.

IDELCHIK, I.E. (1994) *Handbook of Hydraulic Resistance,* 3rd edition, CRC Press, Boca Raton, Florida.

JORGENSEN, R. (ed.) (1983) *Fan Engineering,* 8th edition, Buffalo Forge Company, Buffalo, New York.

Joy Manufacturing Company (1990) *Joy Fan Catalogue,* New Philadelphia, Ohio.

MCCUTCHEN, W.R. (1974) "Ventilation Considerations for the Trans-Bay Tube," *Symposium Proceedings, Ventilation Aspects of Hostile Environments,* American Society of Heating, Refrigerating and Air Conditioning Engineers, Inc. (ASHRAE), New York.

Occupational Safety and Health Act of 1970, Public Law 91-596, December 1970, as amended by Public Law 101-552, November 1970.

OSBORNE, W.C., and TURNER, C.G. (eds.) (1964) *Woods Practical Guide to Fan Engineering,* 2nd edition, Woods of Colchester, Ltd., Colchester, England.

Parsons Brinckerhoff Quade and Douglas, Inc. (PBQD) (1966) *Libby Dam Project Railroad Tunnel Ventilation Research and Test Program at Cascade Tunnel,* New York, August.

Permanent International Association of Road Congress (PIARC) (1991) Report of the Technical Committee on Road Tunnels, XIXth World Road Congress, Marakesh, September.

Public Health Service (PHS) (1970) *Air Quality Criteria for Carbon Monoxide,* AP-62, U.S. Dept. of Health, Education and Welfare, Washington, D.C., pp. 2–3.

ROGERS, S.J., ROELICH, F., JR., and PALLADINO, C.A. (1970) *Tunnel Ventilation and Air Pollution Treatment,* Mine Safety Appliances Research Corp. for Federal Highway Administration, Washington, D.C.

Safety and Health Regulations for Construction (1992) Code of Federal Regulations, Title 29, Part 1926, Subpart S, Section 1926.800, Washington, D.C.

SINGSTAD, O. (1929) "Ventilation of Vehicular Tunnels," World Engineering Congress, Tokyo.

SOLOMON, I.M. (1973) "Subway Station Air Conditioning for the Washington Metropolitan Area Transit Authority," Paper 73-RT-7, Rail Transportation Division, American Society of Mechanical Engineers (ASME), New York.

STAHEL, M., ACKERT, J., and HAERTER, A. (1961) *Die Luftung der Autotunnel,* Mittelung Inst. fur Stassenbau, Eidg. Thechn. Hochschule, Zurich, Switzerland.

Subway Environment Design Handbook (SEDH) (1975) Vol. 2: Subway Environment Simulation (SES) computer program, NTIS No. PB-254-789, Part 1-3, Washington, D.C.

Subway Environment Design Handbook (SEDH) (1976) Vol. 1: *Principles and Applications,* 2nd edition, U.S. Dept. of Transportation, Washington, D.C.

Tippetts-Abbett-McCarthy-Strattor (TAMS) (1965) *Research Investigations and Studies to Establish the Ventilation Requirements for the Straight Creek Highway Tunnels,* report for the Colo. Dept. of Highway and U.S. Dept. of Commerce, Bureau of Public Roads, Colorado.

Transportation Research Board (TRB) (1985) *Highway Capacity Manual,* Special Report 209, National Research Council, Washington, D.C.

U.S. Corps of Engineers (USCOE) (1966) *Libby Dam Project Tunnel Ventilation System,* Dept. of the Army, Seattle District, Seattle, Washington.

OTHER RECOMMENDED READING

Associated Engineers, Inc., (1976) *Subway Environmental Handbook, Vol. 1, Principles and Applications,* 2nd edition.

BABA, T., OHASHI, H., NAKAMICHI, F., and AKASHI, H. (1979) "A New Longitudinal Ventilation System Using Electrostatic Precipitator for Long Vehicular Traffic Tunnel," 3rd International Symposium on the Aerodynamics and Ventilation of Vehicle Tunnels, Sheffield, England, March 1979, British Hydromechanics Research Association, Cranfield, England.

BABA, T., OHASHI, H., NAKAMICHI, F., and UESALO, K. (1982) "Aerodynamics Specialties in Connection with the Tsuruca Tunnel," 4th International Symposium on the Aerodynamics and Ventilation of Vehicle Tunnels, Sheffield, England, March 1982, British Hydromechanics Research Association, Cranfield England.

BABA, T., OHASHI, H., and JYOUZKA, S. (1988) "Specialties of Ventilation System of the Kan-Esty Tunnels," 6th International Symposium on the Aerodynamics and Ventilation of Vehicle Tunnels, Durham, England, September 1988, British Hydromechanics Research Association, Cranfield, England.

BENDELIUS, A.G., and HETTINGER, J.C. (1988) "Ventilation of the Sydney Harbour Tunnel," Proceedings: International Symposium on the Aerodynamics and Ventilation of Vehicle Tunnels, BHRA Fluid Engineering, Durham, England.

BENDELIUS, A.G., and MEEKS, K.R. (1989) "Impact of the Development of Tunnel Service Systems on the Planning and Design of the Immersed Tunnel for the Sydney Harbour Crossing," *Immersed Tunnel Techniques,* Thomas Telford, London.

CHARWOOD, R.G., HUGGETT, J.R., and SALT, P.E. (1982) "Tunnel Ventilation Systems for the British Columbia Railway," Proceedings: International Symposium on the Aerodynamics and Ventilation of Vehicle Tunnels, BHRA Fluid Engineering, York, England.

CONRAD, C.W. (1990) "System Effect—A Design Fundamental," *ASHRAE Journal,* American Society of Heating, Refrigerating, and Air Conditioning Engineers, Inc. (ASHRAE), Atlanta, Georgia.

CORY, W.T.W. (19xx) *A Survey of Fan Performance Test Methods and How They Compare with ISO Recommendations,* Woods of Colchester, Ltd., Colchester, England,.

GARAY, P.N. (1990) *Pump Application Desk Book,* The Fairmont Press, Inc., Lilburn, Georgia.

HOBBS, J.R., WALTER, R.A., HARD, T., and DEVOE, D. (1977) *Train Generated Air Contaminants in the Train Crew's Working Environment,* U.S. Department of Transportation, Transportation Systems Center, Cambridge, Massachusetts.

KATZ, M., RENNIE, R.P., and JEGIER, Z. (1959) *Air Pollution Hazards from Diesel Locomotive Traffic in a Railway Tunnel,* A.M.A. Archives of Industrial Health.

KONDA, T., and MITZUTANI, T. (1976) "Application of the Longitudinal Ventilation System for Long Tunnels," 2nd International Symposium on the Aerodynamics and Ventilation of Vehicle Tunnels, March 1976, Cambridge, England, British Hydromechanics Research Association, Cranfield, England.

LEVY, S.S. (xxxx) *Ventilation System for Mount Macdonald Tunnel,* American Railway Engineering Association (AREA).

LONDOLFA, A., MENEGHINI, G.B., BARBETTA, C., and CORY, W.T.W. (1988) "The Application of Jet Fans to Longitudinal Tunnel Ventilation," 6th International Symposium on the Aerodynamics and Ventilation of Vehicle Tunnels, September 1988, British Hydromechanics Research Association, Cranfield, England.

Methane Control in Tunneling (1989) A Bureau of Mines Handbook, U.S. Dept. of the Interior, Washington, D.C.

NAKAMUJRA, S., SHIMOTORI, K., MIZUMO, A., and SASAMOTO, T. "Development of Ceiling Type Electrostatic Precipitation System," 7th International Symposium on the Aerodynamics and Ventilation of Vehicle Tunnels, Brighton, England, British Hydromechanics Research Group, Cranfield, England.

National Fire Protection Association (1992) *Recommended Practice on Fire Protection for Limited Access Highway Tunnels, Bridges, Elevated Roadways, and Air Right Structures,* NFPA 502, Quincy, Massachusettes.

Parsons Brinckerhoff Quade &, Douglas Inc. (1982) *Subway Environmental Design Handbook, Vol. 2, Subway Environment Simulation (SES) Computer Program Version 3,* Part 1: User's Manual, prepared in draft for the U.S. Department of Transportation, October.

Parsons Brinckerhoff Quade & Douglas, Inc. (1982) *Subway Environmental Design Handbook, Vol. 2, Subway Environment Simulation (SES) Computer Program Version 3,* Part 2: Programmer's Manual, prepared in draft for the U.S. Department of Transportation, October.

Permanent International Association of Road Congresses (1987) Technical Committee on Road Tunnels—Report, Technical Committee Report No. 85, XVIIIth World Road Congress, Brussels, September.

WEST, E.M. (1988) "Selecting and Designing a Ventilation System for the Sydney Harbour Tunnel," *AIRAH Journal,* The Australian Institute of Refrigeration, Air Conditioning and Heating (Inc.) (AIRAH), Surrey Hills, Australia.

CONCLUSIONS

At press time, October 1995, the following conclusions were reported as part of the Memorial Tunnel Fire Ventilation Test Program.

The Memorial Tunnel Fire Ventilation Test Program (MTFVTP) represented a unique opportunity to evaluate and develop design methods and operational strategies leading to safe underground transportation facilities. This comprehensive test program, which commenced in September 1993 and concluded in October 1995, resulted in the gathering of much needed data acquired in a full-size facility under controlled conditions over a wide range of system parameters. The results of the program will be processed and made available to the professional community for the use in the development of tunnel ventilation design and emergency operation procedures.

The resulting data summaries were published in a comprehensive Test Report in late 1995 and resulted in the following conclusions:

1. Longitudinal ventilation using jet fans was shown to be capable of controlling the movement of smoke and heat resulting from fire sizes up to 100 MW.
2. Jet fans positioned downstream of, and close to, the fire were subjected to temperatures high enough to cause their failure.
3. Single zone balanced (equal flow rates for supply and exhaust air) full transverse systems indicated limited smoke and temperature management capability. Ventilation rates of 100 cfm per lane-foot exhaust capacity did not satisfactorily manage conditions resulting from heat release rates of 20 MW and greater, unless the system was operated in an unbalanced mode (reduced supply airflow), in which case there was improved smoke and temperature management capability. However, even 100 cfm per lane-foot exhaust capacity provided only limited temperature and smoke management for a 20 MW heat release rate.
4. Single zone semi-transverse systems capable of only supplying air (no possible reversal of fans to exhaust air) evidenced no smoke or temperature management. Single zone semi-transverse systems which can be operated in the exhaust mode provided a degree of smoke and temperature management.
5. Longitudinal airflow within the roadway area was shown to be a considerable factor in the management and control of smoke and heat movement generated in a fire. Ventilation systems which effectively combine extraction and longitudinal airflow can limit the spread of smoke and heat. Multiple zone ventilation systems can allow control over the direction and magnitude of longitudinal airflow, and therefore can effectively limit the migration of smoke and high temperatures in the tunnel.
6. If longitudinal airflow movement is generated in transverse systems, a measure of smoke movement management can be achieved. Installing multiple zones or operating full transverse systems in an unbalanced mode induces longitudinal airflow to varying degrees.
7. Point extraction is a system configuration capable of extracting large volumes of smoke near the fire, through large controlled openings in a ceiling exhaust duct, thus preventing extensive migration of smoke. The single point extraction (SPE) transverse type system effectively managed smoke and temperature conditions for a 20 MW fire with airflow rates lower than 100 cfm per lane-ft.
8. Natural ventilation resulted in extensive spread of heat and smoke upgrade of the fire.
9. Fan response time, the interval between the onset of a fire and ventilation system activation, should be minimized, since hot smoke layers can spread quickly.
10. The restriction to visibility caused by smoke occurs more quickly than does a temperature high enough to be debilitating. Carbon monoxide (CO) levels never exceeded the *guidelines established for the test program.*

The database generated by the MTFVTP should permit professionals engaged in the design, operations and fire fighting fields to develop accepted analytical procedures, standards and operating procedures which appropriately account for fire emergencies. The entire database is archived on magnetic and video tape. The Test Report published in late 1995 includes the resolution of this data into diagrammatic and pictorial form, a degree of analysis, and the findings drawn from that analysis.

The implication of these conclusions for the tunnel ventilation engineer are:

1. Longitudinal ventilation utilizing jet fans can now be accepted and installed where appropriate. Such application is limited to tunnels with uni-directional traffic and where discharge of pollutants at the portals is not an environmental issue.
2. Future full transverse systems should be designed with the consideration of creating longitudinal flow by multi-zoning and/or unbalanced supply and exhaust airflows. In addition, the use of single point extraction and oversized exhaust ports should be considered.
3. A single criteria value for the supply air ventilation airflow rate is not appropriate. Any such criteria should be dependent on tunnel configuration and on the type of ventilation system.
4. Single zone semi-transverse supply air ventilation systems are not appropriate for smoke management if fans are not reversible.
5. The impact of high temperature on fans should be considered in ventilation system design. The row or rows of jet fans closest to the fire could be lost due to the impact of the fire and heat. The central fans in a ventilation system could be exposed to temperatures of 350°F (177°C).

21

Tunnel Lighting

Peter A. Mowczan
Supervising Engineer, Parsons Brinckerhoff Quade & Douglas, Inc.

LIGHTING OF HIGHWAY TUNNELS

The maintenance of traffic volume at design roadway speeds through a tunnel on a bright sunlit day depends on the ability of the motorist to see the interior of the tunnel and objects on the roadway for a safe stopping distance. The geographic location, orientation, and portal surroundings influence the ability of the human eye to adapt from the bright ambient roadway to the "black hole" of the tunnel interior. The lighting of the tunnel interior to eliminate and or diminish the effect of the black hole is achieved through varied lighting concepts. The most prominent lighting concepts employed are the symmetrical and the asymmetrical, of which there are two types: the counter beam and line-of-sight. Linear or point source luminaires or a combination of types of sources are employed to provide for specific illumination requirements for unidirectional or bidirectional traffic tunnels as appropriate to the system.

DEFINITION OF TERMS

Adaptation: The process by which the sensitivity of the retina of the eye adjusts to increases or decreases light from a level it was adjusted to. The resulting sensitivity of the retina is termed the state of adaptation. The luminance causing this state of adaptation is termed the state of adaptation level.

Adaptation Point: The point on the road where the adaptation of the eye of a motorist approaching a tunnel begins to be influenced by the presence of the dark tunnel entrance.

Brightness: The luminous flux per unit of project area and unit solid angle, either leaving a surface at a given point in a given direction or arriving at a given point from a given direction; the luminous intensity of a surface in a given direction per unit of projected area of the surface as viewed from that direction.

Candela: The International System of Units (SI) unit of luminous intensity. One candela (cd) is one lumen per square meter.

Candela per Square Meter: The SI unit of photometric brightness. A perfectly diffused surface emitting or reflecting light at the rate of one lumen per square meter.

Counter Beam: A technique whereby a directional light control is used opposite to the direction of traffic to provide negative contrast (object luminance is less than background luminance).

Flicker: The result of periodic luminance changes in the field of vision, due to the spacing of the lighting fixtures.

Foot-Candle: The incident illumination on a surface 1 ft^2 in area on which is distributed a uniform light output of 1 lumen; 1 lumen/ft^2.

Lux: The incident illumination on a surface 1 m^2 in area on which is distributed a uniform light output of 1 lumen; 1 lumen/m^2.

Foot Lambert: The unit of photometric brightness (luminance). A perfectly diffused surface emitting or reflecting light at the rate of 1 lumen/ft^2 would have an equivalent brightness of 1 foot Lambert (fL).

Glare: The sensation produced by brightness within the visual field that is sufficiently more intense than the luminance to which the eyes are adapted to cause discomfort or loss in visual performance and visibility.

Lamp Lumen Depreciation Factor: One of the many factors used in the evaluation of a maintenance factor causing a reduction of lumen output of a lamp during its life cycle.

Louver: A series of baffles used to shield a source from view at certain angles or to absorb unwanted light.

Lumen: The unit of measure of the quantity of light. The amount of light that falls on an area of 1 ft^2, every point of which is 1 ft from a source of 1 candela (candle).

Luminaire: A complete lighting device consisting of a light source together with its direct appurtenances.

Luminaire Ambient Temperature: The lumen output of fluorescent lamps is affected by the ambient temperature in which they operate; the luminaire ambient temperature must be considered when obtaining the maintenance factor (Figure 21-1).

Luminaire Dirt Depreciation Factor: Another maintenance factor, causing loss of light on the work plane due to the accumulation of dirt on the luminaire.

Luminaire Surface Depreciation: This depreciation of initial luminaire lumen output results from the adverse changes in metal, paint, and plastic components. Glass, porcelain, and processed aluminum depreciate negligibly and are easily restored to their original reflectance. Baked enamels and other painted surfaces have a permanent depreciation because all paints are porous to some degree. For plastics, acrylic is least susceptible to change, but in certain atmospheres, transmittance may be reduced by usage over a period of 15 to 20 years. For the same usage, polystyrene will have lower transmittance than acrylic and will exhibit a faster depreciation. Because of the complex relationship between the light-controlling elements of luminaires using more than one type of material (such as a lensed troffer), it is difficult to predict losses due to deterioration of materials. In addition, some materials are more or less adversely affected depending on the atmosphere in which they are installed.

Luminaire Voltage: High or low voltage at the luminaire will affect the output of most lamps. For incandescent units, deviations from rated lamp voltage cause approximately 3% change in lumen output for each 1% change in primary voltage deviation from rated ballast voltage. When regulated output ballasts are used, the lamp lumen output is relatively independent of the primary voltage within the design range. Fluorescent luminaire output changes approximately 1% for each 2-1/2% change in primary voltage.

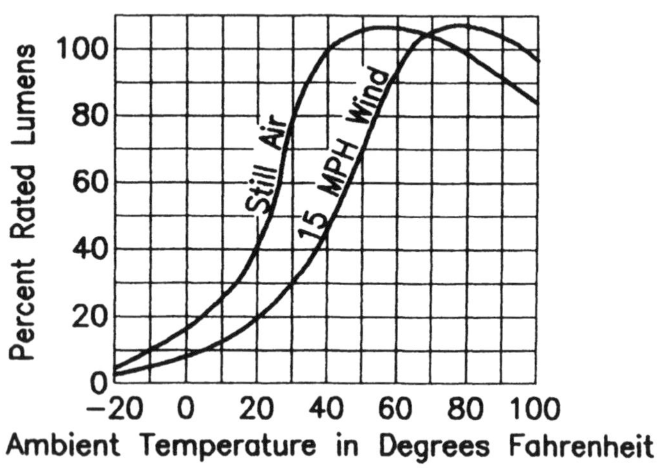

Fig. 21-1. Light output versus ambient temperature (°F) (IES, 1990).

Maintenance Factor: The ratio of the in-service lumens of a lighting system to the initial lumens, which can be determined with reasonable accuracy by the evaluation of various light loss factors that eventually contribute, in varying amounts, to the in-service illumination levels. Some of these light loss factors are luminaire ambient temperature, voltage to luminaire, ballast factor, luminaire refractor depreciation, room surface dirt depreciation, burnouts, lamp lumens depreciation, and luminaire dirt depreciation.

Matte Surface: A surface from which the reflection is predominantly diffused. The ANSI/IES-RP8 road surface classifications are given in Table 21-1.

Negative Contrast: A dark vertical face of an object toward the viewer relative to a light background.

Optical Guidance: The means by which visible aids indicate the course of the road direction. In tunnel applications, this aid is derived from the light sources as well as curb delineation and lane marker lines.

Positive Contrast: A light vertical face of an object toward the viewer relative to a dark background.

Reflectance: The ratio of the flux reflected by a surface or medium to the incident flux. This general definition may be further modified by the use of one or more of these terms: specular, spectral, diffuse.

Room Surface Dirt Depreciation: The accumulation of dirt on room surface reduces the amount of lumens reflected and interreflected to the work plane; it is one of the many factors to be considered in the evaluation of the design maintenance factor.

Utilization Factor: Also known as the *coefficient of utilization.* The ratio of the lumens that reach the work plane to the total lumen output of the bare lamps. It accounts for the luminaire photometric characteristics, mounting heights, room dimensions, and surface reflectances.

TUNNEL LIGHTING NOMENCLATURE

See Figure 21-2 for graphic representation of the following terminology.

Access Zone: That portion of the open approach of the highway immediately preceding the tunnel facade or portal. Also referred to as the approach zone. It is a function of the safe stopping distance for the design speed.

Threshold Zone: The first section at the entrance end of a tunnel, where the first decrease in daylighting takes place. This decrease can be accomplished by a reduction in daylighting with the use of screening or by the use of artificial lighting, or by a combination of these.

Transition Zone: The section at the entrance end that immediately follows the threshold zone and contains diminish-

Table 21-1. ANSI/IES-RP8 Road Surface Classifications

Classification	Description	Q_O
R1	Mostly diffuse	0.10
R2	Mixed diffuse and specular	0.07
R3	Slightly specular	0.07
R4	Mostly specular	0.08
Q_O is the representative mean luminance coefficient.		

ing light levels until the interior zone levels are reached. It is sufficiently long to provide for adequate eye adaptation time from open load brightnesses to interior tunnel brightnesses.

Interior Zone: (Sometimes known as the normal day zone.) The length of tunnel between a point just beyond the entrance transition zone, where eye adaptation is no longer a consideration for visual perception, and the exit portal.

Exit Zone: That portion at the end of the tunnel that during daytime appears brilliant when a motorist has driven for several minutes in the tunnel interior.

TUNNEL CLASSIFICATION

Figure 21-3 shows profiles of the various types of vehicular tunnels.

Underpasses

From the point of view of adequate supplementary lighting, there is no definite line of demarcation between an underpass and a short tunnel. The American Association of State Highway and Transportation Officials (AASHTO) defines an underpass as a portion of roadway extending through and beneath some natural or manmade structure, which because of its limited length-to-height ratio requires no supplementary daytime lighting. Length to height ratios of approximately 10:1 or lower will not, under normal conditions, require daytime underpass lighting. Other authorities, such as the Illuminating Engineering Society (IES) and the International Commission on Illumination (CIE), generally recognize all covered highways as tunnels and do not recognize an underpass as a separate and distinct structure. These authorities indicate that for lighting purposes every artificial or natural covering of a road, irrespective of the length and nature of the covering, is considered to be a tunnel.

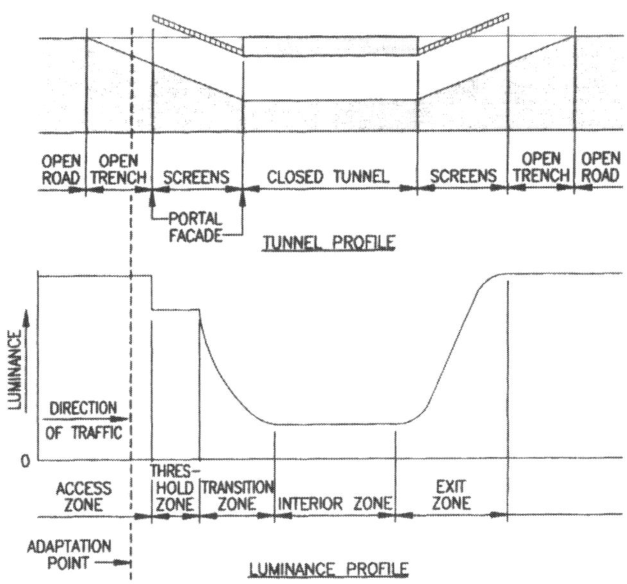

Fig. 21-2. Tunnel lighting nomenclaure (Schreuder, 1964).

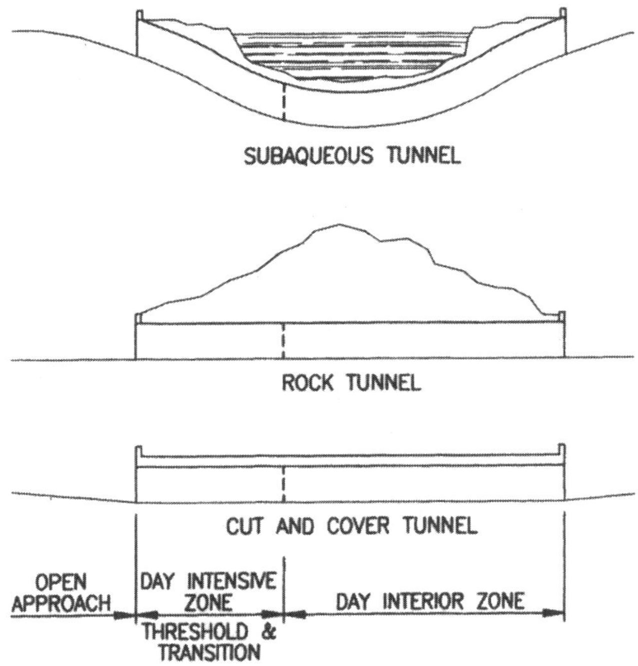

Fig. 21-3. Tunnel types (Thompson and Fanslor, 1968).

Short Tunnels

IES and CIE define a short tunnel as one where, in the absence of traffic, the exit and the area behind the exit can be clearly visible from a point ahead of the entrance portal. For lighting purposes, the length of a short tunnel is usually limited to approximately 150 ft. Some tunnels up to 400 ft long may be classified as short if they are straight, level, and have a high width and/or height to length ratio.

Long Tunnels

For lighting purposes, IES defines a long tunnel as one with an overall length greater than the safe stopping sight distance. CIE terminology defines every cover over a road as included in the concept "tunnel" irrespective of the length and nature of the covering.

PHYSIOLOGICAL CONSIDERATIONS IN TUNNEL LIGHTING DESIGN

The speed with which a motor vehicle can be safely operated depends entirely on the information that the motorist can quickly obtain about his environment. In gathering this information, the most important sense is vision. It is the only means by which the motorist can detect and discern objects and obstacles in the path of the vehicle. Therefore, the amount of information obtained depends on the lighting conditions. Obstacles and objects must be visible at such a distance that the necessary maneuvers can be executed in an effective manner.

Since the ability to see depends on the lighting conditions, which are subject to sudden brightness changes, especially during daytime hours, a great deal depends on the sensitivity of the eye. Continuing research on the physiology of the visual system has yielded important refinements to the calculation of adaptation luminance L_a. Previous luminance relationships between outside intensity, L_1, and interior intensity, L_2, expressed as the L_1/L_2 ratio, yielded a lower L_a in actual conditions than laboratory experiments. An improved method for determining L_a by means of a weighted average of the luminance measured over a 20° cone from the safe stopping distance is termed L_{20} and is presently included in the CIE technical report, *Guide for the Lighting of Road Tunnels and Underpasses.* The great variations encountered in the orientation, geographic, and seasonal conditions suggest careful examination of the 20° field for weighting. For example, in snow conditions the periphery of the 20° field has the same weight as equally large central areas, with the effect of higher entrance luminances. A more practical and precise method currently employed to determine L_a is expressed by

$$L_a = L_{2^\circ} + L_{seq} \qquad (21\text{-}1)$$

where

L_a = adaptation luminance

L_{2° = the average luminance in the center of a field of 2° in size

L_{seq} = equivalent veiling luminance

Table 21-2 presents reference data for sky, road, and environmental luminances in kcd/m² suggested by the CIE. These data may be used if no accurate local values are available.

Table 21-2. Reference Data for Sky, Road, and Environmental Luminances

Driving directions (Northern Hemisphere)	L_C (sky)	L_R (road)	L_E (environment)			
			Rocks	Buildings	Snow [a]	Meadows
N	8	3	3	8	15 (V) 15 (H)	2
E-W	12	4	2	6	10 (V) 15 (H)	2
S	16	5	1	4	5 (V) 15 (H)	2

[a] V = mountainous country with mainly steep surfaces facing drivers
H = flat, more or less horizontal country

Equation (21-1) is based on the physiology of the visual system and that the essential foveal adaptation is composed of the luminance in a 2° field of fixation and the equivalent veiling luminance L_{seq} caused by the environmental field luminance.

Two methods are available to determine L_{seq}, by calculation and by direct measurement. Equation (21-1) together with a polar diagram (Figure 21-4) superimposed and centered on a graphic of the tunnel portal and field, may be used to calculate the equivalent veiling luminance:

$$L_{seq} = K \sum_{i=1}^{n} \frac{E_{Gli}}{\theta_i^2} \quad (\text{cd/m}^2) \qquad (21\text{-}2)$$

where:

K = a constant (e.g., for the 20 to 30 year-old age group, K = 9.2)

E_{Gli} = illuminance at the eye produced by the glare source i in lux

θ = the angle between the fixation line and the glare source i. Valid for θ between 1.5° and 40° (the central foveal area of 2° should not be included in the evaluation).

Where no measured luminances of the field are available, Table 21-2 may be used.

The equivalent veiling luminance may be measured directly with a luminance meter fitted with a glare lens and entering measured data on a polar diagram superimposed and centered on a photographic or graphic representation of the tunnel portal and field adjusted for perspective. The total of L_{seq} is obtained from

$$L_{seq} = 0.5131 \times 10^{-3} \sum L_i \quad (\text{cd/m}^2) \qquad (21\text{-}2)$$

Important advances in understanding adaptance processes have been made by W. Adrian (1987, 1990), and fur-

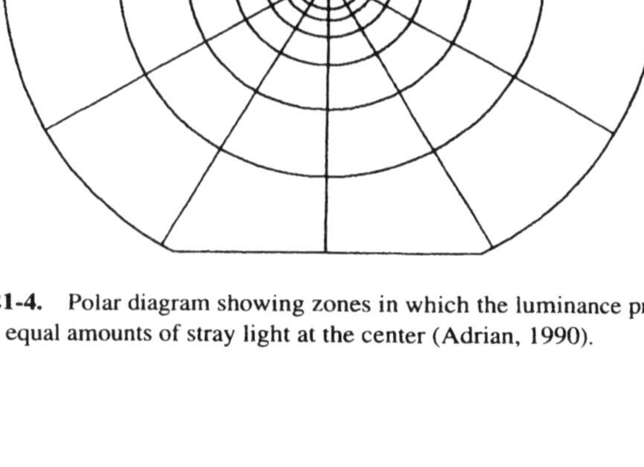

Fig. 21-4. Polar diagram showing zones in which the luminance produces equal amounts of stray light at the center (Adrian, 1990).

ther reading is suggested for derivation of the preceding equations.

The luminance required in the threshold zone, L_{th}, is a function of the adaptation luminance L_a and the threshold visibility level VL, where VL $= \Delta L/\Delta L_0$ for ΔL_0 equal to the threshold of luminance necessary for perception and for a target having photometric contrast, $c = \Delta c/L_{th}$ or $\Delta L = c \times L_{th}$.

Since L_a is composed of $L_{2°} + L_{seq}$, calculation of L_{seq} is required to reduce the visibility level VL to 1, thereby bringing the target to its perception threshold ΔL_0. Figure 21-5 shows various target contrasts that allow values of equivalent veiling luminance L_{seq} to determine entrance luminance L_{th} above the threshold of target perception.

Values of the L_{th} falling below the contrast line would cause a tunnel entrance to appear as a black hole, with loss of target visibility. Target visibility may be by positive or negative contrast to background luminance.

Symmetrical and asymmetrical concepts (asymmetrical consisting of positive and negative contrast for line-of-sight and counter beam, respectively) are illustrated in Figures 21-6–21-8. Selection of concept and subsequent design of the threshold lighting system requires careful examination to ensure visibility for specific tunnel applications. It should be noted that contrast of object to background luminance is diminished in the first 40 m of the tunnel by daylight penetration, which may influence selection of symmetrical or asymmetrical systems.

Figure 21-5 presents results of the calculation of the threshold visibility of the 10-min angle target of various photometric contrasts sitting in the entrance zone. The targets are seen against L_{th} as background, and their contrast is reduced to the threshold by the equivalent veiling luminance, L_{seq}, produced by the tunnel environment (solid curves). The dashed curves show results of subjective rat-

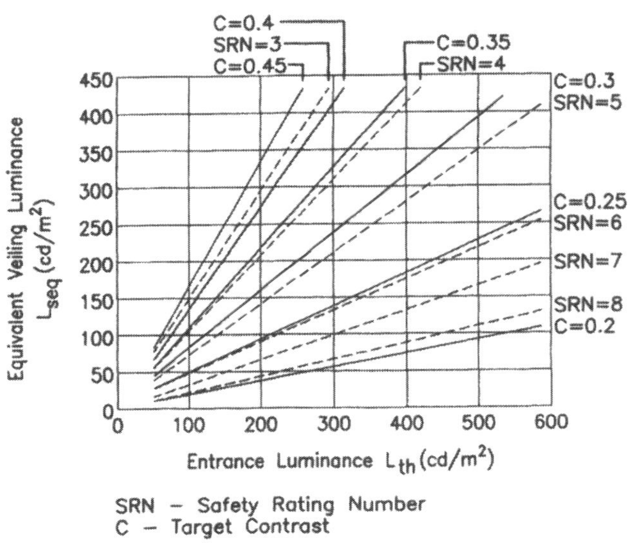

Fig. 21-5. Target contrasts (Adrian, 1990).

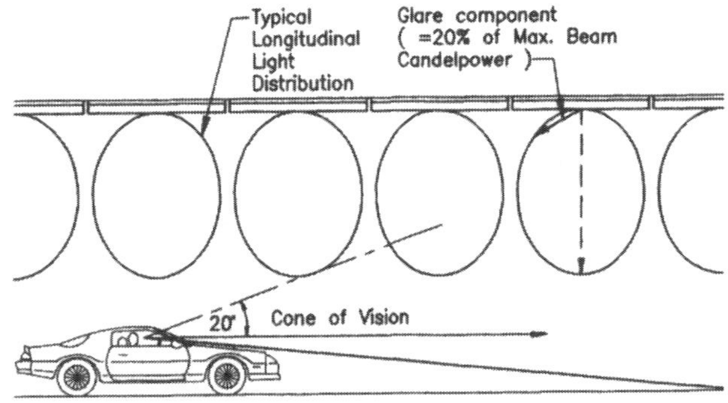

Fig. 21-6. Symmetrical lighting system.

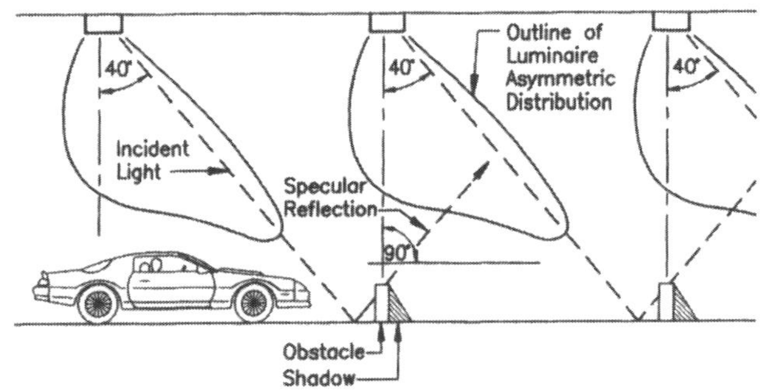

Fig. 21-7. Directional lighting system "line-of-sight" or "pro-beam lighting."

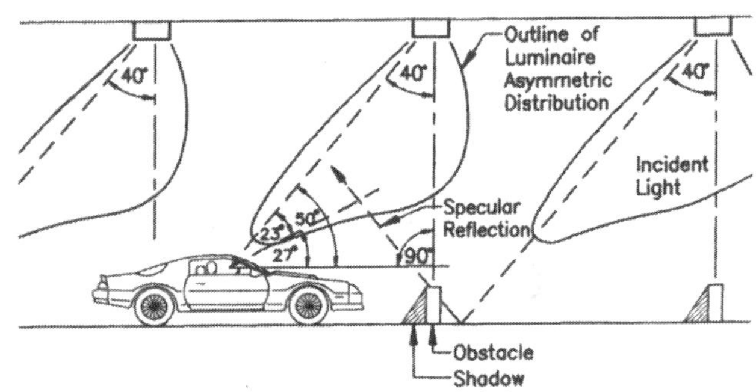

Fig. 21-8. Directional lighting system "counter beam lighting."

ings of the entrance luminance according to an ordinal scale as in Table 21-2.

ENTRANCE LIGHTING

The most critical section of tunnel lighting is the entrance section, which comprises the threshold and transition zones (Figure 21-2). Recommended illumination and luminance levels at the threshold and transition zones vary somewhat in different countries. Figures 21-9–21-11 illustrate these variations. To produce luminance levels solely by artificial

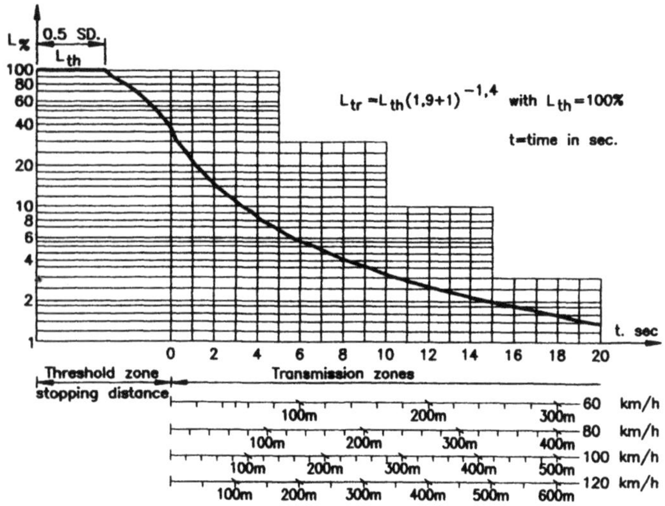

Fig. 21-9. Schematic represetation of lighting level in the various zones (CIE, 1990).

means using fluorescent lamps requires many luminaires on the walls and ceiling. Use of low-pressure sodium and high-intensity point sources permits reduction in the number of units. Attention to luminaire type selection, location, and

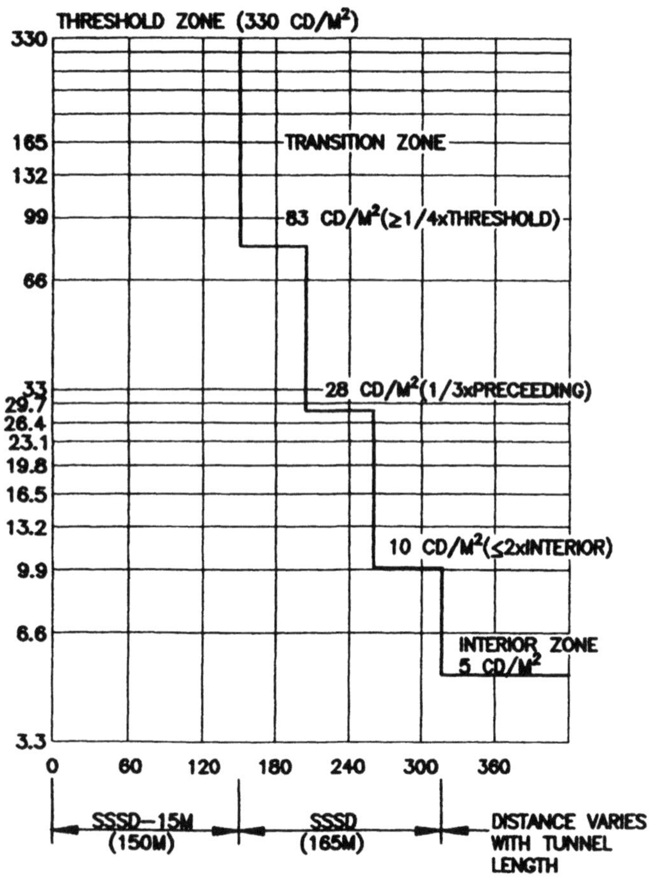

Fig. 21-10. Example of tunnel lighting luminance profile (ANSI/IES, 1987).

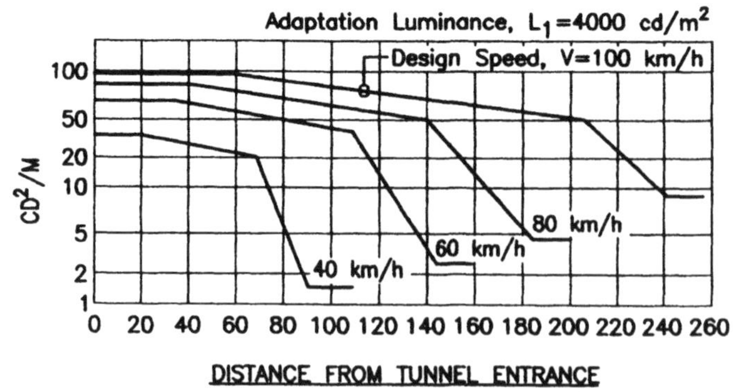

Fig. 21-11. Tunnel entrance lighting curves (Ketvertis, 1971).

spacing for reduction of glare and flicker is a primary design requisite in the entrance zone and throughout the tunnel.

To reduce the number of luminaires, some tunnel lighting designers have incorporated a grid at the tunnel entrance to screen out natural daylight in the proper amounts to assist in the eye adaptation process. The subject of sun screens will be treated later in this chapter, but as presently designed, they have not proven to be the ultimate answer to the first step in reduction of natural daylight.

Other factors to be considered in the evaluation of the lighting levels, as well as transition time calculated from the 20° field to the portal, are tunnel orientation, latitude, geographical location, approach grades, terrain, and conditions where the tunnel lighting problem can be readily solved using conventional equipment—for example, is a tunnel located in the higher latitudes where the sun or high sky brightnesses never come into the field of view; the orientation is east–west; the location is through a mountain; and the approach is on an upgrade, which allows the tunnel ceiling (as well as the walls) to act as an appropriate backdrop. An example of unfavorable conditions is a location of a subaqueous tunnel in the lower latitudes having a southeast or southwest orientation, where the sun or its reflection over open water would be directly in the field of view during morning or afternoon rush hours, on a downgrade approach to the entrance portal, where the roadway surface (an eventually soiled and usually poorly reflecting surface) becomes the principal backdrop for silhouette discernment of objects or stalled cars. A good example is the Hampton Roads Tunnel and Bridge Crossing between Hampton and Norfolk, Virginia, where additional rows of luminaires on the tunnel walls provide increased illumination levels.

LUMINANCE LEVEL IN THE TUNNEL INTERIOR

Recommendations for tunnel interior lighting levels, as well as methods of measurement, presently vary among authoritative sources. The CIE recommends average luminance of the road surface to vary with safe sight stopping distance (SSSD) and traffic flow between 1 and 15 cd/m² in heavy traffic conditions. The American National Standard Institute practice for tunnel lighting, ANSI/IES RP-22 (1987), suggests at least 5 cd/m^2 maintained luminance on the roadway. AASHTO also recommends an average maintained luminance on roadway 5 cd/m^2. These values are based on a wall reflectance factor of at least 50%. When the reflectance factor is less than 50%, lighting should be increased to compensate. This suggests that the wall luminance is the parameter that should be kept constant. These values are minimum maintained values measured on the road surface. The walls should have at least the same level of luminance up to 2 m (6.2 ft) above the roadway.

Based on a reflectance factor of 50% for the walls and ceilings and a reflectance factor of 20% for the roadway, Table 21-3 summarizes recommended values for luminaire levels.

While there is disparity in these recommendations, there are conditions under which the extremes can be shown to be adequate for daytime interior lighting. Where there is sufficient time for complete adaptation, such as in long tunnels, the interior lighting can approach the criteria for nighttime street lighting. On the other hand, if only partial adaptation can be effected, then the higher values appear to be the better choice during daytime hours. In many cases, economic factors, as well as the availability of the proper lighting equipment, will play an important role in the final interior lighting level.

EXIT LIGHTING

During the daytime, the exit of the tunnel appears as a bright hole to the motorist. Usually, all obstacles will be discernible by silhouette against the bright exit and thus will be clearly visible. This visibility by silhouette can be further improved by lining the walls with tile or panels having a high reflectance and thus permitting a greater daylight penetration into the tunnel. This effect is shown in Figure 21-12.

When the exit is not entirely clear but partially screened by a large object such as a truck, then a different visibility requirement is present. In this instance, if a smaller object is following the large object, the smaller object may not be readily visible. Whereas the large object screening the exit would be clearly visible, the following small object would have equivalent brightness to the large one and, therefore, would not be readily discerned by silhouette.

Under certain conditions and during certain periods of the day, sunlight may penetrate directly into the tunnel exit,

Table 21-3. Summary of Recommended Day Interior Maintained Luminarie Levels in Candela Per Square Meter.

Authority	Walls (up to 2 meters above roadway)	Roadway
IES	5	5
AASHTO	5+	5+
CIE	1-15	1-15

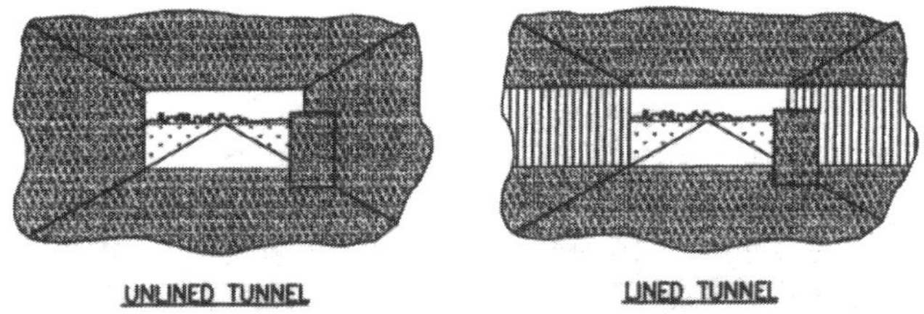

Fig. 21-12. Effect of natural light penetration on walls at tunnel exit (Thompson and Fanslor, 1968).

creating an extremely difficult visual condition. Coupled with the use of glazed walls and glossy paints, the resulting specular reflection may create considerable interreflections, thus preventing the motorist from identifying the tunnel outlines clearly, as well as creating a situation where discomfort and disability glare would be experienced. Under these conditions, discernment of objects by silhouette may be difficult, and a hazardous traffic situation may result. Even sun screens will not offer much help, and so the situation should be avoided. Reorientation of the tunnel exit or the use of materials with diffuse reflectances (matte or flat) should prove helpful.

LIGHTING OF SHORT TUNNELS

In most cases, a lighting system is not required inside short tunnels for adequate driver visibility. Daylight penetration from each end, plus the silhouette effect of the opposite end brightness, generally will assure satisfactory visibility. On the other hand, tunnels between 75 and 150 ft in length may require supplementary daytime lighting if the daylight is restricted due to roadway depression, tunnel curvature, or the proximity of tall buildings in urban areas. Short tunnels appear to the approaching driver as a black frame, as opposed to the black hole usually experienced in long tunnels. The effect, illustrated in Figure 21-13, indicates that adaptation to the lower level inside the tunnel will not have an opportunity to occur. After entering the tunnel, the central part of the field of view is taken up by the brightness of the exit, so that the brightness of the walls and ceiling has insignificant adaptation stimulus.

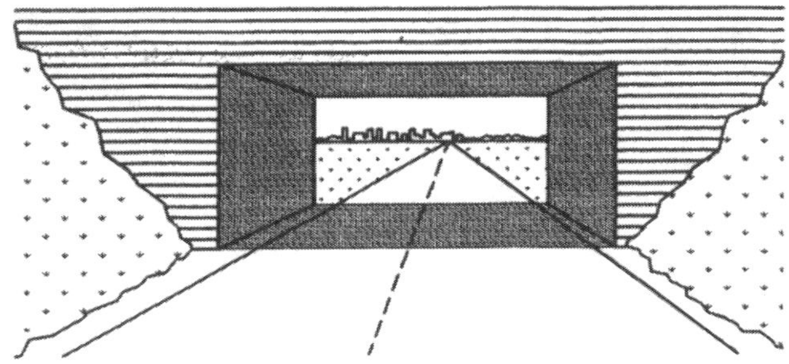

Fig. 21-13. A short tunnel appears as a dark frame (Schreuder, 1964).

Obstructions in the short tunnel can be seen if they are high enough to silhouette against the bright environment of the exit. If the vertically projected dimension of the dark frame across the roadway is less than the height of the smallest object that must be seen, then the object will be silhouetted against the exit brightness and will be visible. Figure 21-14 illustrates this for various tunnel lengths on a straight and level roadway with the driver's eye at a height of 3.75 ft (1.14 m) and at a distance of 475 ft (145 m) (stopping sight distance at 60 mph) from the entrance portal.

Due to daylight penetration from both portals, a short tunnel on a straight and level roadway can be as much as 75 ft long (23 m) and not require daytime lighting for an obstacle having a height of 6 in. (152 mm). Silhouetting can be enhanced by lining the structure wall with a light-colored material to reduce the darkness within the tunnel.

Where local conditions warrant the installation of an artificial lighting system for daytime operation, the level of luminance within the entire length of the tunnel should be at least 0.1 of the expected maximum open approach luminance. This requirement may mean that a luminance level of 300–350 fL is needed for satisfactory visibility. Where it is not practical to achieve these levels, visibility can be improved by providing an opening in the ceiling about midway through the structure to permit daylight penetration. In effect, the opening creates two structures and may satisfy the requirement for lighting during the day. Walls and ceilings lined with a light-colored, matte-finished material will enhance visibility. A design with a funneled-up ceiling at the portals will permit additional daylight penetration.

LIGHTING OF LONG TUNNELS

Lighting in long tunnels should generally follow the luminance profile illustrated in Figure 21-10 for satisfactory visibility during daylight hours. The system should be flexible enough to permit its operation at night at a reduced level.

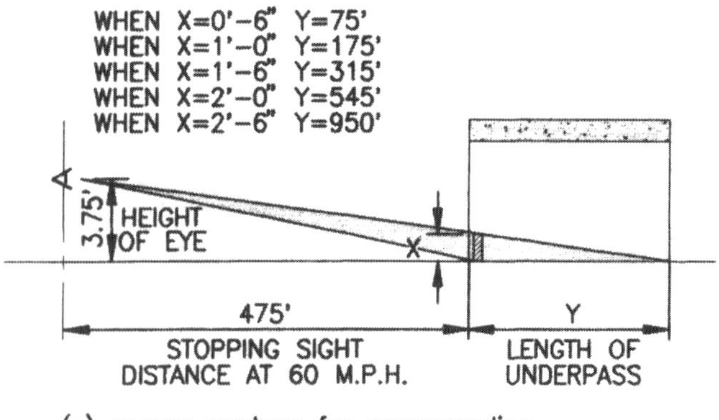

Fig. 21-14. Vertical dimension of dark frame (Schreuder, 1964).

The long tunnel requires two daytime lighting levels, one for the intensive zone (the entrance zone comprising the threshold and transition sections) and another for the normal day zone (interior zone).

The Entrance Zone

The entrance zone is the most critical area, because without sufficient portal brightness, the entrance will appear to the approaching driver as a black hole. The most severe visual task is not when the driver is passing through the plane of the portal shadow, but when he is outside of it and is trying to see within the portal shadow. The point in front of the portal at which must be discerned objects within the tunnel is dictated by the safe stopping sight distance at the posted speed. At 60 mph (87 kph), this would be 650 ft (198 m) in front of the portal and, in most cases, will be at a point ahead of the adaptation point. This means that the motorist's eye is

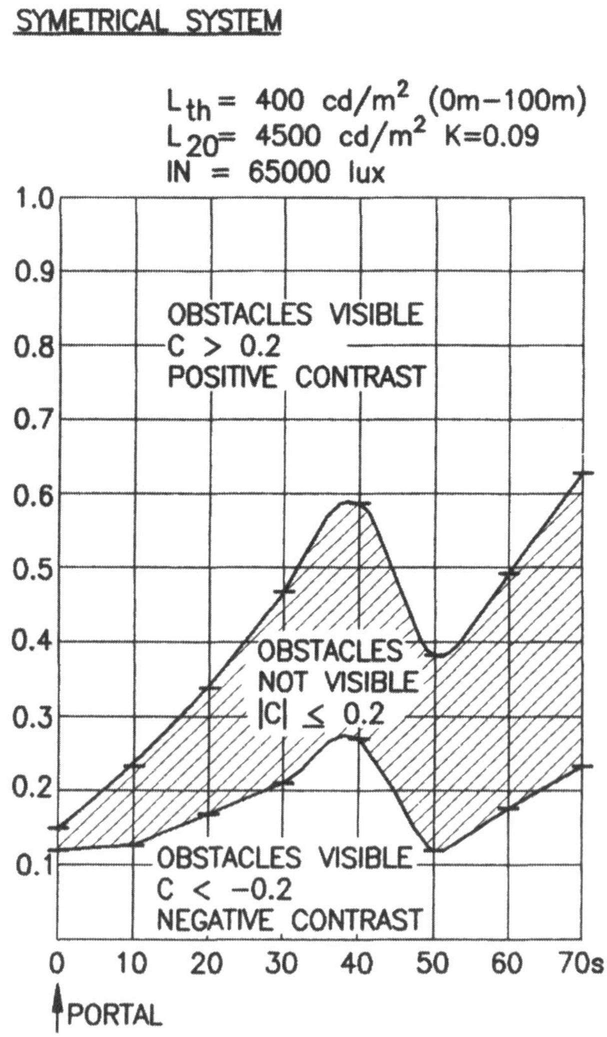

Fig. 21-15(a). Zones of visibility of obstacles (luminance meter 160 m in front of the portal): (a) symetrical system; (b) counter beam lighting system; (c) pro-beam lighting system.

at that time adapted to the ambient luminance level of the open road. According to the recommendations of IES, CIE, and Schreuder, the portal luminance level should be 1/10, and preferably 1/8, of the maximum expected open approach luminance. This portal luminance should be displayed on the walls, ceiling, and roadway so that these surfaces can serve as an effective background for discernment of objects by contrast and silhouette. The effects of the so-called black hole would then be drastically reduced or even eliminated. The first reduction in luminance should occur in the threshold zone and should be at a constant level. The length of the threshold zone depends on the driving speed and the position of the adaptation point, which, in turn, depends on the geometry and dimensions of the access zone.

The symmetrical lighting system utilizes luminaires having photometric characteristics that direct light transverse to the longitudinal axis of the tunnel roadway of sufficient magnitude to enable objects to be observed by positive or negative contrast.

The asymmetrical lighting system utilizes luminaires mounted above the road with photometric distribution to direct the light toward the driver for the counter beam system and away from the driver for the pro-beam system. The counter beam system produces a negative contrast between an object and the motorist, which enhances visibility. A common concern of the counter beam system is that there is an increase in glare. Designs currently employ beam angles of 40 from the vertical to minimize the glare. The pro-beam system produces a positive contrast to enhance object visibility; however, there are concerns that reflected glare from vehicle windows ahead will produce diminished visibility. Experiments at Wevelgen Tunnel in Belgium were carried out to compare these three systems. Each system has some advantages and disadvantages in its visibility zones, as shown in Figures 21-15a–c. Selection by the designer of a tunnel lighting system will require evaluation of proposed ceiling and wall finishes, roadway reflectance, roadway treatment, and proposed maintenance and cleaning measures.

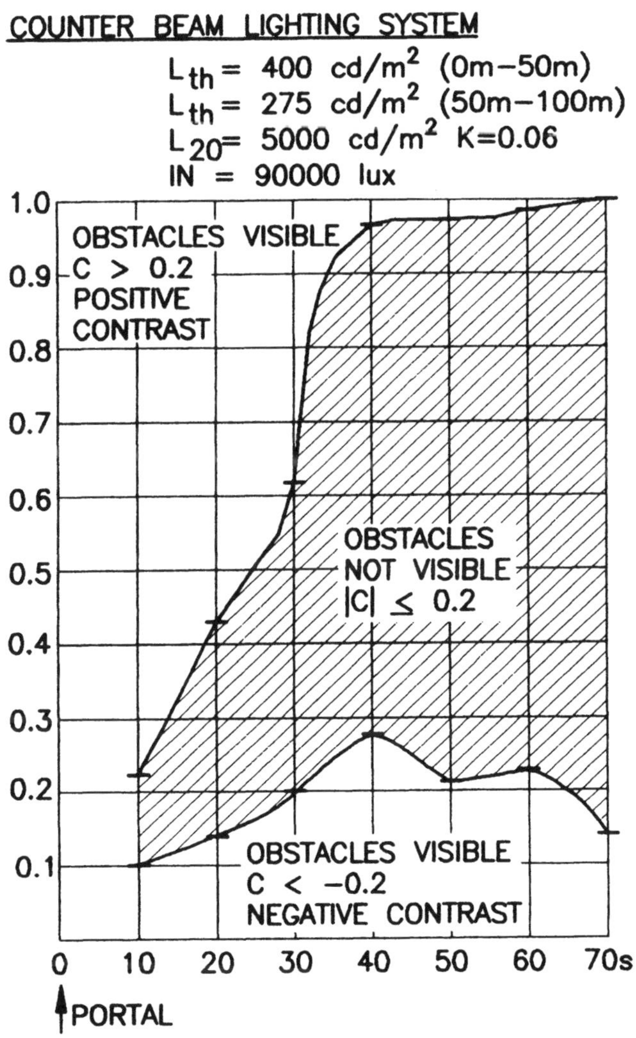

Fig. 21-15(b).

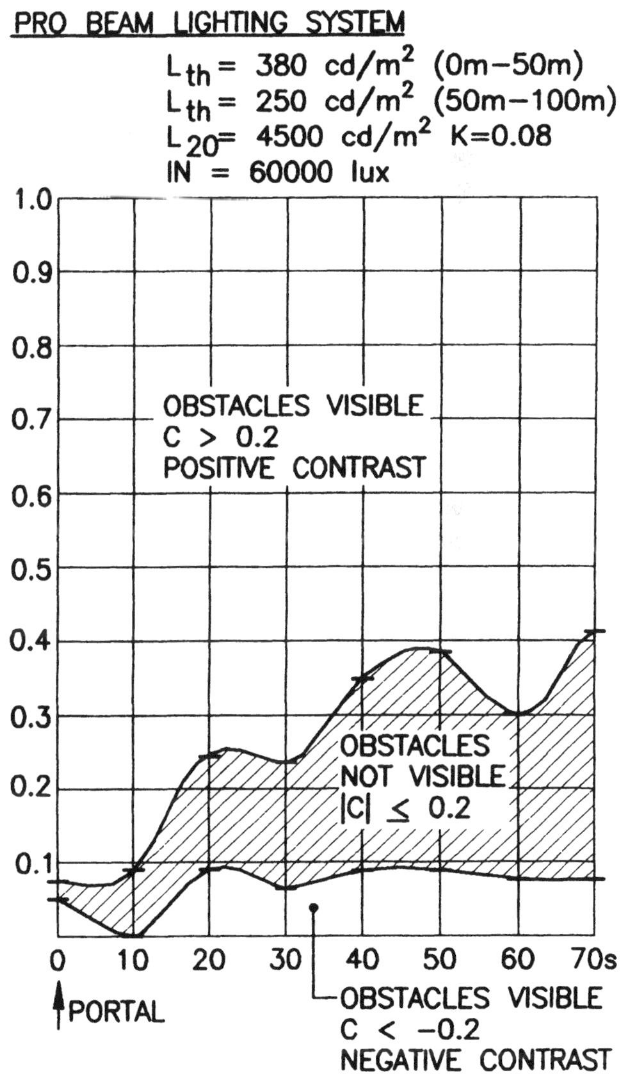

Fig. 21-15(c).

Stopping Sight Distance

The distance at which a critical object can be seen in the threshold must correlate with the driving speed and the safe stopping sight distance. If a test distance of 325 ft is to be safe stopping sight distance on a wet road, the corresponding vehicle speed would be about 42 mph. In a similar manner, the safe stopping sight distance should be used as the distance to the critical object for other speeds in determining the length of the threshold zone. From the geometric relation shown in Figure 21-16, an adaptation point can be calculated from

$$d_a = (H - h)\cot I \qquad (21\text{-}4)$$

where

d_a = adaptation point
H = portal height, in ft
h = height of driver's eye above pavement (usually taken as 3.75–4 ft (1.143–1.22 m)
I = vertical angle in degrees (Schreuder (1964) suggested 7°; Ketvertis (1989) suggests 25°)

Knowing the location ahead of the portal of the adaptation point, Ketvertis (1989) suggests that the total length of the supplementary zone (threshold plus transition) can be computed from

$$L_s = (V_s \times t_a) - d_a \qquad (21\text{-}5)$$

where

V_s = speed (ft/sec)
t_a = adaptation time (sec, assumed to be reasonable at 8 sec)
d_a = adaptation point (ft)

The length of the threshold zone can be calculated from

$$L_t = \frac{a}{h - a} \times d_s \qquad (21\text{-}6)$$

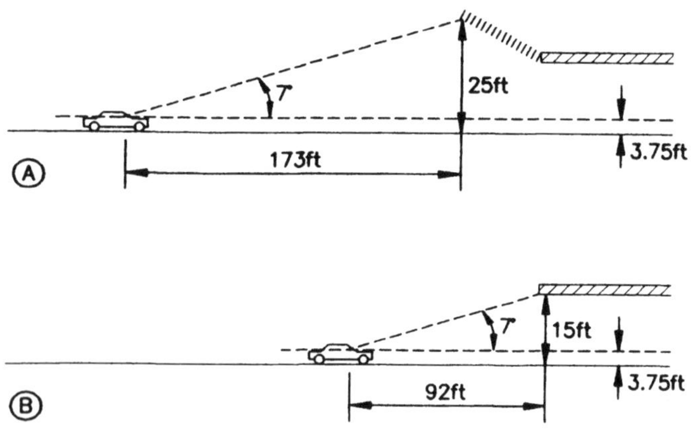

Fig. 21-16. Location of adaptation point.

where

a = hazardous object height (1 ft)
h = height of driver's eye position (4 ft)
d_s = safe stopping sight distance

Table 21-4 shows safe stopping sight distances for various design speeds.

Adaptation Point

Figure 21-16 illustrates the method by which the adaptation point can be located. The vertical angle shown is a value taken as an average for the shielding angle of vehicle windshields. The two profiles demonstrate that, as the portal height increases, the distance of the adaptation point in front of the portal also increases. The raising of the facade, consequently, is a very effective expedient to start the adaptation process sooner and, as a result, to reduce the length of the threshold zone. Further aids are dark finish of the facade and other neighboring surfaces.

When the motorist has passed the adaptation point, his eyes begin to adapt to the dark area of the tunnel entrance. If the luminance within the transition zone of the tunnel decreases, inconvenience caused by afterimages is prevented. These luminances exist at points equal in distance to the safe stopping sight distance in front of the observer.

Haze

At the entrance of tunnels, the presence of haze due to the exhaust emissions of vehicles may create a visual problem, particularly when there is no open-grid construction for screening of daylight through which the fumes can escape. Whenever this haze is illuminated by the sun, the luminance can become very high, and it may prove difficult to see through it because of the veiling effect.

The outside luminance L_1, the background luminance L_2, and the luminance of an object L_3 define contrast as

$$C = \frac{L_2 - L_3}{L_2} \qquad (21\text{-}7)$$

A veiling luminance L_v between the object and the observer increases the luminance of the object to $L_{3'} = L_3 +$

Table 21-4. Minimum Stopping Sight Distance.

Design Speed, mph (km/hr)	Stopping Sight Distance, ft (m)	
	Wet Pavement [a]	Dry Pavement [b]
30 (48)	200 (61)	158 (48)
40 (64)	300 (91)	236 (72)
50 (80)	450 (137)	236 (99)
60 (96)	650 (198)	434 (132)

[a] Rounded for design
[b] Computed

L_v and that of the tunnel to $L_{2'} = L_2 + L_v$. The new apparent contrast, C', between the object and the tunnel background becomes

$$C' = \frac{L_{2'} - L_{3'}}{L_{2'}} = \frac{(L_2 + L_v) - (L_3 + L_v)}{L_2 + L_v} = \frac{L_2 - L_3}{L_2 + L_v} \quad (21\text{-}8)$$

or

$$C' = C\frac{L_2}{L_2 + L_v} \quad \text{and} \quad \frac{C'}{C} = \frac{L_2}{L_2 + L_v} \quad (21\text{-}9)$$

The presence of the interposing veiling brightness acts on the eye the same way as a source of stray light that, within the eye, produces a superimposed veiling luminance upon the retinal image of the object to be seen. This alters the luminance of the image and its background; hence, the contrast. The ratio of the apparent contrast to the original threshold contrast, shown in the equation above, will always be less than 1 because of the presence of L_v in the denominator. This means that the object, which was previously seen at the threshold level, will not be seen under conditions of a lower contrast, with the same background luminance, L_2. To bring the object back to threshold visibility, the background luminance L_2 would have to be increased. Figure 21-17 shows a curve of the visual performance criterion function indicating the variation of the background luminance with the task contrast; illustrated is an example of the additional background luminance required for a decrease in task contrast to bring the object to threshold visibility. Another manifestation of a veiling brightness is that produced by the reflection of the top deck of a car's dashboard through the inclined windshield on a bright, sunlit day. This also appears to the motorist as a bright screen between the object and the viewer. A dirty and dusty windshield will produce a similar effect.

Theoretically, it should be possible to "see through" all of these veiling brightnesses and sources of stray light, whether they occur individually or collectively, by providing sufficient background luminance. However, in practice, overcoming these effects by providing brightnesses using artificial means is neither practical nor economical.

Night Lighting

The requirements for lighting at night in short and long tunnels are identical. The lighting must be designed to avoid flicker and glare, and the illumination level should be the same throughout the entire length. The ratio of the roadway lighting inside the tunnel to that of the roadway outside the tunnel should not be greater than 3:1 and preferably should not exceed 2:1.

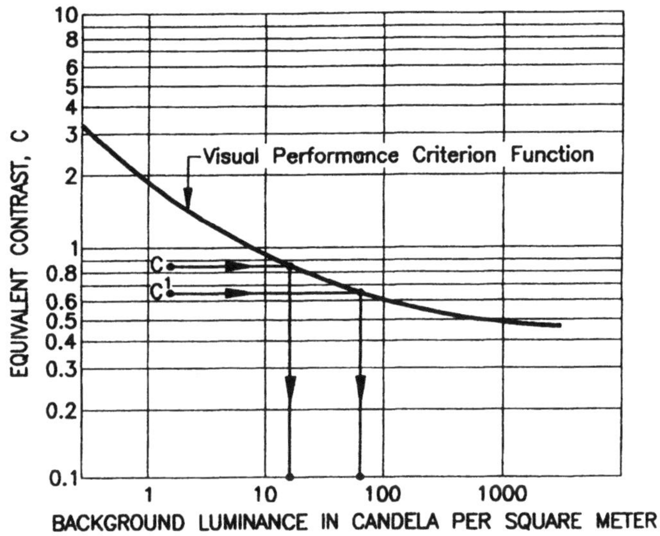

Fig. 21-17. Visual performance criterion.

At night, the most severe visual task occurs at the exit of the tunnel when the roadway outside the tunnel has no lighting or has a lower level of illumination than inside the tunnel, and the exit appears as a dark hole to the motorist. The common cause for this situation is excessive lighting inside the tunnel. Many tunnels have the same level of illumination for night lighting as for daytime lighting in the interior zone. Inasmuch as the normal daylighting system usually must provide substantially more brightness than is required for night lighting, a single lighting system, without provisions for varying the amount of illumination, will not satisfy both operating requirements. Lighting in excess of the optimum amount is detrimental and may create a hazardous condition at the exit.

In instances when it is not desirable to reduce the night lighting level to a ratio of, at most, 3:1 with the approach roadway lighting, consideration should be given to providing transition lighting. The zone for the transition on the approach roadway should be about 500 to 600 ft long (150 to 180 m).

Approach roadways to many tunnels are of the depressed type, and these sections of the highway should be provided with roadway lighting. For the comfort of the motorist, the nighttime lighting level in this type of depressed highway should be 50% greater than that required on the open road. If none is provided on the open road, then the lighting level in this section should be 50% of the level within the tunnel.

When sun screens are applied at the entrance and/or exit of the tunnel, the nighttime lighting system must also be extended under and to the end of the sun screens. These sections of the lighting installation must also operate at dusk and at low levels of outside illumination during the daytime.

Very short tunnels that are straight and on a flat grade, and that are not provided with lighting for daytime operation, can usually be provided with satisfactory nighttime visibility by the proper positioning and mounting height of pole-mounted street lighting luminaires at each end of the tunnel. This principle of obtaining artificial light penetration into the tunnel is illustrated in Figure 21-18. When higher-than-usual mounting heights are used for the luminaires adjacent to the tunnel portals, care must be exercised not to exceed the uniformity ratio; otherwise, some tunnel lighting may be necessary.

Use of Daylight in Tunnels

Natural daylight can be employed in lighting the interior of tunnels in three ways:

1. The use of subdued daylight at the entrance portal of a long tunnel by the use of daylight screening.
2. The application of light slits in the roof of a short tunnel.
3. The use of the light that provides a positive target contrast at a tunnel exit.

Sun Screens

Daylight screening is used as a means of reducing roadway luminance in many European tunnels and, to a limited extent, in this country (Chesapeake Bay Bridge–Tunnel and Harbor Tunnel, Baltimore). All of these tunnels use a form of louver spanning over the roadway and are constructed with vertical or inclined slats and in various geometric shapes and patterns.

Since louvers must be designed with an upper cut-off angle sufficient to prevent direct sunlight penetration to the roadway, they do not perform with a constant transmission factor when the ambient access illumination level varies from sunrise to sunset and from summer to winter as the sun's altitude gets progressively lower in the sky. As a result, there will be times, under certain atmospheric conditions, when artificial lighting will be required under the louvers to supplement daylight so that the portal shadow, in effect, will not be transferred to the beginning of the louver.

The important characteristic for a louver to operate successfully is that the transmission factor must remain constant under all outdoor illumination conditions. It appears, however, that additional research is required in this field before a satisfactory solution is reached. For these reasons, sun screens are not recommended as a solution to the tunnel entrance visibility problem.

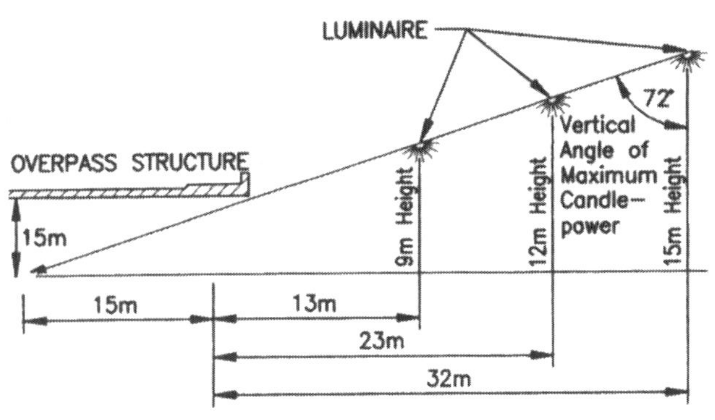

Fig. 21-18. Location of roadway lighting luminaires adjacent to underpass portals.

Daylight Slits

These can sometimes be used in tunnels of medium length. An estimate can be made of the effect of such slits by the following formula if we assume that the slit is very long and that it can be considered a diffuse light source:

$$E = \frac{\pi B}{2}(\sin \gamma_2 - \sin \lambda_1) \tag{21-10}$$

where

- E = the illumination at a point P (Figure 21-19)
- B = the luminance of the slit
- γ_1, γ_2 = the angles at which the edges of the slit are seen

Figure 21-20 illustrates some of the results obtained using this formula.

Daylight Entering a Tunnel Portal

Figure 21-21 illustrates the amount of horizontal (E_H) and vertical (E_V) illumination from penetration of daylight in terms of the percentage of outside illumination. Figure 21-20 is plotted from measurements and from calculated data. It shows that the contribution to tunnel illumination by the penetrating daylight can be neglected after about 20 ft (6 m) from the portal.

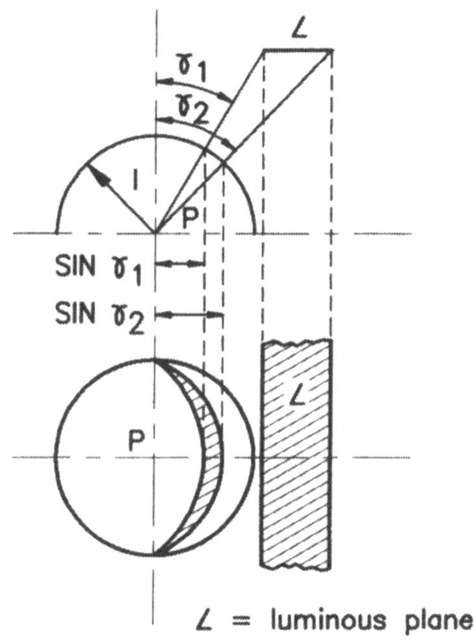

Fig. 21-19. Determining transmission of daylight slits.

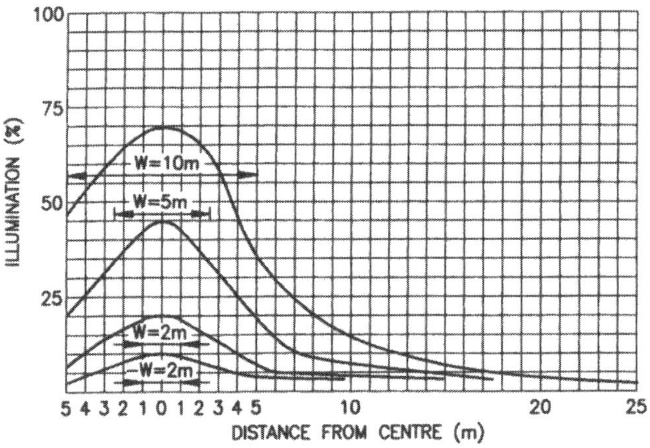

Fig. 21-20. Transmission of daylight slits.

TUNNEL LINING

The interior walls and ceiling of the tunnel are important adjuncts of the lighting system, and their brightnesses and uniformity depend on the reflectance quality of the surface. While the ceiling may not play an important part as a background against which objects may be silhouetted, except in tunnels with an upgrade approach, its light color and high reflectance is desirable because of the higher wall and roadway brightnesses that will result. The surfaces soil and can be easily cleaned. A light-colored matte (nonspecular) finish surface with a reflectance factor of 70% or greater is recommended.

The ceiling should also consist of a durable and reflective surface. Epoxy paint coatings have been applied with mixed success.

The tunnel roadway surface should have as high a reflectance factor as possible. To enhance reflective qualities, natural baked flint is used in England in conjunction with black asphalt. Some European tunnels have used equivalent artificial materials. Concrete, with the same additives, would perhaps produce almost ideal tunnel pavement material.

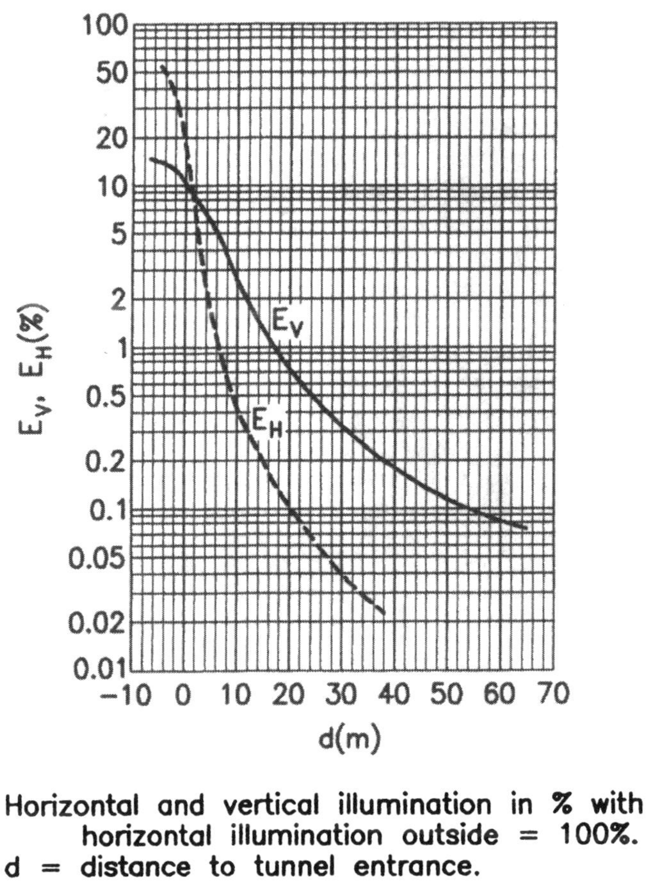

Fig. 21-21. Daylight entering a tunnel portal (Schreuder, 1964).

TUNNEL LIGHTING LUMINAIRES

Development of higher efficacy lamps with significantly greater life has eliminated the use of the incandescent lamp for tunnel lighting. Fluorescent, low-pressure sodium, high-pressure sodium, and metal halide offer the tunnel lighting designer options for color and the application of symmetrical and asymmetrical systems.

Multilamp Enclosed Fluorescent Lamp Luminaire

This luminaire with internal reflector is placed either on the walls or in the angle between the ceiling and the walls and is one way of adding new lighting to an old tunnel. This type of luminaire has been installed in some recently completed tunnels. It offers considerable flexibility and must be watertight, bug-tight, and dust-tight for efficient lighting and maintenance. This is illustrated in Figure 21-22. The

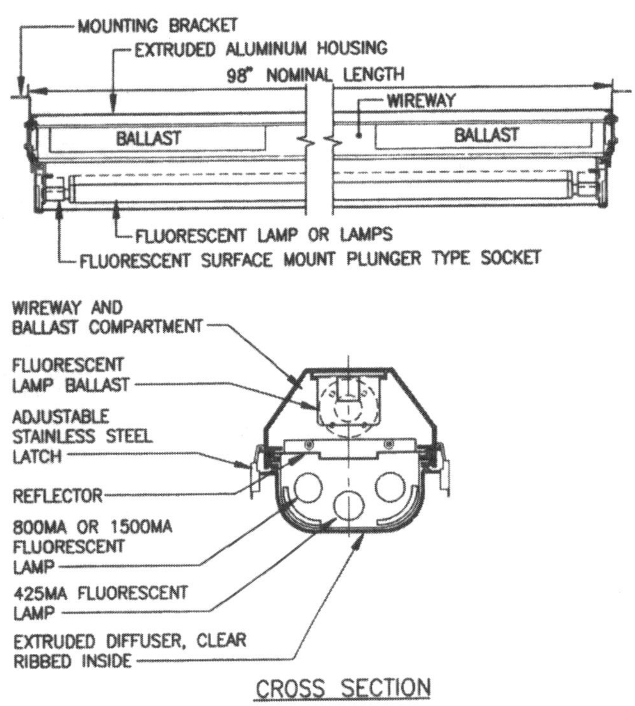

Fig. 21-22. Typical multilamp fluorescent luminaire.

particular advantage of this fixture is the control of the photometric distribution by the design of the reflector in conjunction with the enclosing diffuser and the higher lumen output per unit length. Multilamp luminaires have been used in the Hampton Roads Tunnels and Elizabeth River Tunnels in Virginia.

Low-Pressure Sodium Luminaire

The low-pressure sodium lamp has the highest efficacy of lamps applicable for tunnel lighting. The lamp's output is monochromatic yellow, which can be objectionable for color recognition of vehicles in a tunnel. A mix of fluorescent luminaires with low-pressure sodium luminaires has been utilized in the Fort McHenry Tunnel in Baltimore, Maryland, and also the Elizabeth River Tunnels in Norfolk, Virginia. This combination enables the viewing of a broader spectrum of color. Low-pressure sodium lamps have been installed in tandem in 8-ft-long luminaries as illustrated in Figure 21-23 to match the configuration of multilamp fluorescent luminaires.

High-Intensity Discharge Luminaires

High-pressure sodium and metal halide lamps have been utilized for symmetrical and asymmetrical tunnel lighting systems. These lamps are very versatile and enable precise luminaire photometric control for both wall and ceiling mounting applications. These lamps are preferred for counter beam and pro-beam systems. Wall-mounted and ceiling-mounted luminaires are illustrated in Figures 21-24 and 21-25. These fixtures must be watertight and dust-tight to operate in a tunnel environment for many years.

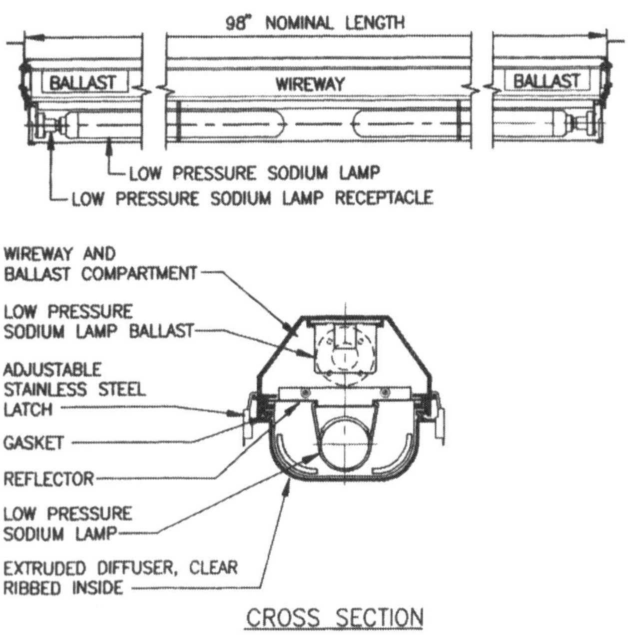

Fig. 21-23. Typical low-pressure sodium luminaire.

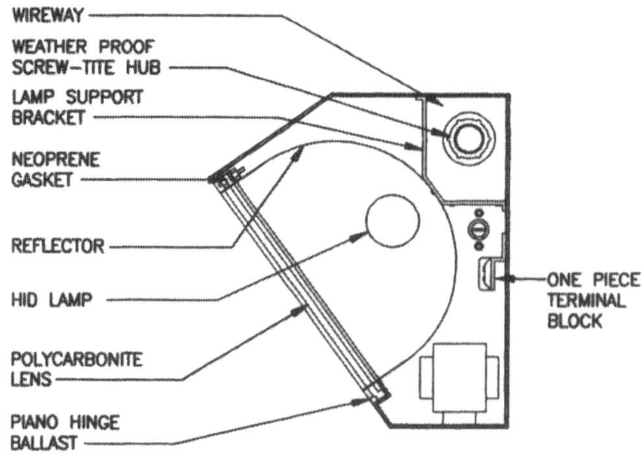

Fig. 21-24. High-intensity discharge luminaire (wall-mounted).

High-Intensity Discharge Lamp and Pipe System

The lighted pipe system uses high-pressure sodium or metal halide lamps in a luminaire coupled to a light guide consisting of an acrylic tube having an internal optical film and reflector. The lighted pipe system is used for symmetrical lighting systems. The system uses a 6-in.-diameter tube in lengths up to 44 ft. Luminance of the tube can be varied by varying the length of the tubes and lamp size. Emitting sectors or aperture from 90° to 180° are used to suit the photometric requirements of the tunnel.

MAINTENANCE

Reliability of the tunnel lighting system is crucial, and its continuous operation without interruption must be assured. The lighting system depends on many factors that should be considered at the design stage. Maintenance of lighting is a most important factor, and the proposed maintenance program must be reflected in the initial design.

The amount of maintenance required depends on the location, type, and volume of the traffic; the type and capacity of the ventilation system; the tunnel cross section and shape; the ceiling and wall finish; operating speeds in the tunnel;

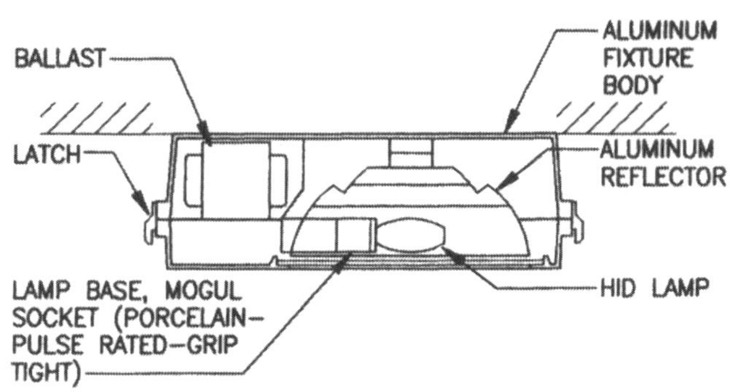

Fig. 21-25. High-intensity discharge luminaire (ceiling-mounted).

the type and location of luminaires; grades and alignment within the tunnel; and the electrical system and supply.

A tunnel must have sufficiently high initial illumination to compensate for the many factors that will reduce it while it is in service. These factors fall into two categories: light loss factors not to be recovered and light loss factors to be recovered. In the first category are luminaire ambient temperature, voltage to luminaire, ballast factor, and luminaire surface depreciation. In the latter category are room lumen depreciation and luminaire dirt depreciation.

The most important factor, with the exception of the luminaires, is the surface treatment of the walls and ceiling. Reducing the reflection factor of these surfaces may be more detrimental to lighting effectiveness than any other single factor. Vehicular tunnels must be finished with an interior surface that will not deteriorate as time progresses and as chemicals attack it, that will not readily soil, and that can be easily cleaned. These attributes are characteristic of a light-colored matte-finished tile or porcelain-enameled panel with an initial reflectance of 70% or higher.

Luminaires should be sealed to prevent the entry of dust and water when the tunnel is being cleaned by high-pressure water spray. They should be designed to permit quick and easy internal cleaning. The materials used should be resistant to alkaline deposits, to concentrated exhaust fumes, and especially to cleaning solutions that must be used to thoroughly clean the tunnel walls and ceiling.

Lamps should be replaced on a group replacement program, which not only will help to maintain the light output at the desired design level, but will also provide for a balanced maintenance work load. The magnitude of the relamping task will generally dictate that group relamping must be done by sections. To replace lamps on a burnout basis may be acceptable in certain cases, but this is usually false economy from the standpoint of lighting output and equipment maintenance cost.

The inside of the luminaire will require cleaning more often than the lamps are replaced, but the maintenance schedules should still prescribe cleaning each time a lamp is replaced. Some tunnels require a program of cleaning on a weekly basis, whereas others may not require cleaning for several weeks. Locations in the more northern latitudes, where freezing conditions may prevail for long periods, make cleaning operations difficult and frequently impractical. All of these factors must be recognized in the initial design of the lighting system.

EMERGENCY LIGHTING

Complete interruption of the tunnel lighting, even for an instant, cannot be tolerated, which requires a reliable power supply (for detailed information, see Chapter 22).

For dual-utility power sources, one-half of the tunnel lighting is connected to each supply, so that in case of failure at least one-half of the system remains energized until transfer of the entire load to the remaining source.

For single-utility service and standby diesel generator, one-sixth of the tunnel lighting is connected to an emergency circuit, which in case of power failure is immediately transferred to a central emergency battery system until the diesel generator picks up to carry one-half of the tunnel lighting. An alternative to a central battery system is to utilize individual battery packs within one-sixth of the fixtures to maintain illumination from portal to portal.

LIGHTING OF TRANSIT TUNNELS

The lighting of transit tunnels (rapid transit or subway tunnels) shares similar design features with the lighting of vehicular tunnels. The portal, in bright daylight conditions, presents a "black hole" and diminished visibility for the motorman or train operator. The tunnel interior also requires illumination for both normal and emergency operations, which applies to nighttime conditions from portal to portal. Emergency operations include lighting for passenger egress to exits.

Train operations differ from vehicle operations in a highway tunnel through strict regulatory requirements for approach and through speed, signal systems, occupancy control, and public access.

Individual transit operating agencies have criteria defining levels of illumination for both portal and interior zones for the design engineer. For applications where no criteria exist, criteria for vehicular tunnels may be used, with consideration for operational constraints on approach, safe stopping sight distance, and through speed. Recommendations of ANSI/IES RP-22 (1987) suggest pavement luminance values based on traffic volume and speed, and characteristics of the tunnel. These values require adjustment for the actual use since the traffic volumes and pavement luminance are not applicable. Therefore calculation procedures using an illuminance method should be applied.

LIGHTING OF RAILWAY TUNNELS

The design of lighting for railway tunnels is different from the design of highway tunnels and transit tunnels. Highway and transit tunnels typically have intensive lighting for 200–400 m from the portal to the interior, for eye adaptation from a high external to low interior illumination level. The railway tunnel lighting design can have an intensive lighting zone greater than 1 km, owing to the great mass of the train and stopping distance.

Close coordination with railroad operation agencies to establish intensive zone criteria will be required. Issues such as mandatory train entry speed, grade, alignment, and train mass will provide a basis for engineering judgments. Economic considerations must include capital costs, and energy and maintenance costs.

Portal-to-portal illumination will require assessment of tunnel lining reflectance. Low reflectance resulting from the characteristics of the material and minimal maintenance will require greater lamp lumens than highway tunnels with highly reflective surfaces. This suggests that an asymmetrical lighting concept be employed for positive and negative visibility. Location of luminaries to coincide with refuge niches, to indicate their location as well as providing visibility, is recommended.

DESIGN COMPUTATIONS

Improvement in tunnel lighting has generated the need for taking a closer look at the accuracy of predicting illuminance and luminance levels. For nearly 70 years, lumen or flux methods for calculation of the illumination level using coefficients of utilization factors have been employed. These original methods used coefficients of utilization that were determined by empirical methods for lighting equipment available at the time. Later developments used mathematical analysis as a method for computing coefficients of utilization data. All of these methods were based on the theory that average illumination is equal to lumens divided by the work area over which they are distributed. Later mathematical methods of analysis took into account the concept of interreflection of light, and they have led to progressively more accurate coefficients of utilization data. The Zonal Cavity Method improved older systems by providing increased flexibility in lighting calculations as well as greater accuracy.

For the detailed procedures to be followed in the use of the Zonal Cavity Method, refer to ANSI/IES RP-22, American National Standard Practice for Tunnel Lighting, latest edition. The designer may utilize a computer to aid in calculating illuminance and luminance. Computer software is readily available for this application.

To complement the accuracy in predicting the initial average illuminance and luminance levels in tunnels, the designer should also use the new procedure for evaluating the maintenance factor for the determination of an equally important accurate prediction of the overall average in-service illumination level. A comprehensive procedure for developing a meaningful maintenance factor will be found in the IES *Handbook* (1990).

REFERENCES

Adrian, W.K. (1990) " A Method for the Design of Tunnel Entrance Lighting," *Journal of the Illuminating Engineering Society,* Winter .

Adrian, W.K. (1987) Adaptation Luminance When Approaching a Tunnel in Daytime, Lighting Research and Technology, *19*(3), 73–79.

American Association of State Highway and Transportation Officials (AASHTO) (1984) *Guide for Roadway Lighting,* Washington, D.C.

ANSI/IES RP-22 (1987) *American National Standard Practice for Tunnel Lighting.*

Commission Internationale de l'Eclainage (CIE) (1990) *Guide for the Lighting of Road Tunnels and Underpasses,* Publication, CIE 53-1990 CIE Central Bureau, Vienna, Austria.

Dijon, J.M., and Winkin, Ph. (1990) *Tunnel Lighting. Symmetrical, Counter Bean and Pro Beam—Comparison and Tests,* Symposium and Visibility, National Research Council, Washington, D.C., July .

Faucett, R.E. (1969) "The Zonal Cavity System Applied to Tunnels," Illuminating Engineering Society National Conference, August.

Illuminating Engineering Society (1990) *IES Lighting Handbook,* New York.

Ketvertis, A. (1971) Foundation of Canada Engineering Corps.

Ketvertis, A. (1989) "Directional Light Application in Vehicular Tunnel Illumination Design," Illuminating Engineering Society National Conference, 1989.

Schreuder, A. (1964) *The Lighting of Vehicular Tunnels,* Phillips Technical Library, Eindhoven, The Netherlands.

Thompson, J.A. and Fanslor, B.L. (1968) "Criteria for Highway Tunnel Lighting Design." *Public Roads, 35*(4) October .

22 | Power Supply and Distribution

Elies Elvove
Senior Supervising Engineer, Parsons Brinckerhoff Quade & Douglas, Inc.

Power supply and distribution systems for tunnels are similar to those of a high-quality industrial facility, where power must be distributed to the associated equipment and systems with a high degree of safety, reliability, voltage quality, and maintainability. Basic requirements of such systems are outlined in various publications, including the IEEE Red Book (IEEE, 1986a).

Safety relates to the protection of life and equipment, with the protection of life being paramount. While tunnel electrical installations must comply with applicable local statutory regulations and codes, compliance with provisions of applicable NFPA standards and codes such as the NEC and the NESC (ANSI, 1993; NFPA, 1991, 1993) should be considered a basic minimum requirement, whether or not such codes are legally mandated. Safety involves the use of quality equipment, factory tested to accepted standards, properly selected and installed, with protective devices of adequate capacity and with a properly designed grounding system, supplemented by a program of carefully supervised operations and maintenance.

Reliability relates to service continuity of all or a specific portion of the electrical systems. Reliability involves both the selection of systems suitable for the application and the use of properly selected high-quality, long-life equipment throughout, i.e., equipment with low failure rates. Life safety systems must include alternative power sources, both utility supplied and from on-site generation, with circuit and equipment redundancy and flexibility to permit equipment to be removed from service for maintenance. Systems selected should be basically simple, avoiding highly complex arrangements that might result in operator confusion during periods of emergency.

Voltage quality relates to the stability of voltage and frequency, particularly the limitation of voltage spread (difference between maximum and minimum voltage) to values appropriate to the associated equipment. While utility supply voltage is generally kept to limits as mandated by appropriate regulatory authorities, system load currents fluctuate, resulting in variations in the voltage delivered to individual customers. For customers located a distance from major utility substations and with systems involving long feeders, this variation can create problems unless adequately considered in the system design. The effect of voltage variations on various equipment is indicated in Table 22-1.

Maintainability of an electrical system pertains to arrangements permitting regular inspection of various components and the removal of major units from service for repair or replacement without interrupting vital operations. Accessibility to equipment and choice of equipment with replaceable components are important aspects of maintainability. Design should permit equipment access and/or removal for servicing or repair by individuals with limited training.

PECULIAR ELECTRICAL REQUIREMENTS OF TUNNELS

The physical location and configuration of tunnels and their specialized power requirements create peculiar problems in the design of associated electrical systems. Service buildings and portal structures are frequently located remotely from existing utility company major facilities, necessitating services with long primary feeders, often of limited capacity. Tunnels are of necessary long linear structures, involving long secondary feeders. Space for equipment within the tunnel is extremely limited (and costly), necessitating maximum use of space in associated service buildings (generally located at the ends of the tunnel) for all major equipment and the careful selection of tunnel space for raceway systems and associated pull boxes.

Electrical equipment within tunnels is subject to harsh normal environmental conditions—moisture, high humidity, cold temperatures, exhaust emissions and pollutants from internal combustion engines, iron dust in rapid transit and railroad tunnels, as well as high temperatures in event of a tunnel fire. Equipment in vehicular tunnels is also subject to tunnel washing operations with high-pressure hoses, caustic detergents, and mechanical scrub brushes. In addition to

Table 22-1. Effects of Voltage and Frequency Variations on Induction Motor Characteristics

Characteristic	Function	Voltage		Function	Frequency	
		110%	90%		105%	95%
Torque Starting and Max. Running	$(\text{Voltage})^2$	Up 21%	Down 19%	$\frac{1}{(\text{Frequency})^2}$	Down 10%	Up 11%
Percent Slip	$\frac{1}{(\text{Voltage})^2}$	Down 17%	Up 23%		Little Change	Little Change
Efficiency Full Load		Up 0.5-1%	Down 2%		Up slightly	Down slightly
3/4 Load		Down slightly	Little change		Up slightly	Down slightly
1/2 Load		Down 0-5%	Up 1-2%		Up slightly	Down slightly
Power Factor Full Load		Down 5-15%	Up 1-7%		Up slightly	Down slightly
3/4 Load		Down 5-15%	Up 2-7%		Up slightly	Down slightly
1/2 Load		Down 10-20%	Up 3-10%		Up slightly	Down slightly
Full-Load Current		Down 7%	Up 11%		Down slightly	Up slightly
Starting current	Voltage	Up 10-12%	Down 10-12%	$\frac{1}{\text{Frequency}}$	Down 5-6%	Up 5%
Full-Load Temperature Rise		Down 3 - 4C	Up 4-7C		Down slightly	Up slightly
Maximum Overload Capacity	$(\text{Voltage})^2$	Up 21%	Down 19%		Down slightly	Up slightly
Magnetic Noise		Up Slightly	Down slightly		Down slightly	Up slightly

power and lighting, auxiliary systems of many types are required—equipment control, communications, signal, fire alarm, closed-circuit TV (CCTV), traffic control, etc.—and associated equipment and raceways must be accommodated in the limited space that must also house the required associated mechanical systems.

TYPES OF TUNNELS

Requirements also vary with tunnel functional use, various types of which are indicated in Chapter 1. Those requiring extensive electrical facilities are highway tunnels, rapid transit tunnels, and railroad tunnels.

The physical arrangement of electrical equipment and raceways also varies with the type of tunnel construction, basic types of which can be classified as indicated in Chapter 1. These are bored tunnels, regardless of method; cut-and-cover tunnels; and sunken tube tunnels Typical cross sections indicating location of electrical facilities are shown in Figures 22-1–22-4.

ELECTRICAL LOADS

Tunnel electrical load include requirements for lighting, auxiliary power for ventilation, cooling, and pumps, power for communications and control, and in the case of transit and railroad tunnels, traction and signal power. Traction and signal power requirements are established by the design criteria of the associated traction system and are based on the proposed consists, grade, speed, and headway. While space requirements for catenary or third rail have a major impact on the tunnel configuration, the associated power supply

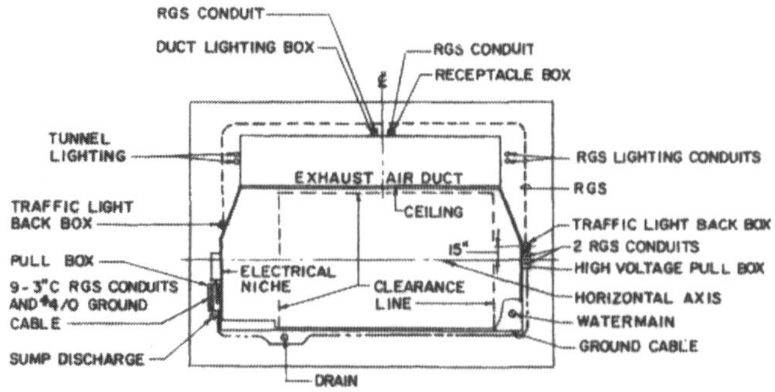

Fig. 22-1. Typical cut-and-cover tunnel (semitransverse ventilation system).

systems are generally independent of the supply of the remaining tunnel loads and are beyond the scope of this handbook. These other tunnel loads are discussed in this and other chapters.

Normal tunnel power supply should be based on the peak coincident system demand of all loads, including lighting, power, and auxiliary systems of the tunnel proper as well as the associated service buildings and any ancillary facilities such as approach roadway lighting. Details of tunnel lighting requirements are discussed in Chapter 21. Mechanical equipment associated with ventilation is discussed in Chapter 20, and that associated with water supply and drainage in Chapter 23. Service buildings are discussed in Chapter 26. Emergency loads include the lighting, control and communications, and other equipment that must be maintained in operation in the event of normal power supply failure, as discussed subsequently.

LIGHTING LOAD

As discussed in Chapter 21, the maximum lighting load for a highway tunnel is that occurring during daylight hours due to the requirement for additional illumination levels at the tunnel entrance to overcome the problems associated with the great difference in ambient sky brightness and tunnel interior luminance ("black hole" effect). The nighttime light-

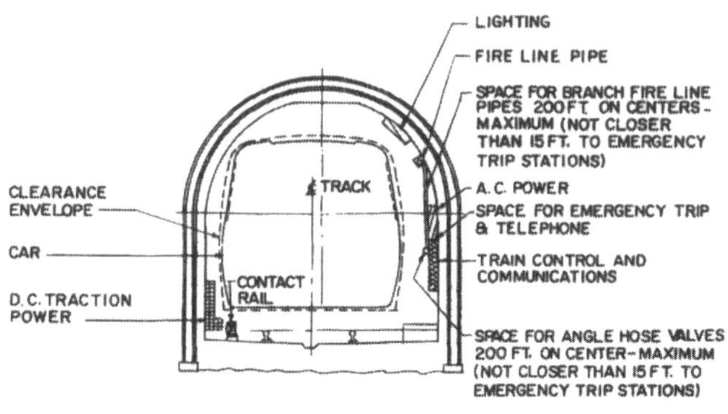

Fig. 22-2. Single-track bored tunnel (horseshoe section).

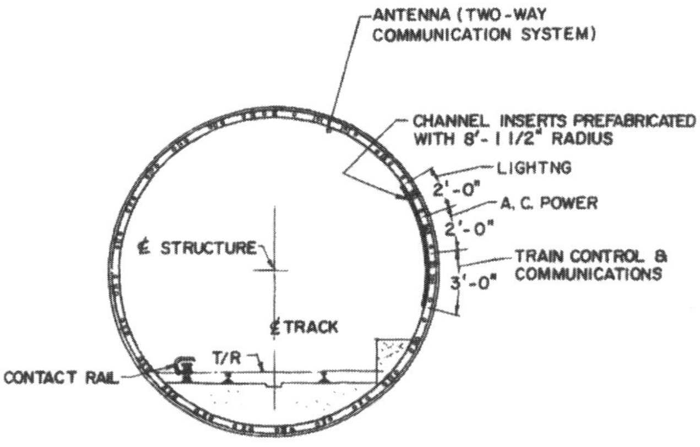

Fig. 22-3. Single-track bored tunnel (circulation section—cast iron).

ing load is substantially less. Since the tunnel power supply normally serves the associated service buildings and access roadway lighting, these loads should be added to the tunnel day-lighting load in the determination of the associated lighting demand. While the lighting requirements of rapid transit and railroad tunnels are substantially less complex, the same principles apply.

POWER LOAD

Pumps

Pumps are provided for water supply and for drainage as indicated in Chapter 23. Since fire pumps are operated only intermittently, their contribution to the system peak coincident demand can be assumed to be zero. Tunnel low-point or mid-river drainage pumps are installed in pump rooms at tunnel low points to discharge tunnel accumulated leakage, rainfall (carryover from portal stormwater systems, if any), tunnel washing operations, vehicle drippings, and fire-fighting operations. Since operation of these pumps, while intermittent, can be at times of peak tunnel use, contribution to system demand should be evaluated for each specific installation.

Tunnel portal drainage pumps are installed in pump rooms at or near the tunnel entrances or ventilation structures to discharge stormwater from the open approach and/or water

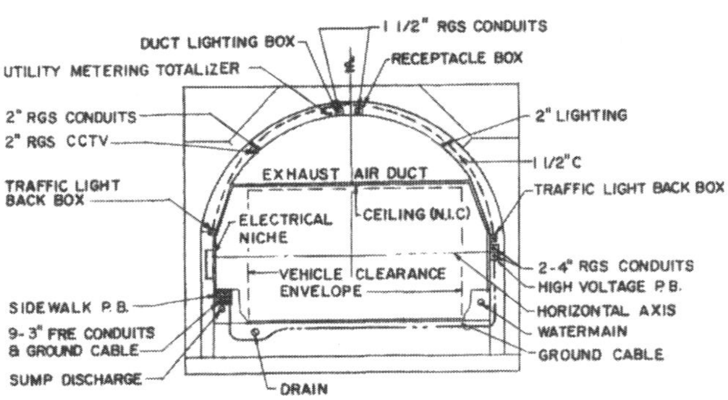

Fig. 22-4. Typical immersed tube (semitransverse ventilation system).

received from the low-point pumping station. Contribution to system demand should be based on the maximum horsepower requirements associated with the design storm, as outlined in Chapter 23.

TUNNEL VENTILATION FAN LOAD

Tunnel ventilation requirements and typical arrangements are outlined in Chapter 20. Associated power requirements depend on fan types and drives and the arrangement selected to provide the required air flow for each ventilation mode in the specific tunnel. In vehicular tunnels, the horsepower requirements for the selected fan arrangements vary with projected traffic volume and traffic conditions. Emergency requirements should be based on a selected worst-case condition. Design power requirements should thus be determined for minimum, intermediate, and maximum normal air flows as well as for the emergency worst-case requirements.

MISCELLANEOUS LOADS

Miscellaneous loads include the requirements of service buildings and the various auxiliary systems. Service building loads, in addition to the previously discussed fan and pump loads, include requirements for building heating, ventilation and air conditioning, elevators, repair shop equipment, etc. Auxiliary equipment loads, although minor, are vital and include the requirements of traffic control and surveillance, fire alarm, communications, CCTV, and heat tracing. These systems are frequently supplied through an uninterruptible power system (UPS) or the emergency power system.

VOLTAGE SELECTION

Utility Service

Service voltage should be selected as the utility's standard, which will permit the most economical arrangement for the load magnitude and service location and which will provide a satisfactory voltage spread under all conditions of tunnel operation. Standard nominal voltages in the United States are indicated in ANSI C84.1 (ANSI, 1992), together with recommended tolerance limits. Utility systems vary widely in their configuration, from strong interconnected systems with major generating plants and regulated transmission and distribution systems to rural systems with limited substation capacity and long overhead primary distribution circuits. Complete, specific information should therefore be obtained from the supplier as to the characteristics of the supply voltages deliverable at points from which the tunnel can be supplied. Information should also be obtained from the utility as to applicable rate schedules for each available service voltage, both with utility-owned transformers and with customer equipment, as well as a history of outages to permit evaluation of the reliability of the supply. The arrangement that will provide the most economical total annual owning costs but with the requisite reliability and voltage quality should be selected. In this regard, care should be taken to obtain complete information as to all special provisions, including charges associated with extending the existing utility facilities to the service location, active and reactive demand charges and method of determination, and contract minimum charges, as well as information as to available system fault capacity at the service location and history of outages.

Design of tunnels involving the installation of very large fan motors—1,000 HP (250 kW) or larger—should also consider the feasibility of the use of medium-voltage motors matching an available utility primary supply voltage, thus eliminating the need of transformer capacity in step-down transformers (or in the case of long motor feeders, step-up transformers to overcome voltage drop). As discussed in Chapter 20, railroad tunnels frequently require fans for ventilation rated 2,000 HP (500 kW) or more. A 13.8-kV utility service supplying motors rated 13.2 kV and controlled from 15-kV switchgear circuit breakers may prove to be an economical installation for this type of tunnel.

Utility service for tunnels located in metropolitan areas may be limited to the voltage of the utility's secondary grid, frequently 208Y/120 V. Under such situations, the design may require the use of multiple supplies to provide the required capacity and reliability, and the use of step-up and step-down transformers as the most economical method of supplying loads with lengthy feeders.

DISTRIBUTION VOLTAGE

Factors affecting the selection of the tunnel distribution voltage include load magnitude, distance power is to be transmitted, available voltage of utilization equipment, and applicable safety codes and standards. Possible physical restrictions affecting the choice are available spaces for the location of transformers, switchgear, and motor control equipment.

Types and sizes of major utilization equipment such as larger fan motors and pumps will also have an impact on the selection of distribution voltage. These motors can frequently be supplied at either low voltage (575 V and below) or medium voltage (2,300–13,200 V). Standard voltages are listed in ANSI/NEMA MG-1 (NEMA ,1993b). Motor supply voltage selection should give special consideration to the specific application, including frequency of starting, type of control (single speed or multiple speed) and load (steady or fluctuating), ambient conditions, and type of enclosure. Motors with higher voltage ratings may prove to be more economical, but they should never be selected on the basis of cost alone and at the expense of winding mechanical strength for those units subject to frequent starting and/or highly fluctuating loads.

A cost comparison of motor drives at different motor voltages must consider the cost of the motor, associated control equipment, wiring, and that portion of the system distributed equipment affected by the choice of motor voltage (transformers, switchgear, etc.). This study assumed squirrel-cage-type induction motors (totally enclosed fan-cooled, 1.0 Service Factor, 1,200 rpm), full voltage combination magnetic starter for 200-HP (150-kW) motors, electrically operated breakers for 300-HP (225-kW) and larger at 480 V, and draw-out-type full voltage combination starters at 2,400 V, and 200-foot feeders in conduit. This study indicated that a nominal 480-V supply was more economical for motors rated up to 400 HP (300 kW), but for larger motors, the 2,400-V system appeared to be more economical.

For typical tunnel installations with maximum motor size 400 HP (300 kW) or smaller, the use of a system voltage of 480 V to supply motors 1/2 HP and larger, 277Y/480 V for fluorescent and HID lighting, and 208Y/120 V for smaller motors, incandescent lighting, and general purpose outlets may prove to be the most desirable arrangement. Utilization equipment is readily available for use at these voltages. The use of a 600-V system (with 575-V motors) provides relatively little decrease in costs for systems with substation ratings of 1,000 kVA or less, and equipment is not as readily available at this voltage.

PRIMARY DISTRIBUTION SYSTEMS

The basic function of a tunnel primary distribution system is to supply load center unit substations located strategically to permit relatively short low-voltage circuits, the entire system conforming to the previously stated general requirements. While many possible types and arrangements of systems are possible, the following three basic systems have been found to be most satisfactory for supplying tunnel power.

Radial Systems

The radial system consists of a single primary feeder supplying the various unit substation transformers, as shown in Figure 22-5. Since no duplication in equipment is provided, cost is a minimum and reliability is limited by the reliability of individual components. Use of the simple radial system should be limited to nonessential loads involving relatively short feeders, such as service building loads.

Primary Selective Systems

In the primary selective system, each unit substation is supplied by two independent feeders with suitable interlocked breakers or load interrupter switches as shown in Figure 22-6. This arrangement has the advantage of improved primary circuit reliability and flexibility due to the availability of an alternate circuit in the event of primary cable failure or outage for maintenance. Transfer can be either manual or automatic. The reliability of the individual

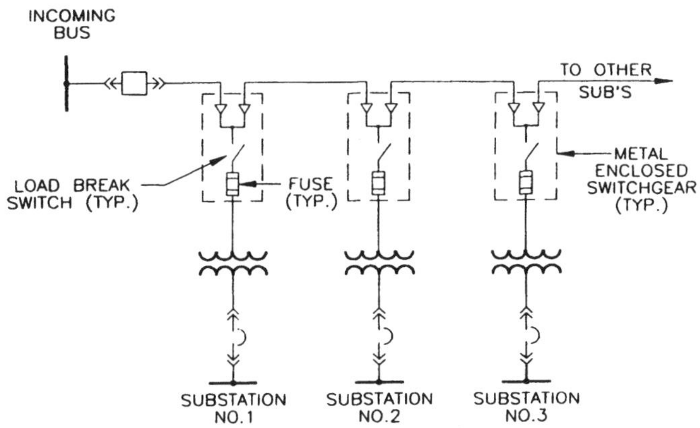

Fig. 22-5. Radial system.

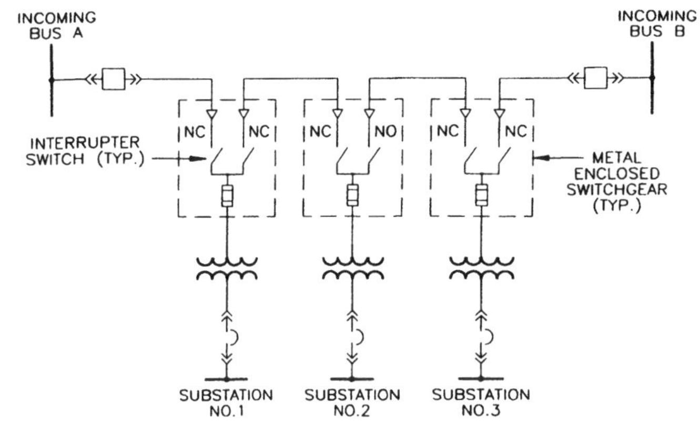

Fig. 22-7. Looped primary system.

primary feeders can be further improved by being connected to separate utility substations, and, in the case of a tunnel, from opposite sides of the tunnel. In many locations, separate utility companies may be the suppliers.

Metal-enclosed load interrupter switchgear is more commonly used for primary selective systems rather than circuit breakers, due to lower cost. Equipment should be carefully selected for the application to provide complete isolation of the individual feeders and protection of the associated transformer substations. Operating procedures should be such as to preclude the possibility of transferring a faulty transformer to an energized feeder. While individual interrupter switches should be of the load break type, opening of the main secondary breaker prior to opening the primary switch provides a safer operating practice.

Looped Primary System

The looped primary system as shown in Figure 22-7 may permit somewhat lower installation costs for systems with widely separated load centers fed from separated primary

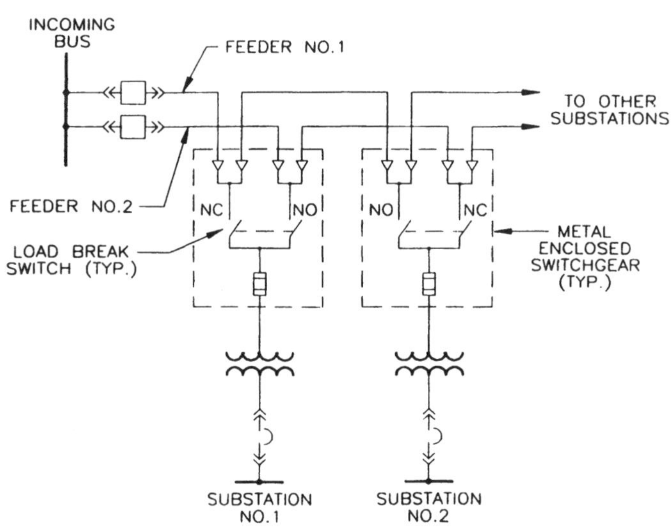

Fig. 22-6. Primary selective system.

buses, due to lower feeder costs. However, reliability and safety are greatly reduced. There is no practical way that the individual switches can be interlocked to prevent accidents/closing into a fault. Thus it becomes necessary to deenergize the feeder whenever a fault occurs to permit locating the fault. The use of the radial or primary selective systems thus appears preferable for operating reliability.

SERVICE BUS ARRANGEMENTS

The arrangement of the incoming switchgear and buses supplying primary power to the tunnel has a major impact on the reliability and maintainability of the electrical system. While many arrangements are possible and have been used to supply tunnel systems, several types that have been used most frequently are discussed here, with their advantages and disadvantages. Each arrangement presumes one or more utility primary supply feeders, preferably from opposite tunnel portals, and a primary distribution system supplying a number of individual unit substations.

Single Bus

With the single bus arrangement as shown in Figure 22-8, each utility primary supply feeder supplies a dedicated switchgear bus with feeder breakers to each associated unit substation and with a cable tie to the remote service bus. This arrangement has the advantages of lowest initial cost and operational simplicity. Disadvantages include loss of power to all associated feeders and to the tie connections upon a power loss of the incoming line, and a total shutdown in the event of bus failure or its removal from service for maintenance.

Double Bus, Single Breaker

A typical double bus, single breaker arrangement, with twin primary service feeders at both portals, is shown in Figure 22-9. This arrangement provides improved service reliability, because the bus tie breaker permits service to be restored from the adjacent bus in the event of failure of one of

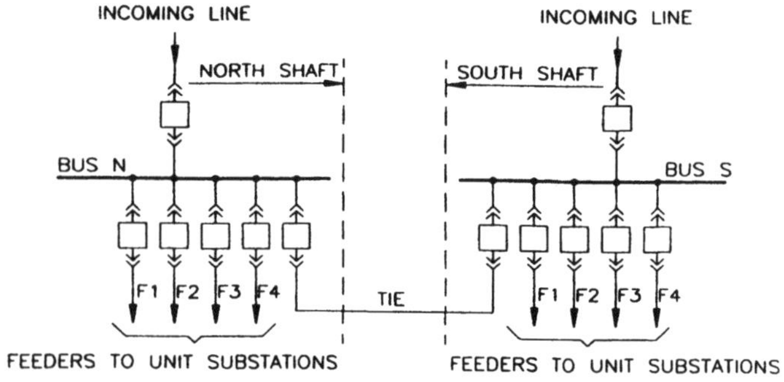

Fig. 22-8. Typical single bus arrangement.

the incoming lines. Tie feeders from the opposite portal permit additional means of restoring power in the event of failure of both service feeders. Disadvantages include (1) a bus fault will interrupt service to all feeders and ties associated with that bus and (2) with normal metal-clad construction, a fault in the bus tie breaker unit would cause a loss of both buses. It should be noted that the use of double-ended unit substations (secondary selective system, as discussed later) would permit the restoration of power to feeders from a faulted bus by closing the secondary tie breaker.

Double Bus, Double Breakers

Figure 22-10 shows a double bus, double breaker arrangement, with two breakers associated with each incoming feeder and tie feeder. This arrangement provides great reliability, permitting the complete isolation of any bus in the event of a bus fault or for maintenance. Any breaker can be removed for maintenance without interrupting service. Protection would be provided by a bus differential scheme for each individual bus so that a bus fault would result in the interruption of service to the faulted bus only. This scheme requires additional breakers and is therefore more costly, but it can be justified when extreme reliability is required.

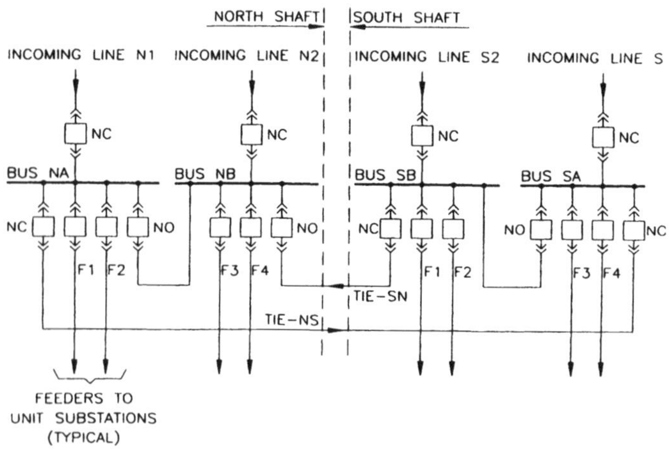

Fig. 22-9. Typical double bus, single breaker arrangement.

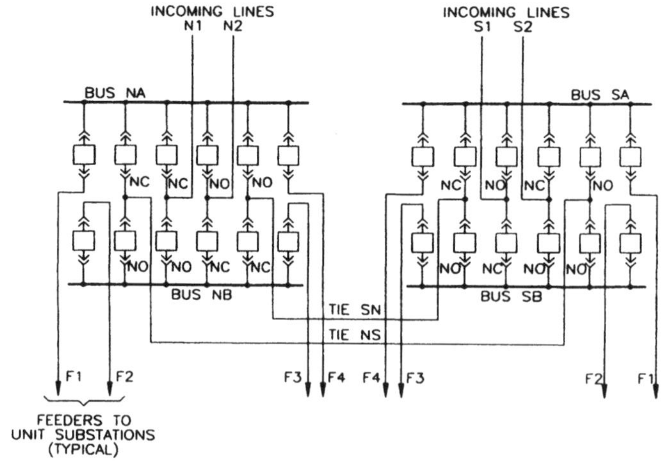

Fig. 22-10. Typical double bus, double breaker arrangement.

SECONDARY DISTRIBUTION SYSTEM

As previously stated, the load center unit substations for the supply of utilization voltage to the various items of tunnel equipment should be located as close to the load as possible to keep the secondary feeders short. Secondary distribution can be configured in numerous arrangements, but the radial and secondary selective systems have been most commonly used in tunnels.

Radial System

The radial system consists of single transformer unit substations supplying distribution switchgear with feeders to individual panels and/or equipment loads. While failure of the associated transformer or feeder would interrupt the load, the quality of modern transformers, low-voltage switchgear, and cable are relatively high, so that failures are uncommon. Operational reliability can be increased by the use of redundant equipment with individual feeders.

Secondary Selective System

With the secondary selective system, utilization equipment is supplied from double-ended unit substations consisting of two separate transformers supplying individual secondary buses, with a secondary tie breaker to permit restoration of service in the event of a single transformer of primary feeder failure. The two main breakers are interlocked with the tie breaker so that only two of the three breakers can be closed at any time. By providing redundant equipment with associated feeders connected to separate secondary buses, a high degree of reliability is obtained. Tie breakers can be either arranged for automatic closure on bus power failure or manually controlled. Transformer capacity should be selected to permit the supply of the total load that occurs with the tie breaker closed.

STANDBY POWER SUPPLY

Certain tunnel loads are essential to normal operation of the facility and/or to life safety and frequently warrant an alternate highly reliable source of power in the event of failure or the normal utility supply. Such loads include drainage pumps, fire pumps, alarm and communication systems, traffic control and surveillance, security and control systems, and egress lighting. The need for standby tunnel lighting is discussed in Chapter 21. While multiple utility supplies from alternate substations can sometimes provide a reliable source of standby power, particularly those utilities in metropolitan areas that operate highly reliable network systems, the most common source of standby power is from engine-generators or uninterruptible power supply (UPS) systems.

Generators

The most common standby power supply for tunnels is an engine-driven generator fueled by oil, gasoline, or natural gas. Generation is commonly at the utilization voltage, and connection to the emergency power distribution system is made through an automatic transfer switch. Combustion turbines are also available, but they have the disadvantage of difficult heat rejection requirements and generally higher cost for the common size of units required for tunnel standby systems.

Equipment ratings should be based on the calculated standby demand load with a reasonable margin for growth (on the order of 25%), and conforming to Diesel Engine Manufacturers Association (DEMA) standards for standby service. Minimum fuel supply for 24 hours of continuous operation is recommended. Units should be provided with complete systems for automatic starting under the most adverse conditions and control panels with all devices required for monitoring operation and stopping the equipment upon malfunction. If two or more units are provided, equipment should permit parallel operation. Controls should include provisions for testing the unit under load and for continued operation after dropping load to permit engine cooldown. Installation should provide for proper ventilation (combustion intake and cooling air) and for the safe discharge of the exhaust.

UNINTERRUPTIBLE POWER SYSTEMS (UPS)

Uninterruptible power systems (UPS) are provided for those loads that require a constant uninterrupted supply of power for proper operation. Typical systems include those associated with life safety, and they utilize computer-based equipment, such as monitoring and control systems (SCADA), fire protection, security, traffic control and surveillance, communications systems, and variable-message sign control. UPS may also be used to supply emergency egress lighting.

A typical UPS system consists of four major components:

- A rectifier-battery charger, which converts normal AC power to DC and both supplies the investor and charges the batteries
- Batteries, which supply power to the investor when normal power fails.
- An investor, which changes the DC power from the battery to AC, which in turn supplies to load.
- A bypass switch, which ensures continuous supply of AC power to the load by transferring the load from investor output to the normal supply in the event of output overload or the unlikely occurrence of UPS system failure. The switch can also be used for maintenance.

UPS equipment also has the added benefit of providing "clean" output power, eliminating transients and noise that may be in the normal supply and adversely affect solid state equipment.

Since UPS units are relatively expensive, the selected output capacity and operating time should be carefully determined. The battery ampere-hour capacity is directly associated with operating time, and the batteries are a major cost of an UPS system. Power will normally be supplied from the utility source, and upon power failure, from the standby generator; UPS battery supply energy is thus only required during the simultaneous failure of both these sources.

STANDBY POWER DISTRIBUTION SYSTEM

The standby power distribution system consists of automatic transfer switches, panelboards, and all associated wiring. Transfer switches of the mechanically held, electrically operated single solenoid should be used because of their reliability and inherent two-position (no "off" position) construction. Four pole units (with switched neutral) are recommended to prevent permanent connection of the standby generator-grounded neutral to the utility system. The neutral poles should be of "make-before-break" design. Switches should have adjustable pickup and dropout voltage sensing relays and adjustable timers for both normal-to-standby and standby-to-normal operation. Cooldown control, permitting the continued operation of the standby generator for a period of time after retransfer of the load to the normal source, is frequently incorporated in the automatic transfer switch. Panelboards and wiring of the standby power systems should be completely separate from the normal supply system.

SUPERVISORY CONTROL AND DATA ACQUISITION (SCADA)

Efficient operation of tunnel electrical and mechanical systems requires a control center with a system for monitoring and controlling the functioning of the various systems. The control center can be located either in a contiguous structure or in a separate ancillary building and generally contains facilities for displaying information concerning the key operating parameters and for controlling designated operations. In rapid transit and railroad tunnels, the control center may be designed to control all or part of the complete rail operation, including traction power, train control, and passenger station auxiliary equipment. The data transmission system (DTS) linking the various elements of the SCADA system may serve as the communication link of the various other systems, as described in Chapter 24.

The SCADA system provides facilities for monitoring the status of electrical equipment necessary for the continuous operation of the tunnel and for controlling certain functions. Typical functions monitored include utility supply breaker position, primary bus tie and feeder breaker position, secondary main and tie breaker position, standby generator condition, automatic transfer switch position, transformer overtemperature, control battery voltage, battery charger failure, and UPS transfer switch position. Controls may include primary supply and feeder breakers, and secondary main and bus tie breakers. The system should provide immediate alarm and visual indication of faults, status changes, and abnormal conditions. It also may include facilities for the remote monitoring of power and energy consumption.

DATA TRANSMISSION SYSTEM (DTS)

The data transmission system links all field control devices with the central control room and should be suitable for either digital or analog signals. The DTS may be common to the various other system as described in Chapter 24, provided suitable interface facilities are provided. The DTS may consist of coaxial cable or fiber optics.

AUXILIARY SYSTEMS

Power supply and wiring systems for tunnels must include provisions for the numerous auxiliary systems required for tunnel operation and, in the case of rapid transit and railroad tunnels, systems involved in traction power, train control, and communications for the rail line. Fire alarm systems are discussed in Chapter 19. Communication, control, security, closed-circuit TV, public address, and signage are described in Chapter 24, with emphasis on vehicular tunnel requirements. The physical space requirement for the associated equipment and raceways in the limited tunnel profile has a major impact on the tunnel design. Full advantage must be taken of the possibility of installing major components in the associated structures such as ventilation buildings and service buildings; this is discussed in Chapter 26.

GROUNDING AND BONDING

Grounding comprises both system and equipment grounding. System grounding is the connection to ground of one of the current-carrying conductors of the electrical system and is provided to control overvoltages associated with certain faults and to permit protection of the system by overcurrent protective devices and relays. Equipment grounding refers to the solid connection to ground of non-current-carrying metal parts of electrical equipment and adjacent structures to provide safety to personnel. Bonding refers to the electrical connection between electrical conductors and other metallic items to maintain a common electrical potential and to establish desired paths for fault and stray current flow. Basic grounding information is contained in the IEEE Green Book (IEEE, 1982; Tencza and Billmyer, 1993).

SYSTEM GROUNDING

The system ground should be designed so that under either normal or abnormal conditions there is no danger to personnel. It should be capable of passing the maximum ground fault current back to the system neutral without establishing a dangerous potential gradient in the earth (step potential) or a dangerous potential difference between exposed parts of equipment and ground (touch potential), and without resulting in thermal or mechanical damage to the insulation of connected apparatus.

The grounding requirements of medium voltage utility service to a tunnel are generally established by the utility company and should be unchanged if extended without transformation as the tunnel primary distribution. If transformation to medium voltage for larger motors is contemplated, consideration should be given to 2,400-V wye connected transformer secondaries, with low-resistance grounding to limit the ground fault current to a value suitable for available protective devices to remove faulty equipment from service. Low-voltage systems (277/480 V and 120/208 V) are generally solidly grounded, breakers being selected with ratings permitting safe interruption of maximum available ground fault current. Use of impedance grounding (high or low), while permitting continued operation in the event of a single-phase ground fault (alarm only), creates the possibility of equipment damage due to transient overvoltages. Resistance grounding of larger standby generators may be considered to limit the ground fault current to the rated three phase fault level to protect the generator windings.

EQUIPMENT GROUNDING

The equipment grounding system consists of grounding conductors, ground bus, and grounding electrodes intended to prevent electrical shock hazard; the requirements for these installations are covered in detail in the NEC and NESC. In addition to providing for the grounding of all electrical raceways and enclosures of the electrical system, the grounding system should provide a low impedance path to ground for all exposed metallic structures in the tunnel, including railings, stairs, fencing, and the like. Removable covers of manholes and removable gratings should have flexible bonding cables to ensure positive connection to the grounding system when the covers are in place. A grounding conductor (green wire) should be provided with each branch circuit feeder, sized in accordance with the NEC. Design of all components should be such as to limit the potential to ground of the grounded equipment to a safe value (on order of 50 V) and to carry the maximum fault current safely (without a temperature rise that could create a fire hazard).

GROUNDING ELECTRODES

Grounding electrodes are conductors embedded in earth that are used to maintain ground potential on conductors connected to it and to conduct current to earth. Electrodes consist of noncorrosive rods, pipes, plates, or conductors embedded in earth as well as connections to underground metallic piping systems. Resistance to ground should be as low as economically possible, and not more than 2 ohms for a typical tunnel utility service. Design should be based on actual test data on associated soil resistivity, if available. Ground buses to provide a uniform potential in areas with major electrical and/or mechanical equipment are generally provided in electrical and mechanical equipment rooms. These ground buses must be continuous, of low resistance, and solidly connected to the grounding electrodes and the grounding conductors. Ground buses are typically 1/4 in. by 2 in. (6.4 mm by 5.0 mm) copper bar, supported from the adjacent wall by suitable insulators.

STRAY CURRENT AND CATHODIC PROTECTION

Grounding systems should be physically isolated from the structural rebars, stray current systems, and cathodic protection systems. Steel rebars to tunnels should be bonded together to create a electrically continuous path, which is in turn connected to the return bus of the electrical supply system (rail return of rapid transit and electrified railroad tunnels, and utility supply ground bus of vehicular tunnels). Cathodic protection system (either impressed current or corrodible anode systems) should be isolated from the protective grounding system.

RACEWAY SYSTEMS

Separate raceway systems are generally provided in the tunnel for the various electrical systems, including medium-voltage feeders and bus ties, low-voltage power feeders, lighting, control, and auxiliary systems fire alarm, public address, CCTV, security, traffic control and surveillance, signs, telephone, etc. In addition, rapid transit and rail tunnels must provide raceways for traction power and train control. Each raceway system should be complete with accessible pull points (manholes, hand holes, or pull boxes) and supports and should be installed so as to minimize possible accidental damage from vehicular mishaps.

Figures 22-1–22-4 show typical tunnel cross sections with locations of raceways. Low-voltage power runs are frequently located under safety walks, with medium-voltage power feeders and tie lines under the opposite benches or in supply air ducts. Auxiliary systems are generally embedded in tunnel walls above the walks. Locations must be coordinated with mechanical piping such as drains and fire protection that must also run the length of the tunnel.

DESIGN

Raceway systems should be designed to accommodate the associated conductors, with pulling tensions and sidewall pressures of embedded installations well within cable manufacturers' recommendations and with consideration given to future requirements. Routing of exposed raceways should, where possible, avoid spaces subject to exhaust air flow in the event of a tunnel fire. Redundant system with generous spares (50% minimum) are highly desirable to permit rapid restoration of service in the event of failure of any run and to provide for future requirements. Raceway sizes should conform with applicable codes for maximum anticipated number and size of conductors, frequent practice utilizing the sizes for embedded construction given in Table 22-2.

For exposed runs, 3/4 in. (19 mm) is a commonly used minimum. All runs should be as straight as possible, unavoidable bends being of the long radius type and pitched to drain to an associated pull point.

MATERIALS

Concrete encased raceways 3 in. and larger are frequently made of nonmetallic, fiberglass-reinforced epoxy (FRE),

Table 22-2. Common Raceway Sizes for Embedded Construction

Type	Size	
	in.	mm
Medium voltage system	4 or 5	102 or 127
Low voltage distribution	3	76
Lighting feeders	3	76
Auxiliary systems	2	51
Minimum size	1	25

which is preferable to PVC due to its nontoxic properties when subjected to arcing or fire. Exposed runs are generally rigid galvanized steel (RGS), with epoxy coating when installed in damp or wet locations or subject to corrosive or outdoor atmospheres, and covered with fire-resistant materials when installed in tunnel exhaust ducts. Raceway systems installed in areas where fire or explosion hazards may exist due to flammable gases or vapors should comply with applicable code provisions for hazardous locations.

Use of electrical metallic tubing (EMT) is generally limited to smaller sizes (1-1/2 in. [38 mm] and smaller) installed above hung ceilings in ancillary structures. Liquid tight flexible metallic conduit is used for equipment connections subject to vibration, such as motors and generators.

Cable and Wire

Cable and wire should be carefully selected on the basis of both electrical and physical characteristics. Electrical characteristics include system voltage, maximum operating temperature, current-carrying capacity (ampacity), voltage drop, and short circuit capability. Ampacity is selected on the basis that the operating temperature will not exceed the designated value for the selected insulation when carrying the design load current when installed in the specified raceway system, taking full account of the ambient conditions and adjacent cables. Conductor impedance is selected to limit voltage drop to the design values, commonly 5% for power feeder and branch circuit to the most distant outlet and 3% for lighting. Minimum conductor size should be such as to withstand the maximum available short circuit current for the time required for the protective device to clear as calculated from ICEA P-32-382 (1994). The minimum size of low-voltage conductor for general light and power is generally limited to #12 AWG, with #14 minimum for control wiring.

Physical factors to be considered include ambient temperature, location (damp, dry, or wet), suitability for associated raceway system, resistance to corrosion (cable armor), and flame retardant and low smoke characteristics. Insulation and jacket materials with little or no hologen content are highly desirable in vehicular and transit tunnels to avoid the generation of toxic fumes during tunnel fires. A limitation in the otherwise excellent properties of PVC is its hologen content.

Conductors in tunnels are frequently specified to be copper to avoid difficulties associated with the need for separate types of devices for aluminum and copper pressure splicing connectors and terminals and special requirements for dissimilar metals. Multiconductor cables are generally used for exposed runs and for runs in cable tray, while single-conductor cables are used in conduits. Cables for exposed runs are generally of types enclosed in a metallic sheath (Type MC). Cables in tray may have a nonmetallic sheath (Type TC). Medium-voltage cables should be shielded and terminated in suitable potheads or terminators.

MAJOR EQUIPMENT

As previously stated, tunnel equipment must be selected for reliability and maintainability. In addition, equipment selection must consider the limited space generally available and the anticipated environment. Space considerations involve not only the assembled physical dimensions of the equipment, but the component requirements for initial installation and future removal for replacement and/or repair, clearance for safety and maintenance, and accommodation of associated raceways and cable terminations, particularly for higher voltage equipment. The operating environment in which the equipment will be installed should be carefully evaluated. Enclosures and all components should be specified to withstand exposure to the anticipated ambient temperatures, dust, humidity, moisture, and other special environments. The requirements of the NEC, NESC, and NEMA standards (ANSI, 1993; NEMA ,1991; NFPA, 1993) should be considered minimum guidelines, actual selection being made to minimize the effect of adverse environmental condition on sustained operation of the equipment.

Transformers

Main supply transformers for vehicular tunnels are normally located in the associated ventilation buildings and can be either isolated units or integrated (throat connected) with the associated low-voltage switchgear. Transformers may be either fluid-filled (flame-resistant, non-PCB) or ventilated dry types. Dry-type equipment is more common. Since transformers are continuously energized and have inherent no load and load losses, economy generally justifies the use of units with lower full-load temperature rise (55°C for liquid-filled units and 80 or 115°C for dry types). Use of supplemental fan cooling for peak load may reduce initial costs at the expense of increased operating costs. Fluid-filled transformers require special provisions for the containment of spillage.

Dry-type transformers are available as conventional design, resin encapsulated, or solid cast construction, with cost increasing in that order. Choice of equipment should be based on the anticipated environment. Where noise criteria require quieter operation, it should be noted that liquid-filled transformers are generally quieter than dry types of the same rating.

Distribution transformers should be of the energy efficient, dry self-cooled types with 220° insulation and 115° or 80° temperature rise. Transformers associated with variable-speed drives should be specifically designed for the application. Loads associated with UPS supply and other solid state equipment should be specifically designed to supply circuits with nonsinusoidal current loads of the anticipated harmonic profile.

Switchgear

Main medium-voltage switchgear for tunnels is generally of the metal-clad type with draw-out air or vacuum circuit breakers. The equipment is usually factory assembled and wired and contains all associated instrumentation, protective devices, and ancillary equipment. Primary connections to transformers of unit substations are generally made through metal-enclosed-type fused or nonfused load break switches. Interrupting and withstand rating must be within the requirements of the available supply system capacity and motor contribution. Since the equipment is normally located in unattended locations, complete interface devices for supervisory control and data acquisition (SCADA) systems should be provided.

Main low-voltage switchgear is generally of the metal enclosed type with draw-out or encased power circuit breakers, with solid state type trip devices. Use of draw-out equipment simplifies maintenance by permitting rapid interchanging of a defective removable unit with a spare operational unit. Use of solid state trip devices permits the selection of short time and long time delay for system coordination and ground fault sensing where desired. Switchgear should contain all associated sensors, instruments, instrument transformers, and supervisory control devices including remote control of main and tie breakers. Construction should permit throat connection to the associated power transformer.

Motor Control Equipment

Operation of the various motors powering tunnel equipment requires suitable motor control equipment for starting, protection, and where needed, speed control. Associated feeders require short circuit and overcurrent protection. These functions are commonly provided with suitable combination-type motor starters, containing fuses or circuit breakers for feeder protection, contactors or solid state devices for motor starting, thermal or solid state devices for motor protection, and where required, control devices. Tunnel emergency ventilating fan motors are usually provided with short circuit protection only (no overload protection). Low-voltage equipment for multiple motors such as fans or pumps is commonly grouped together in factory-wired motor control centers. Each individual motor is associated with a replaceable cell-mounted assembly (draw-out or bolted). Alternative classes of construction and wiring types are available per NEMA ICS 1 (NEMA, 1993a), but Class II, Type B with interlocking wiring and provisions for remotely mounted devices and terminal strips for all external wiring has proven most suitable for tunnel applications. Bus connecting individual medium-voltage starters results in reliable medium-voltage motor control centers for larger motors such as those used for railroad tunnel fans.

Full-voltage motor starters are commonly used for tunnel equipment, with reduced-voltage starters limited to larger motors with long feeders or where mandated by the supplying utility company. Auto-transformer-type reduced-voltage starters are preferable to resistance types. Solid state "soft start" control can also be used where excessive voltage drops on starting are anticipated. Electrically operated circuit breakers can be used for starting infrequently operated medium-voltage motors.

Direct variable starting speed control of tunnel ventilating fans often provides an attractive means of energy savings and cost efficiency with varying air flow requirements, as pointed out in Chapter 20. A cost-effective method of achieving this variable speed control is with solid state AC variable-frequency controls used with conventional AC induction motors. Such controllers also provide "soft start" operation but result in harmonic distortion reflected into the installation distribution system. Equipment should therefore be specified to limit such distortion to levels as defined by IEEE-519 (1992) for general system applications and should include suitable devices for harmonic attenuation.

Motors

Motor-driven tunnel equipment consist primarily of ventilation fans (Chapter 20) and pumps (Chapter 23), with motors generally furnished with the driven equipment. Motors must be carefully selected for mechanical and electrical characteristics as well as environmental considerations. Of particular importance is factory testing of motors both at the plant where manufactured and again when assembled with the driven equipment. Motor tests should be in accordance with NEMA MG-1 (1993b) and documented with performance curves, including speed–torque curves. Driven equipment performance curves should be compared with these motor curves to determine if the application is proper. Test reports should include all relevant physical parameters, including inertia. Assembled equipment tests should include as a minimum current, voltage, kW, and speed versus time during starting.

Since some ventilation fan motors operate continuously for extended periods of time, consideration should be given to use of energy efficient motors meeting Table 12-6C of NEMA MG-1 (1993b) to take advantage of the associated possible savings in energy charges. These motors are becoming more readily available as a result of the Energy Policy Act of 1992, which mandates manufacturers to provide them by October 24, 1997. However, since these motors have higher in-rush or starting current than standard, care must be taken in the selection of motor starters and the design of the associated power supply system to provide for this characteristic (Tencza and Billmyer, 1993).

Location of motors in tunnels should permit cooling to limit temperature rise over anticipated ambient temperature to the manufacturer's rating for continuous duty at the applicable service factor for the design. Totally enclosed fan-cooled motors (TEFC) are generally used, with air over-cooled (TEAO) units on vane axial or tube-type ventilation fans (Chapter 20). Emergency fans are to be serviceable for operations at 300°F (148.8°C) for a period of at least one hour per NFPA 130 (NFPA ,1990).

REFERENCES

American National Standards Institute (ANSI) (1992) *Power Systems,* C84.1-1982, New York.

American National Standards Institute (ANSI) (1993) *National Electrical Safety Code* (NESC), C2-1993, New York.

Federal Register (1981) 296FR Part 1910, Subpart S, (Effective April 16, 1981) Design Safety Standards for Electrical Systems, Washington, D.C.

Institute of Electrical and Electronics Engineers (IEEE) (1974) *Standard Safety Tests of Class IE Electrical Cables; Field Splices and Generators for Nuclear Power Generating Stations,* 383-1974, Piscataway, New Jersey.

Institute of Electrical and Electronics Engineers (IEEE) (1980) *Industrial and Commercial Power Systems* (Gold Book), 493-1980, Piscataway, New Jersey.

Institute of Electrical and Electronics Engineers (IEEE) (1982) *Grounding Industrial and Commercial Power Systems* (Green Book), 142-1982, Piscataway, New Jersey.

Institute of Electrical and Electronics Engineers (IEEE) (1986a) *Recommended Practice for Electrical Power Distribution for Industrial Plants* (Red Book), 141-1986, Piscataway, New Jersey.

Institute of Electrical and Electronics Engineers (IEEE) (1986b) *Protection and Coordination of Industrial and Commercial Power* (Buff Book), 242-1986, Piscataway, New Jersey.

Institute of Electrical and Electronics Engineers (IEEE) (1987) *Industrial and Commercial Applications* (Orange Book), 446-1987, Piscataway, New Jersey.

Institute of Electrical and Electronics Engineers (IEEE) (1990) *Industrial and Commercial Power System Analysis* (Brown Book), 399-1990, Piscataway, New Jersey.

Institute of Electrical and Electronics Engineers (IEEE) (1992) *Recommended Practices and Requirements for Harmonic Control in Electric Power Systems,* 519-1992, Piscataway, New Jersey.

Insulated Power Cable Engineers Association (IPCEA) (1994) *Short Circuit Characteristics of Insulated Cables,* P32-382-1994, South Yarmouth, Massachusetts.

National Electrical Manufacturers Association (NEMA) (1991) *Enclosures for Electrical Equipment (1,000 Volts Maximum),* 250 -1991, Washington, D.C.

National Electrical Manufacturers Association (NEMA) (1993a) *General Standards for Industrial Control and Systems,* ICS1-1993, Washington, D.C.

National Electrical Manufacturers Association (NEMA) (1993b) *Motors and Generators,* MGI -1993, Washington, D.C.

National Fire Protection Association (NFPA) (1990) *Fixed Guideway Transit Systems,* 130-1990, Quincy, Massachusetts.

National Fire Protection Association (NFPA) (1991) *Life Safety Code,* 101-1991, Quincy, Massachusetts.

National Fire Protection Association (NFPA) (1993) *National Electrical Code* (NEC), 70-1993, Quincy, Massachusetts.

TENCZA, J.W. and BILLMYER, C.M. (1993) *Consulting-Specifying Engineer,* December, pp. 32–35.

23 | Water Supply and Drainage Systems

Arthur G. Bendelius
Senior Vice President, Principal Professional Associate,
Technical Director, Parsons Brinckerhoff Quade &
Douglas, Inc.

Most tunnels will require systems to provide water to the tunnel and remove wastewater from the tunnel. A water supply system is required primarily for fire protection and possibly for wall-washing operations. A drainage system is necessary to collect, treat, and discharge the wastewater resulting from fire-fighting operations, washing operations, and leakage.

WATER SUPPLY SYSTEM

Highway, subway, and rapid transit tunnels, as outlined in Chapter 19, are usually provided with water mains, hose valves, and fire extinguishers and, in some instances, with water deluge systems on the exhaust fans. Railroad tunnels, because of their extreme length, are usually not provided with water mains.

The most significant aspect of the water supply requirement is the tunnel fire protection system. The water for fire protection is transported through a water main located within the tunnel. The minimum flow and pressure criteria must be established prior to implementation of design.

WATER SUPPLY DESIGN CRITERIA

Flow Rate

The minimum available water flow rate that should be provided for highway tunnels is stated by the National Fire Protection Association (NFPA) as 1,000 gpm (3,800 L/min) at adequate pressure (NFPA, 1991). This will provide sufficient flow for four 250 gpm (950 L/min) hose streams using the standard 2-1/2-in. (63.5-mm) fire hose connection. Such a water flow quantity will most likely be sufficient for any auxiliary use of the system, such as wall washing.

Pressure

Sufficient pressure must be available at each hose station to create an adequate hose stream. To discharge 250 gpm (950 L/min) from a 2-1/2-in. (63.5-mm) hose outlet equipped with 100 ft (31 m) of hose and a standard tapered hose nozzle, the residual pressure at the hose valve must be 60 psi (420 kPa).

In a subaqueous tunnel, the fire hose valve located at the low point of the tunnel may, because of the difference in elevations, have a greater available pressure than a fire hose valve located near the portal. For the same reason, in a mountain tunnel of fixed grade, the available pressure at opposing ends of the tunnel may vary greatly due to the large difference in elevation. In either case, the fire hose valve located at the higher elevation may often be the most critical from the standpoint of maintaining a minimum pressure requirement at all valve locations. This fact must be considered when hydraulic calculations for a tunnel water supply system are being performed.

Should the supply pressure be insufficient to provide the required pressure at the hose valve outlet, fire pumps must be considered at the point of water supply, as outlined later in this chapter.

WATER SOURCE

The selection of a source of water for a tunnel must be made early in the design process. In the case of a tunnel located in an urban area, the choice is relatively simple, since large quantities of water are normally readily available from a municipal water supply system. Elsewhere, a dedicated water supply may be necessary, using well or surface water.

Municipal Water Supply

A municipal water supply system should be the prime choice as a water source for any tunnel when it is available. This is a valid approach provided that the municipal water lines are relatively close and have a flow rate and pressure adequate to meet the flow and pressure criteria required at the tunnel. If, however, the municipal water lines are not within a reasonable distance from the tunnel, it may be less costly to provide another means of water supply. This could be a dedicated system using an underground water supply, a river, or a lake, together with a storage tank and fire pumps.

As indicated above, a municipal water system may supply the tunnel fire protection system if it is capable of providing at least 60 psi (420 kPa) at any hose outlet while supplying full fire protection demand. The full fire protection demand is a sum of 1,000 gpm (3,800 L/min) for hose outlets, cooling water for fans, and sprinkler demand (if any sprinklers are installed).

According to NFPA 502 (NFPA, 1991) a city main may be used for direct connection to the tunnel's fire protection system only if the pressure in the main would not drop below 20 psi (140 kPa) at the 1,000 gpm (3,800 L/min) load applied in addition to its normal nonfire flow.

Therefore, when considering a municipal water system as a source of water supply for a tunnel, the pressure and capacity of the city main at the point of proposed connection must be taken into account. If both are satisfactory, the tunnel system should be connected directly to the municipal water system.

If the municipal main satisfies the NFPA's minimum flow requirement of 1,000 gpm/20 psi (3,800 L/min/140 kPa) but does not have adequate pressure to provide 60 psi (420 kPa) at any hose outlet in the tunnel at full fire demand, booster fire pumps should be used. The pumps should be installed at the point of water supply to boost the pressure to the required level.

If the municipal main does not satisfy the NFPA's minimum requirement, the city water must be accumulated in a special storage tank and then supplied to the tunnel system by fire pumps having the required capacity and pressure. This arrangement must be used regardless of the pressure that the city mains can provide at any other fire flow rates.

The storage tank reserve capacity must be equal to at least 60 min (NFPA, 1991) of the full fire protection demand. This 60-min reserve is considered a bare minimum; a greater capacity is recommended. Where the municipal water supply is reliable, that is, the storage tank is connected to a city loop, makeup water accumulated in the tank during the fire pump operation may justify selection of a tank with minimum capacity.

Dedicated Water Supply

If a municipal water supply system cannot be used, a dedicated system using wells or surface waters must be considered. If wells are used, a storage tank should be used to accumulate the required water reserve and allow reducing the capacity of the well supply system, thus reducing its cost. The tank capacity should be selected to accommodate the full fire protection demand of the tunnel for a minimum of one hour (NFPA, 1991). Using the recommended flow rate of 1,000 gpm (3,800 L/min), a tank of 60,000-gal (228,000-L) minimum capacity would be required. A thorough analysis should be made prior to selecting the water supply system for each tunnel to determine which supply system is the most reliable and economical given the specific tunnel configuration.

Where possible, it is advisable to have a dual source of water supply for the tunnel fire protection system; that is, one line from each end of the tunnel. This is usually only practical in a subaqueous or urban tunnel, where the municipal water supplies are close at hand. Figure 23-1 shows typical schematic diagrams of three basic water supply system configurations used in highway tunnels.

WATER MAINS

The supply of water for tunnel fire protection is transported through a water main installed in a location that offers protection from both physical and high-temperature damage, such as the location shown in the subaqueous tunnel cross section in Figures 23-2 and 23-3.

Consideration must be given to whether the water main should be wet or dry. The wet main contains water at all times, whereas the dry main is filled only when water is required.

Wet Main

The wet main is ordinarily more advantageous for fire protection, since the water is available immediately when required during a fire.

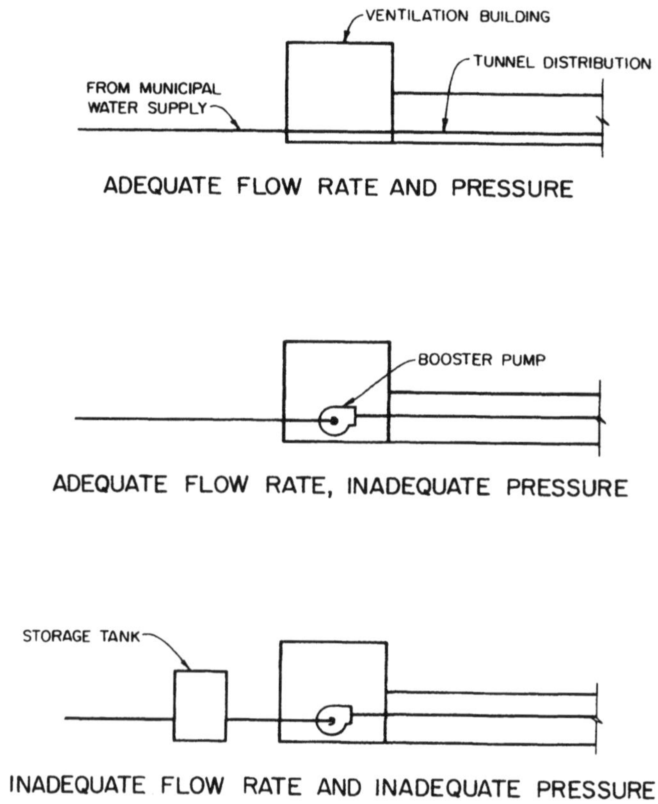

Fig. 23-1. Typical highway tunnel water supply configurations.

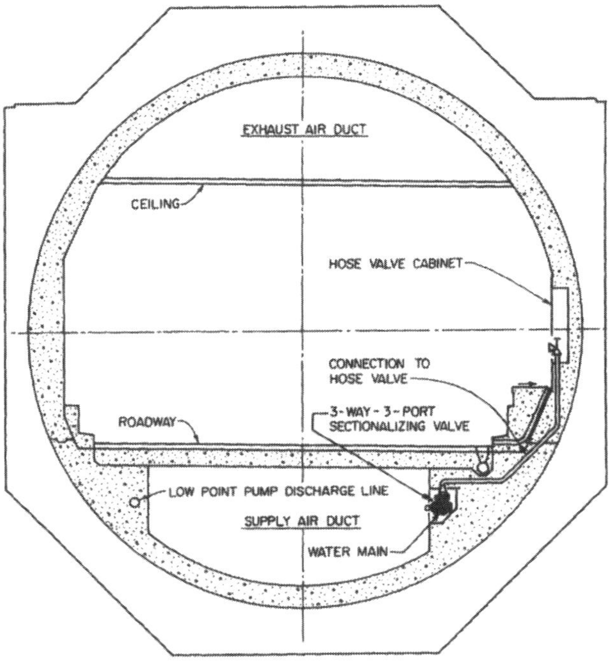

Fig. 23-2. Main and hose valves in a subaqueous highway tunnel.

The water in a wet main, however, cannot be allowed to freeze, and in some locations freeze protection will have to be provided. This can be accomplished in several ways. The most effective method involves recirculating heated water in the main to keep its temperature above freezing. Another method is to embed the main in concrete within the tunnel structure, with sufficient cover to prevent freezing. A third method is to keep a continuous movement of water in the pipe, thus reducing the possibility of freezing. This can be accomplished by a simple bleed system, controlled by a thermostatically controlled valve such as that shown in Figure 23-4. This system uses the available municipal water pressure to keep the water flowing. This approach will, however, result in a waste of potable water if a municipal water supply system is used.

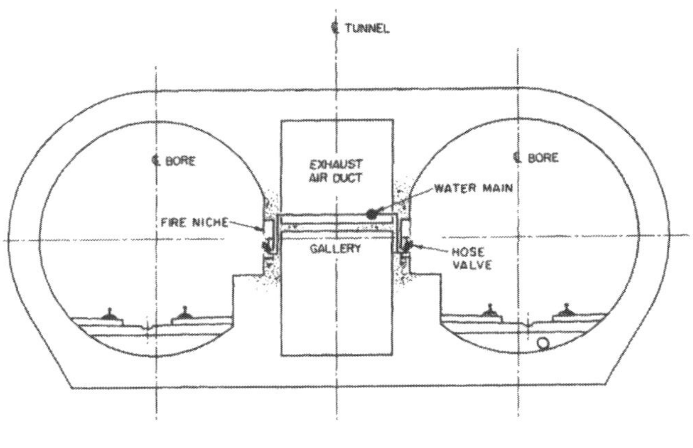

Fig. 23-3. Main and hose valves in a twin-tube subaqueous highway tunnel.

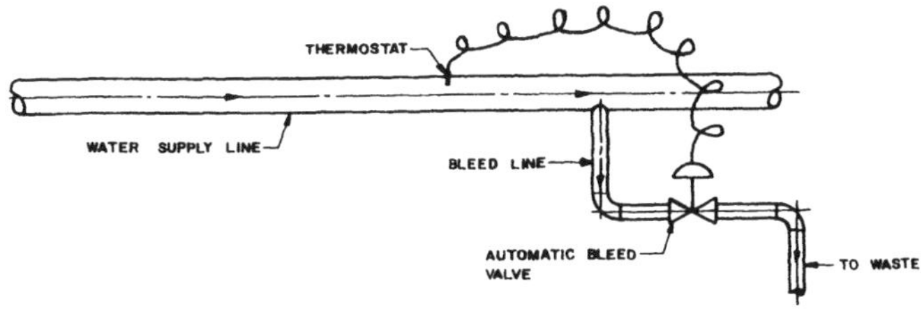

Fig. 23-4. Automatic bleed valve in wet main.

Dry Main

The time it takes to fill a dry main before the water becomes available for fire fighting becomes a serious disadvantage in longer tunnels. For example, using normal municipal water pressure, and a flow rate of 1,000 gpm (3,800 L/min), an 8-in. (203.2-mm) main in a 19,000-ft (5,791-m) tunnel would require more than 30 min to fill. This is not acceptable since it exceeds the 10-min limit set by NFPA 502 (NFPA, 1991) and would create a severe handicap during an emergency. Another disadvantage of a dry main is that it must be drained after each use.

Supply Arrangement

The water main in a tunnel, where practical, should be provided with two sources of supply, preferably one at each end of the tunnel. The main should be suitably valved and automatically controlled to provide water from either supply when required and to prevent flow from one water source to another. A typical method of valving is shown in Figure 23-5.

The water main within the tunnel should also be valved to separate sections to permit repair of a damaged section while minimizing any reduction in fire protection. This can be accomplished by installing sectionalizing valves, provided two sources of water supply are available.

Open Approach

The open approach to the tunnel, where the roadway extends beyond the portal, should be considered an extension of the tunnel and, therefore, protected. This will provide *available* fire protection for a portion of the roadway with limited access. In areas without freezing conditions and in all cases of a tunnel equipped with a dry pipe fire main, the outlets within the open approach should be connected to the tunnel water main.

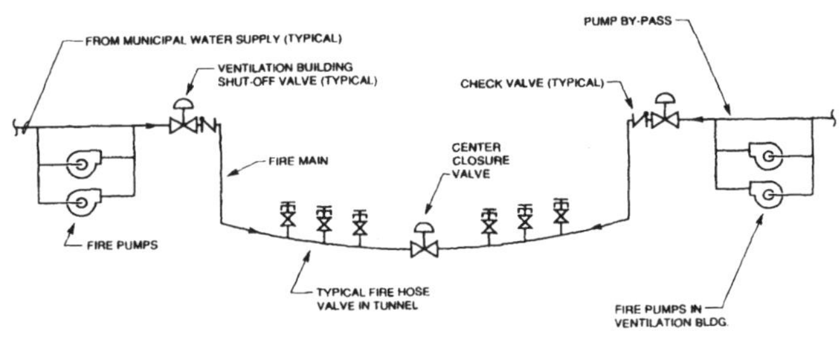

Fig. 23-5. Main with booster pumps for dual municipal supply.

In areas subject to freezing temperatures and where the tunnel is equipped with a wet pipe main, the approach should be protected by a dry pipe main. A fire protection Siamese connection should be provided at the street level so that water could be pumped by the fire department from the nearest street hydrant into the dry pipe system of the open approach.

An alternative method would be to protect the open approach from fire hydrants located above, at the surface of the road behind the approach retaining walls. The road must be accessible to fire-fighting equipment at all times.

Water Supply to Buildings

Water supply for the buildings associated with the tunnel should be provided for standpipes and/or sprinklers as required by local codes, NFPA Standards, and the local authorities having jurisdiction, in accordance with the construction group, area and height of the building, and occupancy classification.

Size of Main

The water main should be sized to minimize the pressure drop through the system and to provide sufficient pressure at the hose valves as required by code. The hydraulic design considerations should be balanced against the space available within the tunnel cross section. This should be verified by hydraulic calculations whether the system is supplied by a fire pump or directly by a municipal water supply.

Material

The piping material used for the fire main can be cast iron, ductile iron, or steel of sufficient strength to withstand the maximum pressure of the system. Sufficient means of compensation for temperature expansion and contraction and projected structural movements must be provided.

HOSE STATIONS

Hose Connections in the Tunnel

Hose connections are required within the tunnel roadway or trackway area. These should be conspicuously marked and located at a maximum spacing of 150 ft (45.7 m) (NFPA, 1991). In wide highway tunnels, a spacing of 300 ft (91.4 m) on each side of the tunnel roadway, with a staggered arrangement to achieve an equivalent longitudinal spacing of 150 ft (45.7 m), would be acceptable.

Hose stations in a highway tunnel should consist of a hose valve and a fire extinguisher (see Chapter 18). In most tunnels, fire hose is not installed at the hose station; it is carried on the emergency response vehicle, as noted in Chapter 25. Elimination of the hose from the hose station minimizes the amount of hose to be maintained, since unused fire hose deteriorates rapidly. A typical location of the fire hose station in the tunnel cross section is shown in Figures 23-2 and 23-3.

The hose valve should be approved for fire service. The valve thread should be compatible with the standard used by the local fire department(s). Most fire departments in the United States use the 2-1/2-in. (63.5-mm) standard.

The enclosure cabinet for the hose station should be provided with a door to protect the equipment. This door should be marked to permit easy identification of the station. In a number of tunnels, a detection system is installed that will provide a signal to the control room indicating the location of any opened hose station door.

Hose Stations for the Open Approach

The same type of hose station used in the tunnel should be installed in the open approach. Another method would be to install Siamese connections on the road above the open approach retaining walls. This method provides for greater access to hose connections with fire-fighting equipment to fight a fire from the service roads above, and it does eliminate the depth at which water piping would be installed to supply water to the hose stations in the open approach.

Hose Stations for Buildings

The hose stations located within the buildings associated with the tunnel should be installed as required by the NFPA standards, local codes, and with the authorities having jurisdiction.

PROTECTION OF EXHAUST FANS

An important part of the fire protection system of any tunnel is the capability to purge the hot gases generated during a fire. This is usually accomplished by the use of exhaust fans capable of removing the smoke and heat to facilitate the evacuation of tunnel occupants and provide clear access for fire-fighting equipment and personnel.

In highway tunnels, this is usually accomplished by either the normal exhaust ventilation system or by a reversible supply system, whereas in rapid transit systems a unique set of fans is installed for emergency ventilation (see Chapter 19).

Designing for Fan Protection

To support continued operation of the exhaust fans during emergency conditions, the critical components such as the power supply, motor, fan controls, bearings, and impeller must be protected.

The power supply should be protected against fire, flooding, and impact by placing it in protective conduits, possibly imbedded in concrete. The experience in the Montreal Metro fire of December 1971 highlights the need for such protection (Donato, 1972). As an added precaution, fans located in rapid transit systems should be provided with power from two sources (NFPA, 1993). For further discussion of power supplies, see Chapter 23.

High-temperature insulation should be provided for all fan motors that could possibly be exposed to hot gases from a fire. In a direct-connected axial-flow fan, this is a must, while, in a centrifugal fan, the motor can be separated from the hot gases (if you have ducted inlets). Elimination of the normal thermal protective devices in a motor should be considered in the case of an emergency fan that would be near the intense heat of a potential fire. This should not be considered for a fan that is used only for normal ventilation.

The fan controls and control wiring should be located so as to be protected from the stream of hot combustion gases and from all possible damage, during either normal operation or emergency operation of the system.

The fan itself (namely the impeller and the bearings) is also susceptible to damage from excessive heat. During the Holland Tunnel fire of 1949 (NBFU, 1949), the hot gases heated the fan wheel shaft to a temperature sufficiently high to soften the babbitt bushings of the bearings and thus cause the shutdown of three centrifugal fans. In the case of axial-flow fans, the required close tolerance between the impeller and the housing is a cause for concern when hot gases pass through the fans, causing serious distortion of the housing and interrupting the fan operation, thus creating a serious condition during a fire.

Deluge Systems

As protection against the possibility of fan damage, deluge or spray systems have been installed in several tunnels, which can provide the cooling effect of a water spray on the hot gases in the duct or on the critical surfaces of the fan itself in the event the temperature of the hot gases becomes excessive.

The tunnels operated by the Port Authority of New York and New Jersey have been provided with such a water deluge system on the fan equipment. These systems were installed after the 1949 Holland Tunnel fire (NBFU, 1949) and consist of two nozzles located so that their spray is directed onto the fan shaft and bearings, thus cooling the critical portions of these fans.

On the Hampton Roads and the Elizabeth River Tunnels in Virginia, a deluge system has been installed to protect the axial-flow fans in the exhaust ventilation system. The deluge system consists of eight water spray nozzles arranged on each fan as shown in Figure 23-6. Each nozzle can deliver 50 gpm (190 L/min) of water at 100 psi (690 kPa) nozzle pressure. The water flow is controlled by a deluge valve, which is activated by heat detectors located in the exhaust air duct.

There is a limited amount of design or test data available regarding fan deluge systems. Most of these systems in the United States were installed after the Holland Tunnel fire, and little, if any, research was done prior to their installation. The report regarding the Montreal Metro fire (Donato, 1972) has suggested the incorporation of such spray systems on fans for rapid transit facilities.

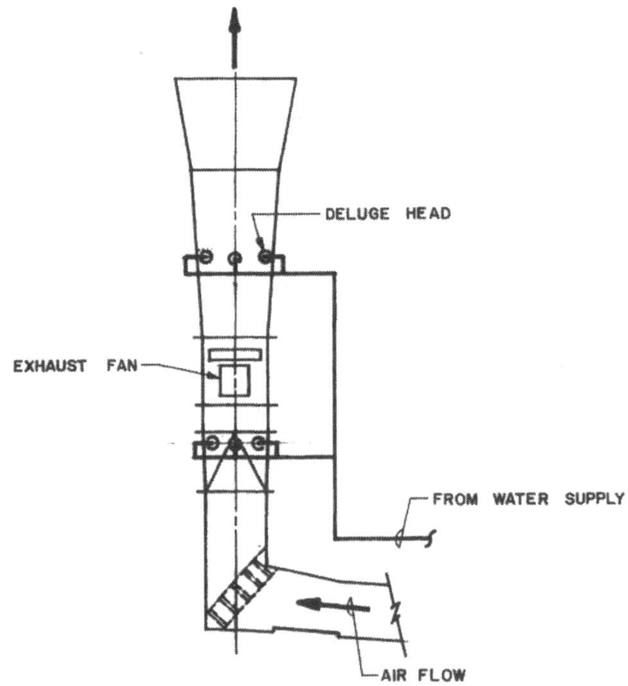

Fig. 23-6. Axial exhaust fan deluge system arrangement.

FIRE PUMPS

Centrifugal Fire Pump

A standard fire pump is of the centrifugal type, which has been specifically rated and listed for fire protection service by the Underwriters Laboratory (UL) and, where applicable, by Factory Mutual (FM). The major advantage of the centrifugal pump is that the discharge flow is reduced when the head increases, thus preventing a buildup of pressure in the piping system.

Standard fire pump sizes, by rated capacity, range from 25 gpm (95 L/min) through 5,000 gpm (19,000 L/min) with pressure ratings ranging from 40 psi (280 kPa) to 400 psi (2,760 kPa).

Horizontal Type

The horizontal, single-stage, split-case, double-suction volute pump is the one most often used in tunnel fire protection systems. In this pump, the suction inlet water flow separates and enters the impeller through both sides. Single-stage end-suction and in-line pumps may be used but are limited to capacities under 500 gpm (1,900 L/min).

Pump Characteristics

A typical set of performance characteristic curves for a fire pump is shown in Figure 23-7. It includes a plot of total head, brake horsepower, efficiency, and required net positive suction head (NPSH) versus discharge water flow rate and is based on constant-speed operation.

The total *dynamic* head of a horizontal centrifugal pump can be defined as the difference between the total heads on

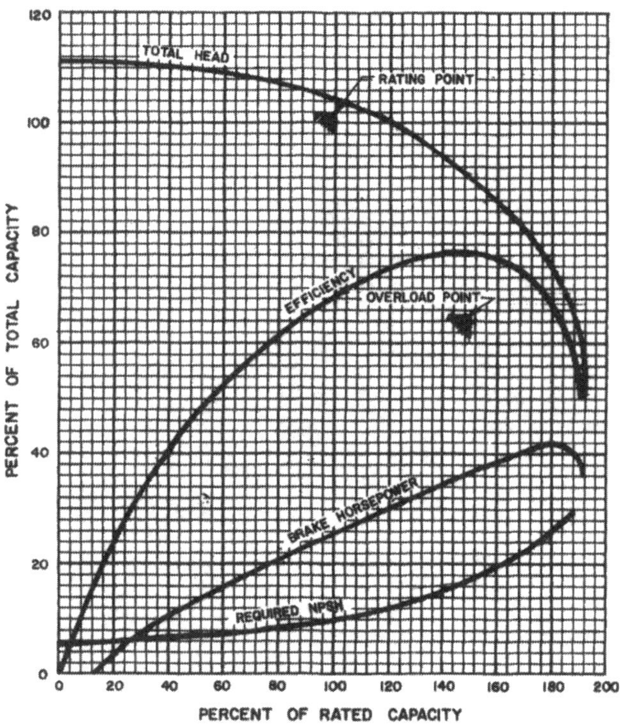

Fig. 23-7. Typical fire pump performance curves (NFPA, 1991).

the discharge and suction sides of the pump. This should take into account the gauge pressure and the static velocity pressures at the pump suction and discharge, as shown in Figure 23-8.

Performance Curves

The acceptable shape of a pump curve is affected by three limiting factors: shut-off head, overload, and rated capacity (Figure 23-7):

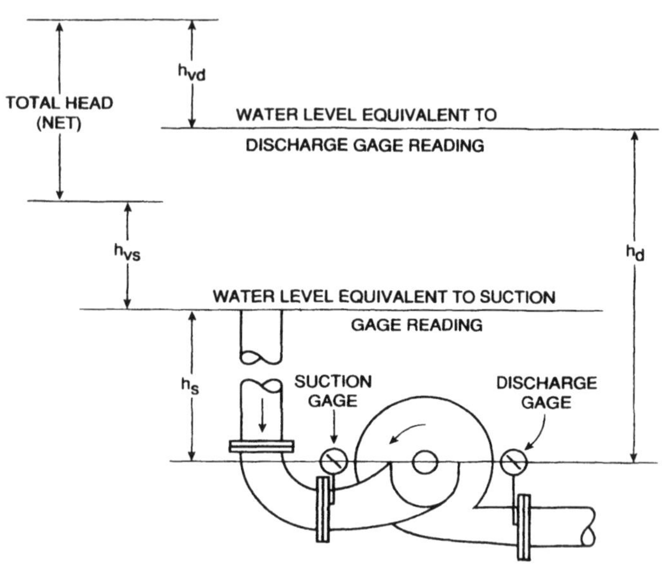

Fig. 23-8. Typical horizontal centrifugal fire pump head (NFPA, 1991).

- *Shut-off head.* The total head of a horizontal centrifugal fire pump operating at rated speed with the discharge line blocked should not exceed 140% of the total head at rated capacity.
- *Overload.* The total head at 150% of the rated capacity should not be less than 65% of the total head at rated capacity.
- *Rating.* The capacity/head curve must pass through or above the point corresponding to the rated capacity and head listed for the approved pump in the applicable listing (UL or FM).

All of these factors must be considered when selecting a fire pump.

Pump Selection

Fire pumps must be carefully selected based on the system resistance characteristics curve, the pump performance curve, and the factors outlined above, as shown in Figure 23-9. The pump capacity and pressure must meet the design and code requirements. For evaluation of the system pressure losses, an appropriate hydraulic handbook should be consulted.

The brake horsepower for a fire pump can be estimated by using the following formula:

$$\text{Bhp} = \frac{Q \times \text{HD}}{1{,}710 \times \text{EFF}} \tag{23-1}$$

where

Bhp = brake horsepower
Q = discharge flow rate (gpm)
HD = total head (psi)
1,710 = conversion factor
EFF = efficiency (percent/100)

In SI units:

$$\text{kW} = \frac{Q \times \text{HD}}{\text{EFF}} \tag{23-2}$$

where

kW = power (kW)
Q = discharge flow rate (L/sec)
HD = total head pressure (kPa)
EFF = efficiency (percent/100)

Approvals

According to the NFPA Standard (NFPA, 1990a), only approved equipment may be used for fire pump installations. The pump manufacturer should be responsible for shop testing the pump equipment, obtaining equipment approval, and providing certified pump performance curves.

Fire Pump Drives

Most fire pumps are driven by electric motors, since a high degree of reliability is usually developed by the entire

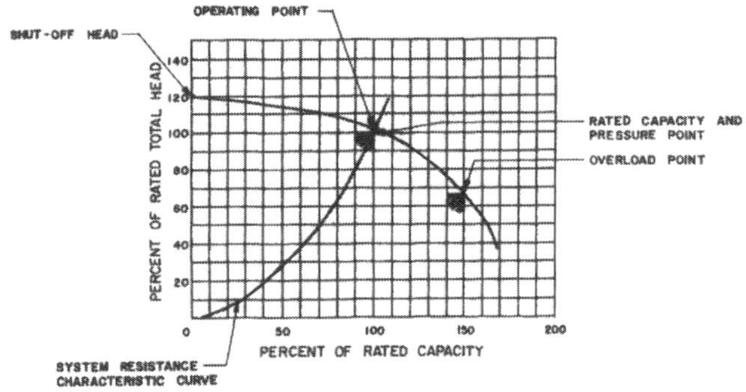

Fig. 23-9. Standard performance and system curves for fire pumps (NFPA, 1991).

tunnel power system, and the chance of power loss is remote. Also, in many of these tunnels, a secondary source of power, such as a standby emergency generator driven by a diesel engine, should be considered for critical services such as the fire pump.

The fire pump should be rated for continuous duty and should comply with all National Electrical Manufacturers Association (NEMA) specifications. All electrical equipment and wiring in a fire pump installation should, as a minimum, comply with the National Electric Code (NFPA, 1990b), except where modified by NFPA standard. (NFPA, 1990a)

The diesel engine or steam turbine engine used to drive a fire pump must be listed for fire pump service by a testing laboratory and should meet the requirements of NFPA 20 (NFPA, 1990a).

Motor and engine controllers for fire pumps must be specifically approved and listed as fire pump controllers. An approved controller is a completely assembled unit, wired and tested by the manufacturer prior to shipment, and ready for service. Listed motor controllers may be used for any configuration of automatic or manual operation. Where emergency power is available, the fire pump controller must be furnished with a listed automatic transfer switch, installed in a separate compartment.

Auxiliary Devices

A number of auxiliary devices are required for the proper functioning of an approved fire pump installation (NFPA, 1990a):

- A listed outside stem and yoke (OS&Y) gate valve installed on the suction side of the pump
- A listed spring-loaded check valve on the pump discharge
- A listed indicating gate or butterfly valve on the system side of the check valve
- A listed relief valve on the pump discharge (where required)
- Hose valves for test purposes
- An automatic air release valve on the top of the pump casing
- A circulating relief valve

Pump Arrangement

It is recommended that 100% standby pump facilities be installed in the tunnel fire protection system. With a single source of electric power, the standby pump should be diesel-driven.

Pump Control

Most tunnel fire pump installations are arranged for automatic or remote manual control. Fire pumps should be installed with a positive suction head to avoid priming problems.

Pump Rooms

The fire pump should be located in a dry enclosure, protected from fire, explosion, flood, dirt, corrosion, and unauthorized access. The enclosure should be provided with light, heat, ventilation, and floor drainage. Enclosing structures should have at least a 2-hour fire rating with self-closing fire doors rated at 1-1/2 hours, and they should be accessible from the outside or from an enclosed fire stair directly or via a 2-hour rated corridor. Figure 23-10 shows a typical tunnel fire pump room.

Acceptance Tests

Tests should be conducted on each completed tunnel fire pump installation, in accordance with applicable NFPA/FM Standards. These tests must demonstrate the adequacy of the pump suction head and the ability of the pump to deliver the quantity of water and pressure in accordance with the certified shop test data and satisfactory operation of the pump controller and all auxiliary devices.

DRAINAGE SYSTEM

A drainage system is required in all tunnels to remove water that could accumulate from rainfall, tunnel washing operations, tunnel seepage, vehicle drippings, fire-fighting operations, or any combination of these sources. Drainage of a tunnel can be accomplished either by a gravity flow system or a pumped system. A gravity flow system will suffice for

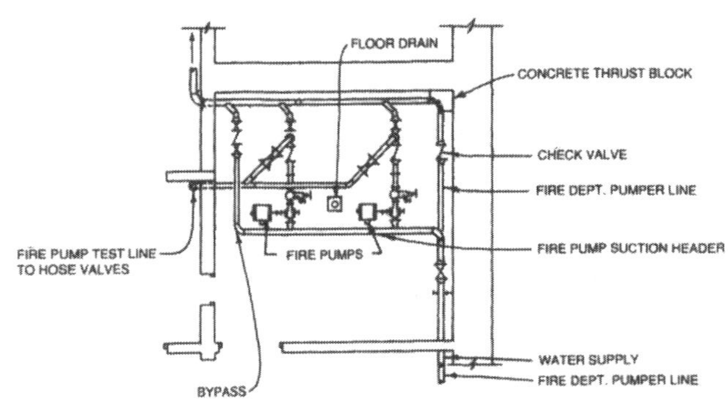

Fig. 23-10. Typical fire pump room arrangement.

tunnels with continuous grades, provided that the collected water can be properly disposed of at the lower end of the tunnel. It may be necessary to have a pumped system to dispose of collected water when a low point occurs within the tunnel.

DRAINAGE DESIGN CRITERIA

The drainage system design must be predicated on a proper determination of the anticipated flow rate, that is, the peak discharge rate of the water to be drained.

Drainage of Rainfall

To design an adequate drainage system, criteria regarding the amount of expected rainfall in the area to be drained must be available. Three key factors in determining the amount of water to be drained from rainfall are intensity, frequency, and time of concentration.

- *Intensity.* This is the amount of rainfall within a specific period of time, usually given in in./hour. Intensity–duration curves for many areas of the United States are published by the U.S. Weather Bureau (U.S. Dept. Commerce, 1961).
- *Frequency.* The frequency of a storm is the average number of years between occurrences of a storm of a given or greater intensity. For normal drainage design a 10-year storm frequency is used; however, for a tunnel where flooding is a serious concern, such as a subaqueous tunnel, a 100-year storm frequency should be considered.
- *Time of Concentration.* The time of concentration is defined as the time required for run-off from the most remote point of the drainage area to arrive at the point where the entire area exposed to the rain contributes to the flow rate in the drainage system.
- *Run-off.* The amount of water left from a rainfall after the losses from evaporation, transpiration, and infiltration is the run-off.

The method commonly used when computing run-off for small areas is the Rational Method (Asphalt Institute, 1984), as shown in the following formula:

$$Q = C \times A \times I \tag{23-3}$$

where

Q = peak run-off rate (ft^3/sec)
C = run-off coefficient
A = drainage area (acres)
I = rainfall intensity for the time of concentration and storm frequency selected (in./hour)

While this formula is not dimensionally correct, it is satisfactory for most tunnel applications since one cubic foot per second is approximately equal to one acre-inch per hour. Application of the Rational Method of design to tunnels allows modification to the above formula to reflect the size of the drainage areas encountered in tunnel design and the applicable units (gpm and ft^2), which results in

$$\text{QG} = 0.010 \times C \times \text{AF} \times I \tag{23-4}$$

where

QG = peak run-off flow rate (gpm)
0.010 = conversion factor (gal-hour/ft^2-min-in.)
AF = drainage area (ft^2)

In SI units,

$$\text{QL} = 30.40 \times C \times \text{AM} \times \text{IM} \tag{23-5}$$

where

QL = peak run-off rate (L/sec)
30.40 = conversion factor (L-hour/m^2-sec-mm)
C = run-off coefficient
AM = drainage area (m^2)
IM = rainfall intensity for the time of concentration and storm frequency selected (mm/hour)

The run-off coefficient is the ratio of the rate of run-off to the rate of rainfall at an average intensity when the entire drainage area is contributing. A list of run-off coefficients is shown in Table 23-1. For most tunnel applications, the value of the run-off coefficient for surface drainage will be about 0.90 (C = 0.90), since most of the drainage areas will be sloped pavement. Therefore, we can further modify the formula for tunnel applications to

$$Q = 0.000207 \times \text{AF} \times I \tag{23-6}$$

where

0.000207 = combined conversion factor (gal-hour/ft^2-min-in.)
AF = drainage area (ft^2)

In SI units,

$$QL = 27.36 \times \text{AM} \times \text{IM} \tag{23-7}$$

Table 23-1. Selected Values of Run-off Coefficients[a]

Surface Type	Runoff Coefficient, *C*
Asphaltic/concrete pavement	0.70–0.95
Roofs	0.75–0.95
Gravel roadway	0.40–0.60
Grassed surface	0.10–0.35
Bare earth	0.20–0.90

[a] For flat slopes, use the lower values; for steep slopes, use the higher values (Asphalt Institute 1984, Water Pollution Control Federation 1970).

where

27.36 = combined conversion factor (L-hour/m^2-sec-mm)
AM = drainage area (m^2)

For open areas, such as vertical vent shafts and fan discharges, the following would apply, since all of the rainfall is assumed to be falling into the opening ($C = 1.0$):

$$QC = 0.0094 \times \text{AF} \times I \qquad (23\text{-}8)$$

where

0.0094 = combined conversion factor (gal-hour/sq ft-min-in.)

In SI units,

$$QL = 30.4 \times \text{AM} \times \text{IM} \qquad (23\text{-}9)$$

where

30.40 = combined conversion factor (L-hour/m^2-sec-mm)

Tunnel Washing Operations

In most highway tunnels, the walls and ceilings are washed periodically, usually in a two-stage procedure. First, a detergent is applied using a wash truck with rotating brushes; second, the detergent and the dirt are rinsed off with water by a separate flush/rinse truck. In a typical tunnel of 7,500 ft (2,286 m) in length, the wash process includes a quantity of water and detergent roughly equal to 1,500 gal (56,780 L) and a rinse quantity of 15,000 gal (56,780 L); the process takes approximately one hour. Thus, an average washwater flow rate for this typical tunnel would be approximately 275 gpm (17.4 L/sec).

Drainage from Fire-Fighting Operations

Fire-fighting operations can also contribute a sizable quantity of water to the drainage system. This quantity can be determined by estimating the maximum flow rate of water that could be pumped into the tunnel during a fire emergency within the capacity of the existing fire protection system.

Drainage of Vehicle Drippings

Vehicle drippings have been shown to be of minimal consequence and, if the system is designed to handle all the above quantities, the water from vehicle drippings will be adequately handled.

Drainage of Seepage

Most tunnels through hills and mountains have water seepage problems. Surface water penetrates through fissures and percolates through permeable soils. Concrete liners are not completely watertight, and water may find its way through cracks in the lining.

Attempts to seal off the rock by grouting with either cement or chemicals usually are not successful. Concrete linings are not completely watertight. Water will find its way through shrinkage cracks in the linings into the interior tunnels. There, it can freeze and cause an unsightly appearance in highway tunnels.

If water appears in considerable quantity during tunneling operations, longitudinal drainpipes should be installed behind the sidewalls, with laterals at regular intervals to the main tunnel drain lines.

Cut-and-cover tunnels can be waterproofed, and with good control, the number of leaks in such a tunnel can be minimized. Seepage in underwater tunnels, either the shield driven or the immersed tube type, is usually limited and can be controlled by caulking joints where leaks do appear in segmented liners.

OPEN APPROACH DRAINAGE

The portions of the tunnel roadway that extend beyond the portals are classed as the tunnel approaches. In cases where the approach road slopes down into the tunnel and cannot be drained by gravity external to the tunnel, the approach drainage system must be included in the tunnel drainage system. This is especially true in a subaqueous tunnel where the open approaches are below the surrounding grade. The open approach drainage system should be designed to minimize the influx of water from the open approach roadway into the tunnel.

The quantity of drainage water on the open approach can be computed from rainfall alone, since this value, in most instances, will be the greatest.

Straight Open Approaches

On straight open approaches without superelevation, transverse interceptors placed approximately 300 ft (91 m) on centers, with the first one located immediately outside the tunnel portal, are most effective in preventing the run-off from entering the tunnel. The actual interceptor spacing will depend on grade, inlet capacity, and pavement type. The interceptors are approximately 18 in. (457.2 mm) in width, extend from curb to curb, and are covered with heavy cast iron gratings.

Superelevated Open Approaches

When the approach is provided with a superelevation, the approach drainage inlets must be placed at regular intervals along the low curb. For comparison of the two arrangements, with and without superelevation, see Figure 23-11. The drainage water flows by gravity from the approach inlets into portal pump stations or into the tunnel drainage system to discharge into a pump station located at a low point in the tunnel (see Figure 23-14, later in the chapter). The gravity line should be a minimum of 8 in. (203.2 mm) in diameter, with cleanouts located at required intervals.

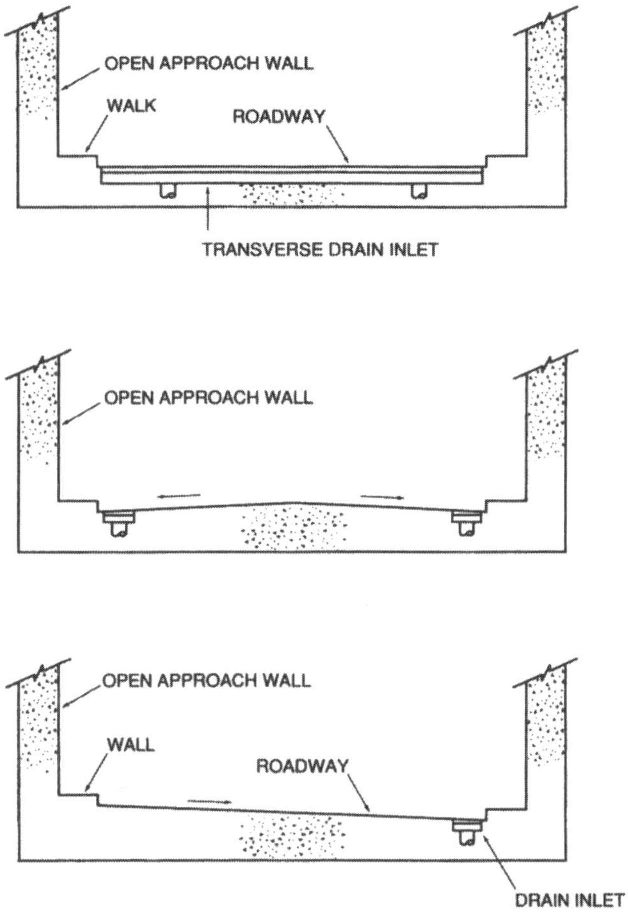

Fig. 23-11. Open approach roadway drainage arrangements.

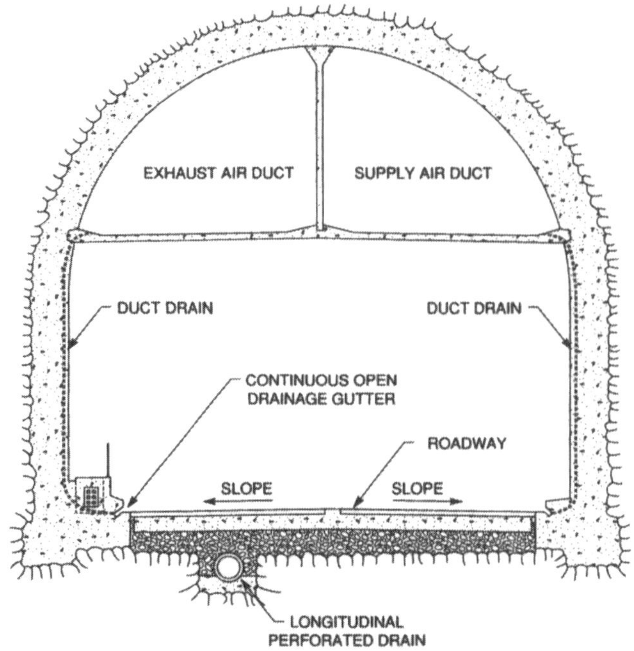

Fig. 23-12. Open gutter drainage system in mountain highway tunnel.

TUNNEL DRAINAGE

Roadway Drainage

The roadway drainage system for a tunnel can be either open or closed.

Open Drainage System. The open type of roadway drainage consists of a continuous gutter recessed into the curb and has been used in many tunnels (Figure 23-12). This system, however, may permit propagation of a fire of burning fuel, in the event of a serious accident, due to a continuous source of air to support combustion.

Closed Drainage System. The closed system, on the other hand, will minimize such fire propagation, since the drainage liquid enters the inlets located at the curb lines, then passes through a closed gravity flow system to a pump station. For this reason, the closed system should be used in all situations. The drainage inlets should be spaced 50–75 ft (15–23 m) apart on both sides of a level roadway and on the low side of a superelevated roadway.

Drainage Inlet. The drainage inlet design is important, since it must remain clear of debris that would prevent influx of water. The drainage inlet must be a key maintenance item to prevent clogging.

Gravity Drain Line. The gravity drain line carrying the drainage water from the inlets to the pump station should be a minimum of 8 in. (203.2 mm) in diameter. Line cleanouts should be located every 100 ft (31 m), with suitable access.

Miscellaneous Tunnel Drainage

There are several areas in the tunnel where water from wall-washing, seepage, or fire-fighting operations can collect and should be drained to maintain proper condition of the tunnel. Figure 23-13 shows some of these for a typical subaqueous immersed tube tunnel cross section. The diagram includes drains from the overhead air duct, niche drains, sidewalk gutter drains, and electrical pull box drains. All of these can be drained by gravity to the roadway. The size of the drain lines from these areas is significant only from the standpoint of possible blockage by debris, since the flow is extremely small. A minimum 2-in. (50.8-mm) diameter pipe is recommended.

DRAINAGE PUMP STATIONS

Any tunnel from which the drainage water cannot be properly removed by gravity must be provided with one or more pump stations having the necessary capacity to remove the maximum drainage demand. There are two basic locations for tunnel pump stations: at the low point of a tunnel, in particular a subaqueous type, and at the portals of any tunnel (Figure 23-14).

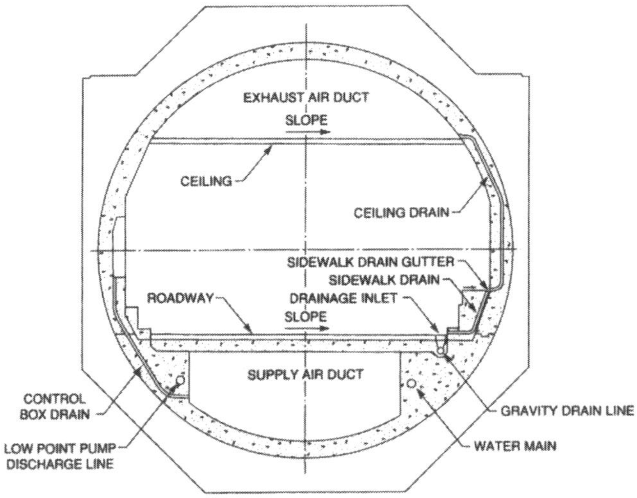

Fig. 23-13. Miscellaneous drains in subaqueous highway tunnel.

Low-Point Pump Station

The purpose of the low-point pump station is to collect all the drainage water within the tunnel and pump it out to the portal pump station or to a designated system.

The low-point pump station in a subaqueous tunnel could have a configuration similar to that shown in Figure 23-15. This pump station is located within the supply air duct, usually at a point in the duct where the air velocity is sufficiently low to minimize the pressure losses.

The pump type most appropriate for the low-point pump station is a vertical dry pit or horizontal centrifugal type because of the usual limited available headroom at such locations.

Portal Pump Station

The portal pump station in a subaqueous tunnel collects water from the open approach and water pumped from the tunnel low point and the ventilation structures, as shown in Figure 23-14.

A portal pump station would be required for a mountain tunnel if the water cannot be properly drained by gravity.

A typical arrangement for the portal pump station in a subaqueous tunnel is shown in Figure 23-16. Almost any type of pump can be used in the portal pump station since headroom is usually not a a serious problem.

The portal pump station may be constructed by either of two methods. The first is in-place construction, where the structure at the portal is formed and constructed to create the necessary settling and holding chambers and pump room.

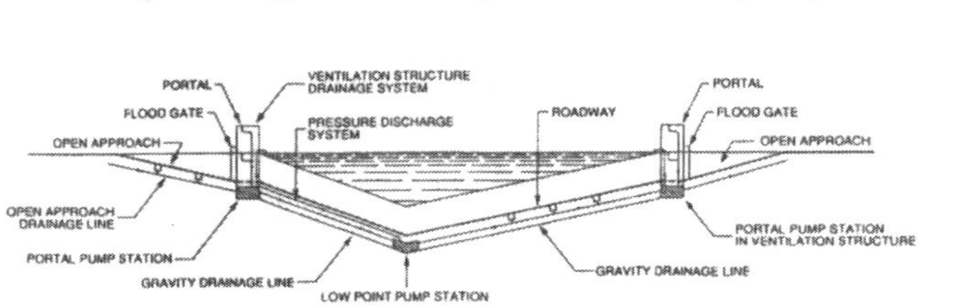

Fig. 23-14. Typical profile of subaqueous highway tunnel.

The pumps and associated piping are then installed in the field. A typical arrangement is shown in Figure 23-16. The second method employs a prefabricated package pump station similar to that shown in Figure 23-17. Package pump stations are most appropriate in locations where they can be installed from the surface directly into the structural enclosure. They usually are not suitable for installations within a tunnel, such as at the low point, due to the difficulties arising in transporting and installing them within confined spaces.

Chambers

All pump stations require one or several chambers designed to hold and treat the drainage water. A full evaluation of the drainage water treatment is presented in the next section of this chapter.

One such system is shown in Figures 23-23 and 23-24. The settling well, or sump, is the first line of water treatment. It should be sized to provide adequate time for solids to settle out of the water, based on normal inflow rates. The settling well could be equipped with a skimming weir, as shown in Figures 23-23 and 23-24, or a bar screen across the width of the well, to prevent floating materials from clogging the pumping system or being discharged out of the system.

The holding or storage sump, which receives the water from the settling well, should be sized to prevent a rapid cycling of the drainage pumps and to allow a minimum of 4 min of running time for each pump. If submersible pumps are used, the need for a separate pump room is eliminated.

Provision must be made in all installation for periodic inspection and removal of sludge from the sumps. Access manholes and proper drainage facilities are necessary, as shown in Figure 23-23. Portable housekeeping pumps can be used to empty the chambers for maintenance purposes.

DRAINAGE PUMPS

Three broad categories of pumps are available: reciprocating, rotary, and centrifugal. However, since only the centrifugal pump is used in tunnel drainage applications, only this type of pump will be discussed in this section.

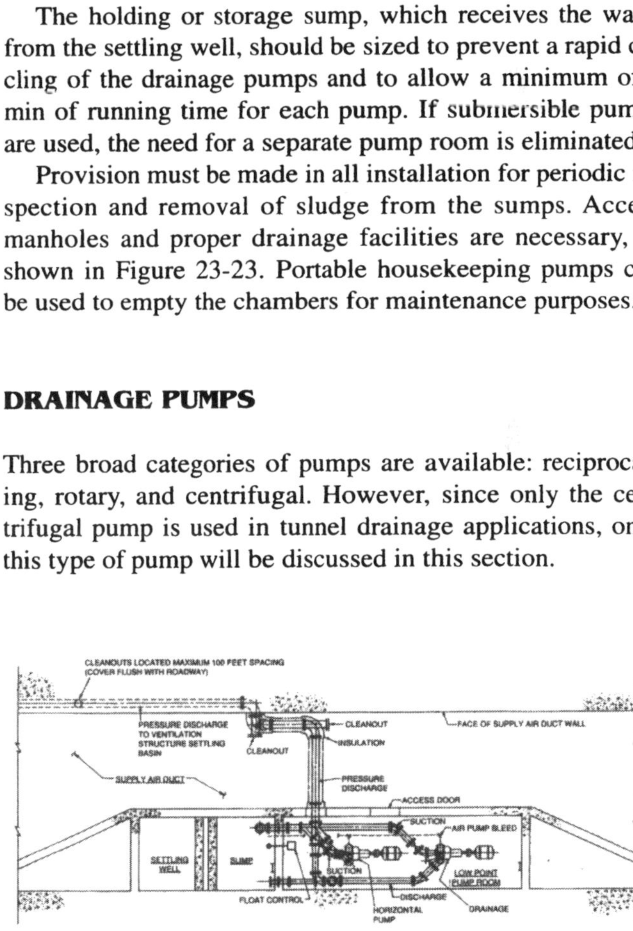

Fig. 23-15. Typical low-point drainage pump station in subaqueous highway tunnel.

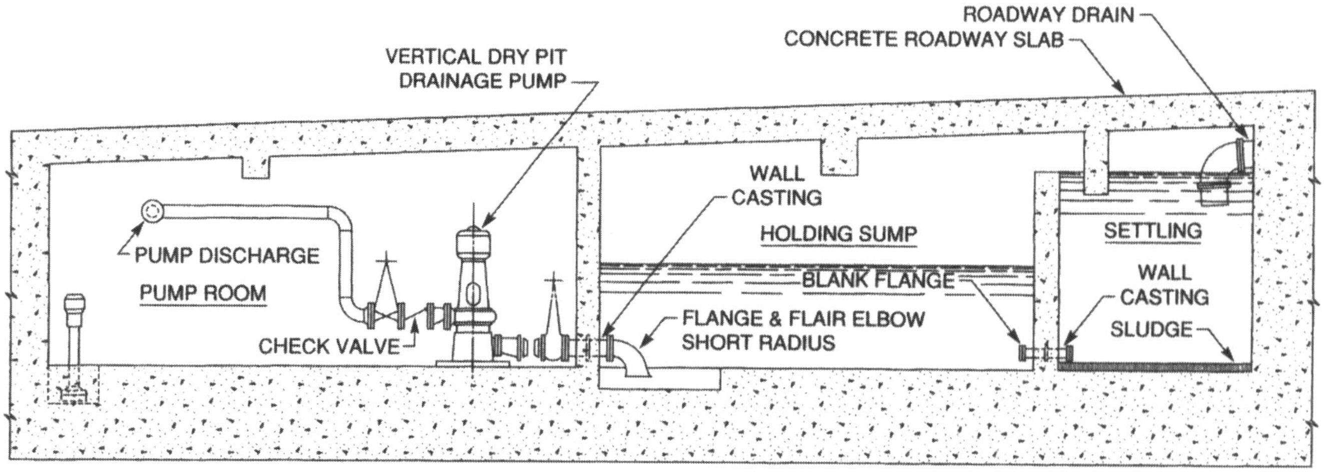

Fig. 23-16. Typical portal pump station subaqueous highway tunnel.

The centrifugal pump can be operated against a completely blocked discharge without overloading the drive, and it is the most suitable pump for tunnel drainage service. Centrifugal pumps can be obtained in several arrangements: horizontal, vertical dry pit, submersible, and vertical wet sump. Each of these is suitable for a particular situation or situations.

The *horizontal pump* is suitable where the headroom is limited and plan space is available, or when high flow capacity and pressures are required. The horizontal split casing pump is easy to maintain because half of the casing can be removed for inspection and maintenance without removing the piping or bearings.

The *vertical dry pit pump* can be used where the vertical space is limited but plan space is available adjacent to the sumps. This pump requires less floor space than does the horizontal type. The entire pump is accessible for ease of maintenance. The vertical dry pit pump is a type most often used in a package pump station.

The *vertical sump pump* is appropriate where floor space is limited, but vertical space is available above the sumps.

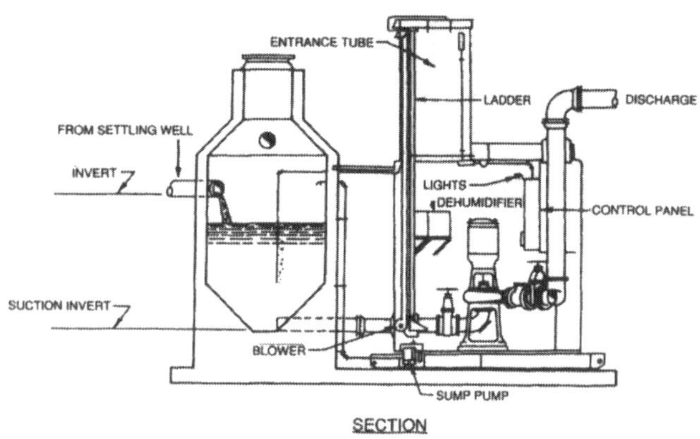

Fig. 23-17. Typical prefabricated package pump station.

The impeller and the bearings of this type of pump are located below the water, thus creating some maintenance problems. When the pump must be removed, considerable space is required to lift the pump.

The *submersible pump* is installed completely below the surface of the water in the sump. The only connection from the pump to the space above the sump is the discharge pipe, the power cable, and the removal cable. These pumps can be mounted on vertical rails for lifting the pumps to the surface for maintenance. It is not recommended where traffic in the tunnel must be interrupted to service the pump. This pump requires a minimum of floor area and vertical space.

The centrifugal pumps outlined above should, if possible, be installed in tunnel drainage systems with a flooded suction. If a suction lift is unavoidable, a reliable priming system must be installed.

Pump Drives

Most tunnel drainage pumps are driven by electric motors. However, there are instances where an internal combustion engine drive would be appropriate. This could be during a power failure when emergency pumping is absolutely necessary.

Pump Arrangement

In most tunnel drainage systems, more than one pump should be installed to provide an adequate factor of safety in what is a critical system, particularly in a subaqueous tunnel. Other benefits of multiple pump installation include smaller individual installations and servicing loads, reduced electrical starting loads and cable sizes, smoother pumping, and overall installation economies. The use of two pumps in each pumping station, each having 100% capacity, implies full spare capacity, and would give too large a pumping increment. Normal inflow is much less than peak capacity, making it undesirable to run one high-capacity pump. If three pumps are used, each having 50% capacity, then as inflow increases the second and third pumps would start sequentially; such a system will have a margin of safety and a smaller pumping increment.

Pump Selection

The performance of a centrifugal pump can be defined by the water flow rate and the total head and can be graphically displayed in a characteristic curves such as that shown in Figure 23-18. A pump should be selected to operate at a point where the pump curve intersects the system resistance curve with an attempt to maximize the pump efficiency (Figure 23-19).

In tunnel drainage systems, the pumps are installed in parallel arrangement so that each pump can operate singly or in parallel. Pump selection must take this parallel arrangement into account. When two pumps operate in parallel, they will not deliver twice the water flow rate of one pump operating on the same system. It will be necessary to draw the system head curve and the system capacity curve to determine the actual flow rate (Figure 23-20).

The *total head,* or pressure, against which a drainage pump must operate is made up of two components: static head, and friction and dynamic losses.

Static head is the maximum total height to which the water must be raised by the pump. In a low-point pump station, this would be the vertical distance from the surface of the water in the low-point sump to the highest point of the discharge piping.

Friction and dynamic losses are due to the water velocity in the piping system. These losses can be calculated by referring to a hydraulic data book.

The *required net positive suction head* (NPSH) of the pump is the minimum head of water required at the pump suction when the pump operates at any given capacity. The pump NPSH must be less than the available NPSH in any water pumping system, to provide sufficient pressure at the pump suction to prevent pump cavitation and, sometimes, loss of prime. The minimum required NPSH is shown on thc pump performance characteristic curves (Figure 23-18).

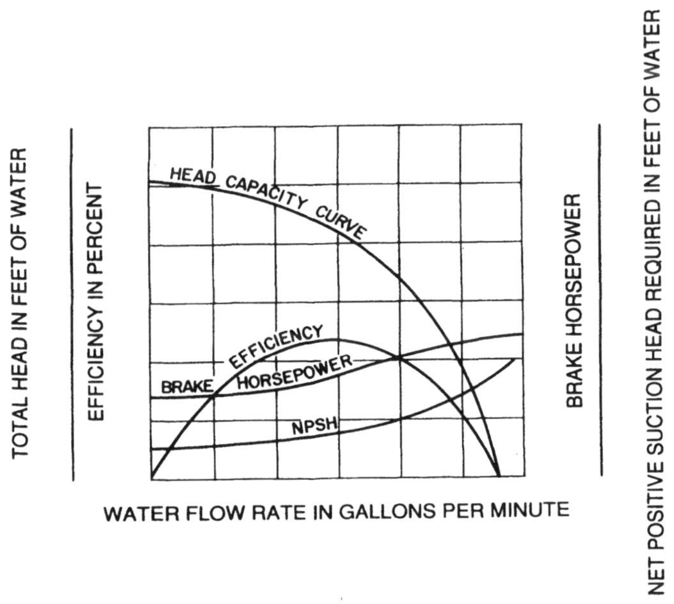

Fig. 23-18. Typical centrifugal pump performance curves.

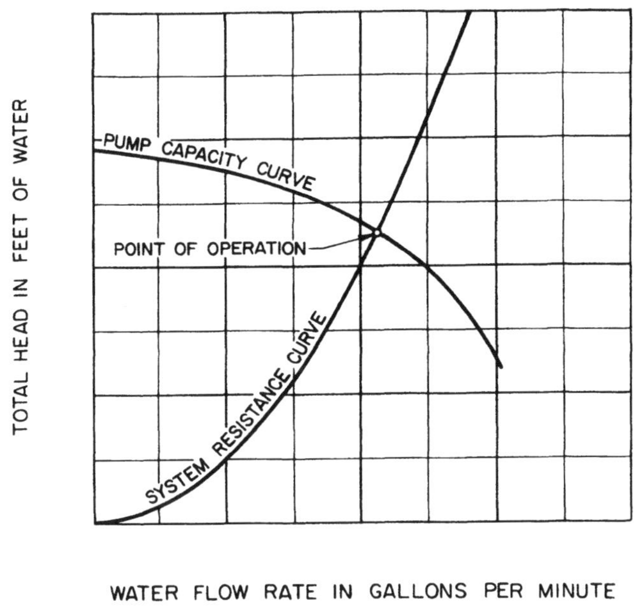

Fig. 23-19. Pump selection curves.

Brake horsepower is the rate of work required as input to the pump:

$$\text{Bhp} = \frac{Q \times \text{TH} \times S}{3{,}960 \times \text{EFF}} \tag{23-10}$$

where

BhP = brake horsepower
Q = water flow rate (gpm)
TH = total head (ft of water)
EFF = pump efficiency (percent/100)
3,960 = conversion factor
S = specific gravity (water at 60°F: S = 1)

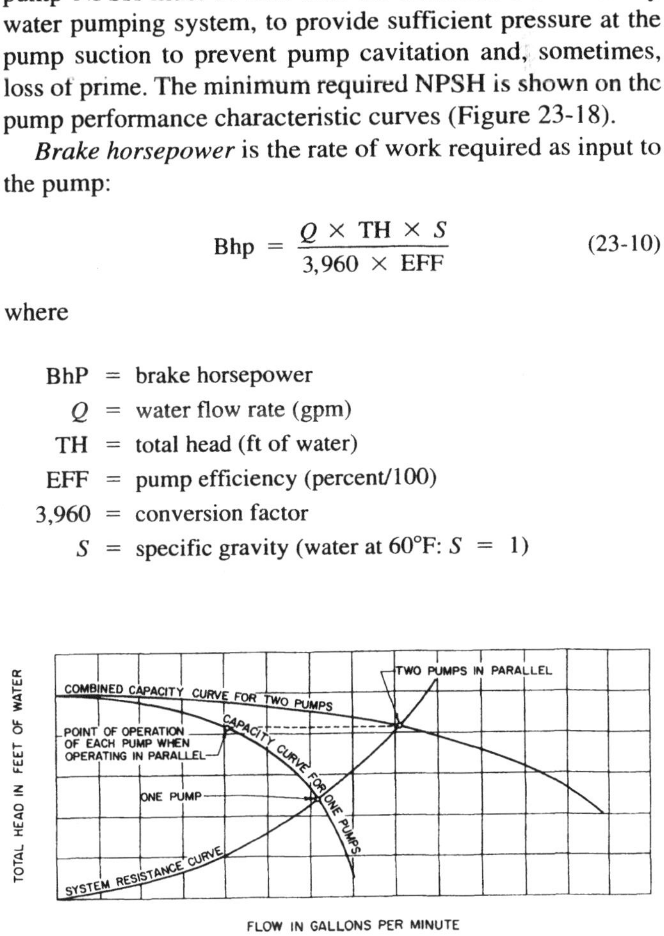

Fig. 23-20. Typical parallel pump operation performance curves.

In SI units,

$$\text{kW} = \frac{O \times \text{TH} \times S}{20{,}747 \times \text{EFF}} \qquad (23\text{-}11)$$

where

kW = power (kW)
O = water flow rate (L/sec)
TH = total head (kPa)
EFF = pump efficiency (percent/100)
20,747 = conversion factor
S = specific gravity (water at 37°C; S = 1)

Pump Control

All tunnel drainage pump stations must be provided with automatic pump control systems that will permit automatic starting and stopping of pumps based on the level of water in the sump. The pump operation should be sequenced and alternated to obtain equal operating time on each pump.

These systems usually consist of some form of water level detection and associated pump controls. There are several types of level detection and control devices suitable for the tunnel drainage system.

The *float type* has probably been used more than any other type in tunnel drainage systems. There are, however, several disadvantages to both the mechanically linked ball float type and the chain- or tape-operated float. A single float is usually used to detect a number of water levels and control points; and if the mechanical portion of the system is jammed, no additional pumps will start, thus possibly creating a flooding condition. The float-type level detector in a tunnel drainage system usually requires a complicated installation.

The *electrical conduction electrode* or probe type of water level detection is easy and inexpensive to install. Although the effectiveness of these probes is affected by foreign material in the drainage water, they remain the most effective in the tunnel environment in reliability and maintainability. Separate probes are required for each control point, as shown in Figure 23-21. An advantage is that the control relays can be located remotely from the probe location.

The *mercury float switch* is a simple and inexpensive water level detection device. The main disadvantages are its wide detection range and that it is seriously affected by turbulent water. A typical installation is shown in Figure 23-22.

There are other types of water level detection devices, such as the newly developed sonic type. However, there is at this time limited experience in tunnel drainage systems with this equipment.

The pump control points should be established to allow a minimum pumping time of 4 min for each pump.

Gas Detection

Each pump station should be outfitted with a hydrocarbon gas detector and analyzer system to provide an alarm to central control when hydrocarbon gas or other combustible or explosive gases are present in the drainage flow stream. This intrusion can come from a fuel spill in the tunnel due to a leak or an accident.

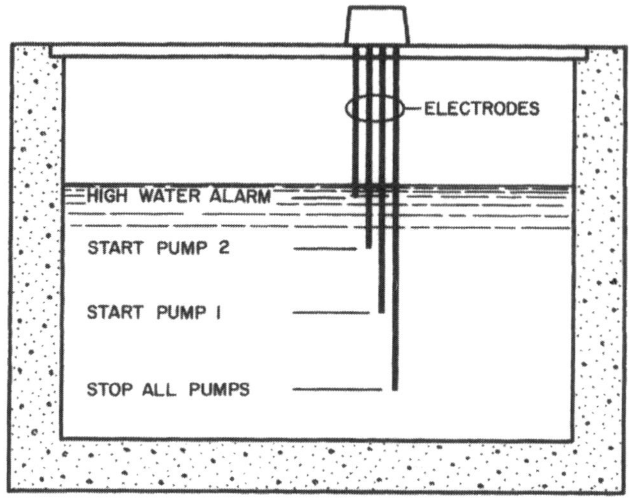

Fig. 23-21. Electrode level control.

WATER TREATMENT

The drainage water from a vehicular tunnel will have contaminants from vehicle drippings and from tunnel washing operations, the washing operations producing the severest problem. The effects of the detergents used in the washing operation on the drainage water quality must be evaluated. The detergent, which is usually diluted by water, must be low sudsing, biodegradable, and a nonemulsifier.

It has been shown that, for many tunnel installations, a treatment system, as shown in Figure 24-24, will be adequate to mitigate the effect of the tunnel operation on water

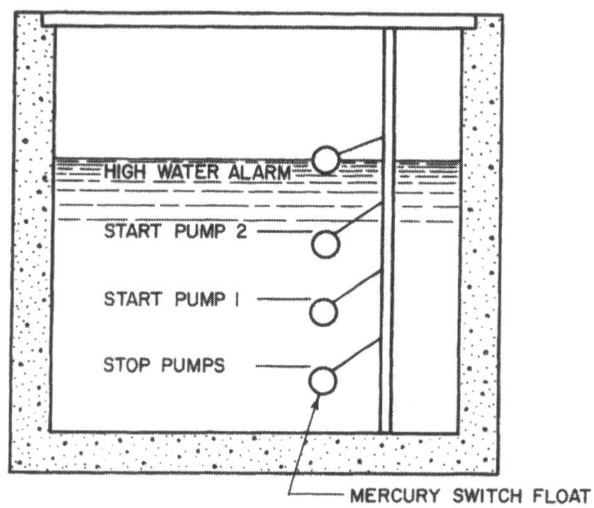

Fig. 23-22. Mercury float switch level control.

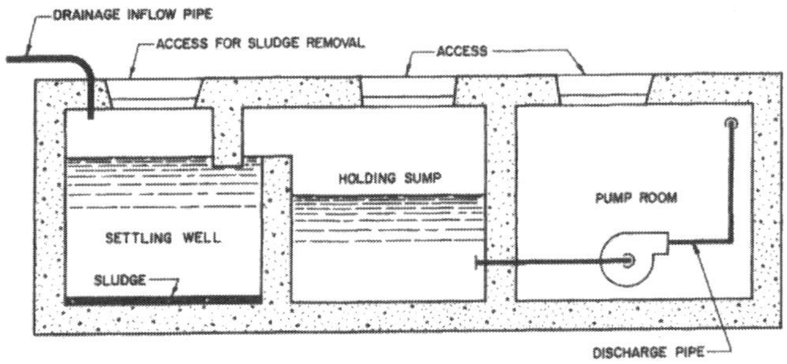

Fig. 23-23. Typical drainage pump station chamber arrangement.

quality. There must be sufficient settling time to permit sedimentation of solids and to permit skimming of floating materials. Periodic removal and proper disposal of sludge, which accumulates at the bottom of the settling well, must be included in the operation of the system.

The critical pump station in a subaqueous tunnel, from a standpoint of settling for contaminants, is at the low point, where all the drainage water is collected and either pumped to a portal pump station or to a designated system. The water will enter this sump at a maximum rate of 275 gpm (1,045 L/min) during washing operations.

The type and location of the drainage discharge must be carefully considered lest the receiving body of water be seriously affected by the discharge. Subsurface discharge should be considered for open bodies of water.

Treatment of drainage from a rail tunnel is normally not necessary, since this type of tunnel is not washed, and therefore detergents are not introduced into the drainage system.

Fire Fighting

The drainage water from fire fighting can carry abundant contaminants, especially if there is an oil or gasoline spill coincident with the fire. These contaminants will be discharged into the low-point pump station with the drainage water. Environmental and local agencies will require these containments to be properly treated and separated prior to discharge of drainage water into sanitary or storm sewer systems.

Oil/Water Separators

The oily waste resulting from an oil spill due to an accident in a tunnel will require separation from the wastewater

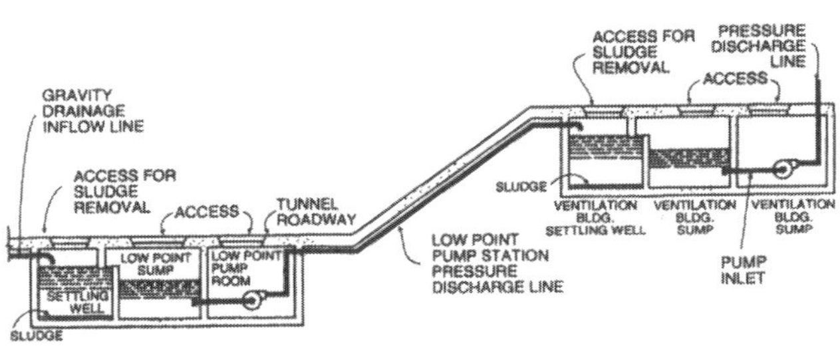

Fig. 23-24. Tunnel drainage discharge treatment system with settling basins.

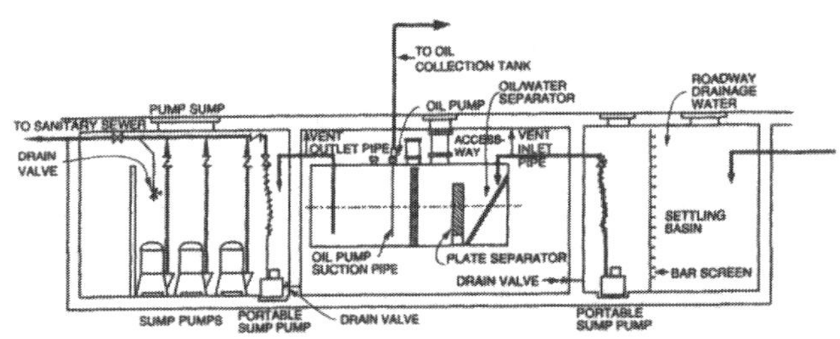

Fig. 23-25. Tunnel drainage discharge treatment system with oil/water separator.

stream. Oil/water separators may be used for this purpose by breaking up the oily influent and causing the oil or lighter-density substances to rise and coalesce at the water surface, from which the oil can be removed and disposed of properly (Figure 23-25). The drainage water can then be discharged to a sanitary or storm sewer system or other designated system.

FLOOD PROTECTION

The only tunnels subjected to serious flooding are subaqueous, either in tidal areas or in flood plains of rivers. Where possible, flooding should be prevented by raising the elevation of the approaches above maximum flood levels. Where this is too expensive or impractical, flood gates must be installed (Figure 23-26).

It may be possible to raise the approach walls above the flood level and install flood barriers at the upper end of the open approaches. This would be less expensive than raising the approach elevation and would prevent flooding of the tunnel and the open approaches.

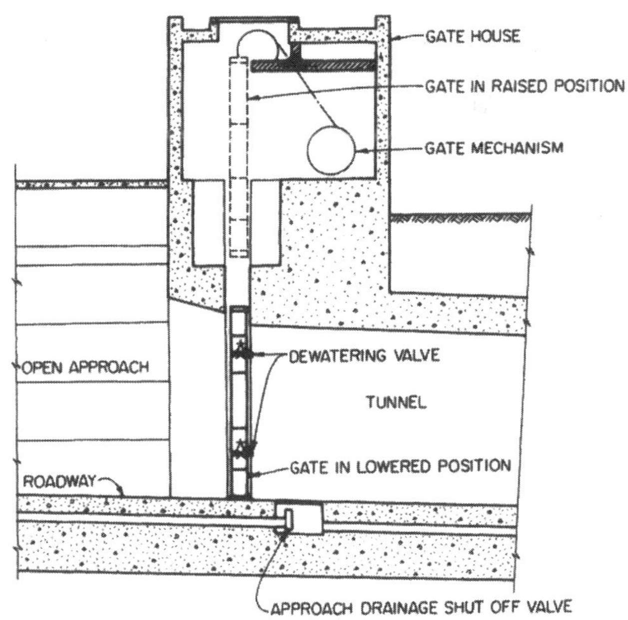

Fig. 23-26. Typical flood gate installation at portal of subaqueous highway tunnel.

Flood Gates

Flood gates installed at the tunnel portals will provide a closure to restrain rising flood waters from entering the tunnel and permit collection of flood water at the portals. A typical installation appears in Figure 23-26. These gates are constructed of steel and designed to withstand the hydraulic forces present during maximum flood conditions. The gate travels in vertical frames and seals against a seat built into the roadway to minimize the leakage into the tunnel. An enclosure may be provided above the gate while it is in its raised or stored position (Figure 23-26).

Leakage criteria should be established for each portal flood gate installation and should be considered in the selection of pumps for the tunnel pumping stations.

Valves are required on the flood gate (Figure 23-26) to permit rapid drainage of the water collected in front of the gates prior to raising the gates. This water can be drained and permitted to enter the tunnel drainage system prior to raising the gates. Gate valves of a minimum 2-1/2 in. (63.5 mm) size with threaded connection for 2-1/2-in. (63.5-mm) hose are most appropriate for this application.

In subaqueous tunnels where there are no portal pump stations, the open approach drain lines are connected to the tunnel drainage system. A means must be provided to isolate these two systems during flooding to prevent the ingress of flood waters through the open drainpipes, which could cause flooding of the tunnel. A method of providing such isolation is to place a shut-off valve in a valve box on the tunnel side of each portal flood gate, as shown in Figure 23-26. This will permit inspection of the valve to confirm that the drainpipe leading from the open approach is clear of debris and can be sealed tightly against the flood waters.

Testing. The flood gate should be tested against a head of water equal to the maximum flood level anticipated. This test is required to assure that the leakage criteria are not exceeded. Construction of a watertight bulkhead on the open approach side of the flood gate will permit development of such a head and testing of the flood gate.

Operation. When flood conditions are imminent, the flood gates should be lowered to seal the tunnel portals. The drainage system isolation valves, along with the dewatering valves on the flood gates, must be in the closed position at this time.

DRAINAGE OF RAIL TUNNELS

Rail tunnels include both those carrying railroad trains and those carrying rapid transit trains, with both ballast and concrete roadbeds. A railroad tunnel with ballast is usually drained by installing a perforated pipe below the track ballast, as shown in Figure 23-27. Where there is no ballast used, such as in a rapid transit tunnel, an open channel or drainage trough is often used, as shown in Figure 23-28. The drainage water is carried through this channel to inlets located at specified intervals. These inlets permit the water to enter the gravity flow line, thus transporting the water either to a low-point sump, in the case of a subaqueous tunnel, or to the low portal in the case of a mountain tunnel.

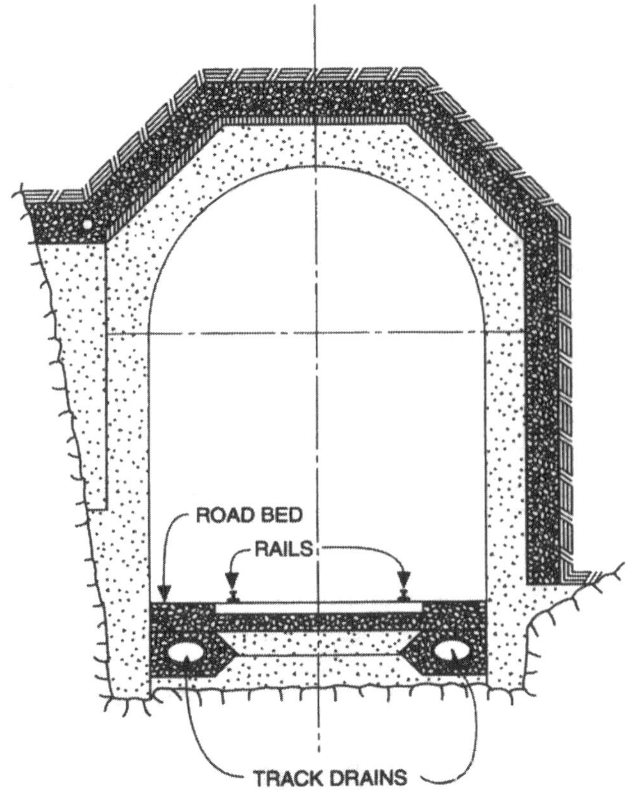

Fig. 23-27. Drainage of railway tunnel.

The water quantity anticipated in the rail tunnel drainage system will consist of water from either fire-fighting operations, vehicle drippings, seepage, or rainfall on the approach tracks and on openings to the surface.

Subway

Drainage in the subway portion of a rail rapid transit system is accomplished by use of a center channel to collect

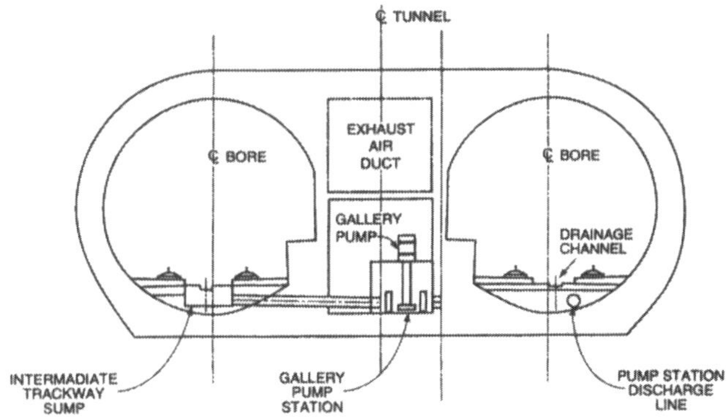

Fig. 23-28. Typical drainage of subaqueous rapid transit tunnel.

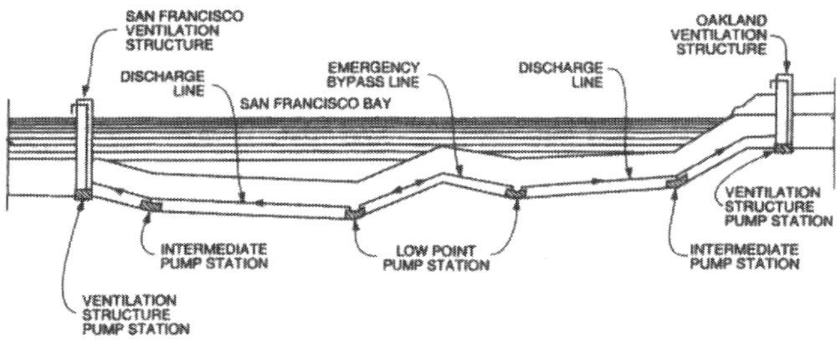

Fig. 23-29. Profile of Trans-Bay Tube.

and transport water, which is then piped to pump stations located at each low point.

System Descriptions

Drainage System with Two Low Points. The Trans-Bay Tube is a 4-1/2-mi (7-1/4-km) subaqueous rapid transit crossing of the San Francisco Bay in California. It is a part of the Bay Area Rapid Transit system. This tunnel has two low points, as shown in Figure 23-29. The drainage system consists of two low-point pump stations, four intermediate-gallery pump stations located at changes in grades, and two ventilation-structure pump stations. The four intermediate-gallery pump stations, as shown in Figure 23-30, are arranged so that two comprise the intermediate pump stations at grade changes and two are included in the low-point pump stations.

The drainage water is collected and transported in the open channel located between the rails in the concrete roadbed and then collected either at the intermediate points or at the low points. At the intermediate point, the water is collected in a trackway sump and then flows by gravity into the intermediate sumps, from which it is pumped through the main discharge line to the pump station located in the ventilation structure at the rear end of the tunnel, either in Oakland or in San Francisco. At the tunnel low point, the water is collected in a trackway sump, drained by gravity to the gallery sump, then pumped into the main sump of the low-point pump station. From the low-point sump station, the water is pumped through the main discharge line to the vent structure pump station and then discharged to the surface.

This system is equipped with an emergency bypass arrangement, whereby either low-point pump station can pump its effluent to either ventilation structure. This provides a method of removing water should there be a break in either end of the discharge line. A recirculating arrangement has also been built into this system to provide the means to exercise the pumps with a minimum of water in the sumps.

Drainage System with One Low Point. The Potomac River Crossing, on the Huntington Route of the Washington Metro, is a 6,000-ft (1,830-m) subaqueous rail tunnel. It is a

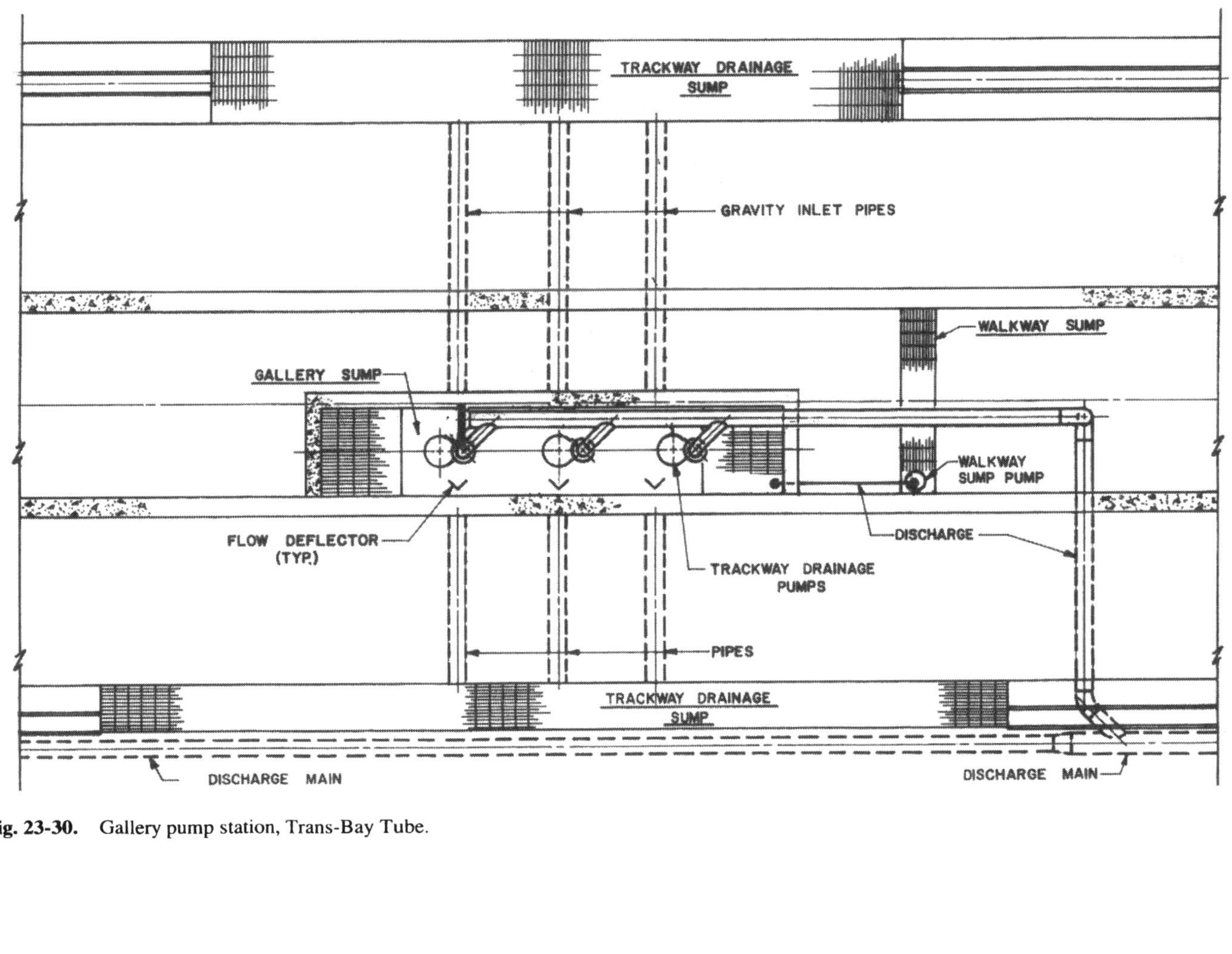

Fig. 23-30. Gallery pump station, Trans-Bay Tube.

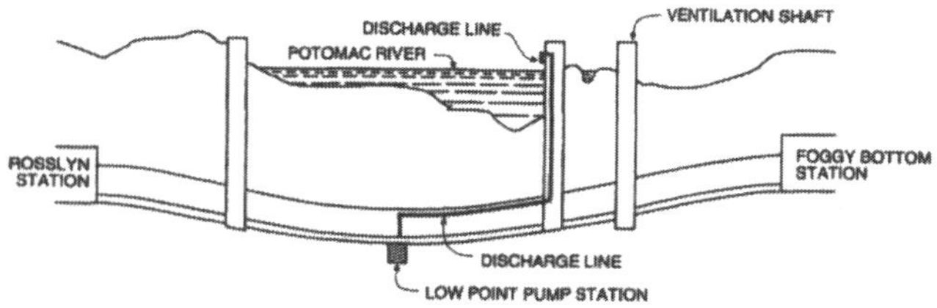

Fig. 23-31. Profile of Potomac River Crossing.

double-track, two-bore tunnel. The crossing profile is shown in Figure 23-31 and has one low point. The water is drained from the trackway by an imbedded drainage pipe. The water will flow in an open channel to drain inlet sumps, which are spaced approximately 300 ft (90 m) on centers. Through these inlets, the water enters the imbedded gravity drain line and then flows to the low-point pump station, which is of a package construction with two pumps. The water is then pumped to the surface through a pressure discharge line.

REFERENCES

Asphalt Institute (1984) *Drainage of Asphalt Pavement Structures,* Manual Series No. 15 (MS-15), Maryland.

DONATO, G. (1972) "Montreal Metro Fire, December 9, 1971," International Metropolitan Railway Committee, International Union of Public Transport, New York.

FREEMAN, J.R. ASCE Transactions XII, American Society of Civil Engineers, New York.

National Board of Fire Underwriters (NBFU) (1949) *The Holland Tunnel Chemical Fire,* May 13, 1949, National Board of Fire Underwriters report, New York.

National Fire Protection Association (NFPA) (1990a) *Centrifugal Fire Pumps,* NFPA 20, Quincy, Massachusetts.

National Fire Protection Association (NFPA) (1990b) *National Electric Code,* NFPA 70, Quincy, Massachusetts.

National Fire Protection Association (NFPA) (1991) *NFPA Fire Protection Handbook,* 17th edition, NFPA 502, Quincy, Massachusetts.

National Fire Protection Association (NFPA) (1992) *Fixed Guideway Transit Systems,* NFPA 130, Quincy, Massachusetts.

U.S. Dept. of Commerce (1961) *Rainfall Intensity Atlas of the United States,* Weather Bureau, Technical Paper No. 25, Washington, D.C.

Water Pollution Control Federation (1970) *Design and Construction of Sanitary and Storm Sewers,* WPCF Manual of Practice No. 9 (also available as ASCE Manual and Report on Engineering Practice No. 37), Washington, D.C.

RECOMMENDED READING

BENJAMIN, T.J. (1972) Pumping Services for the Second Mersey Tunnel, *Tunnels and Tunnelling,* March, p. 12.

Crane Co. (1957) *Flow of Fluids Through Valves, Fittings and Pipes,* Technical Paper No. 410, Chicago.

GARAY, P.N. (1990) *Pump Application Desk Book,* The Fairmont Press, Inc., Lilburn, Georgia.

SHAW G.V., and LOOMIS, A.W. (eds.) (1962) *Cameron Hydraulic Data,* Ingersoll-Rand Co., New York.

24 | Surveillance and Control Systems for Highway Tunnels

Richard J. Naish
Senior Supervising Engineer, Parsons Brinckerhoff Quade & Douglas, Inc.

Unlike open highways, tunnels require special attention to maintain safety under normal and abnormal traffic conditions. Most modern tunnels, including their approach roads, require a centralized control system to meet these goals. Although not all tunnels require the same attention or installed features, they all have the following general features. [Note: The dimensions shown in this chapter are indicative guidelines. All English unit equivalents are soft (rounded) conversions of metric.]

SURVEILLANCE AND CONTROL SYSTEMS

These systems provide means to

1. Monitor traffic flow and identify impending congestion or stoppages caused by breakdowns or accidents
2. Maintain a safe tunnel environment, responsive to traffic density and travel speed
3. Communicate travel restrictions to motorists approaching and passing through the tunnel
4. Mobilize emergency response to clear incidents within the tunnel
5. Initiate, when appropriate, the necessary systems operation for emergency conditions
6. Monitor the status of tunnel service equipment to ensure continued operation and availability when needed.

Central Control

The control center is usually located in a tunnel ventilation or administration building and is manned 24 hours a day. For low-volume rural tunnels or short-length tunnels that are really extended underpasses, an alternative is part-time remote monitoring with tunnel equipment operating automatically.

Human Control

The common control element is the need for human intervention to judge the extent and severity of incidents, followed by initiation and supervision of corrective measures and emergency response until conditions have returned to normal.

OVERVIEW OF AVAILABLE TECHNOLOGY

Tunnel Configuration

Most road tunnels are dual-bored or immersed tubes each carrying two or three full-width traffic lanes with an offset or minishoulder on both sides to safety barriers (see Figure 24-1). Approach roadway shoulders are carried through short tunnels but are eliminated or reduced to offsets on high-construction-cost tunnels. A rule of thumb states the cost of tunnel construction will directly increase on an algorithmic scale with the tunnel diameter. Full-time traffic monitoring has been shown to be more cost effective than providing full shoulders. Tunnels are normally designed for directional operation with provisions to operate bidirectionally should the adjacent bore/tube be closed for emergencies or maintenance. Cross-passages between the tunnels are provided for emergency evacuation, access to fight tunnel fires, a location to install equipment control centers, and in a number of long tunnels, vehicle turnaround.

Tunnel Surveillance and Control Equipment

The following traffic control devices, used to monitor vehicle flow, stop traffic, close individual lanes and/or a tunnel, and provide visual instructions to motorists passing through the tunnel, are installed in the tunnel roadway on the walls or suspended from the ceiling.

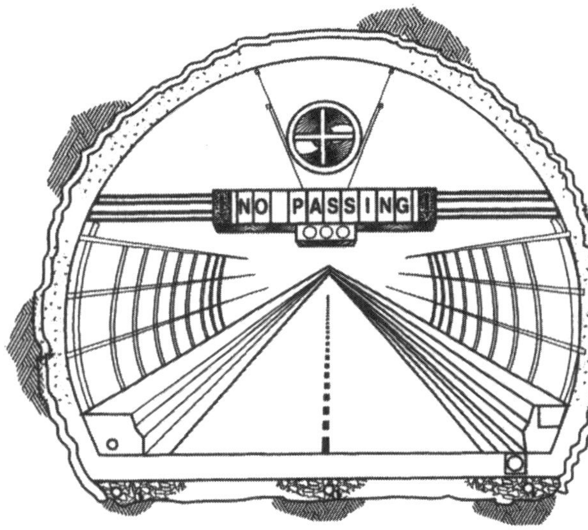

Fig. 24-1. Typical tunnel cross section.

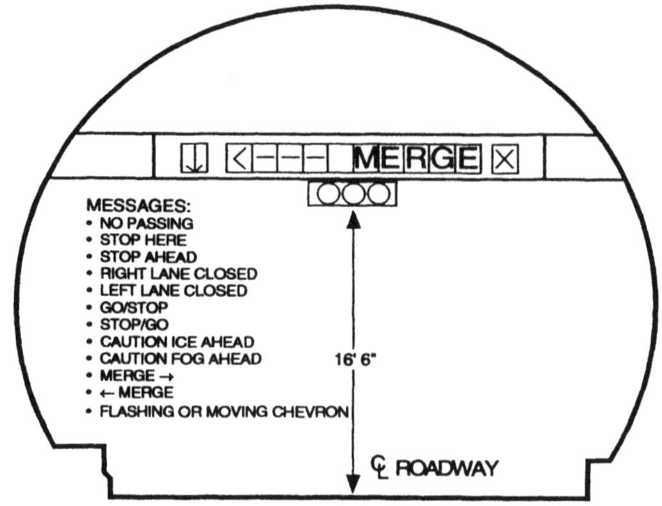

Fig. 24-2. Tunnel variable-message sign at cross-passageway.

- *Vehicle detectors.* Loop detectors embedded in each traffic lane with video radar/microwave detectors mounted overhead to support or replace the loop detectors are used to monitor vehicle volume, speed, and occupancy and, by processing this data, identify the probability of stoppages.
- *Signs and signals.* Spaced at regular intervals over the roadway are 10- or 12-character variable-message signs (VMS) with lane-use signals at both ends. The signs are spaced at regular intervals so that two units are visible ahead. Below the VMS and mounted horizontally is a standard three-head traffic signal. These units should be double-faced to provide a front to back display when the tunnel is operating reversed or bidirectional. This combination of sign and signals provides traffic control continuously through the tunnel (see Figure 24-2). Alternatives to this arrangement, particularly where space is at a premium are
  - The VMS sign centered between a dual-purpose lane-use/traffic signal units
  - Interval location of VMS/signals/VMS/etc.
  - Smaller (vertical) units at closer spacing
- *Television cameras.* Closed-circuit television (CCTV) cameras, either ceiling- or wall-mounted, are spaced throughout the tunnel to provide full coverage. They are coupled with the vehicle detector to provide the visual verification that an incident has occurred.
- *Emergency exits.* The locations of cross-passage doors are marked with strobe lights mounted above or at walking level together with a flashing arrow and the word exit included on the wall side of the overhead sign and signal displays.

Safety and Environmental Monitoring

The following monitors are used to detect fires, measure air pollution levels, and sense dangerous spills in the tunnel:

- *Heat detectors.* Ceiling-mounted thermal detectors or infrared area detectors provided as a backup alert in support of the CCTV system to detect a fire. See Chapter 19 for an in-depth discussion of detectors.
- *Carbon monoxide monitors.* Tunnel air is sampled so units such as infrared absorption analyzers can measure carbon monoxide (CO) levels. Improvements allow monitoring of NO_x to be coupled to CO monitoring. Readings from these sensors can be used for ventilation control (see Chapter 20).
- *Visibility monitoring.* Reductions of visual range in the tunnel are measured with a wall-mounted light transmitter and receiver that record the degree of obscurity caused by particles and diesel smoke. In tunnels with a high percentage of diesel-powered vehicles, ventilation is more often controlled by visibility requirements than CO.
- *Air velocity monitors.* Anemometers or ultrasonic transducers are mounted above the traffic stream to measure tunnel air velocity and direction of flow. This data is fed into the ventilation control system.
- *Hydrocarbon monitors.* Gas sensors in the tunnel drainage system detect quantities of potentially explosive and/or hazardous materials. See Chapter 23.

Voice Communication. A combination of radio and telephones is used to communicate directly with motorists through their car radios and maintain communication between tunnel and emergency response personnel within the tunnels and to units outside the tunnel. These systems and their role in fire safety are addressed in Chapter 19. They include

- *Telephones.* Throughout the tunnels and ventilation buildings, there is a telephone system built around an electronic private automatic branch exchange (EPABX) for internal and external calling.
- *Call boxes.* Motorist and call boxes are located in the cross-passages to provide direct contact with the tunnel operator.
- *Facility radio.* Operations and maintenance personnel use a two-way FM and/or VHF radio system for communication between the control center, vehicles with mobile units, and personnel using handheld transceivers.

- *Miscellaneous radio.* Separate radio systems and/or additional channels linked with the facility radio are provided for police, fire department, and emergency medical response.
- *AM/FM radio rebroadcast.* The rebroadcast system provides continuous broadcast band reception for the motorist while passing through the tunnel. With this system the operator can interrupt commercial broadcasts to relay information directly to the motorists over their car radios.

Approach Road Configuration

A plaza immediately in front of the tunnel portal or between the approach roads is used to transfer traffic from one roadway to the other, for bidirectional tunnel operations, to turn back oversize and hazardous materials carriers, and to provide access to the portal buildings. A typical arrangement for the plaza including the location and type of traffic control devices is shown in Figure 24-3.

Oversize Vehicles. The identification, stopping, and diverting of oversize and overheight vehicles requires a five-station layout that will extend one or more miles before the tunnel plaza and tunnel portal. The location of these installations is centered around station No. 4, the last exit road before the tunnel. If there is a long gap from the last exit to the tunnel, the crossover plaza is then used to turn back vehicles. The linear arrangement of these installations are

1. *Overheight Detector.* Detection is made using an infrared light beam projected across the roadway just below clearance height (5 m/12.5 ft) with a transmitter on one side and a receiver on the other. When the beam is broken, a signal is sent to stations 2 and 3. The distance between stations 1 and 2 should be sufficient to trigger No. 2 flashing lights and gain driver recognition (5-sec minimum, or 135 m [440 ft] at 100 kph [60 mph]).
2. *Flashing Overheight Sign.* Overheight detector signal activates two wig-wig flashing lights on a fixed panel sign that is downstream with the message, "When Flashing All Trucks Exit to Inspection Station Ahead." The flashing time is usually set for 15 sec and may catch several trucks.
3. *Inspection Station.* The inspection station is configured similar to a weighting station. The No. 1 detector alarm alerts the station staff to be prepared to sort out the errant vehicle and direct it to the exit (see No. 4) or turn it back.
4. *Last Roadway Exit.* It is unlikely that once told to leave the roadway at the next exit the driver will continue and enter the tunnel; however, if this does happen or the driver does not stop at the inspection station, the vehicle will be detected again at the next checkpoint (No. 5).
5. *Fail-Safe Overheight Detector.* The fail-safe detector is similar to the first detector, but when triggered all signs and signals ahead leading into the tunnel are turned to stop/red, thus bringing the total traffic stream to a halt and preventing the errant truck from entering the tunnel. The police or tunnel staff will then sort out the traffic jam and apprehend the violator. The linear arrangement of the above installation in urban settings may be severely truncated or replaced by an overpass structure with minimum clearance immediately before the tunnel portal.

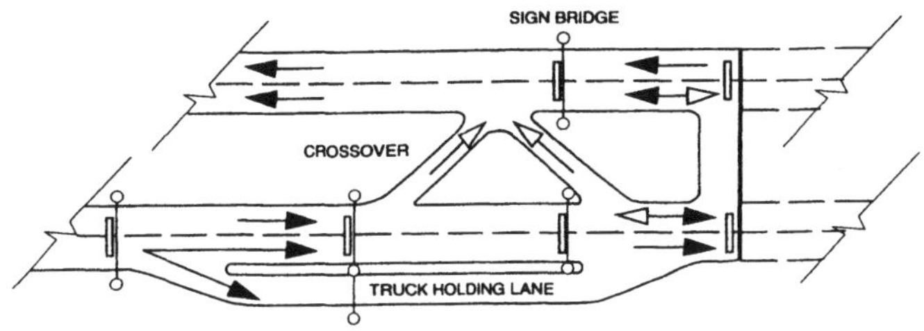

Fig. 24-3. Tunnel portal plaza.

In urban tunnels, where detection and stopping overheight vehicles occurs in a plaza immediately before the portal, a crash curtain consisting of weighted-blocks chain suspended at clearance height across the detector location gives a sound/impact alert to the driver.

Hazardous Materials. Some classes of materials—explosive, toxic, and poisonous—are prohibited passage in tunnels. There are circumstances, however, where flammable and other hazardous materials are allowed passage because alternative routes are long or will pass through local streets, which is considered more of a hazard. To allow for this passage, vehicles are directed to a holding area as shown on Figure 24-3 and at scheduled times are escorted through the tunnel.

Approach Road Traffic Control. The combination of fixed panel signs, variable-message signs, and traffic and lane signals is used to familiarize motorists with the sign and signal arrangements they will see in the tunnel. These control devices will be used to stop and maneuver traffic to single lanes or direct traffic to cross over to the other tunnel.

- *Approach road signs and signals.* The overhead signs and signals are larger than the tunnel units so as to be visible from a greater distance and attract the attention needed. Normally, three sets of sign/signal units are needed to advise action ahead, initiate this action, and confirm the travel restriction (see Figure 24-4). The traffic signals should be paired with the signing to replicate standard intersection arrangements.
- *Approach road CCTV.* The outdoor cameras are fitted with a remotely controlled mechanism to pan and tilt the camera and zoom the lens for area scanning and close-up viewing. A minimum of two cameras, one looking into the portal and the other looking away from the portal down the approach road, is required to supplement the series of in-tunnel cameras and view the approach road (see Figure 24-5).

TRAFFIC CONTROL CONCEPTS

Modes of Operation

The tunnel and approach roads should be equipped to make all changes in traffic operation using the sign and signal units described previously. As there is usually no time to call for police or tunnel staff to manually assist in traffic control for accidents, the control devices and their use must

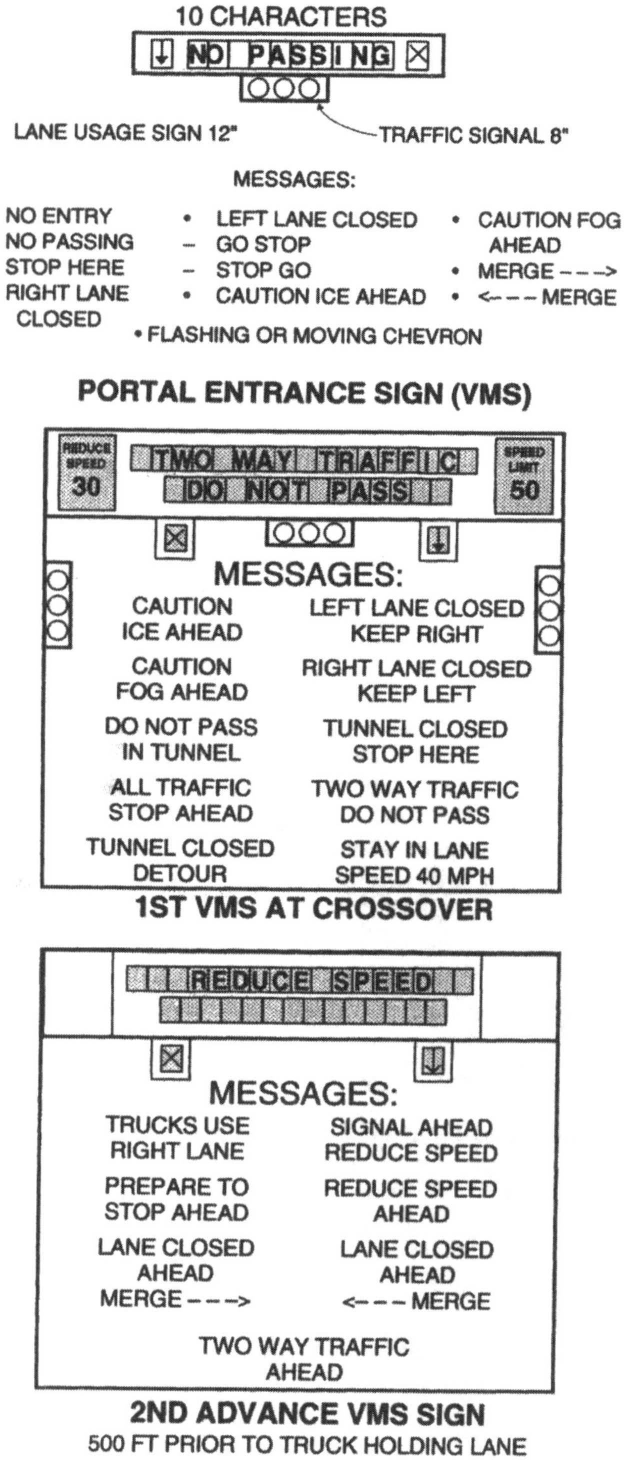

Fig. 24-4. Approach road signing.

be foolproof and easy to understand. Traffic changes fall into two major categories: scheduled changes and incident response. The former is a timed action with manual support starting out on the approach road. Incident response is a reaction requiring quick sign/signal response at the incident site and tunnel portal and then travel back to intercept approaching traffic to close lanes or redirect tunnel entry. For

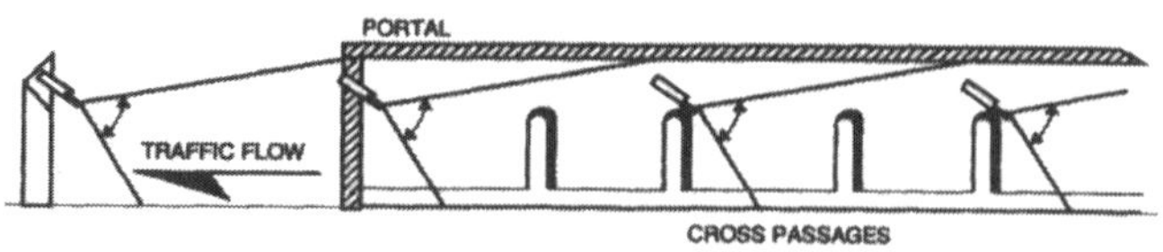

Fig. 24-5. CCTV camera arrangement.

incident response, the first objective is to stop entry of traffic into the tunnel. The holding point is the approach crossover (see Figure 24-3). Inside the tunnel there are always some motorists trapped behind the incident, thus creating a need for dual response to deal with the stopped vehicles and to clear the traffic stream behind.

Sign/Signal Displays

The operator's tools to orchestrate operations are packaged sign/signal displays, known as *sign/signal plans,* resident in his computer. The plans, using messages shown in Figure 24-4 with associated signals, can be prepared using the following operating procedures to organize the displays.

Basic Signing Plans

Five basic sign/signal plans (displays) are needed to control traffic, with several supplemental plans to modify these five basic plans. The following two-letter designations used to develop sign/signal plans are shown for a typical two-lane bidirectional tunnel port of a dual two-lane facility.

- NN—Both lanes operating in the normal mode at posted speed.
- NC or CN—One lane open, the other closed. This plan is used for scheduled closure of one lane, left or right, with the adjacent lane operating normally at reduced speed.
- CC—Both lanes closed. This plan is used to clear the tunnel for maintenance with the adjacent tunnel in two-way operation. Two or more steps are needed to reach the final closure state.
- NE or EN—One lane open, one closed. This plan is used for response to a minor incident in which traffic is allowed to continue flow past the incident, or the first stage of a tunnel closure to allow trapped vehicles to exit the tunnel.
- EE—Major incident plan. This plan, like the minor incident plan, reacts from the incident site closing the tunnel upstream while the lanes downstream remain open to allow traffic to clear.

All signing/signal plans must include transition stages from an existing state to any of the other states or back to normal (NN) (see Figure 24-6).

Condition Change

Most changes to traffic flow can be accomplished by using a two-step sequence of individual and coordinated sign/signal displays. The initial display should be advisory or warning (e.g., Right Lane Closed Ahead), followed by the

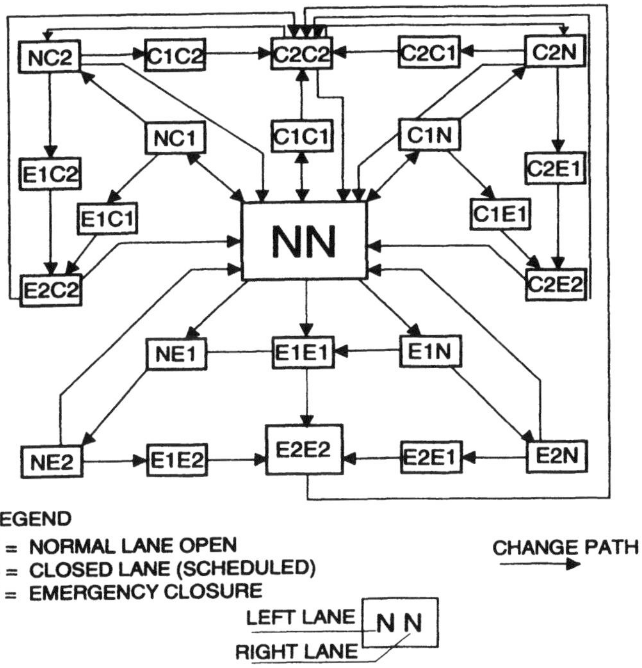

Fig. 24-6. Response plan logic diagram.

action or restriction display (e.g., Keep Left, Right Lane Closed).

Supplemental Plans

These plans call for action by flashing command messages and then returning to the basic plan. Displays can be flashed on and off or kept in a momentary steady state to stop traffic.

Caution. This plan will alert drivers to reduce speed and set in line the next plan. The plan can be used as an automatic alert with incident detection to warn motorists and give the operator time to follow with the appropriate action plan.

Speed Change. Reduction in speed on the approach roadways as a result of fog, ice, or snow.

Flow Control. Driving through a tunnel is a traumatic experience for many motorists, causing overcaution and/or a complete disorientation, and resulting in the need for signing plans to stabilize traffic flow. A common occurrence in subaqueous tunnels is the loss of speed on the downgrade caused by overreaction to brake lights. This is followed by a loss of orientation and failure to accelerate on the upgrade, resulting in a sluggish crawl that greatly reduces throughput of the tunnel. This condition can only be broken up by platooning traffic to break the inertia or, in this case, lack of inertia. Platooning is accomplished by stopping entry at the tunnel portal much like a signalized intersection. When released, traffic will flow freely through the tunnel until it reaches the end of the crawl. Traffic should again be stopped one or more times until smooth flow exists throughout the tunnel length. Stopping time should be limited to a maximum of 3 min, which is about the extent of motorists' patience. The tunnel signs (Figure 24-2) should also include messages such as "Maintain XX kph" or "Upgrade, Accelerate."

Stop. This command is used to momentarily halt traffic approaching the tunnel to allow the turn back of restricted vehicles or escort of vehicles through the tunnel, or as the initial stage in establishing bidirectional flow in the other tunnel.

Action/Advisory. Action message include "Turn on Radio," "Evacuate the Tunnel," "Stay in Lane," etc.

Normal Operation

Most urban tunnels have more traffic lanes on the approaches than in the tunnel; thus, the demand exceeds the tunnel roadway capacity. After entering the tunnel, traffic becomes unstable and slow, which further reduces the flow. The dividing line between free flow and unstable flow can be related to traffic density. As traffic density increases to the optimal flow rate (veh/hour), free flow exists. Beyond this point, as the density increases, flow and speed decrease (see Figure 24-7). A means of gauging these traffic changes is the monitoring of lane occupancy by providing occupancy readouts per lane in the control center. A green display for a traffic density of 0–20% indicates uncongested flow, a yellow of 20–30% indicates unstable or impeding congestion, and red for over 30% indicates congestion. Once notified of unstable flow, the operator has the following options or combinations of options to use:

- *Traffic monitoring.* Flow control or reducing the number of approach lanes.
- *Motorist advisory.* Flash tunnel signs to read "Maintain XX kph," or prompt the motorist to regain free flow over the radio rebroadcast.

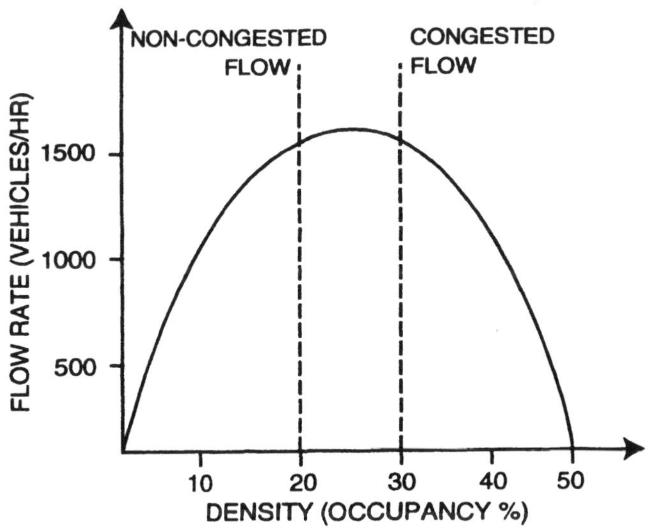

Fig. 24-7. Traffic flow diagram.

Incident Response

The first alert of a possible incident will be received from the lane sensing of a disruption in flow (breakdown or accident). (See Figure 24-8.) The alert will automatically switch the console CCTV monitors to display the incident site. After examining the situation, the operator can verify the incident or reject the alarm as a false alarm. Should the operator not respond or delay his response, the sign/signal program will be triggered, initiating the Caution Plan and a prompt waiting for further instructions from the operator. The usual sequence after verification is for the operator to select the appropriate response plan. As the display changes are sent to the field devices, the feedback status is compared with the change command to verify proper execution. While calling up the sign/signal plans the operator will take the following steps:

- *Emergency response.* Alert the emergency response crew to the location and extent of the incident. Select the access routes to reach the site either by counter flow in the incident tunnel or through the adjacent tunnel and cross-passages.
- *Radio broadcast.* Flash tunnel signs to read "Turn on Radio" and then begin broadcasting instructions.
- *Off-site assistance.* Notify police, fire, and medical assistance with routing instructions to reach the incident site.
- *Equipment operation.* Change and activate tunnel service equipment—set ventilation mode/levels, pressurize water mains, open drainage holding tasks, start standby power generator.
- *Response supervision.* Maintain overall control of response activities using CCTV, phones, and radio.
- *Normalize.* After incident clearance, return traffic control devices and tunnel equipment to normal operating mode.

Tunnel Evacuation

Some accidents may require evacuation actions prompted by sign displays, radio instruction, and tunnel staff, which direct motorists to leave their vehicles and exit the tunnel through cross-passage or portal. In a fire emergency, evacuation of stranded motorists is critical. Tunnel fires are usually identified by CCTV when viewing incidents; however, traffic queues can block the view or secondary incidents in the queue may result in a fire, which establishes the need for backup detectors. Operator action is essential as motorists may attempt to fight the fire with portable extinguishers available from tunnel niches or in vehicles. They usually will not flee the area until there is a flare-up and smoke. When smoke buildup occurs, quick, forceful instructions for direction to flee should be issued using the rebroadcast radio, strobe lights, and signs at the cross-passages to prompt this action.

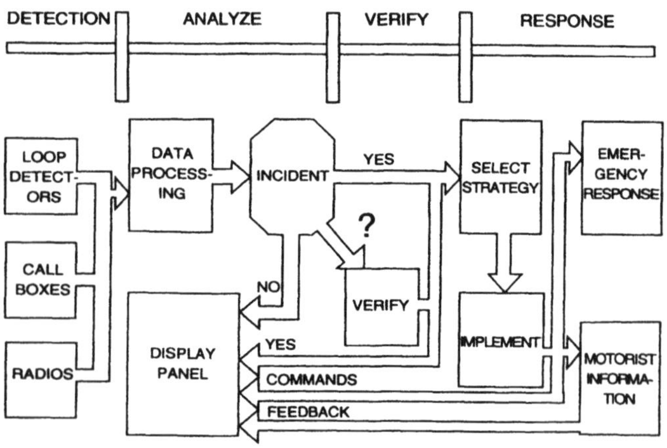

Fig. 24-8. Incident management diagram.

FIELD HARDWARE

Detectors

The efficiency of incident detection depends on the reliability of the vehicle sensing unit. Present practice employs detector loops cut into the approach road and tunnel pavement. Standard traffic controllers like the 170 with programmable logic and communication medium can be used with several detector amplifiers to collect and process traffic data. Since most tunnels prohibit lane changing in the tunnel, the sensing of a stoppage can be accomplished in seconds using developed algorithms such as the queuing and flow-discontinuity programs commonly known as the California algorithm. With a 180-m (400-ft) spacing of detector loops in each lane, the traffic controller will poll the loop amplifier at 1/60-sec intervals to summate 1-sec measurements of lane occupancy over the loop. This data is then processed locally or at the control center to develop 1-min averages of lane occupancy that are updated every 20 sec and fed into the algorithm. The sensitivity parameter built into the program will usually be set to accept a high (2:1) false alarm rate in order to generate quick alarms. This detection procedure is very successful when the traffic flow exceeds 400–600 vph per lane. Detection is uncertain during periods of low volume and at night when lane changing enforcement is improbable, or it can occur when a vehicle pulls off onto a partial shoulder offset. A second algorithm based on vehicle accounting can be used to warn the operator when there appears to be a stalled vehicle somewhere in the tunnel.

Freeway traffic management systems are now using video/radar/microwave detectors to measure flow as a means to identify congestion and incidents. These units could furnish the vehicle speed input to the algorithm and eliminate the need for dual occupancy loops. However, until tested and proven in tunnel use, detector loops are recommended for queuing and counting detection.

Television Cameras

It is important that the tunnel be fully covered by the CCTV system to provide clear images to the operator. The tunnel cameras are fixed-focus and mounted on the ceiling or high on the sidewall to view approaching traffic. Positioning cameras to face oncoming traffic prevents trucks, buses, etc., blocking the camera sight lines. It is possible to add remotely controlled pan and tilt mechanisms to provide double camera coverage of an incident site, but this requires higher tunnel headroom.

Solid State Cameras. The cameras are compact charged-couple devices (CCD) fitted with 12.5-mm (0.5-in.) lenses. They are free from image lag, blurring of images, blooming, image burn, and damage from direct light. Although each camera with a fixed focal lens can cover more than a 300-m (990-ft) length of tunnel, the cross section restraint limits the effective range to 200 m (660 ft). Horizontal or vertical changes in alignment may reduce this range significantly, possibly to 75 m (250 ft).

Camera Spacing. A practical method for spacing the cameras in the tunnel begins with a camera looking in at the exit portal, then going back against the flow locate a camera at 200 m (660 ft) spacing (see Figure 24-5). Following this layout, it is possible to plot the camera sight lines on the tunnel plan and profile to see if full coverage is available. If not, cameras are added and spaced at less than 660 ft (200 m) to obtain full coverage. It should be noted the camera vision cone will not produce wall-to-wall, pavement-to-ceiling coverage immediately in front of the camera, and thus a 25 m (80 ft) sight overlap between cameras is needed.

When selecting the remaining components of the CCTV system, a balance and assurance of quality is needed. The camera should have high resolution and be housed in a weatherproof case. Fiber optic or coaxial cables should be used to transmit the video signals directly back to the control center. Solid state monitors with high screen resolution are standard for control room display. Since the CCTV system is so important to operations, its configuration should use quality components and be simple, easily maintained, and free from unneeded fixtures such as transmission multiplexing, split-screen monitors, etc.

Outdoor Cameras. The outdoor cameras should be fitted with a remotely controlled motorized zoom lens (1:10 or 1:16) enclosed in a weatherproof case with window washer/wiper. The camera is mounted on a pan (350°) and tilt (90°) unit. Two outdoor cameras are recommended at each portal. One is located over the exit roadway to view the immediate portal area and normally is set looking into the tunnel to complete the tunnel coverage. The second camera is mounted high on the portal and will view the approach roads. Additional approach road cameras may be needed if the surveillance length is extensive or road alignment and terrain obstructs the view. The camera height above the roadway also determines the distance viewing is effective; increasing height increases depth of view.

Traffic and Lane Signals

The traffic and lane signals each have separate use. They are mounted separately and not mixed together, so as to avoid conflict in meaning. The traffic signals are used to stop traffic and should have 200-mm (8-in.) diameter lenses using standard three-signal heads with fiber optic light source. The outdoor signals are similar, but should have 300-mm (12-in.) lenses. The lane signal is used for lane control and is mounted over the centerline of each lane using a 300-mm (12-in.) fiber optic light point for green arrows, yellow slanted down arrows, and red crosses.

Approach Road and Tunnel Signing

Directional Signing. Directional signing should be avoided if possible at the tunnel entrance, within the tunnel, and just outside the tunnel exit. However, in many urban facilities and where there are entrance and exit roads in the tunnel, this signing is necessary. Logic shows that the outdoor signing standards are not suitable nor really necessary in a tunnel. With the tunnel confinement, it is difficult to see more than 75–150 m (250–500 ft) ahead, and therefore the signing panel and letter size can be greatly reduced. A second consideration is that the motorist has little time to read and comprehend lengthy texts; therefore, the message must be concise. Using standard 300-mm (12-in.) letter heights for two lines of text arranged to read left to right will result in a compact 900-mm (36-in.) height sign pane (see Figure 24-9). If there are several such signs in the tunnel, the ceiling height can be set to accommodate 900-mm (36-in.) signs. If there are only a few, all or a portion of the sign height may be fitted into a ceiling notch. If this type of signing is used extensively within the tunnel, similar arrangements with larger letters should be used on the approach road to familiarize the motorist.

Regulatory Signing. Regulatory signing is used on the approach roads and in the tunnels to support the traffic and lane signals. For example, they are used to warn ("Left Lane Closed"), to impose restrictions ("Stay in Lane"), to provide motorist advice ("Turn on Radio"), or to command action ("Evacuate Tunnel"). There are several types of variable-message signing units (blank-out, rotating drum, fiber optic, or LED matrix). The matrix type can provide almost limitless numbers of messages and has the capacity to illuminate the letters/symbol points, which is preferred for use in tunnels and approach roads (see Figure 24-10).

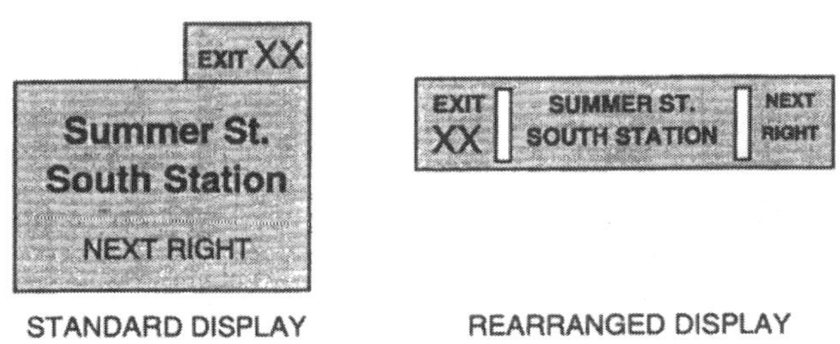

Fig. 24-9. Directional signing panels.

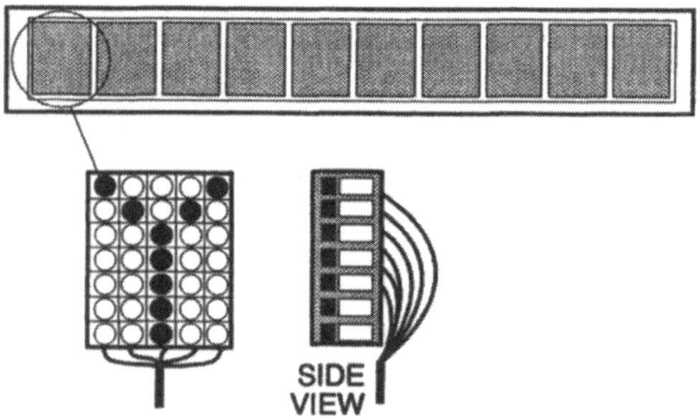

Fig. 24-10. Matrix variable-message signs.

Overheight Vehicle. The infrared beam transmitter and receiver may be mounted on an overhead sign bridge, bridge abutment/pier, or on their own poles. To eliminate false alarms caused by vehicle aerials or birds flying through the beam, a minimum break time is needed to trigger an alarm. The beam can also be coupled with detector loops to prove the presence of a vehicle. For bidirectional roadways where it is not possible to erect poles between opposing travel lanes, two sets of beams are needed with a breaking logic to determine the direction of the vehicle.

Fire Detection and Equipment

The following discussion of fire detection and equipment relates primarily to surveillance and control aspects. For a discussion of the fire protection aspects, see Chapters 19, 20, and 23.

The first line of fire and smoke detection should be the CCTV system, but standard detectors are needed to provide backup and to identify hidden fires. Heat detectors should be used throughout the length of the tunnel, subdivided into alarm zones. The detectors, high-wall- or ceiling-mounted, can be individual point detectors set at a temperature level and rate of rise, linear thermo cables, or infrared area scanners.

Smoke Detection. Alarms can be generated by the visibility monitors when levels fall below set levels. These units are effective for small fires, which can generate large and dangerous quantities of smoke. To provide effective sensing, the number of monitors and their spacing should be increased.

Fire Extinguishers. Small fires can be easily controlled with powder or foam extinguishers, which most motorists know how to use. Extinguishers should be located in tunnel niches at each entrance door. A control center alarm should be activated when the extinguisher is removed from its holder.

Air Quality Monitors

Carbon Monoxide Monitors. The most effective carbon monoxide (CO) monitor in terms of reliability, ability to produce accurate measurements, and need/ease of maintenance and calibration is the pumped sample infrared absorption type. Each unit is equipped to handle six to eight sampling ports. The ideal location for the analyzer unit is at each portal, with sampling ports some 15–30 m (50–100 ft) inside the tunnel on both tunnel walls. This arrangement will provide dual measurements in each tunnel, and the short sample tube lengths will enable minimum time readings. Other similar installations should be at the mid-tunnel and/or quarter point locations.

Visibility Monitors. The visibility monitor consists of a light transmitter and receiver, usually mounted on the tunnel walls, to measure the smoke/particle content of the tunnel. The measurement of obscurity is directly related to the degree of visibility in the tunnel, and when concentrations reach set values fresh air is supplied to dilute this concentration. These units should be located in similar positions to the CO monitors, where the expected concentration is highest and where dual readings can be obtained for comparison.

Air Velocity Monitors. The measurement of air movement in the tunnel—velocity and direction—is needed for control and/or measurement of the ventilation system efficiency. This is accomplished by installing, inside each portal of both tunnels, ultrasonic transducers on the walls to point a signal at approximately 3.65 m (12 ft) above the roadway angled across the roadway at 45° to the receiver on the opposite wall. The analyzer unit can be housed together with the CO and visibility units.

Communication Equipment

The rebroadcast system contains an outdoor receiving antenna, broadband amplifiers for the standard AM and FM bands (530–1620 kHz and 88–108 mHz), a radiating tunnel antenna, and an operator switching/microphone/VHS cassette recorder.

Two-Way Radio. This system will be used by tunnel staff in vehicles or with handheld transceivers for communication between individuals. It is coordinated by the control center operator. A dedicated VHF channel must be obtained for this two-way FM communication system. Local police, fire, and medical assistance may also be included in the system.

Telephone. A direct dialing telephone system for communications within the portal building, between cross-passages, and externally is provided by the EPABX.

Motorist Call Boxes. The motorist call boxes provide direct communication with the control center operator, who can organize assistance as required. The call boxes should be installed at each cross-passage and on pedestals on the immediate approach road. They should be hands-free speaker-microphones with a call button. The operator will receive an alarm tone coded to indicate the location of the call box and will then activate the unit.

Cellular Telephone. Provision should be included in the tunnel for the installation of a 800-MHz cellular telephone system. There is an increasing use of mobile telephones to alert tunnel operators of conditions in the tunnel.

Equipment Locations

The layout of the tunnel services can best be accomplished by using a modular arrangement or spacing to unify power feeds, remote terminal units, and maintenance areas in the cross-passage. Using a 100-m (300-ft) spacing, equipment can be installed cross-passage. This matrix arrangement shown in Table 24-1 illustrates modular spacing and preferred mounting locations in the tunnel cross section.

CONTROL CENTER

The control center is designed for the man–machine interface requirements, for without the need for human intervention, the computers could happily run alone in a dark room. Much of the tunnel operation can run automatically using programmed schedules based on the time of day and limiting levels for changes. It is only when there is an abnormality that the operator is needed.

The control center is therefore configured to first provide displays of operating status for all equipment and then provide means to change operation status (see Figure 24-11).

There may be some need for manual controls and hardwired connections to field equipment, but the basic configuration and communication to remove equipment is electronic code and computer-aided.

Status Displays

Means should be available to status of monitor all display devices, sensing monitors, and operating equipment, but not all at one time or even continuously. The priority level of status addressing should be arranged in the following order.

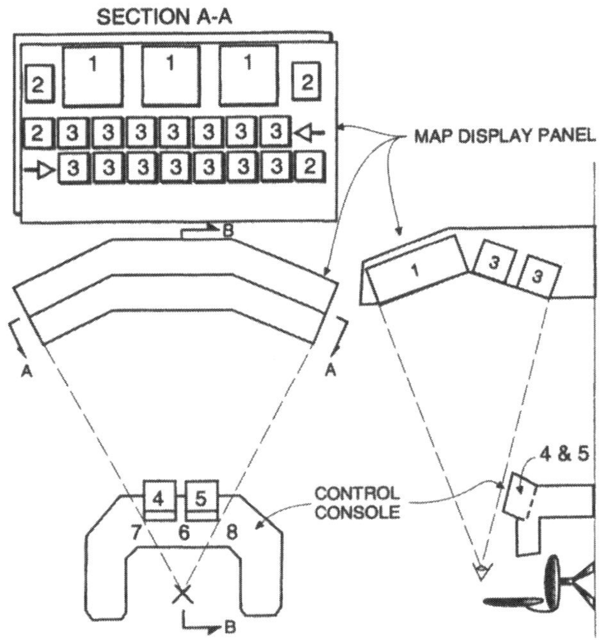

1. STATUS DISPLAYS COMPUTER GRAPHICS
2. APPROACH ROAD REMOTE CONTROLLED CCTV MONITORS
3. TUNNEL CCTV MONITORS
4. MASTER CCTV MONITOR
5. OPERATOR COMMAND MONITOR
6. TERMINAL AND MOUSE
7. CCTV CONTROLS
8. COMMUNICATION CONTROLS/SETS

Fig. 24-11. Control center.

Table 24-1. Tunnel Control Arrangement

System	Unit	Location	Approximate Spacing
Monitors	Carbon Monoxide	I H/L	Portals and 1/4 points
	Visibility	I H/L	1/4 points
	Air velocity	I H/L to O H/L	Portals
	Heat detector	I H/L or C	Continuous
	Vehicle detector	P each lane	600 ft
	CCTV camera	c/l C or I H/L	600 ft (maximum)
Traffic Control	Variable message sign	c/l C	300 ft
	Lane signal	C over each lane	300 ft
	Traffic signals	c/l C	300 ft
	Lighted lane markers	P between lanes	Continuous
Ventilation	Supply air outlet	I L/L	5 to 25 ft
	Exhaust air inlets	I and O C	Continuous
	Smoke exhaust damper	c/l C	150 to 300 ft
Fire fighting	Hydrant niche	X and IW	150 ft
	Portable extinguisher	X	300 ft
Evacuation	Cross passage	IW	300 ft
	Exit sign	I H/L @ X	300 ft
	Strobe light	I H/L @ X	300 ft
Electrical	Mini-power center	X	300 or 600 ft
	Lighting	I and O H/L	Continuous
	High voltage	OW	Continuous
	Low voltage/control	IW	Continuous
Communication	Radio antenna	I H/L	Continuous
	Motorist call box	X	
Drainage	Curb inlet	I and/or O P	100 ft (maximum)
	Low-point sump	Tunnel low point	
	Portal intercept	Tunnel portal	

First Priority. Traffic incidents or major equipment events that require urgent operator response include

- Traffic accidents
- Fire/smoke detection
- Power failure
- Dangerous levels of CO
- Hydrocarbon spillage
- High water levels in sumps

Second Priority. This level includes condition alerts such as detection of a possible incident, equipment failure, or communication shutdown. These alarms require operator acknowledgment and are then recorded. The computer simplifies recordkeeping both by one-line printout of a hardcopy (paper) record and by logging in computer file storage.

Third Priority. Every change in traffic control and equipment operation is logged for record purposes.

Control Procedures

Three means of control should be available to the operator and tunnel staff.

Computer Control. The primary control system allows the operator to send commands to the computer using the keyboard or a mouse and computer graphics to call up preprogrammed traffic management plans, ventilation plans, etc. These plans will be conflict-proof and can be run concurrently with each other, but not layered.

Manual Control. Using the computer terminal, manual switches, or both, the operator can make individual changes to any device or piece of equipment. Changes made manually may not be conflict-proof and may change again with the introduction of a command using preprogrammed plans. Manual control is usually used for testing and maintenance.

Local Control. At the remote location of the device or piece of equipment, control is accomplished by using its local intelligent or manual control (PC or switches). Changes can be made here in the event of a communication failure from the control center or for testing and maintenance.

Map Display Panel

This floor-to-ceiling display is arranged in a semicircle to provide a panoramic view of the status displays to the operator. Centered in the panel are three large color video terminals to display computer graphic text, macro/micro line diagrams of device/equipment configurations, and their operating status. The central or primary screen will usually show a minimap of the tunnel and approach roads with the current traffic control plan in place. The two flanking screens are used for backup and concurrent status call-up as pages from the status menu. Running just above or below are the TV monitors arranged in direction to traffic flow (top right to left, bottom left to right). If there is room to spare on each end of the panel, it is used as a special enunciating display.

CCTV Monitors. For short tunnels with a few CCTV cameras, the display panel will contain one monitor for each camera. With a larger number of tunnel cameras, monitor sequencing is recommended with 3–10 tunnel cameras coupled to each monitor in the map display panel. The sequence should not be a rotation on individual monitors but a rolling sequence through the tunnel. For example, with a 3:1 ratio of cameras to monitors, 1/3 of the tunnel would be displayed moving through the tunnel. In this way a continuous section of the tunnel can be shown. Two separate monitors, one on each end of the panel, are for the portal approach road cameras. The first sequencing monitor can also be dedicated to the second outdoor camera.

Incident Viewing. Upon alert of an incident, the tunnel section display can be centered on the incident site to show conditions upstream and downstream. The operator can pull down to the console monitor the camera showing the incident and return the panel monitors to tunnel sequencing.

Control Console

The operator's position will enable the viewing of the entire map display panel, two master video display units (VDUs) for CCTV displays, and computer dialogue, each built into the console. Arranged around this operating position are switching panels and radio/telephone handsets. The console is U-shaped to provide desk working areas.

Annunciating or Switching Panel

The type or need for annunciating or switching panels depends on code or operating practice particular to the location of the tunnel. A fire annunciating panel may be required here and at the tunnel portals to conform with local regulations. The panel may also contain secondary or backup manual switches for some or all of the tunnel equipment.

Supervisor or Dispatch Desk

A second control/supervisor/communication desk can be located in the control center or at a secondary or remote location. The use of computer control and digital communication allows this operating freedom. For large facilities with one or more tunnels, bridges, toll, or maintenance facilities to manage, the communication needs will require this second desk. With dual VDUs, switching, and communication devices, this subcenter can provide parallel and backup control with the prime center subject to operations protocol.

Computers and Peripherals

The system is built around two industrial-grade minicomputers or PCs to provide 100% redundancy (see Figure 24-12). Each computer (CPU) is sized and equipped with dual disk drives, clocks, a watchdog unit, and a communications unit, so that individually each CPU can support all operation/alert/recording requirements. Both CPUs are supplied with operating programs, but they are not expected to perform parallel processing. The operating software is configured to allow the backup to assume command by simply polling status of displays and operating levels. Included in the computer room is a program desk with a terminal and keyboard to be used for testing and maintenance. This station, as with all computer input terminals, should be tied

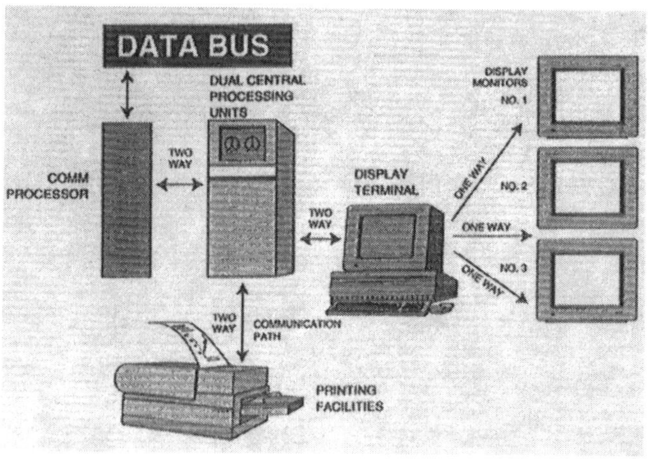

Fig. 24-12. Central control layout.

into access passwords to protect access and use of the computer systems.

Communication Network

The most expensive element of the control system is the communication network. Cost and limitations of a hard-wired system have led to the use of single cable for multiplexing data transmission. Typical single cables of twisted-pair, coaxial and/or fiber optic are now used with time-division (TDM) and/or frequency-division (FDM) multiplexing.

Advances in the use of programmable controllers (PLCs) has also relieved the data processing load in the control center CPUs and the volume of data interchanged between field units and the control center. This concept of distributed intelligence with multiplexing transmissions is the basis of a supervisory control and data acquisition (SCADA) system (see Figure 24-13).

SCADA Configuration

The communication system is made reliable by using two techniques. The first technique is distributed processing, which involves the spreading of control processing throughout the network to minimize the severity of a single failure. Coupled with this is network redundancy. If the primary route of communication has failed, then communication is transferred to a secondary route. In simple terms, the control center is usually in a state of waiting, receiving device and equipment status reports from remote terminal units (RTUs). When called into action, they send execution commands to the RTUs and then verify that the proper change has been carried out. The RTUs control the various downline devices and equipment, which include

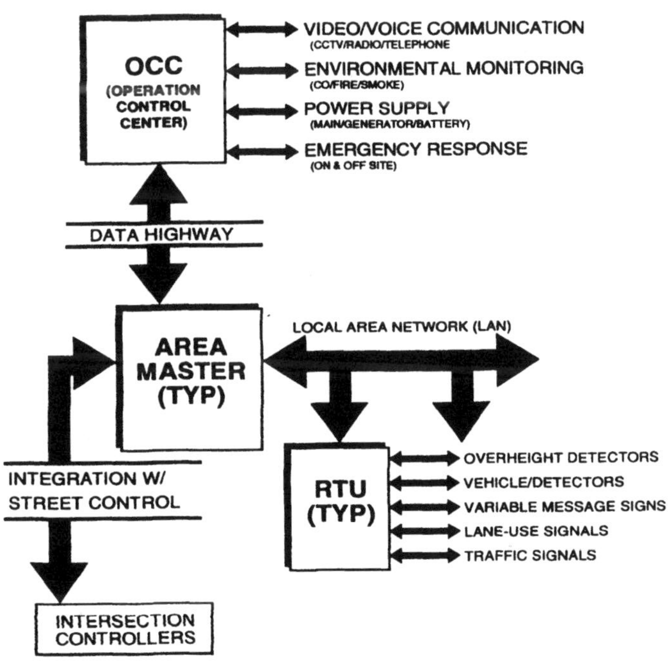

Fig. 24-13. Network.

1. Traffic detectors sending data back to the control center
2. Traffic control equipment on hold waiting for commands, which send basic status to the control center to confirm availability

Distributed Intelligence

The control center will have the capability to interrupt and when necessary take over the duties of any downline RTU using reserve capacity built into the communications network. Depending on the size and number of RTUs, there may be a need for two levels of downline data processing. An example of this would be a local master RTU controlling several subsystems that have RTUs at each piece of equipment. One or more of these master RTUs would feed information and receive commands from the control center, and supervise the RTUs to create a self-contained operating unit. These master RTUs can assume command in the event of a communications break from control center.

Cable Network

The communication cable network should be configured on a semi- or, preferably, fully duplex loop to transmit and receive data simultaneously. Should a break occur within the loop, it will automatically switch to semiduplex operation. The use of independent communication loops can allow the grouping of devices and equipment having similar priority and need for high-speed transmission rate and data refreshment. Other pieces may only need periodic contact at slow speeds and can be grouped on separate cabling loops. Codes of practice may also require separate cabling as for fire detection or alarms.

Software

Recent developments of computer control for industrial applications have made real-time multitasking systems available in several software packages that can be used for the general-purpose software. By using these general application packages, the amount of purpose-written software is reduced and made easier to prepare. The tunnel system package loaded into the central computers and downline in the programmable controllers should meet the following requirements.

Input/Output (I/O). Measurements from field monitors and change commands form this link. This data is usually translated to digital code. The I/O processing speed of this data is critical and must be optimized to handle the number of I/Os to be scanned while maintaining an acceptable level of responsiveness.

Man–Machine Interface (MMI). The preferred interface is computer graphics with simulated network, control panels, and logic diagrams for visual displays. Data may be input via keyboard commands, although using a mouse or trackball to manipulate a graphical user interface is more common. Programmed sequencing of group commands is essential for critical action, particularly when a trained operator is not available. Using these devices, the operator should be able to window in quickly for status, operating, or diagnostic plans that give an overview or point display of all systems. Sound alarms including voice simulation are gaining use for MMI and are recommended.

Operating Plans and Algorithms. Within the software will be routines ranging from complex data manipulation at the central computer to relay-ladder-logic at the PLCs. These operating plans and modifications of parameter are resident in the system's operating plans.

Database. The organization setup is housed here to assign locations, sequencing, alarms, timing, and reporting for all functional subroutines. Included are historical backgrounds of actual operating experience for input into the operating plan.

Communications Network. The SCADA network should conform to an industry standard for a local area network (LAN).

Purpose-Written Software

Title headings for tasks to be included in the application software are

- *Monitoring.* Incident detection, CO levels, visibility, heat and smoke detection, sump water levels, power and equipment failures
- *Man–machine interface.* Computer graphics, alarm priority
- *Operating plans.* Traffic control, emergency response management, ventilation control, plant management
- *Security.* Entry surveillance, computer usage, watchdog, emergency power, system shutdown
- *Recordkeeping.* Event logs, operating timing, traffic and ventilation histograms
- *Maintenance.* Operating logs, servicing alarms
- *Training.* Operation simulation

SYSTEM SELECTION

There are concerns for cost cutting and the question of whether certain features are really necessary. The basic system requirements for a short tunnel in the country and an urban high-volume tunnel are very much the same: attention to tunnel user needs. An answer to the cost question is nonquantifiable, but how well a tunnel performs is its most visible feature. Since there is such a large investment in building a tunnel, why compromise with its operational capabilities?

Basic Requirements

The basic components that should be included in the surveillance and control system are

- *Traffic service.* A full-time means to identify stopped or disabled vehicles in the tunnel, their verification, and availability of on-call emergency response
- *Fire service.* A proven means to identify or detect fire or smoke, and the means to fight fires and evacuate trapped motorists in the tunnel
- *Environment.* Continual monitoring of levels of tunnel pollution and a means to dilute excessive amounts
- *Lighting.* Full-time tunnel illumination with battery backup to prevent total darkness
- *Flooding.* A drainage system including sumps, pumps, and outfalls/storage
- *Power.* Dual sources of power supply or a built-in standby diesel electric power unit

DESIGN AND IMPLEMENTATION

System design employs engineering techniques from traffic engineering, computer/communication design, and software development. To produce a successful end product, their combined input is required from design inception to final acceptance testing and hand over to a client.

Traditional Design

There are two basic design and contracting methods used to implement the system—the traditional preparation of design plans and specifications for contractor construction, or the system manager approach. For the former, the designer prepares either a materials/installation specification or a performance specification for typical contract bidding to furnish/install or design/purchase/install. This method can be successful only by contracting directly with prequalified control system contractors. When this specialty work is lost within a large tunnel civil contract, it is difficult to stop this element from being slighted by the prime contractor and viewed as a nuisance to be passed piecemeal to subcontractors. Success is seldom certain.

System Manager

The system manager is a selected firm working under an engineering service contract to design, prepare procurement and installation contracts, and be responsible for system integration, documentation, and training. Under this method there is freedom to make changes as the system is being developed without the responsibility of claims for extras. The system manager provides the application software, which is the key element in a successful operating system.

The complete services package associated with a control system should include the following:

- *Operating manual.* The design and installation reflect specific operating procedures to define goals that should be spelled out in this manual. This document should be first organized in the system inception stage and then refined and updated throughout the design and installation process.
- *Maintenance manual.* This is an organized reference of original designs, shop drawings, manufacturers' specifications, and maintenance procedures with parts listed for all hardware

items. The software manual should include descriptions of source programs and programming instructions for diagnostics and parameter changing.

- *Training.* Formal book and classroom training may familiarize staff with the system, but the opportunity for hands-on involvement with the contractor/system manager during installation, testing, and commissions is far superior.
- *Initial operation.* Provisions to supply a management staff during start-up for a specified period to further train the takeover staff, debug the software, and apply corrective maintenance is a sound investment. It also gives greater assurance that the warranty/guarantee response will be prompt and complete.
- *Warranty/guarantee.* This provision ensures responsibility for a specified period for all components including manufacturer product in-house warranties that may have expired.

OPERATION AND MAINTENANCE

There are two types of tunnel facilities: tunnels with tolling and those without. The former tend to become kingdoms unto themselves, while the later will contain essential staff and equipment and thus better illustrate the basic needs for operating and maintaining a tunnel.

Organization

The tunnel organization is made up of day staff working a regular 8-hour day, 5 days a week, and shift workers assigned to the day shift from 6 A.M. to 2 P.M., a night shift from 2 P.M. to 10 P.M. or the graveyard shift from 10 P.M. to 6 A.M., on a 7 days a week operation. The number of employees needed to man the shifts will be 40% more than the actual number to cover a 7-day week, holidays, vacations, etc. The day staff perform routine administrative and maintenance tasks. The shift workers monitor traffic, inspect and perform routine maintenance, and are on call for emergencies. During the graveyard shift, most of the major maintenance work is carried out. The permanent tunnel staff supervise temporary or contracted laborers and specialists.

Permanent Key Staff

The permanent key staff is divided into three divisions having the following duties (see Table 24-2):

Management.

- *Tunnel manager.* Provides overall facility management, maintains dealings with government agencies, the community and contracted services.
- *Supervisors.* One supervisor is assigned to each of the three shifts to supervise the day-to-day operation and management of the tunnel, working staff, and contracted services. Assumes command for emergency response.
- *Administration.* An administrator, secretary, and clerk handle correspondence, budget, finances, and purchasing.

Table 24-2. Permanent Key Staff

Division/Key Staff	Number of Personnel				
	Day	Morning	Afternoon	Graveyard	Total employees
Management					
Tunnel Manager	1				1
Supervisors		1	1	1	5
Administration	3				3
Operations					
Control Center		1	1	1	5
Responsible		2	2	2	7
Maintenance					
Technical Specialties	3			1	5
				Total	26

Operations.

- *Control center operators.* The center is manned continually with an operator who monitors traffic and equipment operation. The shift supervisor provides his relief.
- *Response crew.* This crew is manned by two members on both the day and night shift and one for the graveyard shift. They are the manual arm for the control center operator to man the emergency response wreckers, check prohibited vehicles, assist motorists, enforce traffic control, and keep the roadways clear.

Maintenance.

- *Technical specialists.* This four-member crew, three on day shift and one on graveyard shift, perform routine inspection and equipment maintenance and supervise contracted services work. Included in this crew are technical specialists in mechanical, electrical, and electronic equipment.
- *Maintenance crew.* Included with the shift workers is a labor force to assist the technical specialists to perform general janitorial, cleanup, painting, patching, replacement, and repair work. This crew can be permanent staff, part-time drawn from a larger Highway Department, or included in contracted services (i.e., janitorial, tunnel washing).

Contracted Services. Many large transportation authorities serving a number of facilities including tunnels have their own workshops, staff, and equipment to be almost 100% self-maintaining. However, for most individual tunnels it has been found beneficial to contract out all maintenance work except for the day-to-day caretaking. Included below are the professional and technical services suitable for on-call and off-site service:

- *Professional.* Legal, labor relation, employment service, engineering, facility inspection, accounting
- *Site maintenance.* Structural repairs, paving, lighting, lamp replacing, signing, pavement marking, painting, tunnel washing, janitorial
- *Equipment repair.* Fan motors, pumps, switchgear, electronic equipment
- *Equipment servicing.* Computers, radios, telephones, data transmission, office equipment

Tolling Facilities. There are few tunnel facilities where tolls are not needed: no tolls, no tunnel. For tunnel operation, a toll plaza is a godsend where oversize or hazardous cargo vehicles are easily dealt with. It provides a built-in traffic crossover or turn-back area and can be the excuse for traffic delays.

Toll Plaza Layout. The usual transition and number of toll booths in the plaza are three booths per through roadway traffic lane arranged with a truck lane(s). The booth should be arranged to gap automatic collection with manual collection. There is a strong move to introduce automatic vehicle identification (AVI) for toll collection.

Plaza Location. The tolling facility may be located immediately in front of the tunnel portal to consolidate tunnel operations and tolling.

- *Toll supervisor.* Stationed in the support building (administration or toll building) with an overview of the plaza operation; the supervisor maintains supervision of toll plaza operations.
- *Toll collectors.* Shift workers in the collection booths to handle manual and truck booths. A plaza supervisor is included to oversee operations and provide relief.
- *Revenue security.* The collection of manual and automatic collection booths revenue is handled by this group together with the assembly of change packages and the accounting of return packages. Coin and token counting is also handled by this group.

25 | Tunnel Finish

Stanley Lorch
Senior Supervising Architect, Parsons Brinckerhoff Quade & Douglas, Inc.

The elements or materials that constitute the exposed surfaces of a tunnel interior can be described as the finish of a tunnel. In tunnel work, the same term refers to elements installed after completion of the structural lining that constitute an integral part of the complete tunnel interior, often in a separate "finish" contract. Tunnel finish work can include wall and ceiling finish materials and support systems; tunnel roadway pavement, barrier curbs, sidewalks/safety walks, and railings; utility niche frames and doors, doors and frames for cross-passageway utility closets, used increasingly in licu of sidewall niches; and police booths or sidewalk patrol cars. All can be considered finish work, and are discussed in this chapter.

Other systems installed in tunnel finish work contracts can include lighting, power, traffic control, safety and communications devices such as variable-message and static signage, closed-circuit television (CCTV) cameras, signal lights, fire alarms, fire-fighting and detection devices, telephones, and radio antennas. These facilities are described in other chapters.

All tunnels, except those mined through sound, competent rock, require an internal lining. This lining can consist of rock bolt reinforcement with wire mesh, shotcrete, segmental steel or precast concrete liners, a steel envelope, cast-in-place reinforced or unreinforced concrete, or combinations thereof. In all but highway and pedestrian tunnels, the interior surfaces of the structural liner are customarily left exposed without applied finishes.

Railroad, rapid transit, and highway tunnels usually require mechanical ventilation. To minimize air turbulence and consequent frictional losses, the interior surfaces of these tunnels should be relatively smooth. Water conveyance tunnels also require smooth interior surfaces,and sewer tunnels frequently require special corrosion-resistant inverts, as discussed in Chapter 15.

Highway tunnels of considerable length must be ventilated to dilute pollutants from vehicle emissions to acceptable levels. This ventilation is provided by introducing outside air into the tunnel and exhausting polluted air. Where required by the ventilation design, the air is transported in ducts located outside the roadway clearance envelope. For circular, arch, and horseshoe-shaped tunnels, these ducts can be located below the tunnel roadway and/or over a suspended ceiling above the roadway clearance envelope, separated by a suspended ceiling. This ceiling is generally constructed as part of the tunnel finish work.

Some highway tunnels use a ventilation system with longitudinal jet fans suspended from the tunnel roof within the roadway compartment, but above the vehicle clearance diagram. In this case the suspended ceiling is eliminated. The structural roof above the fan area may or may not receive applied finish depending on lighting and operational considerations.

Increasingly, wider box-shaped tunnels are being constructed with duct or utility chambers between or alongside the roadway spaces, often in lieu of overhead or below-roadway ducts.

The selection and incorporation of appropriate finishes and finish support systems must be based on functional, operational, aesthetic, economic, and public safety considerations.

Public perception is generally the measure of a highway tunnel's success. Regardless of its design, economics, and construction innovations, acceptance will be measured by ride comfort, convenience, aesthetics, and safety.

The light-reflecting characteristics of the finish materials used in the tunnel ceiling and wall surfaces bear directly on the effectiveness and efficiency of the tunnel lighting system and, in turn, on the tunnel's safety and the aesthetic impression it leaves upon users (Chapter 21). The presence of moisture and engine exhaust products in the tunnel—especially emissions from diesel-powered trucks—creates an atmosphere that not only can corrode metal components, but darken finish surfaces, detracting from their light-reflecting qualities.

Frequent, high-quality maintenance to maintain reflectivity and appearance is desirable, but cannot always be depended upon.

The use of materials with impervious, soil-resistant surfaces will help to compensate for maintenance deficiencies, but surfaces must be cleaned and should be resistant to long-term deterioration from cleaning agents, brushing, and high-pressure water jets.

Tunnel finishes and support systems are also subject to vibration, dampness, and extremes of temperature and may be exposed to fire and explosion.

The evaluation and selection process for tunnel finish materials must therefore consider reflectivity, driver orientation, adaptability, cleanability, durability, fire resistance, repairability, replaceability, inspectability, and noise reduction. Further, it is vital to evaluate these factors within the framework of public safety, acceptable aesthetic standards, and cost. Economic evaluation should include initial, maintenance, and life-cycle cost comparisons.

SUSPENDED CEILING SYSTEMS

Cast-In-Place Concrete Ceiling

This type of suspended ceiling construction (Figure 25-1) was used in the Holland Tunnel, the first subaqueous highway tunnel, constructed during the years 1920 to 1927. To date, approximately 90% of all highway tunnel ceilings have been built of cast-in-place concrete.

The ceiling is a reinforced concrete slab 4–6 in. (100–150 mm) thick transversely spanning the tunnel sidewalls and interior supports. These interior supports vary from 1 to 3 (or more) in number and consist of composite concrete and structural steel beams or structural steel stringers that span longitudinally between ceiling hangers. Ceiling hangers vary in longitudinal spacing from 4 to 12 ft (1.2 to 3.7 m) on centers and transfer the ceiling load to the tunnel structural lining. Transverse expansion joints are provided in the ceiling slab and vary in spacing from 20 to 50 ft (6.1 to 15.2 m).

The ceiling soffit is located 13 ft (4.0 m) to more than 16 ft, 6 in. (5.0 m) above the tunnel roadway, in accordance with AASHTO Standards for primary federal interstate and defense highway clearances. The longitudinal profile of the ceiling is parallel to the roadway grade, and the transverse slope of the ceiling is usually parallel to the pitch of the roadway surface to provide transverse drainage of water from condensation or leakage above the ceiling surface. Run-off is removed by sidewall drains located on the up-grade side of each expansion joint and discharged into the roadway drainage system. These drains are usually 1-1/2–2 in. (38–50 mm) in diameter and made of hard copper with bends turned to permit cleaning.

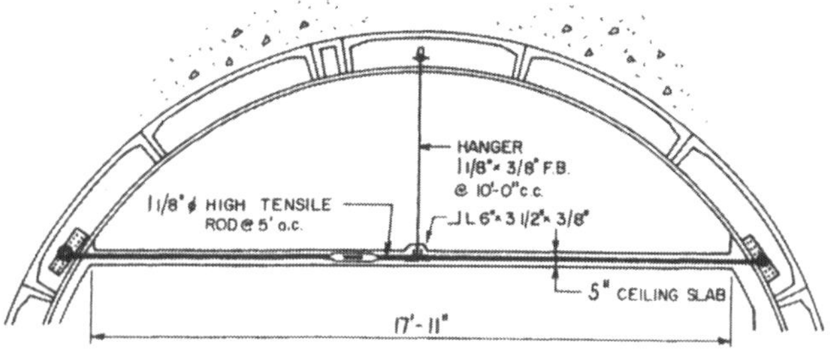

Fig. 25-1. Cast-in-place ceiling.

Design loads are the dead load of slab and ceiling finish and a 20 psf (1 kPa) live load, which provides for maintenance equipment and personnel. Where the ceiling forms the floor of an exhaust air duct, suction loads must be considered. The method for determining total suction pressure in the exhaust air duct is described in Chapter 19 and usually does not exceed 50 psf (2.4 kPa) (10 in. of water).

The stresses for cast-in-place ceilings should be in accordance with ACI Building Code Requirements for Reinforced Concrete. The use of this code is preferable to that of the AASHTO Code for Highway Bridges since the live loads are essentially static rather than repetitive, as provided for in AASHTO. Reinforcing bars are deformed and in accordance with the requirements of ASTM A305. Design strength of the concrete is usually 4,000 psi (27.6 MPa) using Type II Portland cement with a maximum aggregate size of 3/4 in. (19 mm).

Deflection of the structural lining caused by variations in earth, rock, or water pressure can result in the transmission of axial loads (tension or compression) to the ceiling after it has been installed. Consideration should be given to this possibility in the design of any ceilings where keyed and doweled supports are used.

A typical sidewall ceiling support used for tunnels of circular cross section employs a continuous box-out in the interior structural concrete liner centered on tension tie rods connected to the primary lining during excavation. The box-out is filled in when the ceiling concrete is placed. Sidewall support details have also included use of a loop anchor insert and threaded dowel insert within a shallow key formed in the structural liner with or without a haunch.

Because the dead loads of the cast-in-place ceiling are usually greater than the suction loads, the hanger assembly is in tension. A hanger assembly consists of an adjustable rod or flat-bar hanger anchored at the top in or to the structural lining and connected at the bottom to the ceiling slab's interior longitudinal support. Where the interior ceiling support is a reinforced concrete beam, a concrete insert is used for the bottom connection to the hanger. Where the interior support is a composite concrete and steel beam, the hanger is connected to the longitudinal structural steel stringer, which consists of a tee or double angle.

The hanger assembly is fabricated from structural steel that meets the requirements of ASTM A36 or stainless steel that conforms to ASTM A276 Type 304. Protection of the hangers against corrosive exhaust gases and dampness has ranged from painting with red lead to shop and field finish paint to coal-tar enamel coatings. Recently, corrosion-resistant finishes, such as epoxies, ceramic and metal compounds, and metallic zinc paint have been used. In several ceiling installations, bare stainless steel hangers have provided excellent service.

The hanger assembly can be adjusted vertically to compensate for construction and fabrication tolerances of the hanger and tunnel lining and to maintain ceiling-to-roadway clearance. Since the tunnel's structural lining is usually constructed on chords rather than following the roadway profile vertical curve, hangers of varying length are required throughout the length of the vertical curve. To minimize the number of variable-length hangers, an adjustable-length hanger is desirable. Further, to prevent bending in the hanger, the top insert should be set plumb, or a hinged connection should be provided at the top of the hanger. These adjustments are accommodated by threaded top and bottom hanger connections, field-drilled holes, slotted holes, or combinations thereof.

The soffit of the cast-in-place concrete ceiling is adaptable to various finishes, such as unfinished concrete, painted concrete, ceramic or glass tile, and veneers of ceramic or metallic materials. (Selection of the type and color of the finish is discussed later.)

Forms for placing the cast-in-place concrete ceiling are usually steel faced with plywood. They are supported by a laterally and vertically adjustable traveler designed to permit passage of construction equipment and mounted on pneumatic tires or steel-flanged wheels running on rails.

Form length varies from 150 to 175 ft (45 to 53 m), depending on the spacing of construction and expansion joints and the length of the pour. The construction sequence is usually threefold: one form section supports the previously placed concrete, another supports the concrete being placed, and the final form supports the reinforcing steel for the section about to be placed.

Travelers have been designed to permit lowering and bypassing the self-supporting forms. In other cases, each section of form has been supported by its own traveler. Choice of methods is a question of economy and individual preference. The number of three-phase form units depends on the construction schedule and form-stripping time. Most specifications require a minimum strength of 1,500 psi (10.342 Mpa) in the concrete before form removal, which is usually determined by the breaking strength of cylinders cured in the tunnel.

When ceramic or glass tiles are used as a finish, they are cast into the concrete slabs. A layer of gummed paper, sticky side up, is placed on the ceiling form. This is dampened and tiles are placed face down on the paper and accurately spaced. The joints between tiles are then filled with fine sand or, in some cases, a lean cement–sand grout. Next, a 2-in. (50-mm) deep hat-shaped wire mesh is set in place on top of the tile, and a sand–cement grout is placed to a depth of 1 in. (25 mm). After the grout has set, the reinforcing steel and structural steel stringers are installed. Finally, the ceiling concrete is placed using pneumatic pumping equipment. The average cycle of advance is one week. When a veneered or unfinished concrete finish is used on the ceiling, the gummed paper and plywood are omitted, and the ceiling concrete is placed directly in the form. Exhaust and/or supply-air port castings, access hatches, lighting, signal, and other ceiling-mounted system provisions are also placed in the forms prior to placing concrete.

Advantages and Disadvantages. The cast-in-place concrete ceiling offers many advantages. It is suitable for all types of ceiling finishes and requires a minimum number of fabricated metal parts for mechanical adjustment to compensate for construction and fabrication tolerances. Horizontal curves can be accommodated by an adjustable form shoe that will compensate for variations in width. This adjustable shoe also eliminates concrete chipping where sidewall tights occur.

The few disadvantages are of major consequence. Repair or replacement of the ceiling and/or finish is a slow, expensive operation that requires special equipment and usually necessitates closing the tunnel to traffic. Care must be taken in the design and placement of the concrete mix to minimize shrinkage and subsequent cracking of the ceiling. Infiltration of drainage water through these cracks will damage the finish and can cause structural disintegration of the concrete slab. In addition, tile or veneered finishes can be hazardous; during a fire these finish materials can explode off the ceiling, endangering firefighters with falling debris and making footing treacherous.

Metal Panel Ceiling

This type of suspended ceiling construction was first used in the Downtown Elizabeth River Tunnel, built between 1950 and 1952. Support details are shown in Figure 25-2.

A metal panel ceiling consists of cold-formed steel or extruded aluminum panels filled with concrete. The panels vary from 2 to 4 in. (50 to 100 mm) in thickness, 6 to 13 ft (1.8 to 4.0 m) in length and 1 ft to 2 ft, 6 in. (31 to 76 mm) in width. The soffit and sides of the panel are finished with a white or light-colored porcelain enamel. The panels span transversely between longitudinal structural steel stringers, which are supported by brackets at the tunnel sidewalls and by interior hangers suspended from the tunnel's structural lining. In the Downtown Elizabeth River Tunnel, ceiling hangers were located on 12-ft, 7-in. (3.8-m) centers longitudinally. The stringer is continuous over the hanger supports

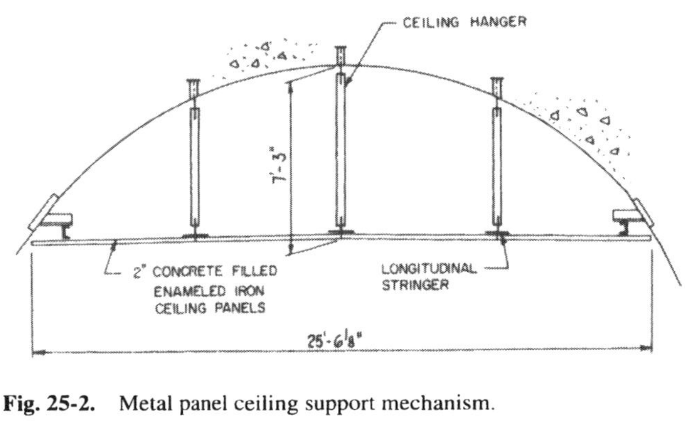

Fig. 25-2. Metal panel ceiling support mechanism.

and has cantilever end spans, which vary from 1 to 3 ft (31 to 91 mm). Transverse expansion joints were located at 25 ft, 2 in. (7.7 m) on centers.

Vertical clearances, drainage, and design loads are similar to those used for the cast-in-place ceiling.

Design stresses are also similar to those used for the cast-in-place ceiling, except that the design of the metal panels conforms to the Light Gage Cold Formed Steel Design Manual published by the AISI. Structural steel design is in accordance with the requirements of the AISC Manual of Steel Construction. Design strength of the concrete is 3,000 psi (20.7 MPa) with a maximum aggregate size of 3/4 in. (19 mm). Where extruded aluminum panels are used, a lightweight (110 pcf) (1762 kg/m^3), 3,000 psi (20.7 Mpa) concrete with expanded shale aggregate is used for panel concrete. Structural steel stringers and brackets conform to the requirements for ASTM A36 steel.

Material for steel panels is usually 14-gauge (1.9 mm) commercial grade enameling iron. Extruded aluminum panels fabricated from Alloy 6063-T42 have been used in one tunnel (Ricker and Manley, 1957). After installation of the ceiling, spalling of the porcelain enamel finish occurred. This spalling may have been caused by the use of an unsuitable alloy or by improper enameling techniques. Studies by the aluminum industry indicate that a controlled 6061 Alloy might eliminate spalling. Aluminum panels for tunnel ceilings have not been used in any other U.S. tunnel. Typical cross sections of steel and aluminum panels are shown in Figure 25-3.

A typical sidewall support for the metal panel ceiling, shown in Figure 25-4, consists of a steel bracket permitting a 2-in. (50-mm) vertical and a 1-1/2-in. (38-mm) lateral adjustment of the ceiling. These brackets are located on the same centers as the hanger inserts.

Figure 25-5 illustrates a typical tension hanger composed of a stainless rod with hinged top and bottom connections. When the dead load of the ceiling approaches the suction load, the hanger assembly is designed as a pipe column, as shown in Figure 25-6. Both types of hangers provide installation adjustment, but the column hanger is fixed at the bottom by welding after the ceiling is installed.

Hanger support has been accomplished using cast-in-place inserts, expansion bolts and resin anchors. Difficulties in accurate placement of cast-in-place inserts, particularly when the structural liner is completed by other contractors prior to ceiling construction, have led to the preferred practice of using drilled-in anchors.

The ceiling hanger assembly for the Second Hampton Roads Tunnel is similar to the one shown in Figure 25-6. However, in lieu of the top insert, a percussion-drilled hole 6 in. (150 mm) deep by 1-3/8 in. (35 mm) in diameter was provided. A polyester resin cartridge was placed in the hole, the upset bolt was spun into the structural lining and, after setup, each bolt was tested to twice the design load. This method of installation permits compensation for construction tolerances. It has been demonstrated that resin anchors have performed as well as mechanical anchors in fire exposure tests.

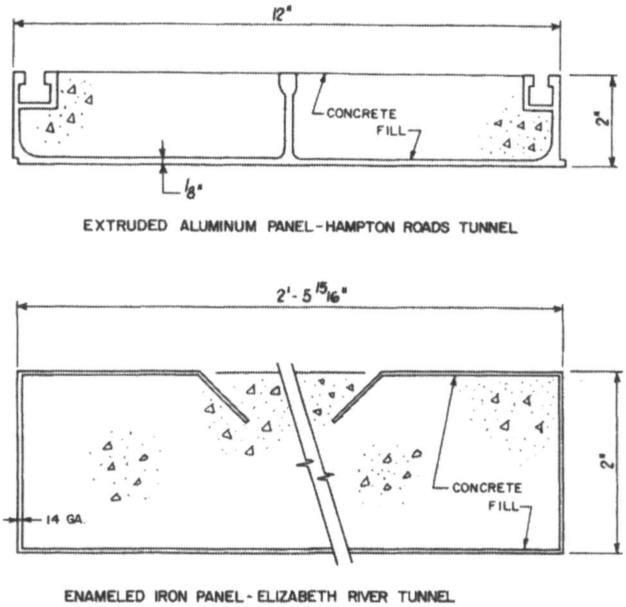

Fig. 25-3. Typical cross sections of steel and aluminum panels.

Vertical adjustment of the panels is provided for at the center of the ceiling, which eliminates the need for fabrication of special panels where sidewall tights occur. Joints between panels are gasketed to provide airtight and watertight connections.

Panels are supported at each corner by stainless steel bolts, which are connected to the stringers. Adjacent panels are connected longitudinally by four stainless steel bolts. The stainless steel conforms to ASTM A276 Type 304 and gaskets are of a fire-resistant (self-extinguishing) neoprene.

Gaskets are cemented to the panels during erection of the ceiling; their shapes prevent them from falling through the joints. Alignment of joints between panels is maintained by slotted holes in the supporting stringers. The bulb shape of the transverse gaskets ensures a watertight joint while providing for longitudinal adjustment of joints necessary for

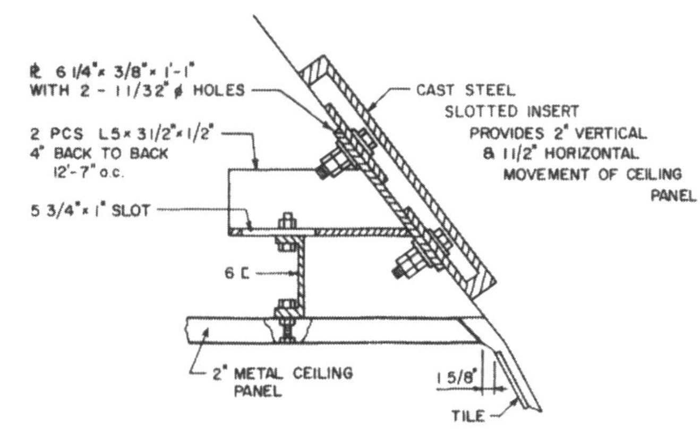

Fig. 25-4. Typical sidewall support for metal panel ceiling.

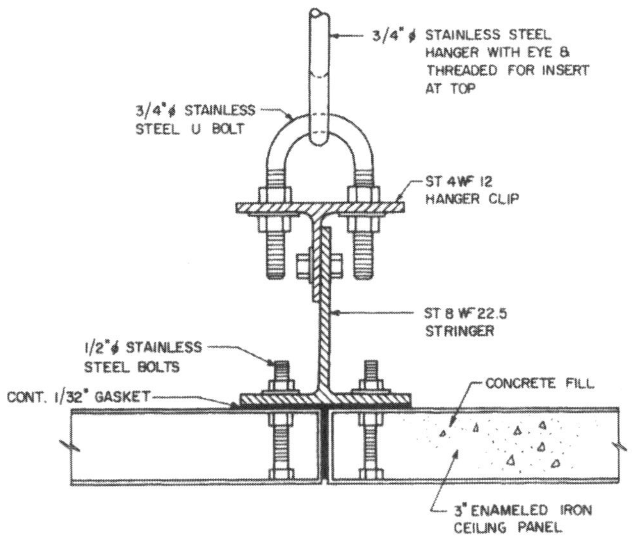

Fig. 25-5. Typical tension hanger.

erection tolerances and maintenance of the expansion joint width.

All structural steel brackets, stringers, and hangers are coated with corrosion-resistant finishes similar to those used for the cast-in-place ceilings. Metal panels are temporarily protected against shipping and installation damage by a covering of adhesive paper or plastic film that is removed after installing the ceiling.

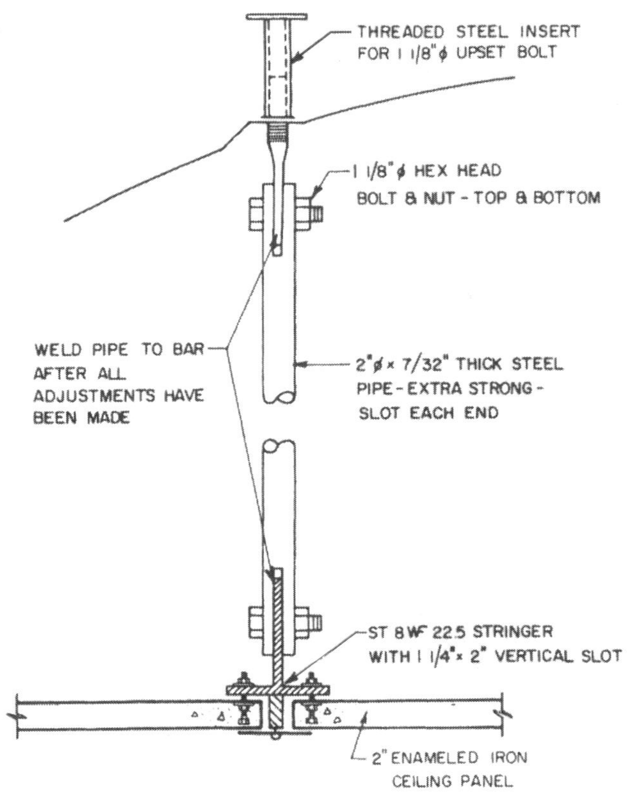

Fig. 25-6. Hanger assembly designed as a pipe column.

Metal panel ceiling installation is a three-phase operation. The initial operation consists of installing the hangers and erecting the stringers to line and grade. Panels are assembled on a traveler similar to those used for the cast-in-place ceiling. The traveler is usually 25 ft (7.6 m) long and adjustable in height. Concrete is usually hoisted to the traveler platform for placing in the panels. Pneumatic placement can be used, but in the past, it has not been considered economical. On the Second Hampton Roads Bridge Tunnel, completed in 1976, concrete was placed in the panels off-site. After curing, the panels were shipped to the site and lifted to the traveler using a cherry picker with a vacuum lifting device. The final stage involves raising the traveler platform to the level of the stringers and attaching the panels to them. Since the metal panels have been designed as forms, the panels are customarily fastened to the stringers as soon as the concrete fill has set (usually the next morning). The number of traveler units depends on the construction schedule. After making final adjustments, all bolts are tightened and connections are welded.

Advantages and Disadvantages. The major advantages of the suspended metal panel ceiling are maximum off-site fabrication, ease of replacement, factory control of fabrication and finishes, and speed of erection. If the ceiling is damaged during tunnel operation, the metal panels can be replaced quickly by unbolting the damaged ceiling panels and inserting new ones. Extra panels, complete with concrete fill, can be stored at the completed tunnel for this purpose. In addition, during a fire there are no tiles or veneered finishes to explode off the soffit as a result of superheated steam created by free water in the concrete. The concrete fill in the metal pans is contained and can only explode upward into the air duct. Mechanical adjustments of all parts of the ceiling permit close tolerances and result in high aesthetic standards of finish. Also, the high corrosion resistance of the porcelainized metal panels minimizes the effect of leakage in the structural lining.

Disadvantages of the metal panel ceiling include difficulty in achieving consistently flat panels with the concrete fill, and porosity of the fill that can permit water to penetrate and pond in back of the pans, where voids are possible because of the lack of bond between concrete and the coated metal and where, ultimately, corrosion may occur. Consideration is now being given to the use of preformed cementitious cores, and to the capping or encapsulation of the core to reduce potential sources of corrosion.

Precast Concrete Panel Ceiling

This type of ceiling construction was first used for the Hong Kong Tunnel, opened to traffic in 1972. The ceiling is similar in concept and design to the suspended metal panel type, except that a precast concrete panel is used in lieu of the metal panel. A precast concrete ceiling was installed in the Holland Tunnel in 1988 to replace the original cast-in-place system. As in the original, ceramic tile was cast into panels as a finish surface. Unique features included the width of the panels that spanned between tunnel sidewalls, and detailing that incorporated a fully prefabricated tiled ceiling surface to serve as centering for accommodation of existing tie rods and permit casting of sidewall keys.

A replacement ceiling of precast concrete structural panels is being installed in the Brooklyn Battery Tunnel. A veneer of light gauge porcelain-on-steel laminated panels will be applied to the new ceiling.

Preformed Laminated Metal Panel Ceilings

This ceiling construction was used as a replacement for a deteriorated tile-faced cast-in-place concrete ceiling in the Callahan Tunnel in Boston in 1991 and has been used extensively in Europe.

Description. The ceiling system consists of a lamination of 24- or 28-gauge (0.61 or 0.38 mm) porcelain-coated steel facing and steel backer sheets and preformed cement core encapsulated in porcelain-coated extruded aluminum frames. The panels used in Boston were 2 in. (50 mm) thick, 6 to 13 ft (1.8 to 4.0 m) long, and between 1 ft and 3 ft, 6 in. (0.3 to 1.1 m) wide. Panels span transversely between longitudinal structural steel hangers, which are supported by brackets at the tunnel sidewalls and by interior hangers suspended from the tunnel's structural lining, similar to the metal panel ceiling previously described.

Panel cores have been built up of laminated layers of preformed fiber-reinforced cement board and, most recently, of mesh-reinforced precast concrete slabs cast to size. Supporting anchors and box-outs are incorporated into the core.

Advantages and Disadvantages. As with suspended metal panels, advantages include maximum off-site fabrication, ease of replacement, factory control of fabrication and finishes, speed of erection, and a nearly dead flat surface because of the factory prefabricated core. The full encapsulation makes the panels impervious to water and thereby highly resistant to deterioration and freeze–thaw damage. These assemblies are generally less costly than heavy gauge metal porcelain panels.

A disadvantage is the use of aluminum frames, which are required to encapsulate the edges of the porcelain veneers and are less corrosion-resistant than the porcelain, and reliance on adhesives for multiple laminations. All surfaces of the aluminum frames, especially the cut edges, must be protected with suitable coatings to increase corrosion resistance.

CEILING VENEERS

This construction approach involves attaching the ceiling finish directly to a structural substrate. The total veneer system thickness can be as little as 1/2 in (12.7 mm).

Veneered ceilings have advantages similar to the various types of suspended ceilings. Their principal disadvantage is that the veneer is susceptible to separation from the structural lining, the result of the formation of ice caused by low temperatures and leakage. It can also become a hazard in the event of a tunnel fire.

Tile Veneer

The most common example of ceiling veneer is ceramic tile. Tile can be cast directly into precast or cast-in-place concrete ceilings, or installed after concrete placement by bonding with mortar.

Heavy Gauge Metal Panels

Another kind of ceiling veneer involves attaching porcelain-enameled heavy gauge metal panels to the structural lining or unfinished ceiling slab. This technique was first used in the cut-and-cover section of the Downtown Elizabeth River Tunnel. There, 1-1/2-in. (38-mm) thick panels measuring 2-ft, 6-in. (760 mm) by 7-ft, 6-in. (2.28 m) were attached by tee bolts after placing the structural cut-and-cover roof slab. Typical details are shown in Figure 25-7. Lightweight vermiculite concrete fill was used to prevent "tin canning" and to minimize noise from "drumming." A similar type of porcelain enamel metal pan veneer ceiling was installed in the Tyne Tunnel.

Ceramic Glass Veneer

A test section of Pyram panels was installed in the Brooklyn Battery Tunnel in 1969. These aluminum-framed, ceramic glass panels were manufactured by Corning Glass Works. The 3 ft by 8 ft (0.9 m by 2.4 m) panels were attached 1-1/2 in. (38 mm) below the existing cast-in-place concrete ceiling, which had a glass tile finish that had been installed during construction in the 1940s. Low temperatures and leakage through the structural cast iron and concrete lining caused large areas of the tile to fall off. Pyram panels were selected to repair the existing ceiling because they had minimum thickness, were a fail-safe material, had a durable and washable fire-resistant finish, and because large panels could be rapidly installed. The test installation made an improvement in the appearance of the ceiling. The

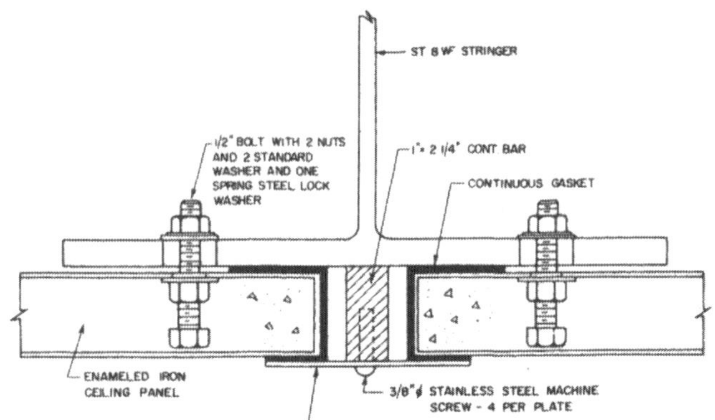

Fig. 25-7. Typical attachments for heavy-gauge metal panels.

panels have endured well, but the material is no longer available. Replacement of the entire ceiling finish and much of the structural ceiling was begun in 1994.

Light Gauge Porcelain-on-Steel Laminated Panels

This system, described previously under "Preformed Laminated Metal Panel Ceilings," can also be utilized as a veneer attached directly to a supporting slab, structural liner, or structural framing.

Attachment methods have included extruded aluminum tracks interlocking with the panel frames, or with sliding clips, that provide for a nonprogressive installation that permits removal of individual units when required. Fasteners are located in joints between panels and concealed from view by use of resilient gaskets or snap-in metal covers.

Coated Cementboard Panels

A thin fiber-reinforced cementboard material with a baked enamel coating has been used as a veneer cladding in vehicular tunnels in Europe and Asia and has been installed in the United States. Produced in Belgium and marketed in the United States under the name of Glasweld, it has also been used extensively as building cladding.

Fastening systems have ranged from coated sheet metal channels and hat-shaped sections to adhesives. Fasteners are often exposed to view.

Stainless Steel Veneer Panels

Stainless steel has had limited use as a ceiling cladding in vehicular tunnels. Series 300 alloys, while highly resistant to corrosion and abuse, are relatively dark in color, difficult to maintain in appearance, and must be quite flat to avoid unsightly reflections.

Vitreous Ceramic Panels

Recent advances in high fired ceramic technology have led to the development of very large, thin ceramic panels that have been successfully used as exterior building cladding in Europe. The full concealment of fastenings has been achieved by use of ceramic disks fused to the panel body incorporating captive stainless steel bolts. These bolts are attached to an aluminum framework that is attached to the structural shell. Tunnel applications, although not yet in place, are being explored.

TUNNEL SIDEWALL FINISHES

With the exception of highway tunnels, the interior surface of the structural lining of tunnels constitutes the tunnel sidewall finish. Paint or an epoxy coating applied directly to the structural lining, ceramic tile veneer, and furred-out walls with various types of finish have been used for highway tunnel sidewalls.

Paint and Epoxy Coating

These finishes are applied directly to the structural lining. They have the lowest initial cost and require no space allowance, resulting in minimum interior tunnel dimensions. (The advantages and disadvantages of these finishes are discussed later.)

Ceramic Tile Veneer

This veneer is, and has been, the most prevalent finish for tunnel sidewalls. The thickness of the finish is approximately 1 in. (25 mm) and consists of a mortar scratch coat, a float coat, and tile applied successively to the structural lining, which has been roughened by 1/4-in. (6-mm) steel wires welded to the sidewall forms on approximately 2-in. (50-mm) centers. When the forms are stripped, the edges of the wire-formed grooves are slightly fractured and provide an additional bonding surface for the scratch coat. The scratch and float coats serve as a leveling course to compensate for forming tolerances required for placing the structural sidewalls. Before the scratch and float coats have cured, the ceramic tiles are cemented in place using a sand–cement bonding mortar. After setting, the joints between tiles are raked and then filled with a neat cement grout which can be pigmented or, if the joints are greater than 1/4 in. (6 mm), with a sanded cement grout. Current practice has added the use of a bonding agent on the substrate prior to application of the scratch coat, as well as use of acrylic latex additives to the scratch, float, and bonding mortar, to improve flexibility and bond.

The average rate of advance is 2,000 ft^2 of wall per 8-hour shift. The tile work starts after the tunnel has been holed through and sidewalks placed and is usually concurrent with electrical and mechanical finish work in the tunnel.

Porcelain-Enameled Steel Tile

This material was used for sidewall finish in the Detroit–Windsor Tunnel in 1928. This installation was probably the first use of porcelain finish in a tunnel. These tiles were replaced after 49 years of continuous service with 12 in. by 24 in. (305 mm by 610 mm) mortar-set porcelain steel tiles.

Metal Panels

The IJ Tunnel in Amsterdam, completed in 1968, has a metal panel sidewall finish. These panels are a greenish-gray porcelain-enameled steel, which can be easily removed and replaced in the event of damage. The wall panels are slightly inclined from the vertical such that sound is reflected upward, where much of it is absorbed by an acoustic ceiling. This ceiling consists of perforated aluminum panels enclosing rock wool batts. Both wall and ceiling panels are isolated from the structural lining by special attachments that incorporate blocks of neoprene to aid in sound reduction.

Furred-out Walls

Recently, furred-out wall construction has been used in tunnels, including the Mersey 2 and Tyne Tunnels in England. The finish sidewalls consist of coated steel panels supported by a structural steel framework, which is attached to the structural lining.

The wall construction for the Tyne Tunnel is 20-gauge (0.91 mm) porcelain-enameled steel panels approximately 12 ft (3.7 m) wide, bent to follow the contour of the tunnel. On the reverse side of the panel, a 1/2-in. (13-mm) thick asbestos insulation board is bonded with a neoprene base adhesive, and the back face of the board is sealed with a chlorinated rubber. The asbestos board increased the rigidity of the panel and acted as a sound deadener. The panels were attached to the structural steel frame with Z clips at the side flanges. Vertical joints between panels were enclosed with a full-height stainless steel cover strip.

A furred-out porcelainized enameled steel wall panel system is shown in Figure 25-8.

The steel sidewall panels for the Mersey 2 Tunnel are similar to those used for the Tyne Tunnel. The steel panels were finished with a field applied epoxy paint.

It should be noted that current Environmental Protection Agency (EPA) regulations prevent the use of asbestos board in the United States. Suitable substitutes are available.

The panels are 8 ft, 6 in. (2.6 m) long by 2 ft, 7 in. (79 cm) high and 1-1/2 in. (38 mm) thick. The face panel is 14 gauge (1.9 mm). To provide rigidity, the panels are filled with fiberglass, which is bonded to the panel after porcelainizing. A 16-gauge porcelainzed back cover plate is then pop riveted in place. The support system consists of vertical structural steel tees spaced 8 ft, 6 in. (2.6 m) on centers, to which the panels are attached. The tees are attached to the concrete tunnel lining with adjustable clips, which are expansion-bolted into place, thus providing for construction tolerances. All structural steel members are coated with corrosion-resistant materials. All hardware is stainless steel.

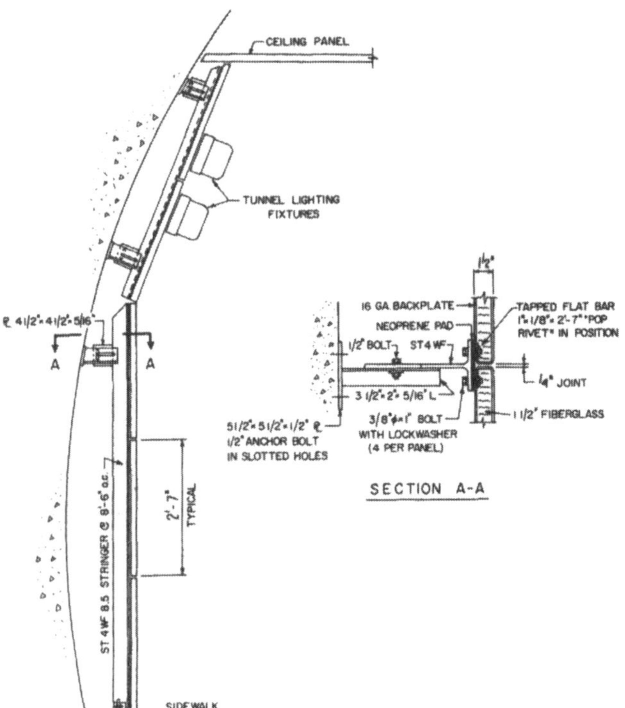

Fig. 25-8. Furred-out porcelainized enamelled steel wall panel system.

Light gauge porcelain-coated steel panels laminated to reinforced cement cores have also been effectively used as a sidewall finish system.

Other types of furred-out walls have been utilized, such as precast concrete panels, ceramic tiled precast panels, and structural glazed tile panels. Furred-out walls require a structural support system, which should be designed for lateral loads resulting from air pressure created by moving traffic and ventilation. The finish wall and the structural support system usually require a space of 4–6 in. (100–150 mm) inside the structural tunnel lining. This air space between walls permits installation of branch ducts for distribution of ventilation air from the air ducts to the tunnel roadway and provides free drainage of seepage water through the structural lining without damage to or discoloration of the interior finished wall. In addition, utility conduits can be surface-mounted on the wall of the tunnel structural lining. Since the furred-out wall panels can be readily removed, access for repair of leaks and conduits and replacement of damaged wall panels is relatively easy. This type of wall construction permits compensation for construction tolerances required for alignment and placing of the concrete lining and is ideally suited to rock tunnels without structural linings.

SIDEWALKS

Sidewalks in the tunnel provide a means of emergency exit for motorists in the event of fire or accident, access for maintenance personnel, and a pathway for traffic control police. In addition, utility conduits are enclosed within the sidewalk and the sidewalk protects the structural lining and sidewall finish of the tunnel from vehicular damage.

Figure 25-9 shows a sidewalk, level with the top of the curb and ranging in width from 2 ft, 6 in. to 4 ft (0.8 to 1.2 m), and on the opposite side of the tunnel, frequently called the ledge side, is a safety barrier. This type of construction is common in tunnels where intermittent vehicular police patrol is provided. A low sidewalk configuration is more compatible with current requirements in fire life safety provisions calling for regularly spaced cross-passageways between parallel tunnels, with ready access from the roadway.

For subaqueous vehicular tunnels, a raised sidewalk and ledge have usually been provided, as shown in Figure 25-10. The type of construction shown is for a cast-in-place concrete sidewalk and ledge faced with ceramic tile. Exposed concrete and architectural terra cotta facings have also been

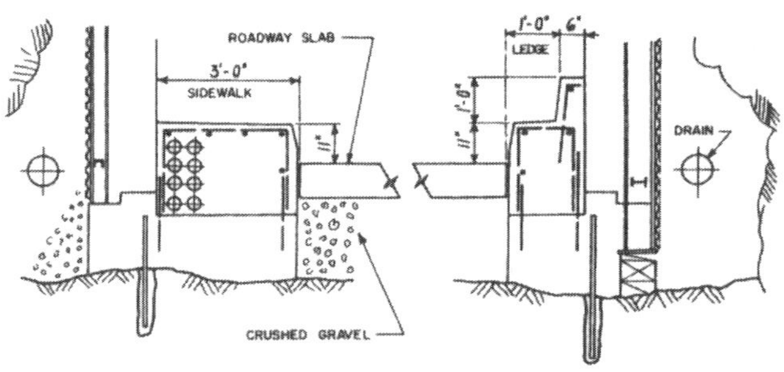

Fig. 25-9. Typical tunnel sidewalk arrangement.

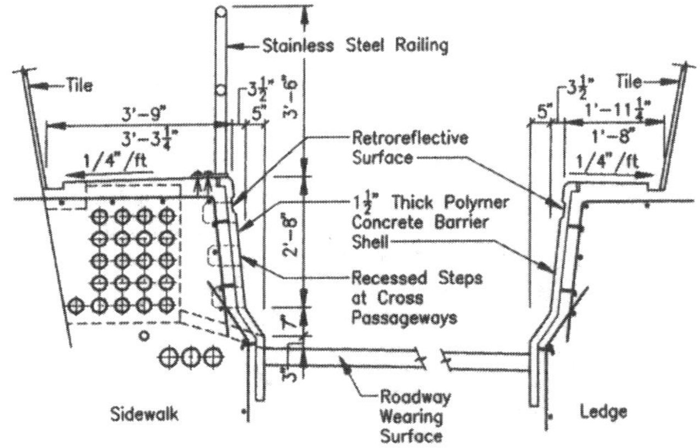

Fig. 25-11. Sidewalk incorporating safety barrier shape to redirect impacting vehicles.

used. Recent developments in removable precast polymer concrete facings have proven more resistant to vehicular damage and easier to replace. The raised sidewalk is ideally suited to a circular or arch tunnel configuration and has the advantage of greater room for utility conduits and pull boxes. It keeps the tunnel patrolman or maintenance personnel in clear view of motorists, while protecting them from traffic in the event of accident, and alleviates somewhat the effects of suction air blast from passing vehicles. It has the significant disadvantage of being more difficult to access from the roadway than low sidewalks.

Another configuration being utilized, as shown in Figure 25-11, incorporates a safety barrier shape to redirect impacting vehicles. One of the drawbacks to this arrangement is the limited accessibility from the roadway. Access is provided by footholes and interruptions in the railing at regular intervals along the length of the tunnel. Present trends are away from in-tunnel foot patrols and toward remote monitoring, with emergency response personnel entering the tunnel by vehicle as needed.

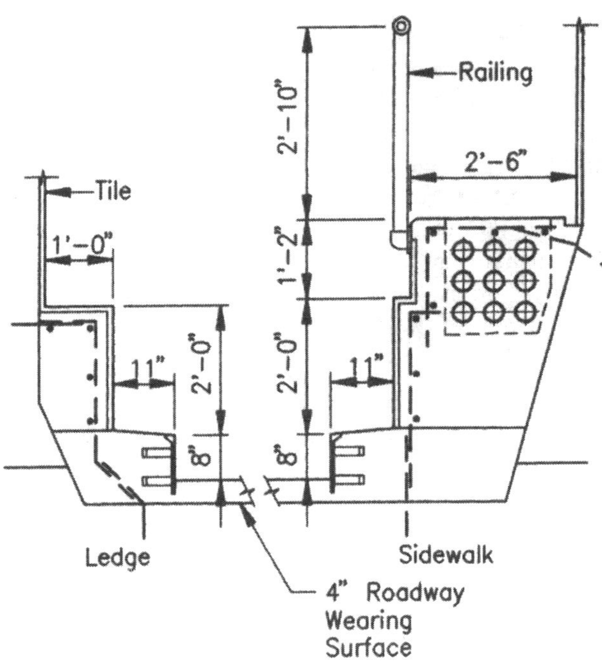

Fig. 25-10. Raised sidewalk and ledge usually provided for subaqueous vehicular tunnels.

In several tunnels, sidewalk patrol cars have been utilized to provide motorized police surveillance of traffic. These vehicles have gasoline engines or battery-operated electric motors and travel on the sidewalk at speeds up to 30 mph (48 kph). The cars are guided by a continuous rail anchored to the sidewalk.

With the growing concerns about long-term exposure to vehicular pollutants, foot patrolmen are less likely to be stationed in tunnels today. There is, instead, increasing reliance upon CCTV surveillance, electronic monitoring, and/or motorized sidewalk patrol vehicles

Sidewalk Railings

These are usually provided where raised sidewalks are used (Figure 25-11). They should comply with OSHA requirements and be configured to permit access to the sidewalk from the roadway. This is generally accomplished by providing interruptions in the railings at appropriate intervals along the length of the tunnel. Railings are generally constructed of tubular steel that is painted or coated, or aluminum alloy, porcelainized aluminum or steel, or stainless steel.

Police Booths

Historically, where sidewalk police foot patrols were utilized for traffic control, police booths were provided within the tunnel. These booths were located at intervals that permitted surveillance of the entire tunnel. The booth provided shelter from air blast, noise, and temperature for patrol personnel during their tour of duty and were provided with an outside air source, heating and air conditioning, appropriate lighting, and convenient access to communications systems. Booths have been constructed of aluminum or stainless steel, with glazed panels arranged to maximize visibility of traffic, a folding or movable seat, and pass-through doors on each end to permit through circulation.

EQUIPMENT NICHES AND DOORS

In-tunnel electrical control equipment for traffic, power, communications, and fire-fighting equipment have traditionally been mounted within tiled niches in the tunnel sidewalls, exposed for easy access. Generally, power conduits were embedded in the concrete lining on the ledge side of the roadway, and control and communication equipment located on the sidewalk side of the tunnel. This separation ensured electrical isolation.

Where practical, equipment recessed in a niche should be provided with a gasketed framed door to protect it from dirt and make regular cleaning easier. To facilitate maintenance, minimize traffic interruption, improve the safety of maintenance personnel, and reduce the risk of theft and vandalism, current practice locates panels and equipment in utility rooms or cross-passageways, separated from the tunnel roadway by fire rated doors.

Door and frame materials can be bronze, aluminum alloy, stainless steel, or painted carbon steel. Selection of material should be based on initial, maintenance, and replacement costs. Current preference is for the use of stainless steel where economics permit.

ROADWAY DESIGN

Vehicular tunnel roadways in the United States should be designed in accordance with AASHTO requirements for an HS20 loading with a check for military highway loading on interstate highways. While there are no published geometric standards for lane widths in tunnels, recent practice has been to provide a 12-ft (3.7-m) width for traveled lanes with a 1-ft (0.3-m) shoulder on each side. Where traffic density is high and design has provided for speeds of 60 mph (96 kph), 8–10-ft (2.4–3.0-m) shoulders are preferred. Design speeds, stopping sight distances, horizontal and vertical curvature and superelevation are discussed in Chapter 2.

Structural roadway slabs are usually reinforced concrete with a 4-in. wearing surface. In circular tunnels that have air ducts beneath the roadway, these slabs span transversely to support roadway loads. In most older tunnels with air ducts beneath the roadway, the roadway slab consists of structural steel beams spaced from 30–60 in. (0.8–1.5 m) on centers and embedded in concrete. The relative merits of asphalt versus concrete wearing surfaces are controversial. Except where ice and snow can extend well into the tunnel, the preferred wearing surface is generally asphalt. Roadway wearing surfaces are predominantly asphalt concrete or concrete.

Selection of the roadway surface requires evaluation of initial costs, maintenance and replacement costs, traffic density, ride comfort, quietness, and performance records for safety. These factors must be evaluated for each tunnel.

TUNNEL FINISH MATERIALS

For the public, the interior finish of a highway tunnel is generally the measure of its success. Regardless of the design, economics, and construction innovations, acceptance will be measured by ride comfort, convenience, aesthetics and safety. However, in addition to these factors, the interior finishes have to satisfy the criteria of adequate lighting, good visibility, fire resistance, noise attenuation, minimal maintenance, and durability.

Visibility

Visibility within a tunnel affects the ability to see the limits of the roadway, objects within the roadway, and informational warnings. There is also an impact on driver comfort level relating to lighting types and placement, color, glare, and patterns of light and shadow or light distribution on objects within the driver's line of sight as well as wall, ceiling, and roadway surfaces.

The selection of appropriate finish material properties must be coordinated with the design of the lighting system, for they are directly interrelated.

The designers of early highway tunnels for many installations used unfinished concrete ceilings and, in some cases, unfinished concrete sidewalls. These surfaces soon darkened, providing poor visibility. Following this trend, sidewalls were surfaced with white tile, at least to the springline. Soon after, ceilings were tiled as well. This increased and allowed for maintaining of surface reflectance, enhancing the effectiveness and efficiency of tunnel lighting, and driver visibility. Thus, prior to 1950, it was almost an axiom that highway tunnels required light-colored tiled walls and ceilings.

While this trend continues in the United States, the concept of light-colored reflective finishes on walls and ceilings is no longer universally accepted. Many designers, particularly in Europe and Asia, rely less on the reflective qualities of the ceiling in the design of lighting systems, finding economic trade-offs between the cost of ceiling washing, tunnel aesthetics, and the increased cost of higher lighting output.

The comfort level of tunnel users must also be factored into the decision on finish type and colors. Designers are becoming more adventurous in the selection of materials, colors, and patterns to enhance the ambiance of the tunnel interior. Care must be given to avoid distractions to the drivers attention and compromising safety and comfort, but the experience of driving through a well-designed vehicular tunnel can be pleasant as well as safe.

Paints and Epoxy Finishes

Increased costs of labor and material and new paint technology in the 1950s and 1960s led to experimentation with paints and epoxy finishes that were applied directly to the concrete structural lining. Under laboratory conditions, these surface finishes appeared to be durable, light-reflective, and economical. Field application of these finishes required extensive preparation of the structural concrete lining, including removal of fins and air bubbles, and thorough cleaning with steam, water, or sandblasting. Most of these surface coatings have to be applied under conditions of controlled humidity and only to dry, clean surfaces.

Although low in initial cost, most of these surfaces have had a very short life (five years or less). To this date, they have not established a satisfactory performance record for durability, low maintenance, and aesthetic acceptance when compared with more durable, longer-lived surfaces.

Ceramic Tile

Traditionally, light-colored ceramic tile has been the accepted finish for walls and ceilings in highway tunnels. The usual tile size has been 4-1/4 in. by 4-1/4 in. (108 mm by 108 mm) but 6 in. by 6 in. (152 mm by 152 mm), 6 in. by 12 in. (152 mm by 305 mm), 8 in. by 8 in. (203 mm by 203 mm), and other sizes have been used. Historically, general manufacturing tolerances for size and warpage versus the labor costs for a reduced number of tile units have favored the 4-1/4 in. by 4-1/4 in. tile, but recent advances in tile manufacturing have reduced the incidence of warpage, thereby permitting increased use of the larger, more economical sizes.

Ceramic tile for tunnels should be vitreous glazed and of uniform thickness, not varying from the nominal size by more than 3/64 in. (1.2 mm). Tiles should have a cushioned edge, and unglazed backs; warpage should not exceed 0.02 in. (0.5 mm) in a length of 4-1/4 in. from a plane surface, wedging or crooked edges should not exceed 0.70 of 1%, and the tiles should be free of imperfections such as pressing cracks, dents, swelling, and chipping. Because these tiles are generally heavier than standard commercial tile and subjected to extreme conditions, and because bond failure of the tile in a tunnel could cause accidents, provisions have been incorporated into their configuration to enhance bond strength by providing keys or lugs on the tile backs to mechanically interlock with the setting mortar. These tiles are specially made for tunnel use and their minimum performance characteristics for absorption, crazing, reflectivity, thermal shock, weathering, glaze hardness, and bond can be found in most tunnel tile specifications.

Although their hardness and smooth finish are not conducive to noise attenuation, those very qualities—along with colorfastness and lack of craze—make ceramic tiles the most widely accepted and thoroughly proven tunnel finish material. They have a successful record of resistance to maintenance abuse, having withstood erosion and loss of glaze and color from both mechanical abrasion and chemical reactions resulting from washing and scrubbing. They are highly fire-resistant, reflective, and economical. Further, they are available in all colors and sizes. Because they are specially manufactured materials, replacements (10% of total tiled surface) are usually provided for at the time of manufacture and stored at the tunnel site. Despite the ease of replacement, new tiling is difficult to conceal, and the tiles are subject to displacement as a result of vibration and temperature changes. The joints between tiles using cement grout become dirty as well, although the introduction of epoxy grout has reduced that problem.

Porcelain Enamel

Porcelain enamel, a combination of glass and inorganic color oxides, when used as a tunnel finish should meet the Porcelain Enamel Institute's "Class A" requirements for acid resistance cited in the current edition of "Test for Acid Resistance of Porcelain Enamel."

Enamel coatings are applied to the front, sides, and back of heavy gauge metal panels. Total coating thickness varies between 0.003 and 0.006 in. (0.08 to 0.15 mm). The soffit side of the panel, including the outside flanges, is clad with a ground and finish coat of porcelain enamel. The back of the panel and the inside of the flanges receive a ground coat.

The porcelain enamel finish is usually medium gloss so that its light-reflecting quality will neither produce undue glare nor absorb enough light to reduce illumination to unsafe levels. Several tunnels have used a "ripple" surface finish. This type of finish is not recommended, since it tends to increase specular reflection. Also, the textured surface harbors dirt, requiring increased washing.

The porcelain finish must be factory applied, and it requires reasonable care and protection during shipping and installation of the panels. More than 25 years of low-maintenance experience have clearly established the durability, colorfastness, and abrasion and impact resistance of this finish under all traffic conditions. It is highly resistant to fire and available in a wide range of colors. It is ideally suited for ceiling and curtain-wall finish. Unfortunately, like ceramic tile, its hard, durable surface is not conducive to noise attenuation.

Effects of Tunnel Finishes on Lighting

To effectively optimize artificial lighting in a tunnel, the wall surfaces should have an easily maintained, highly reflective, nonspecular finish. Ceiling surfaces can also contribute to the effectiveness of the lighting system, if they have similar light reflective properties.

Relatively narrow tunnels, where the width-to-height ratios are approximately 3, or less, to 1 will normally exhibit good visibility as a result of light reflected from the walls. Tunnels having greater width-to-height ratios will normally require supplemental lighting.

As described in Chapter 21, three basic types of lighting systems are commonly used in tunnels: symmetrical—generally continuous rows of light fixtures parallel to or 90° to traffic; line-of-sight—individual fixtures spaced some distance apart with light aimed away from the drivers' eyes; and counter beam—individual separated fixtures aimed toward the drivers. Each has its advantages and disadvantages, and each interacts differently with the tunnel finishes.

In narrow tunnels symmetrically lit by rows of fluorescent or low-pressure sodium fixtures parallel to the roadway, the fixtures direct light in a spread distribution that illuminates the walls and ceilings as well as the roadway. The reflectance of the finish surfaces therefore has a considerable effect on overall tunnel illumination.

Similarly, benefit can be derived from other symmetrical systems with individual point-source fixtures spaced some distance apart, with distribution oriented 90° to the roadway.

Where ceiling-mounted systems are used that direct light mainly at the roadway, with only a small portion of the light cast downward along the walls, ceiling reflectance is of little importance. In all instances attention should be given in the selection and design of lighting to shadow effects on the sidewalls and the interaction of the illumination type with colors and reflective characteristics of the finish surfaces.

Temperature

Because of the high volumes of air introduced to maintain environmental air quality, tunnel surfaces are routinely subject to temperature extremes when outside air is below freezing.

Occasionally there are fires in tunnels, of varying intensity. Finishes must be resistant to fire damage, should not contribute to the fire's spread, and must not generate toxic fumes or smoke that would be harmful to tunnel occupants.

Fire Resistance

All materials used for interior tunnel finish should be noncombustible, should preferably remain undamaged by minor fires, and must not emit smoke or toxic gases during major fires. Particular attention should be given to coatings, gaskets, cements, and other finish materials to ensure that they do not create fire, loss of structural support, environmental, or operational hazards during construction or normal functioning of the tunnel.

Minor fires within highway tunnels are not uncommon. For example, estimates suggest that prior to 1949 more than 50 such fires occurred every year in the Holland Tunnel. Damage to the unfinished concrete ceiling and wall tiles from these minor fires was inconsequential.

Shortly after the major Holland Tunnel fire in 1949, the concept of a suspended metal ceiling with a concrete fill was developed. The objective of this innovation was to limit structural damage in the event of a major fire or explosion and to facilitate repair or replacement to restore ventilation for resumption of traffic. This ceiling design, first installed in the Downtown Elizabeth River Tunnel in 1952, has subsequently (with some modification) been used elsewhere, including the Second Downtown Elizabeth River Tunnel and the Fort McHenry Tunnel.

Utilizing a porcelain-enameled finish on the heavy gauge, concrete-filled metal panels, the initial design was based on fire tests (Ricker and Manley, 1957). These tests indicated that in a minor fire (1,200°F [649°C] for 6 min), no damage to the porcelain enamel finish or metal panel occurred. Minor damage, consisting of slight permanent deflection of the panel, was sustained from a 1,200°F (649°C), 31-min fire. Further tests with a 1,575° (857°C) to 1,600°F (871°C) fire showed that the panel conducted the heat and dissipated it over a large area such that the metal did not approach its melting point. Resulting permanent distortions, however, would require that the panels be replaced. As a cost-saving measure, fusion-bonded epoxy has been used in lieu of the porcelain-enameled finish on heavy gauge metal panels in more recent installations.

Noise Attenuation

Two types of noise are associated with tunnels: noise within the tunnel generated principally by vehicular traffic and reverberating off interior surfaces, and noise produced by exhaust and/or air supply fans located either in ventilation buildings or within the tunnel itself. The former affects principally motorists and tunnel personnel; the latter can affect not only tunnel users but also the general environment in the vicinity of the ventilation building(s) and portals.

Fan-generated noise can be reduced to acceptable levels by sound energy absorbing materials and devices directly at or in close proximity to the source. This approach does, however, adversely affect the operating efficiency of the fans (see Chapter 20).

Reduction of traffic-generated noise within the tunnel is another matter. The level of noise generated by traffic inside a tunnel is generally much higher than that found at an open road site (e.g., 4–10 dB(A) depending on the noise absorption characteristics of the tunnel interior surfaces), as a result of the combination of road noise, aerodynamic noise, and vehicle mechanical noise. Problem noise levels are generally those in the low-frequency range not efficiently absorbed by conventional acoustical materials and applications, let alone those suitable for utilization in a tunnel. Sound energy must be reduced by more than 80% to produce a noticeable change in noise level.

Attentuating tunnel sound to levels that comply with Occupational Safety and Health Administration (OSHA) standards is important to reduce the risk of hearing loss from prolonged exposure. Both motorists and tunnel workers are exposed to ambient tunnel noise, but tunnel personnel are more at risk; the presence of air conditioning and effective internal sound isolation in the modern automobile allows drivers to exert a measure of independent sound control by merely shutting their windows.

Tunnel finishes that are light-reflective, readily maintainable, durable and that meet other criteria described in this chapter do not, by their reflective nature, effectively attenuate noise.

Tunnel noise attenuation has been achieved using cellular materials with intercommunicating cells such as glass wool, mineral compounds, or synthetic materials, which, in combination with corrosion-resistant metals fabricated in eggcrate or perforated sandwiches, will provide a washable finish surface. Eventually, though, the cellular structure becomes clogged with residue from exhaust gases and airborne dust, and its ability to reduce noise is diminished. The use of baffle-type surfaces without cellular sound absorbing material is less effective than sandwich construction, without being any easier to clean. Acoustical ceilings, grooved wall surfaces, and other attenuation methods have also been tried, but with only marginal success.

One system that has been more effective in providing a measure of low-frequency sound absorption in tunnels consists of lightweight resonant panels like sintered aluminum acting as a diaphragm, in combination with a cavity between the panel and the tunnel structural liner, thereby acting as a low-frequency absorber and a panel resonator. The sound absorption curve for the resonant panels depends on the panel fixing method and the physical dimensions of the panel and the air space.

Site-specific operational conditions and cost effectiveness are major determinants of the type and extent of special sound attenuation measures that will need to be provided.

Cost Considerations

While finish systems can be compared solely on the basis of initial installation costs, a more equitable basis for comparison should encompass capital, operational, maintenance, and life-cycle replacement cost considerations as well as corresponding lighting costs. The most useful results can be derived if systems are compared over the presumed service life of the most durable system being considered, using applicable annual rates for cost escalation and including replacement cost at the end of the useful service of each subject system.

Capital costs are those associated with furnishing and installing the finish system, and do not account for the life expectancy of the finish. Operational costs are those related to energizing the utilities and equipment for a tunnel facility. Interior lighting represents a significant portion of the load demand for tunnel power consumption. Factors that affect lighting are reflectivity of the finish, its cleanability, and its durability under repeated cleaning and other forms of degradation. From the standpoint of operational costs, highly reflective durable materials perform better than shorter-lived materials.

Replacement costs are those attributable to the removal and replacement of finishes when they have achieved their average useful service life.

Maintenance of Finishes

The recommended practice for cleaning a vehicular tunnel is to wash the walls (and ceilings) once a week with a cleaning agent, supplemented by a spray rinse directed from high-pressure jets, and selectively scrubbing and rinsing hard-to-reach areas every six months. In cold climates, regular washing must frequently be curtailed in winter to preclude freezing wash water on the roadway. The degradation of reflectivity of finish materials under such circumstances should be considered.

The same relative effort is necessary to maintain regularly cleaned surfaces regardless of type, provided textures are comparable, but unfinished surfaces are not routinely cleaned. Repairs are required for all finishes; however, there is considerable variation in type, frequency, and cost of repairs—some systems can be touched up, others require in-place refinishing, and yet others partial or total replacement.

UNDERGROUND TRANSIT STATION FINISHES

This chapter has dealt primarily with highway tunnel finishes. While underground rail transit stations share some of the same characteristics as highway tunnels, they have many functional differences, especially relating to the accommodation of dense pedestrian flows in confined spaces, that are unique to each station and have significant impact on finishes. General requirements for stations and ancillary facilities are included in Chapter 26, but station organization, accommodation of pedestrian movements, and station finshes are very broad and specialized subjects that could not be included in this handbook.

REFERENCES

National Board of Fire Underwriters (NBFU) (1949) "The Holland Tunnel Chemical Fire, New Jersey–New York," report, May 13.

Permanent International Association of Road Congresses (PIARC) (1967) *XIII Congress,* p. 29.

Ricker, R.W., and Manley, C.R. (1957) "Aluminum Panels Pass Fire Tests," *Ceramic Industry,* June.

Tunnel Paneling (1971) *Tunnels and Tunneling,* September, p. 368.

26 | Service Buildings and Ancillary Spaces

Stanley Lorch
Senior Supervising Architect, Parsons Brinckerhoff Quade & Douglas, Inc.
Hanan Kivett
Senior Architectural Manager, Senior Principal Associate, Parsons Brinckerhoff Quade & Douglas, Inc.

The number, type, and specifics of tunnel ancillary facilities and service buildings depend on the functional requirements of each tunnel. Broadly speaking, ventilation equipment, electrical switchgear, and other mechanical and electrical equipment required for the operation of a roadway tunnel are housed in one or more below- or above-grade buildings. There must be space to house and maintain service vehicles and emergency trucks, shops for electrical and mechanical repairs, and adequate storage for the material and equipment needed to operate and maintain the tunnel and related roadways. Traffic control and surveillance, communications and equipment monitoring functions must be accommodated, and facilities for operating personnel have to be provided. In addition, facilities for the administrative operation of the tunnel must be provided, either as part of a broader district operational facility or as stand-alone quarters.

Ancillary spaces for rapid transit underground stations must provide similar functions. Space for ventilation equipment, electrical switchgear, other mechanical and electrical equipment, train control and communications facilities, and station maintenance provisions are as necessary as patron access provisions, the attendant's booth, and so on.

This chapter addresses general programming requirements for tunnel building facilities and describes the configuration alternatives associated with the various tunnel construction types.

VENTILATION BUILDINGS FOR DUCTED TUNNELS

Depending on the length of a tunnel and the type of ducted ventilation system used, one or more ventilation buildings may be required. Their positioning along the length of the tunnel is related not only to the most efficient arrangement of the ventilation ducts, but to site constraints and economic considerations. They are usually located directly over the tunnel, but may be offset to either side of the tunnel with transition ducting over and/or under the tunnel. In some subaqueous or rugged-terrain tunnels, the buildings may be located some distance above the roadway, connected to the roadway by vertical shaftways. Ventilation buildings are most commonly incorporated into the portal structures (Figure 26-1), but longer tunnels can require supplemental intermediate ventilation stations on the surface connected by shaftway to the tunnel.

Building Functions

The buildings house the fans and appurtenances, fire pumps, emergency power plant, switchgear and controls, workshops for electrical and mechanical repairs, and personnel toilet facilities. The main control room and associated facilities are usually located in one of the ventilation buildings (Figure 26-2).

Depending on availability and site access, there can be economic advantages to utilizing surplus space and substructure foundation support capacity to accommodate additional functions in ventilation buildings. Functions that have been successfully incorporated into ventilation structures, particularly at portal locations, include emergency response, garaging, personnel, and police offices and administrative facilities. Depending on location, access/egress provisions can include corridors, stairways, hoistways, elevators, and parking (Figure 26-3).

Building Facilities for Nonducted Tunnels

Longitudinally ventilated tunnels, depending on length, may need accommodations for the functions described earlier, save the fan equipment space. But in these cases, location is constrained by electrical distribution, access, convenience, and site availability rather than duct efficiency. These facilities are often incorporated into portal structures.

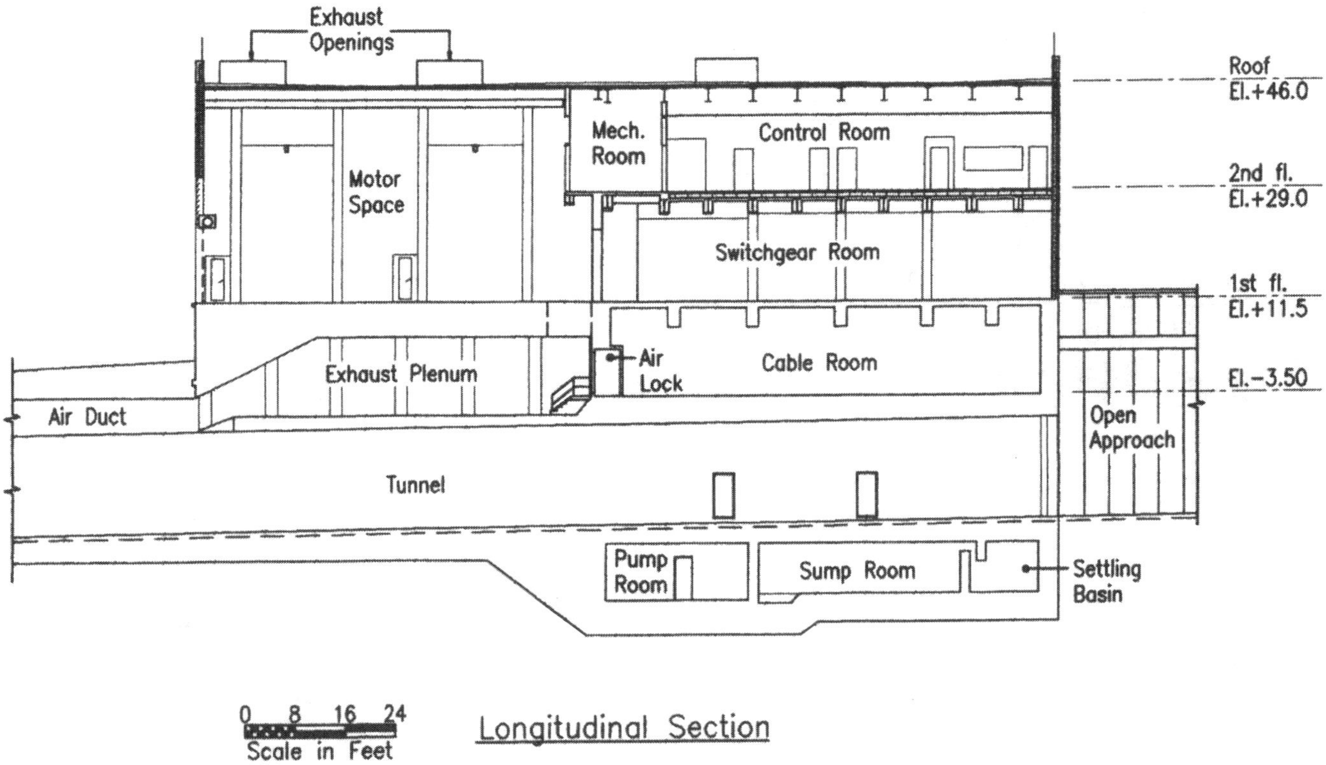

Fig. 26-1. Portal ventilation building, Second Downtown Elizabeth River Tunnel.

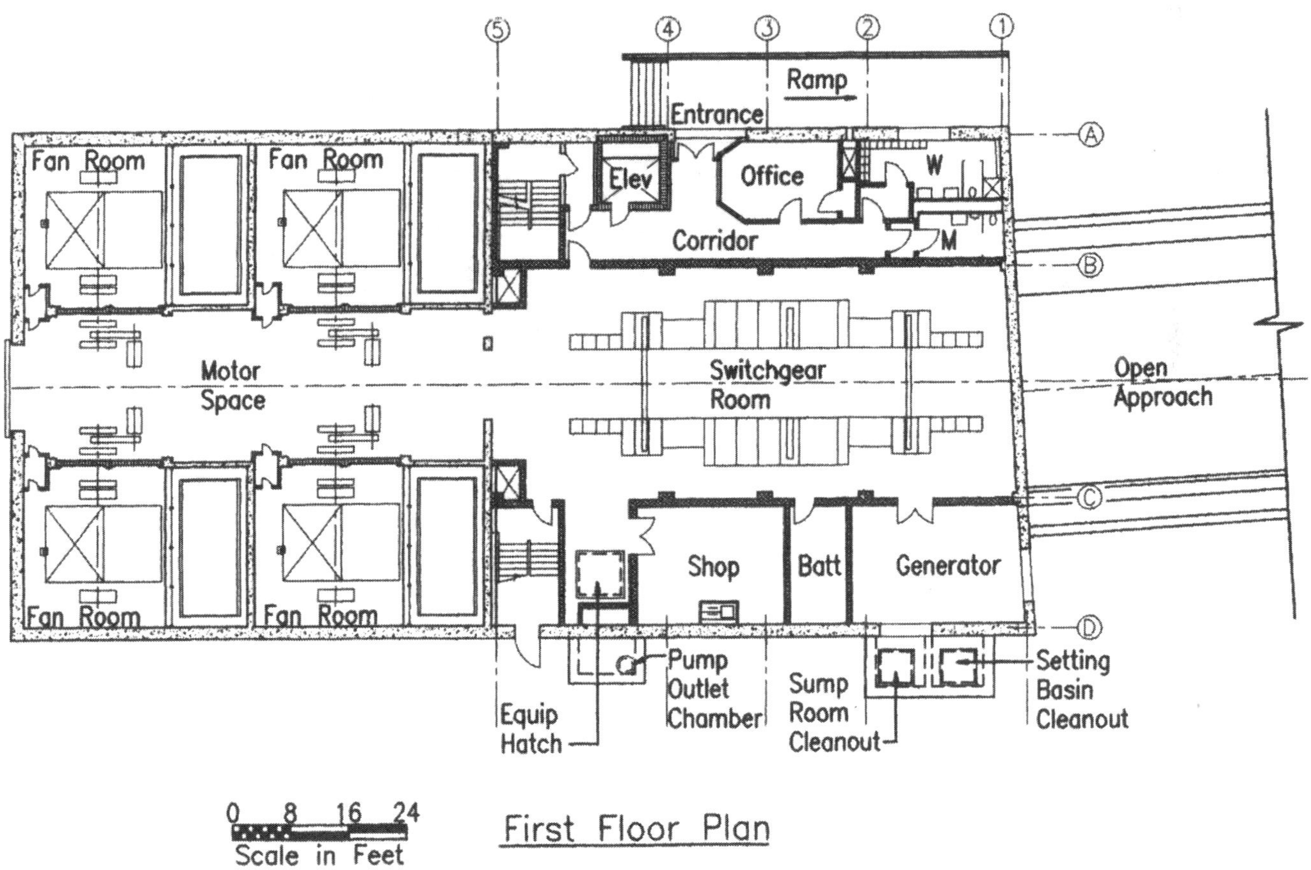

Fig. 26-2. Ventilation building with centrifugal fans, Second Downtown Elizabeth River Tunnel.

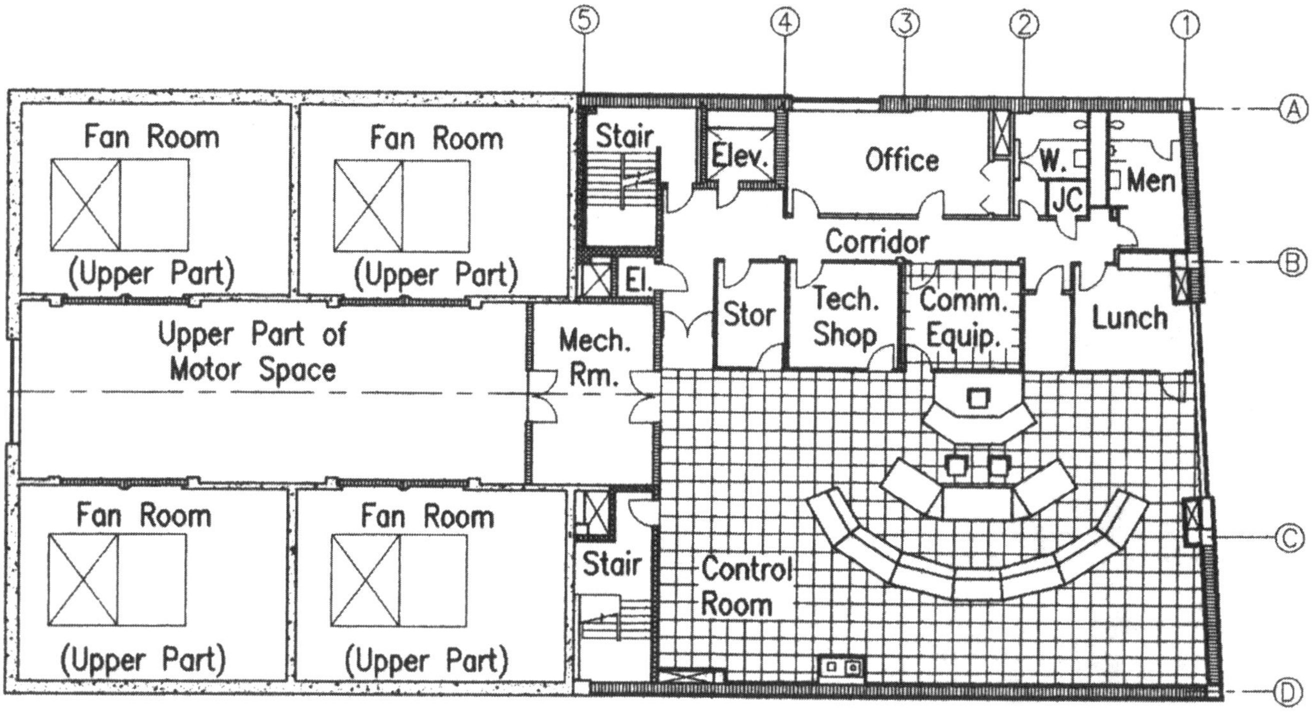

Fig. 26-2. (continued)

Subaqueous Tunnels

In addition to the facilities outlined earlier, subaqueous tunnels require space for a drainage system and associated equipment. This generally consists of a low-point or mid-river sump and pump facility, as well as portal sump and pump facilities.

Where appropriate, flood gates are provided to prevent inundation of the tunnels, with necessary gate storage space, machinery rooms, equipment, and personnel access (Figures 26-4 and 26-5).

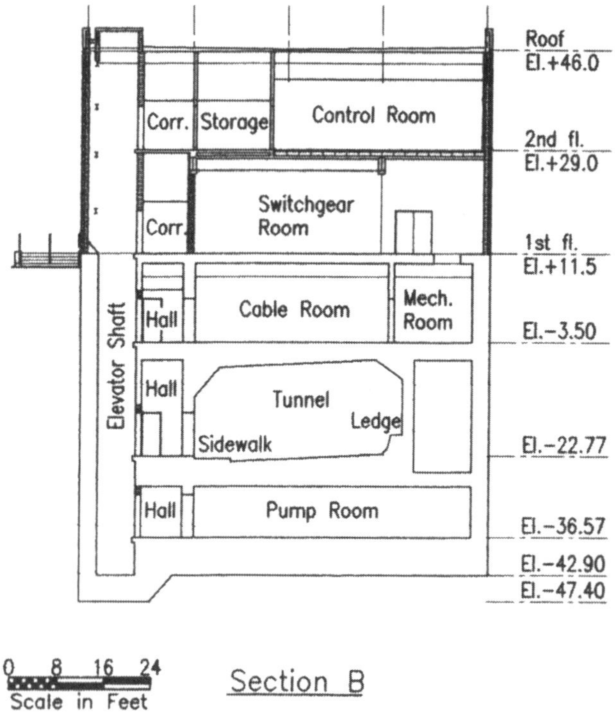

Fig. 26-3. Elevator access to all levels, Second Downtown Elizabeth River Tunnel.

Building Construction and Design

The foundations are incorporated into the cut-and-cover portion of a subaqueous tube tunnel at or near the interface between construction types. In shield-driven tunnels, the vent structures are usually placed over, and form a part of, the shafts used for starting the shields.

The superstructure of a ventilation building may be steel framed or of concrete construction. Walls may be of brick masonry, precast concrete, or some other kind of preformed architectural wall system (Figures 26-6 and 26-7). The architectural design should be appropriate for the context of its site and may have to be approved by local design review boards and comply with local codes. There may be local restrictions on the height of the building and noise restrictions that will affect not only building configuration and location, but the type of ventilation fans selected.

PROGRAM REQUIREMENTS

General program requirements for ancillary spaces are outlined in Table 26-1. In the United States, certain facilities

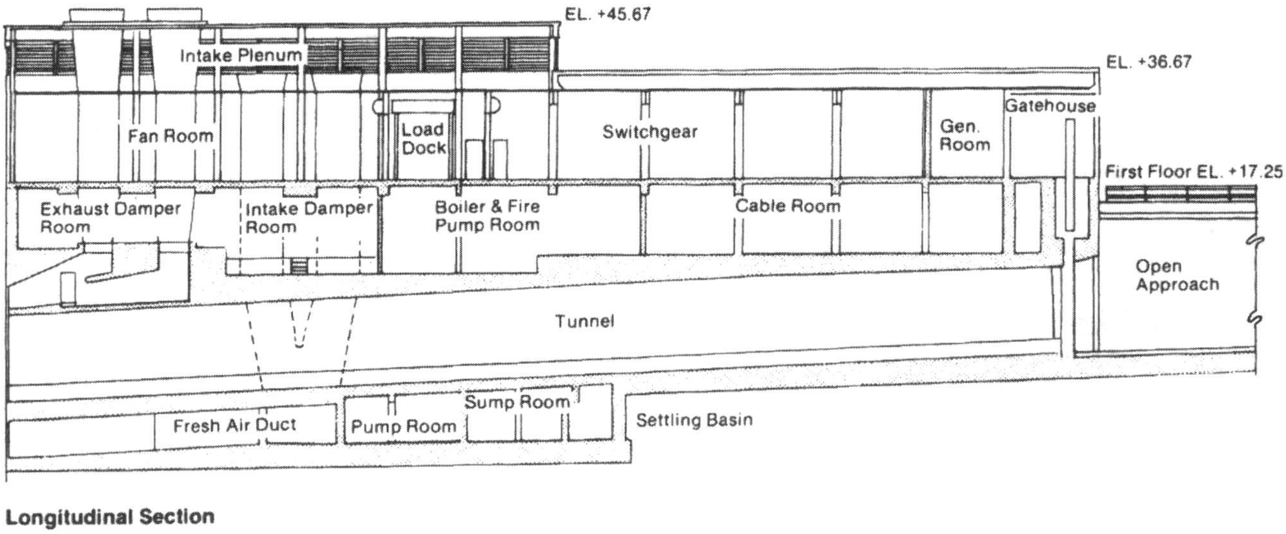

Fig. 26-4. Portal ventilation building with flood gates, Second Hampton Roads Tunnel.

must be built in compliance with Americans with Disabilities Act.

Garage and Service Functions

Buildings may be required to house emergency response vehicles and personnel, to service maintenance and repair vehicles and equipment, and to accommodate associated personnel. Specific requirements are outlined in Table 26-2.

Emergency Vehicle Garage. The emergency vehicle garage can be situated in a separate building(s) or integrated into the portal buildings. Separate facilities will require a room for HVAC equipment. Multiple bays, to allow independent vehicle operation, are preferred (Figure 26-8). Large operations should include an office for the operations supervisor, and first aid facilities should be provided if the tunnel is in a remote location.

Service Vehicle Garage and Maintenance Facilities. Open sheds or fully enclosed buildings can house maintenance and operating vehicles and equipment. Either individual vehicle bays or internal circulation may be used. Maintenance facilities can be combined with the garage or housed

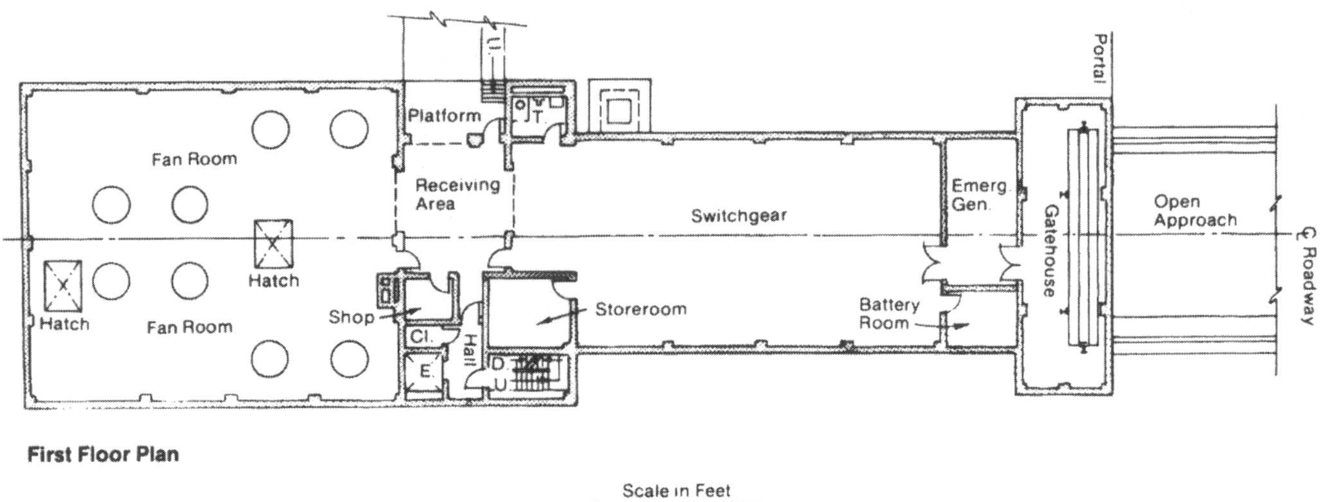

Fig. 26-5. Plan, fan and switchgear rooms, vane axial fans and flood gates, Second Hampton Roads Tunnel.

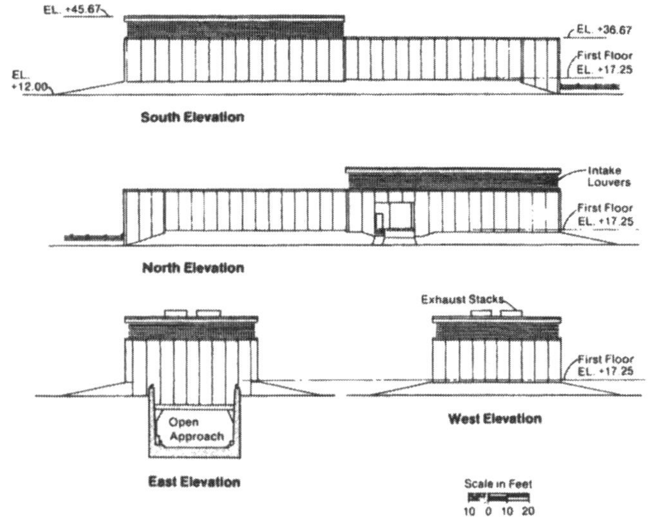

Fig. 26-6. Elevations, ventilation building, precast concrete on steel framing, Second Hampton Roads Tunnel.

separately. These facilities should include wash and service bays, auto parts and tool storage, and crew areas.

Administrative Facilities. The administrative facilities required depend on the structure of the operating organization. In some tunnels, the administrative facilities will be very limited and can be housed among the service and maintenance facilities. Other tunnels, particularly vehicular tunnels that carry heavy traffic loads and require considerable traffic management and toll collection, may require extensive administrative facilities.

Toll Facilities. Toll operations can be housed in separate buildings or partially integrated into an administration building. Typically, three toll lanes are required for each lane of traffic. A sergeant's office with full view of each toll booth is essential. A secure money vault is necessary as well as a secured money counting room. Direct but fully controlled access for authorized toll personnel and armed guards is also required. Parking space for employees, toll payers short of money, and general visitors should be provided.

Table 26-1. Program Requirements

Facility	Requirements
Fan Room and Associated Spaces	Vane axial- Fan room Centrifugal - Motor space with airlocks
Electrical Switchgear Room and Associated Spaces	Switchgear and Transformers Cable room Emergency generator room Battery room Uninterrupted Power Source (U.P.S.)
Workshops and Associated Spaces	Mechanical Electrical Electronics Personnel lockers and tools
Control Room and Associated Spaces [a]	Control room Computer room with UPS Communications equipment room Supervisor's office Technician's office(s) Lunchroom Toilets for men and women Storage room/supplies
Miscellaneous Rooms	Fire pumps and valves Sump room Janitor's closet Staff lockers Toilets for maintenance personnel HVAC equipment space Boiler room
Vertical Access/Egress	Stairways Elevator(s) Equipment shaftways and hoistways
Police Facilities	Day room/training room Sergeant's desk Supervisor's office(s) with weapons safe Locker rooms with toilets and showers for men and women Uniform storage room

[a] Must be handicap accessible.

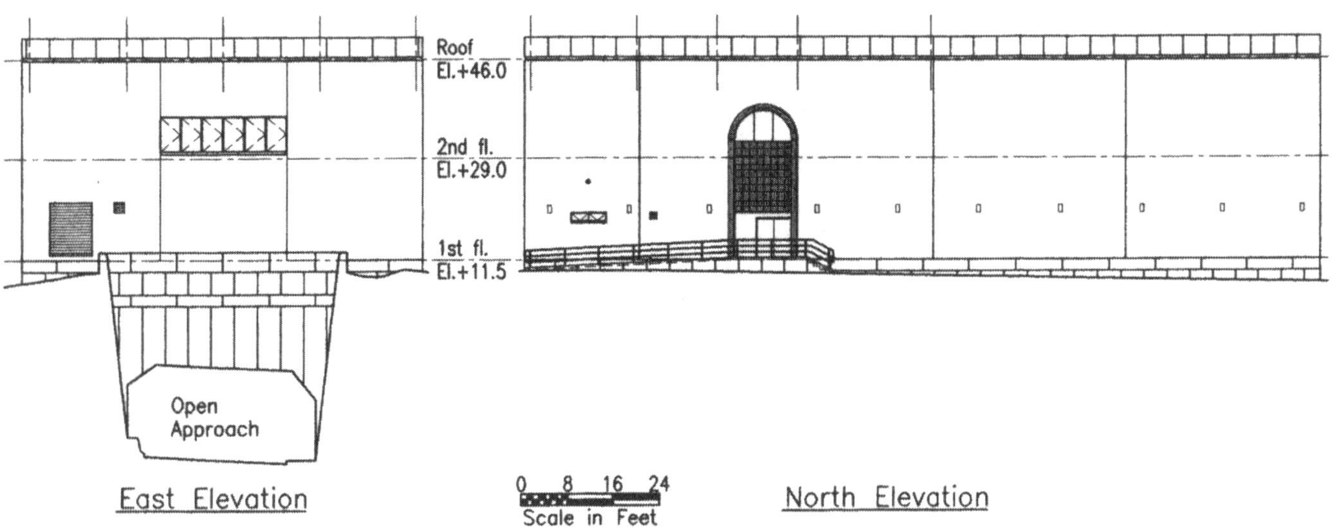

Fig. 26-7. Elevations, brick masonry on steel and concrete framing, Second Downtown Elizabeth River Tunnel.

Table 26-2. Requirements for Garage and Service Functions

Facility	Requirements
Emergency Vehicle Garage	Multiple vehicle bays Crew lunchroom with kitchenette (can double as training room Lockers and toilets for men and women Operations supervisor's office (for large operations) First-aid room (in remote locations)
Service Vehicle Garage and Maintenance Facilities	Wash bay Grease pit bay(s) Body shop/workshop bay(s) Auto parts storage and tool crib Foreman's office(s) Toilets and lockers for men and women Janitor's closet Multi-use training/lunch room
Miscellaneous Facilities	Maintenance shops Carpentry shop Electrical shop with storage Sign shop Maintenance of way with storage Spare parts/consumables warehouse with clerk station Fuel pump(s) Salt storage Personnel facilities Multi-use training/lunch room Maintenance foreman's office(s) Assistant Maintenance superintendent's office Changing rooms, lockers, and toilets for men and women Janitor's closet Uniform storage
Administrative Facilities[a]	Executive Reception area Executive director's office Executive secretary's area Deputy's office General manager's office Toll facility supervisor's office Maintenance supervisor's office Secretary's area Conference room Toilets for men and women Kitchenette Personnel Personnel manager's office Benefits administrator's office Labor relations office Interview room Secretary's area with file storage Purchasing Purchasing manager's office Secretary's area with file storage Computer services Computer room Computer Supervisor's office Technician's office(s) General office/accounting General office personnel work area Accountant's office Comptroller's office Secretary's area File room Vault Audit room (separate from general office area) Employee lunchroom with kitchenette/ vending machines Men's and women's toilets Mailroom Supply room

[a] Must be handicap accessible.

Table 26-3. Ancillary Space Requirements for Underground Rail Transit Stations

Functional Element	Dimensions (width × length × height), in feet	Proximity/ Adjacency
Power		
Traction Power Substation	50 × 80 × 14	Mezzanine level
Traction Power Fan Room	25 × 40 × 14	Substation
Incoming Service Room	40 × 40 × 14	Substation
Auxiliary Power (at each end)	27 × 44 × 14	
Gap Breaker Station	16 × 22 × 12	
Battery Room at each end	11 × 17 × 12	Auxiliary power
Cable Room	—[a]	Platform level
Communications/Control		
Train Control/ Communications	45 × 54 × 14	Mezzanine level
Mechanical Room	15 × 16 × 22	Train control/ communications
Battery Room	15 × 20 × 14	Train control/ communications
HVAC and Exhaust		
Emergency Fan Rooms		
Platform Level (at each end)	45 × 55 × 14	
Mezzanine Level (at each end)	18 × 80 × 14	
Air Supply Unit Room		
Mezzanine Level (at each end--includes FAI plenum)	44 × 48 × 14	Emergency Fan Room
Chiller Room	30 × 44 × 14	Air supply unit room (one side only)
Cooling Tower	—[b]	Entry portal area
Under-Platform Exhaust Room	15 × 32 × 14	
Under-Platform Exhaust Plenum	42 sq ft/track	
Smoke Exhaust Room		Under air supply or platform exhaust room
Station Operations and Maintenance		
Staff Security Room	8 × 12 × 8	Mezzanine level
Toilet	6-6 × 8 × 8	Mezzanine level
Trash Room	8 × 10 × 8	Mezzanine level
Custodial Room	8-6 × 16 × 8	
Custodial Closet	3-6 × 8 × 8	
Emergency Equipment Room	4-6 × 8 × 8	Platform level
Electrical Room (at each end)	8 × 10 × 8	Platform level
Tunnel Valve Room (at each end)	6 × 18 × 10	Platform level
Sprinkler Valve Room (at each end)	8 × 14 × 10	Mezzanine level
Ejector Room	8 × 12 × 11	
Sump Pump	10 × 10 × 8	
Elevator Equipment	8 × 17 × 8	
Emergency Stairs to Surface	Specified by codes	Platform and mezzanine levels (both ends)
Circulation Corridors	Vary	Personnel and equipment access (all levels, both ends)
Storage Rooms	As available	Mezzanine and platform

[a] Use space under emergency stairs
[b] Size varies depending on cooling load requirements; located away from underground structure

UNDERGROUND RAIL TRANSIT STATIONS

Many functional areas are required to support underground operations of rail transit stations on behalf of transit operations as well as patrons. These spaces are generally located at the ends of platforms at the platform and mezzanine levels in cut-and-cover stations. To reduce the overall length of cut-and-cover construction, some ancillary rooms can be located away from the trainway; that is, they are incorporated as part of the station entry portal or as part of joint development in below-grade space.

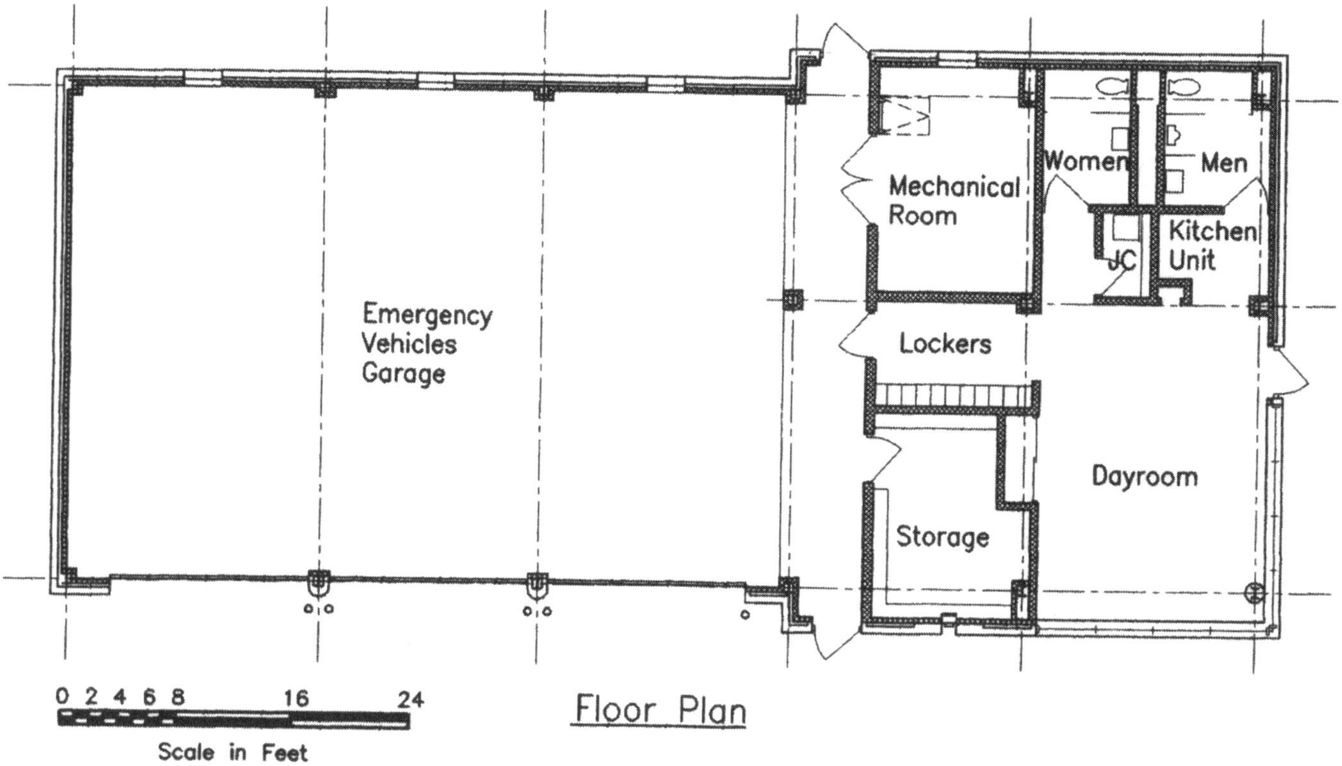

Fig. 26-8. Freestanding emergency vehicle garage, Second Downtown Elizabeth River Tunnel.

In mined stations, ancillary space can be located within the cavern beyond the ends of the platforms and in adjacent caverns with access provided by narrow passageways perpendicular to the trainway itself.

Special consideration must be given to vertical shaft and surface penetration requirements in streets, sidewalks, and private property for air intake and exhaust, equipment replacement, emergency egress, and utility connections. Space for cooling towers must also be provided if air conditioning is required for major public areas.

Specific Room Requirements

The ancillary facilities can be organized into the following functional categories for normal as well as emergency operations:

- Power for train operations and station functions conditions
- HVAC and exhaust for public spaces and ancillary spaces
- Communications and control (trains, patrons, equipment/installations)
- Station operations and maintenance

Specific space requirements (length, width, and height) and relationships between functional elements horizontally and vertically will of course vary depending on train lengths, station depth, mezzanine configurations, available space associated with entry portals, and the type of construction. Table 26-3 represents the full range of ancillary rooms required for an underground heavy rail station. The functional elements have been grouped to indicate the preferred proximity, and adjacency requirements have been indicated in the table.

Additional rooms are required for end-of-line station turnaround and storage facilities and crossovers or pocket tracks built as part of the station complex. It should be noted that no provisions have been made for concessions/shops or readily accessible public restrooms as part of this space tabulation.

Technical Design Issues Checklist

The following list indicates the range of specific design issues that must be addressed during the course of designing ancillary facilities for underground rail transit stations.

- Functional Requirements
- Capacities (train lengths and types, fans, equipment, etc.)
- Adjacency/Proximity (platform, mezzanine, public, operators)
- Room Sizes and Ceiling Heights (related to construction type, station depth, and configuration of entries and mezzanines)

- Surface Penetrations (intake, exhaust, access/egress)
- Circulation/Flow (patrons, operators, emergency personnel)
- Equipment Installation, Maintenance, and Replacement
- Code Compliance (Americans with Disabilities Act, fire life safety, etc.)
- Communications Systems (phones, alarms, sensors)
- Ventilation, Exhaust, and AC Requirements
- Noise and Vibration
- Power Requirements
- Architectural Finishes (floors, walls, ceilings, special support systems, doors, hardware, signing/graphics, etc.)
- Special Requirements (floor drains, eye wash shower, etc.)

In conclusion, underground stations require an extensive array of ancillary support areas not unlike those associated with high-rise office buildings. In fact, that overall square footage of nonpublic space can approach 50% of public space in long-platform stations (greater than 450 ft), and 25% in short-platform stations.

27 | Tunnel Rehabilitation

Henry A. Russell
Principal Professional Associate, Parsons, Brinckerhoff, Quade & Douglas, Inc.

Over the last few years, the concern for infrastructure has created a new interest in the rehabilitation of existing tunnel facilities. Many of the tunnels in the United States and abroad were constructed prior to World War II and are approaching their design-life expectations. However, due to the high cost of replacing these facilities, many tunnel systems are being rehabilitated to extend their useful lives well into the next century.

Rehabilitation of tunnels is rapidly becoming one of the most significant aspects of our profession. The transport of water, vehicles, commuters, and other utility services are necessary for the maintenance and continued development of our metropolitan areas. The cost of developing new tunnel systems is often prohibitive. Therefore, the rehabilitation of existing systems has become a necessity for the future operation of metropolitan areas.

This chapter is intended to illustrate the methods for inspecting lined tunnels while developing a systematic procedure for recording their existing condition. The repairs are based on state-of-the-art products that provide suitable reconstruction while limiting interference with the tunnel operation.

TUNNEL REHABILITATION INSPECTION METHODS

Background Information and Design Data

A key element in the rehabilitation of a tunnel system is a working knowledge of the original construction and any subsequent modifications to the tunnel system. This working knowledge is only obtained through exhaustive review of all existing plans and specifications. The information obtained from the original plans and subsequent modifications to the tunnel system provides the basis for the interpretation of the inspection data as it pertains to the observed tunnel defects.

These plans and specifications are usually obtained from the facilities engineering department. The original plans are often supplemented by photographs, narratives, and other memorabilia from local historical societies. All information obtained should be cataloged in a systematic method allowing rapid retrieval.

Aside from the historical data available on the subject tunnel system, it is also necessary to collect all available design calculations, as-built plans, descriptions of any tunnel instrumentation, and any materials testing information that was done during construction and in subsequent years.

Inspection Purpose

The purpose of a tunnel condition survey is to observe and document current physical conditions and to establish basic criteria for determining the quality and performance of the tunnel system. The condition survey also provides a data baseline for the long-term observation of the tunnel and the prevention of catastrophic structural events that would cause the tunnel system to be shut down for long periods of time and cause unnecessary financial burden to the users of the system.

Tunnel condition surveys are usually conducted during periods of inactivity in the use of the tunnel system. Transit and vehicular tunnels are generally inspected during nonrevenue hours, those hours when the system is shut down for maintenance, or nonpeak traffic times. Fluid transit tunnels are inspected by two distinctly different means: use of remote operated vehicles (ROVs) and/or physical inspection by survey crews during periods of shutdown.

The use of ROVs usually occurs in small-diameter tunnels, say less than 60 in. (1.5 m). The ROV is controlled from the surface and is lowered into the tunnel through existing access points. The ROV advances through the tunnel under its own power; and by the use of video or other means, it generally documents the tunnel condition. The advantage of the use of ROV surveys is that they can be performed in very small tunnels where a physical inspection is difficult.

In fluid transfer tunnels, a survey utilizing an inspection team is difficult to schedule since the system must be shut down, drained, and made safe for the entrance of inspection personnel. The inspection of the tunnel is then performed in the same manner as for transit and vehicular tunnels.

Inspection Staffing and Equipment

The inspection of transit tunnels is best performed by a three-man technical team. The team's personnel are assigned quadrants or limits of the tunnel cross section for inspection, as shown in Figures 27-1–27-4. As the inspection progresses, each member is responsible for the inspection of his/her assigned area of the tunnel, with the extra person utilized as the recorder. It is not always necessary to assign a third member of the team to act as a recorder; however, it is a more efficient way to inspect more tunnel per actual hours of inspection. Unless otherwise required, all of the elements of the tunnel inspection should be inspected simultaneously and coded as shown in Table 27-1. This simultaneous inspection allows for rapid inspection and efficient use of the limited hours for the tunnel inspection. It is helpful, but not required, that plans and cross sections of the tunnel be included in the information carried by the inspection personnel.

Equipment Requirements. The equipment required for the inspection of tunnel systems is in the following categories:

- Individual equipment
- Team equipment
- Safety equipment

The assignment of equipment is based on the usage and the individual who will maintain the equipment for the duration of the project. The designation of safety equipment is that which is solely used for the protection of the personnel and will be listed as part of the team equipment.

Individual Equipment.

Item:

- Flashlight with batteries
- Folding ruler, wood

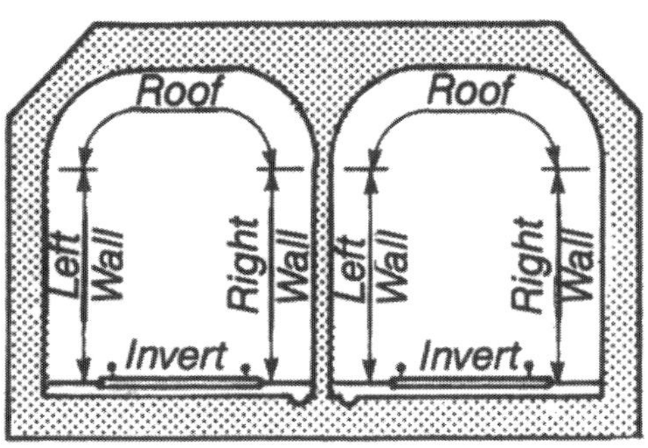

Fig. 27-1. Typical box tunnel, double cell.

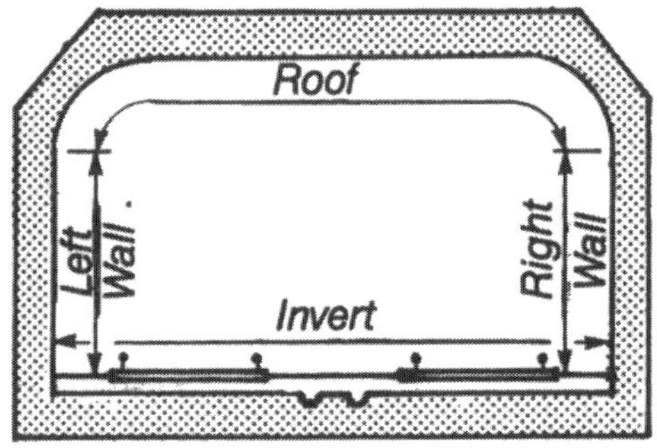

Fig. 27-2. Typical box tunnel, single cell.

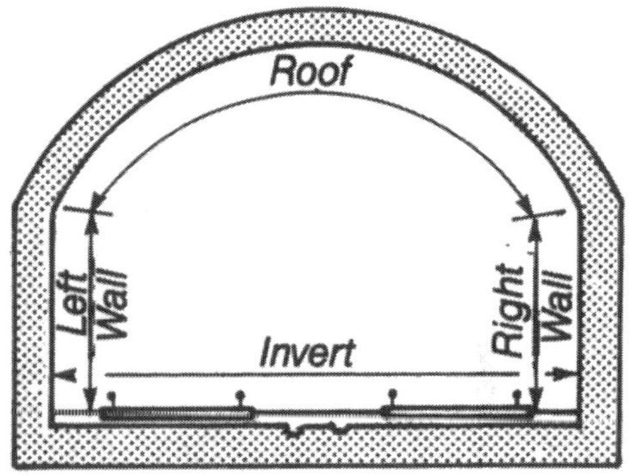

Fig. 27-3. Typical arch roof tunnel, single cell.

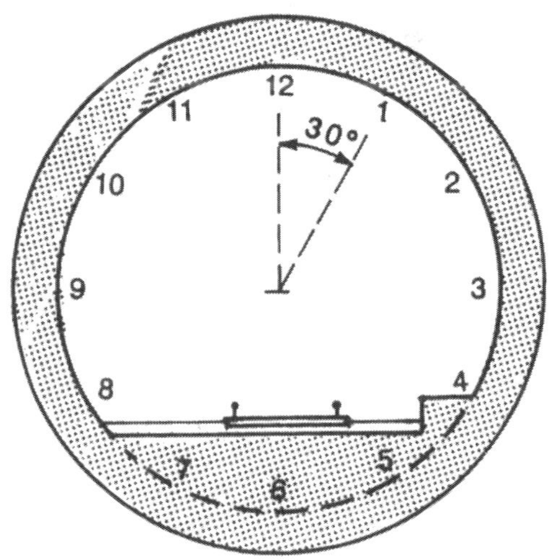

Fig. 27-4. Typical circular cell tunnel.

Table 27-1. Standard Identifier Parameters

Symbol	Identification
Concrete Spalls	
S-1	Concrete spalls less than 2"
S-2	Concrete spall to re-rod
S-3	Concrete spall behind re-rod
S-4	Special concrete spall
Concrete Delaminations	
D	Delamination
Tunnel Cracks	
C-1	Less than 1/8 inch (3 mm)
C-2	1/8 inch to 1/4 inch (3-6 mm)
C-3	1/4 inch to 1/2 inch (6-12 mm)
C-4	Greater than 1/2 inch (12 mm)
Tunnel Joints	
J-1	Joint less than 1/8 inch (3 mm)
J-2	Joint 1/8 inch to 1/4 inch (3-6 mm)
J-3	Joint 1/4 inch to 1/2 inch (6-12 mm)
J-4	Joint greater than 1/2 inch (12 mm)
J-5	Special joint -requires extensive notes
Reinforcing Steel	
R-1	Surface rust
R-2	Loss of section -%
R-3	Out of plane -H=Horz., V=Vert.
R-4	Broken
R-5	Buckled
R-6	Other -requires extensive notes
Framing Steel	
F-1	Surface rust
F-2	Loss of section -%
F-3	Out of plane -H=Horz., V=Vert.
F-4	Broken
F-5	Buckled
F-6	Other -requires extensive notes
Steel Liner Plate Segments	
P-1	Surface rust
P-2	Loss of section -%
P-3	Out of plane -H=Horz., V=Vert.
P-4	Broken
P-5	Buckled
P-6	Other -requires extensive notes; sketches
Steel Liner Plate Flanges	
FL-1	Surface rust
FL-2	Loss of section -%
FL-3	Out of plane -H=Horz., V=Vert.
FL-4	Broken
FL-5	Buckled
FL-6	Other -requires extensive notes
Precast Concrete Liner Segments	
SC-1	Concrete spalls less than 2" (5.0 CM)
SC-2	Concrete spall to re-rod
SC-3	Concrete spall behind re-rod
SC-4	Special concrete spall
SC-5	Other -requires extensive notes
SCD	Delamination
Bolt Connections	
B-1	Surface rust
B-2	Loss of section -%
B-3	Out of plane -H=Horz., V=Vert.
B-4	Broken
B-5	Buckled
B-6	Other -requires extensive notes
B-7	Bolt missing
Tunnel Brick	
BR-1	Loss of section, depth=in., (cm) (M^2) area=S.F.
BR-2	Loss of mortar -FT^2. (M^2)
BR-3	Other -requires extensive notes
Rock Bolts	
RB-1	Surface rust
RB-2	Loss of section -%
RB-3	Broken
RB-4	Buckled
RB-5	Loss of tension
RB-5	Other -requires extensive notes
Tunnel Moisture	
M	Moist
PM	Past Moisture
GS	Glistening Surface
F	Flowing
D	Dry

Notes:
1. For steel sets in rock tunnels, use same terminology as for framing steel.
2. Use terminology from concrete section.

- Safety vest
- Safety glasses
- 20 oz. (0.56 kg) masonry hammer
- Field book
- Hard hat (cap)
- VHF portable radio
- Radio battery charger
- Specialized breathing apparatus (if required)
- Individual air quality monitoring equipment (as required)

Team Equipment.

Item:

- Fiberglass tape, 100 ft (33.48 m)
- 6V hand lantern with batteries
- Schmidt hammer
- Wire brush
- Calipers
- Lumber crayons
- Felt-tip markers
- Marking paint
- Tool bag
- Gloves, dielectric (if required)
- 35 mm camera
- 35 mm film (slide and print)
- First aid kit
- Electronic data collector (if required)
- Air quality monitoring equipment

In addition to the basic equipment listed above, additional specialized equipment may be required by the project as the inspection progresses. This generally includes survey equipment, special testing equipment, and other ancillary equipment required to perform specialized functions. The sources and types of specialty equipment will vary from project to project and must be selected on need and specific project requirements.

Project Safety. The most important elements in the inspection of tunnels and other underground facilities are the safety and protection of inspection personnel. For the inspection and testing personnel to perform their tasks effectively, all personnel must be familiar with a standardized safety program and the implementation of the project safety plan. The project safety plan must contain procedures for protecting personnel without affecting safe operations of the existing tunnel system. The project safety plan should be coordinated with the owner's safety department and, where applicable, should include instruction from that department.

The project safety plan must contain the following elements:

- Clearly defined areas of responsibility
- Description of required safety equipment

- Operational plan for safety training
- Standardized hand, flag, and lantern signals (as required)
- Procedures for the locations of flagmen or special safety personnel
- Evacuation procedures
- Emergency procedures

Project Safety Responsibility. Safety is the responsibility of all personnel, regardless of position on the project. Individual awareness is the single most important element of project safety, but it is also necessary to assign project safety responsibility as follows:

- Project manager: Develop project safety plan
- Deputy project manager: Provide safety training
- Senior inspectors: Implement safety plan and coordinate with tunnel operations
- Staff inspectors: Locate personnel properly and implement safety plan
- All staff members: Have working knowledge of project safety plan

Safety Equipment. The safety equipment required for the inspection of tunnels must

- Provide suitable protection for personnel
- Notify and alert tunnel system operators of the presence of inspection personnel in the tunnel

The safety equipment required for the inspection of tunnels must be capable of providing suitable protection for personnel. It also serves to notify and alert the tunnel system operators of the presence of inspection personnel in the tunnel. To achieve this objective, the following equipment use is suggested for all personnel working in the tunnels:

- Hard hat
- Signaling device (horn or whistle)
- Flashlight (permissible)
- UHF radio (optional)
- Special flagging equipment (as required)
- Safety vest with reflective material (optional)
- Backup breathing equipment (as required)
- Explosive gas detectors (as required)
- Other specialized safety equipment to provide a safe working environment.

Inspection Procedures and Documentation

Standardized inspection parameters are necessary for the expeditious processing and evaluation of the observed data. Coding of the information is necessary for consistency of reporting and incorporation into the database system. This coding is helpful in assuring quality control by providing guidelines for inspection personnel and standardizing visual observations. Standard coding for cataloguing tunnel defects has been developed and is given in Table 27-1.

Preliminary Tunnel Inspection. A key element in the effective inspection of the tunnel structure is the preliminary inspection of the tunnel by senior personnel. Typically, this inspection team should consist of a structural engineer, geotechnical engineer, materials specialist, and any other special discipline required by the scope of work. The purpose of this inspection is to observe the condition of the tunnel and its subsystems, make recommendations for prioritization of elements to be inspected, and provide guidance for the development of the inspection schedule.

Based on the preliminary inspection of the tunnel and the scope of work, the selection of inspection parameters are chosen. Particular emphasis should be given to determining the presence of special or unique structures requiring the addition of special inspection parameters to the program.

Survey Control

All condition surveys require a definitive baseline for location (survey) purposes. This requirement is the same for the inspection of tunnels. Generally, most tunnel systems have an established survey baseline. These baseline systems are usually well defined and are in active use by the operator of the tunnel system.

The tunnel inspection must be tied into the existing baseline stationing system in use by the owner. The use of the established survey baseline

- Allows the inspection data to be used for long-term monitoring of the tunnel structure by the owner's engineering/maintenance staff
- Allows rapid start-up of inspection teams
- Reduces project costs and confusion

In addition to the location along the alignment, it is also necessary to delineate the location of the tunnel defect in relation to its position in the structure. To accomplish this location in the tunnel, the limits of the walls, roof, and invert must be delineated for conformity. The delineation of the limits of walls, roof, and invert are performed as shown in Figures 27-1–27-3. Circular tunnels are divided into 30° segments clockwise from the high point of the tunnel crown as shown in Figure 27-4.

Calibration of Inspection Crews

It is necessary to standardize observations of the inspection teams. This standardization/calibration of the crews is to be performed at the outset of the field inspection. The calibration is performed by having a "dry run" through a typical 100-ft (30.48-m) section of the tunnel. This "dry run" is to be an independent inspection of the tunnel section by an experienced inspector and by the inspection team(s) who will perform the actual collection of data. The two inspections will then be compared for accuracy, and additional training will be performed in areas of discrepancy. Additional training, if required by the field personnel, is to be performed at this time, and an additional check section for inspection calibration may be required. This method of calibrating the inspection team(s) has proven to be successful in assuring similar observations among all personnel involved in the collection of field data.

Project Quality Assurance

The purpose of the quality control plan is to ensure the completeness, clarity, accuracy, and documentation of the inspection data. The development of standard inspection parameters as described earlier and the associated calibration of inspection crews eliminates many of the typical errors and omissions that occur when the work is performed by numerous separate teams.

Effective quality control of a tunnel inspection project requires the development of specific procedures for control of the inspection. These procedures are identified as follows:

- Inspection and identification of tunnel defects
- Tabulation and management of inspection data
- Accurate reporting of inspection results
- Development of contract documents (as required)

Routine Inspection

Specific equipment needs for the inspection of the tunnel vary with the type of tunnel construction and the material composition of that section. A typical inspection is performed in 100-ft (30.48-m) increments for each cell of the tunnel. A 100-ft fiberglass tape is stretched along the tunnel wall with 0 and the 100-ft (30.48-m) marks corresponding to the tunnel stationing. This tape is now used to provide horizontal location of the tunnel defect being cataloged. The location in the tunnel cross section is identified as described earlier. At this time, a description of the tunnel defect is recorded. Utilizing the standard parameters for the identification of defects as outlined in Table 27-1, the following tabular format is used to describe the major defect with other elements included as subdefects:

Example: A concrete spall, located in the crown at station 25 + 31 of the tunnel, with an area of 10 ft^2 at an average depth of 3 in.:

Station	Location	Defect Identification	Remarks
25 = 31	Crown	S-1	10 ft^2, 3 in. (0.92 m^2) (7.62 cm)

The remarks column is used to identify any important fact that must be included in the description.

The inspection/documentation of the various elements of the tunnel system must be performed simultaneously. This inspection of all the elements of the tunnel must be performed in this manner since these elements of the tunnel function in conjunction with each other, and an evaluation of the tunnel system requires analysis as an all inclusive unit rather than a series of small independent elements.

Special Structural Elements

Many tunnels, particularly older rail tunnels, have special/unique structural steel elements that do not easily fall into the standardized inspection format. These special structures are usually unique to a specific transit system or area and require special inspection procedures. These procedures must be tailored to that specific area and, where possible, be incorporated into the database system.

Special Structures

Special structures consist of those types of underground structures that make up the tunnel system; these structures include entrance structures, vent shafts, electromechanical facilities, and other ancillary structures. These structures, while being a part of the tunnel system, are often constructed by different means than that of the mainline tunnels and, therefore, require a different approach to their inspection. In general, though, these structures may be inspected utilizing the same standardized parameters and procedures.

The major element of the inspection process in which these structures differ from the mainline tunnels is in the location of the defects. Since most of these structures are rooms/chambers that are attached to the tunnel system, it is necessary to establish independent survey control for each structure to be inspected. This independent survey for the special structure must be tied into the tunnel baseline survey at one location, typically where the structure meets the mainline tunnel. Once the special room/chamber is identified, an individual condition survey is performed at a later date.

Condition surveys of these ancillary structures are best performed using existing plans and developing elevations of each wall and plans of the floors and roof. These special drawings are used in the field to locate the structural defects. Based on these drawings, specialized location identifiers are developed and used for the data management. Depending on the type of construction, the standardized inspection symbols for defects are used for the inspection.

Data Management Procedures. The procedure for the management of the inspection data requires the development of a systematic approach for the flow of information from the field to the office and back to the field for verification. This transfer of inspection data within the management system is essential to efficient documentation of the data and the required quality assurance necessary for the inspection.

Field Notes. There are three types of field notes required for effective inspection of transit tunnels. They are:

- General notes in field books
- Documentation of defects on field data forms or electronic data collectors (Figure 27-5)
- Documentation of defects by photographs

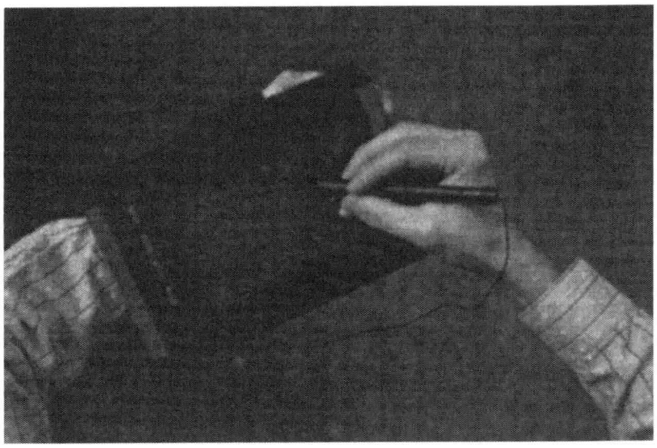

Fig. 27-5. Typical electronic data collector.

General Notes in Field Book. All general notes should be kept in a field book. These general notes consist of a chronology of events. The information contained in the field book should include: safety issues and discussions with contractors, operations personnel, and other interested parties (Figure 27-6).

Field Data Forms. Field data forms (Figure 27-7) are developed for each project and reflect the information required for that project. In general, the data forms are a project standard for the tabulation of the data obtained from the inspection. This information is incorporated into the inspection database.

The use of tabular format for the documentation of inspection data provides excellent quality control by establishing a standard for the visual identification of the various tunnel elements. Over the last few years, computerized field data collectors have been introduced to replace field data forms. These electronic data collectors are basically handheld computers that allow direct input of field data into a database in the field at the time of inspection. These data collectors transfer the data via modem or disk into mainframe computers in the office for data reduction.

Photographic Documentation. The documentation of tunnel defects is best supplemented by the use of 35 mm photographs. These photographs should be taken of typical and atypical conditions, as well as documentation for special or unique conditions. All photographs should exhibit the date, time, location, photographer, and a general description of the item photographed. All photographs must be cataloged and stored in a systematic manner for future recall and use.

Special Testing

In locations where structural analysis is required, detailed information as to the makeup of the construction material or presence of certain elements in the structural unit will require special testing. Special testing comprises those tests used to determine certain characteristics of the material that makes up the tunnel system. Table 27-2 is a listing of typical special tests, which are often used in the determination of tunnel condition.

Geophysical Nondestructive Testing. Numerous nondestructive methods have been applied to tunnel testing

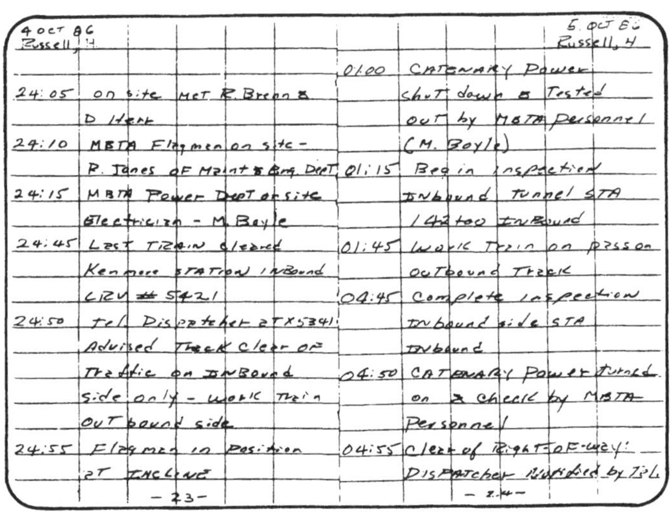

4 OCT 86
Russell, H

24:05 on site Met R. Breen &
D Hert
24:10 MBTA Flagmen on site -
R. Jones of Maint & Eng. Dept
24:15 MBTA Power Dept on site
Electrician - M. Boyle
24:45 Last Train cleared
Kenmore Station Inbound
CRV # 5421
24:50 tell Dispatcher at X5341
Advised Track clear of
Traffic on Inbound
side only - work Train
Out bound side
24:55 Flagmen in Position
at Incline

- 23 -

5 OCT 86
Russell, H

01:00 Catenary Power
shut down & Tested
out by MBTA Personnel
(M. Boyle)
01:15 Begin inspection
Inbound Tunnel STA
142+00 Inbound
01:45 Work Train on pass on
Outbound Track
04:45 Complete inspection
Inbound side STA
Inbound
04:50 Catenary Power turned
on & check by MBTA
Personnel
04:55 Clear of Right-of-way:
Dispatcher Notified by Tel.

- 24 -

Fig. 27-6. Format for general notes.

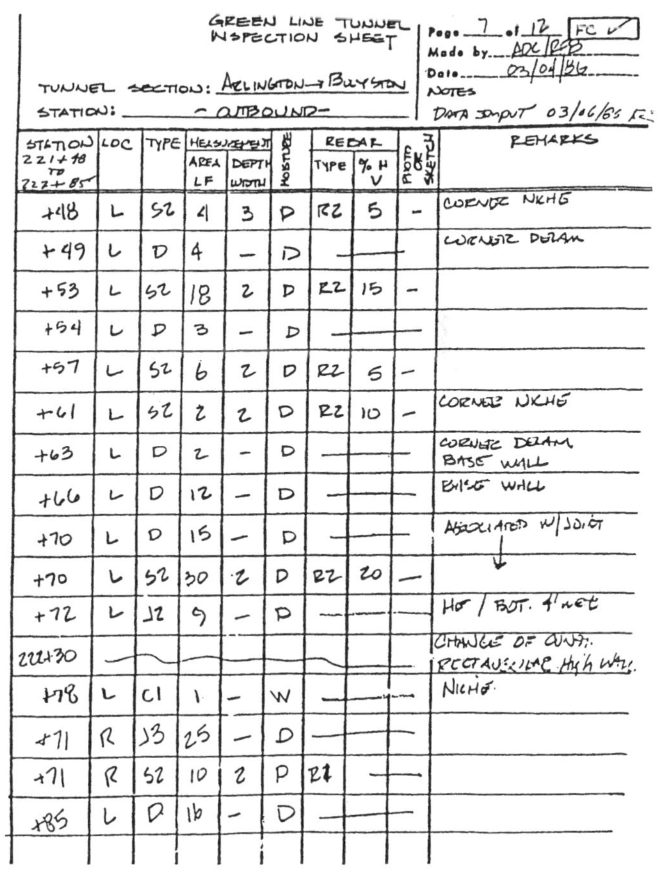

GREEN LINE TUNNEL INSPECTION SHEET

TUNNEL SECTION: ARLINGTON → BOYLSTON
STATION: - OUTBOUND -

Page 7 of 12 FC ✓
Made by ADC/RFB
Date 03/04/86
NOTES
DATA INPUT 03/06/86

STATION 221+48 TO 222+85	LOC	TYPE	MEASUREMENT AREA LF	MEASUREMENT DEPTH WIDTH	MOISTURE	REBAR TYPE	REBAR % H V	PHOTO OR SKETCH	REMARKS
+48	L	S2	4	3	D	R2	5	-	CORNER NICHE
+49	L	D	4	–	D				CORNER DRAIN
+53	L	S2	18	2	D	R2	15	-	
+54	L	D	3	–	D				
+57	L	S2	6	2	D	R2	5	–	
+61	L	S2	2	2	D	R2	10	–	CORNER NICHE
+63	L	D	2	-	D				CORNER DRAIN BASE WALL
+66	L	D	12	–	D				BASE WALL
+70	L	D	15	–	D				ASSOCIATED W/JOINT ↓
+70	L	S2	30	2	D	R2	20	–	
+72	L	J2	9	–	D				H / BOT. 4' wet
222+30									CHANGE OF CONST.: RECTANGULAR High Wall
+78	L	C1	1	–	W				NICHE
+71	R	J3	25	–	D				
+71	R	S2	10	2	D	R2			
+85	L	D	16	–	D				

Fig. 27-7. Field data form.

Table 27-2. Special Tests

In Situ Concrete Strength			
Type	**Test Method**	**Property Measured**	**Region Tested**
Semi-Destructive	Pull Out Test	Indirect Shear/ Tensile Strength	Surface Zone
	Break-Off Test	Flexural Strength	Surface Zone
	Windsor Probe	Penetration Resistance	Surface Zone
	Tescon Probe	Stress-Strain Relationship	Internal Zone
	Cores	Strength	Internal Zone
	Carbonization	Depth CO_2	Surface Zone
	Petrographic Analysis	Composition	Internal Zone
Nondestructive	Rebound Hammer	Resilience	Internal Zone
	Ground Penetrating Radar	General Condition	Internal Zone
	Refraction Survey	Micro Cracking	Internal Zone
	Resistivity	Steel Corrosion	Internal Zone
	Ultrasonic	Elastic Modulus	Surface and
	Pulse Velocity	(If the density and Poisson's Ratio are known)	Internal Zone

Steel Tunnel Elements		
Type	**Test Method**	**Property Measured**
Destructive	ASTM E8	Strength
Destructive	ASTM E10	Hardness
Destructive	ASTM E18	Hardness
Destructive	ASTM A370	Strength of Steel Element
Destructive	ASTM E390	Quality of Welds
Nondestructive	ASTM A325	Torque for Bolt Connections
Nondestructive	ASTM 490	Bolt Connections

(Figure 27-8). These techniques are based on the measurement of different material characteristics such as thermal conductivity (thermography), electrical conductivity and dielectric constant (galvanic electrical, electromagnetic, and ground penetrating radar), the velocity of a stress wave through the material (sonic/ultrasonic). The thermal and electrical properties of concrete (and other materials) are largely controlled by moisture content in the host material. Interpretation of the data relative to a change in moisture content is related to the host material but in a secondary fashion that makes interpretation somewhat subjective. For example, moisture-laden cracks will be distinguished by ground penetrating radar (GPR) because they reflect the electromagnetic wave, whereas dry cracks will not. Cracks, however, weaken the concrete and reduce its modulus value, which is measurable by the sonic/ultrasonic methods whether the cracks are wet or dry; the sonic/ultrasonic changes are directly related to the characteristics of the concrete itself and quantify the changes in the material. This is important because the identification of cracks or delamination alone by GPR or thermography does not evaluate their effect on the concrete.

Nondestructive sonic/ultrasonic wave measurements provide a comprehensive method to evaluate tunnel liner, liner boundary, and behind-liner conditions. The sonic/ultrasonic methods directly determine the deformation moduli values (Young's shear, bulk) and unconfined compressive strength. An important aspect of the sonic/ultrasonic measurements is that while they are affected by macro-cracking, they are also affected by micro-cracking well below the visual level. They are a very early indicator of future macro-cracking, delamination, etc., and in this sense are a valuable management tool for repairs. This method identifies distress conditions of the concrete: cracking, voiding, delamination, etc. GPR can be used to distinguish particular conditions such as the presence of metal rebar, water fill voids, water penetration into the liner, moisture concentration in cracks, and fracture zones. The combination of sonic/ultrasonic methods and GPR provides a complete suite of nondestructive data to evaluate the following critical tunnel parameters:

- Tunnel liner strength
- Identification and mapping of weak, deteriorated, cracked, or delaminated concrete liner areas (sonic/ultrasonic)

Test Method	Cracks		Chemical Alteration	Delamination	Rebar	Moduli Values	Strength
	Macro	Micro					
Sonic/ultrasonic compressional, shear velocity & attenuation	■	■	■	■		■	■
Sonic/Ultrasonic Reflection/Resonanaces				■		□	■
Radar		□	□	□	■		
Thermography		□	□	□			

■ Best use of method

□ Dependent on favorable conditions

Fig. 27-8. Nondestructive testing methods for concrete.

- Identification of metal structural members and rebar (GPR)
- Identification of thinner or thicker tunnel liner areas (sonic/ultrasonic, GPR)
- Identification and mapping of voiding (sonic/ultrasonic, GPR)
- Identification and mapping of water infiltration into the liner and water-filled voiding and cracking behind the liner (GPR, sonic/ultrasonic)

Sonic/Ultrasonic Measurements. A series of highly sensitive sensors (generally accelerometer type) are placed on the surface of the tunnel liner to measure the transit time of a stress wave induced by an impulsive energy source, which may be a pulsed electromagnetic transducer, piezoceramic element, or a high-velocity projectile. Energy transmitted through the concrete is received by the sensors (see Figure 27-9). The detected signals are used to extract several simultaneous measurements of the transmitted wave, which yield important tunnel liner characteristics:

- Velocity measurements of the full thickness of the liner to determine its elastic deformation moduli values and a reasonable determination of its strength
- Reflections (pulse-echo, resonance) from various conditions in the concrete to distinguish such distress conditions as delaminations, voids, joints, cracks, etc., equivalent to the so-called pulse-echo testing of concrete.

The combination of the sonic/ultrasonic reflections and velocities detect the distress conditions of concrete and quantify their effect as a degradation of moduli values and loss of strength.

Velocity Measurements. The compressional and shear wave velocity of the tunnel liner materials are determined by observing the arrival times of direct and refracted waves over a known distance; if the thickness is known, reflected waves are also used. This information is used to determine the elastic moduli values (Young's, bulk, and shear) as well as Poisson's Ratio and the approximate unconfined compressive strength (Figure 27-10). Cracking, voiding, delamination, and deterioration conditions weaken tunnel liner materials and are associated with areas of low strength (Figure 27-10).

Reflected (Pulse-Echo). When a seismic wave impinges on a boundary with contrasting physical parameters, (velocity and density) part of the energy is reflected (see Figure 27-10). The resultant reflection can be measured as a single event, a distinct signal from the back of the liner (pulse-echo), or as a combination of signals representing a resonant frequency created by the continuous reflection "trapped" in the concrete layer. This is similar to, but out of the range of hearing of, the distinct audible signal created by a "chain drag" or hammer impact on a piece of delaminated concrete.

Ground Penetrating Radar. The GPR is similar to airborne radar in that it is based on the reflection of an electromagnetic wave from a "target." The difference in penetrating the ground is that the earth is not as friendly a media as air, where the velocity remains fairly constant. The GPR depends on the electrical conductivity and the dielectric constant of the material. Since in most earth materials the conductivity and dielectric constant is controlled by moisture (conductivity through these materials is almost entirely

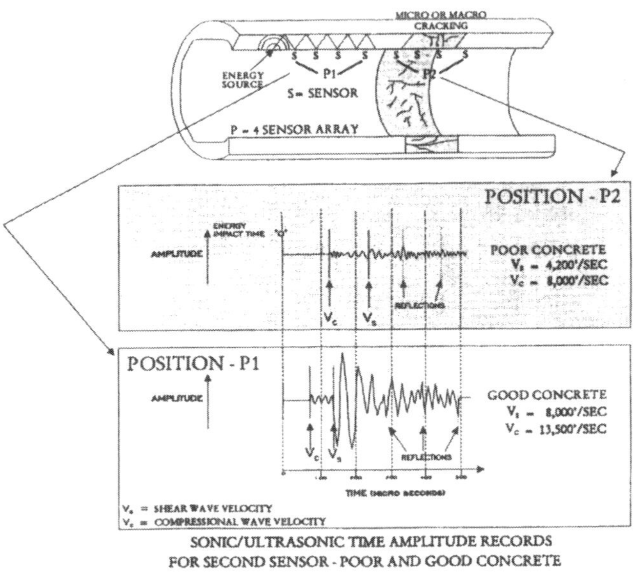

Fig. 27-9. Sonic/ultrasonic tunnel and pipe line testing.

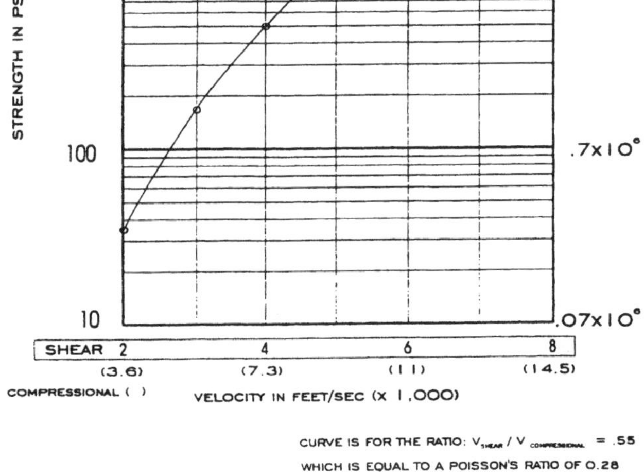

Fig. 27-10. Strength of concrete versus velocity.

ionic), the GPR measurements are highly influenced by moisture content; the electromagnetic wave velocity can change substantially due to the presence of water. The dielectric constant of water is 80, while the dielectric constants of concrete and many earth materials are 10 or less. Since the velocity of the electromagnetic wave is inversely proportional to square root of the dielectric constant, moisture can cause GPR velocity to change by a factor of 3.

The GPR is best used to augment the sonic/ultrasonic measurements for detection of rebar or other metal, the presence of moisture and consequently the loss of integrity of a waterproof barrier, and tunnel boundary conditions (moisture entrapment, drainage, etc.)

Presentation of Data

The inspection on tunnel systems is by nature a complex, extensive collection of data. To systematically catalog and expand the data, one must consider the use of a personal computer and a comprehensive spreadsheet program to sort and manipulate the information collected in a timely manner and produce easily understood information. Recently, many inspection programs in the United States have used this format to develop contract documents with the assistance of computer-aided drafting (CAD).

The most commonly used programs allow for the sorting, tabulation, and summary of information by category, subcategory, and element. This sorting may be performed by survey station, standardized identifier, tunnel location, and by time. These database programs provide a systematic method for comparing current surveys with future surveys and, therefore, allow for rapid comparisons of tunnel performance.

During the performance of the survey, the use of computers and the spreadsheet concept have allowed for extremely fast surveys and for the rapid field check of the data for a quality control check and verification.

Figures 27-11 and 27-12 illustrates a typical printout from a database program for tunnel inspection illustrating sorted data and a complete printout of all of the data for a tunnel section. Figure 27-13 illustrates a compilation of data as developed by CAD showing the tunnel cross section, stationing, and defect summary.

Evaluation of Data

The culmination of a tunnel condition survey is the determination of the status of the tunnel system and its associated subsystems. This determination is based on the review of the original design and the performance of that design. The condition survey also will provide valuable information on what repairs and products were suited for use in the tunnel environment.

The inspection data must be preserved in a format for future use and comparisons. This is best done in the format of a summary data report, which should be prepared in two volumes. The first should be a factual document describing the tunnel system's history and the observed conditions of the tunnel inspection, including any special tests performed and their results. This volume should also include a description of any software used in sorting the inspection data, as well as copies of all magnetic media files. The second volume is an interpretive report that discusses the cause and effect of any tunnel defects and recommends any special repairs or special testing. The summary report should also indicate a schedule for periodic inspection of the tunnel system and develop estimates for repairs and future inspections.

TUNNEL REHABILITATION REPAIRS

A successful rehabilitation project culminates in the economical implementation of repairs to the tunnel system. These repairs must be durable, easy to perform, and capable of being implemented rapidly during nonoperating hours. In addition, they must not pose a safety hazard to the operations. This section presents various techniques that have been used successfully in rehabilitation projects at numerous locations throughout the United States. These repair descriptions are from projects of a similar nature. However, no two sites are exactly the same, and attention to specific site requirements and conditions may reveal a repair technique to be unsuitable.

The following sections will address the various type of repair procedures for concrete restoration, control of groundwater, repairs to structural steel elements, and other specialized repair techniques required for concrete-lined tunnels.

CONCRETE REPAIR

Concrete repairs to existing tunnels depend on type of construction, tunnel operations, and severity of the tunnel defects. The elements that most directly affect the selection of a repair technique are

- Strength of repair material, generally dictated by the strength of the original construction material.
- Durability of the repair material, a function of the strength and the long-term suitability of the repair.
- Environmental setting of the repair location, of prime importance due to the varied environments that exist within the tunnel setting. (The presence of water, atmospheric/temperature changes at vent shaft, station, and portal locations, and the chemical composition of the groundwater and soil that surround the tunnel have a strong influence on the selection of the proper repair product.)
- The allowable time frame for the implementation of the repair, probably the most individualized element in the repair selection process, and dependent on the owner's operations (including the hours of shutdown for repair construction and the maintenance of clearance envelopes for revenue service during the rehabilitation construction period).

Page No. 1
08/18/86

GREEN LINE TUNNEL INSPECTION
HIGHLAND BRANCH
INBOUND TRACK

STATION	LOCA	TYP	AREA SF	LENGTH LF	DPH IN	M	RE BAR	PCT	PHOTO	REMARKS
07+17	L	C1	0.0	6.0	0.0	D				PAST MOISTURE
07+19	L	D	4.0	0.0	0.0	D				
07+19	L	C2	0.0	17.0	0.0	D				SHOWS SIGNS OF PAST MOISTURE
07+80	L	C1	0.0	14.0	0.0	F			B8,9,10	GROUTED JOINT /FAILED /RUSTY
08+53	L-C	C1	0.0	30.0	0.0	W				WET CRACK, HEAVY RUST, FAILED PATCH
08+58	R	C1	0.0	3.0	0.0	D				
08+82	L	C1	0.0	6.0	0.0	D				PAST MOISTURE
08+95	R	C1	0.0	3.0	0.0	D				
09+01	L-C	C1	0.0	30.0	0.0	W				PAST MOISTURE
09+41	L-C	C1	0.0	30.0	0.0	D			B11-B12	FLOWING 10 FEET UP ON LEFT WALL
09+80	L-C	C1	0.0	30.0	0.0	W				WET 10 FEET UP FROM BASE
10+18	L-C	C2	0.0	30.0	0.0	F			B13-B14	FLOW LEFT WALL
10+58	L-C	C2	0.0	30.0	0.0	F			B15	FLOW FROM CROWN /RUST STAIN CROWN
10+84	L-C	C1	0.0	25.0	0.0	D				PAST MOISTURE
11+00	L	C1	0.0	14.0	0.0	D				PAST MOISTURE
11+00	L	S2	2.0	0.0	1.0	D	R-2	10		AT BASE
11+14	L-C-R	C2	0.0	44.0	0.0	W				
11+40	L-C	C1	0.0	30.0	0.0	W				WET 10 FEET UP LEFT WALL
11+57	L-C-R	C2	0.0	44.0	0.0	M				MOIST LEFT WALL
12+00	L-C-R	C1	0.0	44.0	0.0	M				MOIST AT BASE OF LEFT WALL
12+49	L-C-R	C2	0.0	44.0	0.0	M				MOIST LEFT WALL AT NICHE
12+82	C	D	15.0	0.0	0.0	D			B16	AT CATENARY SUPPORT
13+50	L	C2	0.0	15.0	0.0	M				OLD PORTAL, END CENTER WALL

Fig. 27-11. Typical printout of complete inspection data.

Concrete Restoration Methods

Concrete restoration is a process of replacing loose, spalled, or crumbling concrete with new material. Properly done, this process will restore the structural integrity, be compatible with the surrounding concrete, and last as long as the structure. There are three basic techniques for applying the new materials:

- Concrete restoration by replacement or patching
- Shotcreting
- Grouting

Suitable materials to apply by such methods include

- Portland cement concrete (PCC)
- Polymer-modified PCC
- Epoxy-modified PCC
- Polymer mortar
- Epoxy mortar
- Special cement

The selection of the right combinations of method and material depends on such factors as "breathability" (ability to let vapor pass through), shrinkage, thermal coefficient of expansion, thickness, chemical resistance, application, and cost. Limited site access within the tunnel creates additional problems in delivery and installation of the repair.

GREEN LINE INSPCTION
COPLEY STATION TO ARLINGTON STATION
INBOUND REHAB SUMMARY

STATION	LOC	TYPE	AREA IN SF	LENGTH IN LF	DEPTH IN IN	M
** ********************************BEGINNING @ STATION 199						
199+32	R	S2	20.0	0.0	1.5	D
199+57	R	S2	8.0	0.0	2.0	D
199+62	R	S2	8.0	0.0	2.0	D
199+68	R	S2	6.0	0.0	1.5	D
199+81	R	S2	3.0	0.0	2.0	D
** Subtotal **						
			45.0	0.0		
** ********************************BEGINNING @ STATION 200						
200+15	C	S2	12.0	0.0	2.0	D
200+15	R	S2	60.0	0.0	3.0	W
200+30	R	S2	10.0	0.0	3.0	D
200+40	R	S2	20.0	0.0	2.0	D
200+57	R	S2	2.0	0.0	1.0	D
200+62	R	S2	30.0	0.0	3.0	D
200+85	R	S2	4.0	0.0	2.0	D
200+93	R	S2	20.0	0.0	2.0	D
** Subtotal **						
			158.0	0.0		
** ********************************BEGINNING @ STATION 201						
201+67	R	S2	45.0	0.0	3.0	D
201+90	C	S2	3.0	0.0	1.5	D
** Subtotal **						
			48.0	0.0		
** ********************************BEGINNING @ STATION 202						
202+45	R	S2	40.0	0.0	2.0	D
202+61	R	S2	12.0	0.0	1.5	D
202+78	C	S2	8.0	0.0	1.5	D
202+89	R	S2	45.0	0.0	1.5	D
202+98	R	S2	20.0	0.0	3.0	D
** Subtotal **						
			125.0	0.0		
** ********************************BEGINNING @ STATION 203						
203+16	R	S2	20.0	0.0	2.0	D
203+99		S2	3.0	0.0	3.0	D
** Subtotal **						
			23.0	0.0		
** ********************************BEGINNING @ STATION 204						
204+12	R	S2	40.0	0.0	1.5	D
204+57	R	S2	10.0	0.0	1.5	D
204+90	C	S2	15.0	0.0	2.0	D

Fig. 27-12. Typical printout of sorted data summary.

Regardless of which product or method is used, concrete restoration of tunnels requires good surface preparation and repairs to the reinforcing steel.

Surface Preparation

The surface preparation prior to the application of the appropriate material is common to the application of any of the

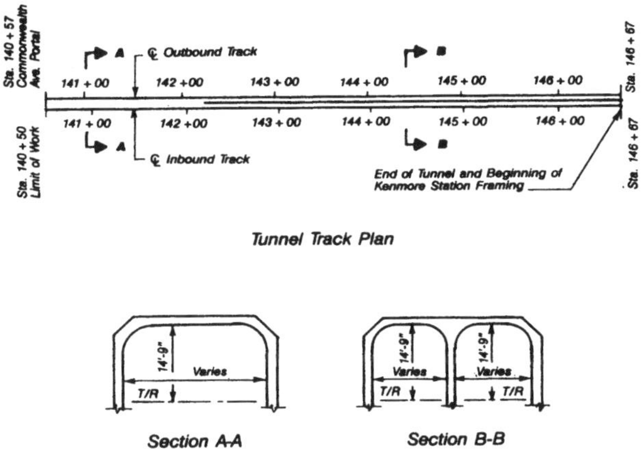

STATION		SPALLS				DELAM	CRACKS				EXP. JOINTS		
		S-1	S-2	S-3	S-4	D	C-1	C-2	C-3	C-4	J-1	J-2	J-3
FROM	TO	SF(D)	SF(D)	SF(D)	SF	LF	LF	LF	LF	LF	LF	LF	LF
Portal	14+00	6(2)	25(2)	44(3)		16	64						
41+01	42+00	50(2)	78(2)	36(3)		83	20	70	27				
42+01	43+00	74(9)	24(2)	8(1)		221	75	53	60	6			
43+01	44+00	51(2)	207(2)	40(3)		193	42	42					
44+01	45+00	33(2)	31(2)		135(8)	104	66	40	30	5			
45+01	46+00	52(2)	45(2)	34(2)		190	65	40		50			
46+01	46+67		47(2)			77	43	41	39				

Note: (D) Average Depth in Inches

Fig. 27-13. Contract document display of data.

concrete products and methods. Surface preparation consists of

- Removal of all loose, unsound existing concrete
- Cleaning of all corrosion from reinforcing steel
- Repairing or replacement of reinforcing steel (as required)

The removal of all loose, unsound concrete is best performed by the use of air-powered chipping hammers, sized so as not to remove excessive quantities of sound concrete. The best method for sizing chipping hammers is to specify the total weight of the hammer. The accepted hammer size for concrete removal in tunnels is a hammer weight not to exceed 30 lb (13.6 kg), not including the bit (Figure 27-14).

The unsound concrete must be removed by a chipping process starting at the middle of the identified defect and moving horizontally to the edges of the defect. The hammers will seek their own depth of penetration and be refused when the hammer's force is repelled by sound concrete. The edges of the area to be patched must not have a feather edge but rather a shoulder of not less than 1/8 in. (20 mm). See Figure 27-15.

Cleaning of Reinforcing Steel. After the removal of all loose concrete, and prior to the placement of any cementitious product, the reinforcing steel must be cleaned of all surface corrosion. The cleaning of reinforcing steel is best

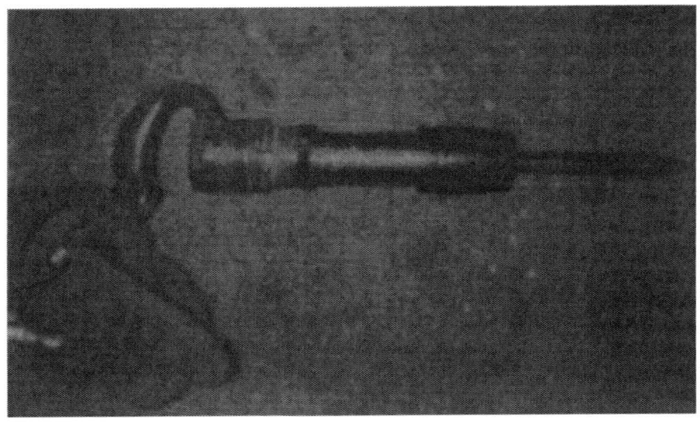

Fig. 27-14. Concrete chipping hammer and bit.

performed by sandblasting, which is a process of cleaning materials by blasting the surface of the item to be cleaned with a sharp sand or synthetic grit. The most common material used for sandblasting steel is Black Beauty, a synthetic carbide grit. The grit size depends on the type of equipment used for the application. In some cases the steel is cleaned by the use of wire brushes or chipping hammers to remove the heavy scale, followed by a sandblast.

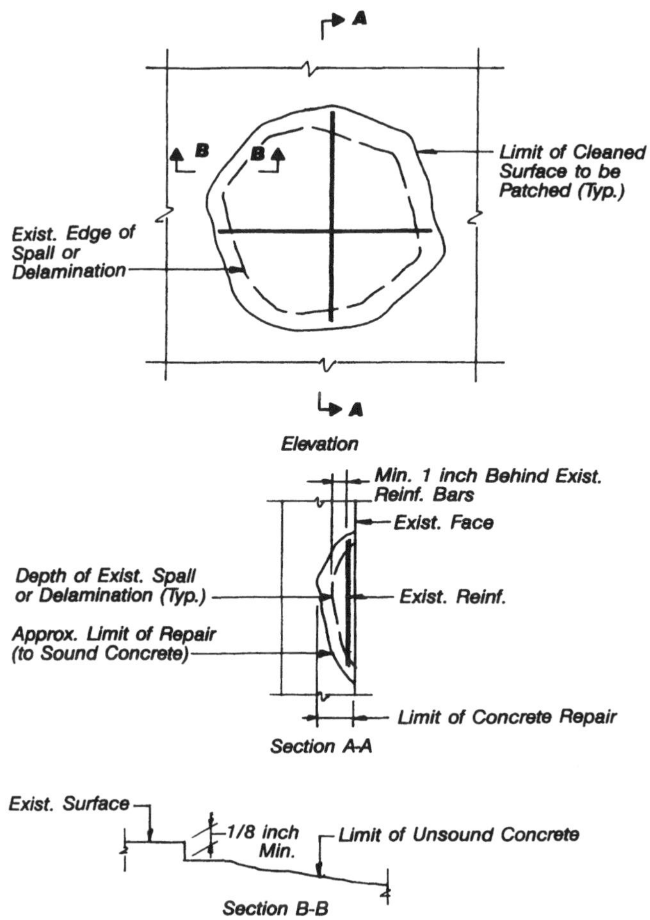

Fig. 27-15. Typical section at concrete repair.

High-pressure waterblasting for cleaning the reinforcing steel is not recommended for transit tunnels due to the presence of numerous electrical cables and traction power systems. In addition, high-pressure waterblasting does not provide acceptable cleaning of the reinforcing steel.

All steel must be cleaned to a "white metal" condition (or commercial grade finish; e.g., SSPC10). This "white metal" condition is required to provide a suitable bond between the steel and the new concrete repair (Figure 27-16). In areas of potentially high corrosion such as tunnel roadway slabs, the reinforcing steel must be painted with a zinc-rich primer.

After the steel located within the repair area is cleaned, the entire patch area must be cleaned of all dust, rust scale, and other debris, using compressed air to flush the area of the repair. Note: If the repair area is left for long periods prior to the placement of the concrete material, the area must be recleaned with compressed air immediately prior to placing the cementitious repair material.

Reinforcing Steel Repairs

Repairs to reinforcing steel generally are limited to the cleaning and restoration of the existing steel to its original configuration. Reinforcing steel that is out of plane, bent, or buckled is restored to its original location by cold bending. However, reinforcing steel that has a loss of section greater than 50% must be analyzed and, if required, replaced.

The replacement of the reinforcing steel must be performed in accordance with current American Concrete Institute (ACI) codes and have a minimum bar lap of 30 diameters (Figure 27-17). In some instances, if only an occasional bar requires replacement and the structural analysis indicates that the bar is not critical to the structural design, the bar need not be replaced.

Once the reinforcing steel is returned to the "white metal" condition, it is of a similar condition to the original construction, and therefore, further special coating protection is not required. The use of epoxy coating for reinforcing steel is not recommended unless the specific site conditions indicate unusual corrosion problems.

Fig. 27-16. Reinforcing steel ("white metal" condition).

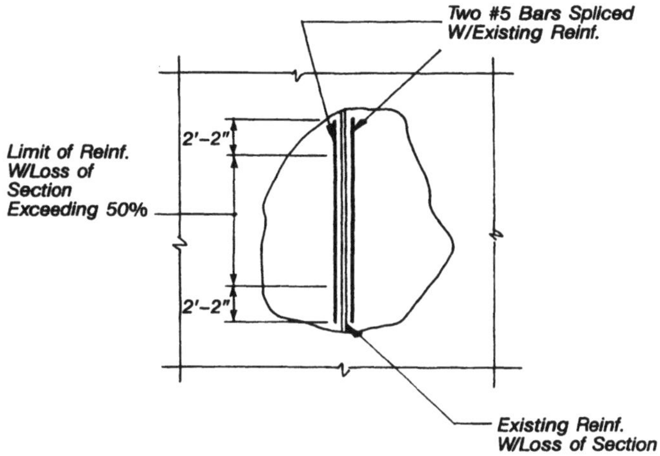

Fig. 27-17. Reinforcing steel splice detail.

Cast-in-Place Concrete

Limited Applications in Tunnels. Cast-in-place concrete for repairs of the tunnel structure is of limited use in the repair of transit tunnels. This limited use is allowable only when the repair is at a location where good site access is available. The two common methods for placing concrete in the tunnel are

- Transporting the material by rail-mounted/rubber-tired equipment
- Supplying the material by use of a "slick line," and pumping the concrete to the area of the repair

The problems with these methods of concrete supply involve the potential for loss of continuity of supply and high delivery costs.

In addition, cast-in-place concrete repairs in rail tunnels requires form work, which can only be constructed in areas where sufficient space prevents encroachment into the operating envelope of the revenue equipment and allows the safe operation of the trains. However, cast-in-place concrete has been successfully used in rehabilitation projects where the tunnel has been shut down for rehabilitation for a period of weeks or months for the construction of the repairs.

The use of form work for cast-in-place concrete poses additional problems in placing the concrete at wall and roof locations. Special construction procedures and/or special forms are required for wall repairs. Cast-in-place concrete should not be used for repairs to the underside of roof slabs due to the potential risk to construction personnel and revenue traffic.

The installation of cast-in-place Portland cement concrete required the use of a bonding agent to provide a strong bond between the existing sound concrete and the new concrete. The bonding agent is usually a latex or epoxy compound that is painted on the existing concrete surface as directed by the manufacturer. Once the bonding agent is placed, the forms are buttoned up, and the concrete is cast within the forms. The forms are removed after curing, according to accepted construction practices and as outlined by the ACI Standards for Placement of Concrete.

In actuality, the proper application of the bonding agent is seldom done, and the quality of the concrete bond is questionable. In response to this problem, specifications for a special polymer-modified Portland cement concrete mix have been developed incorporating the bonding agent as part of the concrete mix as a substitution for 10–30% of the water in the mix. This mix has provided a positive method of ensuring a good bond to the existing concrete surface while increasing the strength of the design mix. Cast-in-place polymer-modified Portland cement concrete is particularly advantageous in areas where a smooth or other special architectural finish is required.

Polymer Cementitious Mortars

Portland cementitious mortars with the addition of polymers create a high-strength, two-component, rapid-curing material for the restoration of structural concrete. Polymers used in concrete should meet the following criteria:

- Improve water resistance
- Improve freeze/thaw resistance
- Improve bond/adhesion
- Reduce shrinkage
- Improve set time

The acrylic copolymers (ACs) are the polymers that best fulfill these criteria. Incorporated in numerous premixed two-component cementitious products, ACs are best suited for the restoration of underground concrete structures due to their "breathability." This unique characteristic, which allows the passage of water vapor through the repair, prevents the buildup of moisture at the interface between the existing concrete surface and the repair, thereby eliminating freeze/thaw problems and extensive corrosion to the reinforcing steel within the repair area.

There are other products that use polymers and epoxy resins in the development of cementitious repair products. However, caution must be exercised in the use of high-alumina cementitious products or products containing polymer resins that create vapor barriers. Some of these products, in the cured state, will support combustion and are not recommended for tunnel use.

The two-component cementitious mortars are usually mixed at the repair site in small quantities and built up in layers of approximately 1–2 in. (2.5–5.0 cm). Common total thickness of these repairs is on the order of 4–6 in. (10–12.5 cm). Application must be in accordance with the manufacturer's recommendations.

Shotcrete

Shotcrete, a pneumatically placed Portland cement concrete mixture, is projected by compressing air against the area of the tunnel structure to be repaired. The concrete mixture of sand and cement is prepared dry and pumped through hoses to a nozzle where the water and any additives are mixed in (Figures 27-18 and 27-19).

Shotcrete can be categorized into three types:

- Cementitious shotcrete
- Latex/acrylic-modified shotcrete
- Two-component epoxy shotcrete

The latex/acrylic-modified and two-component epoxy shotcrete are the most suitable for use in operating tunnel systems. Unlike the cementitious shotcrete, both of these mixes have additives that enhance the bond to the existing concrete, limit the loss of material due to rebound, and allow for rapid cleanup after the repair work is performed. In addition, the use of these additives increases set time and allows higher tensile strengths, as shown in Table 27-3.

Fig. 27-18. Shotcrete mixing equipment.

Fig. 27-19. Shotcrete application.

Table 27-3. Shotcrete Comparisons

Shotcrete Type	Compressive Strength (28 Day) (psi)	Ultimate Bond Stress (psi)	Particle Rebound (%)	Set Time (min)
Cementitious	4,500	90	20-50	10-30
Latex-Modified	7,000	155	5-10	15-20
Epoxy-Two Component	6,500	129	10-25	2-5

Shotcrete is normally applied by rail-mounted equipment. After the existing concrete surface is prepared and the reinforcing steel is cleaned, as discussed previously, wire mesh is applied to areas of deep repairs of more than 9 ft^2 (2.85 m^2). This wire mesh is attached to the wall in increments of 3 in. (7.5 cm) by masonry anchors (Figure 27-20). The mesh is used to prevent sagging and to promote rapid application of the shotcrete. In areas of less than 9 ft^2 (2.85 m^2), the shotcrete is applied directly to the wall. Sagging is controlled by the applicator regulating the moisture in the mix. The surface finish of tunnel shotcrete should be a "gun" finish, a reasonably smooth finish, as placed by the equipment, with no additional troweling required.

Grouting

Grouting is a process of placing materials through ports in the tunnel structure to fill voids on the outside of the tunnel system. Grouting can also be used to seal off the tunnel structure from groundwater infiltration.

Grouts are liquids that are pumped at various pressures and solidify after placement. There are many types of grouts available for the densification of soils and the filling of voids around tunnel structures. Two types of grouts are used for filling voids outside of the tunnel structure:

- *Particle grouts.* The cement-based grouts, the most frequently used product for the filling of voids outside of tunnel structures, are a mixture of Portland cement, fly ash, sand, and water. Often, special fluidizers are added to assist in pumping the material.
- *Single/multiple-component chemical grouts.* The single/multiple-component chemical grouts, relatively new to the market, have been used more extensively in Europe, with mixed results. In the United States the cementitious grouts are used due to their low cost and the relative ease in training personnel in their application.

The results of grouting programs for the filling of voids outside of tunnels are difficult to assess, due to the difficulty and extensive cost of inspecting and verifying the filling process. In addition, the pressure injection of these positive displacement grouts can have a detrimental effect on the tunnel by hydraulically fracturing the tunnel structure. These grouts also have the potential for damaging nearby utilities, foundations, and other structures in an urban environment.

Therefore, pressure grouting is not recommended for filling voids outside existing transit tunnels unless no other alternative is available. In projects requiring pressure injection of voids, caution must be exercised not to exceed injection pressures of 40 psi (275.8 kPa), as measured at the tunnel structure, since pressures in excess of that limit will deform most segmental tunnels liners.

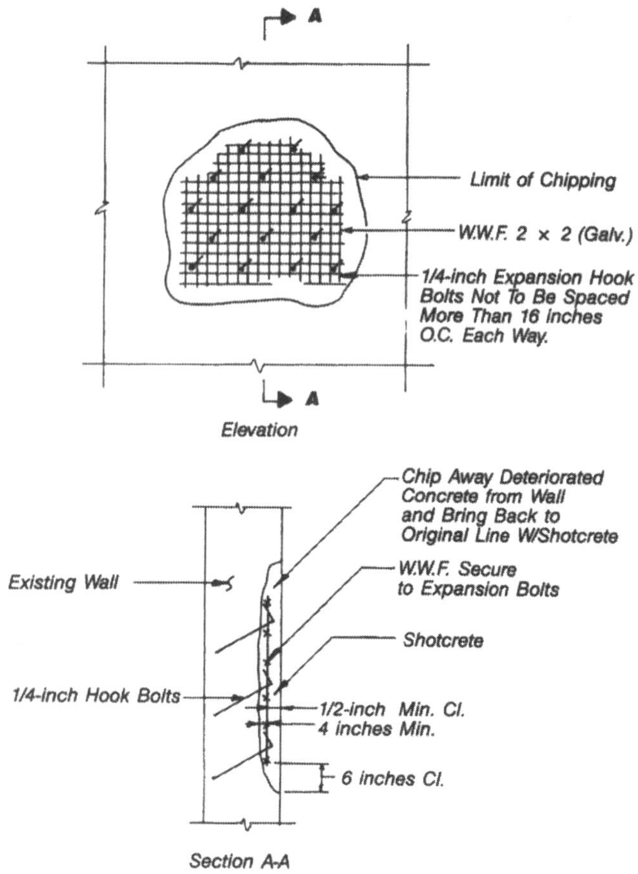

Fig. 27-20. Shotcrete repair detail.

CRACK REPAIR

Structural Repair of Cracks

The sealing/bonding of structural cracks in tunnels is performed to restore the monolithic condition of the structure. Cracking may be a result of concrete shrinkage or of movements of the structure caused by settlement or other external forces acting on the tunnel system. Cracks are often the cause of water inflow into the structure and the subsequent corrosion of reinforcing steel and the deterioration of the concrete structure. Therefore, it is desirable to restore the bond of the sides of these cracks and restore the structural integrity of the tunnel system as originally designed.

Rigid epoxy injection systems have been developed to repair these structural cracks successfully, provided that the crack is dry or merely moist, not saturated, and without flowing water. Special polyester resin grouts have been developed to rebond saturated concrete.

The structural repair of cracks in concrete tunnels is performed as follows:

- Seal the surface of the crack to be injected with a paste gel, and install injection ports over the crack (Figure 27-21).
- Install injection ports a horizontal distance equal to the thickness of the wall to be injected.
- After the initial set of the gel, inject high-solid two-component, moisture-insensitive epoxy resin into the crack (Figure 27-22).

The injection of the resin into the joint, performed using a special high-pressure rotary mixing pump, starts at the lowest point on vertical surfaces. When the resin is observed flowing from the next port up the crack, the injection nozzle is moved up to the next port and the lower port is plugged with a wooden plug. This process is repeated until all of the ports have been pumped with resin.

Injection pressure must not exceed 100 psi (0.69 MPa), and a recommended working pressure is 40 psi (275.8 kPa). Specialized products have been developed for use in horizontal, overhead, and vertical locations.

Tunnel Leakage Control by Crack Injection

The nonstructural sealing of cracks that carry groundwater into the tunnel system has long been one of the most serious problems confronting the operators of tunnel systems. Such cracks are the primary cause of corrosion of the steel elements of the tunnel. In addition, the inflow of groundwater through various tunnel defects causes deterioration and significant maintenance costs. The epoxy resins for structural bonding of cracks are not particularly suitable for groundwater control because, while they are injectable in a moist environment, they cannot be injected in a flowing environment.

Two types of products have been used to control or eliminate tunnel leakage: particle grouts and chemical grouts.

Particle Grouts. The particle grouts consist of cementitious grouts using Type I and Type III Portland cements, microfine cement, and microfine cement silicates. All of the particle grouts have low toxicity and are nonflexible after curing. This lack of flexibility creates a potential for repairs to be short term, and they will leak if any future structural movement occurs.

The installation of the particle grouts is essentially the same as that for grouting exterior tunnel voids. The microfine grouts are also injected into the structure and, due to their fine particle size, are able to penetrate very fine cracks.

Chemical Grouts. The chemical grouts vary greatly in regard and viscosity, and include the following:

- Acrylamides, a highly toxic, low-viscosity grout banned in Europe and Japan due to its toxicity
- Acrylates, a nontoxic equivalent to the acrylamides, with low viscosity
- Sodium silicate, a low-toxicity, medium-viscosity grout with limited usage due to its syneresis and high shrinkage rate
- Lignosulfonates and aminoplasts, not generally used due to their very high toxicity and costs
- Polyurethane, a high-viscosity, medium-toxicity, water-reactive grout

Application. The chemical grouts that have had the most success in the sealing of leaking tunnel cracks are single-component polyurethane grouts. These single-component grouts are flexible, water-reactive, and relatively easy to install.

Fig. 27-21. Sealing of crack surface.

Fig. 27-22. Injection of structural crack.

The injection of chemical grouts is performed in the same manner as epoxy resins. The major change in the installation is the type and location of the injection ports. In areas of high water flow, a surface sealant is placed to control water flow and retain grout. Relief ports must be retained, however, to allow full penetration of the grout.

- The injection ports are placed in holes drilled diagonally from the sides of the crack and intersect the crack within the wall (Figure 27-23).
- A mechanical packer is installed that has a "cirque" fitting, which allows a tight connection and prevents spillage of the grout. (Various types of packers are available for chemical grouting; Figure 27-24 illustrates the most commonly used packers.)
- The surface of the crack is sealed with low-modulus gel. The crack is flushed with clean water, and the crack is injected with the chemical grout using a hydraulic "grease gun" pump (Figure 27-25).

The mechanical packers are left in place, and regrouting is possible for up to approximately 7 days. Successful sealing of cracks must be performed when the surface of the wall to be injected is above 45°F (7°C). Caution must be exercised to ensure the injected polyurethane is capable of sufficient elongation and/or flexibility for potential crack movements.

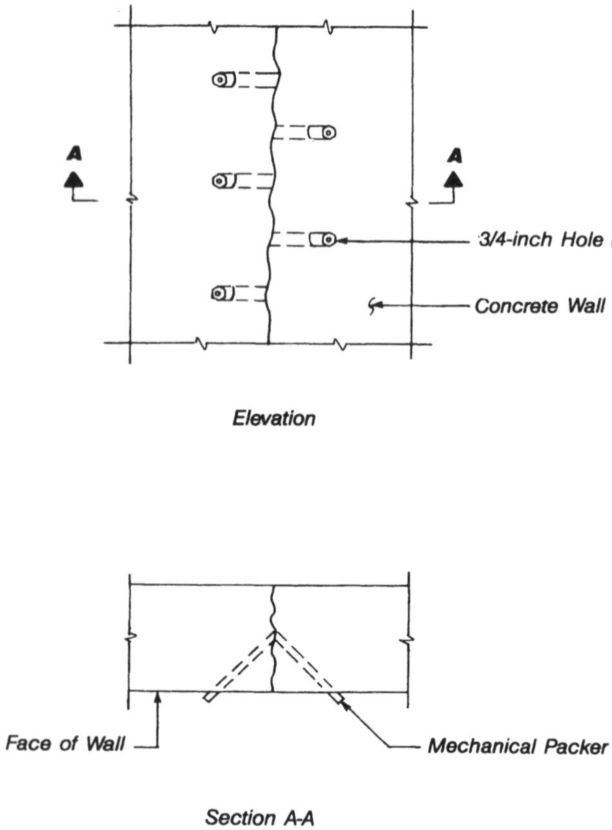

Fig. 27-23. Typical installation of chemical grout ports.

Fig. 27-24. Mechanical packers.

METAL REPAIRS

The restoration of structural metal elements in tunnels includes the repair of structural steel elements and often, in the case of older tunnels, of cast iron elements. These elements are generally incorporated in the tunnel systems as framing steel, structural columns, and miscellaneous metal construction. The use of cast iron for structural elements is generally confined to bored tunnels where the segmental liner is cast iron. Therefore, the repairs to cast iron will be discussed below in "Cast Iron Segmental Liners."

The structural repair of steel in transit tunnels is performed in the same manner as steel repair for bridges and other aboveground structures. Steel elements having a loss of section in excess of 20% must be analyzed and repaired, if required, using accepted practices outlined in the *AISC Manual of Steel Construction* (1981) and the Lincoln Arc Welding Foundation's *Design of Welded Structure* (1966).

The steel unit to be repaired is either built up by the addition of steel plates (either bolted or welded to sound existing

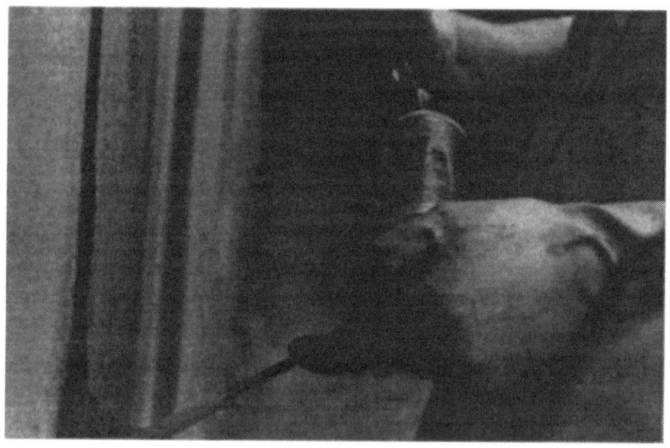

Fig. 27-25. Crack injection of chemical grout.

steel) or completely replaced by a new structural element (Figures 27-26 and 27-27). The bolting of replacement steel elements is to be performed with the use of high-strength bolts in accordance with ASTM Standard A325 or A490. The reuse of existing high-strength bolts is prohibited.

The painting of steel elements is the best protection against future corrosion caused by moisture. An effective painting program requires the thorough cleaning (sandblasting) of the steel elements. Epoxy paints should be applied to the cleaned surfaces as recommended by the manufacturer. The owner must be consulted in the selection of the paint color, and the color must be carefully specified in the contract documents.

BRICK MASONRY REPAIR

The restoration of masonry structures composed of clay brick consists of repointing of mortar joints and replacement of defective brick. The repointing of masonry joints requires cleaning the joints to be repaired to a depth of approximately 1 in. (2.54 cm). Once these joints have been raked clean of old mortar they are repointed with cementitious mortar (Figure 27-28) or with a cementitious mortar that has been fortified with acrylic bonding agents.

Replacement of broken, slaked, or crushed brick requires detailed analysis to determine the cause and extent of the problem; once these are determined, repairs may be performed.

The use of brick in transit tunnels is limited, and the repairs are unique to the individual structure. The removal of more than the occasional brick in the structure requires detailed procedures. The design of the repairs must be performed by specialists familiar with brick masonry construction.

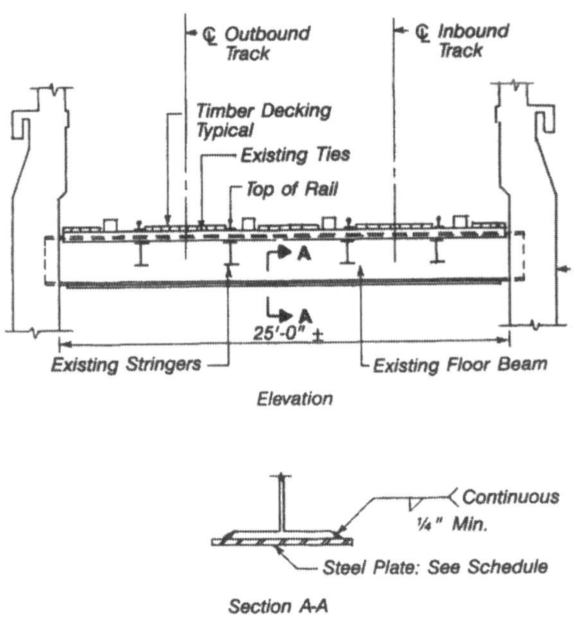

Fig. 27-26. Typical repair of steel framing.

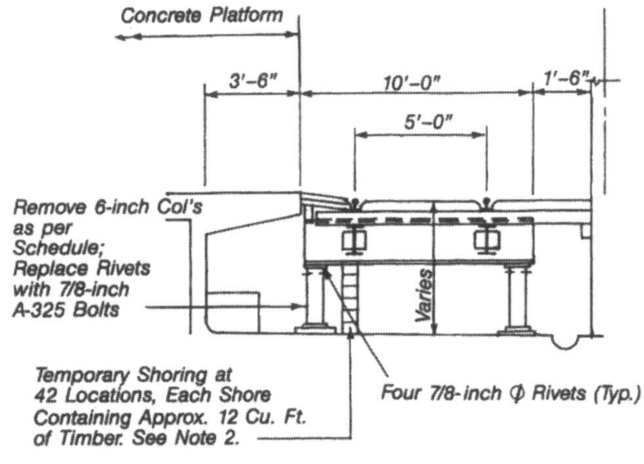

Fig. 27-27. Replacement of steel element.

SEGMENTAL TUNNEL LINERS

Repairs to segmental liners combine the processes discussed earlier with repairs based on the composition of the original construction. These repair processes are adaptable to the repair of segmental liners that exhibit the same construction composition and require the restoration of structural integrity or watertightness. Segmental tunnel liners consist of three basic types:

- Precast concrete
- Steel
- Cast iron

Common defects in segmental liners are

- Cracking of segmental liner plates/precast elements
- Leakage through joints/bolt holes at the interface between plates
- Corrosion caused by leakage or electrical currents
- Distortion of liner segments due to impact/changes in stress.

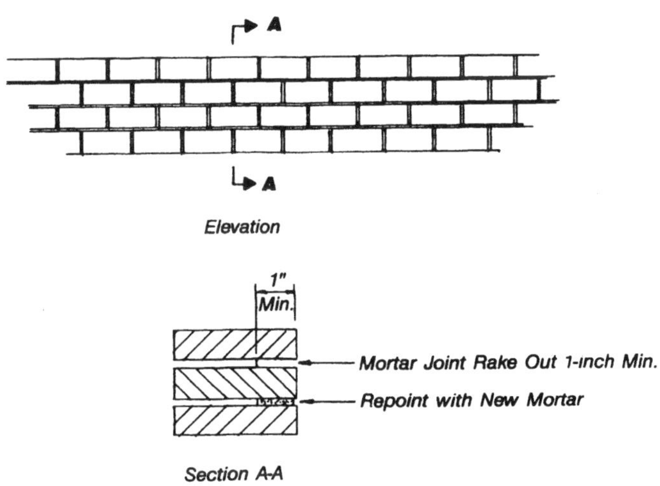

Fig. 27-28. Repair to brick masonry.

Precast Concrete Segmental Liners

The structural bonding of cracks in precast tunnel liners is performed by injecting epoxy resins as described earlier for crack repair. Seepage through the liner is controlled by

- Injection of chemical grout in segmental joints and flowing cracks in the precast panels
- Restoration of the original seals at the panel joints
- Sealing of bolt-hole leakage by installation of new gaskets or chemical grouting

Electrical corrosion in precast tunnel liners has not been a major concern. If evidence of electrical corrosion is found, however, site-specific repairs must be performed.

Impact to precast concrete tunnel liners causes the precast panels to crack and spall. These spalls are usually repaired by the use of two-component cementitious mortars described earlier.

Steel Segmental Liners

Leakage to steel segmental liners is controlled by the injection of cementitious and/or chemical grouts. Leakage at bolt holes and between liner segments is usually fixed by repairing the original lead caulking by repacking the material into the liner joint groove.

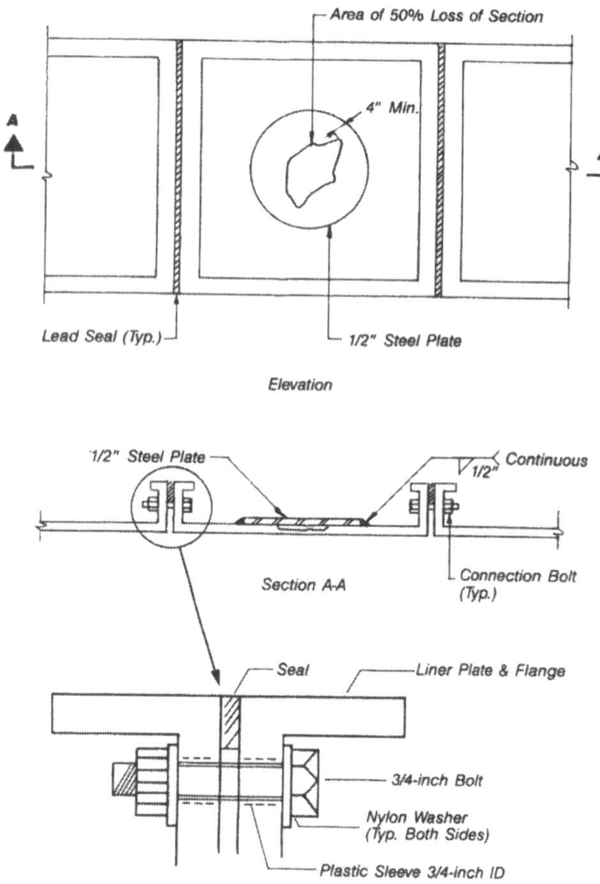

Fig. 27-29. Typical repair to steel liner plate.

Electrical corrosion in segmental steel tunnels is first observed in areas of dissimilar metals. The prescribed method for control is to isolate the bolts from the liner by an insulating sleeve or jacket (Figure 27-29). Corrosion to the liner plate and subsequent loss of section is repaired by welding steel plates to the affected area in accordance with AISC Manual of Steel Construction.

Cast Iron Segmental Liners

Seepage in cast iron liners is stopped by injecting cementitious or chemical grouts and repacking existing lead joint seals. Chemical grouts may also be used to seal the panel joints. Electrical corrosion of the connection bolts can be eliminated by insulating sleeves or jackets (Figure 27-29).

Loss of section and/or replacement of a defective cast iron structural element is more difficult since cast iron liners cannot be welded. Therefore, the replacement or repair of defective cast iron panels requires extensive site-specific information and analysis, which must be tailored to the specific repairs.

CONSTRUCTION COSTS

Table 27-4 illustrates the type and suitability of materials for concrete repairs in the restoration of tunnels. The associated construction costs shown in Table 27-5 are based upon material costs and actual bid prices in the Boston area in 1994 dollars.

Table 27-4. Comparison of Concrete Repair Products

Application	Two-Component Self-Leveling Mortar	Latex-Improved Shotcrete	Two-Component Mortar	Epoxy Shotcrete	Polymer Masonry Mortar
On grade, above, below	X	X	X	X	X
On horizontal	X	X	X	X	X
On vertical			X	X	X
Overlay system	X	X			X
Structural material	X	X	X	X	X
Leveling product	X	X			X
Bolt holes filler --voids and cavities			X		X
Minimum depth	1/2 inch	1/8 inch	1/8 inch	1/8 inch	
Maximum depth					1/4 inch
Extend with aggregate	X	X		X	
High abrasion	X	X	X	X	X
Good bond strength	X	X	X	X	X
Compatible coefficient of expansion -- concrete	X	X	X	X	X
Resistant to salts	X	X	X	X	X
High early strength	X	X	X	X	X
High compressive	X	X	X	X	X
High flexural	X	X	X	X	X
Good freeze thaw	X	X	X	X	X
Not a vapor barrier	X	X	X	X	X
Not flammable	X	X	X	X	X
OK potable water	X	X	X		X
Traffic 1-2 hours	X				X
Low rebound and dust		X		X	

Table 27-5. Comparison of Tunnel Rehabilitation Bid Prices

Repair product	Material cost	Bid price
Portland Cement Concrete (4,000 psi)	$ 60.00 cu yd.	$ 800.00 cu yd
Polymer-Modified Portland Cement (4,000 psi)	$ 72.00 cu yd	$ 800.00 cu yd
Cementitious Shotcrete	$ 60.00 cu yd	$2,100.00 cu yd
Latex-Modified Shotcrete	$ 34.00 cu ft	$ 150.00 cu ft
Epoxy Shotcrete	$300.00 cu yd	$4,000.00 cu yd
Two Component Mortar (Vertical)	$ 50.00 cu ft	$ 150.00 cu ft
Two-Component Mortar (Horizontal)	$ 55.00 cu ft	$ 150.00 cu ft
Polymer Mortar	$ 55.00 cu ft	$ 150.00 cu ft
Crack Injection (Epoxy Resin)	$ 1.20 linear ft	$ 24.00 linear ft
Crack Injection (Polyurethane Foam)	$ 85.00 gal	$ 70.00 linear ft
Painting of Framing Steel	$ 1.00 sq ft	$ 3.00 sq ft

NOTES:
1. All bid prices include demolition and preparation
2. All work based on 8-hour working day (starting at 10 PM)
3. All prices in 1994 dollars for work performed in Boston
4. All work performed during operations shutdown

REFERENCES

ACI Committee 201 (1968) "Guide for Making a Condition Survey of Concrete in Service," *ACI Journal, Proceedings, 65*(11), 908–918.

ACI Committee 318 (1983) *Building Code Requirements for Reinforced Concrete,* ACI 318-83, American Concrete Institute.

ACI Committee 504 (1982) "State-of-the-Art Report on Sealing Joints in Tunnels," Report ACI 504.IR-82, *Concrete International,* pp. 17–29.

ACI Seminar (1985) *Structural Repair, Corrosion Damage and Control,* American Concrete Institute SCM (85).

American Cyanamid Co. (1965) *AM-9 Chemical Grout,* Technical Data, Wayne, IN.

American Institute of Steel Construction (1981) *Manual of Steel Construction.*

American Public Works Association (1971) "Control of Infiltration and Inflow into Sewer Systems," *Report for the Environmental Protection Agency,* Program 11022EFF, January.

American Society for Testing and Materials (1986) *Annual Book of Standards, Part 2, 4, 10, 19.*

BAKER, W.H. (1983) *Design and Control of Precast Tunnel Liners,* Report FHWA/RD-82/039, U.S. Dept. of Transportation.

Bechtel, Inc. (1975) *System Study of Precast Tunnel Liners,* Report TST-75-102, U.S. Dept. of Transportation.

BIRKMEYER, J.A. (1978) *Rapid Transit Subways—Maintenance Guidelines,* Report UMTA-MA-06-0025-78-3, August.

BREKKE, T.L., EINSTEIN, HH., and MASTON, R.E. (1976) *State-of-the-Art Review on Shotcrete,* Contract DACA 39-74-0371.

CARON, C. (1963) "The Development of Grout for the Injection of Fine Sands," *Grouts and Drilling Muds in Engineering Project Description Practice.*

CLOUGH, G.W. (1976) "European Stabilization Practice Using Chemical Systems in Urban Tunneling," *Proceedings, Rapid Excavation and Tunneling Conference,* Las Vegas.

CLOUGH, G.W., et al. (1983) *Development of a Design Technology for Ground Support for Tunnels in Soil, Vol. 1, 2, 3,* U.S. Dept. of Transportation, Report DOT-TSC-UMTA-82-54.1, 2, 3.

Construction Specifications Institute (1975) *Manual of Practice, Vol. 1.*

DEERE, D.U., PECK, R.B., MONSEES, J.E., and SCHMIDT, B. (1969) *Design of Tunnel Liners and Support Systems,* Final Report, Contract 3-0152, U.S. Dept. of Transportation.

DEMPSEY, J.A., and MOLLER, K. (1970) "Grouting in Ground Engineering," *Proceedings of the Conference on Ground Engineering,* Paper 1.

Department of the Army (1966) *Engineering and Design—Foundation Grouting: Planning,* Engineering Manual No. 1110-2-3501, Dept. of the Army.

EINSTEIN, H.H., and BARVENIK, M.J. (1975) *Grouting Applications in Civil Engineering,* Vols. 1 and 2, MIT Research Report R75-8, January.

ERAUD, J. (1974) *Policy in the Field of Inspecting, Maintenance of Tunnels,* SNCF.

GRAHAM, J.R. (1973) *Use of Shotcrete for Underground Structural Support,* ACI Publications SP 45.

HAMPTON, D., MCCUSKER, T.G., and Essex, R.J. (1980) *Representative Ground Parameters for Structural Analysis of Tunnels,* Report FHWA/RD-80/013, Vol. 2, U.S. Dept. of Transportation.

HAUGE, P., and HOFFMAN, W. (1970) "Epoxy Resin Grouting in Large Underground Openings," *Proceedings, International Symposium on Large Permanent Underground Openings, 1969,* Universiteforlaget, Oslo.

JACKSON, D.C. (1984) "Corrosion of Iron Pipes by Action of Electric Railway Currents," *Journal of Association of Engineering Societies,* September.

JANIN, J.J., et al. (1970) "Chemical Grouting for Paris Rapid Transit Tunnels," *Proceedings, ASCE Construction Division, 96.*

JONES, R. (1948) *The Application of Ultrasonics to the Testing of Concrete,* Research.

KOPF, R.J., COOPER, C.G., and WILLIAMS, F.W. (1981) "In Situ Strength Evaluation of Concrete Case Histories and Laboratory Investigations," *Concrete International,* American Concrete Institute, March.

KUESEL, T.R. (1983) "The Structural Behavior of Tunnel Linings," *Proceedings,* American Society of Civil Engineers, Metropolitan Section Seminar on Tunneling and Underground Construction.

LESLIE, J.R., and CHEESMAN, W.J. (1949) "An Ultrasonic Method of Studying Deterioration and Cracking in Concrete Structures," *Journal of American Concrete Institute, 21*(1) (September), 17–36.

Lincoln Arc Welding Foundation (1966) *Design of Welded Structures.*

MALHOTRA, V.M. (1976) "Testing Hardened Concrete: Nondestructive Methods," *ACI Monograph No. 9,* American Concrete Institute.

MALHOTRA, V.M., et al. (1986) *In Situ Nondestructive Testing of Concrete,* ACI Publication SP 82, American Concrete Institute.

MATTHEWS, R. (1975) *Tunnel Manual,* Report DOT-TST-76-40, Monitor 18.

MENAY, J.J. (1974) *A History of Stray Traction Current Corrosion in the United States,* Boston Society of Civil Engineers.

NAIK, T.R. (1979) "Ultrasonic Testing of Concrete," Presented at the 1979 Fall Convention, American Concrete Institute.

New York City Transit Authority (1975) *New York City Transit Authority Design Guidelines,* Project UMTA-IT-09-0014-TS-C160, Report UMTA-IT-09-0014-75-4.

PERROT, W.E. (1965) "British Practice for Grouting Granular Soils," *Journal of Soil Mechanics and Foundation Division, ASCE, 91,* SM6.

POWERS, J.P. (1972) "Groundwater Control in Tunnel Construction," *Proceedings, Rapid Excavation and Tunneling Conference.*

PROCTOR, R.V., and WHITE, R.L. (1977) *Earth Tunneling with Steel Supports,* Commercial Shearing and Stamping Co.

RAMAKRISHNAN, V. (1976) "Past and Future of Concrete Quality Evaluation," *Journal of the Construction Division, ASCE, 102,* CO2.

ROMANOFF, M. (1957) "Underground Corrosion," U.S. Dept. of the Interior, National Bureau of Standards, *Cir 575.*

SAMARIN, A., and DHIR, R. (1984) *Determination of In Situ Concrete Strength Rapidly and Confidently by Nondestructive Tests,* ACI SP 82, American Concrete Institute.

SCHMIDT, B., and DUNNICLIFF, C.J. (1975) *Construction Monitoring of Soft Ground Rapid Transit Tunnels,* Vols. 1 and 2, Report DOT-TSC-UMTA-75-9-VI, -V2.

SCHWARTZ, C.W., and EINSTEIN, H.H. (1980) *Improved Design of Tunnel Supports: Vol. 1—Simplified Analysis for Ground-Structure Interaction in Tunneling,* Report UMTA-MA-06-0100-80-4, U.S. Dept. of Transportation.

SELMER-OLSON, R., and BROCH, E. (1976) "Some Reservations on the Use of Shotcrete," *Tunnels and Tunneling, 8*(3), March–April 1976.

TALLARD, G.R., and CARON, C. (1977a) *Chemical Grouts for Soils—Vol. 1, Available Materials,* FHWA Report FHWA-RD-77-50.

TALLARD, G.R., and CARON, C. (1977b) *Chemical Grouts for Soils—Vol. 2, Engineering Evaluation of Available Materials,* FHWA Report FHWA-RD-77-51.

VOORT, H.B. (1976) "Tunnel Instrumentation," Shotcrete for Ground Support, *Proceedings, Engineered Foundation Conference,* Tidewater Inn, October 4–8.

WELCH, J.P. (1984) *Control of Water Infiltration by Injection Techniques for Underground Transportation Structures,* American Public Transit Association Conference.

WHITEHURST, E.A. (1966) "Evaluation of Concrete Properties from Sonic Tests," *ACI Monograph No. 2,* American Concrete Institute.

WINKERTON, H.F. and FANG, H. (1975) *Foundation Engineering Handbook,* Van Nostrand Reinhold.

28 Tunnel Construction Contracting

Thomas R. Kuesel
Consulting Engineer, Chairman Emeritus,
Parsons Brinckerhoff Quade & Douglas, Inc.

A whole separate book (or several) could be written about tunnel construction contracting. The purpose of this brief chapter is to explain why tunnel construction is markedly different from aboveground construction, and why and how the philosophic basis for modern tunnel construction contracting has evolved.

In English law, from which American practice is derived, a contract is an agreement between two parties in which one party undertakes to perform for the other a definite, mutually understood task for a defined compensation. Construction contract law is generally based on this underlying premise. The American system of competitive bid contracting is based on the assumption that the work to be performed, and the conditions under which it is to be performed, are perfectly and unambiguously defined in the contract documents, so that a simple comparison of bid prices is sufficient to determine a fair selection, and contract award is made to the lowest responsive bidder.

This ideal is rarely reached in any construction contract, but it is especially illusory for tunnel (or any underground) construction, owing to several peculiarities inherent in underground work:

- Tunnels are invariably lengthy structures, and over extended lengths the characteristics of the ground (and groundwater) may vary widely and unpredictably. Despite the most comprehensive geotechnical investigations and the most astute interpretations (neither of which are universally attained), the exact nature of the ground is never completely disclosed until it is exposed by excavation at the tunnel heading (and, indeed, this nature may change subsequently owing to time-dependent effects). As a result, the work that is actually performed may differ in small or large measure from that expected by either or both parties at the time of contract award.
- The methods, equipment, and skills required for safe and economical tunnel construction depend on the nature of the ground, and may be disproportionately sensitive to small changes in ground characteristics. In particular, small changes in groundwater content or permeability, which are especially difficult to predict accurately, may have large effects on how the work must be performed.
- Preconstruction investigations can determine the characteristics of (more or less representative samples of) the existing ground. The processes of construction may change these characteristics (e.g., destressing rock joints by excavation may cause them to open, and convert a dry tunnel into a waterfall). The contractor's choice of equipment and construction methods, and the skill of his workers, may increase or decrease deleterious changes in ground characteristics during construction.
- In urban tunnel work, the existence and location of unknown buried obstructions and hazardous conditions (such as gasoline derived from abandoned leaking underground storage tanks) is difficult to determine beforehand and can have major effects on the work.
- In urban work, it is not uncommon that the cost of dealing with third parties (government regulations, adjacent structure protection, support and relocation of utilities, hauling and disposal of excavated materials, traffic maintenance, site clearance and preparation, and restoration of the surface after completion of construction) may equal or exceed the cost of constructing the desired facility. The best of agreements between two parties regarding how third parties will act or behave is considerably less than infallible.

In traditional construction contracting practice, the owner allocated all risks to the contractor, saying in effect, "you deal with all construction problems and all third parties, and don't bother me." The contractor's bid price was supposed to cover the costs of mitigating all problems related to unknown site or geotechnical conditions, as well as all delays or difficulties introduced by the actions or omissions of third parties. These were essentially gambles, which the owner asked the contractor to make.

But at the same time, the owner told all bidders that the contract would be awarded to the lowest responsive bid. A bid that included reservations or exclusions regarding any risk was deemed "nonresponsive."

This practice produced two results:

1. Experienced and prudent contractors, who included in their bids substantial contingencies to cover perceived risks, found that their bids were rarely low. In the minority of contracts that they did win (generally the more difficult ones that scared inexperienced contractors away), they found that if the risks did not materialize they made a windfall, and if they did materialize they went broke. In this boom-and-bust climate, if the boom came first the contractor survived, and if the bust came first, he went out of business. Tunnel construction contracting was a short-lived occupation.
2. Less experienced, or more adventurous, contractors took an optimistic view of risks of underground construction, and included little or no contingencies in their bids. They won more contracts, and when the risks materialized they mounted a vigorous campaign of claims and litigation, generally on the assertion that the contract was defective, in that the owner knew, or should have known, or failed to take adequate measures to discover, conditions of which the contractor should have been advised in the solicitation of bids.

This was the situation in the 1950s, when a great surge in urban underground construction occurred across the United States. Construction litigation became an increasingly popular and lucrative (for the lawyers) occupation, and tunnel construction developed a bad name among owners and the general public, as prone to large, unexpected overruns of projected costs and schedules.

DIFFERING SITE CONDITIONS CLAUSE

The greatest source of conflict was the allegation of "changed conditions," or more specifically, "differing site conditions." The first attempt to mitigate this problem was the development by the Corps of Engineers of a standard contract clause, which has evolved into the following, which is now almost universally included in tunnel contracts:

> (a) The Contractor shall promptly, and before such conditions are disturbed, notify the Owner in writing of (1) subsurface or latent physical conditions at the site differing materially from those indicated in the contract, or (2) unknown physical conditions at the site, of an unusual nature, differing materially from those ordinarily encountered and generally recognized as inherent in work of the character provided for in this contract. The Owner shall promptly investigate the conditions, and if he finds that such conditions do materially so differ and cause an increase or decrease in the Contractor's cost of, or time required for, performance of any part of the work under this contract, whether or not changed as a result of such conditions, an equitable adjustment shall be made and the contract modified in writing accordingly.
>
> (b) No claim of the Contractor under this clause shall be allowed unless the Contractor has given the notice required in (a) above; provided, however, the time prescribed therefor may be extended by the Owner.
>
> (c) No claim by the Contractor for an equitable adjustment hereunder shall be allowed if asserted after final payment under this contract.

One valuable attribute of this particular clause is that it has been through the courts for a long enough time that there is general agreement among judges and lawyers as to what the words mean, and on that account it is desirable to follow this wording scrupulously.

The Differing Site Conditions clause enabled the owner's contracting officer to make equitable contract adjustments for the more egregious cases, where he could in good conscience find that site conditions did indeed differ so clearly that he could not, by accepting the contractor's claim, be held to breach his duty to the owner, to protect his interests. General acceptance of this clause by the courts eased the strain of loosening the rigid contract bonds, and the climate for tunnel construction improved. Prudent contractors found that they were protected from the more outrageous risks, their contingencies came down, and they started to gain more contracts. Owners found that the quality of their tunnel contractors improved, and their litigation became less raucous.

Nonetheless, many cases remained in which the owner and the contractor did not agree on whether the conditions differed, and these cases still ended up, after the contract appeals process was exhausted, in court.

It became apparent that tunnel construction is an arcane art to the legal profession, and much of the time and cost of litigation was being expended on educating the lawyers and judges (to say nothing of juries) on the technical terms and practicalities of tunnel construction. Since this education was perforce rudimentary, and the field remained obscure and mysterious to the lawyers and judges, they tended to try to find grounds for deciding the case on the basis of fine points of law and alleged legal precedents (which they understood), rather then what was practicable, realistic, and effective in the tunnel heading (which they did not understand). Jury cases tended to be decided by the relative persuasiveness of the opposing lawyers rather then by any consideration of the relative equity of the parties.

This situation satisfied neither the owners nor the contractors. A particular burden was the length of time the litigation process consumed, which meant that final settlement was frequently delayed until years (sometimes many years) after construction was completed.

One of the most prominent tunnel contract disputes of this time concerned the first bore of the Eisenhower (Straight Creek) Tunnel in Colorado. The final settlement doubled the original contract price, long after the tunnel was completed. The Colorado Highway Department determined that on the second bore there should be a better way to resolve contract disputes. Mr. A.A. Mathews, a construction engineering consultant, recommended a Mediation Board to resolve any disagreements between the owner and the contractor that could not be settled through the claims process of the contract. This was so successful that it became a model for what has evolved into the Disputes Review Board process, which is a standard feature of modern tunnel construction contracts.

A Disputes Review Board is composed of three members, all of whom are experts in tunnel construction or tunnel engineering. One member is selected by the owner, one by the contractor, and the third by the first two. They provide informal mediation of technical and contractual issues on which the owner and contractor are unable to reach agreement under the provisions of the contract. The Board provides a written report and recommendation regarding each dispute, which is not binding but carries great weight because the members are chosen, and recognized, for their professional experience and perspective. If either party rejects the Board's recommendation, it is with the knowledge that in subsequent litigation the courts will value the Board's recommendation highly.

The Disputes Review Board (DRB) procedure has been found to have a number of advantages for both the owner and the contractor:

- The DRB procedure is much less costly and time consuming than formal litigation.
- Recommendations of professional experts are more likely to be based on practical considerations than on abstruse points of law.
- The saving in senior management time devoted to contract dispute resolution is significant.
- Disputes are settled promptly while the construction continues to go forward, and consequential delays and costs are reduced.
- The process is much less adversarial than litigation, and the climate of contract administration is improved.

The DRB process has also been found to have some unexpected benefits:

- By its very existence, the Board reduces the incidence of claims and fosters settlement between the two parties, because both parties know that the Board cannot be bluffed and that insecure claims (or arbitrary disallowances) are likely to be rejected by the Board.
- Since the contracting officer is relieved of the onus of being both the owner's representative and the judge of the contractor's claims, he is enabled to be more flexible in dealing with unanticipated developments during construction. Since he can pass ambiguous issues to the Board for equitable resolution, he does not have to find black and white solutions to gray problems.

With respect to the largest classification of tunnel construction contract disputes, Differing Site Conditions, an important companion to the DRB procedure is the Geotechnical Design Summary Report (GDSR), which is discussed in detail in Chapter 4. The GDSR, which is a part of the contract documents, is a record of the engineer's preconstruction site investigations and laboratory tests, as well as a discussion of the engineer's interpretation of how this information has affected the design and may affect construction. In particular, the GDSR is intended to illuminate any restrictions on construction methods, equipment, or sequences that may be included in the specifications.

In broad terms, the GDSR is intended to define a basis for the solicitation of bids. It provides the assumptions with respect to site geotechnical conditions that are to be used by both the contractor and the owner in performance of their respective duties under the contract. The GDSR defines a box—if the actual conditions disclosed during construction fall within this box, they are covered by the contract; if they fall outside the box, they require modification to the contract.

In cases of disagreement about whether a condition falls inside or outside, the matter may be settled directly by negotiation between the owner and contractor, through the stipulated claims process, and if this fails, the DRB mediates the dispute and recommends a settlement.

More than 25 years of experience with the DRB process has now been accumulated. The reduction in the cost and time required for dispute resolution has been so substantial that the process has spread from tunnel and underground construction to general heavy construction and even to complex commercial building projects. The scope of DRB activities has similarly been extended from its original focus on differing site conditions to include all forms of technical and contractual disputes between the owner and contractor.

A comprehensive discussion of the history, operation, and effects of the DRB process is given in the *Construction Disputes Review Board Manual,* by Mathews, Matyas, Smith, and Sperry, 1995 (McGraw-Hill).

A few words are in order on the use and misuse of the Geotechnical Design Summary Report. It is intended to clarify, insofar as the imperfect state of engineering art permits, the inscrutable vagaries of nature, and thereby to reduce the incidence of surprises during construction. It is not a promise or warranty, to either the owner or the contractor, that conditions different from those described will not be encountered. It is, rather, a definition of what is to be covered by the contract (and the bid price), and a notice that excursions beyond what is described in the GDSR will be legitimate grounds for a contract modification. (Note, however, that disputes about what is an inclusion and what is an excursion are contemplated, and these are to be settled through the DRB process.)

The preparation of a GDSR is perhaps the greatest moral challenge faced by a tunnel engineer. It is possible to draw up a GDSR that predicts such great hazards that all contractors will raise their prices and the likelihood of differing site conditions claims and disputes is minimized. This is not likely to serve the owner's best interests.

It is also possible to draw up a GDSR that predicts such benign conditions that all bids will be low, but the risk of disputes, changes, and cost overruns is sharply increased. This is also not in the owner's (or engineer's) interest.

If the owner does not choose to fund an adequate geotechnical exploration and testing program (as discussed in Chapter 4), the engineer's basis for predicting construction conditions is reduced. But even the most comprehensive preconstruction investigation is never perfect. No engineer, owner, or contractor has yet been found who can outsmart nature all the time.

The engineer must therefore draw his "box" with what information he has at hand, and use his experience and intuition to illuminate the dark crevices where nature has squirreled away her surprises for contractors, engineers, and owners. In the end, he must choose to draw his box where, in his professional judgment, it will yield the best balance of risk and price for his client, the owner.

Tunnel engineering and construction remain an arcane art (the lawyers are right about that!). It will always remain a field in which honest professionals can and will disagree, as well as one in which an experienced party can and will take advantage of the inexperience of other parties. Over the past 25 years the concepts of the Disputes Review Board and the Geotechnical Design Summary Report have evolved to provide the means for resolving honest disagreements and for protecting the innocent, and to cast a spell of civilization over the jungle that has been the venue of tunnel construction contracting for so long.